侯伯宇论著集

《侯伯宇论著集》编委会　选编

西北大学名师大家学术文库

西北大学出版社
·西安·

图书在版编目(CIP)数据

侯伯宇论著集 / 侯伯宇论著集编委会选编. -- 西安:
西北大学出版社, 2022. 2
ISBN 978 - 7 - 5604 - 4517 - 5

Ⅰ. ①侯… Ⅱ. ①侯… Ⅲ. ①物理学—文集
Ⅳ. ①O4 - 53

中国版本图书馆 CIP 数据核字(2020)第 057559 号

侯伯宇论著集
HOU BOYU LUNZHUJI

编　　者　《侯伯宇论著集》编委会
出版发行　西北大学出版社
地　　址　西安市太白北路 229 号
网　　址　http://nwupress.nwu.edu.cn
E - mail　xdpress@nwu.edu.cn
邮　　编　710069
电　　话　029-88302590
经　　销　全国新华书店
印　　装　陕西博文印务有限责任公司
开　　本　787 毫米 × 1092 毫米　1/16
印　　张　43
字　　数　618 千字
版　　次　2022 年 2 月第 1 版　2022 年 2 月第 1 次印刷
书　　号　ISBN 978 - 7 - 5604 - 4517 - 5
定　　价　218.00 元

《西北大学名师大家学术文库》
编辑出版委员会

序　言

西北大学是一所具有丰厚文化底蕴和卓越学术声望的综合性大学。在近120年的发展历程中，学校始终秉承“公诚勤朴”的校训，形成了“发扬民族精神，融合世界思想，肩负建设西北之重任”的办学理念，致力于传承中华灿烂文明，融汇中外优秀文化，追踪世界科学前沿。学校在人才培养、科学研究、文化传承创新等方面成绩卓著，特别是在中国大陆构造、早期生命起源、西部生物资源、理论物理、中国思想文化、周秦汉唐文明、考古与文化遗产保护、中东历史，以及西部大开发中的经济发展、资源环境与社会管理等专业领域，形成了雄厚的学术积累，产生了中国思想史学派、“地壳波浪状镶嵌构造学说”“侯氏变换”“王氏定理”等重大理论创新，涌现出了一批蜚声中外的学术巨匠，如民国最大水利模范灌溉区的创建者李仪祉，第一座钢筋混凝土连拱坝的设计者汪胡桢，第一部探讨古代方言音系著作的著者罗常培，中国函数论的主要开拓者熊庆来，五四著名诗人吴芳吉，中国病理学的创立者徐诵明，第一个将数理逻辑及西方数学基础研究引入中国的傅种孙，“曾定理”和“曾层次”的创立者并将我国抽象代数推向国际前沿的曾炯，我国“汉语拼音之父”黎锦熙，丝路考古和我国西北考古的开启者黄文弼，第一部清史著者萧一山，甲骨文概念的提出者陆懋德，我国最早系统和科学地研究“迷信”的民俗学家江绍原，《辩证唯物主义和历史唯物主义》的最早译者、第一部马克思主义哲学辞典编著者沈志远，首部《中国国民经济史》的著者罗章龙，我国现代地理学的奠基者黄国璋，接收南海诸岛和划定十一段海疆国界的郑资约、傅角今，我国古脊椎动物学的开拓者和奠基人杨钟健，我国秦汉史学的开拓者陈直，我国西北民族学的开拓者马长寿，《资本论》的首译者侯外庐，“地壳波浪状镶嵌构造学说”的创立者张伯声，“侯氏变换”的创立者侯伯宇等。这些活跃在西北大学百余年发展历程中的前辈先贤们，深刻彰显着西北大学“艰苦创业、自强不息”的精神光辉和“士以弘道、立德立言”

的价值追求，筑铸了学术研究的高度和厚度，为推动人类文明进步、国家发展和民族复兴做出了不可磨灭的贡献。

在长期的发展历程中，西北大学秉持“严谨求实、团结创新”的校风，致力于培养有文化理想、善于融会贯通、敢于创新的综合型人才，构建了文理并重、学科交叉、特色鲜明的专业布局，培养了数十万优秀学子，涌现出大批的精英才俊，赢得了“中华石油英才之母”“经济学家的摇篮”“作家摇篮”等美誉。

2022年，西北大学甲子逢双，组织编纂出版《西北大学名师大家学术文库》，以汇聚百余年来做出重大贡献、产生重要影响的名师大家的学术力作，充分展示因之构筑的学术面貌与学人精神风骨。这不仅是对学校悠久历史传承的整理和再现，也是对学校深厚文化传统的发掘与弘扬。

文化的未来取决于思想的高度。渐渐远去的学者们留给我们的不只是一叠叠尘封已久的文字、符号或图表，更是弥足珍贵的学术遗产和精神瑰宝。温故才能知新，站在巨人的肩膀上才能领略更美的风景。认真体悟这些学术成果的魅力和价值，进而将其转化成直面现实、走向未来的“新能源”“新动力”和“新航向”，是我们后辈学人应当肩负的使命和追求。编辑出版《西北大学名师大家学术文库》正是西北大学新一代学人践行“不忘本来、面向未来”的文化价值观，坚定文化自信、铸就新辉煌的具体体现。

编辑出版《西北大学名师大家学术文库》，不仅有助于挖掘历史文化资源、把握学术延展脉动、推动文明交流互动，为西北大学综合改革和“双一流”建设提供强大的精神动力，也必将为推动整个高等教育事业发展提供有益借鉴。

是为序。

《西北大学名师大家学术文库》编辑出版委员会

纪念侯伯宇先生

侯伯宇先生是我多年的老朋友，是我主持的中国高等科学技术中心的顾问委员会委员，为中心的发展做出过重要、积极的贡献。

伯宇先生是极有成就的物理学家、教育家，一生致力于理论物理、数学物理研究。在群表示论、量子反常、二维可积场、规范场、共形场论、统计模型、量子群等领域做出过众多很出色的创新工作。他以研究物理为自己的人生理想，勤奋万分，工作到生命的最后一刻。伯宇先生献身于祖国的科学与教育事业，培养了大量优秀人才；同时，为促进高层次研究人员培养，伯宇先生曾积极参与中国博士后研究制度的建立。

伯宇先生在半个世纪的研究生涯中，发表论文二百余篇。这本论著集收录了他的部分代表性论文，使人们对伯宇先生在物理学领域的贡献有初步的了解。文集的出版，也是我们对伯宇先生一个永久的纪念。

李政道

2011年8月15日

前 言

侯伯宇与我们都是张宗燧先生的学生。自1963年他以33岁的年龄在中国科学院数学研究所当研究生开始(据说他是当年年龄最大的研究生),在将近半个世纪的经历中,我们见证了他对理论物理事业的执着。只要有可能,他就会拿起理论物理的文献。甚至在东北参加“四清”,以及在“文化大革命”中在小站部队农场接受“再教育”时,他也是常常捧着书本,这在当时显得与环境气氛很不协调,很可能会受到批判,但他好像也不在乎。他将这种对于理论物理研究孜孜不倦的精神坚持一生,始终不渝。听说,直至最后在医院已病重不起时,他还在读书,还一心想着正在研究的课题,令人敬佩万分。

侯伯宇潜心于量子场论和数学物理的研究,工作于这些方面的最前沿,在群表示论、磁单极子、可积场论模型、量子反常、量子群、统计模型和共形场论等方面取得了一系列极为重要的成果,得到了国内外理论物理界的高度评价。他是我国理论物理事业的优秀学术带头人之一,为我国理论物理队伍培养了一大批人才。特别是到西北大学后,对于得到广泛认可的我国理论物理“西北军”的形成,做出了重大贡献,他的学生遍布全国乃至国外。

《侯伯宇论著集》选编了他的部分论文。在他逝世一周年之际出版这本论著集是很有意义的:一方面,这是他的学术成就和学术风格的记录;另一方面,也将给他的同时代人提供一个永远的纪念,后辈青年也将从中得到教益。

戴元本　朱重远

2011年7月29日

侯伯宇简介

侯伯宇(1930—2010 年),北京人。西北大学终身教授,全国优秀共产党员,五一劳动奖章获得者,不幸于2010 年10 月6 日病逝于西安。

侯伯宇,中国著名的理论物理学家和数学物理学家,他在粒子物理与场论、引力理论、统计物理、群表示论、数学物理等领域进行了广泛深刻而持久的研究,取得了突出的研究成果,其中相当一部分具有原创性、开拓性和前驱性,在国际学术界产生了重要影响,为我国的理论物理和数学物理的发展做出了不可替代的重要贡献。他学术思想活跃,学风严谨求实,工作勤奋刻苦,堪称年轻人的楷模。他勤勤恳恳、默默耕耘、乐于奉献,为培养理论物理、数学物理的青年人才呕心沥血。

侯伯宇教授一生坎坷,历经磨难,但热爱祖国的信念始终如一。无论是在人生的低谷,还是在改革开放的年代,他总是将对祖国的赤诚情怀,融入自己的平凡工作中,成就了"中国人的骄傲"。

侯伯宇教授一生发表论文约200 篇,我们从中选择出50 篇,编撰成集,纪念这位杰出的同人、亲密的朋友和可敬的老师,弘扬他献身科学、潜心探索、淡泊名利、追求真理的科学精神。

侯伯宇教授一生从事的研究领域十分广阔,成果丰富。我们就选集的部分重要内容做一简单的介绍,但编委会才疏学浅,难以对其研究工作进行全面、深入、准确地介绍,尤其对其晚年的工作知之甚少。在对其学术贡献的评价中,如有差错遗漏,敬请读者、相关专家,以及侯伯宇教授生前的合作者和学生指正。

一、坎坷人生

侯伯宇出生于1930 年9 月11 日,其父亲侯镜如是著名爱国人士,黄埔军校第一期毕业生,在东征时经周恩来和郭俊介绍秘密加入中国共产党。十四年抗战中,侯伯宇跟随父母辗转各地,在动荡中完成了中小学学业。苦难中国的屈辱在其幼小的心灵中埋下了忠贞爱国的种子。他从小热爱科学,喜欢音乐,学会的第一首歌就是八十九旅的军歌:"只有铁,只有血,只有铁血才能救中国。救我同胞誓把倭奴灭,醒我国魂誓把奇耻雪……"终生不忘。

1947 年，侯伯宇考上清华大学，保留学籍。1948 年，燕京大学举行优秀中学生保送考试，他考分第一，获得全额奖学金。但由于对物理的酷爱，他最终选定清华攻读物理。因北平战事，侯伯宇在清华仅读 4 个月后，就随家迁往香港，转读台湾大学。不久，其父侯镜如经香港秘密渠道告知“不可停留台湾，立即去香港”。这样，在台湾大学物理系、化工系才读了 3 个月的侯伯宇，赶乘最后一班直通航轮奔赴香港。在港等待赴美签证时，他获悉清华大学已复课，立即回到北京再次就读清华大学物理系（1950.2—1951.1）。此后，抗美援朝战争爆发，侯伯宇毅然报名参加军干校去东北，被安排突击学习俄文，准备为苏联军事顾问做随军翻译。停战后，他给高教部写信，得到了返校继续学习的机会。因经济建设缺俄文翻译，于是服从国家需要，奔赴鞍钢。其间各项工作都名列前茅。1955 年“肃反”，侯伯宇因家庭出身的“复杂性”和在台湾的“历史疑点”，被鞍山市委列为“重点”清查对象隔离审查，直至侯镜如找到“全国肃反小组”说明情况后，才得以解脱。此时他再次萌生回大学学习的意愿，但未获批准。后经北京市委副书记刘仁的直接干预，才将他的人事档案调回北京。可惜，此时侯伯宇已经错过北京高考，无奈之下他随即飞赴西安，考取当时全国唯一招收插班生的西北大学，开始了第三次大学生活。

侯伯宇自学能力极强。入读西北大学时，已自学完大学的主要课程。1958 年，他倦于政治运动，申请提前毕业，被分配到西安交通大学加速器专业，但再次因政治原因被交大转分到西安矿业学院（1958.9—1963.7）。自此，侯伯宇开始了他的物理数学研究，发表了学术研究的处女作。1963 年，他以几乎满分的成绩考取中国科学院数学所的研究生，师从张宗燧教授，与陈景润成为舍友。在“文革”初期，张宗燧被“批斗”，“白专分子”侯伯宇则成为“陪斗”，直至张“畏罪自杀”。“文革”后期，侯伯宇毕业留任数学所工作，被安排转向实用科研领域。当他得知家属的户口“绝不可能”调到北京，便于 1973 年 6 月回到西北大学，开始了在西北大学长达 37 年的科研、教学工作。

侯伯宇 1978 年晋升副教授，1980 年晋升教授。1980 年 2 月创建西北大学现代物理研究所，并任所长。他曾多次应邀赴美国耶鲁大学、纽约市立大学、纽约州立大学石溪分校、意大利国际理论物理中心、加州大学戴维斯分校、京都大学数理解析研究所、墨尔本大学、东京都立大学等访问。1988—1995 年，侯伯宇被选为国际纯粹与应用物理学联合会（IUPAP）委员。

侯伯宇为西部理论物理的人才培养付出了大量心血。1981 年，他受聘为国务院第一批博士生导师，他领导的研究所也是全国第一批被授权的物理学科博士后流动站。几十年来，他指导、培养博士后 7 名、博士研究生 20 多名。他们之中有国家级突出贡献专家 1 人，获得中国科协、国家人事部、中组部的中国青年科技奖 3 人，获得日本学术振兴会(JSPS)科学基金 6 人，获得洪堡基金 7 人。侯伯宇以他杰出的贡献，于 1985 年获得“全国优秀教育工作者”称号和五一劳动奖章。

二、学术成就

1. 群论、格林函数

在 20 世纪 60 年代，侯伯宇与合作者一起从事有关群论等数学物理的研究，创造性地提出利用产生湮灭算子构造 SU(N)群正则基降级的变换算子，进而获得完整的正则基，纠正了前人的错误，解决了这一延续十余年的难题；创新地使用三次贝塞尔函数的沿不同回路积分解决赫施菲尔德(J. Hirschfelder)的交叠区同类项合并的问题，以及构造局域坐标系中的旋量球函数，极大地简化了计算。这些工作都为当时国际先进水平，他本人也受到国内理论物理开创者之一张宗燧教授的注意。

(1)分子链的相互作用 Green 函数

1960 年，与同事文振翼注意到美国科学院院士赫施菲尔德等未能求得交叠区同类项合并的问题。他们创新地使用同一被积函数的不同回路积分给出各分区的表达式，将交叠区的三个贝塞尔函数按幂次级数递推表出，得到简化的递推式；同时，运用边界条件的对称性限定并项为偶次，得到收敛级数，解决了这一问题。

(2)局域坐标系中的旋量球函数及算子

中心辐射(散射)以及从质心系观察的散射与辐射涉及辐射源、散射中心的极化，而极化依赖于方向角与径向距离的关系。通常人们按沿径向极化等量子数写的波函数的各分量沿固定观察系投影，分类合并后，再按径向投影，其计算甚为繁杂。侯伯宇建议采用始终按径向运算的方法，而固定系标架到球面局域三足标架的转动矩阵，可直接表达旋量波函数(helicity 幅)，将微商的横向作 Hodge 对偶，直接表出梯度、旋度等公式；将前人著作中各径向、横向多极辐射的繁杂结果，直截了当地表示为用 Clebsch – Gordan 系数耦合到一起的已知的径向积分式。

(3)SU(N)群不可约表示的正交归一基底与代数

1950 年，盖尔芳(I. Gelfand)和泽特林(M. Tsetlin)给出 SU(N)代数表示式，如何证明却一直没有成功。1965 年，侯伯宇利用正负素根算子与钩算子的玻色实现，依次构成了从各盖尔芳正则基降到与它相邻的各低阶相邻正则基的降(升)算子多项式，成功地完成各正则基的归一化，从而获得正交归一、完整的正则基；证明各正负素根算子表达为各子群内的钩因子及相邻子群间的不可约张量因子的乘积，改正了以往文献中从盖尔芳沿袭下来的相角符号错误。

2. 规范场论、量子场论及大范围拓扑性质

(1)规范场论

侯伯宇教授是我国较早研究规范场、量子场及大范围拓扑性质的理论物理学家之一。早在20世纪70年代，他利用微扰论证明同阶费曼图内非物理粒子的规范贡献全部抵消、物理过程的规范无关性及任意规范的幺正性；利用纤维丛理论研究磁涡流、单极的整体宏观量子数及其规范协变量、电磁荷流等的定域分布，导出SU(2)规范场对称破缺所剩U(1)磁荷的表达式，对应于陪集的第二同伦群。他还和合作者一起研究荷电粒子与单极作用体系的物理规律和效应。这些工作处于当时国际先进水平，获得1978年陕西省科技成果一等奖，作为合作者获得1982年国家自然科学奖三等奖。

(2)量子规范场的反常与大范围拓扑性质研究

20世纪80年代中期，量子反常及其大范围拓扑性质为当时国际理论物理的研究热点，作为主要承担者，侯伯宇参加了量子反常及其大范围拓扑性质的研究。他独立提出和分析3上同调链，给出反常系列整体特征数的继承性；证明流守恒反常、对易反常和Jacobi反常的微分继承式积分满足Cech双上同调关系，江口徹(Eguchi)在国际光子与轻子大会场论总结时评价这些工作"非常漂亮地运用了Cech方法，给出反常的整体特征数继承相等"；侯伯宇还发现结合律反常满足五角恒等式，Jacobi反常的自洽条件是四重对易式，而不是Malcev代数，澄清了捷奇夫与朱米诺(B. Zumino)的争论。这项工作获得1989年国家自然科学二等奖。

3. 非线性可积系统

20世纪80年代初，孤立子的非局域对称性及相应守恒量为研究热点。1981年底，侯伯宇在耶鲁大学访问期间利用协变分解方法，得到了含谱参数的Noether守恒流，形成了有名的耶鲁预印本，谱参数的展开可以给出无穷多守恒流，满足Kac－Moody代数，曾被一些同行称为"H－变换"；与合作者一起，利用拓扑磁荷流的生成元与Hodge对偶的电荷流的生成元扭转，合成得到非定域守恒流，并得到Virasoro型变换；与学生李卫一起得到可用来生成引力场由真空平庸解变到静轴对称及柱对称一切解的变换，被金奈斯雷(W. Kinnersley)等称之为"Hou－Li变换"；特别是可以增减Weyl多级矩，从而实现了杰拉奇(R. Geroch)猜想，被金奈斯雷与朱里亚(B. Julia)先后用来构成四维引力及十一维伸缩超引力(M理论)的渐进几乎处处平坦的整体解；侯伯宇还指出四维自对偶Yang－Mills场表为复化反自偶面的平联络后不满足复二维主手征型“守恒律”，而是两个互对偶二维面上的 WZW 型方程，创新地给出自对偶 Yang－Mills 场的含双拓扑 WZW 项的作用量。

侯伯宇在该方面的工作具有鲜明的特色和原创性，其研究成果被称为“H－变换”和“Hou－Li 变换”。20 世纪 80 年代，“H－变换”入选新华社编发的二十项《中国的骄傲——以中国人姓氏命名的现代科技成果》。该方面的研究于 1987 年获得国家教委科学技术进步奖一等奖。

4. 可积统计模型与量子代数研究

可积模型是可以解析求解的高度非线性和强关联的物理系统，它们在精确揭示物理体系的特性和临界行为等方面扮演着不可替代的角色。自 1987 年开始，侯伯宇转向了这一竞争激烈的研究领域，取得了一系列的研究成果。

（1）有限格点与多体长程作用

椭圆函数型格点模型的求解及对称性研究是该领域多年悬而待决的基础难题。1982 年，斯克良宁（E. Sklyanin）做出了 $n=2$ 椭圆量子代数；1985 年，谢雷德尼克（I. Cherednik）试图推广到 $n>2$，但未成功。侯伯宇等人发现了 sl(n) 椭圆型转移矩阵的差分表示、建立其本征函数和本征谱；通过分离谱参量，成功构成 $n>2$ 椭圆函数型经典及量子代数，创先构成其循环表示及差分算子表示。随后，又成功地将此方法运用于各种统计模型，获得了一系列成果。侯伯宇等人还创新性地研究了与可积系统相匹配的各种边界条件及相关的物理效应，形成了具有特色的研究团队，成为了国内在国际上具有一定影响力的唯一研究组。

（2）量子群研究

20 世纪 80 年代末，量子群研究成为数学物理的热点，侯伯宇、侯伯元、马中骐合作将二子表示递降各顺序间所差 q 相角求和，得到由 q 系数构成的拉卡型 Clebsch－Gordan 系数以及对称关系，按照不同于原角动量理论的聚合顺序，成功推得聚合型系数的 q 乘积式和 144 个系数的全部对称关系，证明拉卡系数满足辫子群与五角结合律。该项工作独立于基里洛夫（A. Kirillov）与雷切迪金（N. Reshetikhin），明显而具体地推得各系数及其对称性与求和律，杰维斯（J. Gervais）等指出这正是他们研究二维引力等聚合时所需的。

此方向的研究成果获得 1994 年陕西省科技进步一等奖（周玉魁）及 2005 年陕西省科技进步二等奖。

5. 共形场与可积场论研究

（1）共形场的对称性

20 世纪 80 年代末，阿瓦雷兹－高默（L. Alvarez-Gaume）讨论了极小共形场与量子群的关系,但实际上并未证明。侯伯宇等最先利用 Dotsenko 库仑气积分中的单值行为计算极小共形场的辫子矩阵,证明其对成群为两个 q - SU(2)的直积,随后又用积分表示直接推出SU(2) WZW模型共形块的聚合与辫子矩阵,证明它们分别用量子群的两种 $q-6j$ 系数表示,同时聚合法则受到共形场阶数限制的截断行为,这完全等同于 q - SU(2)的表示理论。

该方面的工作当时具有国际领先水平,在国内外具有鲜明的特色,其成果与其他工作一起,获得 1994 年陕西省科技进步奖一等奖。

(2)有质量量子代数的可积

如何将量子流代数从共形场推广到无穷多格点及有质量场是数学物理的难点之一。侯伯宇采用德里费尔德(V. Drinfeld)的扭曲 Hopf 代数可从 XYZ 格点的量子代数得到 RSOS 模型的量子代数,发现面模型在热力学极限下,满足既有动力学扭曲又有中心扩张的量子代数对称性,并明显求出该流。通过取双标度极限,得到一种新代数,它描述有质量场的散射关系;利用 Miura 变换获得有质量场的 W 代数及其顶角算子的双畸变玻色振子表达式。这些工作被京都学派及圣彼得堡学派等引为最初文献。

侯伯宇教授对理论物理学和数学物理学的贡献是多方面的。他时刻关注国际理论物理的研究前沿,直至去世前一个月,仍在阅读最新文献,探讨几何 Langlands 纲领与拓扑场论方面的工作。他思维敏锐、洞察力强,善于发现和探索重要科学问题、新的研究领域和方向,做出杰出工作,研究新的理论基础和技术方法,富于创新精神;他勤奋刻苦,为同辈人中的佼佼者,堪称后辈的楷模。侯伯宇教授一生淡泊名利,两袖清风,自由穿行于理论物理的乐园……

侯伯宇教授2010 年 10 月 6 日不幸因病逝世。侯先生病逝后,西北大学做出了《关于开展向侯伯宇同志学习的决定》。2011 年 6 月,陕西省委教育工委、省教育厅做出了《关于在全省教育系统开展向优秀共产党员侯伯宇同志学习的决定》,杨战营教授在陕西高校庆祝中国共产党成立 90 周年大会上做了侯伯宇先进事迹报告。2011 年 11 月,中共陕西省委做出了《关于追授侯伯宇同志"优秀共产党员"称号 开展向侯伯宇同志学习宣传活动的决定》,并组织省内媒体开展了一系列宣传活动。

2011 年 12 月,时任中央政治局委员、中组部部长、中央创先争优活动领导小组组长李源潮同志在侯伯宇同志先进事迹专报《把一切献给教育科学事业的时代先锋》上做出重要批示:"侯伯宇同志的先进思想和先进事迹很感人,应广泛宣传这位在创先争优活动中涌现出的先进典型,倡导侯伯宇同志学高为师,身正为范的教授理想。"2012 年元月,原中央政治局委员、国务委员刘延东同志在第二十次全国高校党建会上号召全国教育战线深入开展学习侯伯宇同志先进事迹活动。3 月，人力资源和社会保障部、教育部联合做出了《关于追授侯伯宇同志全国模范教师荣誉称号的决定》。在 2012 年 6 月建党 91 周年前夕召开的全国创先争优表彰大会上，侯伯宇教授被追授为“全国创先争优优秀共产党员”，是这次活动中我省唯一的、也是全国教育系统唯一的入选者。

2012 年 6 月，时任中共中央政治局常委的李长春同志在新华社的内参上做出重要批示：要求浓墨重彩地宣传侯伯宇同志的先进事迹。时任中共中央政治局委员刘云山、刘延东等也就学习宣传侯伯宇教授先进事迹分别做出重要批示。时任中央政治局委员、书记处书记、中组部部长，陕西省委书记赵乐际同志十余次做出重要批示，并亲自安排部署有关工作。随后，侯先生被中央确定为全国重大先进典型，这是全国高等教育系统和西北地区十余年来的唯一一个全国重大先进典型。按照中央安排，2012 年教师节启动了全国性的学习宣传活动。2012 年 9 月 5 日，由人民日报、新华社、光明日报、经济日报、中国青年报、中国教育报、科技日报、中央人民广播电台、中央电视台等十余家新闻媒体组成的采访团来到西北大学，对侯先生的先进事迹进行了集中采访。9 月 9 日，中央电视台新闻联播和焦点访谈对侯伯宇教授的先进事迹做了报道。从 9 月 10 日开始，所有中央媒体在重要版面或者重要时段，集中宣传报道侯先生的先进事迹。9 月 27 日，侯伯宇同志先进事迹报告会在人民大会堂举行，刘延东同志代表党中央、国务院接见了报告团并发表重要讲话，中央主要媒体进行了报道，中央电视台播出了报告会实况录像。人民日报、光明日报和陕西日报整版刊登了报告稿。从 10 月 14 日开始，由中宣部、教育部和陕西省委共同组织的侯伯宇先进事迹报告团先后在黑龙江、浙江、湖南、四川、河南、陕西开展巡回报告，听众达到近万人。侯先生的感人事迹、高尚品格传遍大江南北，在干部、群众和广大师生中产生了强烈反响。

《十一维空间——物理学家侯伯宇教授的多维人生》荣获陕西省 2012 年度重大文化精品项目。陕西省委以侯伯宇事迹为原型，拍摄了电影《爱的帕斯卡》。2015 年陕西省高教工委将设立在西北大学博物馆中的侯伯宇先进事迹展览馆确立为全省高校师德教育基地，全面向社会公众开放，现已接待省内外专家学者、干部培训班学员及广大师生 9000 余人。2017 年 6 月，教育部部长陈宝生参观西北大学侯伯宇先进事迹展览馆后说：“我们一路寻找伟大，找到的多是昙花，今天我找到了平凡，欣喜地偶遇伟大。”2019 年侯伯宇教授荣获“最美劳动者——新中国成立以来陕西省最具影响的劳动模范”。

Contents

* 为尊重侯伯宇先生原作,书中论文均采用影印版,故目录标题格式与书中论文格式保持一致。

第 19 卷 第 6 期 物 理 学 报 Vol. 19, No. 6
1963 年 6 月 ACTA PHYSICA SINICA June, 1963

局部坐标系中的旋量球函数及算子*

侯 伯 宇
(西安矿业学院)

提 要

本文給出了各种表象的旋量球函数在局部坐标系中的表达式及其与固定坐标式的联系，并用此式討論了光子、电子、双光子系波函数的各种性质. 利用梯度算子及无穷小轉动算子在局部系中的联系将旋量方程分离变量，得到了梯度公式. 最后在局部系中探討了散射問題，計算了 γ 跃迁的角向积分.

研究标量場的中心对称問題时，最方便的是采用球极坐标系. 研究旋量場时，通常仍然对固定极軸(量子化軸)取自旋矩的投影. 这种不考虑具体地点的取分量方法往往掩盖了問題固有的对称性，使旋量場的表达式及計算过程不必要地复杂化. 例如最簡单的旋量——径向单位矢量 $\mathbf{r}_0$ 在球面上各点 (θ, φ) 的表达式为 $-\sqrt{4\pi}\psi^{0;\,0}_{1;\,1}(\theta, \varphi) = -\sqrt{4\pi}\cdot\sum_m C^{0;\,0}_{1;\,m;\,1,-m} Y_{1m}(\theta, \varphi) u_{1,-m}(K)$[1)]，分成的三个分量 $Y_{1m}(\theta, \varphi)$ 是随点而变的，也不易看出它是纵向的. 因此有些作者[2,3]引进了局部坐标系，但他們并沒有詳細討論局部坐标系与固定球极坐标系間的明显变换式，一般文献中也不大采用局部坐标系. 本文在第一节中导出了普遍变换式，引进了具有一定宇称及极化的旋量球函数. 在三、四、五节中指出了通常的“电”及“磁”复极場，有确定宇称及动量矩的电子波函数以及汪容[4]的双光子系波函数都可以由第一节的統一表达式得到. 并且通过这些实例說明了在局部坐标系中旋量球函数的各种性质較为明显，构成符合某些选择律的波函数較为方便. 在这种坐标系中容易理解过去关于双光子系的各工作[4,5,6,7,8]，并发現有些工作中不够明确或錯誤的地方.

Гельфанд 等[2]及 Любарский[3] 在局部坐标系中将旋量的一阶微分方程或二阶的 Maxwell 方程分离了变量求解. 但由于或者是采用了固定系中的微分算子，或者是虽然采用了局部系中的微分算子，而最后仍旧用固定系中的无穷小轉动算子来表示，演算过程較为繁复. 本文第二节引进了局部系中的无穷小轉动算子，找到了它与微分算子 ∇ 的横向分量的連系，从而得到了計算旋量球函数的微商的簡单方法，这时只需利用广义球函数作为轉动羣不可約表示的正則基的特性，用不着具体的微分递推式. 用此法将方程分离了变量，与文献[2]，[3]比較，指出了其中的錯誤. 还可以直接求得梯度公式，避免了过去(例如[9])求梯度时从固定极軸投影到局部軸再投影回固定极軸这种往返周折的过程.

第六节簡单討論了用局部坐标式計算散射問題的方法，与現有的从固定极軸球坐标

* 1961 年 8 月 14 日收到；1962年 7 月 21 日收到修改稿.

1) 式中各符号的含义請参看下节 (1.5) 式. 本文中函数 Y、D 系数 C、W 的定义及位相及 Euler 角均依照文献[1].

式出发的計算結果[10,11,12,13]符合一致．在局部坐标系中討論散射幅的有些性质（例如南氏变換[14,15,16,17]）比較方便．

最后一节在局部系中分离径向及角向变量，計算了与电子在复极場中跃迁有关的一些矩陣的角向积分．

一、旋量球函数的各种表象及其在局部坐标系中的表达式

采用通常的球极坐标系，旋量場在球面上的基底可以选择为

$$\psi_{lm;\,sm_s}(\theta,\,\varphi)\equiv Y_{lm}(\theta,\,\varphi)u_{sm_s}(K), \tag{1.1}$$

式中旋量球函数 $\psi_{lm;\,sm_s}(\theta,\varphi)$ 由两部分的积构成，自旋部分 $u_{sm_s}(K)$ 是 s 阶旋量，s、m_s 分别表示自旋矩及其在固定极軸 K 上的投影；空間分布部分 $Y_{lm}(\theta,\varphi)$ 为立体角 θ,φ 的球函数，l、m 表示軌道矩及其在极軸 K 上的投影． $\psi_{lm;\,sm_s}(\theta,\varphi)$ 是 l、m；s、m_s 的本征态．但在空間轉动时，它属于不同阶的不可約表示，并不具有确定的总动量矩 j．变換到局部坐标系就可以更明显地看出这一点．

經过轉动 g 将以 K 为固定极軸的坐标系轉为以 κ 为极軸的局部坐标系，就得到

$$u_{s\mu_s}(\kappa)=\sum_{m_s}u_{sm_s}(K)D^{s}_{m_s\mu_s}(\varphi,\,\theta,\,\gamma),$$

式中 $u_{s\mu_s}(\kappa)$ 为 s 阶旋量，其自旋在局部軸 κ 上的投影为 μ_s，$D^{(s)}_{m_s\mu_s}(\varphi,\theta,\gamma)$ 是轉动 g 在 s 阶不可約表示中的矩陣元，φ、θ、γ 是轉动 g 的 Euler 参数，这儿 φ、θ 正好是 κ 軸在 K 系中的立体角．利用 D 矩陣的 U 性质，易得逆变換 $u_{sm_s}(K)=\sum_{\mu_s}D^{s*}_{m_s\mu_s}(\varphi,\theta,\gamma)u_{s\mu_s}(\kappa)$，式中 $*$ 表示取复共軛．将此式代入(1.1)式，用 $\sqrt{\dfrac{2l+1}{4\pi}}D^{l*}_{m0}(\varphi,\theta,\gamma)$ 表示 $Y_{lm}(\theta,\varphi)$，用 Clebsch-Gordan 級数展开后，再考虑到 Clebsch-Gordan 系数的对称性，就得到局部坐标系中的表达式：

$$\psi_{lm;\,sm_s}(\theta,\,\varphi)=\sum_{j,\mu_s}(-)^{s+\mu_s}C^{l,\,0}_{j,-\mu_s;\,s,\,\mu_s}C^{j,\,m+m_s}_{l,\,m;\,s,\,m_s}\sqrt{\frac{2j+1}{4\pi}}D^{j*}_{m+m_s,\,\mu_s}(\varphi,\,\theta,\,\gamma)u_{s\mu_s}(\kappa). \tag{1.2}$$

容易証明式中各項在轉动时各自独立地变化．可見在局部坐标系中，旋量場在球面上最自然的基底是

$$\psi^{jM\mu}_{s}(\theta,\,\varphi)\equiv\sqrt{\frac{2j+1}{4\pi}}D^{j*}_{M\mu}(\varphi,\theta,\,\gamma)u_{s\mu}(\kappa). \tag{1.3}$$

它是总动量矩 j、j 在固定极軸 K 上的投影 M、在局部极軸 κ 上的投影 μ（$=s$ 在此軸上的投影 μ_s）及自旋矩 s 的本征态．利用系数 C 的正交性，易由(1.2)得 $\psi^{jM\mu}_s$ 在固定系中的式子：

$$\psi^{jM\mu}_{s}(\theta,\,\varphi)=\sum_{l,m,m_s}(-)^{s+\mu}C^{l,\,0}_{j,\,-\mu;\,s,\,\mu}C^{j,\,M}_{l,\,m;\,sm_s}\psi_{lm;\,sm_s}(\theta,\,\varphi). \tag{1.4}$$

常采用的是具有确定 j、M、s、l 的旋量球函数[18]：

$$\psi^{jM}_{ls}(\theta,\,\varphi)=\sum_{mm_s}C^{j,\,M}_{l,\,m;\,s,\,m_s}Y_{lm}(\theta,\,\varphi)u_{sm_s}(K). \tag{1.5}$$

将 (1.2) 代入卽得局部系中的式子：

$$\psi_{ls}^{jM}(\theta, \varphi) = \sum_{\mu} (-)^{s+\mu} C^{l,\,0}_{j,\,-\mu;\,s,\,\mu} \sqrt{\frac{2j+1}{4\pi}}\, D^{j*}_{M\mu}(\varphi, \theta, \gamma) u_{s\mu}(\kappa) =$$

$$= \sum_{\mu} (-)^{s+\mu} C^{l,\,0}_{j,\,-\mu;\,s,\,\mu} \psi_s^{jM\mu}(\theta, \varphi), \tag{1.6}$$

逆之得

$$\psi_s^{jM\mu}(\theta, \varphi) = \sum_{l} (-)^{s+\mu} C^{l,\,0}_{j,\,-\mu;\,s,\,\mu} \psi_{ls}^{jM}(\theta, \varphi). \tag{1.7}$$

局部坐标式(1.3)的自旋部分 $u_{s\mu}(\kappa)$ 与球极坐标系 K 的轉动无关，对于 K 的轉动来說是零阶 L 张量；空間分布部分 $D^{j*}_{M\mu}(\varphi, \theta, \gamma)$ 则依照 j 阶不可約表示变化．这种写法便于分别研究波函数的空間分布与轉动变換性 D 及粒子的自旋极化状态 u；也便于根据极化条件 u 选择相应的波函数 ψ．虽然 $\psi_s^{jM\mu}$ 不是 l 的本征态，但 l 并非守恆量，因此这并不妨碍用它来描述粒子的状态．不过它也不具有确定的宇称．的确，在原坐标系中在 φ，θ，γ 处的右手局部系变为在反映坐标系中在 $\varphi \pm \pi$，$\pi - \theta$，$-\gamma$ 处的左手系 κ_L（局部极軸方向不变我們約定 φ 軸不变，θ 軸反向），故当 j 为整数时，

$$PD^{j*}_{M\mu}(\varphi, \theta, \gamma) = D^{j*}_{M\mu}(\varphi \pm \pi, \pi - \theta, -\gamma) = (-)^j D^{j*}_{M,-\mu}(\varphi, \theta, \gamma).$$

由于反映时局部系的 θ 軸反向，故磁量子数 μ 的符号改变：

$$Pu_{s\mu}(\kappa) = p_s u_{s,\,\mu}(\kappa_L) = p_s u_{s,-\mu}(\kappa)$$

式中 $p_s = \pm 1$ 为粒子的內禀宇称．整个波函数經反映变为 $P\psi_s^{jM\mu}(\theta, \varphi) = p_s(-)^j \psi_s^{jM,-\mu}(\theta, \varphi)$，并非宇称的本征态．可以构成有确定宇称的函数 $\psi_s^{jM\mu p}$ 如下：

$$\psi_s^{jM\mu+}(\theta, \varphi) = \frac{1}{\sqrt{2}}(\psi_s^{jM,-\mu}(\theta, \varphi) + \psi_s^{jM\mu}(\theta, \varphi)) = \sum_l {}^{+}C^{l,0}_{j,\mu;\,s,-\mu}\psi_{ls}^{jM}(\theta, \varphi),$$

$$\psi_s^{jM\mu-}(\theta, \varphi) = \frac{1}{\sqrt{2}}(\psi_s^{jM,-\mu}(\theta, \varphi) - \psi_s^{jM\mu}(\theta, \varphi)) = \sum_l {}^{-}C^{l,0}_{j,\mu;\,s,-\mu}\psi_{ls}^{jM}(\theta, \varphi), \tag{1.8}$$

式中

$${}^{\pm}C^{l,\,0}_{j,\,\mu;\,s,-\mu} \equiv \frac{1}{\sqrt{2}}[(-)^{s-\mu}C^{l,\,0}_{j,\,\mu;\,s,-\mu} \pm (-)^{s+\mu}C^{l,\,0}_{j,\,-\mu;\,s,\mu}] =$$

$$= \frac{1}{\sqrt{2}}(-)^{s-\mu}C^{l,\,0}_{j,\,\mu;\,s,-\mu}(1 \pm (-)^{l+s+j+2\mu}),$$

$\psi_s^{jM\mu p}$ 的宇称为 $p \cdot p_s(-)^j$．

根据羣表示的矩陣元 $D^j_{M\mu}$ 在羣上积分的正交性及其完备性[19,2,3]以及系数 C 的正交性，不难証明本节列举的各种表象的旋量球函数(1.1)、(1.3)、(1.5)、(1.8)均构成球面上旋量場的正交归一完备基，可用来展开各种旋量函数．例如旋量平面波

$$u_{sm_s}(K)e^{iK\cdot r} = \sum_l \sqrt{4\pi(2l+1)}\,\zeta_l(Kr)\psi_{l0;\,sm_s}(\theta, \varphi) =$$

$$= \sum_{lj} \sqrt{4\pi(2l+1)}\,\zeta_l(Kr) C^{j,\,m_s}_{l,\,0;\,s,\,m_s}\psi_{ls}^{jm_s}(\theta, \varphi) =$$

$$= \sum_{j\mu} \sqrt{4\pi(2j+1)}\left(\sum_l (-)^{\mu-m_s} C^{l,0}_{j,-m_s;s,m_s} C^{l,0}_{j,-\mu;s,\mu}\zeta_l(Kr)\right)\psi_s^{jm_s\mu}(\theta, \varphi) \equiv$$

$$\equiv \sum_{j\mu} \sqrt{4\pi(2j+1)}\,\zeta^{m_s\mu}_{j,\,s}(Kr)\psi_s^{jm_s\mu}(\theta, \varphi), \tag{1.9}$$

当 $r\to\infty$ 时的漸近式为

$$u_{sm_s}(K)e^{iK\cdot r}=\sum_j\frac{\sqrt{4\pi(2j+1)}}{2i}[\psi_s^{jm_sm_s}e^{iKr}+(-)^{j+s+2m_s+1}\psi_s^{jm_s,-m_s}e^{-iKr}]/Kr,\tag{1.9a}$$

式中 $\zeta_l(Kr)$ 为 l 阶球 Bessel 函数[1]的 i^l 倍.

用局部坐标式易求旋量球函数的积:

$$\psi_{s_1}^{j_1M_1\mu_1}(\theta,\varphi)\psi_{s_2}^{j_2M_2\mu_2}(\theta,\varphi)^*=\sum_{\substack{JM\\s\mu}}(-i)^{2M_2}C_{j_1,M_1;j_2,-M_2}^{j,M}C_{j_1,\mu_1;j_2,-\mu_2}^{j,\mu}C_{s_1,\mu_1;s_2,-\mu_2}^{s\mu}\sqrt{\frac{(2j_1+1)(2j_2+1)}{4\pi(2j+1)}}\psi_s^{jM\mu}(\theta,\varphi).\tag{1.10}$$

求上式时，空間部分与自旋部分分别求积. 空間部分利用了 $D_{M\mu}^{j*}(\varphi,\theta,\gamma)=(-)^{M-\mu}\cdot D_{-M,-\mu}^{j}(\varphi,\theta,\gamma)$[19]，再用 Clebsch-Gordan 級数展开;自旋部分約定了 $u_{s\mu}^*$ 的相位为 $u_{s\mu}^*=(-i)^{2\mu}u_{s,-\mu}$，再将直接积 $u_{s_1\mu_1}u_{s_2\mu_2}$ 按 s 阶不可約表示展开.

例如令 $s_1=s_2=s'$, $s=0$ 就得到自旋空間中的不变量:

$$\psi_{s'}^{j_1M_1\mu_1}(\theta,\varphi)\psi_{s'}^{j_2M_2\mu_2}(\theta,\varphi)^*=\sum_{jM}i^{2(s'-M_2-\mu_2)}C_{j_1,M_1;j_2,-M_2}^{j,M}C_{j_1,\mu_1;j_2,-\mu_2}^{j,0}\sqrt{\frac{(2j_1+1)(2j_2+1)}{4\pi(2s'+1)(2j+1)}}Y_{jM}(\theta,\varphi),\tag{1.10a}$$

在 $s'=1$ 时，乘以系数 $-\sqrt{3}$ 即标积. 考虑到 Racah 系数 W 与系数 C 的关系式 $C_{j,0;j,0}^{J,0}W(j,j,j,j;J,1)=C_{j,1;j,-1}^{J,0}$ 等，容易看出与通常的結果是等价的(例如[8]). 令 $s_1=s_2=s=1$，并乘以系数 $\sqrt{2}\,i$ 就得到矢积. 用(1.6)、(1.10a)并采用 Racah[20] 的对磁量子数求和法，即得常見的結果[8]:

$$\psi_{l_1s}^{j_1M_1}(\theta,\varphi)\psi_{l_2s}^{j_2M_2*}(\theta,\varphi)=\sum_{jM}(-i)^{2M_2+4j_2}C_{j_1,M_1;j_2,-M_2}^{j,M}C_{l_1,0;l_2,0}^{j,0}W(j_1j_2l_1l_2;js)\sqrt{\frac{(2j_1+1)(2j_2+1)(2l_1+1)(2l_2+1)}{4\pi(2s+1)(2j+1)}}Y_{jM}(\theta,\varphi).\tag{1.11}$$

利用 D 的 U 性質及 D 为羣的表示 $D(g_1)D(g_2)=D(g_1g_2)$，易求得下式

$$\sum_M\psi_{s_1}^{jM\mu_1}(\theta_1,\varphi_1)\psi_{s_2}^{jM\mu_2}(\theta_2,\varphi_2)^*=\frac{2j+1}{4\pi}D_{\mu_2\mu_1}^{j*}(\Phi,\Theta,\Gamma)u_{s_1\mu_1}(\kappa_1)u_{s_2\mu_2}(\kappa_2)^*,\tag{1.12}$$

式中 Φ,Θ,Γ 为 κ_1 系在 κ_2 系中的 Euler 角. 在二系重合时,

$$\sum_M\psi_{s_1}^{jM\mu_1}(\theta,\varphi)\psi_{s_2}^{jM\mu_2}(\theta,\varphi)^*=\frac{2j+1}{4\pi}\delta_{\mu_1\mu_2}u_{s_1\mu_1}(\kappa)u_{s_2\mu_2}(\kappa)^*,\tag{1.12a}$$

由此式及(1.6)易得[18]中的对 M 求和諸式.

二、局部坐标系中的微分算子

如所周知[2,20]，对于繞固定系的轉动来説，矩陣 $D_{M\mu}^{j*}$ 的第 M 列各元构成 j 阶不可約表示的正則基，固定系中总动量矩算子 $\mathscr{J}$ 的各分量即以 Euler 角为参数时繞相应軸的无穷小轉动算子 A 乘以負 i1)，利用 A 在正則基中的熟知矩陣元 $\|A\|_{ij}$，就可求得

1) 本文采用 $\hbar=C=1$ 的自然单位制.

$$\mathscr{J}_{+}D^{j*}_{M\mu}(\varphi,\theta,\gamma)=\sum_{M'}(-i)\|A_{+}\|_{MM'}D^{j*}_{M'\mu}(\varphi,\theta,\gamma)=\alpha^{j}_{M+1}D^{j*}_{M+1,\mu}(\varphi,\theta,\gamma),$$

同理

$$\mathscr{J}_{-}D^{j*}_{M\mu}(\varphi,\theta,\gamma)=\alpha^{j}_{M}D^{j*}_{M-1,\mu}(\varphi,\theta,\gamma),$$
$$\mathscr{J}_{0}D^{j*}_{M\mu}(\varphi,\theta,\gamma)=MD^{j*}_{M\mu}(\varphi,\theta,\gamma),$$

式中

$$\alpha^{j}_{M}=\sqrt{(j+M)(j-M+1)},$$
$$\mathscr{J}_{0}=\mathscr{J}_{z}=-i\frac{\partial}{\partial\varphi},$$
$$\mathscr{J}_{\pm}=\mathscr{J}_{x}\pm i\mathscr{J}_{y}=-ie^{\pm i\varphi}\left\{\frac{1}{\sin\theta}\left(\frac{\partial}{\partial\gamma}-\cos\theta\frac{\partial}{\partial\varphi}\right)\pm i\frac{\partial}{\partial\theta}\right\}.$$

对于繞局部系的轉动来說，矩陣 $D^{j*}_{M\mu}$ 的第 M 行各元构成正則基，总动量矩算子 J 的各分量是繞局部軸的无穷小算子 A 乘以負 i，这时 A 在正則基中的矩陣元不变，只是改为作用于磁量子数 μ，故有

$$J_{+}D^{j*}_{M,\mu}(\varphi,\theta,\gamma)=\sum_{\mu'}(-i)D^{j*}_{M,\mu'}(\varphi,\theta,\gamma)\|A_{+}\|_{\mu'\mu}=\alpha^{j}_{\mu}D^{j*}_{M,\mu-1}(\varphi,\theta,\gamma),$$

同理

$$J_{-}D^{j*}_{M\mu}(\varphi,\theta,\gamma)=\alpha^{j}_{\mu+1}D^{j*}_{M,\mu+1}(\varphi,\theta,\gamma),$$
$$J_{0}D^{j*}_{M\mu}(\varphi,\theta,\gamma)=\mu D^{j*}_{M,\mu}(\varphi,\theta,\gamma),\tag{2.1}$$

式中

$$J_{0}=J_{r}=-i\frac{\partial}{\partial\gamma},$$
$$J_{\pm}=J_{\theta}\pm iJ_{\varphi}=-ie^{\mp i\gamma}\left\{\frac{1}{\sin\theta}\left(\cos\theta\frac{\partial}{\partial\gamma}-\frac{\partial}{\partial\varphi}\right)\pm i\frac{\partial}{\partial\theta}\right\},$$

本式也可由将写成微分形式的 J 作用于广义球函数而直接驗証，这就是 D 对于指标 μ 的递推关系.

在局部系中 ∇ 算子的第三分量卽径向微商 $\nabla^{0}=\frac{\partial}{\partial r}$，横向部分可求得如下，将軌道矩算子 $\mathbf{L}=-i\mathbf{r}\times\nabla$ 在局部系中写成球基矢分量式

$$\mathbf{L}=\sum_{\mu}u_{1\mu}(\kappa)L^{\mu}=-iru_{1,0}(\kappa)\times\nabla=-ru_{1,1}(\kappa)\nabla^{1}+ru_{1,-1}(\kappa)\nabla^{-1},\tag{2.2}$$

式中 L^{μ}、∇^{μ} 为 L、∇ 在局部系中的共变球矢分量[9]. 比較左右側就求得

$$\nabla^{1}=\frac{-L^{1}}{r}=\frac{L_{-}}{\sqrt{2}r}=\frac{J_{-}-S_{-}}{\sqrt{2}r}$$
$$\nabla^{-1}=\frac{L^{-1}}{r}=\frac{L_{+}}{\sqrt{2}r}=\frac{J_{+}-S_{+}}{\sqrt{2}r}\tag{2.3}$$

式中 $L_{\pm}=L_{\theta}\pm iL_{\varphi}$，为局部系中軌道矩产生及湮沒算子，

$S_{\pm}=S_{\theta}\pm iS_{\varphi}$，为局部系中自旋矩产生及湮沒算子.

从 (2.2) 还得到 $L_0=0$. 这是很自然的，局部系就是 l 的磁量子数为零，j 的磁量子数 μ 与 S 的磁量子数 μ_s 相等的坐标系. 所以算子 L 在此系是横的，$L_0=0$，而算子 J 与 S 的縱向分量恆相等 $J_0=S_0$.

由(2.3)式再利用(2.1)及

$$S_+u_{s\mu}(\kappa)=\alpha^s_{\mu+1}u_{s,\,\mu+1}(\kappa),\quad S_-u_{s\mu}(\kappa)=\alpha^s_\mu u_{s,\,\mu-1}(\kappa),\tag{2.4}$$

易求 $\psi^{jM}_s(r;\theta,\varphi)\equiv\sum\limits_\mu f^\mu_j(r)\psi^{jM\mu}_s(\theta,\varphi)$ 的微商如下:

$$\begin{aligned}\nabla\psi^{jM}_s(r;\theta,\varphi)&=\sum_{\mu\nu}u_{1\nu}\nabla^\nu f^\mu_j(r)\sqrt{\frac{2j+1}{4\pi}}D^{j*}_{M\mu}(\varphi,\theta,\gamma)u_{s\mu}(\kappa)=\\&=\sum_{s'\mu}\Big\{C^{s',\mu}_{1,0;s,\mu}\frac{df^\mu_j}{dr}+C^{s',\mu}_{1,1;s,\mu-1}\frac{1}{\sqrt{2}\,r}(\alpha^j_\mu f^{\mu-1}_j-\alpha^s_\mu f^\mu_j)+\\&+C^{s',\mu}_{1,-1;s,\mu+1}(\alpha^j_{\mu+1}f^{\mu+1}_j-\alpha^s_{\mu+1}f^\mu_j)\Big\}\psi^{jM\mu}_{s'}(\theta,\varphi),\end{aligned}\tag{2.5}$$

此式包含了通常的梯度等公式[1,9,21].

例如取 $S=0$, 则 $S'=1$, 就得到梯度公式:

$$\nabla\psi^{jM}_0(r;\theta,\varphi)=\nabla f(r)Y_{jM}(\theta,\varphi)=\frac{df}{dr}\psi^{jM0}_1+\frac{\alpha^j_0}{r}f\psi^{jM1+}_1,\tag{2.5a}$$

取 $S=1$, $S'=0$, 并乘以系数 $-\sqrt{3}$ 就得到散度公式:

$$\nabla\cdot\psi^{jM}_1(r;\theta,\varphi)=\left(\frac{df^0}{dr}+\frac{2}{r}f^0-\alpha^j_0\frac{f^{1+}}{r}\right)Y_{jM},\tag{2.5b}$$

取 $S=1$, $S'=1$, 并乘以系数 $\sqrt{2}\,i$ 就得到旋度公式:

$$\begin{aligned}\nabla\times\psi^{jM}_1(r;\theta,\varphi)&=i\left(\frac{d}{dr}+\frac{1}{r}\right)(f^{1+}\psi^{jM1-}_1+f^{1-}\psi^{jM1+}_1)-\\&-i\alpha^j_0\frac{f^0}{r}\psi^{jM1-}_1+i\alpha^j_0\frac{f^{1-}}{r}\psi^{jM0}_1,\end{aligned}\tag{2.5c}$$

式中 $f^{1\pm}\equiv\frac{1}{\sqrt{2}}(f^{-1}\pm f^{+1})$. 这三式也可以直接算出. 由(2.1)、(2.3)立即得梯度公式. 求散度与旋度时可以利用

$$\nabla^{\pm1}u_{1\mu}=-\frac{S_\mp}{\sqrt{2}\,r}u_{1\mu}=-\frac{1}{r}u_{1,\,\mu\mp1}.$$

因为 $\alpha^1_1=\alpha^1_0=\sqrt{2}$,而 $\alpha^1_{-1}=\alpha^1_2=0$ 则可以由 $u_{1,2}=u_{1,-2}=0$ 考虑在内.

由(2.1)(2.4)易得

$$\begin{aligned}L\psi^{jM}_s(r;\theta,\varphi)&=\sum_{s'\mu}\Big\{-\frac{1}{\sqrt{2}}C^{s',\mu}_{1,1;s,\mu-1}(\alpha^j_\mu f^{\mu-1}_j-\alpha^s_\mu f^\mu_j)+\\&+\frac{1}{\sqrt{2}}C^{s',\mu}_{1,-1;s,\mu+1}(\alpha^j_{\mu+1}f^{\mu+1}_j-\alpha^s_{\mu+1}f^\mu_j)\Big\}\psi^{jM\mu}_{s'}(\theta,\varphi),\end{aligned}\tag{2.6}$$

例如, $S=0$ 时,

$$LfY_{jM}=Lf\psi^{jM}_0=\sqrt{j(j+1)}\,f\psi^{jM1-}_1,\tag{2.6a}$$

此法易推广到二次微商. 例如 $(\nabla\nabla)\psi$ 前的算子可化簡如下:

$$\begin{aligned}\nabla\nabla&=\sum_{\mu\nu}u_{1\mu}\nabla^\mu u_{1\nu}\nabla^\nu=\sum_{s'\nu}\Big[\sum_\mu(C^{s',\mu+\nu}_{1,\mu;1,\nu}u_{s',\mu+\nu}\nabla^\mu\nabla^\nu)-\\&-(C^{s',\nu}_{1,1;1,\nu-1}+C^{s',\nu}_{1,-1;1,\nu+1})\frac{u_{s',\nu}}{r}\nabla^\nu\Big],\end{aligned}$$

取 $S'=0$，并乘以系数 $-\sqrt{3}$，就得到 Laplace 算子：

$$\nabla\cdot\nabla=\sum_{\mu}(-)^{\mu}\nabla^{\mu}\nabla^{-\mu}+\frac{2}{r}\nabla^{0}=\frac{d^2}{dr^2}+\frac{2}{r}\frac{d}{dr}-\frac{1}{2r^2}(L_+L_-+L_-L_+)=$$

$$=\frac{d^2}{dr^2}+\frac{2}{r}\frac{d}{dr}-\frac{1}{r^2}(J^2-J_0^2+S^2-S_0^2-J_+S_--J_-S_+).$$

作用于 ψ 得到

$$(\nabla\cdot\nabla)\psi_s^{jM}(r;\theta,\varphi)=\sum_{\mu}\left[\left\{\frac{d^2}{dr^2}+\frac{2}{r}\frac{d}{dr}-\frac{1}{r^2}[j(j+1)+S(S+1)-2\mu^2]\right\}f_j^{\mu}+\right.$$

$$\left.+\frac{1}{r^2}\alpha_{\mu+1}^{j}\alpha_{\mu+1}^{s}f_j^{\mu+1}+\frac{1}{r^2}\alpha_{\mu}^{j}\alpha_{\mu}^{s}f_j^{\mu-1}\right]\psi_s^{jM\mu},\tag{2.7}$$

$S=0$ 就是通常的 Laplace 算子径向与横向分离变量的形式：

$$\left(\frac{d^2}{dr^2}+\frac{2}{r}\frac{d}{dr}-\frac{j(j+1)}{r^2}\right)f_jY_{jM}$$

$S=1$，则有

$$\nabla\cdot\nabla\psi_1^{jM}(r;\theta,\varphi)=\sum_{\mu}\left(\frac{d^2}{dr^2}+\frac{2}{r}\frac{d}{dr}-\frac{j(j+1)+2-2\mu^2}{r^2}\right)f_j^{\mu}\psi_1^{jM\mu}+$$

$$+\frac{2}{r^2}\alpha_1^jf_j^{1+}\psi_1^{jM0}+\frac{2}{r^2}\alpha_1^jf_j^0\psi_1^{jM1+},\tag{2.7a}$$

$\nabla(\nabla\psi)$ 类型的一般表达式需将三个表示的积展开。也可以直接选用(2.5)式求得。例如选用(2.5c)，即得

$$\nabla\times(\nabla\times\psi_1^{jM}(r;\theta,\varphi))=\left[(\alpha_0^j)^2\frac{f^0}{r^2}-\alpha_0^j\frac{1}{r^2}\frac{d}{dr}(rf^{1+})\right]\psi_1^{jM0}+$$

$$+\left[\alpha_0^j\frac{d}{rdr}f^0-\frac{1}{r}\frac{d^2}{dr^2}(rf^{1+})\right]\psi_1^{jM1+}+\left[(\alpha_0^j)^2\frac{f^{1-}}{r^2}-\frac{1}{r}\frac{d^2}{dr^2}(rf^{1-})\right]\psi_1^{jM1-},\tag{2.8}$$

用此式可将方程

$$\nabla\times(\nabla\times E)-K^2E=0.\tag{2.9}$$

直接分离变量求解，用不着化为附加有无散条件的 Gordan 方程。設 $f^{1+}=f^0=0$，得解

$$E_{jM}(磁)=\frac{K^{3/2}}{R^{1/2}}\zeta_j(Kr)\psi_1^{jM1-}(\theta,\varphi),\tag{2.9a}$$

即 j 阶磁复极輻射，式中 R 为归一化球体的半径。由(2.9)易見 $\nabla\times E$（磁）也为解，用(2.5c)取(2.9a)的旋度后，得

$$E_{jM}(电)=\frac{1}{iK}\nabla\times E_{jM}(磁)=\left(\frac{K}{R}\right)^{\frac{1}{2}}\left[\alpha_0^j\frac{\zeta_j(Kr)}{r}\psi_1^{jM0}(\theta,\varphi)+\right.$$

$$\left.+\frac{1}{r}\frac{d}{dr}(r\zeta_j(Kr))\psi_1^{jM1+}(\theta,\varphi)\right],\tag{2.9b}$$

代入(2.8)易驗証为(2.9)的解。

本文(2.7a)式与 Гельфанд 等的結果([2]，123 頁．(21))一样，只是后两項的符号相反，原因是他們的章动角 θ 是对 x 軸的旋轉，故轉至局部系的轉动角与本文不同，表

示矩陣 D 的相也与本文不同，結果在相邻二磁量子数之間与我們差一正負号. 由于他的散度式(22)与本文(2.5b)最后一項符号也相反，故并不影响求得的解. 順便指出，利用(1.9)式可看出

$$(\nabla\cdot\nabla+K^2)\left(\sum_{\mu} f_j^{\mu}(r)\psi_s^{jM\mu}(\theta,\varphi)\right)=0$$

有 $2S+1$ 組独立无关的解:

$$\sum_{\mu}\zeta_{js}^{m_s\mu}(Kr)\psi_s^{jM\mu}.$$

文献[2]中的两个解及本文的(2.9a),(2.9b)均为 $S=1$ 时 $m_s=\pm1$ 的两解的和及差.

本文(2.5)式与文献[2] 141 頁上(18)式 $f_j^{\mu\pm1}$(該文符号应为 $f^j_{s,\mu\pm1,r}$)前符号相反，理由同前.但該文 $\frac{m}{r}f^{l_0}_{lm\tau'}$ 前漏掉了 $\frac{1}{r}$，此外 $d^{\tau'/\tau}_{l+1,l,m,m}$ 求錯了，应更正为 $-C^{\tau'/\tau}_{l+1,l}(l+2)\cdot\sqrt{(l+1)^2-m^2}$，故其(18)式应作相应的修改. 否則，例如求散度就是錯的.

文献[3]內 Любарский[3] 的(60.12)式中 $\alpha^t u^p$ 項前的符号与本文相反，因为他用固定系的无穷小轉动算子表示局部系中微分算符的式子錯了，球基矢分量的指标也应改变正負号，作相应修改后，除前述相位的不同外与本文一致. 其例中 Maxwell 方程的径向部分及解都是錯誤的. 用更正符号的(60.12)式就能得到正确的方程及解.

三、光子的波函数[21]

由前两节易得光子波函数的各种表示式及性質見表 1.

由动量空間的表可見: $jM\mu ps$ 表象(1)(及其局部坐标式(3))与宇称(4)及偏振(5)有自然的連系. 根据宇称及偏振选择了 $jM\mu ps$ 表象的波函数后就易由(1.8)式得到 $jMls$ 表象的式子(2)，这样就統一解释了縱波、电复极輻射場及磁复极輻射場前的系数，与(例[13])的形式相仿，并指出了其来源. 采用局部坐标式(3)，很容易利用(2.5a)、(2.5b)及球基矢的标积、矢积式得到(6)(7)(8)栏. 由(2)中的系数 C 或(3)的矩陣 D 可看出 $j=0$ 时只有縱波.

位形空間表中 $\Phi_1^{jM\mu}(r;\theta,\varphi)$ 表示 $\psi_1^{jM\mu}(\theta,\varphi)$ 在位形空間中的像. 其径向部分可由动量空間 $jMls$ 表象的固定坐标式求 Fourier 像而得(2)，再用(1.6)式求得

$$\Phi_1^{jM\mu}(r;\theta,\varphi)=\sum_{\mu'}\zeta_{j1}^{\mu\mu'}\psi_1^{jM\mu'},$$

最后用 Bessel 函数的遞推关系写如(3). 也可以先由动量空間的球函数式求像而得(6)，由(6)用(2.5a)、(2.6a)、(2.5c)可得(3)，再与(2)比較就相当于由 Legendre 函数的遞推关系(包含在梯度等公式內)导出了 Bessel 函数(平面波的球面展式的径向函数)的遞推关系. 根据 Bessel 函数的冪級数展式可見 Φ_1^{jM0} 及 Φ_1^{jM1+} 在靜区是縱横波的混合. 但由 Bessel 函数的漸近式可見他們仍有一定的偏振. 由于 r 空間的散度与旋度对应于 K 空間与 $iKu_{10}(\kappa)$ 的标积与矢积，故 K 空間的横波在 r 空間散度为零，K 空間的縱波在 r 空間旋度为零，K 空間二横波与 $u_{10}(\kappa)$ 构成正交系，在 r 空間則二者互为旋度. 由(8)还可見到后二函数为 $\nabla\times\nabla\times A-K^2A=0$ 的解.

表 1

	$jM\mu ps$ 表象 1	$jMls$ 表象 2	局 部 坐 标 式 3	宇称 4	偏振 5	球函数式 6	$u_{1,0}(\kappa)\cdot\psi$ 7	$u_{1,0}(\kappa)\times\psi$ 8
动量空間	$\psi_1^{jM0}(\theta,\varphi)$	$\sum_l (-)C_{j,0;1,0}^{l,0}\psi_{l1}^{jM}(\theta,\varphi)$	$\sqrt{\frac{2j+1}{4\pi}}D_{M0}^{j*}(\varphi,\theta,\gamma)u_{1,0}(\kappa)$	$(-)^j$	縱	$Y_{jM}(\theta,\varphi)u_{1,0}(\kappa)$	$Y_{jM}(\theta,\varphi)$	0
	$\psi_1^{jM1+}(\theta,\varphi)$	$\sum_l {}^{+}C_{j,1;1,-1}^{l,0}\psi_{l1}^{jM}(\theta,\varphi)$	$\sqrt{\frac{2j+1}{8\pi}}\left(D_{M,1}^{j*}(\varphi,\theta,\gamma)u_{1,1}(\kappa)+D_{M,1}^{j*}(\varphi,\theta,\gamma)u_{1,1}(\kappa)\right)$	$(-)^j$	横	$\frac{K\nabla_K Y_{jM}(\theta,\varphi)}{\alpha_0^j}$	0	$i\psi_1^{jM1-}$
	$\psi_1^{jM1-}(\theta,\varphi)$	$\sum_l {}^{-}C_{j,1;1,-1}^{l,0}\psi_{l1}^{jM}(\theta,\varphi)$	$\sqrt{\frac{2j+1}{8\pi}}\left(D_{M,-1}^{j*}(\varphi,\theta,\gamma)u_{1,-1}(\kappa)-D_{M,1}^{j*}(\varphi,\theta,\gamma)u_{1,1}(\kappa)\right)$	$-(-)^j$	横	$\frac{LY_{jM}(\theta,\varphi)}{\alpha_0^j}$	0	$i\psi_1^{jM1+}$
位形空間	位形空間中的像	$jMls$ 表象	$jM\mu ps$ 表 象	宇称	偏振	球函数式	$\nabla\cdot\bar{\psi}$	$\nabla\times\bar{\psi}$
	$\bar{\psi}_1^{jM0}(r;\theta,\varphi)$	$\sum_l (-)C_{j,0;1,0}^{l,0}\xi_l(Kr)\psi_{l1}^{jM}(\theta,\varphi)$	$-i\frac{d\xi_j(Kr)}{Kdr}\psi_1^{jM0}(\theta,\varphi)-i\alpha_0^j\frac{\xi_j(Kr)}{Kr}\psi_1^{jM1+}(\theta,\varphi)$	$(-)^j$	縱	$\frac{1}{iK}\nabla\xi_j(Kr)Y_{jM}(\theta,\varphi)$	$iK\xi_j(Kr)Y_{jM}(\theta,\varphi)$	0
	$\bar{\psi}_1^{jM1+}(r;\theta,\varphi)$	$\sum_l {}^{+}C_{j,1;1,-1}^{l,0}\xi_l(Kr)\psi_{l1}^{jM}(\theta,\varphi)$	$-i\frac{dr\xi_j(Kr)}{Krdr}\psi_1^{jM1+}(\theta,\varphi)-i\alpha_0^j\frac{\xi_j(Kr)}{Kr}\psi_1^{jM0}(\theta,\varphi)$	$(-)^j$	横	$\frac{-1}{K\alpha_0^j}\nabla\times L\xi_j(Kr)Y_{jM}(\theta,\varphi)$	0	$-K\bar{\psi}_1^{jM1-}$
	$\bar{\psi}_1^{jM1-}(r;\theta,\varphi)$	$\sum_l {}^{-}C_{j,1;1,-1}^{l,0}\xi_l(Kr)\psi_{l1}^{jM}(\theta,\varphi)$	$\xi_j(Kr)\psi_1^{jM1-}$	$-(-)^j$	横	$\frac{1}{\alpha_0^j}L\xi_j(Kr)Y_{jM}(\theta,\varphi)$	0	$-K\bar{\psi}_1^{jM1+}$

\# 位形空間各栏省略了共同的因子$\sqrt{\frac{K}{R}}$

四、双光子系的波函数[4]

双光子系的偏振旋量为二光子旋量的直接积 $u_{1\mu_1}(\kappa)u_{1\mu_2}(\kappa)$,(我們选定以质心系中第一光子的运动方向为局部系的极軸 κ, 如分别以各光子的运动方向为局部軸,则此积等于 $-u_{1\mu_1}(\kappa_1)u_{1,-\mu_2}(\kappa_2)$,这里 $\kappa_1=\kappa$; κ_2 与 κ 的极軸、φ 軸反向,θ 軸同向,簡写为 $\kappa_2=-\kappa$),可分解为不可約表示 $u_{1\mu_1}u_{1\mu_2}=\sum\limits_s C^{s,\mu_1+\mu_2}_{1,\mu_1;1,\mu_2}\omega_{s,\mu_1+\mu_2}$. 取其有一定对称性的組合(1), 则从系数 C 的对称性,易見对称态的 s 为 2 或 2,0 的混合,反对称态的 s 为 1(2). 由于在空間反映时 $Pu_{1,\mu}(\kappa)=u_{1,-\mu}(\kappa)$, 则由系数 C 的奇偶性可見 $P\omega_{s\mu}(\kappa)=(-)^s\omega_{s,-\mu}(\kappa)$(5) 只有 $\mu=0$ 的态有确定的內禀宇称,$\mu\neq 0$ 的态反映互变. 将自旋部分乘以相应的空間部分(同行(6)栏),用(1.8)式构成有一定宇称及 j, M, μ 的波函数如(7),或再用(1.7)即得(8). 根据 $u_{1,\pm 1}(\kappa)$ 为横的,$u_{1,0}(\kappa)$ 为纵的,由(1)得(9). 由(5)及(1.8)得宇称(12). Bose 条件要求偏振对称态宇称为正,反对称态宇称为負,由(4)及(12)得(13). j 的最小值

表

二光子的偏振旋量 1	总　旋　量 2	直角坐标式 3	对称性 4	$P\omega$ 5	空　間　部　分 6	
$u_{1,1}u_{1,1}$	$\omega_{2,2}$	$\frac{1}{2}\begin{pmatrix}1&i&0\\i&-1&0\\0&0&0\end{pmatrix}$	对称	$\omega_{2,-2}$	$\sqrt{\frac{2j+1}{4\pi}}D^{j*}_{M,2}(\varphi,\theta,\gamma)$	+
$u_{1,-1}u_{1,-1}$	$\omega_{2,-2}$	$\frac{1}{2}\begin{pmatrix}1&-i&0\\-i&-1&0\\0&0&0\end{pmatrix}$	对称	$\omega_{2,2}$	$\sqrt{\frac{2j+1}{4\pi}}D^{j*}_{M,-2}(\varphi,\theta,\gamma)$	−
$\frac{1}{\sqrt{2}}(u_{1,1}u_{1,-1}+u_{1,-1}u_{1,1})$	$\frac{1}{\sqrt{3}}\omega_{2,0}+\frac{\sqrt{2}}{\sqrt{3}}\omega_{0,0}$	$\frac{-1}{\sqrt{2}}\begin{pmatrix}1&0&0\\0&1&0\\0&0&0\end{pmatrix}$	对称	+	$\sqrt{\frac{2j+1}{4\pi}}D^{j*}_{M,0}(\varphi,\theta,\gamma)$	
$\frac{1}{\sqrt{2}}(u_{1,1}u_{1,-1}-u_{1,-1}u_{1,1})$	$\omega_{1,0}$	$\frac{1}{\sqrt{2}}\begin{pmatrix}0&i&0\\-i&0&0\\0&0&0\end{pmatrix}$	反对称	−	$\sqrt{\frac{2j+1}{4\pi}}D^{j*}_{M,0}(\varphi,\theta,\gamma)$	
$\frac{1}{\sqrt{2}}(u_{1,1}u_{1,0}+u_{1,0}u_{1,1})$	$\omega_{2,1}$	$-\frac{1}{2}\begin{pmatrix}0&0&1\\0&0&i\\1&i&0\end{pmatrix}$	对称	$\omega_{2,-1}$	$\sqrt{\frac{2j+1}{4\pi}}D^{j*}_{M,1}(\varphi,\theta,\gamma)$	+
$\frac{1}{\sqrt{2}}(u_{1,0}u_{1,-1}+u_{1,-1}u_{1,0})$	$\omega_{2,-1}$	$\frac{1}{2}\begin{pmatrix}0&0&1\\0&0&-i\\1&-i&0\end{pmatrix}$	对称	$\omega_{2,1}$	$\sqrt{\frac{2j+1}{4\pi}}D^{j*}_{M,-1}(\varphi,\theta,\gamma)$	−
$\frac{1}{\sqrt{2}}(u_{1,1}u_{1,0}-u_{1,0}u_{1,1})$	$\omega_{1,1}$	$\frac{1}{2}\begin{pmatrix}0&0&-1\\0&0&-i\\1&i&0\end{pmatrix}$	反对称	$-\omega_{1,-1}$	$\sqrt{\frac{2j+1}{4\pi}}D^{j*}_{M,1}(\varphi,\theta,\gamma)$	+
$\frac{1}{\sqrt{2}}(u_{1,0}u_{1,-1}-u_{1,-1}u_{1,0})$	$\omega_{1,-1}$	$\frac{1}{2}\begin{pmatrix}0&0&-i\\0&0&i\\1&-i&0\end{pmatrix}$	反对称	$-\omega_{1,1}$	$\sqrt{\frac{2j+1}{4\pi}}D^{j*}_{M,-1}(\varphi,\theta,\gamma)$	−
$u_{1,0}u_{1,0}$	$\frac{\sqrt{2}}{\sqrt{3}}\omega_{2,0}-\frac{1}{\sqrt{3}}\omega_{0,0}$	$\begin{pmatrix}0&&\\&0&\\&&1\end{pmatrix}$	对称	+	$\sqrt{\frac{2j+1}{4\pi}}D^{j*}_{M,0}(\varphi,\theta,\gamma)$	

(14)从(6)中的D或(8)中的C均易見．綜合(9)、(13)、(14)就得到在一定jM之下满足横波条件及 Bose 条件的态数，与以前各工作[4,5,6,7,8]相符．自旋张量的直角坐标式(3)見文献[4](本文是在局部系中的)．对角綫部分对应于綫偏振平行态；对角綫外部分对应于綫偏振垂直态，故有(10)．由 $u_{1\mu_1}\cdot u_{1\mu_2}=(-)^{\mu_2}\delta_{\mu_1,-\mu_2}$ 可得(11)．

用局部系各式便于学习、理解过去关于双光子系的工作，在个别細节上可进一步澄清．

1．Шапиро[7] 的反对称张量 A_{iK}，相当于本文 $s=1$ 各态(4)，自旋部分为膺矢量 $\omega_{1\mu}$(2)、(5)．其中满足横波条件的 ψ_1^{jM0}，如該文指出，对应的膺矢量是纵的($\omega_{1\mu}$ 的第三分量 ω_{10})．該文根据横波条件选择迹为零的对称张量 S_{iK}(自旋部分为 $\omega_{2\mu}$)态数的討論不够妥当．对称张量 P_{iK} 的自旋部分为 $\frac{1}{2}(u_{10}u_{1\mu}+u_{1\mu}u_{10})$ 的綫性組合．其中含有迹非零的部分 $u_0u_0=\begin{pmatrix}0&0&0\\0&0&0\\0&0&1\end{pmatrix}$．因此，舍去 P_{iK} 后剩下的 $S_{iK}=S'_{iK}-P_{iK}$ 并不再符合迹为零

2

$jM\mu ps$ 表象 $(j,M;s_1,\mu_1;s_2,\mu_2)$ 7	$jMls$ 表象 8	偏振 9	偏振关系 綫 10	偏振关系 圓 11	宇称 12	Bose条件允許的 j 13	j 可以取的最小值 14
ψ_2^{jM2+}	$\sum_l +C^{l,0}_{j,2;2,-2}\ \psi^{jM}_{l2}$	横	∥及⊥	正交,反向	$(-)^j$	偶	2
ψ_2^{jM2-}	$\sum_l -C^{l,0}_{j,2;2,-2}\ \psi^{jM}_{l2}$	横	∥及⊥	正交,反向	$-(-)^j$	奇	2
$\sqrt{\frac{2}{3}}\psi_2^{jM0}+\sqrt{\frac{2}{3}}\psi_0^{jM0}$ $(\psi^{jM}_{1,\pm1;1,\mp1})$	$\frac{\sqrt{2}}{\sqrt{3}}\psi^{jM}_{j0}+\frac{1}{\sqrt{3}}\sum_l C^{l,0}_{j,0;2\,0}\psi^{jM}_{l2}$	横	∥	同 向	$(-)^j$	偶	0
ψ_1^{jM0}	$\sum_l C^{l,0}_{j,0;1,0}\ \psi^{jM}_{l1}$	横	⊥	同 向	$-(-)^j$	偶	0
ψ_2^{jM1+}	$\sum_l +C^{l,0}_{j,1;2,-1}\psi^{jM}_{l2}$	纵横	⊥	正 交	$(-)^j$	偶	1
ψ_2^{jM1-}	$\sum_l -C^{l,0}_{j,1;2,-1}\psi^{jM}_{l2}$	纵横	⊥	正 交	$-(-)^j$	奇	1
ψ_1^{jM1+}	$\sum_l +C^{l,0}_{j,1;1,-1}\psi^{jM}_{l1}$	纵横	⊥	正 交	$-(-)^j$	偶	1
ψ_1^{jM1-}	$\sum_l -C^{l,0}_{j,1;1,-1}\psi^{jM}_{l1}$	纵横	⊥	正 交	$(-)^j$	奇	1
$\sqrt{\frac{2}{3}}\psi_2^{jM0}-\sqrt{\frac{1}{3}}\psi_0^{jM0}$ $(\psi^{jM}_{1,0;1,0})$	$\sqrt{\frac{2}{3}}\sum_l C^{l,0}_{j,0;2,0}\ \psi^{jM}_{l2}-\sqrt{\frac{1}{3}}\psi^{jM}_{j0}$	纵	∥	同 向	$(-)^j$	偶	0

的条件. 此外,他仅指出标量 S 是偏振平行的偶态,未説明該文表 3 中的偏振平行的偶态究竟是純 S 态,还是与 S_{iK} 的混合态,也沒討論純 S 是否横的. 我們認为較严格的方法,是从六个对称张量 S_{iK} 与 S 中,先选择偶态如文献[8]($j = 2n$ 时为 $\psi_2^{jM2}+$, $\psi_2^{jM1}+$, ψ_2^{jM0}, ψ_0^{jM0}; $j = 2n + 1$ 时为 $\psi_2^{jM2}-$, $\psi_2^{jM1}-$. (4), (13)), 再去掉不符合横波条件的 P_{iK} (含 $u_{1,0}$ 的各态(1), 偶 j 时为 $\psi_2^{jM1}+$, $\sqrt{\frac{2}{3}}\psi_2^{jM0} - \sqrt{\frac{1}{3}}\psi_0^{jM0}$, 奇 j 时为 $\psi_2^{jM1}-$), 就可得他的表中的偶态数($j = 0, 1$ 的討論相仿, 請参照(14)). 文献[4]中已指出 Шапиро S_{iK} (純) 态綫偏振垂直的結論是錯的. Шапиро 認为 S 态光子圓偏振方向相同, S_{iK} 态方向相反也不够严密. 应更正为:純 S_{iK} 横波的二光子圓偏振方向相反而正交, $E_j^1 \cdot E_j^2 = 0$. 但 S_{iK} 中的 ψ_2^{jM0} 与 S 共同构成的横波的圓偏振相同 ($E_j^1 \cdot E_j^2 \neq 0$)(11).

2. Ахиезер 及 Берестецкий[8] 的反对称张量贋矢量形式: $\mathbf{F} = Y_{jM}^{(-1)}(n)$ 即本文 ψ_1^{jM0}. 用对称张量 $f_{\alpha_1\alpha_2}$ 构成的矢量 $G = f_{\alpha\beta}n_\beta$ 中三个不为零的即 $\psi_2^{jM1\pm} \cdot u_{1,0} = \psi_1^{jM1\pm}$ 及 $\psi_{1,0;1,0}^{jM0} \cdot u_{1,0} = \psi_1^{jM0}$, 式中左側的 ψ 是双光子系波函数, 右側为单光子波函数. 在 $j = 2n + 1$ 时, G 中仅 $\psi_1^{jM1}-$ 的 l 是偶的,舍去相应的含縱分量的双光子 $\psi_2^{jM1}-$,剩下符合横波条件的偶态(参看上节的討論)是 $\psi_2^{jM2}-$, $j = 2n$, 1, 0 时的討論相仿.

3. Широков[6] 得的解 $\psi^{(0)}$, $\psi^{(1)}$ 并分离了自旋部分与空間部分,与本文 $\psi_{1,\pm1;1\mp1}^{jM}$ 及 ψ_1^{jM0} 的(4)乘(6)的形式相当. 最后求迹为零的 $\psi^{(2)}$ 时, 利用横波条件引进的 η、η_1 实質上即相当于变到局部系分离出的空間部分. 的确, 由我們的 $\psi_2^{jM2\pm}$ 的局部系式, 将 $\omega_{2,\pm2}(\kappa)$ 变为 $\sum\limits_{m_s}\omega_{2,m_s}(K)D_{m_s,\pm2}^2(\varphi, \theta, \gamma)$, 依次求出他的 $\psi_{ij}^{(2)}$; x_1、$x_m^\pm$、$x_{3m}^\pm$; η, η_1 就可看出 η, η_1 分别与 $D_{m,-2}^j \pm D_{m,+2}^j$ 的偶奇部对应. 如果一上来就用局部系, 也可以簡明地看出这点, 这时 $\psi_{3i}^{(2)} = \psi_{i3}^{(2)} = 0$, $x_m^\pm$ 分别与 $D_{m\mp2}^{j*}$ 对应,在他的横波条件及其推論中[他的横波条件的第二、三式应更正为 $(x_m^+ - x_m^-) + (x_{3m}^+ - x_{3m}^-)\operatorname{ctg}\theta = 0$ 及 $(x_m^+ + x_m^- - 2x) + (x_{3m}^+ + x_{3m}^-)\operatorname{ctg}\theta = 0$, 并且要用到迹为零的条件] 取 $\theta = 0$ 立即得到 η、η_1 分别相当于 $D_{m,-2}^{j*} \pm D_{m,2}^{j*}$. 这样, 他設 $m = 0$ 时得到的方程 $-\Delta_\theta\eta + \frac{4}{\sin^2\theta}\eta = M_2'^2\eta$ 却是磁量子数为 2 型的方程就容易理解了,这正表示在局部系中的量子数为 ±2. (即 $-\Delta_\theta\eta - \frac{1}{\sin^2\theta}\left(\frac{\partial^2}{\partial\varphi^2} - 2\cos\theta\frac{\partial^2}{\partial\varphi\,\partial r} + \frac{\partial^2}{\partial r^2}\right)\eta = j(j+1)\eta$ 在 $\frac{\partial}{\partial\varphi}\eta = 0$ $\frac{\partial}{\partial r}\eta = \pm 2i\eta$ 时的形式). 其解即 $D_{0,\pm2}^{j*}$ 的綫性組合.

4. 楊振宁[5] 的 $a_+^{R*}\psi_{00\cdots}$、$a_+^{L*}\psi_{00\cdots}$、$a_-^{R*}\psi_{00\cdots}$、$a_-^{L*}\psi_{00\cdots}$ 的自旋部分, 分别与我們的 $u_{1,1}(K_1) = u_{1,1}(K)$, $u_{1,-1}(K_1) = u_{1,-1}(K)$, $u_{1,1}(K_2) = u_{1,1}(-K) = -u_{1,-1}(K)$ 及 $u_{1,-1}(K_2) = -u_{11}(K)$ 对应. 故如認为二光子可区别,则 ψ^{RL}、ψ^{LR}、$\psi^{RR} + \psi^{LL}$、$\psi^{RR} - \psi^{LL}$ 依次与本文 ψ_2^{j22}、$\psi_2^{j,-2,-2}$、$\psi_{1\pm1;1\mp1}^{j,0,0}$、$\psi_1^{j00}$ 对应(1). 易見这时繞 K 軸的轉动性 R_φ 如該文. 但楊振宁的反映变换包括粒子交换在內. 的确 $P\psi^{RL} = Pa_+^{R*}a_-^{L*}\psi_{00\cdots} = a_-^{L*}a_+^{R*}\psi_{00\cdots}$ 必須利用 Bose 子的对易性才能化为 $a_+^{R*}a_-^{L*}\psi_{00\cdots} = \psi^{RL}$. 如不考虑光子的全同性, 则与 ψ^{RL} 对应的 ψ_2^{j22} 經反映变为 $(-)^j\psi_2^{j2,-2}$ 不是宇称本征态. 必須再交换粒子[这时 $u_{1,\mu_1}(K)u_{1,\mu_2}(K) \to u_{1,\mu_2}(-K)\cdot u_{1,\mu_1}(-K) = u_{1,-\mu_2}(K)u_{1,-\mu_1}(K)$, 因粒子的自旋 μ_1、μ_2 交换, 且軸 K 的方向也改变为指向新的粒子 1 (原来的粒子 2)的运动方向故要反号. 空間部分則由于K的改变而由 $D_{M\mu}^{j*}$ 变为 $(-)^jD_{M,-\mu}^{j*}$] 才又变为 ψ_2^{j22}. 粒子交换相当于空間反映与自旋交换的积. 故楊振宁的

包含粒子交换的反映变换实质上与自旋交换等价，而楊的結論应改为 ψ^{RL} 及 ψ^{LR} 是对称态，但无确定宇称. 必須将 ψ^{RL} 及 ψ^{LR} 相混构成对二粒子对称的波函数，例如 $\sqrt{\frac{2j+1}{8\pi}}$ $[D^{j*}_{2,2}(\varphi_{12}, \theta_{12}, \gamma_{12})u_{1,1}(K_1)u_{1,-1}(K_2) + D^{j*}_{2,2}(\varphi_{21}, \theta_{21}, \gamma_{21})u_{1,1}(K_2)u_{1,-1}(K_1)]$，这里 $\varphi_{12}, \theta_{12}, \gamma_{12}$ 及 $\varphi_{21}, \theta_{21}, \gamma_{21}$ 分别为从 K_1 系及 K_2 系观察的立体角. 取 $K = K_1 = -K_2$ 卽化为 $-\sqrt{\frac{2j+1}{8\pi}}[D^{j}_{2,2}(\varphi, \theta, \gamma)\omega_{2,2}(K) + (-)^{j}D^{j*}_{2,-2}(\varphi, \theta, \gamma)\omega_{2,-2}(K)]$ 才是宇称有确定值 (+) 的 $\psi^{j,2,2\pm}_{2}$. 轉到其他坐标系就化为 $\psi^{j,M,2\pm}_{2}$. 同样，未考慮光子全同性时平面波 $\psi^{LL} \pm \psi^{RR}$ 的各球面分波宇称为 $\pm(-)^{j}$，对二光子对称化后，奇 j 分波为零，只剩下偶 j 分波，才有了确定的宇称.

III 节的选择律 ii 也必須考慮 Bose 条件后方成立. 如二光子可区别，則 $\psi^{RR} \pm \psi^{LL}$ 态繞 x 軸轉 π 角后二光子对换，并非不变的. 的确这时 $D^{j*}_{00}(\varphi,\theta,\gamma)$ 变为 $\sum_{M'} D^{j*}_{0M'}\left(-\frac{\pi}{2}, \pi, \frac{\pi}{2}\right) D^{j*}_{M'0}(\varphi, \theta, \gamma) = (-)^{j}D^{j*}_{00}(\varphi, \theta, \gamma)$，自旋的局部式不变. 故 ψ^{j00}_{s} 变为 $(-)^{j}\psi^{j00}_{s}$. 用前一段的方法可見与交换粒子后的結果相同. 可見 R_ξ 不一定为 1. 楊的結論应改为 R_ξ 与光子的交换等价，只有考慮了 Bose 条件，构成对二粒子对称的 $\psi^{RR} \pm \psi^{LL}$ 禁戒了奇 j 的存在后 R_ξ 才等于 1.

5. 本文卽在汪容的[4]启发下得到的，将 (8) 栏 $jMls$ 表象中符合选择律的前四个用 (1.5) 写为固定极軸式卽与之相同.

五、半自旋粒子的波函数[9,22]

$\psi^{jM\frac{1}{2}}_{\frac{1}{2}}$ 及 $\psi^{jM,-\frac{1}{2}}_{\frac{1}{2}}$ 具有一定的螺旋性，可作为双分量反中微子及中微子的波函数. 二者依 Lorentz 羣的互为共軛的二表示变换[3]，在空間反映时互变，沒有一定的宇称. 通常分别将二者对称及反对称組合作为 Dirac 4 分量波函数的“大”“小”分量. 这时螺旋性算子 γ_5 不再是对角形的，但內禀宇称算子 β 是对角形的. 表 3 中列的是空間部分的宇称.

在局部坐标系中易求得常用的式子：

$$\sigma_r \equiv \frac{\sigma \cdot \mathbf{r}}{r} = 2S_0, \tag{5.1}$$

$$\sigma \cdot L = \sigma \cdot (J - S) = S_+J_- + S_-J_+ - 1, \tag{5.2}$$

$$r\sigma \cdot \nabla = \sigma_r(r\nabla_0 + 1) + S_+J_- - S_-J_+, \tag{5.3}$$

式中 σ 为 Pauli 矩陣，$\sigma = 2S$. $S_{\pm,0}J_\pm$ 都是局部系中的. 由 (5.2) 可見 $\sigma \cdot L + 1$ 是横的，因此易由 σ 的反对易性得到

$$\sigma_r(\sigma \cdot L + 1)\sigma_r = -(\sigma \cdot L + 1). \tag{5.4}$$

(5.2)与(5.1)相乘代入(5.3)卽得

$$\sigma \cdot \nabla = \sigma_r \frac{\partial}{\partial r} - \frac{1}{r}\sigma_r\sigma \cdot L, \tag{5.5}$$

由前三式得表中后三栏.

$\sigma_r(\frac{\partial}{\partial r} + \frac{1}{r}) - \frac{1}{r}\sigma_r(\sigma \cdot L + 1)$

表

$jM\mu ps$ 表象	局部坐标式	$jMls$ 表象
$\psi_{\frac{1}{2}}^{jM\frac{1}{2}}$	$\sqrt{\frac{2j+1}{4\pi}}\,D^{j*}_{M,\frac{1}{2}}u_{\frac{1}{2}}$	$\frac{1}{\sqrt{2}}(\psi^{jM}_{j-\frac{1}{2},\frac{1}{2}}-\psi^{jM}_{j+\frac{1}{2},\frac{1}{2}})$
$\psi_{\frac{1}{2}}^{jM,-\frac{1}{2}}$	$\sqrt{\frac{2j+1}{4\pi}}\,D^{j*}_{M,-\frac{1}{2}}u_{-\frac{1}{2}}$	$\frac{1}{\sqrt{2}}(\psi^{jM}_{j-\frac{1}{2},\frac{1}{2}}+\psi^{jM}_{j+\frac{1}{2},\frac{1}{2}})$
$\psi_{\frac{1}{2}}^{jM\frac{1}{2}+}$	$\sqrt{\frac{2j+1}{8\pi}}\,(D^{j*}_{M,-\frac{1}{2}}u_{-\frac{1}{2}}+D^{j*}_{M,\frac{1}{2}}u_{\frac{1}{2}})$	$\psi^{jM}_{j-\frac{1}{2},\frac{1}{2}}$
$\psi_{\frac{1}{2}}^{jM\frac{1}{2}-}$	$\sqrt{\frac{2j+1}{8\pi}}\,(D^{j*}_{M,-\frac{1}{2}}u_{-\frac{1}{2}}-D^{j*}_{M,\frac{1}{2}}u_{\frac{1}{2}})$	$\psi^{jM}_{j+\frac{1}{2},\frac{1}{2}}$

六、局部坐标系中的散射幅

将散射幅依 $jMls$ 表象的旋量球函数展开

$$M=\sum_{jM}\frac{2\pi}{iK}\sum_{\substack{ls\\l's'}}\psi^{jM}_{l's'}(\theta',\varphi')R^{j}_{l's';ls}\psi^{jM*}_{ls}(\theta,\varphi), \tag{6.1}$$

式中带撇的量属于終态,无撇的量属于始态. 取极軸沿入射粒子方向,用(1.5)表 ψ^{jM}_{ls} 即通常在固定系中的散射幅. 也可以用 (1.6) 式求局部系中的展式. 但这时最方便的是采用 j、M、μ、p、s 表象:

$$M=\sum_{jM}\frac{2\pi}{iK}\sum_{\substack{|\mu|ps\\|\mu'|p's'}}\psi^{jM\mu'p'}_{s'}(\theta',\varphi')R^{j}_{s'\mu'p';s\mu p}\psi^{jM\mu p}_{s}(\theta,\varphi), \tag{6.2}$$

用(1.8)可得

$$R^{j}_{l's';ls}=\sum_{\substack{|\mu|p\\|\mu'|p'}}{}^{p'}C^{l',0}_{j,\mu';s',-\mu'}R^{j}_{s'\mu'p';s\mu p}{}^{p}C^{l,0}_{j,\mu;s,-\mu}, \tag{6.3}$$

宇称守恆要求:

$$\begin{aligned} &p=p', \quad R_{p,-p}=0, \quad \text{如果} \quad p_s=p'_s,\\ &p=-p', \quad R_{p,p}=0, \quad \text{如果} \quad p_s=-p'_s. \end{aligned} \tag{6.4}$$

代入(6.3),易見(6.4)即通常限制 l 的宇称选择律: $l+l'$ 为偶(奇)数,如果散射道的宇称相同(反). 选球极軸沿入射粒子方向,可由(6.2)求得終态的密度矩陣:

$$\rho'_{s'_2\mu'_2;s'_1\mu'_1}S_p\rho'=S_pM\rho M^+=$$

$$=\sum_{\substack{L,j_1,j_2\\\mu_1,\mu_2\\s_1,s_2}}\frac{(2j_1+1)(2j_2+1)}{4K^2}(-)^{\mu'_1-\mu_1}C^{L,\mu'_2-\mu'_1}_{j_1,-\mu'_1;j_2,\mu'_2}C^{L,\mu_2-\mu_1}_{j_1,-\mu_1;j_2,\mu_2}R^{j_2}_{s'_2\mu'_2;s_2\mu_2}$$

$$\rho_{s_2\mu_2;s_1\mu_1}R^{j_1*}_{s_1\mu_1;s'_1\mu'_1}D^{L*}_{\mu_2-\mu_1;\mu'_2-\mu'_1}(\varphi',\theta',r'), \tag{6.5}$$

設始态是非极化的,終态未观察极化,则可求得角分布:

3

螺 旋 性	空 間 反 映	σ_r	$\frac{1}{\alpha^j_{1/2}}(\sigma\cdot L+1)$	$\frac{1}{\alpha^j_{1/2}}(r\sigma\cdot\nabla-\sigma_r)$
右		$+$	$\psi^{jM,-\frac{1}{2}}_{\frac{1}{2}}$	$\psi^{jM,-\frac{1}{2}}_{\frac{1}{2}}$
左		$-$	$\psi^{jM\frac{1}{2}}_{\frac{1}{2}}$	$-\psi^{jM\frac{1}{2}}_{\frac{1}{2}}$
	$(-)^{j-\frac{1}{2}}$	$-\psi^{jM\frac{1}{2}-}_{\frac{1}{2}}$	$+$	$\psi^{jM\frac{1}{2}-}_{\frac{1}{2}}$
	$(-)^{j+\frac{1}{2}}$	$-\psi^{jM\frac{1}{2}+}_{\frac{1}{2}}$	$-$	$-\psi^{jM\frac{1}{2}+}_{\frac{1}{2}}$

$$\frac{d\sigma}{d\Omega}=\sum_{Lj_1j_2|\mu||\mu'|}\frac{(2j_1+1)(2j_2+1)}{4K^2(2S_a+1)(2S_b+1)}(-)^{\mu'-\mu}\frac{1}{2}\left(C^{L,0}_{j_1,-\mu';j_2,\mu'}+\pi C^{L,0}_{j_1,\mu';j_2,-\mu'}\right)$$
$$\frac{1}{2}\left(C^{L,0}_{j_1,-\mu;j_2,\mu}+\pi C^{L,0}_{j_1,\mu;j_2,-\mu}\right)P_L(\cos\theta')\sum_{p_1p_2ss'}\left(R^{j_2}_{s'\mu'p_2';s\mu p_2}R^{j_1*}_{s\mu p_1;s'\mu'p_1'}\right),\tag{6.6}$$

式中

$$\pi=p_1p_2(-)^{2\mu}=p_1'p_2'(-)^{2\mu'}.\tag{6.7}$$

符号法则(6.7)包含选择律：l_1+l_2+L 及 $l_1'+l_2'+L$ 为偶数．前式中 S_a，S_b 为入射粒子 a，b 的自旋．

如果变换到 $jMls$ 表象，对磁量子数 μ 求和，将 C 系数积的和用 Z 系数表之，結果与寻常从固定系出发得到的式子相同[10]，不过我們是从局部系出发的．

例：$S_a=0$，$S=S'=S_b=\frac{1}{2}$；这时 jp 即已为完整量子数，故有

$$\frac{d\sigma}{d\Omega}=\sum_{j_1j_2}\frac{(2j_1+1)(2j_2+1)}{8K^2}\left(C^{L,0}_{j_1,-\frac{1}{2};j_2,\frac{1}{2}}\right)^2\Big[\frac{1}{2}(1-(-)^{j_1+j_2+L})(R^{j_1+}R^{j_2+*}+$$
$$+R^{j_1-}R^{j_2-*})+\frac{1}{2}(1+(-)^{j_1+j_2+L})(R^{j_1+}R^{j_2-*}+R^{j_1-}R^{j_2+*})\Big]P_2(\cos\theta').\tag{6.6a}$$

变换到 $jMls$ 表象的系数 $^{\pm}C^{l,0}_{j,\frac{1}{2};\frac{1}{2},-\frac{1}{2}}$ 等于零或 1，故 $R^{j\pm}$ 分别与 $l=j\mp\frac{1}{2}$ 的 R^j_l 相等．而 $\sqrt{(2j_1+1)(2j_2+1)}\ C^{L,0}_{j_1,-\frac{1}{2};j_2,\frac{1}{2}}$ 在 j_1+j_2+L 为奇数时，即 $Z\left(j_1-\frac{1}{2},j_1,j_2-\frac{1}{2},j_2;\frac{1}{2},L\right)=Z\left(j_1+\frac{1}{2},j_1,j_2+\frac{1}{2},j_2;\frac{1}{2},L\right)$，在 j_1+j_2+L 为偶数时即 $Z\left(j_1-\frac{1}{2},j_1,j_2+\frac{1}{2},j_2;\frac{1}{2},L\right)=-Z\left(j_1+\frac{1}{2},j_1,j_2-\frac{1}{2},j_2;\frac{1}{2},L\right)$．

自旋 S 粒子衰变为半自旋粒子[23] $S=S_a$（无 b 粒子）．$S'=\frac{1}{2}$．质心系中 $J=S$，μ 即总磁量子数 M，R 与之无关．故只有 R^s_+ 及 R^s_- 分别为宇称守恆或不守恆部分：

$$\frac{d\sigma}{d\Omega}=\sum_{L,\mu_2,\mu_1}\frac{2S+1}{4K^2}C^{L,\mu_2-\mu_1}_{s,-\mu_1;s,\mu_2}C^{L,0}_{s,-\frac{1}{2};s,\frac{1}{2}}\left[\frac{1}{2}(1-(-)^{L+2S})(|R^s_+|^2+|R^s_-|^2)+\right.$$

$$\left.+\frac{1}{2}(1+(-)^{L+2S})(R^s_+R^{s*}_-+R^s_-R^{s*}_+)\right]\rho_{\mu_2\mu_1}D^{L*}_{\mu_2-\mu_1,0}(\varphi',\theta',\gamma'). \qquad (6.6b)$$

光子的散射，这时散射系的总动量矩 j 及其在固定系的磁量子数 M、局部系的磁量子数 μ、与光子的字称 p_γ 构成完整量子数，故有

$$\frac{d\sigma}{d\Omega}=\sum_{\substack{L,j_1,\mu_1,p\gamma_1\\ j_2,\mu_2,p\gamma_2}}\frac{(2j_1+1)(2j_2+1)}{8K^2(2S_a+1)}(-)^{\mu'-\mu}C^{L,0}_{j_1,-\mu';j_2,\mu'}C^{L,0}_{j_1,-\mu;j_2,\mu}\times$$

$$\times R^{j_1}_{\mu_1'\mu_1p\gamma_1}R^{j_2*}_{\mu_2'\mu_2p\gamma_2}P_L(\cos\theta'). \qquad (6.6c)$$

用 $C^{j\gamma,1}_{j,\mu;sa,1-\mu}+p_\gamma C^{j\gamma,-1}_{j,\mu;sa,1-\mu}$（这里 j、S_a、j_γ 为散射系总动量矩，核自旋矩及光子总动量矩．磁量子数都是局部系里的）作为变换系数就可化为[13]中的形式．

由(6.6)式可以看出 R_p 与 R_{-p} 变换并不改变角分布．这就是南氏相变换在 $jM\mu sp$ 表象中的形式．存在此种不确定性的原因是：只测角分布、不测极化时不能察覚繞局部軸的轉动．在 μ 为半整数时，取此轉角 $\gamma=\pi$，则 $\psi^{jM\mu}_s$ 变为 $\psi^{xjM\mu}_s=e^{-i\pi\mu}\psi^{jM\mu}_s$（用附标^{x}表示南氏变换后的量)，而 $\psi^{xjM\mu p}_s=e^{-i\pi\mu}\psi^{xjM\mu,-p}_s$，故 $R^{xp}_{\mu'\mu}=e^{-i\pi(\mu'-\mu)}R^{-p}_{\mu'\mu}$．

由(6.6)式还可以看出，只有最后的括号与 S、S' 有关．故在非弹性散射 S 非守恆量时，还适于采用 $\psi^{jM}_{s_a\mu_a;s_b\mu_b}$ 表象． 这时可以分别繞局部軸轉动粒子 a、b 而不改变角分布．同上有 $\psi^{xjM}_{s_a\mu_a;s_b\mu_b}=e^{-i\gamma_a\mu_a-i\gamma_b\mu_b}\psi^{jM}_{s_a\mu_a;s_b\mu_b}$，$\gamma_a$ 或(和) γ_b 取值 π 为有物理意义的变换．将此式沿有一定 S 的态展开化为 $\psi^{jM\mu}_s$ 表象，再用(1.6)化为 ψ^{jM}_{ls} 表象，就有

$$\psi^{xjM}_{l^xs^x}=\sum_{ls}\left(\sum_{\mu_a\mu_b}(-)^{s^x-s}C^{l^x0}_{j,-\mu;s^x,\mu}C^{s^x\mu}_{s_a,\mu_a;s_b,\mu_b}e^{-i(\gamma_a\mu_a+\gamma_b\mu_b)}C^{s\mu}_{s_a,\mu_a;s_b,\mu_b}C^{l0}_{j,-\mu;s,\mu}\right)\psi^{jM}_{ls},$$

于是我們得到不改变角分布的相变换 $R^x=LRL^+$ 的变换矩陣 L 的普遍式：

$$L_{l^xs^xls}=\sum_{\mu_a\mu_b}(-)^{s^x-s}C^{l^x,0}_{j,-\mu;s^x,\mu}C^{s^x,\mu}_{s_a,\mu_a;s_b;\mu_b}e^{-i(\gamma_a\mu_a+\gamma_b\mu_b)}C^{s,\mu}_{s_a,\mu_a;s_b,\mu_b}C^{l,0}_{j,-\mu;s,\mu}, \qquad (6.8)$$

在 $\gamma_a=\gamma_b=\gamma$ 时，簡化为

$$L_{l^xs^xls}=\sum_{\mu}C^{l^x,0}_{j,-\mu;s^x,\mu}e^{-i\gamma\mu}C^{l,0}_{j,-\mu;s,\mu}\delta_{ss^x}. \qquad (6.8a)$$

例：π-N 散射，$S_a=0$，$S=S^x=S_b=\frac{1}{2}$．取 $\gamma=\pi$ 即南氏变换[14]．

N-N 散射，$S_a=S_b=\frac{1}{2}$，取 $\gamma_a=\pi$，$\gamma_b=0$；或 $\gamma_a=0$，$\gamma_b=\pi$；或 $\gamma=\gamma_a=\gamma_b=\pi$ 即[16]中的变换．

N-D 散射，$S_a=\frac{1}{2}$，$S_b=1$．取 $\gamma=\gamma_a=\gamma_b=\pi$ 即文献[17]中弹性散射的情况．取 $\gamma_a=0$，$\gamma_b=\pi$ 或 $\gamma_a=\pi$，$\gamma_b=0$ 即非弹性散射的情况．

π-D 散射，$S_a=0$，$S=S^x=S_b=1$．取 $\gamma=\pi$，这时在 $jM\mu s$ 表象中只是改变 $\psi^{jM,\pm1}_s$ 相对于 ψ^{jM0}_s 的符号，但 $jMls$ 表象中得到另一組新相移．

在有光子参与的散射过程中，取 $\gamma=\pi$，光子的宇称不改变，而且又沒有 $\mu=0$ 的縱分量，因此得不到新的相移．但若取 $\gamma=\frac{\pi}{2}$就可以改变光子的宇称，得到有物理意义（因无 $\mu=0$ 态）的新相移：即电、磁复极場的相彼此交換．

在粒子轉变过程中，可以采用同时改变終态和始态的宇称的变換而不致改变角分布．

文献[15]中指出在南氏变換下，反冲核的垂直于反应面的极化要改变符号，现在也易理解，这正是繞局部軸轉动 π 角的結果．

七、电子跃迁矩陣的分离变量及其角部分的計算

用第二节的方法分离算子的径向及横向部分，再将三个 D 函数的积的积分表示为两个 C 系数[19]，就可得到

$$\left(j_2\frac{1}{2}\,m_2p_2\right|\psi_1^{JM0}(r;\theta,\varphi)\cdot\mathbf{r}_0\left|j_1\frac{1}{2}\,m_1p_1\right)$$

$$\equiv\left(f_{j_2}^{p_2}(r)\psi_{\frac{1}{2}}^{j_2m_2\frac{1}{2}p_2^*}(\theta,\varphi)\middle|f_J^0(r)Y_{JM}(\theta,\varphi)\middle|f_{j_1}^{p_1}(r)\psi_{\frac{1}{2}}^{j_1m_1\frac{1}{2}p_1}(\theta,\varphi)\right)$$

$$=(-)^{M+m_2+\frac{1}{2}}\sqrt{\frac{(2j_1+1)(2j_2+1)}{4\pi(2J+1)}}\,C^{J,-M}_{j_1,m_1;j_2,-m_2}\frac{1}{2}[1-p_1p_2(-)^{j_1+j_2+J}]\times$$

$$\times\, C^{J,0}_{j_1,-\frac{1}{2};j_2,\frac{1}{2}}\int f_{j_2}^{p_2}f_J^0f_{j_1}^{p_1}r^2dr, \tag{7.1}$$

求角变量积分时避免了通常[1]先写作簡化矩陣 $\left(j_2l_2\,\frac{1}{2}\,\|Y_J\|j_1l_1\,\frac{1}{2}\right)$ 再表为 Racah 系数与 $(l_2\|Y_J\|l_1)$，或与之相同的由 j_1、j_2 的耦合为 J 改为由 l_1l_2 耦合为 J[9] 各方法的周折．

$$\left(j_2\frac{1}{2}\,m_2p_2\right|\psi_1^{JM0}(r;\theta,\varphi)\cdot\sigma\left|j_1\frac{1}{2}\,m_1p_1\right)=$$

$$=(-)^{M+m_2-\frac{1}{2}}\sqrt{\frac{(2j_1+1)(2j_2+1)}{4\pi(2J+1)}}\,C^{J,-M}_{j_1,m_1;j_2,-m_2}\frac{1}{2}[1+p_1p_2(-)^{j_1+j_2+J}]\times$$

$$\times\, C^{J,0}_{j_1,-\frac{1}{2};j_2,\frac{1}{2}}\int f_{j_2}^{p_2}f_J^0f_{j_1}^{p_1}r^2dr, \tag{7.2}$$

$$\left(j_2\frac{1}{2}\,m_2p_2\right|\psi_1^{JM1p}(r;\theta,\varphi)\cdot\sigma\left|j_1\frac{1}{2}\,m_1p_1\right)=$$

$$=(-)^{M+m_2-\frac{1}{2}}\sqrt{\frac{(2j_1+1)(2j_2+1)}{4\pi(2J+1)}}\,C^{J,-M}_{j_1,m_1;j_2,-m_2}\frac{1}{2}\,[p_1+pp_2(-)^{j_1+j_2+J}]\times$$

$$\times\, C^{J,1}_{j_1,\frac{1}{2};j_2,\frac{1}{2}}\int f_{j_2}^{p_2}f_J^pf_{j_1}^{p_1}r^2dr, \tag{7.3}$$

$$\left(j_2\frac{1}{2}\,m_2p_2\right|\psi_1^{JM1p}(r;\theta,\varphi)\cdot L\left|j_1\frac{1}{2}\,m_1p_1\right)=$$

$$=(-)^{M+m_2-\frac{1}{2}}\sqrt{\frac{(2j_1+1)(2j_2+1)}{4\pi(2J+1)}}\,C^{J,-M}_{j_1,m_1;j_2,-m_2}\frac{1}{4}(1+p_1p_2p(-)^{j_1+j_2+J})\times$$

$$\times\,[(p_1\cdot\alpha_{\frac{1}{2}}^{j_1})\;C^{J,1}_{j_1,\frac{1}{2};j_2,\frac{1}{2}}-p_1p_2\alpha_{\frac{3}{2}}^{j_1}C^{J,1}_{j_1,\frac{3}{2};j_2,-\frac{1}{2}}]\int f_{j_2}^{p_2}f_J^pf_{j_1}^{p_1}r^2dr, \tag{7.4}$$

$$\left(j_2\frac{1}{2}m_2p_2\middle|\psi_1^{JM1p}(r;\theta,\varphi)\cdot r\nabla\middle|j_1\frac{1}{2}m_1p_1\right)=$$

$$=(-)^{M+m_2+\frac12}\sqrt{\frac{(2j_1+1)(2j_2+1)}{4\pi(2J+1)}}\,C^{J,-M}_{j_1,m_1;j_2,-m_2}\frac{1}{4}(1-p_1p_2p(-)^{j_1+j_2+J})\times$$

$$\times\left[(p_1\cdot\alpha^{j_1}_{\frac12})C^{J,1}_{j_1,\frac12;j_2,\frac12}-p_1p_2\alpha^{j_1}_{\frac32}C^{J,1}_{j_1,\frac32;j_2,-\frac12}\right]\int f^{p_2}_{j_2}f^{p}_{J}f^{p_1}_{j_1}r^2dr,\qquad(7.5)$$

与 (7.4) 式只有 J 的奇偶性不同，因为 $r\nabla$ 与 L 的横向分量宇称相反.

$$\left(j_2\frac{1}{2}m_2p_2\middle|\psi_1^{JM0}(r;\theta,\varphi)\cdot\nabla\middle|j_1\frac{1}{2}m_1p_1\right)=$$

$$=(-)^{M+m_2+\frac12}\sqrt{\frac{(2j_1+1)(2j_2+1)}{4\pi(2J+1)}}\,C^{J,-M}_{j_1,m_1;j_2,-m_2}\frac{1}{2}(1-p_1p_2(-)^{j_1+j_2+J})\times$$

$$\times C^{J,0}_{j_1,-\frac12;j_2,\frac12}\int f^{p_2}_{j_2}f^{p}_{J}\frac{d}{dr}f^{p_1}_{j_1}r^2dr,\qquad(7.6)$$

除 (7.6) 式外甚至用不着求电子径向函数的微商. 这样就避免了通常求 γ 輻射几率[21]时曲折的計算. 例如求磁复极輻射几率时，各項的径向积分都是文献 [21] 中的 M_{L-1} 型的，用不着根据径向方程作繁复的变化.

利用 C 系数的递推关系：

$$\alpha^J_0(C^{J,1}_{j_1,\frac12;j_2,\frac12}\pm C^{J,-1}_{j_1,-\frac12;j_2,-\frac12})=(\alpha^{j_1}_{\frac12}\pm\alpha^{j_2}_{\frac12})(C^{J,0}_{j_1,-\frac12;j_2,\frac12}\pm C^{J,0}_{j_1,\frac12;j_2,-\frac12})$$

$$\alpha^J_0\alpha^{j_1}_{\frac32}(C^{J,1}_{j_1,-\frac12;j_2,\frac32}\pm C^{J,-1}_{j_1,\frac12;j_2,-\frac32})=\Big(\alpha^{j_1}_{\frac32}\alpha^{j_2}_{\frac32}+\Big[J(J+1)-j_1(j_1+1)-$$

$$-j_2(j_2+1)+\frac{1}{2}\Big]\mp\alpha^{j_1}_{\frac12}\alpha^{j_2}_{\frac12}\Big)(C^{J,0}_{j_1,-\frac12;j_2,\frac12}\pm C^{J,0}_{j_1,-\frac12;j_2,\frac12})$$

及上节所述用 Racah 系数表示 $C^{J,0}_{j_1,-\frac12;j_2,\frac12}$ 的式子，即可将本节各式化成常见的形式.

参 考 文 献

[1] Edmonds, A. R., Angular Momentum in Quantum Mechanics (1957).

[2] Гельфанд, И. М., Минлос, Р. А. и Шапиро, З. Я., Представления группы вращений и группы Лоренца (1958).

[3] Любарский, Теория групп и её применений в физике (1957). 有中譯本.

[4] 汪容，物理学报，**15** (1959)，55.

[5] 楊振宁，*Phys. Rev.*, **77** (1950), 242.

[6] Широков, *Ж.Э.Т.Ф.*, **24** (1953), 14; **26** (1954), 128.

[7] Шапиро, *У.Ф.Н.*, **53** (1954), 7. 中譯文：物理譯报，**5** (1958)，351.

[8] Ахиезер и Берестецкий, Квантовая электродинамика. 1959.

[9] Rose, Elementary Theory of Angular Momentum.

[10] Blatt & Biedenharn, *Rev. Mod. Phys.*, **24** (1952), 258.

[11] Biedenharn, Rose, *Rev. Mod. Phys.*, **25** (1953), 735.

[12] Simon & Welton, *Phys. Rev.*, **90** (1953), 1037.
Simon, *Phys. Rev.*, **92** (1953), 1050.

[13] Morita, Sugie & Yoshido, *Prog. Theor. Phys.*, **12** (1954), 713.

[14] Minami, *Prog. Theor. Phys.*, **11** (1954), 213.

[15] Hayakawa, Kawaguchi, Minami, *Prog. Theor. Phys.*, **12** (1954), 355.

[16] Заставенко, Рынаин, 周光召，*Ж.Э.Т.Ф.*, **34** (1958), 526.

[17] 时学丹，物理学报，**18** (1962)，184.

[18] Берестецкий, Долгинов, Тер-Мартиросян, *Ж.Э.Т.Ф.*, **20** (1950), 527.
[19] Wigner, Group Theory and Its Application to the Quantum Mechanics of Atomic Spectra (1959).
[20] Fano, Racah, Irreducible Tensorial Sets (1959).
[21] Rose, Multipole Fields (1955).
[22] Rose, Relativistic Electron Theory (1961).
[23] 陈中謨、何祚庥、冼鼎昌、朱洪元，物理学报 **15** (1959), 254.

THE SPINOR WAVE FUNCTIONS AND OPERATORS IN THE LOCAL COORDINATE SYSTEM

Hou Bai-yü

Abstract

Spinor wave functions are expressed in terms of variables in the coordinate system carried with the particle. The properties of various simple systems are discussed with the help of wave functions or scattering functions expressed in this form. The differential operators are expressed in terms of the infinitesimal rotation operators in the coordinate system carried with the particle. A simple way of separating the variables of various spinor wave equations is given.

第 20 卷 第 1 期 物 理 学 报 Vol. 20, No. 1
1964 年 1 月 ACTA PHYSICA SINICA Jan., 1964

Green 函数及 δ 函数的三方向球函数展开式*

侯 伯 宇

提 要

将 $\frac{1}{R}$ 及 δ 函数的积分表达式分离径向与角向变量得到了用三个分矢的球函数及冪級数表达的展开式.

在計算分子間的作用力及多原子分子的化学鍵时，常用到 $\frac{1}{R}$ 的双中心球极坐标展开式(例文献[4—6]). 在計算核及基本粒子的相互作用时，如果考虑到其电磁結构，也有可能会用到这种展式. Carlson 及 Rushbrook[1] 首先引进了相离区域 ($r_3 \geqslant r_1 + r_2$) 內的展式, Buehler 及 Hirschfelder[2] 进一步找到了在相迭区域 ($r_1 + r_2 \geqslant r_3 \geqslant |r_1 - r_2|$) 內的展式，并且造了一个表. 唐敖庆及江元生[3]也曾用遞推法得到了与文献[1]相同的結果. 文献[2]虽得到了相迭域的式子，但并未将冪級数的同类項归并在一起写出明显的展式，其中因子 $1/(r_{ab} + r_j)^{2n_a-2t-1}$ 的冪次比 $V(v)$ 項中 $(r_{ab}^2 - r_j^2)^p$ 的冪次还要高, 因此必須証明分母中的 $r_{ab} + r_j$ 可以与其他因子消去，才确能展为冪級数(利用二項式系数的一些性質，經較繁的运算可以証明这一点)，文献[2]并未指出这一問題.

我們利用 Green 函数的积分表达式分离了展式的角向部分及径向部分. 角向部分的系数用矢量耦合 $3j$ 系数表示，这样旣統一了双中心連綫卽量子化軸时及非量子化軸时的系数，又可以用現成的 $3j$ 系数表代替繁杂的計算或造表，也便于利用 $3j$ 系数的熟知性質簡化計算的結果 (例如文献 [5] 中的 (ab, ab, ab) 項及 (ab, bc, ca) 項均可进一步用 Racah 的方法对磁量子数 m 求和，将結果表为 W, X 系数，該文中关于系数 ϕ, θ 的一些等式也就显而易見)，展式的轉动不变性也較明显. 我們用同一个函数在不同域內的冪級数来表示径向部分. 統一了文献[2]中四个域內的式子，而且容易看出在各区域边界上展式的連續性及其相互联系. 在重迭域得到的明显冪級数式較文献[2]旣簡明而有对称性，且各項系数間有簡单的遞推关系.

径向部分的計算要用到三个球 Bessel 函数的乘积的积分. 虽然过去已經有过不少作者对此进行了計算 (例如文献[7, 8])，但所得結果均限于相离域，由于他們在計算中先处理两个 Bessel 函数，故不易推广到三个函数的变量具有对称性的重迭域. 我們用围綫积分法与遞推法求出了在此区域內展式的对称表达式. 所用方法也适于求三个以上的球 Bessel 函数乘积的积分.

本文还作了将 δ 函数依三方向展开的初次尝試.

* 1962 年 8 月 13 日收到; 1963 年 3 月 18 日收到修改稿.

一

本节先用围綫法再换用递推法計算三个球 Bessel 函数乘积的积分：

$$ {}_{\lambda}j_{l_1l_2l_3}(r_1r_2r_3) \equiv \int_0^{\infty} j_{l_1}(Kr_1)j_{l_2}(Kr_2)j_{l_3}(Kr_3)K^{\lambda}dK, $$

式中 l, λ 为整数，且 $l_1+l_2+l_3+\lambda \geqslant 0$, $\lambda \leqslant 2$（这就保証了积分的收斂性）；$j_l(Kr) = \sqrt{\frac{\pi}{2Kr}} J_{l+\frac{1}{2}}(Kr)$ 为 l 阶球 Bessel 函数[1)]，r 取非負值，l 取負值时 $j_l = (-)^l n_{-l-1}$ 为球 Neumann 函数（以后省略 $j_{l_i}(Kr_i)$ 的变量 Kr_i）.

在 $l_1+l_2+l_3+\lambda$ 为偶数时，利用 j_l 的宇称为 $(-)^l$ 可得 ${}_{\lambda}j_{l_1l_2l_3} = \frac{1}{2}\int_{-\infty}^{\infty} j_{l_1}j_{l_2}j_{l_3}K^{\lambda}dK$，由于 j_l 的渐近展式中旣有含因子 e^{iz} 的項，也有含 e^{-iz} 的項，不能用在无穷远处的上半圓或下半圓构成围綫. 故用在上半圓趋于零的第一类球 Hankel 函数 h_l 代替 j_l. 在 $S_1 \leqslant 0 \left(S_1 \equiv S - r_1 \equiv \frac{r_1+r_2+r_3}{2} - r_1 = \frac{1}{2}(r_2+r_3-r_1)\right)$ 的区域內可以化上式如下：

$$ {}_{\lambda}j_{l_1l_2l_3} = \frac{1}{2}\int_{c_1} j_{l_1}j_{l_2}j_{l_3}K^{\lambda}dK = \frac{1}{2}\int_{c_1}(h_{l_1} - in_{l_1})j_{l_2}j_{l_3}K^{\lambda}dK = $$

$$ = \frac{1}{2}\int_{c_2} h_{l_1}j_{l_2}j_{l_3}K^{\lambda}dK - \frac{i}{2}\int_{c_3} n_{l_1}j_{l_2}j_{l_3}K^{\lambda}dK, $$

式中积分路綫 c_1 包括：負半軸 $(-\rho, -\varepsilon)$，繞原点的上半小圓 $\varepsilon e^{i\varphi}$（φ 由 π 至 0），正半軸 $(+\varepsilon, +\rho)$ $(\rho \to \infty, \varepsilon \to 0)$；$c_2$ 则包括：負半軸、上半小圓、正半軸及上半面的大半圓 $\rho e^{i\varphi}$（φ 由 0 至 π）；c_3 为繞原点的上半小圓 $\varepsilon e^{i\varphi}$（φ 由 π 至 0）. 第一个等号是利用被积函数在原点附近一致趋于零而得到的. 最后一个等号則考虑了前一个积分在上半大圓上的值为零；而 $\int_{\varepsilon}^{\rho} + \int_{-\rho}^{-\varepsilon} n_{l_1}j_{l_2}j_{l_3}K^{\lambda}dK = \int_{\varepsilon}^{\rho} - \int_{\varepsilon}^{\rho} n_{l_1}j_{l_2}j_{l_3}K^{\lambda}dK = 0$（$\because n_{l_1}$ 的宇称与 j_{l_1} 相反）.

c_2 內无极点，沿它的积分为零，沿 c_3 的积分为在原点留数的 $-i\pi$ 倍. 将球 Bessel 函数的冪級数展式代入求 K^{-1} 項的系数，并經簡单变化，得

$$ {}_{\lambda}j_{l_1,l_2,l_3}(r_1r_2r_3) = (-)^{\frac{l_1+l_2+l_3+\lambda}{2}} \pi 2^{\lambda-3} \sum_{L_i} \prod_{i=1,2,3} \frac{\Gamma\left(\frac{1}{2}\right) r_i^{L_i}}{\Gamma\left(\frac{L_i+l_i+1}{2}+1\right)\Gamma\left(\frac{L_i-l_i}{2}+1\right)}. \quad (1) $$

式中的冪指数 L_i 需满足条件 $L_1+L_2+L_3 = -\lambda-1$，且 $L_i - l_i$ 中仅 $L_1 - l_1$ 为奇数. 与分母中 Γ 函数的极点对应的項为零，故实际上只有有限項.

在 $S_2 \leqslant 0$ 及 $S_3 \leqslant 0$ 区域內的結果相似，只是 $L_i - l_i$ 的奇偶应作相应的改变.

---

1) 本文中各种球 Bessel 函数的定义請参看文献[9].

在 $S_1S_2S_3 \geqslant 0$ 的区域内则有

$$
{}_\lambda j_{l_1,l_2,l_3} = \frac{1}{4}\int_{c_2}(h_{l_1}h_{l_2}j_{l_3} + j_{l_1}h_{l_2}h_{l_3} + h_{l_1}j_{l_2}h_{l_3} - h_{l_1}h_{l_2}h_{l_3})K^\lambda dK -
$$

$$
-\frac{i}{4}\int_{c_3}(n_{l_1}j_{l_2}j_{l_3} + j_{l_1}n_{l_2}j_{l_3} + j_{l_1}j_{l_2}n_{l_3} + n_{l_1}n_{l_2}n_{l_3})K^\lambda dK =
$$

$$
= (-)^{\frac{l_1+l_2+l_3+\lambda}{2}}\pi 2^{\lambda-4}\sum_{L_i}\alpha(L_i, l_i)\prod_{i=1,2,3}\frac{\Gamma\left(\frac{1}{2}\right)r_i^{L_i}}{\Gamma\left(\frac{L_i+l_i+1}{2}+1\right)\Gamma\left(\frac{L_i-l_i}{2}+1\right)}, \tag{2}
$$

式中 $L_1 + L_2 + L_3 = -\lambda - 1$． $\alpha(L_i, l_i) = \begin{cases} -1 & \text{当 } L_i - l_i \text{ 均为奇数} \\ 1 & \text{当 } L_i - l_i \text{ 之一为奇数.} \end{cases}$
易見其項数也是有限的．

也可以用递推法来求 ${}_\lambda j_{l_1,l_2,l_3}$ 的冪級数展式(2)．为此，設

$$
{}_\lambda j_{l_1,l_2,l_3} = \sum_{L_i}\Gamma^{L_1,L_2,L_3}_{l_1,l_2,l_3}r_1^{L_1}r_2^{L_2}r_3^{L_3}. \tag{2'}
$$

式中 $\Gamma^{L_1,L_2,L_3}_{l_1,l_2,l_3}$ 为待定的系数．在积分号下求 ${}_\lambda j_{l_1,l_2,l_3}$ 对参变量 r_i 的微商，可得等式

$$
\frac{1}{r_i^2}\left[\frac{d}{dr_i}r_i^2\frac{d}{dr_i} - l_i(l_i+1)\right]{}_\lambda j_{l_1,l_2,l_3} = \frac{1}{r_j^2}\left[\frac{d}{dr_j}r_j^2\frac{d}{dr_j} - l_j(l_j+1)\right]{}_\lambda j_{l_1,l_2,l_3}.
$$

将(2′)代入上式，就可求得递推关系

$$
(L_1+l_1+1)(L_1-l_1)\Gamma^{L_1,L_2-2,L_3}_{l_1,l_2,l_3} = (L_2+l_2+1)(L_2-l_2)\Gamma^{L_1-2,L_2,L_3}_{l_1,l_2,l_3}. \tag{3a}
$$

由之可得 ${}_\lambda j_{l_1,l_2,l_3}$ 展式中 L_i 奇偶性相同之各項間的关系，从而可見展式必取(2)式連乘符号后的形式．

此外，再根据

$$
\left(\frac{d}{dr_1} + \frac{l_1+2}{r_1}\right){}_\lambda j_{l_1+1,l_2,l_3} = {}_{\lambda+1}j_{l_1,l_2,l_3} = -\left(\frac{d}{dr_1} - \frac{l_1-1}{r_1}\right){}_\lambda j_{l_1-1,l_2,l_3}
$$

可得递推关系

$$
(L_1+l_1+2)\Gamma^{L_1,L_2,L_3}_{l_1+1,l_2,l_3} = \Gamma^{L_1-1,L_2,L_3}_{l_1,l_2,l_3} = -(L_1-l_1+1)\Gamma^{L_1,L_2,L_3}_{l_1-1,l_2,l_3}. \tag{3b}
$$

由此式及类似的式子可得 $l_1 + l_2 + l_3 + \lambda$ 同为偶数(或同为奇数)的各个函数 ${}_\lambda j_{l_1,l_2,l_3}$ 間的递推关系，从而易見 ${}_\lambda j_{l_1,l_2,l_3}$ 的展式形式如(2)，只要在偶(或奇) ${}_\lambda j_{l_1,l_2,l_3}$ 中各利用一个已知的非零收斂积分求出 $\alpha(L, l)$ 就可以了．

例：当 $l_1 + l_2 + l_3 + \lambda =$ 偶数时，利用

$$
{}_0 j_{0,0,0} \equiv \int_0^\infty j_0 j_0 j_0\, dK = \begin{cases} \dfrac{\pi}{2}\dfrac{1}{r_i} & \text{在 } S_i \leqslant 0 \text{ 域,} \\ \dfrac{\pi}{2}\dfrac{1}{r_1r_2r_3}\sum\limits_{ij}S_iS_j \quad i<j & \text{在 } S_1S_2S_3 \geqslant 0 \text{ 域,} \end{cases} \tag{4}
$$

代入(2)，求得 $\alpha(L_il_i)$ 結果同前．

当 $l_1+l_2+l_3+\lambda=$ 奇数时，利用

$$_2j_{-1,0,0}\equiv-\int_0^\infty n_0j_0j_0K^2dK=\frac{1}{4}\frac{1}{r_1r_2r_3}\ln\frac{|S||S_1|}{|S_2||S_3|},$$

易得在 $S_1<0$ 域内

$$_2j_{-1,0,0}=\sum_{L_i}\frac{-\Gamma\left(\frac{1}{2}\right)\Gamma\left(\frac{-1-L_1}{2}\right)\Gamma\left(\frac{-L_1}{2}\right)}{2\Gamma\left(\frac{L_2+1}{2}+1\right)\Gamma\left(\frac{L_2}{2}+1\right)\Gamma\left(\frac{L_3+1}{2}+1\right)\Gamma\left(\frac{L_3}{2}+1\right)}.$$

式中 $L_1+L_2+L_3=-3$; $-1-L_1-2$ 及 L_2, L_3 为非負偶数.

在 $S_2<0$ 域内，

$$_2j_{-1,0,0}=\sum_{L_i}\frac{\Gamma\left(\frac{1}{2}\right)\Gamma\left(\frac{-1-L_2}{2}\right)\Gamma\left(-\frac{L_2}{2}\right)r_1^{L_1}r_2^{L_2}r_3^{L_3}}{2\Gamma\left(\frac{L_3+1}{2}+1\right)\Gamma\left(\frac{L_3}{2}+1\right)\Gamma\left(\frac{L_1}{2}+1\right)\Gamma\left(\frac{L_1+1}{2}+1\right)}.$$

式中 $L_1+L_2+L_3=-3$; L_1-1, $-L_2-2$ 及 L_3 为非負偶数.

$S_3<0$ 域内的展式相彷. 故可得在 $l_1+l_2+l_3+\lambda$ 为奇数时，在 $S_i<0$ 的区域内 $\alpha(L,l)=(-)^{3/2}\,2\csc\frac{L_i-l_i}{2}\pi\csc\frac{L_i+l_i+1}{2}\pi$，当 l_i-L_i-2 及其余的 $L_j-l_j(j\neq i)$ 为非負偶数，而且 $L_1+L_2+L_3=-\lambda-1$; L 不符合上述条件各項前的 $\alpha(Ll)=0$. 这里的余割函数是形式上引进的，表示应当移相应的 Γ 函数到分子上. 易見，这时展式有无穷个項.

在 $S_1S_2S_3>0$ 域内可化 $_2j_{-1,0,0}$ 为 $\sum_n\frac{1}{2(2n+1)}\left[\frac{r_2^2+r_3^2-r_1^2}{2r_2r_3}\right]^{2n+1}\frac{1}{r_1r_2r_3}$，这是一个条件收斂級数. 将 r_i 的同次項归并后形式上可写成

$$\sum_{L_i}(-)\frac{1}{4\sqrt{\pi}^7}\prod_i\Gamma\left(-\frac{L_i+1}{2}+1\right)\Gamma\left(-\frac{L_i}{2}+1\right)r_i^{L_i},$$

$$L_1+L_2+L_3=-3,$$

L_2、L_3 为偶数，但实质上 Γ 函数是发散的. 故 $l_1+l_2+l_3+\lambda$ 为奇数的 $_\lambda j_{l_1,l_2,l_3}$ 在交迭域不能展为单冪級数展式.

二

本节求 $\frac{1}{R}$ $(\mathbf{R}=\mathbf{r}_1+\mathbf{r}_2+\mathbf{r}_3)$ 的球函数展开式.

$\frac{1}{R}$ 作为 Laplace 算子的 Green 函数可用其本征函数——平面波 $e^{i\mathbf{K}\cdot\mathbf{R}}$ 展开:

$$\frac{1}{R}=\frac{1}{2\pi^2}\int\frac{e^{i\mathbf{K}\cdot\mathbf{R}}}{K^2}d^3K=\frac{1}{2\pi^2}\int\frac{e^{i\mathbf{K}\cdot\mathbf{r}_1}e^{i\mathbf{K}\cdot\mathbf{r}_2}e^{i\mathbf{K}\cdot\mathbf{r}_3}}{K^2}d^3K.$$

将各分矢的平面波用熟知的展式分解为球面波（为了以下运算的方便，用广义球函数代替常用的 Legendre 函数）就得到

$$\frac{1}{R}=\frac{1}{2\pi^2}\sum_{l_i}(-)^{\frac{l_1+l_2+l_3}{2}}(2l_1+1)(2l_2+1)(2l_3+1)\times$$
$$\times\int D^{l_1}_{0,0}(\Omega^{-1}\omega_1)D^{l_2}_{0,0}(\Omega^{-1}\omega_2)D^{l_3}_{0,0}(\Omega^{-1}\omega_3)j_{l_1}(Kr_1)j_{l_2}(Kr_2)j_{l_3}(Kr_3)\frac{d^3K}{K^2},$$

式中 $D^l_{0,0}$ 为广义球函数1)，ω_i 为将固定坐标系的极轴变到矢径 r_i 方向的轉动，Ω 则为将极轴变到 K 的轉动. $\Omega^{-1}\omega_i$ 的章动角就是 K、r_i 的夹角 Θ_i，故 $D^l_{0,0}(\Omega^{-1}\omega_i)$ 卽 Legendre 函数 $P^l(\cos\Theta_i)$. 将 $D^l_{0,0}(\Omega^{-1}\omega)=\sum_m D^l_{0,m}(\Omega^{-1})D^l_{m,0}(\omega)$ 代入上式，并将三个 $D^l_{0,m}$ 的乘积在羣上的积分表为 $3j$ 系数(例如参看文献[10])，就可算得角向积分，径向积分見上节.

$$\frac{1}{R}=\frac{2}{\pi}\sum_{l_i,m_i}(-)^{\frac{l_1+l_2+l_3}{2}}(2l_1+1)(2l_2+1)(2l_3+1)\begin{pmatrix}l_1\,l_2\,l_3\\0\,0\,0\end{pmatrix}\begin{pmatrix}l_1&l_2&l_3\\m_1&m_2&m_3\end{pmatrix}\times$$
$$\times D^{l_1}_{m_1,0}(\varphi_1\theta_10)D^{l_2}_{m_2,0}(\varphi_2\theta_20)D^{l_3}_{m_3,0}(\varphi_3\theta_30)_0j_{l_1,l_2,l_3}(r_1r_2r_3).\tag{5}$$

式中 $\begin{pmatrix}l_1&l_2&l_3\\m_1&m_2&m_3\end{pmatrix}$ 为对称化的 $3j$ 系数，θ_i，φ_i 为 r_i 的方向角. 由 $3j$ 系数的性质易見，(5)式各項的 l 满足动量矩选择律 ($l_2+l_3\geqslant l_1\geqslant|l_2-l_3|$) 及宇称选择律 ($l_1+l_2+l_3$ 为偶数). 故由(1)式可知，在 $S_1\leqslant0$ 域內只有 $l_2+l_3=l_1$ 的項非零. 这时展式可簡化为

$$\frac{1}{R}=\sum_{l_i,m_i}\delta_{l_1,l_2+l_3}\sqrt{\frac{(2l_1+1)!}{(2l_2)!(2l_3)!}}\begin{pmatrix}l_1&l_2&l_3\\m_1&m_2&m_3\end{pmatrix}D^{l_1}_{m_1,0}(\varphi_1\theta_10)D^{l_2}_{m_2,0}(\varphi_2\theta_20)D^{l_3}_{m_3,0}(\varphi_3\theta_30)\frac{r_2^{l_2}r_3^{l_3}}{r_1^{l_1+1}}.\tag{6}$$

沿 r_1 的方向取极軸，则有

$$\frac{1}{R}=\sum_{l_i,m}\delta_{l_1,l_2+l_3}\sqrt{\frac{(2l_1+1)!}{(2l_2)!(2l_3)!}}\begin{pmatrix}l_1&l_2&l_3\\0&m&-m\end{pmatrix}D^{l_2}_{m0}(\varphi_2\theta_20)D^{l_3}_{-m0}(\varphi_3\theta_30)\frac{r_2^{l_2}r_3^{l_3}}{r_1^{l_1+1}}.\tag{7}$$

卽通常[1-3]的双中心球坐标展式，不过为了使得在各相离域展式的形式完全相仿，同时为了使得重迭域內的展式更有对称性，我們采用了 $\mathbf{R}=\mathbf{r}_1+\mathbf{r}_2+\mathbf{r}_3$，而通常用的是 $\mathbf{R}=\mathbf{r}_1+\mathbf{r}_2-\mathbf{r}_3$. 故我們的 φ_3，θ_3 相当于通常的 $\pi+\varphi_3$ 及 $\pi-\theta_3$，而展式与常用展式有相位的差别.

为了与文献[2]在重迭域內的結果比較，在(5)式中取极軸沿 $\mathbf{r}_1$ 方向，有

$$\frac{1}{R}=\sum_{l_i,m}(-)^{\frac{l_1+l_2+l_3}{2}}\frac{2}{\pi}\begin{pmatrix}l_1\,l_2\,l_3\\0\,0\,0\end{pmatrix}\begin{pmatrix}l_1&l_2&l_3\\0&m&-m\end{pmatrix}D^{l_2}_{m,0}(\varphi_2\theta_20)D^{l_3}_{-m,0}(\varphi_3\theta_30)_0j_{l_1,l_2,l_3}(r_1r_2r_3).$$

在計算冪級数各項的系数时，請注意下列事項：

1. 只有 l_1，l_2，l_3 的和为偶数且遵守动量矩选择律的項.

2. 仅当 $l_1=l_2+l_3$ 时，才有 L_1-l_1 为奇数，而 L_2-l_2，L_3-l_3 为偶数的項. 且只有一項，卽 $L_1=-l_1-1$，$L_2=l_2$，$L_3=l_3$ 的項，其系数为 (7) 式相应系数的 $\frac{1}{2}$. 仅 L_2-l_2 或 L_3-l_3 为奇的項的出現規律与之相同.

3. L_i-l_i 均为奇数的項可由 (3a) 式从 $L_1=-l_1-1$，$L_2=-l_2-1$，$L_3=l_1+$

1) 本文中的函数 $D^l_{m,m'}$，P^l 及 $3j$ 系数的定义及位相請参看文献[10].

$+ l_2 + 1$ 的項出发递推到 $L_1 = -l_1 - 1$，$L_2 = l_3 + l_1 + 1$，$L_3 = -l_3 + 1$ 的項，或 $L_1 = l_2 + l_3 + 1$，$L_2 = -l_2 - 1$，$L_3 = -l_3 - 1$ 的項而逐个求得.

4. 还可以用递推关系 (3b) 式及类似的式子从 l 較小的項求 l 較大的項.

5. 当对于同一組 l_2, l_3 的值有若干个符合条件(1)的 l_1 值 ($l_2 + l_3$, $l_2 + l_3 - 2$, $l_2 + l_3 - 4$, $\cdots$, $|l_2 - l_3|$) 时，需对这些 l_1 求和才是文献[2]的結果.

例如：注意到 $(2l_1+1)(2l_2+1)(2l_3+1)\begin{pmatrix} l_1 & l_2 & l_3 \\ 0 & 0 & 0\end{pmatrix}\begin{pmatrix} l_1 & l_2 & l_3 \\ 0 & m & -m\end{pmatrix}$ 当 $l_2 = l_3 = 0$ 时取值 $\delta_{l_1,0}\delta_{m,0}$，而当 $l_2 = 1$，$l_3 = 0$ 时取值 $3\delta_{l_1,1}\delta_{m,0}$；再利用

$$ {}_0j_{0,0,0} = \frac{\pi}{2}\cdot\frac{1}{4}\left(\frac{-r_3}{r_1r_2} - \frac{r_1}{r_2r_3} - \frac{r_2}{r_3r_1} + \frac{2}{r_1} + \frac{2}{r_2} + \frac{2}{r_3}\right) \quad [\text{参看 4 (式)}], $$

$$ {}_0j_{1,1,0} = \frac{\pi}{2}\cdot\frac{1}{48}\left(-\frac{8r_1}{r_2^2} - \frac{8r_2}{r_1^2} - \frac{r_3^3}{r_1^2r_2^2} + \frac{6r_3}{r_1^2} + \frac{6r_3}{r_2^2} + \frac{3r_2^2}{r_1^2r_3} + \frac{3r_1^2}{r_2^2r_3} - \frac{6}{r_3}\right) $$

[由(2)式直接求得，或利用法則(2)，(3)，(4)]，卽得文献[2]的表中第一及第二列.

由(4)式易証，在相离域与重迭域的边界 $S_i = 0$ 处，${}_0j_{0,0,0}$ 的二展式相等. 故由 ${}_0j_{0,0,0}$ 递推出其余 ${}_0j_{l_1,l_2,l_3}$ 的过程可看出 $\frac{1}{R}$ 展式各項的径向部分 ${}_0j_{l_1,l_2,l_3}$ 都是連續的.

三

本节将 δ 函数依三方向球函数展开：

$$ \begin{aligned} \delta(\mathbf{R}) &= \frac{1}{(2\pi)^3}\int e^{i\mathbf{K}\cdot\mathbf{R}}d^3K = \\ &= \frac{1}{2\pi^2}\sum_{l_im_i}(-)^{\frac{l_1+l_2+l_3}{2}}(2l_1+1)(2l_2+1)(2l_3+1)\begin{pmatrix} l_1 & l_2 & l_3 \\ m_1 & m_2 & m_3\end{pmatrix}\begin{pmatrix} l_1 & l_2 & l_3 \\ 0 & 0 & 0\end{pmatrix}\times \\ &\quad\times D^{l_1}_{m_1,0}D^{l_2}_{m_2,0}D^{l_3}_{m_3,0}\,{}_2j_{l_1,l_2,l_3}. \end{aligned} \tag{8} $$

由(1)式易見，此展式各項在 $S_i \leqslant 0$ 的域內为零.

利用 δ 函数的特性 $\int f(\mathbf{r}_1)\delta(\mathbf{R})d^3r_1 = f(-\mathbf{r}_2 - \mathbf{r}_3) \equiv f(\mathbf{r}_{23})$，可由(8)得到一些有用的式子，例如

$$ \begin{aligned} r_{23}^{l_1}D^{l_1}_{m_1,0}(\varphi_{23},\theta_{23},0) &= \sum_{\substack{l_2m_2\\ l_3m_3}}\delta_{l_1,l_2+l_3}\sqrt{\frac{(2l_1+1)!}{(2l_2)!(2l_3)!}}\begin{pmatrix} l_1 & l_2 & l_3 \\ m_1 & m_2 & m_3\end{pmatrix}\times \\ &\quad\times D^{l_2}_{m_2,0}(\varphi_2,\theta_2,0)D^{l_3}_{m_3,0}(\varphi_3,\theta_3,0)\,r_2^{l_2}r_3^{l_3}. \end{aligned} \tag{9} $$

$$ \begin{aligned} r_{23}^{-l_1-1}D^{l_1}_{m_1,0}(\varphi_{23},\theta_{23},0) &= \sum_{\substack{l_2m_2\\ l_3m_3}}\delta_{l_1,l_>-l_<}\sqrt{\frac{(2l_>+1)!}{(2l_1)!(2l_<)!}}\begin{pmatrix} l_1 & l_2 & l_3 \\ m_1 & m_2 & m_3\end{pmatrix}\times \\ &\quad\times D^{l_2}_{m_2,0}(\varphi_2,\theta_2,0)D^{l_3}_{m_3,0}(\varphi_3,\theta_3,0)\frac{r_<^{l_<}r_1^{l_1}}{r_>^{l_>+1}}. \end{aligned} \tag{10} $$

式中 $r_>$, $r_<$ 表示 r_2, r_3 中較大較小者；$l_>$, $l_<$ 是与之对应的 l. 求上二式时利用了在积分

号下对参变量积分而得的

$$\int_{|r_2-r_3|}^{r_2+r_3} {}_2\dot{j}_{l_1,l_2,l_3} r_1^{l_1+2} dr_1 = r_1^{l_1+2} {}_1\dot{j}_{l_1+1,l_2,l_3} \Big|_{r_1=|r_2-r_3|}^{r_1=r_2+r_3}$$

及

$$\int {}_\lambda \dot{j}_{l_1,l_2,l_3} r_1^{1-l_1} dr_1 = - r_1^{1-l_1} {}_{\lambda-2}\dot{j}_{l_1+2,l_2,l_3}$$

二式.

(2),(3)二式即球体函数的 Hobson 公式的推广,也可以由(6)式与 $\frac{1}{R}$ 的寻常二方向 Neumann 展式比較而得(9),(10)式.

結　束　語

过去，δ 函数的展式(正交函数系的封閉性公式)及 Green 函数的 Mercer 展式都是由正交函数两两乘积的和构成的. 我們作了用三重积来展开的尝試.

Gordan 方程的 Green 函数 $\frac{e^{-K_0R}}{R} = \frac{1}{2\pi^2}\int \frac{e^{i\mathbf{K}\cdot\mathbf{R}}}{K^2+K_0^2} d^3K$ 也可用第二节的方法分离变量,再用围綫法求径向积分,从而得到其三方向展式.

利用 $e^{i\mathbf{K}\cdot\mathbf{r}_i}$ 的 Jacobin 函数展式或 Gegenbauer 函数展式,还可以得到 Green 函数及 δ 函数在球极坐标系中的其他三方向展式. 仿此可得 Laplace 算子的其余可分离变量的本征函数的展式.

用本文的方法也可以得到通常的两方向 Neumann 展式,而且还适用于求更多方向的展式. 比較这些展式可以得到更高次的球体函数 Hobson 展式, 以及 Laplace 算子的各种本征函数用本征函数的积展开的式子.

文振翼同志借到了唐敖庆等同志的論文介紹給我学习,并帮我核对一些結果,特此志謝.

参　考　文　献

[1] Carlson Rushbrook, *Proc. Cambridge Phil. Soc.*, **46** (1950), 626.
[2] Buehler Hirschfelder, *Phys. Rev.*, **83** (1951), 628; **85** (1952), 149.
Hirschfelder & others, Molecular theory of gases and liquids (1954).
[3] 唐敖庆、江元生，东北人民大学自然科学学报，**1** (1955)，208.
[4] 孙家鍾、蔣栋成、周木易，物理学报，**18** (1962)，117.
[5] 唐敖庆、孙家鍾，*Science Record*, **1** (1957), 219; **2** (1958), 154.
[6] 孙家鍾、蔣栋成，物理学报，**17** (1961)，559.
[7] Watson, *Jour. London Math. Soc.*, **9** (1934), 16.
[8] Bailey, *Proc. London. Math. Soc.*, (2) **40** (1935), 37.
Jour. London Math. Soc., **11** (1936), 16.
[9] Morse & Feshbach, Methods of Theoretical Physics.
[10] Edmonds, Angular Momentum in Quantum Mechanics.

THREE DIRECTIONAL EXPANSION OF GREEN FUNCTIONS & δ FUNCTION

HOU PEI-YU

ABSTRACT

By using the integral representation of Green function and δ function, it is very easy to separate the angular and radial variables and obtain an explicit expansion.

SCIENTIA SINICA
Vol. XV, No. 6, 1966

MATHEMATICS

ORTHONORMAL BASES AND INFINITESIMAL OPERATORS OF THE IRREDUCIBLE REPRESENTATIONS OF GROUP U_n*

Hou Pei-yu (侯伯宇)
(*Institute of Mathematics, Academia Sinica*)

Abstract

By utilizing the commutation properties of the infinitesimal operators, we constructed the orthonormal bases of the irreducible representations of group U_n explicitly without using any knowledge about the symmetric group. Moreover, by using lowering operators, a simple derivation of the representation of the infinitesimal operators is given.

Introduction

Until now there is no self-contained pure algebraic theory about explicit expressions of the orthonormal bases and infinitesimal operators in an irreducible representation (hereinafter abbreviated as I.R.) of group U_n.

The Weyl bases[1] in their polynomial form[2-4] have not been orthogonlized. The Gelfand bases[5] are orthonormal, but no general explicit forms have been given. Only a few particular cases have been studied by Biedenharn[3] in connexion with the Weyl bases.

We note that if one expresses the bases, which have been suggested by Желобенко ([6], p. 117), with the help of infinitesimal operators, one may achieve all the ortho-bases by letting some polynomial of lowering operators act on the highest vector. The bases obtained in this way still remain unnormalized. However, by studying the various properties of these lowering polynomials, we may obtain the normalizing coefficients, and verify the known I.R. of infinitesimal operators (i.e., the Lie algebra A_{n-1}).

After our result had been obtained, we happened to find that in a paper by Nagel and Moshinsky[9] a part of the results is the same as ours. Their lowering polynomials are similar to ours (such polynomials first came to our notice from a paper by Желобенко). They did not derive the normalizing coefficients by working directly on the lowering operators, but used the known representations of A_{n-1} instead. In this sense their work is not self-contained.

We put all the definitions of the adopted symbols and some mathematical relations in the appendices. It may be more convenient if the appendices are read first.

I

An I.R. of group U_n is determined uniquely by its highest weight $(m_{1n}, \cdots, m_{nn})$. Owing to irreducibility, any vector in the representation space (e.g., the canonical basis

* Received Dec. 10, 1965.

$|m\rangle$ in Appendix I (A1.2)) may be obtained by letting some polynomials of E_{ij}, H_i operate on the highest vector $|m_{1,n}, \cdots, m_{n,n}\rangle \equiv |\Lambda_n\rangle$. By noting the properties (A1.3) and (A1.5) of $|\Lambda_n\rangle$ and the commutation rule (A1.1), one may use only the lowering operator $E_{ij}(i > j)$ in these polynominals. But lowering operators are still connected by nonzero commutation rules (A1.1), hence products of lowering operators are linearly dependent. Obviously, terms with unequal root sums (the sum of all the roots corresponding to the operators E_{ij} in one term) are linearly independent. If we arrange the infinitesimal operators in any term according to definite orders (e.g., with the roots increasing from left to right) and call this term an ortho-order term, then the ortho-order terms constitute a complete set of independent bases in the polynomial space of lowering operators. But vectors, which are obtained by letting ortho-terms act on the highest vector, are not orthogonal, and moreover, not always independent. Hence, it is necessary to orthogonlize these vectors. Here, let us aim at the set of Gelfand's bases. We start from the highest vector $|\Lambda_n\rangle$ of U_n, and by using lowering polynomials try to lower it successively to the highest vector $|\Lambda_n, \Lambda_{n-1}\rangle$ of the induced representation $(m_{1,n-1}, \cdots, m_{n-1,n-1})$ of subgroup U_{n-1}. Then, we start from $|\Lambda_n, \Lambda_{n-1}\rangle$, and by using operator polynomials in U_{n-1} try to obtain $|\Lambda_n, \Lambda_{n-1}, \Lambda_{n-2}\rangle$, etc. This problem is solved by finding the following lowering polynomial:

$$\varepsilon_{ni}|m_{1,n-1}, \cdots, m_{i,n-1}, \cdots m_{n-1,n-1}\rangle \simeq |m_{1,n-1}, \cdots, m_{i,n-1} - 1, \cdots, m_{n-1,n-1}\rangle. \quad (1)$$

Here we use $\simeq$ instead of $=$ to denote the presence of an undetermined normalizing coefficient. Since the right-hand side of (1) is a highest vector of U_{n-1}, we have (cf. (I.4,5))

$$\begin{aligned} H_j\varepsilon_{ni}|m_{1,n-1}, \cdots, m_{i,n-1}, \cdots, m_{n-1,n-1}\rangle = \\ = (m_{j,n-1} - \delta_{ij} + \delta_{nj})|m_{1,n-1}, \cdots, m_{i,n-1}, \cdots, m_{n-1,n-1}\rangle, \end{aligned} \quad (2)$$

$$E_{j,j+1}\varepsilon_{ni}|m_{1,n-1}, \cdots, m_{i,n-1}, \cdots, m_{n-1,n-1}\rangle = 0. \quad j < n-1. \quad (3)$$

But since the vector acted on by ε_{ni} is also a highest vector of U_{n-1}, we have the "highest" condition, which is equivalent to (2),

$$[H_i, \varepsilon_{ni}] = (\delta_{nj} - \delta_{ij})\varepsilon_{ni}. \quad (4)$$

From (4) we see that the root sum of any term in ε_{ni} equals $e_n - e_i$, hence (3) is equivalent to

$$E_{j,j+1}\varepsilon_{ni} - \varepsilon'_{ni}E_{j,j+1} = 0, \quad (5)$$

where ε'_{ni} is some polynomial with root sum $\mathbf{e}_n - \mathbf{e}_j$.

Formula (1) of Appendix III shows that expressions (A2.3) satisfy conditions (3) and (4). We may use (A3.1), (A3.2) to derive another form of ε_{ni} (A2.8)[6]. In the course of deduction of the latter, we see that these polynomials of lowering operators in the ortho-order form are determined uniquely up to a normalizing factor.

By using ε_{ni} successively we have

$$|m\rangle \simeq \varepsilon_{21}^{m_{12}-m_{11}}\varepsilon_{31}^{m_{13}-m_{12}}\varepsilon_{32}^{m_{23}-m_{22}} \cdots \varepsilon_{ij}^{m_{ij}-m_{i,j-1}} \cdots \varepsilon_{n1}^{m_{1,n}-m_{1,n-1}} \cdots \varepsilon_{n,n-1}^{m_{n-1,n}-m_{n-1,n-1}}|\Lambda_n\rangle,$$

It is well known that the Gelfand bases $|m\rangle$ are complete and orthogonal. We may prove the orthogonality directly by (A3.5). The completeness may be proved by a simple argument but with lengthy calculations.

II

In this section, we find the normalization factors

$$\left\langle \Lambda_n \left| \prod_{i=1}^{n-1} \varepsilon_{in}^{m_{i,n}-m_{i,n-1}} \cdot \prod_{i=1}^{n-1} \varepsilon_{ni}^{m_{i,n}-m_{i,n-1}} \right| \Lambda_n \right\rangle, \tag{1}$$

where $\langle \Lambda_n |$ is the lowest vector of the conjugate representation, and $\varepsilon_{in} (i < n)$ is the transpose of ε_{ni}. Since ε_{ni}, ε_{nj} are mutually commutating (A3.8a), their orders in (1) are arbitrary.

We find (1) by induction. Assume that

$$\left\langle \Lambda_n \left| \prod_{i=l}^{n-1} \varepsilon_{in}^{m_{i,n}-m_{i,n-1}} \prod_{i=l}^{n-1} \varepsilon_{ni}^{m_{i,n}-m_{i,n-1}} \right| \Lambda_n \right\rangle \equiv \langle M | M \rangle$$

is already known. It is to be asked what the value of

$$\langle M | \varepsilon_{ln}, \varepsilon_{nl} | M \rangle \equiv \langle M' | M' \rangle = ?\ \langle M | M \rangle$$

is.

Expand ε_{ln} in this expression according to (A2.5). Since $E_{li} | M' \rangle = 0\ (i < n)$, only E_{ln} remains:

$$\varepsilon_{ln} | M' \rangle = L_{l,l+1} L_{l,l+2} \cdots L_{l,n-1} E_{ln} | M' \rangle. \tag{2}$$

By commuting ε_{nl} and by using the relation (A3.4),

$$E_{ln} | M' \rangle = E_{ln} \prod_{i=l+1}^{n-1} \varepsilon_{ni}^{m_{i,n}-m_{i,n-1}} \varepsilon_{nl}^{m_{l,n}-m_{l,n-1}+1} | \Lambda_n \rangle =$$

$$= \prod_{i=l+1}^{n-1} \varepsilon_{n-1}^{m_{i,n}-m_{i,n-1}} E_{l,n} \varepsilon_{nl}^{m_{l,n}-m_{l,n-1}+1} | \Lambda_n \rangle. \tag{3}$$

By using (A1.1), (A3.1b), (A3.2b), we have

$$E_{ln} \varepsilon_{nl}^{\alpha} | \Lambda_n \rangle = E_{l,n-1} E_{n-1,n} \varepsilon_{nl}^{\alpha} | \Lambda_n \rangle =$$

$$= E_{l,n-1} \sum_{\beta=0}^{\alpha-1} \varepsilon_{[H_{n-1}-1]}^{\alpha-\beta-1} \varepsilon_{n-1,l} (L_{l,n} - 1) \varepsilon_{nl}^{\beta} | \Lambda_n \rangle, \tag{4}$$

where $\varepsilon_{nl[H_{n-1}-1]}$ is the same as ε_{nl} but with H_{n-1} substituted by $H_{n-1} - 1$. Transposing $\varepsilon_{n-1,l}$ to the left by (A3.9a), we have

$$E_{l,n-1} \varepsilon_{n-1,l} \sum_{\beta=0}^{\alpha-1} \varepsilon_{n,l}^{\alpha-\beta-1} (L_{l,n} - 1) \varepsilon_{n,l}^{\beta} | \Lambda_n \rangle.$$

Displacing $L_{l,n}$ to the right and taking the eigenvalue and the sum, we obtain $E_{l,n-1} \varepsilon_{n-1,l} \varepsilon_{nl}^{\alpha-1} \alpha (l_{l,n} - \alpha) | \Lambda_n \rangle$. Furthermore, by using (A3.3c), we obtain

$$E_{ln} \varepsilon_{nl}^{\alpha} | \Lambda_n \rangle = (l_{l,l+1} - \alpha)(l_{l,l+2} - \alpha) \cdots (l_{ln} - \alpha) \alpha \varepsilon_{nl}^{\alpha-1} | \Lambda_n \rangle. \tag{5}$$

Substituting (5) into (3) and then into (2), and taking the eigenvalue of L by displacing it to the right, we have

$$\varepsilon_{ln}|M'\rangle = \prod_{i=l+1}^{n-1}(m_{l,n-1} - m_{i,n-1} + i - l)\prod_{i=l+1}^{n}(m_{l,n-1} - m_{i,n} + \\ + i - l - 1)(m_{l,n} - m_{l,n-1} + 1)|M\rangle. \tag{6}$$

The coefficient before $|M\rangle$ on the right-hand side of (6) is just the desired normalization coefficient $\dfrac{\langle M'|M'\rangle}{\langle M|M\rangle}$.

Using (6) successively, we get the modulus of (1):

$$\prod_{l=1}^{n-1}\left[\prod_{i=l}^{n-1}\frac{(m_{l,n} - m_{i,n-1} + i - l)!}{(m_{l,n-1} - m_{i,n-1} + i - l)!}\cdot\prod_{i=l+1}^{n}\frac{(m_{l,n} - m_{i,n} + i - l - 1)!}{(m_{l,n-1} - m_{i,n} + i - l - 1)!}\right]. \tag{7}$$

Comparing the moduli of $\prod_{i=1}^{n-1}\varepsilon_{n,i}^{m_{i,n}-m_{i,n-1}}|A_n\rangle$ and $\varepsilon_{nj}\prod_{i=1}^{n-1}\varepsilon_{n,i}^{m_{i,n}-m_{i,n-1}}|A_n\rangle$, we get

$$\left[\prod_{i=1}^{n}(m_{j,n-1} - m_{i,n} + i - j - 1)\prod_{i=j+1}^{n-1}(m_{j,n-1} - m_{i,n-1} + i - j)\cdot\right. \\ \left.\cdot\left|\prod_{i=1}^{j}(m_{j,n-1} - m_{i,n-1} + i - j - 1)\right]^{\frac{1}{2}}. \tag{8}$$

Expression (6) is a special case of (8). From the fact that there are no factors in the denominator of (6), we see that it is more convenient to derive (6) first. In deriving (6) our trick lies in separating out the "hook factors"[3] in U_{n-1} (2) from the factors joining U_{n-1} and U_n (5).

Finally, the explicit form of Gelfand's orthonormal bases is

$$|m\rangle = \prod_{k=2}^{n}\left\{\prod_{l=1}^{k-1}\left[\left(\prod_{i=l}^{k-1}\frac{(m_{lk} - m_{i,k-1} + i - l)!}{(m_{lk} - m_{i,k-1} + i - l)!}\prod_{i=l+1}^{k}\frac{(m_{l,k} - m_{i,k} + i - l - 1)!}{(m_{l,k-1} - m_{i,k} + i - l - 1)!}\right)^{-\frac{1}{2}}\cdot\right.\right. \\ \left.\left.\cdot\,\varepsilon_{kl}^{m_{lk}-m_{l,k-1}}\right]\right\}|A_n\rangle. \tag{9}$$

Here ε_{kl} stands before ε_{ij} if $k < i$.

III

In this section we verify the known representation matrix of $E_{n,n-1}$:

$$\langle A_n|\prod_{i=1}^{n-1}\varepsilon_{i,n}^{m_{i,n}-m_{i,n-1}}\varepsilon_{q,n}\prod_{j=1}^{n-2}\varepsilon_{j,n-1}^{m_{j,n-1}-m_{j,n-2}}\varepsilon_{q,n-1}^{-1}\prod_{k=2}^{n-2}\prod_{l=1}^{k-1}\varepsilon_{l,k}^{m_{l,k}-m_{l,k-1}}N'\cdot \\ \cdot|E_{n,n-1}|\prod_{k=2}^{n}\prod_{l=1}^{k-1}\varepsilon_{kl}^{m_{lk}-m_{l,k-1}}N|A_n\rangle, \tag{1}$$

where N, N' are normalization factors.

By letting $E_{n,n-1}$ act on the left, it may be commuted with $E_{jk}(k \leqslant n-1)$ until it meets the operator ε_{qn}. Therefore, it is only necessary to find the coefficient x_q in the expansion:

$$\langle \Lambda_n | \prod_{i=1}^{n-1} \varepsilon_{in}^{m_{i,n}-m_{i,n-1}} \varepsilon_{qn} E_{n,n-1} = \sum_{j=1}^{n-1} x_j \langle \Lambda_n | \prod_{i=1}^{n-1} \varepsilon_{in}^{m_{i,n}-m_{i,n-1}} \varepsilon_{qn} \varepsilon_{j,n}^{-1} \varepsilon_{j,n-1}. \tag{2}$$

Noting that the terms on the right-hand side are mutually orthogonal, we get

$$x_q = \frac{\langle \Lambda_n | \prod_{i=1}^{n-1} \varepsilon_{in}^{m_{i,n}-m_{i,n-1}} \varepsilon_{q,n} E_{n,n-1} \varepsilon_{n-1,q} \prod_{i=1}^{n-1} \varepsilon_{ni}^{m_{i,n}-m_{i,n-1}} | \Lambda_n \rangle}{\langle \Lambda_n | \prod_{i=1}^{n-1} \varepsilon_{in}^{m_{i,n}-m_{i,n-1}} \prod_{i=1}^{n-1} \varepsilon_{ni}^{m_{i,n}-m_{i,n-1}} | \Lambda_n \rangle}. \tag{3}$$

The denominator in (3) is already known in previous sections.

Let $E_{n,n-1}$ in the numerator act on the right. From (A2.11) we get

$$x_q = \frac{\langle \Lambda_n | \prod_{i=1}^{n-1} \varepsilon_{i,n}^{m_{i,n}-m_{i,n-1}} \varepsilon_{q,n} \cdot \varepsilon_{n,q} \prod_{i=1}^{n-1} \varepsilon_{n,i}^{m_{i,n}-m_{i,n-1}} | \Lambda_n \rangle / (m_{q,n-1} - m_{n-1,n-1} + n - 1 - q)}{\langle \Lambda_n | \prod_{i=1}^{n-1} \varepsilon_{in}^{m_{i,n}-m_{i,n-1}} \prod_{i=1}^{n-1} \varepsilon_{ni}^{m_{i,n}-m_{i,n-1}} | \Lambda_n \rangle},$$

where $q \neq n-1$, (4a)

$$x_{n-1} = \frac{\langle \Lambda_n | \prod_{i=1}^{n-1} \varepsilon_{in}^{m_{i,n}-m_{i,n-1}} E_{n-1,n} E_{n,n-1} \prod_{i=1}^{n-1} \varepsilon_{ni}^{m_{in}-m_{i,n-1}} | \Lambda_n \rangle}{\langle \Lambda_n | \prod_{i=1}^{n-1} \varepsilon_{i,n}^{m_{i,n}-m_{i,n-1}} \prod_{i=1}^{n-1} \varepsilon_{n,i}^{m_{i,n}-m_{i,n-1}} | \Lambda_n \rangle}. \tag{4b}$$

The ratio of the two moduli in (4) is just expression (8) in the last section.

In conclusion, letting $E_{n,n-1}$ in (1) act on the left and using (2), we have

$$x_q \langle \Lambda_n | \prod_{k=2}^{n} \prod_{l=1}^{k-1} \varepsilon_{lk}^{m_{lk}-m_{l,k-1}} N' | N \prod_{k=2}^{n} \prod_{l=1}^{k-1} \varepsilon_{kl}^{m_{lk}-m_{l,k-1}} | \Lambda_n \rangle.$$

Here the right vector has already its own normalization factor, but the normalization factor of the left vector must be changed from N' to N. Therefore, the value of the last expression is $x_q \cdot \dfrac{N'}{N}$. Substituting the known value of x_q, N, N' into it, we obtain

$$\left\langle \begin{matrix} m_{1,n} \quad m_{2,n} \cdots\cdots\cdots\cdots m_{n,n} \\ m_{1,n-1} \cdots m_{q,n-1} - 1 \cdots m_{n-1,n-1} \\ \cdots\cdots \\ m_{11} \end{matrix} \middle| E_{n,n-1} \middle| \begin{matrix} m_{1,n} \quad m_{2,n} \cdots\cdots\cdots\cdots m_{n,n} \\ m_{1,n-1} \cdots m_{q,n-1} \cdots m_{n-1,n-1} \\ \cdots\cdots \\ m_{11} \end{matrix} \right\rangle =$$

$$= \left[(-) \frac{\prod_{i=1}^{n} (m_{i,n} - m_{q,n-1} + q - i + 1)}{\prod_{i=1}^{n-1} (m_{i,n-1} - m_{q,n-1} + q - i + 1)} \cdot \frac{\prod_{i=1}^{n-2} (m_{i,n-2} - m_{q,n-1} - q + i)}{\prod_{\substack{i=1 \\ i \neq q}}^{n-1} (m_{i,n-1} - m_{q,n-1} - q + i)} \right]^{\frac{1}{2}}, \tag{5}$$

which is the same as the results in [5] and [3].

We find x_q first, and then the entire matrix. These methods are similar to that of Biedenharn, who found the reduced coefficient of U_n and the Wigner coefficient of U_{n-1} separately. But our manipulation is carried out with the whole basis, while they use only the U_{n-1} part of this basis. However, for the part of U_{n-2} they have made out another basis in some relevant representation of U_n.

The auther wishes to express his gratitude to Professor T. S. Chang for his encouragement.

Appendix I

Infinitesimal operators $E_{ij}(i, j = 1, \cdots, n)$ of U_n satisfy the following commutation relations:

$$[E_{ij}, E_{kl}] = \delta_{jk}E_{il} - \delta_{il}E_{jk}. \tag{1}$$

In the natural representation, E_{ij} is a matrix which has unit matrix element at the intersection of the i-th row with the j-th column but zeros elsewhere. Here $E_{ii} \equiv H_i$ commute mutually. Obviously E_{ij} and $H_{ij} \equiv H_i - H_j$ $(i \neq j)$ generate the simple Lie algebra A_{n-1}, whose Cartan subalgebra is H_{ij}.

The root corresponding to E_{ij} is $\mathbf{e}_i - \mathbf{e}_j$[7], where $\mathbf{e}_i$ denotes a unit vector in the root space along the i-th direction. The rising operator $E_{ij}(i < j)$ corresponds to the positive root (cf. [8] for the lexical order of root). The primitive rising operators $E_{i,i+1}$ correspond to primitive roots.

The I.R. of U_n is characterized by its highest weight $(\Lambda_n) \equiv (m_1, \cdots, m_n)$. If we choose a canonical subgroup decomposition $U_n \supset U_1 \times U_{n-1} \supset \cdots \supset U_1 \times U_2 \supset U_1$, where U_i is generated by $E_{jk}(j, k \leqslant i)$, then we may label the canonical basis[5] by

$$\left|\begin{matrix} m_{1,n}, & m_{2,n}, & \cdots, & m_{n,n} \\ m_{1,n-1}, & m_{2,n-1}, & \cdots, & m_{n-1,n-1} \\ m_{1,n-2}, & m_{2,n-2}, & \cdots, & m_{n-2,n-2} \\ & \cdots\cdots & & \\ & \cdots & & \\ & m_{1,1} & & \end{matrix}\right\rangle \equiv \left|\begin{matrix} \Lambda_n \\ \Lambda_{n-1} \\ \Lambda_{n-2} \\ \vdots \\ \Lambda_1 \end{matrix}\right\rangle \equiv |m\rangle, \tag{2}$$

where $m_{i,j+1} \geqslant m_{i,j} \geqslant m_{i+1,j+1}$.

It belongs to the I.R. $(\Lambda_i) \equiv (m_{1,i}, m_{2,i}, \cdots, m_{i,i})$ of subgroup $U_i (n \geqslant i \geqslant 1)$.

Occasionally in certain particular cases we adopt the following abbreviations:

If $m_{j,j} = \cdots = m_{j,k} = \cdots = m_{j,n}$ $(1 \leqslant j \leqslant n-1; j \leqslant k \leqslant n)$, i.e., if it denotes the highest vector with respect to U_n, we write it as $|\Lambda_n\rangle$ or $|m_{1,n}, \cdots, m_{n,n}\rangle$.

If for fixed i, $m_{j,j} = \cdots = m_{j,k} = \cdots = m_{j,i}$ $(1 \leqslant j \leqslant i-1; j \leqslant k \leqslant i)$, i.e., if it is the highest vector with respect to U_i, we omit the lowest i-1 row in (2), and write it as $|\Lambda_n; \cdots; \Lambda_i\rangle$. Frequently, we consider only the transformation properties within subgroup U_i; at that moment we write out only the lowest i rows, and the above-mentioned highest vector is denoted by $|\Lambda_i\rangle$ or $|m_{1,i}, \cdots, m_{i,i}\rangle$.

Obviously, $|\Lambda_i\rangle$ satisfies the "highest" condition:

$$E_{kj}|\Lambda_i\rangle = 0, \quad k < j \leqslant i, \tag{3}$$

or equivalently

$$E_{k,k+1}|\Lambda_i\rangle = 0, \quad k < i, \tag{4}$$

and the "weight" condition

$$H_j|m_{1,i}, \cdots, m_{j,i}, \cdots, m_{i,i}\rangle = |m_{1,i}, \cdots, m_{j,i}, \cdots, m_{i,i}\rangle. \tag{5}$$

APPENDIX II

In order to facilitate calculations, we introduce the following symbolic convention. Let the Bose operators $a_{is}, a^*_{jr}(i, j; s, r = 1, \cdots, n)$ satisfy the commutation rule[2]:

$$[a_{is}, a^*_{jr}] = \delta_{ij}\delta_{sr}, \quad [a_{is}, a_{jr}] = 0, \quad [a^*_{is}a^*_{jr}] = 0. \tag{1}$$

Here i, j are indices of the ordinary space (e.g., the natural representation space of group U_n), and r, s are indices of the permutation space. If we take the "scalar product" with respect to permutation indices, we get

$$E_{ij} = \sum_s a^*_{is}a_{js} \equiv \|a^*_i \cdot a_j\|, \tag{2a}$$

which is a "permutation scalar", but an "ordinary operator". Obviously, operator (2) satisfies the commutation rule (A1.1).

When we denote a product of several infinitesimal operators by a product of the corresponding Bose operators, our conventions are: (i) The construction rule. In the latter product, the first creation operator forms the "scalar product" with the first destruction operator and makes an infinitesimal operator; the second creation operator with the second destruction operator, and so forth. (ii) The rule of orders. We put E_{ij} before (after) E_{lk} if a_j stands before (after) a^*_l. If $j \neq l$, $i \neq k$, we do not bother about whether E_{ij} is before or after E_{lk}. Rule (i) concerns only the orders within the destruction (or within the creation) operators and it determines which infinitesimal operator occurs, while rule (ii) concerns only the orders of a pair of destruction and creation operators which have an index in common and it determines the order of the corresponding infinitesimal operators. Since we use only the lowering operators, these conventions are self-consistent and remain true in the process of commuting the Bose operators according to (1) (cf. Ex. (2c)).

Ex.: $E_{32}E_{21} = \|a^*_3a_2a^*_2a_1\|$, (2a) $\qquad E_{21}E_{32} = \|a^*_3a^*_2a_2a_1\|$, (2b)

$$E_{32}E_{21} - E_{21}E_{32} - E_{31} = 0 \Longleftrightarrow \|a^*_3(a_2a^*_2 - a^*_2a_2 - 1)a_1\| = 0, \tag{2c}$$

$$E_{32}E_{21} + E_{31}H_{1,2} \Longleftrightarrow \|a^*_3a_2a^*_2a_1 + a^*_3a_1H_{1,2}\| \equiv \|a^*_3(a_2a^*_2 + H_{1,2} + 1)a_1\|. \tag{2d}$$

Here $H_{i,j}$ are "permutation scalars" and have nothing to do with rule (i); and we put $E_{kl}(E_{il})$ before (after) it when $a_i(a^*_i)$ is before (after) it. In the following $H_{l,j}$ never occurs with $E_{kj}(E_{jl})$ together in a single term.

Keeping these conventions in mind, we define the lowering polynomials ε_{nl} and discuss their properties:

$$\varepsilon_{nl} \equiv \|a^*_n(a_{n-1}a^*_{n-1} + L_{l,n-1})(a_{n-2}a^*_{n-2} + L_{l,n-2})\cdots(a_{l+1}a^*_{l+1} + L_{l,l+1})a^*_l\|, \tag{3}$$

where

$$L_{l,j} \equiv H_l - H_j + j - l. \tag{4}$$

There are 2^{n-l-1} terms in expression (3), with each term understood as a product of E_{lj} according to the above-mentioned convention. (Sometimes we use ${}_k\varepsilon_{nl}$ to denote varieties of (3), where $L_{l,j}$ is substituted by $L_{k,j}$, and use $\varepsilon_{nl}[H_j - 1]$ to denote that ε_{nl}, for which H_j is substituted by $H_j - 1$). In each term of ε_{nl} the sum of the roots corresponding to all the factors E_{lj} equals $\mathbf{e}_n - \mathbf{e}_l$. In expression (3), E_{lj} is arranged according to increasing orders of roots. From (1), it is easy to change it into an expression with decreasing orders:

$$\varepsilon_{nl} = \|a_n^*(a_{n-1}^*a_{n-1} + L_{l,n-1} + 1)\cdots(a_{l+1}^*a_{l+1} + L_{l,l+1} + 1)a_l\| = \\ = \|a_l(a_{l+1}^*a_{l+1} + L_{l,l+1})\cdots(a_{n-1}^*a_{n-1} + L_{l,n-1})a_n^*\|, \tag{5}$$

or expressions with other orders:

$$\varepsilon_{ni} = \|a_n^*(a_{n-1}a_{n-1}^* + L_{i,n-1})\cdots(a_j^*a_j + L_{i,j} + 1)\cdots(a_{i+1}a_{i+1}^* + L_{i,i+1})a_i^*\|. \tag{6}$$

We have also the factorization:

$$\begin{aligned}\varepsilon_{nl} = {} & \|a_n^*(a_{n-1}a_{n-1}^* + L_{l,n-1})\cdots(a_{m+1}a_{m+1}^* + L_{l,m+1})a_m a_m^*(a_{m-1}a_{m-1}^* + \\ & + L_{l,m-1})\cdots(a_{l+1}a_{l+1}^* + L_{l,l+1})a_l\| + \|a_n^*(a_{n-1}a_{n-1}^* + L_{l,n-1})\cdots(a_{m+1}a_{m+1}^* + \\ & + L_{l,m+1})L_{l,m}(a_{m-1}a_{m-1}^* + L_{l,m-1})\cdots(a_{l+1}a_{l+1}^* + L_{l,l+1})a_l\| = \\ = {} & {}_l\varepsilon_{nm}\varepsilon_{ml} + \sum_{i=m+1}^{n}\sum_{j=l}^{m-1} {}_l\varepsilon_{ni}L_{l,i-1}L_{l,i-2}\cdots L_{l,j+1}E_{ij}\varepsilon_{jl} = \\ = {} & \sum_{i=m+1}^{n}\sum_{j=l}^{m} {}_l\varepsilon_{ni}E_{ij}L_{l,i-1}L_{l,i-2}\cdots L_{l,j+1}\varepsilon_{jl},\end{aligned} \tag{7}$$

and as special cases of (7):

$$\varepsilon_{nl} = \sum_{j=l}^{n-1} L_{l,n-1}L_{l,n-2}\cdots L_{l,j+1}E_{nj}\varepsilon_{jl}, \tag{8}$$

where $m = n - 1$; and

$$\varepsilon_{nl} = \sum_{i=l+1}^{n} {}_l\varepsilon_{ni}L_{l,i-1}L_{l,i-2}\cdots L_{l,l+1}E_{il}, \tag{9}$$

where $m = l$.

All the above relations follow from the structure of (3), and are independent of relations (4).

With relation (4) we have also

$$\varepsilon_{nl} = {}_l\varepsilon_{nm}\varepsilon_{ml}L_{lm} - L_{lm}\varepsilon_{ml}\,{}_l\varepsilon_{nm}[H_l - 1], \tag{10}$$

and, when $m = l$,

$$\varepsilon_{nl} = E_{n,n-1}\varepsilon_{n-1\,l}L_{l,n-1} - L_{l,n-1}\varepsilon_{n-1,l}E_{n,n-1}. \tag{11}$$

Appendix III

In this appendix we introduce a series of multiplication and commutation relations. All proofs are omitted since they are apparent under our convention, but they may be sometimes rather lengthy. Details of proof are to appear in our Chinese paper.

1. *Formulae Which Are Useful When ε_{nl} Are Acted On by a Rising Operator*

$$E_{j,j+1}\varepsilon_{nl}-\varepsilon_{nl}\begin{bmatrix}H_j-1\\H_{j+1}+1\end{bmatrix}E_{j,j+1}=0 \quad \text{if} \quad n-1>j\geqslant l,$$

$$[E_{j,j+1},\varepsilon_{nl}]=0 \quad \text{if} \quad j<l \text{ or } j\geqslant n, \tag{1a}$$

$$E_{j,j+1}\varepsilon_{nl}|\Lambda_{n-1}\rangle=0 \quad \text{if} \quad j<n-1. \tag{1b}$$

$$E_{n-1,n}\varepsilon_{nl}-\varepsilon_{nl}[H_{n-1}-1]E_{n-1,n}=L_{l,n}\varepsilon_{n-1,l},$$

$$[E_{n-1,n},\varepsilon_{nl}]=[E_{n,n-1},\varepsilon_{n-1,l}]E_{n-1,n}+L_{l,n}\varepsilon_{n-1,l}, \tag{2a}$$

$$E_{n-1,n}\varepsilon_{nl}|\Lambda_{n-1}\rangle=L_{l,n}\varepsilon_{n-1,l}|\Lambda_{n-1}\rangle=\varepsilon_{n-1,l}(l_{l,n}-1)|\Lambda_{n-1}\rangle= \\ =\varepsilon_{n-1,l}(l_{l,n}-1)|\Lambda_{n-1}\rangle, \tag{2b}$$

where l_{ln} is the eigenvalue of operator L_{ln} on $|\Lambda_{n-1}\rangle$.

$$E_{ln}\varepsilon_{nl}|\Lambda_{n-1}\rangle=E_{l,n-1}E_{n-1,n}\varepsilon_{nl}|\Lambda_{n-1}\rangle=\cdots= \\ =(l_{l,n}-1)(l_{l,n-1}-1)\cdots(l_{l,l+1}-1)|\Lambda_{n-1}\rangle, \tag{3b}$$

$$[E_{ln},\varepsilon_{nm}]=\sum_{i=m}^{n-1}L_{m,i+1}\cdots L_{m,n-1}E_{l,i}\varepsilon_{im}= \\ =\sum_{i=m}^{n-1}\varepsilon_{im}L_{m,i+1}\cdots L_{m,n-1}E_{il} \quad \text{if} \quad l>n\geqslant m, \text{ or } l<m, \tag{4a}$$

$$E_{ln}\varepsilon_{nm}|\Lambda_{n-1}\rangle=0 \quad \text{if} \quad l<m, \tag{4b}$$

$$\tilde{\varepsilon}_{jl}\varepsilon_{nl}|\Lambda_{n-1}\rangle=0 \text{ if } i<j<n>l, \tag{5b}$$

where $\tilde{\varepsilon}_{jl}$ are the transpose of ε_{jl}, sometimes denoted by ε_{lj}.

2. *Some Cases of the Multiplication or Commutation of Lowering Polynomials ε_{nl}*

$$\varepsilon_{n1}^2=\|a_n^{*2}(a_{n-1}a_{n-1}^*+L_{1,n-1}+1)(a_{n-1}a_{n-1}^*+L_{1,n-1})\cdots(a_2a_2^*+L_{1,2}+1)\cdot \\ \cdot(a_2a_2^*+L_{1,2})a_1^2\|= \\ =\|a_n^{*2}[a_{n-1}^2a_{n-1}^{*2}+2L_{1,n-1}a_{n-1}a_{n-1}^*+(L_{1,n-1}+1)L_{1,n-1}]\cdots[\ ]a_1^2\|, \tag{6}$$

$$\varepsilon_{n1}^{\alpha}=\|a_n^{*\alpha}\left[\sum_{\beta=0}^{\alpha}a_{n-1}^{\beta}a_{n-1}^{*\beta}(L_{1,n-1}+\alpha-\beta-1)\cdots(L_{1,n-1}+1)L_{1,n-1}\frac{\alpha!}{\beta!(\alpha-\beta)!}\right]\cdots \\ \cdots\left[\sum_{\beta=0}^{\alpha}a_2^{\beta}a_2^{*\beta}(L_{1,2}+\alpha-\beta-1)\cdots(L_{1,2}+1)L_{1,2}\frac{\alpha!}{\beta!(\alpha-\beta)!}\right]a_1^{\alpha}\|, \tag{7}$$

$$\varepsilon_{nl}\varepsilon_{nm}=\|a_n^{*2}[a_{n-1}^2a_{n-1}^{*2}+(L_{l,n-1}+L_{m,n-1}-1)a_{n-1}a_{n-1}^*+L_{l,n-1}L_{m,n-1}]\cdots \\ \cdots[a_{m+1}^2a_{m+1}^{*2}+(L_{l,m+1}+L_{m,m+1}-1)a_{m+1}a_{m+1}^*+ \\ +L_{l,m+1}L_{m,m+1}](a_ma_m^*+L_{l,m})a_m\times \\ \times(a_{m-1}a_{m-1}^*+L_{l,m-1})\cdots(a_{l+1}a_{l+1}^*+L_{l\,l+1})a_l\|= \\ =\varepsilon_{nm}\varepsilon_{nl}, \quad n\geqslant m\geqslant l. \tag{8}$$

Hence

$$[\varepsilon_{nl},\varepsilon_{nm}]=0, \tag{8a}$$

$$\begin{aligned}\varepsilon_{nl}\varepsilon_{ml} = \|a_n^*(a_{n-1}a_{n-1}^* + L_{1,n-1})\cdots(a_m a_m^* + L_{1,m})a_m^*[a_{m-1}^2 a_{m-1}^{*2} + \\ + 2L_{1,m-1}a_{m-1}a_{m-1}^* + (L_{1,m-1}+1)L_{1,m-1}]\cdots \\ \cdots[a_2^2 a_2^{*2} + 2L_{1,2}a_2 a_2^* + (L_{1,2}+1)L_{1,2}]a_1^2\| \neq \\ \neq \varepsilon_{ml}\varepsilon_{nl} = \|a_n^*(a_{n-1}a_{n-1}^* + L_{1,n-1}+1)\cdots(a_m a_m^* + L_{1,m}+1)a_m^*[a_{m-1}a_{m-1}^{*2} + \\ + 2L_{1,m-1}a_{m-1}a_{m-1}^* + (L_{1,m-1}+1)L_{1,m-1}]\cdots \\ \cdots[a_2^2 a_2^{*2} + 2L_{1,2}a_2 a_2^* + (L_{1,2}+1)L_{1,2}]a_1^2\|. \end{aligned} \tag{9}$$

Hence

$$\varepsilon_{ml}\varepsilon_{nl} - \varepsilon_{nl}\begin{bmatrix} H_{n-1}-1 \\ \vdots \\ H_{m+1}-1 \\ H_m - 1\end{bmatrix}\varepsilon_{ml} = 0, \quad n \geq m \geq l. \tag{9a}$$

References

[1] Weyl, H. 1946 *Classical Groups*, Princeton.
[2] Moshinsky, M. 1963 *J. Math. Phys.*, 4, 1128.
[3] Baird, G. E. & Biedenharn, L. C. 1963 *J. Math. Phys.*, 4, 1449.
[4] Hou, Pei-yu 1964 *Sci. Sin.*, 14, 368.
[5] Гельфанд И. М. и Цейтлин М. Л. 1950. *ДАН СССР*, 71, 825; 1958. *Представления группы вращений и группы Лоренца*, Москва.
[6] Желобенко Д. П. 1962. *У.М.Н.*, 17, 27.
[7] Racah, G. 1951 *Lecture Notes*, Princeton.
[8] Дынкин Е. Б. 1947. *У.М.Н.*, 2, 59.
[9] Nagel, J. G. & Moshinsky, M. 1965 *J. Math. Phys.*, 6, 682.

研究简报

在磁单极附近荷电粒子的波函数

冼鼎昌 侯伯宇
（中国科学院高能物理研究所） （西北大学物理系）

1. 在 Dirac 提出磁单极的概念[1]后，不少作者讨论了磁单极附近的荷电粒子波函数的问题[2]，但是他们的讨论，都是基于 Dirac 的奇异弦的概念上的，所以在波函数中出现了复杂的奇异性质. 吴大峻和杨振宁[3]曾重新讨论了这个问题，他们引用电磁势是 U_1 纤维从上的联络[4]的概念，指出在磁单极旁的荷电粒子波函数不是通常的函数，而是纤维从理论中的截面，并且引入磁单极球谐函数以讨论波函数与角度有关的部份. 在本文中我们引入规范作为沿电荷——磁单极轴转动自由度的概念，指出吴大峻和杨振宁引入的磁单极球谐函数其实就是转动群有限转动表示所常用的广义球谐函数[5]，而且电荷量子化条件就是角动量量子化条件.

2. 在我们的讨论中，不限定规范的选取. 静止磁单极所产生的势的一般形式为：

$$\mathbf{A}=\mathbf{A}^{(0)}+\nabla\gamma, \tag{1}$$

其中 γ 是决定规范的一个标量函数. $\mathbf{A}^{(0)}$ 可以选择为：

$$A_r^{(0)}=A_\theta^{(0)}=0,\quad A_\phi^{(0)}=-\frac{g\cos\theta}{r\sin\theta}, \tag{2}$$

其中 g 是磁荷.

形状为式(2)的电磁势，曾经在(i)电磁场方向是 SO(3) 对称群的一个特定方向；(ii)对不可积相因子而言，磁单极周围的时空具有弯曲空间的特征，规范势与平移的 christoffel 符号有关这两个观点下推出[6,7]. 如果把资料[6]的图 2 中的同位旋单位球面上每点的标架（经纬度标架）沿着法线方向转一角度 ϕ（ϕ 也就是此点在球面上的经度），则 $\mathbf{A}$ 变成

$$\mathrm{A}_r=\mathrm{A}_\theta=0,$$

$$\mathrm{A}_\phi=\frac{-g}{r\sin\theta}(1+\cos\theta), \tag{3}$$

亦即电磁势 $\mathbf{A}$ 从 Schwinger 势(2)变成 Dirac 势(3). 如果在比北半球稍大一点的区域 R_a 中的每点上的标架都是由经纬度标架转一角度 $-\phi$ 所得到的，而在比南半球稍大一点的区域 R_b 中的每点上的标架都是由经纬度标架转一角度 ϕ 所得到的，用这样的标架，就可得到吴大峻-杨振宁势[4]：

$$\begin{aligned}&A_r=A_\theta=0,\\&A_\phi=-\frac{g}{r\sin\theta}(\cos\theta-1),\ \text{在 } R_a,\\&A_r=A_\theta=0,\\&A_\phi=-\frac{g}{r\sin\theta}(\cos\theta+1),\ \text{在 } R_b.\end{aligned} \tag{4}$$

在一般的情况下，转动角 γ 是 θ 和 ϕ 的函数，每一个转动相应于一个规范变换. 通过这个特殊的例子（U_1 规范群作为 SO(3) 群的子群），我们第一次看到了规范变换和转动的密切关系.

3. 对于有电磁势(1)存在的情况下的角动量算符 $\mathbf{L}$，仍可定义如下（我们取单位 $\hbar=c=1$）：

$$\mathbf{L}=\mathbf{r}\times(\mathbf{p}-e\mathbf{A})-q\frac{\mathbf{r}}{r}, \tag{5}$$

只要满足关系

本文 1976 年 4 月 30 日收到.

$$q = eg, \tag{6}$$

$\mathbf{L}$ 就满足角动量的对易关系式

$$[L_a, L_b] = i\varepsilon_{abc}L_c. \tag{7}$$

注意到角动量在电荷——磁单极联轴方向上的分量 $\mathbf{L}\cdot\mathbf{r}/r = -q$，根据角动量理论，它应与沿此方向转动的无穷小算符联系起来。如果定义沿此方向的转角为 γ，那么有

$$-q = -i\frac{\partial}{\partial\gamma}. \tag{8}$$

现在来论证，$\phi, \theta, \gamma(\theta, \phi)$ 和欧拉角 α, β, γ 可以如下的方式对应起来：$\alpha = \phi$, $\theta = \beta$, $\gamma = \gamma(\theta, \phi)$. 这是由于

$$\begin{aligned}\frac{\partial}{\partial\phi} &= \frac{\partial}{\partial\alpha} + \frac{\partial\gamma}{\partial\phi}\frac{\partial}{\partial\gamma},\\ \frac{\partial}{\partial\theta} &= \frac{\partial}{\partial\beta} + \frac{\partial\gamma}{\partial\theta}\frac{\partial}{\partial\gamma},\end{aligned} \tag{9}$$

以式(9)代入式(5)并注意到式(8)，角动量算符(5)便有用欧拉角写出的标准形式：

$$\begin{aligned}L_x &= i\left\{\sin\alpha\frac{\partial}{\partial\beta} + \cot\beta\cos\alpha\frac{\partial}{\partial\alpha} - \frac{\cos\alpha}{\sin\beta}\frac{\partial}{\partial\gamma}\right\},\\ L_y &= i\left\{-\cos\alpha\frac{\partial}{\partial\beta} + \cot\beta\sin\alpha\frac{\partial}{\partial\alpha} - \frac{\sin\alpha}{\sin\beta}\frac{\partial}{\partial\gamma}\right\},\\ L_z &= -i\frac{\partial}{\partial\alpha},\\ \mathbf{L}^2 &= -\left\{\frac{\partial^2}{\partial\beta^2} + \cot\beta\frac{\partial}{\partial\beta} + \frac{1}{\sin^2\beta}\times\left(\frac{\partial^2}{\partial\alpha^2} + \frac{\partial^2}{\partial\gamma^2}\right) - \frac{2\cos\beta}{\sin^2\beta}\frac{\partial^2}{\partial\alpha\partial\gamma}\right\}.\end{aligned} \tag{10}$$

可见通过式(8)可以把规范作为沿电荷——磁单极轴转动的自由度的概念引进来。

4. 荷电粒子在磁单极附近的波函数 $\psi(\mathbf{r}, \gamma)$ 应是空间(流形)坐标 $\mathbf{r}$ 的函数，同时又是规范 γ (纤维上的点)的函数。正如资料[6]所指出，波函数的奇异性是由坐标架的选取引入的，可以适当选取规范加以避免。在资料[4]中避免奇异弦的引入的方法是采取一种特殊规范(4). 在这里我们讨论一般的情况.

首先注意到 r^2, $\mathbf{L}^2$, L_z 和 $-i\dfrac{\partial}{\partial\gamma}$ 可以同时对角线化，我们先讨论 $\mathbf{L}^2$、L_z 和 $-i\dfrac{\partial}{\partial\gamma}$ 的共同本征函数 $D^l_{m,q}(\alpha, \beta, \gamma)$：

$$\begin{aligned}\mathbf{L}^2 D^l_{m,q} &= l(l+1)D^l_{m,q},\\ L_z D^l_{m,q} &= mD^l_{m,q},\\ -i\frac{\partial}{\partial\gamma}D^l_{m,q} &= qD^l_{m,q}.\end{aligned} \tag{11}$$

由式(11)及(10)不难看出，$D^l_{m,q}(\alpha, \beta, \gamma)$ 就是角动量理论中的广义球谐函数[5]：

$$D^l_{m,q}(\alpha, \beta, \gamma) = e^{im\alpha}d^l_{m,q}(\beta)e^{iq\gamma}, \tag{12}$$

式(12)中的函数 $d^l_{m,q}(\beta)$ 满足陀螺方程：

$$\left\{\frac{d^2}{d\beta^2} + \cot\beta\frac{d}{d\beta} - \frac{m^2 + q^2 - 2mq\cos\beta}{\sin^2\beta} + l(l+1)\right\}d^l_{m,q}(\beta) = 0. \tag{13}$$

这是个与规范无关的方程，其解的表达式是熟知的. 事实上，资料[4]引入的磁单极球谐函数 $\Theta_{q,l,m}$ 就是函数 $d^l_{m,-q}$ 乘上一个常数因子：

$$\Theta_{q,l,m}(\theta) = \sqrt{\frac{2l+1}{4\pi}}(-)^{q+m}d^l_{m,-q}(\theta). \tag{14}$$

5. 方程(13)容许 q 为整数或半整数的解[8]，这时，系统的角动量沿电荷-磁单极轴分量的量子化条件

$$eg = q = \frac{\text{整数}}{2} \tag{15}$$

就是电荷量子化条件.

由于 l, m 和 q 应有相同的奇偶性，于是当 q 为半整数时，就会发生统计性的问题：自旋为零的电荷及磁单极与周围的电磁场能够合成一个半整数角动量的系统！解决这个问题的一个办法就是赋与这个系统新的内部自由度，就象引入 $SU_3\times SU_3$ 对称性以解决层子模型中的统计问题那样[9]. 我们猜测，这意味着有磁单极的复合系统不是单纯意义下的电磁相互作用粒子和可能与基本粒子的构造

有关[10]的基础.

6. 关于函数 $d^l_{m,q}$ 及 $D^l_{m,q}$ 的解析性，零点位置、正交完备性等性质，可以在有关的专著[5,11]中查出. 例如，对于有相同规范的两个波函数的 Clebsch-Gordon 系数关系式是

$$\sum_{m_1m_2} D^{l_1}_{m_1,q_1}(\alpha,\beta,\gamma)\, D^{l_2}_{m_2,q_2}(\alpha,\beta,\gamma)\begin{pmatrix} l_1 & l_2 & L \\ m_1 & m_2 & M\end{pmatrix}$$

$$= D^L_{M,-q_1-q_2}(\alpha,\beta,\gamma)\begin{pmatrix} l_1 & l_2 & L \\ q_1 & q_2 & -q_1-q_2\end{pmatrix}. \quad (16)$$

在这里我们想指出，q 是一个与规范选取无关的量子数，但是在整体规范（Global Gauge）的观点[4]下，它和规范类型有关.

在 q 确定的情况下，选定规范后，由于函数 $D^l_{m,q}$ 的正交完备性，在磁单极附近的荷电粒子波函数的角向部分可以之展开. 我们注意到，波函数中与规范有关的因子是一个相因子 $e^{-iq\gamma}$，它只与 q 有关，与 l, m 无关. 规范变换只是把这个相因子变成另一个相因子 $e^{-iq\gamma'}$，不影响波函数的模和与空间有关部份的正交完备性，因此对物理结果不产生任何影响.

7. 在空间转动（由欧拉角 α_1, β_1, γ_1 所表征）下，波函数 $D^l_{m,q}(\alpha,\beta,\gamma)$ 经受如下的变换：

$$D^l_{m,q}(\alpha',\beta',\gamma')$$

$$=\sum_{m_1=-l}^{l} D^l_{m,m_1}(\alpha_1,\beta_1,\gamma_1)D^l_{m_1,q}(\alpha,\beta,\gamma), \quad (17)$$

转动后的规范 γ'，完全由 $(\alpha_1, \beta_1, \gamma_1)$ 和 (α, β, γ) 所确定，一般而言不再是原来的规范 γ 了. 例如，原来的规范 $\gamma=0$，但转动后的规范 γ' 一般不再是零. 不过由于波函数与规范有关的因子只是一个相因子，转动后的波函数如果用原来的规范写出来所差的必然是一个只与 q 有关，与 l, m 无关的相因子，因为，设原来的规范 $\gamma=\Gamma(\theta,\phi)$，则

$$D^l_{m,q}(\theta',\phi',\gamma')$$

$$= D^l_{m,q}(\theta',\phi',\Gamma(\theta',\phi'))e^{i[\gamma'-\Gamma(\theta',\phi')]}. \quad (18)$$

参 考 资 料

[1] Dirac, P. A. M., *Proc. Roy. Soc.* **A133** (1931), 60.

[2] 例如 Tamm, Ig., *Z. Physik*, **71** (1931), 141. Fierz, M., *Helv. Phys. Acta*, **17** (1944), 27.

[3] 吴大峻、杨振宁, *Dirac Monopole Without Strings: Monopole Harmonics*, Preprint ITP-SB-76-5.

[4] 吴大峻、杨振宁, *Phys. Rev.*, **D12** (1975), 3845.

[5] Edmonds, A. R., *Angular Momentum in Quantum Mechanics*, Princeton University Press, 1957.

[6] 李华钟、冼鼎昌、郭硕鸿，中山大学学报（自然科学版）1975, 3, 7.

[7] 侯伯宇、段一士、葛墨林，兰州大学学报(自然科学版), 1975, 2, 26.

[8] Bopp, F. & Haag, R., *Z. Naturforchg.*, **5A** (1950), 644.

[9] 刘耀阳，原子能，1966, 3, 232; Han, H. Y. & Nambu, Y., *Phys. Rev.*, **139B** (1965), 1006.

[10] Schwinger, J., *Science*, **165** (1969), 757.

[11] 王竹溪、郭敦仁，特殊函数，科学出版社.

第 26 卷 第 5 期 物 理 学 报 Vol. 26, No. 5
1977 年 9 月 ACTA PHYSICA SINICA Sept., 1977

研究简报

不可易规范场的规范不变守恒流*

侯 伯 宇
（西北大学物理系）

不可易规范场 $\boldsymbol{F}^{\mu\nu}$ 与费密子 $\boldsymbol{\phi}$ 的拉氏函数给定如下[1]：

$$\mathscr{L} = \mathscr{L}_F + \mathscr{L}_I + \mathscr{L}_m, \quad \mathscr{L}_F = -\frac{1}{4}\boldsymbol{F}^{\mu\nu}\cdot\boldsymbol{F}_{\mu\nu},$$

$$\mathscr{L}_I = -\boldsymbol{W}_\mu\cdot\boldsymbol{j}^\mu, \quad \mathscr{L}_m = \bar{\boldsymbol{\phi}}(i\partial_\mu\gamma^\mu - m)\boldsymbol{\phi},$$

式中 $\boldsymbol{j}^\mu = ie\bar{\boldsymbol{\phi}}\gamma^\mu\boldsymbol{X}\boldsymbol{\phi}$，$F^{\mu\nu} = \partial^\mu\boldsymbol{W}^\nu - \partial^\nu\boldsymbol{W}^\mu + e\boldsymbol{W}^\mu\times\boldsymbol{W}^\nu$. 这里，场强 $\boldsymbol{F}^{\mu\nu}$ 流 $\boldsymbol{j}^\mu$ 是伴随空间矢量，$\boldsymbol{X}$ 是产生子的表示矩阵. 矢积与标积是伴随空间的. 希腊字母为时空指标，$\boldsymbol{\phi}$ 为普通空间旋量、同位旋空间表示矩阵 $\boldsymbol{X}$ 的作用对象. 在无穷小规范变换下，

$$\boldsymbol{W}^\mu(x)' = \boldsymbol{W}^\mu_{(x)} + e\boldsymbol{W}^\mu_{(x)}\times\boldsymbol{\epsilon}(x) + \partial^\mu\boldsymbol{\epsilon}(x), \quad \boldsymbol{\phi}(x)' = \exp(\boldsymbol{X}\cdot\boldsymbol{\epsilon}(x))\boldsymbol{\phi}(x),$$

式中 $\boldsymbol{\epsilon}(x)$ 是无穷小变换参数.

$\boldsymbol{j}^\mu$ 本为无规范场（$\mathscr{L} = \mathscr{L}_m$）时第一类规范变换（变换参数 $\boldsymbol{\epsilon}(x)$ 取与 x 无关的常量 $\boldsymbol{\epsilon}(x) = \boldsymbol{\delta\theta}$）所对应的守恒流[2]：

$$\boldsymbol{j}^\mu = \frac{\delta\mathscr{L}_m}{\delta(\partial_\mu\boldsymbol{\phi})}\cdot\frac{\delta\boldsymbol{\phi}}{\boldsymbol{\delta\theta}} = ie\bar{\boldsymbol{\phi}}\gamma^\mu\boldsymbol{X}\boldsymbol{\phi}, \quad \partial_\mu\boldsymbol{j}^\mu = \frac{\delta\mathscr{L}_m}{\boldsymbol{\delta\theta}} = 0.$$

但是在 $\boldsymbol{\epsilon}(x)$ 非常量的第二类规范变换下，$\mathscr{L}_m$ 不再是不变的. 为了保持不变性，必须在 $\mathscr{L}$ 中引进规范场，如 $\mathscr{L}_F$ 和 $\mathscr{L}_I$[1]. 由含规范场的 $\mathscr{L}$ 变分得到的运动方程中 $\boldsymbol{j}^\mu$ 是协变散度的源，

$$\nabla_\mu\boldsymbol{F}^{\mu\nu} \equiv (\partial_\mu + e\boldsymbol{W}_\mu\times)\boldsymbol{F}^{\mu\nu} = \boldsymbol{j}^\nu$$

不再是普通散度的源. 与此相应，易见 $\boldsymbol{j}^\mu$ 的协变散度为零，

$$\nabla_\mu\boldsymbol{j}^\mu \equiv \partial_\mu\boldsymbol{j}^\mu + e\boldsymbol{W}_\mu\times\boldsymbol{j}^\mu = 0$$

一般不再是守恒流了. 实际上这时仍然存在有守恒流[1]，我们注意到该流仍然可由整个 $\mathscr{L}$ 在第一类规范变换下的性质求得，只是要考虑到，在引进不可易规范场后，即使在常规范变换 $\boldsymbol{\epsilon}(x) = \boldsymbol{\delta\theta}$ 下，$\boldsymbol{W}^\mu$ 也要有变更：$\delta\boldsymbol{W}^\mu = \boldsymbol{W}^\mu\times\boldsymbol{\delta\theta}$. 考虑到此变更对流的贡献 $\frac{\delta\mathscr{L}}{\delta(\partial_\mu\boldsymbol{W}^\nu)}\cdot\frac{\delta\boldsymbol{W}^\nu}{\boldsymbol{\delta\theta}} = -e\boldsymbol{F}^{\mu\nu}\times\boldsymbol{W}_\nu$. 整个流应是 $\boldsymbol{J}^\mu = \boldsymbol{j}^\mu - e\boldsymbol{F}^{\mu\nu}\times\boldsymbol{W}_\nu$. 由 $\mathscr{L}$ 的第一类规范不变性及运动方程可证[2]，此 $\boldsymbol{J}^\mu$ 是守恒的，$\partial_\mu\boldsymbol{J}^\mu = \frac{\delta\mathscr{L}}{\boldsymbol{\delta\theta}} = 0$. 而且 $\boldsymbol{J}^\mu$ 可表为某一反对称张量的普通散度（因为 $\boldsymbol{J}^\mu\cdot\boldsymbol{\delta\theta}$ 是"动力学流"）. 的确，$\boldsymbol{J}^\nu = \partial_\mu\boldsymbol{F}^{\mu\nu}$. 不过由于 $\boldsymbol{J}^\mu$ 中有非物理量 $\boldsymbol{W}^\nu$，$\boldsymbol{J}^\mu$ 不是规范协变的，其物理意义不清楚.

在文献[3]中提出了物理量应是规范不变的. 该文已发现在无中间矢粒子的情况下，存在规范不变的守恒流 j^μ，它是规范不变的场 $F^{\mu\nu}$ 的普通散度. 本文将这一结果推广到

* 1976 年 2 月 28 日收到.

含中间矢粒子的一般情况．为此，将上一段的第一类规范变换推广为 $\boldsymbol{\epsilon}(x)=\boldsymbol{n}(x)\delta\theta$，式中 $\boldsymbol{n}(x)\cdot\boldsymbol{n}(x)=1$．即各点的规范转角是常量 $\delta\theta$，但转轴的方向则可以不同．按照常规变分得到对应的流为

$$J^{\mu}=\frac{\delta\mathscr{L}}{\delta(\partial_{\mu}\boldsymbol{\phi})}\frac{\delta\boldsymbol{\phi}}{\delta\theta}+\frac{\delta\mathscr{L}}{\delta(\partial_{\mu}\boldsymbol{W}^{\nu})}\frac{\delta\boldsymbol{W}^{\nu}}{\delta\theta}=\boldsymbol{n}\cdot\boldsymbol{j}^{\mu}-\boldsymbol{F}^{\mu\nu}\cdot(\partial_{\nu}\boldsymbol{n}+e\boldsymbol{W}_{\nu}\times\boldsymbol{n})$$
$$=\boldsymbol{n}\cdot\boldsymbol{j}^{\mu}-\boldsymbol{F}^{\mu\nu}\cdot\nabla_{\nu}\boldsymbol{n}=\boldsymbol{n}\cdot(\boldsymbol{j}^{\mu}-e\boldsymbol{F}^{\mu\nu}\times\boldsymbol{B}_{\nu}),$$

式中的中间矢粒子 $\boldsymbol{B}_{\nu}\equiv\frac{1}{e}\boldsymbol{n}\times\nabla_{\nu}\boldsymbol{n}$[3]，在 $\boldsymbol{n}$ 为常矢(转轴方向处处相同)的特殊情况下，上式末端圆括号内的量退化为前一段的 $\boldsymbol{J}^{\mu}$，但一般情况下宜采用此处括号内的明显协变量，整个 J^{μ} 甚至是规范不变的．由 $\mathscr{L}$ 的规范不变性和运动方程可得，$\partial_{\mu}J_{\mu}=\frac{\delta\mathscr{L}}{\delta\theta}=0$，即 J^{μ} 是守恒流．J^{μ} 又是规范不变的反对称张量 $\boldsymbol{n}\cdot\boldsymbol{F}^{\mu\nu}$ 的普通散度源．的确

$$\partial_{\mu}(\boldsymbol{n}\cdot\boldsymbol{F}^{\mu\nu})=\nabla_{\mu}(\boldsymbol{n}\cdot\boldsymbol{F}^{\mu\nu})=\boldsymbol{n}\cdot(\nabla_{\mu}\boldsymbol{F}^{\mu\nu})+(\nabla_{\mu}\boldsymbol{n})\cdot\boldsymbol{F}^{\mu\nu}$$
$$=\boldsymbol{n}\cdot\boldsymbol{j}^{\nu}+(\nabla_{\mu}\boldsymbol{n})\cdot\boldsymbol{F}^{\mu\nu}=J^{\mu}.$$

考虑到 $F^{\mu\nu}$ 是反对称的，也可由此式看出 J^{μ} 的守恒性．守恒荷 $Q=\int J^{0}d^{3}x$ 是规范变换 $\boldsymbol{n}(x)\delta\theta$ 的产生子．

为了阐明此荷流的物理意义，讨论 $\boldsymbol{n}(x)$ 的各种情况如下：

$(\nabla_{\mu}\boldsymbol{n})\cdot\boldsymbol{j}^{\mu}=0$ 的情况　这时 J^{μ} 可分为两个各自单独守恒的流．场的“外源流” $\boldsymbol{n}\cdot\boldsymbol{j}^{\mu}$ 以及场的“自源流” $(\nabla_{\nu}\boldsymbol{n})\cdot\boldsymbol{F}^{\mu\nu}$．这时有

$$\partial_{\mu}(\boldsymbol{n}\cdot\boldsymbol{j}^{\mu})=(\partial_{\mu}\boldsymbol{n})\cdot\boldsymbol{j}^{\mu}+\boldsymbol{n}\cdot(\partial_{\mu}\boldsymbol{j}^{\mu})=(\partial_{\mu}\boldsymbol{n})\cdot\boldsymbol{j}^{\mu}-\boldsymbol{n}\cdot(e\boldsymbol{W}_{\mu}\times\boldsymbol{j}^{\mu})$$
$$=\boldsymbol{j}^{\mu}\cdot(\partial_{\mu}\boldsymbol{n}-e\boldsymbol{n}\times\boldsymbol{W}_{\mu})=\boldsymbol{j}^{\mu}\cdot\nabla_{\mu}\boldsymbol{n}=0,$$
$$\partial_{\mu}[(\nabla_{\nu}\boldsymbol{n})\cdot\boldsymbol{F}^{\mu\nu}]=\nabla_{\mu}[(\nabla_{\nu}\boldsymbol{n})\cdot\boldsymbol{F}^{\mu\nu}]=(\nabla_{\mu}\nabla_{\nu}\boldsymbol{n})\cdot\boldsymbol{F}^{\mu\nu}+(\nabla_{\nu}\boldsymbol{n})\cdot(\nabla_{\mu}\boldsymbol{F}^{\mu\nu})$$
$$=\frac{1}{2}[(\nabla_{\mu}\nabla_{\nu}-\nabla_{\nu}\nabla_{\mu})\boldsymbol{n}]\cdot\boldsymbol{F}^{\mu\nu}+(\nabla_{\nu}\boldsymbol{n})\cdot\boldsymbol{j}^{\nu}=\frac{1}{2}e(\boldsymbol{F}_{\mu\nu}\times\boldsymbol{n})\cdot\boldsymbol{F}^{\mu\nu}=0.$$

将以上两式的推导过程反过来易证，如可选择 $\boldsymbol{n}$ 使相应的外源流或自源流之一守恒，则必有 $(\nabla_{\mu}\boldsymbol{n})\cdot\boldsymbol{j}^{\mu}=0$，且另一流也必守恒．此处，自源流

$$\nabla_{\mu}\boldsymbol{n}\cdot\boldsymbol{F}^{\mu\nu}=\boldsymbol{n}\cdot[\boldsymbol{B}^{\nu}\times(\nabla_{\mu}^{(A)}\boldsymbol{B}_{\nu}-\nabla_{\nu}^{(A)}\boldsymbol{B}_{\mu})]$$

即带电矢粒子流，式中 $\nabla_{\mu}^{(A)}\equiv\partial_{\mu}+e\boldsymbol{A}_{\mu}\times$，$\boldsymbol{A}_{\mu}\equiv\boldsymbol{W}_{\mu}-\boldsymbol{B}_{\mu}$．

如果 $\boldsymbol{j}^{\mu}(x)$ 的四个时空分量的同位旋方向是相同的(对于有质量粒子的流 $\boldsymbol{j}^{\mu}$ 这总是成立的)，就可以选 $\boldsymbol{n}(x)\times\boldsymbol{j}^{\mu}(x)=0$．于是，易证 $(\nabla_{\mu}\boldsymbol{n})\cdot\boldsymbol{j}^{\mu}=0$，为上段的特例．这时守恒的外源流 $\boldsymbol{n}\cdot\boldsymbol{j}^{\mu}=|\boldsymbol{j}^{\mu}|\equiv j^{\mu}$ 正好是 $\boldsymbol{j}^{\mu}$ 在同位旋空间的模．

如果存在 $\boldsymbol{n}(x)$ 满足 $\nabla_{\mu}\boldsymbol{n}=0$，则守恒流 J^{μ} 中的自源流分量恒为零，只剩下外源流 $\boldsymbol{n}\cdot\boldsymbol{j}^{\mu}$．文献 [3] 中证明了这时 $\boldsymbol{F}_{\mu\nu}\times\boldsymbol{n}=0$，此 $\boldsymbol{F}^{\mu\nu}$ 可以 Abel 化．Abel 化场 $F^{\mu\nu}\equiv|\boldsymbol{F}^{\mu\nu}|=\boldsymbol{n}\cdot\boldsymbol{F}^{\mu\nu}$ 满足麦克斯韦方程．如果外源流 $j^{\mu}(x)$ 的荷是电(磁，…)荷，则 $\boldsymbol{n}(x)$ 的同伦性质[3-5]1)决定其对偶荷流矢 ${}^{*}j^{\mu}(x)$ 磁(电，…)荷的分布．

由上述可见，如选 $\boldsymbol{n}(x)$ 处处沿某一荷的产生算符的同位旋方向，则 J^{μ} 即该荷的荷流

1) 在文献 [3] 印出后，承中国科学院物理研究所朱重远同志转寄来文献 [4]．

矢. $\boldsymbol{n}(x)$ 的方向可能是由自发破缺决定的，例如是 Higgs 粒子真空平均值的同位旋方向，也可能就是 $\boldsymbol{j}^{\mu}$ 的方向（如 $\boldsymbol{n}\times\boldsymbol{j}^{\mu}=0$），还可能是 $\boldsymbol{F}^{\mu\nu}$ 的固有方向（如 $\boldsymbol{F}^{\mu\nu}(x)$ 的六时空分量的同位旋方向是相同的）.

文献 [6] 中的无源情况 $\nabla_{\mu}\boldsymbol{n}=0$，$\boldsymbol{j}^{\mu}=0$（取 $\boldsymbol{n}$ 为 $\boldsymbol{F}^{\mu\nu}(x)$ 的同位旋方向单位矢），外源流、自源流均为零. 有源情况 $\nabla_{\mu}\boldsymbol{n}\neq 0$，$\boldsymbol{j}^{\mu}\neq 0$，但是 $\boldsymbol{n}$ 方向的守恒流（电荷流矢）$J^{\mu}=\boldsymbol{n}\cdot\boldsymbol{j}^{\mu}-\boldsymbol{F}^{\mu\nu}\cdot\nabla_{\nu}\boldsymbol{n}=0$，而且 $\boldsymbol{n}\cdot\boldsymbol{j}^{\mu}$ 及 $\boldsymbol{F}^{\mu\nu}\cdot\nabla_{\nu}\boldsymbol{n}$ 均等于零.

参 考 文 献

[1] 杨振宁, R. L. Mills, *Phys. Rev.*, **96** (1954), 191.
[2] S. L. Adler and R. F. Dashen, Current Algebra (1968).
[3] 侯伯宇、 段一士、 葛墨林,兰州大学学报(自然科学版),第 2 期(1975), 26.
[4] J. Arafune, P. G. O. Freund and J. Goebel, *J. Math. Phys.*, **16** (1975), 433.
[5] 吴大峻、 杨振宁, *Phys. Rev.*, **D12** (1975), 3845.
[6] 吴大峻、 杨振宁, *Phys. Rev.*, **D12** (1975), 3843.

GAUGE-INVARIANT CONSERVED CURRENT OF A NON-ABELIAN GAUGE FIELD

Hou Bo-yu
(Department of Physics, Northwest University)

第26卷 第4期 物理学报 Vol. 26, No. 4
1977年7月 ACTA PHYSICA SINICA July, 1977

各级微扰展开下自发破缺的规范无关性、么正性及可重整性* 1)

侯伯宇
(西北大学物理系)

提　　要

用微扰论展开明显地讨论了标粒子与规范场矢粒子自发破缺 Abel 模型，发现在各种可重整规范下同阶各费曼图的内线非物理分量贡献的规范有关部分只与外线的质壳外部分互相依存．在质壳上只剩下物理分量的贡献，亦即转化成了么正规范．这样就明显地验证了么正规范与可重整规范在质壳上的全同，从而说明可重整规范是么正的，以及么正规范下怎样会出现剩余发散，为何剩余发散必然相消，揭穿了么正规范的隐藏可重整性．

规范场的自发破缺理论近几年取得了许多进展[1,2]．它有两种相反而相成的方案．其中一种方案明显可重整，但因含非物理粒子，表观上似违反么正性．另一种方案只含物理粒子，它明显地么正，但是表观上却不可重整．现已证明了可重整方案的么正性[3,4]及两种方案的 S 矩阵元的等价[5]．这些证明采用泛函积分研究全格林函数，优点是简单明瞭，但是泛函积分法换部积分时有可能出毛病，而且笼统地处理整个格林函数，一些问题研究得还不明显，需要用微扰论展开认真地深入探讨．例如：1. 整个 S 矩阵元规范无关，而各级微扰各费曼图却与规范有关．怎样从各自的“有关”抵消为整体的“无关”呢？虽然某些过程的具体计算[6-8]表明，低次图求和时规范有关部分的确抵消掉，但是抵消的普遍规律还未得到．2. 含非物理粒子的可重整方案如何转化为只含物理粒子的么正方案还不够明显．虽然已证明了两种方案 S 矩阵元相等[5]（单环线具体验证见文献[9]），也证明了可重整方案中非物理极点相消[4]，以及可重整的 R_ξ 规范[6]当 $\xi\to 0$ 时，只剩下物理粒子的贡献，亦即么正规范[10]．但是，由于非物理粒子对于质壳外格林函数（虚粒子线间的广义顶角）的贡献不为零，故两种方案的质壳外格林函数并不相等．怎样从内部虚粒子过程广义顶角的“不相等”转化成整个真实过程质壳上 S 矩阵元的“相等”，还不够清楚．3. 么正方案各阶不可约发散费曼图用通常的重整化方法只能消去一部分发散，还可能有剩余[11]．而低阶图的具体计算表明，同阶所有图的剩余发散在质壳上正好相抵消[11,12]．为什么由无剩余发散的可重整规范转化为等价的么正规范时会使某些费曼图有剩余发散，而同阶图的和恰好抵消而无剩余发散呢？上述各种问题的解决就是本文要探讨的内容．

本文采用微扰论，由简入繁地逐步讨论可重整规范下内线非物理粒子的贡献在同阶

* 1976年6月15日收到．

1）本工作于1974年3月完成．

图间的各种相消机构，明显地证明互相抵消后只剩下与标粒子"外"线的质壳外动量互相依存的部分（除与环线对应有 $\delta^4(0)$ 项以外）．当有些"外"线不在质壳上时，这些"外"线实际上是虚粒子内线，应与更外一层的线相连．如此类推，直到所有的外线都是质壳上的实粒子线为止，于是非物理粒子的规范有关贡献正好全部抵消掉（发散项是整个被积式抵消掉，与正常化及重整化无关，也无需取极限 $\xi\to 0$），只剩下物理粒子的贡献，也就是幺正规范．这也就明显地证明了物理过程的规范无关性及任意规范的幺正性．

一、拉氏函数与费曼法则

自发破缺的拉氏函数[5]

$$\mathscr{L} = -\frac{1}{4}(\partial^\mu W^\nu - \partial^\nu W^\mu)^2 + \frac{1}{2}(eV)^2(W^\mu)^2 + \frac{1}{2}(\partial^\mu\varphi + eW^\mu\chi)^2 + \frac{1}{2}(\partial^\mu\chi - eW^\mu\varphi)^2 - eVW_\mu(\partial^\mu\chi - eW^\mu\varphi) - \lambda V^2\varphi^2 - \lambda V\varphi(\varphi^2+\chi^2) - \frac{1}{4}\lambda(\varphi^2+\chi^2)^2. \quad (1)$$

选定规范项 $-\frac{1}{2}\left(\xi\partial^\mu W_\mu - \frac{eV}{\xi}\chi\right)^2$ 及规范补偿项 $\phi^*\left(\partial^2 + \frac{e^2V^2}{\xi} + \frac{e^2V}{\xi}\varphi\right)\phi$（$\phi$ 为虚拟粒子，构成环线，且服从费米统计）．以上为 R_ξ 规范，其费曼法则见图 1．

幺正规范时，

$$\mathscr{L} = -\frac{1}{4}(\partial^\mu\mathscr{W}^\nu - \partial^\nu\mathscr{W}^\mu)^2 + \frac{1}{2}(eV)^2(\mathscr{W}^\mu)^2 + \frac{1}{2}(\partial^\mu\varphi)^2 - \lambda V^2\varphi^2 + \frac{1}{2}e^2(\varphi^2+2V\varphi)\mathscr{W}_\mu^2 - \lambda V\varphi^3 - \frac{1}{4}\lambda\varphi^4. \quad (2)$$

等效哈密顿中还有 $i\delta^4(0)\ln\left(1+\frac{\varphi}{V}\right)$ 以保证协变性[13]．费曼法则中 $\mathscr{W}$，φ 顶角与图 1 W，φ 顶角相同，φ 传播子也不变，但 $\mathscr{W}$ 传播子只含 $U^{\mu\nu}(K) = \dfrac{g^{\mu\nu} - \dfrac{K^\mu K^\nu}{(eV)^2}}{K^2-(eV)^2+i\epsilon}$.

注意到图 1 中，只有 W 的非物理分量 $\frac{K^\mu K^\nu}{(eV)^2}G(K)$，$\chi$，$\phi$ 的传播子 $G(K)$ 及顶点 f 与 ξ 有关，且各顶点都含有偶数根 W，χ 线．可见含 ξ 有关因子 G 的 W，χ 线必然构成连续线；或者两端是横极化的 $\mathscr{W}$ 外线，或者构成闭合环线．ϕ 线则已知是闭合的．因此可以分开考虑各 W-χ 线或 ϕ 环线与 ξ 的关系．

二、$2\mathscr{W}\,n\varphi$ 树枝图

如采用幺正规范，则此树枝图的一般形式是两端 $\mathscr{W}$ 外线间由若干（0 到 $(n-1)$ 根）首尾相连的 $\mathscr{W}$ 传播子 U 连接，各传播子间的 a 或 v 型（图 1）顶点与支状的 φ 线相连（最外端有 n 个 φ 线）．R_ξ 规范时，分支形状与幺正规范相同，只是 $\mathscr{W}$ 内线换为 W-χ

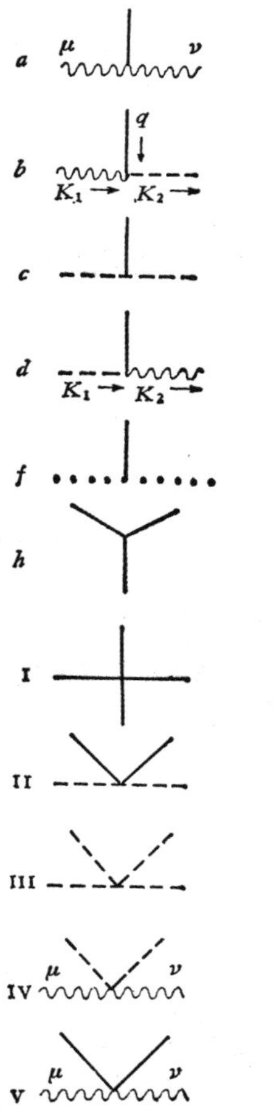

x	$i\frac{1}{K^2-(eV)^2/\xi+i\epsilon}\equiv iG(K)$
W^μ	$\frac{1}{i}\frac{g^{\mu\nu}-K^\mu K^\nu/(eV)^2}{K^2-(eV)^2+i\epsilon}+\frac{1}{i}\frac{K^\mu K^\nu}{(eV)^2}\frac{1}{K^2-(eV)^2/\xi+i\epsilon}$
$\equiv$ U + G	$iU^{\mu\nu}(K)+\frac{1}{i}\frac{K^\mu K^\nu}{(eV)^2}G(K)$
ψ	$iG(K)$
φ	$i\frac{1}{q^2-2\lambda V^2+i\epsilon}\equiv i\Delta(q)$
a	$i2e^2Vg^{\mu\nu}$
b	$-e(q+K_2)=e(K_1-2K_2)$
c	$-i2\lambda V$
d	$e(2K_1-K_2)$
f	$-i\frac{e^2V}{\xi}$
h	$-i6\lambda V$
I	$-i6\lambda$
II	$-i2\lambda$
III	$-i6\lambda$
IV	$i2e^2g^{\mu\nu}$
V	$i2e^2g^{\mu\nu}$

图 1 费曼法则

线的各种组合．本节证明，如将 W 的传播子分为 U，G 二项(图1)后展开，并对 W-χ 各种组合求和，这时如果把因子(传播子 G,U,Δ) 相同的项归并在一起，则任何含 G(非物理粒子内线)的项(亦即规范 ξ 有关项)都将抵消掉，只剩下且必剩下一切只含 U,Δ 的项，这正好与只含物理粒子内线的么正规范全同．先分析最简单的图．

$2\mathscr{W}2\varphi$ 的二阶图(如图 2 及其置换)．将图 2II 的 W 传播子分为 G 项(纵 G)及 U 项，

纵 G 的贡献为 $ie^2 2\epsilon_i \cdot KG(K)2K \cdot \epsilon_f$，而图 2III χ 传播子 G(标 G) 则给出 $-ie^2 2\epsilon_i \cdot KG(K)2K \cdot \epsilon_f$（这里用了 $\mathscr{W}$ 外线是横极化的，$\epsilon_i \cdot p_i = \epsilon_f \cdot p_f = 0$，式中 $\epsilon_{i,f}$ 为初、末态极化矢，$p_{i,f}$ 为其动量）. 这两部分恰好相消（如图 3，以后图中 W 传播子下附以 U, G 时分别代表 U, G 项）. 结果剩下与 ξ 无关的图 2II 中的 U 项及图 2I, 2IV 正好与么正规范相同.

$2\mathscr{W}3\varphi$ 的三阶图(图 4)，其中前三个图与图 2 类似. 图 $4a,b,c,d$ 含有 $U(K_1)U(K_2)$, $U(K_1)G(K_2)$, $G(K_1)U(K_2)$, $G(K_1)G(K_2)$ 等类型的项，其中，含于图 a 的 $U(K_1)U(K_2)$ 项与 ξ 无关. 以下分别讨论 GU 及 GG 项.

含传播子 $G(K_1)U(K_2)$ 的项(图 $5a$, $5d$, $5a'$, $5d'$) 求和后给出 $\frac{ie^2}{V} 2\epsilon_i \cdot K_1 G(K_1) 2K_1 \cdot \epsilon_f$ 正好与少一个 U 传播子的图 4IV 的 G 项相消.

含传播子 $G(K_1)G(K_2)$ 的项来自图 $4a,b,c,d$，其和为

$$ie^2 2\epsilon_i \cdot K_1 G(K_1) \frac{1}{V} [2K_1 \cdot K_2 + (-2K_1 \cdot K_2 + K_1^2) - 2\lambda V^2 - (2K_1 \cdot K_2 - K_2^2)] G(K_2) 2K_2 \cdot \epsilon_f = ie^2 2\epsilon_i \cdot K_1 G(K_1) \frac{1}{V} (q_2^2 - 2\lambda V^2) G(K_2) 2K_2 \cdot \epsilon_f. \tag{1}$$

上式左方方括号内的四项分别与图 $4a$, b, c, d 的贡献相应. 今后将 W 传播子非物理项 G(纵 G) 的动量因子 $\frac{K^\mu}{eV} \cdot \frac{K^\nu}{eV}$ 分别纳入左右侧的顶点，将 χ 传播子 $\frac{-1}{i} G$ (标 G) 的负号分为两个 i 分别给左右侧，于是图 $4a$ 中顶角包括两边传播子 G 中的动量因子成为

$$\frac{K_1^\nu}{eV} i2e^2 V g_{\nu\lambda} \frac{K_2^\lambda}{eV} = i \frac{2K_1 \cdot K_2}{V} \equiv ia(K_1, K_2), \tag{2a}$$

简写为 a. 图 $4b$ 中顶角为

$$\frac{i}{V}(-2K_1 \cdot K_2 + K_1^2) \equiv ib, \tag{2b}$$

图 $4c$ 中顶角为

$$\frac{i}{V} 2\lambda V^2 \equiv ic, \tag{2c}$$

图 $4d$ 中顶角为

$$\frac{i}{V}(2K_1 \cdot K_2 - K_2^2) \equiv id. \tag{2d}$$

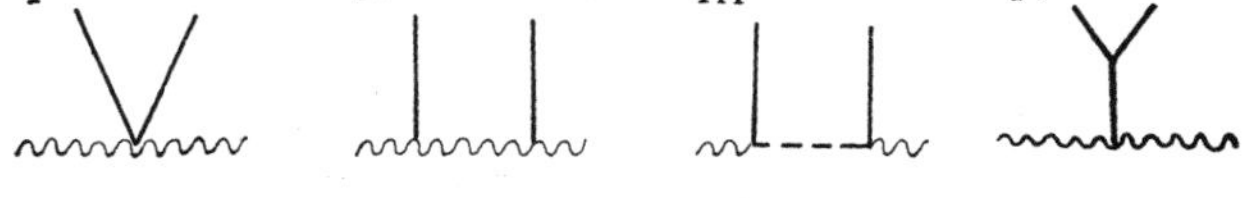

图 2 $2\mathscr{W}2\phi$ 二阶图

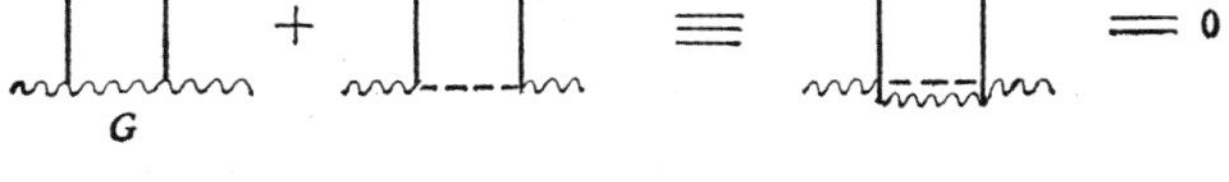

图 3 纵 G 与标 G 相消

（双线表示 χ 及 W 传播子中 G 项的和）

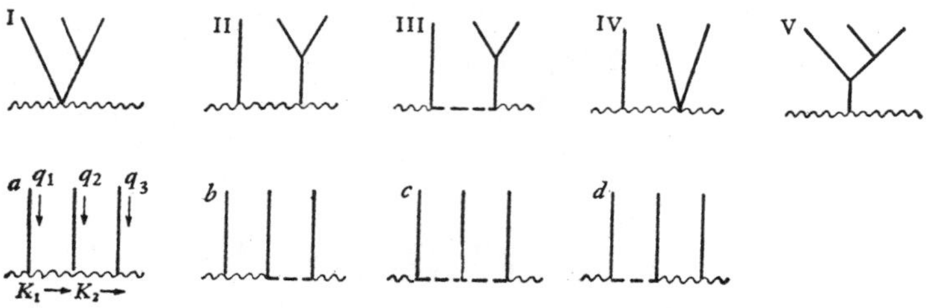

图 4

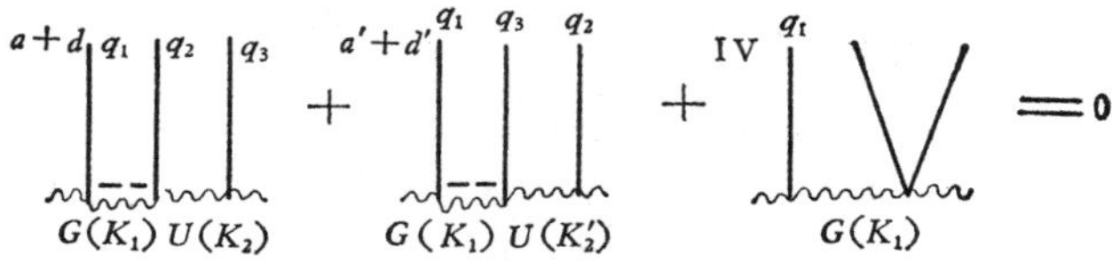

图 5

于是(1)式方括号内四项可写为

$$a+b-c-d=\frac{1}{V}(q_2^2-2\lambda V^2)\equiv\Gamma(q_2)\ \text{（见图 6）.} \tag{3}$$

如 $\varphi(q_2)$ 线在质壳上，则其值为零．可见四个图中 GG 项之和(1)式相消．

综上所述，$2\mathscr{W}3\varphi$ 三阶图（图 4）求和后消去了所有含 G 的 ξ 有关项，剩下图 4I 及 II，IV，a 中的 U 项，正好与么正规范相同．

如 $\varphi(q_2)$ 线不在质壳上，例如与另二外 $\mathscr{W}$ 线耦合，如图 7a,b,c,d，则注意到 $q_2^2-2\lambda V^2=\Delta(q_2)^{-1}$，易见它正好与缩去 $\varphi(q_2)$ 线而得的图 7IV 相消．

以上是 W-χ 单线的最基本相消机构：1. 纵 G 及标 G 与外 $\mathscr{W}$ 线耦合符号相反而相消（图 3）；2. 与纵 G 及标 G 耦合的 U 被约去（图 5）；3. 与纵 G 及标 G 耦合的 φ 线为零（质壳上时，图 6）或被约去（图 7）．

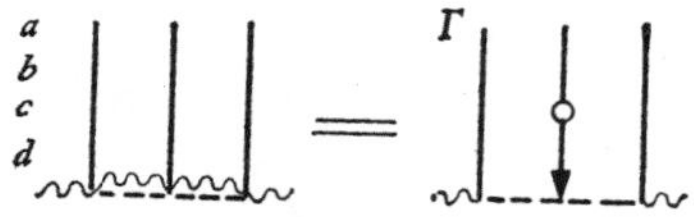

图 6

⊥ 代表 $\frac{-1}{V}$； ○ 代表 $\frac{1}{i}(q^2-2\lambda V^2)$

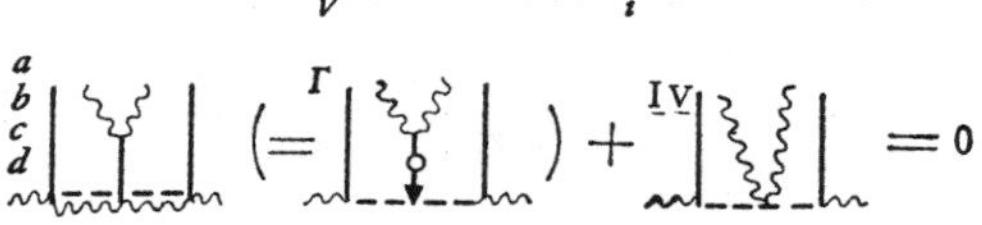

图 7 $4\mathscr{W}2\varphi$ 四阶图中含 $G(K_1)G(K_2)$ 的项

以下逐步推广到 $2\mathscr{W}(n+2)\varphi$ 的各种含 G 同类项，先考虑图 8 所示只含 $G(K_i)$ 不含 U,Δ 的 $\prod_{j=1}^{n+1} G(K_j)$ 同类项的和

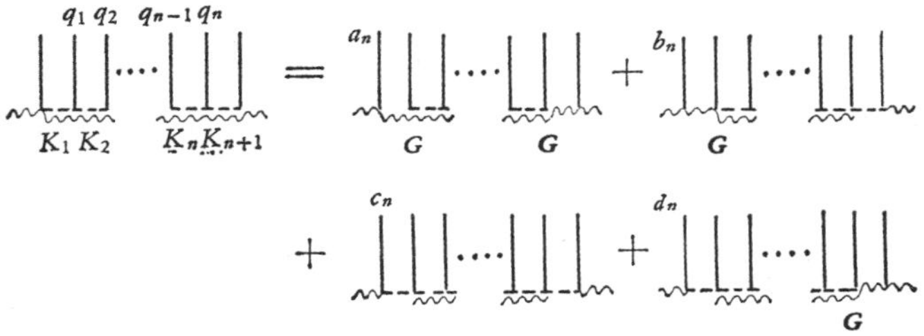

图 8

$$ie^2 2e_i \cdot K_1 G(K_1)[a_n + b_n - c_n - d_n]G(K_{n+1})2K_{n+1}\cdot e_f, \tag{4}$$

式中 a_n, b_n, c_n, d_n 分别为图 8a, b, c, d 内的 $2Gn\varphi$ 内夹 $(n-1)$ 双G的 n 阶顶点,各含 2^{n-1} 项(一般项内因子的顺序为 d 或者 a 后可接 Ga 或 Gb, b 或 c 后接 Gc 或 Gd)可由

$$a_n = a_m G a_{n-m} + b_m G d_{n-m}, \tag{4a}$$

$$b_n = a_m G b_{n-m} + b_m G c_{n-m}, \tag{4b}$$

$$c_n = d_m G b_{n-m} + c_m G c_{n-m}, \tag{4c}$$

$$d_n = d_m G a_{n-m} + c_m G d_{n-m}. \tag{4d}$$

递推构成(各式第一、二项分别是 $\varphi(q_m)$, $\varphi(q_{m+1})$ 间为纵 $G(K_{m+1})$ 或者标时的贡献).可证

$$a_n + b_n - c_n - d_n = \sum_{r+s+t=n}(-b_r + c_r)GX_sG(c_t + d_t) \quad r, t \geqslant 0, s \geqslant 1, \tag{4Γ}$$

式中 $X_s \equiv \sum \Gamma Gc_{j_1} G\Gamma Gc_{j_2} G\Gamma \cdots Gc_{jl} G\Gamma$. 此处求和遍及 $j_1 + j_2 + \cdots + j_l + l + 1 = s$, $l = 0, 1, \cdots, s-1$ 的各项,且

$$a_1 \equiv a, \quad b_1 \equiv b, \quad c_1 \equiv c, \quad d_1 \equiv d; \quad a_0 = c_0 \equiv \frac{1}{G}, \ b_0 = d_0 \equiv 0,$$

此式的归纳证明如下:用(4a—d)式可得

$$a_{n+1} + b_{n+1} - c_{n+1} - d_{n+1} = (a_n - d_n)G(a + b) + (b_n - c_n)G(c + d)$$
$$= (a_n + b_n - c_n - d_n)G(a + b) - (b_n - c_n)G\Gamma.$$

将(4Γ)式代入上式右方第一项,再利用由(4c,d)式易得 $(c_t + d_t)G(a + b) = c_t G\Gamma + c_{t+1} + d_{t+1}$, 即可证 $n+1$ 时的(4Γ)式.

注意,(4Γ)式右方各项为含顶点 $\Gamma(q_i)$ 1 至 n 个的各种组合,它当 $\varphi(q_i)$ 为实粒子外线时为零;而当 $\varphi(q_i)$ 为虚粒子内线时,则与缩去 $\Delta(q_i)$ 的图相关.

以下考虑既含G又含 U,Δ 的一般项.定义顶角 A_l, B_l, C_l, D_l 如图 9,图中阴影代表一切有 l 个 φ 外线且只含物理传播子 U,Δ 的 l 阶树枝图的和 $(A_1 \equiv a, B_1 \equiv b\cdots)$.则含有给定的$G$因子的同类项的和(如图 10),其结构和图 8 相似,只是对应的一阶顶点 a, b, c, d 需换为 l 阶树枝顶点 A_l, B_l, C_l, D_l. 因此,求和后也可化为(4Γ)式的形式. 只需注意这时右方出现因子 $A_{l_i} + B_{l_i} - C_{l_i} - D_{l_i} \equiv \Gamma_{l_i}$ 的各种组合. 以上已得 $\Gamma_1(\equiv \Gamma$, (3)式,如图 6),可得 Γ_2(如图 11),Γ_3,Γ_4(如图 12)如下式:

$$iT_2 = i\frac{-1}{V^2}\sum_{j=1}^{2}(q_j^2 - 2\lambda V^2),$$

$iT_3 = \frac{i}{V^3}(q_1^2 - 2\lambda V^2)[2! - \Delta(q_2 + q_3)\cdot 3\cdot 2\lambda V^2] +$ 含 $(q_2^2 - 2\lambda V^2),(q_3^2 - 2\lambda V^2)$ 的置换类似项，

$$iT_4 = \frac{-i}{V^4}(q_1^2 - 2\lambda V^2)\{3! - 2![\Delta(q_2 + q_3) + \Delta(q_3 + q_4) + \Delta(q_4 + q_2)]\cdot 3\cdot 2\lambda V^2 + \Delta(q_2 + q_3 + q_4)\cdot 3\cdot 2\lambda V^2 + \Delta(q_2 + q_3 + q_4)3\cdot 2\lambda V^2[\Delta(q_2 + q_3) + \Delta(q_3 + q_4) + \Delta(q_4 + q_2)]\cdot 3\cdot 2\lambda V^2\} + \text{置换类似项}. \tag{5}$$

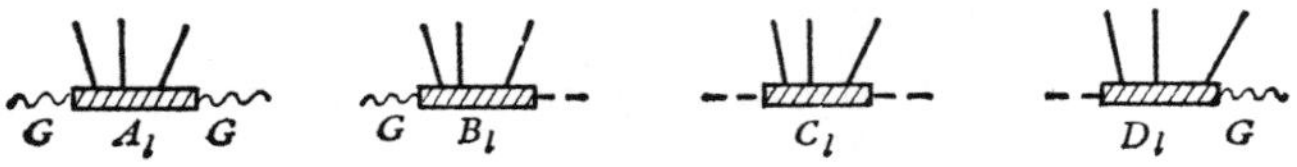

图 9

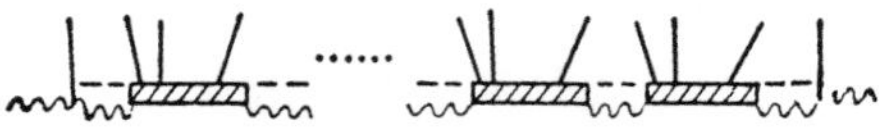

图 10

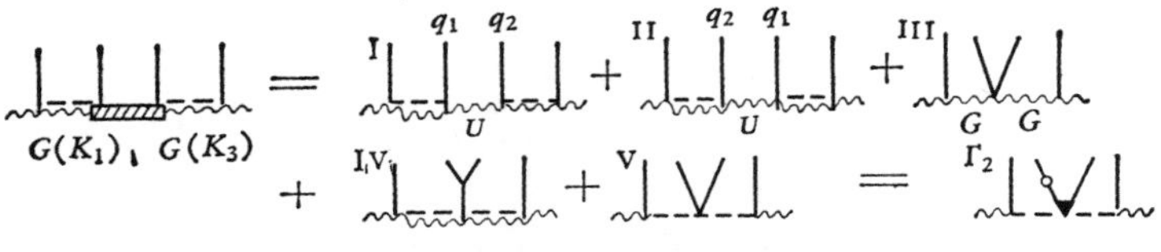

图 11 2𝒲4φ 四阶图中含 $G(K_1)G(K_3)$ 项

一般可证，Γ_l 为 l 个含因子 $q_j^2 - 2\lambda V^2(j = 1,\cdots,l)$ 项的和，各项结构如图 12 的推广．在两侧的 iG 之间夹有共顶点的 m 根 φ 线 $(l \geqslant m \geqslant 2)$，其一为“带○”线，其余 $(m-1)$ 根线通过顶点 h,I（图 1）的各种组合逐步分支为 $(l-1)$ 根 φ 线．此图的对应法则是：$2\,G\,m\,\varphi$ 汇合处对应 $(-V)^{-m}(m-1)!$ 带○线格外有因子 $\frac{1}{i}(q^2 - 2\lambda V^2)$，其余的 Δ 及 3，4 φ 顶角按通常的费曼法则(图1)，且注意需按 l 个 $\varphi(q_i)$ 的各种次序求和：

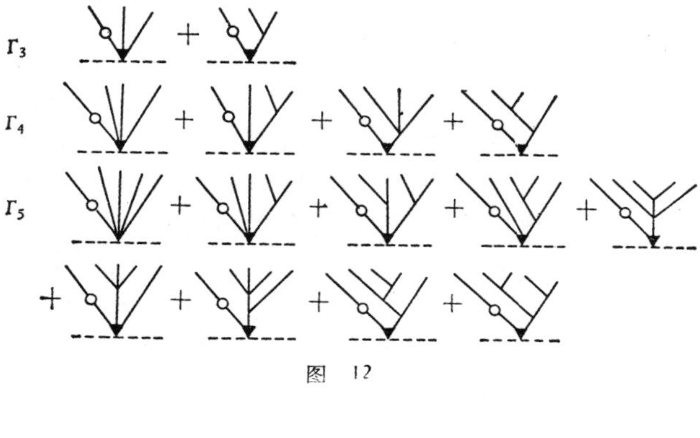

图 12

$$i\Gamma_l = i\frac{(-)^{l-1}}{V^l}(l-1)!\sum_{j=1}^{l}(q_j^2 - 2\lambda V^2) + i\frac{(-)^{l-2}(l-2)!}{V^{l-1}}\left\{\sum_{j=3}^{l}(q_j^2 - 2\lambda V^2)6\lambda V\cdot \Delta(q_1+q_2) + \text{置换对称项}\right\}$$

$$+ i\frac{(-)^{l-3}}{V^{l-2}}(l-3)!\left\{\sum_{j=4}^{l}(q_j^2 - 2\lambda V^2)6\lambda\Delta(q_1+q_2+q_3) + \text{置换对称项}\right\}$$

$$+ i\frac{(-)^{l-3}}{V^{l-2}}(l-3)!\left\{\sum_{j=5}^{l}(q_j^2 - 2\lambda V^2)6\lambda V\Delta(q_1+q_2)6\lambda V\Delta(q_3+q_4) + \text{置换对称项}\right\}$$

$$+\cdots$$

$$\equiv \sum_{j=1}^{l}\left[\frac{i}{V}(q_j^2 - 2\lambda V^2)\cdot Y_{l-1}(q_1\cdots q_{j-1}, q_{j+1}, \cdots q_l)\right]. \tag{6}$$

如再将各种含G的一切ξ有关项求和,并将含同样的 Γ_l 因子的项归并到一起,则得

$$\sum_{l+m+r=n} iE_lG(K_l)(\mathscr{A}_m + \mathscr{B}_m + \mathscr{C}_m + \mathscr{D}_m)G(K_{l+m})E_r + \varphi(q_i)\ \text{置换项}$$

$$= i\sum_{l+s+\Sigma l+\Sigma j+t+r=0} E_lG(-\mathscr{B}_s+\mathscr{C}_s)G\Gamma_{l_1}G\mathscr{C}_{j_1}\Gamma_{l_2}G\mathscr{C}_{j_2}\cdots\Gamma_{l_m}G(\mathscr{C}_t+\mathscr{D}_t)GE_r$$

$$n \geqslant m \geqslant 1, \tag{7}$$

式中 $\mathscr{A}_m, \mathscr{B}_m, \mathscr{C}_m, \mathscr{D}_m$ 代表完全的(含各种 G, U, Δ 的) $2G$ (纵纵、纵标、标标、标纵) $m\varphi$ 树枝顶点, E_l 代表 1 外 $\mathscr{W}$ 1 内纵 $G\left(\text{提出因子 } \frac{K_\mu}{eV}\right)$ l 根 φ 线只含 U, Δ 的树枝图. 上式各项至少含 1 Γ_l, 故当 φ 在质壳上时为零, 即 ξ 有关项全消掉. 只剩不含G的项与么正规范同.

三、n 根 φ 外线的单环线

$n=1$ 即蝌蚪图 (图 13). 图 13a 分为 G, U 二项, G项与图 13f 求和得的与 ξ 无关的 $\frac{1}{V}\delta^4(0)$, 即等效哈密顿 $\delta^4(0)\ \ln\left(1+\frac{\varphi}{V}\right)$ 的 φ 一次项. 此项和图 13a 中剩下的U项以及图 13h 的和已是么正规范的结果. 但多出一个图 13c, $\frac{1}{2V}2\lambda V^2\int\frac{d^4K}{K^2 - e^2V^2/\xi}$, 似乎破坏了 ξ 无关性. 原因是蝌蚪的尾线 $\varphi(q)$ 不在质壳上 (除非 $2\lambda V^2 = 0$), 这正说明蝌蚪图不代表物理过程, 因此可与 ξ 有关[10]. 但以下将看到, 作为物理过程费曼图的一部分时,正是它的 ξ 有关保证了同阶各图之和与 ξ 无关.

$n=2$ 即 φ 的二阶自能图(图 14). 其中 aa, bd, cc, db 图中的 $G(K_1)G(K_2)$ 项求和得(暂不积分)

$$\frac{1}{2}[a(K_1,K_2)G(K_2)a(K_2,K_1)G(K_1) + b(K_1,K_2)G(K_2)d(K_2,K_1)G(K_1)$$

$$+ d(K_1,K_2)G(K_2)b(K_2,K_1)G(K_1)$$

$$+ c(K_1,K_2)G(K_2)c(K_2,K_1)G(K_1)] \equiv \frac{1}{2}(a_2+c_2)G. \tag{1}$$

利用第二节中 (4Γ) 式及

$$a(K_1,K_2)+b(K_1,K_2)=\frac{1}{V}K_1^2,\quad a(K_1,K_2)-d(K_1,K_2)=\frac{1}{V}K_2^2,\tag{2}$$

可得

$$\frac{1}{2}(a_2+c_2)G=-\frac{1}{2}[\Gamma G\Gamma G+cG\Gamma G+\Gamma GcG]+\frac{1}{2}(a+b)G(a+b)G$$
$$+\frac{1}{2}(a-d)G(a-d)G=-\frac{1}{2}[\Gamma G\Gamma G+cG\Gamma G+\Gamma GcG]$$
$$+\frac{K_1^2G(K_1)K_2^2G(K_2)}{V^2}.\tag{3}$$

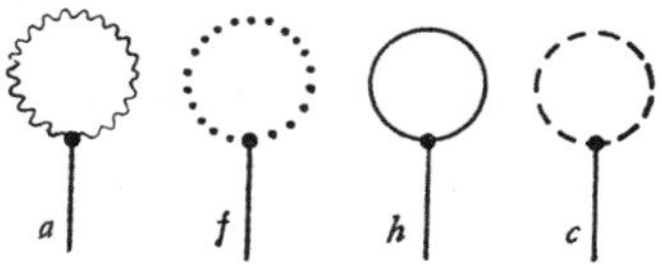

图 13

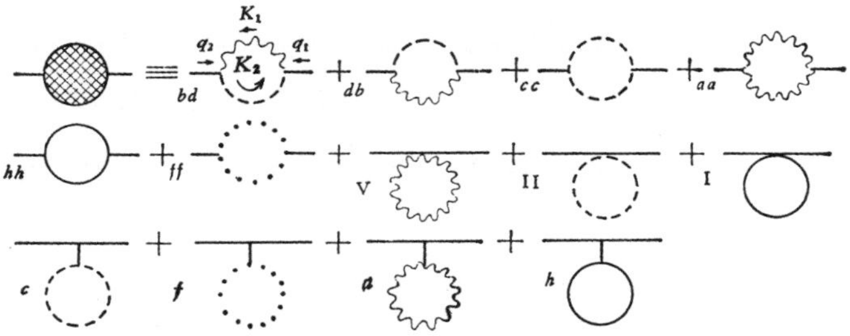

图 14

前一项依存于质壳外动量，与第二节中 (4Γ) 式相似，但多出含循环动量的后一项，这是单环线的特点.

图 14aa, bd 中的 $G(K_2)U(K_1)$ 项求和，并在固定 $G(K_2)$ 的变量 K_2 的条件下，对 $q_1(=q)$, $q_2(=-q)$ 的置换求平均，得 $\frac{1}{2V^2}(-3K_2^2+q^2)G(K_2)$ 约去了 $U(K_1)$：仿此，db, aa 的 $G(K_1)U(K_2)$ 项给出 $\frac{1}{2V^2}(-3K_1^2+q^2)G(K_1)$，再将图 14v 的 G 项 $\frac{1}{V^2}K^2G(K)$ 写作 $\frac{1}{2V^2}(K_1^2G(K_1)+K_2^2G(K_2))$ (相当于积分变量的变换)；图 14II 及 c 的和写作 $\frac{-1}{2V^2}2\lambda V^2[G(K_1)+G(K_2)]$. 以上四项求和得

$$\frac{1}{2V}\Gamma(q)[G(K_1)+G(K_2)]-\frac{1}{V^2}[K_1^2G(K_1)+K_2^2G(K_2)].\tag{4}$$

第一项的构成与图 8 相似，但又多出与循环动量有关的后一项. 环线时矛盾的特殊性即在于与(3),(4)式中存在循环动量项同时还有 ϕ 环线项(图 14ff). 求和得

$$\frac{1}{V^2}\left[K_1^2G(K_1)K_2^2G(K_2)-K_1^2G(K_1)-K_2^2G(K_2)-\frac{e^2V^4}{\xi^2}G(K_1)G(K_2)\right]-\frac{1}{V^2} \qquad (5)$$

与 ξ 无关. 对 K 积分后正好是幺正规范补偿哈密顿的 φ 二次项 $-\frac{1}{V^2}\delta^4(0)$. 总之,质壳上只剩: 这一 $\delta(0)$ 项: 图 14aa 中的 UU,V 中的 U,a 中的 U,a 中 G 与 f 的和 ($\delta(0)$),及含 Δ 环线的 hh, i, h, 与幺正规范全同.

以下推广到 $n\varphi$ 外线的单环线 n 阶图,且只考虑与 ξ 有关的 W-χ 及 ϕ 环线.

W-χ 环线各图中 n 个 G 传播子夹有 n 个一阶顶点的项为

$$\frac{1}{2}[a_n+c_n]G=-\frac{1}{2}\sum_{r+s+t=n}(c_rGX_sGc_tG+b_rGX_sGd_tG)+\frac{1}{V^n}\prod_{j=1}^{n}[K_j^2G(K_j)]. \qquad (6)$$

上式中所用符号同前节. 如利用 $c_tGc_r+d_tGb_r=c_{t+r}$, 则(6)式求和号内并为 $\Sigma\cdots c_{l_1}G\Gamma Gc_{l_2}G\Gamma Gc_{l_3}G\Gamma G\cdots$ 的 n 阶轮换对称和,此项与质壳外动量有关,它的意义和结构与第二节中(4Γ)式同. (6)式末项是环线特有的循环动量项.

对于在 $m(n\geqslant m\geqslant 1)$ 个给定的 G 传播子间夹有 m 个树枝顶点 $A_l\cdots(l\geqslant 1)$ 的 G 同类项,求和后可得相似的结果: 质壳外动量项中的一阶顶点 a, b, c, d, Γ 需换为对应的 l 阶顶点 A_l, B_l, C_l, D_l,Γ_l, 循环动量项中只含给定的 m 个 $G(K_j)$ 与对应的 K_j (不含与 $A_l\cdots$ 的内线对应的 $G(K_j)$ 及 K_j), 还可能含传播子 Δ 与顶角 h,i (见图 1). 可证各循环动量项与 ϕ 环线项求和后约去含 ξ 的因子,构成各种 $\delta(0)\cdot\Delta$ 项与幺正规范补偿哈密顿的贡献相同.

最后,单环线的和可表为

$$\Sigma\cdots\mathscr{C}_{i_1}G\Gamma_{l_1}G\mathscr{C}_{i_2}G\Gamma_{l_2}G\cdots\ +\ 幺正规范. \qquad (7)$$

四、各 W-χ 线的互连

以上已证单 W-χ 线的 ξ 有关项中的 G 同类项可归并为含顶点 Γ_l 的各种组合的项(第二节(7)式,第三节(7)式),而各 Γ_l 又可表为各项均含一因子$(q_i^2-2\lambda V^2)$ 的形式(第二节(6)式). 故当 $\varphi(q_i)$ 均在质壳上时,ξ 有关项为零. 当某些 $\varphi(q_i)$非外线而与其它 W-χ 连接时,对应的 $q_i^2-2\lambda V^2\neq 0$, 故 ξ 有关项中含此因子的项不为零. 本节将证明,考虑到与另外的 W-χ 线相连时出现的以上未曾计算到的图的贡献,就可将此非零项抵消掉.

例如,图 7 中 $\varphi(q)$ 非外线,故图 7Γ 非零,但考虑到另接 $2\mathscr{W}$ 线时应有的IV型顶点(图 7IV) 正好抵消.

再如图 15, 含 Γ 的三种组合的三项分别对应有含顶点 IV 的三种组合的三项与之相消.

仿此易见,同阶单 W-χ 线 ξ 有关项被 $G\mathscr{C}G$ 因子分隔成顶点 Γ_l 的各种组合, 只需其中每个含非外线 $\varphi(q_i)$ 的 $\Gamma_l(\cdots q_i\cdots)$ 均分别对应有抵消项, 即可用抵消项的相应组合消掉此 ξ 有关项. 因此,以下对 Γ_l 顶端连接其它 W-χ 线的各种情况分别寻找

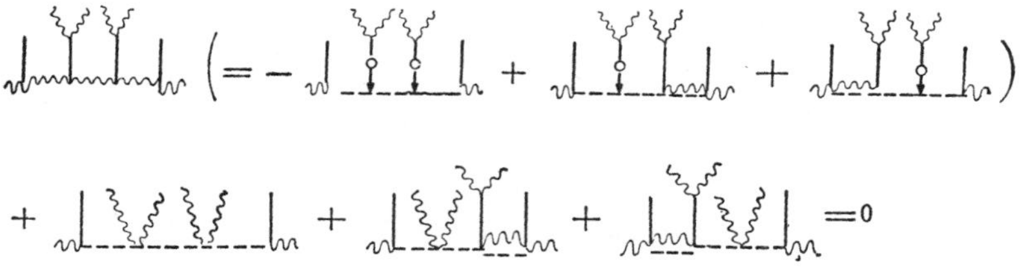

图 15

抵消项.

1. Γ_l 顶端接 $2\mathscr{W}$ 横极化外线　Γ_1 如图 7，讨论如上. Γ_2 连以 $2\mathscr{W}$ 如图 16 第一项(括号内是它按因子 $(q_i^2 - 2\lambda V^2)$ 归并成的两项，见第二节(6)式)，第二项为其抵消项. Γ_3 连以 $2\mathscr{W}$ 如图 17 首项，后二项为抵消项，将第二节(6)式(图 12，8，6)代入左方各项求和即得右方. 所剩各项均含因子 $(q_2^2 - 2\lambda V^2)$ 或 $(q_3^2 - 2\lambda V^2)$，它或为零(当 $\varphi(q_2)$，$\varphi(q_3)$ 为外线)或另有与此因子相关的抵消项使之消去，可以分别考虑之. 故我们舍去左方含因子 $(q_{2,3}^2 - 2\lambda V^2)$ 的各项，简写如图 16，17 下方二式.

一般由 Γ_l 的结构(图 12，第二节(6)式)可见如 $\varphi(q_1)$ 顶端连 $2\mathscr{W}$，则需抵消的是含因子$(q_1^2 - 2\lambda V^2)$的各项，于是，只需将 $2\mathscr{W}$ 线接到图 12 (其 l 阶推广)中含○的 $\varphi(q_1)$ 线上，然后缩去此带○线，滑到共汇合顶点的另一支的最底线中间，构成 II 型(图 1)顶点，再给 II 型顶点下方新出现的传播子 $\Delta(q_1 + \cdots)$ 带上○，乘以 $\Delta^{-1}(q_1 + \cdots)$，即得对应的抵消项. 这样的抵消项必有对应的树枝顶点包含它，如图 16，17 的上方.

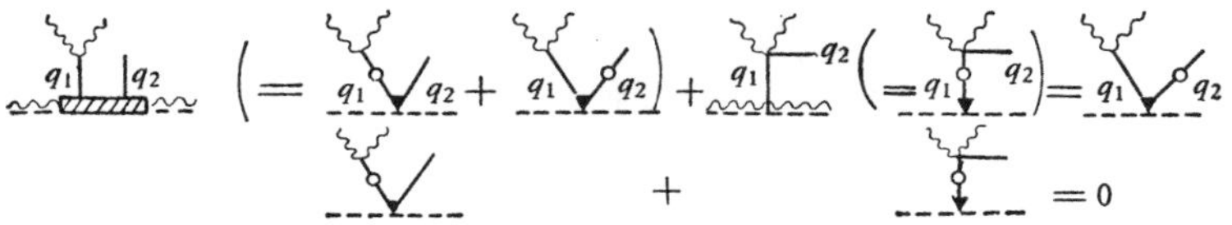

图 16

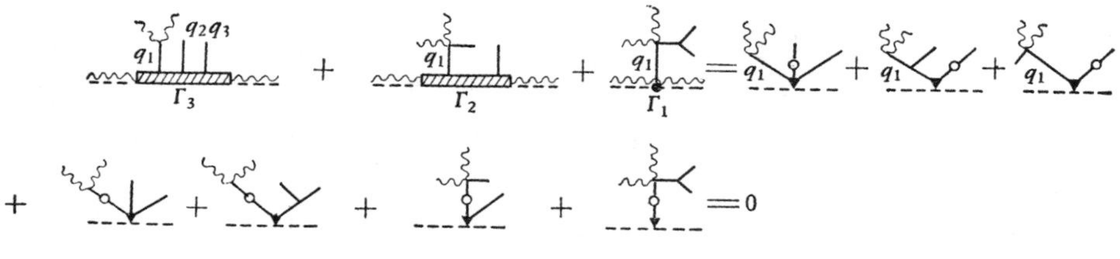

图 17

2. Γ_l 顶端与一横 $\mathscr{W}$ 外线及一 G 内线衔接　例如 Γ_1 (图 18). 一个新特点是要考虑到三道互为抵消:

$$2i\frac{e}{V}\,\epsilon_i \cdot (K_1 + K_2 + K_3) = 2i\,\frac{e}{V}\,\epsilon_i \cdot (-p_i) = 0.$$

本节以下规定 K 的方向朝内为正. 一般情况证明仿下节，从略.

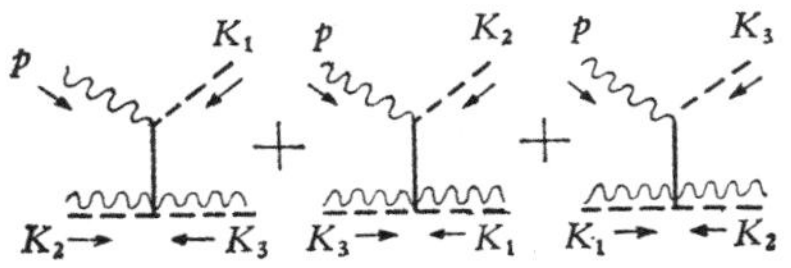

图 18

3. Γ_l 顶端与 2G 内线连接 Γ_1 如图 19，这里后二项为两根 $W-\chi$ 线衔接时特有的 III, IV 型顶点项，其和为

$$-\frac{i}{V^2}\{[(-)(K_1+K_2)(K_3+K_4)-2\lambda V^2]+2K_1K_2+2K_3K_3\}+t,u \text{ 道 } -6i\lambda=0.$$

一般，Γ_{n+1} 上接 2G 需考虑 $4Gn\varphi$ 的全部 $n+2$ 阶树枝图．如图 20．为了避免算重复，把连系上下两根 $W-\chi$ 线的那根内 φ 线上的分支全部归入下面的 Γ_{t+1}（上面的无阴影框代表一切内部只含 U 传播子及 a, V 型顶点的 $2\,G\,u\,\varphi\,u$ 阶图，其和为 Γ_u^0，$\Gamma_1^0=a+b-c=\frac{1}{V}q^2$，$\Gamma_2^0$ 为图 8 内前三项 $\Gamma_u^0=\frac{(-)^{u-1}}{V^u}(u-1)!\sum_{j=1}^{u}q_j^2$）．可证，图 20 各道求和后，$\Gamma_{t+1}$ 中含因子 $Q_t^2-2\lambda V^2$ 的项的贡献之和与含顶点 IV 的最后一项互消．最后只剩下含其它“外”线 $\varphi(q_i)$ 的相应因子 $(q_i^2-2\lambda V^2)$ 的项．

以上证明了 Γ_l 顶端某一线 $\varphi(Q)$ 接其它 $W-\chi$ 线时必有抵消项将含因子 $(Q^2-2\lambda V^2)$ 的项消去(注意：此项的消去，与其他 $\varphi(q_i)$ 是否为外线无关)．情况 2,3 中 IV 型顶点抵消项全部用掉，结果只剩下原 Γ_l 中含其它 $(q_i^2-2\lambda V^2)$ 的项，故如某些 $\varphi(q_i)$ 非外线，又可独立地选择缩去此 $\varphi(q_i)$ 的抵消项使之消去．情况 1 的 II 型顶点项有时不完全用掉(如图 17Γ_2)，但剩余项在 II 型顶点下方的结构仍与图 12 同．易见，可为它找到相应的独立抵消项同上．

以上只讨论了树枝图．但是，假如 $\varphi(Q)$ 所连接的上下两条 $W-\chi$ 线已通过另外的 $\varphi(Q')$ 线相连，或已在左或(和)右方连为一条 $\chi-W$ 线，实质上仍可沿用上述的 $(Q^2-2\lambda V^2)$ 项的抵消机构．细节在此暂不详述，只需注意还要将循环动量项与对应的 ϕ 环线项合并．

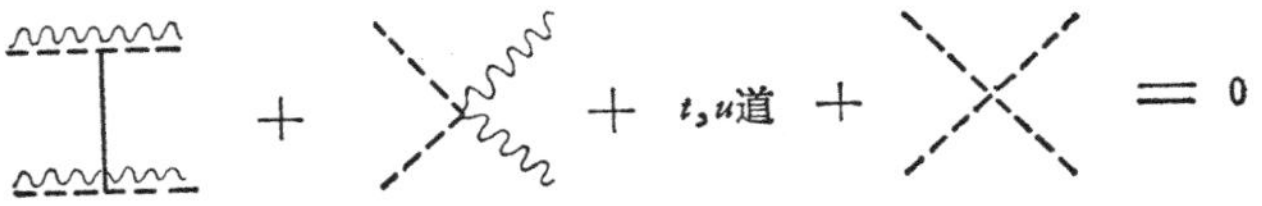

图 19

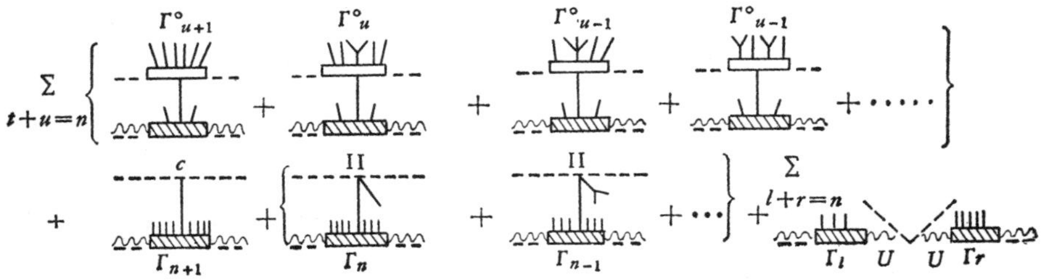

图 20

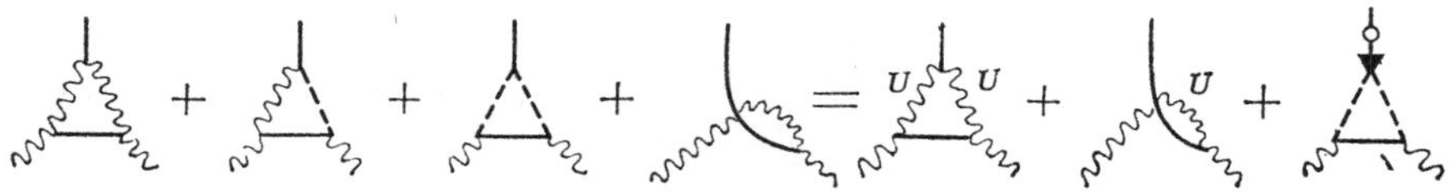

图 21

五、剩 余 发 散

本节以 R_ξ 规范可重整为出发点，利用规范不变性将它与么正规范连系起来，简单讨论么正规范的剩余发散的出现与消除.

以上已证 R_ξ 规范与么正规范只差非物理粒子贡献的质壳外动量项. 例如标粒子二阶自能图(图 14)(第三节(3),(4)式).

$$\Pi_\xi^{(2)} = \Pi_U^{(2)} - \frac{(q^2-2\lambda V^2)^2 + 4\lambda V^2(q^2-2\lambda V^2)}{2V^2}\int G(K)G(K+q)d^4K + \frac{1}{V^2}(q^2-2\lambda V^2)\int G(K)d^4K, \tag{1}$$

式中 $\Pi_\xi^{(2)}$, $\Pi_U^{(2)}$ 分别为 R_ξ, 么正规范(包括 $\delta(0)$ 项)的二阶自能,右方后二项为质壳外动量项. 式中各项为发散的,作质量及波函数重整(亦即各项在 $(q^2-2\lambda V^2)$ 点作二次减除),得

$$\Delta\Pi_\xi^{(2)} = \Delta\Pi_U^{(2)} - \frac{1}{2V^2}[(q^2-2\lambda V^2)^2 + 4\lambda V^2(q^2-2\lambda V^2)]\int G(K)G(K+q)d^4K + \frac{4\lambda V^2}{2V^2}(q^2-2\lambda V^2)[\int G(K)G(K+q)d^4K]|_{q^2=2\lambda V^2}. \tag{2}$$

R_ξ 可重整,故(2)式左方 $\Delta\Pi_\xi$ 有限,但右方减除后的质壳外动量项是对数发散的,其发散部分为与 ξ 无关的 $-\frac{1}{2V^2}(q^2-2\lambda V^2)^2\frac{i\pi^2}{2-\frac{n}{2}}\Big|_{n=4}$[9]. 可见 $\Delta\Pi_U$ 项必须含对数发散

$$\frac{1}{2V^2}(q^2-2\lambda V^2)^2\frac{i\pi^2}{2-\frac{n}{2}}\Big|_{n=4} \tag{3}$$

与之相消.

仿此可见,由于 $W^\mu W^\nu\phi$ 三阶不可约顶角中图 21 所示的部分是 ξ 相关的,可见么正规范的 UU,U 在一次减除(与顶角重整有关)后还要有剩余发散(即右方第三项的发散部分反号):

$$(q^2-2\lambda V^2)g^{\mu\nu}\frac{e^2}{V}i\pi^2\left(2-\frac{n}{2}\right)\Big|_{n=4}. \tag{4}$$

一般,由质壳外动量项在减除后有剩余发散,可见么正规范在减除后也有剩余发散. 但它在实粒子线(质壳上)时为零,故没有妨碍. 它仅在虚粒子内线时残留. 以下将看到,正是虚粒子线的此种剩余保证内含它的高阶图在质壳上无剩余.

以 $\mathscr{W}$ $\mathscr{W}$ 散射四阶不可约图(图 22) 为例. R_ξ 时它非原始发散,各图和应是有限的,的确这时仅图22中 3,5,6,7, 8 图(包括各道)对数发散,其发散部正好相消. 幺正规范时,则只剩下图 22 中 1, 4, 6 图,这时均对数发散,发散部的和非零[11]. 亦即幺正规范此四阶不可约图的发散属于"不可重整"类型. 但易见图 22 中 1,2,3 图之和与 ξ 无关(参考图 3),故幺正规范时,虽不含发散的图 22 中 3 图,但图 22 中 1 图升为发散来代替它. 同理, R_ξ 时图 22 中 5 图的发散项转移给了幺正时的图 22 中 4 图. 于是,幺正规范表观上"不可重整"问题在于未考虑图 22 中 7, 8 图二项的 ξ 相关项. 亦即同时应考虑图 23,它与图 22 中 7, 8 图之和是 ξ 无关的. R_ξ 规范时图 23 中的自能与顶角可重整而无剩余发散,只有图 22 中 7, 8 图给出发散:

$$2e^4 g_{\mu\nu} g_{\lambda\rho} \int G(K)G(K+q)d^4K = 2e^4 g_{\mu\nu} g_{\lambda\rho} \frac{i\pi^2}{2-\frac{n}{2}}\bigg|_{n=4} + \text{有限}, \tag{5}$$

$$-2e^4 g_{\lambda\rho} \int 4K^\mu K^\nu G(K+p_i)G(K-p_f)\Delta(K)d^4K = -2e^4 g_{\mu\nu} g_{\lambda\rho} \frac{i\pi^2}{2-\frac{n}{2}}\bigg|_{n=4} + \text{有限}. \tag{6}$$

幺正时不存在图 22 中 7, 8 图,代替它的是图 23 中自能与顶角的幺正规范部分,重整后它含有剩余发散如本节(3),(4)式,其贡献正好与本节(5),(6)式的发散部相等.

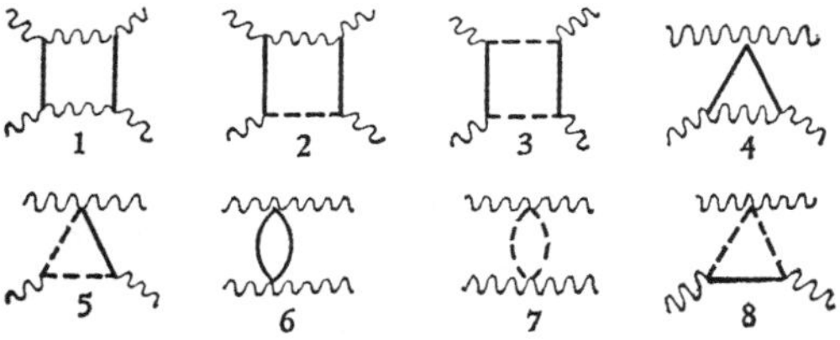

图 22

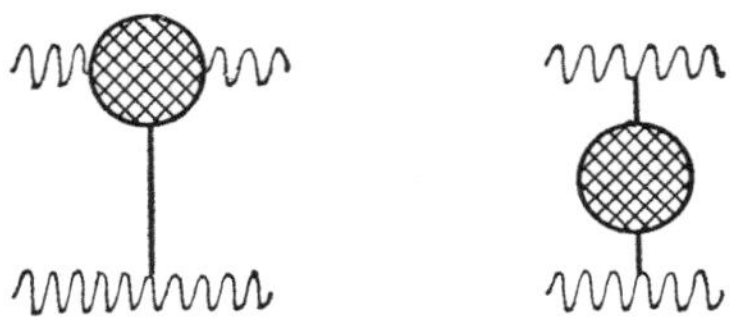

图 23 阴影为三阶不可约顶角、二阶不可约自能(包括蝌蚪)

一般,规范不变性保证发散在 ξ 相关图间转移,因此质壳上幺正规范应与 R_ξ 同样可重整. 但如只考虑"不可约"图,则幺正规范即使在质壳上也有不可重整的发散. 原因是 R_ξ 的有些发散项转移给了 ξ 相关的同阶但"可约"的图,必须由此"可约"图所含低阶不可约图的质壳外剩余发散补充此项以保证可重整性. 细节及重整化方案在此不述.

参 考 文 献

[1] S. Weinberg, *Phys. Rev. Letters,* **19** (1967), 1264; *ibid.*, **27** (1971), 1688.
[2] A. Salam, Elementary Particle Theory (1969), Wiley.
[3] G. t'Hooft, *Nucl. Phys.,* **B33** (1971), 173. **B35** (1971), 127.
[4] B. W. Lee, *Phys. Rev.,* **D5** (1972), 823.
[5] Justin, J. Zinn and B. W. Lee, *Phys. Rev.,* **D5** (1972), 3155.
[6] K. Fujikawa, B. W. Lee, and Sanda, *Phys. Rev.,* **D6** (1972), 2923.
[7] Y. P. Yao, *Phys. Rev.,* **D7** (1973), 1647.
[8] S. Weinberg, *Phys. Rev.,* **D7** (1973), 2887.
[9] K. Shizuya, *Prog. Theor. Phys.,* **49** (1973), 2106.
[10] T. Appelquist, J. Carazzone, T. Goldman, and H. R. Quinn, *Phys. Rev.,* **D8** (1973), 1747.
[11] T. Appelquist, and H. R. Quinn, *Phys. Letters,* **39B** (1972), 229.
[12] S. Y. Lee, *Phys. Rev.,* **D6** (1972), 1701, 1803.
[13] T. D. Lee, and C. N. Yang, *Phys. Rev.,* **128** (1962), 885.

GAUGE INDEPENDENCE, UNITARITY AND RENORMALIZABILITY OF A SPONTANEOUSLY BROKEN MODEL IN THE PERTURBATION THEORY

Hou Bo-yu
(Department of Physics, Northwestern University)

Abstract

This paper studies explicitly the Abelian spontaneously broken gauge model for scalar and gauge bosons by using perturbation expansions . It is discovered that in various renormalizable gauges (R gauge) the gauge dependent part contributed by unphysical components of inner lines in the sum of all Feynman diagrams of the same order coexists only with the outer mass shell components of outer lines. On the mass shell only the contribution by physical components is left, just as in the unitary gauge (U gauge). Thus, the identitical nature of the R gauge and U gauge on the mass shell has been shown explicitly, hence may be seen the unitarity of the R gauge. It is also shown why residue divergences appear in the U gauge, and why these residue divergences will necessarily cancal each other out, thus revealing the crypto-renormalizability of the U gauge.

Vol. XXI No. 4 SCIENTIA SINICA July - August 1978

THE DECOMPOSITION AND REDUCTION OF GAUGE FIELD AND DUAL CHARGED SOLUTION OF ABELIANIZABLE FIELD

HOU BO-YU (侯伯宇)
(*Northwest University, Sian*)

DUAN YI-SHI (段一士) AND GE MO-LIN (葛墨林)
(*Lanchou University*)

Received March 18, 1977.

ABSTRACT

In accordance with the isospin direction of some charge operator, we decompose the non-Abelian gauge potential and field strength explicitly and gauge covariantly into "neutral" fields, whose sources are these charges and their dual charges, and "vector particle" fields which are charged. The decomposition expressions of $SU(N)/U(1)^{N-1}$, $SU(N)/SU(P) \otimes SU(N-P) \otimes U(1)$, $SO(N)/SO(N-1)$ are given explicitly, including various generalized 't Hooft field strength expressions.

By separating variables of the equation of motion of the abelianizable gauge field, solution with static dual charge system and retarded solution of linearly accelerated dual charge have been given.

There have been many arguments which favour the possibility of realization of various interactions between elementary particles by a unified gauge field with broken symmetry. Recently, the nontrivial topological properties[1-4] of such gauge field around dual charge have been noticed, — a fact that explains not only the quantization of charge, but also the origin and quantization of the spin of particles by internal construction and symmetry, and at the same time, the topological stability supplies a hopeful measure for searching stable particle models. Therefore, it is meaningful to analyse the constitution of non-Abelian gauge field and to find out new solutions with dual charges.

In current literature, the field with dual charge has been discussed separately in the two cases: (i) $U(1)$ gauge field with nontrivial fibre bundle; (ii) non-Abelian gauge field but with nontrivial $U(1)$ sub-bundle. We have discovered that both of the two cases are connected[4]. This paper, taking such a connection as the starting point, discusses the structure of gauge field and consequently removes some misunderstandings in some literature, thus providing a series of solutions with given physical properties.

In Section I, we generalize the results of [1] and [4], (the decomposition and reduction of $SU(2)$ into $U(1)$) into the following cases: $SU(N) \to U(1)^{N-1}$; $SU(N) \to SU(P) \otimes SU(N-P)$; $SO(N) \to SO(N-1)$. Firstly, the reducible condition is given explicitly, then by using this condition, the reducible part in gauge potential and gauge field strength are separated out explicitly and gauge covariantly. At the same time it becomes clear how each of the terms in the expression of reducible strength ('t Hoofts' expression)[5] is contributed by the corresponding terms in that of potential.

Similar as the Gauss equation in differential geometry, 't Hooft' expression comes out naturally from reduction. Therefore, it always satisfies the Bianchi identity restricted in the subgroup (when the subgroup is $U(1)$, this identity is just the first class Maxwell equation). In Section II, we search the solution of field equation. As all the known solutions with dual charge are static monocentrical synchronous sphero-symmetrical solution, these solutions have been obtained by using the ansatz with variables separated according to such symmetry. Since the equation is multicomponent and nonlinear, it would be difficult to find the multicentric solutions with no guiding clue. We notice that the reducible part of the sphero-symmetrical solution is the canonical invariant connection of the homogeneous symmetry space. Hence we may induce nonsynchronous connections from this universal one, whose equation in the high rank group may be simplified into equation of a simpler subgroup. If the subgroup is Abelian, the equation becomes linear. Then it is easy to solve it and obtain solutions with system of static magnetic charges and arbitrary electric charges, and solutions with linearly accelerated charges.

In this paper, when we discuss the condition of reducibility, the decomposition and reduction of potential and strength, the topological property, the separation of variable in equation and the expression of solution, all of these are manipulated with the help of charge operator $n(x)$ (mapping from space-time x onto sphere in isospace), which realizes the connection and distinction between space-time and isospace.

I. The Reducible Part of Gauge Potential and Field Strength

1. *Condition of reducibility.* The necessary and sufficient condition for the gauge field with group G defined on space-time manifold M to be reducible into gauge field with subgroup H, may be formulated as follows: A connection on the principal bundle $P(M, G)$ can be reduced into the connection of subbundle $Q(M, H)$, when and only when the associated bundle $E(M, G/H, G)$ have a section $n(x): M \to G/H$, which is invariant under parallel displacement. In order to employ this condition, we must find out proper expression for G/H, consequently we decompose the left invariant algebra $\mathfrak{g}$ of group G canonically into $\mathfrak{g} = \mathfrak{h} + \mathfrak{m}$, where $\mathfrak{h}$ is the subalgebra corresponding to the stationary subgroup of the element on G/H. Observing the natural correspondence between G/H and the subspace spanned by $\mathfrak{h}$ in the left invariant Lie algebra, we may perform the reduction as follows.

2. *Gauge field with group $G = SU(2)$ which may be abelianized into $H = U(1)$.* In this case, $\mathfrak{h}$ is one dimensional everywhere, and its normalized base is taken as $\hat{\boldsymbol{n}}$, $\hat{\boldsymbol{n}} \cdot \hat{\boldsymbol{n}} \equiv -2\,\mathrm{tr}(\hat{\boldsymbol{n}} \cdot \hat{\boldsymbol{n}}) = 1$. Then the set of $\hat{\boldsymbol{n}}$ makes a unit sphere $S^2 \sim SU(2)/U(1)$ in the space of adjoint representation. The section $\hat{\boldsymbol{n}}(x)$ is the mapping of space-time M (except the singular point) onto S^2. This unit isospin field is invariant under the parallel displacement by gauge potential $\boldsymbol{A}_\mu(x)$,

$$\tilde{\nabla}_\mu \hat{\boldsymbol{n}}(x) \equiv \partial_\mu \hat{\boldsymbol{n}}(x) + e[\boldsymbol{A}_\mu(x), \hat{\boldsymbol{n}}(x)] = 0,$$
$$\boldsymbol{A}_\mu(x) \in \mathfrak{g}, \quad \mu = 0, 1, 2, 3 \tag{1}$$

where, under infinitesimal gauge transformation

$$\boldsymbol{A}'_\mu(x) = \boldsymbol{A}_\mu(x) + e[\boldsymbol{A}_\mu(x), \boldsymbol{\alpha}(x)] + e\partial_\mu \boldsymbol{\alpha}(x), \qquad \boldsymbol{\alpha}(x) \in \mathfrak{g}. \tag{2}$$

Using identity $\boldsymbol{V} = (\boldsymbol{V} \cdot \boldsymbol{N})\boldsymbol{N} + [\boldsymbol{N}, [\boldsymbol{V}, \boldsymbol{N}]]$, we can easily see that the necessary and sufficient condition of (1) is

$$\boldsymbol{A}_\mu = (\boldsymbol{A}_\mu \cdot \hat{\boldsymbol{n}})\hat{\boldsymbol{n}} - \frac{1}{e}[\hat{\boldsymbol{n}}, \partial_\mu \hat{\boldsymbol{n}}]. \tag{3}$$

Here, the potential is $SU(2)$ formally, but in reality it may be tranformed at least locally into $U(1)$ potential with a constant $\hat{\boldsymbol{n}}(x)$ as the generator, i.e. the gauge could be chosen to turn $\hat{\boldsymbol{n}}(x)$ into the same direcion in some region of x, $\partial_\mu \boldsymbol{n}(x) = 0$. Then $A_\mu'(x)$ becomes explicit Abelian, i.e., it equals $(\boldsymbol{A}_\mu' \cdot \boldsymbol{n})\boldsymbol{n}$ in this region. If we fix the direction of $\hat{\boldsymbol{n}}(x)$, but rotate a gauge angle $\Gamma(x)$ around $\hat{\boldsymbol{n}}(x)$, then we obtain the $U(1)$ transform generated by $e\hat{\boldsymbol{n}}$: $\boldsymbol{A}'_\mu = \boldsymbol{A}_\mu + e\hat{\boldsymbol{n}}\,\partial_\mu \Gamma$.

Substituting (3) into

$$\boldsymbol{F}_{\mu\nu} = \partial_\mu \boldsymbol{A}_\nu - \partial_\nu \boldsymbol{A}_\mu + e[\boldsymbol{A}_\mu, \boldsymbol{A}_\nu], \tag{4}$$

we get

$$\boldsymbol{F}_{\mu\nu} = [\partial_\mu(\boldsymbol{A}_\nu \cdot \hat{\boldsymbol{n}}) - \partial_\nu(\boldsymbol{A}_\mu \cdot \hat{\boldsymbol{n}})]\hat{\boldsymbol{n}} - \frac{1}{e}[\partial_\mu \hat{\boldsymbol{n}}, \partial_\nu \hat{\boldsymbol{n}}] = (\boldsymbol{F}_{\mu\nu} \cdot \hat{\boldsymbol{n}})\hat{\boldsymbol{n}}. \tag{5}$$

In the region where $\hat{\boldsymbol{n}}(x)$ is well defined, $\boldsymbol{F}_{\mu\nu} \cdot \hat{\boldsymbol{n}}$ satisfies

$$\partial^\mu({}^*\boldsymbol{F}_{\mu\nu} \cdot \hat{\boldsymbol{n}}) = \tilde{\nabla}^\mu({}^*\boldsymbol{F}_{\mu\nu} \cdot \hat{\boldsymbol{n}}) = (\tilde{\nabla}^{\mu\,*}\boldsymbol{F}_{\mu\nu}) \cdot \hat{\boldsymbol{n}} + {}^*\boldsymbol{F}_{\mu\nu} \tilde{\nabla}_\mu \hat{\boldsymbol{n}} = 0, \tag{6}$$

where ${}^*\boldsymbol{F}_{\mu\nu} \equiv \frac{1}{2}\varepsilon_{\mu\nu\lambda\rho}\boldsymbol{F}^{\lambda\rho}$. Locally $\boldsymbol{F}_{\mu\nu} \cdot \hat{\boldsymbol{n}}$ is the same as the ordinary electromagnetic field without magnetic charge, and in explicit Abelian gauge it may be expressed by the $U(1)$ potential $\boldsymbol{A}_\mu \cdot \hat{\boldsymbol{n}}$, $\boldsymbol{F}_{\mu\nu} \cdot \hat{\boldsymbol{n}} = \partial_\mu(\boldsymbol{A}_\nu \cdot \hat{\boldsymbol{n}}) - \partial_\nu(\boldsymbol{A}_\mu \cdot \hat{\boldsymbol{n}})$. But, globally its magnetic flux through some two dimensional space like close surface M' may be non-zero,

$$\frac{1}{2}\iint_{M'} \boldsymbol{F}_{\mu\nu} \cdot \hat{\boldsymbol{n}} dx^\mu \wedge dx^\nu = -\frac{l}{e}\iint_{S^2} \hat{\boldsymbol{n}} \cdot [d\hat{\boldsymbol{n}}, \delta\hat{\boldsymbol{n}}] = -\frac{4\pi l}{e}. \tag{7}$$

Here integer l is the times by which the surface M' covers the isospin sphere S^2 through the mapping $\hat{\boldsymbol{n}}(x)$. Physically it is the quantum number of the magnetic charge surrounded by the surface M′. If $l \neq 0$, it is impossible to turn $\hat{\boldsymbol{n}}(x)$ into one and the same direction globally by non-singular single-valued gauge transformation. Then there must be either singularity or overlapping regions with transition function, the corresponding Abelian potential being the Dirac-Schwinger potential with string or the Wu-Yang global potential. Above all, the characteristic $\pi_1(S^1)$ of bundle $Q(M, H)$ of $U(1)$ gauge field corresponds one to one to $\pi_2(S^2)$ of the section $\boldsymbol{n}(x)$ on the associated coset bundle of $SU(2)$, $\pi_1(S^1) \sim \pi_2(S^2)$. Their common characteristic number is determined physically by the dual charge. Mathematically $\hat{\boldsymbol{n}}(x)$ is the generator of the holonomy group of $P(M, G)$ under given connection; physically $e\boldsymbol{n}(x)$ is the charge operator; abelianizable $\boldsymbol{A}_\mu$ is the potential of pure electromagnetic field $\boldsymbol{F}_{\mu\nu}$; and meantime, $\hat{\boldsymbol{n}}$ is the common isodirection of the six space-time components of $\boldsymbol{F}_{\mu\nu}$.

3. *Non-abelianizable $SU(N)$ potential $\boldsymbol{W}_\mu(x)$ and field strength $\boldsymbol{G}_{\mu\nu}(x)$*. Now, the holonomy group are whole $SU(2)$, thus it is impossible to choose from its generators some $\hat{\boldsymbol{n}}(x)$ which remains invariant under parallel displacement. The non-invariant section $\hat{\boldsymbol{n}}(x)$ must be given otherwise. Physically, as the charge operator, $\hat{\boldsymbol{n}}(x)$ is the phase axis of wave functions of charged particles, or the isodirection of its current vector, or $\hat{\boldsymbol{n}}(x)=\boldsymbol{\varphi}(x)/|\boldsymbol{\varphi}(x)|$, where $\boldsymbol{\varphi}(x)$ is the Higgs particle. In sole gauge field without other particles, $\hat{\boldsymbol{n}}(x)$ may be the privileged direction determined by the intrinsic symmetry of the field, e.g. the generator of the stationary subgroup H for G invariant connection. (In case of synchronous spherical symmetry field, the privileged direction $\hat{\boldsymbol{n}}(x)$ is synchronous with the vector radius.)

Once a section $\hat{\boldsymbol{n}}(x)$ is given, it determines a corresponding subbundle $Q(M, H)$, whose characteristic class is decided by the homotopic property of $\hat{\boldsymbol{n}}(x)$. Physically, as soon as the charge operator $\hat{\boldsymbol{n}}(x)$ is given, one can separate the electromagentic component $\boldsymbol{A}_\mu(x)$ from the $SU(2)$ potential $\boldsymbol{W}_\mu(x)$ as follows: Here $\boldsymbol{A}_\mu(x)$ is the $U(1)$ gauge potential with $\hat{\boldsymbol{n}}(x)$ as the generator. Now from $\tilde{\nabla}_\mu\hat{\boldsymbol{n}}\equiv\partial_\mu\boldsymbol{n}+e[\boldsymbol{W}_\mu,\hat{\boldsymbol{n}}]$ which is not vanishing now, we get

$$\boldsymbol{W}_\mu=(\boldsymbol{W}_\mu\cdot\hat{\boldsymbol{n}})\hat{\boldsymbol{n}}-\frac{1}{e}[\hat{\boldsymbol{n}},\partial_\mu\hat{\boldsymbol{n}}]+\frac{1}{e}[\hat{\boldsymbol{n}},\nabla_\mu\hat{\boldsymbol{n}}]\equiv\boldsymbol{A}_\mu+\boldsymbol{B}_\mu. \tag{8}$$

Here we have set $\boldsymbol{A}_\mu\equiv(\boldsymbol{W}_\mu\cdot\hat{\boldsymbol{n}})\hat{\boldsymbol{n}}-\frac{1}{e}[\hat{\boldsymbol{n}},\partial_\mu\boldsymbol{n}]$. It is easy to prove that $\boldsymbol{A}_\mu$ satisfies (1)—(3), hence it is the electromagnetic $U(1)$ part in $\boldsymbol{W}_\mu$, the remainder $\frac{1}{e}[\hat{\boldsymbol{n}},\nabla_\mu\hat{\boldsymbol{n}}]\equiv$ $\boldsymbol{B}_\mu$ is gauge covariant and represents the charged vector particles. Geometrically $e\boldsymbol{B}_\mu$ corresponds to the second fundamental form, e.g. $\nabla_\mu\hat{\boldsymbol{n}}=[e\boldsymbol{B}_\mu,\hat{\boldsymbol{n}}]$ is the generalized Weingarten formula (since $|\hat{\boldsymbol{n}}|=1$, the "normal" component is absent). Substituting (8) into $\boldsymbol{G}_{\mu\nu}=\partial_\mu\boldsymbol{W}_\nu-\partial_\nu\boldsymbol{W}_\mu+e[\boldsymbol{W}_\mu,\boldsymbol{W}_\nu]$ and making comparison with (5), we get

$$\begin{aligned}\boldsymbol{G}_{\mu\nu}\cdot\hat{\boldsymbol{n}}\hat{\boldsymbol{n}}-[B_\mu,B_\nu]&=\boldsymbol{G}_{\mu\nu}\cdot\hat{\boldsymbol{n}}\hat{\boldsymbol{n}}-[\nabla_\mu\hat{\boldsymbol{n}},\nabla_\nu\hat{\boldsymbol{n}}]\\&=\partial_\mu(\boldsymbol{A}_\nu\cdot\hat{\boldsymbol{n}})\hat{\boldsymbol{n}}-\partial_\nu(\boldsymbol{A}_\mu\cdot\hat{\boldsymbol{n}})\hat{\boldsymbol{n}}-\frac{1}{e}[\partial_\mu\hat{\boldsymbol{n}},\partial_\nu\hat{\boldsymbol{n}}]\\&=\boldsymbol{F}_{\mu\nu}.\end{aligned}\tag{9}$$

This is just the t'Hooft expression. Here $\boldsymbol{F}_{\mu\nu}$ is the electromagnetic field part in $\boldsymbol{G}_{\mu\nu}$ contributed by $\boldsymbol{A}_\mu$. (The Higgs particle does not contribute the electromagnetic field, only its isodirection coincides with that of charge operator.) Geometrically (9) is the generalized Gauss equation. $\boldsymbol{F}_{\mu\nu}\cdot\hat{\boldsymbol{n}}$, as the $U(1)$ "subcurvature" of the total curvature $\boldsymbol{G}_{\mu\nu}$, satisfies the Bianchi identity on subbundle. At the same time we get the generalized Codazzi equation,

$$[\hat{\boldsymbol{n}},[\boldsymbol{G}_{\mu\nu},\hat{\boldsymbol{n}}]]=\tilde{\nabla}_\mu\boldsymbol{B}_\nu-\tilde{\nabla}_\nu\boldsymbol{B}_\mu. \tag{10}$$

4. $G = SU(N)$, $H = (U(1))^{N-1}$[6]. Now $\mathfrak{h}$ is the Carton subalgebra of left invariant Lie algebra g. Let a complete set of left invariant orthonormal base of $\mathfrak{h}$ be $\hat{\boldsymbol{n}}_i$ $(i = 1, \ldots, N-1)$, $\hat{\boldsymbol{n}}_i \cdot \hat{\boldsymbol{n}}_j = \delta_{ij}$. Since $\mathfrak{h}$ is Abelian, $[\hat{\boldsymbol{n}}_i, \hat{\boldsymbol{n}}_j] = 0$. By the orthonormality and the left invariance of $\hat{\boldsymbol{n}}_i$ (consequently, its eigenvalues are constants) we get $\partial_\mu \hat{\boldsymbol{n}}_i \cdot \hat{\boldsymbol{n}}_j = 0$. From the abelianizable condition $\tilde{\nabla}_\mu \hat{\boldsymbol{n}}_i = 0$, and the decomposition expression of $\boldsymbol{v}$ in $su(N)$,

$$\boldsymbol{v} = [\hat{\boldsymbol{n}}_i, [\boldsymbol{v}, \hat{\boldsymbol{n}}_i]] + (\boldsymbol{v} \cdot \hat{\boldsymbol{n}}_i)\hat{\boldsymbol{n}}_i \equiv \boldsymbol{v}^{(m)} + \boldsymbol{v}^{(h)}, \tag{11}$$

we get

$$\boldsymbol{A}_\mu = \boldsymbol{A}_\mu \cdot \hat{\boldsymbol{n}}_i \hat{\boldsymbol{n}}_i - \frac{1}{e}[\hat{\boldsymbol{n}}_i, \partial_\mu \hat{\boldsymbol{n}}_i]. \tag{12}$$

Locally we may choose the gauge $\partial_\mu \hat{\boldsymbol{n}}_i = 0$, then $\boldsymbol{A}_\mu$ becomes explicitly Abelian $\boldsymbol{A}_\mu = (\boldsymbol{A}_\mu \cdot \hat{\boldsymbol{n}}_i)\hat{\boldsymbol{n}}_i$.

Substituting $\boldsymbol{A}_\mu$(12) into $\boldsymbol{F}_{\mu\nu}$(4), and utilizing (11) and

$$[\boldsymbol{v}^{(m)}, \boldsymbol{u}^{(h)}] = [\boldsymbol{v}, \hat{\boldsymbol{n}}_i](\boldsymbol{u} \cdot \hat{\boldsymbol{n}}_i),$$

$$[[\boldsymbol{v}, \hat{\boldsymbol{n}}_i], [\boldsymbol{u}, \hat{\boldsymbol{n}}_i]] = \frac{1}{2}[\boldsymbol{v}^{(m)}, \boldsymbol{u}^{(m)}]^{(m)} + [\boldsymbol{v}^{(m)}, \boldsymbol{u}^{(m)}]^{(h)},$$

we get

$$\begin{aligned}\boldsymbol{F}_{\mu\nu} &= \partial_\mu(\boldsymbol{A}_\nu \cdot \hat{\boldsymbol{n}}_i)\hat{\boldsymbol{n}}_i - \partial_\nu(\boldsymbol{A}_\mu \cdot \hat{\boldsymbol{n}}_i)\hat{\boldsymbol{n}}_i - \frac{1}{e}[\partial_\mu \hat{\boldsymbol{n}}_j, \partial_\nu \hat{\boldsymbol{n}}_j] \cdot \hat{\boldsymbol{n}}_i \hat{\boldsymbol{n}}_i \\ &= \boldsymbol{F}_{\mu\nu} \cdot \hat{\boldsymbol{n}}_i \hat{\boldsymbol{n}}_i,\end{aligned} \tag{13}$$

which in the explicit Abelian gauge becomes $\hat{\boldsymbol{n}}_i \cdot \boldsymbol{F}_{\mu\nu} = \partial_\mu(\boldsymbol{A}_\nu \cdot \boldsymbol{n}_i) - \partial_\nu(\boldsymbol{A}_\mu \cdot \boldsymbol{n}_i)$. Globally $\boldsymbol{F}_{\mu\nu}$ has N—1 independent, mutually commuting dual charges which in the closed surface M' correspond to $\pi_2(SU(N)/U(1)^{N-1})$; meanwhile the ith dual charge is given by $\iint_{M'} \boldsymbol{F} \cdot \hat{\boldsymbol{n}}_i$.

Similarly, a general $SU(N)$ potential $\boldsymbol{W}_\mu$ may be decomposed into

$$\boldsymbol{W}_\mu = \boldsymbol{W}_\mu \cdot \hat{\boldsymbol{n}}_i \hat{\boldsymbol{n}}_i - \frac{1}{e}[\hat{\boldsymbol{n}}_j, \partial_\nu \hat{\boldsymbol{n}}_j] + \frac{1}{e}[\hat{\boldsymbol{n}}_j, \nabla_\nu \hat{\boldsymbol{n}}_j] = \boldsymbol{A}_\mu + \boldsymbol{B}_\mu. \tag{14}$$

The generalized Weingarten formula is $(\nabla_\mu \hat{\boldsymbol{n}}_i) \cdot \hat{\boldsymbol{n}}_i = 0$ (no normal component, because of orthonormality and left invariance). The generalized 't Hooft-Gauss equation is

$$\boldsymbol{F}_{\mu\nu} = (\boldsymbol{G}_{\mu\nu} - [\nabla_\mu \hat{\boldsymbol{n}}_j, \nabla_\nu \hat{\boldsymbol{n}}_j]) \cdot \hat{\boldsymbol{n}}_i \hat{\boldsymbol{n}}_i = (\boldsymbol{G}_{\mu\nu} - [\boldsymbol{B}_\mu, \boldsymbol{B}_\nu]) \cdot \hat{\boldsymbol{n}}_i \hat{\boldsymbol{n}}_i. \tag{15}$$

The generalized Coddazi equation is

$$\begin{aligned}[\hat{\boldsymbol{n}}_i, [\boldsymbol{G}_{\mu\nu}, \hat{\boldsymbol{n}}_i]] - \tilde{\nabla}_\mu \boldsymbol{B}_\nu + \tilde{\nabla}_\nu \boldsymbol{B}_\mu &= [\hat{\boldsymbol{n}}_i, [[\boldsymbol{B}_\mu, \boldsymbol{B}_\nu], \hat{\boldsymbol{n}}_i]] \\ &= \frac{1}{2}[\hat{\boldsymbol{n}}_i, [[\nabla_\mu \hat{\boldsymbol{n}}_j, \nabla_\nu \hat{\boldsymbol{n}}_j], \hat{\boldsymbol{n}}_i]],\end{aligned} \tag{16}$$

whose right-hand side is not vanishing; since $[\mathfrak{m}, \mathfrak{m}] \cap \mathfrak{m} \neq 0$, G/H is not a symmetric space.

Until now, we have used a complete base $\hat{\boldsymbol{n}}_i$ to determine the Cartan subalgebra $\mathfrak{h}$, but in order to determine the subspace spanned by $\mathfrak{h}$, it is sufficient to adopt only one left invariant $\hat{\boldsymbol{n}}$ which has N different eigenvalue. Since such $\hat{\boldsymbol{n}}$ is stationary under H, $[\hat{\boldsymbol{n}}, \mathfrak{h}] = 0$ and transitive under G, i.e. $[\hat{\boldsymbol{n}}, \mathfrak{m}]$ spans the whole $\mathfrak{m}$, which is naturally isomorphic with the tangent space of G/H, therefore this $\hat{\boldsymbol{n}}$ only is sufficient to represent G/H. For example, in $SU(3)$, $\hat{\boldsymbol{n}}$ may be chosen as $-i(Q - Y/2)$, whose eigenvalue are $i/2, -i/2, 0$. Indeed, another base $\hat{\boldsymbol{n}}'$ of $\mathfrak{h}$ may be expressed by $\hat{\boldsymbol{n}}$ such that

$$\hat{\boldsymbol{n}}'^a = \sqrt{3}\, d_{abc} \hat{\boldsymbol{n}}^b \hat{\boldsymbol{n}}^c. \tag{17}$$

It is easy to prove that if $\nabla_\mu \hat{\boldsymbol{n}} = 0$, then $\nabla_\mu \hat{\boldsymbol{n}}' = 0$. Thus, $\nabla_\mu \hat{\boldsymbol{n}} = 0$ has already beome the sufficient condition for abelianizing. It is also easy to see that if $\partial_\mu \hat{\boldsymbol{n}} = 0$, then $\partial_\mu \hat{\boldsymbol{n}}' = 0$. Therefore once the homotopic property of $\hat{\boldsymbol{n}}$ (T_3 Charge) is trivial, there would not be any kind of dual charge ($*Q$, $*Y$ charge).

From (15), (13) and the relation

$$[[\boldsymbol{v}, \hat{\boldsymbol{n}}'], [\boldsymbol{u}, \hat{\boldsymbol{n}}']] \cdot \hat{\boldsymbol{n}}' = 3[[\boldsymbol{v}, \hat{\boldsymbol{n}}], [\boldsymbol{u}, \hat{\boldsymbol{n}}]] \cdot \hat{\boldsymbol{n}}',$$

we may separate the Abelian part $F_{\mu\nu} \cdot \hat{\boldsymbol{n}}'$ from $\boldsymbol{G}_{\mu\nu}$ by using only one term in $[\nabla_\mu \hat{\boldsymbol{n}}_i, \nabla_\nu \hat{\boldsymbol{n}}_i]$ with proper numerical coefficient[7,8],

$$\begin{aligned}\left(\boldsymbol{G}_{\mu\nu} - \frac{4}{e}[\nabla_\mu \hat{\boldsymbol{n}}, \nabla_\nu \hat{\boldsymbol{n}}]\right) \cdot \hat{\boldsymbol{n}}' &= \left(\boldsymbol{G}_{\mu\nu} - \frac{4}{3e}[\nabla_\mu \hat{\boldsymbol{n}}', \nabla_\nu \hat{\boldsymbol{n}}']\right) \cdot \hat{\boldsymbol{n}}' = \boldsymbol{F}_{\mu\nu} \cdot \hat{\boldsymbol{n}}' \\ &= \partial_\mu(\boldsymbol{W}_\nu \cdot \hat{\boldsymbol{n}}') - \partial_\nu(\boldsymbol{W}_\mu \cdot \hat{\boldsymbol{n}}') - \frac{4}{e}[\partial_\mu \hat{\boldsymbol{n}}, \partial_\nu \hat{\boldsymbol{n}}] \cdot \hat{\boldsymbol{n}}' \\ &= \partial_\mu(\boldsymbol{W}_\nu \cdot \hat{\boldsymbol{n}}') - \partial_\nu(\boldsymbol{W}_\mu \cdot \hat{\boldsymbol{n}}') - \frac{4}{3e}[\partial_\mu \hat{\boldsymbol{n}}', \partial_\nu \hat{\boldsymbol{n}}'] \cdot \hat{\boldsymbol{n}}'.\end{aligned} \tag{18}$$

As for $F_{\mu\nu} \cdot \hat{\boldsymbol{n}}$, generally, it cannot be gained by subtraction of only $[\nabla_\mu \hat{\boldsymbol{n}}, \nabla_\nu \hat{\boldsymbol{n}}]$. Reference [10] has adopted the original 't Hooft expression (9), which is improper in this case. Only when $\nabla_\mu \hat{\boldsymbol{n}}' = 0$ (i.e. $[\nabla_\mu \hat{\boldsymbol{n}}, \hat{\boldsymbol{n}}'] = 0$, or when it is already known that the field is reducible to $SU(2) \otimes U(1)$) can we further use (9) to get the fully abelianized component.

5. $G = SU(N)$, $H = SU(p) \otimes SU(q) \otimes U(1)$, $p + q = N$. In this case one must use the left invariant subalgebra spanned by $\mathfrak{h} = su(p) \oplus su(q) \oplus u(1)$ to represent G/H, herein the knowledge of the last term $u(1)$ alone is sufficient. In the N dimensional representation $\frac{i}{2}\hat{\boldsymbol{n}}$ is a matrix conjugated with $\left(\frac{2}{pqN}\right)^{1/2} \begin{pmatrix} qI_p & 0 \\ 0 & -pI_q \end{pmatrix}$. It is easy to see that this $\hat{\boldsymbol{n}}$ satisfies $[\hat{\boldsymbol{n}}, \mathfrak{h}] = 0$, and $[\hat{\boldsymbol{n}}, \mathfrak{m}]$ spans the whole $\mathfrak{m}$. From the condition $\tilde{\nabla}_\mu \hat{\boldsymbol{n}} = 0$ and the relation,

$$\boldsymbol{v} = \boldsymbol{v}^{(p)} + \boldsymbol{v}^{(q)} + \boldsymbol{v} \cdot \hat{\boldsymbol{n}}\hat{\boldsymbol{n}} - \frac{2qp}{N}[[\boldsymbol{v}, \hat{\boldsymbol{n}}], \hat{\boldsymbol{n}}], \tag{19}$$

we get the reducible potential,

$$\boldsymbol{A}_\mu = \boldsymbol{A}_\mu \cdot \hat{\boldsymbol{n}}\hat{\boldsymbol{n}} + \boldsymbol{A}_\mu^{(p)} + \boldsymbol{A}_\mu^{(q)} - \frac{2pq}{eN}[\hat{\boldsymbol{n}}, \partial_\mu \hat{\boldsymbol{n}}], \tag{20}$$

where $\boldsymbol{A}_\mu^{(p)} \equiv \boldsymbol{A}_\mu \cap su(p)$. From this potential we can obtain the field strength by using $[[\boldsymbol{v}, \hat{\boldsymbol{n}}], [\boldsymbol{u}, \hat{\boldsymbol{n}}]]^{(\mathfrak{h})} = \frac{N}{2pq}[\boldsymbol{v}, \boldsymbol{u}]^{(\mathfrak{h})}$ and $\partial_\mu \boldsymbol{v}^{(p)} + \frac{2pq}{N}[\boldsymbol{v}^{(p)}, [\hat{\boldsymbol{n}}, \partial_\mu \hat{\boldsymbol{n}}]] = (\partial_\mu \boldsymbol{v}^{(p)})^{(p)}$ which are derived from (19) and $[\boldsymbol{v}^{(p)}, \hat{\boldsymbol{n}}] = 0$,

$$\boldsymbol{F}_{\mu\nu} = \boldsymbol{F}_{\mu\nu}^{(p)} + \boldsymbol{F}_{\mu\nu}^{(q)} + \boldsymbol{F}_{\mu\nu} \cdot \hat{\boldsymbol{n}}\hat{\boldsymbol{n}},$$

where

$$\boldsymbol{F}_{\mu\nu}^{(p)} \equiv \boldsymbol{F}_{\mu\nu} \cap su(p) = (\partial_\mu \boldsymbol{A}_\nu^{(p)})^{(p)} - (\partial_\nu \boldsymbol{A}_\mu^{(p)})^{(p)} + e[\boldsymbol{A}_\mu^{(p)}, \boldsymbol{A}_\nu^{(p)}] - \frac{2pq}{eN}[\partial_\mu \hat{\boldsymbol{n}}, \partial_\nu \hat{\boldsymbol{n}}]^{(p)},$$

$$\boldsymbol{F}_{\mu\nu} \cdot \hat{\boldsymbol{n}} = \partial_\mu(\boldsymbol{A}_\nu \cdot \hat{\boldsymbol{n}}) - \partial_\nu(\boldsymbol{A}_\mu \cdot \hat{\boldsymbol{n}}) - \frac{2pq}{eN}[\partial_\mu \hat{\boldsymbol{n}}, \partial_\nu \hat{\boldsymbol{n}}] \cdot \hat{\boldsymbol{n}}. \tag{21}$$

Similarly, we have the irreducible potential,

$$\boldsymbol{W}_\mu = (\boldsymbol{W}_\mu \cdot \hat{\boldsymbol{n}})\hat{\boldsymbol{n}} - \frac{2pq}{eN}[\hat{\boldsymbol{n}}, \partial_\mu \hat{\boldsymbol{n}}] + \boldsymbol{W}_\mu^{(p)} + \boldsymbol{W}_\mu^{(q)} + \frac{2pq}{eN}[\hat{\boldsymbol{n}}, \nabla_\mu \hat{\boldsymbol{n}}]^{(p)} + \frac{2pq}{eN}[\hat{\boldsymbol{n}}, \nabla_\mu \hat{\boldsymbol{n}}]^{(q)} = \boldsymbol{A}_\mu + \boldsymbol{B}_\mu, \tag{22}$$

the generalized Gauss 't Hooft equation,

$$\boldsymbol{G}_{\mu\nu}^{(\mathfrak{h})} - \frac{2pq}{eN}[\nabla_\mu \hat{\boldsymbol{n}}, \nabla_\nu \hat{\boldsymbol{n}}] = \boldsymbol{F}_{\mu\nu}, \tag{23}$$

the Guo[9] 't Hooft expression,

$$\boldsymbol{G}_{\mu\nu} \cdot \hat{\boldsymbol{n}} - \frac{2pq}{eN}[\nabla_\mu \hat{\boldsymbol{n}}, \nabla_\nu \hat{\boldsymbol{n}}] \cdot \hat{\boldsymbol{n}} = \boldsymbol{F}_{\mu\nu} \cdot \hat{\boldsymbol{n}}, \tag{24}$$

and the generalized Coddzi equation,

$$\boldsymbol{G}_{\mu\nu}^{(m)} = \tilde{\nabla}_\mu \boldsymbol{B}_\nu - \tilde{\nabla}_\nu \boldsymbol{B}_\mu. \tag{25}$$

6. $G = SO(N)$, $H = SO(N-1)$. In this case, the symmetric homogeneous space $G/H \sim S^{N-1}$ corresponds to the $N-1$ dimensional unit sphere, whose elements are the unit vector $\vec{n}$, $n^a n^a = 1$ $(a = 1, \ldots, N)$, which is invariant under H, $\mathfrak{h}\vec{n} = 0$. Now, the potential $\boldsymbol{W}_\mu \in O(N)$ is an $N \times N$ antisymmetric matrix; from $\boldsymbol{V} = \boldsymbol{V}^{(\mathfrak{h})} + \vec{n} \times (\boldsymbol{V}\vec{n})$, here $\boldsymbol{V} \in o(N)$, $\boldsymbol{V}^{(\mathfrak{h})}$ denote the projection of $\boldsymbol{V}$ on $\mathfrak{h}$, and from $(\vec{l} \times \vec{n})^{ab} \equiv l^a n^b - l^b n^a$, we get

$$\boldsymbol{W}_\mu = \boldsymbol{W}_\mu^{(\mathfrak{h})} - \frac{1}{e}\vec{n} \times \partial_\mu \vec{n} + \frac{1}{e}\vec{n} \times \nabla_\mu \vec{n} \equiv \boldsymbol{A}_\mu + \boldsymbol{B}_\mu, \tag{26}$$

where

$$\boldsymbol{A}_\mu = \boldsymbol{W}_\mu^{(\mathfrak{h})} - \frac{1}{e}\vec{n} \times \partial_\mu \vec{n} = \boldsymbol{A}_\mu^{(\mathfrak{h})} - \frac{1}{e}\vec{n} \times \partial_\mu \vec{n}. \tag{27}$$

It is easy to see that $\tilde{\nabla}_\mu \vec{n} \equiv \partial_\mu \vec{n} + e\boldsymbol{A}_\mu \vec{n} = 0$. Then we have the generalized Gauss 't Hooft equation,

$$G_{\mu\nu}^{(\mathfrak{h})} - e[B_\mu, B_\nu] = G_{\mu\nu}^{(\mathfrak{h})} - \frac{1}{e}[\nabla_\mu \vec{n} \times \nabla_\nu \vec{n}] = F_{\mu\nu} \equiv \partial_\mu A_\nu - \partial_\nu A_\mu$$

$$+ e[A_\mu, A_\nu] = (\partial_\mu A_\nu^{(\mathfrak{h})} - \partial_\nu A_\mu^{(\mathfrak{h})})^{(\mathfrak{h})} + e[A_\mu^{(\mathfrak{h})}, A_\nu^{(\mathfrak{h})}]$$

$$- \frac{1}{e}(\partial_\mu \vec{n} \times \partial_\nu \vec{n}), \tag{28}$$

where we have used $\partial_\mu v^{(\mathfrak{h})} + [v^{(\mathfrak{h})}, \vec{n} \times \partial_\mu \vec{n}] = (\partial_\mu \dot{v}^{(\mathfrak{h})})^{(\mathfrak{h})}$, and the generalized Coddazi equation,

$$\vec{n} \times G_{\mu\nu}\vec{n} = \tilde{\nabla}_\mu B_\nu - \tilde{\nabla}_\nu B_\mu. \tag{29}$$

II. Solution of Gauge Field Equation

1. *Synchro-symmetric reducible gauge field —— canonical connection.* If the reducible gauge field satisfies the orthogonal gauge condition such that $A_\mu \perp \mathfrak{h}$, $A_\mu \in \mathfrak{m}$, then all of the expressions for the potential for the strength in the above section may be simplified with only one term remaining as follows:

$$SU(2)/U(1):\ A_\mu = -\frac{1}{e}[\hat{n}, \partial_\mu \hat{n}],\quad F_{\mu\nu} = -\frac{1}{e}[\partial_\mu \hat{n}, \partial_\nu \hat{n}], \tag{30}$$

$$SU(N)/[U(1)]^{N-1}:\ A_\mu = -\frac{1}{e}[\hat{n}_i, \partial_\mu \hat{n}_i],\quad F_{\mu\nu} = -\frac{1}{e}[\partial_\mu \hat{n}_i, \partial_\nu \hat{n}_i], \tag{31}$$

$$SU(N)/SU(p)\otimes SU(q)\otimes U(1):\ A_\mu = -\frac{2pq}{eN}[\hat{n}, \partial_\mu \hat{n}],$$

$$F_{\mu\nu} = \frac{2pq}{eN}[\partial_\mu \hat{n}, \partial_\nu \hat{n}], \tag{32}$$

$$SO(N)/SO(N-1):\ A_\mu = -\frac{1}{e}\vec{n} \times \partial_\mu \vec{n},\quad F_{\mu\nu} = -\frac{1}{e}\partial_\mu \vec{n} \times \partial_\nu \vec{n}. \tag{33}$$

Here we always have

$$F_{\mu\nu} = -e[A_\mu, A_\nu]. \tag{34}$$

Now, the potential, the strength, the source (charge current) and the dual charge current are all determined by the mapping $n(x)$.

All the known reducible solutions in orthogonal gauge have synchro-symmetric $n(x)$. Take for example the following case:

(i) $SU(2)/U(1)$: Let

$$\hat{n} = \frac{x^a}{|x|}\frac{i}{2}\sigma^a,\quad |x| = (x_1^2 + x_2^2 + x_3^2)^{1/2}$$

be substituted into (30), and then we get the asympototic part of Ref. [5, 11, 12].

(ii) $SO(N)/SO(N-1)$: Let $n^a = x^a/|x|$, (here it is assumed that the base manifold M is R_N—[0], $x^a \in M$,) and substitute it into (33), then we get the expression in Ref. [13, 14, 15].

(iii) $SU(3)/U(1)\otimes U(1)$[6]:

$$\hat{\boldsymbol{n}} = \frac{-i}{2|x|}(-x_2\boldsymbol{\lambda}_5 + x_1\boldsymbol{\lambda}_7 + x_3\boldsymbol{\lambda}_2), \tag{35}$$

$$\hat{\boldsymbol{n}}' = \sqrt{3}(d\hat{\boldsymbol{n}}\hat{\boldsymbol{n}}) = \frac{-\sqrt{3}\,i}{4|x|^2}\Big[(2x_1x_2\boldsymbol{\lambda}_1 + 2x_3x_1\boldsymbol{\lambda}_4 + 2x_2x_3\boldsymbol{\lambda}_6) + (x_1^2 - x_2^2)\boldsymbol{\lambda}_3 + (|x|^2 - 3x_3^2)\frac{\boldsymbol{\lambda}_8}{\sqrt{3}}\Big]. \tag{36}$$

Substituting these into (31), or substituting (36) into (32), we get the results of [16, 17]. Here $\hat{\boldsymbol{n}}'$ is determined by $\hat{\boldsymbol{n}}$; actually, the mapping is determined by $\hat{\boldsymbol{n}}$ alone, which is a synchronous covering from $M=R^3-0$ onto $SO(3)/SO(2)\sim S^2$ (here $SO(3)$ is a maximum subgroup of $SU(3)$). Hence it may be put in the same class as the last two examples.

All the above-mentioned synchro-symmetric reducible potentials in orthogonal gauge are G invariant connection on G/H, which is obtained when the flat connection of trivial bundle $P(G/H, G)$ is reduced naturally to the subbundle[18]. Originally, the flat connection form in the product bundle runs over all directions in $\mathfrak{g}$, but when it is restricted toward the subbundle $G(G/H, H)$, then along the direction tangent to M, the connection form lies wholly in $\mathfrak{m}$ of $\mathfrak{g}$. This is just the above-mentioned potential in orthogonal gauge. The distinguished points of orthogonal gauge are: uniqueness in form with no uncertainty remaining, explicit synchro-symmetry, no singularity and no transition function in overlapping region, but it is expressed in group of a higher rank with superfluous freedom, just as in the research of the sphere S^2 in a three dimensional space.

As for the corresponding potential on $Q(M, H)$ with explicitly reduced gauge, since it is the connection form on G/H of the invariant connection for the homogeneous space $G(G/H, H)$, it may be obtained by the following various equivalent methods: (i) introducing a natural metric in the bundle $G(G/H, H)$, so that the cross term of the fibre and the base would give the potential (the case of fibering of $SU(2)$ by $U(1)$[19]); (ii) giving the sphere G/H a natural Riemann metric from the Christoffel so as to get the potential[13,20]; (iii) getting the connection form on M from the Cartan structure equation on sphere G/H[4]; (iv) getting it from the orthogonal gauge by proper gauge transformation[2,13,21]. In the reduced gauge, the rank of group is lower, and the freedom is lesser, but the cost is that there would be no such explicit symmetry as under higher group; it becomes gauge-dependent without unique canonical form, and must be expressed either with gauge-dependent singularity or with gauge-dependent overlapping region.

2. *The abelianizable $SU(2)$ gauge field induced by canonical connection——static point charges system.* Given any mapping $\hat{\boldsymbol{n}}(x)$ from spacetime manifold to G/H, there would be a corresponding reducible G gauge field induced by using (30) and (33). Given $\hat{\boldsymbol{n}}(x)$, it is also possible to induce the reduced H gauge potential directly from the known connection form on G/H of the canonical connection on $G(G/H, H)$. For example, in the case $SU(2)/U(1)$, let the spherical coordinate of $\hat{\boldsymbol{n}}$ on S^2 be Θ, Φ,

then from the canonical connection form $-\cos\Theta d\Phi - d\Gamma(\Theta, \Phi)$ (where $\Gamma(\Theta, \Phi)$ is the angle between the local frame used on S^2 and the natural frame of the coordinate mesh), one may get the $U(1)$ potential induced by $\hat{\boldsymbol{n}}(x)$,

$$e\boldsymbol{A}_\mu \cdot \hat{\boldsymbol{n}} = -\cos\Theta(x)\partial_\mu\Phi(x) - \partial_\mu\Gamma(x). \tag{37}$$

When $\Gamma(x) = 0$, it is the Schwinger gauge. These are the well-known induced bundle and induced connection in mathematics. But the connection induced by arbitrary $\hat{\boldsymbol{n}}(x)$ may have no well-defined physical meaning, for it may not satisfy the sourceless equation, i.e. $\nabla^\mu F_{\mu\nu} = 0$, which in the abelianizable case may be simplified into

$$\begin{aligned} 0 = \nabla^\mu \boldsymbol{F}_{\mu\nu} &= \nabla^\mu((\boldsymbol{F}_{\mu\nu} \cdot \hat{\boldsymbol{n}})\hat{\boldsymbol{n}}) = \partial^\mu(\boldsymbol{F}_{\mu\nu} \cdot \hat{\boldsymbol{n}})\hat{\boldsymbol{n}} + (\boldsymbol{F}_{\mu\nu} \cdot \hat{\boldsymbol{n}})\nabla^\mu\hat{\boldsymbol{n}} \\ &= \partial^\mu(\boldsymbol{F}_{\mu\nu} \cdot \hat{\boldsymbol{n}})\hat{\boldsymbol{n}}. \end{aligned} \tag{38}$$

Thus the original nonlinear sourceless equation of $\boldsymbol{F}_{\mu\nu} \in g$ is simplified into the linear Maxwell equation satisfied by $\boldsymbol{F}_{\mu\nu} \cdot \hat{\boldsymbol{n}} \in R_1$. Keeping this in mind we may easily show that given any system of magnetic point charges one can get corresponding $SU(2)$ potential with non-trivial bundle as follows[23]: (i) Take synchronous mapping (if magnetic charge is one in Dirac unit) or l-multisynchronous mapping (i.e. let $\Phi = l\varphi$, $\Theta = \theta$, if magnetic charge is l) from the infinitesimal sphere S_ε surrounding each magnetic pole with coordinate φ, θ onto the isospin sphere S^2 with coordinate Φ, Θ. (ii) All the space points along the same magnetic force line are mapped on one and the same points on isospin sphere S^2. Thus, using the known equation of magnetic force line (e.g. produced by magnetic poles distributed on one axis[23]) the expression of $\hat{\boldsymbol{n}}(x)$ and gauge field is obtained[22], and the paradox[2] about the nonexistence of $SU(2)$ potential with two positive poles is clarified. Here the string of singularity after meeting at the middle point is turned into a surface of discontinuity.

If together with the static poles, there are also static electric charges, which may be distributed continuously and may coincide with the pole, it is sufficient to find out the corresponding $U(1)$ electric polential $\varphi_0(x)$ at first, and then combine with the $\hat{\boldsymbol{n}}(x)$ which has been obtained along magnetic force line by the method just related,

$$\boldsymbol{A}_0(x) = \varphi_0(x)\hat{\boldsymbol{n}}(x), \ A_i(x) = \frac{-1}{e}[\hat{\boldsymbol{n}}(x), \partial_i\hat{\boldsymbol{n}}(x)].$$

Solutions of other abelianizable field may be obtained similarly. For instance, substituting the spherical coordinate $\Theta(x), \Phi(x)$ of $\hat{\boldsymbol{n}}(x)$ into $\hat{\boldsymbol{n}}_1 = \sin\Theta(x)(\cos\Phi(x)\cdot \boldsymbol{\lambda}_7 - \sin\Phi(x)\boldsymbol{\lambda}_5) + \cos\Theta(x)\boldsymbol{\lambda}_2$ and further substituting this $\hat{\boldsymbol{n}}_1$, into (17), then from (31) we obtain the $SU(3)$ potential.

3. *The retarded $SU(2)$ and $U(1)$ potentials of dual charge accelerated along a straight line.* Take the synchronous mapping in the optical coordinate of accelerated charge, then $\hat{\boldsymbol{n}}(x)$ $\boldsymbol{A}_\mu(x)$ and $\boldsymbol{F}_{\mu\nu}(x)$ can be found as follows: Let the world line of the pole be $\xi^\mu(\tau)$, and the lower half light cone from field point x intersect $\xi^\mu(\tau)$ at $\xi^\mu(x)$; Further, put $r^\mu = x^\mu - \xi^\mu(x)$, and let $n^a(x)$ be the radial direction in the instaneous rest system of $\xi^\mu(x)$, i.e.,

$$n^a(x) = \frac{\Lambda^a_\nu r^\nu}{u_\mu r^\mu} \equiv \frac{\Lambda^a_\nu r^\nu}{R},$$

where

$$u^\mu(x) = \frac{d\xi^\mu}{d\tau}.$$

Here $\Lambda_\mu^{\ \nu}$ is the special Lorentz transformation which transforms u^μ into (1, 0, 0, 0). Substitute this $n^a(x)$ into (30) together with the retarded $U(1)$ electric potential of electric charge Ze, $\varphi_\mu(x) = \frac{Ze\, u_\mu(x)}{u_\nu r^\nu}$, we obtain $\boldsymbol{A}_\mu(x) = -\frac{1}{e}[\hat{\boldsymbol{n}}, \partial_\mu \hat{\boldsymbol{n}}] + \varphi_\mu \hat{\boldsymbol{n}}$. Then, we may show that

$$A_\lambda^c(x) = \epsilon_{abc} n^a \Lambda_\mu^b (\delta_\lambda^\mu - a^\mu r_\lambda)/e + Zeu_\lambda(x)\frac{n^c}{R}.$$

$$(a^\mu = du^\mu/d\tau) \tag{39}$$

Here we have utilized

$$\partial_\lambda n^a = \Lambda_\mu^b(\delta^{ba} - n^b n^a)(\delta_\lambda^\mu - a^\mu r_\lambda)/R, \tag{40}$$

which is valid only when the pole moves along a straight line. From (39) and (40), we get

$$F_{\lambda\mu} = \left\{\left(Ze\epsilon_{\lambda\mu\alpha\beta} + \frac{1}{e}g_{\alpha\lambda}g_{\beta\mu}\right)\epsilon_{\sigma\nu\rho\tau} r^\rho u^\tau (g^{\sigma\alpha}g^{\nu\beta} - a^\sigma r^\alpha g^{\beta\nu} - a^\nu g^{\beta\sigma} r^\beta)/R^3\right\}\hat{\boldsymbol{n}}. \tag{41}$$

The Abelian potential is the sum of (37) and φ_μ

$$A_\mu \cdot \hat{\boldsymbol{n}} = \frac{n^3(n^2\Lambda_\lambda^1 - n^1\Lambda_\lambda^2)(\delta_\mu^\lambda - a^\lambda r_\mu)}{eR(n_1^2 + n_2^2)} + Ze\frac{u_\mu}{R}. \tag{42}$$

REFERENCES

[1] Hou, B. Y., Duan, Y. S. & Ge, M. L.: *Acta Physica Sinica*, **25**(1976), 514.
[2] Arafune, J., Freund, P. G. O. & Goebel, C. J.: *J. Math. Phys.*, **16**(1975), 433.
[3] Wu, T. T. & Yang, C. N.: *Phys. Rev.*, **D12**(1975), 3845.
[4] Hou, B. Y.: *Acta Physica Sinica*, **26**(1977), 83.
[5] 't Hooft, G.: *Nuc. Phys.*, **B79**(1974), 276.
[6] Hou, B. Y.: *Journal of Fudan Univ.*, (1977), (3), 23.
[7] Marciano, W. J. & Pagels, H.: *Phys. Rev.*, **D12**(1975), 1093.
[8] Sinha, A.: *Phys. Rev.*, **D14**(1976), 2016.
[9] Gu, C. H.: *Journal of Fudan Univ.*, (1976) (3—4), 161.
[10] Patrascioiu, A.: *Phys. Rev.*, **D12**(1975), 523.
[11] Wu, T. T. & Yang, C. N.: in *Properties of Matter Under Unusual Conditions*, p. 349, (eds. Fernbach, S. & Mack, H.), Intersience.
[12] Wu, T. T. & Yang, C. N.: *Phys. Rev.*, **D13**(1976), 3233; Prasad, M. K. & Sommerfield, C. M.: *Phys. Rev. Lett.*, **35**(1975), 760; Hsu, J. P.: *Lett. Nuov. Cim.*, **14**(1975) 189; Hsu, J. P.: *Phys. Rev. Lett.*, **36**(1976), 646.
[13] Li, H. Z., Xian, D. C. & Guo, S. H.: *Acta Physica Sinica*, **25**(1976), 507.
[14] Li, H. Z., Xian, D. C., & Guo, S. H.: (to be Published).
[15] Jackiw, R. & Rebbi, C.: *Phys. Rev.*, **D14**(1976), 517.
[16] Wu, A. C. T. & Wu, T. T.: *J. Math. Phys.*, **15**(1974), 53.
[17] Chakrabarti, A.: *Ann. Inst. Henri Poincare*, **23**(1975), 235.
[18] Kobayashi, S. & Nomizu, K.: *Fundations of Differential Geometry*. p. 103. Interscience.
[19] Kuo, H. Y. & Wu, Y. S.: *Scientia Sinica*, **20**(1977), 408.
[20] Li, H. Z., Xian, D. C. & Guo, S. H.: (to be published).
[21] Chakrabarti, A.: *Nucl. Phys.*, **B101**(1975), 159.
[22] Wang, Y. K., Chang, K. Y., & Hou, B. Y.: (to be published in *Physica Energiae et Physica Nuclearis*).
[23] Jeans, J.: *Mathematical Theory of Electricity and Magnetism*, Cambridge.

第3卷 第6期　　高 能 物 理 与 核 物 理　　Vol. 3, No. 6
1979 年 11 月　　PHYSICA ENERGIAE FORTIS ET PHYSICA NUCLEARIS　　Nov., 1979

闵空间拓扑非平庸球对称场中零能费米子

侯伯宇　　侯伯元
（西北大学）（内蒙古大学）

摘　　要

本文讨论了 $U(1)$ 点磁荷场中零能费米子解的个数及其物理性质，特别是轴矢流部分守恒的反常源为磁荷，及电荷的有效分布集中在磁荷点的现象．本文又采用同步规范局部坐标系明显地将任意同位旋自旋 $\frac{1}{2}$ 粒子在球对称 $SU(2)$ 无源场中的方程分离变量．证明了零能解仅有 $J=0$，$|q|=\frac{1}{2}$；$J=\frac{1}{2}$，$|q|=1$，0 的情况，并且明显地求出了同位旋 I 为任意半整数及 1 时的解．

规范场中零本征值费米子解与 Adler 反常、CP 破缺，瞬子及其微扰解数，真空隧道效应以及费米子数 1/2 等问题有关[1,2]，近来受到了注意．此种解的存在条件涉及非平庸的拓扑性质，故一般底空间需是可紧致化的欧空间，而欧空间中零本征值费米子的物理解释要联系到闵空间的虚时间隧道效应．

本文讨论闵空间拓扑非平庸球对称场中零能费米子．第一节讨论 $U(1)$ 点磁荷场中带电费米子．阐明了四维闵空间的解数如何取决于角向方程，以及球面上的场、方程、波函数分别与二维欧空间对应式的关系．分析了零能解的物理量：特别是轴矢流部分守恒而在原点（磁场的源）破缺，及电荷的有效分布集中在原点等有意思现象．第二节中讨论 $SU(2)$ 无源球对称场中的费米子，运用球对称同步规范标架将此背景场中任意同位旋的自旋1/2 费米子的方程分离变量，利用背景场的边界行为证明了仅有 $J=0$，$|q|=\frac{1}{2}$；及 $J=\frac{1}{2}$，$|q|=1$，0 的零能解．并且明显地求出了同位旋 I 为半整数及 1 时的解．

一、在点磁单极附近的零能费米子

设在强度为 g/e 的点磁单极的磁场中，有一电荷为 νe，质量为M的 Dirac 粒子，e 为电子的电荷值，ν，g 为量子化的整数（半整数）．则它的定态波函数满足下方程：

$$\boldsymbol{\gamma}\cdot\boldsymbol{\nabla}\psi+M\psi=\gamma_4E\psi, \tag{1}$$

其中 $\boldsymbol{\nabla}$ 是规范协变微商

$$\boldsymbol{\nabla}=\boldsymbol{\partial}-i\nu e\boldsymbol{A}. \tag{2}$$

本文 1978 年 8 月 1 日收到．

而规范势 $\boldsymbol{A}=A_\varphi \boldsymbol{e}_\varphi$:

$$A_\varphi=\frac{g}{e}\frac{1-\cos\theta}{r\sin\theta},\quad(\text{在北区})$$
$$A_\varphi=\frac{g}{e}\frac{-1-\cos\theta}{r\sin\theta},\quad(\text{在南区})\tag{3}$$

这时存在守恒的总角动量

$$\boldsymbol{J}=\boldsymbol{r}\times(\boldsymbol{P}-\nu e\boldsymbol{A})-q\frac{\boldsymbol{r}}{r}+\frac{1}{2}\boldsymbol{\Sigma}.\tag{4}$$

其中 $\boldsymbol{\Sigma}$ 为 Dirac 矩阵，$q=g\nu$ 取量子化整数或半整数值，$-q\frac{\boldsymbol{r}}{r}$ 项代表电磁场内禀角动量.

以下求 $E=0$ 的解，为此选 $\gamma_4\gamma_5$ 为对角的表象：

$$\boldsymbol{\gamma}=\begin{pmatrix}-\boldsymbol{\sigma} & 0\\ 0 & \boldsymbol{\sigma}\end{pmatrix},\quad \gamma_4=\begin{pmatrix}0 & -i\\ i & 0\end{pmatrix},\quad \gamma_4\gamma_5=\begin{pmatrix}1 & 0\\ 0 & -1\end{pmatrix},\tag{5}$$

其中 $\boldsymbol{\sigma}$、0、i、1 等均代表 2×2 矩阵. 并将波函数 ψ 表为

$$\psi=\begin{pmatrix}\chi^+\\ \chi^-\end{pmatrix},\tag{6}$$

其中 $\chi^\pm$ 都是两分量的.

为分离变量采用推广的两分量 K 算子：

$$K\equiv -i\varepsilon_{ijK}\sigma_i x_j\nabla_K+1,\tag{7}$$

即将通常 K 算子中的普通微商 ∂_K 改为规范协变微商 ∇_K.

再利用易证的

$$\boldsymbol{\sigma}\cdot\boldsymbol{\nabla}=\sigma_r\left(\frac{\partial}{\partial r}+\frac{1}{r}\right)-\frac{1}{r}\sigma_r K,\tag{8}$$

其中

$$\sigma_r=\frac{1}{r}\boldsymbol{\sigma}\cdot\boldsymbol{r}.\tag{9}$$

(8)式代入(1)式后得（当 $E=0$ 时）：

$$\left[\sigma_r\left(\frac{\partial}{\partial r}+\frac{1}{r}\right)-\frac{1}{r}\sigma_r K\mp M\right]K^\pm=0.\tag{10}$$

解(10)式最方便是用 J^2, J_Z, 与 σ_r 的共同本征态为基展开，即如下式选完备基：

$$\eta^j_{m,\mu}(\varphi,\theta,\gamma)=\sqrt{\frac{2j+1}{4\pi}}D^j_{m,\mu-q}(\varphi,\theta,\gamma)s_\mu(\varphi,\theta,\gamma),\tag{11}$$

其中

$$s_\mu(\varphi,\theta,\gamma)=\sum_{\mu'}S_{\mu'}D^{\frac{1}{2}}_{\mu',\mu}(\varphi,\theta,\gamma).\tag{12}$$

而 $S_{\mu'}$ 为通常 σ_Z 的本征态，表为：

$$s_{\frac{1}{2}}=\begin{pmatrix}1\\0\end{pmatrix},\quad s_{-\frac{1}{2}}=\begin{pmatrix}0\\1\end{pmatrix}.$$

于是 s_μ 为 σ_r 的本征态，例如：

$$
\begin{aligned}
s_{\frac{1}{2}}(\varphi, \theta, -\varphi) &= \begin{pmatrix} \cos\frac{\theta}{2} \\ \sin\frac{\theta}{2} e^{i\varphi} \end{pmatrix}, \quad \text{（在北区）} \\
s_{\frac{1}{2}}(\varphi, \theta, \varphi) &= \begin{pmatrix} \cos\frac{\theta}{2} e^{i\varphi} \\ \sin\frac{\theta}{2} \end{pmatrix}. \quad \text{（在南区）}
\end{aligned} \tag{13}
$$

易见此组完备基(11)具有如下性质：

$$
\begin{aligned}
J^2\eta^j_{m,\mu} &= j(j+1)\eta^j_{m,\mu}, \\
J_Z\eta^j_{m,\mu} &= m\eta^j_{m,\mu}, \\
\sigma_r\eta^j_{m,\mu} &= 2\mu\eta^j_{m,\mu}, \\
K\eta^j_{m,\mu} &= K_q\eta^j_{m,-\mu},
\end{aligned} \tag{14}
$$

其中

$$
K_q = \sqrt{\left(j+\frac{1}{2}\right)^2 - q^2}.
$$

令 $K^{\pm} = \sum\limits_{\mu} f^{\pm}_{\mu}(r)\eta^j_{m,\mu}$ 代入(9)并利用 $\eta^j_{m,\mu}$ 的正交完备性得径向函数 $f^{\pm}_{\mu}(r)$ 满足方程：

$$
\left(\frac{\partial}{\partial r} + \frac{1}{r}\right) f^{\pm}_{\mu}(r) - \frac{1}{r} K_q f^{\pm}_{-\mu}(r) \mp 2\mu M f^{\pm}_{\mu}(r) = 0. \tag{15}
$$

考查在∞的渐近行为，可见仅当有关各方程第一项与最后一项系数比均为正时始有相容的束缚态解．当$K_q \neq 0$时，$f_{\mu}(r)$ 与 $f_{-\mu}(r)$ 两函数有关而其方程组最后一项符号相反，可见无束缚态解．仅当 $K_q = 0\left(\text{即 } j = |q| - \frac{1}{2}\text{，由(11)知还要求 } \mu = \frac{1}{2}\frac{q}{|q|}\right)$ 时，才可能有如下束缚态解：

$$
\begin{aligned}
&\text{当 } q > 0(\mu > 0) \text{ 时：} f^{+}(r) = 0, \quad f^{-}(r) = F(r). \\
&\text{当 } q < 0(\mu < 0) \text{ 时：} f^{-}(r) = 0, \quad f^{+}(r) = F(r).
\end{aligned} \tag{16}
$$

其中 $F(r)$ 满足：

$$
\left(\frac{\partial}{\partial r} + \frac{1}{r}\right) F(r) + MF(r) = 0,
$$

解得

$$
F(r) = \frac{\sqrt{2M}}{r} e^{-Mr}. \tag{17}
$$

虽然 $F(r)$ 在原点奇异，但是

$$
\int_0^{\infty} F^2(r) r^2 dr = 1,
$$

即解可归一．

当 $K_q = 0$ 时基矢(11)为 J^2, J_Z, σ_r 以及K算子的共同本征态，其 j, μ 指标全由 q 值决定，可表为

$$\eta_m(\varphi,\theta,\gamma)=\sqrt{\frac{2j+1}{4\pi}}D^j_{m,\mu-q}(\varphi,\theta,\gamma)s_\mu(\varphi,\theta,\gamma),$$

$$\left(j=|q|-\frac{1}{2},\quad \mu=\frac{1}{2}\frac{q}{|q|}\right),\tag{11a}$$

此式实质上与文献[3]所用 η_m 相同.

总之,在点磁单极附近,经典 Dirac 方程存在可归一零能解 $F(r)\eta_m(\varphi,\theta,\gamma)$,它是 $\gamma_4\gamma_5$ 与 σ_r 的本征态,本征值分别为 $-\frac{q}{|q|}$ 与 $\frac{q}{|q|}$,解是 $2j+1$ 维简并的 $(j\geqslant m\geqslant -j)$,即解数

$$n=2j+1=2|q|.$$

以 Dim ψ^- (Dim ψ^+) 表示 $\gamma_4\gamma_5$ 本征值为负(正)的本征态数,则上式可写为:

$$\text{第一陈类示性数}\equiv 2q=\text{Dim }\psi^- - \text{Dim }\psi^+\tag{18}$$

此式与文献[4]所引 index 定理相似,但该文相应式中左方是二维欧空间涡线势的缠绕数,右方是二维"γ_5"取负正本征值的零能解数差. 本文右方则指 $\gamma_4\gamma_5$ 本征值为负正的零能解数差. 二者的联系是:因 $E=0$ 使 Dirac 方程(1)可分块对角,径向边界条件进一步要求仅当 $K_q=0$ 时有可归一解(于是波函数可同时为 σ_r 的本征态),且解必在径向微商前的 $\gamma_r=-\gamma_4\gamma_5\sigma_r$ 取正值的子空间内,亦即 $\gamma_4\gamma_5$ 的本征值符号与 σ_r 的本征值符号相反,于是问题化为角向方程 $K\psi=0$ 的解 η_m 中 σ_r 的本征值为正负的个数差,即磁量子数 m 的取值范围,有意思的是这里将解的个数与群表示的维数相联系. 为清楚说明问题,如下列表对比:

在二维欧空间的量	在四维闵空间角向部分对应量
$i\gamma_\mu\nabla_\mu$ (此处 $\mu=1,2$, 求和)	$K=-i\varepsilon_{ijK}\sigma_i x_j\nabla_K+1$
$A_\mu(x)=\frac{2g}{e}\frac{1}{1+x^2}\varepsilon_{\lambda\mu}x_\mu$ (Vortex)	$A_r=A_\theta=0,\ A_\varphi=\frac{g}{e}\frac{1-\cos\theta}{\sin\theta}$
$E=0$	$K_q=0$
"γ_5" $=\sigma_z$ 本征态	σ_r 本征态
$u=c\begin{pmatrix}(1+x^2)^{-\frac{1}{2}}\\0\end{pmatrix}$	$\hat{u}=\frac{1}{\sqrt{2}}\begin{pmatrix}\cos\frac{\theta}{2}\\ \sin\frac{\theta}{2}e^{i\varphi}\end{pmatrix}=s_{\frac{1}{2}}(\varphi,\theta,-\varphi)$
$u^{2j+1}=(e^{i\varphi}\sqrt{x^2})^{j+m}\begin{pmatrix}(1+x^2)^{-j-\frac{1}{2}}\\0\end{pmatrix}$	$\hat{u}^{2j+1}=D^j_{m,-q+\frac{1}{2}}(\varphi,\theta,-\varphi)s_{\frac{1}{2}}(\varphi,\theta,-\varphi)$

二者通过下列关系相联系:

$$i\gamma_\mu\nabla_\mu=UKU^{-1},$$

$$u=U\hat{u}.$$

其中 $U=\frac{2}{1+x^2}\frac{1}{\sqrt{2}}(1-i\gamma_\mu x_\mu)=\sqrt{2}\cos\frac{\theta}{2}\begin{pmatrix}\cos\frac{\theta}{2} & \sin\frac{\theta}{2}e^{-i\varphi}\\ -\sin\frac{\theta}{2}e^{i\varphi} & \cos\frac{\theta}{2}\end{pmatrix}$,即二者经

保形变换相连系．其中 $j = q - \frac{1}{2}$，即 $2j + 1 = 2q = 2\nu g > 0$，文献[4]相应于本文的 $\nu = 1$，$2g = n > 0$ 的情况．当 $q < 0$ 时应得含 $S_{-\frac{1}{2}}$的解．

比较中微子理论由于 $M = 0$，波函数可按 γ_5 本征值分为两分量函数，它们是 $\boldsymbol{\sigma} \cdot \boldsymbol{P}$ 的本征态，无确定宇称以光速运动．现在 $E = 0$ 解可按 $\gamma_4\gamma_5$ 本征值分解为两分量函数，它们是 $\boldsymbol{\sigma} \cdot \boldsymbol{r}$ 的本征态，也无确定宇称，但为静态．

以下分析零能束缚态的荷流等物理量的分布(略去了共同因子 νe)，

$$J_K = i\bar{\Phi}\gamma_K\psi = 0,$$

$$\rho = \psi^*\psi = \frac{2M}{r^2} e^{-2Mr}\eta^*\eta,$$

$$J_K^5 = \bar{\Phi}\gamma_5\gamma_K\psi = -\frac{i2M}{r^2} e^{-2Mr}\eta^*\sigma_K\eta,$$

$$J_4^5 = \bar{\Phi}\gamma_5\gamma_4\psi = 0,$$

$$J^5 = \bar{\Phi}\gamma_5\psi = -\frac{q}{|q|}\frac{i2M}{r^2} e^{-2Mr}\eta^*\eta.$$

磁矩

$$\frac{1}{2M}\bar{\Phi}\sigma\psi = 0.$$

电矩

$$P_K = -\frac{1}{2M}\bar{\Phi}\gamma_4\gamma_K\psi = -\frac{q}{|q|}\frac{1}{r^2} e^{-2Mr}\eta^*\sigma_K\eta.$$

注意到 η 为 σ_r 的本征态，$\eta^*\sigma_K\eta$ 中仅径向非零 $\eta^*\sigma_r\eta = \frac{q}{|q|}\eta^*\eta$．所以 J^5 流仅径向部分非零，除原点外，此流是部分守恒的、“正常”的：

$$\partial_\mu J_\mu^5 = \partial_i J_i^5 = -2MJ^5, \tag{19}$$

仅在原点(磁场的散度非零处)存在反常．

由于 ψ 是 $\gamma_4\gamma_5$ 本征态，使

$$J^5 = -i\frac{q}{|q|}\rho,$$

$$J_K^5 = i\frac{q}{|q|}2MP_K.$$

再比较(19)式易见由电极化 P_K 产生的束缚电荷 $-\partial_K P_K$ 与自由电荷 ρ 除原点外处处相消，结果如同电荷全集中在原点．总之，在点磁单极周围有一 Dirac 粒子零能束缚态，不仅能量、动量等都不改变，且除原点外电荷分布也不改变．

由于势 A 在原点奇异(这是规范无关的，因为场强也奇异)，故此方程的算子在原点奇异，与此相关存在有 Lipkin 困难．我们所求解在原点也存在有奇异．严格些应该说本文采用原点的边界条件：$f^+(0) = 0$，当 $q > 0$ (或 $f^-(0) = 0$，当 $q < 0$)，如文献[5]所讨论．或者为避免 Lipkin 困难，文献[6]设 Dirac 粒子还有小反常磁矩 $\frac{\nu e}{2M}(1+\kappa)$ 可证仅当 $\kappa > 0$ 时才有 $\chi^\pm = f^\pm(r)\eta_m$ 型解，但这时

$$f(r)=\frac{1}{r}e^{-Mr-\frac{|\kappa q|}{2Mr}},\quad (\kappa>0) \tag{20}$$

当 $\kappa<0$ 时无此解．此解的角动量仍限制为 $j=|q|-\frac{1}{2}$，其性质与我们前面的讨论完全类似，但它在 $r\to 0$ 与 $r\to\infty$ 时都趋于零，无原点奇异，避免了 Lipkin 困难．有意思的是，这时除了上述类型解外还存在有 $j\geqslant|q|+\frac{1}{2}$ 的零能解，它们仍为 $\gamma_4\gamma_5$ 的本征态，但不再是 $\boldsymbol{\sigma}\cdot\boldsymbol{r}$ 的本征态，也非宇称的本征态．当阵类示性数一定时，它们的数目可以有无穷多，但是当 $\kappa>0$ 与 $\kappa<0$ 时，$\gamma_4\gamma_5$ 本征值分别为 ± 值的 j, m, q 相等的解一一对应，解数相等，而仅(20)型的解数不一样，将 $\kappa\gtrless 0$ 在一块考虑，仍有关系如(18)．

二、SU(2) 规范场中零能费米子

静球对称不可约化 $SU(2)$ 无源规范场必为同步的，可选同步规范表为：

$$W_i^\sigma=\varepsilon_{i\sigma\tau}[r\varphi(r)-1]x^\tau/er^2$$

$$W_4^\sigma=iG(r)x^\sigma/er.\quad [r=(x_1^2+x_2^2+x_3^2)^{\frac{1}{2}},\quad i、\sigma、\tau=1,2,3] \tag{21}$$

它可分解为可 Abel 化部分 A 与不可约化部分 B：

$$W_i^\sigma=A_i^\sigma+B_i^\sigma,$$

$$A_i^\sigma=-\varepsilon_{i\sigma\tau}x^\tau/er^2, \tag{22a}$$

$$B_i^\sigma=\varphi(r)\varepsilon_{i\sigma\tau}x^\tau/er. \tag{22b}$$

为使总能量有限，还要求它们满足如下渐近条件：

$$\begin{aligned}&r\varphi(r)\to 0,\quad G(r)\to i\beta,\quad (|\beta|>0)\quad (当\ r\to\infty\ 时)\\ &r\varphi(r)\to 1,\quad G(r)\to 0.\qquad\qquad\qquad (当\ r\to 0\ 时)\end{aligned} \tag{23}$$

满足上条件的无源解已知形式为：

$$\begin{aligned}\varphi(r)&=\frac{\beta}{\mathrm{sh}\beta r},\\ G(r)&=-i\left(\frac{1}{r}-\beta\,\mathrm{cth}\,\beta r\right).\end{aligned} \tag{24}$$

在此规范场中 $M=0$ 的自旋 1/2 同位旋 I 的粒子满足如下经典 Dirac 方程：

$$\gamma_\mu\nabla^\mu\psi=0. \tag{25}$$

其中

$$\nabla^\mu=\partial^\mu-ieT^aW_a^\mu, \tag{26}$$

T^a 为 $SU(2)$ 产生算符．以下选 γ_5 对角表象：

$$\gamma_K=\begin{pmatrix}0 & -i\sigma_K\\ i\sigma_K & 0\end{pmatrix},\quad \gamma_4=\begin{pmatrix}0 & I\\ I & 0\end{pmatrix},\quad \gamma_5=\begin{pmatrix}I & 0\\ 0 & -I\end{pmatrix}, \tag{27}$$

而波函数表为：

$$\psi=\begin{pmatrix}x^+\\ x^-\end{pmatrix}. \tag{28}$$

其中 $x^\pm$ 都是自旋二维、同位旋 $2I+1$ 维旋量．由(25)得定态方程：

$$(i\sigma_K\nabla_K \pm G(r)T^\sigma x^\sigma/r)K^\pm = \pm EK^\pm. \tag{29}$$

由于同步球对称性，存在守恒的总角动量：

$$\boldsymbol{J} = \boldsymbol{r}\times\boldsymbol{P} + \boldsymbol{T} + \frac{1}{2}\boldsymbol{\sigma}$$

为便于分离变量，仍采用推广的K算子，这时仍有如上节(7)、(8)式的关系，但其中规范协变微商$\boldsymbol{\nabla}$应改为仅仅以规范势可约部分A_i^a((22a)式)为联络的规范协变微商$\boldsymbol{\nabla}^{(A)}$方可与$J$对易，即：

$$\nabla_i^{(A)} = \partial_i - ieA_i^aT^a, \tag{26a}$$

$$K = -i\varepsilon_{ijK}\sigma_i x_j\nabla_K^{(A)} + 1, \tag{7a}$$

$$\sigma_i\nabla_i^{(A)} = \sigma_r\left(\frac{\partial}{\partial r} + \frac{1}{r}\right) - \frac{1}{r}\sigma_r K. \tag{8a}$$

于是可将(29)式化为：

$$\left[\sigma_r\left(\frac{\partial}{\partial r} + \frac{1}{r}\right) - \frac{1}{r}\sigma r K - i\varphi(r)\,\frac{\boldsymbol{r}}{r}\cdot(\boldsymbol{\sigma}\times\boldsymbol{T}) \mp iG(r)\times\left(\boldsymbol{T}\cdot\frac{\boldsymbol{r}}{r}\right)\right]x^\pm = \mp iEx^\pm. \tag{30}$$

在具体对上式分离径向与角向变量以便求解时，通常采取对固定极轴投影的本征态为基矢展开，使计算非常麻烦．现将$x^\pm$的角向部分按反映问题对称性的局部坐标系的完备基展开，即令：

$$x^\pm = \sum_{\mu\nu} f^\pm_{\mu,\nu}(r)\zeta_{\mu,\nu}(\varphi, \theta, \gamma). \tag{31}$$

其中

$$\zeta_{\mu,\nu} = \sqrt{\frac{2j+1}{4\pi}}\,D^j_{m,\mu+\nu}s_\mu i_\nu. \tag{32}$$

而i_ν为电荷算符$\left(\text{同位旋}\,\boldsymbol{T}\,\text{在局部极轴上投影}\ T_r = \boldsymbol{T}\cdot\frac{\boldsymbol{r}}{r}\right)$的本征态，它与同位旋在固定极轴上的投影$T_Z$的本征态$I_\nu$可由规范变换联系如下：

$$i_\nu(\varphi, \theta, \gamma) = \sum_{\nu'} I_{\nu'}D^I_{\nu'\nu}(\varphi, \theta, \gamma).$$

而上节中$\eta^j_{m,\mu}$乘以I_3再经上规范变换，可变为本节所选的$\zeta_{\mu\nu}$(注意与本节A^a_μ对应的磁荷$g=-1$，故$q=-\nu$)．

$\zeta_{\mu,\nu}$是J^2, J_Z, σ_r, T_r的共同本征态，有如下性质：

$$\begin{aligned}
J^2\zeta_{\mu,\nu} &= j(j+1)\zeta_{\mu,\nu},\\
J_Z\zeta_{\mu,\nu} &= m\zeta_{\mu,\nu},\\
\sigma_r\zeta_{\mu,\nu} &= 2\mu\zeta_{\mu,\nu},\\
T_r\zeta_{\mu,\nu} &= \nu\zeta_{\mu,\nu},\\
K\zeta_{\mu,\nu} &= K_\nu\zeta_{-\mu,\nu},\\
\frac{\boldsymbol{r}}{r}\cdot(\boldsymbol{\sigma}\times\boldsymbol{T})\zeta_{\mu,\nu} &= -i2\mu\alpha^I_{\nu+\frac{1}{2}+\mu}\zeta_{-\mu,\nu+2\mu}.
\end{aligned} \tag{33}$$

其中

$$\alpha_\nu^I = (I+\nu)^{\frac{1}{2}}(I-\nu+1)^{\frac{1}{2}},$$

$$K_\nu = \alpha_{|\nu|+\frac{1}{2}}^j = \sqrt{\left(j+\frac{1}{2}\right)^2 - \nu^2}\,.$$

将(31)(32)代入(30),并利用(33)各式,取 $E=0$,再利用 $\zeta_{\mu,\nu}$ 的正交完备性,得零能解径向部分满足下方程组:

$$\left(\frac{\partial}{\partial r}+\frac{1}{r}\right)f_{\mu,\nu}^{\pm}(r) - \frac{K_\nu}{r} f_{-\mu,\nu}^{\pm}(r) + \alpha_{\nu+\frac{1}{2}+\mu}^I \varphi(r) f_{-\mu,\nu+2\mu}^{\pm}(r) \mp 2\mu\nu iG(r) f_{\mu,\nu}^{\pm}(r) = 0,$$

(34)

首先分析此方程组在无穷远处的渐近性质,由(23)知

$$iG(r)\to -\beta,\quad (\text{不失一般性以下假设 } R_e\beta > 0)$$

由要求方程(34)最后一项系数均为正看出如 $\mu\nu>0$ 只可能有 γ_5 的正本征值的解 $f^+(f^-=0)$,如 $\mu\nu<0$,只可能有 γ_5 的负本征值的解 $f^-(f^+=0)$。

由方程(34)第二项使 $f_{\mu,\nu}(r)$ 与 $f_{-\mu,\nu}(r)$ 相关,如 $\nu\neq 0$,与方程最后一项必取正号要求矛盾,故必须

$$K_\nu = 0. \quad (\text{当 } \nu \neq 0)$$

由方程(34)第三项使 $f_{\mu\nu}$ 与 $f_{-\mu,\nu+2\mu}$ 相关,为满足联立方程组各方程最后一项符号相同的要求,$\mu\nu$ 与 $-\mu(\nu+2\mu)$ 必须同号,即 ν 必须满足下三条件之一:

$$\nu=0,\quad \text{或}\quad \nu=-2\mu,\quad \text{或}\quad \nu=-\mu.$$

总之有两种情况:

1. 当 Dirac 粒子同位旋 I 为任意半整数时,可归一的静零能解必满足:

$$\nu=-\mu=\pm\frac{1}{2},\quad K_\nu=0,\quad \text{即}\quad j=0; \tag{35a}$$

2. 当 Dirac 粒子同位旋 I 为任意整数时,必符合下条件:

$$\nu=-2\mu=\pm 1,\quad K_\nu=0,\quad \text{即}\quad j=\frac{1}{2};$$

$$\text{及}\quad \nu=0,\quad \text{而由}\ j=\frac{1}{2}\ \text{知}\quad K_0=1. \tag{35b}$$

二种情况均只含 γ_5 负本征值的 $f^-(f^+=0)$,以下为符号简化,均略去 f^- 的上标而记为 f。

先讨论粒子同位旋 I 为半整数时的情况,将条件(35a)代入(34)得:

$$\left(\frac{\partial}{\partial r}+\frac{1}{r}\right)f_{\mu,-\mu}(r) + \alpha_{\frac{1}{2}}^I \varphi(r) f_{-\mu,\mu}(r) - \frac{1}{2} iG(r) f_{\mu,-\mu}(r) = 0. \tag{36}$$

令

$$F^{\pm}(r) = f_{\frac{1}{2},-\frac{1}{2}}(r) \pm f_{-\frac{1}{2},\frac{1}{2}}(r), \tag{37}$$

得

$$\left(\frac{\partial}{\partial r}+\frac{1}{r}\pm\alpha_{\frac{1}{2}}^I\varphi(r)-\frac{1}{2}iG(r)\right)F^{\pm}(r)=0. \tag{38}$$

其中 $\alpha^I_{\frac{1}{2}} = I + \frac{1}{2}$，再由规范场的边界条件知 $F^+(r)$ 只有平庸解：

$$F^+(r) = f_{\frac{1}{2},-\frac{1}{2}}(r) + f_{-\frac{1}{2},\frac{1}{2}}(r) = 0, \tag{39}$$

而 $F^-(r) = 2f_{\frac{1}{2},-\frac{1}{2}}(r)$ 有满足边界条件的唯一非平庸解．它是电荷数 $\nu = \pm\frac{1}{2}$ 的混合态，同时也是自旋在局部极轴上投影 σ_r 的不同本征态的混合，且处处保持粒子自旋与电磁场的内禀旋方向相反，使得总角动量为零．此解是 γ_5 本征值为 -1 的本征态，由于 $J = 0$，解不简并，解数为 1．这因对静球对称不可约化 $SU(2)$ 规范场只能有同步解，故限为 $g = -1$，而边界条件又限制为 (35a)，$|\nu| = \frac{1}{2}$，故第一阵类示性数只能为 $1(2|q| = 2|\nu| = 1)$，即当 Dirac 粒子同位旋为任意半整数时，零能解都是非简并的．

利用规范势的明显解析表达式(24)，还可具体积分得 Dirac 粒子零能解径向部分的解析表达式为：

$$f(r) = cr^{-\frac{1}{2}}(\operatorname{sh}\beta r)^{-\frac{1}{2}}\left(\operatorname{th}\frac{\beta r}{2}\right)^{I+\frac{1}{2}}\left(I = N + \frac{1}{2},\quad N = 0, 1, 2, \cdots\right) \tag{40}$$

由此式更明显看出，当 $r \to \infty$ 时，解按指数衰减，分布半径由规范势第 4 分量在无穷远处的渐近值 β 决定（它相当于 Higgs 场的真空平均值）．此解在原点的性质与同位旋 I 有关，当 $I = \frac{1}{2}$ 时在原点趋于有限值，而当 I 为大于 $\frac{1}{2}$ 的半整数时，在原点趋于零．

下面再讨论当同位旋为任意整数时的情况，将条件 (35b) 代入(34)得：

$$\begin{cases} \left(\dfrac{\partial}{\partial r} + \dfrac{1}{r}\right) f_{\mu,0}(r) - \dfrac{1}{r} f_{-\mu,0}(r) + \alpha\varphi(r) f_{-\mu,2\mu}(r) = 0, \\ \left(\dfrac{\partial}{\partial r} + \dfrac{1}{r}\right) f_{\mu,-2\mu}(r) + \alpha\varphi(r) f_{-\mu,0}(r) - iG(r) f_{\mu,-2\mu}(r) = 0, \end{cases} \tag{41}$$

其中

$$\alpha = \alpha^I_0 = \alpha^I_1 = \sqrt{I(I+1)}.$$

令

$$F^\pm_0(r) = f_{\frac{1}{2},0}(r) \pm f_{-\frac{1}{2},0}(r), \quad F^\pm_1(r) = f_{\frac{1}{2},-1}(r) \pm f_{-\frac{1}{2},1}(r) \tag{42}$$

则方程组(41)可化为两组方程：

$$\begin{cases} \dfrac{\partial}{\partial r} F^+_0(r) + \alpha\varphi(r) F^+_1(r) = 0, \\ \left(\dfrac{\partial}{\partial r} + \dfrac{1}{r} - iG(r)\right) F^+_1(r) + \alpha\varphi(r) F^+_0(r) = 0, \end{cases} \tag{43a}$$

$$\begin{cases} \left(\dfrac{\partial}{\partial r} + \dfrac{2}{r}\right) F^-_0(r) - \alpha\varphi(r) F^-_1(r) = 0, \\ \left(\dfrac{\partial}{\partial r} + \dfrac{1}{r} - iG(r)\right) F^-_1(r) - \alpha\varphi(r) F^-_0(r) = 0. \end{cases} \tag{43b}$$

在原点这些方程的解常为一个正则函数与一个奇异不可归一函数的线性组合．而它们在无穷远处的渐近行为：(43a) 的解 $\sim c_1 + c_2 e^{-\beta r}$，故一般在原点正则的解在 $r \to \infty$ 时不可归一，即只有平庸解：

$$F_0^+(r) \equiv f_{\frac{1}{2},0}(r) + f_{-\frac{1}{2},0}(r) = 0, \quad F_1^+(r) \equiv f_{\frac{1}{2},-1}(r) + f_{-\frac{1}{2},1}(r) = 0, \tag{44}$$

而（43b）的解在无穷远处的渐近行为 $\sim c_1 r^{-2} + c_2 e^{-\beta r}$，故可得唯一可归一解满足（43b）及条件(44)．它是电荷数 $\nu = \pm 1, 0$ 的混合态，同时也是自旋在局部极轴上不同投影的混合态，其中电磁场内禀旋非零部分处处保持与粒子自旋方向相反，使总角动量 $j = \frac{1}{2}$．当 $r \to \infty$ 时的渐近行为由两部分组成，其中 $\nu = \pm 1$ 部分按指数衰减，而 $\nu = 0$ 部分则按 r^{-2} 衰减．此解是 γ_5 的本征值为 -1 的本征态，但由于 $J = \frac{1}{2}$，故双重简并，解数也满足：

$$n = 2|\nu_{\max}| = 2|q_{\max}|.$$

利用规范势的明显解析表达式（24）式，对同位旋矢量 Dirac 粒子，还可具体积分（43b）得：

$$\begin{aligned} f_{\frac{1}{2},0}(r) &= \frac{1}{\sqrt{2}\,\beta r^2} - \frac{\beta}{\sqrt{2}\,\mathrm{sh}^2 \beta r}, \\ f_{\frac{1}{2},-1}(r) &= \beta \frac{\mathrm{ch}\,\beta r}{\mathrm{sh}^2 \beta r} - \frac{1}{r\,\mathrm{sh}\,\beta r}. \end{aligned} \tag{45}$$

在无穷远处 $f_{\frac{1}{2},-1}(r)$ 按指数衰减，而 $f_{\frac{1}{2},0}(r)$ 按 r^{-2} 缓慢衰减，而在原点它们都趋于有限值．

总之，在静球对称 $SU(2)$ 规范场中的可归一零能费米子解的数目仍等于第一陈类示性数，且它们都是 γ_5 本征值 -1 的本征态，而无 γ_5 正本征值的本征态，即零能解数仍由(18)式决定．

参考文献

[1] R. Jackiw, *Rev. Mod. Phys.*, **49**(1977), 681.
[2] R. Jackiw, C. Rebbi, *Phys. Rev.*, **D13**(1976), 3398.
[3] Y. Kazama, C. N. Yang, A. S. Goldhaber, *Phys. Rev.*, **D15**(1977), 2387.
[4] N. K. Nielsen, B. Schroer, *Nucl. Phys.*, **B127**(1977), 493.
[5] C. J. Callias, *Phys. Rev.*, **D16**(1977), 3068.
[6] Y. Kazama, C. N. Yang, *Phys. Rev.*, **D15**(1977), 2300.

ZERO ENERGY FERMION IN TOPOLOGICAL NONTRIVIAL SPHERICAL SYMMETRICAL FIELD ON MINKOWSKI SPACE

HOU BO-YU

(*Northwest University*)

HOU BO-YUAN

(*Inner Mongolian University*)

ABSTRACT

This paper discusses the number of solution and physical property of zero energy fermions in the field of $U(1)$ pointwise monopole. It is interesting to point out that the anomaly source of PCAC lies on the monopole, and that the effective electric charge is concentrated at the point monopole.

This paper has separated explicitly the variables in the equation of the half spin particles with arbitrary isospins moving in the spherical symmetric sourceless $SU(2)$ gauge field. It is shown that the zero energy solution exists only when the total angular momentum $J=0$ and charge number $|\nu|=1/2$; or when $J=1/2$, $|\nu|=1,0$. The solutions with isospin equals to one or arbitrary half integrals are given explicitly.

Volume 93B, number 4 PHYSICS LETTERS 30 June 1980

NON-SELFDUAL INSTANTONS OF TWO DIMENSIONAL CP(N) ($N \geqslant 2$) CHIRAL THEORY

HOU BO-YU and WANG YU-BIN
Northwest University, Xian, People's China

HOU BO-YUAN
Inner Mongonia University, Huhehot, People's China

and

WANG PEI
Szichuan University, Chengtu, People's China

Received 1 April 1980

We give maps onto a SO(3) subgroup manifold of the SU(3) group manifold, which is reducible to a CP(1) and a CP(2) chiral field. The map on CP(1) is holomorphic, while that on CP(2) may be either holomorphic or not. We also discuss the related gauge field and cohomology properties.

Recently many papers [1–5] discussed the CP(N) chiral theory. All known instanton solutions are induced by holomorphic maps, so they obey the self-duality condition. In this paper we give non-self-dual solutions induced by a conformal but nonholomorphic map. For simplicity we show explicitly the case of CP(2) only. Generalization for CP(N) may be given similarly. Usually in the action and equation of the CP(2) chiral model the elements of coset space G/H = SU(3)/SU(2) × U(1) are expressed by the involution operator N or projection operator P as follows

$$S = \frac{1}{8\pi} \operatorname{tr} \int d^2x \left(\frac{E-N}{2} \partial_\mu N \partial_\mu N \right),$$

or

$$S = \frac{1}{2\pi} \operatorname{tr} \int d^2x \, P \partial_\mu P \partial_\mu P \,, \tag{1}$$

$$[\partial^2 N, N] = 0 \,, \quad \text{or} \quad [\partial^2 P, P] = 0 \,,$$

where:

$$N(g(x)) = g(x)\, N_0 g(x)^{-1} \,, \quad N_0 \equiv \begin{pmatrix} 1 & & \\ & 1 & \\ & & -1 \end{pmatrix},$$

$$P(g(x)) = g(x)\, P_0 g(x)^{-1} \,, \quad P_0 \equiv \begin{pmatrix} 0 & & \\ & 0 & \\ & & 1 \end{pmatrix},$$

$$N = E - 2P \,, \quad g \in \mathrm{SU}(3) \,. \tag{2}$$

Equivalently this coset space may be represented by the "8th" element λ_8 in Lie algebra.

$$\Lambda_8(g(x)) = g(x)\, \lambda_8 g(x)^{-1} \,,$$

$$\lambda_8 \equiv (3N_0 - E)/2\sqrt{3} = (E - 3P_0)/\sqrt{3} \,, \tag{3}$$

and eq. (2) becomes

$$[\partial^2 \Lambda_8, \Lambda_8] = 0 \,. \tag{4}$$

The self-dual solution of eq. (2) may be obtained by taking the following map $g(x)$;

$$g(x) = \exp(\mathrm{i}\alpha(x)\, j_3) \exp(\mathrm{i}\beta(x)\, j_2) \exp(\mathrm{i}\gamma(x)\, j_3) \,, \tag{5}$$

which covers a subgroup manifold J = SO(3) in G = SU(3), where the generators of this J are

$$j_1 = \lambda_6/2 \,, \quad j_2 = \lambda_7/2 \,,$$

$$j_3 = -\mathrm{i}[j_1, j_2] = (\sqrt{3}\,\lambda_8 - \lambda_3)/4 \,, \tag{6}$$

and

Volume 93B, number 4 PHYSICS LETTERS 30 June 1980

$$\mathrm{tg}\,\tfrac{1}{2}\alpha(x)\,\mathrm{e}^{-\mathrm{i}\beta(x)} = w(x) \equiv u(x) + \mathrm{i}v(x)\,, \tag{7}$$

$$w(x_1, x_2) = f(x_1 + \mathrm{i}x_2)\,, \tag{8}$$

or

$$w(x_1, x_2) = f(x_1 - \mathrm{i}x_2)\,, \tag{$\bar{8}$}$$

where f is an arbitrary rational function.

We explain shortly that eqs. (3), (5)–(8) (($\bar{8}$)) really give the self-dual (anti) solution of eq. (4). From eqs. (3), (5), (6)

$$\Lambda_8 = D^0_{00}(\alpha,\beta,\gamma)(\lambda_8 + \sqrt{3}\lambda_3)/4 + \sqrt{3}\{D^1_{00}(\alpha,\beta,\gamma) \times(\sqrt{3}\lambda_8 - \lambda_3)/4 + D^1_{10}(\alpha,\beta,\gamma)(-\lambda_6 - \mathrm{i}\lambda_7)/2\sqrt{2} + D^1_{-10}(\alpha,\beta,\gamma)(\lambda_6 - \mathrm{i}\lambda_7)/2\sqrt{2}\}\,, \tag{9}$$

where $D^j_{m,\mu}(\alpha,\beta,\gamma)$ are the familiar rotation matrices of SU(2). Notice that in fact $\Lambda_8(x)$ is independent of $\gamma(x)$ (since $[j_3, \lambda_8] = 0$). Therefore only the map onto subcoset space SO(3)/SO(2) $\sim S^2$ is relevant. This S^2 is parameterized by geodesic projection (7) onto CP(1). At last, $f(x)$ [eq. (8)] gives a holomorphic map from euclidean space $\mathrm{R}^2\{x\}$ onto this CP(1), and determines the function $\Lambda_8(x)$. Then we can calculate successively:

$$\partial^2 \equiv \partial_1^2 + \partial_2^2 = |f'|^2(\partial_u^2 + \partial_v^2)\,, \tag{10}$$

$$\frac{1}{4(\cos^2\frac{1}{2}\beta)}(\partial_u^2 + \partial_v^2) = \frac{1}{\sin\beta}\frac{\partial}{\partial\beta}\left(\sin\beta\frac{\partial}{\partial\beta}\right) + \frac{1}{\sin\beta}\frac{\partial^2}{\partial\alpha^2}\,, \tag{11}$$

which is just the Casimir operator of SO(3) acting on sphere S^2, finally we get

$$\partial^2\Lambda_8(x) = 8\sqrt{3}\cos^2\tfrac{1}{2}\beta\,|f'|^2 J_3(x)\,, \tag{12}$$

where $J_3(\alpha,\beta,\gamma) \equiv g(\alpha,\beta,\gamma)\,j_3 g(\alpha,\beta,\gamma)^{-1}$, is the SO(3) generator in moving frames [6], now it is

$$g(x)(\sqrt{3}\lambda_8 - \lambda_3)\,g(x)^{-1}/4 \equiv (\sqrt{3}\,\Lambda_8(x) - \Lambda_3(x))/4$$

and equals the expression in the figure bracket of eq. (9), which is commutable with the total Λ_8, so eq. (4) is fulfilled. Furthermore, from eqs. (8), ($\bar{8}$), (7):

$$f'(\partial_u - \mathrm{i}\partial_v) = \partial_1 - \mathrm{i}\partial_2\,, \quad \bar{f}'(\partial_u + \mathrm{i}\partial_v) = \partial_1 + \mathrm{i}\partial_2\,, \tag{13}$$

$$f'(\partial_u - \mathrm{i}\partial_v) = \partial_1 + \mathrm{i}\partial_2\,, \quad \bar{f}'(\partial_u + \mathrm{i}\partial_v) = \partial_1 - \mathrm{i}\partial_2\,, \tag{$\overline{13}$}$$

$$\partial_u \mp \mathrm{i}\partial_v = 2\cos^2\tfrac{1}{2}\beta\,\mathrm{e}^{\mp\mathrm{i}\alpha} \times\left(\frac{\partial}{\partial\beta} \mp \mathrm{i}\frac{1}{\sin\beta}\frac{\partial}{\partial\alpha}\right), \tag{14}$$

$$\mathrm{e}^{\mp\mathrm{i}\gamma}\left(\mp\frac{\partial}{\partial\beta} + \frac{\mathrm{i}}{\sin\beta}\frac{\partial}{\partial\alpha} - \mathrm{i}\,\mathrm{ctg}\beta\frac{\partial}{\partial\gamma}\right)\Lambda_i(\alpha,\beta,\gamma) = [J_1(\alpha,\beta,\gamma) \pm \mathrm{i}J_2(\alpha,\beta,\gamma), \Lambda_i(\alpha,\beta,\gamma)]\,, \tag{15}$$

$$-\mathrm{i}\frac{\partial}{\partial\gamma}\Lambda_i(\alpha,\beta,\gamma) = [J_3(\alpha,\beta,\gamma), \Lambda_i(\alpha,\beta,\gamma)]\,, \tag{16}$$

here $\Lambda_i(\alpha,\beta,\gamma) = g(\alpha,\beta,\gamma)\,\lambda_i g(\alpha,\beta,\gamma)^{-1}$ and we have used the differential expression of J in moving frames [6]. In summary

$$(\partial_1 \mp \mathrm{i}\partial_2)\,\Lambda_i = -2\,|f'|\cos^2\tfrac{1}{2}\beta \times\exp[\pm\mathrm{i}(\arg f' - \alpha + \gamma)] \times[J_1 - \mathrm{ctg}\,\beta J_3 \mp \mathrm{i}J_2, \Lambda_i]\,, \tag{17}$$

$$(\partial_1 \pm \mathrm{i}\partial_2)\,\Lambda_i = -2\,|f'|\cos^2\tfrac{1}{2}\beta \times\exp[\pm\mathrm{i}(\arg f' - \alpha + \gamma)] \times[J_1 - \mathrm{ctg}\,\beta J_3 \mp \mathrm{i}J_2, \Lambda_i]\,, \tag{$\overline{17}$}$$

for the particular j_i in eq. (6) besides

$$[J_3, \Lambda_8] = 0\,, \tag{18}$$

we also have

$$J_1 = \tfrac{1}{2}\Lambda_6\,, \quad J_2 = \tfrac{1}{2}\Lambda_7\,, \tag{19}$$

so that

$$[J_\mu, \Lambda_8/\sqrt{3}] = -\mathrm{i}\epsilon_{\mu\nu}J_\nu\,, \tag{20}$$

here: $\epsilon_{12} = -\epsilon_{21} = 1$. Thus it is easy to check the self-duality condition

$$\mathrm{i}[\partial_\mu\Lambda_8, \Lambda_8/\sqrt{3}] = \epsilon_{\mu\nu}\partial_\nu\Lambda_8\,. \tag{21}$$

To get the non-self-dual solution we choose

$$j_3 = \lambda_2\,, \quad j_1 = \lambda_7\,, \quad j_2 = -\lambda_5\,. \tag{22}$$

Hence

$$\Lambda_8(x) = D^2_{00}(\alpha,\beta,\gamma)\,\lambda_8 + D^2_{10}(\alpha,\beta,\gamma)(\lambda_4 + \mathrm{i}\lambda_6)/\sqrt{2} - D^2_{-10}(\alpha,\beta,\gamma)(\lambda_4 - \mathrm{i}\lambda_6)/\sqrt{2} - D^2_{20}(\alpha,\beta,\gamma)(\lambda_3 + \mathrm{i}\lambda_1)/\sqrt{2} - D^2_{-20}(\alpha,\beta,\gamma)(\lambda_3 - \mathrm{i}\lambda_1)/\sqrt{2}\,. \tag{23}$$

It is an eigenvector of the SO(3) Casimir operator

Volume 93B, number 4 PHYSICS LETTERS 30 June 1980

(11) with eigenvalue 2(2 + 1), so from eqs. (10), (11)

$$\partial^2 \Lambda_8 = 24\,|f'|^2 \cos^2 \tfrac{1}{2}\beta\, \Lambda_8 \,, \tag{24}$$

therefore

$$[\partial^2 \Lambda_8, \Lambda_8] = 0 \,. \tag{25}$$

In the same way as in the previous case, from eqs. (13)–(18) we also have the sequence of complex operators $\partial_1 - \mathrm{i}\partial_2 \to \partial_u - \mathrm{i}\partial_v \to J_1 - \mathrm{i}J_2$, but instead of eq. (19) where the pair Λ_6 and Λ_7 is mutual dual under commutation with respect to $\Lambda_8/\sqrt{3}$ (this operation is the endomorphism on coset space SU(3)/SU(2) $\times$ U(1) $\sim$ CP(2), which defines its complex structure), now we have the non-dual pair $\Lambda_7, -\Lambda_5$ (cf. eq. (22)), so the operation ∂ gives $(\partial_1 - \mathrm{i}\partial_2)\,\Lambda_8 \to [(\Lambda_7 + \mathrm{i}\Lambda_5), \Lambda_8] \sim \Lambda_6 + \mathrm{i}\Lambda_4$ whose dual image with respect to $\Lambda_8/\sqrt{3}$ is

$$\mathrm{i}\,[(\partial_1 - \mathrm{i}\partial_2)\,\Lambda_8, \Lambda_8/\sqrt{3}]$$

$$\to \mathrm{i}\,[\Lambda_6 + \mathrm{i}\Lambda_4, \Lambda_8/\sqrt{3}] = \Lambda_7 + \mathrm{i}\Lambda_5 \,.$$

Thus the non-self-dualness is obvious. When we embed the subgroup $J \sim$ SO(3) into group $G \sim$ SU(3), the complex structure of CP(1) $\sim$ SO(3)/SO(2) does not match into the complex structure of CP(2) $\sim$ SU(3)/SU(2) $\times$ U(1).

To understand this result more apparently, we introduce the SU(3) gauge potential [7] induced by map $\Lambda_8(x)$

$$A_\mu = \tfrac{1}{3}[\Lambda_8, \partial_\mu \Lambda_8] = \tfrac{1}{4}[N, \partial_\mu N] = [P, \partial_\mu P] \,, \tag{26}$$

which is reducible to SU(2) $\times$ U(1) gauge, since by eq. (26) $N(x)$ are invariant under parallel displacement:

$$D_\mu N \equiv \partial_\mu N + [A_\mu, N] = 0 \,,$$

$$D_\mu P = 0 \,, \quad D_\mu \Lambda_8 = 0 \,. \tag{27}$$

From eq. (26) we have $F_{\mu\nu} = -[A_\mu, A_\nu]$ or rewrite it as

$$\epsilon_{\mu\nu} D_\mu A_\nu = 0 \,. \tag{28}$$

Eq. (4) becomes

$$\partial_\mu A_\mu = 0 \,. \tag{29}$$

Moreover from ansatz (3), (5), (6) (or (22)), the A_μ in eq. (26) also equals

$$A_\mu = [J_3, \partial_\mu J_3] \,. \tag{30}$$

(Therefore

$$D_\mu J_3 = 0 \,, \tag{31}$$

Λ_8, J_3 span a Cartan subalgebra invariant under displacement, so the gauge is abelianizable) i.e. it is induced by the map $J_3(x)$ so $A_\mu(x)$ is the pull back of $\mathrm{i}J_\mu$. Thus from $J_\mu = -\mathrm{i}\epsilon_{\mu\nu}[J_\nu, J_3]$ we have

$$A_\mu = -\mathrm{i}\epsilon_{\mu\nu}[A_\nu, J_3] \,, \tag{32}$$

or

$$A_\mu = \mathrm{i}\epsilon_{\mu\nu}[A_\nu, J_3] \,. \tag{$\overline{32}$}$$

Similarly in case (6) (or (22)) the pull back of fulfillment (or unfulfillment) of $J_\mu = -\mathrm{i}\epsilon_{\mu\nu}[J_\nu, \Lambda_8/\sqrt{3}]$ gives the self-dualness (non-self-dualness)

$$A_\mu = -\mathrm{i}\epsilon_{\mu\nu}[A_\nu, \Lambda_8/\sqrt{3}] \tag{33}$$

(or unequal). We call the dual relation (32) in CP(1) $\sim$ SO(3)/SO(2) an "outer" duality for CP(2) $\sim$ SU(3)/SU(2) $\times$ U(1). The "outer" duality together with eq. (31), is sufficient to derive eq. (29) from eq. (28):

$$\partial_\mu A_\mu = D_\mu A_\mu = -\mathrm{i}D_\mu \epsilon_{\mu\nu}[A_\nu, J_3] = -\mathrm{i}\,[\epsilon_{\mu\nu} D_\mu A_\nu, J_3] = 0$$

The inner duality (33) is not necessary.

It is not difficult to find the topo-charge Q: By definition: for the map $x \to$ SU(3)/SU(2) $\times$ U(1) $\sim$ CP(2)

$$4\pi Q_2 = \frac{1}{4}\int \mathrm{d}^2x \,\operatorname{tr}(\dot{N}[\partial_\mu N, \partial_\nu N]) \,,$$

for the map $x \to$ SO(3)/SO(2) $\sim$ CP(1)

$$4\pi Q_1 = \int \mathrm{d}^2x \,\operatorname{tr}([\partial_\mu J_3, \partial_\nu J_3]\, J_3)/\operatorname{tr}(J_3 J_3) \,.$$

From eq. (26)

$$F_{\mu\nu} = \tfrac{1}{4}[\partial_\mu N, \partial_\nu N] \,.$$

From eq. (30)

$$F_{\mu\nu} = [\partial_\mu J_3, \partial_\nu J_3] = \operatorname{tr}(F_{\mu\nu} J_3)\, J_3/\operatorname{tr}(J_3 J_3) \,.$$

Notice that $\operatorname{tr}(J_3 N) = \operatorname{tr}(j_3 N_0) =$ constant. It is easy to obtain the relation $Q_2 = \operatorname{tr}(j_3 N_0)\, Q_1$. So in case (6) $Q_2 = Q_1$, in case (22) $Q_2 = 0$. Although in both cases Q_1 equals the number of poles in eqs. (8), ($\overline{8}$).

Our result may be easily extended to CP(N) $\sim$ SU(N + 1)/SU(N) $\times$ U(1). If we choose j_3, j_μ such that $[j_3, \lambda_{N^2+N}] = 0$, $[j_\mu, \lambda_{N^2+N}] \neq 0$, then we can induce solutions of $[\partial^2\Lambda_{N^2+N}, \Lambda_{N^2+N}] = 0$ similarly. Meanwhile if and only if $[[j_\mu, \lambda_{N^2+N}], \lambda_{N^2+N}] \propto \pm i\epsilon_{\mu\nu} \times [j_\mu, \lambda_{N^2+N}]$ this solution is self-dual.

References

[1] F. Gürsey and H.C. Tze, YTP 79-02.
[2] V.L. Golo and A.M. Perelomov, Lett. Math. Phys. 2 (1978) 477.
[3] A. D'Adda, P. di Vecchia and M. Lüscher, Nucl. Phys. 146B (1978) 63;
M. Lüscher, Phys. Lett. 78B (1978) 465.
[4] H. Eichenherr, Nucl. Phys. 135B (1978) 1.
[5] W.E. Zakharov and A.W. Mikhailov, JETP 74 (1978) 1953.
[6] U. Fano and G. Racah, Irreducible tensorial sets (Academic Press, 1959) p. 146.
[7] Hou Bo-yu, Duan I-shi and Ge Mo-lin, Scientia Sinica 21 (1978) 446.

PHYSICAL REVIEW D VOLUME 24, NUMBER 8 15 OCTOBER 1981

Noether analysis for the hidden symmetry responsible for an infinite set of nonlocal currents

Hou Bo-yu,* Ge Mo-lin,† and Wu Yong-shi‡

Institute for Theoretical Physics, State University of New York at Stony Brook, Stony Brook, New York 11794

(Received 17 November 1980; revised manuscript received 17 June 1981)

In general two-dimensional principal chiral models we exhibit a parametric infinitesimal transformation which is defined for all field configurations and leaves the action invariant. This "hidden" symmetry (i.e., the invariance of action) leads, through a Noether-type analysis, to a parametric conservation law. Expanding in the parameter, we find a systematic procedure to write down the infinitesimal transformations responsible for higher nonlocal currents and thus complete the derivation of the infinite set of nonlocal currents as Noether currents.

I. INTRODUCTION

In the last few years, an infinite set of nonlocal conservation laws has been found in classical two-dimensional chiral models.[1-9] The existence of these nonlocal charges is a signal for the presence of a highly nontrivial hidden symmetry in these models.

From earlier derivations of these nonlocal currents which made use of either the equations of motion[2] or the solution-generating "dual symmetry"[1,4,8] or the "linearized" (i.e., inverse-scattering) equations,[3,5,6] the hidden symmetry was shown to exist in the solution subset of all field configurations.[7] Naturally there arises the following question: Does such a symmetry exist in the entire space of field configurations which leads, via Noether's theorem, to the infinite set of nonlocal currents?

Dolan and Roos[10] have given a partial answer to the question. They wrote down explicitly two nonlocal infinitesimal transformations, from which the first two nonlocal currents were derived as Noether currents. However, they were not able to simply generalize their results for arbitrary higher nonlocal currents.

In this paper we give a complete answer to this question for principal chiral models. In the spirit of using a parametric conservation law to summarize the infinite set of nonlocal currents,[3-8] we use a parametric infinitesimal transformation to summarize the infinite set of infinitesimal transformations responsible for the nonlocal currents. To find an appropriate form for the parametric transformation, we first present in Sec. II a new derivation for the conservation laws starting from a parametric symmetry transformation in the solution set. Then by generalizing it to arbitrary field configurations, we obtain the desired transformation which is displayed in Sec. III. In the same section, we also give a proof for the invariance of the action, thus showing that the parametric infinitesimal transformation is indeed a symmetry of the whole space of field configurations. In the subsequent sections (Secs. IV and V), using different methods, we show how to obtain nontrivial nonlocal conservation laws from the hidden symmetry. In particular, we find a systematic expansion for obtaining the infinitesimal transformations responsible for arbitrary higher nonlocal currents, so that we can claim that we have completed the derivation of the infinite set of nonlocal currents as Noether currents.

Finally, conclusions are summarized in Sec. VI, and a discussion about the interpretation of our transformation and possible generalizations of our results are also presented.

Our calculations are performed in Minkowski space-time. The metric used is $\eta_{00}=-\eta_{11}=1$, $\epsilon_{01}=-\epsilon_{10}=1$, $\epsilon^{01}=-\epsilon^{10}=-1$. All formulas in this paper may be extended to Euclidean space without difficulty.

II. A NEW ON-SHELL DERIVATION

To exhibit more explicitly the connection between nonlocal conserved charges and a hidden symmetry which exists at least on-shell, we present here a new derivation of nonlocal currents starting from an on-shell symmetry transformation, which will give us some hints to generalize the symmetry off-shell.

For the principal chiral models, the Lagrangian density is

$$\mathcal{L}(x)=\tfrac{1}{16}\operatorname{tr}\{\partial_\mu g(x)\partial^\mu g^{-1}(x)\}, \tag{2.1}$$

where $g(x)\in G$, a matrix Lie group. Defining

$$A_\mu(x)=g^{-1}\partial_\mu g, \tag{2.2}$$

the equations of motion obtained from Eq. (2.1) are

$$\partial^\mu A_\mu=0. \tag{2.3}$$

As a pure gauge potential, $A_\mu(x)$ satisfies

$$\partial_\mu A_\nu-\partial_\nu A_\mu+[A_\mu,A_\nu]=0. \tag{2.4}$$

It is easy to see that under the global transformation

$$\delta g=-gT\,, \tag{2.5}$$

where $T=\alpha^a T_a$ (with infinitesimal constants α^a) belongs to the Lie algebra of the group G, the Lagrangian density (2.1) is invariant:

$$\delta\mathcal{L}=-\tfrac{1}{8}\mathrm{tr}\{A_\mu\partial^\mu(g^{-1}\delta g)\}=0\,. \tag{2.6}$$

This invariance gives rise to the conservation of A_μ. If the constant generator T in Eq. (2.5) is replaced by a space-time-dependent $T(x)$, the Lagrangian density is generally not invariant, as seen from Eq. (2.6).

However, if we assume $T(x)$ to be the following particular matrix function of space-time

$$T(x)=U(x;l)T\,U(x;l)^{-1}\,, \tag{2.7}$$

with $U(x;l)\in G$ satisfying the so-called "inverse-scattering equations" (l being a complex parameter)[6]

$$\begin{aligned}\partial_0 U&=\frac{l}{1-l^2}(lA_0-A_1)U\,,\\ \partial_1 U&=\frac{l}{1-l^2}(lA_1-A_0)U\,,\end{aligned} \tag{2.8}$$

then we can show that for those field configurations which satisfy Eqs. (2.3) and (2.4), the Lagrangian density (2.1) is changed by a total divergence under the infinitesimal transformation

$$\delta g=-gT(x)\equiv -gU(x)TU(x)^{-1}\,. \tag{2.9}$$

In fact, in this case we have

$$\begin{aligned}\delta\mathcal{L}&=-\tfrac{1}{8}\mathrm{tr}\{A_\mu\partial^\mu(g^{-1}\delta g)\}\\ &=\tfrac{1}{8}\mathrm{tr}\{A_\mu\partial^\mu(UTU^{-1})\}\\ &=\tfrac{1}{8}\mathrm{tr}\{[U^{-1}A_\mu U,U^{-1}\partial^\mu U]T\}\,.\end{aligned} \tag{2.10}$$

From Eq. (2.8) we can express $U^{-1}A_\mu U$ in terms of $U^{-1}\partial_\mu U$,

$$U^{-1}A_\mu U=\frac{1}{l}\epsilon_{\mu\nu}U^{-1}\partial^\nu U-U^{-1}\partial_\mu U\,. \tag{2.11}$$

Upon substituting Eq. (2.11) into Eq. (2.10), $\delta\mathcal{L}$ can be expressed as a total divergence

$$\delta\mathcal{L}=\tfrac{1}{8}\partial^\mu\,\mathrm{tr}\left\{\frac{2}{l}\epsilon_{\mu\nu}U^{-1}\partial_\nu UT\right\}. \tag{2.12}$$

This means that the action is invariant under the transformation (2.9).

It should be pointed out that since the integrability conditions of Eq. (2.8) are just Eq. (2.4) and the equations of motion Eq. (2.3), the transformation (2.9) is defined only for those field configurations $g(x)$, which satisfy Eqs. (2.3) and (2.4), and only for them is $\delta\mathcal{L}$ given by Eq. (2.12). Because of this, we call this symmetry, Eq. (2.9), an on-shell symmetry.

The derivation of nonlocal conserved currents from this on-shell symmetry is straightforward. Using the equations of motion Eq. (2.3), we easily obtain another expression for $\delta\mathcal{L}$:

$$\delta\mathcal{L}=-\tfrac{1}{8}\partial^\mu\,\mathrm{tr}\{A_\mu g^{-1}\delta g\}=\tfrac{1}{8}\partial^\mu\,\mathrm{tr}\{U^{-1}A_\mu UT\}\,. \tag{2.13}$$

Now from Eqs. (2.12) and (2.13) we get the conservation law

$$\partial^\mu J_\mu(x;l)=0\,, \tag{2.14}$$

where the parametric conserved current J_μ is

$$\begin{aligned}J_\mu(x;l)&=U(x;l)^{-1}A_\mu U(x;l)\\ &\quad-\frac{2}{l}\epsilon_{\mu\nu}U(x;l)^{-1}\partial^\nu U(x;l)\,,\end{aligned} \tag{2.15}$$

which summarizes the usual infinite number of nonlocal conserved currents. Performing a Taylor expansion in l around $l=0$, we obtain the desired infinite set of nonlocal charges given in Ref. 1, the first two being

$$\begin{aligned}Q_1&=\int_{-\infty}^{+\infty}dx\,A_0(x,t)\,,\\ Q_2&=\int_{-\infty}^{+\infty}dx\int_{-\infty}^{x}dx'\,A_0(x,t)A_0(x',t)\\ &\quad-\int_{-\infty}^{+\infty}dx\,A_1(x,t)\,.\end{aligned} \tag{2.16}$$

III. THE OFF-SHELL HIDDEN SYMMETRY

The above derivation is not in the spirit of Noether's theorem, because in a derivation of the conservation law in the manner of Noether one needs an off-shell symmetry, i.e., a symmetry of the whole space of field configurations. We will show in this section that we can indeed improve our derivation to exhibit the existence of an off-shell symmetry which is responsible for the infinite set of nonlocal charges.

To generalize the transformation (2.9) off-shell, it is sufficient to require that $U(x)$ should satisfy one of the inverse-scattering equations (2.8), e.g.,

$$\partial_1 U=\frac{l}{1-l^2}(lA_1-A_0)U\,. \tag{3.1}$$

Thus, we are led to consider the following nonlocal infinitesimal transformation defined for all field configurations:

$$\delta g=-gU(x;l)TU(x;l)^{-1} \tag{3.2}$$

with the space-time function $U(x)$ given as follows:

$$U(x;l)=P\exp\left(\frac{l}{1-l^2}\int_{-\infty}^{x'}dy[lA_1(y,t)-A_0(y,t)]\right). \tag{3.3}$$

To show that (3.2) is really an off-shell symmetry, we need to prove that for arbitrary $g(x)$, the change in $\mathcal{L}$ under (3.2) can still be expressed as a total divergence without using the equations of motion. Hence in this derivation, use of only Eqs. (3.1) and (2.4) is allowed. We start with Eq. (2.10), i.e.,

$$\delta\mathcal{L}=\tfrac{1}{8}\mathrm{tr}\{[U^{-1}A_0U,U^{-1}\partial_0U]T-[U^{-1}A_1U,U^{-1}\partial_1U]T\}. \tag{3.4}$$

From Eq. (3.1) we have

$$U^{-1}\partial_1U=\frac{l}{1-l^2}(lU^{-1}A_1U-U^{-1}A_0U). \tag{3.5}$$

Making use of this equation, we can rewrite $\delta\mathcal{L}$ as follows:

$$\begin{aligned}\delta\mathcal{L}&=\tfrac{1}{8}\mathrm{tr}\left\{\frac{1-l^2}{l}[-U^{-1}\partial_1U,U^{-1}\partial_0U]T+l[U^{-1}A_1U,U^{-1}\partial_0U]T+\frac{l}{1-l^2}[U^{-1}A_1U,U^{-1}A_0U]T\right\}\\&=\tfrac{1}{8}\mathrm{tr}\left\{\frac{1-l^2}{l}[U^{-1}\partial_0U,U^{-1}\partial_1U]T+l[U^{-1}A_1U,U^{-1}\partial_0U]T+l[U^{-1}A_1U,U^{-1}A_0U]T+l[U^{-1}\partial_1U,U^{-1}A_0U]T\right\}.\end{aligned} \tag{3.6}$$

Observe that, by using Eq. (2.4), we have

$$[U^{-1}A_1U,U^{-1}A_0U]=U^{-1}(\partial_0A_1-\partial_1A_0)U. \tag{3.7}$$

Then the sum of the last three terms in Eq. (3.6) can be recast into the form $-l\epsilon^{\mu\nu}\partial_\mu(U^{-1}A_\nu U)T$. Finally we get the desired form for $\delta\mathcal{L}$,

$$\delta\mathcal{L}=\tfrac{1}{8}\partial^\mu\,\mathrm{tr}\left\{\left(\frac{1-l^2}{l}\epsilon_{\mu\nu}U^{-1}\partial^\nu U-l\epsilon_{\mu\nu}U^{-1}A^\nu U\right)T\right\}, \tag{3.8}$$

which shows that the action is invariant under the off-shell infinitesimal transformation (3.2), with U given by Eq. (3.3).

In addition to the off-shell invariance of the action, the transformation (3.2) with (3.3) leads to the following two on-shell invariances:

(1) The equations of motion are invariant under δg.

(2) The variation of the energy-momentum density due to δg vanishes on-shell.

The proof follows from the fact that the function U defined by Eq. (3.3) satisfies both inverse-scattering equations (2.8), not just Eq. (3.1), for on-shell field configurations $g(x)$. In fact, writing $\delta g=-gT(x)$, the invariance of the equations of motion (2.3) requires

$$D_\mu\partial^\mu T(x)\equiv\partial_\mu\partial^\mu T(x)+[A_\mu,\partial^\mu T(x)]=0, \tag{3.9}$$

while for the energy-momentum density

$$T_{\mu\nu}=\tfrac{1}{8}\mathrm{tr}\{\tfrac{1}{2}g_{\mu\nu}A^\lambda A_\lambda-A_\mu A_\nu\},$$

the vanishing of its variation requires

$$\begin{aligned}\delta T_{00}=\delta T_{11}&=-\tfrac{1}{8}\mathrm{tr}\{A_0D_0T(x)+A_1D_1T(x)\}=0,\\\delta T_{01}=\delta T_{10}&=-\tfrac{1}{8}\mathrm{tr}\{A_0D_1T(x)+A_1D_0T(x)\}=0.\end{aligned} \tag{3.10}$$

For on-shell $g(x)$, using Eq. (2.8), we have

$$\partial_\mu T(x)\equiv\partial_\mu[U(x)TU(x)^{-1}]=l\epsilon_{\mu\nu}D^\nu T(x),$$

$$D_\mu T(x)=\left[\frac{1}{1-l^2}A_\mu-\frac{l}{1-l^2}\epsilon_{\mu\nu}A^\nu,T(x)\right],$$

from which it is easy to check Eqs. (3.9) and (3.10).

IV. NONLOCAL CURRENTS AS NOETHER CURRENTS

Now we turn to seeing how the off-shell hidden symmetry transformation (3.2) with (3.3) gives rise to the infinite set of nonlocal currents as Noether currents.

The simplest way to derive nonlocal conserved currents seems to be the following. For on-shell $g(x)$, Eq. (2.13) still holds. By equating Eq. (2.13) still holds. By equating Eq. (2.13) and Eq. (3.8) we are led to the "conservation law"

$$\partial^\mu\tilde{J}_\mu\equiv\partial^\mu\left\{U^{-1}A_\mu U+l\epsilon_{\mu\nu}U^{-1}A^\nu U-\frac{1-l^2}{l}\epsilon_{\mu\nu}U^{-1}\partial^\nu U\right\}=0. \tag{4.1}$$

However, from Eq. (3.1) we see that $\tilde{J}_0$ is identically zero even off-shell; similarly J_1 is equal to zero on-shell by Eq. (3.8). Thus the "conserved current" $\tilde{J}_\mu$ turns out to be a trivial current.

Nonetheless, this does not mean that we can obtain only a trivial current from the off-shell hidden symmetry (3.2) with (3.3). Actually, as shown below in this section, we can obtain the usual infinite set of nonlocal conserved currents by dropping some terms, which are total divergences and identically equal to zero, from the expression (3.8) for $\delta\mathcal{L}$.

To this end, we expand the function $U(x;l)$, defined by Eq. (3.3) in powers of the parameter l,

$$U(x;l)=1+\sum_{n=1}^{\infty}l^n\psi^{(n)}(x), \tag{4.2}$$

in which, for convenience of comparison with the known results in the literature, we put

$$\begin{aligned}
\psi^{(1)} &= \chi^{(1)}\,,\\
\psi^{(2)} &= \chi^{(2)} + \tfrac{1}{2}(\chi^{(1)})^2\,,\\
\psi^{(3)} &= \chi^{(3)} + \tfrac{1}{2}(\chi^{(1)}\chi^{(2)} + \chi^{(2)}\chi^{(1)})\,,\\
\psi^{(4)} &= \chi^{(4)} + \tfrac{1}{2}[(\chi^{(2)})^2 + \chi^{(2)}(\chi^{(1)})^2 - \tfrac{1}{4}(\chi^{(1)})^4]\\
&\quad + \tfrac{1}{2}(\chi^{(1)}\chi^{(3)} + \chi^{(3)}\chi^{(1)}),\ \ldots
\end{aligned} \tag{4.3}$$

From $UU^{-1}=1$ we obtain the expansion of U^{-1} in l as follows:

$$U(x;l)^{-1} = 1 - \sum_{n=1}^{\infty} l^n \psi^{(-n)}(x)\,, \tag{4.4}$$

where

$$\begin{aligned}
\psi^{(-1)} &= \chi^{(1)}\,,\\
\psi^{(-2)} &= \chi^{(2)} - \tfrac{1}{2}(\chi^{(1)})^2\,,\\
\psi^{(-3)} &= \chi^{(3)} - \tfrac{1}{2}(\chi^{(1)}\chi^{(2)} + \chi^{(2)}\chi^{(1)})\,,\\
\psi^{(-4)} &= \chi^{(4)} - \tfrac{1}{2}[(\chi^{(2)})^2 - \chi^{(2)}(\chi^{(1)})^2 - \tfrac{1}{4}(\chi^{(1)})^4]\\
&\quad - \tfrac{1}{2}(\chi^{(1)}\chi^{(3)} + \chi^{(3)}\chi^{(1)}),\ \ldots
\end{aligned} \tag{4.5}$$

Substituting Eq. (4.2) with (4.3) into Eq. (3.1) and comparing the coefficients of the terms in l^n on both sides, the following equations satisfied by $\chi^{(n)}(x)$ are obtained recursively order by order:

$$\begin{aligned}
\partial_1\chi^{(1)} &= A_0\,,\\
\partial_1\chi^{(2)} &= -A_1 + \tfrac{1}{2}[\chi^{(1)}, A_0]\,,\\
\partial_1\chi^{(3)} &= \tfrac{1}{2}[\chi^{(2)}, A_0] - \tfrac{1}{2}[\chi^{(1)}, A_1]\\
&\quad + \tfrac{1}{4}[[A_0, \chi^{(1)}], \chi^{(1)}] + \tfrac{1}{2}\chi^{(1)}A_0\chi^{(1)} + A_0,\ \ldots
\end{aligned} \tag{4.6}$$

It is easy to obtain $\chi^{(n)}(x)$ by integrating the right-hand sides of the above equations. Using these $\chi^{(n)}(x)$'s, we define the following infinite set of infinitesimal transformations obtained by expanding the parametric transformation (3.2):

$$\delta^{(n)}g = -g\lambda^{(n)} \quad (n=1,2,3,\ldots) \tag{4.7}$$

$$\begin{aligned}
\lambda^{(1)} &= [\chi^{(1)}, T]\,,\\
\lambda^{(2)} &= [\chi^{(2)}, T] + \tfrac{1}{2}[\chi^{(1)}, [\chi^{(1)}, T]]\,,\\
\lambda^{(3)} &= [\chi^{(3)}, T] + \tfrac{1}{2}[\chi^{(1)}T\chi^{(1)}, \chi^{(1)}] + \tfrac{1}{2}[\chi^{(2)}[\chi^{(1)}, T]]\\
&\quad + \tfrac{1}{2}[\chi^{(1)}, [\chi^{(2)}, T]]\,.
\end{aligned} \tag{4.8}$$

By a direct but very tedious calculation, the variation of $\mathcal{L}$ for $\delta^{(n)}g$

$$\delta^{(n)}\mathcal{L} = \tfrac{1}{8}\,\mathrm{tr}(A_\mu \partial^\mu \lambda^{(n)}) \tag{4.9}$$

can be expressed as a total divergence order by order without using the equations of motion:

$$\delta^{(1)}\mathcal{L} = \tfrac{1}{8}\partial_\mu \mathrm{tr}\{(\tfrac{1}{2}\epsilon^{\mu\nu}[\partial_\nu\chi^{(1)}, \chi^{(1)}] - \epsilon^{\mu\nu}A_\nu)T\}\,, \tag{4.10a}$$

$$\delta^{(2)}\mathcal{L} = \tfrac{1}{8}\partial_\mu \mathrm{tr}\{(\epsilon^{\mu\nu}[\partial_\nu\chi^{(1)}, \chi^{(2)}] + \tfrac{1}{6}\epsilon^{\mu\nu}[[\partial_\nu\chi^{(1)}, \chi^{(1)}], \chi^{(1)}] - \epsilon^{\mu\nu}[A_\nu, \chi^{(1)}])T\}\,, \tag{4.10b}$$

$$\begin{aligned}
\delta^{(3)}\mathcal{L} = \tfrac{1}{8}\partial_\mu \mathrm{tr}\{&\epsilon^{\mu\nu}([\partial_\nu\chi^{(1)}, \chi^{(3)}] + \tfrac{1}{2}[\chi^{(2)}, \partial_\nu\chi^{(2)}] + \tfrac{1}{2}[\chi^{(1)}, \partial_\nu\chi^{(1)}] + \tfrac{1}{4}[[\partial_\nu\chi^{(1)}, \chi^{(1)}], \chi^{(2)}]\\
&+ \tfrac{1}{4}[[\partial_\nu\chi^{(2)}, \chi^{(1)}], \chi^{(1)}] + \tfrac{1}{8}[\partial_\nu(\chi^{(1)})^2, (\chi^{(1)})^2] + [\chi^{(2)}, A_\nu] - \tfrac{1}{2}[\chi^{(1)}, [\chi^{(1)}, A_\nu]])T\}\,.
\end{aligned} \tag{4.10c}$$

Using the equations of motion (2.3), from (4.9) $\delta^{(n)}\mathcal{L}$ can be directly written as

$$\delta^{(n)}\mathcal{L} = \tfrac{1}{8}\partial^\mu\, \mathrm{tr}\{A_\mu \lambda^{(n)}\}\,. \tag{4.11}$$

Equating Eqs. (4.11) and (4.10a)–(4.10c), respectively, we obtain the conserved currents

$$\partial^\mu J_\mu^{(n)} = 0\,, \tag{4.12}$$

where

$$J_\mu^{(1)} = [A_\mu, \chi^{(1)}] + \epsilon_{\mu\nu}A^\nu - \tfrac{1}{2}\epsilon_{\mu\nu}[\partial^\nu\chi^{(1)}, \chi^{(1)}]\,, \tag{4.13a}$$

$$\begin{aligned}
J_\mu^{(2)} &= [A_\mu, \chi^{(2)}] + \tfrac{1}{2}[[A_\mu, \chi^{(1)}], \chi^{(1)}] - \epsilon_{\mu\nu}[\partial_\nu\chi^{(1)}, \chi^{(2)}]\\
&\quad - \tfrac{1}{6}\epsilon_{\mu\nu}[[\partial^\nu\chi^{(1)}, \chi^{(1)}], \chi^{(1)}] + \epsilon_{\mu\nu}[A^\nu, \chi^{(1)}]\,,
\end{aligned} \tag{4.13b}$$

$$\begin{aligned}
J_\mu^{(3)} &= [A_\mu, \chi^{(3)}] - \tfrac{1}{2}[\chi^{(1)}A_\mu\chi^{(1)}, \chi^{(1)}] + \tfrac{1}{2}[[A_\mu, \chi^{(1)}], \chi^{(2)}] + \tfrac{1}{2}[[A_\mu, \chi^{(2)}], \chi^{(1)}]\\
&\quad - \epsilon_{\mu\nu}([\partial^\nu\chi^{(1)}, \chi^{(3)}] + \tfrac{1}{2}[\chi^{(2)}, \partial^\nu\chi^{(2)}] + \tfrac{1}{2}[\chi^{(1)}, \partial^\nu\chi^{(1)}]\\
&\quad + \tfrac{1}{4}[[\partial^\nu\chi^{(1)}, \chi^{(1)}], \chi^{(2)}] + \tfrac{1}{4}[[\partial^\nu\chi^{(2)}, \chi^{(1)}], \chi^{(1)}]\\
&\quad + \tfrac{1}{8}[\partial^\nu(\chi^{(1)})^2, (\chi^{(1)})^2] + [\chi^{(2)}, A^\nu] - \tfrac{1}{2}[\chi^{(1)}, [\chi^{(1)}, A^\nu]])\,.
\end{aligned} \tag{4.13c}$$

These currents are obviously not trivial even on-shell. It can be easily checked that the corresponding nonlocal charges

$$Q^{(n)} = \int_{-\infty}^{+\infty} dx'\, J_0^{(n)}(x', t) \tag{4.14}$$

obtained from Eq. (4.13) are just the usual ones in literature [see also Eq. (2.16)]. In Eq. (4.13) the currents $J_\mu^{(n)}$ are defined for off-shell field configurations, and they reduce to the standard form when the field $g(x)$ are restricted on-shell.

For $n=1,2$, the above formulas Eqs. (4.6)–(4.13) are identical to corresponding ones, appearing in Ref. 9 [Eqs. (2.2)–(2.9)], except for some differences in sign arising from the difference in the signature of space-time. However, by a systematic expansion, we can obtain the infinitesimal transformations and higher nonlocal currents for $n>2$; for example, we have already worked out explicitly the case $n=3$ in detail, in which both the expression for $\lambda^{(3)}$, Eq. (4.8), and the equation for $\chi^{(3)}$, Eq. (4.6), are too complicated to be guessed.

It is interesting to compare Eq. (4.10) with Eq. (3.8). We find that Eqs. (4.10a)–(4.10c) are obtained if we expand the right-hand side of Eq. (3.8) in powers of l and drop terms which are identically zero such as, $\epsilon^{\mu\nu}\partial_\mu\partial_\nu\chi^{(n)}$, $\epsilon^{\mu\nu}(\partial_\mu\chi^{(1)}\chi^{(1)}\partial_\nu\chi^{(1)}+\partial_\nu\chi^{(1)}\chi^{(1)}\partial_\mu\chi^{(1)})$, $\epsilon^{\mu\nu}(\partial_\mu\chi^{(2)}\partial_\nu\chi^{(1)}+\partial_\nu\chi^{(2)}\partial_\mu\chi^{(1)})$, and so on. So in effect, the nontrivial currents $J_\mu^{(n)}$ or Eqs. (4.13a)–(4.13c) can also be obtained from the trivial current J_μ [see Eq. (4.1)] by expanding the latter in powers of l and then dropping terms whose divergence is identically zero.

We point out that we have the freedom to redefine the function $\chi^{(n)}(x)(n\geq 2)$ in terms of which $\delta^{(n)}g$ and $\delta^{(n)}\mathcal{L}$ are expressed. We choose Eq. (4.3) to define $\chi^{(n)}(x)$ only for convenience in comparing with literature and to write all terms in $\delta^{(n)}\mathcal{L}$ in the form of a series of commutators. Actually, the simplest way is to define $\chi^{(n)}(x)=\psi^{(n)}(x)$ [see Eq. (4.2)]. The above procedure can be applied to the redefined $\chi^{(n)}$; Eqs. (4.5)–(4.13) will change their appearance, but the physics is the same. Especially, the off-shell current $J_\mu^{(n)}$ may change by divergenceless terms, while the nth nonlocal charges $Q^{(n)}$ does not change on-shell.

V. PARAMETRIC NONLOCAL CURRENT AS NOETHER CURRENT

The merit of the derivation in the last section lies in the fact that it allows one to express the change in $\mathcal{L}$ as a total divergence order by order, even for off-shell configurations. However, by means of this method one cannot obtain a parametric conserved current which summarizes an infinite set of nonlocal currents.

In order to derive a parametric current from the hidden symmetry (3.2) with (3.3), we observe that to get a conservation law from the symmetry it is sufficient to recast the $\delta\mathcal{L}$ in Eq. (3.4) into a form which is a total divergence of a vector different from $U^{-1}A_\mu U$ only when on-shell. To this end, let us introduce the function

$$\chi(x)=\int_{-\infty}^{x^1}dx^1\left(\frac{1}{l}U^{-1}\partial_1U-\frac{l}{1-l^2}U^{-1}(lA_0-A_1)U\right) \quad (5.1)$$

which satisfies

$$\partial_1\chi=\frac{1}{l}U^{-1}\partial_1U-\frac{l}{1-l^2}U^{-1}(lA_0-A_1)U\,. \quad (5.2)$$

By a direct and somewhat lengthy calculation, in which only Eqs. (3.1) and (5.2) are used, we can recast the off-shell $\delta\mathcal{L}$ in Eq. (3.4) into the form

$$\delta\mathcal{L}=\tfrac{1}{8}\,\mathrm{tr}\left\{\left(\frac{2}{l}\epsilon^{\mu\nu}\partial_\mu(U^{-1}\partial_\nu U)+\frac{l}{1-l^2}[\chi,U^{-1}(\partial^\mu A_\mu)U]+\partial_\mu\left[\chi,\epsilon^{\mu\nu}U^{-1}\partial_\nu U-\frac{l}{1-l^2}U^{-1}(A^\mu+l\epsilon^{\mu\nu}A_\nu)U\right]\right)T\right\}. \quad (5.3)$$

It is easily seen that the last two terms vanish while on-shell so that $\delta\mathcal{L}$ then becomes a total divergence. Then equating Eqs. (5.3) and (2.13) gives also the conservation law (2.14), so in some generalized sense, the parametric current (2.15) can also be viewed as a Noether current derived from the hidden symmetry transformation (3.2) with (3.3).

Incidentally, we observe that the total-divergence form for the on-shell $\delta\mathcal{L}$ is not unique so that we can have several different forms for the parametric conserved current. For instance, on-shell we have

$$\delta\mathcal{L}=\tfrac{1}{8}\partial_\mu\,\mathrm{tr}(U^{-1}A^\mu UT) \quad (5.4)$$

$$=\frac{1}{4l}\partial_\mu\,\mathrm{tr}(\epsilon^{\mu\nu}U^{-1}\partial_\nu UT) \quad (5.5)$$

$$=\frac{-l}{4(1+l^2)}\partial_\mu\,\mathrm{tr}(\epsilon^{\mu\nu}U^{-1}A_\nu UT) \quad (5.6)$$

$$=-\tfrac{1}{4}\partial_\mu\,\mathrm{tr}(U^{-1}\partial^\mu UT)\,. \quad (5.7)$$

They are equivalent to each other through the inverse-scattering equation (2.8), i.e.,

$$\left(\partial_\mu+A_\mu-\frac{1}{l}\epsilon_{\mu\nu}\partial^\nu\right)U=0\,. \quad (5.8)$$

By choosing different pairs from these equations we may get different forms of parametric conservation laws. For example, the J_μ in Eq. (2.15) is obtained by choosing Eqs. (5.4) and (5.5); equating Eq. (5.4) and Eq. (5.6) gives the current as given in Ref. 5; and the conservation law as given in Ref. 3 can be recovered by combining Eqs. (5.5) and (5.7). Other combinations may be used to give some formally new (but essentially old) parametric conservation laws, e.g.,

$$\partial^\mu(U^{-1}\partial_\mu U+\tfrac{1}{2}U^{-1}A_\mu U)=0\,, \quad (5.9)$$

$$\partial^\mu\left(U^{-1}\partial_\mu U-\frac{l}{1+l^2}\epsilon_{\mu\nu}U^{-1}A^\nu U\right)=0\,, \tag{5.10}$$

$$\partial^\mu\epsilon_{\mu\nu}\left(U^{-1}\partial^\nu U+\frac{l}{1+l^2}U^{-1}A^\nu U\right)=0\,. \tag{5.11}$$

VI. CONCLUSIONS AND DISCUSSION

We have shown that in two-dimensional principal chiral models there exists a (off-shell) hidden symmetry which shifts the Lagrangian density by a total divergence and is responsible for the existence of an infinite sequence of nonlocal conserved charges. The infinitesimal transformation for the (off-shell) hidden symmetry is exhibited by Eq. (3.2) with (3.3). Since the matrix function U given by Eq. (3.3) depends on A_μ nonlocally, this infintesimal transformation (3.2) transforms the field $g(x)$ nonlinearly and nonlocally. Moreover, because the function U contains a complex parameter l, the infinitesimal transformation is also dependent on l, so that it leads to a parametric Noether current. Expanding the latter in l gives an infinite set of usual nonlocal currents. The derivation of the first two nonlocal currents as Noether currents in Ref. 9 becomes part of our discussion, appearing as the special case corresponding to the first two terms in the expansion of our parametric infinitesimal transformation. Our formalism also provides a systematic way to obtain the infinitesimal transformations for all higher nonlocal currents.

The infinitesimal transformation (3.2) with (3.3) proposed here has an interesting physical interpretation. We know that $\delta g=-gT$ represents a global isospin rotation. So our transformation $\delta g=-gU(x)TU(x)^{-1}$ represents a local isospin rotation, which looks like one and the same isospin rotation if seen at various space-time points from the local isospin frames obtained by doing a gauge rotation $V(x)=U(x)^{-1}$. Equation (3.3) implies that these local isospin frames are parallel to each other[11] along the equal-time lines with respect to the gauge potential $\tilde{A}_\mu=-l(lA_\mu+\epsilon_{\mu\nu}A^\nu)/(1-l^2)$. For on-shell configurations, $\tilde{A}_\mu$ is also curvature-free[3,6] so the local isospin frames at all space-time points are parallel to each other with respect to $\tilde{A}_\mu$. However, we have not succeeded in understanding why just this potential appears in the symmetry transformation. Further implications and the geometrical interpretation of the transformation (3.2) deserve more attention.

We emphasize that the gauge rotation $U(x)^{-1}$ in the transformation is not an arbitrary one. It is a particular function of space-time, nonlocally dependent on the field $g(x)$ by Eq. (3.1). While on-shell it satisfies the inverse-scattering equations (2.8), in which the functions $U(x;l)$ can also be viewed as the dual transformation operators.[4,5] In this way, our off-shell parametric "hidden" symmetry can be viewed as the off-shell generalization of the usual on-shell, dual symmetry. It would be very interesting to generalize the off-shell dual symmetry to four-dimensional non-Abelian gauge theories.

It is easy to generalize the method and results of this paper to the discussion of hidden symmetry in more general cases such as supersymmetric chiral models[6,12] and nonlinear σ models on symmetric spaces.[4,8] The details for these generalizations will be published elsewhere. The discussion concerning the group structure of the symmetry and related problems are in progress.

Note added. After this paper was submitted, we became aware of a paper by T. Curtright and C. Zachos [Phys. Rev. D (to be published)] that parallels some of the discussion of the present paper for the O(N) Gross-Neveu model using a different approach.

ACKNOWLEDGMENTS

The authors are grateful to Professor C. N. Yang, Professor H. T. Nieh, and the Institute for Theoretical Physics, SUNY at Stony Brook for the warm hospitality extended to them. They thank Dr. L.-L. Chau Wang for useful discussions and for reading the manuscript. B. Y. Hou would like to thank the Yale Physics Department for their hospitality where part of the work was done and to thank Professor H. C. Tze for helpful discussions. He also wishes to thank Dr. L. Dolan for mailing her papers to him before publication. This work is supported in part by the National Science Foundation Grant No. PHY-79-06376A01.

---

*On leave from Northwest University, Xi-an, China.
†On leave from Lanzhou University, Lanzhou, China.
‡Permanent address: Institute of Theoretical Physics, Academia Sinica, Beijing, China. Address after September, 1981: Institute for Advanced Study, Princeton, New Jersey 08540.

[1] M. Lüscher and K. Pohlmeyer, Nucl. Phys. B137, 46 (1978).
[2] E. Brézin, C. Itzykson, J. Zinn-Justin, and J. B. Zuber, Phys. Lett. 82B, 442 (1979).
[3] H. J. de Vega, Phys. Lett. 87B, 233 (1979).
[4] H. Eichenherr and M. Forger, Nucl. Phys. B155, 381

(1979).
[5]A. T. Ogielski, Phys. Rev. D 21, 406 (1980).
[6]T. L. Curtright and C. K. Zachos, Phys. Rev. D 21, 411 (1980).
[7]C. Zachos, Phys. Rev. D 21, 3462 (1980).
[8]Chou Kuang-chao and Song Xing-chang, Report No. BUTP 80-003 (unpublished).
[9]For a review on this and related subjects see L.-L. Chau Wang, in *Proceedings of the 1980 Guangzhou Conference on Theoretical Particle Physics* (Science, Beijing, China, 1980); and talk at the International School of Subnuclear Physics, Erice, Italy, 1980 (unpublished).
[10]L. Dolan and A. Roos, Phys. Rev. D 22, 2018 (1980).
[11]For a discussion about the use of parallel-transported local frames in gauge theories, see B. Y. Hou, Y. S. Duan, and M. L. Ge, Sci. Sinica 21, 446 (1978).
[12]P. DiVecchia and S. Ferrara, Nucl. Phys. B130, 93 (1977); E. Witten, Phys. Rev. D 16, 299 (1977); Z. Popowicz and L.-L. Chau Wang, Phys. Lett. 98B, 253 (1981).

Commun. in Theor. Phys.(Beijing, China) Vol. 1, No. 3 (1982) 333-344

NONLOCAL CURRENTS AS NOETHER CURRENTS DUE TO A NONLOCALLY TRANSFORMED GLOBAL SYMMETRY

HOU Bo-yu (侯伯宇)
(Northwest University, Xi'an)

Received November 21, 1981.

Abstract

In this paper we first present the gauge invariant conserved Noether current for the Yang-Mills theory, which is nonlocal in some sense. Then we introduce for the two dimensional chiral model, plain and supersymmetric, the nonlocal one-complex parameter-dependent symmetric generators, which shift the Lagrangian density by a total divergence. Both the equations of motion and the energy momentum density are invariant. The associated Noether current may be expanded analytically into infinite series of nonlocal currents.

I. Introduction

An infinite set of nonlocal conservation laws has been found in two dimensional chiral models[1-4] in their supersymmetric extensions[5-7] and in the self-dual sector of Yang-Mills theory[8,24]. The origin of this conservation must be some hidden symmetry.

Recently, Dolan and Roos[9] found some nonlocal symmetry transformation in the chiral model to show explicitly that the first two nonlocal currents are in fact Noether currents. In this paper we show that the hidden symmetry in total is just the usual isotopic (or involution) symmetry but one transformed nonlocally by the well-known dual transformation[10,11]. Since the latter contains one complex parameter analytically[12], the corresponding Noether current can be expanded around different points in a complex plane into various infinite series of currents.

Usually the Noether currents are generated by a global symmetry, which are constant in the whole space time, e.g. to turn the same infinitesimal angle (or phase) around parallel axes. But in "curved" space (either ordinary physical or internal space) one cannot infer a priori which axes are mutually parallel at different points. There must be some objective curteria for parallelism. For a Yang-Mills field it is natural to define parallelism by using the nonintegrable phase factor, generally it will be path dependent. Only in the case of pure gauge does there exist a complete set of path-independent absolutely parallel axes. Or in the case of reducibility[13], i.e., when the holonomy group is a nontrivial subgroup H of the entire gauge group G, the generators, which commute with H, are absolutely parallel. Otherwise we must choose some privileged direction according to some intrinsic property of the system, e.g. the isotopic direction of Higgs field. In Section II we briefly review[14] the method of decomposing gauge potential and field gauge covariantly with respect to this privileged direction and derive a gauge invariant Noether current. When this direction is determined by gauge parallel transport, it implies that the generators will depend nonlocally on the field (connection), and so will the currents.

In the 2-dimensional chiral model with group G, the Lagrangian remains invariant only under a constant global G transformation, hence it seems useless trying to interpret the known nonlocal current by some local transformation. But in reality the dual symmetry or the integrability condition of the linear scattering equation implies the existence of a whole family of pure gauges. It occurs that the infinitesimal operators of g (or involution projection operator) transported by this pure gauge will generate the desired nonlocal current. In Section IV, we show that under this transformed infinitesimal operation the Lagrangian density changes into its Kahler partner[15] — the topological density. We have tried to express the variation of Lagrangian as a total divergence in different variants, in order to clarify the origin of a nontrivial nonlocal conserved Noether current. We also verify the invariance of equations of motion and energy momentum density. It is interesting that all these properties are the same as the dual transformation in electromagnetic theory and may thus be connected to some hidden symmetry of the 4-dimensional Yang-Mills theory. Finally in Section IV a somewhat new generating expression for two-dimensional supersymmetric nonlocal currents is derived as a Noether current.

II. Gauge invariant conserved currents for the Yang-Mills field

Assume that there exists a privileged isotopic direction $\eta(x)$ at each space time point x

$$\eta(x)=\eta^a(x)T^a\ ,\quad \eta^a(x)\eta^a(x)=1\ , \tag{2.1}$$

where the index a runs over the adjoint space g, T^a is the matrix representation for the basis in g. (For simplicity we shall be restricted to $G=SU(2)$ in this section, for general cases see Ref.[14].) Accordingly we decompose the potential A_μ into h_μ and k_μ.

$$A_\mu=h_\mu+k_\mu\ ,\quad h_\mu\equiv(A_\mu\cdot\eta)\eta-[\eta,\partial_\mu\eta],\quad k_\mu\equiv[\eta,D_\mu\eta]\ ,$$

where

$$A_\mu=A_\mu^a\,\sigma^a/2i\ ,\quad A_\mu\cdot\eta\equiv-2\,\mathrm{tr}(A_\mu\cdot\eta)=A_\mu^a\eta^a\ ,$$

$$D_\mu\equiv\partial_\mu+[A_\mu,\ ]\ . \tag{2.2}$$

The motivation of such a decomposition is to make $\eta(x)$ parallel under transport by h_μ.

$$D_\mu^{(h)}\eta\equiv\partial_\mu\eta+[h_\mu,\eta]=0\ . \tag{2.3}$$

The local transformation

$$A'_\mu = S^{-1} A_\mu S + S^{-1}\partial_\mu S\,,\quad \eta' = S^{-1}\eta S \qquad S\in G \tag{2.4}$$

now implies h_μ transforms as the usual potential while K_μ transforms covariantly

$$h'_\mu = S^{-1} h_\mu S + S^{-1}\partial_\mu S,\quad K'_\mu = S^{-1} K_\mu S\,. \tag{2.5}$$

In the infinitesimal form we have

$$\delta A_\mu = D_\mu \Lambda\,,\quad \delta\eta = [\eta, \Lambda]\,. \qquad \Lambda\in g \tag{2.4a}$$

Thus

$$\delta h_\mu = [h_\mu \cdot \Lambda] + \partial_\mu \Lambda \equiv D_\mu^{(h)}\Lambda\,,\quad \delta K_\mu = [K_\mu \cdot \Lambda] \tag{2.5a}$$

especially if $\Lambda(x) = \theta(x)\eta(x)$, (2.6)

where $\theta(x)$ is a scalar phase function. Then,

$$\delta\eta = 0,\quad \delta h_\mu = (\partial_\mu\theta(x))\eta(x) + \theta(x) D_\mu^{(h)}\eta(x) = (\partial_\mu\theta(x))\eta(x)\,,$$

$$\delta K_\mu = \theta(x)\,[K_\mu, \eta]\,, \tag{2.7}$$

h_μ changes as an Abelian field along η, while K_μ may be divided into differently charged sectors according to its eigenvalue by commutation with η.

Correspondingly, gauge fields $F_{\mu\nu}$ decompose into

$$F^{/\!/}_{\mu\nu} \equiv (F_{\mu\nu}\cdot\eta)\eta = f^{(h)}_{\mu\nu} + [K_\mu, K_\nu]\,, \tag{2.8}$$

$$F^{\perp}_{\mu\nu} \equiv -[[F_{\mu\nu}, \eta], \eta] = D_\mu^{(h)} K_\nu - D_\nu^{(h)} K_\mu\,, \tag{2.9}$$

where

$$f^{(h)}_{\mu\nu} \equiv \partial_\mu h_\nu - \partial_\nu h_\mu + [h_\mu, h_\nu] = -[\partial_\mu\eta, \partial_\nu\eta] + \partial_\mu(h_\nu\cdot\eta)\eta - \partial_\nu(h_\mu\cdot\eta)\eta\,. \tag{2.10}$$

One may recognize Eqs. (2.8) and (2.9) as a generalized Gauss Codazzi equation. When η denotes the isotopic direction of Higgs field $\phi(x)$, $\eta(x) = \phi(x)/|\phi(x)|$ (2.8) gives the t'Hooft[16] expression for electromagnetic field $f^{(h)}_{\mu\nu}\cdot\eta$, whose surface integral gives Kronecker map[17,18] as shown in (2.10).

Now we are prepared for the derivation of Noether current. Given the Lagrangian

$$L = -\frac{1}{4}(F^{\mu\nu}\cdot F_{\mu\nu}) + \bar\Psi(i\gamma^\mu(\partial_\mu + A_\mu) - m)\Psi\,, \tag{2.11}$$

where the Dirac spinor ψ are vectors in a representation space of G, we can gauge transform it into

$$\delta\Psi(x) = -\Lambda(x)\Psi(x)\,. \tag{2.12}$$

Usually it seems natural to assume that the constant variation

$$\Lambda(x) = \delta\theta T^3, \qquad \delta\theta = const. \tag{2.13}$$

that is, everywhere it rotates by the same angle $\delta\theta$ around the same third axis and we get

$$J_\mu = \frac{\delta L}{\delta(\partial^\mu\psi)}\frac{\delta\psi}{\delta\theta} + \frac{\delta L}{\delta(\partial^\mu A_\nu)}\frac{\delta A_\nu}{\delta\theta} = i\bar{\psi}\gamma^\mu T^3\psi + F_{\nu\mu}\cdot[A^\nu, T^3]. \tag{2.14}$$

Since this expression is not gauge covariant, its physical meaning is obscure. The problem is that in the very spirit of local gauge symmetry we cannot simply take the third axis as parallel forever, since the third "axis" in some original gauge becomes a different axis after a local transformation. Therefore instead of (2.13), we assume that

$$\Lambda(x) = \delta\theta\eta(x) \qquad \delta\theta = const. \tag{2.15}$$

Then, just as by ordinary gauge transformation of the first kind

$$\delta h_\mu = 0, \qquad \delta K_\mu = \delta\theta[K_\mu, \eta] \tag{2.16}$$

we obtain

$$J_\mu = i\bar{\psi}\gamma_\mu\eta\psi - F_{\mu\nu}D^\nu\eta, \tag{2.17}$$

which consists of two gauge invariant parts. The first term is the η component $\eta^a j^a_\mu$ for the matter current $j^a_\mu \equiv i\bar{\psi}\gamma_\mu T^a\psi$, which is the source of the field

$$D^\mu F_{\mu\nu} = j_\nu. \tag{2.18}$$

The second term is contributed by the "charged" constituent in the field; in reality it is bilinear in K_μ as the usual charged boson current

$$-F_{\mu\nu}D^\nu\eta = \eta\cdot[K_\nu, D^{(h)}_\mu K_\nu - D^{(h)}_\nu K_\mu]. \tag{2.19}$$

By using E.L. Eq. (2.18), we have

$$\partial_\mu(\eta\cdot F^{\mu\nu}) = \eta\cdot(D_\mu F^{\mu\nu}) + (D_\mu\eta)\cdot F^{\mu\nu} = \eta\cdot j^\nu + D_\mu\eta\cdot F^{\mu\nu} = J^\nu. \tag{2.20}$$

So J_μ is conserved

$$\partial_\mu J^\mu = 0 \tag{2.21}$$

while from (2.18)

$$\partial_\mu j^\mu = -[A_\mu, j^\mu] \neq 0. \tag{2.22}$$

Generally, the two constituents in (2.17) are not conserved separately. Now we consider some special cases:

1. If $$(D_\mu\eta)\cdot j^\mu = 0, \tag{2.23}$$

then both the matter part $\eta\cdot j^\mu$ and charged gauge boson part are conserved:

$$\partial_\mu(\eta\cdot j^\mu)=(\partial_\mu\eta)\cdot j^\mu+\eta\cdot(\partial_\mu j^\mu)=(\partial_\mu\eta)\cdot j^\mu-\eta[A_\mu\cdot j^\mu]=j^\mu\cdot D_\mu\eta=0\,;$$

$$\partial_\mu(F^{\mu\nu}\cdot D_\nu\eta)=(D_\mu D_\nu\eta)\cdot F^{\mu\nu}+(D_\nu\eta)\cdot D_\mu F^{\mu\nu}=$$

$$(D_\mu D_\nu-D_\nu D_\mu)\eta+(D_\nu\eta)\cdot j^\nu=\frac{1}{2}F^{\mu\nu}\cdot[F_{\mu\nu},\eta]+(D_\nu\eta)\cdot j^\nu=j^\nu\cdot D_\nu\eta=0. \tag{2.24}$$

2. If all four space-time components of j_μ point to the same isotopic direction, we may choose this direction as the privileged $\eta(x)$

$$[\eta\,,\,j^\mu]=0. \tag{2.25}$$

Thus

$$j^\mu\cdot D_\mu\eta=j^\mu\cdot[K_\mu,\eta]=[\eta,j^\mu]\cdot K_\mu=0. \tag{2.26}$$

So (2.23) is satisfied and $|j^\mu|=\eta.j^\mu$ is conserved.

3. If furthermore this j_μ is static and timelike, then $\partial_0\eta=0$. Since (2.23) and $[D_\mu\eta,j^\mu]=D_\mu[\eta,j^\mu]=0$, $D_0\eta=0$, we have $[A_0,\eta]=0$, which is the case considered by Jackiw et al.[19]

4. The reducible case

$$D_\mu\eta=0 \tag{2.27}$$

Now, only the matter current $\eta.j_\mu$ remains. We can express $\eta(x)$ nonlocally by A_μ as

$$\eta(x)=U^{-1}(x)\,TU(x),$$

where

$$U\equiv P\exp\int_{-\infty}^{x}A_\mu dx^\mu,\quad T\equiv\eta^a(-\infty)\,T^a. \tag{2.28}$$

Subsequently, the gauge transformation generated by (2.15) becomes a nonlocal operation, and $J_\mu=\eta\cdot j_\mu$ is a nonlocal current.

5. If (2.27) is satisfied only on the space boundary

$$D_\mu\eta(x)=0+0\left(\frac{1}{r}\right), \tag{2.29}$$

then no field-contributed current flows outside and the total charge $Q=\int J_0\,d^3x$ will be conserved and gauge invariant, irrespective of the choice of $\eta(x)$ internal, just as the energy momentum conservation for the asymptotically flat gravitation field.

6. Otherwise $\eta(x)$ may be the symmetry generator for space-time symmetrical solution[20]. Finally $\eta(x)$ may be just the direction of Higgs field[16], which is zero charged with respect to this η.

III. Nonlocal conservation for 2-dimensional chiral model

For simplicity we discuss explicitly in detail the principle model only. We shall point out briefly that the same method and result are also applicable to the reduced chiral model and the functional version of three-dimensional Yang Mills field[21,22].

For the principle chiral model

$$L(x)=\frac{1}{16} tr\, \partial^{\mu} g(x)\, \partial_{\mu} g^{-1}(x). \qquad g\in G \tag{3.1}$$

The equation of motion

$$\partial_{\mu} A^{\mu}(x)=0, \tag{3.2}$$

where

$$A_{\mu}\equiv g^{-1}(x)\, \partial_{\mu} g(x) \tag{3.3}$$

is a pure gauge, so

$$\partial_{\mu} A_{\nu}-\partial_{\nu} A_{\mu}+[A_{\mu}, A_{\nu}]=0. \tag{3.4}$$

We define

$$j_{\mu}=\frac{\delta L}{\delta \partial^{\mu} g}\delta g. \tag{3.5}$$

The Lagrangian is invariant by constant gauge transformation: $\delta L=0$, when $\delta g=-gT$, where $T\epsilon g$ is a constant matrix and $j_{\mu}=\frac{1}{8}tr(A_{\mu}T)$ is conserved by (3.2). The Lagranian is not invariant under any local transformation, but there exist the dual transformation operators $U(x,w)$ [10,11]

$$\partial_{0} U=U(A_{1}-wA_{0})\, w/(1-w^{2}), \tag{3.6}$$

$$\partial_{1} U=U(A_{0}-wA_{1})\, w/(1-w^{2}), \tag{3.7}$$

where w is a complex parameter. The integrability condition of (3.6), (3.7) is just (3.2), (3.4). We may treat $-(wA_{\mu}+\varepsilon_{\mu\nu}A^{\nu})\, w/(1-w^{2})$ as a pure gauge $(\varepsilon_{01}=-\varepsilon_{10}=1)$, U as the phase factor, and introduce a nonlocal transformation

$$\delta g(x)=-g(x)U^{-1}(x)TU(x), \tag{3.8}$$

so

$$\delta L=\frac{1}{8}tr\{A_{\mu}\partial^{\mu}(U^{-1}TU)\}=\frac{1}{8}tr\{[UA_{\mu}U^{-1}, U\partial^{\mu}U^{-1}]T\}. \tag{3.9}$$

The incomplete current (3.5) is not conserved

$$j_{\mu}=\frac{1}{8}tr(AU^{-1}TU), \tag{3.10}$$

$$\partial_{\mu} j^{\mu}=\frac{w}{8(1-w^{2})}tr(\varepsilon^{\mu\nu}[A_{\mu}, A_{\nu}]U^{-1}TU)=\delta L\ (\text{on shell}) \neq 0, \tag{3.11}$$

where (3.2),(3.6) and (3.7) have been used. To get a conserved current, we must express the off shell δL as a total divergence

$$\delta L=-\partial_{\mu} i^{\mu} \tag{3.12}$$

so that the action is not changed, and the current

$$J_{\mu}=i_{\mu}+j_{\mu} \tag{3.13}$$

is conserved. But off shell, the pair (3.6), (3.7) is not consistent, only the nonintegrable phase factor along some path is available. We use the space direction path

$$U(x)=P\exp\int_{-\infty}^{x^1}(A_0-WA_1)Wdx^1/(1-W^2) \tag{3.14}$$

and try to get the total divergent form (3.12) in the following ways:

1. The off shell δL is a total divergence: we have

$$\varepsilon^{\mu\nu}\partial_\mu(U\partial_\nu U^{-1})=[U\partial_0 U^{-1},U\partial_1 U^{-1}], \tag{3.15}$$

$$\varepsilon^{\mu\nu}\partial_\mu(UA_\mu U^{-1})=\varepsilon^{\mu\nu}U(\partial_\mu A_\nu)U^{-1}-\varepsilon^{\mu\nu}[U\partial_\mu U^{-1},UA_\nu U^{-1}]$$
$$=\frac{1}{W}[UA_1U^{-1},U\partial_1 U^{-1}]+\frac{1-W^2}{W^2}[U\partial_0 U^{-1},U\partial_1 U^{-1}]-\frac{1}{W}[UA_0U^{-1},U\partial_0 U-1], \tag{3.16}$$

where (3.4) and (3.14) have been used. Thus from (3.9), we get

$$\partial_\mu i^\mu\equiv\frac{1}{8}tr\{-W\varepsilon^{\mu\nu}\partial_\mu(UA_\nu U^{-1})T+\frac{1-W^2}{W}\varepsilon^{\mu\nu}\partial_\mu(U\partial_\nu U^{-1})T\}=\delta L \tag{3.17}$$

since we also have

$$\partial_\mu j^\mu=\frac{1}{8}tr\,\partial^\mu(UA_\mu U^{-1}T)=\frac{1}{8}tr\,U(\partial^\mu A_\mu)U^{-1}T+\frac{1}{8}tr[UA_\mu U^{-1},$$
$$U\partial^\mu U^{-1}]T=\delta L. \qquad \text{on shell} \tag{3.18}$$

Hence we get the on shell conserved current J_μ

$$J_\mu=j_\mu+i_\mu=\frac{1}{8}tr(UA_\mu U^{-1}+W\varepsilon_{\mu\nu}UA^\nu U^{-1}-\frac{1-W^2}{W}\varepsilon_{\mu\nu}U\partial^\nu U^{-1})T. \tag{3.19}$$

But from (3.14), it happens that J_0 is identically zero even off shell. And the surviving off shell component J_1 also vanishes on shell by using equation of motion (3.2) (more precisely J_1 may be some function of x_0 depending on the boundary value of U at $x_1=-\infty$). Thus J_μ turns out to be a trivial one.

2. A critical parameter gives a nonlocal Noether current in order to construct a new total divergence; we introduce

$$\chi=\int_{-\infty}^{x^1}(PUA_0U^{-1}+(P-1)U\partial_0 U^{-1})dx^1. \quad \text{P arbitrary constant} \tag{3.20}$$

Then for and only for $W\to\infty$,

$$\delta L=\frac{1}{8}tr\,\varepsilon^{\mu\nu}\partial_\mu[\chi,(U\partial_\nu U^{-1}+UA_\nu U^{-1})]T=\frac{1}{8}tr\,\partial[\chi,(U\partial_0 U^{-1}+$$
$$UA_0U^{-1})]T\equiv-i\partial_\mu i^\mu. \tag{3.21}$$

Thus $J_\mu = j_\mu + i_\mu = \frac{1}{8}\mathrm{tr}\, UA_\mu U^{-1}T - \varepsilon^{\mu\nu}\frac{1}{8}\mathrm{tr}[\chi, (UA_\nu U^{-1} + U\partial_\nu U^{-1})]T$ is conserved. Here $i_0 \equiv 0$, $i_1 = 0$ on shell is in accord with $\delta L = 0$ on shell (3.11) when $w \to \infty$. This J_μ is nontrivial, but since the parameter w has been fixed, we cannot expand it into an infinite series of current. By the way, now one possible solution of U is g, then J_μ on shell $= j_\mu = \partial_\mu g g^{-1}$ becomes the right invariant partner with the left invariant A_μ (3.3).

3. Set $$\chi = \int_{-\infty}^{x^1} \left\{ \frac{-w}{1-w^2} U(wA_0 - A_1)U^{-1} + \frac{1}{w}U\partial_1 U^{-1} \right\} dx^1 , \tag{3.22}$$

then

$$\delta L = \frac{1}{8}\mathrm{tr}\Big(\partial_\mu[\chi, \{\frac{-w}{1-w^2}U(w\epsilon^{\mu\nu}A_\nu - A^\mu)U^{-1} + \epsilon^{\mu\nu}U\partial_\nu U^{-1}\}]$$

$$+ \frac{2}{w}\partial_\mu \varepsilon^{\mu\nu}(U\partial_\nu U^{-1}) + \frac{w}{1-w^2}[\chi, U(\partial_\mu A^\mu)U^{-1}]\Big)T \tag{3.23}$$

On shell the last term vanishes, δL is a total divergent $-\partial_\mu i^\mu$ with $-i^\mu$ different from j^μ. Therefore, we obtain a nontrivial conserved current depending analytically on one complex variable:

$$J_\mu = \frac{1}{8}\mathrm{tr}(UA_\mu U^{-1}T - \frac{2}{w}\xi_{\mu\nu}U\partial^\nu U^{-1}T). \tag{3.24}$$

(The figure bracket in (3.23) also vanishes on shell, so we omit it.) Or we can use (3.6), (3.7). in the light cone coordinate.

$$J_\xi = \frac{-1}{8}\mathrm{tr}(\gamma^{-1}UA_\xi U^{-1}T), \quad J_\eta = -\frac{1}{8}\mathrm{tr}(\gamma UA_\eta U^{-1}T). \tag{3.25}$$

One may insist on using such a $-i^\mu$, whose divergence always equals L not merely on shell, and may be more patient and clever to find such an i^μ with nontrivial sum $i_\mu + j_\mu$. But since J^μ never is conserved off shell, it seems not very necessary and a little bit too stringent to restrict the overall off shell behavior of i_μ in order to secure the on shell only conservation of $i_\mu + j_\mu$. It is reasonable to relax it somehow, but we must reinforce it by further requirement in order to ensure the obtained conservation law is not an artifact and is really connected with the symmetry of the system;

1. The variation of Lagrangian density generated by δg is a total divergence (a surface term in action) without use of the equation of motion, consequently the action is invariant. (This condition is not sufficient for the existence of a nontrivial current.)

2. The δL on shell can be expressed as a total divergence by another vector different from $\frac{\delta L}{\delta \partial_\mu g}\delta g$.

3. The equation of motion is invariant by δg[9].

4. The variation of energy momentum density by δg vanishes on shell.

To satisfy 2, we have some freedom of choice since on shell

$$\frac{1}{8}\mathrm{tr}\,\partial_\mu(UA^\mu U^{-1}T) = \delta L \tag{3.26}$$

$$=\frac{1}{4W}\mathrm{tr}\,\varepsilon^{\mu\nu}\partial_\mu(U\partial_\nu U^{-1})T \tag{3.27}$$

$$=-\frac{W}{4(1+W^2)}\mathrm{tr}\,\varepsilon^{\mu\nu}\partial_\mu(UA_\nu U^{-1})T \tag{3.28}$$

$$=-\frac{1}{4}\mathrm{tr}\,\partial_\mu(U\partial^\mu U^{-1})T. \tag{3.29}$$

The J_μ in (3.24) is given by choosing (3.26), (3.27), J_μ in (3.25) by (3.26), (3.28) (it is the same as [4]), and the J_μ in Vega's paper (his equation 19), by (3.27), (3.29). They are all equivalent by on shell relation (3.6), (3.7).

Now we turn to show 3, 4. If $\delta g=-g\Lambda$, then the invariance of equation of motion (3.2) requires[9] that

$$D_\mu\partial_\mu\Lambda=0. \tag{3.30}$$

Now our $\Lambda=U^{-1}TU$, using the on shell relation

$$\partial_\mu(U^{-1}TU)=W\{\varepsilon_{\mu\nu}D^\nu(U^{-1}TU)\}\quad \text{We get}$$

$$D_\mu\partial^\mu(U^{-1}TU)=-D_\mu W\varepsilon^{\mu\nu}D_\nu(U^{-1}TU)=0. \qquad \text{Q.E.D.}$$

The energy momentum density is

$$M_{\mu\nu}=\frac{1}{8}\mathrm{Tr}\left(\frac{1}{2}g_{\mu\nu}A^\lambda A_\lambda-A_\mu A_\nu\right),$$

$$\delta M_{00}=\delta M_{11}=\frac{-1}{8}\mathrm{Tr}\,(A_0D_0\Lambda+A_1D_1\Lambda),$$

$$\delta M_{10}=\delta M_{01}=\frac{-1}{8}\mathrm{Tr}\,(A_0D_1\Lambda+A_1D_0\Lambda).$$

Now

$$D_\mu\Lambda=\left[\frac{1}{1-W^2}A_\mu-\frac{W\varepsilon^{\mu\nu}A_\nu}{1-W^2},\ T\right]\quad \text{so}\quad \delta M_{\mu\nu}=0.$$

For the reduced chiral model, e.g. the O(N) σ model,

$$L=\frac{1}{2}(\partial_\mu\eta^a\partial^\mu\eta^a+\ell(\eta^a\eta^a-1)),\quad a=1,\cdots,N.$$

The equation of motion is $\partial_\mu A^\mu=0$, $A^{ab}_\mu=2\eta^a\overleftrightarrow{\partial_\mu}\eta^b$, $\eta^a\eta^a=1$. The dual operator takes the same form as (3.6), (3.7). Assume that $\delta\eta^a=\eta^b(U^{-1}TU)^{ba}$. Note that T and $U^{-1}TU$ are antisymmetric. We get $\delta L=\mathrm{Tr}(A_\mu\partial^\mu U^{-1}TU)$. Then the derivations of condition 1-4 are all the same as before. Similarly, we use the dual operator for the functional version of three-dimensional Yang-Mills theory and may easily translate the final result into that case as well.

IV. Supersymmetric nonlocal Noether current

We follow the notation of Curtright and Zachos[7] for the O(N) case

$$L=\frac{1}{2}\partial_\mu\eta^a\partial^\mu\eta^a+\frac{1}{2}i\bar{\Psi}^a\partial_\mu\gamma^\mu\psi^a+\frac{1}{8}(\bar{\Psi}^a\Psi^a)^2,\quad \eta^a\eta^a=1,\ \eta^a\gamma^a=0 \tag{4.1}$$

$$A_\mu^{ab} \equiv 2\eta^a \overleftrightarrow{\partial_\mu} \eta^b ,$$

$$B_\mu^{ab} \equiv -i\bar{\Psi}^a \gamma_\mu \Psi^b . \tag{4.2}$$

For the CP(N-1) case, we have

$$L = (D_\mu Z^a)^* D_\mu Z^a + i\bar{\Psi}^a D_\mu \gamma^\mu \Psi^a + \frac{1}{4}\{(\bar{\Psi}^a \Psi^a)^2 -$$

$$-(\bar{\Psi}^a \gamma_5 \Psi^a)^2 - (\bar{\Psi}^a \gamma_\mu \Psi^a)(\bar{\Psi}^b \gamma^\mu \Psi^b)]\} .$$

$$z^{a*} z^a = 1 ,\ z^{a*} \Psi^a = 0 = \bar{\Psi}^a z^a ,\ D_\mu \equiv \partial_\mu - z^{a*} \partial_\mu z^a . \tag{4.3}$$

$$A_\mu^{ab} = 2[z^{a*} D_\mu z^b - (D_\mu z^a)^* z^b] .$$

$$B_\mu^{ab} \equiv -i\bar{\Psi}^a \gamma_\mu \Psi^b . \tag{4.4}$$

On shell we know in both cases[7] that

$$\partial^\mu A_\mu = [A_\mu B^\mu] ,$$

$$\partial^\mu B_\mu = -\frac{1}{2}[A_\mu B^\mu] ,$$

$$\varepsilon^{\mu\nu} \partial_\mu A_\nu = -\varepsilon^{\mu\nu} A_\mu A_\nu , \tag{4.5}$$

$$\varepsilon^{\mu\nu} \partial_\mu B_\nu = -\varepsilon^{\mu\nu} B_\mu B_\nu - \frac{1}{2}\varepsilon^{\mu\nu}[A_\mu , B_\nu] .$$

The generalized dual operators have also been given

$$U_\mu = U[wA_\mu - \varepsilon_{\mu\nu} A^\nu + \frac{4w}{1-w^2} B - \frac{2(1+w^2)}{1-w^2} \varepsilon_{\mu\nu} B_\nu] . \tag{4.6}$$

Let the infinitesimal O(N) (SU(N-1)) rotation be the dual transformed one $U^{-1}TU$,

then

$$\delta L = tr[A_\mu + 2B_\mu ,\ U\partial_\mu U^{-1}]T , \tag{4.7}$$

$$j^\mu = \frac{\delta L}{\delta \partial_\mu \eta} \delta\eta + \frac{\delta L}{\delta \partial_\mu \Psi} \delta\Psi = tr\{(A_\mu + 2B_\mu) U^{-1} T U\} . \tag{4.8}$$

The on shell δL equals not only $\partial_\mu j^\mu$ but also

$$\delta L = tr[-\frac{2}{w}\varepsilon^{\mu\nu}\partial_\mu (U\partial_\nu U^{-1}) + \frac{4w^2}{1-w^2}\{\frac{1+w^2}{1-w^2}\partial_\mu (UB^\mu U^{-1}) -$$

$$-\frac{2w}{1-w^2}\varepsilon^{\mu\nu}\partial_\mu (UB^\nu U^{-1})\}T] \equiv -\partial_\mu i^\mu . \tag{4.9}$$

So we get $J_\mu = j_\mu + i_\mu$, which is conserved.

V. Discussion

The nonlocal conservation laws are nothing else but symmetries generated by nonlocal infinitesimal generators. These infinitesimal operators generate a finite one parameter w group at each space-time point. Expanding with respect to the complicated parameter w, an infinite set of nonlocal currents will be obtained naturally. These infinitesimal operators are invariant under parallel transport, so they satisfy a condition similar to that of the projective operator in Ref.[12]. But our dual operator is not Backlund Transformation. So, when they are substituted into the energy momentum density, they do not yield an infinite series of local conservation laws as in Ref.[23]. Instead, they give only a trivial result. Since the Backlund Transformation in Sine Gordan equation is also related to the shift of total divergence, the method of this paper may be generalized accordingly.

The presence of a dual transformation is crucial to the existence of nonlocal currents in two dimensions. Hence a generalized dual transformation will probably give nonlocal currents for the four-dimensional Yang-Mills theory. The ordinary dual electromagnetic transformation will be a possible successful candidate when made local and extended to a non-Abelian group.

Acknowledgment

I would like to thank Yale Physics Department for their hospitality and Professor H.C. Tze for his helpful discussions and for correcting the English. I also wish to thank Dr. Dolan for mailing her preprints to me before publication and Mary Hoffman for her excellent typewriting.

References

1. M. Luscher and K. Pohlmeyer, Nucl. Phys. B137, 46 (1978).
2. E. Brezin, C. Itzykson, J. Zinn-Justin and J.B. Zuber, Phys. Lett. B82, 442 (1979).
3. H.J. deVega, Phys. Lett. B87, 233 (1979).
4. A.T. Ogielski, Phys. Rev. D21, 406 (1980).
5. E. Corrigan and C.K. Zachos, Phys. Lett. B88, 273 (1979).
6. T.L. Curtright, Phys. Lett. B88, 276 (1979).
7. T.L. Curtright and C.K. Zachos, Phys. Rev. D21, 411 (1980).
8. L.L. Chau Wang, BNL-27617-R.
9. L. Dolan and A. Roos, DOE/EY/2232B-201.
10. K. Pohlmeyer, Commun. Math. Phys. 46, 207 (1976).
11. H. Eichenherr and M. Forger, Nucl. Phys. B155, 381 (1979).
12. V.E. Zakhalov and A.V. Mikhailov, Sov. Phys. JETP 47, 1017 (1978).
13. S. Kobayashi and K. Nomizu, Foundations of Differential Geometry (Interscience, New York).
14. B.Y. Hou, Y.S. Duan and M.L. Ge, Scientia Sinica 21, 446 (1978).
15. F. Gursey and H.C. Tze, to appear in Annals of Physics.
16. G. 't Hooft, Nucl. Phys. B79, 276 (1974).
17. B.Y. Hou, Y.S. Duan and M.L. Ge, J. Northwest University (Xian) No. 1, 11 (1975), J. Lanchow Univ. No. 2, 26 (1975), Acta Physica Sinica, 25, 514 (1976).
18. J. Arafune, P.G.O. Freund and C.J. Goebel, J. Math. Phys. 16, 433 (1975).
19. R. Jackiw, L. Jacobs, and C. Rebbi, Phys. Rev. D20, 474 (1979); R. Jackiw and P. Rossi, Phys. Rev. D21, 426 (1980).
20. P. Forgacs and N.S. Manton, Commun. Math. Phys. 72, 15 (1980).
21. A.M. Polyakov, Nucl. Phys. B164, 171 (1979); Phys. Lett. B82, 247 (1979).
22. L. Dolan, DOE/EY/2232B-202.
23. B. Yoon, Phys. Rev. D13, 3440 (1976).
24. M.K. Prasad, A. Sinha and Ling-Lie Chau Wang, Phys. Lett. B93, 415 (1979).

Volume 125B, number 5 PHYSICS LETTERS 9 June 1983

THE GYROELECTRIC RATIO OF A SUPERSYMMETRIC MONOPOLE DETERMINED BY ITS STRUCTURE

Bo-Yu HOU [1]
International Centre for Theoretical Physics, Trieste, Italy

Received 20 December 1982

This paper analyzes the distribution of the electric dipole produced by the zero energy charged particle around a Dirac monopole and then finds the unbroken U(1) electric polarization inherent by the supersymmetric monopole field so one can obtain its gyroelectric ratio as an intrinsic property.

The classical Maxwell equations behave dual symmetrically with respect to electricity and magnetism, while the quantum theory displays both an intimate connection and a striking contrast between charge and pole. Their quantum values are united cunningly by the famous Dirac condition. But while the electric charges conserve Noether-wise and generate local gauge transformations, the magnetic pole determines the global gauge pattern and maintains its stability topologically. In current field theory electric particles are treated as quantums of local fields while the singularity free 't Hooft monopole realizes a classical extended object. Complementary to this tradition Goddard et al. [1] have made a daring conjecture that the roles of charges and poles are exchangeable. The developments in supersymmetrical Yang–Mills theory [3–6], especially the discovery by Mandelstam [7], on the ultraviolet finiteness of the $N = 4$ model makes this ambitious assumption quite plausible [8,9]. To further explore the mystery of dual symmetry it is worthwhile to study various dual properties more seriously.

Recently, an interesting paper by Osborn [10] shows that the gyroelectric ratio of a supersymmetric monopole carrying its zero mode fermions really assumes the value -2, just as expected by the dual conjecture. His results have been obtained tactically by calculating the energy shift in external electric field. In this paper we shall show that this ratio appears plainly as the intrinsic property of monopole determined by its internal structure, i.e. by the distribution of its inherent electric polarization and spin and by their correlations.

To see the physical origin more intuitively, we show at first how the electric polarization occurs around the Dirac monopole as the preliminaries for the case of supersymmetric monopole.

The zero mode normalizable solutions of a spin $\frac{1}{2}$ fermion with mass M and charge e around a Dirac monopole g situated at the centre are given as [11–14]

$$\psi = f(r)[(2J+1)/4\pi]^{1/2} D^J_{m,\mu-n}(\varphi,\theta,\gamma)\, u_\mu(\varphi,\theta,\gamma)\,, \tag{1}$$

where

$$f(r) \equiv (2M)^{1/2}\, \mathrm{e}^{-Mr}/r\,, \tag{2}$$

$D^J_{m,m'}$ is the usual SU(2) rotation matrix,

$$J = |n| - |\mu| \quad (\therefore\ \mu/|\mu| = n/|n|)\,, \tag{3}$$

$4\pi n = eg$. The Euler angle γ has been put identical with the gauge parameter (e.g. in Dirac's gauge $\gamma = -\varphi$, in Schwinger's gauge $\gamma = 0$).

In order to express the spin distribution more apparently we have introduced the spinor u_μ which is an eigenfunction in local "body axis"

$$u_\mu(\varphi,\theta,\gamma) \equiv \sum_{\mu'} D^{1/2}_{\mu'\mu}(\varphi,\theta,\gamma)\, U_{\mu'}\,, \tag{4}$$

[1] On leave of absence from North West University, Xian, China.

Volume 125B, number 5 PHYSICS LETTERS 9 June 1983

where the constant spinor U_μ are common eigenfunctions of

$$\sigma_3 = -\mathrm{i}\gamma^1\gamma^2 \quad \text{and} \quad \gamma^0\gamma^5 ,$$

$$\tfrac{1}{2}\sigma_3 U_\mu = \mu U_\mu , \quad \tfrac{1}{2}\mathrm{i}\,\gamma^0\gamma^5 U_\mu = \mu U_\mu , \qquad (5\mathrm{a,b})$$

therefore

$$\tfrac{1}{2}\sigma_i \hat{r}_i u_\mu = \mu u_\mu , \quad \tfrac{1}{2}\mathrm{i}\gamma^0\gamma^5 u_\mu = \mu u_\mu , \qquad (6\mathrm{a,b})$$

$$\hat{r}_i \equiv x^i/r .$$

From (1), noticing the orthonormality of U_μ, it is easy to see that the charge density $e\rho \equiv e\bar{\psi}\gamma^0\psi$ is distributed unsymmetrically with respect to the equator ($\theta = \frac{1}{2}\pi$) (except for $|N| = \frac{1}{2}, J = 0$) so there is an electric moment with respect to the centre along the polar direction ($\theta = 0$)

$$P_3 = \int e\rho r \cos\theta\, r^2\, \mathrm{d}r \sin\theta\, \mathrm{d}\theta\, \mathrm{d}\varphi . \qquad (7)$$

Meanwhile the spin density

$$s_3 = \tfrac{1}{2}\psi^+\sigma_3\psi = \tfrac{1}{2}u_\mu^\dagger\sigma_3 u_\mu \rho = \mu \cos\theta\, \rho , \qquad (8)$$

such that totally

$$S_3 = \int \mu\rho \cos\theta\, r^2 \mathrm{d}r \sin\theta\, \mathrm{d}\theta\, \mathrm{d}\varphi . \qquad (9)$$

Since from (2)

$$\int_0^\infty f^2 r^3 \mathrm{d}r = 1/2M = \frac{1}{2M}\int_0^\infty f^2 r^2\, \mathrm{d}r , \qquad (10)$$

so finally we have

$$(P_3/S_3)\, 2M/e = 1/\mu = 2n/|n| , \qquad (11)$$

e.g. when $e > 0, g < 0$, we have $n/|n| = -1$, the ratio is -2. We see the non-vanishingness of the parity changing electric dipole operator originates from the indeterminancy in parity (diagonal in $\gamma^0\gamma^5$ and so in $\sigma\cdot\hat{r}$).

To see the correlation of P and S more delicately, we derive the polarization density p from the charge density ρ as follows. Using the Dirac equation, we may divide the current density $e\bar{\psi}\gamma^\mu\psi$ into two parts:

$$e\bar{\psi}\gamma^\mu\psi = (\mathrm{i}e/2M)[\partial^\nu\bar{\psi}\gamma_\nu\gamma^\mu\psi + \bar{\psi}(\partial^\mu + \mathrm{i}eA^\mu)\psi - (\partial^\mu - \mathrm{i}eA^\mu)\bar{\psi}\psi] . \qquad (12)$$

Since the energy equals zero, so we have $(\partial_0 + \mathrm{i}eA_0)\psi = 0$ therefore

$$e\rho = e\bar{\psi}\gamma^0\psi = (\mathrm{i}e\rho/2M)\partial_i(\bar{\psi}\gamma^0\gamma^i\psi) . \qquad (13)$$

Thus the charge density may be expressed entirely (except at the origin) as the divergence of a polarization density, $\rho = \partial_i p^i = -\nabla\cdot\boldsymbol{p}$, i.e.

$$p^i = (\mathrm{i}e/2M)\bar{\psi}\gamma^0\gamma^i\psi . \qquad (14)$$

Comparing with the spin density $s^i = \frac{1}{2}\psi^+\sigma^i\psi$, using (6) and $\gamma^5 = \mathrm{i}\gamma^0\gamma^1\gamma^2\gamma^3$, we have at last

$$p^i/s^i = 2(n/|n|)\, e/2M . \qquad (15)$$

Once more we see the crucial role played by the diagonality of $\gamma^0\gamma^5$ (6), which in turn is required by the regularity of the asymptotic behaviour for the zero mode function (more generally for the bound state around a pole [11–13], but here we shall not deviate too much to discuss in detail the boundary condition at the centre. And as a consequence of (6) not only does the current vanish $j^i = e\bar{\psi}\gamma^i\psi = 0$ (we emphasize that the zero mode solution is strictly static not only stationary) but also inhibits the magnetic moment $\bar{\psi}\gamma^i\gamma^j\psi = 0$.

It is interesting to point out that the charge density ρ is equivalent to that of a point charge e imposed on the pole with an inhomogeneous polarization $\boldsymbol{p}$ induced by surrounding it such that there is a full compensation of charge in the centre (dielectric constant $\epsilon = \infty$ here), while the charge is smeared outward with density ρ. This phenomenon is consistent with Dirac's veto [15], Lipkin's paradox [16] and behaves like a magnetic superconductor in the centre. By the way, we can obtain the field strength E produced by ρ simply as the sum of the Coulomb field D produced by a point charge e at the centre and with minus the polarization $\boldsymbol{p}$, $\boldsymbol{E} = \boldsymbol{D} - \boldsymbol{p}$.

Now we turn to study the singularity free 't Hooft monopole. The well-known Prasad–Sommerfield gauge potential A_i^a, Higgs field ϕ^a and SU(2) magnetic field B_i^a are

$$A_i^a(x) = \epsilon_{iaj}\hat{r}_j(K(r) - 1)/er ,$$

$$\Phi^a(x) = \Phi(r)\hat{r}^a/e ,$$

$$B_i^a(x) = \{[K^2(r) - 1]/er^2\}\,\hat{r}_i\hat{r}_a + [K\Phi(r)/er](\delta_{ia} - \hat{r}_i\hat{r}_a) , \qquad (16)$$

Volume 125B, number 5 PHYSICS LETTERS 9 June 1983

where $K(r) = evr/\mathrm{sh}(evr)$, $\phi(r) = -1/r + ev\,\mathrm{cth}(evr)$. The supersymmetric solution has in addition a normalized fermion [12,17]

$$\psi^a_\mu = M^{-1/2}\sigma_i B^a_i U_\mu\,, \tag{17}$$

where M is the mass of the monopole, $M = 4\pi v/e = gv$. $\mu = \pm\frac{1}{2}$ denote two independent solutions, respectively [cf. (5a) and (5b)] with accompanying scalar field A_0 and a pseudoscalar G [8,9]

$$A^a_0 = (2M)^{-1}u^\dagger\sigma_i B^a_i u\,, \quad G^a = (2M)^{-1}u^\dagger\sigma_i B^a_i u\,. \tag{18}$$

(We have corrected this expression with a factor $\frac{1}{2}$.)

Now the crucial point is that the external electric field is an abelian U(1) field with generator $\hat\varphi^a \equiv \varphi^a/|\varphi|$. Consequently, we must find the inherent U(1) polarization p_i which constitutes the source of only the corresponding U(1) part among the SU(2) F^a_{i0} and no more, i.e. just produces the 't Hooft electromagnetic field

$$f_{i0} = \hat\varphi^a F^a_{i0} - (\mathrm{D}_i\hat\varphi)^a(\mathrm{D}_0\hat\varphi)^b\hat\varphi^c\epsilon_{abc}\,.$$

Here D_i are gauge covariant derivatives. Since the total U(1) charge in this case equals zero (for a careful analysis of charge distribution, cf. ref. [18]), so we have

$$0 = \boldsymbol{D} = \boldsymbol{E} + \boldsymbol{p}\,, \quad \text{i.e.}\ p_i = -E_i = -f_{i0}\,. \tag{19}$$

Thus to find p_i we need to simply find f_{i0}. As for the latter it is more convenient to use the following formulation [19] of 't Hooft's U(1) field: The SU(2) A^a_μ has to be decomposed into two parts gauge covariantly in accordance with the U(1) generator $\hat\varphi$ as

$$A^a_\mu = H^a_\mu + K^a_\mu\,, \tag{20}$$

where

$$H^a_\mu \equiv (A^b_\mu\hat\varphi^b)\hat\varphi^a + e^{-1}\epsilon_{abc}\hat\varphi^b\partial_\mu\hat\varphi^c\,.$$

Then K^a_μ behaves covariantly and constitutes the charged particles, while H^a_μ is the U(1) potential in an embedding SU(2) gauge and generates the 't Hooft field

$$f_{\mu\nu}\hat\varphi^a = \partial_\mu H^a_\nu - \partial_\nu H^a_\mu + e\epsilon_{abc}H^b_\mu H^c_\nu\,. \tag{21}$$

In the 't Hooft–Prasad–Sommerfield gauge $\partial_0\hat\varphi = 0$, we have

$$H^a_0 = (A_0\cdot\hat{\boldsymbol{\Phi}})\hat\varphi^a\,, \quad H^a_i = e^{-1}\epsilon_{abc}\hat\varphi^b\partial_i\hat\varphi^c\,. \tag{22}$$

Then it is easy to obtain from (21), (22) and (20)

$$f_{i0} = \hat\varphi^a(\partial_i H^a_0 + eH^b_i H^c_0\epsilon_{abc}) = \partial_i(\hat{\boldsymbol{\Phi}}\cdot H_0) = \partial_i(\hat{\boldsymbol{\Phi}}\cdot A_0)\,. \tag{23}$$

With the explicit expression of A_0 (18) we get

$$p_i = -f_{i0} = (-1/2M)[2K^2\Phi\hat r_3\hat r_i/er^2 - (K^2-1)(3\hat r_3\hat r_i - \delta_{i3})/er^3]\mu/|\mu|\,, \tag{24}$$

$$P_i = \int p_i\,\mathrm{d}^3x = \delta_{i3}\,\tfrac{4}{3}(\pi/e)(1/2M)\mu/|\mu|\,. \tag{25}$$

It is straightforward to get

$$S_i = \frac{1}{2}\int\psi^\dagger\sigma_i\psi\,\mathrm{d}^3x = -\delta_{i3}(4\pi v/3e)M^{-1}\mu\,. \tag{26}$$

Finally, the gyroelectrical ratio is actually

$$P/S = -1/v = -2g/2M\,. \tag{27}$$

We may unit (15) and (27) just as

$$P/S = (n/|n|)\,v^{-1}\,. \tag{28}$$

Here v is the universal mass charge (pole) ratio

$$v = M_e/e = M_g/g\,. \tag{29}$$

It is easy to see that around the Dirac point monopole we would have states with 2^{2n+1} electric poles and 2^n magnetic poles. A systematic calculation for the PS monopole would show similar phenomena.

I would like to thank Professor Abdus Salam, the International Atomic Energy Agency and UNESCO for hospitality at the International Centre for Theoretical Physics, Trieste. Thanks are due to Professor I. Jengo, Professor J. Strathdee and Professor A.M. Harun ar Rashid for interesting discussions. It is a pleasure to thank Professor W. Nahm for introducing the recent developments in monopole theory and calling my attention to Osborn's paper [10].

References

[1] P. Goddard, J. Nuyts and D. Olive, Nucl. Phys. B125 (1977) 1;
C. Montonen and D. Olive, Phys. Lett. 72B (1977) 117;
E. Witten and D. Olive, Phys. Lett. 78B (1978) 97;
D. Olive, Monopole in quantum field theory (World Scientia, Singapore, 1982).
[2] N.S. Craigie, P. Goddard and W. Nahm, Monopole in quantum field theory (World Scientia, Singapore, 1982).

Volume 125B, number 5 PHYSICS LETTERS 9 June 1983

[3] L. Brink, J. Schwarz and J. Scherk, Nucl. Phys. B121 (1977) 77;
F. Gliozzi, J. Scherk and D. Olive, Nucl. Phys. B122 (1977) 253.
[4] A. D'Adda, R. Horsley and P. Di Vecchia, Phys. Lett. 76B (1978) 298.
[5] D. Olive, Nucl. Phys. B153 (1979) 1.
[6] H. Osborn, Phys. Lett. 83B (1979) 321; in Monopole in quantum field theory (World Scientia, Singapore, 1982).
[7] S. Mandelstam, UCB-PTH-82/15; TH.3385-CERN.
[8] P. Rossi, Phys. Lett. 99B (1981) 229.
[9] P. Rossi, Phys. Rep. 86 (1982) 318.
[10] H. Osborn, Phys. Lett. 115B (1982) 226.
[11] Y. Kazama and C.N. Yang, Phys. Rev. D15 (1977) 2300;
Y. Kazama, C.N. Yang and A.S. Goldhaber, Phys. Rev. D15 (1977) 2287.
[12] B.-Y. Hou and B.-Y. Hou, Phys. Energ. Fort. Phys. Nucl. 3 (1979) 697.
[13] C.J. Callias, Phys. Rev. D16 (1977) 3068; Commun. Math. Phys. 62 (1978) 213.
[14] P. Rossi, Nucl. Phys. B127 (1977) 518.
[15] P.A.M. Dirac, Phys. Rev. 74 (1948) 817.
[16] H.J. Lipkin, W.I. Weisberger and M. Peshkin, Ann. Phys. 53 (1969) 203.
[17] E. Mottola, Phys. Lett. 79B (1978) 242.
[18] B.-Y. Hou and B.-Y. Hou, Phys. Energ. Fort. Phys. Nucl. 3 (1979) 255.
[19] B.Y. Hou, Y.S. Duan and M.L. Ge, Sci. Sinica 21 (1978) 445.

IL NUOVO CIMENTO VOL. 84 A, N. 4 21 Dicembre 1984

A Correspondence between the σ-Model and the Liouville Model.

H. BOHR
International School for Advanced Studies - Trieste, Italy

B. HOU
North-West University - Xian, People's Republic of China

S. SAITO
Tokyo Metropolitan University - Setagaya-ku, Tokyo, Japan

(ricevuto il 25 Giugno 1984)

Summary. — The σ-model and the Liouville model are both derived from a common geometrical basis. Therefore, solutions to the Liouville equation can be derived from solutions of the σ-model and *vice versa*. The two independent field variables in the σ-model on a coset space correspond then to two similar variables in an extended Liouville model, one being the usual « string variable », the other corresponding to a topological winding parameter. This Liouville model is finally shown to have instanton solutions of fractionally topological charge and finite energy.

PACS. 11.10. – Field theory.

In this paper we should like to introduce a general framework for 2 dimensions from which both the nonlinear σ-model [1] and the Liouville mod-

[1] For a review see H. EICHENHERR: in *Current Topics in Elementary Particle Physics* (Sommer Institute (1980) Bad Honnef, Germany) (Plenum Press, New York, N.Y., 1981).

el ([2]) can be derived and furthermore show that there is a clear correspondence between solutions of both models. Such a correspondence is very desirable since it enables us to establish and understand mathematical properties in one model that has been already realized in the other.

Of course the two models cannot be equivalent since, for example, the σ-model has a topological, conserved charge while the Liouville is known not to possess one. On the other hand, we can start with a solution in the σ-model, and using our correspondence we can generate a similar solution in the ordinary Liouville model, plus an extra field variable that satisfies the Laplace equation and gives rise to a topological number in such an extended Liouville system. Furthermore, this extended system can be made more well defined (*i.e.* having finite-energy configurations) than the ordinary Liouville system.

Let us first briefly introduce both models as gauge theories.

1. – Liouville gauge theory.

The Liouville model is described by the Lagrangian

$$\mathscr{L} = \tfrac{1}{2}\partial_+\varrho\partial_-\varrho - 2\exp[\varrho]\,, \tag{1}$$

where we have used light-cone co-ordinates which in Euclidean space are $x_\pm = \frac{1}{2}(x_1 \pm ix_2)$. The equation of motion (Euler-Lagrange equation) from eq. (1) is then the Liouville equation

$$\partial_+\partial_-\varrho = 2\exp[\varrho]\,. \tag{2}$$

(In Minkowskian co-ordinates $x_\pm = \frac{1}{2}(x_1 \pm x_2)$ eq. (2) becomes the Toda lattice equation.)

The solutions to eq. (2) are in one-to-one correspondence, as shown in ref. ([3]), to that of a two-dimensional, zero-curvature $SL_{2,C}$ Yang-Mills theory. Take the field strength

$$F_{+-} = \partial_+A_- - \partial_-A_+ + [A_+, A_-]\,. \tag{3}$$

([2]) J. Liouville: *J. Math. Pure Appl.*, **18**, 71 (1953); A. Polyakov: *Phys. Lett. B*, **103**, 207, 211 (1981); T. L. Curtright and C. B. Thorn: *Phys. Rev. Lett.*, **48**, 1309 (1982); E. D'Hoker and R. Jackiw: *Phys. Rev. D*, **26**, 3517 (1982). A. Kihlberg and R. Marnelius: Göteborg preprint 82-2 (1982). J. L. Gervais and A. Neveu: *Nucl. Phys. B*, **199**, 59 (1982). B. Durhuus, P. Olesen and J. L. Petersen: *Nucl. Phys. B*, **198**, 157 (1982). H. Bohr and H. B. Nielsen: *Nucl. Phys. B*, **227**, 547 (1983).

([3]) A. Lezhov and M. Saveliev: *Lett. Math. Phys.*, **3**, 489 (1979). G. Bhattacharya and H. Bohr: *Nuovo Cimento A*, **80**, 393 (1984).

Zero curvature, *i.e.* $F_{+-}=0$, implies that

(4) $$A_\pm = g^{-1}\partial_\pm g\,, \qquad g\in SL_{2,C}\,.$$

We can always choose g to be such that the gauge potentials $A_\pm$ can be written as upper and lower triangular matrices

(5) $$A_\pm = u^\pm H + f^\pm E^\pm\,,$$

where H, E^+ and E^- are the generators of the $SL_{2,C}$ Cartan-Weyl basis with the commutation relations

(6) $$[H, E_\pm] = \pm 2E^\pm\,, \qquad [E^+, E^-] = H\,.$$

It is easy to verify that the zero-curvature condition implies

(7*a*) $$\partial_+\partial_-(\ln(f^+f^-)) = 2f^+f^-$$

and

(7*b*) $$\partial_\pm \ln f^\mp = \pm 2u^\pm$$

and defining ϱ to be

(8) $$\varrho = \ln f^+f^-\,,$$

we obtain the Liouville equation in (2).

2. – Chiral gauge theory.

Similarly for the O_3 nonlinear σ-model we can start from the following Lagrangian:

(9) $$\mathscr{L} = \tfrac{1}{2}\,\mathrm{Tr}\,(\partial g^{-1}\partial g)\,.$$

If we choose the vector

$$\boldsymbol{N} = (\sin\theta\cos\varphi, \sin\theta\sin\psi, \cos\theta)\,,$$

we can write g as

(10) $$g = i\boldsymbol{N\sigma} = i\begin{pmatrix}\cos\theta & \sin\theta\exp[-i\psi]\\ \sin\theta\exp[i\psi] & -\cos\theta\end{pmatrix}, \qquad \det g = 1\,,$$

and we recall the constraint $\sum N_i^2 = 1$.

From g we can define the pure gauge potentials as before:

$$A_{\pm} = g^{-1}\partial_{\pm} g\,, \tag{11}$$

so that the curvature is again zero:

$$F_{+-} = \partial_{+}A_{-} - \partial_{+}A_{-} + [A_{+}, A_{-}] = 0\,, \tag{12}$$

but this time the equation of motion from the Lagrangian in eq. (9) is

$$\partial_{+}A_{-} + \partial_{-}A_{+} = 0\,. \tag{13}$$

Because of the constraint it is useful to define field variables $W_{\pm}$ given by a stereographic projection in order to eliminate the nonlinear constraint, so essentially we consider the field on the coset space SU_2/U_1, therefore defined on the two-sphere S^2:

$$W_{\pm} = \frac{N_1 \pm iN_2}{1 - N_3} = \operatorname{ctg}\frac{\theta}{2}\exp[\mp i\varphi] \tag{14}$$

and the equation of motion can be written in the usual form (Belavin-Polyakov) (4):

$$\partial_{-}W_{+} = \partial_{+}W_{-} = 0\,, \qquad W_{-} = W_{+}^{*}\,. \tag{15}$$

Besides the trivial solutions, eq. (15) is also satisfied by the well-known instanton solutions which can be written in the form

$$W_{+} = \prod_{i}^{q} c_i\left(\frac{x_{+} - a_i}{x_{+} - b_i}\right), \qquad W_{-} = \prod_{i}^{q} c_i\left(\frac{x_{-} - a_i}{x_{-} - b_i}\right) \tag{16}$$

or written as in ref. (4):

$$W_{+} = W_{-}^{*} = c\prod_{i}(x_{+} - a_i)^{m_i}(x_{-} - b_i)^{-n_i}\,. \tag{16a}$$

3. – A correspondence on the Lagrangian level.

Before introducing a general geometrical framework for the two models we can already see naively that there is a correspondence. Since the constrained σ-model has two independent field variables W_{+}, W_{-} it will of course corre-

(4) A. BELAVIN and A. POLYAKOV: *JETP Lett.*, **22**, 245 (1975).

spond to an extended Liouville system with two similar field variables which we call ϱ and β. Therefore, we first define the following gauge potential:

$$A_{\pm} = A_1 \mp iA_2 = \pm\tfrac{1}{2}\partial_{\pm}\varphi_{\mp}\sigma_3 + \exp[\varphi_{\pm}]\sigma_{\pm}, \qquad \sigma_{\pm}\sim E_{\pm}, \tag{17}$$

where $\varphi_{\pm}$ is defined as

$$\varphi_{\pm} = \tfrac{1}{2}\varrho \pm 2i\beta . \tag{18}$$

A Lagrangian of the form

$$\mathscr{L} = -\tfrac{1}{2}\,\mathrm{Tr}\,(A_{\mu}^2) = \partial_+\varphi_-\partial_-\varphi_+ - 2\exp[\varphi_+ + \varphi_-] \tag{19}$$

yields the following equation (the Liouville equation in ϱ and the Laplace equation in β):

$$\tfrac{1}{2}\partial_+\partial_-\varrho + 2\exp[\varrho] = 0, \tag{20a}$$

$$\partial_+\partial_-\beta = 0 . \tag{20b}$$

It is well known that the solution to eq. (20*a*) is given by

$$\exp[\varrho] = -\frac{1}{2}\,\frac{\partial_+ M_+ \partial_- M_-}{(1 - M_+ M_-)^2}, \tag{21}$$

where M_+ and M_- are any analytic function satisfying

$$\partial_+ M_- = \partial_- M_+ = 0 . \tag{22}$$

We can already now see that the Lagrangian, $L = -\frac{1}{2}\,\mathrm{Tr}\,A_{\mu}^2$ in eq. (19) can be written as the Lagrangian for the σ-model in eq. (9) since a zero-curvature condition on $A_{\pm}$ implies the existence of g satisfying

$$A_{\pm} = g^{-1}\partial_{\pm} g ,$$

which enables us to write the Lagrangian in eq. (19) as

$$\mathscr{L} = \tfrac{1}{2}\,\mathrm{Tr}\,(\partial g^{-1}\partial g), \tag{23}$$

so the same form of Lagrangian yields the σ-model and an extended Liouville model. Similarly the two analytic functions M_+, M_- in eq. (21) correspond to the two analytic functions W_+, W_- in eq. (14) since both satisfy the same equation.

4. – A general framework for the Liouville and σ-model.

To introduce a general framework we consider a general $SL_{2,C}$ operator g of the form

$$g = i\begin{pmatrix} -(1-YW)\alpha & 2W \\ 2Y & (1-YW)/\alpha \end{pmatrix}(1+YW)^{-1}, \tag{24}$$

where the functions Y, W and α are similar to the three functions N_1, N_2 and N_3 in the O_3 σ-model. The Lagrangian considered in eq. (9) then becomes

$$\begin{aligned} \mathscr{L} &= \tfrac{1}{2}\,\mathrm{Tr}\,(\partial_\mu g^{-1}\partial^\mu g) = \\ &= -\tfrac{1}{2}[(1-YW)^2\alpha^{-2}\partial_\mu\alpha\partial^\mu\alpha - 4\partial_\mu Y\partial^\mu W](1+YW)^{-2}. \end{aligned} \tag{25}$$

The solutions of the equation of motion

$$\begin{cases} \partial_\mu^2\alpha - \alpha^{-1}\partial_\mu\alpha\partial^\mu\alpha - 4(W\partial^\mu\alpha\partial_\mu Y + Y\partial^\mu\alpha\partial_\mu W)(1-Y^2W^2)^{-1} = 0, \\ \partial_\mu^2 W - 2Y(1+YW)^{-1}\partial_\mu W\partial^\mu W - \\ \qquad - (1-YW)(1+YW)^{-1}W\alpha^{-2}\partial_\mu\alpha\partial^\mu\alpha = 0, \\ \partial_\mu^2 Y - 2W(1+YW)^{-1}\partial_\mu Y\partial^\mu Y - \\ \qquad - (1-YW)(1+YW)^{-1}Y\alpha^{-2}\partial_\mu\alpha\partial^\mu\alpha = 0 \end{cases} \tag{26}$$

are those Y and W satisfying either

$$\partial_- Y = \partial_+ W \qquad \text{(self-dual equation)} \tag{27a}$$

or

$$\partial_+ Y = \partial_- W = 0 \qquad \text{(anti-self-dual equation)} \tag{27b}$$

with

$$\partial_\pm\alpha = 0 \qquad (\alpha \text{ constant}) \tag{28}$$

and g being unitary implies $|\alpha| = 1$ and $Y = W^*$. These equations are again just the Belavin-Polyakov equation, and we verify the fact that the solutions of the Euclidean σ-model are either the self-dual or anti-self-dual solutions.

The problem with the metric g in eq. (24) as a mean of deriving the Liouville equation is that it is not easily decomposed into an upper and lower triangular form. We, therefore, write g of eq. (24) in terms of another set of variables

$$g = i\begin{pmatrix} -\alpha & u \\ v & \alpha^{-1} \end{pmatrix}(1+uv)^{-\frac{1}{2}}, \tag{29}$$

where

(30) $$u = 2W(1 - YW)^{-1}, \qquad V = 2Y(1 - YW)^{-1}.$$

The geometrical meaning of g becomes clear when we choose α to some constant value. For $\alpha = 1$, $g = i\sigma_3$ $(u = v = 0)$.

We can easily get the old metric from eq. (24) back by inserting in eq. (29) the expressions for u and v from eq. (30). The Lagrangian in eq. (25) with these new variables then becomes

(31) $$\mathscr{L} = \tfrac{1}{2}\,\mathrm{Tr}\,(\partial_\mu g^{-1}\partial^\mu g) = [(1 + uv)\,\alpha^{-2}\partial_\mu\alpha\partial^\mu\alpha + \tfrac{1}{2}u^2\partial_\mu v\partial^\mu v + \\ + \tfrac{1}{2}v^2\partial_\mu u\partial^\mu u - (2 + uv)\,\partial_\mu u\partial^\mu v](1 + uv)^{-2}.$$

We can now impose similar restrictions on u and v that were derived for Y and W in eqs. (27) and (28) so as to restrict g in eq. (29) to those satisfying

(32) $$\partial_- u = \partial_+ v = -0\,, \qquad \partial_\pm\alpha = 0\,.$$

This restriction could seem artificial here, but it turns out that it is the same (see eq. (36) below) as the previous restriction on $A_\pm$ in eq. (5).

Then the Lagrangian becomes

(33) $$\mathscr{L} = \tfrac{1}{2}(2 + uv)(1 + uv)^{-2}\partial_+ u\partial_- v$$

and the Euler-Lagrange equation derived from eq. (33) is

(34) $$\partial_+\partial_- u = \partial_+\partial_- v = 0\,,$$

which of course is automatically satisfied as long as eq. (32) holds. (In order to avoid confusion we should add that once the conditions in eq. (32) are imposed they imply that Y and W no longer satisfy eq. (27).)

We can now again choose a pure gauge potential

(35) $$A_\pm \equiv g^{-1}\partial_\pm g\,.$$

The constraints of eq. (32) will then restrict this gauge potential into the form

(36) $$A_\pm = \pm\tfrac{1}{2}\partial_\pm\varphi_\mp\,\sigma_3 + \exp\,[\varphi_\pm]\,\sigma_\pm$$

as in eq. (17), and where

(37) $$\exp\,[\varphi_+] = -\frac{\partial_+ u}{1 + uv}\,, \qquad \exp\,[\varphi_-] = \frac{\partial_- v}{1 + uv}\,.$$

We have already shown that the flatness condition of $A_{\pm}$ in eq. (35) guarantees that ϱ defined by

$$\varrho = \varphi_{+} + \varphi_{-} = \ln \frac{-\partial_{+}u\,\partial_{-}v}{(1+uv)^{2}} \tag{38}$$

will satisfy the Liouville equation

$$\partial_{+}\partial_{-}\varrho = 2\exp[\varrho]\,.$$

Moreover, we can derive the Liouville equation from the Lagrangian in eq. (33) by substituting

$$\varphi_{\pm} = \tfrac{1}{2}\varrho \pm 2i\beta \tag{39}$$

into the Lagrangian equivalent to eq. (33)

$$\mathscr{L} = \tfrac{1}{2}\partial_{+}\varphi_{-}\partial_{-}\varphi_{+} - \exp[\varphi_{+}+\varphi_{-}] \tag{40}$$

and we obtain

$$\mathscr{L} = \frac{1}{8}\partial_{+}\varrho\,\partial_{-}\varrho - \frac{i}{2}(\partial_{-}\varrho\,\partial_{+}\beta - \partial_{-}\beta\,\partial_{+}\varrho) + 2\partial_{+}\beta\,\partial_{-}\beta - \exp[\varrho]\,. \tag{41}$$

The Euler-Lagrange equations for ϱ and β from this Lagrangian are

$$\partial_{+}\partial_{-}\varrho + 4\exp[\varrho] = 0\,, \tag{42}$$

$$\partial_{+}\partial_{-}\beta = 0\,, \tag{43}$$

where ϱ and β are independent field variables.

Again we have obtained the Liouville equation together with a Laplace equation for β. (It is easy to absorb the extra factor 2 in eq. (42) by a suitable scale transformation of the space-time variables). Regarding the two equations above, it is well known that there is a unique correspondence between them such that having found a solution β' to the Laplace equation one can find another solution ϱ' to the Liouville equation by considering the Bäcklund transformation

$$\begin{cases} \partial_{+}(\varrho' - \beta') = a\cdot\exp[\tfrac{1}{2}(\varrho'+\beta')]\,, \\ \partial_{-}(\varrho' + \beta') = \dfrac{2}{a}\cdot\exp[\tfrac{1}{2}(\varrho'-\beta')] \end{cases} \qquad (a \text{ is a constant})\,. \tag{44}$$

Altogether we can conclude from this study so far that having found a solution to the σ-model in terms of Y and W we can find the similar solutions for the Liouville system simply by replacing u and v by Y and W in eq. (37) in order to find the corresponding field variables ϱ and β given by

$$\varrho = \ln\left[\frac{-\partial_+ Y \partial_- W}{(1+YW)^2}\right], \qquad \beta = \frac{i}{4}\ln\left[\frac{-\partial_- W}{\partial_+ Y}\right], \tag{45}$$

so this relation can be considered as a kind of Bäcklund transformation between the two models.

5. – Geometrical interpretation.

At this point a few geometrical remarks are in order. We have introduced the σ-model on SU_2, $g \in SU_2$, and with the nonlinear constraint on N in eqs. (9), (10):

$$N^i N_i = 1,$$

where

$$\mathscr{L} = \tfrac{1}{4}\,\mathrm{Tr}\,(\partial_\mu N \partial^\mu N), \qquad N = N^i \sigma_i \quad \text{or} \quad N = \hat{g}\sigma_3\hat{g}^{-1},$$

we have restricted the σ-model to a coset space $SU_2/U_1 \sim$ the two-sphere $S^2(N^i N_i = 1)$. N remains unchanged under a gauge transformation.

Going from the σ-model to the Liouville model, that was previously introduced as a $SL_{2,C}$ gauge theory, we should restrict g from $SL_{2,C}$ to SU_2. Then the ϱ-field becomes real: $\bar{\varrho} = \varrho$ and $f^+ = -f^-$, $A_+ = -A_-^+$ in eq. (5). The $SL_{2,C}$ gauge freedom ($U \sim \exp[\alpha H]$, $\alpha \in \mathbf{C}$) is then restricted to $\alpha \in \mathbf{R}$, *i.e.* a U_1 gauge freedom. Similarly the self-dualness of the σ-model will correspond to triangularness of the pure gauge potential $g^{-1}\partial_\pm g = A_\pm$ in the Liouville case. Since $\varrho = \ln[-\partial_+ u \partial_- v/(1+uv)^2]$, it corresponds to the conformal ratio of the conformal map (harmonic map: σ-model) from the x_μ-plane to the sphere $S^2(N^2 = 1)$.

6. – The instanton solutions and the Liouville model.

An interesting problem would now be to find topological instanton solutions in the Liouville model similar to the ones already known in the σ-model. We do not know of the existence of such topologically stable solitonic solutions in the pure Liouville model, but it is unlikely since the only stable classical solution with minimal energy is $\varrho = -\infty$. However, by appealing to the above-described correspondence we might get similar topological configurations in the extended Liouville system described by the Lagrangian in eq. (41), as we know from the σ-model.

Therefore start with the Belavin-Polyakov instanton solution that satisfy eq. (27):

$$Y = \prod_i^q c_i \frac{x_+ - a_i}{x_+ - b_i}, \qquad W = \prod_i^q \bar{c}_i \frac{x_- - \bar{a}_i}{x_- - \bar{b}_i}. \tag{46}$$

Inserting the one-instanton solution ($q = 1$) in eq. (45) we obtain

$$\varrho = \varphi_+ + \varphi_- = \ln(c\bar{c}) + \ln((a-b)(\bar{a}-\bar{b})) - 2\ln((x_+ - b)(x_- - \bar{b})) - \\ - 2\ln\left(1 + \frac{c\bar{c}(x_+ - a)(x_- - \bar{a})}{(x_+ - b)(x_- - \bar{b})}\right) + \ln(-1) \tag{47}$$

and

$$\beta = \frac{i}{4}[\varphi_- - \varphi_+] = \frac{i}{4}\left[\ln\frac{\bar{c}}{c} + \ln\left(\frac{\bar{a}-\bar{b}}{a-b}\right) - \ln(-1) + 2\ln\left(\frac{x_+ - b}{x_- - \bar{b}}\right)\right]. \tag{48}$$

It is easy to verify that ϱ satisfies the Liouville equation $\partial_+\partial_-\varrho = 2\exp[\varrho]$ and β satisfies the Laplace equation.

By looking at the radial and phase part of

$$\varphi_\pm = \sqrt{\frac{\varrho'^2}{4} + \beta'^2 \cdot 4} \cdot \exp\left[\pm i \operatorname{tg}^{-1}\frac{4\beta'}{\varrho'}\right], \tag{49}$$

where $\varrho' = \varrho - i\pi$, $\beta' = \beta + \pi/4$, one can easily see that the winding parameter is given by

$$\eta = \operatorname{tg}^{-1}\frac{4\beta'}{\varrho'}. \tag{50}$$

Let us calculate the topological charge q for the instantonlike solutions in eqs. (46), (16a).

The calculation of the topological charge q can be generalized and also simplified in the case of SU_2/U_1 as follows: Let

$$\cos\theta = f(|W|), \qquad \psi = \frac{i}{2}\ln\frac{W}{\overline{W}} \qquad \text{and} \qquad W = c\prod_i (x - a_i)^{m_i} \prod_i (x - b_i)^{-n_i}.$$

Then, with $x = x_0 + r\exp[i\varphi]$,

$$q = \frac{1}{4\pi}\int d\cos\theta\, d\psi = \frac{1}{4\pi}\int dr\, d\varphi\left[\frac{\partial}{\partial r}\left(\cos\theta\frac{\partial\psi}{\partial\varphi}\right) - \frac{\partial}{\partial\varphi}\left(\cos\theta\frac{\partial\psi}{\partial r}\right)\right] =$$

$$= \frac{1}{4\pi}\int_0^{2\pi} d\varphi\left[f(|W|)\frac{\partial\psi}{\partial\varphi}\right]_{r=0}^{\infty} - \frac{1}{4\pi}\int_0^{\infty} dr\left[f(|W|)\frac{\partial\psi}{\partial r}\right]_{\varphi=0}^{2\pi} =$$

$$= -\frac{1}{8\pi}\left\{\sum_i m_i \int_0^{2\pi} d\varphi \left[f(|W|)\left(\frac{r\exp[i\varphi]}{r\exp[i\varphi]+x_0-a_i}+\frac{r\exp[-i\varphi]}{r\exp[-i\varphi]+\bar{x}_0-\bar{a}_i}\right)\right]_{r=0}^{\infty} - \right.$$

$$\left. - \sum_i n_i \int_0^{2\pi} d\varphi \left[f(|W|)\left(\frac{r\exp[i\varphi]}{r\exp[i\varphi]+x_0-b_i}+\frac{r\exp[-i\varphi]}{r\exp[-i\varphi]+\bar{x}_0-\bar{b}_i}\right)\right]_{r=0}^{\infty}\right\}.$$

In each term of the summations we can choose $x_0 = a_i$ for the first line and $x_0 = b_i$ for the second line, so that we get

$$q = \tfrac{1}{2}\left\{\sum_i m_i[f(|W(\infty)|) - f(0)] - \sum_i n_i[f(|W(\infty)|) - f(\infty)]\right\} =$$

$$= \begin{cases} -\tfrac{1}{2}\sum_i m_i(f(\infty) - f(0)) & \text{if } W(\infty) = \infty\,, \\ -\tfrac{1}{2}\sum_i n_i(f(\infty) - f(0)) & \text{if } W(\infty) = 0 \end{cases}$$

$$= -\tfrac{1}{2}\max\left(\sum_i m_i, \sum_i n_i\right)(f(\infty) - f(0))\,.$$

In the case of the *σ-model*, $f(|W|) = (1 - |W|^2)/(1 + |W|^2)$ and thus we get

$$q = \max\left(\sum_i m_i, \sum_i n_i\right),$$

while in the case of the *Liouville model*, $f(|W|) = +1/\sqrt{1+|W|^2}$ and we get

$$q = \tfrac{1}{2}\max\left(\sum_i m_i, \sum_i n_i\right).$$

This shows that the instanton solutions in the Liouville model have fractional charge.

Let us finally calculate the energy of a one-instanton solution in the Liouville model. Although the energy of the classical ground-state (excluding $-\infty$) solutions is known to be infinite in the usual Liouville model one could hope that this situation will be improved with the present Lagrangian in eq. (41) (including the β-field) when we insert a one-instanton solution. In fact we can show much more general that the Hamiltonian becomes zero for the classical ground state of the two independent fields u and v from eq. (33).

Starting from the Lagrangian in eq. (33)

$$\mathscr{L} = \frac{2 + uv}{(1 + uv)^2}\,\partial_+ u\partial_- v + \partial_- u\partial_+ v)\,,$$

we calculate $\pi_u = \delta\mathscr{L}/\delta\partial_2 u$, $\pi_v = \delta\mathscr{L}/\delta\partial_2 v$ and construct the Hamiltonian (in the naive approach)

$$H = \pi_u \partial_2 u + \pi_v \partial_2 v - \mathscr{L} = -\frac{1}{2}\frac{(2+uv)}{(1+uv)^2}(\partial_+ u \partial_+ v + \partial_- u \partial_- v) = 0$$

which is zero (*).

To conclude, a desirable connection has been established in this paper between the σ-model and an extended Liouville model, extended by an extra scalar field that satisfies the Laplace equation without altering much the dynamics of the theory. Finally fractionally charged instanton solutions have been found in such a Liouville model.

* * *

We wish to thank Drs. K. NARAIN, O. FODA and H. B. NIELSEN for helpful remarks and Profs. ABDUS SALAM and P. BUDINICH for hospitality.

(*) That the classical energy is zero is also expected from classical field theory since the self-duality conditions implies that $E = iH$, thus making the energy $\int (E^2 + H^2) dx_1 = 0$.

● RIASSUNTO (*)

Si deriva il modello σ e quello di Liouville da una base geometrica comune. Perciò, si possono derivare soluzioni dell'equazione di Liouville da soluzioni del modello σ e vice versa. Le due variabili indipendenti di campo nel modello σ su uno spazio di coinsieme corrispondono quindi a due variabili simili in un modello esteso di Liouville, una delle quali è la consueta variabile stringa, l'altra corrisponde a un parametro topologico di avvolgimento. Si mostra in fine che questo modello di Liouville ha soluzioni istantoniche di carica frazionaria topologica ed energia finita.

(*) *Traduzione a cura della Redazione.*

Соответствие между σ моделью и моделью Лиувилля.

Резюме (*). — σ-модель и модель Лиувилля выводятся из общего геометрического базиса. Следовательно, решения уравнения Лиувилля могут быть выведены из решений σ-модели и, наоборот. Две независимых полевых переменных в σ-модели на пространстве смежного класса соответствуют двум аналогичным переменным в обобщенной модели Лиувилля, причем одна из переменных является обычной «переменной струны», а другая соответствует топологическому параметру «обмотки». В заключение, показывается, что модель Лиувилля имеет инстантонные решения с дробным топологическим зарядом и конечной энергией.

(*) *Переведено редакцией.*

Some series of infinitely many symmetry generators in symmetric space chiral models

Bo-yu Hou[a)]
Institute for Theoretical Physics, State University of New York at Stony Brook, Stony Brook, New York 11794

(Received 4 October 1983; accepted for publication 6 January 1984)

This paper shows that, starting from any conserved current generated by some given infinitesimal symmetry generator, one may use finite dual transformations to induce infinitely many infinitesimal symmetry generators. Thus, besides starting from ordinary isotopic and space-time translation, this paper also discovers the infinitesimal generators for Bäcklund transformation, for dual symmetry itself and other general cases, and then uses them to generate infinitely many local or nonlocal currents, respectively.

PACS numbers: 11.30. – j, 11.10.Lm

I. INTRODUCTION

In this decade, there has been much interest and considerable progress in the nonlinear physical systems and in the nonlinear mathematics. The two dimensional chiral model[1–6] is one of the nonlinear problems under extensive investigation. The chiral model behaves quite similarly with the four-dimensional Yang–Mills field, e.g., both have topologically nontrivial solutions such as instantons and merons[7] both possess some kind of BT (Bäcklund transformation) with similar structures.[1,5–10] It is reasonable to expect that the thorough investigation of this simpler model will be helpful for deeper understanding of the more complicated Yang–Mills field. As a complete integrable system solvable by inverse scattering method, the chiral model possesses a lot of rather interesting and mutually connected properties such as multisoliton solutions, BT, and sets of infinitely many conserved currents, either local or nonlocal.[11–19] What is the hidden symmetry behind so much conservation laws is a crucial question to answer for understanding the structure of the solution space of the chiral model. A lot of work already shows that this phenomenon is closely related with dual symmetry. Results of previous papers[18] show that speaking more exactly the infinitesimal operator generating nonlocal currents is nothing else but the ordinary isospin generator T transformed by DT (dual transformation) $U(x;\gamma)$. The DT with parameter γ is the origin of the existence of infinitely many symmetries. Since a $U(x;\gamma_1)$ with a fixed γ_1 gives one automorphism in solution space, it maps one known explicit symmetry (e.g., constant T) into another hidden symmetry $U^{-1}(x;\gamma)TU(x;\gamma_1)$ generating a conserved current $J_\mu(x;\gamma_1)$ (cf. Sec. III). From the same T but with different parameter γ we get different symmetries $U^{-1}(x;\gamma)TU(x;\gamma)$ generating different currents $J_\mu(x;\gamma)$. In summary, dual symmetry is the symmetry which induces infinitely many symmetries and maps different currents, but itself is not the symmetry which generates the conserved currents J_μ.

Accordingly, this paper tries at first to find out the infinitesimal variations which leave the Lagrangian unchanged, then takes the DT and thus gets the corresponding set of infinitely many symmetry operators generating conserved currents. In this way, after review the results about dual transformed T shortly in Sec. III, we give subsequently in Sec. IV the current which corresponds to the infinitesimal generator of dual symmetry itself. We show that it is a Noether current and a dynamical symmetry of the equation of motion. The infinitesimal BT plays an important role in the soliton equation. In Sec. V we find the infinitesimal BT. For chiral model, it is given by the solution of a matrix Riccati equation; we show also the local current is just the related Noether current. In Sec. VI, we give the infinitesimal generators and Noether currents for more general cases, including the ordinary space-time translation and energy momentum density.

Since the finite dual transformation is quite well known now, the main role of the second section consists in introducing notations. By the way, deviating from the current conventions, which deal with gauge transformations within the isotropic subgroup H only, we discuss somehow in detail the gauge transformations in the whole group G, so that the different formulations may be treated as gauge equivalent expressions and the distinction and relation between the connections, the second fundamental forms, and the invariantly conserved currents are clarified. We use the local involutive operator $N(x) = g(x)ng^{-1}(x)$ of the symmetric space as the dynamical variable, so that our formulation essentially includes the O(N) nonlinear σ-model, the CP($N-1$) model, the Grassman chiral model, and the principle chiral model.

II. CHIRAL MODEL IN VARIOUS GAUGES, DUAL SYMMETRY

A. Symmetric space and canonical variable

The chiral field may be defined as a map from space time x_μ ($\mu = 0,1$) onto a symmetric space (G, H, n), i.e., a coset space G/H with involutive automorphism n,

$$H = \{h \in G; nhn = h\}, \quad n^2 = 1, \tag{2.1}$$

where G is a connected Lie group with Lie algebra $\mathfrak{G}$ and $H \subset G$ is a closed subgroup with Lie algebra $\mathfrak{H}$. In the adjoint representation the same matrix n gives involutive automorphism for the Lie algebra also

$$[n,\mathfrak{H}] = 0, \quad \{n,\kappa\} = 0, \tag{2.2}$$

---

a) Permanent address: Northwest University, Xian, People's Republic of China.

 0022-2488/84/072325-06$02.50

where

$$\mathfrak{H} \oplus \kappa = \mathfrak{G},$$

$$[\mathfrak{H},\mathfrak{H}] \subset \mathfrak{H}, \quad [\mathfrak{H},\kappa] \subset \kappa, \quad [\kappa,\kappa] \subset \mathfrak{H}. \tag{2.3}$$

The elements of G/H are represented by canonical variables

$$N(x) = g(x) n g^{-1}(x), \quad N(x)^2 = 1. \tag{2.4}$$

Then, if g_1 and g_2 are in the same coset class, $g_1 = g_2 h$, hence $g_1 n g_1^{-1} = g_2 n g_2^{-1}$, both correspond to the same N.

B. Gauges with diagonal connections

The left Maurer Cartan form is divided into vertical (connection) and horizontal (second fundamental form) parts and pulled back onto x space

$$a_\mu(x) = g^{-1}(x)\partial_\mu g(x) = h_\mu(x) + k_\mu(x), \tag{2.5}$$

where

$$[h_\mu(x), n] = 0, \tag{2.6}$$

$$\{k_\mu(x), n\} = 0, \tag{2.7}$$

$$h_\mu(x) = \tfrac{1}{2}[g^{-1}(x)\partial_\mu g(x) + n g^{-1}(x)\partial_\mu g(x) n], \tag{2.8}$$

$$k_\mu(x) = \tfrac{1}{2}[g^{-1}(x)\partial_\mu g(x) - n g^{-1}(x)\partial_\mu g(x) n], \tag{2.9}$$

in this gauge h_μ is diagonal with respect to n. The pure gauge a_μ has zero curvature $a_{\mu\nu}(x) = \partial_\mu a_\nu - \partial_\nu a_\mu + [a_\mu, a_\nu] = 0$; it may be divided into the Gauss equation

$$\begin{aligned}\tfrac{1}{2}[a_{\mu\nu}(x) + n a_{\mu\nu}(x) n] &= \partial_\mu h_\nu(x) - \partial_\nu h_\mu(x)\\ &\quad + [h_\mu(x), h_\nu(x)] + [k_\mu(x), k_\nu(x)]\\ &\equiv f_{\mu\nu}(x) + [k_\mu(x), k_\nu(x)] = 0\end{aligned} \tag{2.10}$$

and the Coddazi equation

$$\begin{aligned}\tfrac{1}{2}[a_{\mu\nu}(x) - n a_{\mu\nu}(x) n] &= \partial_\mu k_\nu(x) + [h_\mu(x), k_\nu(x)]\\ &\quad - \partial_\nu k_\mu(x) - [h_\nu(x), k_\mu(x)]\\ &\equiv D_\mu k_\nu(x) - D_\nu k_\mu(x) = 0.\end{aligned} \tag{2.11}$$

C. General gauge transformation

$$h'_\mu(x) = S^{-1}(x) h_\mu(x) S(x) + S^{-1}(x)\partial_\mu S(x),$$

$$k'_\mu(x) = S^{-1}(x) k_\mu(x) S(x). \tag{2.12}$$

Usually S is restricted in H, then all relations (2.6)–(2.11) remains unchanged. If we allow $S(x)$ to be any element in G, then only (2.5), (2.7), (2.10), and (2.11) still remain valid, but the n therein must be replaced by $n'(x) = S^{-1}(x) n S(x)$; meanwhile, instead of the diagonal of h_μ (2.6) and $\partial_\mu n = 0$, we have a covariant condition

$$D'_\mu n'(x) \equiv \partial_\mu n'(x) + [h'_\mu(x), n'(x)] = 0, \tag{2.13}$$

i.e., the reducibility condition for h'_μ [20]: "if there exists on the coset bundle $\{\{x\} G/H, G\}$ a section $n'(x)$ invariant under parallel displacement with respect to h'_μ, then the h'_μ are reducible to a connection in H."

D. Canonical gauge

Choosing $S^{-1}(x) = g(x)$, it occurs that both expressions in (2.12) are expressed solely by the canonical variable $N(x)$ in (2.4); thus

$$H_\mu(x) = \tfrac{1}{2} N(x)\partial_\mu N(x), \tag{2.14}$$

$$K_\mu(x) = -\tfrac{1}{2} N(x)\partial_\mu N(x). \tag{2.15}$$

In summary, we have the flat Gauss Coddazi equation

$$\begin{aligned}F_{\mu\nu}(x) &\equiv \partial_\mu H_\nu(x) - \partial_\nu H_\mu(x) + [H_\mu(x), H_\nu(x)]\\ &= -[K_\mu(x), K_\nu(x)],\end{aligned} \tag{2.16}$$

$$\epsilon^{\mu\nu} D_\mu K_\nu(x) \equiv \epsilon^{\mu\nu}(\partial_\mu K_\nu(x) + [H_\mu(x), K_\nu(x)]) = 0,$$

$$\epsilon^{10} = -\epsilon^{01} = 1, \tag{2.17}$$

the reducibility condition

$$D_\mu N(x) = \partial_\mu N(x) + [H_\mu(x), N(x)] = 0, \quad N(x)^2 = 1, \tag{2.18}$$

and the local involutive condition for K_μ

$$\{K_\mu(x), N(x)\} = 0. \tag{2.19}$$

All equations (2.16)–(2.19) are gauge-covariant under (2.12). In addition we have chosen the canonical gauge condition

$$A_\mu(x) = H_\mu(x) + K_\mu(x) = 0; \tag{2.20}$$

then, from (2.18) and (2.19), we get the expressions of H_μ, K_μ in terms of N as (2.14) and (2.15).

It is interesting to point out that, complementary to the diagonal gauge (1.6), now

$$\{H_\mu(x), N(x)\} = 0. \tag{2.21}$$

Connection $H_\mu(x)$ is fixed by gauge condition (2.21), but we may further change the canonical gauge without breaking (2.21) by using $S = \exp(i\theta N(x))$, where θ is a constant parameter; then $K'_\mu(x) = \tfrac{1}{2}(\cos 2\theta N(x)\partial_\mu N(x) - i\sin 2\theta\partial_\mu N(x))$, e.g., $\theta = \tfrac{1}{4}\pi$, $K'_\mu(x) = H'_\mu(x) = \tfrac{1}{2} N(x)\partial_\mu N(x) = \tfrac{1}{2} A'_\mu(x)$.

E. Dynamics

Let Lagrangian

$$L(x) = \tfrac{1}{8}\mathrm{tr}(\partial_\mu N(x)\partial^\mu N(x)), \quad N(x)^2 = 1 \tag{2.22}$$

and with some further constraints. The Euler–Lagrangian equation

$$[\partial_\mu \partial^\mu N(x), N(x)] = 0 \tag{2.23}$$

may be expressed in K_μ as

$$\partial_\mu K^\mu(x) = 0, \tag{2.24}$$

or rewritten into covariant form

$$D_\mu K^\mu(x) \equiv \partial_\mu K(x) + [H_\mu(x), K^\mu(x)] = 0. \tag{2.25}$$

F. Intermediate DT, $K_\mu(x) \to \tilde{K}_\mu(x;\gamma)$

Since (2.17) and (2.25) are mutually dual in two-dimensional space-time, it is obvious that (2.16)–(2.19) and (2.25) are invariant under DT:

$$\begin{aligned}K_\mu(x) &\to \tilde{K}_\mu(x;\gamma)\\ &= K_\mu(x)(\gamma + \gamma^{-1})/2 + \epsilon_{\mu\nu} K^\nu(x)(\gamma - \gamma^{-1})/2,\\ &\equiv K_\mu(x)\cosh\phi + \epsilon_{\mu\nu} K^\nu(x)\sinh\phi,\end{aligned} \tag{2.26}$$

$$H_\mu(x) \to \widetilde{H}_\mu(x;y) = H_\mu(x). \tag{2.27}$$

Thus, we have $\widetilde{F}_{\mu\nu}(x;\gamma) = \partial_\mu \widetilde{H}_\nu(x;\gamma) - \partial_\nu \widetilde{H}_\mu(x;\gamma) + [\widetilde{H}_\mu(x;\gamma), \widetilde{H}_\nu(x;\gamma)] = -[\widetilde{K}_\mu(x;\gamma), \widetilde{K}_\nu(x,\gamma)]$ as (2.$\tilde{1}$6), and (2.$\tilde{1}$7)–(2.$\tilde{1}$9) and (2.$\tilde{2}$5) by replacing $K_\mu(x)$ in (2.17)–(2.19) and (2.25) in terms of $\widetilde{K}_\mu(x;\gamma)$. Since the explicitly pure condition (2.20) has been broken, $\widetilde{H}_\mu$, $\widetilde{K}_\mu$ could not be expressed directly by some $\widetilde{N}$ as in (2.14) and (2.15). But from Eqs. (2.$\tilde{1}$6) and (2.$\tilde{1}$7), $\widetilde{A}_\mu(x;\gamma) = \widetilde{H}_\mu + \widetilde{K}_\mu$ are pure gauge, so we may discover some new $N(x;\gamma)$ which satisfies the dynamical Eq. (2.23) as follows.

G. Final dual transformation $N(x) \to N(x;\gamma)$

Equations (2.$\tilde{1}$6) and (2.$\tilde{1}$7) show that $\widetilde{A}_\mu(x;\gamma)$ are pure gauge; therefore there exists an $U(x;\gamma)$ such that

$$U^{-1}(x;\gamma)\,\partial_\mu\, U(x;\gamma) = \widetilde{A}_\mu(x;\gamma) \equiv \widetilde{K}_\mu(x;\gamma) + \widetilde{H}_\mu(x;\gamma) \tag{2.28}$$

or

$$\partial_\mu\, U(x;\gamma) = U(x;\gamma)(\widetilde{K}_\mu(x;\gamma) - K_\mu(x)). \tag{2.29}$$

If we gauge transform $\widetilde{H}_\mu(x;\gamma)$ with $S(x) = U^{-1}(x;\gamma)$, i.e., let

$$H_\mu(x;\gamma) = U(x;\gamma)\widetilde{H}(x;\gamma)U^{-1}(x;\gamma) + U(x;\gamma)\,\partial_\mu\, U^{-1}(x;\gamma), \tag{2.30}$$

$$K_\mu(x;\gamma) = U(x;\gamma)\widetilde{K}(x;\gamma)U^{-1}(x;\gamma). \tag{2.31}$$

Then, using (2.28), we get

$$A_\mu(x;\gamma) = H_\mu(x;\gamma) + K_\mu(x;\gamma) = 0. \tag{2.20γ}$$

Now, gauge covariant equations (2.$\tilde{1}$6)–(2.$\tilde{1}$9), (2.$\tilde{2}$5) become (2.16γ)–(2.19γ), (2.25γ) after substituting:

$$\begin{aligned}&\widetilde{H}_\mu(x;\gamma) \to H_\mu(x;\gamma), \quad \widetilde{K}_\mu(x;\gamma) \to K_\mu(x,\gamma),\\ &\widetilde{D}_\mu = D_\mu \to D_\mu(\gamma) \equiv \partial_\mu + [H_\mu(x;\gamma)],\\ &\widetilde{N}(x;\gamma) \equiv N(x) \to N(x;\gamma),\end{aligned} \tag{2.32}$$

where $N(x;\gamma) \equiv U(x;\gamma)N(x)U^{-1}(x;\gamma)$. In gauge (2.20$\gamma$), Eq. (2.25$\gamma$), $D_\mu(\gamma)K^\mu(x;\gamma) = 0$, may be simplified as

$$\partial^\mu K_\mu(x;\gamma) = 0. \tag{2.24γ}$$

From (2.20γ) and (2.18γ) we have

$$H_\mu(x;\gamma) = \tfrac{1}{2} N(x;\gamma)\,\partial_\mu N(x;\gamma), \tag{2.14γ}$$

$$K_\mu(x;\gamma) = -\tfrac{1}{2} N(x;\gamma)\,\partial_\mu N(x;\gamma). \tag{2.15γ}$$

We may check (2.14γ) and (2.15γ) directly by substituting on their right-hand sides (2.32) and then use (2.29), (2.$\tilde{1}$9), (2.19), (2.30), or (2.31) to attain the left-hand side. Compare (2.32) with (2.4); we see that if $g(x;\gamma) = U(x;\gamma)g(x)$, then $N(x;\gamma) = g(x;\gamma)ng^{-1}(x;\gamma)$. Finally from (2.24$\gamma$) and (2.15$\gamma$) we get the dual transformed EL equation (2.23γ). (Equations labeled with γ, are just the same equation, only with N, H_μ, K_μ replaced by $N\langle\gamma\rangle$, $H_\mu\langle\gamma\rangle$, $K_\mu\langle\gamma\rangle$.)

In the latter we shall adopt following abbreviations:

$$\begin{aligned}&H_\mu \equiv H_\mu(x),\quad K_\mu \equiv K_\mu(x),\quad N \equiv N(x),\quad \widetilde{K}_\mu \equiv \widetilde{K}_\mu(x;\gamma),\\ &N\langle\gamma\rangle \equiv N(x;\gamma),\quad H_\mu\langle\gamma\rangle \equiv H_\mu(x;\gamma),\\ &K_\mu\langle\gamma\rangle \equiv K_\mu(x;\gamma),\quad U \equiv U(x;\gamma).\end{aligned}$$

III. DUAL TRANSFORMATION OF ISOTOPIC SYMMETRY OPERATOR

A. Ordinary generator for conserved current

Let

$$\delta N(x) = -[N(x), \Lambda(x)]\delta\epsilon. \tag{3.1}$$

(For simplicity, we omit the infinitesimal constant $\delta\epsilon$ in the future.) Then

$$\delta L = \mathrm{tr}(K_\mu\, \partial^\mu \Lambda). \tag{3.2}$$

Define

$$j^\mu(x) = \frac{\delta L}{\delta \partial_\mu N}\,\delta N = \mathrm{tr}(K^\mu \Lambda). \tag{3.3}$$

Its on-shell (2.24) divergence equals

$$\partial_\mu j^\mu = \mathrm{tr}(K_\mu\, \partial^\mu \Lambda) = \delta L. \tag{3.4}$$

If we have chosen $\Lambda(x)$ such that

$$\mathrm{tr}(K_\mu\, \partial^\mu \Lambda) = 0. \tag{3.5}$$

Then $j_\mu(x)$ is conserved

$$\partial_\mu\, J^\mu(x) = 0. \tag{3.6}$$

For example, let $\Lambda(x) \equiv T$, where T is a constant element in g. Then

$$J_\mu = \mathrm{tr}(K_\mu\, T) \tag{3.7}$$

is a conserved current.

B. Dual transformed current

Heuristically, in the dual transformed functional space with canonical variable $N(x;\gamma)$, let $L(x;\gamma) = \frac{1}{8}\,\mathrm{tr}(N\langle\gamma\rangle N\langle\gamma\rangle)$; take $\delta N\langle\gamma\rangle = -[N\langle\gamma\rangle, T]$. We get $J_\mu\langle\gamma\rangle = \mathrm{tr}(K_\mu\langle\gamma\rangle T)$, (3.7$\gamma$), which is conserved because of (1.24γ). Expanding $J_\mu\langle\gamma\rangle$ into series of γ; we get an infinite series of conserved nonlocal currents.

C. Dual transformed generator

Now, return to the original functional space. Tentatively, neglecting the dependence of $U^{-1}(x;\gamma)$ on $T\delta\epsilon$ via $K_\mu(x;\gamma)$, assume

$$\begin{aligned}\delta N(x) &= U^{-1}(x;\gamma)\delta N(x;\gamma)U(x;\gamma)\\ &= -[N(x), U^{-1}(x;\gamma)TU(x;\gamma)];\end{aligned} \tag{3.8}$$

subsequently, $j_\mu(x;\gamma) = \mathrm{tr}(K_\mu(x)U^{-1}(x;\gamma)TU(x;\gamma))$, but it occurs to us that now its on shell divergence

$$\partial_\mu j^\mu(x;\gamma) = \delta L = \mathrm{tr}(K_\mu\, \partial^\mu(U^{-1}TU)) \neq 0. \tag{3.9}$$

However, using (2.29), (2.26), and (2.17), one may show that

$$\begin{aligned}\delta L &= -\sinh\phi\, \mathrm{tr}(\epsilon_{\mu\nu}\, K^\mu K^\nu T)\\ &= -\tanh\phi\, \partial_\mu\, \mathrm{tr}(\epsilon^{\mu\nu} K_{\mu\nu}\, U^{-1}TU)\\ &\equiv \partial_\mu\, i^\mu(x;\gamma).\end{aligned} \tag{3.10}$$

Put (3.10) together with (3.9); we regain the conserved current (3.7γ) with some coefficient, i.e.,

$$\begin{aligned}J_\mu(x;\gamma) &\equiv j_\mu(x;\gamma) + i_\mu(x;\gamma) = \mathrm{sech}\,\phi\, \mathrm{tr}(\widetilde{K}_\mu\, U^{-1}TU)\\ &= \mathrm{sech}\,\phi\, \mathrm{tr}(K_\mu(\gamma)T),\end{aligned} \tag{3.11}$$

where (2.31) has been used.

Thus, we see that just as $j\mu$ (3.7) is related to the symmetry of rotation $\delta\epsilon$ around the fixed T axis, $J_\mu\langle\gamma\rangle$ (3.7γ) or (3.11) is related to the rotation $\delta\epsilon$ around the transformed axis $U^{-1}TU$.

The variation (3.8) satisfies the condition for invariance of EL equation (2.23) under δN

$$D_\mu D^\mu[N,\delta N] = [K_\mu,[K^\mu,[N,\delta N]]] \quad (3.12)$$

or, using (3.1),

$$D_\mu D^\mu A - [K_\mu,[K^\mu,A]] - N(D_\mu D^\mu A - [K_\mu,[K^\mu,A]])N = 0. \quad (3.13)$$

At last, we emphasize that K_μ does not conserve invariantly with respect to local gauge transformation, as a covariant quantity; it conserves only covariantly (2.25) in general gauge. The true invariantly conserved currents are always gauge-invariant quantities such as projections of K_μ on T or $\widetilde{K}_\mu$ on $U^{-1}TU$, etc., i.e., $\mathrm{tr}(K_\mu T)$ or $\mathrm{tr}(\widetilde{K}_\mu U^{-1}TU)$, etc. (cf. later sections).

IV. INFINITESIMAL DUAL TRANSFORMATION

A. Finite DT

Under finite DT (2.23), the finite variation of L equals zero

$$\Delta L = \tfrac{1}{8}\,\mathrm{tr}(K_\mu\langle\gamma\rangle K^\mu\langle\gamma\rangle - K_\mu K^\mu) = \tfrac{1}{8}\,\mathrm{tr}(\widetilde{K}_\mu \widetilde{K}^\mu - K_\mu K^\mu) = 0. \quad (4.1)$$

B. Infinitesimal DT

But in order to find out the corresponding conserved currents, we must use the infinitesimal DT operator $u(x)$:

$$u(x) \equiv \left(\gamma \frac{dU(x;\gamma)}{d\gamma} U^{-1}(x;\gamma)\right)\Big|_{\gamma=1} = -\int_{-\infty}^{x_1} K_0(x_0, x_1')\,dx^1. \quad (4.2)$$

It satisfies

$$\partial_\mu u(x) = \epsilon_{\mu\nu} K^\nu(x) \quad (4.3)$$

from (4.2) and (2.24). The covariant form of (4.3) is

$$D_\mu u = -[K_\mu, u] + \epsilon_{\mu\nu} K^\nu. \quad (4.4)$$

Now let

$$\delta N = -[N, u]; \quad (4.5)$$

we have

$$\delta L = \mathrm{tr}(K_\mu \partial^\mu u) = \mathrm{tr}(\epsilon^{\mu\nu} K_\mu K_\nu) = 0. \quad (4.6)$$

(Really, K_μ in L has been changed into its dual $\epsilon_{\mu\nu} K^\nu$.) Therefore,

$$J_\mu = \mathrm{tr}(K_\mu u) \text{ is a conserved current.} \quad (4.7)$$

C. Dual transformed infinitesimal DT

Let

$$\tilde{u}(x;\gamma) = U^{-1}(x;\gamma)u(x;\gamma)U(x;\gamma), \quad (4.\tilde{2})$$

where

$$u(x;\gamma) = -\int_{-\infty}^{x_1} K_0(x_0, x_1';\gamma)\,dx_1' = \gamma \frac{dU}{d\gamma} U^{-1}; \quad (4.2\gamma)$$

they satisfy

$$\partial_\mu u(x;\gamma) = \epsilon_{\mu\nu} K^\nu(x;\gamma), \quad (4.3\gamma)$$

$$D_\mu\langle\gamma\rangle u\langle r\rangle = -[K_\mu\langle\gamma\rangle, u\langle\gamma\rangle] + \epsilon_{\mu\nu} K^\nu\langle\gamma\rangle, \quad (4.4\gamma)$$

$$D_\mu \tilde{u} = -[\widetilde{K}_\mu, \tilde{u}] + \epsilon_{\mu\nu} \widetilde{K}^\nu. \quad (4.\tilde{4})$$

Let

$$\delta N = -[N, \tilde{u}]. \quad (4.\tilde{5})$$

Then

$$j_\mu(x;\gamma) = \mathrm{tr}(K_\mu(x)\tilde{u}(x;\gamma)), \quad (4.8)$$

$$\begin{aligned}\partial_\mu j^\mu(x;\gamma) = \delta L &= \mathrm{tr}(K_\mu \partial^\mu \tilde{u}) \\ &= \sinh\phi\, \mathrm{tr}(-\epsilon^{\mu\nu}[K_\mu, K_\nu]\tilde{u} + K_\mu K^\mu) \\ &= -\tanh\phi\, \partial_\mu\, \mathrm{tr}(\epsilon^{\mu\nu} K_\nu \tilde{u}) \equiv -\partial_\mu i^\mu(x;\gamma).\end{aligned} \quad (4.9)$$

$$\begin{aligned}J_\mu(x;\gamma) \equiv j_\mu(x;\gamma) + i_\mu(x;\gamma) &= \mathrm{sech}\,\gamma\, \mathrm{tr}(\widetilde{K}_\mu \tilde{u}) \\ &= \mathrm{sech}\,\phi\, \mathrm{tr}(K_\mu\langle\gamma\rangle u\langle\gamma\rangle)\end{aligned} \quad (4.10)$$

is conserved.

It is easy to check that the EL equation is invariant under (4.5) by substituting it into (3.12).

In the two-dimensional Euclidean space with self-dual (anti-dual) solution $\partial_\mu N(x) = \pm\epsilon_{\mu\nu} N(x)\partial^\nu N(x)$, we get $u(x) = \pm N(x)$. All these currents are trivial.

V. BÄCKLUND TRANSFORMATION

A. Finite BT

It operates on solution $N(x)$ of (2.23); giving a new solution

$$N'(x|\gamma) = N(x)B(x|\gamma) = B^+(x|\gamma)N(x) \quad (5.1)$$

when N, N', B satisfy

$$2K_\mu - 2K_\mu' \equiv N'\partial_\mu N' - N\partial_\mu N = \epsilon_{\mu\nu}\partial^\nu B, \quad (5.2)$$

$$B(x|\gamma) + B^+(x|\gamma) = -2\tanh\phi. \quad (5.3)$$

Let

$$\widetilde{R} \equiv \tfrac{1}{2}\cosh\phi\,(B(x|\gamma) - B^+(x|\gamma)); \quad (5.4)$$

then from (5.1)–(5.3)

$$D_\mu \widetilde{R} \equiv \partial_\mu \widetilde{R} + [H_\mu, \widetilde{R}] = \epsilon^{\mu\nu}(\widetilde{K}_\nu + \widetilde{R}\widetilde{K}_\nu\widetilde{R}). \quad (5.5)$$

It is integrable from (2.16), (2.17), and (2.25). Conversely, from R, satisfying (5.5), let

$$B(x\gamma) = \sec\phi\widetilde{R} - \tanh\phi = \exp(2(\cot^{-1}\phi)\widetilde{R}); \quad (5.6)$$

we obtain from (5.1) the new solution N'. The BT (5.1), (5.2) satisfies variational Bäcklund principle, i.e., δL equals total divergence:

$$\begin{aligned}\Delta L &= \tfrac{1}{8}\,\mathrm{tr}(K_\mu' K^{\mu\prime} - K_\mu K^\mu) \\ &= -\tfrac{1}{4}\,\mathrm{sech}^2\phi\, tr(K_\mu K^\mu + \widetilde{R}K^\mu\widetilde{R}K_\mu) \\ &= \tfrac{1}{8}\,\mathrm{sech}^3\phi\,\partial_\mu\, \mathrm{tr}(\epsilon^{\mu\nu} K_\nu \widetilde{R}) \\ &= \tfrac{1}{8}\,\mathrm{csch}\,\phi\, \mathrm{tr}\,\partial_\mu(\widetilde{K}^\mu\widetilde{R}).\end{aligned} \quad (5.7)$$

B. Infinitesimal BT

Let

$$\delta N = \left[N, B^{-1}\frac{dB}{d\gamma}\Big|_{\gamma=1}\right]$$

$$= 2\left[N, \frac{-1}{(1+\gamma^2)}B + \cot^{-1}\gamma\frac{dR}{\alpha\gamma}\right]\Big|_{\gamma=1}. \quad (5.8)$$

The contribution of the second term in δL equals

$$2\cot^{-1}\gamma\,\mathrm{tr}\left(K_\mu\,\partial^\mu\frac{dR}{d\gamma}\right)$$

$$= 2\gamma^2\cot^{-1}\gamma\,\mathrm{tr}(K_\mu K^\mu + \widetilde{R}K^\mu\widetilde{R}K_\mu)/(1+\gamma^2), \quad (5.9)$$

which is a total divergence as the rhs of (5.7). Hence, we omit this term, keep the first only. Since we need dual transformed operator later, we replace $B(x|1)$ by

$$\widetilde{R}(x;1) = R(x;1) = B(x;1) \equiv R(x),$$

where

$$R(x;\gamma) = U(x;\gamma)\widetilde{R}(x;\gamma)U^{-1}(x;\gamma). \quad (5.10)$$

It satisfies

$$D_\mu\langle\gamma\rangle R(x;\gamma) = \epsilon_{\mu\nu}(K^\nu(x;\gamma) + R(x;\gamma)K^\nu(x;\gamma)R(x;\gamma)); \quad (5.5\gamma)$$

thus, we take

$$\delta N(x) = [N(x), B(x|1)] = [N(x), R(x)]. \quad (5.11)$$

Since now $\delta L = \mathrm{tr}(K_\mu(x)\partial^\mu R(x)) = 0$, the current $\mathrm{tr}(K_\mu(x)R(x))$ are conserved.

C. Dual transformed BT

Let

$$\delta N(x;\gamma) = [N(x), \widetilde{R}(x;\gamma)], \quad (5.11\gamma)$$

we have

$$j_\mu(x;\gamma) = \mathrm{tr}(K_\mu\widetilde{R}), \quad (5.12)$$

$$\partial_\mu j^\mu(x;\gamma) = \delta L = \mathrm{tr}(K_\mu\,\partial^\mu\widetilde{R})$$

$$= -2\sinh\phi\,\mathrm{tr}(K_\mu K^\mu + K_\mu\widetilde{R}K^\mu\widetilde{R}),$$

$$= -\partial_\mu\,\mathrm{tr}(\epsilon^{\mu\nu}\widetilde{K}_\nu\widetilde{R})\tanh\phi \equiv -\partial_\mu\,i^\mu(x;\gamma). \quad (5.13)$$

Finally, we get the conserved current

$$J^\mu(x;\gamma) \equiv j^\mu(x;\gamma) + i^\mu(x;\gamma) = \mathrm{sech}\,\Phi\,\mathrm{tr}(\widetilde{K}_\mu\widetilde{R})$$

$$= \mathrm{sech}\,\phi\,\mathrm{tr}(K^\mu(x;\gamma)R(x;\gamma)). \quad (5.14)$$

The geometrical meaning are rotations around axis $\widetilde{R}$; $B(x|\gamma)$ are finite rotations with angle $\theta = 2\cot^{-1}\gamma$, while the $\delta N(x;|\gamma)$ are generated by rotation with infinitesimal constant angle $\delta\epsilon$.

VI. GENERAL CASE

Generally, we must find $\Lambda(x)$ such that

$$\mathrm{tr}(\widetilde{K}_\mu D^\mu\Lambda) = 0. \quad (6.1)$$

The most general equation for Λ is

$$D_\mu\Lambda = \alpha\epsilon_{\mu\nu}\widetilde{K}^\nu + \beta\epsilon_{\mu\nu}\Lambda\widetilde{K}^\nu\Lambda$$

$$+ a[\Lambda, \widetilde{K}_\mu] + s\epsilon_{\mu\nu}\{\Lambda, \widetilde{K}^\nu\}; \quad (6.2)$$

it is integrable if $a^2 - s^2 + \alpha\beta = 1$. Let

$$\delta N = [N, \Lambda]. \quad (6.3)$$

Then,

$$j_\mu(x;\gamma) = \mathrm{tr}(K_\mu\Lambda), \quad (6.4)$$

$$\partial_\mu j^\mu(x;\gamma)$$

$$= \delta L = \mathrm{tr}(K_\mu\,\partial^\mu\Lambda) = -\tanh\phi\,\mathrm{tr}(\epsilon^{\mu\nu}K_\nu D_\mu\Lambda)$$

$$= -\tanh\phi\,\mathrm{tr}\,\partial_\mu(\epsilon^{\mu\nu}K_\mu\Lambda) \equiv -\partial_\mu i^\mu, \quad (6.5)$$

so

$$J_\mu \equiv j_\mu(x;\gamma) + i_\mu(x;\gamma) = \mathrm{sech}\,\phi\,\mathrm{tr}(\widetilde{K}_\mu\Lambda)$$

are conserved. This includes all currents discussed above:

if $a = 1$, $\alpha = \beta = s = 0$, $\Lambda = U^{-1}(x;\gamma)TU(x;\gamma)$ in Sec. III;
if $a = \alpha = 1$, $\beta = s = 0$, $\Lambda = \widetilde{u}$ in Sec. IV;
if $\alpha = \beta = 1$, $a = s = 0$, $\Lambda = \widetilde{R}$ in Sec. V;

more generally, if $\mathrm{tr}[K^\mu, D_\mu\Lambda] = \partial_\mu\tilde{l}^\mu$, then let $\delta N = [N, \Lambda]$; we get the conserved current

$$J_\mu = \mathrm{sech}\,\phi\,(\mathrm{tr}(\widetilde{K}_\mu\Lambda) - \tilde{l}_\mu). \quad (6.6)$$

For example, under infinitesimal translation, $\delta N(x;\gamma) = \partial_\nu N(x;\gamma)$. Let

$$\delta N(x) = U^{-1}\delta N(x;\gamma)U = U^{-1}\partial_\nu N(x;\gamma)U$$

$$= -U^{-1}[N(x;\gamma), K_\nu(x;\gamma)]U$$

$$= -[N(x), \widetilde{K}_\nu(x;\gamma)], \quad (6.7)$$

i.e.,

$$\Lambda(x,\gamma) = \widetilde{K}_\nu(x;\gamma). \quad (6.8)$$

Then

$$\mathrm{tr}(\widetilde{K}_\mu D^\mu\widetilde{K}_\nu) = -\tfrac{1}{2}\partial_\mu\,\mathrm{tr}(\widetilde{K}_\mu\widetilde{K}^\mu) \equiv -\partial_\mu\tilde{l}_\mu. \quad (6.9)$$

The current (6.6) becomes energy momentum density $M_{\mu\nu}$

$$J_\mu = \mathrm{sech}\,\phi\,\mathrm{tr}(K_\mu K_\nu - \tfrac{1}{2}g_{\mu\nu}K_\lambda K^\lambda).$$

VII. DISCUSSION

Thus, we formulate a general way to get infinitely many Noëther currents from any given Noëther current.

If we expand the generator $U^{-1}\langle\gamma\rangle TU\langle\gamma\rangle$ in series of the parameter $\lambda \equiv (\gamma-1)/(\gamma+1)$, we would obtain the series of generators of the so-called Kac–Moody algebra.[21] Meanwhile, to get the recurrence formulas for each order, one may simply use $\partial_\mu(U^{-1}TU) = \lambda\epsilon_{\mu\nu}(\partial^\mu U^{-1}TU + [H^\nu - K^\nu, U^{-1}TU])$. But the form $U^{-1}\Lambda U$ shows more apparently the origin of symmetry—dual transformed isotopic symmetry T, etc.; and the related current is constructed explicitly from the dual transformed solution $N\langle\gamma\rangle$ in the same way as the original current from original N. All our currents are related to a given symmetry of the action. Almost all of them (except the infinitesimal BT) keep the equation of motion invariant, while each elements of the Kac–Moody algebra (except the zero-order one) does not generate the symmetry of the original equation.

We have found a lot of new Noëther currents and related generators. It is interesting to point out that the infinitesimal generator $u(x)$ of dual transformation (which is the Lie transformation for the related sine–Gordon equation[22]) is just the position vector[23] of the so-called soliton surface,[24] in the case of the O(3) σ-model; it is the well-known pseudospherical surface with $N(x)$ as its normal and $\partial_\xi N$, $\partial_\eta N$ as its asymptotic directions. Then Eq. (5.2) becomes $2du\text{–}2du' = \cosh\phi\, dR$, we can identify the Riccati function $R\langle\gamma\rangle \cosh\phi$ as the common tangent of two pseudospherical surfaces.[23] Using the covariance of our formulation, we can show that $\operatorname{tr}(K_\mu\langle\gamma\rangle R\langle\gamma\rangle)$ gives the series of local conservation current in the ordinary soliton theory and is related to a total geodesic differential along the common tangent direction.

Our formulation is easy to generalize to supersymmetric cases.[25] Then, from the dual similar of the supersymmetric generator, we get infinitely many supersymmetric currents correspondingly obtaining Kac–Moody algebra with both anticommutators and commutators.

ACKNOWLEDGMENTS

The author would like to thank Professor C. N. Yang for his hospitality and for the support of the ITP, SUNY at Stony Brook. This work was supported in part by NSF Grant #PHY-81-09110-A-01.

[1]K. Pohlmeyer, Commun. Math. Phys. **46**, 207 (1976).
[2]A. A. Belavin and A. M. Polyakov, Pis'ma Zh. Eksp. Teor. Fiz. **22**, xxx (1975) [JETP Lett. **22**, 245 (1975)].
[3]H. Eichenherr, Nucl. Phys. B **146**, 215 (1978); B **115**, 544 (1979).
[4]H. Eichenherr and M. Forger, Nucl. Phys. B **155**, 381 (1979); B **164**, 528 (1980); Commun. Math. Phys. **82**, 227 (1981).
[5]V. E. Zakharov and A. V. Mikhailov, Zh. Eksp. Teor. Fiz. **74**, 1953 (1978) [Sov. Phys. JETP **47**, 1017 (1978)].
[6]M. Lüscher and K. Pohlmeyer, Nucl. Phys. B **137**, 46 (1978).
[7]D. Gross, Nucl. Phys. B **132**, 439 (1978); F. Gürsey and X. C. Tze, Ann. Phys. (N.Y.) **128**, 29 (1980); Bo-yu Hou, Y. B. Wang, Bo-yuan Hou, and P. Wang, Phys. Lett. B **93**, 415 (1980).
[8]A. T. Ogielski, M. K. Prasad, A. Sinha, and L. L. Chau Wang, Phys. Lett. B **91**, 387 (1980); M. K. Prasad, A. Sinha, and L. L. Chau Wang, Phys. Lett. B **87**, 237 (1979). L. L. Chau and T. Koikawa, Phys. Lett. B **123**, 47 (1983).
[9]K. Scheler, Phys. Lett. B **93**, 331 (1980); Z. Phys. C **6**, 365 (1980).
[10]K. C. Chou and X. C. Song, Sci. Sin. A **25**, 716 (1982).
[11]B. Brezin, C. Itzykson, J. Zinn-Justin, and J. B. Zuber, Phys. Lett. B **82**, 442 (1979).
[12]H. J. de Vega, Phys. Lett. B **87**, 233 (1979).
[13]A. T. Ogielski, Phys. Rev. D **21**, 406 (1980).
[14]L. Dolan and A. Roos, Phys. Rev. D **22**, 2018 (1980).
[15]E. Corrigan and C. K. Zachos, Phys. Lett. B **88**, 273 (1979); T. L. Cutright and C. K. Zachos, Phys. Rev. D **21**, 441 (1980).
[16]C. K. Zachos, Phys. Rev. D **21**, 3462 (1980).
[17]R. P. Zaikov, Teor. Math. Phys. **53**, 238 (1982).
[18]B. Y. Hou, Comm. Theor. Phys. **1**, 333 (1982); B. Y. Hou, M. L. Ge, and Y. S. Wu, Phys. Rev. D **24**, 2238 (1981); C. Devchand and D. B. Fairlie, Nucl. Phys. B **194**, 232 (1982).
[19]I. V. Cherednik, Theor. Math. Phys. **23**, 120 (1979); **41**, 997 (1979).
[20]S. Kobayashi and K. Nomizu, *Foundations of Differential Geometry* (Interscience, New York, 1969), Vol. 1.
[21]L. Dolan, Phys. Rev. Lett. **47**, 1371 (1981); M. L. Ge and Y. S. Wu, Phys. Lett. B **108**, 411 (1982); K. Ueno and Y. Nakamura, Phys. Lett. B **109**, 273 (1982).
[22]S. S. Chern and C. L. Terng, Rocky Mountain J. Math. **10**, 105 (1980).
[23]A. Sym. Lett. Nuov Cimento **33**, 394 (1982).
[24]B. Y. Hou, B. Y. Hou, and P. Wang, preprint, University of Inner Mongolia, 1982.
[25]E. Witten, Phys. Rev. D **16**, 2991 (1977); P. D. Vecchia and S. Ferrara, Nucl. Phys. B **130**, 93 (1977); E. Cremmer and J. Scherk, Phys. Lett. B **74**, 341 (1978); E. Napolitano and S. Sciuto, Nuov. Cimento **64**, 406 (1981); Commun. Math. Phys. **84**, 171 (1982); Phys. Lett. B **113**, 43 (1982); Z. Popowicz and L. L. Chau Wang, Phys. Lett. B **98**, 253 (1981); L. L. Chau, M. L. Ge, and Y. S. Wu, Phys. Rev. D **25**, 1080 (1982); J. F. Shonfeld, Nucl. Phys. B **109**, 49 (1980).

Volume 145B, number 5,6 PHYSICS LETTERS 27 September 1984

GAUGE COVARIANT FORMULATION OF SYMMETRIC-SPACE FIELDS

Ling-Lie CHAU
Physics Department, Brookhaven National Laboratory, Upton, NY 11973, USA

and

Bo-Yu HOU [1]
Institute of Theoretical Physics, SUNY, Stony Brook, NY 11794, USA

Received 18 June 1984

For fields taking value in the symmetric space G/H, we give a general G gauge covariant formulation, and *explicit* vertical and horizontal decomposition of the Cartan–Maurer relation in arbitrary gauges. From such explicit construction we show that a "zero"-gauge can be chosen for all symmetric-space fields so that the Cartan–Maurer conditions are automatically satisfied and the equations of motion become the only equations for the fields $N(x)$, with $N^2(x) = 1$, Tr $N(x)$ given. Explicit symmetric-space formulations for the principal chiral fields, and for the self-dual Yang–Mills field are given. Our formulation can easily be adapted to the corresponding supersymmetric fields.

1. Introduction. Recently there has been much progress made in the understanding of nonlinear systems [1]. Interestingly, many similarities have been found between the chiral fields and the non-abelian gauge fields, i.e., the self-dual Yang–Mills fields, and much progress has been made on both fields [2,4] [‡1].

Another interesting aspect of the study is the symmetric-space formation of these systems, e.g., the O(3) non-linear σ-model field $N(x)$, 2×2 matrices with $N(x)^2 = 1$, trace $N(x) = 0$ can be viewed as a map from the base x-space to the symmetric space $S^2 \approx SO(3)/SO(2)$ [or $SU(2)/U(1)$] [5,6]. The sine-Gordon field formulated in a specific way [7] can be viewed as mapping from x-space to the symmetric space of $SU(1+1)/S(U(1) \times U(1))$ [8]. Similarly the $O(l)$ σ-model (or the CP^{l-1}) fields $N(x)$, $l \times l$ matrix with $N(x)N^+(x) = 1$ and trace $N(x) = l - 2$, can be viewed as a map from the x-space to the symmetric space $SO(l)/SO(l-1) \approx S^{l-1}$ [or $SU(l)/SU(l-1) \times U(1)$] [5,6,9]. As will be discussed later in section 2, these are the symmetric space formulations in what we call the "zero"-gauge.

The chiral fields in the general symmetric spaces G/H have been formulated by Eihenherr and Forger [5]. In this formulation, only gauge transformation under $S = h \in H$ (see furtheron) was given. We classify this particular formation as the "normal"-gauge formulation.

Recently it was pointed out that the SU(n) self-dual Yang–Mills fields can also be viewed as symmetric space fields [4].

It has been observed that in spaces with two killing vectors (e.g., stationary axially symmetric space) the Einstein equations can be formulated into two-dimensional chiral-like equations [10]. Again their integrability conditions are discussed in what we call the "zero"-gauge. The detail discussions on these subjects will be left for further publications.

In this paper we shall first discuss the formulation of general symmetric-space fields in the "normal"-gauge, section 2. An explicit decomposition for $a_\mu(x) = g^{-1}\partial_\mu g$ into the vertical part $h_\mu(x)$ and horizontal part $k_\mu(x)$ are given. We then define the general G-gauge transformation, section 3,

[1] Permanent address: Department of Physics, Xibei University, Xian, China, resident from February 1984.

[‡1] See also the review in ref. [3].

Volume 145B, number 5,6 PHYSICS LETTERS 27 September 1984

$$a_\mu \to A_\mu = S^{-1}a_\mu S + S^{-1}\partial_\mu S,$$

$$h_\mu \to H_\mu = S^{-1}h_\mu S + S^{-1}\partial_\mu S,$$

$$k_\mu \to K_\mu = S^{-1}k_\mu S\,,$$

$S \in$ G, and again explicit decomposition of A_μ into H_μ and K_μ are given. We then see that the formulation of symmetric space fields with $N(x)$ coincides with the formulation in a particular gauge $S^{-1}(x) = g(x)$ so that $A_\mu^{(0)}(x) = 0$ is satisfied both in the horizontal and vertical directions, section 4. Following our general construction we can explicitly obtain

$$-K_\mu^{(0)}(x) = H_\mu^{(0)}(x) = \tfrac{1}{2}N(x)\partial_\mu N(x)\,,$$

with $N^2(x) = 1$, Tr $N(x) \equiv t$ fixed. In this "zero"-gauge we see that the Cartan–Maurer equation, i.e., both the Gauss and Coddazi equations, are automatically satisfied and the dynamical equation $D^\mu k_\mu = 0$ reduces to $\partial^\mu[N(x)\partial_\mu N(x)] = 0$. From our explicit construction, we see that such "zero"-gauge can be chosen for all symmetric-space fields.

In sections 5 and 6 we give an explicit symmetric-space formulation for the principal chiral fields and for the self-dual Yang–Mills fields.

The advantages of introducing this general G-gauge transformation for the symmetric space fields are many. The connections of different formulations of symmetric space fields are clarified, e.g., between the "normal"-gauge and the "zero"-gauge. The dynamical equations certainly are the simplest in the "zero"-gauge. It is also clear, for any dynamical equations for $N(x)$, that the necessary and sufficient conditions for the $N(x)$ being symmetric-space fields are that the set of $N(x)$ fields can all be diagonalized by a group element $S \in$ G to the same matrix n with only eigenvalues of +1 and −1. More importantly, since the dynamical equations become the simplest in the "zero"-gauge, their integrability properties can be discussed most easily in the "zero"-gauge. Indeed, as it turned out, both the self-dual Yang–Mills equations as well as the principal chiral equations have been essentially discussed in the "zero"-gauge in the literature.

Since we know the general G-gauge transformation properties, all those equations of integrability [2], like the linear system, the Riemann–Hilbert transformations, the Bianchi–Bäcklund transformations, Riccati equations and their interrelations so far given in some particular gauge can all be transformed into covariant form. Specific discussions on integrability properties and their gauge covariant formulation will be discussed in another paper.

All our discussions here can easily be carried over to the corresponding supersymmetric systems.

2. Formulation of symmetric-space chiral fields in the "normal"-gauge. The symmetric-space chiral (s-chiral) fields can be defined as a map from space time x_μ onto a symmetric space (G, H, n), i.e., a coset space G/H with an involute automorphism n, H = $\{h \in G;\ nhn = h\}$, with $n^2 = 1$, where G is a connected Lie group with Lie algebra $\mathcal{G}$, and H ∈ G is a close subgroup with Lie algebra $\mathcal{H}$, and $\mathcal{K} + \mathcal{H} = \mathcal{G}$. In the adjoint representation (and also in the fundamental representation for some particular kind of chiral models such as the models taking values on the Grassman manifold) the same matrix n gives an involute automorphism for the Lie algebra as well,

$$[n, \mathcal{H}] = 0\,, \quad \{n, \mathcal{K}\} = 0\,, \qquad (2.1, 2)$$

thus $[\mathcal{H}, \mathcal{H}] \subset \mathcal{H}$, $[\mathcal{K}, \mathcal{H}] \subset \mathcal{K}$, $[\mathcal{K}, \mathcal{K}] \subset \mathcal{H}$. More explicitly, let p be the projection on the subalgebra $\mathcal{H}$, then $n = 2p - 1$. In the special case of Grassman manifolds, SU($l + m$)/S[U(l) × U(m)], SO($l + m$)/S[O(l) × O(m)], as discussed in refs. [5,6, 9], the simple fundamental ($l + m$)-vector representation can be used. Then n is a matrix with $n^2 = 1$, and with l (+1)-eigenvalues and m (−1)-eigenvalues. But for general symmetric spaces such a fundamental representation does not exist, e.g. SU(3)/SO(3) is such an example. However, we can always use the adjoint representation of the larger dimension $d_G \times d_G$ matrices, where d_G is the dimension of the group manifold. Here n is a matrix with $n^2 = 1$, and d_H (+1)-eigenvalues, and $(d_G - d_H)$ (−1)-eigenvalues, where d_H is the dimension of the subgroup H.

The elements of G/H are represented by canonical variables

$$N(x) = g(x)ng^{-1}(x)\,, \quad N(x)^2 = 1\,, \quad \text{where } g(x) \in G\,. \qquad (2.3)$$

Then if g_1 and g_2 are in the same coset class, $g_1 = g_2 h$, hence $g_1 n g_1^{-1} = g_2 n g_2^{-1}$ both correspond to the same $N(x)$. As discussed in ref. [5], the left Cartan–Maurer form, i.e., curvatureless potential, $g^{-1}(x) \times \partial_\mu g(x)$, which is invariant under a global left transform $g(x) \to g_0 g(x)$, where g_0 is a constant group element of G, can be divided into vertical (connection), and horizontal (second fundamental form) parts, and pulled back onto x-space.

$$a_\mu(x) = g^{-1}(x)\partial_\mu g(x) = h_\mu(x) + k_\mu(x) . \quad (2.4)$$

From (2.1, 2.2),

$$[h_\mu(x), n] = 0 , \quad \{k_\mu(x), n\} = 0 . \quad (2.5, 6)$$

Now we give the explicit construction of h_μ and k_μ, which satisfy (2.5, 2.6):

$$h_\mu(x) = \tfrac{1}{2}(a_\mu + n a_\mu n) , \quad k_\mu(x) = \tfrac{1}{2}(a_\mu - n a_\mu n), \quad (2.7, 8)$$

i.e., in this gauge h_μ is diagonal with respect to n.

From the curvatureless condition of a_μ, i.e., $a_{\mu\nu} \equiv \partial_\mu a_\nu - \partial_\nu a_\mu + [a_\mu, a_\nu] = 0$, it can be divided into two equations, the Gauss (or vertical) equation: $\frac{1}{2}[a_{\mu\nu}(x) + n a_{\mu\nu}(x) n] = 0$, i.e.,

$$\partial_\nu h_\mu(x) - \partial_\mu h_\nu(x) + [h_\mu(x), h_\nu(x)] + [k_\mu(x), k_\nu(x)]$$

$$\equiv f_{\mu\nu}(x) + [k_\mu(x), k_\nu(x)] = 0 , \quad (2.9)$$

and the Codazzi (or horizontal) equation: $\frac{1}{2}[a_{\mu\nu}(x) - n a_{\mu\nu}(x) n] = 0$, i.e.,

$$\partial_\mu k_\nu(x) + [h_\mu(x), k_\nu(x)] - \partial_\nu k_\mu(x) - [h_\nu(x), k_\mu(x)]$$

$$\equiv \mathrm{D}_\mu k_\nu(x) - \mathrm{D}_\nu k_\mu(x) = 0 , \quad (2.10)$$

where $\mathrm{D}_\mu \equiv \partial_\mu + [h_\mu, \ ]$.

The s-chiral equation of motion is

$$\mathrm{D}_\mu k^\mu(x) = 0. \quad (2.11)$$

Eqs. (2.9)–(2.11) are the basic equations for the s-chiral fields.

3. The general G gauge transformation. So far in the literature only the H gauge transformation has been used, i.e., the gauge transformation by a group element belonging to the subgroup H. Here we introduce general gauge transformations by a group element $S(x) \in$ G: $g(x) \to g(x)S(x)$,

$$a_\mu(x) \to A_\mu(x) = S^{-1}(x) a_\mu(x) S(x) + S^{-1}(x)\partial_\mu S(x) , \quad (3.1)$$

$$h_\mu(x) \to H_\mu(x) = S^{-1}(x) h_\mu(x) S(x) + S^{-1}(x)\partial_\mu S(x) , \quad (3.2)$$

$$k_\mu(x) \to K_\mu(x) = S^{-1}(x) k_\mu(x) S(x) . \quad (3.3)$$

Now the corresponding equations as given previously in the "normal"-gauge have to be rederived. The matrix n now is transformed to a new matrix $m(x)$ by $m(x) = S^{-1}(x) n S(x)$, still $m^2(x) = 1$ and eqs. (2.5), (2.6) become

$$\partial_\mu m(x) + [H_\mu(x), m(x)] = 0 ,$$

$$\text{or } [H_\mu(x) - \tfrac{1}{2} m(x)\partial_\mu m(x), m(x)] = 0 , \quad (3.4)$$

and

$$\{k_\mu(x), m(x)\} = 0. \quad (3.5)$$

Eqs. (2.7), (2.8) become

$$H_\mu(x) = \tfrac{1}{2}[m(x)\partial_\mu m(x) + m(x) A_\mu(x) m(x) + A_\mu(x)] , \quad (3.6)$$

$$K_\mu(x) = \tfrac{1}{2}[-m(x)\partial_\mu m(x) - m(x) A_\mu(x) m(x) + A_\mu(x)] . \quad (3.7)$$

Note that eq. (3.4) is the important reducibility condition for H_μ, [11], i.e., if there exists on the coset bundle $\{\{x\}, \mathrm{G/H}, \mathrm{G}\}$ a section $m(x)$ invariant under parallel displacement with respect to H_μ, i.e., eq. (3.4), then the H_μ are reducible to a connection in H.

The s-chiral eqs. (2.9)–(2.11) are gauge covariant, i.e., the equations are correct if $n \to m(x)$, $k_\mu(x) \to K_\mu(x)$, $h_\mu(x) \to H_\mu(x)$. Note that when $S(x) \in H$, or $S(x)$ being a constant matrix, all equations (2.10), (2.11) are unchanged.

4. The "zero"-gauge. Now we want to choose a specific gauge which we call the "zero"-gauge. i.e., $S^{-1}(x) = g(x)$. It is possible to choose such a gauge only because we have introduced a general G gauge covariant formulation, not restricted to H as was done previously. Under this "zero-gauge" eqs. (2.3), (2.4) become

$$m^{(0)}(x) = S^{-1}(x) n S(x) = g(x) n g^{-1}(x) = N(x) ,$$

$$\text{still } N^2(x) = 1 , \quad (4.1)$$

$$a_\mu(x) = h_\mu(x) + k_\mu(x)$$

$$\to A_\mu^{(0)}(x) = H_\mu^{(0)}(x) + K_\mu^{(0)}(x) = 0 . \quad (4.2)$$

Then using eqs. (3.6), (3.7), we can explicitly find out $h_\mu(x), k_\mu(x) \to H_\mu^{(0)}(x), K_\mu^{(0)}(x)$,

$$H_\mu^{(0)}(x) = -K_\mu^{(0)}(x) = \tfrac{1}{2} N(x)\partial_\mu N(x) . \quad (4.3)$$

Among the s-chiral equations, eqs. (2.9), (2.10) are trivially satisfied. The equation of motion eq. (2.11)

Volume 145B, number 5,6 PHYSICS LETTERS 27 September 1984

becomes simply

$$\partial^\mu K_\mu^{(0)}(x) = 0\,, \quad \text{or} \quad \partial^\mu [N(x)\partial_\mu N(x)] = 0\,, \tag{4.4}$$

which is a conserved current equation. Thus in this "zero"-gauge we have introduced $H_\mu^{(0)}(x)$ satisfying $\partial_\mu m^{(0)}(x) + [H_\mu^{(0)}(x), m^{(0)}(x)] = 0$ so that the equation of motion eq. (2.11) and the conservation equation eq. (4.4) are the same equation in different gauge.

For $N(x)$ being an $d \times d$ dimensional matrix, eq. (4.4) is just like the $O(d)$-σ-model $\approx SO(d)/S\{O[\frac{1}{2}(d+t)] \times O[\frac{1}{2}(d-t)]\}$ field, where $t = \mathrm{Tr}\, N(x)$. However, our discussion holds for general symmetric-space fields, based crucially on the introduction of the general G gauge transformation.

We now also see that in the "zero"-gauge, we have reduced the study of s-chiral fields to a single equation for $N(x)$. Once $N(x)$ is solved we can obtain $g(x)$ using eq. (2.3), i.e., to find the similarity transformation that diagonalizes $N(x)$. Then $h_\mu(x)$, $k_\mu(x)$ can be obtained via eqs. (2.7), (2.8). Thus conversely saying, for any set of $N(x)$ satisfying eq. (4.4), a corresponding set of s-chiral fields can be obtained, if such a set of $N(x)$ can be diagonalized to the same matrix n. One example of a set of sufficient conditions for such a set of $N(x)$ is $N^2(x) = 1$, and trace $N(x) = t$, where $t = d_G - d_H$ for adjoint representations; and $t = l - m$ for fundamental representations.

5. An explicit symmetric-space formulation for the principle chiral fields. The principal chiral fields of a group can be viewed as s-chiral fields of the coset space $G/H = [G_1 \times G_1]/G_1$. Here we give an explicit formulation. For $g_L, g_R \in G_1$, we construct

$$g \equiv \begin{pmatrix} g_L & 0 \\ 0 & g_R \end{pmatrix}, \quad n = \begin{pmatrix} 0 & I \\ I & 0 \end{pmatrix},$$

$$g \in H = G_1 \text{ if } g_L = g_R\,. \tag{5.1}$$

Following eqs. (3.6), (3.7), we can construct H_μ, K_μ explicitly

$$H_\mu = \tfrac{1}{2}[g^{-1}(\partial_\mu g) + n g^{-1}(\partial_\mu g) n]$$

$$= \frac{1}{2}\begin{pmatrix} g_L^{-1}\partial_\mu g_L + g_R^{-1}\partial_\mu g_R & 0 \\ 0 & g_L^{-1}\partial_\mu g_L + g_R^{-1}\partial_\mu g_R \end{pmatrix}, \tag{5.2}$$

$$K_\mu = \tfrac{1}{2}[g^{-1}(\partial_\mu g) - n g^{-1}(\partial_\mu g) n]$$

$$= \frac{1}{2}\begin{pmatrix} g_L^{-1}\partial_\mu g_L - g_R^{-1}\partial_\mu g_R & 0 \\ 0 & -g_L^{-1}\partial_\mu g_L + g_R^{-1}\partial_\mu g_R \end{pmatrix}, \tag{5.3}$$

Note that in this construction, the fundamental relations of symmetric-space fields, eqs. (2.6), (2.7): $[n, h] = 0$, $\{n, k\} = 0$ are satisfied. The $N(x)$ field can also be calculated:

$$N = g(x) n g^{-1}(x)$$

$$= \begin{pmatrix} 0 & g_L g_R^{-1} \\ g_R g_L^{-1} & 0 \end{pmatrix} \equiv \begin{pmatrix} 0 & q \\ q^{-1} & 0 \end{pmatrix}. \tag{5.4}$$

Note that $N^2(x) = 1$. The equation of motion becomes

$$\partial_\mu [N(x)\partial_\mu N(x)] = 0$$

$$= \begin{pmatrix} \partial_\mu(q\partial_\mu q^{-1}) & 0 \\ 0 & \partial_\mu(q^{-1}\partial_\mu q) \end{pmatrix}. \tag{5.5}$$

It is nice that both the so-called right-equation $\partial_\mu(q\partial_\mu q^{-1}) = 0$, and the left-equation $\partial_\mu(q^{-1}\partial_\mu q) = 0$ are incorporated nicely into one equation for $N(x)$.

6. Symmetric-space formulation for the self-dual Yang–Mills field (SDYM). The fact that SDYM fields can be viewed as symmetric-space fields was first noted by Chou and Song [4]. Here we give an explicit $G/H = [SL(n,c) \times SL(n,c)]/SL(n,c)$ formulation, which is very similar to the chiral fields.

For $D(x), \bar{D}(x) \in SL(n,c)$, we construct $g(x) \in G$

$$g \equiv \begin{pmatrix} D & 0 \\ 0 & \bar{D} \end{pmatrix}, \quad n = \begin{pmatrix} 0 & I \\ I & 0 \end{pmatrix}, \tag{6.1}$$

and $g \in H = SL(n,c)$, if $D = \bar{D}$. Similar to eqs. (5.2), (5.3), from eqs. (3.6), (3.7) we can construct

$$H_\mu = \frac{1}{2}\begin{pmatrix} D^{-1}\partial_\mu D + \bar{D}^{-1}\partial_\mu \bar{D} & 0 \\ 0 & D^{-1}\partial_\mu D + \bar{D}^{-1}\partial_\mu \bar{D} \end{pmatrix}, \tag{6.2}$$

$$K_\mu = \frac{1}{2}\begin{pmatrix} D^{-1}\partial_\mu D - \bar{D}^{-1}\partial_\mu \bar{D} & 0 \\ 0 & -D^{-1}\partial_\mu D + \bar{D}^{-1}\partial_\mu \bar{D} \end{pmatrix}, \tag{6.3}$$

Volume 145B, number 5,6 PHYSICS LETTERS 27 September 1984

where $\mu = 1, 2, 3, 4$.

The $N(x)$ fields are

$$N(x) = g(x) n g^{-1}(x)$$

$$= \begin{pmatrix} 0 & D\bar{D}^{-1} \\ \bar{D}D^{-1} & 0 \end{pmatrix} \equiv \begin{pmatrix} 0 & J \\ J^{-1} & 0 \end{pmatrix}. \quad (6.4)$$

The equation of motion for SDYM is

$$\begin{pmatrix} \partial_u & 0 \\ 0 & \partial_{\bar{u}} \end{pmatrix} \left[N \begin{pmatrix} \partial_{\bar{u}} & 0 \\ 0 & \partial_u \end{pmatrix} N \right] = 0$$

$$= \begin{pmatrix} \partial_u (J \partial_{\bar{u}} J^{-1}) & 0 \\ 0 & \partial_{\bar{u}} (J^{-1} \partial_u J) \end{pmatrix}, \quad (6.5)$$

here the summation of $u\bar{u} = y\bar{y} + z\bar{z}$ is used.

For real Yang–Mills fields $\bar{D} \doteq D^{+-1}$ for x_μ real, the symmetric space becomes $SL(n, c)/SU(n)$. Note that in this case H_μ is hermitian and K_μ is anti-hermitian. Interestingly the J-formulation [2,12] for the SDYM was already in the "zero"-gauge.

7. Concluding remarks and outlooks. Our gauge covariant formulation can easily be adopted to the corresponding supersymmetric field [‡2].

With our new understanding of the formulation of the symmetric-space fields in the general gauge, many discussions on the symmetric-space fields previously given in any specific gauge can now be given in a gauge covariant way. For example, the integrability properties of the principle chiral fields, and the self-dual Yang–Mills fields previously given in the "zero"-gauge, as reviewed in ref. [2], can be written in a gauge covariant way. Details will be given in another publication [15].

We would like to thank Professor K.-C. Chou and Professor X.-C. Song for enlightening discussions.

[‡2] For supersymmetric chiral fields see ref. [13] for supersymmetric Yang–Mills fields see ref. [14].

References

[1] See M. Serdaroglu and E. Inönü, eds., Proc. Group theoretical methods in physics (Istanbul, 1983), Lecture Notes in Physics (Springer, Berlin);
K.B. Wolf, ed., Proc. Workshop on Nonlinear phenomena (Oaxtepec, Mexico, 1982), Lecture Notes in Physics, Vol. 189 (Springer, Berlin).
[2] L.L. Chau, lectures in: Proc. Group theoretical methods in physics (Istanbul, 1983), eds. M. Serdaroglu and E. Inönü, Lecture Notes in Physics (Springer, Berlin); Proc. Workshop on Nonlinear phenomena (Oaxtepec, Mexico, 1982), ed. K.B. Wolf, Lecture Notes in Physics, Vol. 189 (Springer, Berlin).
[3] A.M. Perelomov, Physica 4D (1981) 1;
F. Gürsey and H.C. Tze, Ann. Phys. 128 (1980) 29.
[4] K.-C. Chou and X.-C. Song, Commun. Theor. Phys. 1 (1982) 185.
[5] H. Eichenherr and M. Forger, Nucl. Phys. B155 (1979) 381; B164 (1980) 528; Commun. Math. Phys. 82 (1981) 227;
E. Brézin, S. Hikami and J. Zinn-Justin, Nucl. Phys. B165 (1980) 381;
K.-C. Chou and X.-C. Song, Commun. Theor. Phys. 1 (1982) 69.
[6] K. Pohlmeyer, Commun. Math. Phys. 46 (1976) 207;
V.E. Zakharo and A.V. Michaihailov, Sov. Phys. JETP 47 (1978) 1017.
[7] L.I. Fadeev, in: Solitons, Springer Topics in Current Physics, Series TCP17, ed. R.K. Bullough and P.J. Caudrey;
R. Sasaki and R.K. Bullough, Proc. R. Soc. London A376 (1981) 401.
[8] L.-L. Chau and T. Koikawa, Phys. Lett. 123B (1983) 47.
[9] S. Helgason, Differential geometry, Lie groups and symmetric spaces (Academic Press, New York, 1978).
[10] E. Ernst, J. Math. Phys. 20 (1978) 871;
V.A. Belinskii and V.E. Zakharov, Zh. Teor. Fiz. 75 (1978) 1953 [Sov. Phys. JETP 50 (1979) 1];
D. Maison, Phys. Rev. Lett. 41 (1978) 521; J. Math. Phys. 20 (1979) 871.
[11] S. Kobayashi and K. Nomizu, Foundation of differential geometry, Vol. 1 (Interscience, New York, 1969).
[12] C.N. Yang, Phys. Rev. Lett. 38 (1977) 1377;
S. Ward, Phys. Lett. 61A (1977) 81;
Y. Brihaye, D.B. Fairlie, J. Nuyts and R.F. Yates, J. Math. Phys. 19 (1978) 2528.
[13] P. DiVecchia and S. Ferrara, Nucl. Phys. B130 (1977) 93;
E. Witten, Phys. Rev. D16 (1977) 2991;
E. Cremmer, J. Scherk and C.K. Zachos, Phys. Rev. Phys. Lett. 74B (1978) 341;
J.F. Shonfeld, Nucl. Phys. B169 (1980) 49;
T.L. Curtright and C.K. Zachos, Phys. Rev. D21 (1980) 411;
Z. Popowicz and L.-L. Chau Wang, Phys. Lett. 98B (1981) 1080;
L.-L. Chau, M.-L. Ge and Y.-S. Wu, Phys. Rev. D25 (1982) 1080.
[14] M. Sohnius, Nucl. Phys. B136 (1978) 461;
E. Witten, Phys. Lett. 77B (1978) 394;
I.V. Volovich, Phys. Lett. 129B (1983) 429; Teor. Mat. Fiz. 57, No. 3 (1983);

C. Devchand, Durham University Report No. DTP-83/15 (1983), to be published;
L.-L. Chau, Ge Mo-Lin and Z. Popowicz, Phys. Rev. Lett. 52 (1984) 1940;
L.-L. Chau, Supersymmetric Yang–Mills fields as an integrable system and connections with other non-linear systems, Invited talk Workshop on Vertex operators in mathematics and physics (Berkeley, November 1983); invited talk Thirteenth Intern. Colloq. on Group theoretical methods in physics (Maryland, May 1984).
[15] L.-L. Chau and B.-Y. Hou, Gauge covariant formulation of integrability properties of symmetric-space fields Phys. Lett., to be published.

CHINESE PHYS. LETT.
Vol.2, No.2(1985)

CHINESE PHYSICS LETTERS
Science Press,Beijing
Springer-Verlag

THE THIRD ORDER COHOMOLOGY CYCLE FOR GAUGE GROUP AND THE ANOMALOUS ASSOCIATIVE LAW FOR GAUGE TRANSFORMATION

HOU Bo-yu
(Northwestern University,Xian)
HOU Bo-yuan
(Inner Mongolian University,Huhehaote)

(Received 31 October 1984)

This note shows that the nontrivial third order cohomology cycle for gauge group implies the existence of anomalous associative law and consequently anomaly for Jacobi identity.

Now is the 30th anniversary of Yang-Mills field bearing local symmetry[1]. In the last decade the global topological properties[2] of gauge field have aroused wide interests. Since last year a series of papers have appeared dealing[3] with the cunning anomaly[4] in quantum field theory by using the cohomology of the infinitesimal gauge transformation operator. Since the measure of a chiral Fermion field in path integral changes under gauge transformation, the formally gauge-independent effective action will change also[5]. By family index theory this variation has been related with a series of characteristic forms ω^m_{2n-m-1}(here 2n equals the sum of dimension of space-time and dimension of the family parameter space; m is the order of cohomology). The m-th order nontrivial cohomology realizes the m-th order topological obstruction for breaking the gauge invariance. Atiyah and Singer[3] put forth the question whether the higher obstructions have any physical significance as well. Soon after, Faddeev[6] showed that the second order cocycle α_2 of gauge group characterizes a projective representation; correspondingly, for the infinitesimal transformations there exists in 2n-3 dimension a Schwinger term given by the second order cohomology form ω^2_{2n-3}. But the role of higher order cocycles for the group representation remains less evident.

This note attempts to show that α_3, which originates from a topological obstruction, measures the associativity dependence of gauge transformations. As in ref.6, let us consider the space of gauge connections a with gauge transformations g. The cohomology operation Δ is defined on the arrays of functions $\alpha_n(a;g_1,\ldots,g_n)$ of connection a and n ordered elements $g_1,\ldots g_n$ in the following way,

$$(\Delta\alpha_n)(a;g_1,\ldots,g_{n+1})=\alpha_n(ag_1;g_2,\ldots,g_{n+1})-\alpha_n(a;g_1g_2,g_3,\ldots,g_{n+1})$$
$$+\alpha_n(a;g_1,g_2g_3,g_4,\ldots,g_{n+1})+\ldots+(-1)^n\alpha_n(a;g_1,\ldots,g_ng_{n+1})+(-1)^{n+1}$$
$$\times\alpha_n(a;g_1,\ldots,g_n)\ .$$

One can easily check $\Delta^2=0$. Now, starting from the case n=0 we will analyse order by order: how the definition of the n-multiple transformation depends on cochain α_n ; how the consistency for the transformations of higher orders requires α_n to be a closed cocycle; and how $\Delta\alpha_n$ measures the degree of nonconsistency. Then it is natural to introduce a new definition of (n+1)-multiple transformation which is characterized by $\Delta\alpha_n$, while the more general and nontrivial character of this new definition is α_{n+1}.

Zero-cocycle, Invariance and Identity Transformation: Let $f_0(a)$ be a gauge invariant function of connection a: $f_0(a)=f_0(ag)$. In reality it is a function defined on the orbits only. If a change in phase factor is admitted $f(a)=f_0(a)\exp[i\alpha_0(a)]$, then the condition for $f(ag)$ remaining independent of g is $\Delta\alpha_0=0$.

Define $V(g)f(a)=f(ag)$. Then $V(g)f(a)=\exp(i\Delta\alpha_0)f(a)$. (1)

Hence when $\Delta\alpha_0=0$, single V(g) and multiple V(g) are all identity operators $V(g)f(a)=f(a)$. Thus, invariants $f_0(a)$ (or $\alpha_0(a)$) and identity transformations constitute a trivial selfconsistent closed structure.

If $\Delta\alpha_0\neq 0$, then $f(ag)$ becomes gauge dependent, i.e. it has been extended to a function of connections in reality. The V(g) also becomes a transformation in the real meaning:

$$V(g)f(a)=f(ag) . \tag{2}$$

The multiple V(g) acts as the regular representation induced by $f(a)$, so it is consistent obviously.

Or from (1) we can show that

$$V(g_1)V(g_2)f(a)=\exp(i\Delta\Delta\alpha_0)V(g_1g_2)f(a)=V(g_1g_2)f(a) , \tag{3}$$

i.e. V(g) constitutes a representation.

One-cocycle α_1 , Functions of Connection and Representation: Generalizing the $\Delta\alpha_0=0$ case of (1), we define

$$V(g)f(a)=\exp[i\alpha_1'(a;g)f(a)\equiv\exp[i\alpha_1(a;g)+i(\Delta\alpha_0)(a;g)]f(a)$$
$$=\exp[i\alpha_1(a;g)]f(ag). \tag{4}$$

Since now

$$V(g_1)V(g_2)f(a)=\exp[i(\Delta\alpha_1)(a;g_1,g_2)]V(g_1g_2)f(a) , \tag{5}$$

so when $\Delta\alpha_1=0$, V(g) constitutes a representation. When α_1' is the zero-coboundary, $\alpha_1'=\Delta\alpha_0$. The rule (4) degenerates into the regular case (2). If $\Delta\alpha_1\neq 0$, the rule of multiplication becomes different from the ordinary representation rule by a modification factor as shown in (5). We will generalize these double product rule by using a general function α_2 of (a,g_1,g_2).

Two-cocycle α_2 and Projective Representations: Now the double product are defined similar to (5)

$$V(g_1)V(g_2)f(a)=\exp[i\alpha_2(a;g_1,g_2)]V(g_1g_2)f(a) . \tag{6}$$

As in ref.6, let $V(g)f(a)=U(a,g)f(ag)$. One obtains the projective representation:

$$U(a,g_1)U(ag_1;g_2)\equiv\exp[i\alpha_2(a;g_1,g_2)]U(a;g_1,g_2)$$
$$=\exp[i\alpha_2(a;g_1,g_2)]U(a;g_1g_2) \tag{7}$$

(When $\alpha_2 = \Delta\alpha_1$, one may redefine $\exp[-i\alpha_2(a;g)]U(a;g) \to U(a;g)$. Then the phase factor $e^{i\alpha_2}$ in (7) disappears, but the dependence of $U(a;g)$ on a remains. Therefore these degenerate cases are not ordinary representation yet (unless α_1 does not depend on a). In order that the projective representation be well defined, we must have $U(a;g_1,g_2)=U(a;g_3,g_4)$. When $g_1g_2=g_3g_4$, this condition is easily shown to be equivalent to associativity. By using (6) we get

$$[V(g_1)V(g_2)]V(g_3)=\exp[-i(\Delta\alpha_2)(a;g_1,g_2,g_3)]V(g_1)[V(g_2)V(g_3)] . \quad (8)$$

So if $\Delta\alpha_2=0$, then associative law is satisfied and all higher product are consistent.

If $\Delta\alpha_2\neq 0$, the $U(a;g)$ cannot be defined consistently. But we will show the products of $V(g)$ with the given association may be defined self-consistently. Meanwhile $\Delta\alpha_2$ will give the ratio factor between different associations. The self-consistency of this assignment will be checked together with more general ratio factors.

Three-cocycle and Associative Dependent "Representation": Let

$$[V(g_1)V(g_2)]V(g_3)=\exp[-i\alpha_3(a;g_1,g_2,g_3)]V(g_1)[V(g_2)V(g_3)] . \quad (9)$$

To check the consistency, we consider the 4-multiple product which has five different association modes: $(\underline{12}3)4$, $(12)(34)$, $1(2\underline{34})$, $1(\underline{23}4)$ and $(1\underline{23})4$. (here: $(\underline{12}3)4\equiv\{[V(g_1)V(g_2)]V(g_3)\}V(g_4)$, etc.). Their difference in phase may be shown as follows:

$(\underline{12}3)4$, $\alpha(a;g_1g_2,g_3,g_4)$; $(12)(34)$, $\alpha(a;g_1,g_2,g_3g_4)$;
$1(2\underline{34})$, $(-)\alpha(ag_1,g_2,g_3,g_4)$; $1(\underline{23}4)$, $(-)\alpha(a;g_1,g_2g_3,g_4)$;
$(1\underline{23})4$, $(-)\alpha(a;g_1,g_2,g_3)$.

Thus, if $\Delta\alpha_3=0$ the 4-multiple products and consequently higher products with given mode of association are all well defined. If $\alpha_3=\Delta\alpha_2$ one may put $\exp[-i\alpha_2(\underset{m}{a};g_1,g_2)]V(g_1)V(g_2)\equiv V(g_1;g_2)$, and so on. Then formally we get $V[(g_1;g_2);g_3]=V[g_1;(g_2;g_3)]$ instead of (9). The factors in (9) disappear, but the result still depends on modes of association.

If $\Delta\alpha_3\neq 0$, then running cyclically over the five 3-simplex corresponding to the five α_3 ($\alpha_3(a;g_1,g_2,g_3)$ related with the simplex: $a, ag_1, ag_1g_2, ag_1g_2g_3$), a dismatch value $\Delta\alpha_3$ appears. These five 3-simplex together constitute a 4-simplex $a, ag_1, ag_1g_2, ag_1g_2g_3, ag_1g_2g_3g_4$. If we use $\Delta\alpha_3$ (or more generally some α_4) to characterize this dismatching, and use the 5-simplex $a, ag_1, \ldots, ag_1g_2g_3g_4g_5$ which includes six 4-simplex, one may show that the consistence condition is $\Delta\alpha_4=0$.

We know that in the infinitesimal gauge transformation case, α_3 is related with the integral of ω^3_{2n-4} [7]. The infinitesimal forms of association are the Jacobi. Thus ω^3_{2n-4} gives the anomaly in Jacobi for 2n-4 dimensional space. It is similar to the L.W.P.[8] parodox, and relate with a lot of interesting phenomena such as the boundary condition of monopole.

It is our pleasure to express our gratitude to Prof.GUO, H.Y, Dr.WU,K. and Prof.SONG,X.C. for helpful discussions and fruitful cooperation. One of the author (Bo-yu,HOU) would like to express his gratitude to Prof. B.Zumino for his excellent report and valuable handwritten manuscript and particularly thanks to Prof.C.N.Yang and K.C.CHOU for drawing his attention to this field of physics. This work is supported by the Science Fund of the Chinese Academy of Sciences.

REFERENCES

1. C.N.Yang, R.L.Mills, Phys.Rev.96(1954)191.
2. I. S.Tuan, M.L.Ge and B.Y.Hou, Journal Xibei Univ.Xian(1975)38 (in Chinese); Acta Physica Sinica, 25(1976)514 (in Chinese); Scientia Sinica, 21(1978)446; J.Arafune, P.G.O.Freund and C.J.Goebel, J.Math.Phys. 16(1975)433; Bo-Yu HOU, Acta Physica Sinica, 26(1977)83 (in Chinese); T.T.Wu and C.N.Yang, Phys.Rev.D12(1975)3845.
3. E.Witten, Nucl. Phys.B23(1983)422;433; L.Bonora and Cotta-Ramusino, Commun.Math.Phys.87(1983)589; K.C.Chou, H.Y.Guo, K.Wu, X.C.Song, Phys. Lett.134B(1984)67; B.Zumino,A.Zee and Y.S.Wu, Nucl.Phys.B239(1984)477; A.Niemi and G.Semenoff, Phys.Rev.Lett.51(1983)2077; B.Zumino, UCB-PTH-83/15; R.Stora, LAPP-TH-94; L.Alvarez-Gaume and P.Ginsparg, HUTP-84/A016; M.F.Atiyah and I.M.Singer, Proc.Natl.Acad.Sci.USA, 81(1984) 2597.
4. S.L.Adler, Phys.Rev.177(1969)2426; J.Bell and R.Jackiw.Nuovo.Cimento. 60A(1969), 47; W.A.Bardeen, Phys.Rev.186(1969)1491.
5. K.Fujikawa, Phys.Rev.Lett.42(1979)1195.
6. D.L.Faddeev, T.M.Ф., TOM. 60, вып.2(1984)206; For Schwinger term cf. R.Jackiw and K.Johnson, Phys.Rev.182(1969)1459.
7. H.Y.Guo, B.Y.Hou, S.K.Wang and K.Wu, AS-ITP-84-0, NWU-IMP-84-5; X.C. Song, private communication; B.Zumino, Santa Sarbara Preprint.
8. H.J.Lipkin, W.I.Weisberger and M.Peshkin, Ann.Phys.53(1969)203.
9. After this note has been submitted, we receive a preprint by R.Jackiw, where the role of 3-cocycle has been given explicitly also, see B.Y. Hou et al. NWU-IMP-84-9 for further discussion.

Commun. in Theor. Phys. (Beijing, China) Vol.4, No.2 (1985) 233-251

COHOMOLOGY OF GAUGE GROUPS AND CHARACTERISTIC CLASSES OF CHERN-SIMONS TYPE

GUO Han-ying (郭汉英)
Institute of Theoretical Physics, Academia Sinica
P.O. Box 2735, Beijing, China
HOU Bo-yu (侯伯宇)
Institute of Morden Physics, Northwestern University, Xian,China
WANG Shi-kun (王世坤)
Institute of Applied Mathematics, Academia Sinica
P.O. Box 300, Beijing, China
and WU Ke (吴 可)
Institute of Theoretical Physics, Academia Sinica
P.O. Box 2735, Beijing, China

Received October 4, 1984

Abstract

We systematically clarify the characteristic classes of Chern-Simons type proposed by some of the present authors (GWW), the cohomologies of gauge groups, and the relations between these classes and cohomologies. We also explain in detail the cohomology of gauge groups realized upon the degenerate forms of those classes and show that the cohomology used in Faddeev's and Zumino's approaches is, in fact, such a degenerate one with a certain local sense. The applications to the analyses on the anomalies in spacetimes of different dimensions are also made by means of these classes and cohomologies of gauge groups.

I. Introduction

In the last few months, some new developments have been made in studying the global aspects of the anomalies. These[1-5] constitute in a certain sense a deepgoing exploration of the previous works on the topological origins of the anomalies by means of the differential geometry methods[6-9], in which the well-known Chern-Simons secondary characteristic classes[10,11] have played an important role.

Some of the authors (GWW)[1] of the present paper have generalized the conception of the Chern-Simons secondary classes associated with certain characteristic polynomials in the curvature[11-13] to introduce a sequence of new characteristic classes, the classes of Chern-Simons type, and shown a remarkable property of this new sequence of classes; that is the coboundary of the k-th classes of this type is exactly equal to the exterior differential of a (k+1)st class of this type. This means that if a certain characteristic polynomial in the curvature is regarded as the zero-th class of this type, then, all corresponding classes of Chern-Simons type can be descended from this zero-th class under the coboundary operation step by step, which implies a certain cohomology meaning. It has been also pointed out in Ref.[1] that these new classes would show not only the topological origins but also the directly topological meaning of the anomalies in spacetimes of different dimensions.

Faddeev[2] has applied, on the other hand, the methods of group theory cohomology[14] to the analysis of the anomalies. He has shown that from the viewpoint of a certain cohomology of gauge groups the Wess-Zumino term[15,16] in 4-dimensional spacetime R^4 is a one-cocycle descended from the 3rd Chern class and the anomalous Schwinger term in the equal time commutator in current algebra[17] is related to a 2-cocycle in R^3 descended from the same Chern class, whereas the center term in the Kac-Moody algebra can be obtained from a 2-cocycle in R^1 descended from the second Chern class.

The relations between Faddeev's approach and the previous differential geometric ones have been studied by SONG[3] and Zumino[4]. SONG has applied H. Cartan's homotopy formula[18,19] widely used in Zumino's and Stora's paper[7,8]. In order to get Faddeev's results, he has introduced some new polynomials descended from the Chern character. Zumino[4] has shown that the gauge group cocycles introduced by Faddeev can be obtained as integrals of the differential forms discussed in Refs.[7,8]. And in this sense the cohomology of the gauge groups can be reduced to that of their Lie algebras.

In a recent note[5], the authors of this paper have further explored the properties of the classes of Chern-Simons type and their applications to the anomalies. The cohomologic meaning of the descent equation for the classes of Chern-Simons type has been analysed and certain cohomologic conditions on the anomalies have been addressed from the viewpoint of the Chern-Simons cochain[5]. It was also pointed out that the cohomology of the gauge groups in both Faddeev's and Zumino's approaches is a degenerate one realized by some degenerate polynomials of Chern-Simons type. And those polynomials introduced by SONG[3] are also the degenerate forms of the classes of Chern-Simons type. As a matter of fact, all results given in Faddeev's paper have been reformulated by means of the degenerate form of the cohomology of gauge group presented in Ref.[5].

In this paper, we will systematically clarify these issues and related ones, especially the cohomologic meaning of the sequence of the characteristic classes of Chern-Simons type, the relations between these classes and the cohomologies of gauge groups, as well as their applications to the course of analyses on the anomalies. We will also explain in some detail the cohomology of gauge group realized upon the degenerate forms of these classes. Since these degenerate forms imply a certain local meaning in some sense, the corresponding cohomology used in Faddeev's and Zumino's approaches is in fact a degenerate one with a local sense as pointed out in Ref.[5].

In Sec. II, we will give a brief review on the concepts and properties concerned with the classes of the Chern-Simons type. Most of the contents in this section have been given in Ref.[1]. In Sec. III, we will first underline some concepts on cohomology and introduce two coboundary operations, then, we show the realizations of the cohomology of gauge groups and that of degenerate one by means of the classes of Chern-Simons type and give rise to the cohomology used in Faddeev's approach. The analyses on anomalies in spacetimes of different dimensions are given in Sec. IV based upon these classes and cohomologies. In Appendices, we sketch out the proof of the fundamental theorems on these classes and show the applications of the theorems to the case with ghosts which includes Zumino's approach.

II. Characteristic Classes of Chern-Simons Type

We consider a gauge theory with gauge group G on a manifold $\mathscr{M}$. The gauge potential specifies a connection 1-form

$$A=A_\mu dx^\mu \quad , \qquad A_\mu=A_\mu^a\lambda_a \quad , \tag{2.1}$$

where A_μ is a Lie algebra-valued potential, λ^a the generators of the Lie algebra of G. The curvature F, defined by

$$F=dA+A^2 \tag{2.2}$$

is a Lie algebra-valued 2-form (the wedge product symbol, $\wedge$, is suppressed). The Bianchi identity

$$DF\equiv dF+[A,F]=0 \tag{2.3}$$

is a simple consequence of the aforementioned definitions. Under a gauge transformation. the change in A, and F is given by

$$\left.\begin{array}{l} {}^gA=g^{-1}Ag+g^{-1}dg \ , \\ {}^gF=g^{-1}Fg \ . \end{array}\right\} \tag{2.4}$$

Let $A^0,\ldots,A^k$ be k+1 connection 1-forms on a submanifold $M^{2n-k+1}\subset\mathscr{M}$, $t^1,\ldots,t^k$, be k parameters with constraints $0\leqslant t^1,\ldots,t^k\leqslant 1$. The interpolation connection 1-form on M^{2n-k+1} is

$$\left.\begin{array}{l} A_{0,t^1\ldots t^k}=A^0+\sum\limits_{j=1}^k t^j\eta^{j,0} \ , \\ \eta^{j,i}=A^j-A^i \ , \end{array}\right\} \tag{2.5}$$

which can be extended as a connection 1-form, $\mathscr{A}_{0,t^1\ldots t^k}$, on $M^{2n-k+1}\times\Delta_k$ with t-components of $\mathscr{A}$ taken to be zeros, where Δ_k is a simplex of dimension k in $R^k(t)$ defined by

$$\Delta_k=\left\{(t^1,\ldots,t^k)\,|\,\sum_{j=1}^k t^j=1,\ t^j\geqslant 0 \qquad j=1,\ldots,k\right\} \ , \tag{2.6}$$

The curvature 2-form $\mathscr{F}_{0,t^1\ldots t^k}$ on $M^{2n-k+1}\times\Delta_k$ is then

$$\mathscr{F}_{0,t^1\ldots t^k}=d\mathscr{A}_{0,t^1\ldots t^k}+\mathscr{A}^2_{0,t^1\ldots t^k} \ , \tag{2.7}$$

where $d=d_{(x)}+d_{(t)}$, d, $d_{(x)}$, and $d_{(t)}$ are the exterior differential operators defined on $M^{2n-k+1}\times R^k(t)$, M^{2n-k+1}, and $R^k(t)$ respectivley. Obviously, if we define

$$\begin{array}{l} F_{0,t^1\ldots t^k}=d_{(x)}A_{0,t^1\ldots t^k}+A^2_{0,t^1\ldots t^k} \ , \\ H_0=\sum\limits_{j=1}^k dt^j\,\eta^{j,0} \ , \end{array} \tag{2.8}$$

Then it follows that

$$\mathscr{F}_{0,t^1\dots t^k}=F_{0,t^1\dots t^k}+H_0 \ . \tag{2.9}$$

Let $P(F^n)$ be a characteristic polynomial of degree n in the curvature which satisfies the invariance property

$$P(g^{-1}Fg)=P(F)$$

We now introduce the density of the k-th characteristic polynomial of Chern-Simons type, $q_n^{(k)}$, associated with the invariant polynomial $P_n(\mathscr{F})$ which is a 2n-form on $M^{2n-k+1}\times R^k(t)$

$$q_n^{(k)}(A^0,\dots,A^k)=(-1)^{\frac{1}{2}k(k-1)}\frac{n!}{k!(n-k)!}P(H_0^k,\mathscr{F}^{n-k}_{0,t^1\dots t^k}),\quad q_n^{(0)}=P_n \ . \tag{2.10}$$

Then the k-th class of Chern-Simons type, $Q_n^{(k)}$, associated with the characteristic class $P_n(\mathscr{F})$ is defined by an integral of the class density $q_n^{(k)}$ over the k-simplex Δ_k

$$Q_n^{(k)}(A^0,\dots,A^k;\Delta_k)=\int_{\Delta_k} q_n^{(k)}(A^0,\dots,A^k),\quad Q_n^{(0)}=P_n \ . \tag{2.11}$$

It is easy to show that the density $q_n^{(k)}$ can be reexpressed as

$$q_n^{(k)}(A^0,\dots,A^k)=(-1)^{\frac{1}{2}k(k-1)}\frac{n!}{k!(n-k)!}P(H_0^k,F^{n-k}_{0,t^1\dots t^k}) \ , \tag{2.12}$$

and both $q_n^{(k)}$ and $Q_n^{(k)}$ satisfy

$$\left.\begin{aligned} q_n^{(k)}(A^0,\dots,A^k)&=\varepsilon^{0,\dots,k}_{i_0,\dots,i_k}q_n^{(k)}(A^{i_0},\dots,A^{i_k}) \ ,\\ Q_n^{(k)}(A^0,\dots,A^k)&=\varepsilon^{0,\dots,k}_{i_0,\dots,i_k}Q_n^{(k)}(A^{i_0},\dots,A^{i_k}) \ ,\end{aligned}\right\} \tag{2.13}$$

where $(i_0,\dots,i_k)$ is a permutation of $(0,\dots,k)$.

The most important properties of $q_n^{(k)}$ and $Q_n^{(k)}$ are discribed by the following two theorems:

Theorem 1. The class densities of Chern-Simons type q's satisfy a hierarchy of equations

$$d_{(x)}q_n^{(k)}=d_{(t)}q_n^{(k-1)} \ . \tag{2.14}$$

Theorem 2. The characteristic polynomials of Chern-Simons type Q's satisfy a descent equation

$$\sum_{\Delta^{(i)}_{k-1}\in\partial\Delta_k} Q_n^{(k-1)}(A^{i_0},\dots,A^{i_{k-1}};\Delta^{(i)}_{k-1})=d_{(x)}Q_n^{(k)}(A^0,\dots,A^k;\Delta_k) \ . \tag{2.15}$$

The proof of these theorems will be sketched out in Appendix A.

Furthermore, we introduce the characteristics of the Chern-Simons type, $\eta_n^{(k)}$, which are the integrals of the polynomials $Q_n^{(k)}$ over $M^{2n-k}\subset\mathscr{M}$

$$\eta_n^{(k)}(A^0,\ldots,A^k;\Delta_k,M^{2n-k})=\int_{M^{2n-k}} Q_n^{(k)}(A^0,\ldots,A^k;\Delta_k) \tag{2.16}$$

Then by Stokes' formula, we have

Theorem 3. The characteristics of Chern-Simons type, η's, satisfy

$$\begin{aligned}\sum_{\Delta_{k-1}^{(i)}\in\partial\Delta_k} &\eta_k^{(k-1)}(A^{i_0},\ldots,A^{i_{k-1}};\Delta_{k-1}^{(i)};M^{2n-k+1})\\ &=\eta_n^{(k)}(A^0,\ldots,A^k;\Delta_k;\partial M^{2n-k+1})\ ,\end{aligned} \tag{2.17}$$

where ∂M^{2n-k+1} is the boundary of the manifold M^{2n-k+1}. If $\partial M^{2n-k+1}=\Phi$, the right-hand side of Eq.(2.17) vanishes.

Let us now illustrate the meaning of these polynomials and theorems by some simple cases.

In the case of k=0, we have only one connection 1-form, A^0, thus from the above formulas it follows that

$$\left.\begin{aligned}&\mathscr{A}_0=A_0=A^0\ ,\\ &Q_n^{(0)}(A^0;\Delta_0)=q_n^{(0)}(A^0)=P(F(A^0)^n)\ ,\\ &d(x)P(F^n)=0\ .\end{aligned}\right\} \tag{2.18}$$

The last equation in Eqs.(2.18) presents an important property of the characteristic classes in the curvature, $P(F^n)$, i.e., it is classed.

In the case of k=1, we have

$$\left.\begin{aligned}&A_{0,t^1}=A^0+t^1\eta^{1,0}\ ,\\ &\eta^{1,0}=A^1-A^0\ ,\\ &Q_n^{(1)}(A^0,A^1;\Delta_1)=n\int_0^1 dt^1 P(A^1-A^0,F_{0,t^1}^{n-1})\ .\end{aligned}\right\} \tag{2.19}$$

The theorem 2 shows that

$$P(F(A^1)^n)-P(F(A^0)^n)=dQ_n^{(1)}(A^0,A^1;\Delta_1)\ . \tag{2.20}$$

This is the well-known transgression formula to link the invariant polynomial $P(F^n)$ and the Chern-Simons secondary classes defined by the second expression in Eqs.(2.19). And the theorem 3 shows that if $\partial M^{2n}=\Phi$, the integral of $P(F^n)$ over M^{2n} is topologically invariant (i.e., it is invariant under deformations of the gauge potential or connection). This is the second important property of $P(F^n)$. If $\partial M^{2n}\neq\Phi$, the integral

$$\eta_n^{(1)}(A^0,A^1;\Delta_1,\partial M^{2n})=\int_{\partial M^{2n}} Q_n^{(1)}(A^0,A^1;\Delta_1) \tag{2.21}$$

is not necessarily zero. This means that the polynomials $Q_n^{(1)}(A^0,A^1;\Delta_1)$ are characteristic classes in their own right and are of independently topological importance, especially they specify the contributions from the boundary.

In the case of k=2, we have

$$A_{0,t^1t^2}=A^0+t^1\eta^{1,0}+t^2\eta^{2,0},$$
$$Q_n^{(2)}(A^0,A^1,A^2;\Delta_2)=n(n-1)\int_0^1 dt^1\int_0^{1-t^1}dt^2P(\eta^{1,0},\eta^{2,0},F_{0,t^1t^2}^{n-2}). \tag{2.22}$$

The theorem 2 reads

$$Q_n^{(1)}(A^1,A^2;\Delta_1)-Q_n^{(1)}(A^0,A^2,\Delta_1)+Q_n^{(1)}(A^0,A^1;\Delta_1)=dQ_n^{(2)}(A^0,A^1,A^2;\Delta_2), \tag{2.23}$$

then if $\partial M^{2n-1}=\emptyset$, we get

$$\eta_n^{(1)}(A^1,A^2;\Delta_1,M^{2n-1})-\eta_n^{(1)}(A^0,A^2;\Delta_1,M^{2n-1})+\eta_n^{(1)}(A^0,A^1;\Delta_1,M^{2n-1})=0, \tag{2.24}$$

which implies the topological meaning of $\eta_n^{(1)}$, i.e., an $\eta_n^{(1)}$ of arbitrary two connections among A^0,A^1, and A^2, each of them can be regarded as a representative of an equivalence class under the gauge transformations, is equal to the signed sum of the other two $\eta_n^{(1)}$'s and the sum is independent of the deformation of the third connection. If $\partial M^{2n-1}\neq\emptyset$, then we have

$$\eta_n^{(1)}(A^1,A^2;\Delta_1,M^{2n-1})-\eta_n^{(1)}(A^0,A^2;\Delta_1,M^{2n-1})+\eta_n^{(1)}(A^0,A^1;\Delta_1,M^{2n-1})$$
$$=\eta_n^{(2)}(A^0,A^1,A^2;\Delta_2,\partial M^{2n-1}). \tag{2.25}$$

As if the Chern-Simons secondary classes, in this case the polynomials $Q_n^{(2)}(A^0,A^1,A^2;\Delta_2)$ have certain meaning of characteristic classes in their own right and present the boundary contribution from ∂M^{2n-1} and so on.

Generally speaking, the classes of Chern-Simons type $Q_n^{(k)}(A^0,\ldots,A^k;\Delta_k)$ on a series of submanifolds $M^{2n-k}\subset\mathscr{M}$ (k=0,1,...,n) construct a hierarchy of classes descended from the same characteristic classes in the curvature on M^{2n}. On each submanifold M^{2n-k+1}, the descent equation (2.15) gives the relation between the (k-1)-st classes of Chern-Simons type and the k-th one. And for each class of Chern-Simons type it not only has topological meaning in their own right but also shares some common topological properties with those classes of this type with higher degree of curvatures, especially the leading characteristic class in the curvature.

Finally, it should be pointed out that the above theorems can be easily applied to the case in which some ghosts are involved and Zumino's approach[4] is also included (See Appendix B).

III. Cohomology of Gauge Groups

In this section, we will first underline some abstract formalism on cohomology algebra[14,19]. This and the characteristic classes of Chern-Simons type constitute the bases of our analyses on the cohomology of gauge groups. Then we will show that these classes present certain realizations of the cohomology of gauge groups as well as that of the degenerate one. Those realizations will play important roles in the course of analyses on the anomalies in the next section.

1. Coboundary Operators and Their Cohomologies

Let $R^{(k)}$ be a set of arrays of functions with respect to arbitrary (k+1) ordered objects $(A^0, A^1, \ldots, A^k)$. We define an operator

$$\Delta: R^{(k)} \to R^{(k+1)} \tag{3.1}$$

acting on the sequence

$$0 \xrightarrow{\Delta} R^{(0)} \xrightarrow{\Delta} R^{(1)} \xrightarrow{\Delta} \cdots \tag{3.2}$$

in the following way:

$$(\Delta r^{(k)})(A^0, \ldots, A^{k+1}) = \sum_{j=0}^{k+1} (-1)^j r^{(k)}(A^0, \ldots \hat{A}^j, \ldots, A^{k+1}) , \quad \forall r^{(k)} \in R^{(k)} , \tag{3.3}$$

where the caret denotes omission.

It is easy to check that

$$\Delta^2 = 0 , \tag{3.4}$$

which allows for the introduction of the cohomology of $R^{(k)}$ with respect to the operator Δ defined as a coboundary operator. We define $\Delta r^{(k)} \in R^{(k+1)}$ to be a (k+1)-cochain. A k-cochain $r^{(k)}$ is a k-cocycle if $\Delta r^{(k)}=0$ whereas a k-cochain $r^{(k)}=\Delta s^{(k-1)}$, for some $s^{(k-1)} \in R^{(k-1)}$, is a coboundary. Let $Z_\Delta^{(k)}=\{r^{(k)} \mid \Delta r^{(k)}=0\}$ be the set of cocycles (i.e., k-cochains with no coboundaries), $B_\Delta^{(k)}=\{r^{(k)} \mid r^{(k)} =\Delta s^{(k-1)}$, $s^{(k-1)} \in R^{(k-1)}\}$ be the set of coboundaries. Since the coboundary of a coboundary is always empty, $B_\Delta^{(k)}$ is a subset of $Z_\Delta^{(k)}$. The cohomology of $R^{(k)}$ with respect to the operator Δ is defined by

$$H_\Delta^{(k)} = Z_\Delta^{(k)} / B_\Delta^{(k)} , \qquad k=0,1,2,\ldots \tag{3.5}$$

Obviously, $H_\Delta^{(k)}$ is the set of equivalence classes of cocycles $z^{(k)} \in Z_\Delta^{(k)}$ which differ only from coboundaries. The nontrivial element of $H_\Delta^{(k)}$, of course, is a cocycle which is not a coboundary.

It should be pointed out that one can introduce another operator $\bar{\Delta}$ relevant to the coboundary operator Δ as follows:

$$\bar{\Delta} r^{(k)} = \Delta r^{(k)} - r^{(k)} \cdot \pi , \qquad \forall r^{(k)} \in R^{(k)} , \tag{3.6}$$

where π is an exclusion operator which excludes the first object from the ordered ones, for instance

$$\pi : (A^0, A^1, \ldots, A^k) \to (A^1, \ldots, A^k) . \tag{3.7}$$

The operation rule of $\bar{\Delta}$ on $R^{(k)}$ is defined as the same as Eq.(3.3), then $\bar{\Delta}$ is also a coboundary operator:

$$\bar{\Delta}^2=0 \ . \tag{3.8}$$

It is easy to check that the operators Δ, $\bar{\Delta}$, and π satisfy the following equation:

$$(\bar{\Delta}r^{(k)})\cdot\pi=-\Delta\cdot(r^{(k)}\pi) \ . \tag{3.9}$$

An important point is that $\bar{\Delta}r^{(k)}(A^0,\ldots,A^k)$ has one less than $\Delta r^{(k)}(A^0,\ldots,A^k)$ in number of terms, because of the relation (3.6), thus the number of terms in $\bar{\Delta}r^{(k)}$ is equal to that in $\Delta r^{(k-1)}$. This means that the operation of $\bar{\Delta}$ on $r^{(k)}$ is equivalent to the one of Δ on a (k-1)-cochain rather than a k-cochain and in this sense the k-cochain $r^{(k)}$ with respect to the operator Δ is degenerated under the operation of $\bar{\Delta}$ which could be called degenerate coboundary operator. It is obvious that one can introduce another cohomology of $R^{(k)}$ with respect to this operator $\bar{\Delta}$ in the sense of degeneration of the cohomology with respect to the operator Δ.

2. Cohomology of Chern-Simons Type Classes

Let us now consider the case in which these objects, A's, (or some of them) are the connection 1-forms. Thus the cohomologies with respect to the operators Δ and $\bar{\Delta}$ become the ones of the gauge potentials. If $R^{(k)}$'s are taken to be some characteristic polynomials which are invariant under the gauge transformations, then, we can obtain some cohomologies of gauge groups.

The cohomologic meaning of the characteristic classes of Chern-Simons type is very clear and definite. In fact, by means of the coboundary operator Δ, the descent formula (2.15) can be rewritten as

$$\Delta\omega_n^{(k-1)}(A^0,\ldots,A^{k-1};\Delta_{k-1})=d_x\omega_n^{(k)}(A^0,\ldots,A^k;\Delta_k) \ , \tag{3.10}$$

which means that the coboundary of the (k-1)-st classes of Chern-Simons type is exactly equal to the exterior differential of a k-th one as pointed out in Ref.[1]. Furthermore, Eq.(2.17) satisfied by the characteristics of Chern-Simons type, η_n's, can also be rewritten as

$$\Delta\eta_n^{(k-1)}(A^0,\ldots,A^{k-1};\Delta_{k-1},M^{2n-k+1})=\eta_n^{(k)}(A^0,\ldots,A^k;\Delta_k,\partial M^{2n-k+1}) \ . \tag{3.11}$$

For the sake of simplicity, we abbreviate Eq.(3.11) to

$$\Delta\eta_n^{(k-1)}(M^{2n-k+1})=\eta_n^{(k)}(\partial M^{2n-k+1}) \ . \tag{3.11'}$$

It is clear that we can define $\eta_n^{(k)}(M^{2n-k})$ as a k-cochain with respect to the coboundary operator Δ. As explained above, a k-cochain $\eta_n^{(k)}$ is a k-cocycle if $\Delta\eta_n^{(k)}=0$ whereas a k-cochain $\eta_n^{(k)}$ is a coboundary if there exists a (k-1)-cochain $\eta_n^{(k-1)}(M^{2n-k+1})$ such that $\eta_n^{(k)}=\Delta\eta_n^{(k-1)}$. And a k-cocycle $\eta_n^{(k)}$ is nontrivial if it is not a coboundary. Therefore, one would introduce such a cohomology

$$\left.\begin{aligned} Z^k(\eta_n^{(k)},\Delta)&=\left\{\eta_n^{(k)}\,|\,\Delta\eta_n^{(k)}=0\right\} ,\\ B^k(\eta_n^{(k)},\Delta)&=\left\{\eta_n^{(k)}\,|\,\eta_n^{(k)}=\Delta\eta_n^{(k-1)}\right\} ,\\ H^k(\eta_n^{(k)},\Delta)&=Z^k(\eta_n^{(k)},\Delta)/B(\eta_n^{(k)},\Delta) . \end{aligned}\right\} \tag{3.12}$$

The theorem 3 in the preceding section or the formula (3.11) shows that a k-cochain $\eta_n^{(k)}(M^{2n-k})$ is a cocycle if and only if the submanifold $M^{2n-k}\subset\mathfrak{M}$ has no boundary, i.e., $\partial M^{2n-k}=\emptyset$. A k-cochain $\eta_n^{(k)}(M^{2n-k})$ is a coboundary if and only if the submanifold $M^{2n-k}\subset\mathfrak{M}$ is the boundary of a submanifold of one more dimensions, i.e., $M^{2n-k}=\partial M^{2n-k+1}$. And a k-cocycle $\eta_n^{(k)}(M^{2n-k})$ is a nontrivial element of the cohomology group $H^k(\eta_n^{(k)},\Delta)$ if and only if the boundaryless submanifold M^{2n-k} cannot be the boundary of a submanifold M^{2n-k+1}.

As a matter of fact, it is easy to show that there exists a homomorphic relation between the cohomology group $H^k(\eta_n^{(k)},\Delta)$ and the homology of the manifold[13,19] which is used to distinguish topologically inequivalent manifolds; that is

$$\begin{aligned} f:\ H_k(M,\partial)&\to H^k(\eta,\Delta) ,\\ [M^{2n-k}]&\to[\eta_n^{(k)}(M^{2n-k})] , \end{aligned} \tag{3.13}$$

where $[\cdots]$ denotes a representative of an equivalence class, ∂ the boundary operator of homology of manifolds, $H_k(M,\partial)$.

In the simplest case of $k=0$, for instance, $H^0(\eta_n^{(0)},\Delta)$ is the zero-th cohomology group. If M^{2n} is compact,

$$\eta_n^{(0)}(M^{2n})=\int_{M^{2n}} P(F^n) . \tag{3.14}$$

If the characteristic polynomial in the curvature, $P(F^n)$, is taken to be the n-th Chern class, then the zero-th cohomology group $H^0(\eta_n^{(0)},\Delta)$ is the Chern number on the compact manifold M^{2n}.

3. Cohomology of Degenerate Cochains

We now consider the relation between the cohomology of degenerate cochains with respect to the coboundary operator $\bar{\Delta}$ and the classes of Chern-Simons type. We will show that one can find some polynomials, denoted by $\bar{Q}^{(k)}$, consisting of the polynomials Chern-Simons type such that they solve the descent equation for the opeator:

$$\bar{\Delta}\bar{Q}_n^{(k-1)}=d\bar{Q}_n^{(k)} \tag{3.15}$$

with a condition

$$\bar{Q}_n^{(0)}=Q_n^{(0)}=P(F^n) . \tag{3.16}$$

Naturally, if we find such kind of Q's, we would have the following sequence

$$0\xrightarrow{d^{-1}\bar{\Delta}}\bar{Q}^{(0)}\xrightarrow{d^{-1}\bar{\Delta}}\bar{Q}^{(1)}\xrightarrow{d^{-1}\bar{\Delta}}\cdots . \tag{3.17}$$

And the condition (3.16) implies that there should be some relations between the polynomials $\bar{Q}_n^{(k)}$ and $Q_n^{(k)}$ since they are descended from the same polynomial in the curvature, $P(F^n)$.

In order to show how to find such kind of polynomials, we start from Eq. (3.16) to construct $\bar{Q}^{(1)}$. Since $P(F^n)$ is closed, $dQ^{(0)}\pi(A^0,A^1)=dQ^{(0)}(A^1)=dP(A^1)=0$, by the Poincaré Lemma, $Q^{(0)}\pi(A^0,A^1)$ can locally be written as an exterior differential of something denoted by $d^{-1}Q^{(0)}\pi(A^0,A^1)$, where d^{-1} is an anti-differentiation operator acting only on some closed forms. From the relation (3.6), it follows that

$$(\bar{\Delta}\bar{Q}^{(0)})(A^0,A^1)=(\Delta Q^{(0)}-Q^{(0)}\pi)(A^0,A^1)$$
$$=d(Q^{(1)}(A^0,A^1)-d^{-1}Q^{(0)}\pi(A^0,A^1)) \ .$$

This shows that as long as we take

$$\bar{Q}^{(1)}=Q^{(1)}-d^{-1}Q^{(0)}\pi \ ,$$

the descent equation (3.15) can automatically be satisfied. As a matter of fact, Eq.(3.15) can be solved by taking

$$\bar{Q}^{(k)}=Q^{(k)}-d^{-1}\{Q^{(k-1)}\pi+\bar{\Delta}d^{-1}[Q^{(k-2)}\pi+\bar{\Delta}d^{-1}(\cdots+\bar{\Delta}d^{-1}Q^{(0)}\pi)]\} \ . \qquad (3.18)$$

In order to prove that Eq.(3.18) solves Eq.(3.15), we have made use of some relations

$$d\bar{\Delta}=\bar{\Delta}d \ , \qquad (3.19)$$

$$d(r^{(k)}\cdot\pi)=(dr^{(k)})\cdot\pi \ . \qquad (3.20)$$

However, it should be emphasized that since the local exact expression of $Q^{(0)}\pi$ is not unique, the solution (3.18) is of course not unique. We would explain this in some detail. For convenience, we introduce an inclusion operator, which puts zero at the first place of some ordered objects on which it acts, for instance,

$$\iota:(A^1,A^2,\ldots,A^k)\to(0,A^1,A^2,\ldots,A^k) \ . \qquad (3.21)$$

Then the local exact expression for $Q^{(0)}$ can be written as

$$Q_n^{(0)}(A^1)=dQ_n^{(1)}(0,A^1)=dQ_n^{(1)}\iota\pi(A^0,A^1) \ . \qquad (3.22)$$

Obviously, $\bar{Q}_n^{(1)}=Q_n^{(1)}-Q_n^{(1)}\iota\pi$ can also solve Eq.(3.15). And in general, we can prove that

$$\bar{Q}^{(k)}=Q^{(k)}-Q^{(k)}\iota\pi \ , \qquad k=1,2,\ldots \qquad (3.23)$$

are also the solutions of Eq.(3.15).

The differences between the two sequences of degenerate cochains (3.18) and (3.23) stem from the ununiqueness of the local exact expressions. For instance, although $Q^{(0)}$ can be locally written as Eq.(3.22), globally it is true only up to addition of (2n-1) form which has integervalued integral over R^{2n-1}. In fact, we should write

$$d^{-1}Q_n^{(0)}\pi(A^0,A^1)=Q_n^{(0)}i\pi(A^0,A^1)\ ,\quad (\mathrm{mod}\ \mathbb{Z})\ . \tag{3.24}$$

It should be noted that the operator d^{-1} can only be defined in the local sense which leads to the ununiqueness.

Notice that d^{-1} satisfies

$$dd^{-1}=I\ ,\qquad \text{but}\qquad d^{-1}d\neq I \tag{3.25}$$

and the ununiqueness of this kind exists for each value of k.

Finally, we would show the cohomology in Faddeev's approach[2] is a degenerate one with a certain local sense. To see this, we take[5]

$$A^0=A,\quad A^1=0,\quad A^2=g_1dg_1^{-1},\ldots,A^k=(g_1\ldots g_{k-1})d(g_1\cdots g_{k-1})^{-1},\ldots \tag{3.26}$$

then, the solution (3.18) of descent equation (3.15) with the condition (3.16) gives rise to the cohomology in Refs.[2,4].

IV. The Hierarchy of Anomalies

As pointed out before[1,5], the classes of Chern-Simons type and relevant cohomologies of the gauge group would play an important role in the course of analyses on the anomalies. As a matter of fact, from the viewpoint of these topology and cohomology we are given a hierarchy of anomalous objects in quantum field theories with gauge interactions on a series of spacetimes of different dimensions as submanifolds of $\mathscr{M}$ as long as there exists the Abelian or singlet anomalies as the leading ones of the hierarchy in the highest dimensional spacetime, such as $M^{2k}\subseteq\mathscr{M}$.

We have shown that the classes of Chern-Simons type in differently dimensional submanifolds of $\mathscr{M}$ are cohomologically related to each other. On the one hand, these classes stem from a common characteristic polynomial in the curvature; on the other hand, they have definitely topological meaning in their own right. Since the Abelian or singlet anomalies are exactly certain curvature invariant polynomials of the gauge field, there must exist anomalous objects due to proper classes of Chern-Simons type in a series of submanifolds of $\mathscr{M}$ if the singlet anomalies do exist in the highest even dimensional submanifold of $\mathscr{M}$. And these anomalous objects have definitely topological and cohomological meaning as that of corresponding classes of Chern-Simons type as well.

We will start from the analysis on the singlet anomalies in $M^{2n}\subseteq\mathscr{M}$. Let U(x) be an element of a Lie group or its nonsingular matrix representation which represents some background field. Under the action of the gauge group G, U(x) is transformed into

$$U(x)\to g^{-1}(x)U(x),\quad g(x)\in G,\quad \forall x\in M^{2n}\subseteq\mathscr{M}\ , \tag{4.1}$$

and its covariant derivative is defined by

$$DU=dU+AU\ , \tag{4.2}$$

where A is gauge potential 1-form valued on the Lie algebra of G. Let J(U,A) be the expectation value of a singlet anomalous current 1-form of some quantized field on the classical background U(x) and external gauge field A. The J(U,A) always satisfies the singlet anomalous divergence equation[6]

$$d{*}J(U,A)=P(F^n)\ , \tag{4.3}$$

where "*" is the Hodge star operator. This equation can be easily solved by means of the descent formula (3.10) in the cohomology of the gauge group. To see this, we take

$$A_0=UdU^{-1},\qquad A^1=A\ .$$

Then

$$A_t=UdU^{-1}+tDUU^{-1}\ ,$$
$$F_t=tF-t(1-t)(DUU^{-1})^2\ . \tag{4.4}$$

The descent formula gives

$$(\Delta Q^{(0)})(UdU^{-1},A)=dQ^{(1)}(UdU^{-1},A)=P(F(A))\ , \tag{4.5}$$

$$Q^{(1)}(UdU^{-1},A)=n\int_0^1 dtP(DUU^{-1},F_t^{n-1})\ . \tag{4.6}$$

Thus, we get the dual of minimum of J:

$$*J_{min}(U,A)=Q^{(1)}(UdU^{-1},A)=n\int_0^1 dtP(DUU^{-1},F_t^{n-1})\ . \tag{4.7}$$

That is $*J_{min}(U,A)$ is a 1-cochain density with respect to Δ whereas the singlet anomalous divergent is a zero-cochain density, which will become the density of a zero-cocycle if M^{2n} has no boundary.

Let us now consider a submanifold $M^{2n-1}\subset\mathcal{M}$, which may have no relation with the submanifold $M^{2n}\subseteq\mathcal{M}$. For some gauge invariant field theory on M^{2n-1} with both U(x) and A fields, we can define the 1-cochain density $Q^{(1)}(UdU^{-1},A)$ as a gauge invariant Euler-Heisenberg effective Lagrangian which presents the parity violation anomalies on odd dimensions[20-22]. This Euler-Heisenberg Lagrangian can be decomposed into a gauge field part, the U-field part and the contribution from the boundary of M^{2n-1} by means of the descent formula

$$(\Delta Q^{(1)})(UdU^{-1},A,0)=dQ^{(2)}(UdU^{-1},A,0)\ , \tag{4.8}$$

$$Q^{(2)}(UdU^{-1},A,0)=n(n-1)\int_0^1 dt^1\int_0^{1-t^1} dt^2P(DUU^{-1},dUU^{-1},F^{n-2}_{t^1t^2})\ , \tag{4.9}$$

where

$$\left.\begin{aligned}&A_{t^1t^2}=UdU^{-1}+t^1DUU^{-1}+t^2dUU^{-1}\ ,\\&F_{t^1t^2}=dA_{t^1t^2}+A^2_{t^1t^2}\ .\end{aligned}\right\} \tag{4.10}$$

From Eq.(4.8), it follows that

$$Q^{(1)}(UdU^{-1},A)=Q^{(1)}(UdU^{-1},0)-Q^{(1)}(A,0)+dQ^{(2)}(UdU^{-1},A,0)\ . \qquad (4.11)$$

This is just the decomposition formula for $Q^{(1)}(UdU^{-1},A)$[6]. The terms on the right-hand side of Eq.(4.11) present the contributions from the U(x)-field, the gauge field A and the fields on the boundary. It is also reasonable to take $(\Delta Q^{(1)})(UdU^{-1}\ A,0)$ as a part of Lagrangian, then $Q^{(2)}$ discribes the boundary effect of the spacetime M^{2n-1}.

It should be pointed out that

$$\tilde{\Gamma}(U,A)\equiv 2\pi\int_{M^{2n-1}} Q^{(1)}(UdU^{-1},0)+2\pi\int_{M^{2n-2}=\partial M^{2n-1}} Q^{(2)}(UdU^{-1},A,0) \qquad (4.12)$$

is just the Wess-Zumino-Witten effective anomalous action[6] in M^{2n-2}, which has no boundary and can be dealt with as the boundary of M^{2n-1}, if some anomaly-free condition holds. Otherwise

$$\tilde{\Gamma}(U,A) \qquad \text{or} \qquad 2\pi\int_{M^{2n-1}} Q^{(1)}(A,0) \qquad (4.13)$$

gives rise to the generating functional of non-Abelian anomaly on $M^{2n-2}=\partial M^{2n-1}$ and under the gauge transformation (4.1) they generate the finite form of such an anomaly, which contains some global aspects about the anomaly, such as so-called non-purturbative anomaly,[6,16,22,23]

$$\mathcal{O}\!\!\mathcal{C}(g(x),A)=2\pi\int_{M^{2n-1}} Q^{(1)}(gdg^{-1},0)+2\pi\int_{M^{2n-2}} Q^{(2)}(gdg^{-1},A,0)\ , \qquad (4.14)$$

whose infinitesimal form corresponds to the perturbative one[24]

$$\mathcal{O}\!\!\mathcal{C}(\varepsilon(x),A)=2\pi\int_{M^{2n-2}} \varepsilon^a\ Q_a(A,T^a)\ . \qquad (4.15)$$

Note that each term in Eqs.(4.12) and (4.14) has topological meaning from the viewpoint of the Chern-Simons type classes and relevant cohomology of the gauge group, both $\tilde{\Gamma}(U,A)$ and $\mathcal{O}\!\!\mathcal{C}(g,A)$ have also certain cohomologic meaning which we will explain later.

However, if M^{2n-2} has boundary or cannot be the boundary of some M^{2n-1}, the problem is how to write down the Wess-Zumino-Witten effective action, the generating functional of non-Abelian anomaly and the finite form of the anomaly on such kind of manifolds with which we have to deal in the cases of some gravitational fields[13], hybrid bag models[25] and so on. From the viewpoint of the classes of Chern-Simons type and relevant cohomology, this problem could be solved automatically by taking into account $Q^{(2)}$'s directly. Let A^3 be some connection appropriately taken to be relevant to the boundary of M^{2n-2} or the nontrivial topology of M^{2n-2} Then by means of the descent equation, we have

$$(\Delta Q_n^{(2)})(UdU^{-1},A,0,A^3)=dQ_n^{(3)}(UdU^{-1},A,0,A^3) \quad . \tag{4.16}$$

The last term in the left-hand side is just $Q_n^{(2)}(UdU^{-1},A,0)$ whose infinitesimal form is the purturbative anomaly, and the right-hand side gives rise to the contribution from the boundary of M^{2n-2}. If the submanifold M^{2n-2} is a nontrivial one in the homology of the manifolds $H_2(\mathscr{M},\partial)$, then the second characteristic of Chern-Simons type, $\eta_n^{(2)}$, is correspondingly a nontrivial element of the cohomology group $H^2(\eta,\Delta)$; that is

$$\Delta\eta_n^{(2)}(UdU^{-1},A,0,A^3;M^{2n-2})=0 \ , \quad \eta_n^{(2)}\in H^2(\eta,\Delta) \ . \tag{4.17}$$

If we go one step further, by taking $A^0=g_1dg_1^{-1}$ instead of UdU^{-1} and $A^3=g_1g_2d(g_1g_2)^{-1}$, $A^4=g_1g_2g_3d(g_1g_2g_3)^{-1}$ and repermuting A's, we have

$$A^0=A,\ A^1=0,\ A^2=g_1dg_1^{-1},\ A^3=(g_1g_2)d(g_1g_2)^{-1},\ A^4=g_1g_2g_3d(g_1g_2g_3)^{-1}\ , \tag{4.18}$$

then, the descent formula gives

$$(\Delta Q_n^{(3)})(A,0,g_1dg_1^{-1},(g_1g_2)d(g_1g_2)^{-1},g_1g_2g_3d(g_1g_2g_3)^{-1})$$
$$=dQ_n^{(4)}(A,0,g_1dg_1^{-1},(g_1g_2)d(g_1g_2)^{-1},g_1g_2g_3d(g_1g_2g_3)^{-1}) \ . \tag{4.19}$$

By performing integral of both sides over a submanifold $M^{2n-3}\subset\mathscr{M}$, we get

$$(\Delta\eta_n^{(3)})(A,0,g_1dg_1^{-1},(g_1g_2)d(g_1g_2)^{-1},g_1g_2g_3d(g_1g_2g_3)^{-1};M^{2n-3})$$
$$=\eta_n^{(4)}(A,0,g_1dg_1^{-1},(g_1g_2)d(g_1g_2)^{-1};g_1g_2g_3d(g_1g_2g_3)^{-1},\partial M^{2n-3}) \ . \tag{4.20}$$

We will show that these relations are cohomologic conditions for the anomalous Schwinger term in equal time commutator on space $M^{2n-3}(n\geqslant3)$, or the center term in Kac-Moody algebra on space, $M^1\simeq R^1$(See below).

In what follows, we consider the cohomologic meaning of these anomalous objects from the viewpoint of the cohomology of the degenerate coboundary operator $\bar\Delta$ and the degenerate polynomials $\bar Q_n^{(k)}$ consisting of the classes of Chern-Simons type.

In the case of k=0, we have already assumed that

$$\bar Q^{(0)}=Q^{(0)} \tag{4.21}$$

so the singlet anomaly on $M^{2n}\subseteq\mathscr{M}$ is also a density of zero-cochain or of (-1)-degenerate cochain.

In the case of k=1, corresponding to that of 1-cochain density $Q_n^{(1)}$, we now have

$$\bar\Delta\bar Q_n^{(1)}(A,0,UdU^{-1})=d\bar Q^{(2)}(A,0,UdU^{-1}) \ , \tag{4.22}$$

by means of the relation between $Q_n^{(k)}$'s and $\bar{Q}_n^{(k)}$'s, we get

$$Q_n^{(1)}(UdU^{-1},A)+Q^{(1)}(A,0)=d(Q^{(2)}(UdU^{-1},A,0)-d^{-1}Q^{(1)}(0,UdU^{-1})). \quad (4.23)$$

The integral of Eq.(4.23) over $M^{2n-1}\subset\mathcal{M}$ gives rise to a kind of Euler-Heisenberg effective action on M^{2n-1}. And if $\partial M^{2n-1}=0$, M^{2n-1} can be covered by one coordinate neighbourhood, then this action is a 1-degenerate cocycle on such an M^{2n-1}.

On the other hand, the integral of the right-hand side of Eq.(4.23) over a submanifold M^{2n-1}, whose boundary is taken to be a 2n-2 dimensional spacetime, gives the Wess-Zumino-Witten type effective action, $\tilde{\Gamma}(U,A)$, i.e.,

$$\tilde{\Gamma}(U,A)=2\pi\int_{x\in M^{2n-2}}(Q_n^{(2)}(A,0,UdU^{-1})-d^{-1}Q_n^{(1)}(0,UdU^{-1})), \quad (4.24)$$

and $\tilde{\Gamma}(U,A)$ is a 1-degenerate cocycle:

$$\bar{\Delta}\tilde{\Gamma}(U,A)=0. \quad (4.25)$$

Similarly, the finite form of non-Abelian anomaly, $\mathcal{O}\!\!\mathcal{C}(g,A)$, is also a 1-degenerate cocycle

$$\mathcal{O}\!\!\mathcal{C}(g,A)=2\pi\int_{x\in M^{2n-2}}(Q_n^{(2)}(A,0,gdg^{-1})-d^{-1}Q_n^{(1)}(0,gdg^{-1})),$$
$$\bar{\Delta}\mathcal{O}\!\!\mathcal{C}(g,A)=0. \quad (4.26)$$

It should be noted that we must take care of the integration domain because of the local sense of the degenerate cochains.

Furthermore, we have for k=2 and k=3

$$(\bar{\Delta}\bar{Q}_n^{(2)})(A,0,g_1dg_1^{-1},g_1g_2d(g_1g_2)^{-1})=d\bar{Q}_n^{(3)}(A,0,g_1dg_1^{-1},g_1g_2d(g_1g_2)^{-1}), \quad (4.27)$$

$$(\bar{\Delta}\bar{Q}_n^{(3)})(A,0,g_1dg_1^{-1},\dots,g_1g_2g_3d(g_1g_2g_3)^{-1})=d\bar{Q}_n^{(4)}(A,0,\dots,g_1g_2g_3d(g_1g_2g_3)^{-1}). \quad (4.28)$$

If the submanifold $M^{2n-3}\subset\mathcal{M}$ has no boundary and can be covered by one coordinate neighbourhood, the integral of Eq.(4.28) over such an M^{2n-3} gives a 2-degenerate cocycle $\bar{\eta}_n^{(3)}$

$$\bar{\Delta}\bar{\eta}_n^{(3)}(M^{2n-3})=0. \quad (4.29)$$

For the case in which the characteristic polynomial P(F) is taken to be the Chern class and n=3, $\bar{\eta}_n^{(3)}$ is just the 2-cocycle α_2 on R^3 introduced by Faddeev[2] to get the anomalous Schwinger term. For n=2, $\bar{\eta}_2^{(3)}$ is Faddeev's 2-cocycle α_2 on R^1 relevant to the center of the Kac-Moody algebra[2].

As mentioned in Ref.[5], one of the advantages in our approach is that we can deal with the cases in which both left- and right-hand gauge fields are involved. Let us now take Wess-Zumino term[6,8,16] as an example which corresponds to k=1. By taking

$$A^0=A_L,\ A^1=A_R,\ A^2={}^{U^{-1}}A_R=UA_RU^{-1}+UdU^{-1}\ , \qquad (4.30)$$

then from the descent formula (3.15) we have

$$(\bar{\Delta}\bar{Q}^{(1)})(A_L,A_R,{}^{U^{-1}}A_R)=d(Q^{(2)}-d^{-1}Q^{(1)}\pi)(A_L,A_R,{}^{U^{-1}}A_R)\ . \qquad (4.31)$$

From which it follows that

$$Q^{(1)}(A_L,A_R)-Q^{(1)}(A_L,{}^{U^{-1}}A_R)=d(Q^{(2)}-d^{-1}Q^{(1)}\pi)(A_L,A_R,{}^{U^{-1}}A_R)\ . \qquad (4.32)$$

If the submanifold $M^{2n-2}\subset\mathcal{M}$ can be regarded as the boundary of some submanifold M^{2n-1}, then by the Stoke's formula, the integral of right-hand side of Eq.(4.32) gives rise to the Wess-Zumino term with both left- and right-hand gauge field on M^{2n-2}, $\tilde{\Gamma}(U,A_L,A_R)$, i.e.,

$$\tilde{\Gamma}(U,A_L,A_R)=2\pi\int_{x\in M^{2n-2}}(Q^{(2)}-d^{-1}Q^{(1)}\pi)(A_L,A_R,{}^{U^{-1}}A_R)\ . \qquad (4.33)$$

Since

$$2\pi\int_{x\in M^{2n-1}}Q^{(1)}(A_L,A_R)$$

gives rise to Bardeen's counterterm R^3 to transfer the anomaly from the (L,R)-form to the (V-A)-form in 4-dimensional spacetime, therefore, Eq.(4.33) is the Wess-Zumino term in (V-A) form. From Eq.(4.31), this term is a 1-degenerate cocycle with respect to the degenerate coboundary operator $\bar{\Delta}$,

$$\bar{\Delta}\tilde{\Gamma}(U,A_L,A_R)=0\ . \qquad (4.34)$$

Finally, we would like to point out that the formalism proposed in this paper can also be used to deal with the gravitational anomaly and the Lorentz anomaly[22,16,27,28] as long as the characteristic polynomial in the curvature, P(F), is suitably taken. For instance, in the case of Dirac particle in the gravitational field, the polynomial P should be taken as

$$\hat{A}(M)ch(F)\ , \qquad (4.35)$$

where $\hat{A}$ is the A roof genus and ch the Chern cheracter[13]. Then we would have a hierarchy of Lorentz anomalies as well.

Appendix A

We now sketch out how to prove the hierarchy of Eq.(2.14) and the descent formula (2.15) which describe the most important properties of the polynomials of Chern-Sinons type.

From Eq.(2.8), it is easy to see that

$$DH_0=d_{(x)}H_0+A_{0\,t^1\ldots t^k}H_0+H_0A_{0\,t^1\ldots t^k}=d_{(t)}F_{0\,t^1\ldots t^k}\ ,\quad d_{(t)}H_0=0\ . \qquad (A1)$$

By taking the exterior derivatives, $d_{(x)}$, of both sides of Eq.(2.12), we get

$$
\begin{aligned}
d_{(x)} q_n^{(k)} &= (-1)^{\frac{1}{2}k(k-1)} \frac{n!}{k!(n-k)!} d_{(x)} P(H_0^k,\ F^{n-k}_{0,t^1\dots t^k}) \\
&= (-1)^{\frac{1}{2}k(k-1)} \frac{n!}{(k-1)!(n-k)!} P(H_0^{k-1}, d_{(t)} F_{0,t^1\dots t^k}, F^{n-k}_{0,t^1\dots t^k}) \\
&= (-1)^{\frac{1}{2}k(k-1)} \frac{n!}{(k-1)!(n-k+1)!} d_{(t)} P(H_0^{k-1}, F^{n-k}_{0,t^1\dots t^k}) \\
&= d_{(t)} q_n^{(k-1)} \ .
\end{aligned}
\tag{A2}
$$

Thus we complete the proof of the theorem 1 in Sec. II for the densities of the Chern-Simons type classes.

Then by taking the integral of (A2) over a k-simplex Δ_k whose orientation is taken as $[\partial/\partial t^1,\dots,\partial/\partial t^k]$, we get

$$d_{(x)} Q_n^{(k)} = \int_{\Delta_k} d_{(t)} q_n^{(k-1)} \ . \tag{A3}$$

By Stokes' theorem, the right-hand side of Eq.(A3) turns out to be

$$\int_{\partial\Delta_k} q_n^{(k-1)} = \sum_{j=0}^{k-1} (-1)^j Q_n^{(k-1)} \ . \tag{A4}$$

Here we made use of the relation between the orientation of Δ_k and that of Δ_{k-1}'s which construct the boundary of Δ_k, $\partial\Delta_k$[1]. Then by definition of the coboundary operation of Δ, we have

$$d_{(x)} Q_n^{(k)} = \sum_{j=0}^{k-1} (-1)^j Q_n^{(k-1)} = \Delta Q_n^{(k-1)} \ . \tag{A5}$$

Thus we get the descent formula for the classes of Chern-Simons type (2.15) or (3.10).

The other important property, the theorem 3, has been already proved in Sec. II.

Appendix B

Let $A^0(x),\dots,A^k(x)$ be the connection 1-form on a principal bundle $P(\mathscr{M},G)$ with base space $\mathscr{M}$, structure group G, $d_{(u)}$, $d_{(x)}$, $d_{(s)}$ be the exterior differential operators with respect to the coordinates of $u\in P$, $x\in\mathscr{M}$, $s\in F\simeq G$ respectively, where F is the standard fibre of the bundle which is isomorphic to the structure group. Obviously, we have

$$d_{(u)} = d_{(x)} + d_{(s)} \ , \tag{B1}$$

$$d_{(x)}^2 = d_{(s)}^2 = d_{(u)}^2 = 0 \ , \qquad d_{(x)} d_{(s)} + d_{(s)} d_{(x)} = 0 \ . \tag{B2}$$

Denoting that

$$\overset{(i)}{\mathscr{A}}(x,s) = g^{-1}(x,s) \overset{(i)}{A}(x) g(x,s) + g^{-1}(x,s) d_{(u)} g(x,s), \quad i=1,\dots,k. \tag{B3}$$

$$v(x,s)=g^{-1}(x,s)d_{(s)}g(x,s) \ , \tag{B4}$$

then, it is easy to see that

$$d_{(s)}v=-v^2 \ , \qquad d_{(u)}v+v^2=d_{(x)}v=F(v) \ , \tag{B5}$$

$$d_{(s)}\mathscr{A}(x,s)=-dv-v\mathscr{A}-\mathscr{A}v-v^2 \ . \tag{B6}$$

By taking ${}^gA^0=0$; the interpolation then is

$$\mathscr{A}_{0,t^1\dots t^k}=\mathscr{A}^0+\sum_{j=1}^{k}t^j(\mathscr{A}^j-\mathscr{A}^0)=v+\sum_{j=1}^{k}t^j\,{}^gA^j(x,s) \ , \tag{B7}$$

where ${}^gA=g^{-1}Ag+g^{-1}d_{(x)}g$. Introduce the polynomials with respect to $\mathscr{A}^0,\dots,\mathscr{A}^k$

$$Q_n^{(k)}(\mathscr{A}^0,\dots,\mathscr{A}^k;\Delta_k)=\frac{n!}{k!(n-k)!}\int_{\Delta_k}P(\mathscr{K}^k,\mathscr{F}^{n-k}_{0,t^1\dots t^k}) \ , \tag{B8}$$

where

$$\mathscr{K}=\sum_{j=1}^{k}dt^j(\mathscr{A}^j-\mathscr{A}^0), \quad \mathscr{F}_{0,t^1\dots t^k}=d_{(u)}\mathscr{A}_{0,t^1\dots t^k}+\mathscr{A}^2_{0,t^1\dots t^k} \ . \tag{B9}$$

It is easy to see that $Q_n^{(k)}$'s satisfy a descent equation

$$\Delta Q_n^{(k-1)}=d_{(u)}Q_n^{(k)} \ , \qquad k=1,2,\dots \ . \tag{B10}$$

For k=1, having in mind that $\mathscr{A}^0=v$, we have

$$P(F^n(A^1))-P(dv^n)=(d_{(x)}+d_{(s)})n\int_0^1 dt P(A^1,(td\,{}^gA^1+t^2({}^gA^1)^2+(1-t)dU)^{n-1}) \ . \tag{B11}$$

The expansions of Eq.(B11) with respect to the power of dv give rise to Zumino's results with v regarded as ghost[4].

References

[1] H.Y. GUO, K. WU and S.K. WANG, Commun. in Theor. Phys. 4(1985) 113.

[2] L.D. Faddeev, Phys. Lett. 145B(1984)81.

[3] X.C. SONG, Private Communication, Aug. 1984.

[4] B. Zumino, "Cohomology of Gauge Groups", to appear as NSF-ITP preprint.

[5] H.Y. GUO, B.Y. HOU, S.K. WANG and K. WU, Commun. in Theor. Phys. 4(1985) 145.

[6] K.C. CHOU, H.Y. GUO, K. WU and X.C. SONG, Phys. Lett. 134B(1984) 67; Commun. in Theor. Phys. 3(1984) 73; K.C. CHOU, H.Y. GUO, X.Y. LI, K. WU and X.C. SONG, ibid. 3(1984) 491.

[7] B. Zumino, Less Houches Lectures 1983, to be published by North-Holland, Ed. R. Stora and B. De Witt; B. Zumino. Y.S. WU and A. Zee, Nucl. Phys. B239(1984) 477.

[8] R. Stora, Cargese lectures 1983.

[9] L. Alvarez-Gaumê and P. Ginsparg, "The Structure of Gauge and Gravitational Anomalies", preprint HUTP-84/A016; and the references therein.

[10] S.S. Chern and J. Simons, Ann. Math. 99(1974) 48.

[11] S.S. Chern, "Complex Manifolds without Potential Theory", Springer, New York, 1979.

[12] S. Kobayashi and K. Nomizu, "Foundations of Differential Geometry", Vol. II, Interscience Publ. 1969.

[13] T. Eguchi, P.B. Gilkey and A.J. Hanson, Phys. Reports 66(1980) 213.

[14] H. Cartan and S. Eilenberg, "Homological Algebra", Princeton Univ. Press, Princeton 1956.

[15] J. Wess and B. Zumino, Phys. Lett. 37B(1971) 95.

[16] E. Witten, Nucl. Phys. B223(1983) 422.

[17] R. Jackiw, "In Lectures on Current Algebra and Its Applications", Princeton Univ. Press, Princeton, 1972.

[18] H. Cartan, "In Colloque de Topologie" (Espaces fibre's) Bruxelles (1950).

[19] R. Bott and L. TU, "Differential Forms in Algebraic Geometry", Springer, New York, 1982.

[20] S. Deser, R. Jackiw and S. Templeton, Phys. Rev. Lett. 48(1982) 975; Ann. Phys. (N,Y) 140(1982) 372.

[21] A.J. Niemi and G.W. Semenoff, Phys. Rev. Lett. 51(1983) 2077; A.N. Redlich, ibid. 52(1984) 18.

[22] K.C. CHOU, H.Y. GUO and K. WU, Commun. in Theor. Phys. 4(1985) 91.

[23] E. Witten, Phys. Lett. 117B(1982) 324.

[24] S. Adler, Phys. Rev. 177(1969) 2426; J. Bell and R. Jackiw, Nuovo Cim. 60A(1969) 47.

[25] See, for example, J. Goldstone and R.L. Jaffe, Phys. Rev. Lett. 51(1983) 1518.

[26] L. Alvarez-Gaumê and E. Witten, Nucl. Phys. B234(1984) 269; L.N. CHANG and H.T. Nieh, Stony Brook preprint ITP-SB-84-25; W.A. Bardeen and B. Zumino, Lawrence Berkeley Lab. preprint LBL-1763, 1984.

[27] M.F. Atiyah and I.M. Singer, Proc. Nat. Acad. Sci. USA 81(1984) 2597.

[28] O. Alvarez, I.M. Singer and B. Zumino, "Gravitational Anomalies and the Family's Index Theorem", preprint, UCB-PTH-84/9, LBL-17672 (April 1984).

Letters in Mathematical Physics **11** (1986) 179–187.

The Global Topological Meaning of Cocycles in a Gauge Group

BO-YU HOU
Institute of Modern Physics, Northwest University, Xian, China

BO-YUAN HOU
Physics Department, Neimonggu University, Huhehaote, China

and

PEI WANG
Physics Department, Sichuan University, Chengdu, China

(Received: 31 January 1985; revised version: 30 November 1985)

Abstract. This Letter shows explicitly that the descent sequence of cocycles in a gauge group realizes the Čech–de Rham double complex. Meanwhile the Čech complex corresponds to finite gauge transformations. This Letter also shows how the indices of each order of cocycles characterizes the obstructions on overlapping spheres in different dimensions and how these indices are equal to a common one.

1. Introduction

Recently, much interesting progress has been made in studying the topological origin of anomalies. Zumino [1] and Stora [2] introduced the descent sequence of infinitesimal cocycles ω_m^k, $\delta\omega_m^k = \mathrm{d}\omega_{m-1}^{k+1}$ (here δ denotes the infinitesimal gauge transformation, i.e., BRS transformation [3], and d denotes the exterior differential), then Atiyah *et al.* [4–5] related it with the family index theorem, and discussed the topological obstructions. Furthermore, Faddeev [6] introduced the gauge-group cohomology and the sequence of forms Ω_m^k, $\Delta\Omega_m^k = \mathrm{d}\Omega_{m-1}^{k+1}$ (here Δ denotes a finite gauge transformation), he also analyzed the structure of the representation of 1- and 2-cocycles of a group, and pointed out that the anomalous Schwinger terms are infinitesimal 2-cocycles. Jackiw [7] and the present authors [8] pointed out independently that the 3-cocycle in group cohomology implies the nonassociativity of the representation of a group and the invalidation of the Jacobi identity of infinitesimal generators. In addition, Zumino [9] discussed the relationship between the Δ–d sequence Ω and the δ–d sequence ω. Guo *et al.* [10] introduced the simplex coboundary operator Δ and descent sequence cocycle Q on the connection space $\{A\}$. And Guo *et al.* [11], Song [11], and Faddeev *et al.* [11] independently found the relationship between sequences Q and Ω (in fact, the definition of the coboundary operator Δ operating on Q or Ω is different, but we use the same notation for simplicity).

In this Letter, we explain the global topological meaning of the Δ–d sequence Ω, the cocycles of gauge-group transformations. As is well known, in the nontrivial case ($\Delta\alpha^k = z \neq 0$), the Poincaré lemma is only valid locally, so there exist obstructions to obtaining the Δ–d sequence globally. In order to overcome these, we introduce an open cover $\{U_\alpha\}$ of S^{2n}, in which each of the cover patches corresponds to different gauges, with consistent transition functions in the intersections of patches, so we have constructed a global bundle. Since there is a duality between the simplex, whose vertices are equivalent connections related by the aforementioned transition functions, and the nerve of patches which cover the base manifold, the Čech cohomology difference operator turns out to be exactly the coboundary operator Δ on the cochain of the gauge group. Thus, by using the generalized Mayer–Vietoris sequence, we can globally define the operator d^{-1} and resolve the ambiguity in mod 2π. We have also shown that each order of topological obstructions $\Delta\alpha^k/2\pi$ can be realized by the same Čech–de Rham double cohomology index which is equal to the Chern number Z of the principle bundle on S^{2n}. This explicitly describes the global topological meaning of the cocycles of the gauge group.

2. The Čech–De Rham Double Complex

Let $A = A^a_\mu(x)\lambda_a\,\mathrm{d}x^\mu$ be a Lie algebra-valued gauge potential 1-form and $\mathscr{A} = \{A\}$ be a topologically-trivial affine space. Let $\mathscr{G}$ be the gauge group

$$\begin{aligned} &M \xrightarrow{g} G\,, \quad g(x) \in \mathscr{G}\,, \\ &A(x) \to A(x)^g = g(x)^{-1}A(x)g(x) + g(x)^{-1}\,\mathrm{d}g(x)\,. \end{aligned} \tag{1}$$

$\mathscr{G} = \{g(x) \mid x \in M\}$ is a group of infinite dimensions and may be topologically nontrivial, so that the orbital space $\mathscr{A}/\mathscr{G}$ may also be topologically nontrivial ($\pi_N(\mathscr{G}) = \pi_{N+1}(\mathscr{A}/\mathscr{G})$). This causes a topological obstruction to the existence of the covariant Dirac operator

$$\mathrm{Ind}\, i\not{D}_{1/2} = \int_M \hat{A}(M)\,\mathrm{Ch}(F)\,. \tag{2}$$

where $\mathrm{Ch}(F)$ is known as the Chern character, and $\hat{A}(M)$ the A-roof genus of the manifold. Without loss of generality, we consider $M = S^6$ as an example, then $\hat{A}(M) = 1$, and $\mathrm{Ind}\, i\not{D}_{1/2}$ can be expressed as

$$\mathrm{Ind}\, i\not{D}_{1/2} = -\frac{i}{48\pi^3}\int_{S^6} \mathrm{Tr}(F^3) \equiv \int_{S^6} \Omega_6^{-1}(F) = Z\,. \tag{3}$$

Here $F = \mathrm{d}A + A^2 = \frac{1}{2}F_{\mu\nu}\,\mathrm{d}x^\mu\,\mathrm{d}x^\nu$, integer Z is the third Chern number, and $\Omega_6^{-1}(F)$ is the third Chern class, which is a closed 6-form on the base space. However, it cannot be expressed as an exact form on the whole S^6 when $Z \neq 0$. But if we cover the manifold $M = S^6$ by two patches U_i $(i = 0, 1)$ such that $U_0 \cup U_1 = S^6$, each U_i is a contractible six-dimensional disc. Then we can find a form $\Omega_5^0(A_i)$ on each U_i such that

$$\Omega_6^{-1}(F) = -\mathrm{d}\Omega_5^0(A_i)\,, \tag{4}$$

where $\Omega_5^0(A_i)$ is the well-known Chern–Simons form and can be expressed as

$$\Omega_5^0(A) = \frac{3i}{48\pi^3}\int_0^1 \mathrm{d}t\,\mathrm{Tr}(AF_t^2)$$

$$= \frac{i}{48\pi^3}\,\mathrm{Tr}\left(F^2A - \tfrac{1}{2}FA^3 + \tfrac{1}{10}A^5\right), \tag{5}$$

where $F_t = tF + t(t-1)A^2$. Note that on the intersection $U_0 \cap U_1 = U_{01}$, the two-gauge potentials A_i, defined on U_i, respectively, are related by a gauge transformation. Combining the de Rham cohomology and the Čech cohomology, we obtain the following Δ–d double complex

$$\begin{array}{ccccc} M & & U_0 & & U_{01} \\ \Omega_6^{-1}(F) & \xrightarrow{r} & \Omega_6^{-1}(F) & \xrightarrow{\Delta} & 0 \\ & & \mathrm{d}\uparrow & & \mathrm{d}\uparrow \\ & & \Omega_5^0(A_i) & \xrightarrow{\Delta} & \Omega_5^0(A^g) - \Omega_5^0(A), \end{array}$$

where r denotes restriction, d is the exterior differential, and Δ is the difference map $\Omega(1) - \Omega(0)$. Here $\Omega(i)$ is defined on U_i. It is easy to show that

$$\Delta\Omega_5^0(A, A^g) \equiv \Omega_5^0(A^g) - \Omega_5^0(A) \tag{6}$$

is closed, i.e., $\mathrm{d}\Delta\Omega_5^0 = 0$. Since the intersection $U_{01} = U_0 \cap U_1$ is the homotopic equivalent to S^5 which is not contractible, so $\Delta\Omega_5^0$ is not generally exact. In fact, if we choose some 'S^5' in U_{01} to divide S^6 into two discs $D^6(i)$, i.e., $D^6(1) \cup D^6(0) = S^6$ and $\partial D^6(i) = S^5(i) = (-1)^{i+1}S^5$, then we have

$$-\int_{S^5}\Delta\Omega_5^0(A, A^g) = -\sum_{i=0,1}\int_{D_i^6}\mathrm{d}\Omega_5^0(A_i) = \sum_{i=0,1}\int_{D_i^6}\Omega_6^{-1}(F) = \int_{S^6}\Omega_6^{-1}(F) = Z. \tag{7}$$

Therefore, it is clear that, when index $Z \neq 0$, $\Delta\Omega_5^0$ is a nontrivial cohomology form and it has the same characteristic number as the Chern form.

Let

$$\alpha^0(A) = 2\pi\int_{S^5}\Omega_5^0(A),$$

then

$$\Delta\alpha^0(A; g) = \alpha^0(A^g) - \alpha^0(A) = 2\pi Z, \tag{8}$$

and the coboundary operation Δ of the gauge group appears naturally and α^0 on S^5 represents a cohomology class. It is obvious that the one-simplex $A \to A^g$ in the gauge-potential space is the dual of the common boundary S^5 of $D^6(i)$ on the base space. By using the aforementioned open cover, we can individually obtain $\mathrm{d}^{-1}\Omega_6^{-1} = \Omega_5^0$ in each patch of the cover.

Repeating this division procedure, we obtain the higher 1-cocycle α^1 in the lower-dimensional space S^4 (more precisely, a submanifold homeomorphic to S^4). For this purpose we divide S^6 into three discs $D^6(i)$, $i = 0, 1, 2$

$$\begin{aligned} Z &= \int_{S^6} \Omega_6^{-1}(F) = \sum_i \int_{D^6(i)} \Omega_6^{-1}(F) = -\sum_i \int_{S^5(i)} \Omega_5^0(A_i) = -\sum_{i\neq j} \int_{D^5(i,j)} \Omega_5^0(A_i) \\ &= -\sum_{i<j} \int_{D^5(i,j)} \Delta\Omega_5^0(A_i, A_j) = -\sum_{i<j} \int_{D^5(i,j)} \mathrm{d}\Omega_4^1(A_i, A_j) = -\int_{S^4(i,j)} \Omega_4^1(A_i, A_j) \\ &= \int_{S^4} \Delta\Omega_4^1(A_0, A_1, A_2) \end{aligned} \tag{9}$$

where $A_0 \equiv A$, $A_1 \equiv A^{g_1}$, $A_2 \equiv A^{g_1 g_2}$. Now topologically nontrivial S^5 is divided into sums of D^5, which is topologically trivial, and the Poincaré lemma is valid, hence there exists a Ω_4^1 such that $\mathrm{d}\Omega_4^1 = \Delta\Omega_5^0$, and

$$\partial D^6(i) = S^5(i) = \sum_{j\,(\neq i)} D^5(i, j)\,, \qquad \partial D^5(i, j) = S^4(i, j) = (-1)^{i+j} S^4\,, \quad (i<j)\,,$$

i.e., discs $D^5(i, j)$ have a common boundary S^4. Let

$$\alpha^1(A; g) = 2\pi \int_{S^4} \Omega_4^1(A, A^g)\,. \tag{10}$$

then

$$\Delta\alpha^1(A; g_1, g_2) = \alpha^1(A^{g_1}, g_2) - \alpha^1(A; g_1 g_2) + \alpha^1(A; g_1) = 2\pi Z\,. \tag{11}$$

$\alpha^1(A; g)$ gives the 1-cocycle of the gauge group acting on the manifold of the Yang–Mills fields on S^4. Here, in terms of Equation (9) we have related the character of the topological obstruction to the index Z and interpreted the topological meaning of 'mod 2π'. The reader will find that the open cover now includes three patches U_i ($i = 0, 1, 2$). All intersections U_{01}, U_{02}, and U_{12} of each pair of patches are nonempty and homotopically equivalent to the topologically trivial discs D^5, but the last common intersection $U_0 \cap U_1 \cap U_2$ of the patches is the homotopic equivalent to the topologically nontrivial sphere S^4 – we call this an 'almost good cover'. Generally, we can cover S^6 by $m + 2$ ($m \leqslant 5$) patches such that the intersections of all $k + 2$ patches ($0 \leqslant k < m$) are nonempty and homotopically equivalent to the disc D^{6-k-1}, and only the common intersection of all the $m + 2$ patches is a homotopic equivalent to a sphere S^{6-m-1}.

Correspondingly, we can divide the manifold S^6 into $m + 2$ discs D^6 and obtain the Δ–d double sequence

$$\Delta\Omega_{2n-k-1}^k(A_0, A_1, \ldots, A_{k+1}) = (-1)^k \mathrm{d}\Omega_{2n-k-2}^{k+1}(A_0, A_1, \ldots, A_{k+1})\,, \tag{12}$$

$$\Delta\Omega_{2n-k-1}^k(A_0, A_1, \ldots, A_{k+1}) \equiv \sum_{i=0}^{k+1} (-1)^i \Omega_{2n-k-1}^k(A_0, A_1, \ldots, \hat{A}_i, \ldots, A_{k+1})\,, \tag{13}$$

where $k \leqslant m$, $\hat{A}_i$ denotes that A_i is excised, and

$$A_i \equiv A^{g_1, g_2, \ldots, g_i} \equiv (g_1 \cdots g_i)^{-1} A (g_1 \cdots g_i) + (g_1 \cdots g_i)^{-1} \mathrm{d}(g_1 \cdots g_i) . \tag{14}$$

It is easy to prove that

$$\Delta^2 = 0 . \tag{15}$$

Let

$$\alpha^m(A; g_1, g_2, \ldots, g_m) = 2\pi \int_{S^{5-m}} \Omega^m_{5-m}(A_0, A_1, \ldots, A_m) , \tag{16}$$

then

$$\Delta\alpha^m(A; g_1, g_2, \ldots, g_{m+1}) = 2\pi \int_{S^{5-m}} \Delta\Omega^m_{5-m}(A_0, A_1, \ldots, A_{m+1}) = 2\pi Z , \tag{17}$$

where

$$\begin{aligned} &\Delta\alpha^m(A; g_1, g_2, \ldots, g_{m+1}) \\ &\quad = \alpha^m(A^{g_1}; g_2, g_3, \ldots, g_{m+1}) - \alpha^m(A; g_1, g_2, g_3, \ldots, g_{m+1}) + \cdots \\ &\quad\quad + (-1)^i \alpha^m(A; g_1, \ldots, g_i g_{i+1}, \ldots, g_{m+1}) + \cdots \\ &\quad\quad + (-1)^{m+1} \alpha^m(A; g_1, g_2, \ldots, g_m) . \end{aligned} \tag{18}$$

When the number of patches are more than 8, we will get a good cover and obtain the whole descent sequence (Table I), which can be described by the Čech cohomology. Notice that $\Delta^2 = 0$, $\mathrm{d}^2 = 0$, and $\Delta \mathrm{d} = \mathrm{d}\Delta$, so we can introduce the de Rham–Čech cohomology operator $D = \Delta + (-1)^p \mathrm{d}$ for the intersections of the p patches. It satisfies

$$D^2 = 0 . \tag{19}$$

Table I. The Čech–de Rham complex. Here $\cap_p U$ denotes the intersection of p patches, and $\Omega^k_{5-k} \equiv \Omega^k_{5-k}(A_{i_0}, \ldots, A_{i_k})$, $C^p(U, R)$ consists of the locally-constant real functions on the $(p+1)$-fold intersection $\cap_{p+1} U$.

M	U_i	$\bigcap_2 U$	$\bigcap_3 U$	$\bigcap_4 U$	$\bigcap_5 U$	$\bigcap_6 U$	$\bigcap_7 U$	
$0 \to \Omega_6^{-1} \to$	Ω_6^{-1}	0						
	Ω_5^0	$\Delta\Omega_5^0$	0					
		Ω_4^1	$\Delta\Omega_4^1$	0				
			Ω_3^2	$\Delta\Omega_3^2$	0			
				Ω_2^3	$\Delta\Omega_2^3$	0		
					Ω_1^4	$\Delta\Omega_1^4$	0	
						Ω_0^5	$\Delta\Omega_0^5$	0

$\uparrow$ inclusion

$C^0(U, R) \to C^1(U, R)$ - — — — — — — — — — — — — — — —$C^6(U, R) \to 0$

$\uparrow$

0

Let

$$\Omega^* = \sum_{p=-1}^{6} \Omega_p^{5-p}, \tag{20}$$

in which $\Omega_{-1}^6 \equiv C^6(U, R)$. Furthermore, let

$$\begin{aligned} r\Omega^{-1} &\equiv \text{`}\Delta\text{'}\Omega^{-1} \quad (r \text{ denotes restriction}), \\ i\Omega_{-1} &\equiv \text{`d'}\Omega_{-1} \quad (i \text{ denotes inclusion}). \end{aligned} \tag{21}$$

Then we have

$$D\Omega^* = 0. \tag{22}$$

Here we have used Equation (12) and $d\Omega_6^{-1} = 0$, $\Delta C^6(U, R) = 0$. The latter equation holds because $\{U_i\}$ is a good cover of the sphere S^6.

The sequence Ω_{2n-k}^{k-1} can be expressed in terms of the simplex cocycle form Q_{2n-k}^k on the connection space $\{A\}$. The definition and the explicit expression of Q_{2n-k}^k will be given in the next section.

3. Group Cocycles Ω are Expressed by Known Simplex Cocycles Q on $\{A\}$

In the connection space $\mathscr{A} = \{A\}$, we are given a k-simplex with vertices $A_0, A_1, \ldots, A_k$. Let $P(F^n)$ be a characteristic polynomial of degree n, then

$$\begin{aligned} &Q_{2n-k}^k(A_0, A_1, \ldots, A_k) \\ &= (-1)^{[k/2]} \frac{i^n}{(2\pi)^n (n-k)!} \int_{\Delta_k} dt_1 \cdots dt_k S \operatorname{Tr}((A_1 - A_0) \cdots (A_k - A_0) F_{t_1 \cdots t_k}^{n-k}) \end{aligned}$$

in which (23)

$$\begin{aligned} F_{t_1 \cdots t_k} &= dA_{t_1 \cdots t_k} + A_{t_1 \cdots t_k}^2, \\ A_{t_1 \cdots t_k} &= A_0 + \sum_{j=1}^{k} t_j (A_j - A_0). \end{aligned} \tag{24}$$

It is easy to see that Q_{2n-k}^k has the following property

$$Q_{2n-k}^k(A_0, \ldots, A_k) = \varepsilon_{i_0 \cdots i_k}^{0 \cdots k} Q_{2n-k}^k(A_{i_0}, \ldots, A_{i_k}) \tag{25}$$

which is invariant under the simultaneous gauge transformation of all A's

$$Q_{2n-k}^k(A_0^g, A_1^g, \ldots, A_k^g) = Q_{2n-k}^k(A_0, A_1, \ldots, A_k), \tag{26}$$

and satisfies

$$\Delta Q_{2n-k}^k(A_0, \ldots, A_{k+1}) = (-1) = (-1)^k dQ_{2n-k-1}^{k+1}(A_0, \ldots, A_{k+1}), \tag{27}$$

where

$$\Delta Q_{2n-k}^k(A_0, \ldots, A_{k+1}) \equiv \sum_{i=0}^{k+1} (-1)^i Q_{2n-k}^k(A_0, \ldots, \hat{A}_i, \ldots, A_{k+1}). \tag{28}$$

Assume that

$$Q^* = \sum_{k=0}^{n} Q^k_{2n-k}, \tag{29}$$

in which we denote

$$Q^0_{2n} \equiv P(F^n) \equiv \Omega^{-1}_{2n}, \tag{30}$$

and let

$$D = \Delta + (-1)^p \, \mathrm{d} \tag{31}$$

where p is the order of the simplex, then we get (here $\Delta Q^n_n = 0$ is used)

$$DQ^* = 0. \tag{32}$$

It was proved in [11] that

$$\Omega^k_{2n-k-1}(A, A_1, \ldots, A_k) = (-1)^{k+1} Q^{k+1}_{2n-k-1}(0, v_1, \ldots, v_k, A) + \\ + R^k_{2n-k-1}(0, v_1, \ldots, v_k), \tag{33}$$

in which $v_1 = g_1 \, \mathrm{d}g_1^{-1}$, $v_2 = (g_1 g_2) \, \mathrm{d}(g_1 g_2)^{-1}, \ldots$. On the right-hand side of Equation (33), the first term depends on A and has a global meaning without a topological obstruction because $\mathscr{A}$ is topologically trivial. The second term R^k_{2n-k-1}, however, can be expressed by means of the recurrence method. Let

$$\bar{\Delta} Q^k_{2n-k}(0, v_1, \ldots, v_k, A) \\ \equiv \Delta Q^k_{2n-k}(0, v_1, \ldots, v_k, A) + (-1)^k Q^k_{2n-k}(0, v_1, \ldots, v_k). \tag{34}$$

It can be proved [11] that

$$\Delta \Omega^{k-1}_{2n-k}(A, A_1, \ldots, A_k) \\ = (-1)^k \bar{\Delta} Q^k_{2n-k}(0, v_1, \ldots, v_k, A) + \Delta R^{k-1}_{2n-k}(0, v_1, \ldots, v_k). \tag{35}$$

From Equations (12), (27), (33), (34), and (35) we get

$$\Delta R^{k-1}_{2n-k}(0, v_1, \ldots, v_k) + Q^k_{2n-k}(0, v_1, \ldots, v_k) \\ = (-1)^{k+1} \, \mathrm{d} R^k_{2n-k-1}(0, v_1, \ldots, v_k). \tag{36}$$

Let

$$R^* \equiv \sum_{k=1}^{2n-1} R^k_{2n-k-1} \qquad (R^0_{2n-1} = 0), \tag{37}$$

$$\mathring{Q}^* = \sum_{k=1}^{n} \mathring{Q}^k_{2n-k} \qquad (\mathring{Q}^k_{2n-k} \equiv Q^k_{2n-k}(0, v_1, \ldots, v_k)). \tag{38}$$

Then we get

$$DR^* = -\mathring{Q}^*, \tag{39}$$

so R^* is a cochain in the Čech–de Rham D cohomology and $\mathring{Q}^*$ is a coboundary.

The explicit expression of Q^k_{2n-k} (23) can be derived recurrently from formula (27), because ΔQ^k_{2n-k} is closed and $\mathscr{A}$ is topologically trivial. We can, therefore, construct the operator d^{-1} by means of the homotopic operator (see [10]). Using the same method, we can construct d^{-1} for $\Delta R^k_{2n-k-1} + Q^{k+1}_{2n-k-1}$ on D^5 and then get R^{k+1}_{2n-k-2} because of its closeness.

When the dimension of the base space $q = 4$, the WZW action is a 1-cocycle α^1 [12]. The leading term (the first-order infinitesimal quantity when $v_i \to 0$) of this 1-cocycle gives the nonabelian anomaly of four-dimensional divergence of the current (4-form). For $q = 3$, the leading term (the second-order infinitesimal) of the 2-cocycle gives the anomalous Schwinger term (3-form) of the equal-time three-dimensional commutator (two 3-forms) of the infinitesimal gauge algebra. And for $q = 2$, the leading term (the third-order infinitesimal) of the 3-cocycle gives the 2-dimensional Jacobi identity anomaly (2-form) of three 2-forms.

Recently, an interesting discussion was introduced by Jackiw for the cohomology of a translation group [7], and we also gave a short comment about that [13]. The Čech–de Rham cohomology of the cocycles of a translation group can also be obtained in the same way, but in this case the base manifold coincides with the group manifold and the cohomology does not generally relate with the family index theorem.

Acknowledgements

This work was supported by the Science Fund of the Chinese Academy of Sciences. We are grateful to Profs. R. Jackiw and B. Zumino for showing us their handwritten manuscripts. We would also like to express our thanks to Profs. H. Y. Guo and X. C. Song for helpful information. We are also grateful to the referees for their kind help in correcting the English and telling us where Faddeev's paper was published [14].

References

1. Zumino, B., in B. S. De Witt and R. Stora (eds.), *Relativity, Groups, and Topology II*, Les Houches, 1983, North Holland, Amsterdam, 1984.
2. Stora, R., LAPP-TH-94 (Cargese Lectures (1983)).
3. Borona, L. and Cotta-Ramusino, P., *Commun. Math. Phys.* **87**, 589 (1983).
4. Atiyah, M. F. and Singer, I. M., *Proc. Natl. Acad. Sci. (USA)* **81**, 2597 (1984).
5. Alvarez-Gaume, L. and Ginsparg, P., Harvard preprint HUTP-83/A081.
6. Faddeev, L. D., *Phys. Lett.* **145B**, 81 (1984);
   Faddeev, L. D. and Shatashvili, S. L., *Theor. Math. Phys.* **60**, 206 (1984).
7. Jackiw, R., *Phys. Rev. Lett.* **54**, 159 (1985).
8. Hou, B. Y. and Hou, B. Y., *Chinese Phys. Lett.* **2** (1985).
9. Zumino, B., Santa Barbara preprint NSF-ITP-84-150.
10. Guo, H. Y., Wu, K., and Wang, S. K., Beijing preprint AS-ITP-84-035.

11. Guo, H. Y., Hou, B. Y., Wang, S. K., and Wu, K., Beijing preprints AS-ITP-84-039; AS-ITP-84-041; Xian preprints NWU-IMP-84-4; NWU-IMP-84-5;
Song, X. C., BNL preprint;
In October 1984, L. D. Faddeev told H. Y. Guo that they had obtained similar results (see [14]).
12. Wess, J. and Zumino, B., *Phys. Lett.* **37B**, 95 (1971);
Witten, E., *Nucl. Phys.* **223**, 22 (1983).
13. Hou, B. Y., Xian preprint NWU-IMP-84-9; see also
Hou, B. Y., Hou, B. Y., and Wang, P., Xian preprint NWU-IMP-85-4.
14. Reyman, A. G., Semenov-Tian-Shansky, M. A., and Faddeev, L. D., *Funkz. Analiz i ego Priloz.* **18**, 64 (1984).

PHYSICAL REVIEW D VOLUME 34, NUMBER 6 15 SEPTEMBER 1986

Integrability properties of symmetric-space fields reduced from axially symmetric Einstein and Yang-Mills equations

Ling-Lie Chau
Physics Department, Brookhaven National Laboratory, Upton, New York 11973

Kuang-Chao Chou
Institute of Theoretical Physics, Academia Sinica, Beijing, China

Bo-Yu Hou
Department of Physics, Xibei University, Xian, China

Xing-Chang Song*
Physics Department, Brookhaven National Laboratory, Upton, New York 11973
and Institute for Theoretical Physics, State University of New York at Stony Brook, Stony Brook, New York 11794

(Received 3 May 1985)

A gauge-covariant symmetric-space formulation and integrability properties are given for symmetric-space fields reduced from the stationary axially symmetric Einstein equation and static axially symmetric self-dual Yang-Mills equations.

I. INTRODUCTION

The gauge-covariant formulation of the symmetric-space fields proposed in recent papers[1-3] has the advantage of uniformity and simplicity. It has been shown that the various formalisms[4] for the nonlinear equation of motions, the linear system, and relevant relations adopted by different authors are the same equations in different gauges.[1-3] This is true not only for the ordinary nonlinear σ models taking values on some symmetries space [e.g., the O(n) vector models, the CP_n models on complex projective space, and models on the Grassmann manifold] but also for the principal chiral fields and even for the self-dual Yang-Mills (SDYM) fields in four-dimensional Euclidean space. In this paper we shall use this formulation to discuss the properties of the two-dimensional symmetric-space chiral fields reduced from static axially symmetric Einstein and Yang-Mills equations.

In the stationary axially symmetric case, the Einstein equation was shown to reduce[5,6] to the two-dimensional SL($2R$) chiral equations with an axial factor ρ. In the static axially symmetric case, the SDYM fields with group G were known to become[7-10] the $\widetilde{G}/G$ symmetric-space chiral fields with an axial factor ρ ($\widetilde{G}$ is the complexified G). Afterwards, such chiral fields with an axial factor were formulated with arbitrary groups.[8] Some of the integrability properties of such systems were discussed, e.g., the existence of linear equations[5,6,8-12] and the Bianchi-Bäcklund transformations (BBT's).[7,8,11,12]

Here we shall give an overall discussion of the integrabilities of such chiral fields with an axial factor in the view of Refs. 1—3. Interestingly, although the equations become much more complicated, they possess all the integrability properties of the ordinary chiral equations. This is an interesting addition to the integrable nonlinear systems. Our results will further provide a useful basis for future quantum development of the system.

This paper is organized as follows. After recalling some important results for the ordinary (flat-space) chiral fields in the Introduction, we formulate in Sec. II the static axially symmetric SDYM field into the symmetric-space formulation which is appropriate for our later discussions. This part overlaps with the first part of Ref. 8, and we include it for completeness and for introducing notations.

In Sec. III, we introduce dual-like transformations and show that such transformations actually provide a way to construct the linear equations. The whole family of the one-parameter dual-like transformations form an Abelian group U(1), and the solutions $U(x,\lambda)$ of the linear system form a nonlinear realization of the group U(1). Section IV is devoted to constructing the BBT's and their corresponding Riccati equations. We then construct the generalized Riccati equations and classify them according to their corresponding linear systems. These different classes of Riccati equations lead to different kinds of symmetry transformations, BBT's, Lie transform, and Riemann-Hilbert transforms.

In previous papers[1,3] we start with three basic equations for symmetric-space fields $A_\mu(x)=g^{-1}(x)\partial_\mu g(x)=H_\mu(x)+K_\mu(x)$ in an arbitrary gauge: the Gauss equation

$$\partial_\mu H_\nu(x)-\partial_\nu H_\mu(x)+[H_\mu(x),H_\nu(x)]+[K_\mu(x),K_\nu(x)]=0\ ; \quad (1.1a)$$

the Codazzi equation

$$D_\mu K_\nu(x)-D_\nu K_\mu(x)=0,\ (D_\mu\equiv\partial_\mu+[H_\mu,\ ])\ ; \quad (1.1b)$$

and the equation of motion

$$D_\mu K^\mu(x)=0 \,. \tag{1.1c}$$

In the "zero" gauge, $A_\mu(x)\to 0$ and

$$H_\mu(x)\to H_\mu^{(0)}(x)=\tfrac{1}{2}N(X)\partial_\mu N(x) \,, \tag{1.2a}$$

$$K_\mu(x)\to K_\mu^{(0)}(x)=-\tfrac{1}{2}N(x)\partial_\mu N(x) \,, \tag{1.2b}$$

where $N(x)$ is the canonical variable for the symmetric-space field satisfying $N^2(x)=1$. Then the equation of motion (1.1c) turns out to be the conservation equation

$$\partial_\mu[N(x)\partial^\mu N(x)]=0 \,, \tag{1.3}$$

and the other two basic equations, (1.1a) and (1.1b), take the same form and reduce to the curvatureless condition for $N(x)\partial_\mu N(x)$.

It can easily be seen that the dual transformed[13,14] fields $\widetilde{A}_\mu(x)=H_\mu(x)+\widetilde{K}_\mu(x)$, with $\widetilde{K}_\mu$ defined by

$$\begin{aligned}&\widetilde{K}_\mu(x)=K_\mu(x)\cos\phi+{}^*K_\mu(x)\sin\phi \quad \text{for the } E^2 \text{ case} \,,\\ &\widetilde{K}_\mu(x)=K_\mu(x)\cosh\phi+{}^*K_\mu(x)\sinh\phi \quad \text{for the } M^2 \text{ case} \,,\end{aligned} \tag{1.4}$$

satisfy the same set of basic equations. Here the dual vector of K_μ is defined as ${}^*K_\mu=\epsilon_{\mu\nu}K^\nu$, $\epsilon_{12}=-\epsilon_{21}=-1$, and

$$\cos\phi=\tfrac{1}{2}(\gamma+\gamma^{-1})=(1-\lambda^2)/(1+\lambda^2) \,,$$

$$\sin\phi=(1/2i)(\gamma-\gamma^{-1})=2\lambda/(1+\lambda^2)$$

for the E^2 case; $\epsilon_{01}=-\epsilon_{10}=1$ and

$$\cosh\phi=\tfrac{1}{2}(\gamma+\gamma^{-1})=(1+\lambda^2)/(1-\lambda^2) \,,$$

$$\sinh\phi=\tfrac{1}{2}(\gamma-\gamma^{-1})=2\lambda/(1-\lambda^2)$$

for the M^2 case. This means that $\widetilde{A}_\mu(x)$ is a pure gauge so that it can be set into $\widetilde{A}_\mu(x)=U\partial_\mu U^{-1}$. Thus

$$\begin{aligned}D_\mu U(x,\lambda)&\equiv[\partial_\mu+H_\mu(x)]U(x,\lambda)\\ &=-\widetilde{K}_\mu U(x,\lambda)\\ &=-[K_\mu c(\lambda)+{}^*K_\mu s(\lambda)]U(x,\lambda) \,,\end{aligned} \tag{1.5}$$

where $c(\lambda)=\cos\phi$ ($\cosh\phi$) and $s(\lambda)=\sin\phi$ ($\sinh\phi$) for the E^2 (M^2) case. Equation (1.5) is nothing but the linear system for the chiral field whose integrability conditions are the basic equations (1.1). In the "zero" gauge, this linear system turns out to take the following familiar form:

$$\partial_\mu U(x,\lambda)=\frac{\lambda}{1\pm\lambda^2}(\epsilon_{\mu\nu}N\partial^\nu N\mp\lambda N\partial_\mu N)U(x,\lambda) \,, \tag{1.6}$$

and its integrability conditions are just the curvatureless condition and conservation equation (1.3) for $N\partial_\mu N$. In Eq. (1.6), the upper signs stand for the E^2 case and the lower signs for the M^2 case. From the linear system (1.5) or (1.6), we have discussed the Bianchi-Bäcklund transformation, the corresponding Riccati equation, and the generalized Riccati equation.[1,3]

For definiteness, let us consider the E^2 case and introduce the conjugate coordinates

$$\xi=(x_1+ix_2), \quad \eta=(x_1-ix_2) \,. \tag{1.7}$$

Thus

$$A_\xi=\tfrac{1}{2}(A_1-iA_2), \quad A_\eta=\tfrac{1}{2}(A_1+iA_2) \,, \tag{1.8}$$

for any vector A_μ and

$${}^*A_\xi=-iA_\xi, \quad {}^*A_\eta=+iA_\eta \,. \tag{1.9}$$

Then the basic equations (1.1) can be written as the following: the Gauss equation

$$\partial_\xi H_\eta-\partial_\eta H_\xi+[H_\xi,H_\eta]+[K_\xi,K_\eta]=0 \,; \tag{1.10a}$$

the Codazzi equation

$$D_\xi K_\eta-D_\eta K_\xi=0 \,; \tag{1.10b}$$

and the equation of motion

$$D_\xi K_\eta+D_\eta K_\xi=0 \,. \tag{1.10c}$$

Equations (1.10b) and (1.10c) can be combined to write

$$D_\xi K_\eta=D_\eta K_\xi=0 \,, \tag{1.11}$$

Then the dual transformation (1.4) takes the simpler form[14]

$$\begin{aligned}&H_\xi\to\widetilde{H}_\xi=H_\xi, \quad H_\eta\to\widetilde{H}_\eta=H_\eta \,,\\ &K_\xi\to\widetilde{K}_\xi=\gamma^{-1}K_\xi, \quad K_\eta\to\widetilde{K}_\eta=\gamma K_\eta \,.\end{aligned} \tag{1.12}$$

The transformed fields $\widetilde{H}_\mu$ and $\widetilde{K}_\mu$ satisfy the same set of equations as in Eq. (1.10).

II. THE SYMMETRIC-SPACE FORMULATION FOR THE STATIC AXIALLY SYMMETRIC SDYM

First we will set the self-dual equations for the static axially symmetric Yang-Mills field into the symmetric-space field formulation, which is more convenient for later discussion.

Consider the four-dimensional Euclidean space with the coordinates x_1, x_2, $x_3=z$, and $x_4=t=$ Euclidean time, and introduce the cylindrical coordinates ρ, θ by $x_1=\rho\cos\theta$ and $x_2=\rho\sin\theta$. The cylindrical components of an arbitrary vector A_μ are defined as[8]

$$A_\rho=\frac{1}{\rho}x^a A_a, \quad A_\theta=-\frac{1}{\rho}\epsilon_{ab}x^a A^b \,, \tag{2.1}$$

where $x^a=x_a$, $A^a=A_a$, $a=1,2$. Consider the Yang-Mills potential A_μ and the strength $F_{\mu\nu}$ taking values on the Lie algebra of the gauge group G:

$$A_\mu=A_\mu^i\lambda^i, \quad F_{\mu\nu}=\partial_\mu A_\nu-\partial_\nu A_\mu+[A_\mu,A_\nu] \,, \tag{2.2}$$

with λ^i the anti-Hermitian generators of the gauge group, satisfying the commutation relation

$$[\lambda^i,\lambda^j]=f^{ij}{}_k\lambda^k \,. \tag{2.3}$$

The components of the "electric field" $E_k=F_{k4}$, ($k=1,2,3$) can be expressed as

$$\begin{aligned}&F_z=D_z A_4-\partial_4 A_z, \quad E_\rho=D_\rho A_4-\partial_4 A_\rho \,,\\ &E_\theta=D_\theta A_4-\partial_4 A_\theta \,,\end{aligned} \tag{2.4}$$

where D_k denotes the covariant derivative:

$$D_k=\partial_k+[A_k,\ ]\quad \text{for } k=z,\rho\ ,\qquad D_\theta=\frac{1}{\rho}\partial_\theta+[A_\theta,\ ]\ . \tag{2.5}$$

The components of the "magnetic field" $B_k=\frac{1}{2}\epsilon_{kij}F_{ij}$ take the form

$$\rho B_z=D_\rho(\rho A_\theta)-\partial_\theta A_\rho,\quad \rho B_\rho=\partial_\theta A_z-D_z(\rho A\theta)\ ,\qquad B_\theta=\partial_z(A_\rho)-D_\rho A_z\ . \tag{2.6}$$

Then the self-dual conditions

$$F_{\mu\nu}=\frac{1}{2}\epsilon_{\mu\nu\alpha\beta}F_{\alpha\beta} \tag{2.7}$$

can be expressed as $B_k=E_k$ ($k=z,\rho,\theta$), i.e.,

$$B_\theta=E_\theta:\ \partial_z A_\rho-D_\rho A_z=D_\theta A_4-\partial_4 A_\theta\ ; \tag{2.8a}$$

$$B_\rho=E_\rho:\ \frac{1}{\rho}\partial_\theta A_z-D_z A_\theta=D_\rho A_4-\partial_4 A_\rho\ ; \tag{2.8b}$$

$$B_z=E_z:\ -\frac{1}{\rho}\partial_\theta A_\rho+\frac{1}{\rho}\partial_\rho(\rho A_\theta)+[A_\rho,A_\theta]=D_z A_4-\partial_4 A_z\ . \tag{2.8c}$$

For static axially symmetric fields, Eqs. (2.8) are reduced since A_μ are independent of $x_4=t$ and θ. Then by introducing the new notations

$$z=y_1,\quad \rho=y_2\ , \tag{2.9a}$$

$$A_z=h_1,\quad A_\rho=h_2,\quad A_4=ik_1,\quad A_\theta=ik_2\ , \tag{2.9b}$$

the self-dual conditions (2.8) can be rewritten as

$$\partial_1 h_2-\partial_2 h_1+[h_1,h_2]+[k_1,k_2]=0\ , \tag{2.10a}$$

$$D_1 k_2-D_1 k_1=0\ , \tag{2.10b}$$

$$D_\mu(\rho k_\mu)=0\quad (\mu=1,2)\ . \tag{2.10c}$$

This is the symmetric-space formulation for the static axially symmetric SDYM. For the SU(n) gauge field, A_μ ($\mu=1,2,34$) are all anti-Hermitian $n\times n$ matrices. Then h_μ ($\mu=1,2$) are anti-Hermitian while k_μ ($\mu=1,2$) are Hermitian. This gives the symmetric space SL(n,c)/SU(n) which has been discussed in Ref. 10 and is named as a full ansatz in Ref. 8. In Ref. 8 special attention has been given to the so-called reduced ansatz in which h_μ (A_z and A_ρ) are assumed to take values on the subalgebra L_2 corresponding to a compact subgroup H of the gauge group G, whereas A_4 and A_θ are assumed to take values on the complementary subspace L_1+L_3 corresponding to the coset G/H.[8] Then k_μ ($-iA_4$ and iA_θ) take values on iL_i+iL_3 corresponding to the coset G^*/H, with G^* being a particular noncompact group associated with G. In the case of the SU(2) gauge field, $H=$SO(2) and $G^*=$SL(2,R).

As indicated in the introduction, in the case of the symmetric space being SL(2,R)/SO(2), the same set of equations (2.10) can be obtained from the reduction of the stationary axially symmetric Einstein equation.

Equations (2.10) are indeed the basic equations for a σ model over the symmetric coset space G^*/H (reduced ansatz) or $\tilde{G}/G$ (full ansatz, $\tilde{G}$ being the complexification of G) in a curved two-dimensional space.[8] From the Gauss equation (2.10a) and Codazzi equation (2.10b), one knows immediately that $h_\mu+k_\mu$ is pure gauge and then can be set into the form[14]

$$h_\mu+k_\mu=g^{-1}\partial_\mu g,\quad g\in G^*\ (\text{or }\tilde{G})\ ; \tag{2.11}$$

then inversely

$$h_\mu=\frac{1}{2}g^{-1}\partial_\mu g+\frac{1}{2}ng^{-1}\partial_\mu gn\ ,\qquad k_\mu=\frac{1}{2}g^{-1}\partial_\mu g-\frac{1}{2}ng^{-1}\partial_\mu gn\ , \tag{2.12}$$

where n is the involution matrix with respect to which the coset space is defined: $n^2=1$, $nhn=h$ for any $h\in H$, $n\phi n=\phi^{-1}$ for any $d\in G^*/H$. In the case of $G=$SU(2), $n=\sigma_2$, under the gauge transform of subgroup $h\in H$,

$$g\to gh,\quad g^{-1}\to h^{-1}g^{-1}\ ; \tag{2.13a}$$

then

$$h_\mu\to h^{-1}(h_\mu+\partial_\mu)h,\quad k_\mu\to h^{-1}k_\mu h\ . \tag{2.13b}$$

This transform is generalized[1-3] to any $S(x)\in G^*$:

$$g\to gS,\quad g^{-1}\to S^{-1}g^{-1},\quad n\to S^{-1}nS\ ; \tag{2.14a}$$

$$h_\mu\to S^{-1}(h_\mu+\partial_\mu)S=H_\mu,\quad k_\mu\to S^{-1}k_\mu S=K_\mu\ . \tag{2.14b}$$

For the particular choice $S=g^{-1}$, $g\to 1$, $g^{-1}\to 1$, then

$$n\to gng^{-1}=N(x),\quad N(x)^2=1\ ,\qquad H_\mu^{(0)}=-K_\mu^{(0)}=\frac{1}{2}N\partial_\mu N\ . \tag{2.15}$$

In this "zero" gauge, the equation of motion (2.10c) takes the form of a conservation law:

$$\partial_\mu(\rho N\partial_\mu N)=0\ . \tag{2.16}$$

Comparing with Eqs. (1.2) and (1.3), one sees immediately that the only difference is the appearance of the axial factor ρ in the equation of motion. The curvatureless quantity is $M_\mu=N\partial_\mu N$: $\partial_\mu M_\nu-\partial_\nu M_\mu+[M_\mu,M_\nu]=0$; while the quantity being conserved is $J_u=\rho M_\mu$: $\partial_\mu J_\mu=0$.

III. DUAL-LIKE TRANSFORMATION AND THE LINEAR SYSTEM

In an arbitrary gauge, the basic equations for the σ model with an axial factor or, equivalently, the self-dual conditions for the static axially symmetric[8] Yang-Mills field take the form

$$\partial_\mu H_\nu-\partial_\nu H_\mu+[H_\mu,H_\nu]+[K_\mu,K_\nu]=0\ , \tag{3.1a}$$

$$D_\mu K_\nu-D_\nu K_\mu=0,\quad D_\mu(\rho K_\mu)=0\ . \tag{3.1b}$$

By using the complex-conjugate coordinates ξ,η defined as

$$z=\frac{1}{2}(\xi+\eta),\quad \rho=\frac{i}{2}(\eta-\xi)\ , \tag{3.2}$$

the basic equations can be rewritten as

$$\partial_\xi H_\eta-\partial_\eta H_\xi+[H_\xi,H_\eta]+[K_\xi,K_\eta]=0\ , \tag{3.3a}$$

$$D_\xi K_\eta = D_\eta K_\xi, \quad D_\xi K_\eta + D_\eta K_\xi = -\frac{1}{\rho}(K_\eta \rho_\xi + K_\xi \rho_\eta), \tag{3.3b}$$

where we have used the notation $\rho_\xi = \partial_\xi \rho$, etc. Comparing Eq. (3.3) with Eq. (1.11), we see that the only difference between the two cases is the appearance of ρ and ρ_ξ, ρ_η in the equation of motion. But from this essential difference we see immediately that the dual transformations defined in Eqs.(1.4) or (1.12) are no longer the symmetry transformations for Eqs. (3.3) with the factor ρ.

To obtain an appropriate symmetry transformation for Eqs. (3.3), we consider the following form of the dual-like transformation:[12]

$$\begin{aligned} &\partial_\xi \to f(\xi)\partial_\xi, \quad \partial_\eta \to g(\eta)\partial_\eta, \\ &\rho \to \tilde{\rho}(\xi,\eta), \\ &H_\xi \to f(\xi) H_\xi, \quad H_\eta \to g(\eta) H_\eta, \\ &K_\xi \to \gamma_1(\xi,\eta) K_\xi, \quad K_\eta \to \gamma_2(\xi,\eta) K_\eta, \end{aligned} \tag{3.4}$$

where the coordinate-dependent functions $f(\xi)$, $g(\eta)$, $\tilde{\rho}(\xi,\eta)$, $\gamma_1(\xi,\eta)$, and $\gamma_2(\xi,\eta)$, which may also depend on some spectrum parameter, are all to be determined later.

Firstly, it is evident that the transformed Gauss equation gives a restriction among the as-yet unspecified functions, i.e., $\gamma_1\gamma_2 = fg$. Secondly, inserting the Codazzi equation (3.3b) and the equation of motion (3.3c) into the transformed Codazzi equation and comparing the coefficients of K_ξ and K_η on both sides, respectively, we obtain

$$\left[\frac{f^2-\gamma_1^2}{\rho}\right]_\eta = 0, \quad \left[\frac{g^2-\gamma_2^2}{\rho}\right]_\xi = 0, \tag{3.5}$$

from which it follows that

$$\begin{aligned} \gamma_1(\xi,\eta)^2 &= f(\xi)^2 - \rho F(\xi)^{-1}, \\ \gamma_2(\xi,\eta)^2 &= g(\eta)^2 + \rho G(\eta)^{-1}, \end{aligned} \tag{3.6}$$

where $F(\xi)$ $[G(\eta)]$ is only a function of ξ $[\eta]$. Substituting (3.6) into $\gamma_1\gamma_2 = fg$, we get

$$\rho = F(\xi) f(\xi)^2 - G(\eta) g(\eta)^2,$$

from which we obtain

$$F(\xi) = -\frac{i}{2}(\xi+\tau) f(\xi)^{-2}, \quad G(\eta) = -\frac{i}{2}(\eta+\tau) g(\eta)^{-2} \tag{3.7}$$

with τ as an arbitrary parameter. Then (3.6) leads to

$$\begin{aligned} \gamma_1(\xi,\eta) &= f(\xi)\left[\frac{\eta+\tau}{\xi+\tau}\right]^{1/2}, \\ \gamma_2(\xi,\eta) &= g(\eta)\left[\frac{\xi+\tau}{\eta+\tau}\right]^{1/2}. \end{aligned} \tag{3.8}$$

Similarly, when we insert (3.3) into the transformed equation of motion, we obtain the relation

$$\frac{1}{\tilde{\rho}} = \frac{f}{\rho}\frac{\eta+\tau}{\xi+\tau} = \frac{g}{\rho}\frac{\xi+\tau}{\eta+\tau} \tag{3.9}$$

by comparing the coefficients of K_ξ and K_η on both sides. This implies

$$\frac{f(\xi)}{(\xi+\tau)^2} = \frac{g(\eta)}{(\eta+\tau)^2} = \text{const}. \tag{3.10}$$

The constant can be taken as τ^{-2} by imposing the normalization condition that the dual-like transform becomes an identity when the parameter τ tends to infinity.[12] By introducing the new parameter $\kappa = \tau^{-1}$, we obtain, at last,

$$f(\xi) = (1+\kappa\xi)^2, \quad g(\eta) = (1+\kappa\eta)^2, \tag{3.11a}$$

$$\begin{aligned} \gamma_1(\xi,\eta)^2 &= (1+\kappa\xi)^3(1+\kappa\eta), \\ \gamma_2(\xi,\eta)^2 &= (1+\kappa\xi)(1+\kappa\eta)^3, \end{aligned} \tag{3.11b}$$

and

$$\begin{aligned} &\tilde{\rho}(\xi,\eta) = \frac{i}{2}(\tilde{\eta}-\tilde{\xi}), \\ &\tilde{\eta} = \eta(1+\kappa\eta)^{-1}, \quad \tilde{\xi} = \xi(1+\kappa\xi)^{-1}. \end{aligned} \tag{3.12}$$

Now it is easy to check $\partial_{\tilde{\eta}} = g(\eta)\partial_\eta$, $\partial_{\tilde{\xi}} = f(\xi)\partial_\xi$ as demanded at the first line of Eq. (3.4).

So, for the two-dimensional chiral field with the ρ factor (or, equivalently, the static axially symmetric SDYM field), there exists indeed a set of transformations we call the dual-like transformations, which are the generalization of the dual transformations for the flat two-dimensional chiral field and which are coordinate dependent:

$$\begin{aligned} &\xi \to \tilde{\xi} = \xi(1+\kappa\xi)^{-1}, \quad \eta \to \tilde{\eta} = \eta(1+\kappa\eta)^{-1}, \\ &H_\xi \to \tilde{H}_{\tilde{\xi}} = (1+\kappa\xi)^2 H_\xi, \quad H_\eta \to \tilde{H}_{\tilde{\eta}} = (1+\kappa\eta)^2 H_\eta, \\ &K_\xi \to \tilde{K}_{\tilde{\xi}} = [(1+\kappa\xi)^3(1+\kappa\eta)]^{1/2} K_\xi, \\ &K_\eta \to \tilde{K}_{\tilde{\eta}} = [(1+\kappa\xi)(1+\kappa\eta)^3]^{1/2} K_\eta. \end{aligned} \tag{3.13}$$

Sometimes it is more convenient to use the differential forms for the two-dimensional space. Then

$$d = \partial_\xi d\xi + \partial_\eta d\eta, \quad \tilde{d} = \partial_{\tilde{\xi}} d\tilde{\xi} + \partial_{\tilde{\eta}} d\tilde{\eta} = d, \tag{3.14a}$$

$$H = H_\xi d\xi + H_\eta d\eta, \quad \tilde{H} = \tilde{H}_{\tilde{\xi}} d\tilde{\xi} + \tilde{H}_{\tilde{\eta}} d\tilde{\eta} = H, \tag{3.14b}$$

and

$$\begin{aligned} K &= K_\xi d\xi + K_\eta d\eta, \\ \tilde{K} &= K_{\tilde{\xi}} d\tilde{\xi} + K_{\tilde{\eta}} d\tilde{\eta} = \left[\frac{1+\kappa\eta}{1+\kappa u}\right]^{1/2} K_\xi du \\ &\quad + \left[\frac{1+\kappa\xi}{1+\kappa\eta}\right]^{1/2} K_\eta d\eta \\ &\equiv \tilde{K}_\xi d\xi + \tilde{K}_\eta d\eta. \end{aligned} \tag{3.14c}$$

The basic equations (3.3) can be rewritten in forms

$$dH + H^2 + K^2 = 0, \quad DK = 0, \quad D(\rho^* K) = 0, \tag{3.15}$$

where $DK = dK + HK + KH$, and the transformed equations can also be expressed as

$$\tilde{d}\tilde{H} + \tilde{H}^2 + \tilde{K}^2 \equiv dH + H^2 + \tilde{K}^2 = 0, \tag{3.16a}$$

$$\tilde{D}\tilde{K} \equiv D\tilde{K} = 0, \quad \tilde{D}(\tilde{\rho}^*\tilde{K}) \equiv D(\tilde{\rho}^*\tilde{K}) = 0 . \tag{3.16b}$$

Now, similar to the case in flat space, the transformed Gauss equation (3.16a) and the Codazzi equation (3.16b) imply that $H+\tilde{K}$ is a pure gauge so that

$$H+\tilde{K} = UdU^{-1} = -(dU)U^{-1} ,$$

and then

$$dU = -(H+\tilde{K})U, \text{ or } DU \equiv dU + HU = -\tilde{K}U , \tag{3.17}$$

whose compatibility gives the basic equations (3.15). So Eq. (3.17) is just the linear system for the curved two-dimensional chiral field or static axially symmetric SDYM field.

In the arbitrary coordinate system, $\tilde{K}$ defined in (3.14c) and (3.13) can be written similarly to $\tilde{K}$ in Eq. (1.4) for the flat case, i.e.,

$$\tilde{K} = K\cos\phi + {}^*K\sin\phi ,$$
$${}^*K = {}^*K_\mu dy^\mu = \epsilon_{\mu\nu}K^\nu dy^\mu \quad (\mu = 1.2) , \tag{3.18}$$

The parameters ϕ or γ or λ are all coordinate dependent and can be specified by the relation ($\gamma_1 = \gamma^{-1}f$, $\gamma_2 = \gamma g$)

$$\gamma \equiv \gamma(\xi,\eta;\kappa) = \left[\frac{1+\kappa\xi}{1+\kappa\eta}\right]^{1/2} , \tag{3.19}$$

which can be obtained by comparing the expression in Eqs. (3.18) and (3.14c) in conjugate coordinates ξ and η.

In "zero" gauge, $H = -K = \frac{1}{2}NdN$, the Gauss and the Codazzi equations (3.15) reduce to the curvatureless condition for $M \equiv NdN$, and the equation of motion becomes

$$d(\rho N^*dN) \equiv d(\rho^*M) = 0 . \tag{3.20}$$

Then the linear system in coordinates ξ,η takes the form

$$\partial_\xi U(\kappa) = \frac{\gamma^{-1}(\kappa)-1}{2} M_\xi U(\kappa) = \frac{-\lambda}{\lambda - i} M_\xi U(\kappa) ,$$
$$\partial_\eta U(\kappa) = \frac{\gamma(\kappa)-1}{2} M_\eta U(\kappa) = \frac{-\lambda}{\lambda + i} M_\eta U(\kappa) , \tag{3.21}$$

or in other coordinates

$$dU(\kappa) = \tfrac{1}{2}[{}^*M\sin\phi + M(\cos\phi - 1)]U(\kappa)$$
$$= \frac{\lambda}{1+\lambda^2}({}^*M - \lambda M)U(\kappa) , \tag{3.22}$$

where $d \equiv \partial_\mu dy^\mu$, ${}^*d = \epsilon_{\mu\nu}\partial^\nu dy^\mu$, and

$$\lambda = \frac{[(1+\kappa z)^2 + \kappa^2\rho^2]^{1/2} - (1+\kappa z)}{\kappa\rho} , \tag{3.23}$$

from which an important relation follows:

$$\kappa\left[\rho\frac{\lambda^{-1}-\lambda}{2} - z\right] = 1 . \tag{3.24}$$

The linear systems given in Eqs. (3.21) and (3.22) are equivalent to the ones given by Maison[5] for the stationary axially symmetric gravitation field, and to the one given by Bais and Sasaki[8] for the static axially symmetric SDYM field. It is evident that, apart from the important feature λ being coordinate dependent as shown in Eq. (3.23), the linear system in Eqs. (3.21) or (3.22) has the same form as the one in Eq. (1.8) for the flat case. It can be shown by direct calculation that there is an important identity for the parameter ϕ, i.e.,

$$d\cos\phi - \rho^* d(\rho^{-1}\sin\phi) = 0 , \tag{3.25}$$

from which we can prove the transformed equations (3.16) directly from the original ones in (3.15), e.g.,

$$D\tilde{K} \equiv D(K\cos\phi + {}^*K\sin\phi)$$
$$= \cos\phi DK + \rho^{-1}\sin\phi D(\rho^*K)$$
$$- K[d\cos\phi - \rho^* d(\rho^{-1}\sin\phi)] = 0 ,$$

and show more directly that the compatibility condition for the linear system (3.22) leads to the equation of motion (3.20), i.e.,

$$d^2U = \tfrac{1}{2}[\rho^{-1}\sin\phi d(\rho^*M) + (\cos\phi - 1)(dM + M^2)]U$$
$$+ \tfrac{1}{2}[d\cos\phi - \rho^* d(\rho^{-1}\sin\phi)]MU = 0 .$$

As pointed out in Ref. 12, if, instead of the parameter λ, we introduce another coordinate-dependent parameter $\omega = \rho\lambda^{-1}$, then

$$\omega = (\tau + x) + [(\tau + z)^2 + \rho^2]^{1/2} , \tag{3.26}$$

and the linear system (3.21) becomes

$$-\partial_\xi U = \frac{\rho}{\rho - i\omega} M_\xi U = \frac{\rho + i\omega}{\rho^2 + \omega^2}\rho M_\xi U ,$$
$$-\partial_\eta U = \frac{\rho}{\rho + i\omega} M_\eta U = \frac{\rho - i\omega}{\rho^2 + \omega^2}\rho M_\eta U , \tag{3.27}$$

which is the form given by Mikhailov and Yaremchuk[11] in constructing the solitonlike solutions. Now passing to the original variables (z,ρ), Eq. (3.27) can be rewritten in the form

$$-\partial_z U = \frac{\rho M_z + \omega M_\rho}{\rho^2 + \omega^2}\rho U, \quad -\partial_\rho U = \frac{\rho M_\rho - \omega M_z}{\rho^2 + \omega^2}\rho U . \tag{3.28}$$

We can consider the solution of (3.28) as being a function of $\omega(z,\rho)$ as well as a function of z and ρ: $U = U(\omega(z,\rho); z,\rho)$. Then the differentiation with respect to the variable z,ρ must be a compound one:

$$\frac{\partial}{\partial z} = \left[\frac{\partial}{\partial z}\right]_\omega + \frac{\partial\omega}{\partial z}\frac{\partial}{\partial\omega} = \left[\frac{\partial}{\partial z}\right]_\omega + \frac{2\omega^2}{\omega^2 + z^2}\frac{\partial}{\partial\omega} ,$$
$$\frac{\partial}{\partial\rho} = \left[\frac{\partial}{\partial\rho}\right]_\omega + \frac{\partial\omega}{\partial\rho}\frac{\partial}{\partial\omega} = \left[\frac{\partial}{\partial\rho}\right]_\omega + \frac{2\omega\rho}{\omega^2 + \rho^2}\frac{\partial}{\partial\omega} , \tag{3.29}$$

where we have used (3.25) to calculate the explicit forms of $\partial\omega/\partial z$ and $\partial\omega/\partial\rho$. On account of the relations (3.29), the pair of linear equations in (3.28) is just the one given by Belinsky and Zakharov.[6] In their paper, the differentiation operators in (3.29) were denoted as $D_1 = D_z$ and $D_2 = D_\rho$.

At the end of this section we would like to indicate two

important features of the dual-like transforms we have discussed above. First, starting from the "zero" gauge field N, $N^2=1, H=K=\frac{1}{2}N\,dN$, we can construct the transformed $\tilde{H}=H$ and $\tilde{K}$ as in (3.14). H and $\tilde{K}$ themselves are not in zero gauge, but can be made to be in zero gauge by a gauge transformation specified by the solution of the linear system, i.e.,

$$K(\kappa;N)=U^{-1}(\kappa;N)\tilde{K}(N)U(\kappa;N) = -\tfrac{1}{2}N(\kappa;N)dN(\kappa;N)\,, \quad (3.30a)$$

$$H(\kappa;N)=U^{-1}(\kappa;N)[H(N)+d]U(\kappa;N) = \tfrac{1}{2}N(\kappa;N)dN(\kappa;N)\,, \quad (3.30b)$$

where

$$N(\kappa;N)=U^{-1}(\kappa;N)NU(\kappa;N)\,. \quad (3.30c)$$

The combined transformation $N\to N(\kappa;N)$, $K\to K(\kappa;N)$, $H\to H(\kappa;N)$ is called a final dual-like transformation[2] and the solution of the linear system is the appropriate amplitude for this transformation, as shown in (3.30).

The second feature is the group property for the one-parameter family of the dual-like transformations. If κ_1 is the parameter that specifies the transform from $(\xi,\eta;H,K)$ to $(\tilde{\xi},\tilde{\eta};H,\tilde{K})$:

$$\tilde{\xi}=\xi(1+\kappa_1\xi)^{-1},\quad \tilde{\eta}=\eta(1+\kappa_1\eta)^{-1}\,,$$

$$\gamma(\xi,\eta;\kappa_1)=\left[\frac{1+\kappa_1\eta}{1+\kappa_1\xi}\right]^{1/2},$$

and κ_2 is the parameter that specifies the transform from $(\tilde{\xi},\tilde{\eta};H,\tilde{K})$ to $(\tilde{\tilde{\xi}},\tilde{\tilde{\eta}};H,\tilde{\tilde{K}})$, then $(\xi,\eta;H,K)\to(\tilde{\tilde{\xi}},\tilde{\tilde{\eta}};H,\tilde{\tilde{K}})$ is still a dual-like transform specified by the parameter $\kappa_1+\kappa_2$. This can easily be checked by

$$\tilde{\tilde{\xi}}=\tilde{\xi}(1+\kappa_2+\xi)^{-1}=\frac{\xi(1+\kappa_1\eta)^{-1}}{1+\kappa_2\xi(1+\kappa_1\xi)^{-1}} = \frac{\xi}{1+\kappa_1\xi+\kappa_2\xi}\,, \quad (3.31)$$

$$\gamma(\tilde{\xi},\tilde{\eta};\kappa_2)\gamma(\xi,\eta;\kappa_1)=\left[\frac{1+\kappa_2\tilde{\eta}}{1+\kappa_2\tilde{\xi}}\,\frac{1+\kappa_1\eta}{1+\kappa_1\xi}\right]^{1/2} = \left[\frac{1+\kappa_1\eta+\kappa_2\eta}{1+\kappa_1\xi+\kappa_2\xi}\right]^{1/2} = \gamma(\xi,\eta;\kappa_1+\kappa_2)\,. \quad (3.32)$$

This means that the dual-like transformations form an Abelian group U(1) with κ as the group parameter. Then it is not difficult to prove that, by choosing the boundary condition properly, the set of solutions of the linear system gives a nonlinear realization of this group:[12]

$$U^{-1}(\kappa_2;N(\kappa_1))U^{-1}(\kappa_1;N)=U^{-1}(\kappa_1+\kappa_2;N) = U^{-1}(\kappa_1;N(\kappa_2))U^{-1}(\kappa_2;N)\,. \quad (3.33)$$

More precisely, the set of solutions of the linear system constructs a nonlinear realization of $G\wedge U(1)/G$, where $G\wedge U(1)$ is the semidirect product of the isotropic group G with the U(1). That is to say, if $U(\kappa_1;N)$ is the solution of the linear system with κ_1 as the spectrum parameter,

$$DU(\kappa_1;N)=-\tilde{K}(\kappa_1;N)U(\kappa_1;N)\,,$$

and $U(\kappa_2;N(\kappa_1))$ is the solution of a similar system with κ_2 as spectrum parameter and N replaced by the transformed field $N(\kappa_1)$,

$$DU(\kappa_2;N(\kappa_1))=-\tilde{K}(\kappa_2;N(\kappa_1))U(\kappa_2;N(\kappa_1))\,,$$

then the product $U(\kappa_1;N)\,U(\kappa_2;N(\kappa_1))$ solves the equation as $U(\kappa_1+\kappa_2;N)$ solves

$$DU(\kappa_1;\kappa_2;N)=-\tilde{K}(\kappa_1+\kappa_2;N)U(\kappa_1+\kappa_2;N)\,.$$

So they are equivalent to each other up to a constant (coordinate-independent, but may be parameter-dependent) matrix in G.

IV. BÄCKLUND TRANSFORM AND ASSOCIATED RICCATI EQUATION

Now we turn to a discussion of the Bianchi-Bäcklund transformation for our curved chiral field. Starting from a given solution $N(y)$, we use the Riemann-Hilbert (RH) method to find a new solution $N'(y)$. First, consider the transform $R(y;\kappa)=U(y;\kappa)U'^{-1}(y;\kappa)$, where U,U' are, respectively, solutions to the linear system (3.22) for the field $N(y),N'(y)$ with the same spectrum parameter κ. From the linear system, $R(y;\kappa)$ can be shown to satisfy the following equation:

$$dR=\frac{2\lambda}{1+\lambda^2}(\lambda K-{}^*K)R-\frac{2\lambda}{1+\lambda^2}R(\lambda K'-{}^*K')\,, \quad (4.1)$$

where $K=-\frac{1}{2}NdN$, $K'=-\frac{1}{2}N'dN'$, and λ is given by (3.23). Then following the procedure used in Refs. 4 and 15, we assume that R takes the form

$$R=\tfrac{1}{2}(1+\kappa\eta)^{1/2}(1-i\lambda_0B)+\tfrac{1}{2}(1+\kappa\xi)^{1/2}(1+i\lambda_0B)\,, \quad (4.2)$$

with λ_0 a parameter and

$$B=N^{-1}N'\,. \quad (4.3)$$

Substituting (4.2) and (4.3) into (4.1), and completing a series of algebraic calculations, we obtain an equation for B:

$$d(\rho\lambda_0B-z)=2\rho({}^*K-{}^*K')\,. \quad (4.4)$$

It is easy to show that this is a BBT by taking the exterior derivative d on both sides of the equation. Then $d(\rho{}^*K)-d(\rho{}^*K')=0$. So N' is a solution to the basic equations if N is. Obviously, Eq. (4.4) is the BBT formula corresponding to the one in the flat case, as given in Refs. 3 and 16. In the case of the symmetry group being a unitary group, by using an argument similar to that presented in Ref. 16 for the flat case, we can also show further algebraic constraint imposed by BBT itself, i.e.,

$$\rho(\lambda_0 B+\bar{\lambda}_0 B^+)-2z=\text{const.} \tag{4.5}$$

Then it can be shown that, starting from a given N with $N^2=1$, the only consistent new solution N' is restricted to satisfy the condition $N'^2=-1$. From Eq. (4.5) we can also see that $\lambda_0 B+\bar{\lambda}_0 B^+$ must be coordinate dependent in the present (curved) case, whereas in the flat case the constraint requires that it must be a constant. Then by comparing (4.5) to (3.24) we see that the only reasonable ansatz is

$$\tfrac{1}{2}(\lambda_0 B+\bar{\lambda}_0 B^+)=\frac{\lambda^{-1}-\lambda}{2}=\cot\phi\ . \tag{4.6}$$

Now it is easy to show from the definition of K,K', and B in Eq. (4.3), that

$$2(K-K')=B^{-1}(dB-2[K,B])\ . \tag{4.7}$$

Substituting this into the BBT formula (4.4), we obtain the Riccati equation for B:

$$dB=B^{-1}({}^*dB-2[{}^*K,B])\ . \tag{4.8}$$

To simplify this equation, we choose $\lambda_0=1$ and make the assumption

$$B=\cot\phi+\tilde{\Lambda}\csc\phi,\quad B^{-1}=-\cot\phi+\tilde{\Lambda}\csc\phi\ , \tag{4.9}$$

such that

$$\tilde{\Lambda}^2=1\ . \tag{4.10}$$

Then by making use of the identity (3.25), we can show from Eq. (4.8) that

$$d\tilde{\Lambda}=\frac{1}{\sin\phi}(\tilde{\Lambda}-\cos\phi)({}^*d\tilde{\Lambda}-2[{}^*K,\tilde{\Lambda}])\ , \tag{4.11}$$

from which we see

$$-{}^*d\tilde{\Lambda}=\frac{1}{\sin\phi}(\tilde{\Lambda}-\cos\phi)(d\tilde{\Lambda}-2[K,\tilde{\Lambda}])\ ,$$

$$-\tilde{\Lambda}{}^*d\tilde{\Lambda}=\frac{1}{\sin\phi}(1-\tilde{\Lambda}\cos\phi)(d\tilde{\Lambda}-2[K,\tilde{\Lambda}])\ .$$

By taking the linear combination of the above three equations to eliminate terms containing ${}^*d\tilde{\Lambda}$ and $\tilde{\Lambda}{}^*d\tilde{\Lambda}$, we obtain

$$d\tilde{\Lambda}=[K,\tilde{\Lambda}]-\tilde{\Lambda}\tilde{K}\tilde{\Lambda}+\tilde{K}\ , \tag{4.12a}$$

or covariantly

$$D\tilde{\Lambda}\equiv d\tilde{\Lambda}+[H,\tilde{\Lambda}]=-\tilde{\Lambda}\tilde{K}\tilde{\Lambda}+\tilde{K}\ , \tag{4.12b}$$

with $\tilde{K}=K\cos\phi+{}^*K\sin\phi$. This is the gauge-covariant Riccati equation for the curved chiral field. It is not difficult to show that the compatibility condition for the Riccati equation (4.12) is just the set of three basic equations (3.15). As a matter of fact, by taking the covariant derivative to Eq. (4.12), we have

$$\text{LHS}=D^2\tilde{\Lambda}=d^2\tilde{\Lambda}+d[H,\tilde{\Lambda}]+\{H,d\tilde{\Lambda}+[H,\tilde{\Lambda}]\}$$
$$=d^2\tilde{\Lambda}+[dH+H^2,\tilde{\Lambda}]\ ,$$
$$\text{RHS}=-(D\tilde{\Lambda})\tilde{K}\tilde{\Lambda}+\tilde{\Lambda}\tilde{K}D\tilde{\Lambda}-\tilde{\Lambda}(D\tilde{K})\tilde{\Lambda}+D\tilde{K}$$
$$=-[\tilde{K}^2,\tilde{\Lambda}]-\tilde{\Lambda}(D\tilde{K})\tilde{\Lambda}+D\tilde{K}\ .$$

where LHS and RHS stand for the left- and right-hand sides, respectively. Therefore,

$$d^2\tilde{\Lambda}=-[dH+H^2+\tilde{K}^2,\tilde{\Lambda}]-\tilde{\Lambda}(D\tilde{K})\tilde{\Lambda}+D\tilde{K}=0$$

follows from Eq. (3.16) or equivalently from Eq. (3.15). From the Riccati equation (4.12) and the basic equations (3.16), we can show that the parametric current ${}^*J\equiv\text{Tr}\tilde{K}\tilde{\Lambda}$ is conserved,[14] i.e.,

$$d{}^*J=d\,\text{Tr}\tilde{K}\tilde{\Lambda}=\text{Tr}(D\tilde{K})\tilde{\Lambda}-\text{Tr}\tilde{K}D\tilde{\Lambda}=0\ , \tag{4.13}$$

where we have used the property that $\text{Tr}A^2=0$ for any one-form matrix A. Making proper expansion in parameter κ, we can obtain an infinite number of conserved currents.

In contrast with the usual case,[14,16] the currents thus obtained are still nonlocal ones. This can be seen in the following way. In the conjugate coordinate, the Riccati equation (4.12) becomes

$$D_\xi\tilde{\Lambda}=-\tilde{\Lambda}\tilde{K}_\xi\tilde{\Lambda}+\tilde{K}_\xi=-\gamma^{-1}\tilde{\Lambda}K_\xi\tilde{\Lambda}+\gamma^{-1}K_\xi\ , \tag{4.14a}$$

$$D_\eta\tilde{\Lambda}=-\tilde{\Lambda}\tilde{K}_\eta\tilde{\Lambda}+\tilde{K}_\eta=-\gamma\tilde{\Lambda}K_\eta\tilde{\Lambda}+\gamma K_\eta\ . \tag{4.14b}$$

For the case without the ρ factor, γ is the essential parameter. We make a formal expansion $\tilde{\Lambda}=\sum_{n=0}^{\infty}\gamma^n\Lambda^{(n)}$, substitute it into (4.14a), and compare the coefficients on both sides, terms in the RHS always contain lower powers in γ, e.g.,

$$0=-\Lambda(0)K_\xi\Lambda(0)+K_\xi\ ,$$

$$D_\xi\Lambda(0)=-\Lambda(0)K_\xi\Lambda(1)-\Lambda(1)K_\xi\Lambda(0)\ ,$$

etc. Then we can get $\Lambda^{(n)}$ step by step from $\Lambda^{(i)}$ $(i\le n-1)$ by solving the matrix algebraic equation so that $\Lambda^{(n)}$ depends only on the local expression of K_ξ and $D_\xi K_\xi$, etc. Thus (4.13) gives the infinite set of local conservation laws. But in the case with the ρ factor, γ is coordinate dependent, and we must assume the formal expansion of $\tilde{\Lambda}$ in terms of the essential parameter κ, $\tilde{\Lambda}=\sum_{n=0}^{\infty}\kappa^n\Lambda^{(n)}$. Then the LHS and RHS are in the same order and we have to solve the matrix differentiation equation step by step. Thus $\Lambda^{(n)}$ contains integration of K_ξ, and then the currents obtained from (4.13) are nonlocal. These nonlocal conservation laws may be especially significant when one considers the corresponding quantum system.

V. GENERAL RICCATI EQUATION AND LINEAR EQUATIONS

As discussed for the flat case,[3] the concepts of the Riccati equation (4.12) and the current (4.13) attached to it can be generalized by introducing a more general form of $\tilde{\Lambda}$. From the basic equations

$$dH+H^2+\tilde{K}^2=0,\quad D\tilde{K}=0,\quad D(\rho{}^*\tilde{K})=0\ , \tag{5.1}$$

we can construct two different kinds of conserved currents.

(a) The current

$$*J^{(1)}=\mathrm{Tr}\tilde{\rho}^*\tilde{K}\tilde{\Lambda} \tag{5.2}$$

will be conserved, i.e.,

$$d^*J^{(1)}=\mathrm{Tr}\tilde{\Lambda}D(\tilde{\rho}^*\tilde{K})-\mathrm{Tr}\tilde{\rho}^*\tilde{K}D\tilde{\Lambda}=0 \tag{5.3}$$

if $D\tilde{\Lambda}$ is trace orthogonal to $\tilde{\rho}^*\tilde{K}$, e.g., $D\tilde{\Lambda}$ takes the form

$$D\tilde{\Lambda}=\alpha_1\tilde{\rho}^*\tilde{K}+\beta_1\tilde{\Lambda}\tilde{\rho}^*\tilde{K}\tilde{\Lambda}+a_1[\tilde{\Lambda},\tilde{K}]+s_1\{\tilde{\Lambda},\tilde{\rho}^*\tilde{K}\} . \tag{5.4}$$

It is not difficult to check that under the constraint among the parameters

$$a_1^2=1, \quad s_1^2-\alpha_1\beta_1=0 , \tag{5.5}$$

the integrability condition of (5.4) gives basic equations (5.1). Therefore, Eq. (5.4) together with the constraint (5.5) is a set of general Riccati equations.

(b) The current

$$*J^{(2)}=\mathrm{Tr}\tilde{K}\tilde{\Lambda} \tag{5.6}$$

will be conserved, i.e.,

$$d^*J^{(2)}=\mathrm{Tr}(D\tilde{K})\tilde{\Lambda}-\mathrm{Tr}\tilde{K}D\tilde{\Lambda}=0 , \tag{5.7}$$

if $D\tilde{\Lambda}$ is trace orthogonal to $\tilde{K}$, e.g., $D\tilde{\Lambda}$ takes the form

$$D\tilde{\Lambda}=\alpha_2\tilde{K}+\beta_2\tilde{\Lambda}\tilde{K}\tilde{\Lambda}+a_2[\tilde{\Lambda},\tilde{\rho}^*\tilde{K}]+s_2\{\tilde{\Lambda},\tilde{K}\} . \tag{5.8}$$

It is not difficult to check that under the constraint

$$a_2=0, \quad s_2^2-\alpha_2\beta_2=1 , \tag{5.9}$$

the integrability condition of (5.8) gives basic equations (5.1). Therefore, Eq. (5.8) together with the constraint (5.9) is another set of general Riccati equations.

For the usual case,[3] we can also construct two sets of general Riccati equations. They are dual to each other and equivalent to each other when the spectrum parameter (say ϕ) is taken to be complex. But now for the case with a factor ρ, we see that these two sets of general Riccati equations are completely different, and the constraints among the parameters are also completely different. We will see later that they are in charge of different kinds of symmetries for the theory.

The general form of the Riccati equation in two dimensions and its relation to the corresponding linear system have been discussed in Refs. 4 and 17, and the covariant form is used in Ref. 3. Here we give some comments on it. Consider an $n\times n$ matrix $\bar{\Lambda}$ satisfying the equation

$$d\bar{\Lambda}=\bar{G}^{(1)}\bar{\Lambda}+\bar{G}^{(2)}-\bar{\Lambda}\bar{G}^{(3)}\bar{\Lambda}-\bar{\Lambda}\bar{G}^{(4)} . \tag{5.10}$$

One can show that the sufficient condition for the integrability of this equation

$$d^2\bar{\Lambda}=0 \tag{5.11}$$

implies

$$d\bar{G}-\bar{G}^2=0, \quad \bar{G}=\begin{bmatrix}\bar{G}^{(1)} & \bar{G}^{(2)}\\ \bar{G}^{(3)} & \bar{G}^{(4)}\end{bmatrix} , \tag{5.12}$$

which is the same as the integrability condition for the following linear system:

$$d\bar{U}=\bar{G}\bar{U}, \quad \bar{U}=\begin{bmatrix}\bar{U}_{11},\bar{U}_{12}\\ \bar{U}_{21},\bar{U}_{22}\end{bmatrix} . \tag{5.13}$$

When (5.12) gives three basic equations for the chiral field, (5.10) is the Riccati equation for the chiral field and (5.13) is the corresponding linear system. Conversely, as discussed in Ref. 17, given the linear system (5.13), one can easily show that

$$\bar{\Lambda}\equiv(\bar{U}_{11}a+\bar{U}_{12})(\bar{U}_{21}a+\bar{U}_{22})^{-1} , \tag{5.14}$$

with a being a constant matrix, satisfies the Riccati equation (5.10). Linear equations (5.13) and the relation (5.14) can be generalized to $2n\times mn$ form:

$$\bar{U}=\begin{bmatrix}\bar{U}_{11},\bar{U}_{12},\ldots,\bar{U}_{1m}\\ \bar{U}_{21},\bar{U}_{22},\ldots,\bar{U}_{2m}\end{bmatrix} ,$$
$$\bar{\Lambda}=\left[\sum_{j=1}^{m}\bar{U}_{1j}a_j\right]\left[\sum_{k=1}^{m}\bar{U}_{2k}a_k\right]^{-1} , \tag{5.15}$$

with a_j $(j=1,2,\ldots,m)$ being constant matrices.

All the formulas in Eqs. (5.10)–(5.15) can be easily taken over to the covariant form. The covariant form of the general Riccati equation reads

$$D\Lambda\equiv d\Lambda+[H,\Lambda]=G^{(1)}\Lambda+G^{(2)}-\Lambda G^{(3)}\Lambda-\Lambda G^{(4)} , \tag{5.10$'$}$$

and the integrability condition

$$D^2-DG+G^2=0 \tag{5.11$'$}$$

implies that

$$d^2=-(dH+H^2)+DG-G^2=0 . \tag{5.12$'$}$$

Equations (5.13) to (5.15) remain unchanged for unbarred quantities. The integrability condition takes its simplest form in the complex-conjugate coordinates for the Euclidean case where we can set

$$G_\xi=K_\xi\times L^{(\xi)}, \quad G_\eta=K_\eta\times L^{(\eta)} . \tag{5.16}$$

In the flat case, $L^{(\xi)}$ and $L^{(\eta)}$ being 2×2 constant matrices, then Eq. (5.12) implies that

$$L^{(\xi)}L^{(\eta)}=L^{(\eta)}L^{(\xi)}=1 . \tag{5.17}$$

The conservation property of the current $J=\mathrm{Tr}K\Lambda$, i.e., $d^*J=0$, requires that $D\Lambda$ must trace orthogonal to $^*K$. This further restricts the constant matrices $L^{(\xi)}$ and $L^{(\eta)}$, namely,

$$\det L^{(\xi)}=\det L^{(\eta)}=1 . \tag{5.18}$$

The general Riccati equation and the corresponding linear system for the flat chiral field have several interesting features. (i) The $2n\times2n$ linear equations for both ξ and η must be diagonalized simultaneously because of (5.17), if they are diagonalizable. (ii) Any Riccati equation is connected with some conservation law since $L^{(\xi)}$ and $L^{(\eta)}$ are nonsingular from (5.17) and then can be made unit determinant by a proper rescaling. (iii) For a given Riccati equation, $L^{(\xi)}$ and $L^{(\eta)}$ characterize its properties since the restrictions (5.17) and (5.18) are invariant under any similar matrix transform The concrete properties of the general Riccati equation have been discussed in Ref. 3 in some detail.

This becomes much more complicated for the case with the factor ρ where $L^{(\xi)}$ and $L^{(\eta)}$ may contain some coordinate-dependent factor, say ρ. For choice (a) in Eq. (5.4),

$$G_1=-a_1K-M_1\hat{\rho}^*K,M_1=\begin{bmatrix}-s_1, & -\alpha_1\\ \beta_1 & s_1\end{bmatrix}, \quad (5.19)$$

Eq. (5.12) gives the constraint (5.5) exactly. For choice (b) in (5.8),

$$G_2=-a_2\rho^*K-M_2K,\ M_2=\begin{bmatrix}-s_2, & -\alpha_2\\ \beta_2, & s_2\end{bmatrix}, \quad (5.20)$$

Eq. (5.12) gives the restriction (5.9). For a more general case, we can choose

$$G=G_1+G_2=-(a_1+M_2)K-(a_2+M_1)\rho^*K\ . \quad (5.21)$$

Equation (5.21) will be the Riccati equation for the chiral field with a factor ρ, only when the parameters are subject to the constraints

$$a_1=a_2=0,\ s_1{}^2-\alpha_1\beta_1=0,\ s_2{}^2-\alpha_2\beta_2=1\ . \quad (5.22)$$

Now we turn to a discussion of the classification of the general Riccati equation according to the type of linear equation $DU=GU$ it corresponds to. The central problem is to analyze the properties of M:

$$M=\begin{bmatrix}-s, & -\alpha\\ \beta & s\end{bmatrix}.$$

For a 2×2 traceless non-Hermitian matrix M, its eigenvalues are

$$\mu_\pm=\pm(s^2-\alpha\beta)^{1/2}=\pm\sqrt{-\Delta},\Delta=\det M\ . \quad (5.23)$$

The conclusion is that M is diagonalizable in the case $\Delta\neq0$; M is nondiagonalizable in the case $\Delta=0$ but $M\neq0$ (nilpotent).

To deal with the diagonal problem for an arbitrary 2×2 traceless non-Hermitian matrix M, the following procedure is proposed.

(i) To bring M into a totally off-diagonal form by using an orthogonal (generally complex) matrix S,

$$M\to M'=S^{-1}MS,\ S=\begin{bmatrix}\cos\theta, & -\sin\theta\\ \sin\theta, & \cos\theta\end{bmatrix}, \quad (5.24)$$

then

$$M'=\begin{bmatrix}-s\cos2\theta+\frac{\beta-\alpha}{2}\sin2\theta, & \frac{\beta-\alpha}{2}\cos2\theta-\frac{\beta+\alpha}{2}+s\sin2\theta\\ \frac{\beta-\alpha}{2}\cos2\theta+\frac{\beta+\alpha}{2}+s\sin2\theta, & s\cos2\theta-\frac{\beta-\alpha}{2}\sin2\theta\end{bmatrix}.$$

By choosing $\tan2\theta=2s/(\beta-\alpha)$, M' takes the totally off-diagonal form

$$M'=\begin{bmatrix}0 & -\alpha'\\ \beta' & 0\end{bmatrix},$$

$$-\alpha'=\left[\left[\frac{\beta+\alpha}{2}\right]^2+\mu^2\right]^{1/2}-\frac{\beta+\alpha}{2}, \quad (5.25)$$

$$\beta'=\left[\left[\frac{\beta+\alpha}{2}\right]^2+\mu^2\right]^{1/2}+\frac{\beta+\alpha}{2}.$$

In the case $\Delta=0$, $\mu^2=0$, either α' or β' must be zero. This gives the standard form for the nondiagonalizable case.

(ii) In the case $\Delta\neq0$,M' can be cast to the diagonal form M'' by a non-Hermitian similar matrix transform:

$$M'\to M''=T^{-1}M'T=\begin{bmatrix}\mu & 0\\ 0 & -\mu\end{bmatrix}.$$

$$T^{-1}=T=\frac{1}{\sqrt{2}\mu}(M'+M'')\ . \quad (5.26)$$

This is the standard form for the diagonalizable case.

Now we apply this procedure to the equations we considered above. For choice (a) in (5.4) and (5.5), $\Delta=0$, we have only two different cases.

Case (A) $M=0$, the linear system has already been diagonalized. For $a=1$, it takes the form

$$DU=-\tilde{K}U,\ U=\begin{bmatrix}U_1\\ U_2\end{bmatrix}. \quad (5.27)$$

Now the upper and lower components satisfy the same equation, which is indeed the original linear system (3.17). So we obtain $U_2=U(\kappa)$, $U_1=U(\kappa)T$. According to the general argument in (5.15), $\tilde{\Lambda}$ can be expressed as

$$\tilde{\Lambda}\equiv\Lambda^i_{\rm RH}=U(y;\kappa)T^iU^{-1}(y;\kappa)\ , \quad (5.28)$$

which is the infinitesimal RH transform satisfying the same algebra of T^i, i.e.,

$$[\Lambda^i_{\rm RH},\Lambda^j_{RH}]=f^{ij}{}_k\Lambda^k_{RH}\ . \quad (5.29)$$

Then the Kac-Moody algebra follows,[18] and the conserved current corresponding to the transform is

$$j={\rm Tr}(\tilde{K}UTU^{-1})\ , \quad (5.30)$$

which generates the infinite set of the nonlocal conserved currents. Comparing (5.30) with (3.30a), we see immediately that this current is the dual transformed isotopic current.[2] The corresponding finite transformation for Eq. (5.29) is related to the Geroch group discussed by Geroch, Hauser, and Ernst.[19]

Case (B) $M\neq0$, then $M^2=0$. So M is nilpotent and can be brought to the standard form $M'=\left(\begin{smallmatrix}0 & -\alpha'\\ 0 & 0\end{smallmatrix}\right)$. And $-\alpha'$ can be further chosen as 1 by a proper rescaling. For $a=1$, $-\alpha'=1$, the linear system becomes

$$DU = -\begin{bmatrix} \tilde{K} & \tilde{\rho}^*\tilde{K} \\ 0 & \tilde{K} \end{bmatrix} U$$

or

$$DU_1 = -\tilde{K}U_1 - \tilde{\rho}^*\tilde{K}U_2, \quad DU_2 = -\tilde{K}U_2 . \qquad (5.31)$$

We see that U_2 satisfies the same equation as $U(\kappa)$ in (3.17). So $U_2 = U(\kappa)$. Differentiating (5.31) with respect to the parameter κ and using the relation

$$\frac{\partial}{\partial\kappa}\tilde{K} = \tilde{\rho}\frac{\partial}{\partial\phi}\tilde{K} = \tilde{\rho}^*\tilde{K} , \qquad (5.32)$$

which can be obtained by a direct calculation from Eqs. (3.18) and (3.23), we obtain

$$D\frac{\partial U(\kappa)}{\partial\kappa} = -\tilde{\rho}^*\tilde{K}U(\kappa) - \tilde{K}\frac{\partial U(\kappa)}{\partial\kappa} .$$

This means

$$U_1 = \frac{\partial U(\kappa)}{\partial\kappa} . \qquad (5.33a)$$

Thus in this case,

$$\tilde{\Lambda} \equiv \Lambda_{\text{Lie}} = \frac{\partial U(\kappa)}{\partial\kappa}U^{-1}(\kappa) , \qquad (5.33b)$$

which is a Lie transform just as in the flat case.[2,3]

For the choice (b), given in Eqs. (5.8) and (5.9), $\Delta \neq 0$; then M can always be brought to the completely off-diagonal form

$$M \rightarrow M' = \begin{bmatrix} 0 & -\alpha' \\ \beta' & 0 \end{bmatrix}, \quad \alpha'\beta' = \Delta . \qquad (5.34)$$

We can set $\alpha' = -\beta' = 1$ by a proper rescaling and then obtain the BBT's characterized by (4.12b). M' can be further diagonalized and the linear system then takes the form

$$DU = -\begin{bmatrix} \tilde{K} & 0 \\ 0 & -\tilde{K} \end{bmatrix} U . \qquad (5.35)$$

Then

$$U = \begin{bmatrix} U(y;\gamma(\kappa)) \\ U(y;-\gamma(\kappa)) \end{bmatrix} , \qquad (5.36)$$

and the Riccati equation for BBT's can now be solved by

$$\tilde{\Lambda} = [U(y;-\gamma(\kappa))c - U(y;\gamma(\kappa))]$$
$$\times [U(y;-\gamma(\kappa))c + U(y;\gamma(\kappa))]^{-1} , \qquad (5.37)$$

where c is an arbitrary constant matrix.

ACKNOWLEDGMENTS

The authors would like to express their thanks to Professor C. N. Yang for his hospitality and inspiration during the visit to the Institute for Theoretical Physics, Stony Brook. One of the authors (X.-C.S) is supported in part by the Feng King-Hey Foundation through the Committee for Educational Exchange with China at Stony Brook. This work was supported by the U.S. Department of Energy under Contract No. DE-AC02-76CH00016.

*Permanent address: Institute of Theoretical Physics, Peking University, Beijing, China.

[1]L.-L. Chau and B.-Y. Hou, Phys. Lett. **145B**, 347 (1984).
[2]B.-Y. Hou, J. Math. Phys. **25**, 2325 (1984).
[3]L.-L. Chau, B.-Y. Hou, and X.-C. Song, Phys. Lett. **151B**, 421 (1985).
[4]See L. L. Chau, in *Group Theoretical Methods in Physics*, proceedings of the 11th International Colloquium, Istanbul, 1982, edited by M. Serdaroğlu and E. İnönü (Lecture Notes in Physics, Vol. 180) (Springer, Berlin, 1983); in *Nonlinear Phenomena*, proceedings of the Workshop, Oaxtepec, Mexico, 1982, edited by K. B. Wolf (Lecture Notes in Physics, Vol. 189) (Springer, Berlin, 1983).
[5]D. Maison, Phys. Rev. Lett. **41**, 521 (1978); J. Math. Phys. **20**, 871 (1979).
[6]V. A. Belinsky and V. E. Zakharov, Zh. Eksp. Teor. Fiz. **77**, 3 (1979) [Sov. Phys. JETP **50**, 1 (1979)].
[7]L. Witten, Phys. Rev. D **19**, 718 (1979); P. Forgács, Z. Horváth, and L. Palla, Phys. Rev. Lett. **45**, 505 (1980).
[8]F. A. Bais and R. Sasaki, Nucl. Phys. **B195**, 522 (1982).
[9]K. Pohlmeyer, Commun. Math. Phys. **72**, 317 (1980).
[10]K.-C. Chou and X.-C. Song, Commun. Theor. Phys. **1**, 69 (1982); **1**, 185 (1982).
[11]A. V. Mikhailov and A. I. Yaremchuk, Nucl. Phys. **B202**, 508 (1982).
[12]K.-C. Chou and X.-C. Song, Commun. Theor. Phys. **2**, 971 (1983).
[13]H. Eichenherr and M. Forger, Nucl. Phys. **B155**, 381 (1979); H. Eichenherr, Phys. Lett., **B90**, 121 (1980); K. Scheler, *ibid.* **B93**, 331 (1980).
[14]K.-C. Chou and X.-C. Song, Sci. Sin. **25**, 716 (1982); **25**, 825 (1982).
[15]D. Levi, O. Ragnisco, and A. Sym, Lett. Nuovo Cimento **33**, 401 (1982); D. Levi and O. Ragnisco, Phys. Lett. **87A**, 381 (1982).
[16]A. T. Ogielski, M. K. Prasad, A. Sinha, and L.-L. Chau Wang, Phys. Lett. **91B**, 387 (1980).
[17]J. J. Levin, Bull. Am. Math. **10**, 519 (1959).
[18]L. Dolan and A. Roos, Phys. Rev. D **22**, 20 (1980); B.-Y. Hou, Commun. Theor. Phys. **1**, 333 (1982); L. Dolan, Phys. Rev. Lett. **47**, 1371 (1981); M.-L. Ge and Y.-S. Wu, Phys. Lett. **108B**, 411 (1982); C. Devchand and D. B. Fairlie, Nucl. Phys. **B194**, 232 (1982); K. Ueno and Y. Nakamura, Phys. Lett. **117**, 208 (1982).
[19]R. Geroch, J. Math. Phys. **12**, 918 (1971); **13**, 394 (1972); I. Hauser and F. J. Ernst, *ibid.* **22**, 1051 (1981).

Cohomology in connection space, family index theorem, and Abelian gauge structure

Bo-Yu Hou and Yao-Zhong Zhang
Institute of Modern Physics, Northwest University, Xian, The People's Republic of China

(Received 11 July 1986; accepted for publication 8 April 1987)

Using the natural connection form on a principal bundle $P(M,G)$ and the bundle $\mathfrak{A}(\mathfrak{A}/\mathscr{G},\mathscr{G})$ a systematic derivation of the double-cohomological series constituted by the exterior differential d on space-time M and arbitrary, horizontal, and vertical variations in connection space is given. The relationship between these cohomologies and the family index theorem is clarified. The formalism is then used to analyze Abelian gauge structure inside non-Abelian gauge theory. The pertinent functional $U(1)$ connection form, curvature form, and three-form "curvature" are identified and computed, and are related to the θ vacuum, anomalous commutation relation, and Jacobi identity, respectively. Some of the results differ from those obtained by Wu and Zee [Nucl. Phys. B **258**, 157 (1985)] and Niemi and Semenoff [Phys. Rev. Lett. **55**, 227 (1985)] and the results of this paper recover theirs under certain conditions. Finally the generalization of the formalism to a nontrivial principal bundle by introduction of a fixed background connection form is discussed.

I. INTRODUCTION

Recently there has been active interest in anomalies[1-5] in quantum theory. The cohomology of Lie algebras and Lie groups has been discussed by a number of authors. Bonora *et al.*,[6] Stora,[7] and Zumino[8] give the descent equation of the gauge algebra δ_v and the exterior differential d. Faddeev[9] discusses the double-cohomological series of the transformation Δ_v of the gauge group and d. Zumino[10] shows that the cohomology of the gauge group can be reduced to that of its Lie algebra.

If we describe the connection space $\mathfrak{A}$ as a fiber bundle with $\mathfrak{A}/\mathscr{G}$ as the base and $\mathscr{G}$ as the structure group, the variation $\delta_v(\Delta_v)$ is the one along the fiber $\mathscr{G}$. The corresponding cohomology is called the vertical one. On the other hand, Gelfand[11] and Faddeev *et al.*[12] introduce the horizontal variation $\delta_h(\Delta_h)$ along the base $\mathfrak{A}/\mathscr{G}$, and establish the descent equation of $\delta_h(\Delta_h)$ and d. We call the cohomology associated with $\delta_h(\Delta_h)$ the horizontal one. In addition to these cohomologies, Guo *et al.*[13] first discuss the cohomological series of d and the interpolation variation in the connection space $\mathfrak{A}$.

However, few attempts have been made to study the relationship between the horizontal and vertical cohomologies, which seem to be different. Faddeev[12] and Hou *et al.*[14] claim that they have given the relationship between these cohomologies. Since the operations of Δ_v and Δ_h do not match each other, the relation they obtain in their paper is very complicated and not obvious. A deep understanding of the relationship between horizontal and vertical cohomologies is still lacking. In addition, one may ask the following question: What is the relationship between these cohomologies and the family index theorem?

This paper is a modest attempt to try to answer these questions. Using the natural connection form on a principal bundle $P(M,G)$ and bundle $\mathfrak{A}(\mathfrak{A}/\mathscr{G},\mathscr{G})$ we deal with an arbitrary variation δ, a horizontal one δ_h, and vertical one δ_v in a unified point of view. We also give a systematical derivation of the generalized finite double-cohomological series in the space of all connection forms. We show very simple and obvious relations between horizontal and vertical cohomologies and, in particular, between these cohomologies and the family index theorem.[15,16]

As an application of our formalism, we examine the Abelian gauge structure inside non-Abelian gauge theory. This has been discussed first by Jackiw[17] and then by Wu *et al.*[18] and Niemi *et al.*[19] They compute the functional $U(1)$ connection form and curvature form on connection space by using the path integral formulation, the Hamiltonian formulation, and the η invariant of Dirac operator, respectively. In this paper we obtain the results in their works in a different way by using the family index theorem and cohomology in connection space. We also obtain some new results that reduce to those in Refs. 18 and 19 under certain conditions.

This paper will be organized as follows: In Sec. II we discuss three kinds of cohomological series on connection space, the family index theorem, and the relations among them. In Secs. III–V, we present the application of our formalism to examining Abelian gauge structure inside non-Abelian gauge theory. Section VI is devoted to the generalization of our formalism to nontrivial principal bundle. In Sec. VII we give some conclusions and discussions.

II. THE FORMALISM

Consider connection forms on compactified Euclidean space-time M. We shall first assume the principal bundle $P(M,G)$ has trivial topology, i.e., $P=M\times G$, where G is a semisimple Lie group. (The case that principal bundle is nontrivial will be discussed in Sec. VI.) Let $\mathfrak{A}$ be the set of all such connection forms, $\mathfrak{A}=\{A(x)\}$. Let $\mathscr{G}$ be the gauge group, i.e., the set of maps $g: M\to G\,(x\mapsto g(x))$, $\mathscr{G}=\{g(x)\}$. Let $\mathfrak{A}/\mathscr{G}$ be the corresponding orbit space, that is, the set of different orbits. Because $\mathscr{G}$ is nontrivial, $\mathfrak{A}/\mathscr{G}$ has a topology. In fiber bundle language, the connection space $\mathfrak{A}$ may be described as a fiber bundle with $\mathfrak{A}/\mathscr{G}$ as base and $\mathscr{G}$ as struc-

 0022-2488/87/081709-07$02.50

ture group. Following Stora,[7] we can locally parametrize $\mathfrak{A}$ by

$$A = g^{-1} a g + g^{-1} dg, \tag{2.1}$$

where a represents an orbit. Then a cotangent vector in $\mathfrak{A}$ is

$$\delta A = g^{-1} \delta a g - D_A (g^{-1} \delta g) \quad \text{with } D_A \equiv d + [A, \ ]. \tag{2.2}$$

We can introduce on $\mathfrak{A}$ the connection form

$$\mathscr{A} = -G_A D_A^+ \delta A \quad \text{with } G_A \equiv (D_A^+ D_A)^{-1} \tag{2.3}$$

so that

$$\mathscr{A}|_{\text{fiber}} = g^{-1} \delta_v g. \tag{2.4}$$

Here δA can be decomposed into a horizontal component $\delta_h A$ and a vertical one $\delta_v A$,

$$\begin{aligned} \delta_h A &= (1 - D_A G_A D_A^+) \delta A, \\ \delta_v A &= D_A G_A D_A^+ \delta A = -D_A \mathscr{A}. \end{aligned} \tag{2.5}$$

The total connection form over $M \times \mathfrak{A}/\mathscr{G}$ at $(g(x), g(\cdot); x, A)$ is

$$A + \mathscr{A} = A - G_A D_A^+ \delta A. \tag{2.6}$$

The corresponding curvature form is

$$\begin{aligned} \mathscr{F} &= (d + \delta)(A + \mathscr{A}) + \tfrac{1}{2}[A + \mathscr{A}, A + \mathscr{A}] \\ &= F + \delta_h A + (\delta \mathscr{A} + \mathscr{A}^2) \\ &= F + (\mathscr{F})_1^1 + (\mathscr{F})_0^2, \end{aligned} \tag{2.7}$$

where $F = dA + \frac{1}{2}[A, A]$ and $(\ )_j^i$ stands for a j-form on P and i-form on $\mathfrak{A}$.

The bundle and connections can be pulled back onto $M \times \mathfrak{A}/\mathscr{G}$. According to the family index theorem,[15] the characteristic classes on $\mathfrak{A}/\mathscr{G}$ are expressible as integral of the higher classes on $M \times \mathfrak{A}/\mathscr{G}$ over the base M:

$$Q^k(M) = \int_M P_{2n}(\mathscr{F}^n), \tag{2.8}$$

forms of degree k on $\mathfrak{A}/\mathscr{G}$, where $P_{2n}(\mathscr{F}^n)$ is an n-rank invariant polynomial in $\mathscr{F}$ and the dimension of M is $(2n - k)$. We can choose locally horizontal gauge[12] $D_A^+ \delta A = 0$ such that $\mathscr{A} = 0$. It is convenient to define space in which $\mathscr{A} = 0$ as $\mathfrak{A}/\mathscr{G}$. So, from Eq. (2.7) we obtain $\mathscr{F}$ in horizontal gauge and we denote it by $\hat{\mathscr{F}}$,

$$\hat{\mathscr{F}} = F + \delta_h A = (d + \delta_h) A + \tfrac{1}{2}[A, A]. \tag{2.9}$$

The Bianchi identity $D_h \hat{\mathscr{F}} \equiv (d + \delta_h) \hat{\mathscr{F}} + [A, \hat{\mathscr{F}}] = 0$ implies that

$$(d + \delta_h) P_{2n}(\hat{\mathscr{F}}^n) = 0. \tag{2.10}$$

Expanding the $P_{2n}(\hat{\mathscr{F}}^n)$,

$$P_{2n}(\hat{\mathscr{F}}^n) = \sum_k q_{2n-k}^k (F, \delta_h A), \tag{2.11}$$

we have the following descent equation:

$$\begin{aligned} & dq_{2n}^0 (F, \delta_h A) = 0, \quad \delta_h q_0^{2n} (F, \delta_h A) = 0, \\ & dq_{2n-k}^k (F, \delta_h A) = -\delta_h q_{2n-k+1}^{k-1} (F, \delta_h A). \end{aligned} \tag{2.12}$$

Introducing integration over an arbitrary nonclosed k-chain $\hat{\Delta}^k$ in $\mathfrak{A}/\mathscr{G}$,

$$\hat{Q}_{2n-k}(\hat{\Delta}^k) = \int_{\hat{\Delta}^k} q_{2n-k}^k (F, \delta_h A) \tag{2.13}$$

and over an arbitrary $(2n - k)$-chain C_{2n-k} in space-time M,

$$\hat{Q}^k (C_{2n-k}) = \int_{C_{2n-k}} q_{2n-k}^k (F, \delta_h A), \tag{2.14}$$

we obtain the following finite double-cohomological series from Eq. (2.12) and the Stokes theorem:

$$d\hat{Q}_{2n-k-1}(\hat{\Delta}^{k+1}) = -\hat{Q}_{2n-k}(\partial_h \hat{\Delta}^{k+1}), \tag{2.15a}$$

$$\delta_h \hat{Q}^{k-1}(C_{2n-k+1}) = -\hat{Q}^k (\partial C_{2n-k+1}). \tag{2.15b}$$

Here ∂_h denotes the boundary operation in $\mathfrak{A}/\mathscr{G}$. The Chern classes associated with the family index, when a locally horizontal gauge is chosen, are given by Eq. (2.14) and satisfy the descent equation (2.15b) for horizontal variation δ_h and exterior differential d. Note that when $\hat{\Delta}$ is a simplex, Eq. (2.15b) reduces to the double-cohomological series given by Faddeev.[12]

It follows from Eq. (2.15b) that

$$\delta_h \hat{Q}^k (M_{2n-k}) = 0, \quad \text{when } \partial M_{2n-k} = 0. \tag{2.16}$$

However, $\hat{Q}^k(M_{2n-k}) \neq \delta_h$ (something) because of nontriviality of the base space $\mathfrak{A}/\mathscr{G}$.

The Chern classes $\hat{Q}^k(M_{2n-k})$, which are k-forms on $\mathfrak{A}/\mathscr{G}$, can be lifted to k-forms $Q^k(M_{2n-k})$ on $\mathfrak{A}$. In order to do this, we first lift the $q_{2n-k}^k(F, \delta_h A)$ to $\mathfrak{A}$ and obtain

$$q_{2n-k}^k (\mathscr{F}^n) = \sum_{\substack{l_1, \ldots, l_n = 0 \\ l_1 + \cdots + l_n = k}}^{2} P_{2n}((\mathscr{F})_{2-l_1}^{l_1}, \ldots, (\mathscr{F})_{2-l_n}^{l_n}). \tag{2.17}$$

Invariance of the $P_{2n}(\mathscr{F}^n)$, i.e., $(d + \delta) P_{2n}(\mathscr{F}^n) = 0$, gives

$$\begin{aligned} & dq_{2n}^0 (\mathscr{F}^n) = 0, \quad \delta q_0^{2n} (\mathscr{F}^n) = 0, \\ & dq_{2n-k}^k (\mathscr{F}^n) = -\delta q_{2n-k+1}^{k-1} (\mathscr{F}^n). \end{aligned} \tag{2.18}$$

We can thus construct k-forms on $\mathfrak{A}$,

$$Q^k (C_{2n-k}) = \int_{C_{2n-k}} q_{2n-k}^k (\mathscr{F}^n), \tag{2.19}$$

and $(2n - k)$-forms on M,

$$Q_{2n-k}(\Delta^k) = \int_{\Delta^k} q_{2n-k}^k (\mathscr{F}^n), \tag{2.20}$$

where the C_{2n-k} are arbitrary $(2n - k)$-chains in space-time M and Δ^k k-chains in $\mathfrak{A}$. We can easily show from Eq. (2.18) and the Stokes theorem that

$$dQ_{2n-k-1}(\Delta^{k+1}) = -Q_{2n-k}(\partial \Delta^{k+1}), \tag{2.21a}$$

$$\delta Q^{k-1}(C_{2n-k+1}) = -Q^k (\partial C_{2n-k+1}). \tag{2.21b}$$

These are double-cohomological series in the space $\mathfrak{A}$.

According to the family index theorem,[15] the $Q^k(C_{2n-k})$, defined on $\mathfrak{A}$, defines a k-form on $\mathfrak{A}/\mathscr{G}$. This can be proved[7] by checking that the term $-D_A(g^{-1}\delta g)$ in δA does not contribute to the result (up to d of some form) if we look at the nonintegrated $q_{2n-k}^k(\mathscr{F}^n)$. This observation is important when we exmine the global property of horizontal cohomology.

We obviously have from Eq. (2.21)

$$\delta Q^k (M_{2n-k}) = 0 \quad \text{when } \partial M_{2n-k} = 0, \tag{2.22}$$

which shows that the $Q^k(M_{2n-k})$ are cocycles on $\mathfrak{A}$. The

$Q^k(M_{2n-k})$ are also exact since $\mathfrak{A}$ has no topology, that is,

$$Q^k(M_{2n-k}) = \delta\Omega^{k-1}(M_{2n-k}). \qquad (2.23)$$

Here $\Omega^{k-1}(M_{2n-k})$ can be determined as follows: Since the principal bundle has trivial topology, we obtain from the relation that

$$(d+\delta)P_{2n}(\mathscr{F}^n) = 0$$

that

$$P_{2n}(\mathscr{F}^n) = (d+\delta)\omega_{2n-1}(A+\mathscr{A}). \qquad (2.24)$$

Here $\omega_{2n-1}(A+\mathscr{A})$ is given by the well-known transgression formula due to Chern,

$$\omega_{2n-1}(A+\mathscr{A}) = n\int_0^1 dt\, P_{2n}(A+\mathscr{A},\mathscr{F}_t^{n-1}) \qquad (2.25)$$

with $\mathscr{F}_t = t\mathscr{F} + (t^2-t)(A+\mathscr{A})^2$.

Integrating Eq. (2.24) over M_{2n-k} without boundary and then comparing it with Eq. (2.23) we obtain

$$Q^k(M_{2n-k}) = \int_{M_{2n-k}} q^k_{2n-k}(\mathscr{F}^n), \qquad (2.26)$$

$$\Omega^{k-1}(M_{2n-k}) = \int_{M_{2n-k}} \omega^{k-1}_{2n-k}(A;\mathscr{A}), \qquad (2.27)$$

where $\omega^k_{2n-k}(A;\mathscr{A})$ stands for the $(2n-k,k-1)$ component of the $\omega_{2n-1}(A+\mathscr{A})$.

When limited to fiber, Eq. (2.24) becomes

$$P_{2n}(\check{\mathscr{F}}^n) = P_{2n}(F^n) = (d+\delta v)\check{\omega}_{2n-1}(A+v), \qquad (2.28)$$

where use has been made of the following facts:

$$\mathscr{A}|_{\text{fiber}} \equiv \check{\mathscr{A}} = g^{-1}\delta_v g \equiv v, \quad \mathscr{F}|_{\text{fiber}} \equiv \check{\mathscr{F}} = F,$$
$$\omega_{2n-1}(A+\mathscr{A})|_{\text{fiber}} \equiv \check{\omega}_{2n-1}(A+v). \qquad (2.29)$$

Expanding the $\check{\omega}_{2n-1}(A+v)$ in v:

$$\check{\omega}_{2n-1}(A+v) = \sum_k \check{\omega}^k_{2n-k}(v;A). \qquad (2.30)$$

We obtain from Eq. (2.28):

$$d\check{\omega}^0_{2n-1}(v;A) = q^0_{2n}(F^n) = P_{2n}(F^n),$$
$$\delta_v\check{\omega}_0^{2n-1}(v;A) = 0, \qquad (2.31)$$
$$\delta_v\check{\omega}^{k-1}_{2n-k}(v;A) = -d\check{\omega}^k_{2n-k-1}(v;A).$$

Introducing the integrations

$$\check{\Omega}^k(C_{2n-k-1}) = \int_{C_{2n-k-1}} \check{\omega}^k_{2n-k-1}(v;A), \qquad (2.32)$$

$$\check{\Omega}_{2n-k-1}(\Gamma^k) = \int_{\Gamma^k} \check{\omega}^k_{2n-k-1}(v;A), \qquad (2.33)$$

where the Γ^k are arbitrary k-chains in $\mathscr{G}$, we have the following finite double-cohomological series from Eq. (2.31) and the Stokes theorem:

$$d\check{\Omega}_{2n-k-1}(\Gamma^k) = -\check{\Omega}_{2n-k}(\partial_v\Gamma^k), \qquad (2.34a)$$
$$\delta_v\check{\Omega}^{k-1}(C_{2n-k}) = -\check{\Omega}^k(\partial C_{2n-k}), \qquad (2.34b)$$

where ∂_v stands for the boundary operation in $\mathscr{G}$. When Γ^k is a simplex, Eq. (2.34a) reduces to the $\Delta_v - d$ series given by Faddeev.[9]

Now the relation between the horizontal and the vertical cohomologies, and in particular between these cohomologies and the family index theorem, may be formulated: Introduce integrations

$$\alpha(\Gamma^k,C_{2n-k-1}) = \int_{\Gamma^k}\Omega^k(C_{2n-k-1}), \qquad (2.35)$$

$$Q(\Delta^{k+1},C_{2n-k-1}) = \int_{\Delta^{k+1}} Q^{k+1}(C_{2n-k-1}). \qquad (2.36)$$

We choose the $(k+1)$-chains Δ^{k+1} in such a way that their boundaries belong to $\mathscr{G}$, i.e., $\partial\Delta^{k+1} = \Gamma^k\in\mathscr{G}$, and the projection of Δ^{k+1} into $\mathfrak{A}/\mathscr{G}$ is a k-dimensional sphere S^k in it. Noticing that

$$\Omega^k(M_{2n-k-1})|_{\text{fiber}} = \check{\Omega}^k(M_{2n-k-1}), \qquad (2.37)$$

we can easily show by means of Eq. (2.23) that

$$Q(\Delta^{k+1},C_{2n-k-1}) = \alpha(\Gamma^k,C_{2n-k-1}), \qquad (2.38)$$

and, in particular,

$$\pi^{k+1}(\mathfrak{A}/\mathscr{G}) = Q(\Delta^{k+1},M_{2n-k-1})$$
$$= \alpha(\Gamma^k,M_{2n-k-1}) = \pi^k(\mathscr{G}). \qquad (2.39)$$

We have thus linked the horizontal and vertical cohomologies with the family index theorem.

The above analyses are general and abstract. To further understand the formalism requires a concrete calculation, which we will carry out in the following sections.

III. U(1) CONNECTION FORM ON $\mathfrak{A}$ AND θ VACUUM

As a first application of the above formalism, we consider the (3 + 1)-dimensional gauge theory without non-Abelian anomaly.[20] Fix gauge $A_0 = 0$ and consider the space $\mathfrak{A}^3$ of all static gauge field configurations $A_i^a(\vec{x})$. In the Schrödinger formulation the wave functional is $\psi[A]$. The Gauss law

$$D_i(\delta/\delta A_i^a)\psi[A] = 0 \qquad (3.1)$$

can be used to eliminate the residual static gauge freedom, $D_i(\delta/\delta A_i^a)$ is an infinitesimal generator of the gauge transformation. In the homotopically trivial case a finite gauge transformation can be obtained from an infinitesimal one. Therefore the Gauss law means that the wave functional $\psi[A]$ is invariant under "small" gauge transformations. Here the so-called small gauge transformations are those that can be obtained from the identity by infinitesimal ones. We know that there are "large" gauge transformations that cannot be obtained in this way in the homotopically nontrivial case. The wave functional is not invariant under large transformations

$$\psi[A^g] = e^{in\theta}\psi[A] \quad (A^g = g^{-1}Ag + g^{-1}dg). \qquad (3.2)$$

From the discussions in the above section we can easily compute the one-form on $\mathfrak{A}^3/\mathscr{G}^3$:

$$\hat{Q}^1(S^3) = \int_{S^3} q_3^1(F,\delta_h A)$$
$$= -\frac{1}{8\pi^2}\int_{S^3}(\mathrm{tr}(F+\delta_h A)^2)_3^1$$
$$= -\frac{1}{4\pi^2}\int_{S^3}\mathrm{tr}(F\delta_h A), \qquad (3.3)$$

which is closed, i.e., $\delta_h \hat{Q}^1(S^3) = 0$, but not exact. We can lift $\hat{Q}^1(S^3)$ and obtain a one-form on $\mathfrak{A}^3$:

$$\begin{aligned} Q^1(S^3) &= \int_{S^3} q_3^1(\mathcal{F}^2) \\ &= -\frac{1}{8\pi^2}\int_{S^3} \operatorname{tr}((\mathcal{F})_1^1 F) + \operatorname{tr}(F(\mathcal{F})_1^1) \\ &= -\frac{1}{4\pi^2}\int_{S^3} \operatorname{tr}(F(\mathcal{F})_1^1) \\ &= -\frac{1}{4\pi^2}\int_{S^3} \operatorname{tr}(F\delta A). \end{aligned} \tag{3.4}$$

We can explain $Q^1(S^3)$ as a $U(1)$ connection form on $\mathfrak{A}^3$. The corresponding curvature two-form is

$$\begin{aligned} \delta Q^1(S^3) &= -\frac{1}{4\pi^2}\int_{S^3} \operatorname{tr}(\delta A\delta A) \\ &= \frac{1}{4\pi^2}\int_{S^3} \operatorname{tr}(\delta A D_A(\delta A)) \\ &= \frac{1}{2\pi^2}\int_{S^3} d(\operatorname{tr}(\delta A\delta A)) = 0. \end{aligned} \tag{3.5}$$

Therefore the field strength form of the $U(1)$ potential form is zero. Equations (3.4) and (3.5) are the $U(1)$ connection and the curvature form given by Wu and Zee,[18] respectively. We have obtained these anew using our formalism.

Since base $\mathfrak{A}^3$ has the trivial topology, Eq. (3.5) means that $Q^1(S^3)$ is also exact on $\mathfrak{A}^3$, i.e.,

$$Q^1(S^3) = \delta\Omega^0(S^3) \tag{3.6a}$$

with $\Omega^0(S^3)$ given by Eq. (2.27):

$$\begin{aligned} \Omega^0(S^3) &= \int_{S^3} \omega_3^0(A + \mathcal{A}) \\ &= -\frac{1}{4\pi^2}\int_{S^3}\int_0^1 dt(\operatorname{tr}((A+\mathcal{A})\mathcal{F}_t))_3^0 \\ &= -\frac{1}{8\pi^2}\int_{S^3} \operatorname{tr}(A\,dA + \tfrac{2}{3}A^3). \end{aligned} \tag{3.6b}$$

Thus the connection form $Q^1(S^3)$ is a pure gauge.

Now we consider integration of the potential one-form $Q^1(S^3)$ over an open path Δ^1 in $\mathfrak{A}^3$. Let its boundary be two points A^g and A on the same gauge orbit. Then from Eq. (3.6),

$$\begin{aligned} \int_{\Delta^1} Q^1(S^3) &= \int_{\Delta^1} \delta\Omega^0(S^3) = \int_{\partial\Delta^1} \Omega^0(S^3) = \int_{\Gamma^0} \Omega^0(S^3) \\ &= -\frac{1}{8\pi^2}\int_{S^3} (\operatorname{tr}(A^g\,dA^g + \tfrac{2}{3}(A^g)^3) \\ &\quad - \operatorname{tr}(A\,dA + \tfrac{2}{3}A^3)) \\ &= -\frac{1}{24\pi^2}\int_{S^3} \operatorname{tr}(g^{-1}\,dg)^3 = Z. \end{aligned} \tag{3.7}$$

The above equation is an explicit manifestation of the relation $\pi^1(\mathfrak{A}^3/\mathcal{G}^3) = \pi^0(\mathcal{G}^3) = Z$, which shows that there exist noncontractible loops in $\mathfrak{A}^3/\mathcal{G}^3$. We thus have vortex field in $\mathfrak{A}^3/\mathcal{G}^3$, in agreement with the conclusion in Ref. 18.

If we define

$$\Phi[A] = \exp\left(\frac{1}{8\pi^2}\int_{S^3} \operatorname{tr}\left(A\,dA + \frac{2}{3}A^3\right)\theta\right)\chi[A], \tag{3.8}$$

$\Phi[A]$ is gauge invariant from Eqs. (3.2) and (3.7). However, in quantum theory a phase change in the wave functional corresponds to a canonical transformation,[17] which will induce a change in the Lagrangian by a total derivative, and therefore leads to the vacuum θ angle in the Yang–Mills Lagrangian.

IV. U(1) CURVATURE FORM AND ANOMALOUS COMMUTATOR ON $\mathfrak{A}$

In this section, we shall consider non-Abelian gauge theory[21] defined on space-time manifold S^3. We shall fix $A_0 = 0$ and consider the infinite-dimensional affine space $\mathfrak{A}^2$ of all static gauge field configurations $A_i^a(\vec{x})$. The two-form on $\mathfrak{A}^2/\mathcal{G}^2$ is

$$\begin{aligned} \hat{Q}^2(S^2) &= \int_{S^2} q_2^2(F,\delta_h A) = -\frac{1}{8\pi^2}\int_{S^2} (\operatorname{tr}(F + \delta_h A)^2)_2^2 \\ &= -\frac{1}{8\pi^2}\int_{S^2} \operatorname{tr}(\delta_h A\delta_h A). \end{aligned} \tag{4.1}$$

Lifting $\hat{Q}^2(S^2)$, we obtain a two-form on $\mathfrak{A}^2$,

$$\begin{aligned} Q^2(S^2) &= \int_{S^2} q_2^2(\mathcal{F}^2) \\ &= -\frac{1}{8\pi^2}\int_{S^2} (\operatorname{tr}(\mathcal{F}^2))_2^2 = -\frac{1}{4\pi^2}\int_{S^2} \operatorname{tr}((\mathcal{F})_0^2 F) \\ &\quad -\frac{1}{8\pi^2}\int_{S^2} \operatorname{tr}((\mathcal{F})_1^1(\mathcal{F})_1^1) \\ &= -\frac{1}{8\pi^2}\int_{S^2} \operatorname{tr}(\delta A\delta A) - \frac{1}{4\pi^2}\delta\int_{S^2} \operatorname{tr}(\mathcal{A}F), \end{aligned} \tag{4.2}$$

which is obviously closed: $\delta Q^2(S^2) = 0$. Here $Q^2(S^2)$ can be identified as the $U(1)$ curvature two-form on $\mathfrak{A}^2$. The $Q^2(S^2)$ is also exact because of topological triviality of the space $\mathfrak{A}^2$:

$$Q^2(S^2) = \delta\Omega^1(S^2), \tag{4.3a}$$

where

$$\begin{aligned} \Omega^1(S^2) &= \int_{S^2} \omega_2^1(A + \mathcal{A}) \\ &= -\frac{1}{4\pi^2}\int_{S^2}\int_0^1 \mathrm{dt}(\operatorname{tr}((A+\mathcal{A})\mathcal{F}_t))_2^1 \\ &= -\frac{1}{8\pi^2}\int_{S^2} \operatorname{tr}(A\delta A) - \frac{1}{4\pi^2}\int_{S^2} \operatorname{tr}(\mathcal{A}F). \end{aligned} \tag{4.3b}$$

Thus we have obtained the connection one-form $\Omega^1(S^2)$ on $\mathfrak{A}^2$. Note that when $\mathcal{A} = 0$, only the first term in Eq. (4.2) and (4.3b) survives: This corresponds to the expressions obtained by Wu and Zee.[18]

Now let us examine the integration of $Q^2(S^2)$ over a two-dimensional disk Δ^2 in $\mathfrak{A}^2$. Assume its boundary $\partial\Delta^2 = \Gamma^1$ is a one-dimensional loop in $\mathcal{G}^2$. We then have from Eq. (4.3)

$$\int_{\Delta^2} Q^2(S^2) = \int_{\Delta^2} \delta\Omega^1(S^2) = \int_{\partial\Delta^2} \Omega^1(S^2) = \int_{\Gamma^1} \Omega^1(S^2)$$
$$= \int_{\Gamma^1} \check{\Omega}^1(S^2) = \int_{\Gamma^1}\int_{S^2} \check{\omega}_2^1$$
$$= \int_{\Gamma^1_+}\int_{S^2} \check{\omega}_2^1 - \int_{\Gamma^1_-}\int_{S^2} \check{\omega}_2^1$$
$$= \int_{\Gamma^0}\int_{D^3_+} \check{\omega}_3^0 - \int_{\Gamma^0}\int_{D^3_-} \check{\omega}_3^0 = \int_{\Gamma^0}\int_{S^3} \check{\omega}_3^0 = Z. \tag{4.4}$$

Therefore the Chern number associated with a functional $U(1)$ connection form $\Omega^1(S^2)$ is an integer n, in contrast to the conclusion in Ref. 18. Note that Eq. (4.4) means that $\pi^2(\mathfrak{A}^2/\mathcal{G}^2) = \pi^1(\mathcal{G}^2) = Z$, which shows that there exists a noncontractible two-dimensional sphere in $\mathfrak{A}^2/\mathcal{G}^2$. So, we have a monopole in the orbit space $\mathfrak{A}^2/\mathcal{G}^2$.[18]

The above system, quantized in Schrödinger formulation, is the infinite-dimensional version of quantum electrodynamics in ordinary space, and as in the finite case, the field velocity operator

$$V_X = \delta_X + \Omega^1(S^2), \quad \delta_X = \int d^2x\, \delta A_i^a(x) \frac{\delta}{\delta A_i^a(x)}, \tag{4.5}$$

on $\mathfrak{A}^2$ describes a gauge field (where X stands for sum over a, i, and integration over $\vec{x}$) for an external $U(1)$ connection form $\Omega^1(S^2)$. The commutation relation leads to the nonzero curvature two-form

$$[V_X, V_Y] = \delta\Omega^1(S^2) = Q^2(S^2)$$
$$= -\frac{1}{8\pi^2}\int_{S^2} \mathrm{tr}(\delta A\delta A) - \frac{1}{4\pi^2}\delta\int_{S^2}\mathrm{tr}(\mathscr{A}F), \tag{4.6}$$

which, being independent of the dynamical variable, satisfies the Jacobi identity

$$[V_X,[V_Y,V_Z]] + (\text{perm.}) = \delta Q^2(S^2) = 0. \tag{4.7}$$

Thus the translations (i.e., field velocity operators) on $\mathfrak{A}^2$ constitute a Lie group. However, the Gauss law fails because V_X satisfies anomalous commutation relation.

V. FUNCTIONAL U(1) GAUGE THEORY WITHOUT A CONNECTION ONE-FORM AND THE ANOMALOUS JACOBI IDENTITY ON $\mathfrak{A}$

We now consider (3 + 1)-dimensional gauge theory with non-Abelian anomaly.[1] Assume a non-Abelian gauge field is minimally coupled to Weyl fermions in a complex representation of some gauge group. We once again fix $A_0 = 0$ and consider the infinite-dimensional affine space $\mathfrak{A}^3$ of all static gauge field configurations $A_i^a(x)$. Form $\mathfrak{A}^3/\mathcal{G}^3$, the space of three-dimensional gauge fields modulo three-dimensional gauge transformations. Since we are dealing with anomalous gauge theory, we have to consider a two-form on $\mathfrak{A}^3$. To this end, we first compute a three-form on $\mathfrak{A}^3/\mathcal{G}^3$. We have, from Eq. (2.14),

$$\hat{Q}^3(S^3) = \int_{S^3} q_3^3(F,\delta_h A) = -\frac{i}{48\pi^3}\int_{S^3}(\mathrm{tr}(F+\delta_h A)^3)_3^3$$
$$= -\frac{i}{48\pi^3}\int_{S^3}\mathrm{tr}(\delta_h A\delta_h A\delta_h A). \tag{5.1}$$

This form is closed, i.e., $\delta_h \hat{Q}^3(S^3) = 0$, but not exact. In order to relate to anomaly, we lift the $\hat{Q}^3(S^3)$ and derive a three-form on $\mathfrak{A}^3$:

$$Q^3(S^3) = \int_{S^3} q_3^3(\mathscr{F}^3) = -\frac{i}{48\pi^3}\int_{S^3}(\mathrm{tr}\,\mathscr{F}^3)_3^3$$
$$= -\frac{i}{48\pi^3}\int_{S^3}\mathrm{tr}((\mathscr{F})_1^1)^3 - \frac{i}{16\pi^3}\int_{S^3}\mathrm{tr}((\mathscr{F})_1^1(\mathscr{F})_0^2 F) - \frac{i}{16\pi^3}\int_{S^3}\mathrm{tr}((\mathscr{F})_0^2(\mathscr{F})_1^1 F)$$
$$= -\frac{i}{48\pi^3}\int_{S^3}\mathrm{tr}(\delta A\delta A\delta A) - \frac{i}{16\pi^3}\delta\int_{S^3}\mathrm{tr}(F\delta A\mathscr{A} + F\mathscr{A}\delta A + F\mathscr{A}D_A\mathscr{A}). \tag{5.2}$$

This is also closed: $\delta Q^3(S^3) = 0$. Notice that the $Q^3(S^3)$ is exact on $\mathfrak{A}^3$ because of its topological triviality

$$Q^3(S^3) = \delta\Omega^2(S^3). \tag{5.3a}$$

The two-form $\Omega^2(S^3)$ is given by Eq. (2.27),

$$\Omega^2(S^3) = \int_{S^3}\omega_3^2(A+\mathscr{A}) = -\frac{i}{16\pi^3}\int_{S^3}\int_0^1 dt(\mathrm{tr}((A+\mathscr{A})\mathscr{F}_t^2))_3^2$$
$$= -\frac{i}{48\pi^3}\int_{S^3}\mathrm{tr}(A\delta A\delta A) - \frac{i}{16\pi^3}\int_{S^3}\mathrm{tr}(F\delta A\mathscr{A} + F\mathscr{A}\delta A + F\mathscr{A}D_A\mathscr{A}) - \frac{i}{16\pi^3}\delta\int_{S^3}\mathrm{tr}\Big(\mathscr{A}\Big(AF + FA - \frac{1}{2}A^3\Big)\Big). \tag{5.3b}$$

We can identify $\Omega^2(S^3)$ and $Q^3(S^3)$ with the generalized functional $U(1)$ "potential" two-form and "field strength" three-form on $\mathfrak{A}^3$, respectively. Therefore, the system, quantized in the Schrödinger formulation, is an infinite-dimensional version of a quantum mechanical point particle that moves in a background field without potential one-form as discussed by Hou *et al.* in Ref. 22. We have to deal with the functional $U(1)$ antisymmetrical tensor gauge theory[23] without potential one-form. Such a system shares many features with quantum electrodynamics without potential in ordinary space-time: First there is no smooth $U(1)$ potential one-form on $\mathfrak{A}^3$. Second the representation space of the translation group acting on $\mathfrak{A}^3$ is not ordinary functional Hilbert space, and we have to introduce a membrane-dependent wave functional[22] on $\mathfrak{A}^3$. So, we have the situation as in Mandelstam.[24] In analogy with the finite-dimensional case covariant translations on $\mathfrak{A}^3$ are generated by the "velocity" operator (or so-called electric field) E_X,[19] although it is not a smooth functional. The commutation relation gives the "connection" two-form

$$[E_X,E_Y]=\Omega^2(S^3)=-\frac{i}{48\pi^3}\int_{S^3}\mathrm{tr}(A\delta A\delta A)-\frac{i}{16\pi^3}\int_{S^3}\mathrm{tr}(F\delta A\mathscr{A}+F\mathscr{A}\delta A+F\mathscr{A}D_A\mathscr{A})-\frac{i}{16\pi^3}\delta\int_{S^3}\mathrm{tr}\Big(\mathscr{A}\Big(AF+FA-\frac{1}{2}A^3\Big)\Big), \tag{5.4}$$

and the Jacobi identity gives the "curvature" three-form

$$[E_X,[E_Y,E_Z]]+(\text{perm.})=\delta\Omega^2(S^3)=Q^3(S^3)=-\frac{i}{48\pi^3}\int_{S^3}\mathrm{tr}(\delta A\delta A\delta A)-\frac{i}{48\pi^3}\int_{S^3}\mathrm{tr}(F\delta A\mathscr{A}+F\mathscr{A}\delta A+F\mathscr{A}D_A\mathscr{A}). \tag{5.5}$$

Note that the above anomalous terms in the commutation relation and the Jacobi identity differ from ones given by Niemi and Semenoff in Ref. 19. Our results reduce to theirs for $\mathscr{A}=0$.

Since the Jacobi identity fails, the covariant translations on $\mathfrak{A}^3$ are not associative and cannot form a Lie group.[25] However, since $\delta Q^3(S^3)=0$, the electric field trivially satisfies the algebra identity given by the present authors[26]

$$[E_X,[E_Y,[E_Z,E_W]]]+(\text{perm.})=\delta Q^3(S^3)=0. \tag{5.6}$$

As pointed out in Ref. 26, they do not form the so-called Malcev identity.[27]

It remains to examine the global properties of the $Q^3(S^3)$ and of the $\Omega^2(S^3)$. To this end, we integrate $Q^3(S^3)$ over the three-dimensional disk Δ^3 in $\mathfrak{A}^3$. Select Δ^3 so that its boundary $\partial\Delta^3=\Gamma^2$ is a two-dimensional sphere in $\mathscr{G}^3$, and its projection onto $\mathfrak{A}^3/\mathscr{G}$ is also a two-dimensional sphere. Then, from Eq. (5.3),

$$\begin{aligned}\int_{\Delta^3}Q^3(S^3)&=\int_{\Delta^3}\delta\Omega^2(S^3)=\int_{\partial\Delta^3}\Omega^2(S^3)=\int_{\Gamma^2}\Omega^2(S^3)=\int_{\Gamma^2}\check{\Omega}^2(S^3)=\int_{\Gamma^2}\int_{S^3}\check{\omega}_3^2\\&=\int_{\Gamma^2_+}\int_{S^3}\check{\omega}_3^2-\int_{\Gamma^2_-}\int_{S^3}\check{\omega}_3^2=\int_{\Gamma^1}\int_{D^4_+}\check{\omega}_4^1-\int_{\Gamma^1}\int_{D^4_-}\check{\omega}_4^1=\int_{\Gamma^1}\int_{S^4}\check{\omega}_4^1\\&=\int_{\Gamma^1_+}\int_{S^4}\check{\omega}_4^1-\int_{\Gamma^1_-}\int_{S^4}\check{\omega}_4^1=\int_{\Gamma^0}\int_{D^5_+}\check{\omega}_5^0-\int_{\Gamma^0}\int_{D^5_-}\check{\omega}_5^0=\int_{\Gamma^0}\int_{S^5}\check{\omega}_5^0=Z,\end{aligned} \tag{5.7}$$

which means that $\pi^3(\mathfrak{A}^3/\mathscr{G}^3)=\pi^2(\mathscr{G}^3)=Z$. Thus we show that there exists a noncontractible three-dimensional sphere in $\mathfrak{A}^3/\mathscr{G}^3$.

VI. GENERALIZATION TO A NONTRIVIAL PRINCIPAL BUNDLE

When the principal bundle $P(M,G)$ is nontrivial, the theory is slightly complicated. We choose a fixed background connection[15,28] form $\mathring{A}$ on P and extend it to $P\times\mathfrak{A}$. We shall not transform $\mathring{A}$, i.e., $\delta\mathring{A}=0=\delta_h\mathring{A}=0=\delta_v\mathring{A}$. Therefore the extension of $\mathring{A}$ from P to $P\times\mathfrak{A}$ is still $\mathring{A}$. The corresponding curvature form is

$$\mathring{F}=d\mathring{A}+\tfrac{1}{2}[\mathring{A},\mathring{A}]. \tag{6.1}$$

Equation (2.24) used in Sec. II is replaced by

$$P_{2n}(\mathscr{F}^n)-P_{2n}(\mathring{F}^n)=(d+\delta)\omega_{2n-1}(A+\mathscr{A},\mathring{A}). \tag{6.2}$$

This can be shown easily as follows: Let $\widetilde{A}_t=t(A+\mathscr{A})+(1-t)\mathring{A}$. Then

$$\widetilde{F}_t=(d+\delta)\widetilde{A}_t+\tfrac{1}{2}[\widetilde{A}_t,\widetilde{A}_t], \tag{6.3}$$

$$\frac{\partial\widetilde{F}_t}{\partial t}=(d+\delta)\frac{\partial\widetilde{A}_t}{\partial t}+\Big[A_t,\frac{\partial\widetilde{A}_t}{\partial t}\Big]\equiv\widetilde{D}_t\frac{\partial\widetilde{A}_t}{\partial t}. \tag{6.4}$$

The Bianchi identity is

$$\widetilde{D}_t\widetilde{F}_t\equiv(d+\delta)\widetilde{F}_t+[\widetilde{A}_t,\widetilde{F}_t]=0. \tag{6.5}$$

Therefore

$$\begin{aligned}\frac{\partial}{\partial t}P_{2n}(\widetilde{F}_t^n)&=nP_{2n}\Big(\frac{\partial\widetilde{F}_t}{\partial t},\widetilde{F}_t^{n-1}\Big)\\&=nP_{2n}\Big(\widetilde{D}_t\frac{\partial\widetilde{A}_t}{\partial t},\widetilde{F}_t^{n-1}\Big)\\&=(d+\delta)nP_{2n}\Big(\frac{\partial\widetilde{A}_t}{\partial t},\widetilde{F}_t^{n-1}\Big)\\&=n(d+\delta)P_{2n}(A+\mathscr{A}-\mathring{A},\widetilde{F}_t^{n-1}).\end{aligned}$$

Integrating the above equation with respect to t from 0 to 1 gives

$$P_{2n}(\mathscr{F}^n)-P_{2n}(\mathring{F}^n)=(d+\delta)n\int_0^1dt\,P_{2n}(A+\mathscr{A}-\mathring{A},\widetilde{F}_t^{n-1}). \tag{6.6a}$$

Thus

$$\omega_{2n-1}(A+\mathscr{A},\mathring{A})=n\int_0^1dt\,P_{2n}(A+\mathscr{A}-\mathring{A},\widetilde{F}_t^{n-1}). \tag{6.6b}$$

Integrating Eq. (6.2) over M_{2n-k} without boundary, we still have

$$Q^k(M_{2n-k})=\delta\Omega^{k-1}(M_{2n-k}). \tag{6.7}$$

Here

$$Q^k(M_{2n-k})=\int_{M_{2n-k}}P_{2n}(\mathscr{F}^n), \tag{6.8}$$

$$\Omega^{k-1}(M_{2n-k}) = \int_{M_{2n-k}} \omega_{2n-1}(A + \mathscr{A}, \mathring{A}). \tag{6.9}$$

Equation (6.2), when limited to fiber, becomes

$$P_{2n}(F^n) - P_{2n}(\mathring{F}^n) = (d + \delta_v)\omega_{2n-1}(A + v, \mathring{A}). \tag{6.10}$$

Using the expansion of $\omega_{2n-1}(A + v, \mathring{A})$ in v,

$$\omega_{2n-1}(A + v, \mathring{A}) = \sum_k \omega_{2n-k-1}^k(v; A, \mathring{A}), \tag{6.11}$$

we can easily show that

$$\begin{aligned} &P_{2n}(F^n) - P_{2n}(\mathring{F}^n) = d\omega_{2n-1}^0(v; A, \mathring{A}), \\ &\delta_v \omega_0^{2n-1}(v; A, \mathring{A}) = 0, \\ &\delta_v \omega_{2n-k}^{k-1}(v; A, \mathring{A}) = -d\omega_{2n-k-1}^k(v; A, \mathring{A}). \end{aligned} \tag{6.12}$$

We thus obtain the double-cohomological series in the nontrivial case.[28]

Note that all formulas so far written, together with the unchanged expressions in Sec. II, are global on $P(M,G)$ and that only for a trivial bundle one can choose $\mathring{A} = 0$ and recover the local formulas in Sec. II.

VII. CONCLUSION AND DISCUSSION

We have stressed the relation and difference between an arbitrary variation δ, and the horizontal δ_h and the vertical δ_v ones in connection space. By using the natural connection form on the bundle $P(M,G)$ and $\mathfrak{A}(\mathfrak{A}/\mathscr{G}, \mathscr{G})$ and by introducing integration over an arbitrary chain, the generalized finite double-cohomological series can be obtained. Thus the relation between these cohomologies, and the family index theorem can be made simple and obvious. When applying this formalism to analyze Abelian gauge structure inside non-Abelian gauge theory, we reproduce known results in Refs. 18 and 19, and also give some new expressions. The method discussed here has a few important advantages: First, it is very simple. Second, it exposes the mathematical origin of Abelian gauge structures in the sense that these structures can be analyzed by the method. Finally, it can easily be generalized to higher dimensions and to other theories such as gravitational and supersymmetric Yang–Mills theories.

ACKNOWLEDGMENTS

We are grateful to professor Richard P. Goblirsch for carefully reading the manuscript and correcting the English.

[1] S. Adler, Phys. Rev. **177**, 2426 (1969); J. Bell and R. Jackiw, Nuovo Cimento A **60**, 47 (1969); W. A. Bardeen, Phys. Rev. **184**, 1848 (1969); R. Jackiw, in *Lectures on Current Algebra and Its Applications* (Princeton U.P., Princeton, NJ, 1972).
[2] J. Wess and B. Zumino, Phys. Lett. B **37**, 95 (1971); E. Witten, Nucl. Phys. B **223**, 422 (1983).
[3] L. Alvarez-Gaume and E. Witten, Nucl. Phys. B **234**, 422 (1984).
[4] R. Jackiw, in *Relativity, Group and Topology, Les Houches Lectures (1983)*, edited by B. S. De Witt and R. Stora (North-Holland, Amsterdam, 1984).
[5] B. Zumino, Y. S. Wu, and A. Zee, Nucl. Phys. B **329**, 422 (1984); W. A. Bardeen and B. Zumino, Nucl. Phys. B **244**, 477 (1984).
[6] L. Bonora and P. Cotta-Ramusino, Commun. Math. Phys. **87**, 589 (1983).
[7] R. Stora, "Cargese lectures (1983)," in *Progress in Gauge Field Theory*, edited by H. Lehmann (Plenum, New York, 1984).
[8] B. Zumino, in *Relativity, Group and Topology* (North-Holland, Amsterdam, 1984).
[9] L. D. Faddeev, Phys. Lett. B **145**, 81 (1985).
[10] B. Zumino, Nucl. Phys. B **253**, 477 (1985).
[11] A. M. Gabrielov, I. M. Gelfand, and M. Y. Losik, Funct. Anal. Appl. **9**, 12 (1975).
[12] A. G. Reiman, M. A. Semenov-Tian-Shansky, and L. D. Faddeev, Funct. Anal. Appl. **18**, 64 (1984).
[13] Guo Han-Ying, B. Y. Hou, Wang Shi-Kun, and Wu Ke, Commun. Theor. Phys. **4**, 145,233 (1985).
[14] B. Y. Hou, B. Y. Hou, and P. Wang, Lett. Math. Phys. **11**, 179 (1986).
[15] M. F. Atiyah and I. M. Singer, Proc. Natl. Acad. Sci. **81**, 2597 (1984).
[16] O. Alvarez, I. M. Singer, and B. Zumino, Commun. Math. Phys. **96**, 409 (1984); L. Alvarez-Gaume and P. Ginsparg, Nucl. Phys. B **243**, 449 (1984), Ann. Phys. (NY) **161**, 423 (1985).
[17] R. Jackiw, in *The E. S. Fradkin Festschrift* (Hilger, Bristol, 1985).
[18] Y. S. Wu and A. Zee, Nucl. Phys. B **258**, 157 (1985).
[19] A. J. Niemi and G. W. Semenoff, Phys. Rev. Lett. **55**, 227 (1985).
[20] A. A. Belavin, A. M. Polyakov, A. S. Schwarz, and Y. Tyupkin, Phys. Lett. B **59**, 85 (1975); R. Jackiw and C. Rebbi, Phys. Rev. Lett. **37**, 172 (1976); G. t'Hooft, Phys. Rev. Lett. **37**, 8 (1976); C. Callan, R. Dashen, and D. Gross, Phys. Lett. B **63**, 334 (1976).
[21] R. Jackiw and S. Templeton, Phys. Rev. D **23**, 2291 (1981); S. Deser, R. Jackiw, and S. Templeton, Phys. Rev. Lett. **48**, 975 (1982); Ann. Phys. (NY) **223**, 422 (1983).
[22] B. Y. Hou, B. Y. Hou, and P. Wang, Ann. Phys. (NY) **171**, 172 (1986).
[23] D. Z. Freedman and P. K. Townsend, Nucl. Phys. B. **177**, 282 (1981); D. S. Dewitt, Phys. Rev. **125**, 2189 (1962).
[24] S. Mandelstam, Ann. Phys. (NY) **19**, 1 (1962).
[25] R. Jackiw, Phys. Rev. Lett. **54**, 159 (1985); B. Y. Hou and B. Y. Hou, Chin. Phys. Lett. **2**, 49 (1985); B. Grossman, Phys. Lett. B **152**, 93 (1985); Y. S. Wu and A. Zee, *ibid.* B **152**, 98 (1985).
[26] B. Y. Hou and Y.-Z. Zhang, Mod. Phys. Lett. A **1**, 103 (1986).
[27] A. Malcev, Math. Sb. **78**, 569 (1955); A. A. Sagle, Trans. Am. Math. Soc. **101**, 426 (1961); M. Gunaydin and B. Zumino, LBI#19200 (1985).
[28] J. Manes, R. Stora, and B. Zumino, Commun. Math. Phys. **102**, 157 (1986).

Letters in Mathematical Physics **13** (1987) 1–6. 1

Virasoro Algebra in the Solution Space of the Ernst Equation

BO-YU HOU and WEI LEE
Institute of Modern Physics, Northwest University, Xian, China

(Received: 20 May 1986; revised version: 30 September 1986)

Abstract. In this Letter, a new transformation given in the solution space of the Ernst equation is shown to keep the Ernst equation invariant. It is also proved that the transformation enlarges the Cosgrove symmetry and constructs the Virasoro algebra.

1. Introduction

There has been much progress made in applying the Kac–Moody algebra [1] and the Virasoro algebra [2] in theoretical physics [3–11]. For example, the symmetry hidden in the classically principal chiral model [3] was found to be related to the Kac–Moody algebra [4]. Witten [5] proved that the Wess–Zumino model [12] and free fermion theory are equivalent under the identification of their Kac–Moody algebra, and it was later noted [6] that energy momentum tensors must provide the structure of the Virasoro algebra. On the other hand, the existence of the Kac–Moody algebra in the solution space of the Ernst equation [13] was first shown by Kinnersley and Chitre [8]. We would like to indicate that the Virasoro algebra also exists in the same space.

However, the situation discussed in this Letter is different from that in the Wess–Zumino model because their Virasoro algebras are dependent on different circles S'. According to the theory of the infinite-dimensional Lie algebra [14], the Virasoro algebra can be considered as an algebra of a group of smooth maps $S' \to S'$. For the Wess–Zumino model, S' lies in two-dimensional spacetime coordinates, but for the Ernst equation, S' is in a complex parameter plane.

2. Brief Review of the Ernst Equation

As discussed in [8, 9], we briefly review the Ernst equation for our purpose and introduce our conventions in this section. We treat the case of the cylindrically symmetric Einstein space but the case of the stationary, axially-symmetric Einstein space can be discussed in a similar way without difficulty.

In the cylindrically symmetric Einstein space, the metric of spacetime is written

$$ds^2 = -f(t, z)(-dt^2 + dz^2) - g_{ab}(t, z)\, dx^a\, dx^b \tag{1}$$

where $a, b = 1, 2$ and $(x^1, x^2) = (x, y)$. In a vacuum, the Einstein field equations can be reduced to

$$\partial_\xi(\alpha^{-1}g\varepsilon\,\partial_\eta g) + \partial_\eta(\alpha^{-1}g\varepsilon\,\partial_\xi g) = 0 \quad \varepsilon = \begin{pmatrix} 0 & 1 \\ -1 & 0 \end{pmatrix} \tag{2}$$

where $g = (g_{ab})$ is a 2×2 symmetric matrix, $\alpha^2 = \det g$, $\xi = \frac{1}{2}(t + z)$ and $\eta = \frac{1}{2}(t - z)$. The equation that $f(t, z)$ satisfies has not been written out because it is not used in this Letter.

From Equation (2), the twist potential Ψ is defined as

$$\partial_\xi\psi = \alpha^{-1}g\varepsilon\,\partial_\xi g\,, \qquad \partial_\eta\psi = -\alpha^{-1}g\varepsilon\,\partial_\eta g\,. \tag{3}$$

Letting $\Psi - \Psi^T = 2\beta\varepsilon$, we obtain

$$\partial_\xi\beta = \partial_\xi\alpha\,, \qquad \partial_\eta\beta = -\partial_\eta\alpha \tag{4}$$

by taking the trace of Equation (3), so that β and α are the conjugate solutions of a two-dimensional wave equation, i.e., $\partial_\xi\,\partial_\eta\beta = \partial_\xi\,\partial_\eta\alpha = 0$.

Introducing the complex potential $H = g + i\Psi$, we have

$$\begin{aligned} 2(\beta + \alpha)\,\partial_\xi H &= (H + H^+)i\varepsilon\,\partial_\xi H\,, \\ 2(\beta - \alpha)\,\partial_\eta H &= (H + H^+)i\varepsilon\,\partial_\eta H\,, \end{aligned} \tag{5}$$

due to $\frac{1}{2}(H + H^+) = g + i\psi$ (+ represents the Hermitian conjugate) and Equation (3). Following Hauser and Ernst [9], it is not difficult to prove that Equation (5) is the integrability of the following linearization equations

$$\begin{aligned} \partial_\xi F(s) &= \frac{s}{1 - 2s(\beta + \alpha)}\,\partial_\xi H i\varepsilon F(s)\,, \\ \partial_\eta F(s) &= \frac{s}{1 - 2s(\beta - \alpha)}\,\partial_\eta H i\varepsilon F(s)\,. \end{aligned} \tag{6}$$

Here s is a parameter.

For the convenience of discussion, we adopt differential forms to rewrite Equations (5)–(6). Defining the dual transformation $*\mathrm{d}\xi = \mathrm{d}\xi$, $*\mathrm{d}\eta = -\mathrm{d}\eta$, it follows that

$$2(\beta + \alpha *)\,\mathrm{d}H = (H + H^+)i\varepsilon\,\mathrm{d}H \tag{7a}$$

or

$$s\,\mathrm{d}H = A(s)\Gamma(s) \tag{7b}$$

and

$$\mathrm{d}F(s) = \Gamma(s)i\varepsilon F(s) \tag{8}$$

where

$$A(s) = 1 - s(H + H^+)i\varepsilon\cdot, \tag{9}$$

$$\Gamma(s) = s[1 - 2s(\beta + \alpha^*)]^{-1}\,\mathrm{d}H\,. \tag{10}$$

In order to determine the generating function $F(s)$ uniquely, we impose some auxiliary conditions on $F(s)$ as follows

$$F(0) = 1\,, \qquad \dot{F}(0) = Hi\varepsilon\,, \tag{11, 12}$$

$$F^{\times}(s) i\varepsilon A(s) F(s) = i\varepsilon\,, \tag{13}$$

$$\lambda(s) = \det F(s) = [(1 - 2s\beta)^2 - (2s\alpha)^2]^{-1/2}\,, \tag{14}$$

where $\dot{F}(s) = (\partial/\partial s)F(s)$ and $F^{\times}(s) = F^{+}(\bar{s})$ ($\bar{s}$ stands for the complex conjugate of s).

Moreover, since we shall use the two-index potentials $N^{(m,n)}$, introduced by Kinnersley and Chitre [8], in the next section, let us reproduce the definition of the hierarchy of potentials:

$$H^{(0)} = 1\,, \qquad H^{(1)} = Hi\varepsilon\,, \qquad H^{(n)} = -N^{(0,n)}\,, \tag{15, 16, 17}$$

$$N^{(m+k,n)} - N^{(m,n+k)} = \sum_{p=1}^{k} N^{(m,p)} N^{(k+p,n)}\,, \quad k \geqslant 1\,, \tag{18}$$

$$\mathrm{d}N^{(m,n)} = \varepsilon H^{(m)+}\,\varepsilon\,\mathrm{d}H^{(n)} \tag{19}$$

and

$$F(s) = \sum_{n=0}^{\infty} H^{(n)} s^n\,. \tag{20}$$

By definition, it is restricted so that all other quantities of $N^{(m,n)}$, with $m, n < 0$ and $N^{(m,0)}$ $(m \neq 0)$, are assumed to vanish.

3. The Representation of the Virasoro Algebra in the Solution Space of the Ernst Equation

As is well known, the infinitesimal transformations of the Geroch group [7] are simply written

$$\gamma_a^{(k)} N^{(m,n)} = -T_a N^{(m+k,n)} + N^{(m,n+k)} T_a + \sum_{p=1}^{k} N^{(m,p)} T_a N^{(k-p,n)} \quad (m \geq 0, n > 0) \tag{21a}$$

in terms of the hierarchy of the potentials $N^{(m,n)}$, *where* T_a $(a = 1, 2, 3)$ are generators of $SL(2, \mathbb{R})$. The algebra of the Geroch group is the Kac–Moody algebra

$$[\gamma_a^{(k)}, \gamma_b^{(l)}] N^{(m,n)} = f_{ab}^{c} \gamma_c^{(k+l)} N^{(m,n)} \tag{21b}$$

where f_{ab}^{c} is the structure constant of $SL(2, \mathbb{R})$.

Outside the Geroch group, there also exists the Cosgrove group [15] of which the infinitesimal transformations are given by

$$(\mathcal{O})_t N^{(m,n)} = -\delta^{(+1)} N^{(m,n)} = (m+1) N^{(m+1,n)} + n N^{(m,n+1)}\,, \tag{22a}$$

$$(\mathscr{R})_\lambda N^{(m,n)} = -\delta^{(0)} N^{(m,n)} = (m+n) N^{(m,n)}, \tag{22b}$$

$$(\mathscr{L})_\mu N^{(m,n)} = -\delta^{(-1)} N^{(m,n)} = \tfrac{1}{2}(2m-1) N^{(m-1,n)} + \tfrac{1}{2}(2n-1) N^{(m,n-1)}. \tag{22c}$$

Here we have adopted the conventions used by Cosgrove. Obviously, such trans-formations form the generators of the lowest order of the Virasoro algebra

$$[\delta^{(k)}, \delta^{(l)}] N^{(m,n)} = (k-l)\delta^{(k+l)} N^{(m,n)}, \quad k, l = 0, \pm 1. \tag{23}$$

However, we cannot get the generators of the Virasoro algebra for $|k|, |l| \geqslant 2$ from the Cosgrove group.

In order to construct the full Virasoro algebra, it is supposed that there exist the infinitesimal transformation

$$\delta^{(k)} N^{(m,n)} = \tfrac{1}{2}\{(2m+k) N^{(m+k,n)} + (2n+k) N^{(m,n+k)} + \sum_{p=1}^{k} (2p-k) N^{(m,p)} N^{(k-p,n)}, \quad (m \geqslant 0, n > 0).$$

Having calculated simply, we obtain the following commutators

$$[\delta^{(k)}, \delta^{(l)}] N^{(m,n)} = (k-l)\delta^{(k+l)} N^{(m,n)}, \tag{25}$$

$$[\delta^{(k)}, \gamma_a^{(l)}] N^{(m,n)} = -l\gamma_a^{(k+l)} N^{(m,n)}. \tag{26}$$

According to the theory of the infinite-dimensional Lie algebra, the double potentials $N^{(m,n)}$ not only provide us with a representation of the Kac–Moody algebra, but also with a representation of the Virasoro algebra.

It should be pointed out that the symmetry found by us includes the symmetry of the Cosgrove group but it is beyond the symmetry of the Geroch group. For this reason, a new approach is given to generate the new solutions of the Ernst equation by means of the transformation (24). The details can be seen in the next section.

4. Generating the New Solutions of the Ernst Equation from the Old Ones

Now let us discuss the relationship between the solutions of the Ernst equation and our transformations. Because of Equations (15)–(20), the transformation (24) can be expressed in the form

$$\delta(s) H = \dot{F}(s) F^{-1}(s) i\varepsilon \tag{27}$$

where $F(s)$ satisfies the linearization equation (8) and $\delta(s) = \Sigma_{k=0}^{\infty} \delta^{(k)} s^k$. We need to show that Equation (7) is invariant under the transformation (27), i.e.,

$$2(\beta + \alpha^*)\,\mathrm{d}\delta H + 2(\beta + \delta\alpha^*)\,\mathrm{d}H$$
$$= (H + H^+)\,\mathrm{d}\delta H + (\delta H + \delta H^+)\,\mathrm{d}H \tag{28}$$

exists.

First of all it is necessary to derive the transformed forms of β and α under the transformation (27). For a 2×2 arbitrary matrix A, the relations

$$\varepsilon A + A^T \varepsilon = \operatorname{tr} A \varepsilon , \tag{29a}$$

$$A \varepsilon A^T = \det A \varepsilon \tag{29b}$$

hold. Here A^T stands for the transposed matrix of A. Using the definition of β and Equations (13)–(14), (29), we have

$$\begin{aligned}\delta\beta &= \tfrac{1}{2}(\delta H - \delta H^T) i\varepsilon \\ &= \tfrac{1}{2} \operatorname{tr}(\dot{F}(s) F^{-1}(s)) \\ &= -\tfrac{1}{2}\dot{\lambda}^{-1}(s)\dot{\lambda}(s) \\ &= \frac{\beta(1-2s\beta)+2s\alpha^2}{(1-2s\beta)^2-(2s\alpha)^2} .\end{aligned} \tag{30}$$

Then substituting Equation (30) into (4), the solution of Equation (4) is

$$\delta\alpha = \frac{\alpha}{(1-2s\beta)^2-(2s\alpha)^2} . \tag{31}$$

Comparatively, formulas (30)–(31) are similar to those of β and α under Cosgrove's transformation but different from those of β and α under the Geroch transformation because

$$\gamma(s)\beta = \gamma(s)\alpha = 0 \tag{32}$$

where $\gamma(s) = \sum_{k=0}^{\infty} \gamma^{(k)} s^k$.

We turn to prove the identity (28). Equations (7)–(10) yield

$$\begin{aligned}\mathrm{d}\delta H &= \mathrm{d}(\dot{F}(s)F^{-1}(s) i\varepsilon \\ &= \left\{\frac{\partial}{\partial s}(\Gamma(s) i\varepsilon \dot{F}(s))F^{-1}(s) - \dot{F}(s)F^{-1}(s)\Gamma(s) i\varepsilon\right\} i\varepsilon \\ &= [\Gamma(s) i\varepsilon, \dot{F}(s)F^{-1}(s)] i\varepsilon + \dot{\Gamma}(s) \\ &= s[1-2s(\beta+\alpha^*)]^{-1}[\mathrm{d}H i\varepsilon, \dot{F}(s)F^{-1}(s)] i\varepsilon + \dot{\Gamma}(s) .\end{aligned} \tag{33}$$

Putting $2(\beta + \alpha^*)$ on both sides of Equation (33) and using Equation (7a), thus

$$\begin{aligned}&2(\beta+\alpha^*)\,\mathrm{d}\delta H \\ &= s[1-2s(\beta+\alpha^*)]^{-1}[(H+H^+) i\varepsilon\, \mathrm{d}H i\varepsilon, \dot{F}(s)F^{-1}(s)] i\varepsilon + 2(\beta+\alpha^*)\dot{\Gamma}(s) \\ &= (H+H^+) i\varepsilon\, \mathrm{d}\delta H + [(H+H^+) i\varepsilon, \dot{F}(s)F^{-1}(s)]\Gamma(s) .\end{aligned} \tag{34}$$

As the final step, we use $(H + H^+) i\varepsilon\dot{\Gamma}(s) = 2(\beta + \alpha^*)\dot{\Gamma}(s)$ and Equation (33) again.

Moreover, we use Equation (13) to derive

$$(\delta H + \delta H^+) i\varepsilon\, \mathrm{d}H$$

$$= \frac{1}{s}(H + H^+) i\varepsilon \Gamma(s) + [(H + H^+) i\varepsilon, \dot{F}(s) F^{-1}(s)] i\varepsilon\,. \tag{35}$$

Substituting Equation (35) and

$$2(\delta\beta + \delta\alpha^*)\, \mathrm{d}H = \frac{1}{s}(H + H^+) i\varepsilon \Gamma(s) \tag{36}$$

into Equation (34), confirms the identification of Equation (28).

Acknowledgement

This work was partially supported by the Chinese National Science Foundation.

References

1. Kac, V. C., *Matt. USSR-Izv.* **2**, 1271 (1968);
   Moody, R. V., *J. Alg.* **10**, 211 (1968).
2. Virasoro, M. A., *Phys. Rev.* **D1**, 2933 (1970).
3. Hou, B. Y., Yale preprint 80–29 (1980).
   Hou, B. Y., Ge, M. L., and Wu, Y. S., *Phys. Rev.* **D24**, 2238 (1981);
   Dolan, L. and Roos, A., *Phys. Rev.* **D22**, 2018 (1980).
4. Dolan, L., *Phys. Rev. Lett.* **47**, 1371 (1981);
   Ge, M. L. and Wu, Y. S., *Phys. Lett.* **108B**, 411 (1982).
5. Witten, E., *Commun. Math. Phys.* **92**, 455 (1984).
6. Knizhnik, V. G. and Zamolodchikov, A. B., *Nucl. Phys.* **B247**, 83 (1984).
7. Geroch, R., *J. Math. Phys.* **12**, 918 (1971); **13**, 394 (1972).
8. Kinnersley, W. and Chitre, D. M., *J. Math. Phys.* **18**, 1538 (1977); **19**, 1926 (1978).
9. Hauser, I. and Ernst, F. J., *J. Math. Phys.* **21**, 1126, 1418 (1980);
   Wu, Y. S. and Ge, M. L., *J. Math. Phys.* **24**, 1187 (1983).
10. Goddard, P. and Olive, D., *Nucl. Phys.* **B257** [FS14], 226 (1984).
11. Gervais, J. L. and Neveu, A., *Nucl. Phys.* **B209**, 125 (1982).
12. Wess, J. and Zumino, B., *Nucl. Phys.* **B70**, 39 (1974).
13. Ernst, F. J., *Phys. Rev.* **167**, 1175 (1968).
14. Goddard, P., DAMTR 85/7 (1985).
15. Cosgrove, C. M., *J. Math. Phys.* **21**, 2417 (1980).

Class. Quantum Grav. 6 (1989) 163–171. Printed in the UK

Generating functions with new hidden symmetries for cylindrically symmetric gravitational fields

Bo-yu Hou and Wei Li†

Center of Theoretical Physics, CCAST (World Laboratory), Beijing, People's Republic of China, and Institute of Modern Physics, Northwest University, Xian, People's Republic of China

Received 8 October 1987

Abstract. In this paper we present the infinitesimal Virasoro symmetry transformation in terms of the generating function in the vacuum which makes it easy to investigate the algebraic structure of the transformation. Extending this formulation into the electrovac fields, we can find a similar symmetry transformation acting in the solution space of the electrovac field equations. Moreover, we point out that these infinitesimal transformations can be exponentiated by using the Riemann–Hilbert transformation.

1. Introduction

In his pioneer work [1] about twenty years ago, Geroch discovered an infinite-dimensional symmetry in the stationary, axially symmetric vacuum Einstein field equations. Following it, Kinnersley and Chitre [2] derived the explicit expressions of the infinitesimal transformations for the Geroch symmetry in terms of the Kinnersley–Chitre hierarchy of potentials and identified the structure of an infinite-dimensional algebra spanned by these transformations. Hauser and Ernst [3] succeeded in reformulating the Kinnersley–Chitre transformations in terms of the generating function for the Kinnersley–Chitre hierarchy of potentials. They were able to exponentiate the infinitesimal transformations for the Geroch symmetry which have been proved very useful for inducing new solutions from old ones. Since then some progress has been made in this direction. One development [4, 5] is to find the equivalence between the Kinnersley–Chitre algebra and the Kac–Moody algebra with elements of the form $SL(2R)\otimes[t, t^{-1}]$ where t is the parameter in the Hauser–Ernst linearisation equation.

Recently we [6] showed that beyond the Geroch symmetry there exists another infinite-dimensional symmetry known as the Virasoro symmetry. Our work uses the Kinnersley–Chitre hierarchy of potentials to define new infinitesimal transformations and proves these transformations are symmetric to the field equations. We thus enlarged the algebra to the semidirect product of the Kac–Moody algebra and the Virasoro algebra. The purpose of this paper is to use the generating function for another approach to investigate the Virasoro symmetry. We find that it is convenient to express our infinitesimal symmetry transformations as the generating function and to compute their commutators. Based on this formalism, we are able to extend the Virasoro

† Present address: Department of Mathematics and Computer Science, Clarkson University, Potsdam, NY 13676, USA.

0264-9381/89/020163+09$02.50

symmetry in the vacuum to the electrovac fields. We also point out that our infinitesimal transformations are concerned with the Riemann–Hilbert transformations. In this way, it is possible to obtain some useful transformations which can be used to generate new solutions of much interest in physics.

2. Generating functions

We start by considering spacetime possessing two commuting spacelike Killing vectors, for which the metric can be written in the form

$$ds^2 = -f(dz^2 - d\rho^2) - g_{ab}\, dx^a\, dx^b \tag{2.1}$$

where f and g_{ab} $(a, b = 3, 4)$ are functions of ρ and z. We denote a 2×2 matrix g as

$$g = (g_{ab}) \tag{2.2}$$

with the suppressed condition

$$\det g = \alpha^2 \tag{2.3}$$

where α satisfies a two-dimensional wave equation.

It is well known that the vacuum Einstein field equations can be reduced to the Ernst equations which are specified by the following self-dual relation:

$$2(\beta + \alpha *)\, dE_v = (E_v + E_v^\dagger)\Omega_2\, dE_v. \tag{2.4}$$

Here the matrix Ernst potential E_v is defined by

$$g = \mathrm{Re}\, E_v \qquad \mathrm{Tr}(E_v\Omega_2) = 2\beta \qquad \Omega_2 = \begin{pmatrix} 0 & \mathrm{i} \\ -\mathrm{i} & 0 \end{pmatrix} \tag{2.5}$$

and $*$ denotes a two-dimensional dual operation such that

$$*d\rho = dz \qquad *dz = d\rho. \tag{2.6}$$

Using the dual operation, the relation between α and β is given by

$$*d\beta = d\alpha. \tag{2.7}$$

The self-dual relation given above possesses the important property of being integrable in the sense that it has a linearisation formulation. It is specified that the integrability conditions of the linearisation equation are the non-linear equations under study.

Let t be a spectral parameter and $F_v(t) \equiv F_v(t;\rho, z)$ be a 2×2 matrix function satisfying

$$dF_v(t) = \Gamma_v(t)\Omega_2 F_v(t) \tag{2.8}$$

where

$$\Gamma_v(t) = t[1 - 2t(\beta + \alpha *)]^{-1}\, dE_v. \tag{2.9}$$

Once E_v is given, $F_v(t)$ is incompletely determined by the linearisation equation. There are some other constraints imposed on $F_v(t)$:

$$F_v(0) = I_2 \tag{2.10}$$

$$\dot{F}_v(0) = E_v\Omega_2 \tag{2.11}$$

$$\lambda^{-1}(t) \det F_v(t) = 1 \tag{2.12}$$

$$F_v(t)^\dagger \Omega_2 A(t) F_v(t) = \Omega_2 \tag{2.13}$$

where I_2 is the 2×2 unit matrix, the dot represents differentiation with respect to the parameter t and

$$\lambda(t)=[(1-2t\beta)^2-(2t\alpha)^2]^{-1/2} \tag{2.14}$$

$$A(t)=I_2-t(E+E^{\dagger})\Omega_2. \tag{2.15}$$

We should bear in mind that while we take the Hermitian conjugation for $F_v(t)$, the parameter keeps reality.

According to the work of Hauser and Ernst [3], we know that there exist similar formulations for the Einstein-Maxwell fields. Let us introduce some notations. Set

$$\Sigma(t)=\begin{pmatrix}\Omega_2 & 0\\ 0 & \frac{1}{2}t\end{pmatrix} \tag{2.16}$$

and

$$\Sigma=\Sigma(0) \qquad \Omega_3=\Sigma(2). \tag{2.17}$$

The matrix Ernst potential $E_{\rm EM}$ and the generating function $F_{\rm EM}$ in the Einstein-Maxwell fields are 3×3 matrices. In vacuum, they reduce to

$$E_{\rm EM}=\begin{pmatrix}E_v & 0\\ 0 & 0\end{pmatrix} \qquad F_{\rm EM}=\begin{pmatrix}F_v & 0\\ 0 & 1\end{pmatrix}. \tag{2.18}$$

Now the self-dual relation is given as the form

$$2\Sigma(\beta+\alpha *)\,{\rm d}E_{\rm EM}=\Pi\,{\rm d}E_{\rm EM} \tag{2.19}$$

and its linearisation equation corresponds to

$${\rm d}F_{\rm EM}(t)=\Gamma_{\rm EM}(t)\Omega_3\,{\rm d}F_{\rm EM}(t) \tag{2.20}$$

with the auxiliary conditions

$$F_{\rm EM}(0)=I_3 \tag{2.21}$$

$$\dot F_{\rm EM}(0)=E_{\rm EM}\Omega_3 \tag{2.22}$$

$$\lambda^{-1}(t)\det F_{\rm EM}(t)=1 \tag{2.23}$$

$$F_{\rm EM}(t)^{\dagger}\mathscr{H}(t)F_{\rm EM}(t)=\Sigma(t) \tag{2.24}$$

where I_3 is a 3×3 unit matrix, $\lambda(t)$ is defined as before and

$$\Pi=-\dot\Sigma(0)+\Sigma E_{\rm EM}\Omega_3+\Omega_3E^{\dagger}_{\rm EM}\Sigma \qquad \mathscr{H}(t)=\Sigma-t\Pi \tag{2.25}$$

$$\Gamma_{\rm EM}(t)=t[1-2t(\beta+\alpha *)]^{-1}\,{\rm d}E_{\rm EM}. \tag{2.26}$$

In this paper, we concentrate on the spacetime in which two commuting Killing vectors are spacelike, i.e. the cylindrically symmetric spacetime. It is not difficult to extend our treatments to the spacetime in which one of two commuting Killing vectors is timelike. The correspondence between both cases is to change ρ into $\mathrm{i}\rho$ and α into $\mathrm{i}\alpha$.

3. The new symmetry in vacuum

In our previous paper [6], we gave an infinitesimal transformation acting on the Ernst potential

$$\delta(s)E_v=-\dot F_v(t)F_v^{-1}(t)\Omega_2 \tag{3.1}$$

where s is also a spectral parameter and $F_v(t)$ satisfies the linearisation equations. For convenience of discussion, the infinitesimal constant is neglected in the above transformation. One should caution that some terms in the following formulae are infinitesimal although the infinitesimal constant does not appear. It can be verified that the Ernst equation (2.4) remains invariant under this infinitesimal transformation, i.e.

$$2(\beta+\alpha*)\,\mathrm{d}(\delta E_v)+2(\delta\beta+\delta\alpha)\,\mathrm{d}E_v=(E_v+E_v^\dagger)\Omega_2\,\mathrm{d}(\delta E_v)+(\delta E_v+\delta E_v^\dagger)\Omega_2\,\mathrm{d}E_v. \quad (3.2)$$

To prove this equation, we initially derived the variation of β directly

$$\delta\beta=-\frac{\beta(1-2s\beta)+2s\alpha^2}{(1-2s\beta)^2-(2s\alpha)^2} \quad (3.3)$$

and then used (2.7) to find the transform of α. Now we would like to rederive the variation of α directly as that of β without using (2.7). It states that, under transformation, $\alpha+\delta\alpha$ and $\beta+\delta\beta$ still satisfy the two-dimensional wave equation.

In terms of the definition of the matrix Ernst potential, there exist relations

$$g=\tfrac{1}{2}(E_v+E_v^\dagger)-\beta\Omega_2 \qquad (g\Omega_2)^2=\alpha^2 I_2. \quad (3.4)$$

Applying transformation (3.1) to these relations, we have

$$\begin{aligned}\delta\alpha&=(1/\alpha)\,\mathrm{Tr}(\delta g\Omega_2 g\Omega_2)\\&=(1/4\alpha)\,\mathrm{Tr}[(\delta E_v^\dagger+\delta E_v^\dagger)\Omega_2(E_v+E_v^\dagger-2\beta\Omega_2)\Omega_2]\\&=-(1/4s\alpha)\,\mathrm{Tr}[(\delta E_v+\delta E_v^\dagger)\Omega_2 A(s)-(1-2s\beta)(\delta E_v+\delta E_v^\dagger)\Omega_2]\\&=-(1/4s\alpha)\,\mathrm{Tr}[\dot{A}(s)-(1-2s\beta)(\delta E_v+\delta E_v^\dagger)\Omega_2]\\&=(1/2s\alpha)\,\mathrm{Tr}[(E_v\Omega_2)+(1-2s\beta)\delta E_v\Omega_2]\\&=-\alpha/[(1-2s\beta)^2-(2s\alpha)^2].\end{aligned} \quad (3.5)$$

Here we used (2.5), (2.15), (3.3) and the relation

$$(\delta E_v+\delta E_v^\dagger)\Omega_2 A(s)=[A(s),\,\dot{F}_v(s)F_v^{-1}(s)]+\dot{A}(t) \quad (3.6)$$

which results from (2.13). In summary, we thus obtain

$$\delta\beta\pm\delta\alpha=-\frac{\beta\pm\alpha}{1-2s(\beta\pm\alpha)}. \quad (3.7)$$

We now need to investigate the variation of the generating function. Because of the change of the Ernst potential under the infinitesimal transformation, the linearisation equation (2.8) gives rise to the form

$$\delta F_v(t)=\delta\Gamma_v(t)\Omega_2 F_v(t)+\Gamma_v(t)\Omega_2\delta F_v(t). \quad (3.8)$$

Let us now show that a solution of (3.8) can be expressed in the following form:

$$\delta F_v(t)=-\frac{t}{t-s}[t\dot{F}_v(t)F_v^{-1}(t)-s\dot{F}_v(s)F_v^{-1}(s)]F_v(t) \quad (3.9)$$

where $F_v(t)$ and $F_v(s)$ are solutions of the linearisation equation.

To prove this, we differentiate equation (3.9) to obtain

$$\begin{aligned}
\mathrm{d}\delta F_v(t) &= -\frac{t}{t-s}[t\,\mathrm{d}\dot{F}_v(t) - s\,\mathrm{d}(\dot{F}_v(s)F_v^{-1}(s)F_v(t))] \\
&= -\frac{t}{t-s}\{t\dot{\Gamma}_v(t)\Omega_2 F_v(t) + t\Gamma_v(t)\Omega_2\dot{F}_v(t) - s\dot{\Gamma}_v(s)\Omega_2 F_v(t) \\
&\quad - s[\Gamma_v(s)\Omega_2, \dot{F}_v(s)F_v^{-1}(s)]F_v(t) - s\dot{F}_v(s)F_v^{-1}(s)\Gamma_v(t)\Omega_2 F_v(t)\} \\
&= -\frac{t}{t-s}\{t\dot{\Gamma}_v(t)\Omega_2 - s\dot{\Gamma}_v(s)\Omega_2 + [(\Gamma_v(t)-\Gamma_v(s))\Omega_2, \dot{F}_v(s)F_v^{-1}(s)]\}F_v(t) \\
&\quad + \Gamma_v(t)\Omega_2\delta F_v(t).
\end{aligned} \tag{3.10}$$

In terms of the definition of Γ_v, we evaluate the following relations:

$$\begin{aligned}
\Gamma_v(t) - \Gamma_v(s) &= \frac{t-s}{s}[1-2t(\beta+\alpha*)]^{-1}\Gamma_v(s) \\
t\dot{\Gamma}_v(t) - s\dot{\Gamma}_v(s) &= (t-s)\left(\frac{2s(\beta+\alpha*)}{[1-2t(\beta+\alpha*)]^2[1-2s(\beta+\alpha*)]}\mathrm{d}E_v\right. \\
&\quad \left. + \frac{1}{[1+2t(\beta+\alpha*)]}\dot{\Gamma}_v(s)\right).
\end{aligned} \tag{3.11}$$

Then we substitute (3.11) into (3.10) to verify the identification of (3.8). Moreover, it can also be proved that the auxiliary conditions for $F_v(t)$ are preserved under our infinitesimal transformation.

4. The commutations of transformations

As known before, the infinitesimal transformation given in (3.1) constitutes the structure of the Virasoro algebra. To calculate these commutations of the transformation, we had to express it in terms of the Kinnersley-Chitre hierarchy of potentials. Instead we now use the explicit expression for the action of the infinitesimal transformations on the generating function to check directly that it provides the representation of the Virasoro algebra. We shall find that this approach is simpler and more obvious in concept and in calculation.

Let us consider two types of the infinitesimal transformations:

$$\begin{aligned}
\delta' E_v &= -\dot{F}_v(s')F_v^{-1}(s')\Omega_2 \\
\delta'' E_v &= -\dot{F}_v(s'')F_v^{-1}(s'')\Omega_2
\end{aligned} \tag{4.1}$$

where s' and s'' are real parameters. Corresponding to them, we have

$$\begin{aligned}
\delta' F_v(t) &= -\frac{t}{t-s'}[t\dot{F}_v(t)F_v^{-1}(t) - s'\dot{F}_v(s')F_v^{-1}(s')]F_v(t) \\
\delta'' F_v(t) &= -\frac{t}{t-s''}[t\dot{F}_v(t)F_v^{-1}(t) - s''\dot{F}_v(s'')F_v^{-1}(s'')]F_v(t).
\end{aligned} \tag{4.2}$$

Therefore, we can calculate the commutation of δ' and δ'' as follows:

$$
\begin{aligned}
[\delta', \delta'']E_v &= (\delta'\delta'' - \delta'\delta'')E_v \\
&= -\delta'(\dot{F}_v(s'')F_v^{-1}(s''))\Omega_2 + \delta''(\dot{F}_v(s')F_v^{-1}(s'))\Omega_2 \\
&= \frac{s''}{s''-s'}[s''\dot{F}_v(s'')F_v^{-1}(s'') - s'\dot{F}_v(s')F_v^{-1}(s'), \dot{F}_v(s'')F_v^{-1}(s'')]\Omega_2 \\
&\quad + \frac{1}{(s''-s')^2}[s''(s''-2s')\dot{F}_v(s'')F_v^{-1}(s'') + s'^2\dot{F}_v(s')F_v^{-1}(s')]\Omega_2 \\
&\quad + \frac{s''}{s''-s'}\partial_{s''}[\dot{F}_v(s'')F_v^{-1}(s'')]\Omega_2 - (s'' \leftrightarrow s') \\
&= -\frac{s''^2}{s''-s'}\dot{\delta}''E_v + \frac{s'^2}{s'-s''}\dot{\delta}'E_v + \frac{2s's''}{(s''-s')^2}(\delta''E_v - \delta'E_v).
\end{aligned}
\tag{4.3}
$$

Expanding (4.3) in powers of s' and s'' and setting

$$\delta' = \sum_{k=0} \delta^{(k)} s'^k \qquad \delta'' = \sum_{k=0} \delta^{(k)} s''^k \tag{4.4}$$

we get the commutations of the Virasoro algebra

$$[\delta^{(k)}, \delta^{(l)}]E_v = (k-l)\delta^{(k+l)}E_v \qquad k, l \geqslant 0. \tag{4.5}$$

On the other hand, it is well known that the Kinnersley-Chitre transformations form the structure of the Kac-Moody algebra, i.e.

$$[\gamma_a^{(k)}, \gamma_b^{(l)}]E_v = f_{ab}^c \gamma_c^{(k+l)} E_v \tag{4.6}$$

where the transformation is defined by

$$\gamma_a F_v(t) = -\frac{t}{t-s}[F_v(t)T_a F_v^{-1}(t) - F_v(s)T_a F_v^{-1}(s)]F_v(t) \tag{4.7}$$

and where $T_a \in \mathrm{SL}(2R)$ and f_{ab}^c is a structure constant of $\mathrm{SL}(2R)$, and $\gamma_a = \Sigma_{k=0}\, \gamma_a^{(k)} s^k$. With the help of these expressions we can discuss the relation between our transformations and the Kinnersley-Chitre transformations. Similarly to the previous treatment, we get

$$[\delta', \gamma_a'']E_v = -\frac{1}{s'-s''}\left(s'^2\dot{\gamma}_a' + \frac{s's''}{s'-s''}(\gamma_a'' - \gamma_a')\right)E_v. \tag{4.8}$$

Finally, we are led to the infinite set of commutators for $\delta^{(k)}$ and $\gamma_a^{(l)}$:

$$[\delta^{(k)}, \gamma_a^{(l)}]E_v = -l\gamma_a^{(k+l)} \qquad k, l \geqslant 0. \tag{4.9}$$

Thus we have proved that there exists the algebraic structure of the semidirect product of the Kac-Moody and Virasoro algebras in the solution space of the Ernst equation.

5. The new symmetry in the electrovac fields

As pointed out before, many similar properties exist between the vacuum gravitational fields and the electrovac fields; they have similar self-dual relations and similar linearisation equations, and both have Kac-Moody symmetries in the solution spaces of the field equations. So it is natural to conjecture that the Virasoro symmetry must exist in the electrovac fields as in the vacuum. The task in this section is to confirm this.

As discussed before, we propose the following infinitesimal transformation in the electrovac fields:

$$\tilde{\delta} E_{\mathrm{EM}} = -\dot{F}_{\mathrm{EM}}(s) F_{\mathrm{EM}}^{-1}(s) \Omega_3 \tag{5.1}$$

where $F_{\mathrm{EM}}(s)$ satisfies the linearisation equation (2.20). According to the definition of β and α in the electrovac fields, it is not difficult to give their variation as

$$\tilde{\delta}(\beta \pm \alpha) = -\frac{\beta \pm \alpha}{1 - 2s(\beta \pm \alpha)}. \tag{5.2}$$

Applying the transformation to (2.19), we can derive the following equation:

$$\begin{aligned} &2\Sigma(\beta + \alpha *)\, \mathrm{d}\tilde{\delta} E_{\mathrm{EM}} + 2\Sigma(\tilde{\delta}\beta + \tilde{\delta}\alpha *)\, \mathrm{d}E_{\mathrm{EM}} \\ &\quad = 2\tilde{\delta}\Pi\, \mathrm{d}E_{\mathrm{EM}} + 2\Pi\, \mathrm{d}\tilde{\delta} E_{\mathrm{EM}} - (1/s) F_{\mathrm{EM}}^{-1}(s)^{\dagger} \dot{\Sigma}(s) F_{\mathrm{EM}}^{-1}(s) \Gamma_{\mathrm{EM}}(s). \end{aligned} \tag{5.3}$$

Since there exists an infinitesimally additional term in the above equation, we no longer think of $E_{\mathrm{EM}} + \delta E_{\mathrm{EM}}$ being a solution of (2.19).

Proof. First of all, differentiate (5.1) and use the linearisation equation to yield

$$\mathrm{d}\tilde{\delta} E_{\mathrm{EM}} = -\dot{\Gamma}_{\mathrm{EM}}(s) - [\Gamma_{\mathrm{EM}}(s)\Omega_3, \dot{F}_{\mathrm{EM}}(s) F_{\mathrm{EM}}^{-1}(s)]\Omega_3. \tag{5.4}$$

Then act with the operator $2\Sigma(\beta + \alpha *)$ on (5.4) to get

$$\begin{aligned} 2\Sigma(\beta + \alpha *)\, \mathrm{d}\tilde{\delta} E_{\mathrm{EM}} &= -2\Sigma(\beta + \alpha)\dot{\Gamma}_{\mathrm{EM}}(s) - 2\Sigma(\beta + \alpha)[\Gamma_{\mathrm{EM}}(s)\Omega_3, \dot{F}_{\mathrm{EM}}(s) F_{\mathrm{EM}}^{-1}(s)]\Omega_3 \\ &= -\Pi\dot{\Gamma}_{\mathrm{EM}}(s) - [\Pi\Gamma_{\mathrm{EM}}(s)\Omega_3, \dot{F}_{\mathrm{EM}}(s) F_{\mathrm{EM}}^{-1}(s)]\Omega_3 \\ &\quad + [2\Sigma(\beta + \alpha *), \dot{F}_{\mathrm{EM}}(s) F_{\mathrm{EM}}^{-1}(s)]\Gamma_{\mathrm{EM}}(s) \\ &= \Pi\, \mathrm{d}\delta E_{\mathrm{EM}} + [-\Pi + 2\Sigma(\beta + \alpha *)]\dot{F}_{\mathrm{EM}}(s) F_{\mathrm{EM}}^{-1}(s) \Gamma_{\mathrm{EM}}(s) \end{aligned} \tag{5.5}$$

where we have used (2.20) and (5.4).

On the other hand, let us change (2.19) into the form

$$\mathscr{H}(s)\Gamma_{\mathrm{EM}}(s) = s\Sigma\, \mathrm{d}E_{\mathrm{EM}} \tag{5.6}$$

and then use it to derive

$$\begin{aligned} 2\tilde{\delta}\Pi\, \mathrm{d}E_{\mathrm{EM}} &= -\Sigma\dot{F}_{\mathrm{EM}}(s) F_{\mathrm{EM}}^{-1}(s)\, \mathrm{d}E_{\mathrm{EM}} + \dot{F}_{\mathrm{EM}}^{-1\dagger}(s) F_{\mathrm{EM}}(s)^{\dagger}\Sigma\, \mathrm{d}E_{\mathrm{EM}} \\ &= -\Sigma\dot{F}_{\mathrm{EM}}(s) F_{\mathrm{EM}}^{-1}(s)\, \mathrm{d}E_{\mathrm{EM}} + \dot{F}_{\mathrm{EM}}^{-1\dagger}(s) F_{\mathrm{EM}}(s)^{\dagger}(1/s)\mathscr{H}(s)\Gamma_{\mathrm{EM}}(s) \\ &= -\Sigma\dot{F}_{\mathrm{EM}}(s) F_{\mathrm{EM}}^{-1}(s)\, \mathrm{d}E_{\mathrm{EM}} + (1/s)\dot{F}_{\mathrm{EM}}^{-1\dagger}(s)\Sigma(s) F_{\mathrm{EM}}^{-1}(s)\Gamma_{\mathrm{EM}}(s) \\ &= -\Sigma\dot{F}_{\mathrm{EM}}(s) F_{\mathrm{EM}}^{-1}(s)\, \mathrm{d}E_{\mathrm{EM}} + (1/s)[\mathscr{H}(s)\dot{F}_{\mathrm{EM}}(s) + \dot{\mathscr{H}}(s) F_{\mathrm{EM}}(s) \\ &\quad - F_{\mathrm{EM}}^{-1}(s)^{\dagger}\dot{\Sigma}(s)] F_{\mathrm{EM}}^{-1}(s)\Gamma_{\mathrm{EM}}(s) \\ &= [-\Pi + 2\Sigma(\beta + \alpha *)]\dot{F}_{\mathrm{EM}}(s) F_{\mathrm{EM}}^{-1}(s)\Gamma_{\mathrm{EM}}(s) \\ &\quad - (1/s)\Pi\Gamma(s)_{\mathrm{EM}} - (1/s) F_{\mathrm{EM}}^{-1}(s)^{\dagger}\dot{\Sigma}(s) F_{\mathrm{EM}}^{-1}(s)\Gamma_{\mathrm{EM}}(s). \end{aligned} \tag{5.7}$$

From (2.19) and (5.2) we obtain

$$2\Sigma(\tilde{\delta}\beta + \tilde{\delta}\alpha *)\, \mathrm{d}E_{\mathrm{EM}} = -(1/s)\Pi\Gamma_{\mathrm{EM}}(s) \tag{5.8}$$

and combine (5.5), (5.7) and (5.8) to obtain the identity (5.3).

To drop out the additional term in (5.3), which causes the equation (2.19) to change under the transformation (5.1), we have to consider the other transformation. Let us introduce a new infinitesimal transformation

$$\tilde{\gamma} E_{\mathrm{EM}} = -(1/s)(F_{\mathrm{EM}}(s)\tilde{T} F_{\mathrm{EM}}^{-1}(s) - \tilde{T})\Omega_3 \tag{5.9}$$

where $F_{\mathrm{EM}}(s)$ is the generating function in the electrovac fields and $\tilde{T}$ is a constant matrix such that

$$\tilde{T}^{+}\Sigma(s)+\Sigma(s)\tilde{T}^{+}=s\dot{\Sigma}(s). \tag{5.10}$$

In a similar way, it is not difficult to prove that

$$\begin{aligned}&2\Sigma(\beta+\alpha*)\,\mathrm{d}\tilde{\delta}E_{\mathrm{EM}}+2\Sigma(\tilde{\delta}\beta+\tilde{\delta}\alpha*)\,\mathrm{d}E_{\mathrm{EM}}\\&\quad=2\tilde{\delta}\Pi\,\mathrm{d}E_{\mathrm{EM}}+2\Pi\,\mathrm{d}\tilde{\delta}E_{\mathrm{EM}}+(1/s)F_{\mathrm{EM}}^{-1}(s)^{\dagger}\dot{\Sigma}(s)F_{\mathrm{EM}}^{-1}(s)\Gamma_{\mathrm{EM}}(s)\end{aligned} \tag{5.11}$$

and

$$\tilde{\gamma}\beta=\tilde{\gamma}\alpha=0. \tag{5.12}$$

By means of these two transformations, we find that

$$\begin{aligned}\delta E_{\mathrm{EM}}&=\tilde{\delta}E_{\mathrm{EM}}+\tilde{\gamma}E_{\mathrm{EM}}\\&=-\dot{F}_{\mathrm{EM}}(s)F_{\mathrm{EM}}^{-1}(s)\Omega_3-(1/s)[F_{\mathrm{EM}}(s)\tilde{T}F_{\mathrm{EM}}^{-1}(s)-\tilde{T}]\Omega_3\end{aligned} \tag{5.13}$$

will keep equation (2.19) invariant, i.e.

$$2\Sigma(\beta+\alpha*)\,\mathrm{d}\tilde{\delta}E_{\mathrm{EM}}+2\Sigma(\tilde{\delta}\beta+\tilde{\delta}\alpha*)\,\mathrm{d}E_{\mathrm{EM}}=2\tilde{\delta}\Pi\,\mathrm{d}E_{\mathrm{EM}}+2\Sigma\,\mathrm{d}\tilde{\delta}E_{\mathrm{EM}}. \tag{5.14}$$

Corresponding to this new transformation, the transform of the generating function can be easily derived as

$$\begin{aligned}\delta F_{\mathrm{EM}}(t)&=-\frac{t}{t-s}[t\dot{F}_{\mathrm{EM}}(t)F_{\mathrm{EM}}^{-1}(t)-s\dot{F}_{\mathrm{EM}}(s)F_{\mathrm{EM}}^{-1}(s)]F_{\mathrm{EM}}(t)\\&\quad-\frac{t}{t-s}[F_{\mathrm{EM}}(t)\tilde{T}_aF_{\mathrm{EM}}^{-1}(t)-F_{\mathrm{EM}}(s)\tilde{T}_aF_{\mathrm{EM}}^{-1}(s)]F_{\mathrm{EM}}(t).\end{aligned} \tag{5.15}$$

Expanding transformations (5.1) and (5.9) in terms of powers of s we can get the following commutators:

$$\begin{aligned}&[\tilde{\delta}^{(k)},\tilde{\delta}^{(l)}]E_{\mathrm{EM}}=(k-l)\tilde{\delta}^{(k+l)}E_{\mathrm{EM}}\\&[\tilde{\delta}^{(k)},\tilde{\gamma}^{(l)}]E_{\mathrm{EM}}=-l\tilde{\gamma}^{(k+l)}E_{\mathrm{EM}}.\end{aligned} \tag{5.16}$$

Thus we observe

$$[\delta^{(k)},\delta^{(l)}]E_{\mathrm{EM}}=(k-l)\delta^{(k+l)}E_{\mathrm{EM}} \tag{5.17}$$

and so we succeed in generalising the Virasoro symmetry in the vacuum into the electrovac fields.

Let us consider the selection of $\tilde{T}$. Since it is a constant matrix independent of ρ, z and the parameter, and satisfies the relation (5.10), we can choose it as

$$\tilde{T}=\begin{pmatrix}0&0&0\\0&0&0\\0&0&\frac{1}{2}\end{pmatrix}. \tag{5.18}$$

When the electromagnetic fields vanish, the transformations (5.13) and (5.15) correspond to the transformations (3.1) and (3.8) in the vacuum case. Very recently we found a new formulation of the electrovac fields in which the self-dual relation is exactly the same as that of the vacuum case [7] so that we avoid introducing this additional matrix in the Virasoro symmetry transformation.

6. Remarks

In this paper, we have further investigated the Virasoro symmetry in the vacuum in terms of the generating function and have generalised the symmetry to the electrovac fields. We have also discussed the links between these new symmetries and the well known Kac-Moody symmetries. A more important question now is to find the finite form of the new infinitesimal symmetry transformations which will be used to generate new solutions of the vacuum Einstein field equations and the Einstein-Maxwell field equations.

In order to see how to construct the finite form of it, the infinitesimal transformation given in (3.9) can be rewritten as the integral form

$$\delta F_v(t) = -\frac{t}{2\pi \mathrm{i}} \int_{C_{0,s,t}} \frac{w}{(w-s)(w-t)} \dot{F}_v(w) F_v^{-1}(w)\, \mathrm{d}w\, F_v(t) \tag{6.1}$$

where $C_{0,s,t}$ is a circle in the complex w plane around the points 0, s and t. Expanding it in terms of powers of s, we find that for $k \geqslant 0$

$$\delta^{(k)} F_v(t) = -\frac{t}{2\pi \mathrm{i}} \int_{C_{0,t}} \frac{w^{-k}}{w-t} \dot{F}_v(w) F_v^{-1}(w)\, \mathrm{d}w\, F_v(t). \tag{6.2}$$

This is nothing but the infinitesimal Riemann-Hilbert transformation [8] which can be derived from the integral equation

$$\frac{1}{2\pi \mathrm{i}} \int_{C_{0,t}} \frac{F_v(w) F_v^{-1}(p(w))}{w(w-t)} \mathrm{d}w = 0 \tag{6.3}$$

where p is a scalar dependent on the parameter w only. We can give the more general integral equation as follows:

$$\frac{1}{2\pi \mathrm{i}} \int_{C_{0,t}} \frac{F(w) u(w) F^{-1}(p(w))}{w(w-t)} \mathrm{d}w = 0 \tag{6.4}$$

where u is a matrix dependent on the parameter w only and $F(w)$ a generating function in the vacuum gravitational fields or the electrovac fields. We can thus derive the transformation (5.15) as well. We shall use the Riemann-Hilbert transformation developed by Hauser and Ernst [3] to identify the integral equation. The details will be published elsewhere.

References

[1] Geroch R 1971 *J. Math. Phys.* **12** 918; 1972 *J. Math. Phys.* **13** 394

[2] Kinnersley W and Chitre D M 1977 *J. Math. Phys.* **18** 1538; 1978 *J. Math. Phys.* **19** 1926

[3] Hauser I and Ernst F J 1980 *J. Math. Phys.* **21** 1126, 1418

[4] Julia B 1981 *Preprint* LPTEN 81/4 (Invited talk presented at the Johns Hopkins Workshop on Particle Theory)

[5] Wu Y S and Ge M L 1983 *J. Math. Phys.* **24** 1187

[6] Hou B Y and Li W 1987 *Lett. Math. Phys.* **13** 1
Li W 1988 *Phys. Lett.* **129A** 301

[7] Li W 1988 *Phys. Lett.* A to be published

[8] Hou B Y and Li W 1988 *J. Phys. A: Math. Gen.* **20** L897

Algebras connected with the Z_n elliptic solution of the Yang–Baxter equation

Hou Bo-yu
Institute of Modern Physics, P.O. Box 105, Northwest University, Xian 710069, People's Republic of China
Wei Hua
Center of Theoretical Physics, CCAST (World Laboratory), Institute of Modern Physics, P.O. Box 105, Northwest University, Xian 710069, People's Republic of China

(Received 12 May 1989; accepted for publication 2 August 1989)

The quantum and classical algebras connected with the Z_n-symmetric elliptic solution of the Yang–Baxter equation are derived; their structure constants and the relations between the quantum algebra and the classical one are investigated in detail. Moreover, the trigonometric limit of these algebras is worked out.

I. INTRODUCTION

In recent years the Yang–Baxter equation (YBE)[1] and its exact solutions have been studied fruitfully.[2–6] This investigation is extended and associated with quantum algebra (quantum group),[7–9] conformal field theories, completely integrable models, and braid groups etc.[10–15] The solutions of YBE have been classified as rational, trigonometric, and elliptic cases, including their high-spin (fusion) representations.[5,6] The quantum groups for rational and trigonometric cases have been well studied.[7] For the elliptic case, Sklyanin investigated the quantum and classical algebras connected with the eight-vertex model in 1982.[8] In a short article on the representation,[9] Cherednik wrote down an expression for the quantum algebra of Belavin's Z_n-symmetric model in 1985. Since the elliptic case is related naturally to the Kac–Moody–Virasoro characters, and can be generated easily onto a high-genus Riemann surface, and its degeneration gives the trigonometric case, hence the elliptic quantum algebra is more interesting.

In Sec. II we first reduce the Yang–Baxter relation for the Z_n-symmetric elliptic solution (Z_n-SES) to a spectroparameter independent form by means of the Heisenberg group, and we give various explicit expressions for the structure constants of the algebras and their symmetric relations. In Sec III we deduce the quantum algebra. In Sec. IV we treat the classical YBE and find the classical algebra. In Sec. V we exhibit the correspondences between the quantum quantities as well as the equations and the classical ones. In Sec. VI we study the corresponding trigomometric algebras.

II. THE YANG–BAXTER RELATION

The Boltzmann weight in 2D statistical mechanics can be written as

$$R_{jk}(u)=\sum_{\alpha\in Z_n^2} W_\alpha(u) I_\alpha^{(j)} I_\alpha^{\dagger(k)}, \tag{1}$$

where $I_\alpha^{(j)}$ acts on the subspace of the jth site, $I_\alpha=h^{\alpha_1}g^{\alpha_2}$, h and g are $n\times n$ matrices with elements

$$h_{jk}=\delta^{k+1}_{j(\mathrm{mod}\, n)},\quad g_{jk}=\omega^k\delta_{jk}, \tag{2}$$

and $\omega=e^{2\pi i/n}$, $\alpha=(\alpha_1,\alpha_2)$, $\alpha_j=0,1,\dots,n-1$, $W_\alpha(u)$ is the Boltzmann coordinate. The YBE for the Boltzmann weights is

$$R_{12}(u-v)R_{13}(u)R_{23}(v)=R_{23}(v)R_{13}(u)R_{12}(u-v). \tag{3}$$

By means of

$$I_\alpha^{-1}=I_\alpha^\dagger=\omega^{\alpha_1\alpha_2}I_{-\alpha},\quad I_\alpha I_\beta=\omega^{\alpha_2\beta_1}I_{\alpha+\beta},$$
$$\mathrm{tr}(I_\alpha I_\beta^\dagger)=n\delta_\alpha^\beta,\quad \mathrm{tr}(A^{(j)}B^{(k)})=\mathrm{tr}\,A^{(j)}\mathrm{tr}\,B^{(k)}, \tag{4}$$

(3) is reduced to an equivalent YBE for the Boltzmann coordinates

$$\sum_{\gamma\in Z_n^2}(\omega^{\langle\gamma,\alpha\rangle}-\omega^{\langle\alpha,\gamma-\beta\rangle})\times W_{\gamma-\alpha}(u-v)W_{\alpha+\beta-\gamma}(u)\,W_\gamma(v)=0, \tag{5}$$

where $\langle\alpha,\beta\rangle=\alpha_1\beta_2-\alpha_2\beta_1$. Taking $\gamma\to\alpha+\beta-\gamma$ in the latter term, (5) turns to

$$\sum{}' \omega^{\langle\gamma,\alpha\rangle}W_{\alpha\beta\gamma}(u,v)=0, \tag{6}$$

where the summation Σ' means the restriction $\langle\alpha,2\gamma-\beta\rangle\neq0$, and

$$W_{\alpha\beta\gamma}(u,v)=W_{\gamma-\alpha}(u-v)W_{\alpha+\beta-\gamma}(u)W_\gamma(v)-W_{\beta-\gamma}(u-v)W_\gamma(u)W_{\alpha+\beta-\gamma}(v). \tag{7}$$

For the Z_n-SES (Ref. 3) we have

$$W_\alpha(u)=\frac{\sigma_\alpha(u+\eta)}{\sigma_\alpha(\eta)}\frac{\sigma_0(\eta)}{\sigma_0(u+\eta)}, \tag{8}$$

with the Jacob theta function of rational characteristics $(1/2+\alpha_1/n,\ 1/2+\alpha_2/n)$,

$$\sigma_\alpha(u)=\sigma_\alpha(u,\tau)=\sum_{m=-\infty}^{\infty}\exp\left\{i\pi\tau\left(m+\frac12+\frac{\alpha_1}{n}\right)^2+i2\pi\left(m+\frac12+\frac{\alpha_1}{n}\right)\left(u+\frac12+\frac{\alpha_2}{n}\right)\right\}.$$

Now we will construct a spectroparameter independent quantity $C_{\alpha\beta\gamma}$ by dividing (7) with a γ independent factor. Let

$$C_{\alpha\beta\gamma}(u,v)=\frac{\sigma_\beta(2\eta)}{\sigma_0(u-v)\sigma_\beta(u+2\eta)\sigma_\alpha(v)}\left[\frac{\sigma_{\gamma-\alpha}(u-v+\eta)}{\sigma_{\gamma-\alpha}(\eta)}\frac{\sigma_{\alpha+\beta-\gamma}(u+\eta)}{\sigma_{\alpha+\beta-\gamma}(\eta)}\frac{\sigma_\gamma(v+\eta)}{\sigma_\gamma(\eta)}-\frac{\sigma_{\beta-\gamma}(u-v+\eta)}{\sigma_{\beta-\gamma}(\eta)}\frac{\sigma_\gamma(u+\eta)}{\sigma_\gamma(\eta)}\frac{\sigma_{\alpha+\beta-\gamma}(v+\eta)}{\sigma_{\alpha+\beta-\gamma}(\eta)}\right]. \tag{9}$$

By means of the Heisenberg group, we have

$$C_{\alpha\beta\gamma}(u+1,v)=C_{\alpha\beta\gamma}(u+\tau,v)=C_{\alpha\beta\gamma}(u,v).$$

Hence $C_{\alpha\beta\gamma}(u,v)$ is a doubly periodic function with respect to (wrt) u and it has at most two poles on the lattice generated by 1 and τ. On account of the zero of its denominator, $u=v$, cancelling out one zero of its numerator, we confirm that $C_{\alpha\beta\gamma}(u,v)$ is an entire function wrt u, hence it is independent of u. A similar analysis for the spectroparameter v shows that $C_{\alpha\beta\gamma}(u,v)$ is also independent of v. We denote it as $C_{\alpha\beta\gamma}(\eta,\tau)$. Based on the spectroparameter independence of $C_{\alpha\beta\gamma}(\eta,\tau)$, we may get its various expressions by substituting appropriate values of u,v into (9). Taking $u=0$, $v=-\eta-(\gamma_1\tau+\gamma_2)/n$, we obtain

$$C_{\alpha\beta\gamma}(\eta,\tau)=\frac{\omega^{\gamma_1-\alpha_1}\sigma_{\alpha+\beta-2\gamma}(0)\sigma_\beta(2\eta)}{\sigma_{\gamma-\alpha}(\eta)\sigma_{\alpha+\beta-\gamma}(\eta)\sigma_\gamma(\eta)\sigma_{\beta-\gamma}(\eta)}. \tag{10}$$

We see that $C_{\alpha\beta\gamma}(\eta,\tau)$ is a four-order elliptic function of η with periods 1 and τ. Now we rewrite (6) as

$$\sum_{\langle\alpha,2\gamma-\beta\rangle\neq 0}\omega^{\langle\gamma,\alpha\rangle}C_{\alpha\beta\gamma}(\eta,\tau)=\sum_{\gamma\in Z_n^2}\omega^{\langle\gamma,\alpha\rangle}C_{\alpha\beta\gamma}(\eta,\tau)=0. \tag{11}$$

However, we should prove the validity for the latter without the summation-restriction equations. Denote

$$f(\eta)=\sum_{\gamma\in Z_n^2}\omega^{\langle\gamma,\alpha\rangle}C_{\alpha\beta\gamma}(\eta,\tau),$$

with $C_{\alpha\beta\gamma}$ as in (10). It is obvious that $f(\eta)/\sigma_\beta(2\eta)$ is holomorphic for $\eta\neq(\delta_1\tau+\delta_2)/n$, $\delta\in Z_n^2$. Using (10) we evaluate $f(\eta)$ at $\eta=\epsilon+(\delta_1\tau+\delta_2)/n$. For $\epsilon\to 0$ there are four singular terms, $\gamma=\alpha-\delta,\alpha+\beta+\delta,-\delta,$ and $\beta+\delta$, However, the singular parts cancel exactly:

$$f(\epsilon+(\delta_1+\delta_2)/n)=\sigma_\beta(2\eta)\cdot(\text{regular terms at } \varepsilon=0).$$

Hence $f(\eta)/\sigma_\beta(2\eta)$ is an entire function wrt η. As an entire doubly periodic function, $f(\eta)$ is independent of η. Taking $2\eta_0=-(\beta_1\tau+\beta_2)/n$ we get $f(\eta_0)=0$. This proves $f(\eta)=0$. This also gives another explicit proof of Belavin's ansatz[3] as a solution of YBE.

We shall see in the next section that $C_{\alpha\beta\gamma}$'s are quantum algebra structure constants (QSC's). In order to be more convenient for expressing the symmetries of the QSC's and the relations between the QSC's and the classical algebra structure constants (CSC's) we introduce a modified QSC,

$$F_{\alpha\beta\gamma}(\eta,\tau)=\sigma_0'(0)\sigma_\alpha(0)C_{\alpha\beta\gamma}(\eta,\tau),\quad \alpha\neq 0,$$
$$F_{0\beta\gamma}(\eta,\tau)=\sigma_0'^2(0)C_{0\beta\gamma}(\eta,\tau). \tag{12}$$

Equation (10) gives an explicit expression for $F_{\alpha\beta\gamma}$ or $F_{0\beta\gamma}$ Two other expressions come from (9) by setting $v=0$, $u\to 0$ or $u=0$, $v\to 0$:

$$F_{\alpha\beta\gamma}(\eta,\tau)=\frac{\partial}{\partial\eta}\ln\frac{\sigma_{\gamma-\alpha}(\eta)\sigma_{\alpha+\beta-\gamma}(\eta)}{\sigma_{\beta-\gamma}(\eta)\sigma_\gamma(\eta)}$$
$$=\zeta\left(\eta+\frac{\gamma_1-\alpha_1}{n}\tau+\frac{\gamma_2-\alpha_2}{n}\right)+\zeta\left(\eta+\frac{\alpha_1+\beta_1-\gamma_1}{n}\tau+\frac{\alpha_2+\beta_2-\gamma_2}{n}\right)-\zeta\left(\eta+\frac{\beta_1-\gamma_1}{n}\tau+\frac{\beta_2-\gamma_2}{n}\right)$$
$$-\zeta\left(\eta+\frac{\gamma_1}{n}\tau+\frac{\gamma_2}{n}\right),\quad \alpha\neq 0,$$
$$F_{0\beta\gamma}(\eta,\tau)=\frac{\partial^2}{\partial\eta^2}\ln\frac{\sigma_{\beta-\gamma}(\eta)}{\sigma_\gamma(\eta)}$$
$$=\wp\left(\eta+\frac{\gamma_1\tau+\gamma_2}{n}\right)-\wp\left(\eta+\frac{\beta_1-\gamma_1}{n}\tau+\frac{\beta_2-\gamma_2}{n}\right), \tag{13}$$

where $\zeta(x)$ and $\wp(x)$ are the Weierstrass zeta function and elliptic function, respectively

Using (10), (13), and

$$\sigma_{x_1+np,\,\alpha_2+nq}(u)=e^{2i\pi(1/2+\alpha_1/n)q}\sigma_\alpha(u),\quad p,q\in Z, \tag{14}$$

we obtain the following relations:

$$F_{\alpha\beta\gamma}=-F_{-\alpha,\beta,\gamma-\alpha}=F_{2\gamma-\alpha-\beta,\beta,\gamma},\quad \alpha\neq 0, \tag{15}$$
$$=-F_{\alpha,\beta,\alpha+\beta-\gamma}, \tag{16}$$
$$F_{0,\alpha+\beta,\beta}+F_{0,\beta+\gamma,\gamma}+\cdots+F_{0,\delta+\rho,\rho}+F_{0,\rho+\alpha,\alpha}=0, \tag{17}$$

$$\sum_{\gamma\in Z_n^2}F_{0\beta\gamma}=0,$$
$$\sum_{\gamma\in Z_n^2}F_{\alpha\beta\gamma}=\sum_{\beta\in Z_n^2}F_{\alpha\beta\gamma}=0,\quad \alpha\neq 0. \tag{18}$$

Now (11) is rewritten as

$$\sum_{\langle\alpha,2\gamma-\beta\rangle\neq 0}\omega^{\langle\gamma,\alpha\rangle}F_{\alpha\beta\gamma}(\eta,\tau)=\sum_{\gamma\in Z_n^2}\omega^{\langle\gamma,\alpha\rangle}F_{\alpha\beta\gamma}(\eta\tau)=0. \tag{19}$$

III. THE QUATNUM ALGEBRA

The operator representation of the YBE,

$$L_j(u)=\sum_{\alpha\in Z_n^2}W_\alpha(u)I_\alpha^{(j)}S_\alpha, \tag{20}$$

satisfies

$$R_{12}(u-v)L_1(u)L_2(v)=L_2(v)L_1(u)R_{12}(u-v). \tag{21}$$

Using (4), (21) is equivalent to

$$\sum_{\gamma\in Z_n^2}W_{\alpha\beta\gamma}(u,v)\,\omega^{(\beta_1-\gamma_1)(\gamma_2-\alpha_2)}S_{\alpha+\beta-\gamma}S_\gamma=0, \tag{22}$$

where $S_\alpha = S_{\alpha(\mathrm{mod}\, n)}$. For the Z_n-SES (8) using (9), (22) is reduced to a spectroparameter independent form

$$\sum_{\gamma\in Z_n^2} F_{\alpha\beta\gamma}(\eta,\tau)\,\omega^{(\beta_1-\gamma_1)(\gamma_2-\alpha_2)}S_{\alpha+\beta-\gamma}S_\gamma = 0, \quad (23)$$

This is an algebra for the quantum operators S's, and (11) is its compatibility conditions. Define a normalized QSC,

$$F^1_{\alpha\beta\gamma} = F_{\alpha\beta\gamma}/F_{\alpha\beta\alpha}. \quad (24)$$

From (10) we have

$$F^1_{\alpha\beta\gamma}(\eta,\tau) = \omega^{\gamma_1-\alpha_1} \times \frac{\sigma_{\alpha+\beta-2\gamma}(0)\sigma_0(\eta)\sigma_\beta(\eta)\sigma_\alpha(\eta)\sigma_{\beta-\alpha}(\eta)}{\sigma_{\beta-\alpha}(0)\sigma_{\gamma-\alpha}(\eta)\sigma_{\alpha+\beta-\gamma}(\eta)\sigma_\gamma(\eta)\sigma_{\beta-\gamma}(\eta)}. \quad (25)$$

On account of $F_{\alpha\beta\beta} = -F_{\alpha\beta\alpha}$, we rewrite (23) as

$$[S_\alpha, S_\beta] = \sum_{\gamma\neq\alpha,\beta} F^1_{\alpha\beta\gamma}(\eta,\tau) \times\omega^{(\beta_1-\gamma_1)(\gamma_2-\alpha_2)}S_{\alpha+\beta-\gamma}S_\gamma, \quad \alpha\neq\beta, \quad (26a)$$

$$\sum_{\gamma\in Z_n^2} F_{\alpha,\alpha,\alpha+\gamma}(\eta,\tau)\omega^{-\gamma_1\gamma_2}[S_{\alpha-\gamma}, S_{\alpha+\gamma}] = 0 \quad (n\geqslant 3). \quad (26b)$$

For $n=2$, we have

$$F^1_{10,11,01} = F^1_{11,01,10} = F^1_{01,10,11} = 1,$$

$$F^1_{10,11,00} = F^1_{11,01,00} = F^1_{01,10,00} = -1,$$

$$F^1_{0,01,10} = -F^1_{0,01,11} = -\frac{\vartheta_1^2(\eta)\vartheta_2^2(\eta)}{\vartheta_3^2(\eta)\vartheta_4^2(\eta)} \equiv J_{12},$$

$$F^1_{0,10,11} = -F^1_{0,10,01} = -\frac{\vartheta_1^2(\eta)\vartheta_4^2(\eta)}{\vartheta_3^2(\eta)\vartheta_2^2(\eta)} \equiv J_{23},$$

$$F^1_{0,11,01} = -F^1_{0,11,10} = \frac{\vartheta_1^2(\eta)\vartheta_3^2(\eta)}{\vartheta_2^2(\eta)\vartheta_4^2(\eta)} \equiv J_{31}. \quad (27)$$

Note that $I_{10}=\sigma_1$, $I_{11}=-i\sigma_2$, $I_{01}=\sigma_3$, $I_0=1$, and write $S_1=S_{10}$, $iS_2=S_{11}$, $S_3=S_{01}$, $S_0=S_{00}$, we obtain the result of Ref. 8,

$$[S_0, S_a] = iJ_{bc}[S_b, S_c]_+, \quad (28)$$

$$[S_a, S_b] = i[S_c, S_0]_+, \quad \text{abc: cycle of 123}. \quad (29)$$

Now let us see the simplest case $\eta\to 0$ for (26). We get $F^1_{\alpha\beta\gamma}(0,\tau) = 1$ for $\gamma = \alpha+\beta$, -1 for $\gamma = 0$; 0 otherwise. Then (26a) turns to

$$[S_0, S_\alpha] = 0,$$

$$[S_\alpha, S_\beta] = (\omega^{-\alpha_1\beta_2} - \omega^{-\alpha_2\beta_1})\,S_0S_{\alpha+\beta}. \quad (30)$$

By means of (30), (16), and (19), the left-hand side of (26b),

$$\sum_{\gamma\in Z_n^2} F_{\alpha,\alpha,\alpha+\gamma}(0,\tau)\omega^{-\gamma_1\gamma_2}[S_{\alpha-\gamma}, S_{\alpha+\gamma}]$$

$$= \omega^{-\alpha_1\alpha_2}\sum_{\gamma\in Z_n^2} F_{\alpha,\alpha,\alpha+\gamma}(0,\tau)(\omega^{\langle\gamma,\alpha\rangle} - \omega^{\langle\alpha,\gamma\rangle})S_0S_{2\alpha}$$

$$= 2\omega^{-\alpha_1\alpha_2}\sum_{\gamma\in Z_n^2} F_{\alpha,\alpha,\gamma}(0,\tau)\omega^{\langle\gamma,\alpha\rangle}S_0S_{2\alpha},$$

vanishes, hence (26b) does not give more relations except (26a). We conclude that the algebra with $\eta = 0$ is equivalent tp (30) and in another basis it appears as the algebra $u(n)$. Moreover, in this case we have a matrix realization $S_\alpha \cong I^\dagger_\alpha$ for it.

IV. THE CLASSICAL YBE AND THE CLASSICAL ALGEBRA

The classical YBE corresponding to the YBE (3) is

$$[r_{12}(u-v), r_{13}(u) + r_{23}(v)] + [r_{13}(u), r_{23}(v)] = 0, \quad (31)$$

with

$$r_{jk}(u) = \sum_{\alpha\in Z_n^2} w_\alpha(u) I_\alpha^{(j)} I_\alpha^{\dagger(k)}, \quad (32)$$

$$w_\alpha(u) = \frac{\partial}{\partial\eta}W_\alpha(u)|_{\eta=0}, \quad w_0 = 0. \quad (33)$$

By means of (4), (31) is reduced to

$$(\omega^{\langle\alpha,\beta\rangle} - 1)[w_{-\alpha}(u-v)w_{\alpha+\beta}(u) - w_\beta(u-v) \times w_{\alpha+\beta}(v) + w_\beta(u)w_\alpha(v)] = 0, \quad \alpha,\beta\in Z_n^2. \quad (34)$$

The classical representation

$$\mathscr{L}_j(u) = s_0 + i\sum_{\alpha\neq 0} w_\alpha(u) I_\alpha^{(j)} s_\alpha \quad (35)$$

satisfies the fundamental Poisson bracket relations (FPR)

$$\{\mathscr{L}_1(u), \mathscr{L}_2(v)\} = [r_{12}(u-v), \mathscr{L}_1(u)\mathscr{L}_2(v)]. \quad (36)$$

Using (4), the FPR is reduced to

$$\{s_\alpha, s_\beta\} = \sum_{\gamma\in Z_n^2} \frac{w_\gamma(u-v)\bar{w}_{\alpha-\gamma}(u)\bar{w}_{\beta+\gamma}(v)}{\bar{w}_\alpha(u)\bar{w}_\beta(v)} \times[\omega^{(\alpha_1-\beta_1-\gamma_1)\gamma_2} - \omega^{(\alpha_2-\beta_2-\gamma_2)\gamma_1}] \times s_{\alpha-\gamma}s_{\beta+\gamma}, \quad (37)$$

with $\bar{w}_0 = 1$, $\bar{w}_\alpha = iw_\alpha$ $(\alpha\neq 0)$. Letting $\alpha = \beta'$, $\beta = \alpha'$; $\gamma = \gamma' - \alpha'$ for the former term and $\gamma = \beta' - \gamma'$ for the latter term on the right-hand side of (37) we get

$$\{s_\alpha, s_\beta\} = -\sum_{\langle\beta-\gamma,\gamma-\alpha\rangle\neq 0} f_{\alpha\beta\gamma}(u,v)\omega^{(\beta_1-\gamma_1)(\gamma_2-\alpha_2)} \times s_{\alpha+\beta-\gamma}s_\gamma, \quad \alpha\neq\beta, \quad (38)$$

where

$$f_{\alpha\beta\gamma}(u,v) = [\bar{w}_\beta(u)\bar{w}_\alpha(v)]^{-1} \times[w_{\gamma-\alpha}(u-v)\bar{w}_{\alpha+\beta-\gamma}(u)\bar{w}_\gamma(v) - w_{\beta-\gamma}(u-v)\bar{w}_\gamma(u)\bar{w}_{\alpha+\beta-\gamma}(v)]. \quad (39)$$

For the Z_n-SES (8) we have

$$w_\alpha(u) = \frac{\sigma_0'(0)}{\sigma_0(u)}\frac{\sigma_\alpha(u)}{\sigma_\alpha(0)}, \quad \alpha\neq 0, \quad w_0 = 0. \quad (40)$$

Taking a similar analysis as in Sec. II, we conclude that $f_{\alpha\beta\gamma}(u,v)$ is independent of both the spectroparameters u and v. Setting $v=0$, $u = -(\gamma_1\tau+\gamma_2)/n$ for $f_{\alpha\beta\gamma}$ and $f_{\alpha 0\gamma}$; $u=0$, $v = (\gamma_1\tau+\gamma_2)/n$ for $f_{0\beta\gamma}$; $v=0$ for $f_{\alpha\beta 0}$, we obtain the CSC's expressed in terms of the QSC's:

$$f_{\alpha\beta\gamma}(\tau) = F_{\alpha\beta\gamma}(0,\tau), \quad \alpha,\beta,\gamma,\alpha-\gamma,\beta-\gamma,\alpha+\beta-\gamma\neq 0,$$
$$f_{0\beta\gamma}(\tau) = f_{\beta 0\gamma}(\tau) = iF_{0\beta\gamma}(0,\tau), \quad \beta,\gamma,\beta-\gamma\neq 0,$$
$$f_{\alpha\beta 0} = -f_{\alpha,\beta,\alpha+\beta} = i, \quad \alpha,\beta,\alpha+\beta\neq 0. \tag{41}$$

By means of (10), (13), and (41) we get, besides (15)–(18), the following relations for the CSC's:

$$f_{\alpha\beta\gamma} = f_{\beta\alpha\gamma}, \quad \gamma\neq\alpha,\beta, \quad \alpha+\beta,0, \tag{42}$$
$$= f_{-\alpha,-\beta,-\gamma}, \quad \alpha,\beta,\gamma,\alpha-\gamma,\beta-\gamma,\alpha+\beta-\gamma\neq 0, \tag{43}$$
$$f_{0\beta\gamma} = f_{0,-\beta,-\gamma}, \quad \beta,\gamma,\beta-\gamma\neq 0. \tag{44}$$

Moreover, the summation restriction in (38) can be weakened to $\gamma\neq\alpha,\beta$. To see this, using (16) we combine every pair terms of $\gamma=\gamma'$ and $\gamma=\alpha+\beta-\gamma'$ in (38) and denote the summation of all the pairs as Σ'',

$$\{s_\alpha,s_\beta\} = -\sum'' f_{\alpha\beta\gamma}(\tau)[\omega^{(\beta_1-\gamma_1)(\gamma_2-\alpha_2)} - \omega^{(\beta_2-\gamma_2)(\gamma_1-\alpha_1)}]s_{\alpha+\beta-\gamma}s_\gamma;$$

we see that the terms satisfying $\langle\beta-\gamma,\gamma-\alpha\rangle = 0$ cancel. Therefore we write the classical algebra as

$$\{s_\alpha,s_\beta\} = -\sum_{\langle\alpha-\gamma,\beta-\gamma\rangle\neq 0} f_{\alpha\beta\gamma}(\tau)\omega^{(\beta_1-\gamma_1)(\gamma_2-\alpha_2)} \times s_{\alpha+\beta-\gamma}s_\gamma$$
$$= -\sum_{\gamma\neq\alpha,\beta} f_{\alpha\beta\gamma}(\tau)\omega^{(\beta_1-\gamma_1)(\gamma_2-\alpha_2)} \times s_{\alpha+\beta-\gamma}s_\gamma, \tag{45}$$

For the $n=2$ case we have

$$f_{0,01,10} = -f_{0,01,11} = -i\pi^2\vartheta_2^4(0,\tau)\equiv ij_{12},$$
$$f_{0,10,11} = -f_{0,10,01} = -i\pi^2\vartheta_4^4(0,\tau)\equiv ij_{23},$$
$$f_{0,11,01} = -f_{0,11,10} = i\pi^2\vartheta_3^4(0,\tau)\equiv ij_{31},$$
$$f_{\alpha\beta 0} = -f_{\alpha,\beta,\alpha+\beta} = i, \quad \alpha,\beta\neq 0. \tag{46}$$

The summation in (45) is only for two terms. Denoting $s_{11}=is_2$, $s_{10}=s_1$, $s_{01}=s_3$ we get

$$\{s_0,s_a\} = -2j_{bc}s_bs_c, \tag{47}$$
$$\{s_a,s_b\} = -2s_cs_0, \quad abc\text{:cycle of 123}. \tag{48}$$

V. RELATIONS BETWEEN THE QUANTUM AND CLASSICAL CASES

Based on the quantization correspondences

$$[A,B]\sim -i\hbar\{A,B\}, \quad S_0\sim\hbar s_0, \quad S_\alpha\sim s_\alpha \quad (\alpha\neq 0),$$

where the Poisson bracket $\{A,B\} = \Sigma_i(\partial A/\partial p_i\partial B/\partial q_i - \partial A/\partial q_i\partial B/\partial p_i)$ and (33), $w_\alpha(u) = (\partial/\partial\eta)W(u)|_{\eta=0}$, all the classical quantities and equations may be obtained by taking first $\hbar$ terms of $\hbar$-expansions for the corresponding quantum ones. Suppressing the higher-$\hbar$ terms, the relations are

$$R = 1 + i\hbar r, \quad W_\alpha = \delta_\alpha^0 + i\hbar w_\alpha,$$
$$F^1_{\alpha\beta\gamma}(i\hbar,\tau) = i\hbar f_{\alpha\beta\gamma}(\tau), \quad \alpha,\beta\neq 0,$$
$$F^1_{0\beta\gamma}(i\hbar,\tau) = i\hbar^2 f_{0\beta\gamma}(\tau), \quad \beta,\beta-\gamma\neq 0,$$
$$\text{YBE (3)} = -\hbar^2\,\text{CYBE (31)}, \quad (5) = -\hbar^2(34),$$
$$L\sim\hbar\mathscr{L}, \quad (21)\sim -i\hbar^3(36),$$
$$(26\text{a})\sim -i\hbar(45), \quad \alpha\neq 0; \quad -i\hbar^2(45), \quad \alpha\neq 0;$$
$$(28)\sim -i\hbar^2(47),$$
$$(29)\sim -i\hbar^2(48), \quad J_{bc} = \hbar^2 j_{bc}.$$

VI. LIMITING TO THE TRIGONOMETRIC CASE

When the elliptic parameter $\tau\to i\infty$, the Z_n-SES turns to a Z_n-symmetric trigonometric solution. Correspondingly the algebras turn to trigonometric ones. The equations for the trigonometric algebras are the same as (23), (26), and (45). We exhibit the structure constants below. For compactness we use the notations

$$(u\quad p) = \sin\pi\left(u+\frac{p}{n}\right), \quad \begin{pmatrix} u & l \\ v & m\end{pmatrix} = \frac{(u\quad l)}{(v\quad m)},$$
$$\epsilon = \epsilon(\beta_1-\alpha_1) = \begin{cases}1, \beta_1>\alpha_1,\\ -1, \beta_1<\alpha_1,\end{cases}$$
$$\theta_+ = \theta(\beta_1-\alpha_1) = \begin{cases}1, \beta_1>\alpha_1,\\ 0, \beta_1<\alpha_1,\end{cases}$$
$$F_{\alpha\beta\gamma} = F_{\alpha\beta\gamma}(\eta,i\infty), \quad \omega = e^{2\pi i/n}, \quad q(n) = q(\text{mod } n).$$

On account of $S_\alpha = S_{\alpha(\text{mod } n)}$ in (23) we restrict below $0\leqslant\alpha_i,\beta_i,\gamma_i\leqslant n-1$; $F^1_{\alpha\beta\gamma}$ for $\alpha\neq\beta$, $\gamma\neq\alpha,\beta$ only; $\alpha_1\neq\beta_1$ when ϵ or θ_+ appears; $\alpha_1=\beta_1$ when $F^{1'}_{\alpha\beta\gamma}$ is used; $\alpha\neq(0,0)$ in 1- and 2-items. The nonvanishing QSC's are

1-0, $0<\gamma_1<\alpha_1,\beta_1$: $F_{\alpha\beta\gamma} = 2\pi i$, $F^1_{\alpha\beta\gamma} = 2ie^{-i\pi\eta\epsilon}(\eta\quad 0)$, $F^{1'}_{\alpha\beta\gamma} = 2i(\eta\quad 0)\begin{pmatrix}\eta & \beta_2-\alpha_2\\ 0 & \beta_2-\alpha_2\end{pmatrix}$;

1-1, $0<\alpha_1,\beta_1<\gamma_1<\alpha_1+\beta_1$: $F_{\alpha\beta\gamma} = -2\pi i$, $F^1_{\alpha\beta\gamma} = -2ie^{-i\pi\eta\epsilon}(\eta\quad 0)$, $F^{1'}_{\alpha\beta\gamma} = -2i(\eta\quad 0)\begin{pmatrix}\eta & \beta_2-\alpha_2\\ 0 & \beta_2-\alpha_2\end{pmatrix}$;

1-2, $\begin{cases}\gamma_1 = 0,\\ \alpha_1,\beta_1,\alpha_1+\beta_1\neq 0(n):\end{cases}$ $F_{\alpha\beta\gamma} = -\omega^{-\gamma_2/2}e^{-i\pi\eta}\dfrac{\pi}{(\eta\quad\gamma_2)}$,

$$F^1_{\alpha\beta\gamma} = -\omega^{-\gamma_2/2}e^{-2i\pi\eta\theta_+}\begin{pmatrix}\eta & 0\\ \eta & \gamma_2\end{pmatrix},$$
$$F^{1'}_{\alpha\beta\gamma} = -\omega^{-\gamma_2/2}e^{-i\pi\eta}\begin{pmatrix}\eta & 0\\ \eta & \gamma_2\end{pmatrix}\begin{pmatrix}\eta & \beta_2-\alpha_2\\ 0 & \beta_2-\alpha_2\end{pmatrix};$$

1-3 $\begin{cases}\gamma_1 = \alpha_1\neq 0,\\ \beta_1\neq 0,\alpha_1:\end{cases}$ $F_{\alpha\beta\gamma} = \omega^{(\gamma_2-\alpha_2)\epsilon/2}e^{i\pi\eta\epsilon}\dfrac{\pi}{(\eta\quad\gamma_2-\alpha_2)}$, $F^1_{\alpha\beta\gamma} = \omega^{(\gamma_2-\alpha_2)\epsilon/2}\begin{pmatrix}\eta & 0\\ \eta & \gamma_2-\alpha_2\end{pmatrix}$;

1-4 $\begin{cases}\gamma_1=\alpha_1+\beta_1\neq 0(n),\\ \alpha_1,\beta_1\neq 0:\end{cases}$ $F_{\alpha\beta\gamma}=\omega^{-(\alpha_2+\beta_2-\gamma_2)/2}\dfrac{e^{-i\pi\eta}\pi}{(\eta \quad \alpha_2+\beta_2-\gamma_2)}$,

$$F^{1}_{\alpha\beta\gamma}=\omega^{-(\alpha_2+\beta_2-\gamma_2)/2}e^{-2i\pi\eta\theta_+}\begin{pmatrix}\eta & 0\\ \eta & \alpha_2+\beta_2-\gamma_2\end{pmatrix},$$

$$F^{1'}_{\alpha\beta\gamma}=\omega^{-(\alpha_2+\beta_2-\gamma_2)/2}e^{-i\pi\eta}\begin{pmatrix}\eta & 0\\ 0 & \beta_2-\alpha_2\end{pmatrix}\begin{pmatrix}\eta & \beta_2-\alpha_2\\ \eta & \alpha_2+\beta_2-\gamma_2\end{pmatrix};$$

1-5 $\begin{cases}\gamma_1=\beta_1\neq 0,\\ \alpha_1\neq 0,\beta_1:\end{cases}$ $F_{\alpha\beta\gamma}=-\omega^{(\beta_2-\gamma_2)\epsilon/2}\dfrac{\pi e^{i\pi\eta\epsilon}}{(\eta \quad \beta_2-\gamma_2)}$, $F^{1}_{\alpha\beta\gamma}=-\omega^{(\beta_2-\gamma_2)\epsilon/2}\begin{pmatrix}\eta & 0\\ \eta & \beta_2-\gamma_2\end{pmatrix}$;

1-6 $\begin{cases}\gamma_1=\beta_1=0,\\ \alpha_1>0:\end{cases}$ $F_{\alpha\beta\gamma}=-\dfrac{\pi}{(\eta \quad \gamma_2)}\begin{pmatrix}2\eta & \beta_2\\ \eta & \beta_2-\gamma_2\end{pmatrix}$, $F^{1}_{\alpha\beta\gamma}=-\begin{pmatrix}\eta & 0\\ \eta & \gamma_2\end{pmatrix}\begin{pmatrix}\eta & \beta_2\\ \eta & \beta_2-\gamma_2\end{pmatrix}$;

1-7 $\begin{cases}\gamma_1=\alpha_1=0,\\ \beta_1>0:\end{cases}$ $F_{\alpha\beta\gamma}=\dfrac{\pi}{(\eta \quad \gamma_2)}\begin{pmatrix}0 & \alpha_2\\ \eta & \gamma_2-\alpha_2\end{pmatrix}$, $F^{1}_{\alpha\beta\gamma}=\begin{pmatrix}\eta & 0\\ \eta & \gamma_2\end{pmatrix}\begin{pmatrix}\eta & \alpha_2\\ \eta & \gamma_2-\alpha_2\end{pmatrix}$;

1-8 $\begin{cases}\gamma_1=0,\\ \alpha_1+\beta_1=n,\\ \alpha_1,\beta_1\neq 0:\end{cases}$ $F_{\alpha\beta\gamma}=-\dfrac{\pi}{(\eta \quad \gamma_2)}\begin{pmatrix}0 & \alpha_2+\beta_2-2\gamma_2\\ \eta & \alpha_2+\beta_2-\gamma_2\end{pmatrix}$

$$F^{1}_{\alpha\beta\gamma}=-e^{-i\pi\eta\epsilon}\begin{pmatrix}\eta & 0\\ \eta & \gamma_2\end{pmatrix}\begin{pmatrix}0 & \alpha_2+\beta_2-2\gamma_2\\ \eta & \alpha_2+\beta_2-\gamma_2\end{pmatrix},$$

$$F^{1'}_{\alpha\beta\gamma}=-\begin{pmatrix}\eta & 0\\ \eta & \gamma_2\end{pmatrix}\begin{pmatrix}\eta & \beta_2-\alpha_2\\ 0 & \beta_2-\alpha_2\end{pmatrix}\begin{pmatrix}0 & \alpha_2+\beta_2-2\gamma_2\\ \eta & \alpha_2+\beta_2-\gamma_2\end{pmatrix};$$

1-9 $\begin{cases}\beta_1=0,\\ \gamma_1=\alpha_1\neq 0:\end{cases}$ $F_{\alpha\beta\gamma}=\dfrac{\pi}{(\eta \quad \gamma_2-\alpha_2)}\begin{pmatrix}2\eta & \beta_2\\ \eta & \alpha_2+\beta_2-\gamma_2\end{pmatrix}$, $F^{1}_{\alpha\beta\gamma}=\begin{pmatrix}\eta & 0\\ \eta & \gamma_2-\alpha_2\end{pmatrix}\begin{pmatrix}\eta & \beta_2\\ \eta & \alpha_2+\beta_2-\gamma_2\end{pmatrix}$;

1-10 $\begin{cases}\gamma_1=\beta_1\neq 0,\\ \alpha_1=0:\end{cases}$ $F_{\alpha\beta\gamma}=-\dfrac{\pi}{(\eta \quad \beta_2-\gamma_2)}\begin{pmatrix}0 & \alpha_2\\ \eta & \alpha_2+\beta_2-\gamma_2\end{pmatrix}$, $F^{1}_{\alpha\beta\gamma}=-\begin{pmatrix}\eta & 0\\ \eta & \beta_2-\gamma_2\end{pmatrix}\begin{pmatrix}\eta & \alpha_2\\ \eta & \alpha_2+\beta_2-\gamma_2\end{pmatrix}$;

1-11 $\alpha_1=\beta_1=\gamma_1\neq 0$: $F_{\alpha\beta\gamma}=\dfrac{\pi}{(\eta \quad \gamma_2-\alpha_2)}\begin{pmatrix}0 & \alpha_2+\beta_2-2\gamma_2\\ \eta & \beta_2-\gamma_2\end{pmatrix}$,

$$F^{1}_{\alpha\beta\gamma}=\begin{pmatrix}\eta & 0\\ \eta & \gamma_2-\alpha_2\end{pmatrix}\begin{pmatrix}\eta & \beta_2-\alpha_2\\ \eta & \beta_2-\gamma_2\end{pmatrix}\begin{pmatrix}0 & \alpha_2+\beta_2-2\gamma_2\\ 0 & \beta_2-\alpha_2\end{pmatrix};$$

1-12 $\alpha_1=\beta_1=\gamma_1=0$: $F_{\alpha\beta\gamma}=\dfrac{\pi}{(\eta \quad \gamma_2)}\begin{pmatrix}0 & \alpha_2\\ \eta & \gamma_2-\alpha_2\end{pmatrix}\begin{pmatrix}0 & \alpha_2+\beta_2-2\gamma_2\\ \eta & \alpha_2+\beta_2-\gamma_2\end{pmatrix}\begin{pmatrix}2\eta & \beta_2\\ \eta & \beta_2-\gamma_2\end{pmatrix}$,

$$F^{1}_{\alpha\beta\gamma}=\begin{pmatrix}\eta & 0\\ \eta & \gamma_2\end{pmatrix}\begin{pmatrix}\eta & \alpha_2\\ \eta & \gamma_2-\alpha_2\end{pmatrix}\begin{pmatrix}\eta & \beta_2\\ \eta & \alpha_2+\beta_2-\gamma_2\end{pmatrix}\begin{pmatrix}\eta & \beta_2-\alpha_2\\ \eta & \beta_2-\gamma_2\end{pmatrix}\begin{pmatrix}0 & \alpha_2+\beta_2-2\gamma_2\\ 0 & \beta_2-\alpha_2\end{pmatrix};$$

2-1 $\begin{cases}\alpha_1,\beta_1\neq 0,\\ \alpha_1\neq\beta_1:\end{cases}$ $F_{\alpha\beta\alpha}=e^{i\pi\eta\epsilon}\dfrac{\pi}{(\eta \quad 0)}$;

2-2 $\alpha_1=\beta_1\neq 0$:$F_{\alpha\beta\alpha}=\dfrac{\pi}{(\eta \quad 0)}\begin{pmatrix}0 & \beta_2-\alpha_2\\ \eta & \beta_2-\alpha_2\end{pmatrix}$;

2-3 $\begin{cases}\beta_1=0,\\ \alpha_1\neq 0:\end{cases}$ $F_{\alpha\beta\alpha}=\dfrac{\pi}{(\eta \quad 0)}\begin{pmatrix}2\eta & \beta_2\\ \eta & \beta_2\end{pmatrix}$;

2-4 $\begin{cases}\alpha_1=0,\\ \beta_1\neq 0:\end{cases}$ $F_{\alpha\beta\alpha}=\dfrac{\pi}{(\eta \quad 0)}\begin{pmatrix}0 & \alpha_2\\ \eta & \alpha_2\end{pmatrix}$;

2-5 $\alpha_1=\beta_1=0$:$F_{\alpha\beta\alpha}=\dfrac{\pi}{(\eta \quad 0)}\begin{pmatrix}0 & \alpha_2\\ \eta & \alpha_2\end{pmatrix}\begin{pmatrix}2\eta & \beta_2\\ \eta & \beta_2\end{pmatrix}\begin{pmatrix}0 & \beta_2-\alpha_2\\ \eta & \beta_2-\alpha_2\end{pmatrix}$;

3-1 $\gamma_1\neq 0,\beta_1$: $F_{0\beta\gamma}=0$;

3-2 $\begin{cases}\gamma_1=0,\\ \beta_1\neq 0:\end{cases}$ $F_{0\beta\gamma}=\dfrac{\pi^2}{(\eta \quad \gamma_2)^2}$, $F^{1}_{0\beta\gamma}=\begin{pmatrix}\eta & 0\\ \eta & \gamma_2\end{pmatrix}^2$;

3-3 $\gamma_1=\beta_1\neq 0$: $F_{0\beta\gamma}=-\dfrac{\pi^2}{(\eta \quad \beta_2-\gamma_2)^2}$, $F^{1}_{0\beta\gamma}=-\begin{pmatrix}\eta & 0\\ \eta & \beta_2-\gamma_2\end{pmatrix}^2$;

3-4 $\gamma_1=\beta_1=0$: $F_{0\beta\gamma}=\dfrac{\pi^2}{(\eta \quad \gamma_2)^2}\dfrac{(0 \quad \beta_2-2\gamma_2)(2\eta \quad \beta_2)}{(\eta \quad \beta_2-\gamma_2)^2}$, $F^{1}_{0\beta\gamma}\begin{pmatrix}\eta & 0\\ \eta & \gamma_2\end{pmatrix}^2\begin{pmatrix}\eta & \beta_2\\ \eta & \beta_2-\gamma_2\end{pmatrix}^2\begin{pmatrix}0 & \beta_2-2\gamma_2\\ 0 & \beta_2\end{pmatrix}$;

4-1 $\beta_1\neq 0$: $F_{0\beta 0}=\dfrac{\pi^2}{(\eta \quad 0)^2}$;

4-2 $\beta_1=0$: $F_{0\beta 0}=\frac{\pi^2}{(\eta\quad 0)^2}\frac{(0\quad \beta_2)(2\eta\quad \beta_2)}{(\eta\quad \beta_2)^2}$.

For $n=2$ we have

$$F^{1}_{0,10,11}(\eta,i\infty)=J_{23}(i\infty)=-\tan^2\pi\eta,$$

$$F^{1}_{0,11,01}(\eta,i\infty)=J_{31}(i\infty)=\tan^2\pi\eta,$$

$$F^{1}_{0,01,10}(\eta,i\infty)=J_{12}(i\infty)=0.$$

The CSC's $f_{\alpha\beta\gamma}(i\infty)$'s may be written down from above $F_{\alpha\beta\gamma}(\eta,i\infty)$ by means of (41). We see that nonvanishing $f_{\alpha\beta\gamma}(i\infty)$'s exhibit a nontrivial classical trigonometric algebra.

VII. CONCLUDING REMARKS

Recently quantum algebra (quantum group) has become an interesting subject and many approaches to the trigonometric quantum group are appearing. However, the algebra in the elliptic case is more complicated; on that subject, fewer results have been published, and its relation to the trigonometric case is known only for $n=2$. In this article we have, for $n>2$, derived the classical and quantum algebras as well as their compatibility conditions, obtained some explicit expressions for the structure constants and their symmetric relations, given exact relations between the classical algebra and the quantum one, and worked out the corresponding trigonometric algebras.

ACKNOWLEDGMENTS

The authors are indebted to Professor C. N, Yang for his stimulating lectures on statistical mechanics and braid group at Nanki University.

This work was supported in part by the Natural Science Fund of China.

[1]C. N. Yang, Phys. Rev. Lett. **19**, 1312 (1967).
[2]R. J. Baxter, *Exactly Solved Models in Statistical Mechanics* (Academic, London, 1982).
[3]A. A. Belavin, Nucl. Phys. B **180**, 189 (1980).
[4]B. Y. Hou, M. L. Yan, and Y. K. Zhou, Nucl. Phys. B **324**, 715 (1989).
[5]P. P. Kulish, N. Yu. Reshetikhin, and E. K. Sklyanin, Lett. Math. Phys. **5**, 393 (1981).
[6]E. Date, M. Jimbo, A. Kuniba, T. Miwa, and M. Okado, Nucl. Phys. B **290**, 231 (1987).
[7]V. G. Drinfeld, Proc. of the Intl. Congress of Mathematicians, Berkeley, 1986, Vol. 1, p. 798.
[8]E. K. Sklyanin, Funct. Anal. Appl. **16**, 263 (1982); **17**, 273 (1983).
[9]I. V. Cherednik, Func. Anal. Appl. **19**,77 (1985).
[10]V. Jones, Ann. Math. **126**, 335 (1987).
[11]T. Kohno, Ann. Inst. Fourier **37**, 139 (1987).
[12]Y. Akutsu and M. Wadati, J. Phys. Soc. Jpn. **56**, 839 (1987).
[13]A. Tsuchiya and Y. Kanie, Lett. Math. Phys. **13**, 303 (1987).
[14]E. Witten, Proc. of the IAMP Congress, Swansea, July, 1988.
[15]H. Wei and B. Y. Hou, NWU-IMP-89-59.

The Riemann–Hilbert transformation for an approach to a representation of the Virasoro group

Wei Li[a)]
Department of Mathematics and Computer Science, Clarkson University, Potsdam, New York 13676
Bo-yu Hou
Center of the Theoretical Physics, CCAST, and Institute of Modern Physics, Northwest University, Xian, China

(Received 18 May 1988; accepted for publication 25 January 1989)

In this paper, it is the intent to apply the Riemann–Hilbert transformation developed by Hauser and Ernst [J. Math. Phys. **21**, 1126, 1418 (1980)] in providing a new representation of the Virasoro group. It is found that the Geroch group that acts on the solution space of the Einstein field equations is extended to the semidirect product of the Virasoro and Kac–Moody groups; also, the relationship between the infinitesimal transformation given previously [B. Y. Hou and W. Li, Lett. Math. Phys. **13**, 1 (1987); J. Phys. A **20**, L897 (1987); W. Li, Phys. Lett. A **129**, 301 (1988)] and the infinitesimal Riemann–Hilbert transformation is pointed out. Finally, it is shown that the well-known Neugebauer–Backlund transformation can be derived from the Riemann–Hilbert transformation.

I. INTRODUCTION

Several years ago Hauser and Ernst[1] first pointed out that the Riemann–Hilbert transformation is associated with the infinite-dimensional loop group: They were able to give the explicit action of the Geroch group,[2] which is shown to be isomorphic to an affine Kac–Moody group,[3] in the two-dimensional Einstein field equations, the Ernst equation. The work of Hauser and Ernst developed the result of Kinnersley and Chitre,[4] who gave an infinite set of generators of the Geroch group. The Hauser–Ernst method can also offer an effective and powerful technique for generating a new solution of the Ernst equation from the known solution. Later, Ueno and Nakamura[5] applied the Hauser–Ernst method to the principal chiral model and the self-dual Yang–Mills fields and established similar representations of the Kac–Moody groups in these systems. Moreover, Ueno and Nakamura pointed out the link between the Kac–Moody algebra found by Dolan[6] and the so-called hidden symmetry transformations.[7] Thus far much investigation and application has been made toward understanding the Hauser–Ernst method as related to the Riemann–Hilbert transformation for some integrable systems.[8]

In this paper we shall find a way to extend the Hauser–Ernst method to the more general case. We shall indicate that the Riemann–Hilbert transformation for the Hauser–Ernst approach can be related with the Virasoro group as well as the Kac–Moody group and has a richer structure than previously expected. The Riemann–Hilbert transformation gives rise to the construction of the semidirect product of the Kac–Moody and Virasoro groups.

In a series of recent papers,[9,10] the present authors have succeeded in constructing an infinite set of infinitesimal transformations for the Ernst equation. Our transformations are different from those given by Kinnersley and Chitre.[4] Careful calculation shows that these new transformations constitute a representation of the Virasoro algebra which has no central extension and no highest weight, so that the representation is nonunitary. As a result, a new symmetry, like the Kac–Moody symmetry, is confirmed to exist in the solution space of the Ernst equation and the Geroch group is thus extended by the Virasoro and Kac–Moody groups.

We are motivated to find the exponentiation of our infinitesimal transformations and give the representation of the enlarged Geroch group. The problem can be solved by giving an integral equation which will be proved by means of the Riemann–Hilbert transformation in Sec. III; it is given by

$$\frac{1}{2\pi i}\int_{C_{0,t}} \frac{F(s)u(s)F_0(v(s))^{-1}}{s(s-t)}\,ds=0, \tag{1.1}$$

where C represents a circle surrounding the origin and t in the complex s plane; $u(s)$ and $v(s)$ are, respectively, a 2×2 matrix function and a scalar function of s; and $F(s)$ and $F_0(s)$ satisfy the Hauser–Ernst linearization equations. The details of the restrictions to the quantities in (1.1) will be given in Sec. III.

We see that if $v(s)=s$, the integral equation (1.1) is identical with that initially given by Hauser and Ernst[1] and is used to provide the representation of the Kac–Moody group. If $u(t)=I$ (where I is a unit matrix), the successive transform of $F(t)$ for Eq. (1.1) is offered by

$$v_2=v_0\circ v_1 \quad\text{or}\quad v_2(t)=v_0(v_1(t)). \tag{1.2}$$

We know that according to the representation theory of an infinite-dimensional group,[11] Eq. (1.2) is the composition law of the Virasoro group, for which elements consisting of the set of functions $v(t)$ satisfy the conditions given in Sec. III. It is apparent that our work generalizes the application of the Riemann–Hilbert transformation and presents a new

[a)] Permanent address: Institute of Modern Physics, Northwest University, Xian, China.

 0022-2488/89/061198-07$02.50

approach for generating solutions of the Ernst equation.

The structure of this paper is organized as follows. In Sec. II we recall the formulation developed by Hauser and Ernst[1] for the Ernst equation which will be used in the following discussions. In Sec. III, we shall describe how to generalize the Hauser and Ernst approach to the Riemann–Hilbert transformation. Then we shall exploit the Riemann–Hilbert transformation to prove the integral equation (1.1). In Sec. IV, we derive the infinitesimal transformations given in (1.1) and identify the infinitesimal Riemann–Hilbert transformations with those of the Virasoro algebra. Finally, we shall discuss the group structure of the Riemann–Hilbert transformation and apply it to rederive some known Backlund transformations such as the Neugebauer[12] and Maison–Cosgrove transformations.[13]

II. NOTATIONS AND CONVENTIONS

In order to describe our objective more clearly, it is helpful to introduce a few of the notations and conventions that shall often be used in this paper.

We first start with the metric of the space-time, which admits two commuting Killing vectors under the line element

$$ds^2 = g_{ij}\, dx^i\, dx^j + g_{+-}\, dx_+\, dx_-, \tag{2.1}$$

where g_{ij} $(i,j = 1,2)$ and g_{+-} are functions of x_+ and defined by

$$x_+ = \tfrac{1}{2}(x^3 + \Lambda x^4), \quad x_- = \tfrac{1}{2}(x^3 - \Lambda x^4). \tag{2.2}$$

Here we have two cases to be distinguished for the space-time: If one of the Killing vectors is timelike, i.e., the space-time possesses the stationary and axially symmetric fields, we take the value of Λ as $-i$; if both Killing vectors are spacelike, i.e., the space-time has the cylindrically symmetric fields or the gravitational plane-wave fields, we take Λ as 1. The treatments in the following discussion are very similar for these two cases; thus we no longer underline their differences.

The reduction of the vacuum Einstein field equations leads to

$$\partial_+(\alpha^{-1}g\Omega\, \partial_-g) + \partial_-(\alpha^{-1}g\Omega\, \partial_+g) = 0, \tag{2.3}$$

where

$$\partial_+ = \frac{\partial}{\partial x_+}, \quad \partial_- = \frac{\partial}{\partial x_-}, \qquad \Omega = \begin{pmatrix} 0 & i \\ -i & 0 \end{pmatrix}, \tag{2.4}$$

and g is the 2×2 symmetric real matrix whose elements g_{ij} are the metric components in the line element (2.1) satisfying

$$\det g = (\Lambda\alpha)^2. \tag{2.5}$$

We will not consider the remainder of the vacuum Einstein field equations governing the metric component g_{+-} in this paper.

We then introduce the matrix Ernst potential E, which is a 2×2 matrix field and may be defined as a solution of

$$\begin{aligned} 2(\beta + \Lambda^3\alpha)\partial_+E &= (E + E^\dagger)\Omega\, \partial_+E, \\ 2(\beta - \Lambda^3\alpha)\partial_-E &= (E + E^\dagger)\Omega\, \partial_-E \end{aligned} \tag{2.6}$$

such that

$$g = \tfrac{1}{2}(E + E^\dagger) - \beta\Omega, \tag{2.7}$$

$$2\beta = \mathrm{tr}(E\Omega), \tag{2.8}$$

where the dagger stands for the Hermitian conjugation. It is not difficult to show that Eq. (2.3) is equivalent to Eqs. (2.6).

According to the definition of the fields α and β from Eq. (2.3), it is apparent that α and β are solutions of the wave equation in two dimensions and have the relation

$$\partial_+\beta = +\Lambda^3\, \partial_+\alpha, \quad \partial_-\beta = -\Lambda^3\, \partial_-\alpha. \tag{2.9}$$

By defining

$$\eta_+ = \beta + \Lambda^3\alpha, \quad \eta_- = \beta - \Lambda^3\alpha, \tag{2.10}$$

Eq. (2.9) implies that

$$\partial_+\eta_- = \partial_-\eta_+ = 0. \tag{2.11}$$

Following Hauser and Ernst's treatment for the linearization of Eq. (2.6), we define a set of 2×2 matrix functions F of x_+ and x_- and a complex parameter t such that for a given E, $F(t) = F(x_+, x_-; t)$ is any solution of the linearization equations

$$\begin{aligned} \partial_+F(t) &= [t/(1 - 2t\eta_+)]\, \partial_+E\, \Omega F(t), \\ \partial_-F(t) &= [t/(1 - 2t\eta_-)]\, \partial_-E\, \Omega F(t). \end{aligned} \tag{2.12}$$

In the sense of Frobenius the integrable condition of Eqs. (2.12) has to be identical to that of Eq. (2.6). We know that the linearization equations do not define $F(t)$ uniquely; thus we need to suppress some subsidiary conditions consistent with Eqs. (2.12) such that $F(t)$ is holomorphic in a neighborhood $t = 0$ and $F(t)$ satisfies

$$F(0) = I, \tag{2.13}$$

$$\dot F(0) = E\Omega, \tag{2.14}$$

$$\det F(t) = \lambda^{-1}(t), \tag{2.15}$$

$$F(t)^\dagger \Omega A(t) F(t) = \Omega, \tag{2.16}$$

where

$$\dot F(t) = \frac{\partial F(t)}{\partial t},$$

$$F(t)^\dagger = \text{Hermitian conjugate of } F(t^*) \tag{2.17}$$

and

$$\begin{aligned} \lambda(t) &= [(1 - 2t\eta_+)(1 - 2t\eta_-)]^{1/2}, \\ A(t) &= I - t(E + E^\dagger)\Omega. \end{aligned} \tag{2.18}$$

[Also, I denotes the 2×2 unit matrix and the asterisk denotes the complex conjugation.]

On the other hand, Eqs. (2.6) can be written in the form

$$A(t)[t/(1 - 2t\eta_\pm)]\partial_\pm E = t\, \partial_\pm E. \tag{2.19}$$

We then operate $A(t)$ on the linearization equations (2.12) and use Eq. (2.19) to obtain

$$A(t)\partial_\pm F(t) = t\, \partial_\pm E\, \Omega F(t), \tag{2.20}$$

which will be used in the following discussion.

Except for these restrictions on $F(t)$ there still exists the general gauge transformation of the function $F(t)$, i.e., for given any solution $F(t)$,

$$F'(t) = F(t)u(t) \tag{2.21}$$

is a solution of the linearization equations (2.12) also, where $u(t)$ is any 2×2 matrix function of t only and $u(t)$ is holomorphic in a neighborhood $t=0$ satisfying

$$u(0)=I,\quad u(t)^{\dagger}\Omega u(t)=\Omega,\quad \det u(t)=1. \tag{2.22}$$

Since we know that the analytic properties of the function $F(t)$ play an important role in our discussion we thus hope to restrict the gauge to one for which the set of t-plane singularities of the function $F(t)$ is minimized. Hauser and Ernst[1,14] proved that for fixed (x_+,x_-) the function $F(t)$ has t-plane singularities at $t=1/2\eta_+$ and $t=1/2\eta_-$ regardless of the choice of gauge: For this reason we can always choose the gauge such that for fixed (x_+,x_-) $F(t)$ is a holomorphic function of t on the whole t plane except for the points $t=1/2\eta_+$ and $t=1/2\eta_-$.

III. PROOF OF THE INTEGRABLE EQUATION

Before describing the proof of the integrable equation (1.1) given in Sec. I, we would like to briefly recall the formulation of the Riemann–Hilbert transformation, which is an essential tool to our discussion.

First, let us select a circle C surrounding the origin in the complex t plane, in which the interior and exterior regions of C are, respectively, denoted by C_+ and C_-. Then there exist a pair of functions $X_+(t)$ and $X_-(t)$ of x_+,x_- and a parameter t such that $X_+(t)$ and $X_-(t)$ are holomorphic in C_+ and C_-, respectively, and both are continuous on C. If a given function $G(t)$ on C, called the kernel, is analytic and connects $X_+(t)$ with $X_-(t)$ by a relation

$$X_-(t)=X_+(t)G(t)\ \text{ on } C, \tag{3.1}$$

the solutions $X_+(t)$ and $X_-(t)$ to Eq. (3.1) are unique: This is the so-called Riemann–Hilbert transformation. It is well known that the Riemann–Hilbert transformation can be used in connection with the generation of a new solution to some nonlinear equations from prior solutions once the kernel $G(t)$ is explicitly given.

The key to the problem is how to find an explicit expression of $G(t)$ and solve $X_+(t)$ and $X_-(t)$ in the different regions. Hauser and Ernst[1,14] proposed that for a given solution $F_0(t)$, which is assumed to be holomorphic on the whole t plane except at $t_+=1/2\eta_+$ and $t_-=1/2\eta_-$ and lying in C_-, the kernel $G(t)$ can be constructed by the form

$$G(t)=F_0(t)u(t)F_0^{-1}(t), \tag{3.2}$$

where $u(t)$ is defined as a 2×2 matrix function of t, independent of x_+ and x_-, such that $u(t)$ is holomorphic in $C+C_-$ and satisfies the conditions

$$u(t)^{\dagger}\Omega u(t)=\Omega,\quad \det u(t)=1. \tag{3.3}$$

Then Hauser and Ernst were able to verify that with the aid of the solutions $X_+(t)$ and $X_-(t)$ to the Riemann–Hilbert transformation, with the boundary condition

$$X_+(0)=I, \tag{3.4}$$

one can construct a new solution $F(t)$ to Eqs. (2.12)–(2.17) by defining

$$F(t)=X_+(t)F_0(t)\ \text{ in } C_+,\qquad =X_-(t)F_0(t)u(t)^{-1}\ \text{ in } C_-, \tag{3.5}$$

with the new matrix Ernst potential E:

$$E=E_0+\dot{X}_+(0). \tag{3.6}$$

Further investigation shows that the Hauser–Ernst approach to the Riemann–Hilbert transformation establishes a simpler representation of the Geroch group, a group whose elements consist of the set of 2×2 matrix functions $u(t)$ subject to the conditions (3.3) and whose composition law corresponds to the exact form of the Kac–Moody group. However, for our purpose we would like to extend the Hauser–Ernst method to the more general cases in which the Virasoro group, of which an infinite set of infinitesimal generators were found to act on the solution space of the Ernst field equations, will be described in addition to the Kac–Moody group.

Now let us first define a scalar function $v(t)$ such that $v(t)$ is independent of x_+ and x_- and is holomorphic on $C+C_-$ except at infinity, where $v(t)$ tends to linear divergence and such that $v(t)$ is a linear function or has singularities in C_+. We further state the restriction that for fixed x_+ and x_-, t_+ and t_- are single-value solutions to

$$(1-2v(t)\eta_+)(1-2v(t)\eta_-)=0, \tag{3.7}$$

which lie in C_-; $\dot{v}(t_\pm)\neq0$. We introduce the new notations η'_+ and η'_- such that

$$t_+=1/2\eta'_+\ =v^{-1}(1/2\eta_+),\qquad t_-=1/2\eta'_-\ =v^{-1}(1/2\eta_-). \tag{3.8}$$

From Eq. (2.11), we can easily prove that η'_+ and η'_- satisfy

$$\partial_+\eta'_-=\partial_-\eta'_+=0. \tag{3.9}$$

In fact, Eqs. (3.8) can be interpreted as the transforms of variables between different coordinate systems. Thus we define that under the transformations (3.8) the given solution $F_0(t)$ to Eqs. (2.12)–(2.16) is changed into $F'_0(t)$, which satisfies the linearization equations and corresponding subsidiary conditions by replacing η_+ and η_- with η'_+ and η'_-. Thus we have E'_0.

Hence we prefer to select the kernel $G(t)$ as the form

$$G(t)=F'_0(t)u(t)F_0(v(t))^{-1}\ \text{ on } C, \tag{3.10}$$

where $u(t)$ is the same as given above. Thus in our case the Riemann–Hilbert transformation can be written in the form

$$X_-=X_+(t)G(t)=X_+(t)F'_0(t)u(t)F_0(v(t))^{-1}\ \text{ on } C, \tag{3.11}$$

with the boundary condition

$$X_+(0)=I. \tag{3.12}$$

Similar to Hauser and Ernst,[1,14] we shall exploit $X_+(t)$ and $X_-(t)$ in order to construct the new function $F(t)$ by

$$F(t)=X_+(t)F'_0(t),\ \text{ in } C_+,\qquad =X_-(t)F_0(v(t))u^{-1}(t)\ \text{ in } C_-. \tag{3.13}$$

We shall show that $F(t)$ constructed in Eqs. (3.13) is a solution to the following:

$$\partial_+ F(t) = [t/(1-2t\eta'_+)]\partial_+ E\,\Omega F(t),$$
$$\partial_- F(t) = [t/(1-2t\eta'_-)]\partial_- E\,\Omega F(t), \quad (3.14)$$

$$A(t)\partial_+ F(t) = \partial_+ E\,\Omega F(t),$$
$$A(t)\partial_- F(t) = \partial_- E\,\Omega F(t), \quad (3.15)$$

$$F(0) = I, \quad (3.16)$$

$$\dot F(0) = E\Omega, \quad (3.17)$$

$$\det F(t) = \lambda'(t)^{-1}, \quad (3.18)$$

$$F(t)^\dagger \Omega A(t) F(t) = \Omega, \quad (3.19)$$

where

$$E = E'_0 + \dot X'_+(0)\Omega, \quad (3.20)$$

$$A(t) = I - t(E + E^\dagger)\Omega, \quad (3.21)$$

$$\lambda'(t) = [(1-2t\eta'_+)(1-2t\eta'_-)]^{1/2}. \quad (3.22)$$

To prove Eqs. (3.14), we operate Eqs. (3.13) in differentiation with respect to x_+ and obtain

$$\partial_+ F(t)\, F(t)^{-1} = \partial_+ X_+(t) X_+(t)^{-1} + [t/(1-2t\eta'_+)] X_+(t)\partial_+ E'_0\,\Omega X_+^{-1}(t) \text{ in } C_+,$$
$$= \partial_+ X_-(t) X_-(t)^{-1} + \{v(t)/[1-2v(t)\eta_+]\} X_-(t)\partial_+ E_0\,\Omega X_-^{-1}(t) \text{ in } C_-. \quad (3.23)$$

Since $X_+(t)$ and $X_-(t)$ are, respectively, holomorphic in C_+ and C_- and $v(t)$ is holomorphic in C_- we observe Eqs. (3.23) such that there exists only one simple singularity at $t = 1/2\eta'_+$ on the whole t plane. Thus Eqs. (3.23) imply that

$$\partial_+ F(t)\, F(t)^{-1} = P + [t/(1-2t\eta'_+)]Q, \quad (3.24)$$

where P and Q independent of t are undetermined. From Eqs. (3.12) and (3.23) we obtain

$$P = 0, \quad Q = \partial_+(\dot X_+(0) + E'_0\Omega). \quad (3.25)$$

If we set

$$Q = \partial_+ E, \quad (3.26)$$

we obtain Eqs. (3.14) for x_+, together with Eq. (3.20); in a similar way, we can verify a component of Eqs. (3.14) for x_-.

Before we prove Eqs. (3.15), we show

$$X_+^{-1}(t)^\dagger \Omega A'_0(t) X_+^{-1}(t)$$
$$= X_-^{-1}(t)^\dagger \Omega A_0(v(t)) X_-^{-1}(t)$$
$$= \Omega A(t) \quad (3.27)$$

and

$$\Omega A(t)\partial_\pm X_+(t) X_+^{-1}(t)$$
$$+ t X_+^{-1}(t)^\dagger \Omega \partial_\pm E'_0\,\Omega X_+^{-1}(t)^\dagger$$
$$= \Omega A(t)\partial_\pm X_-(t) X_-^{-1}(t) \quad (3.28)$$
$$+ v(t) X_-^{-1}(t)^\dagger \Omega \partial_\pm E_0\,\Omega X_-^{-1}(t)^\dagger$$
$$= t\Omega\,\partial_\pm E.$$

The first equality of Eq. (3.27) is derived from Eqs. (2.16) and (3.13); it can be expressed as a linear function of t since $v(t)$ is linear in t as well as $A(v(t))$ when t tends to infinity. After determining the coefficients of this linear function, we confirm the second equality of Eq. (3.27). Under similar consideration and using Eq. (2.20), it is not difficult to prove Eq. (3.28).

Therefore, it follows from Eqs. (2.16), (3.27), and (3.28) that

$$A(t)\partial_\pm F(t) = A(t)\partial_\pm F'_0(t)$$
$$+ A(t) X_+(t)\partial_\pm F'_0(t)$$
$$= A(t)\partial_\pm F'_0(t) + \Omega X_+(t)\Omega$$
$$\times A'_0(t)\,\partial_\pm F'_0(t) \quad (3.29)$$
$$= A(t)\partial_\pm F'_0(t) + t\Omega X_+(t)\Omega$$
$$\times \partial_\pm E'_0\Omega F'_0(t)$$
$$= t\Omega\,\partial_\pm E\,\Omega X_+(t) F'_0(t)$$
$$= t\Omega\,\partial_\pm E\,\Omega F(t).$$

Thus we complete the proof of Eqs. (3.15).

To prove Eq. (3.18) from Eqs. (3.14), we observe that

$$\det X_+(t) = \det X_-(t)[\lambda(v(t))/\lambda'(t)]. \quad (3.30)$$

Since the singularities at t_+ and t_- on the rhs of Eq. (3.30) can be eliminated, the functions on both sides entice functions for all t. Under the restriction of boundary condition (3.12) we conclude that the function is equal to 1. As a consequence of Eqs. (2.15), (3.13), and (3.30), we can obtain Eq. (3.18).

By using Eqs. (2.16), (3.3), and (3.27), we observe that

$$F(t)^\dagger \Omega A(t) F(t)$$
$$= u^{-1}(t)^\dagger F'_0(t)^\dagger X_+(t)^\dagger \Omega A(t) X_+(t) F'_0(t) u^{-1}(t)$$
$$= u^{-1}(t)^\dagger F'_0(t)^\dagger \Omega A'_0(t) F'_0(t) u^{-1}(t)$$
$$= u^{-1}(t)^\dagger \Omega u^{-1}(t)$$
$$= \Omega. \quad (3.31)$$

Now let us prove that the integral equation (1.1) given in Sec. I is identical to the representation of the Riemann–Hilbert transformation, i.e., both are equivalent. To do this, we note that since $X_-(t)$ is analytic in $C + C_-$ (including $t = \infty$), we have

$$\frac{1}{2\pi i}\int_C \frac{X_-(s)}{s(s-t)}\,ds = 0 \quad (t \text{ in } C_+), \quad (3.32)$$

where we used

$$\frac{1}{2\pi i}\int_C \frac{1}{s(s-t)}\,ds = 0 \quad (t \text{ in } C_+). \quad (3.33)$$

In terms of the Cauchy theorem, we substitute Eq. (3.13) into Eq. (3.32) to obtain

$$\frac{1}{2\pi i}\int_{C_{0,t}}\frac{1}{s(s-t)}F(s)u(s)\,F_0^{-1}(v(s))ds=0 \qquad (3.34)$$

subject to the conditions

$$F(0)=I,\quad \dot F(0)=E\Omega. \qquad (3.35)$$

Therefore, we complete our proof of the integral equation (1.1).

Finally, we should prove that E given in Eq. (3.20) satisfies the Ernst field equations and that its trace is equal to β'; we can obtain the new metric g from this new Ernst potential.

From Eqs. (3.14) and (3.15), we have

$$A(t)\partial_\pm F(t)=(1-2t\eta'_\pm)\partial_\pm F(t). \qquad (3.36)$$

Differentiating (3.36) twice with respect to t and setting $t=0$, we see that E satisfies Eqs. (2.6) with $\eta'_\pm$.

We then take the trace to Eq. (3.20) to obtain

$$\mathrm{tr}(E\Omega)=\mathrm{tr}(E'_0\Omega)=2\beta' \qquad (3.37)$$

because

$$\mathrm{tr}\,\dot X_+(0)=\mathrm{tr}(\dot X_+(0)\,X_+^{-1}(0))$$
$$=(\det X_+(0))^{-1}\frac{\partial}{\partial t}(\det X_+(0))=0, \qquad (3.38)$$

where we used $X_-(0)=I$. Equation (3.37) is equivalent to

$$E-E^T=2\beta'\Omega, \qquad (3.39)$$

where the superscript T stands for the transport operator.

According to the definition of g,

$$g=\tfrac{1}{2}(E+E^+)-\beta'\Omega, \qquad (3.40)$$

it is obvious that g is Hermitian. If we show that g is real, then it must be real symmetric. From Eq. (3.39), it follows that

$$g=\tfrac{1}{2}(E-E^T)+\tfrac{1}{2}(E^T+E^+)-\beta'\Omega$$
$$=\tfrac{1}{2}(E^T+E^+)$$
$$=g^*. \qquad (3.41)$$

It is not difficult to deduce

$$\det g=(\Lambda\alpha')^2, \qquad (3.42)$$

which follows from

$$A(t)^T\Omega A(t)=\det A(t)\Omega=\lambda'^2(t)\Omega \qquad (3.43)$$

and Eqs. (3.19) and (3.40).

IV. DERIVATION OF THE INFINITESIMAL SYMMETRY TRANSFORMATIONS

In this section we shall discuss the relationship between the Reimann–Hilbert transformation and the infinitesimal symmetry transformations given by us previously.[9] We proposed[9] the infinitesimal symmetry transformations

$$\delta F(t)=-[t/(t-t')]\{t\dot F(t)F^{-1}(t)$$
$$-t'\dot F(t')F^{-1}(T')\}F(t) \qquad (4.1)$$

to the linearization equations (2.12) and

$$\delta E=-t'\dot F(t')F^{-1}(t')\Omega \qquad (4.2)$$

to the Ernst field equation, where for the convenience of discussion infinitesimal constants are not written out. We showed[9] that these transformations constitute the Virasoro algebra by expanding the powers of the parameter t'. On the other hand, as indicated above, the Riemann–Hilbert transformation for $u(t)=I$ corresponds to the representation of the Virasoro group; thus we expect to derive the transformations (4.1) and (4.2) from

$$\frac{1}{2\pi i}\int_{C_{0,t}}\frac{F(s)\,F_0^{-1}(v(s))}{s(s-t)}ds=0, \qquad (4.3)$$

with the boundary conditions

$$F(0)=I,\quad \dot F(t)=E\Omega. \qquad (4.4)$$

Let us consider the infinitesimal case for Eq. (4.4). Under the infinitesimal transform, we set

$$\delta t=v(t)-t \qquad (4.5)$$

and

$$\delta F_0(t)=F(t)-F_0(t). \qquad (4.6)$$

Substituting Eqs. (4.4) and (4.5) into Eq. (4.3) we have

$$\frac{1}{2\pi i}\int_{C_{0,t}}\frac{1}{s(s-t)}(F_0(s)+\delta F_0(s))(F_0^{-1}(s)$$
$$-\dot F_0^{-1}(s)\delta s)\,ds=0, \qquad (4.7)$$

where we omitted the higher orders of $(\delta s)^2$. Since $F_0(t)$ is holomorphic in C_+ and $F_0(0)=I$, we can integrate

$$\frac{1}{2\pi i}\int_{C_{0,t}}\frac{\delta F_0(s)\,F_0^{-1}(s)}{s(s-t)}ds=\delta F_0(t)\,F_0^{-1}(t). \qquad (4.8)$$

Hence Eq. (4.6) can be written in the form

$$\delta F_0(t)\,F_0^{-1}(t)=-\frac{1}{2\pi i}\int_{C_{0,t}}\frac{\dot F_0(s)\,F_0^{-1}(s)}{s(s-t)}\delta s\,ds. \qquad (4.9)$$

For convenience, we no longer write the subscript of $F_0(t)$.

According to the definition of $v(t)$ such that $v(t)$ has the singularities of or is a linear function in C_+, without loss of generality, we can select

$$\delta t=\delta^{(k)}t=-t^{-k+1}\quad (k\geqslant 0). \qquad (4.10)$$

Substituting Eq. (4.10) into Eq. (4.9), we have

$$\delta^{(k)}F(t)\,F^{-1}(t)$$
$$=-\frac{1}{2\pi i}\int_{C_{0,t}}\frac{s^{-k+1}\dot F(s)\,F^{-1}(s)}{s(s-t)}ds\quad (k\geqslant 0). \qquad (4.11)$$

Equation (4.11) is the infinitesimal Riemann–Hilbert transformation for the Virasoro symmetry.[9,10]

To derive Eq. (4.1), we obtain

$$\delta F(t)=\sum_{k=0}^{\infty}\delta^{(k)}F(t)t'^k, \qquad (4.12)$$

where t' lies in C_+. Since we always have $|t'/s|\leqslant 1$, by using

$$\frac{1}{s-t}=\sum_{k=0}^{\infty}\frac{t'^k}{s^{k+1}} \qquad (4.13)$$

we thus obtain

$\delta F(t)\, F^{-1}(t)$

$$= -\frac{1}{2\pi i}\int_{C_{0,t}}\sum_{k=0}^{\infty}\frac{t'^k s^{-k+1}}{s(s-t)}\dot F(s)\,F^{-1}(s)\,ds$$

$$= -\frac{1}{2\pi i}\int_{C_{0,t}}\frac{s}{(s-t)(s-t')}\dot F(s)\,F^{-1}(s)\,ds$$

$$= -\frac{t}{t-t'}\{t\dot F(t)F^{-1}(t) - t'\dot F(t')F^{-1}(t')\}. \quad (4.14)$$

Equation (4.14) implies Eq. (4.2) as a result of Eq. (4.3).

Moreover, we need to give the transform of $\eta_\pm$ under the infinitesimal Riemann–Hilbert transformation. In terms of Eqs. (3.8) and (4.10), the transform yields

$$\delta^{(k)}\eta_\pm = \eta'_\pm - \eta_\pm$$
$$= 1/2v^{-1}(t_\pm) - 1/2t_\pm$$
$$= (1/2t_\pm^2)\,\delta^{(k)}t_\pm$$
$$= -\eta_\pm(2\eta_\pm)^k \quad (k\geqslant 0) \quad (4.15)$$

Then we obtain

$$\delta\eta_\pm = \sum_{k=0}\delta^{(k)}\eta_\pm t'^k$$
$$= -\eta_\pm\sum_{k=0}(2\eta_\pm)^k t'^k$$
$$= -\frac{\eta_\pm}{1-2t'\eta_\pm}, \quad (4.16)$$

or equivalently,

$$\delta\alpha = -\alpha/\lambda^2, \quad (4.17)$$

$$\delta\beta = -[\beta(1-2t\beta) + 2\Lambda^2 t\alpha^2]/\lambda^2, \quad (4.18)$$

which are the same as derived from the infinitesimal transformation (4.2) directly.

In order to investigate the structures of the infinitesimal Riemann–Hilbert transformation, we consider the integral equation (1.1) in the case of $v(t) = t$. Parallel to the above treatment, it is known that Kinnesley–Chitre transformations are given by

$\gamma_a^{(k)}\, F(t)\, F^{-1}(t)$

$$= -\frac{1}{2\pi i}\int_{C_{0,t}}\frac{s^{-k}F(s)T_a\,F^{-1}(s)}{s(s-t)}\,ds \quad (k\geqslant 0) \quad (4.19)$$

if we take

$$\gamma_a^{(k)}T_a = u(t) - T_a = -T_a t^{-k} \quad (k\geqslant 0), \quad (4.20)$$

where T_a ($a = 1, 2, 3$) are generators of the Lie algebra SL(2,R) for which the structure constant is denoted by C_{ab}^c. Following the calculations in Ref. 9, we can obtain the following commutations:

$$[\delta^{(k)},\delta^{(l)}]E = (k-1)\delta^{(k+l)}E, \quad (4.21)$$

$$[\delta^{(k)},\gamma_b^{(l)}]E = -l\,\gamma_b^{(k+l)}E, \quad (4.22)$$

$$[\gamma_a^{(k)},\gamma_b^{(l)}]E = C_{ab}^c\,\gamma_c^{(k+l)}E \quad (4.23)$$

for all $k, l\geqslant 0$. Equations (4.21)–(4.23) reveal the fact that the infinitesimal Riemann–Hilbert transformations span the structure of the semidirect product of the Kac–Moody and Virasoro algebras.

According to the previous discussion, it should be emphasized that the infinite-dimensional Lie algebra only has a positive part because k and l are not allowed to be negative. In another paper,[10] we considered the transform $t\to 1/t$ for the linearization equations: Using solutions to these new linearization equations, we proceeded to find another type of infinitesimal transformations constituting the negative part of the infinite-dimensional Lie algebra. We combined these two parts to form the full algebra, which still lacks the central term. In a similar way, in the present paper we can also show that the other type of infinitesimal transformations originate from the infinitesimal Riemann–Hilbert transformation. Therefore, we can drop out the restriction of the positive k and l in Eqs. (4.21)–(4.23).

V. DISCUSSION AND APPLICATION

To examine the structure of the infinite-dimensional group for the new form of the Riemann–Hilbert transformation, let us take the following cases into account.

(i) If $u(t) = I$, Eqs. (3.13) will give rise to

$$F(t) = X_+(t)\,F_0'(t) \text{ in } C_+,$$
$$= X_-(t)F_0(v(t)) \text{ in } C_-. \quad (5.1)$$

Now we take the transformations

$$F_1(t) = X_+^0(t)\,F_0'(t) \text{ in } C_+,$$
$$= X_-^0(t)\,F_0(v_0(t)) \text{ in } C_- \quad (5.2)$$

and

$$F_2(t) = X_-^1(t)\,F_1'(t) \text{ in } C_+,$$
$$= X_-^1(t)F_1(v_1(t)) \text{ in } C_-. \quad (5.3)$$

Then we define the new transformation

$$F_2(t) = X_+^2(t)\,F_0''(t) \text{ in } C_+,$$
$$= X_-^2(t)F_0(v_2(t)) \text{ in } C_-, \quad (5.4)$$

where $F_0''(t)$ denotes the transform of $F_0(t)$ by replacing $1/2\eta_+$ in $v_2^{-1}(1/2\eta_+)$. Using Eqs. (5.2) and (5.3), the successive transform of $F_2(t)$ can be expressed by

$$v_2(t) = v_0(v_1(t)), \quad (5.5)$$

where $v_0(t)$, $v_1(t)$, and $v_2(t)$, respectively, transform $F_0(t)$ into $F_1(t)$, $F_1(t)$ into $F_2(t)$, and $F_0(t)$ into $F_2(t)$ and where $X^2(t)$ will be determined by the forms $X^0(t)$ and $X^1(t)$ in terms of the inside or outside of circle C. As explained above, this formulation will provide us with a representation of the Virasoro group.

(ii) If $u(t) = I$, Eqs. (3.13) will be reduced to the original Riemann–Hilbert transformation proposed by Hauser and Ernst,[1] i.e,

$$F(t) = X_+(t)F_0(t) \text{ in } C_+,$$
$$= X_-(t)F_0(t)u(t)^{-1} \text{ in } C_-. \quad (5.6)$$

Similar to procedure (i), we can express a successive transform of $F(t)$ as the form

$$u_2(t) = u_0(t)u_1(t), \quad (5.7)$$

where $u_0(t)$, $u_1(t)$, and $u_2(t)$, respectively, transform $F_0(t)$ into $F_1(t)$, $F_1(t)$ into $F_2(t)$, and $F_0(t)$ into $F_2(t)$. This corresponds to a representation of the Kac–Moody group.

(iii) In general, neither $v(t)\neq t$ nor $u(t)\neq I$, the Riemann–Hilbert transformation given in Sec. III, admits the representation of the semidirect product of the Kac–Moody and Virasoro groups. This expression of the Riemann–Hilbert transformation is very important and useful because it becomes possible to prove a Geroch conjecture stating that any given stationary axisymmetric vacuum space-time can be generated from Minkowski space by an infinite set of the symmetry transformations.[14]

In order to see how to apply our method to the derivation of some useful Backlund transformations, let us give a simpler example. Under our consideration, we set

$$u(t)=b \tag{5.8}$$

and

$$v(t)=\tau t+\mu, \tag{5.9}$$

where b is an element of SL $(2, R)$ independent of t and τ and μ are parameters. Thus the corresponding Riemann–Hilbert transformation can be written in the form

$$F(t)b\,F_0^{-1}(\tau t+\mu)=X_-(t). \tag{5.10}$$

Since the lhs of Eq. (5.10) is analytic on the whole t plane, we can always set

$$X_-(t)=B, \tag{5.11}$$

where B is independent of t. From $F(0)=I$, it follows that

$$B=b\,F_0^{-1}(\mu). \tag{5.12}$$

Thus we finally obtain that

$$F(t)=b\,F_0^{-1}(\mu)F_0(\tau t+\mu)b^{-1}, \tag{5.13}$$

which is the formulation of the Neugebauer Backlund transformation found by Cosgrove.[13] If we set $b=I$, the transformation (5.13) is reduced to the combination of the Maison–Cosgrove transformation.[13] We would like to emphasize that from the Hauser–Ernst formalism[1] of the Riemann–Hilbert transformation one cannot derive the Neugebauer Backlund transformation because it was proved[13] that the Neugebauer Backlund transformation lies outside of transformations of the Kac–Moody group. However, since we enlarge the Geroch group by the Virasoro group, we can state that the Neugebauer Backlund transformation is still included in the Geroch symmetry transformations.

It is interesting to compare the present formalism of the Riemann–Hilbert transformation with the Belinskii-Zakharov transformation.[15] We notice that in the Belinskii–Zakharov method some scalar functions are put into a solution to the Belinskii–Zakharov linearization equations to obtain a new solution. This treatment is similar to the present paper. However, the new η'_- and η'_- in our case are still solutions of the wave equation (2.11); this is not true for the Belinskii–Zakharov method. This is the reason why we, unlike Belinskii and Zakharov, need not redefine the determinant of the new metric to satisfy the wave equation. On the other hand, it is shown[16] that the Belinskii–Zakharov formalism does not correspond to the Hauser–Ernst formalism[1] completely. Thus we conclude that the Belinskii–Zakharov transformation must be associated with the Virasoro symmetry transformation. We need to further investigate these relationships in the future.

Finally, we point out that the approach to the Riemann–Hilbert transformation introduced in this paper can also be applied in other nonlinear systems such as the two-dimensional Heisenberg model and the nonlinear Schrödinger equation.[17] We are hopeful that we can use this new transformation to generate some new solution of the Einstein field equations, which is of much interest in physics, from the known solution.

ACKNOWLEDGMENTS

One of the authors (WL) should like to thank Professor F. Ernşt for his support and encouragement.

This research was supported in part by Grant No. PHY-8605958 from the National Science Foundation.

[1]I. Hauser and F. J. Ernst, J. Math. Phys. **21**, 1126; **21**, 1418 (1980).
[2]R. Geroch, J. Math. Phys. **12**, 918 (1971); **13**, 394 (1972).
[3]Y. S. Wu and M. L. Ge, J. Math. Phys. **24**, 1187 (1983).
[4]W. Kinnersly and D. M. Chitre, J. Math. Phys. **18**, 1538 (1977); **19**, 1926 (1978).
[5]K. Ueno and Y. Nakamura, Phys. Lett. B **109**, 273; B **117**, 208 (1982).
[6]L. Dolan, Phys. Rev. Lett. **47**, 1371 (1981).
[7]B. Y. Hou, M. L. Ge, and Y. S. Wu, Phys. Rev. D **24**, 2238 (1982); C. Devchand and D. B. Fairlia, Nucl. Phys. B **194**, 1086 (1982); Y. S. Wu, Commun. Math. Phys. **90**, 461 (1983).
[8]H. Eichenherr, Phys. Lett. B **115**, 385 (1982); H. Bohr, M. L. Ge, and T. V. Volovich, Nucl. Phys. B **260**, 701 (1985).
[9]B. Y. Hou and W. Li, Lett. Math. Phys. **13**, 1 (1987); J. Phys. A **20**, L897 (1987).
[10]W. Li, Phys. Lett. A **129**, 301 (1988).
[11]P. Goddard and D. Olive, J. Mod. Phys. A **1**, 303 (1986).
[12]G. Neugebauer, J. Phys. A **12**, L67 (1979).
[13]D. Maison, J. Math. Phys. **20**, 871 (1979); C. M. Cosgrove, *ibid.* **21**, 2417 (1980).
[14]I. Hauser and F. J. Ernst, J. Math. Phys. **22**, 1051 (1981).
[15]V. A. Belinskii and V. E. Zakharov, Sov. Phys. JETP **48**, 985 (1978).
[16]C. M. Cosgrove, J. Math. Phys. **23**, 615 (1982).
[17]W. Li, Phys. Lett. B **214**, 79 (1988).

Nuclear Physics B324 (1989) 715–728
North-Holland, Amsterdam

EXACT SOLUTION OF BELAVIN'S $Z_n \times Z_n$ SYMMETRIC MODEL

Bo-Yu HOU[1], Mu-Lin YAN[2] and Yu-Kui ZHOU[1]

Center of Theoretical Physics, CCAST (World Laboratory)
[1]*Institute of Modern Physics, Xibei University, Xian, P.R. of China*
[2]*Fundamental Physics Center, University of Science and Technology of China, Hefei, P.R. of China*

Received 1 August 1988
(Revised 9 March 1989)

Belavin's $Z_n \times Z_n$ symmetric model is studied. The transfer matrix of this model is exactly diagonalized using the quantum inverse scattering method. The Bethe ansatz equations and the eigenvalues are obtained.

1. Introduction

For most two-dimensional exactly solvable models in statistical mechanics, it is important to find exact solutions for the eigenvalue problem of the transfer matrix of the models in order to find the partition function and thermodynamic properties. These solutions have attracted a lot of research interest [20].

Since the discoveries of the Bethe ansatz method (BAM) [1,21] in early years and the quantum inverse scattering method (QISM) [2–4] recently, much remarkable work has been done on the exactly solvable models. In particular, the eight vertex model was solved exactly by Baxter [5] using BAM and by Takhtadazhan and Faddeev [6] using QISM. The multi-component generalization of the six vertex model was solved by Babelon et al. [7] and Schultz [8] using QISM.

It is the purpose of this paper to diagonalize exactly the transfer matrix of Belavin's $Z_n \times Z_n$ symmetric model by using QISM, which is a natural generalization of Baxter's symmetric eight vertex model and which was first introduced by Belavin [9] and Chudnovsky and Chudnovsky [10]. The partition function of this model has been calculated before [11,12], but, as indicated by Tracy [13], one should note that various assumptions of physical origin have still been made there and these assumptions might be more questionable given negative weights. However, the whole family of transfer matrices has been diagonalized exactly, without any assumptions being made and we find the BA equations of the model. Our results can be used to calculate not only the partition function, but also the whole spectrum of excitations. In our treatment the Jimbo, Miwa and Okado (JMO)

0550-3213/89/$03.50

vertex–IRF (interaction-round-face) correspondence [14, 15], which transforms the $Z_n \times Z_n$ symmetric vertex model of Belavin into an IRF model and reduces to the Baxter vertex–IRF correspondence [5] for $n=2$, plays a fundamental role in the construction of the eigenstates of the transfer matrix of the model.

The contents of the paper are as follows. Firstly, we review Belavin's $Z_n \times Z_n$ symmetric model and introduce the transfer matrix of the model in sect. 2. Then, in sect. 3 we analyze the JMO vertex–IRF correspondence of the model and introduce W-matrices. In sects. 4 and 5 we find the eigenvalues of eigenstates of the transfer matrix and the BA equations of the model. We write out the eigenstates of the transfer matrix explicitly, and show that they are symmetric functions of arguments. This enables us to carry out a relatively simple derivation. Finally, we present our conclusion and discussion in sect. 6. Some complicated calculations are carried out briefly in the appendices.

2. Transfer matrix of Belavin's $Z_n \times Z_n$ symmetric model

Consider a two-dimensional square lattice with M rows and N columns with periodic boundary conditions. Let S_{ij}^{kl} be the Boltzmann weight for a single vertex with bond states $i, j, k, l \in Z_n$. Following Richey and Tracy [12], Belavin's $Z_n \times Z_n$ symmetric model has Boltzmann weight [15]

$$S_{ij}^{kl}(z) = \begin{cases} h(z)\theta^{(i-j)}(z+w)/(\theta^{(i-k)}(w)\theta^{(k-j)}(z)), & \text{for } i+j=k+l \bmod n, \\ 0, & \text{otherwise}, \end{cases} \tag{2.1}$$

where

$$h(z) = \prod_{j=0}^{n-1}\theta^{(j)}(z) \Big/ \prod_{j=1}^{n-1}\theta^{(j)}(0), \qquad \theta^{(i)}(z) = \begin{pmatrix} \frac{1}{2}-i/n \\ \frac{1}{2} \end{pmatrix}(z, n\tau) \tag{2.2}$$

are the theta functions of rational characteristic $(\frac{1}{2}-i/n, \frac{1}{2})$. w is a parameter and z is a spectral parameter. The Boltzmann weights $S(z)$ satisfy the YBR (Yang–Baxter relation) [9, 17–19]

$$\begin{aligned} &S(z_1-z_2)_{i_1i_2}^{k_1k_2}S(z_1-z_3)_{k_1i_3}^{j_1k_3}S(z_2-z_3)_{k_2k_3}^{j_2j_3} \\ &\quad = S(z_2-z_3)_{i_2i_3}^{k_2k_3}S(z_1-z_3)_{i_1k_3}^{k_1j_3}S(z_1-z_2)_{k_1k_2}^{j_1j_2}, \end{aligned} \tag{2.3}$$

where the double indices k imply summations over 0 to $n-1$.

From S we construct the monodromy matrix for a row of N sites

$$T_N(z) = S(z - z_N^0) \dots S(z - z_2^0) S(z - z_1^0), \tag{2.4}$$

where $(1 \leqslant s \leqslant N)$

$$S(z - z_s^0)_{ik} = \sum_{l,j=0}^{n-1} S(z - z_s^0)_{ij}^{kl} E_{jl}.$$

E_{jl} is an $n \times n$ matrix with $(E_{jl})_{ik} = \delta_{ij}\delta_{kl}$. z_s^0 $(1 \leqslant s \leqslant N)$ are a set of constants attached to the square lattices on the row.

This monodromy matrix satisfies the remarkable permutation relation obtained from eq. (2.3)

$$S(z_1 - z_2)_{i_1 i_2}^{k_1 k_2} T_N(z_1)_{k_1 j_1} T_N(z_2)_{k_2 j_2} = T_N(z_2)_{i_2 k_2} T_N(z_1)_{i_1 k_1} S(z_1 - z_2)_{k_1 k_2}^{j_1 j_2} \tag{2.5}$$

and consequently the transfer matrix

$$t(z) = \operatorname{tr} T_N(z) \tag{2.6}$$

satisfies $[t(z), t(z')] = 0$. Thus, $t(z)$ with different spectral parameter can be diagonalized simultaneously.

3. Vertex–IRF correspondence and W-matrices

In this section, we recapitulate the JMO vertex–IRF correspondence [14, 15] for Belavin's $Z_n \times Z_n$ symmetric model and introduce W-matrices in an IRF model for the sake of convenience of further calculations.

To proceed one must prepare several notations briefly.

For details the reader is referred to the JMO papers [14, 15]. Λ_μ $(0 \leqslant \mu \leqslant n-1)$ and $\Lambda_0 = \Lambda_n$ are the fundamental weights of the affine Lie algebra $A_{n-1}^{(1)}$. Set

$$\hat{\mu} = \Lambda_{\mu+1} - \Lambda_\mu = \epsilon_\mu - \frac{1}{n} \sum_{\nu=0}^{n-1} \epsilon_\nu .$$

ϵ_μ, $(0 \leqslant \mu \leqslant n-1)$ form a set of the orthonormal vectors satisfying $\langle \epsilon_\mu, \epsilon_\nu \rangle = \delta_{\mu\nu}$ and $\langle \epsilon_\mu, \Lambda_0 \rangle = 0$. For a generic complex weight $\zeta \in \sum_{\mu=0}^{n-1} \mathbf{C} \Lambda_\mu$ we consider elements of the form

$$a \in P_\zeta = \zeta + \sum_{\mu=0}^{n-1} m_\mu \hat{\mu}, \qquad m_\mu \in \mathbb{Z}, \qquad a^{\mu\nu} = \langle a + \rho, \epsilon_\mu - \epsilon_\nu \rangle = a^\mu - a^\nu,$$

where a^μ, $0 \leqslant \mu \leqslant n-1$, contain fixed arbitrary parameters $s^\mu \in \mathbf{C}$.

We define a column vector

$$\phi_a^\mu(z) = {}^t\left[\theta^{(0)}(z - na^\mu w), \ldots, \theta^{(n-1)}(z - na^\mu w)\right].$$

Then the JMO vertex correspondence has the form [14,15]

$$S(z_1 - z_2)\phi_a^\mu(z_1) \otimes \phi_{a+\hat{\mu}}^\nu(z_2)$$
$$= \sum_\kappa W\left(\begin{matrix} a, & a+\hat{\kappa} \\ a+\hat{\mu}, & a+\hat{\mu}+\hat{\nu} \end{matrix} \middle| \frac{z_1 - z_2}{w}\right) \phi_{a+\hat{\kappa}}^{\mu+\nu-\kappa}(z_1) \otimes \phi_a^\kappa(z_2). \tag{3.1}$$

If $\mu = \nu$, the right-hand side consists of a single term $\kappa = \mu = \nu$, and if $\nu \neq \mu$ two terms $\kappa = \mu$ and ν are contributing. The $W\left(\begin{matrix} \cdot & \cdot \\ \cdot & \cdot \end{matrix} \middle| \frac{z_1 - z_2}{w}\right)$'s are the Boltzmann weights in the IRF model given in refs. [14,15] (also see eqs. (3.8)–(3.11)).

This vertex–IRF correspondence transforms Belavin's $Z_n \times Z_n$ symmetric model into the IRF model. The IRF model can however be thought of as a multi-component generalization, using elliptic functions, of the six vertex model. We expect the model is exactly solvable in the sense that the eigenstates of the transfer matrix $t(z)$ can be exactly constructed. This has been made for the case $n = 2$ [5,6].

For the sake of convenience we define $(0 \leqslant \mu \leqslant n-1)$

$$M_a^\mu(z) = \phi_a^\mu(z) \prod_{i=0}^{n-1} g_{i\mu}^{-1}(a), \tag{3.2}$$

$$g_{i\mu}(a) = \begin{cases} 1, & \text{if } i \geqslant \mu, \\ h(wa^{0\mu}), & \text{if } i = 0, \\ h(wa^{i\mu} - w), & \text{if } 0 < i < \mu. \end{cases} \tag{3.3}$$

It is easy to rewrite the JMO vertex–IRF correspondence (3.1) in the following two forms

$$S(z_1 - z_2) M_a^\nu(z_1) \otimes M_{a+\hat{\mu}}^\mu(z_2) = M_{a+\hat{\mu}}^{\mu'}(z_1) \otimes M_{a+\hat{\mu}-\hat{\mu}'}^{\nu'}(z_2) W_1(a|z_1 - z_2)_{\mu'\nu'}^{\mu\nu}, \tag{3.4}$$

$$S(z_1 - z_2) M_a^\mu(z_1) \otimes M_{a+\hat{\mu}}^\nu(z_2) = M_{a+\hat{\nu}'}^{\mu'}(z_1) \otimes M_a^{\nu'}(z_2) W_2(a|z_1 - z_2)_{\mu'\nu'}^{\mu\nu}, \tag{3.5}$$

where the double indices μ', ν' mean the summations over 0 to $n-1$. We have introduced the W-matrices

$$W_1(a|z)_{\mu'\nu'}^{\mu\nu} = \alpha_1^{\mu\nu}(a|z)\, \delta_{\mu'\nu}\, \delta_{\nu'\mu} + \beta_1^{\mu\nu}(a|z)\, \delta_{\mu\mu'}\, \delta_{\nu\nu'}, \tag{3.6}$$

$$W_2(a|z)_{\mu'\nu'}^{\mu\nu} = \alpha_2^{\mu\nu}(a|z)\, \delta_{\mu\mu'}\, \delta_{\nu\nu'} + \beta_2^{\mu\nu}(a|z)\, \delta_{\mu\nu'}\, \delta_{\mu'\nu} \tag{3.7}$$

with

$$\alpha_1^{\mu\nu}(a|z) = \frac{h(z)}{h(w)} \frac{h(wa^{\mu\nu})}{h(wa^{\mu\nu}+w)} \prod_{i=0}^{n-1} \frac{g_{i\mu}(a+\hat{\mu}-\hat{\nu})}{g_{i\mu}(a+\hat{\mu})} \prod_{i=0}^{n-1} \frac{g_{i\nu}(a+\hat{\mu})}{g_{i\nu}(a)}, \tag{3.8}$$

$$\alpha_2^{\mu\nu}(a|z) = \frac{h(z)}{h(w)} \frac{h\left(w(a-\hat{\nu})^{\nu\mu}\right)}{h\left(w(a-\hat{\nu})^{\nu\mu}+w\right)} \prod_{i=0}^{n-1} \frac{g_{i\nu}(a)}{g_{i\nu}(a+\hat{\mu})} \prod_{i=0}^{n-1} \frac{g_{i\mu}(a+\hat{\nu})}{g_{i\mu}(a)}, \tag{3.9}$$

$$\beta_1^{\mu\nu}(a|z) = \frac{h(wa^{\mu\nu}+w+z)}{h(wa^{\mu\nu}+w)}, \tag{3.10}$$

$$\beta_2^{\mu\nu}(a|z) = \beta_1^{\nu\mu}(a-\hat{\nu}|z). \tag{3.11}$$

Here eqs. (3.4) and (3.5) are valid if we impose $(a-\hat{\nu})^{\mu\mu}=0$ and $a^{\mu\mu}=0$. The vectors $\phi_a^{\mu}(z)$, $0 \leqslant \mu \leqslant n-1$, are linear independent [15]. We can define the row-vectors $M_a^{-\mu}(z)$ $(0 \leqslant \mu \leqslant n-1)$ by

$$M_a^{-\mu}(z) M_a^{\nu}(z) = \delta_{\mu\nu}, \qquad \sum_{\mu=0}^{n-1} M_a^{\mu}(z)_i M_a^{-\mu}(z)_j = \delta_{ij}. \tag{3.12}$$

Thus we can easily find the vertex–IRF correspondence for the $M_a^{-\mu}(z)$'s from eq. (3.4)

$$M_a^{-\mu}(z_1) \otimes M_{a-\hat{\mu}}^{-\nu}(z_2) S(z_1-z_2) = W_1(a-\hat{\mu}'|z_1-z_2)_{\mu\nu}^{\mu'\nu'} M_{a-\hat{\mu}'}^{-\nu'}(z_1) \otimes M_a^{-\mu'}(z_2), \tag{3.13}$$

where the double indices μ', ν' mean the summations over 0 to $n-1$.

The elements of the W-matrices are the Boltzmann weights in the IRF model. With the help of eqs. (3.4) and (3.5) and using (2.3), we can find the YBR for the W-matrices. Here only the YBRs for the W_2 are used and they read (see appendix A)

$$W_2(a+\hat{\mu}_3|z_1-z_2)_{\mu_2\mu_1}^{\alpha_2\alpha_1} W_2(a|z_1-z_3)_{\alpha_2\mu_3}^{\beta_2\alpha_3} W_2\left(a+\hat{\beta}_2|z_2-z_3\right)_{\alpha_1\alpha_3}^{\beta_1\beta_3}$$
$$= W_2(a|z_2-z_3)_{\mu_1\mu_3}^{\alpha_1\alpha_3} W_2(a+\hat{\alpha}_1|z_1-z_3)_{\mu_2\alpha_3}^{\alpha_2\beta_3} W_2(a|z_1-z_2)_{\alpha_2\alpha_1}^{\beta_2\beta_1}, \tag{3.14}$$

$$W_2(a|z_1-z_2)_{\beta_1\beta_2}^{\alpha_1\alpha_2} W_2(a+\hat{\alpha}_1|z_1-z_3)_{\beta_3\alpha_2}^{\alpha_3\mu_2} W_2(a|z_2-z_3)_{\alpha_3\alpha_1}^{\mu_3\mu_1}$$
$$= W_2\left(a+\hat{\beta}_2|z_2-z_3\right)_{\beta_3\beta_1}^{\alpha_3\alpha_1} W_2(a|z_1-z_3)_{\alpha_3\beta_2}^{\mu_3\alpha_2} W_2(a+\hat{\mu}_3|z_1-z_2)_{\alpha_1\alpha_2}^{\mu_1\mu_2}, \tag{3.15}$$

where the double indices α mean the summations over 0 to $n-1$.

4. Vacuum states and permutation relations

Like in the Baxter eight vertex model [6], we cannot obtain a local vacuum, either for $S(z)$ or for a finite product of such matrices, and the permutation relations (2.5) for $T(z)_{ij}$ do not have the simple form so that we do not construct the eigenstates of the transfer matrix $t(z)$. Consequently, the simple method used for solving $Z_n \times Z_n$ symmetric six vertex model in refs. [7,8] are not directly applicable to the case of Belavin's $Z_n \times Z_n$ symmetric model. However, the W-matrices (3.6), (3.7) can be thought of as the Boltzmann weights, by using elliptic functions, of a $Z_n \times Z_n$ symmetric six vertex model in the IRF model. Hence, with the help of the vertex–IRF correspondence (3.4), (3.5) and (3.13) we can get a solution of Belavin's $Z_n \times Z_n$ symmetric model.

Now we find the vacuum states from eq. (3.4). We multiply eq. (3.4) on the left by $M_{a+\hat{\mu}}^{-\kappa}(z_1)$. Using eq. (3.12) and setting $\mu = 0$, we have

$$\begin{aligned} &M_{a+\hat{0}}^{-\kappa}(z_1)S(z_1-z_2)M_a^{\nu}(z_1)M_{a+\hat{0}}^{0}(z_2) \\ &= \frac{h(w+z_1-z_2)}{h(w)}\delta_{\kappa,0}\delta_{\nu,0}M_a^0(z_2) \\ &\quad +(1-\delta_{\nu,0})\left[\frac{h(z_1-z_2)}{h(w)}\delta_{\kappa,\nu}M_{a+\hat{0}-\hat{\nu}}^{0}(z_2)+\beta_1^{0\nu}(a|z_1-z_2)\delta_{\kappa,0}M_a^{\nu}(z_2)\right], \\ &\qquad 0 \leqslant \kappa, \nu \leqslant n-1. \end{aligned} \tag{4.1}$$

Eq. (4.1) shows that $M_{a+\hat{0}}^0(z_2)$ is a local vacuum for $M_{a+\hat{0}}^{-\kappa}(z_1)S(z_1-z_2)M_a^{\nu}(z_1)$, which is gauge-equivalent to $S(z_1-z_2)$. This leads us to consider a new monodromy matrix $T_{b,a}(z)$ which is the gauge-equivalent to $T_N(z)$ as was done in ref. [6]. Its matrix elements are

$$T_{b,a}^{\mu,\nu}(z) = M_b^{-\mu}(z)T_N(z)M_a^{\nu}(z)$$

$$= \begin{cases} A_{b,a}(z), & \text{for} \quad \nu=\mu=0, \\ B_{b,a}^{\nu}(z), & \text{for} \quad \mu=0<\nu\leqslant n-1, \\ C_{b,a}^{\mu}(z), & \text{for} \quad \nu=0<\mu\leqslant n-1, \\ D_{b,a}^{\mu,\nu}(z), & \text{for} \quad 0<\mu,\nu\leqslant n-1. \end{cases} \tag{4.2}$$

From eqs. (2.4) and (4.1), (4.2) we find that the state

$$|0,a\rangle = M_{a+N\hat{0}}^{0}(z_N^0)\otimes\cdots\otimes M_{a+2\hat{0}}^{0}(z_2^0)\otimes M_{a+\hat{0}}^{0}(z_1^0) \tag{4.3}$$

satisfies the following relations

$$A_{a+N\hat{0},a}(z)|0,a\rangle = \prod_{s=1}^{N} \frac{h(w+z-z_s^0)}{h(w)}|0,a-\hat{0}\rangle,$$

$$D^{\mu;\nu}_{a+N\hat{0},a}(z)|0,a\rangle = \delta_{\mu\nu}\prod_{s=1}^{N} \frac{h(z-z_s^0)}{h(w)}|0,a-\hat{\mu}\rangle,$$

$$B^{\nu}_{a+N\hat{0},a}(z)|0,a\rangle \neq 0. \tag{4.4}$$

We call the $|0,a\rangle$ ($a \in P_\zeta$) a family of generating vectors for the transfer matrix of Belavin's $Z_n \times Z_n$ symmetric model. The transfer matrix $t(z)$ of eq. (2.6) can be rewritten as

$$t(z) = A_{a,a}(z) + \operatorname{tr} D_{a,a}(z) \qquad \text{for all } a \in P_\zeta. \tag{4.5}$$

In the following we will find the permutation relations we need for the construction of the eigenstates of the transfer matrix. We multiply the left-hand side of eq. (2.5) by $M_b^{-0}(z_2)_{i_2} M_{b+\hat{0}}^{-0}(z_1)_{i_1}$ and the right-hand side by $M_a^{\lambda_0}(z_1)_{j_1} M_{a+\hat{\lambda}_0}^{\lambda}(z_2)_{j_2}$, and sum i_1, i_2, j_1, j_2 from 0 to $n-1$. Using eqs. (3.5), (3.13) and (4.2), we find for all $a, b \in P_\zeta$

$$B^{\lambda_0}_{b,a}(z_1) B^{\lambda}_{b+\hat{0},a+\hat{\lambda}_0}(z_2) = B^{\lambda'_0}_{b,a}(z_2) B^{\lambda'}_{b+\hat{0},a+\hat{\lambda}'_0}(z_1) W(a|z_1-z_2)^{\lambda_0\lambda}_{\lambda'\lambda'_0}. \tag{4.6}$$

We multiply the left-hand side of eq. (2.5) by $M_{b+\hat{0}-\hat{\alpha}}^{-0}(z_2)_{i_2} M_{b+\hat{0}}^{-\alpha}(z_1)_{i_1}$ and the right-hand side by $M_a^{\lambda_0}(z_1)_{j_1} M_{a+\hat{\lambda}_0}^{\lambda}(z_2)_{j_2}$, and sum i_1, i_2, j_1 and j_2 from 0 to $n-1$. Using eqs. (3.5), (3.13) and (4.2), we now find for all $a, b \in P_\zeta$

$$\begin{aligned} D^{\alpha,\lambda_0}_{b,a}(z_1) B^{\lambda}_{b+\hat{0},a+\hat{\lambda}_0}(z_2) \\ &= \frac{h(w+z_1-z_2)}{h(z_1-z_2)} B^{\lambda'}_{b+\hat{0}-\hat{\alpha},a}(z_2) D^{\alpha,\lambda'_0}_{b+\hat{0},a+\hat{\lambda}'}(z_1) W(a|z_1-z_2)^{\lambda_0\lambda}_{\lambda'_0\lambda'} \\ &\quad - \frac{h(w)\beta_1^{\alpha,0}(b-\hat{\alpha}+\hat{0}|z_1-z_2)}{h(z_1-z_2)} B^{\lambda_0}_{b+\hat{0}-\hat{\alpha},a}(z_1) D^{\alpha,\lambda}_{b+\hat{0},a+\hat{\lambda}_0}(z_2). \end{aligned} \tag{4.7}$$

In eqs. (4.6) and (4.7) the double indices λ'_0 and λ' mean summations from 1 to $n-1$.

We multiply the left-hand side of eq. (2.5) by $M_b^{-0}(z_2)_{i_2} M_{b+\hat{0}}^{-0}(z_1)_{i_1}$ and the right-hand side by $M_a^{0}(z_2)_{j_2} M_{a-\hat{0}}^{\lambda}(z_1)_{j_1}$ and sum i_1, i_2, j_1 and j_2 from 0 to $n-1$. Using eqs. (3.4), (3.13), (4.2) and

$$M^0_{a-\hat{\lambda}}(z) = M^0_{a-\hat{\alpha}}(z), \qquad \text{for } \alpha, \lambda > 0, \tag{4.8}$$

we find for all $a, b \in \mathrm{P}_\zeta$

$$A_{b,a}(z_1) B^{\lambda}_{b+\hat{0},a+\hat{\lambda}_0}(z_2)$$

$$= \frac{h(w+z_2-z_1)}{h(z_2-z_1)} B^{\lambda}_{b,a+\hat{\lambda}_0-\hat{0}}(z_2) A_{b+\hat{0},a+\hat{\lambda}_0}(z_1)$$

$$+ \frac{h(w)\beta_1^{\lambda,0}(a-\hat{\lambda}+\hat{\lambda}_0|z_1-z_2)}{h(z_1-z_2)} B^{\lambda}_{b,a+\hat{\lambda}_0-\hat{0}}(z_1) A_{b+\hat{0},a+\hat{\lambda}_0}(z_2). \quad (4.9)$$

In eqs. (4.6) and (4.7) we have taken

$$W(a|z) = \frac{h(w)}{h(w+z)} W_2(a|z) \quad (4.10)$$

for the sake of convenience. The remaining permutation relations for the operators $T_{b,a}(z)$ can be deduced similarly.

5. The general Bethe ansatz equations

In this section, using eqs. (4.4) and (4.6)–(4.9), we construct an algebraic generalization of the BA method to find the eigenvalues of eigenstates of the transfer matrix $t(z)$.

From eqs. (4.4) and (4.6)–(4.9) it is obvious that we should examine the state

$$|z_1^1, z_2^1, \ldots, z_{p_1}^1; \theta\rangle$$

$$= \sum_{\{m_\mu\}} \exp\left(i \left\langle \hat{\lambda}_0 + \sum_{\mu=0}^{n-1} m_\mu \hat{\mu}, \theta \right\rangle \right) f(a+\hat{\lambda}_0) |a+\hat{\lambda}_0; z_1^1, z_2^1, \ldots, z_{p_1}^1\rangle \quad (5.1)$$

with

$$f(a+\hat{\lambda}_0)|a+\hat{\lambda}_0; z_1^1, z_2^1, \ldots, z_{p_1}^1\rangle$$

$$\equiv B^{\lambda_1}_{a+\hat{0},a+\hat{\lambda}_0}(z_1^1) B^{\lambda_2}_{a+2\hat{0},a+\hat{\lambda}_0+\hat{\lambda}_1}(z_2^1) \cdots$$

$$\times B^{\lambda_{p_1}}_{a+p_1\hat{0},a+\Lambda}(z_{p_1}^1) \left| 0, a+\sum_{s=0}^{p_1-1} \hat{\lambda}_s \right\rangle f(a+\hat{\lambda}_0)_{\lambda_{p_1}\cdots\lambda_2\lambda_1},$$

$$\Lambda = \sum_{s=0}^{p_1-1} \hat{\lambda}_s, \quad (5.2)$$

where the double indices λ_i $(1 \leqslant i \leqslant p_1)$ mean summations from 1 to $n-1$. $\theta = \sum_{\mu=0}^{n-1} \theta_\mu \epsilon_\mu$ is an n-dimensional constant vector. The summation over $\{m_\mu\}$ is over all possible arrangements of the m_μ $(0 \leqslant \mu \leqslant n-1)$ satisfying $m_\mu \in Z$. $f(a)$'s are numerical coefficients defined by the following equation (5.9).

To proceed further one must discuss the properties of the state (5.1). Firstly, using eqs. (3.15) and (4.6), we can verify that the state (5.1) is a symmetric function of arguments z_r^1 $(1 \leqslant r \leqslant p_1)$ (see appendix B.3). Next, since the Boltzmann weight $W(a|z)_{\mu\nu}^{\mu'\nu'}$ given by eqs. (3.7) and (4.10) obeys a conservation law $\mu' + \nu' = \mu + \nu$, we can find that the state (5.1) has a set of the conserved quantities p_i $(1 \leqslant i \leqslant n-1)$ given by the following: there are $N_i = p_i - p_{i-1}$ λ's being equal to i $(1 \leqslant i \leqslant n-2)$ and $N_{n-1} = p_{n-1}$ λ's being equal to $n-1$ in the set $(\lambda_1, \ldots, \lambda_{p_1})$. Similar discussions can be found in refs. [8,16]. The state (5.2) is dependent on $\hat{\lambda}_0$ and the state (5.1) is not due to the summation over $\{m_\mu\}$.

Now we apply to $|a + \hat{\lambda}_0; z_1^1, \ldots, z_{p_1}^1\rangle$ the operator $A_{a,a}(z)$. Using the permutation relation (4.9) for $a=b$, we commute $A_{a,a}(z)$ with B's step by step to $|0, a + \sum_{s=0}^{p_1-1} \hat{\lambda}_s\rangle$. We obtain $A_{a+p_1\hat{0}, a+\sum_{s=0}^{p_1-1}\hat{\lambda}_s}$. From eq. (4.8) we have

$$A_{a+p_1\hat{0}, a+\sum_{s=0}^{p_1-1}\hat{\lambda}_s} = A_{a+p_1\hat{0}, a+\sum_{s=1}^{p_1}\hat{\lambda}_s}, \qquad \left|0, a + \sum_{s=0}^{p_1-1} \hat{\lambda}_s\right\rangle = \left|0, a + \sum_{s=1}^{p_1} \hat{\lambda}_s\right\rangle. \tag{5.3}$$

Taking $N_i = N/n$ $(1 \leqslant i \leqslant n-1)$, we have

$$\sum_{s=1}^{p_1} \hat{\lambda}_s = \frac{N}{n} \sum_{\mu=1}^{n-1} \hat{\mu} = -\frac{N}{n} \hat{0}.$$

Thus $A_{a+p_1\hat{0}, a+\sum_{s=0}^{p_1-1}\hat{\lambda}_s}$ can be applied to $|0, a + \sum_{s=0}^{p_1-1} \hat{\lambda}_s\rangle$ by eq. (4.4) only on the conditions that $p_\mu = (1 - \mu/n)N$, $1 \leqslant \mu \leqslant n-1$. Finally, we have

$$\begin{aligned} A_{a,a}(z)&|a + \hat{\lambda}_0; z_1^1, \ldots, z_{p_1}^1\rangle \\ &= \prod_{s=1}^{p_1} \frac{h(w + z_s^1 - z)}{h(z_s^1 - z)} \prod_{s=1}^{N} \frac{h(w + z - z_s^0)}{h(w)} |a + \hat{\lambda}_0 - \hat{0}; z_1^1, \ldots, z_{p_1}^1\rangle \\ &+ \frac{h(w)\beta_1^{\lambda_1,0}(a - \hat{\lambda}_1 + \hat{\lambda}_0 | z - z_1^1)}{h(z - z_1^1)} \prod_{s=2}^{p_1} \frac{h(w + z_s^1 - z_1^1)}{h(z_s^1 - z_1^1)} \prod_{s=1}^{N} \frac{h(w + z_1^1 - z_s^0)}{h(w)} \\ &\times |a + \hat{\lambda}_0 - \hat{0}; z, z_2^1, \ldots, z_{p_1}^1\rangle + \cdots, \end{aligned} \tag{5.4}$$

where the remaining terms which are not written out explicitly can easily be obtained by the symmetric property of the left-hand side of eq. (5.4) under permutations of z_s^1 $(1 \leqslant s \leqslant p_1)$.

Similarly, we obtain

$$D_{a,a}^{\lambda_0,\lambda_0}(z) f(z+\hat{\lambda}_0)|a+\hat{\lambda}_0; z_1^1, \dots, z_{p_1}^1\rangle$$

$$= \exp(-i\langle\theta, \hat{\lambda}_0\rangle) \prod_{s=1}^{p_1} \frac{h(w+z-z_s^1)}{h(z-z_s^1)} \prod_{s=1}^{N} \frac{h(z-z_s^0)}{h(w)}$$

$$\times |a; z_1^1, \dots, z_{p_1}^1\rangle \left[T_{p_1}^{\lambda_0,\lambda_0}(a|z) f(a+\hat{\lambda}_0)\right]$$

$$- \frac{h(w)\beta_1^{\lambda_1',0}(a-\hat{\lambda}_1'+\hat{0}|z-z_1^1)}{h(z-z_1^1)} \exp(-i\langle\theta, \hat{\lambda}_0\rangle)$$

$$\times \prod_{s=2}^{p_1} \frac{h(w+z_1^1-z_s^1)}{h(z_1^1-z_s^1)} \prod_{s=1}^{N} \frac{h(z_1^1-z_s^0)}{h(w)}$$

$$\times |a; z, z_2^1, \dots, z_{p_1}^1\rangle \left[T_{p_1}^{\lambda_0,\lambda_0}(a|z_1^1) f(a+\hat{\lambda}_0)\right] - \cdots, \tag{5.5}$$

where $T_{p_1}^{\lambda_0,\lambda_0}(a|z)$ is given by $(1 \leqslant \alpha, \alpha' \leqslant n-1)$

$$T_{p_1}^{\alpha,\alpha'}(a|z)_{\lambda_{p_1}' \cdots \lambda_1'}^{\lambda_{p_1} \cdots \lambda_1} = \exp(i\langle\theta, \hat{\alpha}\rangle) W\left(a + \sum_{s=1}^{p_1-1} \hat{\lambda}_s' \middle| z - z_{p_1}^1\right)_{\alpha\lambda_{p_1}'}^{\alpha_{p_1}\lambda_{p_1}} \cdots$$

$$\times W(a+\hat{\lambda}_1'|z-z_2^1)_{\alpha_3\lambda_2'}^{\alpha_2\lambda_2} W(a|z-z_1^1)_{\alpha_2\lambda_1'}^{\alpha'\lambda_1}. \tag{5.6}$$

We multiply eqs. (5.4) and (5.5) by $\exp(i\langle\hat{\lambda}_0 + \sum_{\mu=0}^{n-1} m_\mu \hat{\mu}, \theta\rangle)$, add them and sum the resulting equations over all integers m_μ, $0 \leqslant \mu \leqslant n-1$, from $-\infty$ to ∞ and over λ_0 from 1 to $n-1$. We obtain

$$t(z)|z_1^1, \dots, z_{p_1}^1; \theta\rangle = t_0(z)|z_1^1, \dots, z_{p_1}^1; \theta\rangle \tag{5.7}$$

with the eigenvalue

$$t_0(z) = \prod_{s=1}^{p_1} \frac{h(w+z_s^1-z)}{h(z_s^1-z)} \prod_{s=1}^{N} \frac{h(w+z-z_s^0)}{h(w)} \exp(i\langle\hat{0}, \theta\rangle)$$

$$+ \prod_{s=1}^{p_1} \frac{h(w+z-z_s^1)}{h(z-z_s^1)} \prod_{s=1}^{N} \frac{h(z-z_s^0)}{h(w)} t_1(z), \tag{5.8}$$

provided that

$$\sum_{\lambda_0=1}^{n-1} T_{p_1}^{\lambda_0\lambda_0}(a|z) f(a+\hat{\lambda}_0) = t_1(z) f(a), \tag{5.9}$$

$$f(a+\hat{0}) = f(a), \tag{5.10}$$

and these z_r^1 $(1 \leqslant r \leqslant p_1)$ satisfy

$$\exp(i\langle\hat{0},\theta\rangle) \prod_{s=1}^{N} \frac{h(w+z_r^1-z_s^0)}{h(z_r^1-z_s^0)} = \prod_{s=1}^{p_1} \frac{h(w+z_r^1-z_s^1)}{h(w+z_s^1-z_r^1)} t_1(z_r^1). \tag{5.11}$$

Eq. (5.9) can be solved by using QICM (see eq. (B.1)). The results are that $t_1(z)$ is given by the recurrence relations $(1 \leqslant \mu \leqslant n-2)$:

$$t_\mu(z) = \prod_{s=1}^{p_{\mu+1}} \frac{h(w+z_s^{\mu+1}-z)}{h(z_s^{\mu+1}-z)} \exp(i\langle\hat{\mu},\theta\rangle)$$

$$+ \prod_{s=1}^{p_{\mu+1}} \frac{h(w+z-z_s^{\mu+1})}{h(z-z_s^{\mu+1})} \prod_{s=1}^{p_\mu} \frac{h(z-z_s^\mu)}{h(w+z-z_s^\mu)} t_{\mu+1}(z),$$

$$t_{n-1}(z) = \exp(i < \widehat{n-1}, \theta) \tag{5.12}$$

with

$$p_\mu = (1-\mu/n)N, \qquad \mu = 1,2,\ldots,n. \tag{5.13}$$

These $z_r^{\mu+1}$, $1 \leqslant r \leqslant p_{\mu+1}$, $0 \leqslant \mu \leqslant n-2$, satisfy

$$\exp\left(i\langle\hat{\mu}-(\widehat{\mu+1}),\theta\rangle\right) \prod_{s=1}^{p_\mu} \frac{h(w+z_r^{\mu+1}-z_s^\mu)}{h(z_r^{\mu+1}-z_s^\mu)}$$

$$= \prod_{s=1}^{p_{\mu+1}} \frac{h(w+z_r^{\mu+1}-z_s^{\mu+1})}{h(w+z_s^{\mu+1}-z_r^{\mu+1})} \prod_{s=1}^{p_{\mu+2}} \frac{h(w+z_s^{\mu+2}-z_r^{\mu+1})}{h(z_s^{\mu+2}-z_r^{\mu+1})}, \tag{5.14}$$

where we have used the conventions that $p_0 = N$, $p_n = 0$ and $\prod_{s=1}^{0} \ldots = 1$. Eq. (5.11) is contained in (5.14).

The equations (5.14) are just the BA equations of Belavin's $Z_n \times Z_n$ symmetric model.

6. Conclusion and discussion

In this paper we have presented the QISM for Belavin's $Z_n \times Z_n$ symmetric model. The eigenvalues of the transfer matrix and the BA equations of the model have been found. For a special case of $n=2$ our results coincide with those of Baxter's eight vertex model [5,6].

The convergence of the series (5.1), which θ limits, requires a special discussion. This is not studied here.

There are many solutions z_r^{μ}, $1 \leqslant r \leqslant p_{\mu}$, $1 \leqslant \mu \leqslant n-1$, of the BA equations (5.14), corresponding to the different eigenvalues $t_0(z)$ of the transfer matrix. The BA equations (5.14) are quite complicated but, in order to calculate the partition function and discuss the thermodynamic properties of the model, one must study the solutions of the BA equations (5.14).

We would like to thank Professors M. Jimbo, T. Miwa and M. Okado for their preprints, and Professors B.H. Zhao and F.C. Pu for their interest in this work and discussions. One of us, B.Y. Hou, wishes to thank Professor C.A. Tracy for discussions. This work was supported in part by the science fund of the Chinese Academy of Science.

Appendix A

Here we will show the YBR (3.14) and (3.15).

Firstly, we multiply both sides of eq. (2.3) with $M_a^{\nu_1}(z_1)_{j_1} M_{a+\hat{\nu}_1}^{\nu_2}(z_2)_{j_2} \times M_{a+\hat{\nu}_1+\hat{\nu}_2}^{\nu_3}(z_3)_{j_3}$, then sum j_1, j_2 and j_3 from 0 to $n-1$. Using (3.5), we find

$$M_{a+\hat{\mu}_1+\hat{\mu}_3}^{\mu_2}(z_1)_{i_2} M_{a+\hat{\mu}_3}^{\mu_1}(z_2)_{i_2} M_a^{\mu_3}(z_3)_{i_3} S(\mu_1, \mu_2, \mu_3; a; z_1, z_2, z_3) = 0, \quad \text{(A.1)}$$

where $S(\mu_1, \mu_2, \mu_3; a; z_1, z_2, z_3)$ stand for the difference between the left-hand side and the right-hand side of eq. (3.14). Because the $M_a^{\mu}(z)$'s are linearly independent, we conclude from eq. (A.1) that $S(\mu_1, \mu_2, \mu_3; a; z_1, z_2, z_3) = 0$ as desired. This leads to eq. (3.14).

Next, we replace definition (3.3) by

$$g_{i\mu}(a) = \begin{cases} 1, & \text{if } i \geqslant \mu, \\ h(wa^{0\mu} - w), & \text{if } i = 0, \\ h(wa^{i\mu}), & \text{if } 0 < i < \mu. \end{cases}$$

Similarly, we can obtain another YBR. This is exactly equivalent to eq. (3.15).

Appendix B

(1) Eq. (5.9) can be constructed by using QISM. Using eqs. (3.14), (4.10) and (5.6), we can obtain the YBR for $T_{p_1}(a|z)$

$$W(a|z_1 - z_2)_{\lambda\sigma}^{\lambda'\sigma'} T_{p_1}^{\lambda'\rho}(a|z_1) T_{p_1}^{\sigma'\eta}(a+\hat{\rho}|z_2)$$

$$= T_{p_1}^{\sigma\sigma'}(a|z_2) T_{p_1}^{\lambda\lambda'}(a+\hat{\eta}'|z_1) W(a|z_1 - z_2)_{\lambda'\sigma'}^{\rho\eta}, \quad \text{(B.1)}$$

where the double indices λ', σ' are summed from 1 to $n-1$. In eq. (B.1) the conditions $p_\mu = (1-\mu/n)N$, $1 \leqslant \mu \leqslant n-1$, have been considered.

It is obvious that we should examine the state

$$f(a)_{\lambda_{p_1}\dots\lambda_2\lambda_1}$$

$$= \left[T_{p_1}^{1\alpha_1}(a|z_1^2) T_{p_1}^{1\alpha_2}(a+\hat{\alpha}_1|z_2^2) \dots T_{p_1}^{1\alpha_{p_2}}\left(a + \sum_{s=1}^{p_2-1} \hat{\alpha}_s \middle| z_{p_2}^2\right)\right]_{\lambda_{p_1}\dots\lambda_1}^{1\dots1}$$

$$\times \left[T_{p_2}^{2\beta_1}(a|z_1^3) T_{p_2}^{2\beta_2}(a+\hat{\beta}_1|z_2^3) \dots T_{p_2}^{2\beta_{p_3}}\left(a + \sum_{s=1}^{p_3-1} \hat{\beta}_s \middle| z_{p_3}^3\right)\right]_{\alpha_{p_2}\dots\alpha_1}^{2\dots2} \cdots$$

$$\times \left[T_{p_{n-2}}^{n-2,n-1}(a|z_1^{n-1}) \dots T_{p_{n-2}}^{n-2,n-1}\left(a + (N/n-1)(\widehat{n-1}) | z_{p_{n-1}}^{n-1}\right)\right]_{\rho_{p_{n-2}}\dots\rho_1}^{n-2,\dots,n-2}, \tag{B.2}$$

where the set of double indices are summed for $(\alpha_1 \dots \alpha_{p_2})$ from 2 to $n-1$, for $(\beta_1 \dots \beta_{p_3})$ from 3 to $n-1, \dots,$ for $(\rho_1 \dots \rho_{p_{n-2}})$ from $n-2$ to $n-1$.

We apply to $f(a+\hat{\lambda}_0)$ the operator $T_{p_1}^{\lambda_0\lambda_0}(a|z)$. Using the YBR (B.1), we can obtain (5.9) with $t_1(z)$ given by the recurrence relations (5.12) provided that z_r^μ ($1 \leqslant r \leqslant p_\mu, 2 \leqslant \mu \leqslant n-1$) satisfy eq. (5.14) and $p_\mu = (1-\mu/n)N$, $1 \leqslant \mu \leqslant n$.

2. From eq. (B.2) it can be seen easily that

$$f(a+\hat{0}) = f(a). \tag{B.3}$$

3. From eq. (3.15) we can show that

$$W\left(a + \sum_{s=1}^{i-1} \hat{\lambda}_s \middle| z_i^1 - z_{i+1}^1\right)_{\lambda'_{i+1}\lambda'_i}^{\lambda_i\lambda_{i+1}} f(a)_{\lambda_{p_1}\dots\lambda_{i+1}\lambda_i\dots\lambda_1} = f(a)_{\lambda_{p_1}\dots\lambda'_{i+1}\lambda'_i\dots\lambda_1}. \tag{B.4}$$

From eqs. (B.4) and (4.6) it is obvious that the state (5.1) is a symmetric function of arguments z_r^1, $1 \leqslant r \leqslant p_1$.

References

[1] H.A. Bethe, Z. Phys. 71 (1931) 205
[2] L.D. Faddeev, Sov. Sci. Math. Phys. C1 (1981) 107
[3] H.B. Thacker, Rev. Mod. Phys. 53 (1981) 253
[4] J. Honerkamp, P. Weber and A. Wiesler, Nucl. Phys. B152 (1979) 266
[5] R.J. Baxter, Ann. Phys. (NY) 76 (1973) 1,25,48

[6] L.A. Takhtadzhan and L.D. Faddeev, Russ. Math. Surveys 34 (1979) 11
[7] O. Babelon, H.J. de Vega and C.M. Viallet, Nucl. Phys. B200 [FS4] (1982) 266
[8] C.L. Schultz, Physica A122 (1983) 71
[9] A.A. Belavin, Nucl. Phys. B180 (1981) 189
[10] D.V. Chudnovsky and G.V. Chudnovsky, Phys. Lett. A81 (1981) 105
[11] A.A. Belavin and A.B. Zamolodchikov, Phys. Lett. B116 (1982) 165
[12] M.P. Richey and C.A. Tracy, J. Stat. Phys. 42 (1986) 311
[13] C.A. Tracy, Physica D25 (1987) 1
[14] M. Jimbo, T. Miwa and M. Okado, Lett. Math. Phys. 14 (1987) 123
[15] M. Jimbo, T. Miwa and M. Okado, preprint (1987) RIMS-594
[16] Y.K. Zhou, J. Phys. A (Math Gen.) 21 (1988) 2391
[17] A. Bovier, J. Math. Phys. 24 (1983) 631
[18] I.V. Cherednik, Sov. J. Nucl. Phys. 36 (1982) 320
[19] C.A. Tracy, Physica D14 (1985) 253
[20] R.J. Baxter, Exactly solved models in statistical mechanics (Academic Press, New York, 1982)
[21] C.N. Yang, Phys. Rev. Lett. 19 (1967) 1312

Volume 229, number 1,2 PHYSICS LETTERS B 5 October 1989

QUANTUM GROUP STRUCTURE IN THE UNITARY MINIMAL MODEL

Bo-Yu HOU [a], Ding-Ping LIE [b] and Rui-Hong YUE [a]

[a] *Institute of Modern Physics, Northwest University, Xian, P.R. China*

[b] *Department of Physics, TsiHua University, Beijing, P.R. China*

Received 5 June 1989

We obtain a symmetry algebra for any unitary minimal model by using the representation of conformal field theories. This symmetry algebra can be interpreted as a quantum group. The generalization to non-unitary minimal models is direct.

In recent papers by Moore and Seiberg (MS), Alvarez-Gaumé, Gomez and Sierra (AGS) a new approach has been initiated for the understanding of the classification of rational conformal field theory by using fusion algebra and polynomial equations [1–3]. For any RCFT one can associate a series of polynomial equations which are important for the classification problem. The moduli space is the union over the fusion rule algebra of the solution of the polynomial equations and the classification of RCFTs is the classification of the equations. AGS have shown that to any RCFT we can associate a new quantum symmetry (quantum group) which neatly include both the fusion algebra and polynomial equation, and they have given a simple example (Ising model) [3]. Given a set of fusion rules, we can define an algebra of chiral vertices with sewing rules allowing us to construct conformal blocks. These blocks satisfy the duality of the conformal amplitude and provide representations of the braid and modular groups.

In ref. [4] the integral representation for conformal functions (conformal blocks) in the minimal model is discussed. The correlation function of the general conformal theory in 2D can be represented by averages of vertex operators in a Coulomb-like system with the special boundary condition that there is a fixed charge $-2\alpha_0$ placed at infinity. By use of the screen operators and the monodromy of the physical correlators, we can calculate the multipoint correlation function. In this paper, we discuss the fusion rule in the integral representation of the correlation function for the minimal model, and give the braid matrix (exchange matrix).

For conformal theory, the four-point correlation function is

$$G^{kl}_{nm} = \langle 0 | \phi_k(z_4)\phi_l(z_3)\phi_n(z_2)\phi_m(z_1) | 0 \rangle ,$$

where ϕ_k, ϕ_l, ϕ_n, ϕ_m are conformal fields. Because of the project invariance of G^{kl}_{nm}, we can fix three points at arbitrary values. The standard choice is $z_1 = 0$, $z_2 = z$, $z_3 = 1$, $z_4 = \infty$, we have

$$G^{kl}_{nm}(z, \bar{z}) = G^{mk}_{nl}(1-z, 1-\bar{z}) = G^{lm}_{nk}\left(\frac{1}{z}, \frac{1}{\bar{z}}\right) z^{-2\Delta_n} \bar{z}^{-2\bar{\Delta}_n} . \tag{1}$$

Considering the left–right symmetry of G^{kl}_{nm},

$$G^{kl}_{nm}(z, \bar{z}) = \sum C^{p}_{nm} C_{klp} | \mathscr{F}^{klnm}_{p}(z) |^2 , \tag{2}$$

C^{p}_{nm}, C_{klp} are structure coefficients (or operator product expansion coefficients). The conformal block $\mathscr{F}^{klnm}_{p}$ can be represented by vertex operators which cannot be uniquely determined [5]. So the normalization of $\mathscr{F}^{klnm}_{p}$ may be different, e.g. for $z \to 0$

0370-2693/89/$ 03.50 © Elsevier Science Publishers B.V.
(North-Holland Physics Publishing Division)

$$\mathcal{F}_p^{klnm}(z) \xrightarrow{z\to 0} z^{\Delta_p-\Delta_n-\Delta_m}[A+O(z)]\,. \tag{3}$$

For the minimal model, A can be chosen to be 1, which is consistent with the construction of the vertex operator of the model.

The central charge of the unitary minimal model is [6] $c=1-6/M(M+1)$, $M=3, 4, ...$, and the conformal dimension of the primary field is

$$h_{p,q}=\frac{[(M+1)p-Mq]^2-1}{4M(M+1)}\,, \quad p=1,...,M-1\,, \quad q=1,...,M\,, \quad p+q=\text{even}\,.$$

For convenience, we use the following notation:

$$\alpha_+=\left(\frac{M+1}{M}\right)^{1/2}, \quad \alpha_-=-\left(\frac{M}{M+1}\right)^{1/2}, \quad \alpha_\pm=\alpha_0\pm\sqrt{1+\alpha_0^2}\,, \quad \Delta_{q,p}=\tfrac{1}{4}[(p\alpha_++q\alpha_-)^2-(\alpha_++\alpha_-)^2]=h_{p,q}\,.$$

In the integral representation, the basic conformal operators are represented as exponents of the free field $\phi(z,\bar{z})$: $V_\alpha(z,\bar{z})=\mathrm{e}^{\mathrm{i}\alpha\phi}$, and the dynamics of the field are defined by the action $A[\phi]\sim\int \mathrm{d}\bar{z}\,\mathrm{d}z\,\partial_z\phi\,\partial_{\bar{z}}\phi$ with the special boundary condition that there is a charge $-2\alpha_0$ at infinity. So for the time being, we shall ignore the z dependencies altogether. The energy–momentum tensor is

$$T(z)=-\tfrac{1}{4}:\partial_z\phi\,\partial_z\phi:+\mathrm{i}\alpha_0\partial_z^2\phi\,, \quad \langle\phi(z)\phi(w)\rangle=2\ln(z-w)^{-1}\,. \tag{4}$$

$V_\alpha(z)$ has conformal dimension $\Delta_\alpha=\Delta_{\alpha-2\alpha_0}=\alpha^2-2\alpha_0\alpha$, if $\alpha_{q,p}=\frac{1}{2}(1-q)\alpha_-+\frac{1}{2}(1+p)\alpha_+$, corresponding to the primary field ϕ_{pq}. Making use of the screen operators $J_\pm(z)=V_{\alpha_\pm}(z)$, we find that the three-point correlation function $\langle\phi_{p_1q_1}\phi_{p_2q_2}\phi_{pq}\rangle$ is not vanishing only when $p_1+p_2+p=$odd, $|p_1-p_2|+1\leqslant p\leqslant\min(p_1+p_2-1, 2M-p_1-p_2-1)=p_{\max}$, $|q_1-q_2|+1\leqslant q\leqslant\min(q_1+q_2-1, 2M-q_1-q_2+1)=q_{\max}$. We can state the fusion rule

$$\phi_{p_1q_1}\times\phi_{p_2q_2}=\sum_{p=|p_1-p_2|+1}^{p_{\max}}\ \sum_{q=|q_1-q_2|+1}^{q_{\max}}[\phi_{pq}]\,, \quad p_1+p_2+p=\text{odd}, p_1+q_1=\text{even}, p_2+q_2=\text{even}\,. \tag{5}$$

The fusion algebra can be decided on with modular transformation [7]. For the minimal model, however, the method of derivation of fusion above is more direct and naive.

In RCFT, there exist two basic duality matrices describing the change in the representation content of conformal blocks. One is the fusion matrix N, the other is the braid matrix $B(\varepsilon)$, $\varepsilon=\pm1$ corresponding to the two different analytic continuations [1]. We restrict our further discussion to $\varepsilon=1$.

MS and AGS have shown that there is a relation between the braid and fusion matrices:

$$N_{pp'}\begin{bmatrix} j & k \\ i & l\end{bmatrix}=\varepsilon_{ij}^p\varepsilon_{il}^{p'}\exp[\mathrm{i}\pi(\Delta_p+\Delta_{p'}-\Delta_j-\Delta_l)]B_{pp'}\begin{bmatrix} i & k \\ j & l\end{bmatrix}=\varepsilon_{kl}^p\varepsilon_{il}^{p'}\exp[\mathrm{i}\pi(\Delta_p+\Delta_{p'}-\Delta_i-\Delta_k)]B_{pp'}\begin{bmatrix} j & l \\ i & k\end{bmatrix}. \tag{6}$$

ε_{ij}^p is the (+) or (−) sign depending on whether p appears symmetrically or antisymmetrically in the operator product of i and j. In the minimal model $\varepsilon_{ij}^p=1$, and the conformal block $\mathcal{F}_{\varphi_{pq}}^{\varphi_4\varphi_3\varphi_2\varphi_1}$ can be written as an integration:

Volume 229, number 1,2 PHYSICS LETTERS B 5 October 1989

$$\mathcal{F}^{\varphi_4\varphi_3\varphi_2\varphi_1}_{\varphi_{pq}}=z^{2\alpha_1\alpha_2}(1-z)^{2\alpha_3\alpha_2}I^{(n,m)}_{(k_1k_2)}\begin{pmatrix}a, & b, & c, & \rho \\ a', & b', & c', & \rho'\end{pmatrix}; z\Big) \Big/ M^{(n,m)}_{(k_1k_2)}\begin{pmatrix}a, & b, & c, & \rho \\ a', & b', & c', & \rho'\end{pmatrix},$$

$$I^{(n,m)}_{(k_1k_2)}\left(\begin{matrix}a, & b, & c, & \rho \\ a', & b', & c', & \rho'\end{matrix}; z\right)=\int_1^\infty \mathrm{d}u_1 \int_1^{u_1} \mathrm{d}u_2 \ldots \int_1^{u_{n-k_1-1}} \mathrm{d}u_{n-k_1} \int_0^z \mathrm{d}u_{n-k_1+1} \int_0^{u_{n-k_1+1}} \mathrm{d}u_{n-k_1+2} \ldots \int_0^{u_{n-2}} \mathrm{d}u_{n-1}$$

$$\times \int_1^\infty \mathrm{d}v_1 \ldots \int_1^{v_{m-k_2-1}} \mathrm{d}v_{m-k_2} \int_0^z \mathrm{d}v_{m-k_2+1} \ldots \int_0^{v_{m-2}} \mathrm{d}v_{m-1}$$

$$\times \prod_1^{n-1} u_i^{a'} \prod_1^{n-k_1} (u_i-1)^{b'}(u_i-z)^{c'} \prod_{n-k_1+1}^{n-1} (1-u_i)^{b'}(z-u_i)^{c'} \prod_{i<j}^{n-1} (u_i-u_j)^{2p'}$$

$$\times \prod_1^{m-1} v_i^a \prod_1^{m-k_2} (v_i-1)^b(v_i-z)^c \prod_{m-k_2+1}^{m-1} (1-v_i)^b(z-v_i)^c \prod_{i<j}^{m-1} (v_i-v_j)^{2p} \prod_{i,j} (u_i-v_j)^{-2}, \tag{7}$$

$$\alpha_i=\alpha_{n_im_i}, \quad a'=2\alpha_-\alpha_1, \quad b'=2\alpha_-\alpha_3, \quad c'=2\alpha_-\alpha_2,$$

$$a=2\alpha_+\alpha_1, \quad b=2\alpha_+\alpha_3, \quad c=2\alpha_+\alpha_2, \quad \rho'=\alpha_-^2, \quad \rho=\alpha_+^2, \tag{8}$$

$$\mathcal{F}^{\varphi_4\varphi_3\varphi_2\varphi_1}_{\varphi(p,q)}=f^{n,m}_{k_1k_2}(z)\, z^{(k_1-1)(1+a'+c'+p'(k_1-1))+k_2(1+a+c+p(k_2-1))}, \quad f^{n,m}_{k_1k_2}(z)|_{z=0}=1, \tag{9}$$

where k_1, k_2 are determined by p, q and n_1, m_1, n_2, m_2. With charge neutrality, $\sum\alpha_i=2\alpha_0$, we have $n=\frac{1}{2}(n_1+n_2+n_3-n_4)$, $m=\frac{1}{2}(m_1+m_2+m_3-m_4)$. If the intermediate state $\varphi_{p,q}$ exists, we can prove that $n_1+n_2+n_3-n_4>0$ is even, the same result for m. From the fusion rule we know

$$\mathcal{F}^{n,m}_{k_1k_2}=f^{n,m}_{k_1k_2}\, z^{h_{pq}-h_{m_1n_1}-h_{m_2n_2}}. \tag{10}$$

Comparing the exponents of (9) and (10), we get the unique solution

$$k_1=\tfrac{1}{2}(n_1+n_2+1-q), \quad k_2=\tfrac{1}{2}(m_1+m_2+1-p). \tag{11}$$

The fusion matrix is determined by the following relation:

$$\mathcal{F}^{ijkl}_p(z)=N_{pp'}\begin{bmatrix} j & k \\ i & l\end{bmatrix}\mathcal{F}^{ilkj}_{p'}(1-z). \tag{12}$$

With knowledge of the contour integration, we can verify

$$I^{(n,m)}_{(k_1k_2)}\left(\begin{matrix}a, & b, & c, & \rho \\ a', & b', & c', & \rho'\end{matrix}; z\right)=\alpha^{(n,m)}_{k_1k_2,k'_1k'_2} I^{(n,m)}_{(k'_1k'_2)}\left(\begin{matrix}b, & a, & c, & \rho \\ b', & a', & c', & \rho'\end{matrix}; 1-z\right) \tag{13}$$

and we have from (7) and (13)

$$\mathcal{F}^{\varphi_4\varphi_3\varphi_2\varphi_1}_{\varphi(p,q)}(z)=\left[M^{(n,m)}_{(k'_1k'_2)}\begin{pmatrix}b, & a, & c, & \rho \\ b', & a', & c' & \rho'\end{pmatrix} \Big/ M^{(n,m)}_{(k'_1k_2)}\begin{pmatrix}a, & b, & c, & \rho \\ a', & b', & c' & \rho\end{pmatrix}\right]\alpha^{(n,m)}_{k_1k_2,k'_1k'_2}\mathcal{F}^{\varphi_4\varphi_1\varphi_2\varphi_3}_{\varphi(p'q')}(1-z) \tag{14}$$

and

$$N_{(p,q)(p'q')}\begin{bmatrix} m_3n_3 & m_2n_2 \\ m_4n_4 & m_1n_1\end{bmatrix}=(M^{(n,m)}_{(k'_1k'_2)}/M^{(n,m)}_{(k_1k_2)})\,\alpha^{(n,m)}_{k_1k_2,k'_1k'_2},$$

$$k'_1=\tfrac{1}{2}(n_1+n_2+1-q'), \quad k'_2=\tfrac{1}{2}(m_1+m_2+1-p'), \tag{15}$$

where [4]

Volume 229, number 1,2 PHYSICS LETTERS B 5 October 1989

$$M^{(n,m)}_{(k_1k_2)}\begin{pmatrix} a, & b, & c, & \rho \\ a', & b', & c', & \rho' \end{pmatrix}=j_{n-k_1,m-k_2}\begin{pmatrix} -a-b-c-2\rho(m-2)+2(n-2), b, \rho \\ -a'-b'-c'-2\rho'(m-2)+2(n-2), b, \rho' \end{pmatrix} j_{k_1-1,k_2-1}(a,c,\rho)\,,$$

$$j_{kl}(\alpha,\beta,\rho)=\rho^{2kl}\prod_{i,j=1}^{k,l}\frac{1}{(-i+j\rho)}\prod_{i=1}^{k}\frac{\Gamma(i\rho')}{\Gamma(\rho')}\prod_{i=1}^{l}\frac{\Gamma(i\rho)}{\Gamma(\rho)}$$

$$\times\prod_{i,j=0}^{k-1,l-1}\frac{1}{(\alpha+j\rho-i)(\beta+j\rho-i)[\alpha+\beta+\rho(l-1-j)-(k-1-i)]}\prod_{i=0}^{k-1}\frac{\Gamma(1+\alpha'+i\rho')\Gamma(1+\beta'+i\rho')}{\Gamma(2-2l+\alpha'+\beta'+(k-1+i)\rho')}$$

$$\times\prod_{i=0}^{l-1}\frac{\Gamma(1+\alpha+i\rho)\Gamma(1+\beta+i\rho)}{\Gamma(2-2k+\alpha+\beta+\rho(l-1+i))}\,.$$

The important property of $\alpha^{(n,m)}_{k_1k_2k_1'k_2'}$ is factorization, e.g.

$$\alpha^{(n,m)}_{k_1k_2,k_1'k_2'}(a,b,c,\rho,a',b'c'\rho')=\alpha^{n}_{k_1k_1'}(a'b'c'\rho')\alpha^{m}_{k_2k_2'}(a,b,c,\rho)\,. \tag{16}$$

This can be seen from the formula (7); only $(u_i-v_j)^{-2}$ exists in the formula for the interaction sector of u_i, v_j. In changing integral contours, it does not give a contribution to $\alpha^{(n,m)}_{k_1k_2,k_1'k_2'}$. In order to calculate $\alpha^{n}_{k_1k_1'}$, we introduce the integration

$$J^{n}_{k_1}(a'b'c'\rho',z)=\lambda_{k_1}(\rho')I^{n}_{k_1}(a'b'c'\rho',z)$$

$$=\int_1^\infty du_1\ldots\int_1^\infty du_{n-k_1}\int_0^z du_{n-k_1+1}\ldots\int_0^z du_{n-1}\prod_1^{n-1}u_i^{a'}\prod_{i=1}^{n-k_1}(u_i-1)^{b'}(u_i-z)^{c'}$$

$$\times\prod_{i=n-k_1+1}^{n-1}(1-u_i)^{b'}(z-u_i)^{c'}\prod_{i<j}^{n-1}(u_i-u_j)^{2\rho'}\,,$$

$$\lambda_{k_1}(\rho')=\prod_{i=1}^{n-k_1}\frac{\mathscr{S}(i\rho')}{\mathscr{S}(\rho')}\prod_{i=1}^{k_1-1}\frac{\mathscr{S}(i\rho')}{\mathscr{S}(\rho')},\quad \mathscr{S}(\rho')=\mathscr{S}(\pi\rho')\,, \tag{17}$$

and define another integration

$$J(\mu,\nu,\lambda,\sigma)=\int_0^{-\infty}\ldots\int_0^{-\infty}\prod_1^{\mu}du_i\int_0^z\ldots\int_0^z\prod_{\mu+1}^{\mu+\nu}du_i\int_1^z\ldots\int_1^z\prod_{\mu+\nu+1}^{\mu+\nu+\lambda}du_i\int_1^\infty\ldots\int_1^\infty\prod_{\mu+\nu+\lambda+1}^{\mu+\nu+\lambda+\sigma}du_i$$

$$\times\prod_{i=1}^{\mu}(-u_i)^{a'}(1-u_i)^{b'}(z-u_i)^{c'}\prod_{i=\mu+1}^{\mu+\nu}(u_i)^{a'}(1-u_i)^{b'}(z-u_i)^{c'}\prod_{i=\mu+\nu+1}^{\mu+\nu+\lambda}u_i^{a'}(1-u_i)^{b'}(u_i-z)^{c'}$$

$$\times\prod_{\mu+\nu+\lambda+1}^{\mu+\nu+\lambda+\sigma}u_i^{a'}(u_i-1)^{b'}(u_i-z)^{c'}\prod_{i>j}^{\mu+\nu+\lambda+\sigma}(u_i-u_j)^{2\rho'}\,. \tag{18}$$

Let us continue analytically the integral contour in the two intervals $(0, Z)$ and $(1, \infty)$, we obtain two useful formulas

$$J(\mu,\nu,\lambda,\sigma)=\frac{\mathscr{S}[b'+c'+a'+\rho'(2\nu+\mu+\sigma+2\lambda-2)]}{\mathscr{S}[b'+c'+\rho'(2\lambda+\sigma+\nu-1)]}J(\mu+1,\nu-1,\lambda,\sigma)$$

$$+\frac{\mathscr{S}[b'+\rho'(\lambda+\nu)]}{\mathscr{S}[b'+c'+\rho'(2\lambda+\sigma+\nu-1)]}J(\mu,\nu-1,\lambda+1,\sigma)\,,$$

Volume 229, number 1,2 PHYSICS LETTERS B 5 October 1989

$$J(\mu,\nu,\lambda,\sigma)=\frac{\mathscr{S}[c'+\rho'(\lambda+\nu)]}{\mathscr{S}[b'+c'+\rho'(\sigma+2\lambda+\nu-1)]}J(\mu,\nu,\lambda+1,\sigma-1)$$
$$-\frac{\mathscr{S}[a'+\rho'(\mu+\nu)]}{\mathscr{S}[b'+c'+\rho'(\sigma+2\lambda+\nu-1)]}J(\mu+1,\nu,\lambda,\sigma-1)\,. \tag{19}$$

By direct calculation, we get

$$J(0,\nu-1,0,\sigma)=\sum_{k=1}^{\nu}\sum_{l=1}^{\sigma+1}\beta_{\nu k}^{\nu+\sigma}\gamma_{kl}^{\nu\sigma}J(\nu+\sigma+1-k-l,0,k+l-2,0)\,,$$

$$\beta_{\nu k}^{\nu+\sigma}=\prod_{i=0}^{\nu-k-1}\frac{\mathscr{S}[a'+b'+c'+2\rho'(\nu-2)+\rho'(\sigma-i)]}{\mathscr{S}[b'+c'+\rho'(\nu+k+\sigma-3-i)]}$$
$$\times\prod_{i=0}^{k-2}\frac{\mathscr{S}[b'+\rho'(\sigma+i)]}{\mathscr{S}[b'+c'+\rho'(k+\sigma+i-2)]}\frac{\prod_{i=1}^{\nu-1}\mathscr{S}(i\rho')}{\prod_{i=1}^{k-1}\mathscr{S}(i\rho')\prod_{i=1}^{\nu-k}\mathscr{S}(i\rho)}\,,$$

$$\gamma_{kl}^{\nu\sigma}=\prod_{i=0}^{\sigma-l}\frac{\mathscr{S}[1+a'+\rho'(\nu-k+i)]}{\mathscr{S}[b'+c'+\rho'(\sigma+2k+l-4-i)]}\prod_{i=0}^{l-2}\frac{\mathscr{S}[c'+\rho'(k-1-i)]}{\mathscr{S}[b'+c'+\rho'(2k+l-4+i)]}\frac{\prod_{i=1}^{\sigma}\mathscr{S}(i\rho')}{\prod_{i=1}^{l-1}\mathscr{S}(i\rho')\prod_{i=1}^{\nu-l+1}\mathscr{S}(i\rho')}\,. \tag{20}$$

We can also show that

$$J(\mu,0,\nu,0)(a',b',c',\rho',z)=(-1)^{\mu+\nu}J(0,\nu,0,\mu)(b',a',c',\rho',1-z)\,.$$

Finally we get

$$\alpha_{k_1k_1'}^{n}(a',b',c',\rho')=\sum_{\substack{\mu=1\\ \mu+\nu=k_1'+1}}^{k_1}\sum_{\nu=1}^{n-k_1+1}\beta_{k\mu}^{n}\gamma_{\mu\nu}^{k_1,n-k_1}(-1)^{n-1}\frac{\lambda_{k_1'}(\rho')}{\lambda_{k_1}(\rho')}$$
$$=\sum_{\substack{\mu=1\\ \mu+\nu=k_1'+1}}^{k_1}\sum_{\nu=1}^{n-k_1+1}\prod_{i=0}^{k_1-\mu-1}\frac{\mathscr{S}[1+a'+b'+c'+2\rho'(k_1-2)+\rho'(n-k_1-i)]}{\mathscr{S}[b'+c'+\rho'(n+\mu-3-i)]}$$
$$\times\prod_{i=0}^{\mu-2}\frac{\mathscr{S}[1+b'+\rho'(n-k_1+i)]}{\mathscr{S}[b'+c'+\rho'(\mu+n-k_1-2+i)]}$$
$$\times\prod_{i=0}^{n-k_1-\nu}\frac{\mathscr{S}[2+a'+\rho'(k_1-\mu+i)]}{\mathscr{S}[b'+c'+\rho'(n-k_1+2\mu+\nu-4-i)]}\prod_{i=0}^{\nu-2}\frac{\mathscr{S}[1+c'+\rho'(\mu-1-i)]}{\mathscr{S}[b'+c'+\rho'(2\mu+\nu-4+i)]}$$
$$\times\frac{\prod_{i=1}^{k_1'-1}\mathscr{S}(i\rho')\prod^{n-k_1'}\mathscr{S}(i\rho')}{\prod_{i=1}^{k_1-\mu}\mathscr{S}(i\rho')\prod_{i=1}^{n-k_1-\nu+1}\mathscr{S}(i\rho')\prod_{i=1}^{\nu-1}\mathscr{S}(i\rho')\prod_{i=1}^{\mu-1}\mathscr{S}(i\rho')}\,, \tag{21}$$

and

$$N_{(p,q)(p'q')}\begin{bmatrix} m_3n_3 & m_2n_2\\ m_4n_4 & m_1n_1\end{bmatrix}$$
$$=\left[M_{(k_1'k_2')}^{(n,m)}\begin{pmatrix} b, & a, & c, & \rho\\ b', & a', & c' & \rho'\end{pmatrix}\Big/M_{(k_1k_2)}^{(n,m)}\begin{pmatrix} a, & b, & c, & \rho\\ a', & b', & c', & \rho'\end{pmatrix}\right]\alpha_{k_1k_1'}^{n}(a',b',c',\rho')\alpha_{k_2k_2'}^{m}(a,b,c,\rho)\,. \tag{22}$$

From this formula and (6), we can get any braid matrix B. The matrices N and B must satisfy a series of polynomial equations. For the Wess–Zumino–Witten model, the problem is complicated. It is an extended con-

Volume 229, number 1,2 PHYSICS LETTERS B 5 October 1989

formal field theory and has more symmetries in addition to conformal symmetry. But for SU(2) WZW some special case $B_{pp'}\left[\begin{smallmatrix} 1/2 & j_2 \\ j_4 & j \end{smallmatrix}\right]$ is obtained, which is SU(2, q). It is not clear whether there is a relation between the unitary minimal model and an unknown quantum deformation of some algebra.

We thank Professor Bo-Yuan Hou for useful discussions. One of us (D.P.L.) is grateful to the Institute of Modern Physics Northwest University for the hospitality extended to him, and he would like to thank Professor K.C. Chou for supporting him doing this.

References

[1] G. Moore and N. Seiberg, preprint IASSNS-HEP-88/8.
[2] G. Moore and N. Seiberg, Nucl. Phys. B 313 (1989) 16.
[3] L. Alvarez-Gaumé, C. Gomez and G. Sierra, preprint CERN TH-5129/88; Phys. Lett. B 220 (1989) 142.
[4] V.S. Dotsenko and V.A. Fateev, Nucl. Phys. B 240 (1984) 312; B 251 (1985) 691; Phys. Lett. B 154 (1985) 291.
[5] A. Tsuchiya and Y. Kanie, Lett. Math. Phys. 13 (1987) 303.
[6] D.H. Friedan, Z. Qiu and S. Shenker, Phys. Rev. Lett. 52 (1984) 1575.
[7] E. Verlinde, Nucl. Phys. B 300 (1988) 360.

Nuclear Physics B345 (1990) 659–684
North-Holland

THE CROSSING MATRICES OF WZW SU(2) MODEL AND MINIMAL MODELS WITH THE QUANTUM 6j SYMBOLS

Bo-Yu HOU[1], Kang-Jie SHI[1,2], Pei WANG[1] and Rui-Hong YUE[1,2]

[1]*Institute of Modern Physics, Northwest University, Xian, Shaanxi 710069, China*

[2]*Center of Theoretical Physics, CCAST (World Laboratory), P.O. Box 8730, Beijing 100080, China*

Received 12 January 1990
(Revised 21 May 1990)

We directly derive the crossing (fusion and braiding) matrices for arbitrary isospins of the WZW SU(2) model in the Feigin–Fuchs integral representation and prove explicitly that they are quantum Racah–Wigner 6j matrices under proper normalization conditions. For integer k of the Kac–Moody algebra, the matrices may be truncated and the admissible states are closed under analytic continuation. The non-admissible terms will not appear in the other channels after braiding. The consistency of crossing matrices with the locality condition is also proven. Similar results are obtained for the minimal models, which are in agreement with the results obtained by G. Felder, J. Frölich and G. Keller.

1. Introduction

The crossing matrices (CM) [1] (including fusion and braiding matrices) are important for determining the structure and the classification of the conformal fields [2]. In particular, as a realization of the braid group [3], CM are found [4,5] to be closely related to the quantum groups, Chern–Simons fields [6] and solvable statistical models [7,8], and thus have attracted renewed interest.

Part of the CM results for the WZW SU(2) model [9,10] were obtained by several authors. For the four-point function with one point belonging to the vector representation, the crossing matrix has been shown to be essentially a quantum 6j matrix [3]. Some matrix elements for the general case have also been obtained [11].

Recently, Felder et al. obtained all the braiding matrices for the minimal models. They showed that these matrices are essentially proportional to the direct products of two quantum 6j matrices [12]. However, to our knowledge no such complete results for the WZW SU(2) model have been reported so far.

In principle, one can derive CM of WZW SU(2) model for high representations from the vector representation by polynomial relations [2]. Since the elements of

CM are essentially quantum $6j$ symbols for the vector representations and the fusion rules of conformal fields are similar to that of quantum groups, the elements of CM for high representations should also be essentially quantum $6j$ symbols. But they are different in some coefficients depending on the normalization condition. With such coefficients, CM may not satisfy the polynomial equations. Thus it is interesting to find these coefficients explicitly.

In a bootstrap procedure due to Dotsenko and Fateev [13], one can determine the constants for the locality condition of the four-point functions of WZW model [11]. Similar to the minimal models, the equations of these constants are overdetermined. Thus the self-consistency of the equations also needs to be verified.

Furthermore, due to the integer k (central charge of Kac–Moody algebra), the quantum $6j$ coefficients may diverge. This indicates the decoupling of the Knizhnik–Zamolodchikov equations [14], which results in a truncation of intermediate isospins under the restriction of the Kac–Moody algebra. Therefore it is unclear whether the above process for the CM and the quantum $6j$ symbols still keeps consistence and coincidence.

In this paper, we directly derive the CM for general isospins of WZW SU(2) model in the Feigin–Fuchs integral representation. We study the explicit mechanism for the above truncation and obtain consistent results under the restriction of integer k. We show that the CM of WZW SU(2) model are expressed as $M_{\text{WZW}} = \gamma_1^{-1} S \gamma_2$, where S is the quantum $6j$ matrix and the γ's are diagonal matrices, which reflect the relation between the Feigin–Fuchs integral and the normalized conformal block. Furthermore, our derivation shows that these quantum $6j$ symbols come merely from the Feigin–Fuchs integral of Dotsenko–Fateev type [13] and the results may be applied to a large class of fields. Since the quantum $6j$ matrices are orthogonal, we show that the conditions for locality [11, 13] can be satisfied consistently. For small (compare to the isospins of the fields) integer k of Kac–Moody algebra, the CM are truncated naturally. The admissible states are closed under analytic continuation and the non-admissible terms will not appear in the other channels after braiding. Our results also imply that a properly truncated quantum $6j$ matrix for q being a root of unity may still satisfy polynomial equations, in agreement with refs. [12, 15–17].

In sect. 2, we first briefly review the Feigin–Fuchs integrals for WZW SU(2) model given by Christ and Flume [11] (and equivalently by Fateev and Zamolodchikov [18]; they are also obtained by screened operators [19–22]*). We next use a method of Dotsenko and Fateev [13] to obtain the CM of the four-point functions. We then relate the CM to the quantum $6j$ symbols in sect. 3. We discuss the truncation of the CM due to the restriction of integer k in sect. 4 and the locality of four-point functions in sect. 5. In sect. 6, using our approach we study the CM for the minimal models, which are in agreement with Felder et al. [12].

* Recently, Bilal obtained the crossing matrices for the W-algebra [23]. For W-algebras, see ref. [24].

2. The Feigin–Fuchs integral and the crossing matrices

The conformal block of WZW SU(2) four-point function can be expressed as [11]

$$\langle \varphi_{1j_1}(\eta)\varphi_{2j_2}(0)\varphi_{3j_3}(1)\varphi_{4j_4}(\infty)\rangle \sim \sum_{l=1}^{n} \eta^{2j_1j_2/(k+2)}(1-\eta)^{2j_1j_3/(k+2)} f^{(l)}(\eta)\, T^{(l)}_{j_1j_2j_3j_4},$$

where $T^{(l)}_{j_1j_2j_3j_4}$ denote SU(2)-invariant tensors built out of Clebsch–Gordan coefficients, the total number of $T^{(l)}$ is n, and k is the Kac–Moody central charge.

The functions $f^{(l)}(\eta)$ satisfying the reduced Knizhnik–Zamolodchikov equation (RKZE)

$$\partial_\eta f^{(l)} = -\frac{2}{(k+2)} \sum_{l'} \left(\frac{\Sigma_a t_1^a \otimes t_2^a + (j_1\, j_2)\mathbb{1}}{\eta} + \frac{\Sigma_a t_1^a \otimes t_3^a + (j_1\, j_3)\mathbb{1}}{\eta - 1} \right)_{ll'} f^{(l')}$$
$$\equiv \sum_{l'} G_{ll'} f^{(l')}, \tag{2.1}$$

with t_i^a being the SU(2) generators, are obtained as Feigin–Fuchs integrals,

$$f^{(l)} = \eta^{l-1} \int S\Bigg[\prod_{i=1}^{n-1} \mathrm{d}t_i \prod_{i<j} (t_i - t_j)^{2\rho}$$
$$\times \prod_{i=1}^{l-1} t_i^{\alpha\rho - 1}(t_i - 1)^{\beta\rho}(t_i - \eta)^{\gamma\rho - 1} \prod_{i=l}^{n-1} t_i^{\alpha\rho}(t_i - 1)^{\beta\rho - 1}(t_i - \eta)^{\gamma\rho} \Bigg], \tag{2.2}$$

where S denotes the symmetrizing of the t's. The parameters are

$$\alpha = -2j_2, \quad \beta = -2j_3, \quad \gamma = -2j_1, \quad \rho = 1/(k+2), \tag{2.3a}$$

for

$$j_1 \leqslant j_2 \leqslant j_3 \leqslant j_4 = \sum_{i=1}^{3} j_i + 1 - n, \quad n \leqslant 2j_1 + 1. \tag{2.3b}$$

Define

$$\tilde{j}_1 = \tfrac{1}{2}(j_1 + j_2 + j_3 - j_4), \qquad \tilde{j}_2 = \tfrac{1}{2}(j_1 + j_2 + j_4 - j_3),$$
$$\tilde{j}_3 = \tfrac{1}{2}(j_1 + j_3 + j_4 - j_2), \qquad \tilde{j}_4 = \tfrac{1}{2}(j_2 + j_3 + j_4 - j_1). \tag{2.4}$$

Then eq. (2.3) is expressed as

$$\alpha=-(\tilde{j}_1+\tilde{j}_2+\tilde{j}_4-\tilde{j}_3),\quad \beta=-(\tilde{j}_1+\tilde{j}_3+\tilde{j}_4-\tilde{j}_2),\quad \gamma=-(\tilde{j}_1+\tilde{j}_2+\tilde{j}_3-\tilde{j}_4),\tag{2.5a}$$

with

$$\tilde{j}_1\leqslant\tilde{j}_2\leqslant\tilde{j}_3\leqslant\tilde{j}_4,\qquad \tilde{j}_1+\tilde{j}_4\leqslant\tilde{j}_2+\tilde{j}_3.\tag{2.5b}$$

For a set of properly normalized bases $T^{(l)}_{\tilde{j}_1\tilde{j}_2\tilde{j}_3\tilde{j}_4}$, P. Christe and R. Flume showed that $G_{ll'}$ are unchanged under the transformation (2.4) [11]. Thus we can use

$$\alpha=-(j_1+j_2+j_4-j_3),\quad \beta=-(j_1+j_3+j_4-j_2),\quad \gamma=-(j_1+j_2+j_3-j_4)\tag{2.6a}$$

in the integrals (2.2) for the case

$$j_1\leqslant j_2\leqslant j_3\leqslant j_4,\qquad j_1+j_4\leqslant j_2+j_3.\tag{2.6b}$$

(The second condition implies $\tilde{j}_1+\tilde{j}_4\geqslant\tilde{j}_2+\tilde{j}_3$.)

Considering both (2.5) and (2.6), we change the notations for $\overset{(\sim)}{j_i}$,

$$\overset{(\sim)}{j_1}=d,\quad \overset{(\sim)}{j_2}=c,\quad \overset{(\sim)}{j_3}=f,\quad \overset{(\sim)}{j_4}=a,\tag{2.7}$$

and assign

$$\alpha=-(d+c+a-f),\quad \beta=-(d+f+a-c),\quad \gamma=-(d+c+f-a),\tag{2.8a}$$

with

$$d\leqslant c\leqslant f\leqslant a,\qquad d+a\leqslant c+f.\tag{2.8b}$$

Correspondingly the number of possible SU(2) invariants $T^{(l)}_{j_1j_2j_3j_4}$ is $n=2d+1$.

There are three contour systems for the integration (2.2) as depicted in figs. 1a–c,

$$\text{(i)}\quad \int_0^{\eta}\prod_{i=1}^{\nu}\mathrm{d}t_i\int_1^{\infty}\prod_{i=\nu+1}^{\nu+\sigma=n-1}\mathrm{d}t_i\ \ldots\sim(\vec{f}_\nu)_1,$$

$$\text{(ii)}\quad \int_{-\infty}^{0}\prod_{i=1}^{\mu}\mathrm{d}t_i\int_{\eta}^{1}\prod_{i=\mu+1}^{\mu+\lambda=n-1}\mathrm{d}t_i\ \ldots\sim(\vec{f}_\mu)_2,$$

$$\text{(iii)}\quad \int_0^{1}\prod_{i-1}^{\lambda}\mathrm{d}t_i\int_{\eta}^{\infty}\prod_{i=\lambda+1}^{\mu+\lambda=n-1}\mathrm{d}t_i\ \ldots\sim(\vec{f}_\lambda)_3,\tag{2.9}$$

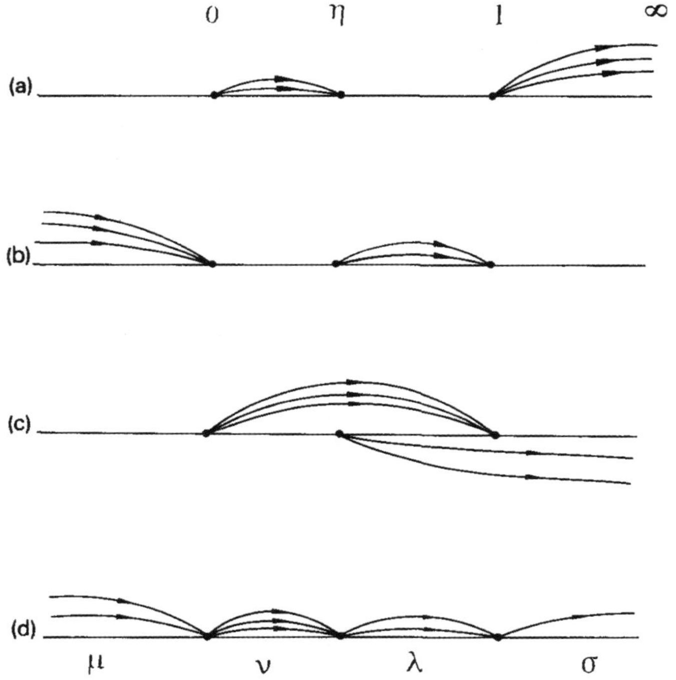

Fig. 1. (a)–(c) Contour systems for the integration in eq. (2.2). (d) Contour system for the calculation of $A(\mu, \nu, \lambda, \sigma)$.

and each system gives a complete set of n solutions of the RKZE. The solutions of system (1) get only a phase factor after analytic continuation around the point $\eta = 0$, and is said to be canonical at $\eta = 0$. System (ii), (iii) is canonical at $\eta = 1$, $\eta = \infty$, respectively. Since the solutions of each system are complete, the solutions of one system can be obtained by linear superposition of that of another system via the crossing matrix,

$$(\vec{f}_\nu)_i = \sum_\mu M^{(ij)}_{\nu\mu} (\vec{f}_\mu)_j,$$

which is the same for all components of $\vec{f}$. For simplicity, we use the first component $f^{(1)}$ to calculate the crossing matrix $M^{(12)}$. We define integrals $A(\mu, \nu, \lambda, \sigma)$ (see fig. 1d) through

$$A(\mu,\nu,\lambda,\sigma) \equiv \int_{-\infty}^{0} \prod_{i=1}^{\mu} \mathrm{d}t_i \int_0^{\eta} \prod_{i=\mu+1}^{\mu+\nu} \mathrm{d}t_i \int_\eta^1 \prod_{i=\mu+\nu+1}^{\mu+\nu+\lambda} \mathrm{d}t_i \int_1^\infty \prod_{i=\mu+\nu+\lambda+1}^{\mu+\nu+\lambda+\sigma=n-1} \mathrm{d}t_i$$

$$\times \prod_{i=1}^{n-1} t_i^{\alpha\rho} (t_i - 1)^{\beta\rho - 1} (t_i - \eta)^{\gamma\rho} \prod_{i<j} (t_i - t_j)^{2\rho},$$

with the same phase convention as that of ref. [13]. We then have

$$\left(f_\nu^{(1)}\right)_1 = A(0,\nu,0,n-\nu-1), \qquad \left(f_\mu^{(1)}\right)_2 = A(\mu,0,n-\mu-1,0).$$

Using the method of changing integration contours of t_i, introduced in ref. [13], we obtain recursion relations for $A(\mu,\nu,\lambda,\sigma)$,

$$A(\mu,\nu,\lambda,\sigma) = A(\mu,\nu-1,\lambda+1,\sigma)\frac{(-1)[\beta+\lambda+\sigma]}{[\beta+\gamma+\nu+2\lambda+\sigma-1]}$$

$$+A(\mu+1,\nu-1,\lambda,\sigma)\frac{(-1)[\alpha+\beta+\gamma+\mu+2\nu+2\lambda+\sigma-2]}{[\beta+\gamma+\nu+2\lambda+\sigma-1]},$$

$$A(\mu,\nu,\lambda,\sigma) = A(\mu,\nu,\lambda+1,\sigma-1)\frac{[\gamma+\nu+\lambda]}{[\beta+\gamma+\nu+2\lambda+\sigma-1]}$$

$$+A(\mu+1,\nu,\lambda,\sigma-1)\frac{(-1)[\alpha+\mu+\nu]}{[\beta+\gamma+\nu+2\lambda+\sigma-1]}, \tag{2.10}$$

where

$$[x] \equiv (e^{i\pi\rho x} - e^{-i\pi\rho x})/(e^{i\pi\rho} - e^{-i\pi\rho}) \equiv \frac{q^x - q^{-x}}{q - q^{-1}}. \tag{2.11}$$

By induction, one can verify that

$$B(m,s) = \prod_{i=0}^{m} \frac{1}{[N+s-i]}[N-m+2s]\frac{[m]!}{[m-s]![s]!} \tag{2.12}$$

is the solution of the recursion relation

$$B(m+1,s+1) = B(m,s)/[N-m+2s] + B(m,s+1)/[N-m+2s+2],$$

with

$$B(m,m) = [N-1]!/[N+m-1]!, \qquad B(m,0) = [N-m]!/[N]!.$$

From this relation we can derive [25]

$$
\begin{aligned}
&A(0,\nu,0,\sigma)\\
&=\sum_s A(s,0,\nu-s,\sigma)\frac{[\nu]!}{[\nu-s]![s]!}\frac{[\beta+\gamma+2\nu+\sigma-2s-1]}{\prod_{i=0}^{\nu}[\beta+\gamma+2\nu+\sigma-1-s-i]}\\
&\quad\times\prod_{i=0}^{s-1}(-1)[\alpha+\beta+\gamma+2\nu+\sigma-2-i]\prod_{i=0}^{\nu-s-1}(-1)[\beta+\sigma+i]\\
&=\sum_\mu A(\mu,0,\nu+\sigma-\mu,0)\sum_s\frac{[\nu]!}{[\nu-s]![s]!}\frac{[\sigma]!}{[\mu-s]![\sigma-\mu+s]!}\\
&\quad\times\prod_{i=0}^{s-1}[\alpha+\beta+\gamma+2\nu+\sigma-2-i]\\
&\quad\times\prod_{i=0}^{\nu-s-1}[\beta+\sigma+i]\frac{[\beta+\gamma+2\nu+\sigma-2s-1]}{\prod_{i=0}^{\nu}[\beta+\gamma+2\nu+\sigma-s-1-i]}\prod_{i=0}^{\mu-s-1}[\alpha+s+i]\\
&\quad\times\prod_{i=0}^{\sigma-\mu+s-1}[\gamma+\nu-s+i]\frac{[\beta+\gamma-2\mu+2\nu+2\sigma-1]}{\prod_{i=0}^{\sigma}[\beta+\gamma-\mu+2\nu+2\sigma-s-1-i]}(-1)^{\mu+\nu+\sigma+s}\\
&\equiv\sum_\mu a_\mu A(\mu,0,\nu+\sigma-\mu,0),
\end{aligned}
\tag{2.13}
$$

where

$$a_\mu=M_{\nu\mu}^{(12)}. \tag{2.14}$$

In sect. 3 we will relate a_μ with the quantum $6j$ symbols.

3. The relation of crossing matrices with the quantum 6*j* symbols

The recursion relation (2.10) can be depicted as in fig. 2, with

$$a'=\frac{(-1)[\beta+\lambda+\sigma]}{[\beta+\gamma+\nu+2\lambda+\sigma-1]},\qquad a''=\frac{(-1)[\alpha+\beta+\gamma+\mu+2\nu+2\lambda+\sigma-2]}{[\beta+\gamma+\nu+2\lambda+\sigma-1]}$$

and

$$b'=\frac{[\gamma+\nu+\lambda]}{[\beta+\gamma+\nu+2\lambda+\sigma-1]},\qquad b''=\frac{(-1)[\alpha+\mu+\nu]}{[\beta+\gamma+\nu+2\lambda+\sigma-1]},$$

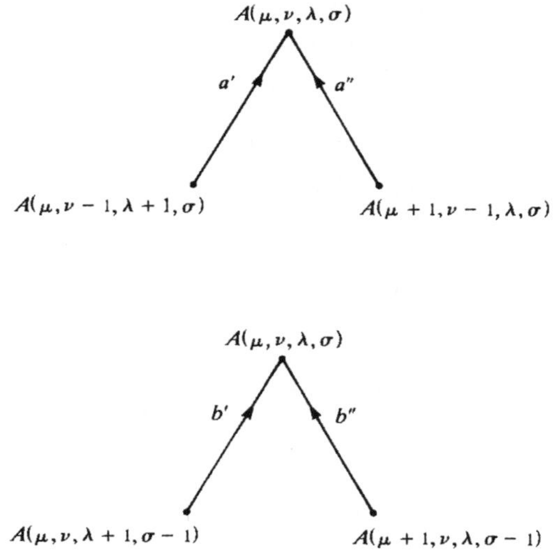

Fig. 2. Representation of the recursion relations (2.10).

which are called the factors of corresponding edges. (They should not be confused with the a and b for isospins.) Using eq. (2.10), we can express $A(0,\nu,0,\sigma)$ as a linear superposition of $A(\mu,0,\lambda,0)$ and therefore get the matrix elements of $M^{(12)}$. We may first subtract ν step-by-step to zero, and then subtract σ to zero as depicted in fig. 3. In this graph every vertex represents an $A(\mu,\nu,\lambda,\sigma)$ and every directed edge together with its factor represents a linear relation between two A's.

We define the factor of a path P as

$$g(\mathrm{P})=\frac{\prod_i a_i}{\prod_j a'_j},$$

where the a_i are factors of edges in the path with directions along the path, the a'_j denote the factors of edges in the path with directions opposite to the path. A path is admissible if all edges in it are along the path.

Now from the graph in fig. 3, we have

$$A(0,\nu,0,\sigma)=\sum_{\mu} a_\mu A(\mu,0,\lambda,0)|_{\mu+\lambda-\nu+\sigma=n-1},$$

where a_μ is the sum of all factors of admissible paths from $A(\mu,0,\lambda,0)$ to $A(0,\nu,0,\sigma)$,

$$a_\mu=M^{(12)}_{\nu\mu}=\sum_i g(\text{admissible path } \mathrm{P}_i)\,, \tag{3.1}$$

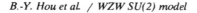

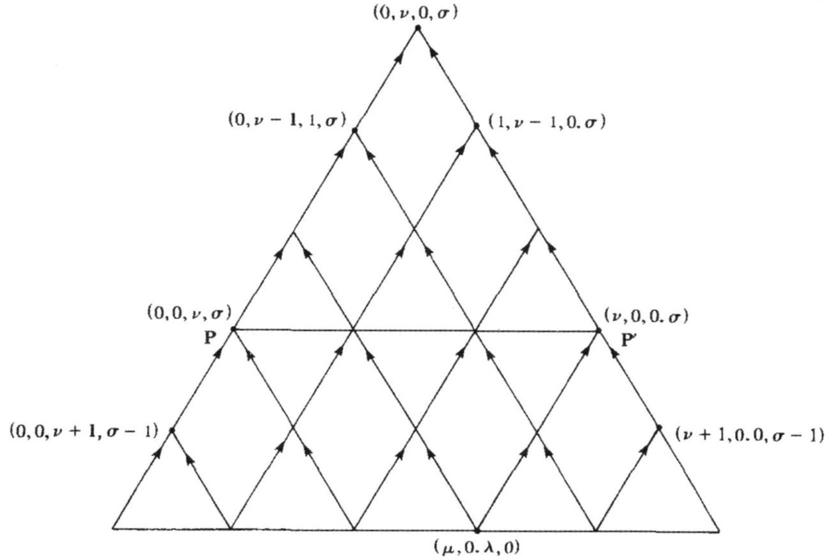

Fig. 3. Graph used for the calculation of $A(\mu,\nu,\lambda,\sigma)$.

where the sum is taken over all admissible paths from $A(\mu,0,\lambda,0)$ to $A(0,\nu,0,\sigma)$ in fig. 3. Let us call a_μ the coupling coefficient between $A(0,\nu,0,\sigma)$ and $A(\mu,0,\lambda,0)$. If a path is closed, we call it a circuit. The factor of that path is called the factor of the circuit. Obviously, if a circuit A is composed of several circuits B, C, ... with correct directions (common edges of two of them have the opposite direction), then the factor of A equals the product of the factors of B, C, ... The factor ratio of two paths having common starting points and ending points equals the factor of the circuit between these two paths.

We now assume that there are two systems A and B. The linear relations of them are represented by two topologically isomorphic graphs (the factors of corresponding edges may be different). Also, the factors of corresponding elementary circuits are equal. We say that these two graphs are equivalent. For these graphs, the factors of any corresponding circuits are equal. We can then express coupling coefficients of two points in one system by that of another system,

$$a_\mu = \sum_i g(\text{admissible path } P_i)_A ,$$

$$b_\mu = \sum_i g(\text{admissible path } P_i)_B ,$$

$$\frac{a_\mu}{b_\mu} = \frac{(g(P_0)_A)(\Sigma_i g(\text{circuit } i)_A}{(g(P_0)_B)(\Sigma_i g(\text{circuit } i)_B} = \frac{g(P_0)_A}{g(P_0)_B}, \tag{3.2}$$

where $g(P_0)$ is the factor of a given path P_0 (it does not need to be admissible) and $g(\text{circuit } i)$ are factors of circuits between P_0 and the admissible path P_i.

Recursion relations of quantum $6j$ symbols [26],

$$\begin{Bmatrix} a & b & c \\ d & e & f \end{Bmatrix}[2f+1]\{[c+d-e][c+e-d+1]\}^{1/2}$$

$$= \begin{Bmatrix} a & b & c \\ d-\frac{1}{2} & e+\frac{1}{2} & f-\frac{1}{2} \end{Bmatrix}\{[a+f-e][a+e-f+1][b+d+f+1][d+f-b]\}^{1/2}$$

$$+ \begin{Bmatrix} a & b & c \\ d-\frac{1}{2} & e+\frac{1}{2} & f+\frac{1}{2} \end{Bmatrix}\{[a+e+f+2][e+f-a+1][b+d-f][b+f-d+1]\}^{1/2},$$

$$\begin{Bmatrix} a & b & c \\ d & e & f \end{Bmatrix}[2f+1]\{[c+d+e+1][d+e-c]\}^{1/2}$$

$$= \begin{Bmatrix} a & b & c \\ d-\frac{1}{2} & e-\frac{1}{2} & f-\frac{1}{2} \end{Bmatrix}(-1)\{[a+e+f+1][e+f-a][b+d+f+1][d+f-b]\}^{1/2}$$

$$+ \begin{Bmatrix} a & b & c \\ d-\frac{1}{2} & e-\frac{1}{2} & f+\frac{1}{2} \end{Bmatrix}\{[a+e-f][a+f-e+1][b+d-f][b+f-d+1]\}^{1/2}, \tag{3.3}$$

enable us to construct a procedure to calculate a q-$6j$ symbol, such that its graph (see fig. 4) is topologically isomorphic to that of $A(\mu,\nu,\lambda,\sigma)$ (fig. 3). From $\begin{Bmatrix} a & b & c \\ d & e & f \end{Bmatrix}$ at vertex B and keeping a,b,c (we omit them in fig. 4) unchanged, we first subtract d and increase e. After ν steps, we get to the line PP′ in fig. 4. We then subtract both d and e, by σ steps and arrive at the line CD. For clarity, we make the following convention. The primed indices in the two graphs are subject to change from point to point, while the indices without prime are fixed in the whole graph. Thus ν,σ,μ,λ in fig. 3 are fixed; they are determined by the chosen $A(0,\nu,0,\sigma)$ and $A(\mu,0,\lambda,0)$. The indices $\mu',\nu',\lambda',\sigma'$ are changing in the graph. In fig. 4, a,b,c,d,e,f denote the six isospins of vertex B and are fixed, while d',e',f' are changing in the graph.

We require that $d'=0$ in the bottom line CD such that only one of the $6j$ symbols in the line CD is nonzero,

$$\begin{Bmatrix} a & b & c \\ d'=0 & e'=c & f'=b \end{Bmatrix} = \begin{Bmatrix} a & b & c \\ 0 & c & b \end{Bmatrix} \neq 0,$$

and further require that this term corresponds to $A(\mu,0,\lambda,0)$ of fig. 3. We thus have

$$d=\tfrac{1}{2}\nu+\tfrac{1}{2}\sigma=\tfrac{1}{2}\mu+\tfrac{1}{2}\lambda,\quad c=e+\tfrac{1}{2}\nu-\tfrac{1}{2}\sigma,\qquad b=f+\tfrac{1}{2}\mu-\tfrac{1}{2}\lambda. \tag{3.4}$$

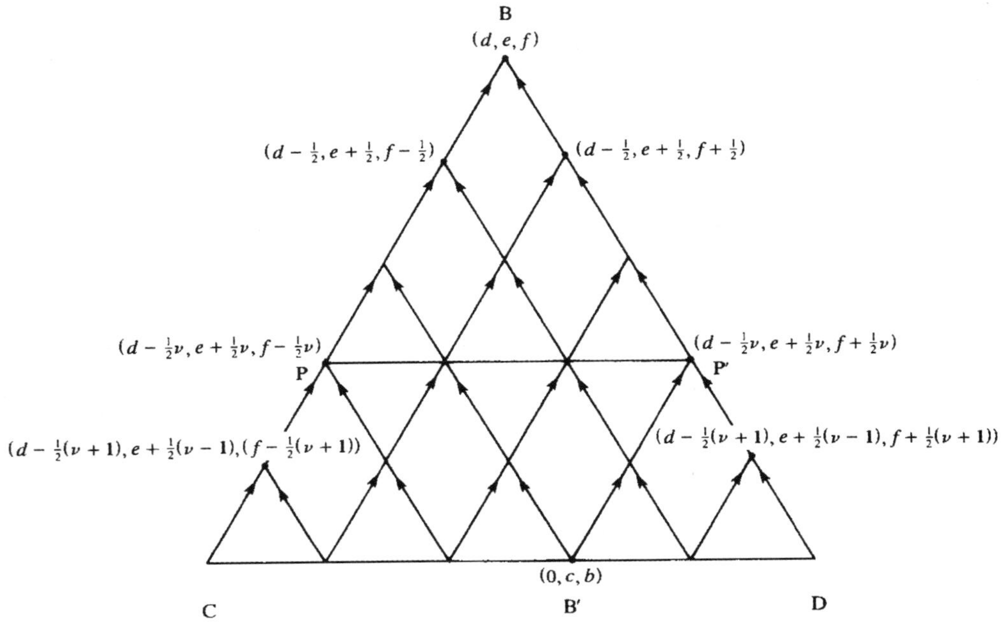

Fig. 4. Graph used for the calculation of a quantum $6j$ symbol.

Since only one term survives in the line CD, we have

$$\begin{Bmatrix} a & b & c \\ d=\frac{1}{2}\nu+\frac{1}{2}\sigma & e=c+\frac{1}{2}\sigma-\frac{1}{2}\nu & f=b+\frac{1}{2}\lambda-\frac{1}{2}\mu \end{Bmatrix} = b_\mu \begin{Bmatrix} a & b & c \\ 0 & c & b \end{Bmatrix},$$

where

$$b_\mu = \sum_i g(\text{admissible path } \mathrm{P}_i) = \begin{Bmatrix} a & b & c \\ d & e & f \end{Bmatrix} \Big/ \begin{Bmatrix} a & b & c \\ 0 & c & b \end{Bmatrix}.$$

with the sum taken over all admissible paths from B′ to B in fig. 4.

Now let the system of $A(\mu', \nu', \lambda', \sigma')$ in fig. 3 be system A and let that of $\begin{Bmatrix} a & b & c \\ d' & e' & f' \end{Bmatrix}$ in fig. 4 be system B. If we can properly choose a, c, f [d, e, b are then determined by eq. (3.4)] such that the factors of every corresponding elementary circuit in the two systems are equal, then from eq. (3.2) we have

$$\frac{a_\mu}{b_\mu} = \frac{g(\mathrm{P}_0)_\mathrm{A}}{g(\mathrm{P}_0)_\mathrm{B}},$$

where $g(\mathrm{P}_0)$'s are factors of the same path P_0 in two graphs. The crossing (fusion) matrix $M^{(12)}$ is obtained through

$$M^{(12)}_{\nu\mu}=a_\mu=\frac{g(\mathrm{P}_0)_\mathrm{A}\begin{Bmatrix}a&b&c\\d&e&f\end{Bmatrix}}{g(\mathrm{P}_0)_\mathrm{B}\begin{Bmatrix}a&b&c\\0&c&b\end{Bmatrix}}$$

$$\equiv\frac{1}{\Gamma}\begin{Bmatrix}a&b&c\\d&e&f\end{Bmatrix}. \qquad (3.5)$$

We now claim that if a, f, c, d are the same quantities as in (2.8a) and d, e, b satisfy eq. (3.4), then the two graphs are equivalent.

The factors of elementary circuits in fig. 3 beyond the crossing line PP′ are obtained from eq. (2.10),

$$(g_1)_\mathrm{A}=\frac{[-\beta-\gamma-\nu'-2\lambda'-\sigma']}{[-\beta-\gamma-\nu'-2\lambda'-\sigma'+2]},$$

$$(g_2)_\mathrm{A}=\frac{[-\beta-\gamma-\nu'-2\lambda'-\sigma']}{[-\beta-\gamma-\nu'-2\lambda'-\sigma'+2]}, \qquad (3.6a)$$

where $(g_1)_\mathrm{A}$ is the factor above PP′ (where ν' changes) and $(g_2)_\mathrm{A}$ the factor under PP′. The indices $\mu', \nu', \lambda', \sigma'$ are for the top vertex of an elementary circuit. The factor of an elementary circuit in the crossing line PP′ is

$$(g_3)_\mathrm{A}=\frac{[-\beta-\gamma-\nu'-2\lambda'-\sigma']}{[-\beta-\gamma-\nu'-2\lambda'-\sigma'+2]}$$
$$\times(-1)\frac{[\gamma+\nu'+\lambda'-1][-\alpha-\beta-\gamma-2\nu'-2\lambda'-\mu'-\sigma'+2]}{[-\beta-\lambda'-\sigma'][\alpha+\mu'+\nu'-1]}. \qquad (3.6b)$$

We can similarly calculate factors of elementary circuits for the graph of $6j$ symbols from eq. (3.3), giving

$$(g_1)_\mathrm{B}=\frac{[2f']}{[2f'+2]}, \qquad (g_2)_\mathrm{B}=\frac{[2f']}{[2f'+2]}, \qquad (3.7a)$$

$$(g_3)_\mathrm{B}=\frac{[2f']}{[2f'+2]}(-1)\frac{[a-e'-f'-1][a+e'+f'+2]}{[a+f'-e'][f'-a-e'-1]}, \qquad (3.7b)$$

where $\begin{Bmatrix}a&b&c\\d'&e'&f'\end{Bmatrix}$ are indices of the top vertex of an elementary circuit (see fig. 4). Note that a, b, c are constants in this graph.

To ensure $(g_i)_A = (g_i)_B$ the sufficient conditions are

$$\begin{aligned} 2f' &= -\beta - \gamma - \nu' - 2\lambda' - \sigma', \\ a - e' - f' - 1 &= \gamma + \nu' + \lambda' - 1, \\ f' - a - e' - 1 &= \alpha + \mu' + \nu' - 1, \\ a + e' + f' + 2 &= -\alpha - \beta - \gamma - 2\nu' - 2\lambda' - \mu' - \sigma' + 2, \\ a + f' - e' &= -\beta - \lambda' - \sigma' \end{aligned} \tag{3.8a}$$

at the top vertices of the elementary circuits. Eqs. (3.8a) are equivalent to

$$\begin{aligned} 2a &= -\alpha - \beta - \mu' - \nu' - \lambda' - \sigma', \\ 2e' &= -\alpha - \gamma - \mu' - 2\nu' - \lambda', \\ 2f' &= -\beta - \gamma - \nu' - 2\lambda' - \sigma'. \end{aligned} \tag{3.8b}$$

We know from eq. (3.4) that the sum of $\mu', \nu', \lambda', \sigma'$ is

$$\mu' + \nu' + \lambda' + \sigma' = n - 1 = 2d.$$

Also e' is related to c by construction of the graph,

$$e' = e'|_{\text{CD line}} + \tfrac{1}{2}\sigma' - \tfrac{1}{2}\nu' = c + \tfrac{1}{2}\sigma' - \tfrac{1}{2}\nu'$$

and f' is related to f by

$$f' = f + \tfrac{1}{2}\mu' - \tfrac{1}{2}\lambda'.$$

Thus eq. (3.8b) is equivalent to

$$2a = -\alpha - \beta - 2d, \quad 2c = -\alpha - \gamma - 2d, \quad 2f = -\beta - \gamma - 2d$$

and these equations are satisfied by the assumptions of α, β, γ appearing in eq. (2.2) [see eq. (2.8a)]. Therefore the graphs of $A(\mu', \nu', \lambda', \sigma')$ and $6j$ symbols are equivalent. The matrix element of $M^{(12)}$ is given by eq. (3.5),

$$M^{(12)}_{\nu\mu} = \frac{1}{\Gamma} \begin{Bmatrix} a & b(\mu) & c \\ d & e(\nu) & f \end{Bmatrix}, \tag{3.9a}$$

where

$$b = f - d + \mu, \qquad e = c + d - \nu, \tag{3.9b}$$

$$\Gamma = \begin{Bmatrix} a & b & c \\ 0 & c & b \end{Bmatrix} \frac{g(\mathrm{P}_0)_B}{g(\mathrm{P}_0)_A}. \tag{3.9c}$$

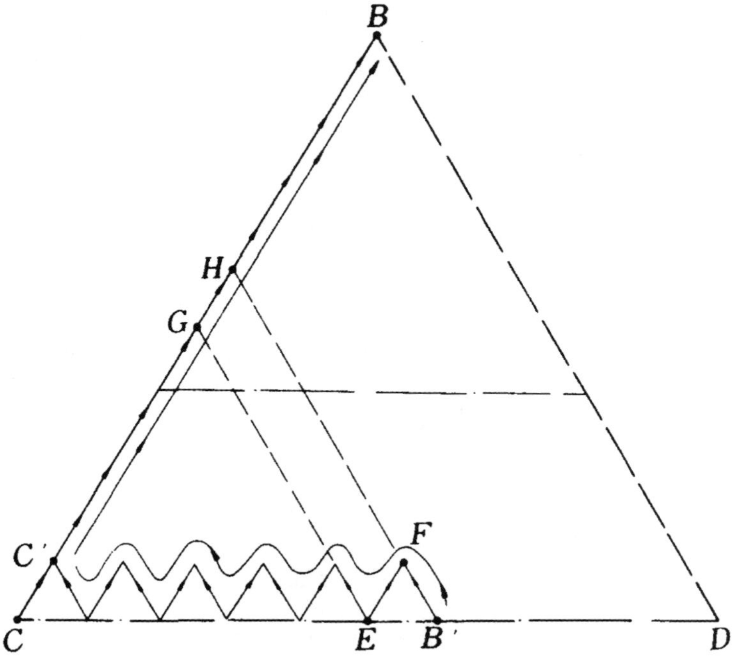

Fig. 5.

$g(\mathrm{P}_0)$ is the factor of any given path from point $A(\mu,0,\lambda,0)$ to $A(0,\nu,0,\sigma)$ (or $\{d'=0,\ e'=c,\ f'=b\}$ to $\{d,e,f\}$, respectively). We choose the path as in fig. 5, obtaining

$$g(\mathrm{P}_0)_{\mathrm{A}} = \left\{ \prod_{\mathrm{P}_0'} \frac{1}{[-\beta-\gamma-\nu'-2\lambda'-\sigma'+1]} \right\}$$

$$\times \prod_{i=0}^{\nu-1} [\beta+\sigma+i] \prod_{i=0}^{\sigma-1} (-1)[\gamma+\nu+i] \prod_{i=0}^{\mu-1} (-1) \frac{[\alpha+i]}{[\gamma+\nu+\sigma-1-i]}$$

$$= \left\{ \prod_{\mathrm{P}_0'} \frac{1}{[2f'+1]} \right\}$$

$$\times (-1)^{\nu+\mu} \frac{[a+f-e]![f+e-a]![a+d+c-f]!}{[a+c-b]![b+c-a]![a+f-d-c]!},$$

where P_0' denotes the path of line C'B in fig. 5, and

$$g(P_0)_B = \left\{ \prod_{P_0'} \frac{1}{[2f'+1]} \right\} (-1)^{\mu+\sigma}$$

$$\times \prod_{i=0}^{\nu-1} \left\{ \frac{[a+f-e-i][a+e-f+1+i] \times [b+d+f+1-i][d+f-b-i]}{[c+d-e-i][c+e-d+1+i]} \right\}^{1/2}$$

$$\times \prod_{i=0}^{\sigma-1} \left\{ \frac{[a+e+f+1-i][e+f-a-i] \times [b+d+f+1-\nu-i][d+f-b-\nu-i]}{[c+d+e+1-i][d+e-c-i]} \right\}^{1/2}$$

$$\times \prod_{i=0}^{\mu-1} \left\{ \frac{[a+e-f+\nu-i][a+f-e+1-\nu+i] \times [b+d-f-i][b+f-d+1+i]}{[a+e+f-\sigma+2+i][e+f-a-\sigma+1+i] \times [b+d+f-\nu-\sigma+2+i][d+f-b-\nu-\sigma+1+i]} \right\}^{1/2}$$

$$= \left\{ \prod_{P_0'} \frac{1}{[2f'+1]} \right\} (-1)^{\mu+\sigma} \left\{ \frac{[2c+1]}{[2b+1]} \right\}^{1/2}$$

$$\times \left\{ \frac{[d+f-b]![b+d-f]![a+f-e]![e+f-a]![a+b-c]!}{[d+e-c]![c+d-e]![b+f-d]![a+c-b]![b+c-a]!} \right.$$

$$\times \left. \frac{[c+e-d]![a+e+f+1]![f+b+d+1]!}{[a+e-f]![c+d+e+1]![a+b+c+1]!} \right\}^{1/2} \frac{[a+c+d-f]!}{[a+f-c-d]!} .$$

It follows from eq. (3.9c) that Γ is given as

$$\Gamma_{\nu\mu} = (-1)^{\nu+\sigma+a+c}$$

$$\times 4m \left\{ \frac{[c+e-d]![a+e+f+1]!}{[d+e-c]![c+d-e]![a+f-e]![e+f-a]! \times [a+e-f]![c+d+e+1]!} \right\}^{1/2}$$

$$\times \frac{(-1)^b}{[2b+1]}$$

$$\times \left\{ \frac{[d+f-b]![b+d-f]![a+c-b]![b+c-a]! \times [a+b-c]![f+b+d+1]!}{[b+f-d]![a+b+c+1]!} \right\}^{1/2}$$

$$\equiv \gamma_1(\nu)/\gamma_2(\mu) . \tag{3.9d}$$

Thus we have from the symmetry of quantum $6j$ symbols and eq. (3.9)

$$M^{(12)}_{\nu\mu}=\frac{1}{\Gamma}\begin{Bmatrix} d & c & e(\nu) \\ a & f & b(\mu) \end{Bmatrix}=\begin{Bmatrix} j_1 & j_2 & e(\nu) \\ j_4 & j_3 & b(\mu) \end{Bmatrix}\left(\frac{\gamma_2(\mu)}{\gamma_1(\nu)}\right) \tag{3.10}$$

for both (2.5) and (2.6), where γ_1 and γ_2 are given in eq. (3.9d) and $e(\nu)=c+d-\nu$, $b(\mu)=f-d+\mu$ [see eq. (3.4)].

Two remarks are in order. The case $\sigma=0$ needs special consideration. The result is consistent with eq. (3.9). We encounter zero factors at two edges EF and GH for $g(P_0)_B$. However, we may treat them as a nonzero number, corresponding to a new graph which is equivalent to that of $A(\mu',\nu',\lambda',\sigma')$.

Properly assigning the phase factors of the integrals in system (iii) of (2.9), we can also obtain the crossing matrix $M^{(13)}$ (which is the braiding matrix for the relation between contour systems a and c in fig. 1. The result is

$$M^{(13)}_{\nu\mu}=\frac{1}{\Gamma'}\begin{Bmatrix} f & a & e=c+d-\nu \\ d & c & b=a-d+\mu \end{Bmatrix}, \tag{3.11a}$$

where

$$\Gamma'=(-1)^{\nu+\sigma+b+c+f}\,e^{i\pi\rho\theta}$$

$$\times\frac{1}{[2b+1]}\left\{\frac{[d+a-b]![b+d-a]![f+c-b]![b+c-f]![f+b-c]!}{[d+e-c]![c+d-e]![b+a-d]![a+f-e]![e+a-f]!}\right.$$

$$\left.\times\frac{[c+e-d]![a+e+f+1]![a+b+d+1]!}{[e+f-a]![c+d+e+1]![b+c+f+1]!}\right\}^{1/2}, \tag{3.11b}$$

$$\theta=f(f+1)+d(d+1)+c(c+1)+a(a+1)-b(b+1)-e(e+1)$$
$$-(f+d)(f+d+1). \tag{3.11c}$$

We can write Γ' as

$$\Gamma'\equiv\gamma_1'(\nu)/\gamma_2'(\mu), \tag{3.11d}$$

which is also factorized.

4. The restriction by integer k

In the above derivation, we considered k as a generic number. That is, in the graph of $A(\mu',\nu',\lambda',\sigma')$, all denominators of edge factors

$$[N]\equiv[-\beta-\gamma-\nu'-2\lambda'-\sigma'+1]$$

are considered to be nonzero. This is not true for $N = mk$, however. We have to discuss this problem in more detail.

Let us rewrite the recursion relations (2.10) of $A(\mu, \nu, \lambda, \sigma)$ as

$$
\begin{aligned}
A(\mu', \nu', \lambda', \sigma') &= A(\mu', \nu' - 1, \lambda' + 1, \sigma') \frac{(-1)[M]}{[N]} \\
&\quad + A(\mu' + 1, \nu' - 1, \lambda', \sigma') \frac{(-1)[Q]}{[N]} \\
&= A(\mu', \nu', \lambda' + 1, \sigma' - 1) \frac{[R]}{[N]} \\
&\quad + A(\mu' + 1, \nu', \lambda', \sigma' - 1) \frac{(-1)[S]}{[N]},
\end{aligned}
$$

where

$$
\begin{aligned}
&N = -\beta - \gamma - \nu' - 2\lambda' - \sigma' + 1 = -\beta - \gamma - 2d + \mu' - \lambda', \\
&M = -\beta - \lambda' - \sigma', \\
&Q = -\alpha - \beta - \gamma - \mu' - 2\nu' - 2\lambda' - \sigma' + 2 = -\alpha - \beta - \gamma - 4d + \mu' + \sigma', \\
&R = -\gamma - \nu' - \lambda', \\
&S = -\alpha - \mu' - \nu'.
\end{aligned}
$$

From eq. (3.8) we can express $N, \ldots, S$ as

$$
\begin{gathered}
N = 2f' + 1, \quad M = a + f' - e', \quad Q = a + f' + e' + 2, \\
R = f' + e' - a, \quad S = a + e' - f'.
\end{gathered}
\tag{4.1}
$$

It is easy to check that if e satisfies the CG conditions in the $6j$ symbols,

$$
a + f \geqslant e \geqslant |a - f| = a - f, \qquad d + c \geqslant e \geqslant |c - d| = c - d, \tag{4.2a}
$$

then $b = f + d - \lambda = f - d + \mu$ also satisfies the CG condition by (2.8b) and the fact $\mu, \lambda \geqslant 0$,

$$
0 \leqslant |a - c| \leqslant |f - d| \leqslant f - d + \mu = b = f + d - \lambda \leqslant f + d \leqslant a + c. \tag{4.2b}
$$

We further assume an "$\not\ell$" condition [3] for e (for ν, respectively),

$$
a + f + e \leqslant k, \qquad c + d + e \leqslant k. \tag{4.3a}
$$

We would like to know whether b also satisfies the "$\not\angle$" condition,

$$a+b+c \leqslant k, \qquad d+b+f \leqslant k. \tag{4.3b}$$

There are two possibilities. First we can have

$$a+c+f+d \leqslant k. \tag{4.4a}$$

One can check that all the "$\not\angle$" conditions for e and b are satisfied, and from

$$0<(2f+1)-(2d-1) \leqslant 2f'+1 \leqslant (2f+1)+(2d-1) \leqslant k$$

one has

$$[N] \equiv (e^{i\pi N/(k+2)} - e^{-i\pi N/(k+2)})/(e^{i\pi/(k+2)} - e^{-i\pi/(k+2)}) \neq 0. \tag{4.4b}$$

Secondly we can have

$$a+c+f+d>k.$$

Assume that the $\not\angle$CG conditions for e are satisfied. Along the line BP′ in fig. 6,

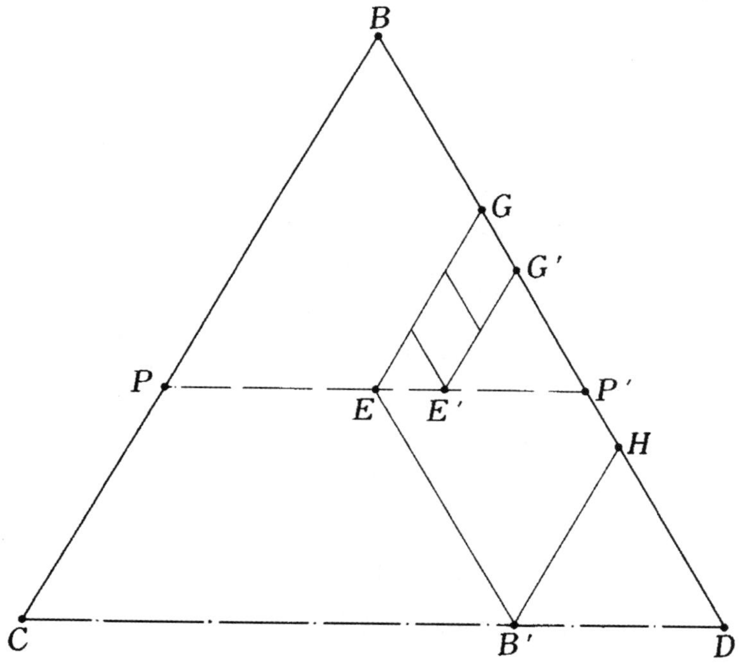

Fig. 6.

where N and Q increase, we have

$$(Q)_{\max}=a+f+e+2+\nu-1=a+f+c+d+1\geqslant k+2.$$

Thus there is a point G in BP′, such that $Q=k+2$, giving $[Q]=0$ for the edge GG′. Let us now draw a parallelogram GEB′H, with GE‖B′H‖BC and EB′‖BD. For the point H,

$$\begin{aligned}N&=2f'+1=2f+1+(k+2-f-e-a-2)+\sigma\\&=k+1+f+\sigma-e-a\\&=k+1+f+d-c-a\leqslant k+1\end{aligned}$$

is the biggest N in the region BCB′HB. Thus $[N]$ is nonzero in this region. On the other hand, all the edges parallel to GG′ and within the area GEE′G′ have the same $[Q]=0$. Thus $A(0,\nu,0,\sigma)$ can be composed by only those $A(\mu',0,\lambda',\sigma)$ which are in the line PE, and therefore by those $A(\mu',0,\lambda',0)$ in the line CB′. Furthermore, all admissible paths from line CB′ to B and going through the line E′G′ must have zero factors. Thus we can still use the generic results (3.9) to calculate the coefficients a_μ.

We now inspect the "$\not\angle$" condition for b in CB′. The maximum μ in the interval CB′ is

$$\begin{aligned}\mu(\mathrm{B}')&=\mu(\mathrm{E})+\sigma=\mu(\mathrm{G})+\sigma\\&=k+2-(a+f+e+2)+\sigma,\end{aligned}\tag{4.5}$$

where $\mu(\mathrm{B}')$ denotes the μ at point B′ etc., and in the last equality we used the fact that $\mu(\mathrm{G})$ is the number of steps from B to G and the condition on the point G. We then have

$$\begin{aligned}f+d+b\leqslant a+b+c&=a+c+f-d+\mu\leqslant a+c+f-d+\mu(\mathrm{B}')\\&=a+c+f-d+(k+2-a-f-e-2)+\sigma=k.\end{aligned}\tag{4.6}$$

Thus the $\not\angle$CG condition for b in CB′ is satisfied. In summary, the integral $A(0,\nu,0,\sigma)$ whose corresponding e satisfies $\not\angle$CG condition can be composed by those $A(\mu,0,\lambda,0)$ whose corresponding b also satisfies $\not\angle$CG condition. For these e and b, we can check that both $(1/\Gamma_{\nu\mu})$ and $\begin{Bmatrix}a&b&c\\d&e&f\end{Bmatrix}$ are well defined and we can always use the generic eq. (3.9) to calculate the matrix elements $M^{(12)}_{\nu\mu}$ for the expansion

$$A(0,\nu,0,\sigma)|_{\not\angle\text{CG admissible}}=\sum_{\not\angle\text{CG admissible}}M^{(12)}_{\nu\mu}A(\mu,0,\lambda,0).\tag{4.7}$$

We are able to analyze the possible values of b and e for the second case. We have from (2.10b), (4.2a), (4.3) and (4.4b)

$$\min(k-a-f, a+f, c+d) = \min(k-a-f, c+d) = k-a-f \geqslant e$$
$$\geqslant \max(a-f, c-d) = c-d,$$

giving $\Delta e = k-a-f-c+d$. On the other hand, we have from (3.4) and (4.5)

$$\Delta e = k-a-f-c+d = k-a-f-e+\sigma = \mu(\mathrm{B}') = \Delta b.$$

All of these equations agree with refs. [12, 14, 28, 29].

5. Locality

To satisfy the locality condition for a four-point function $\langle\varphi_1\varphi_2\varphi_3\varphi_4\rangle$, it is sufficient [13] to find a set of constants $X^{(\nu)}, Y^{(\mu)}$, such that

$$\sum_\nu X^{(\nu)} A(0,\nu,0,\sigma)\bar{A}(0,\nu,0,\sigma) = \sum_\mu Y^{(\mu)} A(\mu,0,\lambda,0)\bar{A}(\mu,0,\lambda,0).$$

Using the crossing matrix $M^{(12)}$, we can write down the equation for $X^{(\nu)}$ and $Y^{(\mu)}$,

$$\sum_\nu X^{(\nu)} M^{(12)}_{\nu\mu} M^{(12)}_{\nu\mu'} = \delta_{\mu\mu'} Y^{(\mu)}, \tag{5.1}$$

which is overdetermined.

Since $\Gamma_{\nu\mu} = \gamma_1(\nu)/\gamma_2(\mu)$ is factorized, we can properly normalize the integrals $A(0,\nu,0,\sigma)$ and $A(\mu,0,\lambda,0)$ [and correspondingly $(\vec{f_p})_1$, and $(\vec{f_p})_2$] such that the transfer matrices of

$$I_\nu = \sqrt{[2e+1]}\,\gamma_1(\nu)A(0,\nu,0,\sigma), \qquad \tilde{I}_\mu = \frac{1}{\sqrt{[2b+1]}}\gamma_2(\mu)A(\mu,0,\lambda,0) \tag{5.2a}$$

can be expressed via

$$I_\nu = \sqrt{[2e+1]}\,\gamma_1(\nu)A(0,\nu,0,\sigma) = \sum_\mu \sqrt{[2e+1]}\,\gamma_1(\nu)M^{(12)}_{\nu\mu}A(\mu,0,\lambda,0)$$
$$= \sum_\mu \left(\sqrt{[2e+1]}\,\gamma_1(\nu)M^{(12)}_{\nu\mu}\sqrt{[2b+1]}/\gamma_2(\mu)\right)\tilde{I}_\mu \equiv \sum_\mu \hat{M}^{(12)}_{\nu\mu}\tilde{I}_\mu$$
$$= \sum_b \sqrt{[2e+1][2b+1]}\begin{Bmatrix} a & f & e \\ d & c & b\end{Bmatrix}\tilde{I}_\mu.$$

That is, the crossing matrix becomes an orthogonal $6j$ matrix. Since in the generic case, the $\hat{M}$ matrix is an orthogonal matrix about e and b [26,27], we can verify

$$\sum_{\nu} I_\nu(\eta)\overline{I_\nu(\eta)} = \sum_{\mu} \tilde{I}_\mu(\eta)\overline{\tilde{I}_\mu(\eta)} , \tag{5.2b}$$

where $\bar{I}_\nu(\bar{\tilde{I}}_\mu)$ is the complex conjugate of $I_\nu(\tilde{I}_\mu)$. Both sides are unchanged after an analytic continuation around $\eta = 0$, 1 and ∞, which ensures the locality condition.

In particular, when the matrix $M^{(12)}$ is restricted by the $\not{\ell}$CG conditions, the truncated part is still subject to the above treatment. We can show that the inverse matrix $\hat{M}^{(21)}$ is similarly truncated (which is the transport matrix of the original matrix $\hat{M}^{(12)}$). Therefore, the locality condition is also satisfied in the same way.

One can also treat the $M^{(13)}$ matrix similarly. The new crossing matrix is the quantum $6j$ matrix of second type [27].

From the fact that the crossing matrices of conformal blocks normalized in this way satisfy polynomial relations such as the Yang–Baxter equation and pentagonal relation, one can show that the truncated q-$6j$ matrix also obey these relations [15–17]. Eq. (5.2) is equivalent to choosing

$$X^{(\nu)} = [2e+1](\gamma_1(\nu))^2, \qquad Y^{(\mu)} = (\gamma_2(\mu))^2/[2b+1]. \tag{5.3}$$

Thus the overdetermined equations for $X^{(\nu)}, Y^{(\mu)}$ have a self-consistent solution for the WZW SU(2) model (and similarly for the minimal model).

6. Crossing matrices of the minimal model

In this section we calculate the crossing (braiding and fusion) matrices of the minimal model, and discuss the relation with SU(2) WZW model.

In the minimal model the four-point function is given by

$$G(z_i, \bar{z}_i) = \langle \phi_1(z_1, \bar{z}_1)\phi_2(z_2, \bar{z}_2)\phi_3(z_3, \bar{z}_3)\phi_4(z_4, \bar{z}_4)\rangle \sim \sum_K X_K I_K(\eta)\overline{I_K(\eta)} .$$

$I_K(\eta), \overline{I_K(\eta)}$ are conformal blocks depending upon $\eta, \bar{\eta}$ (the projective invariant), respectively, which satisfy a differential equation due to the existence of null state. In this section we only discuss one of the conformal blocks, $I_K(\eta)$, for they all have the same properties. The blocks $I_k(\eta)$ can be expressed in terms of the vertex operator [13] $V_\alpha(z) = e^{i\alpha\varphi(z)}$. The field $\varphi(z)$ is a free boson field satisfying $\langle\varphi(z)\varphi(w)\rangle = -2\ln(z-w)$ with a special boundary condition: a charge $-2\alpha_0$ fixed at infinity. The stress–energy tensor $T(z)$ for the field $\varphi(z)$ is

$$T(z) = -\tfrac{1}{4} :\partial_z\varphi(z)\partial_z\varphi(z): + i\alpha_0\partial_z^2\varphi(z).$$

The vertex operator $V_\alpha(z)$ has conformal scale $\Delta_\alpha=\alpha(\alpha-2\alpha_0)$. In this representation the nonzero $I_K(\eta)$ can be obtained with 4 vertex operators $V_{\alpha_i}(z_i)$ and a suitable number of screening operators $V_{\alpha_\pm}(z)$. The parameters $\alpha_\pm$ are associated to the central term C of the minimal model by the relation $C=1-24\alpha_0^2=1-6/M(M+1)$, and $\alpha_\pm=\alpha_0\pm\sqrt{1+\alpha_0^2}$.

The charge conservation of the correlation function implies

$$\alpha_{p,q}=\tfrac{1}{2}[(1-p)\alpha_+ + (1-q)\alpha_-],$$

with integers p and q. We can thus write the $I_K(\eta)$ as

$$I_K^{n,m}\cong\left\langle V_{\alpha_1}V_{\alpha_2}V_{\alpha_3}V_{2\alpha_0-\alpha_4}\int\prod_{i=1}^{n-1}\mathrm{d}u_i V_{\alpha_+}(u_i)\int\prod_{i=1}^{m-1}\mathrm{d}v_i V_{\alpha_-}(v_i)\right\rangle,$$

where $\alpha_i=\alpha_{n_i m_i}$, and n,m are determined by the values of n_i, m_i $(i=1,\dots,4)$. If we assume $|n_1-n_2|\leqslant|n_4-n_3|$, $n_1+n_2\leqslant n_4+n_3$, $n_1+n_2+n_3+n_4\leqslant M$, then we have $n=\frac{1}{2}(n_1+n_2+n_3-n_4)$ and similarly for m.

There are three different contour systems as in the WZW SU(2) model. One of them gives

$$\begin{aligned}
&I_K^{nm}\begin{pmatrix} a & b & c & \rho & \\ a' & b' & c' & \rho' & \eta\end{pmatrix}\\
&=I_{k_1k_2}^{nm}\begin{pmatrix} a & b & c & \rho & \\ a' & b' & c' & \rho & \eta\end{pmatrix}\\
&=\int_1^\infty\prod_{i=1}^{n-k_1}\mathrm{d}u_i\int_0^\eta\prod_{i=n-k_1+1}^{n-1}\mathrm{d}u_i\prod_{i=1}^{n-k_1}(u_i)^a(u_i-1)^b(u_i-\eta)^c\\
&\times\prod_{i=n-k_1+1}^{n-1}u_i^a(1-u_i)^b(\eta-u_i)^c\prod_{i<j}^{n-1}(u_i-u_j)^{2\rho}\\
&\times\int_1^\infty\prod_{i=1}^{m-k_2}\mathrm{d}v_i\int_0^\eta\prod_{i=m-k_2+1}^{m-1}\mathrm{d}v_i\prod_{i=1}^{m-k_2}v_i^{a'}(v_i-1)^{b'}(v_i-\eta)^{c'}\\
&\times\prod_{i=m-k_2+1}^{m-1}v_i^{a'}(1-v_i)^{b'}(\eta-v_i)^{c'}\prod_{i<j}^{m-1}(v_i-v_j)^{2\rho'}\prod_{i,j}^{n-1,m-1}(u_i-v_j)^{-2},
\end{aligned}\tag{6.1}$$

where $a=2\alpha_+\alpha_1$, $b=2\alpha_+\alpha_3$, $c=2\alpha_+\alpha_2$, $\rho=\alpha_+^2$, $a'=2\alpha_-\alpha_1$, $b'=2\alpha_-\alpha_3$, $c'=2\alpha_-\alpha_2$, $\rho'=\rho^{-1}=\alpha_-^2$. This expression is canonical at $\eta=0$.

In the above integral there are n choices for u_i, and m for v_i, so there are nm different possibilities for K. Other contour systems yield $\hat{I}_{k_1 k_2}^{nm}$ and $\hat{\hat{I}}_{k_1 k_2}^{nm}$, which are canonical at $\eta = 1$ and $\eta = \infty$, respectively,

$$\hat{I}_{k_1 k_2}^{nm}\left(\begin{matrix} a & b & c & \rho \\ a' & b' & c' & \rho' \end{matrix}\ \eta\right) = I_{k_1 k_2}^{nm}\left(\begin{matrix} b & a & c & \rho \\ b' & a' & c' & \rho' \end{matrix}\ 1-\eta\right), \quad (6.2)$$

$$\hat{\hat{I}}_{k_1 k_2}^{nm} = \int_0^1 \prod_{i=1}^{n-k_1} du_i \int_\eta^\infty \prod_{i=n-k_1+1}^{n-1} du_i \prod_{i=1}^{n-k_1} u_i^a (1-u_i)^b (\eta - u_i)^c$$

$$\times \prod_{i=n-k_1+1}^{n-1} u_i^a (u_i - 1)^b (u_i - \eta)^c \prod_{i<j}^{n-1} (u_i - u_j)^{2\rho}$$

$$\times \int_0^1 \prod_{i=1}^{m-k_2} dv_i \int_\eta^\infty \prod_{i=m-k_2+1}^{m-1} dv_i \prod_{i=1}^{m-k_2} v_i^{a'} (1-v_i)^{b'} (\eta - v_i)^{c'}$$

$$\times \prod_{i=m-k_2+1}^{m-1} (v_i)^{a'} (v_i - 1)^{b'} (v_i - \eta)^{c'} \prod_{i<j}^{m-1} (v_i - v_j)^{2\rho'} \prod_{i,j}^{n-1,m-1} (u_i - v_j)^{-2}.$$

$$(6.3)$$

The fusion matrix $\alpha_{KK'}^{nm}$ is defined through

$$I_K^{nm}\left(\begin{matrix} a & b & c & \rho \\ a' & b' & c' & \rho' \end{matrix}\ \eta\right) = \sum_{K'} \alpha_{KK'}^{nm} \hat{I}_{K'}^{nm}\left(\begin{matrix} a & b & c & \rho \\ a' & b' & c' & \rho' \end{matrix}\ \eta\right). \quad (6.4)$$

We can perform the same procedure as for WZW SU(2) model. Since the indices $a, b, c, 2\rho, -2$ are proportional to $a', b', c', -2, 2\rho'$, and the index of $(u_i - v_i)$ is -2, the integration contours of u_i can be deformed over through contours of v_j (See refs. [12, 13] for details.) Thus the recursion relations of the corresponding integrals

$$A\left(\begin{matrix} \mu, \nu, \lambda, \sigma \\ \mu', \nu', \lambda', \sigma' \end{matrix}\right) = \int_{-\infty}^0 \prod_{i=1}^\mu du_i \int_0^\eta \prod_{i=\mu+1}^{\mu+\nu} du_i \int_\eta^1 \prod_{i=\mu+\nu+1}^{\mu+\nu+\lambda} du_i \int_1^\infty \prod_{i=\mu+\nu+\lambda}^{\mu+\nu+\lambda+\sigma=n-1} du_i$$

$$\times \prod_{i=1}^{n-1} u_i^a (u_i - 1)^b (u_i - \eta)^c \prod_{i<j}^{n-1} (u_i - u_j)^{2\rho}$$

$$\times \int_{-\infty}^0 \prod_{i=1}^{\mu'} dv_i \int_0^\eta \prod_{i=\mu'+1}^{\mu'+\nu'} dv_i \int_\eta^1 \prod_{i=\mu'+\nu'+1}^{\mu'+\nu'+\lambda'} dv_i \int_1^\infty \prod_{i=\mu'+\nu'+\lambda'}^{\mu'+\nu'+\lambda'+\sigma'=m-1} dv_i$$

$$\times \prod_{i=1}^{m-1} v_i^{a'} (v_i - 1)^{b'} (v_i - \eta)^{c'} \prod_{i<j}^{m-1} (v_i - v_j)^{2\rho'} \prod_{ij} (u_i - v_j)^{-2}, \quad (6.5)$$

with the same phase convention as in ref. [13], are decoupled. Therefore, the crossing (fusion and braiding) matrices are decoupled for parameters k_1 and k_2, as pointed out in ref. [13]. For each of them we can use a similar strategy for deriving eq. (3.9). Finally we have

$$\alpha^{nm}_{KK'} = M^n_{k_1k'_1} M^m_{k_2k'_2}, \tag{6.6}$$

$$M^n_{k_1k'_1} = \frac{1}{\Gamma_{k_1k'_1}} \begin{Bmatrix} r & s & t \\ d & e & f \end{Bmatrix},$$

$$\Gamma_{k_1k'_1} = (-1)^{k_1(m_2-m_3-1)-k'_1(m_1-m_3-1)+\frac{1}{4}(n_1+n_2+n_3+3n_4)-m_2+n(m_1+\frac{3}{2})}$$

$$\times [2s+2d-2k'_1+3]^{-1}$$

$$\times \left(\frac{\begin{matrix}[d+s-f]![f+d-s]![r+e-f]![f+e-r]![r+f-e]! \\ \times [e+t-d]![r+t+s+1]![s+f+d+1]!\end{matrix}}{\begin{matrix}[d+t-e]![e+d-t]![f+s-d]![r+s-t]![t+s-r]! \\ \times [r+t-s]![e+d+t+1]![r+f+e+1]!\end{matrix}} \right)^{1/2},$$

where

$$r = \tfrac{1}{4}(n_1+n_3+n_4-n_2) - \tfrac{1}{2}, \quad s = \tfrac{1}{4}(n_2+n_3+n_4-n_1) - \tfrac{1}{2}, \quad t = \tfrac{1}{2}(n_1+n_2) - k_1,$$

$$d = \tfrac{1}{4}(n_1+n_2+n_3-n_4) - \tfrac{1}{2}, \quad e = \tfrac{1}{4}(n_1+n_2+n_4-n_3) - \tfrac{1}{2}, \quad f = \tfrac{1}{2}(n_2+n_3) - k'_1,$$

and { } stands for the q-6j symbols with $q = \exp(i\pi\rho)$. $M^m_{k_2k'_2}$ is similar to $M^n_{k_1k'_1}$, but m_i and n_i are interchanged, and furthermore the q-6j symbols and the square brackets have a different parameter $q' = \exp(i\pi\rho')$.

From the definition of the braiding matrix $\beta^{nm}_{KK'}$,

$$I^{nm}_K = \sum_{K'} \beta^{nm}_{KK'} \hat{I}^{nm}_{K'},$$

we can obtain

$$\beta^{nm}_{KK'} = \tilde{M}^n_{k_1k'_1} \tilde{M}^m_{k_2k'_2},$$

$$\tilde{M}^n_{k_1k'_1} = q^{s(s+1)+d(d+1)+e(e+1)+r(r+1)-t(t+1)-f(f+1)-(s+d)(s+d+1)}$$

$$\times M^n_{k_1k'_1}(n_1 \leftrightarrow n_2). \tag{6.7}$$

Similar results can be obtained for $\tilde{M}^m_{k_2k'_2}$ (with $q' = e^{i\pi\rho'}$). Thus the CM for the minimal models are essentially direct products of two q-$6j$ matrices, which agrees with Felder et al. [12].

Note that although the CM are decomposed into two parts which depend on (a, b, c, ρ) and (a', b', c', ρ'), respectively, the Dotsenko–Fateev normalization constants of I_K^{nm} are not decomposable. If we define

$$I_{k_1k_2}^{nm} = J_{k_1k_2}^{nm} \hat{\Gamma}_{k_1}^{-1}(n_i m_i) \hat{\Gamma}_{k_2}(n_i \leftrightarrow m_i) , \tag{6.8}$$

with

$$\hat{\Gamma}_{k_1}(n_i m_i) = (-1)^{k_1(m_1 - m_3 - 1) - \frac{1}{2}m_2 + \frac{1}{2}n(m_1+3)} \left\{ \frac{[2e+2d-2k_1+3][d+t-e]!}{[e+t-d]![r+s+t+1]!} \right.$$
$$\left. \times [e+d-t]![r+s-t]![t+s-r]![e+d+t+1]![r+t-s]! \right\}^{1/2} ,$$

then we find that the fusion matrix of J_K^{nm} is a direct product of two orthogonal q-$6j$ matrices satisfying the pentagonal identity. The $G(z_i, \bar{z}_i)$ can be expressed in terms of J's in a simple form,

$$G(z_i, \bar{z}_i) = |\eta|^{2\mu_{12} + 4\alpha_1\alpha_2} |1-\eta|^{2\mu_{23} + 4\alpha_1\alpha_3} \prod_{i<j} |z_i - z_j|^{-2\mu_{ij}} \sum_K J_K^{nm}(\eta) \overline{J_K^{nm}(\eta)} ,$$

$$\mu_{ij} = \Delta_i + \Delta_j - \tfrac{1}{3} \sum_K \Delta_k , \qquad \Delta_i = \Delta_{\alpha_i} = \Delta_{\alpha_{n_i m_i}} ,$$

which satisfies the locality condition.

7. Discussion

The structure constants can be obtained by comparing the γ's in eqs. (3.9d) and (6.8) with the Dotsenko–Fateev normalization constants [13, 30].

The correspondences between the graph of $A(\mu, \nu, \lambda, \sigma)$ and the graph of $6j$ symbols are not unique. They are related to a group. Thus a number of indices (α, β, γ) in the Feigin–Fuchs integral representation can have the same q-$6j$ matrix as the CM. This can be related to the Riemann monodromy problem [31].

We thank Bo-Yuan Hou, Zhong-Qj Ma and Kang Li for helpful discussions. This work was supported in part by the Natural Science Foundation of China through the Nankai Institute of Mathematics.

References

[1] A.A. Belavin, A.M. Polykov and A.B. Zamolodchikov, Nucl. Phys. B241 (1984) 333
[2] G. Moore and N. Seiberg, Phys. Lett. B212 (1988) 451; Nucl. Phys. B313 (1989) 16
[3] A. Tsuchiya and Y. Kanie, Adv. Stud. Pure Math. 16 (1988) 297; Lett. Math. Phys. 13 (1987) 303
[4] L. Alvarez-Gaumé, C. Gomez and G. Sierra, Phys. Lett. B220 (1989) 142; Nucl. Phys. B319 (1989) 155
[5] G. Moore and N. Reshetikhin, Nucl. Phys. B328 (1989) 557
[6] E. Witten, Commun. Math. Phys. 121 (1989) 351; Nucl. Phys. B322 (1989) 629
[7] E. Date, M. Jimbo, T. Miwa and M. Okado, Lett. Math. Phys. 12 (1986) 209
[8] M. Jimbo, A. Kuniba, T. Miwa and M. Okado, Commun. Math. Phys. 119 (1988) 543
[9] V.G. Knizhnik and A.B. Zamolodchikov, Nucl. Phys. B247 (1984) 83
[10] D. Gepner and E. Witten, Nucl. Phys. B278 (1986) 493
[11] P. Christ and R. Flume, Nucl. Phys. B282 (1987) 466
[12] G. Felder, J. Frölich and G. Keller, IAS-ETH preprint, Commun. Math. Phys., to be published
[13] Vl.S. Dotsenko and V.A. Fateev, Nucl. Phys. B240 (1984) 312; B251 (1985) 691
[14] P. Christ and R. Flume, Phys. Lett. B188 (1987) 219
[15] A.C. Ganchev and V.B. Petkova, preprint IC/89/158
[16] Vl.S. Dotsenko, Adv. Stud. Pure Math. 16 (1988) 123
[17] M. Jimbo, T. Miwa and M. Okado, Adv. Stud. Pure Math. 16 (1988) 17; Mod. Phys. Lett. B1 (1987) 73
[18] V.A. Fateev and A.B. Zamolodchikov, Yad. Fiz. 43 (1986) 75
[19] D. Nemeschansky, Phys. Lett. B224 (1989) 121
[20] A.V. Marshakov, Phys. Lett. B224 (1989) 141
[21] D. Bernard and G. Felder, ETH-TH/89-26; SPhT/89/113
[22] K. Ito and Y. Kasama, UT-Komaba 89-22
[23] A. Bilal, Nucl. Phys. B330 (1990) 399
[24] V.A. Fateev and S. Lykyanov, Int. J. Mod. Phys. A3 (1988) 507
[25] B.Y. Hou, D.P. Lie and R.H. Yue, Phys. Lett. B154 (1989) 291
[26] A.N. Kirillov and N.Yu. Reshetikhin, LOMI preprint E-9-88
[27] Bo-Yu Hou, Bo-Yuan Hou and Z.Q. Ma, Commun. Theor. Phys. 13 (1990) 181, 341
[28] B.Y. Hou, K. Li and P. Wang, J. Phys. A: Math. Gen., to be published
[29] J. Fuchs and P.V. Driel, preprint ITFA-90-03
[30] Vl.S. Dotsenko and V.A. Fateev, Phys. Lett. B154 (1985) 291
[31] B. Blok and S. Yankielowicz, Nucl. Phys. B321 (1989) 717; Phys. Lett. B226 (1989) 279

Commun. Theor. Phys. 13(1990)181–198
© China Ocean Press Vol. 13, No. 2

Clebsch-Gordan Coefficients, Racah Coefficients and Braiding Fusion of Quantum sl(2) Enveloping Algebra (I)[1]

Bo-yu HOU
Institute of Modern Physics, Northwest University Xi'an 710069, China
Bo-yuan HOU
Graduate School, Chinese Academy of Sciences, P. O. Box 3908, Beijing 100039, China
Zhong-qi MA
Institute of High Energy Physics, P. O. Box 918(4), Beijing 100039, China

(Received April 1, 1989)
(Accepted December 8, 1989)

Abstract

Quantum Clebsch-Gordan coefficients and the first type quantum Racah coefficients of quantum sl (2) enveloping algebra are given explicitly. The quantum $3-j$ and $6-j$ symbols, similar to those in the theory of angular momentum are also introduced. The solution $\check{R}_q^{j_1 j_2}$ of quantum Yang-Baxter equaton is expressed in terms of the quantum Clebsch-Gordan coefficients. It is shown that when $j_1 = j_2$, $\check{R}_q^{jj}$ is just the same as R^j_{AW} matrix obtained by Akutsu and Wadati for the representation of the braid group. The second type quantum Racah coefficients, which are the solutions of the face models, are also computed explicitly and related to the first type quantum Racah coefficients. The famous pentagonal relation is proved from the formula between two quantum Racah coefficients. The graphical representation of those formulas is discussed.

I. Introduction

For a variety of reasons, there is much interest in Yang-Baxter equation (YBE)[1,2], which plays a crucial role in classical and quantum integrable systems[3,4]. The quantum YBE

$$R_{12}(u,q)R_{13}(u+v,q)R_{23}(v,q) = R_{23}(v,q)R_{13}(u+v,q)R_{12}(u,q) \tag{1}$$

is a highly nonlinear equation, but the classical YBE is a commutative relation

$$R_{ij}(u,q) \sim \mathbb{1} + (q-1)r_{ij}(u)\,, \tag{2}$$

$$[r_{12}(u),\, r_{13}(u+v)] + [r_{12}(u),\, r_{23}(v)] + [r_{13}(u+v),\, r_{23}(v)] = 0\,. \tag{3}$$

From a known solution of the classical YBE, based on a representation of a simple Lie algebra or a Kac-Moody algebra, a set of solutions based on other representations follows. The classical

[1]The project was supported by the National Natural Science Foundation of China through the Nankai Institute of Mathematics.

r matrices associated with a simple Lie algebra are classified[5], and some solutions have been obtained[1,6,7,8]. However, to find out the corresponding solution R of the quantum YBE having the solution r as its classical limit is a quite difficult problem. The method for finding R matrix is called fusion process. A typical fusion process is based on the quantum enveloping algebra[9], which is a direct generalization of the Lie algebra and the Kac-Moody algebra and has drawn the considerable interests of both physicists and mathematicians. It is interesting to notice that the solutions of the classical YBE based on Lie algebras and their quantizations based on the quantum Lie enveloping algebras do not depend on the spectral parameter u explicitly, that is, the solutions R obtained by this method satisfy a quantum YBE without spectral parameter u.

Recently, an unexpected close connection between physics and mathematics has been found by Y. Akutsu *et al.*[4,10,11]. Let $\check{R}_{ij}(u\,,\,q)=R_{ij}(u\,,\,q)P_{ij}$, where P_{ij} is a permutation operator $(P_{ij}:|V_i\times V_j\to V_j\times V_i,\ P_{ij}(\psi_i\times\psi_j)=\psi_j\times\psi_i)$. $\check{R}_{ij}$ satisfy a modified YBE

$$\check{R}_{12}(u,q)\check{R}_{23}(u+v,q)\check{R}_{12}(v,q)=\check{R}_{23}(v,q)\check{R}_{12}(u+v,q)\check{R}_{23}(u,q)\,. \tag{4}$$

Akutsu and Wadati (AW) found that equation (4) is nothing but the multiplication rule of the generators b_i of the braid group

$$b_ib_{i+1}b_i=b_{i+1}b_ib_{i+1}\,, \tag{5}$$

if the spectral parameter u can be removed from $\check{R}$. From the Boltzmann weights of the N-state models which are the solutions of the quantum YBE, AW[10] obtained a set of representations of the braid group in terms of normalizing, the symmetry-breaking transformation and the limit process of $u\to\infty$. Then, AW successfully found a set of new link polynomials to classify knots and links. Jones polynomial[12] is one belonging to this set ($N=2$), and the difficulty of Jones polynomial in distinguishing some links[13] is removed by the else link polynomials (e.g. $N=3$) in this set.

The main progress made by AW is to obtain the representations of the braid group by removing the spectral parameter in $\check{R}$ matrices. An interesting question is raised: what is the relation between R_{AW} matrix obtained by AW and the solution $\check{R}$ of the quantum YBE obtained based on quantum Lie enveloping algebra? Both of them are independent of the spectral parameter u.

Most papers on quantum algebra are devoted to the solutions of the quantum YBE with the spectral parameter, i.e., they used the quantum Kac-Moody algebra without the central term instead of the quantum Lie algebra. An elegant paper by Reshetikhin[14] discussed the solutions of the YBE connected with the quantum universal enveloping algebras of simple Lie algebras generally, and built the link polynomials from the solutions. However, it still has abundant and interesting content to study a simple quantum Lie enveloping algebra explicitly.

In this paper, we study quantum sl(2) enveloping algebra (q-sl(2)) in detail. From the representations of q-sl(2) (Sec. II), we compute the explicit forms of quantum Clebsch-Gordan (q-CG) coefficients (Sec. III) and the first type quantum Racah (q-Racah 1) coefficients (Sec. IV). We discuss their symmetries and introduce quantum $3-j$ and $6-j$ symbols. It will be seen that the theory of q-sl(2) is a straight generalization of the theory of angular momentum in quantum mechanics developed by Wigner, Racah and Biedenharn[15,16]. The $\check{R}_q^{j_1 j_2}$ matrices, which are proved to be both the solutions of quantum YBE and the representations of the braid group, are calculated and expressed by q-CG coefficients. It is interesting that R_{AW}^j matrices are nothing but $\check{R}_q^{jj}$ obtained from q-sl(2) when $j_1 = j_2 = j$.

In the bases of the so-called fusion paths[17], the matrix form of $\check{R}$ is proved to be a solution of the quantum YBE for the face model. We call it the second type quantum Racah (q-Racah 2) coefficient, and express it as a sum of the products of two q-Racah 1 coefficients. Then, we prove a direct relation between q-Racah 1 and q-Racah 2 coefficients. From this important relation we prove the famous pentagonal relation, and then, prove that $\check{R}$ matrices and q-Racah 2 coefficients satisfy the quantum YBE for the vertex model and the face model, respectively. All the equations satisfied by the q-CG and q-Racah coefficients have the corresponding ones in the theory of angular momentum (Sec. VI).

There is a graphical representation for $\check{R}$ matrices, q-CG and q-Racah coefficients, proposed firstly by Reshetikhin[14]. In the representation it is easy to understand the above formulas by the rules of generalized Reidemeister moves[18]. We leave the detailed computation for q-CG and q-Racah coefficients in the next paper (II)[19].

II. Representation of Quantum sl(2) Enveloping Algebra

The general relations of quantum Lie enveloping algebra have been given by Jimbo[9]. For quantum sl(2) enveloping algebra the multiplication relations among the generators e, f and k are as follows:

$$\begin{aligned} &k = q^{h/2}\,, \quad k\,k^{-1} = k^{-1}\,k = 1\,, \quad k\,e = q\,e\,k\,, \quad k\,f = q^{-1}\,f\,k\,, \\ &[e\,,\,f] = \frac{k^2 - k^{-2}}{q - q^{-1}}\,, \end{aligned} \tag{6}$$

where q is a quantum parameter. Throughout this paper, q is not a root of unity, except for the classical case, $q = 1$. In the classical limit, $q \to 1$,

$$e \to J_+ = J_1 + i\,J_2\,, \qquad f \to J_- = J_1 - i\,J_2\,, \qquad h \to 2\,J_3 \tag{7}$$

and relations (6) reduce to those in the sl(2) algebra. The Casimir operator C for q-sl(2) is

$$C = \frac{(k - k^{-1})(kq - k^{-1}q^{-1})}{(q - q^{-1})^2} + f\,e = \frac{(k - k^{-1})(kq^{-1} - k^{-1}q)}{(q - q^{-1})^2} + e\,f\,. \tag{8}$$

In the k diagonal representation, if $k|\,m\rangle = q^m\,|m\rangle$, we have

$$k\{e|m\rangle\} = q^{m+1}\{e|m\rangle\}\,, \qquad k\{f|m\rangle\} = q^{m-1}\{f|m\rangle\}\,. \tag{9}$$

Similar to the discussion in quantum mechanics[16], we obtain the finite dimensional representation of q-sl(2)[9]

$$\begin{aligned} &k|\,j,m\rangle = q^m\,|\,j,m\rangle\,, \qquad &&e|\,j,m-1\rangle = \Gamma_m^j(q)|\,j,m\rangle\,, \\ &f|\,j,m\rangle = \Gamma_m^j(q)|\,j,m-1\rangle\,, \qquad &&C|\,j,m\rangle = [j][j+1]|\,j,m\rangle\,, \end{aligned} \tag{10}$$

where $j = 0, 1/2, 1, 3/2 \ldots, -j \le m \le j$, and

$$\Gamma^j_m(q) = \{[j+m][j-m+1]\}^{1/2}, \qquad [m] = \frac{q^m - q^{-m}}{q - q^{-1}}, \tag{11}$$

$$\begin{gathered}[m] = -[-m] \neq 0 \quad \text{if} \quad m \neq 0, \\ [1] = 1, \qquad [0] = 0,\end{gathered} \tag{12}$$

$\Gamma^j_m(q)$ is a positively real number when q is real.

The representation matrices of generators are

$$D^j_q(k)_{mm'} = \delta_{mm'} q^m, \qquad D^j_q(e)_{mm'} = \delta_{m(m'+1)}\Gamma^j_m(q) = D^j_q(f)_{m'm}, \tag{13}$$

when q is real, the representation D^j_q is called the real orthogonal representation.

III. Quantum Clebsch-Gordan Coefficients

The direct product of representations of a Lie algebra satisfies its commutative relation, but that of a quantum enveloping algebra does not satisfy the multiplication relations of the quantum algebra. So the direct product has to be generalized to co-product[9] for the quantum algebra. For q-sl(2) we have the co-product Δ as follows:

$$\begin{aligned}&\Delta(k) = k \times k, & &\Delta(e) = e \times k^{-1} + k \times e, \\ &\Delta(f) = f \times k^{-1} + k \times f, & &\Delta(I_a I_b) = \Delta(I_a)\Delta(I_b),\end{aligned} \tag{14}$$

which satisfy the multiplication relations of q-sl(2). In the basis $|j_1 m_1\rangle \times |j_2 m_2\rangle$, we have

$$\begin{aligned}&\Delta(k)(|j_1 m_1\rangle \times |j_2 m_2\rangle) = (k|j_1 m_1\rangle) \times (k|j_2 m_2\rangle), \\ &\Delta(e)(|j_1 m_1\rangle \times |j_2 m_2\rangle) = (e|j_1 m_1\rangle) \times (k^{-1}|j_2 m_2\rangle) + (k|j_1 m_1\rangle) \times (e|j_2 m_2\rangle),\end{aligned} \tag{15}$$

and that obtained by replacing $e \leftrightarrow f$. Therefore, we define the quantum direct product of representations

$$\begin{aligned}&D^{j_1 j_2}_q(k) = D^{j_1}_q(k) \times D^{j_2}_q(k), \\ &D^{j_1 j_2}_q(e) = D^{j_1}_q(e) \times D^{j_2}_q(k^{-1}) + D^{j_1}_q(k) \times D^{j_2}_q(e), \\ &D^{j_1 j_2}_q(f) = D^{j_1}_q(f) \times D^{j_2}_q(k^{-1}) \times D^{j_1}_q(k) \times D^{j_2}_q(e).\end{aligned} \tag{16}$$

There is another co-product satisfying the multiplication relations obtained by replacing $k \leftrightarrow k^{-1}$[9]. For definiteness we only discuss the co-product (14).

$D^{j_1 j_2}_q$ is generally a reducible representation and can be transformed to the direct sum of irreducible representations D^J_q by a similarity transformation $(C^{j_1 j_2}_q)$. According to the multiplicity of weights of $|j_1 m_1\rangle \times |j_2 m_2\rangle$, it is easy to see that $J = (j_1 + j_2), (j_1 + j_2 - 1) \ldots, |j_1 - j_2|$:

$$(C^{j_1 j_2}_q)^{-1} D^{j_1 j_2}_q(I)(C^{j_1 j_2}_q) = \bigoplus_{J=|j_1 - j_2|}^{j_1 + j_2} D^J_q(I), \tag{17a}$$

$$|j_1 j_2 JM\rangle = \sum_{m_1 m_2} (|j_1 m_1\rangle \times |j_2 m_2\rangle)(C^{j_1 j_2}_q)_{m_1 m_2 JM}, \tag{17b}$$

$$\Delta(I)|j_1 j_2 JM\rangle = \sum_{M'=-J}^{J} |j_1 j_2 JM'\rangle D^J_q(I)_{M'M}, \tag{17c}$$

where I denotes k, e or f, $(C_q^{j_1j_2})$ is a $(2j_1+1)(2j_2+1)\times(2j_1+1)(2j_2+1)$ matrix with row indices m_1m_2 and column indices JM, and is called quantum Clebsch-Gordan (q-CG) coefficients. Fixing J, we obtain a $(2j_1+1)(2j_2+1)\times(2J+1)$ matrix $(C_q^{j_1j_2})_J$. A factor for each $(C_q^{j_1j_2})_J$ has not been specified. From Eq. (17a) we obtain the recursive relations between q-CG coefficients:

$$(C_q^{j_1j_2})_{m_1m_2JM}=0 \qquad \text{if} \qquad M\neq m_1+m_2\,, \tag{18}$$

$$\begin{aligned}&\Gamma^{j_1}_{\pm m_1}(q)q^{-m_2}(C_q^{j_1j_2})_{(m_1\mp1)m_2JM}+\Gamma^{j_2}_{\pm m_2}(q)q^{m_1}(C_q^{j_1j_2})_{m_1(m_2\mp1)JM}\\&\quad=(C_q^{j_1j_2})_{m_1m_2J(M\pm1)}\Gamma^{J}_{\mp M}(q)\,.\end{aligned} \tag{19}$$

When q is real, we can choose $C_q^{j_1j_2}$ to be real orthogonal. Then, by analytic continuation, $(C_q^{j_1j_2})$ is orthogonal when q is complex

$$\begin{aligned}&\sum_J (C_q^{j_1j_2})_{m(M-m)JM}(C_q^{j_1j_2})_{m'(M-m')JM}=\delta_{mm'}\,,\\&\sum_m (C_q^{j_1j_2})_{m(M-m)JM}(C_q^{j_1j_2})_{m(M-m)J'M}=\delta_{JJ'}\,.\end{aligned} \tag{20}$$

We can firstly calculate the $(C_q^{j_1j_2})_{m(J-m)JJ}$ from equations (18)–(20), and then the general $(C_q^{j_1j_2})_{m_1m_2JM}$. The detailed computation will be written in the next paper (II)[19]. The result can be expressed as two equivalent forms:

1. Quantum Racah's form

$$\begin{aligned}(C_q^{j_1j_2})_{m_1m_2JM}&=\delta_{M(m_1+m_2)}q^{\{j_1(j_1+1)-j_2(j_2+1)+J(J+1)\}/2-m_1(M+1)}\\&\times\left\{\frac{[2J+1][j_1+j_2-J]![j_1-m_1]![j_2-m_2]![J+M]![J-M]!}{[j_1+j_2+J+1]![J+j_1-j_2]![J-j_1+j_2]![j_1+m_1]![j_2+m_2]!}\right\}^{1/2}\\&\times\sum_n(-1)^{n+j_1-m_1}q^{-n(M+J+1)}\\&\times\frac{[j_1+m_1+n]![j_2+J-m_1-n]!}{[n]![j_1-m_1-n]![J-M-n]![j_2-J+m_1+n]!}\,.\end{aligned} \tag{21}$$

2. Quantum Van der Waerden's form

$$\begin{aligned}(C_q^{j_1j_2})_{m_1m_2JM}&=\delta_{M(m_1+m_2)}q^{-(j_1+j_2-J)(j_1+j_2+J+1)/2+m_1j_2-m_2j_1}\\&\times\Delta(j_1j_2J)\{[2J+1][j_1+m_1]![j_1-m_1]!\\&\times[j_2+m_2]![j_2-m_2]![J+M]![J-M]!\}^{1/2}\\&\times\sum_n(-1)^nq^{n(j_1+j_2+J+1)}\{[n]![j_1-m_1-n]![j_2+m_2-n]!\\&\times[j_1+j_2-J-n]![J-j_1-m_2+n]![J-j_2+m_1+n]!\}^{-1}\,,\end{aligned} \tag{22}$$

where

$$\Delta(abc)=\left\{\frac{[a+b-c]![a-b+c]![-a+b+c]!}{[a+b+c+1]!}\right\}^{1/2}, \tag{23}$$

and

$$[n]!=[n][n-1]\cdots[1]\,,\qquad [0]!=1\,,\qquad [-n]!=\infty\,, \tag{24}$$

where n is a positive integer and $[n]$ is given in Eq. (12). It is easy to verify from the quantum Van der Waerden's form (22) that q-CG coefficients have the following symmetries:

$$(C_q^{j_1 j_2})_{m_1 m_2 JM} = (C_q^{j_2 j_1})_{-m_2 -m_1 J -M} \tag{25a}$$

$$= (-1)^{j_1+j_2-J}(C_{q^{-1}}^{j_1 j_2})_{-m_1 -m_2 J -M} \tag{25b}$$

$$= (-1)^{j_1+j_2-J}(C_{q^{-1}}^{j_2 j_1})_{m_2 m_1 JM} \tag{25c}$$

$$= (-1)^{j_1-m_1}\left\{\frac{[2J+1]}{[2j_2+1]}\right\}^{1/2} q^{-m_1}(C_{q^{-1}}^{j_1 J})_{m_1 -M j_2 -m_2} \tag{25d}$$

$$= (-1)^{j_2+m_2}\left\{\frac{[2J+1]}{[2j_1+1]}\right\}^{1/2} q^{m_2}(C_{q^{-1}}^{J j_2})_{-M m_2 j_1 -m_1} \,. \tag{25e}$$

Introduce the quantum $3-j$ symbol:

$$\begin{pmatrix} j_1 & j_2 & j_3 \\ m_1 & m_2 & m_3 \end{pmatrix}_q = (-1)^{j_1-j_2-m_3}\{[2j_3+1]\}^{-1/2} q^{(m_1-m_2)/3}(C_q^{j_1 j_2})_{m_1 m_2 j_3 -m_3}\,,$$
$$|j_1-j_2| \le j_3 \le j_1+j_2\,, \qquad m_1+m_2+m_3=0\,. \tag{26}$$

The quantum $3-j$ symbols have similar symmetries to the classical Wigner $3-j$ symbols:

$$\begin{pmatrix} j_1 & j_2 & j_3 \\ m_1 & m_2 & m_3 \end{pmatrix}_q = \begin{pmatrix} j_2 & j_3 & j_1 \\ m_2 & m_3 & m_1 \end{pmatrix}_q = (-1)^{j_1+j_2+j_3}\begin{pmatrix} j_2 & j_1 & j_3 \\ m_2 & m_1 & m_3 \end{pmatrix}_{q^{-1}}$$
$$= (-1)^{j_1+j_2+j_3}\begin{pmatrix} j_1 & j_2 & j_3 \\ -m_1 & -m_2 & -m_3 \end{pmatrix}_{q^{-1}}\,. \tag{27}$$

The orthogonal relations (20) become

$$\sum_{j_3 m_3}[2j_3+1]\begin{pmatrix} j_1 & j_2 & j_3 \\ m_1 & m_2 & m_3 \end{pmatrix}_q \begin{pmatrix} j_1 & j_2 & j_3 \\ m_1' & m_2' & m_3 \end{pmatrix}_q = q^{2(m_1-m_2)/3}\delta_{m_1 m_1'}\delta_{m_2 m_2'}\,,$$
$$\sum_{m_1 m_2} q^{-2(m_1-m_2)/3}\begin{pmatrix} j_1 & j_2 & j_3 \\ m_1 & m_2 & m_3 \end{pmatrix}_q \begin{pmatrix} j_1 & j_2 & j_3' \\ m_1 & m_2 & m_3' \end{pmatrix}_q$$
$$= [2j_3+1]^{-1}\delta_{j_3 j_3'}\delta_{m_3 m_3'}\,. \tag{28}$$

It is easy to see that in the classical limit $(q \to 1)$ all the quantities and relations reduce to the corresponding ones in the theory of angular momentum.

In the following we list some concrete formulas for the q-CG coefficients and quantum $3-j$ symbols.

$$(C_q^{jj})_{M-m00} = \frac{(-1)^{j-m}q^{-m}}{[2j+1]^{1/2}} = q^{-2m/3}\begin{pmatrix} j & j & 0 \\ m & -m & 0 \end{pmatrix}_q\,, \tag{29a}$$

$$(C_q^{j_1 j_2})_{m_1 m_2 (j_1+j_2)(m_1+m_2)} = A \tag{29b}$$

with

$$
\begin{aligned}
A &= q^{m_1 j_2 - m_2 j_1} \left\{ \frac{[2j_1]![2j_2]![j_1+j_2+m_1+m_2]![j_1+j_2-m_1-m_2]!}{[2j_1+2j_2]![j_1+m_1]![j_1-m_1]![j_2+m_2]![j_2-m_2]!} \right\}^{1/2} \\
&= (-1)^{j_1-j_2+m_1+m_2} q^{-(m_1-m_2)/3} \{[2j_1+2j_2+1]\}^{1/2} \begin{pmatrix} j_1 & j_2 & j_1+j_2 \\ m_1 & m_2 & -(m_1+m_2) \end{pmatrix}_q .
\end{aligned}
$$

$$
\begin{aligned}
(C_q^{j_1 j_2})_{j_1(-j_1-m_3)j_3-m_3} &= q^{\{j_1(j_1+2m_3-1)-j_2(j_2+1)+j_3(j_3+1)\}/2} \\
&\times \left\{ \frac{[2j_3+1][2j_1]![-j_1+j_2+j_3]![j_1+j_2+m_3]![j_3-m_3]!}{[j_1+j_2+j_3+1]![j_1-j_2+j_3]![j_1+j_2-j_3]![-j_1+j_2-m_3]![j_3+m_3]!} \right\}^{1/2} \\
&= (-1)^{j_1-j_2-m_3} [2j_3+1]^{1/2} q^{-(2j_1+m_3)/3} \begin{pmatrix} j_1 & j_2 & j_3 \\ j_1 & -j_1-m_3 & m_3 \end{pmatrix}_q
\end{aligned}
\tag{29c}
$$

IV. The First Type Quantum Racah Coefficients

For the co-product of three representations of the quantum sl(2) enveloping algebra, we have

$$
\begin{aligned}
D_q^{j_1 j_2 j_3}(k) &= D_q^{j_1}(k) \times D_q^{j_2}(k) \times D_q^{j_3}(k) , \\
D_q^{j_1 j_2 j_3}(e) &= D_q^{j_1}(e) \times D_q^{j_2}(k^{-1}) \times D_q^{j_3}(k^{-1}) + D_q^{j_1}(k) \times D_q^{j_2}(e) \times D_q^{j_3}(k^{-1}) \\
&\quad + D_q^{j_1}(k) \times D_q^{j_2}(k) \times D_q^{j_3}(e) ,
\end{aligned}
\tag{30}
$$

and that obtained by replacing $e \leftrightarrow f$. There are different paths to reduce it to be an irreducible one. As usual in the quantum mechanics

$$
\begin{aligned}
&\left\{(C_q^{j_1 j_2})_{J_{12}} (C_q^{J_{12} j_3})_J\right\}^{-1} D_q^{j_1 j_2 j_3} \left\{(C_q^{j_1 j_2})_{J_{12}} (C_q^{J_{12} j_3})_J\right\} = D_q^J \\
&\quad = \left\{(C_q^{j_2 j_3})_{J_{23}} (C_q^{j_1 J_{23}})_J\right\}^{-1} D_q^{j_1 j_2 j_3} \left\{(C_q^{j_2 j_3})_{J_{23}} (C_q^{j_1 J_{23}})_J\right\} .
\end{aligned}
$$

The first type quantum Racah (q-Racah 1) coefficients relates the two similarity transformations

$$
\begin{aligned}
(C_q^{j_2 j_3})_{m_2 m_3 J_{23}(m_2+m_3)} (C_q^{j_1 J_{23}})_{m_1(M-m_1)JM} &= \sum_{J_{12}} (C_q^{j_1 j_2})_{m_1 m_2 J_{12}(m_1+m_2)} \\
&\cdot (C_q^{J_{12} j_3})_{(M-m_3)m_3 JM} \{[2J_{12}+1][2J_{23}+1]\}^{1/2} W[j_1\, j_2\, J\, j_3; J_{12}\, J_{23}]_q ,
\end{aligned}
\tag{31}
$$

where $M = m_1 + m_2 + m_3$ and W is independent of m_i. By making use of the orthogonal relation (20) we have

$$
\begin{aligned}
&(C_q^{J_{12} j_3})_{(M-m_3)m_3 JM} \{[2J_{12}+1][2J_{23}+1]\}^{1/2} W[j_1\, j_2\, J\, j_3; J_{12}\, J_{23}]_q \\
&\quad = \sum_{m_1} (C_q^{j_1 j_2})_{m_1(M-m_1-m_3)J_{12}(M-m_3)} (C_q^{j_2 j_3})_{(M-m_1-m_3)m_3 J_{23}(M-m_1)} \\
&\quad\quad \cdot (C_q^{j_1 J_{23}})_{m_1(M-m_1)JM} ,
\end{aligned}
\tag{32a}
$$

$$
\begin{aligned}
W[j_1 j_2 J j_3; J_{12} J_{23}]_q \delta_{JJ'} &= \{[2J_{12}+1][2J_{23}+1]\}^{-1/2} \sum_{m_1 m_2 m_3} (C_q^{J_{12} j_3})_{(M-m_3)m_3 J'M} \\
&\times (C_q^{j_1 j_2})_{m_1 m_2 J_{12}(M-m_3)} (C_q^{j_2 j_3})_{m_2 m_3 J_{23}(M-m_1)} (C_q^{j_1 J_{23}})_{m_1(M-m_1)JM} ,
\end{aligned}
\tag{33}
$$

and

$$(C_q^{j_1 J_{23}})_{m_1(M-m_1)JM}\{[2J_{12}+1][2J_{23}+1]\}^{1/2}W[j_1j_2Jj_3;J_{12}J_{23}]_q$$
$$=\sum_{m_3}(C_q^{J_{12}j_3})_{(M-m_3)m_3JM}(C_q^{j_1j_2})_{m_1(M-m_1-m_3)J_{12}(M-m_3)} \qquad (32b)$$
$$\cdot(C_q^{j_2j_3})_{(M-m_1-m_3)m_3J_{23}(M-m_1)}\ .$$

We demontrate the q-CG and q-Racah 1 coefficients by graphs[14] in Fig. 1

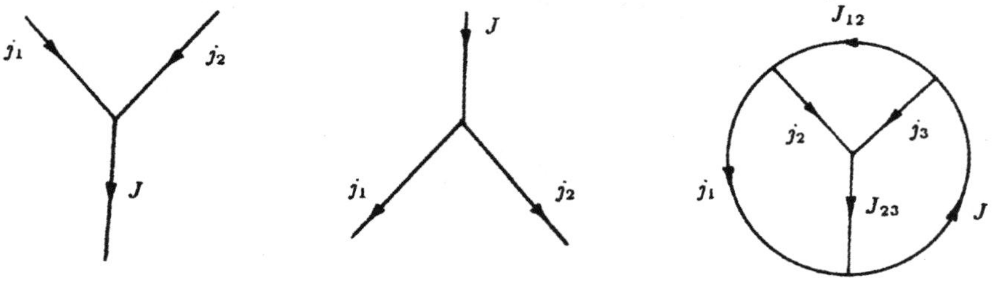

(a) $(C_q^{j_1j_2})_J$ (b) $(\widetilde{C_q^{j_1j_2}})_J$ (c) $\{[2J_{12}+1][2J_{23}+1]\}^{1/2}W[j_1j_2Jj_3;J_{12}J_{23}]_q$

Fig. 1 Graphical representations of q-CG and q-Racah Coefficients.

Through tedious computation[19] we obtain a beautiful expression for q-Racah 1 coefficient which is the same as that for the classical Racah coefficients in quantum mechanics[15] except by replacing round brackets with rectangular ones.

$$\begin{aligned}W[a\,b\,c\,d;e\,f]_q=&(-1)^{a+b+c+d}\Delta(abe)\Delta(dec)\Delta(afc)\Delta(bdf)\\&\times\sum_n(-1)^n[n+1]!\{[n-a-b-e]![n-c-d-e]!\\&\times[n-c-a-f]![n-b-d-f]![a+b+c+d-n]!\\&\times[a+d+e+f-n]![b+c+e+f-n]!\}^{-1}\ .\end{aligned}\qquad(34)$$

Define quantum $6-j$ symbol:

$$\begin{Bmatrix}a & b & e\\ d & c & f\end{Bmatrix}_q=(-1)^{a+b+c+d}W[abcd;ef]_q\ . \qquad (35)$$

There are 144 symmetry relations for q-Racah 1 coefficients

$$\begin{aligned}\begin{Bmatrix}a & b & e\\ d & c & f\end{Bmatrix}_q&=\begin{Bmatrix}b & a & e\\ c & d & f\end{Bmatrix}_q=\begin{Bmatrix}a & e & b\\ d & f & c\end{Bmatrix}_q=\begin{Bmatrix}d & c & e\\ a & b & f\end{Bmatrix}_q\\&=\begin{Bmatrix}(a+b-c+d)/2 & (a+b+c-d)/2 & e\\ (a-b+c+d)/2 & (-a+b+c+d)/2 & f\end{Bmatrix}_q\ .\end{aligned}\qquad(36)$$

V. R-Matrix

In the classical level, the transposition P_{12} commutes with the direct product of two representations of simple Lie algebra

$$D^{j_1j_2}(I)=D^{j_1}(I)\times\mathbb{1}+\mathbb{1}\times D^{j_2}(I)\ ,\qquad D^{j_2j_1}(I)P_{12}=P_{12}D^{j_1j_2}(I)\ . \qquad (37)$$

Because of the symmetry of CG coefficients,

$$P_{12}(C^{j_1j_2})_J = (-1)^{j_1+j_2-J}(C^{j_2j_1})_J\ , \tag{38}$$

r matrix, expressed by the generators of a simple Lie algebra, satisfies the classical YBE:

$$[r_{12}\ , r_{13}] + [r_{12}\ , r_{23}] + [r_{13}\ , r_{23}] = 0\ . \tag{39}$$

For sl(2) algebra, we have[7]

$$r = -\frac{1}{2}h \times h - 2f \times e\ . \tag{40}$$

Note that, the solution to Eq. (39) is not unique, for example, a new solution can be obtained by multiplying a factor, adding a constant and/or transposing. It corresponds to another choice of the quantum parameter q, multiplying a constant to R matrix and/or replacing $R \leftrightarrow R^{-1}$. We choose the solution (40) in order to compare with the R_{AW} matrix obtained by Akutsu-Wadati[10].

In the quantum level, the Lie algebra and the symmetry group are quantized to the quantum enveloping algebra and the braid group, respectively. Then, equation (37) is replaced by the following equation[9]

$$D_q^{j_2j_1}\check{R}_q^{j_1j_2} = \check{R}_q^{j_1j_2}D_q^{j_1j_2}\ , \tag{41}$$

where $D_q^{j_1j_2}$ is given in (16). When $q \sim 1$ we have

$$\check{R}_q = \{1 + (q-1)r\}P\ . \tag{42}$$

In Sec. VI we shall prove that $\check{R}_q^{j_1j_2}$ satisfies the quantum YBE without the spectral parameter u:

$$\check{R}_q^{j_2j_1}\check{R}_q^{j_3j_1}\check{R}_q^{j_3j_2} = \check{R}_q^{j_3j_2}\check{R}_q^{j_3j_1}\check{R}_q^{j_2j_1}\ . \tag{43}$$

Their graphical representations are given in Fig. 2 where the Reidemeister moves II and III in the link theory[18] hold in this case.

From Eqs. (16) and (25c) we have

$$D_q^{j_2j_1}P_{12} = P_{12}D_{q^{-1}}^{j_1j_2}\ , \tag{44a}$$

$$P_{12}(C_q^{j_1j_2})_J = (-1)^{j_1+j_2-J}(C_{q^{-1}}^{j_2j_1})_J\ . \tag{44b}$$

Since

$$D_q^{j_1j_2}(P_{12}\check{R}_{q^{-1}}^{j_1j_2}P_{12}) = (P_{12}\check{R}_{q^{-1}}^{j_1j_2}P_{12})D_q^{j_2j_1}\ ,$$

we define[14]

$$(R_q^{j_1j_2})^{-1} = P_{12}R_{q^{-1}}^{j_1j_2}P_{12}\ , \tag{45}$$

which is a quasi-normalizing condition. Note that

$$(R_q^{j_1j_2})^{-1} \neq R_q^{j_2j_1}\ , \tag{46}$$

so equation (43) is a multiplication rule of a "colored" braid group. When $j_1 = j_2 = j_3$, we have a representation of the usual braid group.

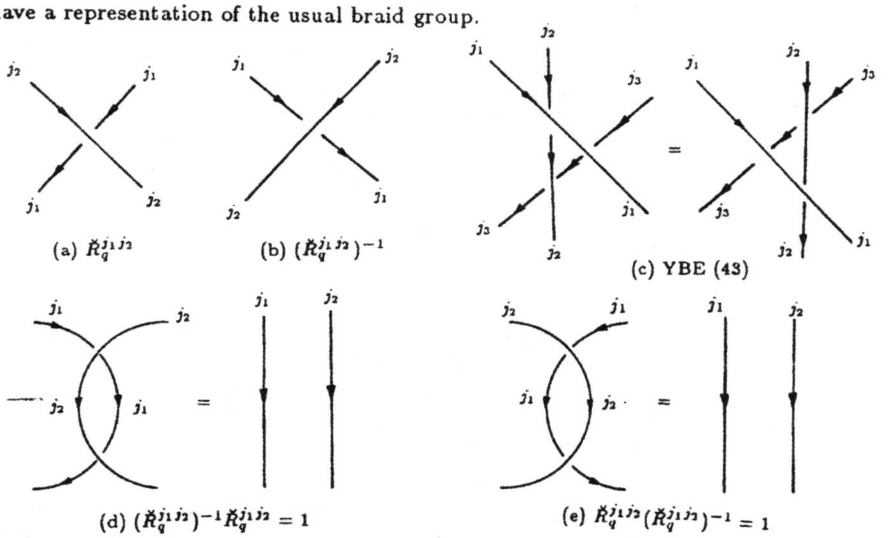

Fig. 2 Graphical representations for R matrices.

Now, we are going to compute the explicit form of $\check{R}$ matrix[14]. From the definitions (41) and (17a) we have

$$(C_q^{j_1 j_2})^{-1}(\check{R}_q^{j_1 j_2})^{-1} D_q^{j_2 j_1} \check{R}_q^{j_1 j_2} C_q^{j_1 j_2} = \bigoplus_J D_q^J = (C_q^{j_2 j_1})^{-1} D_q^{j_2 j_1} C_q^{j_2 j_1} ,$$

so

$$R_q^{j_1 j_2}(C_q^{j_1 j_2})_J = \lambda_J^{j_1 j_2}(q)(C_q^{j_2 j_1})_J . \tag{47}$$

Equation (47) is the quantization form of Eq. (38), and

$$\lambda_J^{j_1 j_2}(q)\Big|_{q=1} = (-1)^{j_1+j_2-J} .$$

Because

$$(C_q^{j_1 j_2})_J = P_{12}\check{R}_{q^{-1}}^{j_1 j_2} P_{12}\check{R}_q^{j_1 j_2}(C_q^{j_1 j_2})_J = \lambda_J^{j_1 j_2}(q)\lambda_J^{j_1 j_2}(q^{-1})(C_q^{j_1 j_2})_J ,$$

we have

$$\lambda_J^{j_1 j_2}(q) = (-1)^{j_1+j_2-J} q^{\alpha} . \tag{48a}$$

where α will be proved to be

$$\alpha = j_1(j_1+1) + j_2(j_2+1) - J(J+1) = -2j_1 j_2 + (j_1+j_2)(j_1+j_2+1) - J(J+1) . \tag{48b}$$

Therefore, we obtain $(2j_2+1)(2j_1+1) \times (2j_1+1)(2j_2+1)$ matrix $\check{R}_q^{j_1 j_2}$

$$\begin{aligned}\check{R}_q^{j_1 j_2} = \widetilde{\check{R}_q^{j_2 j_1}} &= \sum_J \lambda_J^{j_1 j_2}(q)(C_q^{j_2 j_1})_J(\widetilde{C_q^{j_1 j_2}})_J \\ &= q^{-2j_1 j_2}\sum_J (-1)^{j_1+j_2-J} q^{(j_1+j_2)(j_1+j_2+1)-J(J+1)}(C_q^{j_2 j_1})_J(\widetilde{C_q^{j_1 j_2}})_J ,\end{aligned} \tag{49}$$

$$(\check{R}_q^{j_1 j_2})^{-1} = q^{2j_1 j_2}\sum_J (-1)^{j_1+j_2-J} q^{-(j_1+j_2)(j_1+j_2+1)+J(J+1)}(C_q^{j_1 j_2})_J(\widetilde{C_q^{j_2 j_1}})_J ,$$

$$(\check{R}_q^{j_1 j_2})_{m_2 m_1 m_1' m_2'} = 0\,, \qquad \text{if} \qquad m_2 + m_1 \neq m_1' + m_2'\,. \tag{50}$$

Taking the terms in the order $(q-1)$ in Eq. (47), we have

$$(C_q^{j_1 j_2})_J \sim (C^{j_1 j_2})_J + (q-1)(C^{(1)j_1 j_2})_J\,,$$

$$P_{12}(C^{(1)j_1 j_2})_J + rP_{12}(C^{j_1 j_2})_J = (-1)^{j_1+j_2-J}\{\alpha(C^{j_2 j_1})_J + (C^{(1)j_2 j_1})_J\}\,,$$

$$\alpha = (\widetilde{C^{j_2 j_1}})_J r (C^{j_2 j_1})_J + (\widetilde{C^{j_1 j_2}})_J (C^{(1)j_1 j_2})_J - (\widetilde{C^{j_2 j_1}})_J (C^{(1)j_2 j_1})_J\,.$$

The last two terms are vanishing owing to the different permutation parity of $(C^{j_1 j_2})_J$ and $(C^{(1)j_1 j_2})_J$, because

$$P_{12}(C_q^{j_1 j_2})_J = (-1)^{j_1+j_2-J}(C_{q^{-1}}^{j_2 j_1})_J(-1)^{j_1+j_2-J}\{(C^{j_2 j_1})_J - (q-1)(C^{(1)j_2 j_1})_J\}\,.$$

In the representation $D^{j_2 j_1}$, $\widetilde{e \times f} = f \times e$, so

$$(\widetilde{C^{j_2 j_1}})_J (e \times f)(C^{j_2 j_1})_J = (\widetilde{C^{j_2 j_1}})_J (f \times e)(C^{j_2 j_1})_J\,,$$

$$\alpha = -(\widetilde{C^{j_2 j_1}})_J \Big\{\frac{1}{2}h \times h + e \times f + f \times e\Big\}(C^{j_2 j_1})_J\,.$$

Since α is independent of M, from the recursive relation (19) $(q=1)$ we have

$$\begin{aligned}
&\Big\{\Big(\frac{1}{2}h \times h + e \times f + f \times e\Big)C^{j_2 j_1}\Big\}_{m(J-m)JJ} \\
&= 2m(J-m)(C^{j_2 j_1})_{m(J-m)JJ} + \Gamma_m^{j_2}\Gamma_{-J+m}^{j_1}(C^{j_2 j_1})_{(m-1)(J-m+1)JJ} \\
&\quad + \Gamma_{-m}^{j_2}\Gamma_{J-m}^{j_1}(C^{j_2 j_1})_{(m+1)(J-m-1)JJ} \\
&= \{J(J+1) - j_1(j_1+1) - j_2(j_2+1)\}(C^{j_2 j_1})_{m(J-m)JJ}\,,
\end{aligned}$$

so equation (48b) is proved. Equations (47) and (49) can be demonstrated in Fig. 3. The graphical rules in Fig. 3 is a generalization of the Reidemeister move I in link theory[18], because q-CG coefficient denotes a fusion process which does not appear in the usual link theory.

Through calculation we obtain the exact forms of Eq. (49) for $\check{R}_q^{j_1 j_2}$ with $j_1, j_2 = 1/2$, 1, $3/2$, and compare with R_{AW}^j obtained by Akutsu and Wadati[10].

$$\check{R}_q^{\frac{1}{2}\frac{1}{2}} = q^{-1/2}\begin{pmatrix} 1 & & & \\ & 0 & q & \\ & q & 1-q^2 & \\ & & & 1 \end{pmatrix}, \qquad R_{AW}^{\frac{1}{2}} = \begin{pmatrix} 1 & & & \\ & 0 & -\sqrt{t} & \\ & -\sqrt{t} & 1-t & \\ & & & 1 \end{pmatrix},$$

$$(R_q^{11})_{M=\pm 2} = q^{-2}\,, \qquad (R_{AW}^1)_{M=\pm 2} = 1\,,$$

$$(\check{R}_q^{11})_{M=\pm 1} = q^{-2}\begin{pmatrix} 0 & q^2 \\ q^2 & 1-q^4 \end{pmatrix}, \qquad (R_{AW}^1)_{M=\pm 1} = \begin{pmatrix} 1 & t \\ t & 1-t^2 \end{pmatrix},$$

$$(\check{R}_q^{11})_{M=0}=q^{-2}\begin{pmatrix}0 & 0 & q^4\\ 0 & q^2 & -q^3(q^2-q^{-2})\\ q^4 & -q^3(q^2-q^{-2}) & q^6-q^4-q^2+1\end{pmatrix},$$

$$(R_{AW}^1)_{M=0}=\begin{pmatrix}0 & 0 & t^2\\ 0 & t & \sqrt{t}(t^2-1)\\ t^2 & \sqrt{t}(t^2-1) & t^3-t^2-t+1\end{pmatrix},$$

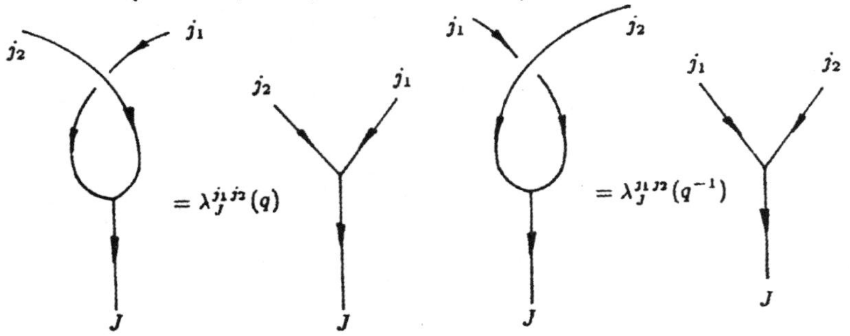

(a)Graphs for Eq. (47)

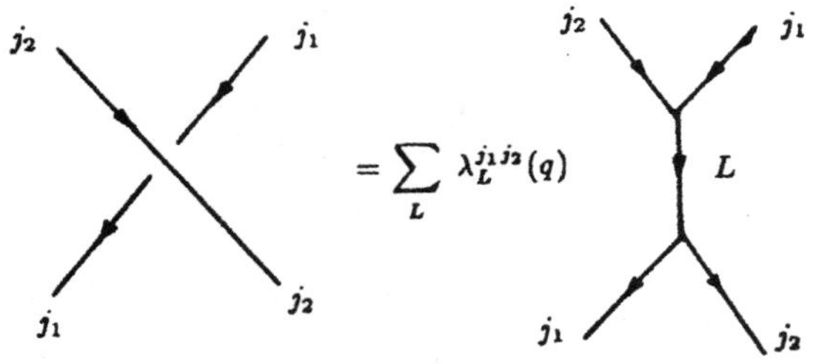

(b) Graph for Eq. (49)

Fig. 3 Graphical representations for the product of $\check{R}$ and q-CG coefficients.

$$(\check{R}_q^{\frac{3}{2}\frac{3}{2}})_{M=\pm 3}=q^{-9/2}, \qquad (\check{R}_q^{\frac{3}{2}\frac{3}{2}})_{M=\pm 2}=q^{-9/2}\begin{pmatrix}0 & q^3\\ q^3 & 1-q^6\end{pmatrix},$$

$$(\check{R}_q^{\frac{3}{2}\frac{3}{2}})_{M=\pm 1}=q^{-9/2}\begin{pmatrix}0 & 0 & q^6\\ 0 & q^4 & \Delta_1\\ q^6 & \Delta_1 & (q^6-1)(q^4-1)\end{pmatrix},$$

$$(\check{R}_q^{\frac{3}{2}\frac{3}{2}})_{M=0}=q^{-9/2}\begin{pmatrix}0 & 0 & 0 & q^9\\ 0 & 0 & q^5 & -q^4(q^6-1)\\ 0 & q^5 & \Delta_2 & \Delta_3\\ q^9 & -q^4(q^6-1) & \Delta_3 & \Delta_4\end{pmatrix},$$

$$(\check{R}_q^{\frac{1}{2}1})_{M=\pm 3/2}=q^{-1},$$

$$(\check{R}_q^{\frac{1}{2}1})_{M=1/2}=(\widetilde{\check{R}_q^{\frac{1}{2}1}})_{M=-1/2}=q^{-1}\begin{pmatrix}0 & q^2\\ q & (1-q^2)(1+q^2)^{1/2}\end{pmatrix},$$

where $\Delta_1 = -q^2(q^2+1)\{(q^6-1)(q^2-1)\}^{1/2}$, $\Delta_2 = -q^2(q^2+1)(q^4-1)$, $\Delta_3 = q(q^6-1)(q^4-1)$, $\Delta_4 = -(q^2-1)(q^4-1)(q^6-1)$.

Except for a factor q^{-2j^2}, $\check{R}_q^{jj}$ coincides with R^j_{AW} by replacing $-\sqrt{t} \leftrightarrow q$. This coincidence is due to the following reasons: i). They satisfy the same YBE (4) without the spectral parameter u. ii). Up to a factor q^{-2j^2}, they have the same eigenvalues (see (48) of this paper and (7.16) of Ref. [4]). iii). Because of the form (40) for r, R takes the form of $A\times B$ where the elements over diagonal line in A and lower diagonal line in B are vanishing, respectively, and $R_{AW}P$ is in the same form. Therefore, equation (49) is the general form of Akutsu-Wadati R^j_{AW} matrices

$$\tilde{R}_q^{jj} = q^{-2j^2} R^j_{AW} \ . \tag{51}$$

VI. The Second Type Quantum Racah Coefficients

We can use q-CG matrix $(C_q^{jj_1})_{J_1}$ to reduce the co-product $D_q^{jj_1}$ of D_q^j and $D_q^{j_1}$, and obtain $D_q^{J_1}$, then, we can use $(C_q^{J_1j_2})_{J_2}$ to reduce the co-product $D_q^{J_1j_2}$ of $D_q^{J_1}$ and $D_q^{j_2}$, and obtain $D_q^{J_2}$. Repeating this process, we can use the product of q-CG matrices

$$(C_q^{jj_1})_{J_1}(C_q^{j_1j_2})_{J_2}\cdots(C^{J_{n-1}j_n})_\Gamma\ , \qquad J_n\equiv\Gamma \tag{52}$$

to reduce the co-product $D_q^{jj_1j_2\cdots j_n}$ of D_q^j and a set of $D_q^{j_1}$, $D_q^{j_2}$, $\cdots D_q^{j_n}$, and to obtain the representation D_q^Γ. For the definite $j, j_1\cdots j_n$, and Γ, there are different sets of J_1, $J_2,\cdots$, and J_{n-1} which satisfy the triangle rule

$$\Delta(J_{i-1}, j_i, J_i): \qquad J_{i-1}+j_i \ge J_i \ge |J_{i-1}-j_i| \tag{53}$$

and are called admissible. An admissible set of J_1, $J_2,\cdots$, J_{n-1} is called a fusion path[17] with the corresponding basis (52). These bases corresponding to the different fusion paths are linearly independent of each other. Note that there is no summation for the matrix indices in the basis (52), in fact, owing to the condition (18a). Multiplying the bases by $\check{R}_q^{j_ij_{i+1}}$ from the left, we obtain a matrix representation W of $\check{R}_q^{j_ij_{i+1}}$ which can also be proved to satisfy the quantum YBE, and is nothing but the solution for the face model. Kohno[17] discussed the simplest case where $j_1 = j_2 = \cdots j_n$ = fundamental representation. Now, we discuss the general one.

Let

$$\sum_{m_1'm_2'}(\check{R}_q^{j_1j_2})_{m_2m_1m_1'm_2'}(C_q^{jj_1})_{mm_1'J(m+m_1')}(C_q^{Jj_2})_{(m+m_1')m_2'\Gamma M}$$

$$=\sum_{J'}(C_q^{jj_2})_{mm_2J'(m+m_2)}(C_q^{J'j_1})_{(m+m_2)m_1\Gamma M}W\begin{bmatrix} j & J \\ J' & \Gamma\end{bmatrix}; j_1j_2q\Big] , \tag{54a}$$

$$M = m+m_1+m_2 = m+m_1'+m_2' ,$$

where $W\left[\begin{matrix} j & J \\ J' & \Gamma\end{matrix}; j_1j_2q\right]$ is called the second type quantum Racah (q-Racah 2) coefficients. In the matrix form, equation (54a) can be written briefly as

$$(\check{R}_q^{j_1j_2})(C_q^{jj_1})_J(C_q^{Jj_2})_\Gamma = \sum_{J'}(C_q^{jj_2})_{J'}(C_q^{J'j_1})_\Gamma W\left[\begin{matrix} j & J \\ J' & \Gamma\end{matrix}; j_1j_2q\right] . \tag{54b}$$

Because of the orthogonal relation (20) for q-CG coefficients, we obtain from (54)

$$\begin{aligned}(C_q^{J'j_1})_{(M-m_1)m_1\Gamma M}W\left[\begin{matrix} j & J \\ J' & \Gamma\end{matrix}; j_1j_2q\right] &= \sum_{mm_1'}(C_q^{jj_2})_{m(M-m-m_1)J'(M-m_1)}\\ &\times(\check{R}_q^{j_1j_2})_{(M-m-m_1)m_1m_1'(M-m-m_1')}(C_q^{jj_1})_{mm_1'J(m+m_1')}\\ &\times(C_q^{Jj_2})_{(m+m_1')(M-m-m_1')\Gamma M}\ ,\end{aligned} \tag{55}$$

$$\begin{aligned}\delta_{\Gamma\Gamma'}W\left[\begin{matrix} j & J \\ J' & \Gamma\end{matrix}; j_1j_2q\right] = &\sum_{mm_1m_2m_1'm_2'}(C_q^{J'j_1})_{(M-m_1)m_1\Gamma' M}(C_q^{jj_2})_{mm_2J'(M-m_1)}\\ &\times(\check{R}_q^{j_1j_2})_{m_2m_1m_1'm_2'}(C_q^{jj_1})_{mm_1'J(m+m_1')}(C_q^{Jj_2})_{(m+m_1')m_2'\Gamma M}\ ,\end{aligned} \tag{56}$$

where $M = m + m_1 + m_2 = m + m_1' + m_2'$. Substituting Eq. (49) into Eq. (56), and using Eqs. (32b) and (20) we obtain the relation between two types of quantum Racah coefficients:

$$\begin{aligned}W\left[\begin{matrix} j & J \\ J' & \Gamma\end{matrix}; j_1j_2q\right] &= \{[2J'+1][2J+1]\}^{1/2}\sum_L(-1)^{j_1+j_2-L}[2L+1]\\ &\times W[jj_2\Gamma j_1\,; J'L]_qW[jj_1\Gamma j_2\,; JL]_q\,q^{j_1(j_1+1)+j_2(j_2+1)-L(L+1)}\ .\end{aligned} \tag{57}$$

We can demonstrate Eqs. (54), (56) and (57) in Fig. 4. Note that figure 4(c) is an application of Fig. 3(b).

In the next paper[19], from Eqs. (57), (32b) and (20) we will prove a quantum Racah sum rule[15] which gives a direct relation between q-Racah 1 and q-Racah 2 coefficients:

$$\sum_L\lambda_L^{j_1j_2}(q)[2L+1]W[jj_2\Gamma j_1; J'L]_qW[jj_1\Gamma j_2; JL]_q = \lambda_{J'}^{j_2j}(q^{-1})\lambda_\Gamma^{j_2J}(q)W[j_2j\Gamma j_1; J'J]_q\ . \tag{58}$$

Then,

$$W\left[\begin{matrix} j & J \\ J' & \Gamma\end{matrix}; j_1j_2q\right] = \{[2J'+1][2J+1]\}^{1/2}\lambda_{J'}^{j_2j}(q^{-1})\lambda_\Gamma^{j_2J}(q)W[j_2j\Gamma j_1;\ J'J]_q\ , \tag{59}$$

where $\lambda_J^{j_1j_2}(q)$ is given in Eq. (48). In terms of this important relation (59), we deduce the famous pentagonal relation from the definitions (31) and (54) for q-Racah coefficients

$$\begin{aligned}(C_q^{j_2j_3})_{J_{23}}\check{R}_q^{J_{23}j_1}(C_q^{J_{23}j_1})_J &= \lambda_J^{j_1J_{23}}(q)(C_q^{j_2j_3})_{J_{23}}(C_q^{j_1J_{23}})_J\\ &= \lambda_J^{j_1J_{23}}(q)\sum_{J_{12}}(C_q^{j_1j_2})_{J_{12}}(C_q^{J_{12}j_3})_J\{[2J_{12}+1][2J_{23}+1]\}^{1/2}W[j_1\,j_2\,J\,j_3; J_{12}\,J_{23}]_q\\ &= \sum_{J_{12}}\lambda_{J_{12}}^{j_1j_2}(q)(C_q^{j_1j_2})_{J_{12}}(C_q^{J_{12}j_3})_JW\left[\begin{matrix} j_2 & J_{23} \\ J_{12} & J\end{matrix}; j_3j_1q\right]\\ &= \check{R}_q^{j_2j_1}\sum_{J_{12}}(C_q^{j_2j_1})_{J_{12}}(C_q^{J_{12}j_3})_JW\left[\begin{matrix} j_2 & J_{23} \\ J_{12} & J\end{matrix}; j_3j_1q\right]\\ &= \check{R}_q^{j_2j_1}\check{R}_q^{j_3j_1}(C_q^{j_2j_3})_{J_{23}}(C_q^{J_{23}j_1})_J\ .\end{aligned}$$

Therefore, we obtain the pentagonal relation

$$\check{R}_q^{j_2j_1}\check{R}_q^{j_3j_1}(C_q^{j_2j_3})_J = (C_q^{j_2j_3})_J\check{R}_q^{Jj_1} \tag{60a}$$

or

$$\sum_{mn_1n_3} (\check{R}_q^{j_2j_1})_{m_1m_2mn_1} (\check{R}_q^{j_3j_1})_{n_1m_3n_3m'_1} (C_q^{j_2j_3})_{mn_3JM}$$
$$= (C_q^{j_2j_3})_{m_2m_3J(m_2+m_3)} (R_q^{Jj_1})_{m_1(m_2+m_3)Mm'_1} , \quad (60b)$$

where $M = m_1 + m_2 + m_3 - m'_1$, $n_1 = m_1 + m_2 - m$ and $n_3 = M - m$.

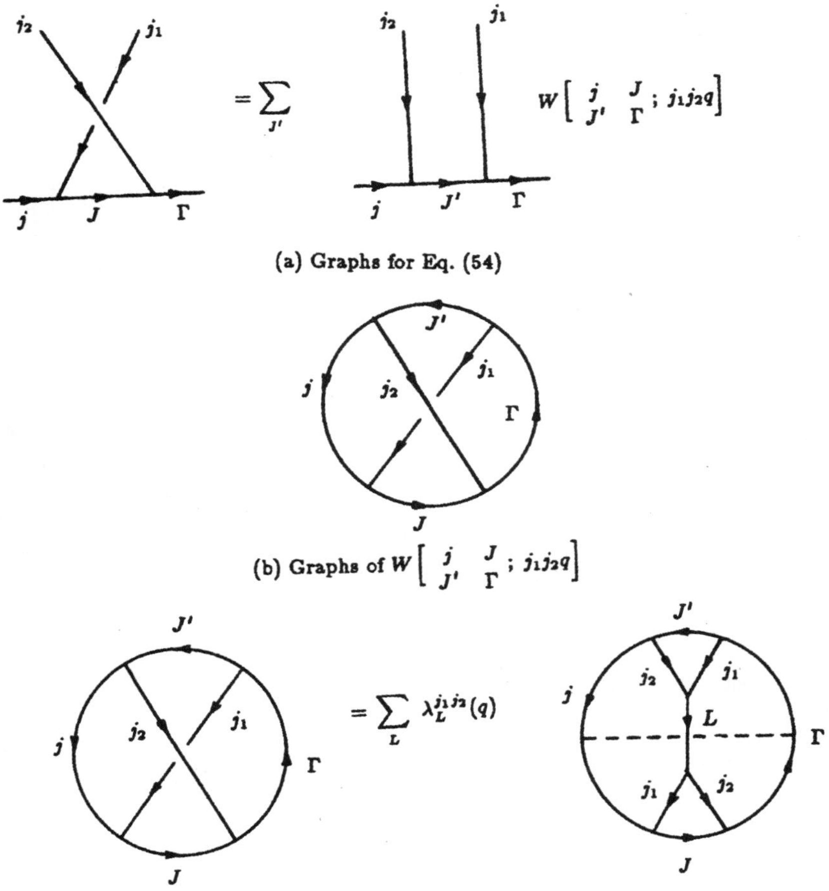

(a) Graphs for Eq. (54)

(b) Graphs of $W\begin{bmatrix} j & J \\ J' & \Gamma \end{bmatrix}; j_1j_2q$

(c) Graph for Eq. (57)

Fig. 4 Graphical representations related to q-Racah 2 coefficients.

In the graphical representation, the pentagonal relation denotes a generalised Reidemeister move III (Fig. 5, (a)), and it is easy to understand (59) by graphical rules (Fig. 5, (b))

Now, it is easy to prove the quantum YBE (43) satisfied by $\check{R}_q^{j_1j_2}$ matrices in terms of the pentagonal relations (60) and (47)

$$\check{R}_q^{j_2j_1}\check{R}_q^{j_3j_1}\check{R}_q^{j_3j_2}(C_q^{j_3j_2})_J = \lambda_J^{j_3j_2}(q)\check{R}_q^{j_2j_1}\check{R}_q^{j_3j_1}(C_q^{j_2j_3})_J$$
$$= \lambda_J^{j_3j_2}(q)(C_q^{j_2j_3})_J\check{R}_q^{Jj_1} = \check{R}_q^{j_3j_2}(C_q^{j_3j_2})_J\check{R}_q^{Jj_1} \quad (61)$$
$$= \check{R}_q^{j_3j_2}\check{R}_q^{j_3j_1}\check{R}_q^{j_2j_1}(C_q^{j_3j_2})_J .$$

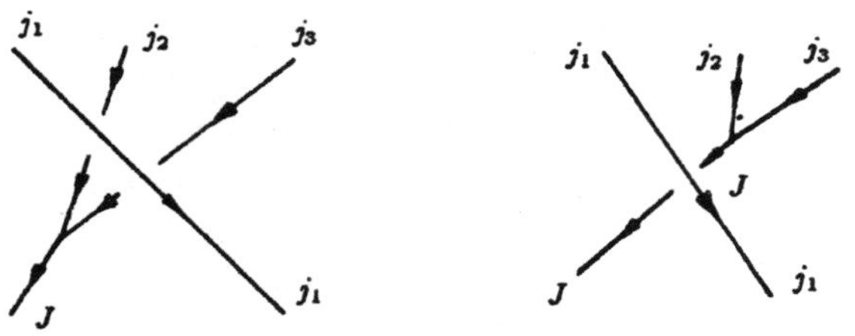

(a) Pentagonal relation

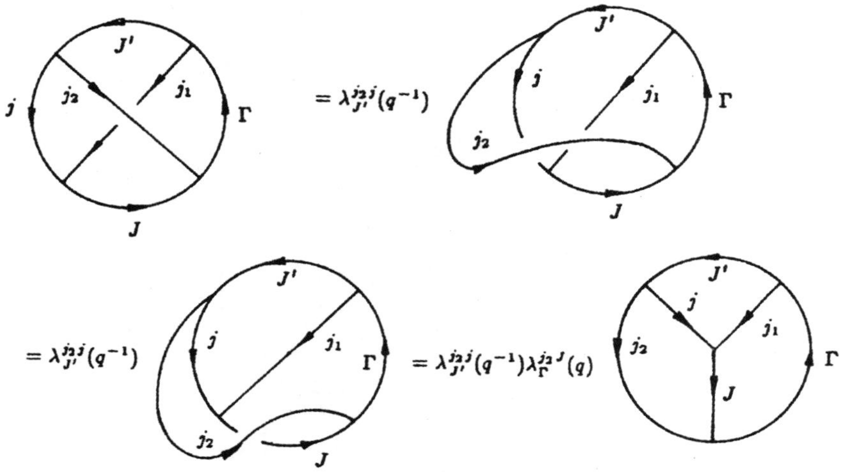

(b) Graphical representaton for Eq. (59)

Fig. 5 The graphical representation for the pentagonal relation.

In the bases (52) we have

$$\sum_{b_3a_2L}(C_q^{a_1j_1})_{b_3}(C_q^{b_3j_2})_{a_2}(C_q^{a_2j_3})_{b_1}W\left[\begin{matrix}a_1 & L\\ b_3 & a_2\end{matrix};j_2j_1q\right]W\left[\begin{matrix}L & a_3\\ a_2 & b_1\end{matrix};j_3j_1q\right]$$

$$\cdot W\left[\begin{matrix}a_1 & b_2\\ L & a_3\end{matrix};j_3j_2q\right]=\check{R}_q^{j_2j_1}\check{R}_q^{j_3j_1}\check{R}_q^{j_3j_2}(C_q^{a_1j_3})_{b_2}(C_q^{b_2j_2})_{a_3}(C_q^{a_3j_1})_{b_1}$$

$$=\check{R}_q^{j_3j_2}\check{R}_q^{j_3j_1}\check{R}_q^{j_2j_1}(C_q^{a_1j_3})_{b_2}(C_q^{b_2j_2})_{a_3}(C_q^{a_3j_1})_{b_1}=\sum_{a_2b_3L}(C_q^{a_1j_1})_{b_3}(C_q^{b_3j_2})_{a_2}(C_q^{a_2j_3})_{b_1}$$

$$\cdot W\left[\begin{matrix}b_3 & L\\ a_2 & b_1\end{matrix};j_3j_2q\right]W\left[\begin{matrix}a_1 & b_2\\ b_3 & L\end{matrix};j_3j_1q\right]W\left[\begin{matrix}b_2 & a_3\\ L & b_1\end{matrix};j_2j_1q\right],$$

that is

$$\sum_L W\left[\begin{matrix}a_1 & L\\ b_3 & a_2\end{matrix};j_2j_1q\right]W\left[\begin{matrix}L & a_3\\ a_2 & b_1\end{matrix};j_3j_1q\right]W\left[\begin{matrix}a_1 & b_2\\ L & a_3\end{matrix};j_3j_2q\right]$$

$$=\sum_L W\left[\begin{matrix}b_3 & L\\ a_2 & b_1\end{matrix};j_3j_2q\right]W\left[\begin{matrix}a_1 & b_2\\ b_3 & L\end{matrix};j_3j_1q\right]W\left[\begin{matrix}b_2 & a_3\\ L & b_1\end{matrix};j_2j_1q\right]. \tag{62}$$

When $j_1 = j_2 = j_3 = j$, equation (62) is nothing but the quantum YBE for the face model[17,20] with the correspondence (see Fig. 6)

$$W\begin{bmatrix} b & c \\ a & d \end{bmatrix}; jjq\Big] \to W[a\,b\,c\,d|u]\Big|_{u=\text{const.}} \tag{63}$$

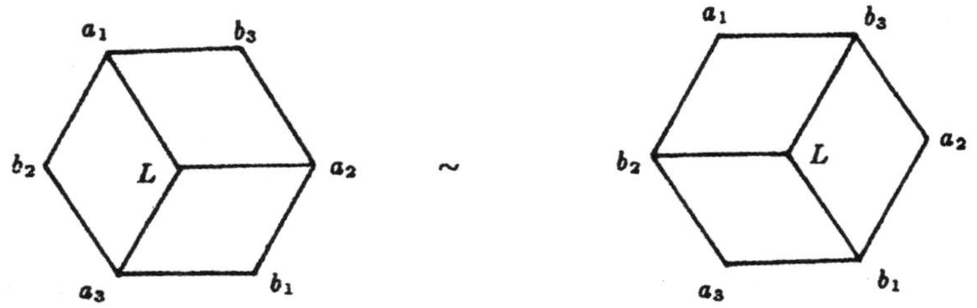

Fig. 6 Graph for the face model.

Substituting Eq. (59) into Eq. (62) we obtain a relation of the quantum $6-j$ symbols which describes the symmetry of the quantum $9-j$ symbol which reduces to that of the classical $9-j$ symbol[15] when $q \to 1$:

$$\begin{aligned}
\begin{Bmatrix} a_1 & j_1 & b_3 \\ j_3 & b_1 & a_2 \\ b_2 & a_3 & j_2 \end{Bmatrix}_q &= \sum_L (-1)^{2L}[2L+1]\begin{Bmatrix} a_1 & j_1 & b_3 \\ a_2 & j_2 & L \end{Bmatrix}_q \\
&\quad \cdot \begin{Bmatrix} j_3 & b_1 & a_2 \\ j_1 & L & a_3 \end{Bmatrix}_q \begin{Bmatrix} b_2 & a_3 & j_2 \\ L & a_1 & j_3 \end{Bmatrix}_q \\
&= \sum_L (-1)^{2L}[2L+1]\begin{Bmatrix} b_3 & a_2 & j_2 \\ b_1 & L & j_3 \end{Bmatrix}_q \\
&\quad \begin{Bmatrix} a_1 & j_3 & b_2 \\ L & j_1 & b_3 \end{Bmatrix}_q \begin{Bmatrix} j_1 & b_1 & a_3 \\ j_2 & b_2 & L \end{Bmatrix}_q .
\end{aligned} \tag{64}$$

Note added in proof

The computation of this paper was completed in January 1989. From the paper of Alvarez-Gaume *et al.* (Phys. Lett. **B220**(1989)142), in April we were aware of a similar computation done by Kirillov and Reshetikhin. In May we read their paper (A.M. Kirillov and N.Yu. Reshetikhin, LOMI preprint E-9-88) and found that there are some differences in the starting point and the derivation. Our paper may be easier to be accepted by physicists because it is a quantum analogue of the theory of angular momentum.

Acknowledgement

The authors would like to thank Prof. Kang-Jie SHI for helpful discussions.

References

[1] C.N. Yang, Phys. Rev. Lett. **19**(1967)1312.

[2] R.J. Baxter, *Exactly Solved Models in Statistical Mechanics*, London, Academic Press (1982), Ann. Phys. **70**(1972)193.

[3] L. Faddeev, *Integrable Models in 1+1-Dimensional Quantum Field Theory*, in Les Houches Lecture (1982), Elsevier, Amsterdam (1984).

[4] M. Wadati and Y. Akutsu, Prog. Theor. Phys. Supp. No.94 (1988).

[5] A.A. Belavin and V.G. Drinfel'd, Funct. Anal. Appl. **17**(1983)220.

[6] P.P. Kulish, N.Yu. Reshetikhin and E.K. Sklyanin, Lett. Math. Phys. **5**(1981)393.

[7] D.I. Olive and N. Turok, Nucl. Phys. **B220**(1983)491.

[8] E.K. Sklyanin, Func. Anal. Supl. **4**(1982)27, (in Russian).

[9] M. Jimbo, Comm. Math. Phys. **102**(1986)537; Lett. Math. Phys. **10**(1985)63; **11**(1986)247.

[10] Y. Akutsu and M. Wadati, J. Phys. Soc. Jpn. **56**(1987)3039; Comm. Math. Phys. **117**(1988)243.

[11] Y. Akutsu, T. Deguchi and M. Wadati, J. Phys. Soc. Jpn. **56**(1987)3464; **57**(1988)1173; T. Deguchi, Y. Akutsu and M. Wadati, J. Phys. Soc. Jpn. **57**(1988)757,1905.

[12] V.F. R. Jones, Bull. Am. Math. Soc. **12**(1985)103.

[13] J.S. Birman, Invent. Math. **81**(1985)287.

[14] N.Yu. Reshetikhin, *Quantized Universal Enveloping Algebras, the Yang-Baxter Equation and Invariants of Links I and II*, LOMI reprints, E-4-87 and E-17-87, Steklov Mathematical Institute, Leningrad Dept. (1988).

[15] L.C. Biedenharn and J.D. Louck, *Angular Momentum in Quantum Physics, Theory and Application, Encyclopedia of Mathematics and Its Applications*, ed. Gian-Carl Rots, Addison-Wisley Publishing Co. Reading, Massachusetts, Vol. 8 (1981).

[16] A.R. Edmonds, *Angular Momentum in Quantum Mechanics*, Princeton University Press (1957).

[17] T. Kohno, *Fusion for the Monodromy of Braid Groups*, Lectures given at Nankai Institute of Mathematics in Tianjin in September, 1988.

[18] L.H. Kauffman, *State Models for Link Polynomials*, IHES/M/88/46, Sept. 1988, Topology, **26**(1987)395.

[19] Bo-Yu HOU, Bo-Yuan HOU and Zhong-Qi MA, *Clebsch-Gordan Coefficients, Racah Coefficients, and Braiding Fusion of Quantum sl (2) Enveloping Algebra (II)*, BIHEP-TH-89-8 (NWU-IMP-89-12), to appear in Commun. Theor. Physics.

[20] M. Jimbo, T. Miwa and M. Okado, Lett. Math. Phys. **14**(1987)123.

Commun. Theor. Phys. 13(1990)341-354
© China Ocean Press
Vol. 13, No. 3

Clebsch–Gordan Coefficients, Racah Coefficients and Braiding Fusion of Quantum sl(2) Enveloping Algebra II[1]

Bo–yu HOU
Institute of Modern Physics, Northwest University, Xi'an 710069, China
Bo–yuan HOU
Graduate School, Chinese Academy of Sciences, P.O. Box 3908, Beijing 100039, China
Zhong–qi MA
Institute of High Energy Physics, P.O. Box 918 (4), Beijing 100039, China

(Received April 1, 1989)
(Accepted December 8, 1989)

Abstract

The explicit forms of the quantum Clebsch–Gordan coefficients and the first–type quantum Racah coefficients for the quantum sl(2) enveloping algebra are computed in detail. Exact values of some coefficients are listed. An important relation, quantum Racah sum rule, is proved.

I. Introduction

Classical Yang–Baxter equation (YBE) is a commutative equation. From its one solution r based on a representation of a simple Lie algebra or a Kac–Moody algebra, a set of solutions based on all representations follows. To find out the corresponding solution R of the quantum YBE[1,2] having the solution r as its classical limit is quite a difficult problem. A typical method is based on the quantum enveloping algebra[3] which has drawn the considerable interests of both physicists and mathematicians. In the preceding paper[4] (paper I), we discussed the properties of the simplest quantum enveloping algebra sl(2) and its representations, but left some detailed computations in this paper, i.e., those on the explicit forms of the quantum Clebsch–Gordan (q-CG) coefficients (Sec. III), the quantum Racah (q–Racah 1) coefficients (Sec. IV), and on a proof for the quantum Racah sum rule (Sec. V). In addition to the other reasons, these computations are interesting because all the results contain but generalise those in the theory of angular momentum, and reduce to them when $q \to 1$. As tools, some fundamental identities were given by Andrews[5], but in order to be completive, we will list them in Sec. II.

II. Fundamental Identities

Andrews[5] has given some identities related to the number of partitions and to the Gaussian polynomials. In applications to our problems, we list some of them, and change their

[1]The project supported by National Natural Science Foundation of China through the Nankai Institute of Mathematics.

forms to be more convenient ones. The readers can find the proof of them in Andrew's book[5], or prove them by induction. Define

$$[n]=\frac{q^n-q^{-n}}{q-q^{-1}}\,,\qquad [-n]=-[n]\,,\qquad [1]=1\,,\qquad [0]=0\,, \tag{1}$$

and

$$[n]!=[n][n-1]\cdots[1]\,,\qquad [0]!=1\,,\qquad [-n]!=\infty \tag{2}$$

where n is a positive integer. Obviously, $[n]$ does not change when q is replaced by q^{-1}. Throughout this paper, q is not a root of one, and

$$\lim_{q\to 1}[n]=n\,. \tag{3}$$

The symbol

$$\begin{bmatrix} n \\ m \end{bmatrix}_q=\begin{bmatrix} n \\ n-m \end{bmatrix}_q \tag{4}$$

is defined by

$$\begin{bmatrix} n \\ m \end{bmatrix}_q=\begin{cases} \dfrac{[n]!}{[m]![n-m]!} & 0\le m\le n\,, \\ 0 & \text{otherwise}\,, \end{cases} \tag{5}$$

which is related to the Gaussian polynomials by

$$\left[\!\left[\begin{matrix} n \\ m \end{matrix}\right]\!\right]_{t=q^2}=\frac{(1-t^n)(1-t^{n-1})\cdots(1-t^{n-m+1})}{(1-t)(1-t^2)\cdots(1-t^m)}=q^{m(n-m)}\begin{bmatrix} n \\ m \end{bmatrix}_q\,. \tag{6}$$

Then, we have the following identities:

$$\begin{bmatrix} n \\ m \end{bmatrix}_q=q^{\pm m}\begin{bmatrix} n-1 \\ m \end{bmatrix}_q+q^{\mp(n-m)}\begin{bmatrix} n-1 \\ m-1 \end{bmatrix}_q\,, \tag{7}$$

$$\begin{bmatrix} n+m+1 \\ m+1 \end{bmatrix}_q=q^{\mp n(m+1)}\sum_{r=0}^{n}q^{\pm r(m+2)}\begin{bmatrix} m+r \\ m \end{bmatrix}_q\,, \tag{8}$$

$$\prod_{r=0}^{n-1}(1-zq^{2r})=\sum_{m=0}^{n}(-1)^m\begin{bmatrix} n \\ m \end{bmatrix}_q z^m q^{m(n-1)}\,, \tag{9}$$

$$\prod_{r=0}^{n-1}(1-zq^{2r})^{-1}=\sum_{m=0}^{\infty}\begin{bmatrix} n+m-1 \\ m \end{bmatrix}_q z^m q^{m(n-1)}\,, \tag{10}$$

$$\begin{bmatrix} u+v \\ r \end{bmatrix}_q=\sum_p q^{\pm\{p(u+v)-ru\}}\begin{bmatrix} u \\ p \end{bmatrix}_q\begin{bmatrix} v \\ r-p \end{bmatrix}_q\,,$$

$$\sum_p q^{\pm\{p(u+v)-ru\}}\{[p]![u-p]![r-p]![v-r+p]!\}^{-1}=[u+v]!\{[u]![v]![r]![u+v-r]!\}^{-1}\,, \tag{11}$$

$$\begin{bmatrix} u+v+r-1 \\ r \end{bmatrix}_q=\sum_p q^{\pm\{p(u+v)-ru\}}\begin{bmatrix} u+p-1 \\ p \end{bmatrix}_q\begin{bmatrix} v+r-p-1 \\ r-p \end{bmatrix}_q\,,$$

$$\sum_p q^{\pm\{p(u+v)-ru\}}\frac{[u+p-1]![v+r-p-1]!}{[p]![r-p]!}=\frac{[u+v+r-1]![u-1]![v-1]!}{[r]![u+v-1]!}\,, \tag{12}$$

$$\begin{bmatrix} v-u \\ r \end{bmatrix}_q = \sum_p (-1)^p q^{\pm\{p(v-u-r+1)+ru\}} \begin{bmatrix} u \\ p \end{bmatrix}_q \begin{bmatrix} v-p \\ r-p \end{bmatrix}_q ,$$
$$\sum_p (-1)^p q^{\pm\{p(v-u-r+1)+ru\}} \frac{[v-p]!}{[p]![u-p]![r-p]!} = \frac{[v-u]![v-r]!}{[u]![r]![v-u-r]!} . \tag{13}$$

At last, let $p(N,M,n)$ denote the number of partitions of n into at most M parts, each $\leq N$, and we have

$$\sum_{n=0}^{NM} p(N,M,n) q^{NM-2n} = \begin{bmatrix} N+M \\ N \end{bmatrix}_q . \tag{14}$$

III. Quantum Clebsch–Gordan Coefficients

Quantum Clebsch–Gordan (q-CG) matrix $C_q^{j_1 j_2}$ is a similarity transformation to reduce the co–product representation $D_q^{j_1 j_2}$ of the quantum sl(2) enveloping algebra (q-sl(2)) to the irreducible one

$$\left(C_q^{j_1 j_2}\right)^{-1} D_q^{j_1 j_2} C_q^{j_1 j_2} = \bigoplus_{J=|j_1-j_2|}^{j_1+j_2} D_q^J . \tag{15}$$

From Eq. (15) we have

$$\left(C_q^{j_1 j_2}\right)_{m_1 m_2 J M} = 0 \quad \text{if} \quad M \neq m_1 + m_2 , \tag{16a}$$

$$\Gamma^{j_1}_{\pm m_1}(q) q^{-m_2} \left(C_q^{j_1 j_2}\right)_{(m_1 \mp 1) m_2 J M} + \Gamma^{j_2}_{\pm m_2}(q) q^{m_1} \left(C_q^{j_1 j_2}\right)_{m_1 (m_2 \mp 1) J M}$$
$$= \left(C_q^{j_1 j_2}\right)_{m_1 m_2 J (M \pm 1)} \Gamma^J_{\mp M}(q) , \tag{16b}$$

where

$$\Gamma^j_m(q) = \{[j+m][j-m+1]\}^{1/2} = \Gamma^j_{-m+1}(q) . \tag{17}$$

When q is real, we can choose q–CG matrix to be unitary and

$$\left(C_q^{j_1 j_2}\right)_{j_1 - j_2 J (j_1 - j_2)} = \text{positive real} , \tag{18}$$

then, it is easy to show from Eq. (16b) that $\left(C_q^{j_1 j_2}\right)_{m_1 m_2 J (m_1 + m_2)}$ is positive real when $m_1 = j_1$ or $m_2 = -j_2$, and real for any m_1 and m_2. Therefore, $C_q^{j_1 j_2}$ is real orthogonal when q is real, and by analytic continuation $C_q^{j_1 j_2}$ is orthogonal when q is complex

$$\sum_J \left(C_q^{j_1 j_2}\right)_{m(M-m)JM} \left(C_q^{j_1 j_2}\right)_{m'(M-m')JM} = \delta_{mm'}$$
$$\sum_m \left(C_q^{j_1 j_2}\right)_{m(M-m)JM} \left(C_q^{j_1 j_2}\right)_{m(M-m)J'M} = \delta_{J'J} . \tag{19}$$

At first, we compute $\left(C_q^{j_1 j_2}\right)_{m(J-m)JJ}$. From Eq. (16b) with the upper signs we have

$$\begin{aligned}\left(C_q^{j_1 j_2}\right)_{m(J-m)JJ} &= -\frac{\Gamma^{j_2}_{J-m}(q)}{\Gamma^{j_1}_{m+1}(q)} q^{J+1}\left(C_q^{j_1 j_2}\right)_{(m+1)(J-m-1)JJ} \\ &= (-1)^{j_1-m} q^{(J+1)(j_1-m)} \frac{\Gamma^{j_2}_{J-m}(q)\Gamma^{j_2}_{J-m-1}(q)\cdots\Gamma^{j_2}_{J-j_1+1}(q)}{\Gamma^{j_1}_{m+1}(q)\Gamma^{j_1}_{m+2}(q)\cdots\Gamma^{j_1}_{j_1}(q)} \\ &\times\left(C_q^{j_1 j_2}\right)_{j_1(J-j_1)JJ} = (-1)^{j_1-m} q^{(J+1)(j_1-m)} \\ &\times\left\{\frac{[j_2+J-m]![j_2-J+j_1]![j_1+m]!}{[j_2+J-j_1]![j_2-J+m]![j_1-m]![2j_1]!}\right\}^{1/2}\left(C_q^{j_1 j_2}\right)_{j_1(J-j_1)JJ} .\end{aligned}$$

From the orthogonal condition (19) and identity (12) with $r=j_1+j_2-J$, $u=j_2+J-j_1+1$, $v=j_1+J-j_2+1$, $p=j_1-m$, we obtain

$$\left(C_q^{j_1 j_2}\right)_{j_1(J-j_1)JJ} = q^{-\{j_2(j_2+1)-(J-j_1)(J-j_1+1)\}/2}\left\{\frac{\begin{bmatrix}2j_1\\ J+j_1-j_2\end{bmatrix}_q}{\begin{bmatrix}j_1+j_2+J+1\\ 2J+1\end{bmatrix}_q}\right\}^{1/2}, \tag{20a}$$

$$\begin{aligned}\left(C_q^{j_1 j_2}\right)_{m(J-m)JJ} &= (-1)^{j_1-m} q^{\{j_1(j_1+1)-j_2(j_2+1)+(J+1)(J-2m)\}/2} \\ &\cdot\left\{\frac{\begin{bmatrix}j_2+J-m\\ j_2+J-j_1\end{bmatrix}_q\begin{bmatrix}j_1+m\\ j_1+J-j_2\end{bmatrix}_q}{\begin{bmatrix}j_1+j_2+J+1\\ 2J+1\end{bmatrix}_q}\right\}^{1/2} .\end{aligned} \tag{20b}$$

By the way, we have from Eq. (20a)

$$\left(C_{q^{-1}}^{j_1 j_2}\right)_{m(J-m)JJ} = q^{-j_1(j_1+1)+j_2(j_2+1)-(J+1)(J-2m)}\left(C_q^{j_1 j_2}\right)_{m(J-m)JJ} . \tag{21}$$

Secondly, we compute $\left(C_q^{j_1 j_2}\right)_{m(M-m)JM}$ by the recursive relation (16b) with the lower sign,

$$\begin{aligned}\left(C_q^{j_1 j_2}\right)_{m(M-m)JM} &= \frac{\Gamma^{j_1}_{-m}(q)}{\Gamma^{J}_{M+1}(q)} q^{-M+m}\left(C_q^{j_1 j_2}\right)_{(m+1)(M-m)J(M+1)} \\ &+\frac{\Gamma^{j_2}_{-M+m}(q)}{\Gamma^{J}_{M+1}(q)} q^{m}\left(C_q^{j_1 j_2}\right)_{m(M-m+1)J(M+1)} .\end{aligned} \tag{22}$$

By making use of Eq. (22) iteratively, we can raise M to J, and express $\left(C_q^{j_1 j_2}\right)_{m(M-m)JM}$ by $\left(C_q^{j_1 j_2}\right)_{s(J-s)JJ}$. There are different paths to raise M. The factors of Γ do not depend on the path, and equal to

$$\begin{aligned}&\frac{\Gamma^{j_1}_{-m}(q)\Gamma^{j_1}_{-m-1}(q)\cdots\Gamma^{j_1}_{-s+1}(q)\Gamma^{j_2}_{-M+m}(q)\Gamma^{j_2}_{-M+m-1}(q)\cdots\Gamma^{j_2}_{-J+s+1}(q)}{\Gamma^{J}_{M+1}(q)\Gamma^{J}_{M+2}(q)\cdots\Gamma^{J}_{J}(q)} \\ &=\left\{\frac{[j_1-m]![j_1+s]![j_2-M+m]![j_2+J-s]![J+M]!}{[j_1-s]![j_1+m]![j_2-J+s]![j_2+M-m]![J-M]![2J]!}\right\}^{1/2} .\end{aligned} \tag{23}$$

But, the power of q depends on the path. For example, the power for the path $(m_1, m_2) \to (m_1+1, m_2) \to (m_1+1, m_2+1)$ is larger than that for the path $(m_1, m_2) \to (m_1, m_2+1) \to (m_1+1, m_2+1)$ by two: $(-m_2+m_1+1)-(m_1-m_2-1)=2$. It can be seen in Fig. 1 that the path I corresponds to the largest power of q,

$$-(M-m)(s-m)+s(J-s-M+m)=(s-m)(J-s-M+m)+mJ-sM\,, \tag{24}$$

and the power of q for the other path II is less than (24) by $2n$ where n is the area between two paths

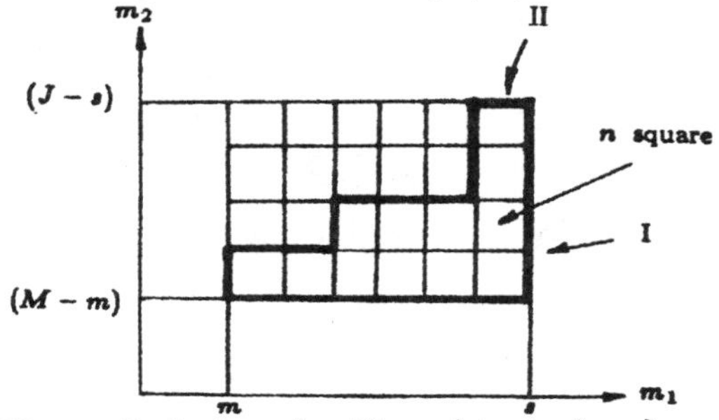

Fig. 1. Paths from $(m, M-m)$ to $(s, J-s)$.

The number of paths under which the area is n is the number of partitions $p(s-m, J-s-M+m, n)$, so all the contributions to the power of q from the different paths are added to be

$$\begin{aligned}&\sum_{n=0}^{(s-m)(J-s-M+m)} p(s-m, J-s-M+m, n)q^{(s-m)(J-s-M+m)+mJ-sM-2n}\\&=\begin{bmatrix}J-M\\s-m\end{bmatrix}_q q^{mJ-sM}\,.\end{aligned} \tag{25}$$

Thus,

$$\begin{aligned}\left(C_q^{j_1j_2}\right)_{m(M-m)JM}&=\sum_s q^{mJ-sM}\begin{bmatrix}J-M\\s-m\end{bmatrix}_q\left\{\frac{[j_1-m]![j_1+s]![j_2-M+m]!}{[j_1-s]![j_1+m]![J_2-J+s]!}\right.\\&\quad\times\left.\frac{[j_2+J-s]![J+M]!}{[j_2+M-m]![J-M]![2J]!}\right\}^{1/2}\left(C_q^{j_1j_2}\right)_{s(J-s)JJ}\\&=q^{\{j_1(j_1+1)-j_2(j_2+1)+J(J+2m+1)\}/2}\\&\quad\times\left\{\frac{[j_1-m]![j_2-M+m]![J+M]![J-M]![j_1+j_2-J]!}{[j_1+m]![j_2+M-m]![j_1+J-j_2]![j_2+J-j_1]!}\right.\\&\quad\times\left.\frac{[2J+1]}{[j_1+j_2+J+1]!}\right\}^{1/2}\sum_s(-1)^{j_1-s}q^{-s(J+M+1)}\\&\quad\times\frac{[j_1+s]![j_2+J-s]!}{[j_1-s]![J-M+m-s]![s-m]![j_2-J+s]!}\,.\end{aligned}$$

Changing the summation index s to $n=s-m$, we have

$$\begin{aligned}\left(C_q^{j_1j_2}\right)_{m(M-m)JM}&=q^{\{j_1(j_1+1)-j_2(j_2+1)+J(J+1)\}/2-m(M+1)}\\&\quad\times\left\{\frac{[2J+1][j_1+j_2-J]![j_1-m]![j_2-M+m]![J+M]![J-M]!}{[J+j_1+j_2+1]![J+j_1-j_2]![J-j_1+j_2]![j_1+m]![j_2+M-m]!}\right\}^{1/2}\end{aligned}$$

$$\times \sum_n (-1)^{n+j_1-m} q^{-n(J+M+1)} \frac{[j_1+m+n]![j_2+J-m-n]!}{[n]![j_2-J+m+n]![j_1-m-n]![J-M-n]!},$$

$$n: \quad \max\left\{\begin{matrix} 0 \\ J-j_2-m \end{matrix}\right\} \cdots \min\left\{\begin{matrix} j_1-m \\ J-M \end{matrix}\right\}. \tag{26}$$

This is the quantum Racah's form for q–CG coefficients. In order to discuss the symmetries of q–CG coefficients, we carry out the following steps towards the quantum Van der Waerden's form:

1. From (11) with $u=j_2+M-m$, $v=J-M-n$, $r=J-j_1+j_2$ and $p=s$, we have:

$$\frac{[j_2+J-m-n]!}{[j_2+M-m]![J-M-n]![J-j_1+j_2]![j_1-m-n]!}$$
$$=\sum_s q^{\ell}\{[s]![j_2+M-m-s]![J-j_1+j_2-s]![j_1-j_2-M-n+s]\}^{-1}$$

with $\ell=-s(j_2+J-m-n)+(j_2+M-m)(J-j_1+j_2)$.

2. Replacing the summation index n by $t=j_1-j_2-M+s-n$, and using Eq. (13) with $v=2j_1-j_2-M+m+s$, $u=j_1-J-M+m+s$, $r=j_1-j_2-M+s$ and $p=t$, we have

$$\sum_n (-1)^{-n+j_1-m} \frac{q^{-n(J+M+1-s)}[j_1+m+n]!}{[n]![j_2-J+m+n]![j_1-j_2-M-n+s]!}$$
$$=\sum_t (-1)^{j_2+M-m-s+t} q^{t(J+M+1-s)-(j_1-j_2-M+s)(J+M+1-s)}$$
$$\times \frac{[2j_1-j_2-M+m+s-t]!}{[t]![j_1-J-M+m+s-t]![j_1-j_2-M+s-t]!}$$
$$=(-1)^{j_2+M-m-s} \frac{q^{-(j_1-j_2-M+s)(j_1+m+1)}[j_1-j_2+J]![j_1+m]!}{[j_1-J-M+m+s]![j_1-j_2-M+s]![J+M-s]!}.$$

3. Replace the summation index s by $n=j_2+M-m-s$

$$\left(C_q^{j_1 j_2}\right)_{m(M-m)JM} = q^{-(j_1+j_2-J)(j_1+j_2+J+1)/2+mj_2-(M-m)j_1}[2J+1]^{1/2}\Delta(j_1 j_2 J)$$
$$\times\{[j_1+m]![j_1-m]![j_2+M-m]![j_2-M+m]![J+M]![J-M]!\}^{1/2}$$
$$\times\sum_n (-1)^n q^{n(J+j_1+j_2+1)}\{[n]![j_1-m-n]![j_2+M-m-n]![j_1+j_2-J-n]! \tag{27}$$
$$\times[J-j_1-M+m+n]![J-j_2+m+n]!\}^{-1},$$

$$n: \quad \max\left\{\begin{matrix} 0 \\ j_1+M-m-J \\ j_2-m-J \end{matrix}\right\} \cdots \quad \min\left\{\begin{matrix} j_1-m \\ j_2+M-m \\ j_1+j_2-J \end{matrix}\right\},$$

where $|m|\le j_1$, $|M-m|\le j_2$, $|M|\le J$, $|j_1-j_2|\le J\le j_1+j_2$ and

$$\Delta(a,b,c)=\left\{\frac{[a+b-c]![b+c-a]![c+a-b]!}{[a+b+c+1]!}\right\}^{1/2}.$$

This is the quantum Van der Waerden's form for q–CG coefficients.

It is easy to check from Eq. (27) that the q–CG coefficients have the following symmetries:

$$\begin{aligned}\left(C_q^{j_1 j_2}\right)_{m_1 m_2 J M} &= \left(C_q^{j_2 j_1}\right)_{-m_2 m_1 J-M} = (-1)^{j_1+j_2-J}\left(C_{q^{-1}}^{j_1 j_2}\right)_{-m_1-m_2 J-M} \\ &= (-1)^{j_1+j_2-J}\left(C_{q^{-1}}^{j_2 j_1}\right)_{m_2 m_1 J M} \\ &= (-1)^{j_1-m_1}\left\{\frac{[2J+1]}{[2j_2+1]}\right\}^{1/2} q^{-m_1}\left(C_{q^{-1}}^{j_1 J}\right)_{m_1-M j_2-m_2} \\ &= (-1)^{j_2+m_2}\left\{\frac{[2J+1]}{[2j_1+1]}\right\}^{1/2} q^{m_2}\left(C_{q^{-1}}^{J j_2}\right)_{-M m_2 j_1-m_1},\end{aligned} \tag{28}$$

where $M = m_1 + m_2$. In Tables 1 and 2 we list the q–CG coefficients with $j_2 = 1/2$ and 1.

Table 1. $\left(C_q^{j\frac{1}{2}}\right)_{(M-m)mJM}$.

	$m = 1/2$	$m = -1/2$
$J = j + 1/2$	$q^{-(j-M+1/2)/2}\left\{\frac{[j+M+1/2]}{[2j+1]}\right\}^{1/2}$	$q^{(j+M+1/2)/2}\left\{\frac{[j-M+1/2]}{[2j+1]}\right\}^{1/2}$
$J = j - 1/2$	$-q^{(j+M+1/2)/2}\left\{\frac{[j-M+1/2]}{[2j+1]}\right\}^{1/2}$	$q^{-(j-M+1/2)/2}\left\{\frac{[j+M+1/2]}{[2j+1]}\right\}^{1/2}$

Table 2. $\left(C_q^{j1}\right)_{(M-m)mJM}$.

	$m = 1$	$m = 0$	$m = -1$
$J = j + 1$	$q^{M-j-1}\left\{\frac{[j+M][j+M+1]}{[2j+1][2j+2]}\right\}^{1/2}$	$q^{M}\left\{\frac{[2][j-M+1][j+M+1]}{[2j+1][2j+2]}\right\}^{1/2}$	$q^{M+j+1}\left\{\frac{[j-M][j-M+1]}{[2j+1][2j+2]}\right\}^{1/2}$
$J = j$	$-q^{M}\left\{\frac{[2][j+M][j-M+1]}{[2j][2j+2]}\right\}^{1/2}$	$\frac{q^{M-j-1}[j+M]-q^{M+j+1}[j-M]}{\{[2j][2j+2]\}^{1/2}}$	$q^{M}\left\{\frac{[2][j-M][j+M+1]}{[2j][2j+2]}\right\}^{1/2}$
$J = j - 1$	$q^{M+j}\left\{\frac{[j-M][j-M+1]}{[2j][2j+1]}\right\}^{1/2}$	$-q^{M}\left\{\frac{[2][j-M][j+M]}{[2j][2j+1]}\right\}^{1/2}$	$q^{M-j}\left\{\frac{[j+M][j+M+1]}{[2j][2j+1]}\right\}^{1/2}$

IV. The First Type Quantum Racah Coefficients

The first type quantum Racah (q–Racah 1) coefficients is defined as

$$\begin{aligned}\left(C_q^{j_2 j_3}\right)_{m_2 m_3 J_{23}(m_2+m_3)}\left(C_q^{j_1 J_{23}}\right)_{m_1(m_2+m_3)JM} &= \sum_{J_{12}}\left(C_q^{j_1 j_2}\right)_{m_1 m_2 J_{12}(m_1+m_2)} \\ \times\left(C_q^{J_{12} j_3}\right)_{(m_1+m_2)m_3 JM} &\{[2J_{12}+1][2J_{23}+1]\}^{1/2} W[j_1 j_2 J j_3; J_{12} J_{23}]_q,\end{aligned} \tag{29}$$

where $M = m_1 + m_2 + m_3$, and $W[j_1 j_2 J j_3; J_{12} J_{23}]_q$ is independent of m_1, m_2 and m_3, and satisfies the orthogonal conditions

$$\begin{aligned}\sum_{J_{12}}[2J_{12}+1]W[j_1 j_2 J j_3; J_{12} J_{23}]W[j_1 j_2 J j_3; J_{12} J'_{23}] &= [2J_{23}+1]^{-1}\delta_{J_{23} J'_{23}}, \\ \sum_{J_{23}}[2J_{23}+1]W[j_1 j_2 J j_3; J_{12} J_{23}]W[j_1 j_2 J j_3; J'_{12} J_{23}] &= [2J_{12}+1]^{-1}\delta_{J_{12} J'_{12}}.\end{aligned} \tag{30}$$

From the orthogonal condition (19) of q–CG coefficients, we have

$$\begin{aligned}&\left(C_q^{J_{12}j_3}\right)_{(M-m_3)m_3JM}\{[2J_{12}+1][2J_{23}+1]\}^{1/2}W[j_1j_2Jj_3;J_{12}J_{23}]_q\\&=\sum_{m_1}\left(C_q^{j_1j_2}\right)_{m_1(M-m_1-m_3)J_{12}(M-m_3)}\left(C_q^{j_2j_3}\right)_{(M-m_1-m_3)m_3J_{23}(M-m_1)}\\&\times\left(C_q^{j_1J_{23}}\right)_{m_1(M-m_1)JM},\end{aligned}\tag{31}$$

$$\begin{aligned}&W[j_1j_2Jj_3;J_{12}J_{23}]_q\delta_{JJ'}=\{[2J_{12}+1][2J_{23}+1]\}^{-1/2}\\&\times\sum_{m_1m_2m_3}\left(C_q^{J_{12}j_3}\right)_{(M-m_3)m_3J'M}\left(C_q^{j_1j_2}\right)_{m_1m_2J_{12}(M-m_3)}\\&\times\left(C_q^{j_2j_3}\right)_{m_2m_3J_{23}(M-m_1)}\left(C_q^{j_1J_{23}}\right)_{m_1(M-m_1)JM},\end{aligned}\tag{32}$$

$$\begin{aligned}&\left(C_q^{j_1J_{23}}\right)_{m_1(M-m_1)JM}\{[2J_{12}+1][2J_{23}+1]\}^{1/2}W[j_1j_2Jj_3;J_{12}J_{23}]_q\\&=\sum_{m_2m_3}\left(C_q^{J_{12}j_3}\right)_{(M-m_3)m_3JM}\left(C_q^{j_1j_2}\right)_{m_1m_2J_{12}(M-m_3)}\left(C_q^{j_2j_3}\right)_{m_2m_3J_{23}(M-m_1)},\end{aligned}\tag{33}$$

where $M=m_1+m_2+m_3$.

Now, we are going to compute the explicit form of q–Racah 1 coefficients. The method is a generalization of that given by Biedenharn and Louck[6]. However, since the complication from the q factors, we have to change the summation order as follows. Let $J=J'=M$ and change notation: $a=j_1, b=j_2, c=J, d=j_3, e=J_{12}, f=J_{23}$, and $\alpha=J-m_2-m_3=m_1$, $\gamma=J-m_3=m_1+m_2$. Substituting Eqs. (20) and (27) into Eq. (32), we have

$$\begin{aligned}W[abcd;ef]_q&=\frac{\Delta(abe)\Delta(edc)\Delta(bdf)\Delta(afc)[2c+1]!}{[c+d-e]![c-d+e]![c+a-f]![c-a+f]!}(-1)^{a-b+c+d}\\&\times q^{c(c+d+f+2)+df-ae}\sum_{\alpha\gamma tn}(-1)^{t+n}q^{-\alpha(c+d+e+2)-\gamma(a+c+f+2)+t(a+b+e+1)-n(b+d+f+1)}\\&\times\Big\{[a+\alpha]![c+f-\alpha]![e+\gamma]![c+d-\gamma]![b+\alpha-\gamma]![b-\alpha+\gamma]!\Big\}\\&\times\Big\{[t]![a-b+e-t]![n]![-b+d+f-n]![a+\alpha-t]![-a+b-\gamma+t]!\\&\times[c+d-\gamma-n]![b-c-d+\alpha+n]![e+\gamma-t]![b-e-\alpha+t]!\\&\times[c+f-\alpha-n]![b-c-f+\gamma+n]!\Big\}^{-1}.\end{aligned}$$

1). From Eq. (11) with $u=-a+b-\gamma+t$, $v=a+\alpha-t$, $r=c+d-\gamma-n$ and $p=m-e-\gamma$, we have

$$\begin{aligned}&[b+\alpha-\gamma]!\Big\{[a+\alpha-t]![-a+b-\gamma+t]![c+d-\gamma-n]![b-c-d+\alpha+n]!\Big\}^{-1}\\&=\sum_m q^{(b+\alpha-\gamma)(m-e-\gamma)-(-a+b-\gamma+t)(c+d-\gamma-n)}\\&\times\Big\{[m-e-\gamma]![-c+b+e-m+t]![c+d+e-n-m]!\\&\times[a-c-d-e+\alpha-t+n+m]!\Big\}^{-1}.\end{aligned}$$

2). From Eq. (11) with $u=b-e-\alpha+t$, $v=e+\gamma-t$, $r=c+f-\alpha-n$ and $p=c+f-\alpha-n-m+t+w$, we have

$$[b-\alpha+\gamma]!\Big\{[e+\gamma-t]![b-e-\alpha+t]![c+f-\alpha-n]![b-c-f+\gamma+n]!\Big\}^{-1}$$
$$=\sum_w q^{-(b-\alpha+\gamma)(c+f-\alpha-n-m+t+w)+(c+f-\alpha-n)(b-e-\alpha+t)}$$
$$\times[c+f-\alpha-n-m+t+w]![b-c-e-f+n+m-w]![m-t-w]!$$
$$\times[e+\gamma-m+w]!\Big\}^{-1}.$$

3). Sum over α by Eq. (12) with $u=n+m-t-w+1$, $v=c+d+e+t-n-m+1$, $r=a-d-e+f+w$, and $p=c+f-\alpha-n-m+t+w$,

$$\sum_\alpha q^{-\alpha(c+d+e-w+2)}[a+\alpha]![c+f-\alpha]!\Big\{[a-c-d-e+\alpha-t+n+m]!$$
$$\times[c+f-\alpha-n-m+t+w]!\Big\}^{-1}$$
$$=q^{(-c-f+n+m-t-w)(c+d+e-w+2)+(a-d-e+f+w)(n+m-t-w+1)}$$
$$\times[a+c+f+1]![n+m-t-w]![c+d+e+t-n-m]!$$
$$\times\Big\{[a-d-e+f+w]![c+d+e-w+1]!\Big\}^{-1}.$$

4). From Eq. (11) with $u=a-b+e-t$, $v=-a+b-e+n+m-w$, $r=n$ and $p=j$, we have

$$[n+m-t-w]!\Big\{[m-w-t]![a-b+e-t]![n]!\Big\}^{-1}=\sum_j q^{(n+m-t-w)j-n(a-b+e-t)}$$
$$\times[-a+b-e+n+m-w]!\Big\{[j]![a-b+e-t-j]![-a+b-e+m-w+j]![n-j]!\Big\}^{-1}.$$

5). From Eq. (11) with $u=-b+d+f-n$, $v=b+c+e-f+t-m$, $r=t$ and $p=k$, we have

$$[c+d+e+t-n-m]!\Big\{[-b+d+f-n]![t]![c+d+e-n-m]!\Big\}^{-1}$$
$$=\sum_k q^{-k(c+d+e+t-n-m)+t(-b+d+f-n)}[b+c+e-f+t-m]!$$
$$\times\Big\{[k]![-b+d+f-n-k]![t-k]![b+c+e-f-m+k]!\Big\}^{-1}.$$

Now, we have

$$W[abcd;ef]=\Delta(abe)\Delta(edc)\Delta(bdf)\Delta(afc)[2c+1]![a+c+f+1]!$$
$$\times\Big\{[c+d-e]![c-d+e]![c+a-f]![c-a+f]!\Big\}^{-1}$$
$$\times(-1)^{a-b+c+d}q^{c(a-b-2e)+e(-a-b-2f-1)+d(a-b-1)+a-f}$$
$$\times\sum_\gamma q^{\gamma(-2a-c+d+e-2f-2)}[e+\gamma]![c+d-\gamma]!\sum_j\Big\{[j]!\Big\}^{-1}$$

$$\times\sum_k q^{-k(c+d+e)}\Big\{[k]!\Big\}^{-1}\sum_m q^{m(a+2b+c+f+j+k+2)}$$

$$\times\Big\{[b+c+e-f+k-m]![-e+m-\gamma]!\Big\}^{-1}\sum_w q^{-w(a+b+j+\gamma+1)}$$

$$\times\Big\{[a-d-e+f+w]![e+c+d-w+1]![-a+b-e-m-w+j]![e+\gamma-m+w]!\Big\}^{-1}$$

$$\times\sum_n(-1)^n q^{n(-a+b+c-d+j+k+1)}[-a+b-e+m+n-w]!$$

$$\times\Big\{[b-c-e-f+m+n-w]![n-j]![-b+d+f-k+n]!\Big\}^{-1}$$

$$\times\sum_t(-1)^t q^{t(-b-c+e+f-j-k-1)}[b+c+e-f-m+t]!$$

$$\times\Big\{[-a+b+e-m+t]![a-b+e-t-j]![t-k]!\Big\}^{-1}.$$

6). Sum over t by Eq. (13) with $u=2e-j-m$, $v=a+c+2e-f-j-m$, $r=a-b+e-j-k$ and $p=a-b+e-j-t$

$$\sum_t(-1)^t q^{t(-b-c+e+f-j-k-1)}[b+c+e+t-f-m]!$$

$$\times\Big\{[-a+b+e-m+t]![a-b+e-j-t]![t-k]!\Big\}^{-1}$$

$$=(-1)^{-a+b-e+j}q^{(a-b+e-j)(-b-c+e+f-j-k-1)-(a-b+e-j-k)(2e-j-m)}$$

$$\times[a+c-f]![b+c+e-f+k-m]!\Big\{[2e-j-m]![a-b+e-j-k]!$$

$$\times[b+c-e-f+j+k]!\Big\}^{-1}.$$

7). Sum over n by Eq. (13) with $u=-b+d+f-j-k$, $v=-a+d-e+f-k+m-w$, $r=-c+d-e-k+m-w$ and $p=-b+d+f-k-n$,

$$\sum_n(-1)^n q^{n(-a+b+c-d+j+k+1)}[-a+b-e+m+n-w]!$$

$$\times\Big\{[b-c-e-f+m+n-w]![n-j]![-b+d+f-k-n]!\Big\}^{-1}$$

$$=(-1)^{b-d-f+k}q^{(-b+d+f-k)(-a+b+c-d+j+k+1)+(-b+d+f-j-k)(-c+d-e-k+m-w)}$$

$$\times[-a+b-e+j+m-w]![-a+c+f]!\Big\{[-b+d+f-j-k]![-c+d-e-k+m-w]!$$

$$\times[-a+b+c-d+j+k]!\Big\}^{-1}.$$

8). Sum over w by Eq. (11) with $u=-c+d-k+\gamma$, $v=a+c+f+1$, $r=c+d+2e-m+\gamma+1$ and $p=e-m+w+\gamma$,

$$\sum_w q^{-w(a+d+f-k+\gamma+1)}\Big\{[a-d-e+f+w]![c+d+e-w+1]![e-m+w+\gamma]!$$

$$\times[-c+d-e-k+m-w]!\Big\}^{-1}$$

$$=q^{(e-m+\gamma)(a+d+f-k+\gamma+1)-(c+d+2e-m+\gamma+1)(-c+d-k+\gamma)}[a+d+f-k+\gamma+1]!$$

$$\times\Big\{[-c+d-k+\gamma]![a+c+f+1]![c+d+2e-m+\gamma+1]![a-c-2e+f-k+m]!\Big\}^{-1}$$

9). Sum over m by Eq. (11) with $u=e-j-\gamma$, $v=a+d+f-k+\gamma+1$, $r=c+d+e+1$ and $p=-e+m-\gamma$

$$\sum_m q^{m(a+d+e+f-j-k+1)}\Big\{[-e+m-\gamma]![c+d+2e-m+\gamma+1]![a-c-2e+f-k+m]!$$
$$\times[2e-j-m]!\Big\}^{-1}$$
$$=q^{(e+\gamma)(a+d+e+f-j-k+1)+(c+d+e+1)(e-j-\gamma)}[a+d+e+f-j-k+1]!$$
$$\times\Big\{[e-j-\gamma]![a+d+f-k+\gamma+1]![c+d+e+1]![a-c+f-j-k]!\Big\}^{-1}.$$

Let $n=j+k$, and we have

$$W[abcd;ef]_q=(-1)^{b+c-e-f}\Delta(abe)\Delta(edc)\Delta(afc)\Delta(bdf)$$
$$\times[2c+1]!\Big\{[c+e-d]![c-e+d]![c+d+e+1]!\Big\}^{-1}\sum_n(-1)^n q^{n(c-d)}$$
$$\times[a+d+e+f-n+1]!\Big\{[a-b+e-n]![b+c-e-f+n]![a-c+f-n]!$$
$$\times[-b+d+f-n]![-a+b+c-d+n]!\Big\}^{-1}\sum_\gamma q^{-\gamma(2c+n+2)}[e+\gamma]![c+d-\gamma]!$$
$$\times\sum_k q^{k(-c+d+e)}\Big\{[n-k]![k]![-c+d+\gamma-k]![e-\gamma-n+k]!\Big\}^{-1}.$$

10). Sum over k by Eq. (11) with $u=n$, $v=-c+d+e-n$, $r=-c+d+\gamma$ and $p=k$,

$$\sum_k q^{k(-c+d+e)}\Big\{[n-k]![k]![-c+d+\gamma-k]![e-\gamma-n+k]!\Big\}^{-1}$$
$$=q^{n(-c+d+\gamma)}[-c+d+e]!\Big\{[n]![-c+d+e-n]![-c+d+\gamma]![e-\gamma]!\Big\}^{-1}.$$

11). Sum over γ by Eq. (12) with $u=c-d+e+1$, $v=c+d-e+1$, $r=-c+d+e$, and $p=-c+d+\gamma$,

$$\sum_\gamma q^{-\gamma(2c+2)+c(c+1)-d(d+1)+e(e+1)}[e+\gamma]![c+d-\gamma]!\Big\{[-c+d+\gamma]![e-\gamma]!\Big\}^{-1}$$
$$=[c+d+e+1]![c-d+e]![c+d-e]!\Big\{[-c+d+e]![2c+1]!\Big\}^{-1}. \tag{34}$$

Let $z=a+d+e+f-n$, and we obtain at last

$$W[abcd;ef]_q=(-1)^{b+c-e-f}\Delta(abe)\Delta(edc)\Delta(bdf)\Delta(afc)$$
$$\times\sum_n(-1)^n[a+d+e+f-n+1]!\Big\{[n]![b+c-e-f+n]![-a+b+c-d+n]!$$
$$\times[a-b+e-n]![-c+d+e-n]![-b+d+f-n]![a-c+f-n]!\Big\}^{-1} \tag{35}$$
$$=(-1)^{a+b+c+d}\Delta(abe)\Delta(dec)\Delta(bdf)\Delta(afc)\sum_z(-1)^z[z+1]!\Big\{[z-a-b-e]!$$

$$\times[z-c-d-e]![z-b-d-f]![z-a-c-f]![a+b+c+d-z]!$$
$$\times[a+d+e+f-z]![b+c+e+f-z]!\Big\}^{-1},$$

$$z:\quad \max\begin{Bmatrix} a+b+e\\ c+d+e\\ a+c+f\\ b+d+f\end{Bmatrix}\quad\cdots\quad \min\begin{Bmatrix} a+b+c+d\\ a+d+e+f\\ b+c+e+f\end{Bmatrix},$$

where a, b, c, d, e and f should satisfy four triangle rules as shown in Fig. 2.

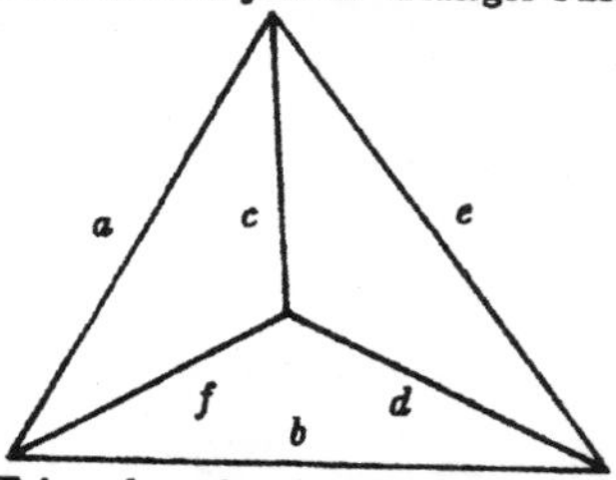

Fig. 2. Triangle rules for the indices in q-Racah 1 coefficients.

It is obvious to see from Eq. (24) that

$$W[abcd;ef]_q = W[abcd;ef]_{q^{-1}}\,. \tag{36}$$

Equation (35) is the same as that for the classical (usual) Racah coefficients except for replacing round brackets (usual number) by rectangular ones, so it is easy to prove that some symmetries of classical Racah coefficients can be generalized to the quantum case. For example, introduce the quantum $6-j$ symbol,

$$\begin{Bmatrix} a & b & e\\ d & c & f\end{Bmatrix}_q = (-1)^{a+b+c+d}W[abcd;ef]_q\,. \tag{37}$$

There are 144 quantum $6-j$ symbols to be equal to each other just like the classical ones[6]

$$\begin{Bmatrix} a & b & e\\ d & c & f\end{Bmatrix}_q = \begin{Bmatrix} b & a & e\\ c & d & f\end{Bmatrix}_q = \begin{Bmatrix} a & e & b\\ d & f & c\end{Bmatrix}_q = \begin{Bmatrix} d & c & e\\ a & b & f\end{Bmatrix}_q$$
$$= \begin{Bmatrix} (a+b-c+d)/2 & (a+b+c-d)/2 & e\\ (a-b+c+d)/2 & (-a+b+c+d)/2 & f\end{Bmatrix}_q\,. \tag{38}$$

In the following, we list the concrete values of those quantum $6-j$ symbols where one index d is equal to 0, 1/2 or 1. Those forms are the straight generalizations of the classical ones[7].

$$\begin{Bmatrix} a & b & c\\ 0 & c & b\end{Bmatrix}_q = (-1)^{a+b+c}\{[2b+1][2c+1]\}^{-1/2}\,,$$

$$\begin{Bmatrix} a & b & c\\ \frac{1}{2} & c-\frac{1}{2} & b-\frac{1}{2}\end{Bmatrix}_q = (-1)^{a+b+c}\left\{\frac{[a+b+c+1][-a+b+c]}{[2b][2b+1][2c][2c+1]}\right\}^{1/2},$$

$$\begin{Bmatrix} a & b & c \\ \frac{1}{2} & c-\frac{1}{2} & b+\frac{1}{2} \end{Bmatrix}_q = (-1)^{a+b+c}\left\{\frac{[a-b+c][a+b-c+1]}{[2b+1][2b+2][2c][2c+1]}\right\}^{1/2},$$

$$\begin{Bmatrix} a & b & c \\ 1 & c-1 & b-1 \end{Bmatrix}_q =$$

$$(-1)^{a+b+c}\left\{\frac{[a+b+c+1][a+b+c][-a+b+c][-a+b+c-1]}{[2b-1][2b][2b+1][2c-1][2c][2c+1]}\right\}^{1/2},$$

$$\begin{Bmatrix} a & b & c \\ 1 & c-1 & b+1 \end{Bmatrix}_q =$$

$$(-1)^{a+b+c}\left\{\frac{[a-b+c-1][a-b+c][a+b-c+1][a+b-c+2]}{[2b][2b+1][2b+2][2c-1][2c][2c+1]}\right\}^{1/2},$$

$$\begin{Bmatrix} a & b & c \\ 1 & c-1 & b \end{Bmatrix}_q =$$

$$(-1)^{a+b+c}\left\{\frac{[a+b+c+1][a+b-c+1][a-b+c][-a+b+c][2]}{[2b][2b+1][2b+2][2c-1][2c][2c+1]}\right\}^{1/2},$$

$$\begin{Bmatrix} a & b & c \\ 1 & c & b \end{Bmatrix}_q = (-1)^{a+b+c+1}\frac{[a+b+c+2][-a+b+c]-[a+b-c][a-b+c]}{\{[2b][2b+1][2b+2][2c][2c+1][2c+2]\}^{1/2}},$$

The last formula shows that it should be careful to use the formulas of the classical Racah coefficients for the quantum ones.

V. Quantum Racah Sum Rule

From Eqs. (33) and (36) we have

$$\{[2e+1][2L+1]\}^{1/2}W[bacd;eL]_q\left(C^{bL}_{q^{-1}}\right)_{m_2(c-m_2)cc}$$

$$=\sum_{m_1m_4}\left(C^{ed}_{q^{-1}}\right)_{(c-m_4)m_4cc}\left(C^{ba}_{q^{-1}}\right)_{m_2m_1e(c-m_4)}\left(C^{ad}_{q^{-1}}\right)_{m_1m_4L(c-m_2)}$$

$$=\sum_{m_1m_4}(-1)^{b+L-d-e}q^{-e(e+1)+d(d+1)+(c+1)(c-2m_4)}$$

$$\times\left(C^{ed}_{q}\right)_{(c-m_4)m_4cc}\left(C^{ab}_{q}\right)_{m_1m_2e(c-m_4)}\left(C^{da}_{q}\right)_{m_4m_1L(c-m_2)}$$

and

$$\{[2f+1][2L+1]\}^{1/2}W[bdca;fL]_q\left(C^{bL}_{q^{-1}}\right)_{m_2(c-m_2)cc}=q^{-b(b+1)+L(L+1)-(c+1)(c-2m_2)}$$

$$\times\sum_{n_1n_4}\left(C^{fa}_{q}\right)_{(c-n_1)n_1cc}\left(C^{bd}_{q}\right)_{m_2n_4f(c-n_1)}\left(C^{da}_{q}\right)_{n_4n_1L(c-m_2)}.$$

Therefore

$$\{[2e+1][2f+1]\}^{1/2}\sum_L \lambda_L^{ad}(q)[2L+1]W[bacd;eL]_q W[bdca;fL]_q$$

$$\times\sum_{m_2}\left(C_{q^{-1}}^{bL}\right)_{m_2(c-m_2)cc}\left(C_{q^{-1}}^{bL}\right)_{m_2(c-m_2)cc}=\lambda_e^{ab}(q^{-1})\lambda_c^{af}(q)$$

$$\times\sum_{m_1m_2m_4n_1n_4} q^{a(a+1)+2d(d+1)-2e(e+1)-f(f+1)+(c+1)(c+2m_2-2m_4)}$$

$$\times\left(C_q^{ab}\right)_{m_1m_2e(c-m_4)}\left(C_q^{ed}\right)_{(c-m_4)m_4cc}q^{-a(a+1)+f(f+1)+(c+1)(2n_1-c)} \quad (39)$$

$$\times\left(C_q^{af}\right)_{n_1(c-n_1)cc}\left(C_q^{bd}\right)_{m_2n_4f(c-n_1)}\sum_L\left(C_q^{da}\right)_{m_4m_1L(c-m_2)}\left(C_q^{da}\right)_{n_4n_1L(c-m_2)}$$

$$=\lambda_e^{ab}(q^{-1})\lambda_c^{af}(q)\sum_{m_1m_2m_4} q^{2d(d+1)-2e(e+1)-2c(c+1)+4(c+1)(m_1+m_2)}$$

$$\times\left(C_q^{ab}\right)_{m_1m_2e(c-m_4)}\left(C_q^{ed}\right)_{(c-m_4)m_4cc}\left(C_q^{af}\right)_{m_1(c-m_1)cc}\left(C_q^{bd}\right)_{m_2m_4f(c-m_1)},$$

where $c=m_1+m_2+m_4$. Let $m_1=\alpha$ and $m_2=\gamma-\alpha$. The sum is the same as Eq. (32) except for the factor of q which is nothing to do with the first ten steps of the computation in Sec. IV. Since there are two signs in Eq. (12), this factor does keep the result of Eq. (34) invariant. Therefore, the sum in the right-hand side of Eq. (39) is equal to $\{[2e+1][2f+1]\}^{1/2}W[abcd;ef]_q$, so the quantum Racah sum rule is proved:

$$\sum_L \lambda_L^{ad}(q)[2L+1]W[bacd;eL]_q W[bdca;fL]_q=\lambda_e^{ab}(q^{-1})\lambda_c^{af}(q)W[abcd;ef]_q\,. \quad (40)$$

Acknowledgement

The authors would like to thank professor Kang-Jie SHI for helpful discussions.

References

[1] C.N. Yang, Phys. Rev. Lett. **19**(1967)1312.
[2] R.J. Baxter, Ann. Phys. **70**(1972)193.
[3] M. Jimbo, Comm. Math. Phys. **102**(1986)537; Lett. Math. Phys. **10**(1985)63; **11**(1986)247.
[4] Bo-yu HOU, Bo-Yuan HOU and Zhong-qi MA, Commun. Theor. Phys. **13**(1990)181.
[5] G.E. Andrews, *The Theory of Partitions*, Chap. 3, Addison-Wesley (1976).
[6] L.C. Biedenharn and J.D. Louck, *Angular Momentum in Quantum Physics, Theory and Application, Encyclopedia of Mathematics and Its Application*, ed. G.C. Rota, Addison-Wesley Publishing Co., Massachusetts (1981), Vol. 8.
[7] A.R. Edmonds, *Angular Momentum in Quantum Mechanics*, Princeton University Press (1957).

J. Phys. A: Math. Gen. 24 (1991) 11-21. Printed in the UK

Modular invariance and the Feigin–Fuch representation of characters for $SU_k(2)$ WZW and minimal models

Bo-Yu Hou† and Rui-Hong Yue†‡

† Institute of Modern Physics, Northwest University, Xian 710069, People's Republic of China

‡ CCAST (World Laboratory), PO Box 8730, Beijing 100080, People's Republic of China

Received 21 May 1990, in final form 18 September 1990

Abstract. In this paper, we have discussed the modular properties of characters in Feigin–Fuch representation for $SU_k(2)$ WZW and minimal models, and we have given the explicit expressions of the S-matrix. The proof is given for a simple case.

1. Introduction

In recent years, a useful method has been developed to discuss the correlation function and quantum group structure for rational conformal field theories (RCFT) [1–5]. In this method, the conformal block can be written as a Feigin–Fuch (FF) contour integral and the correlation function is represented as the product of holomorphic and anti-holomorphic conformal blocks. Based on the monodromy invariance of the correlation function, one can obtain the structure constant (operator product expansion), fusion rule and crossing matrices (fusion and braiding matrices) [3–6].

On the other hand, the classification of conformal field theory is an important problem. Mathur *et al* [7, 8] give an approach to produce in principle all RCFT characters based on the modular invariant differential equation which depends on two integers N, the number of characters, and L, which is proportional to the number of zeroes of the Wronskian determinant of the characters in the interior of moduli space. The characters satisfy an Nth order differential equation. For large values of N it is difficult to solve the equation. Mukhi *et al* [9] propose that the characters for $L=0$ and large N can be written as an FF contour integral, and give some useful results.

In this paper we will discuss the monodromy transformation properties of the FF contour integrals and calculate the modular matrices of characters in FF integral representation for the A–D–E classification of $SU_k(2)$ WZW and minimal models. The organization of this paper is as follows. In section 2 we briefly recall the FF integral representation for characters. In section 3 we generally discuss the monodromy behaviour of FF integrals. The transformation matrices of the characters in the A–D–E classification of the $SU_k(2)$ WZW model are given in section 4. In section 5 we obtain the S-matrix for the minimal model, and give some checks for the identity of the monodromy of FF integrals and the modular property of characters in the appendix.

0305-4470/91/010011+11\$03.50

2. FF contour integral representation for characters

In this section, we will recall the FF integral representation for characters of RCFT with $L=0$.

In RCFT, any partition function has the general form

$$Z(\tau)=\sum_{i,j} N_{ij}\chi_i(\tau)\chi_j(\bar{\tau}) \tag{2.1}$$

where $\chi_i(\tau)$ is the character for the i representation of the chiral algebra A_τ. This character can be thought as a multivalued function on the complex. The variable λ on the complex plan has a power series expansion in half-integer power of the variable $q=e^{i2\pi\tau}$ (τ on the torus):

$$\lambda=16q^{1/2}(1-8q^{1/2}+44q+O(q^{3/2})). \tag{2.2}$$

The corresponding generator of SL(2, 2) of the modular transformation in the term λ are

$$\begin{aligned} &S:\lambda\to 1-\lambda \\ &T:\lambda\to\frac{\lambda}{\lambda-1}. \end{aligned} \tag{2.3}$$

First, we investigate the behaviour of the characters of a conformal field theory with conformal centre c and spectrum h_i, which has N characters. A character χ_i associated with a primary field of conformal weight h_i behaves in the $\lambda\to 0$ limit as

$$\chi_i\to\lambda^{-c/12+2h_i}. \tag{2.4}$$

MPS conjecture is that the characters of RCFT with $L=0$ can be written as an FF integral:

$$\begin{aligned} J_{AB}(\lambda)=(\lambda(1-\lambda))^{\alpha}&\int_0^{\lambda}\prod_{i=1}^{A}\mathrm{d}u_i[u_i(1-u_i)(\lambda-u_i)]^a\int_1^{\infty}\prod_{i=A+1}^{n}\mathrm{d}u_i[u_i(u_i-1)(u_i-\lambda)]^a \\ &\times\int_0^{\lambda}\prod_{i=1}^{B}\mathrm{d}v_i[v_i(1-v_i)(\lambda-v_i)]^b\int_1^{\infty}\prod_{i=B+1}^{m}\mathrm{d}v_i[v_i(v_i-1)(v_i-\lambda)]^b \\ &\times\prod_{i>j=1}^{n}(u_i-u_j)^{-2a/b}\prod_{i>j=1}^{m}(v_i-v_j)^{-2b/a}\prod_{i,j=1}^{n,m}(u_i-v_j)^{-2} \end{aligned} \tag{2.5}$$

where

$$\alpha=\tfrac{1}{3}\left[\frac{a}{b}(n-1)n+\frac{b}{a}(m-1)m-(1+3a)n-(1+3b)m+2nm\right] \tag{2.6}$$

and a, b, n, m are undetermined parameters.

Now, let us consider the behaviour of $J_{AB}(\lambda)$ as $\lambda\to 0$. A simple calculation shows that

$$J_{AB}(\lambda)\to\lambda^{\alpha+\Delta_{AB}} \tag{2.7}$$

and

$$\Delta_{AB}=A(1+2a)+B(1+2b)-\frac{a}{b}A(A-1)-\frac{b}{a}B(B-1)-2AB. \tag{2.8}$$

Since $0 \leqslant A \leqslant n$, $0 \leqslant B \leqslant m$, we know that J_{AB} has $(n+1)(m+1)$ choices. Comparing (2.4) with (2.7), we have three useful equations:

$$N=(n+1)(m+1) \tag{2.9a}$$

$$\frac{a}{b}n(n-1)+\frac{b}{a}m(m-1)-n(1+3a)-m(1+3b)+2nm=\frac{c}{4} \tag{2.9b}$$

$$A(1+2a)+B(1+2b)-\frac{a}{b}A(A-1)-\frac{b}{a}B(B-1)-2AB=2h_i. \tag{2.9c}$$

For RCFT with $L=0$, given the number of characters N, the central term c and conformal scale h_i, we must solve (2.9) and determine a, b, n and m.

Consider, for example, the $SU_k(2)$ WZW model in the diagonal case:

$$C=\frac{3k}{k+2} \qquad h_i=\frac{i(i+1)}{k+2} \qquad \left(i=0, \frac{1}{2}, 1, \ldots, \frac{k}{2}\right). \tag{2.10}$$

Since h_i depends on one parameter i, the LHS of (2.9c) must have only one parameter. Without any loss of generality, we assume $A=2i$, $B=0$, $b=(1+2k)/2$ and $a=-(1+2k)/4(k+2)$, and can show that they satisfy (2.9) if $n=k$ and $m=0$. In the following sections, we will give the solution of (2.9) for the $SU_k(2)$ WZW and minimal models.

3. The monodromy transformation of the FF integral

Since the modular group of the torus has two generators, it is enough to consider the T and S transformations.

In (2.5), we have chosen the phase so that the integral has zero phase. With the knowledge of the contour integral, the function $J_{AB}(\lambda)$ is canonical at $\lambda=0$, and is defined in the interval $0 \leqslant \lambda \leqslant 1$. In order to discuss the T transformation, we must analytically continue $J_{AB}(\lambda)$ from $0 \leqslant \lambda \leqslant 1$ to $\lambda \leqslant 0$:

$$\begin{aligned} J_{AB}\left(\frac{\lambda}{\lambda-1}\right) &= [\lambda(\lambda-1)^2]^{\alpha} \int_0^{\lambda/(\lambda-1)} \prod_{i=1}^{A} du_i \left[u_i(1-u_i)\left(u_i-\frac{\lambda}{\lambda-1}\right)\right]^a \\ &\times \int_1^{\infty} \prod_{i=A+1}^{n} du_i \left[u_i(u_i-1)\left(u_i-\frac{\lambda}{\lambda-1}\right)\right]^a \\ &\times \int_0^{\lambda(\lambda-1)} \prod_{j=1}^{B} dv_j \left[v_j(1-v_j)\left(v_j-\frac{\lambda}{\lambda-1}\right)\right]^b \\ &\times \int_1^{\infty} \prod_{j=B+1}^{m} dv_j \left[v_j(v_j-1)\left(v_j-\frac{\lambda}{\lambda-1}\right)\right]^b \prod_{i>j=1}^{n} (u_i-u_j)^{-2a/b} \\ &\times \prod_{i>j=1}^{m} (v_i-v_j)^{-2b/a} \prod_{i,j=1}^{n,m} (u_i-v_j)^{-2}. \end{aligned} \tag{3.1}$$

Under the transformation T, $J_{AB}(\lambda)$ changes into $J_{AB}(\lambda/(\lambda-1))$. One can show that

$$J_{AB}\left(\frac{\lambda}{\lambda-1}\right)=e^{i\pi(\alpha+\Delta_{AB})}J_{AB}(\lambda). \tag{3.2}$$

From (2.9), we see that $J_{AB}(\lambda)$ has correct transformation properties under the transformation T.

In what follows, we discuss monodromy properties of the FF integral. Generally, under the transformation S, $J_{AB}(\lambda)$ changes as follows:

$$J_{AB}(\lambda)\to J_{AB}(1-\lambda)=\sum_{C,D} S_{AB,CD}J_{CD}(\lambda). \tag{3.3}$$

It is important that the positions of two contours which run through two points can be interchanged without changing the value of the integral. So $S_{AB,CD}$ can be decomposed into two parts: $\alpha_{AC}(n, a, a/b)$ and $\alpha_{BD}(m, b, b/a)$.

In order to calculate $\alpha_{AC}(n, a, a/b)$ we introduce a contour integral:

$$\begin{aligned}F_A(a, b, \lambda)=&\int_0^\lambda \prod_{i=1}^{A} du_i[u_i(1-u_i)(\lambda-u_i)]^a\\&\times\int_1^\infty \prod_{i=A+1}^{n} du_i[u_i(u_i-1)(u_i-\lambda)]^a\prod_{i>j}^{n}(u_i-u_j)^{-2a/b}.\end{aligned} \tag{3.4}$$

Under the monodromy transformation, $F_A(\lambda)$ changes into $F_A(1-\lambda)$, i.e.

$$F_A(\lambda)\to F_A(1-\lambda)=\sum_c \alpha_{AC}\left(n, a, \frac{a}{b}\right)F_c(\lambda). \tag{3.5}$$

According to [1-2], we can write the integration between $(0, \lambda)$ and $(1, \infty)$ as a linear combination in the integrations between $(-\infty, 0)$ and $(\lambda, 1)$. For convenience, we define a contour integral $F(\mu, \nu, \rho, \sigma)$ as follows:

$$\begin{aligned}F(\mu, \nu, \rho, \sigma)=&\int_{-\infty}^0 \prod_{i=1}^{\mu} du_i[(1-u_i)(1-u_i)(\lambda-u_i)]^a\int_0^\lambda \prod_{i=\mu+1}^{\mu+\nu} du_i[u_i(1-u_i)(\lambda-u_i)]^a\\&\times\int_\lambda^1 \prod_{i=\mu+\nu+1}^{\mu+\nu+\rho} du_i[u_i(1-u_i)(u_i-\lambda)]^a\\&\times\int_1^\infty \prod_{i=\mu+\nu+\rho+1}^{\mu+\nu+\rho+\sigma} du_i[u_i(u_i-1)(u_i-\lambda)]^a\\&\times\prod_{i>j}^{\mu+\nu+\rho+\sigma}(u_i-u_j)^{-2a/b}.\end{aligned} \tag{3.6}$$

With the help of contour integration, we can obtain two useful formulae:

$$\begin{aligned}F(\mu, \nu, \rho, \sigma)=&\frac{-\mathscr{S}[3a-(2\mu+\mu+\sigma+2\rho-2)a/b]}{\mathscr{S}[2a-(2\rho+\sigma+\nu-1)a/b]}F(\mu+1, \nu-1, \rho, \sigma)\\&+\frac{-\mathscr{S}[a-(\rho+\sigma)a/b]}{\mathscr{S}[2a-(2\rho+\sigma+\nu-1)a/b]}F(\mu, \nu-1, \rho+1, \sigma)\end{aligned} \tag{3.7}$$

and

$$\begin{aligned}F(\mu, \nu, \rho, \sigma)=&\frac{-\mathscr{S}[a-(\rho+\nu)a/b]}{\mathscr{S}[2a-(\nu+2\rho+\sigma-1)a/b]}F(\mu, \nu, \rho+1, \sigma-1)\\&+\frac{\mathscr{S}[a-(\mu+\nu)a/b]}{\mathscr{S}[2a-(\nu+2\rho+\sigma-1)a/b]}F(\mu+1, \nu, \rho, \sigma-1)\end{aligned} \tag{3.8}$$

where $\mathscr{S}(x)=\sin(\pi x)$.

We first change all integral contours which run through $(0, \lambda)$ into ones which run through $(-\infty, 0)$ and $(\lambda, 1)$ by using (3.7), then remove away all integral contours over $(1, \infty)$ by using (3.8). Finally, we have

$$\alpha_{k_1 k_1'}\left(n, a, \frac{a}{b}\right) = \sum_{\substack{\mu=1 \\ \mu+\nu=k_1'+1}}^{k_1} \sum_{\nu=1}^{n-k_1+2} \prod_{i=0}^{k_1+\mu+1} \frac{\mathscr{S}[1+3a-2(k_1-2)a/b-(n-k_1-i+1)a/b]}{\mathscr{S}[2a-(n-\mu-3-i)a/b]}$$

$$\times \prod_{i=0}^{\mu-2} \frac{\mathscr{S}[1+a-(n-k_1+i)a/b]}{\mathscr{S}[2a-(\mu+n-k_1-2+i)a/b]}$$

$$\times \prod_{i=0}^{n-k_1-\mu+1} \frac{\mathscr{S}[2+a-(k_1-\mu+i)a/b]}{\mathscr{S}[2a-(n-k_1+2\mu+\nu-4-i)a/b]}$$

$$\times \prod_{i=0}^{\nu-2} \frac{\mathscr{S}[1+a-(\mu-1+i)a/b]}{\mathscr{S}[2a-(2\mu+\nu-4+i)a/b]} \left[\prod_{i=1}^{n-k_1'+1} \mathscr{S}(ia/b) \prod_{i=1}^{k_1'-1} \mathscr{S}(ia/b)\right]$$

$$\times \left[\prod_{i=1}^{k_1-\mu} \mathscr{S}(ia/b) \prod_{i=1}^{\nu-1} \mathscr{S}(ia/b) \prod_{i=1}^{\mu-1} \mathscr{S}(ia/b) \prod_{i=1}^{n-k_1-\nu+2} \mathscr{S}(ia/b)\right]^{-1}. \tag{3.9}$$

So far we have discussed the monodromy transformation in general. In the next section, we consider the modular matrix S in the $SU_k(2)$ wzw model.

4. The modular matrix of the $SU_k(2)$ wzw model

The conformal scale of the $SU_k(2)$ wzw model is

$$h_j = \frac{j(j+1)}{k+2} \tag{4.1}$$

and the central charge $C = 3k/(k+2)$.

As shown in section 2, the FF integral J_{AB} has the same modular invariance as the characters of the $SU_k(2)$ wzw model if we set

$$\begin{aligned} &n = k \qquad m = 0 \qquad a = -\frac{1+2k}{4(k+2)} \\ &b = \frac{1+2k}{2} \qquad B = 0 \qquad A = 2j \qquad j = 0, \frac{1}{2}, \ldots, \frac{k}{2}. \end{aligned} \tag{4.2}$$

When q approaches zero, $\chi_j \simeq (2j+1) g^{h_j - c/24}$. So we obtain

$$\chi_j(\lambda) = \frac{2j+1}{N_{2j+1}^{(k+1)}} (16)^{-2} [j(j+1)/(k+2) - k/\delta(k+2)] I_{2j,0}(\lambda) \tag{4.3}$$

where

$$N_{2j+1}^{(k+1)} = \prod_{i=1}^{k-2j} \frac{\Gamma(-ia/b)}{\Gamma(-a/b)} \prod_{i=1}^{2j} \frac{\Gamma(-ia/b)}{\Gamma(-a/b)}$$

$$\times \prod_{i=1}^{k-2j+1} \frac{\Gamma[-\frac{1}{2}-(i+\frac{3}{2})a/b]\Gamma[\frac{1}{2}-(i+\frac{3}{2})a/b]}{\Gamma(1+(4j+3+i)a/b)}$$

$$\times \prod_{i=1}^{2j-1} \frac{\Gamma^2[\frac{1}{2}-(i+\frac{3}{2})a/b]}{\Gamma[1-(2j+i+2)a/b]}. \tag{4.4}$$

Making use of (3.9), we find

$$I_{2j,0}(1-\lambda)=\sum_{j'}\alpha^{k+1}_{2j+1,2j'+1}\left(a,\frac{a}{b}\right)I_{2j',0}(\lambda) \tag{4.5}$$

and

$$\alpha^{k+1}_{2j+1,2j'+1}\left(a,\frac{a}{b}\right)=\sum_{\substack{\mu=1\\ \mu+\nu=2j+2}}^{2i+1}\sum_{\nu=1}^{k-2i+1}\prod_{l=0}^{2i-\mu}\frac{\mathcal{S}[(2i+\frac{1}{2}-l)a/b]}{c[(\mu-1-l)a/b]}$$
$$\times\prod_{l=0}^{\mu-2}\frac{\mathcal{S}[(2i-l+\frac{1}{2})a/b]}{c[(\mu-2i-1+l)a/b]}\prod_{l=0}^{k-2j-\nu}\frac{c[(2i+\frac{5}{2}-\mu+l)a/b]}{\mathcal{S}[(2j-2i-1+\mu-l)a/b]}$$
$$\times\prod_{l=0}^{\nu-2}\frac{c[(\mu+\frac{1}{2}+l)a/b]}{\mathcal{S}[(2j+1+\mu+l)a/b]}\prod_{l=1}^{k-2i+1-\nu}\mathcal{S}^{-1}(la/b)\prod_{l=1}^{2j}\mathcal{S}(la/b)$$
$$\times\prod_{l=1}^{k-2j}\mathcal{S}(la/b)\prod_{l=1}^{2i-\mu+1}\mathcal{S}^{-1}(la/b)\prod_{l=1}^{\nu-1}\mathcal{S}^{-1}(la/b)\prod_{l=1}^{\mu-1}\mathcal{S}^{-1}(la/b) \tag{4.6}$$

where $c(x)=\cos(x\pi)$.

Comparing (4.6) with (4.1), we get the modular transformation matrix S:

$$S_{ij}=\alpha^{k+1}_{2i+1,2j+1}M_{ji} \tag{4.7}$$

and

$$M_{ij}=\begin{cases}\prod\limits_{l=2i+1}^{2j}\dfrac{-c[(l+\frac{1}{2})a/b]\mathcal{S}[(l+1)a/b]}{\mathcal{S}[(l+\frac{1}{2})a/b]c[la/b]} & i<j\\ 1 & i=j\\ \prod\limits_{l=2j+1}^{2i}\dfrac{-c[la/b]\mathcal{S}[(l+\frac{1}{2})a/b]}{\mathcal{S}[(l+1)a/b]c[(l+\frac{1}{2})a/b]} & i>j.\end{cases} \tag{4.8}$$

Equations (3.6) and (3.7) do not appear to be identical to the desired expression for S_{ij} which is given as

$$S_{ij}=\sqrt{\frac{2}{k+2}}\sin\left(\frac{(2i+1)(2j+1)}{k+2}\pi\right). \tag{4.9}$$

The proof of the equality of (4.7) and (4.9) is very difficult. We calculate some simple cases in detail and show their equality in the appendix.

In the $SU_k(2)$ WZW model, there are other modular invariant combinations of the characters, which are D and E series of the A-D-E classification [10]. In FF integral representation, we can write the combinations of characters in the diagonal case (D_{2p+2}, E_6 and E_8) as follows.

For integer $k/4$ ($D_{k/2+2}$), the characters are

$$\hat{\chi}_j=\chi_j+\chi_{k/2-j}\qquad j=0,1,\ldots,\frac{k}{4}. \tag{4.10}$$

One can set

$$n=\frac{k}{4}\qquad m=B=0\qquad a=\frac{2-k}{2(k+2)}$$
$$b=\frac{k-2}{4}\qquad A=j=0,1,\ldots,\frac{k}{4} \tag{4.11}$$

and obtain

$$S_{ij}=\frac{N_{2j+1}^{k/4}(2i+1)}{N_{2i+1}^{k/4}(2j+1)}(16)^{-2}\frac{1}{k+2}[i(i+1)-j(j+1)]\alpha_{2i+1,2j+1}^{k/2+1} \tag{4.12}$$

where

$$N_{2j+1}^{k/4}=\prod_{l=1}^{k/2-2j}\frac{\Gamma(-la/b)}{\Gamma(-a/b)}\prod_{l=0}^{k/2-2j-1}\frac{\Gamma[\frac{1}{2}-(l+\frac{1}{2})a/b]\Gamma[-\frac{3}{2}-(l+\frac{5}{2})a/b]}{\Gamma[1+(4j+1-l)a/b]}$$
$$\times\prod_{l=1}^{2j}\frac{\Gamma(-la/b)}{\Gamma(-a/b)}\prod_{l=0}^{2j-1}\frac{\Gamma^2[\frac{1}{2}-(l+\frac{1}{2})a/b]}{\Gamma[1-(2j+l)a/b]} \tag{4.13}$$

and

$$\alpha_{2i+1,2j+1}^{x/2+1}=\sum_{\substack{\mu=1\\ \mu+\nu=2j+2}}^{2i+1}\sum_{\nu=1}^{k/2+2i+1}\prod_{l=0}^{2i-\mu}\frac{c[(2i-l+\frac{3}{2})a/b]}{\mathscr{S}[(\mu-l-2)a/b]}\prod_{l=0}^{\mu-2}\frac{c[(2i+\frac{1}{2}-l)a/b]}{-\mathscr{S}[(\mu-2i+l-2)a/b]}$$
$$\times\prod_{l=0}^{k/2-2i-\nu}\frac{c[(2i+\frac{3}{2}-\mu+l)a/b]}{\mathscr{S}[(2\mu-l-4+\nu-2i)a/b]}\prod_{l=0}^{\nu-2}\frac{c[(\mu-\frac{1}{2}+l)a/b]}{-\mathscr{S}[(2\mu+\nu-3+l)a/b]}$$
$$\times\prod_{l=1}^{2j}\mathscr{S}(la/b)\prod_{l=1}^{k/2-2j}\mathscr{S}(la/b)\prod_{l=1}^{2i-\mu}\mathscr{S}^{-1}(la/b)\prod_{l=1}^{\mu-1}\mathscr{S}^{-1}(la/b)$$
$$\times\prod_{l=1}^{\nu-1}\mathscr{S}^{-1}(la/b)\prod^{k/2-2i-\nu+1}\mathscr{S}^{-1}(la/b). \tag{4.14}$$

For E_6, $k=10$, there are three characters which have conformal scale 0, $\frac{5}{16}$, $\frac{1}{2}$, and central charge $\frac{5}{2}$. We set

$$n=2 \qquad m=B=0 \qquad A=j=0,1,2 \qquad a=-\tfrac{3}{16} \qquad b=-\tfrac{3}{2} \qquad a/b=\tfrac{1}{8} \tag{4.15}$$

and find

$$S_{ij}=\begin{bmatrix}\frac{1}{2} & \sqrt{\frac{1}{2}} & \frac{1}{2}\\ \sqrt{\frac{1}{2}} & \sigma & -\sqrt{\frac{1}{2}}\\ \frac{1}{2} & -\sqrt{\frac{1}{2}} & \frac{1}{2}\end{bmatrix}. \tag{4.16}$$

Here, the formula $\Gamma(2x)=\pi^{1/2}2^{2x-1}\Gamma(\bar{x})\Gamma(x+\frac{1}{2})$ has been used.

For E_8, $k=28$, two characters can be written as

$$\chi_1=[\lambda(1-\lambda)]^{-7/30}(16)^{7/10}\frac{\Gamma(\frac{1}{15})}{\Gamma(-\frac{7}{10})\Gamma(\frac{9}{10})}\int_1^{\infty}du[u(u-1)(u-\lambda)]^{-1/10}$$
$$\chi_2=[\lambda(1-\lambda)]^{-7/30}(16)^{-1/10}7\frac{\Gamma(1+\frac{4}{5})}{\Gamma^2(\frac{9}{10})}\int_0^{\lambda}du[u(1-u)(\lambda-u)]^{-1/10}. \tag{4.17}$$

We find the 2×2 matrix S as

$$S=\begin{bmatrix}2/\sqrt{5}\cos 3\pi/10 & 2/\sqrt{5}\cos\pi/10\\ 2/\sqrt{5}\cos\pi/10 & -2/\sqrt{5}\cos 3\pi/10\end{bmatrix}. \tag{4.18}$$

For $D_{2\rho+1}$ and E_7, the characters can be obtained from $A_{4\rho-2}$ and D_{10} with an automorphism of the characters. So the characters of $SU_k(2)$ wzw can be written as FF integrals.

5. The modular transformation for the minimal model

The minimal model is labelled by a pair of coprime integers p and t; the Virasoro central charge C and conformal scale $h_{r,s}$ are

$$c=\frac{1-6(p-t)^2}{pt} \tag{5.1}$$

and

$$h_{r,s}=\frac{(sp-rt)^2-(p+t)^2}{4pt}. \tag{5.2}$$

Since p and t are relatively prime, they cannot be both even. Without any loss of generality, we set p even. The FF representation of characters is as (2.5) with given parameters

$$\begin{array}{llll} A=r-1 & 0\leqslant A\leqslant p-2 & B=\frac{(s-1)}{2} & 0\leqslant B\leqslant\frac{(t-3)}{2} \\ a=\frac{3t-4p}{4p} & b=-\frac{3t-4p}{2t} & n=p-2 & m=\frac{(t-3)}{2}. \end{array} \tag{5.3}$$

Now we identify $\chi_{rr'}$ with J_{AB} with (5.3). The normalization is chosen so that $\chi_{rr'}\simeq g^{h_{rr'}-c/24}$ as $q\to 0$. So we can obtain

$$\chi_{r,s}=\frac{1}{N_{rs}}16^{-2(h_{rs}-c/24)}\prod_{i=1}^{n+1-r}\frac{\mathscr{S}(a/b)}{\mathscr{S}(ia/b)}\prod_{i=1}^{r-1}\frac{\mathscr{S}(a/b)}{\mathscr{S}(ia/b)}$$
$$\times\prod_{i=1}^{m+1-5}\frac{\mathscr{S}(b/a)}{\mathscr{S}(ib/a)}\prod_{i=1}^{s-1}\frac{\mathscr{S}(b/a)}{\mathscr{S}(ib/a)}J_{r-1,s-1} \tag{5.4}$$

where

$$N_{r,s}=j_{n+1-r,m+1-s}\begin{pmatrix}-3a+2(m-2)a/b+2n-4,\ a,\ -a/b\\ -3b+2(n-2)b/a+2m-4,\ b,\ -b/a\end{pmatrix}$$
$$\times j_{r-1,s-1}\begin{pmatrix}a,\ b,\ -a/b\\ b,\ a,\ -b/a\end{pmatrix}$$

$$j_{kl}\begin{pmatrix}\alpha,\beta,\rho\\ \alpha',\beta',\rho'\end{pmatrix}=\rho^{2kl}\prod_{i,j=1}^{k,l}\frac{1}{(-i+j\rho)}\prod_{i=1}^{k}\frac{\Gamma(i\rho')}{\Gamma(\rho')}\prod_{i=1}^{l}\frac{\Gamma(i\rho)}{\Gamma(\rho)}$$
$$\times\prod_{i,j=0}^{k-1,l-1}\frac{1}{(\alpha+j\rho-i)(\beta+j\rho-i)[\alpha+\beta+\rho(l-1-j)-k+1+i]}$$
$$\times\prod_{i=0}^{k-1}\frac{\Gamma(1+\alpha'+i\rho')\Gamma(1+\beta'+i\rho')}{\Gamma[2-2l+\alpha'+\beta'+(k-1+i)\rho']}$$
$$\times\prod_{i=0}^{l-1}\frac{\Gamma(1+i\rho+\alpha)\Gamma(1+\beta+i\rho)}{\Gamma[2-2k+\alpha+\beta+(l-1+i)\rho]}. \tag{5.5}$$

Under the transformation S, $\chi_{rs}(\lambda)$ changes into $\chi_{rs}(1-\lambda)$

$$\chi_{rs}(\lambda)\to\chi_{rs}(1-\lambda)=\sum_{r',s'}S_{rs,r's'}\chi_{r's'}(\lambda). \tag{5.6}$$

Making use of (3.9), we can get

$$S_{rs,r's'}=(16)^{2(h_{r's'}-h_{rs})}\frac{N_{r'-1,s'-1}}{N_{r-1,s-1}}\alpha^{n}_{rr'}\left(a,\frac{a}{b}\right)\alpha^{m}_{ss'}\left(b,\frac{b}{a}\right) \tag{5.7}$$

where

$$\begin{aligned}\alpha^{n}_{rr'}\left(a,\frac{a}{b}\right)=&\sum_{\substack{\mu=1\\ \mu+\nu=r'+1}}^{r}\sum_{\nu=1}^{n-r+1}\prod_{i=0}^{r-\mu-1}\\ &\times\frac{\mathcal{S}[1+3a-2(r-2)a/b-(n-r-i)a/b]}{\mathcal{S}[2a-(n+\mu-3-i)a/b]}\\ &\times\prod_{i=0}^{\mu-2}\frac{\mathcal{S}[1+a-(n-r+i)a/b]}{\mathcal{S}[2a-(\mu+n-r-2+i)a/b]}\\ &\times\prod_{i=0}^{n-r-\nu}\frac{\mathcal{S}[2+a-(r-\mu+i)a/b]}{\mathcal{S}[2a-(n-r+2\mu+\nu-4+i)a/b]}\\ &\times\prod_{i=0}^{\nu-2}\frac{\mathcal{S}[1+a-(\mu-1+i)a/b]}{\mathcal{S}[2a-(2\mu+\nu-4+i)a/b]}\\ &\times\left[\prod_{i=1}^{r'-1}\mathcal{S}(ia/b)\prod_{i=1}^{n-r'}\mathcal{S}(ia/b)\right]\\ &\times\left[\prod_{i=1}^{n-r-\nu+1}\mathcal{S}(ia/b)\prod_{i=1}^{\nu-1}\mathcal{S}(ia/b)\prod_{i=1}^{\mu-1}\mathcal{S}(ia/b)\prod_{i=1}^{r-\mu}\mathcal{S}(ia/b)\right]^{-1}.\end{aligned} \tag{5.8}$$

This is the (A_{p-1}, A_{t-1}) series of the minimal model classification. As discussed in the above section, the characters of $(D_{2+(p-1)/2}, A_{t-1})$ can be written as FF integrals. Here, we only give the result for this case:

$$n=\frac{(p-2)}{4}\qquad m=\frac{(t-3)}{2}\qquad A=\frac{r=1}{2}\qquad B=s-1$$
$$a=\frac{4t-3p}{2p}\qquad b=-\frac{4t-3p}{4t} \tag{5.9}$$

and

$$S_{rs,r's'}=(16)^{2(h_{r's'}-h_{rs})}\frac{N_{r'-1,s'-1}}{N_{r-1,s-1}}\alpha^{n}_{rr'}\left(a,\frac{a}{b}\right)\alpha^{m}_{ss'}\left(b,\frac{b}{a}\right). \tag{5.10}$$

Generally, for $(D_{1+(p-1)/2}, A_{t-1})$ and (B_k, A_{t-1}) $(k=6,7,8)$ the characters cannot be directly written as the FF contour integrals, but they can be written as a linear combination of the A-series characters. So the characters of the $SU_k(2)$ wzw and minimal models can be represented as the FF integrals.

6. Conclusion

In this paper, we have discussed the modular property of characters of RCFT in FF contour integral representation and given explicitly the expression of modular transformation matrices. In fact, all characters whose Wronskian determinant is zero can be written as an FF integral. So the FF integral gives a sign to construct characters of conformal field theories (at least ones of rational conformal field theories). This can be generalized to construct the characters of non-unitary $SU_k(2)$ WZW [11] and other models.

Acknowledgment

We are grateful to Professors K J Shi and P Wang for useful discussion. This work was supported in part by the National Science Foundation of China through the Nankai Institute of Mathematics.

Appendix

Here we shall describe the calculation of S_{ij} in some simple cases. For the $SU_k(2)$ WZW model, we obtain S_{ij} from (5.6), (5.7) and (5.8), which can be written as follows:

$$S_{ij} = M_{ji}\alpha^{k+1}_{2i+1,2j+1} \tag{A1}$$

and

$$\alpha^{k+1}_{2i+1,2j+1} = \sum_{\substack{\mu=1 \\ \mu+\nu=2j+2}}^{2i+1}\sum_{\nu=1}^{k-2i+1} \prod_{l=0}^{2i-\mu} \frac{S(2i+\frac{1}{2}-l)}{-C(\mu-1-l)} \prod_{l=0}^{\mu-2} \frac{S(2i+\frac{1}{2}-l)}{-C(\mu-2i-1+l)}$$

$$\times \prod_{l=0}^{k-2i-\nu} \frac{c(2i+\frac{5}{2}-\mu+l)}{-s(2j-2i-1+\mu-l)} \prod_{l=0}^{\nu-2} \frac{c(\mu+\frac{1}{2}+l)}{-s(2j+1+\mu+l)}$$

$$\times \prod_{l=1}^{2j} s(l) \prod_{l=1}^{k-2j} s(l) \prod_{l=1}^{2i-\mu+1} s^{-1}(l) \prod_{l=1}^{\nu-1} s^{-1}(l) \prod_{l=1}^{\mu-1} s^{-1}(l) \prod_{l=1}^{k-2i-\nu+1} s^{-1}(l) \tag{A2}$$

where $s(t)=\sin(t\pi/2(k+2))$, $c(t)=\cos(t\pi/2(k+2)$.

First, setting $j=k/2$ we have

$$S_{ik/2} = \prod_{l=1}^{k-2i} \frac{c(l+1)c(l+\frac{1}{2}+2i)}{c(2i+l-1)s(l)} \prod_{l=1}^{2i} \frac{s(l-2i-\frac{3}{2})}{c(l-1)} \prod_{l=2i+1}^{k} \frac{c(l)s(l+\frac{1}{2})}{c(l+\frac{1}{2})s(l+1)}. \tag{A3}$$

Making use of the formula

$$\prod_{r=1}^{n-1} \sin\frac{\pi r}{n} = n2^{-(n-1)} \tag{A4}$$

one can show that $S_{ik/2}$ in (A3) is equal to $\sqrt{2/(k+2)}\sin(k+1)(2i+1)\pi/(k+2)$.

Secondly, setting $j=0$, we can show

$$S_{i0} = \prod_{l=1}^{k} s(l+\tfrac{1}{2}) \prod_{l=1}^{k+1} \frac{1}{s(l)} \times s(l)s(4i+2)$$

$$= \sqrt{\frac{2}{k+2}}\sin\frac{2i+1}{k+2}\pi. \tag{A5}$$

Similarly

$$S_{0j} = \prod_{l=1}^{k-1} \frac{1}{c(l)} \prod_{l=1}^{k-} s(l+\tfrac{1}{2}) \frac{s(2j+1)c(2j+1)}{c(1)s(1)} = \sqrt{\frac{2}{k+2}}\sin\frac{2j+1}{k+2}\pi. \tag{A6}$$

For $i=\frac{1}{2}$, we find the following identity from (A2):

$$\alpha^{k+1}_{2,2j+1}=-\frac{s(\frac{3}{2})}{s(1)}\prod_{l=1}^{k-1-2j}\frac{c(\frac{3}{2}+l)}{c(2j-l)}\prod_{l=1}^{2j-1}\frac{c(l+\frac{3}{2})}{-s(2j+2+l)}\left[-\frac{c(\frac{3}{2})c(2j+2)}{s(2j+2)}+\frac{c(\frac{3}{2})s(2j)}{c(2j)}\right]$$

$$=\frac{1}{2}\frac{s(3)c(4j+2)}{c(2j)s(1)}\prod_{l=1}^{k-2j}c(l+\tfrac{1}{2})\prod_{l=1}^{2j}c(l+\tfrac{1}{2})\prod_{l=1}^{2j-1}\frac{1}{c(l)}$$

$$\times\prod_{l=1}^{k-2j-1}\frac{1}{c(l)}\prod_{l=2j+2}^{4j+1}\frac{1}{s(l)}. \tag{A7}$$

By a direct calculation, we have shown that

$$S_{1/2j}=M_{j1/2}\alpha^{k+1}_{2,2j+1}=\sqrt{\frac{2}{k+2}}\sin\pi\frac{2(2j+1)}{k+2} \tag{A8}$$

For small k, one can directly calculate S_{ij} and find

$$S_{ij}=\sqrt{\frac{2}{k+2}}\sin\frac{(2i+1)(2j+1)}{k+2}\pi. \tag{A9}$$

We have to point out that for an arbitrary integer k, the proof of (A9) is difficult. For a given k, one can find that (A9) is true by a tedious calculation.

References

[1] Dotsenko S Vl and Fateev V A 1984 *Nucl. Phys.* B **240** 312
[2] Dotsenko S Vl and Fateev V A 1985 *Nucl. Phys.* B **251** 691
[3] Felder G 1988 *Nucl. Phys.* B **317** 215
[4] Felder G, Frohlich J and Keller G 1989 *Commun. Math. Phys.* **124** 647
[5] Hou B Y, Lie D P and Yue R H 1989 *Phys. Lett.* B **229** 45
[6] Hou B Y, Shi K J, Wang P and Yue R H 1989 *Preprint NWU-IMP-1219*
[7] Mathur S D, Mukhi S and Sen A 1988 *Phys. Lett.* B **213** 303
[8] Mathur S D, Mukhi S and Sen A 1989 *Nucl. Phys.* B **312** 15
[9] Mukhi S, Panda S and Sen A 1989 *Preprint TIFR/TH/89-01*
[10] Capelli A, Itzykson C and Zuber J B 1987 *Nucl. Phys.* B **280** 445; 1987 *Commun. Math. Phys.* **113** 1
[11] Mukhi S and Panda S 1989 *Preprint TIFR/TH/89-64*

Physics Letters B 266 (1991) 353–362
North-Holland

PHYSICS LETTERS B

Sine-Gordon and affine Toda fields as non-conformally constrained WZNW model

Bo-Yu Hou, Liu Chao and Huan-Xiong Yang
Institute of Modern Physics, Northwest University, Xian 710 069, China

Received 9 January 1991; revised manuscript received 20 May 1991

The sine-Gordon and affine Toda fields are obtained by imposing a conformal breaking constraint on the WZNW action. It is found that the Drinfeld–Sokolov linear systems for the sine-Gordon system arise naturely as the equations of motion of the restricted WZNW model and are transformed into each other by the restricted WNZW field $g(x)$. The truncation from conformal affine Toda fields to affine ones is given also.

1. Introduction

Over the last few years, conformal field theory, completely integrable systems and their relations have received much attention. For example, the W algebra [1,2] of conformal fields, the completely integrable Liouville–Toda equations [3] and the relation between these two objects [4] have been studied quite well. Zamolodchikov [5] showed that conformal fields can be perturbed to the off-critical states while keeping their integrability unchanged. It is known that the sine-Gordon [6] and affine Toda [7] equations have close relations with off-critical conformal field theories. The crucial point is that one can perturb the free field or WZNW action by a combination of relevent primary fields, usually with Z_2 or Z_n symmetries.

On the other hand, the possibility of deriving soliton equations out of the WZNW model by imposing certain constraints has attracted attention quite recently. This method gives natural explanations for the conformality or inconformality of completely integrable systems. It is shown that the Liouville–Toda [8] and the (m)KdV [9,10] equations can all be derived from the restricted WZNW action, and the Virasoro algebra and W algebra can also be realized. It is even possible [11] to relate this method to the two-dimensional induced gravity [12]. In ref. [10], an integro-differential equation containing the affine Toda equation as a special case is obtained by imposing on the originally two-sided chirally invariant WZNW action only a left-hand-side constraint.

The motivation of the present work is to show that the two-hand-side conformal preserving constraint method of ref. [8] can be generalized to the conformal breaking case with sine-Gordon and affine Toda equations as concrete examples. Such a motivation is clearly different from that of the classic works [13] which derive sine-Gordon and affine Toda equations by reduction of the principal chiral model. Actually, although the present work seems to given no new integrable models and no new method for solving integrable systems, it certainly gives new understanding of the relation between conformal breaking integrable models and conformal invariant theories. By imposing conformal breaking constraints on the SU(2) WZNW action it is found that the only field invariant with respect to the surviving gauge freedom is the sine-Gordon field. The spectral parameter for the constrained system may be introduced invariantly in the action, which ensures the integrability. It is also shown that the Drinfeld–Sokolov linear systems [14] are transformed into each other by $g(x)$. The generalization to

0370-2693/91/$ 03.50

Volume 266, number 3,4 PHYSICS LETTERS B 29 August 1991

the Sl$(n, \mathbb{R})$ affine Toda fields is straightforward. We also show how to obtain the SL$(2, \mathbb{R})$ conformal affine Toda fields and how to truncate from these fields to affine ones.

2. Action and equations of motion

We start from the action [#1]

$$I(g, A_-, A_+) = S(g) + \kappa \int d^2\xi \, \mathrm{Tr}(A_- \, \partial_+ g g^{-1} + g^{-1} \, \partial_- g A_+ + A_- g A_+ g^{-1} - A_- \mu - A_+ \nu) , \tag{1}$$

where the standard WZNW action reads

$$S(g) = \tfrac{1}{2}\kappa \int d^2x \, \eta^{\mu\nu} \, \mathrm{Tr}(\partial_\mu g g^{-1} \, \partial \nu g g^{-1}) - \tfrac{1}{3}\kappa \int d^3x \, \epsilon^{ijk} \, \mathrm{Tr}(\partial_i g g^{-1} \, \partial_j g g^{-1} \, \partial_k g g^{-1}) ,$$

in the case of $g(x) \in \mathrm{SU}(2)$, the constant constraint matrices are taken to be

$$\mu = \tfrac{1}{2}\mathrm{i}(m\sigma_+ + \bar{m}\sigma_-), \quad \nu = \tfrac{1}{2}\mathrm{i}(n\sigma_+ + \bar{n}\sigma_-), \quad m, n = \text{const.} ,$$

and $\sigma_{\pm,3}$ are Pauli matrices. The lagrangian multipliers $A_\pm$ are also taken as antihermitian matrices taking values only in the $\sigma_\pm$ directions,

$$A_\pm(x) = \tfrac{1}{2}\mathrm{i} \sum_j A^j_\pm(x) \sigma_j, \quad j = (+, -) .$$

Taking the variation δg from the left- and right-hand side in the action (1), one gets the following "zero curvature equations":

$$\partial_-(\partial_+ g g^{-1} + g A_+ g^{-1}) + [A_-, \partial_+ g g^{-1} + g A_+ g^{-1}] + \partial_+ A_- = 0 , \tag{2}$$

$$\partial_+(g^{-1} \, \partial_- g + g^{-1} A_- g) - [A_+, g^{-1} \, \partial_- g + g^{-1} A_- g] + \partial_- A_+ = 0 , \tag{3}$$

which are equivalent in the sense that they can be transformed into each other by a similar transformation of the matrix $g(x)$. Taking the variations of A_- and A_+, one gets the "constraint equations"

$$\mathrm{Tr}[\sigma_\pm(\partial_+ g g^{-1} + g A_+ g^{-1} - \mu)] = 0 , \tag{4}$$

$$\mathrm{Tr}[\sigma_\pm(g^{-1} \, \partial_- g + g^{-1} A_- g - \nu)] = 0 . \tag{5}$$

3. Realization of the sine-Gordon equation

On the constraint surface (4),

$$\partial_+ g g^{-1} + g A_+ g^{-1} = \tfrac{1}{2}\mathrm{i} J^3_+(x) \sigma_3 + \mu ,$$

where $J^3_+(x)$ is to be determined. Substituting this equation into (2), the $\sigma_\pm$ component read

$$J^3_+ = -\tfrac{1}{2}\mathrm{i} \, \partial_+ \log(A^+_- / A^-_-), \quad 0 = \partial_+ \log(A^+_- A^-_-) ;$$

therefore one can always choose a pair of functions $\varphi_+(x)$ and $\lambda_-(x_-)$ to express $A^\pm_-$ and J^3_+ as follows:

[#1] Throughout this letter, we adopt the following notations and conventions for our two-dimensional Minkowski spacetime: the metric $\eta^{\mu\nu}$ and the antisymmetric tensor $\epsilon^{\mu\nu}$ are defined by $\eta^{00} = -\eta^{11} = \epsilon^{01} = -\epsilon^{10} = 1$, the light-cone variables are $x_\pm = \tfrac{1}{2}(x_0 \pm x_1)$ and the derivatives with respect to these variables are $\partial_\pm = \partial_0 \pm \partial_1$.

Volume 266, number 3,4 PHYSICS LETTERS B 29 August 1991

$$A^{+}_{-}=\overline{A^{-}_{-}}=\lambda_{-}(x_{-})\exp[i\varphi_{+}(x)], \quad J^{3}_{+}(x)=\partial_{+}\varphi_{+}(x)\,. \tag{6}$$

Similarly, one gets from eqs. (5) and (3) that

$$g^{-1}\partial_{-}g+g^{-1}A_{-}g=\tfrac{1}{2}iJ^{3}_{-}(x)\sigma_{3}+\nu, \quad A^{+}_{+}=\overline{A^{-}_{+}}=\lambda_{+}(x_{+})\exp[-i\varphi_{-}(x)], \quad J^{3}_{-}(x)=\partial_{-}\varphi_{-}(x)\,. \tag{7}$$

From eqs. (6), (7) and the equivalence of (2) and (3) under the similar transformation by $g(x)$, one has

$$|\lambda_{+}(x_{+})|=|m|, \quad |\lambda_{-}(x_{-})|=|n|, \quad \arg[m/\lambda_{+}(x_{+})]=\arg[\lambda_{-}(x_{-})/n]\equiv\delta=\text{const.},$$

$$\varphi_{+}(x)+\delta=\varphi_{-}(x)+\delta,$$

showing that $\varphi_{+}(x)=\varphi_{-}(x)\equiv\varphi(x)$ and that $g(x)=\exp[\frac{1}{2}i\varphi(x)\sigma_{3}]$. The σ_3 component of each of the two zero curvature equations now shows that

$$\partial_{+}\partial_{-}\varphi(x)=|m||n|\sin\varphi(x)\,, \tag{8}$$

which is the well known sine-Gordon equation.

4. Gauge invariance of the constrained WZNW action

In this section we choose without loss of generality $m=n=1$. It is easy to see that the action (1) is invariant up to a total derivative term under the following $U_L(1)\times U_R(1)$ chiral gauge transformations:

$$g(x)\to\alpha(x_{-})g(x)\beta^{-1}(x_{+})\,,$$

$$A_{-}(x)\to\alpha(x_{-})A_{-}(x)\alpha^{-1}(x_{-})+\alpha(x_{-})\partial_{-}\alpha^{-1}(x_{-})\,,$$

$$A_{+}(x)\to\beta(x_{+})A_{+}(x)\beta^{-1}(x_{+})+\partial_{+}\beta(x_{+})\beta^{-1}(x_{+})\,,$$

$$\alpha(x_{-})=\exp[ia(x_{-})\sigma_{1}], \quad \beta(x_{+})=\exp[ib(x_{+})\sigma_{1}]\,,$$

provided the values of A_- and A_+ are not restricted to be zero in the σ_3 direction. Therefore the Lagrange multipliers $A_\pm$ become U(1) gauge fields, each having three components. The σ_1 component acts as a U(1) connection, the other two components form a gauge covariant vector normal to the σ_1 direction. It is a trivial task to verify that the zero curvature equations are invariant under the transformation (9), and the constraint equations are transformed into

$$\mathrm{Tr}[(\alpha^{-1}\sigma_{j}\alpha)(\partial_{+}gg^{-1}+gA_{+}g^{-1}-\mu)]=0\,, \tag{9}$$

$$\mathrm{Tr}[(\beta^{-1}\sigma_{j}\beta)(g^{-1}\partial_{-}g+g^{-1}A_{-}g-\nu)]=0, \quad j=1,2\,. \tag{10}$$

For $j=1$, eqs. (9), (10) remain of the form of (4) and (5); for $j=2$, it changes covariantly. So it is clear that the choice of $A_\pm$ in section 2 is equivalent to a gauge fixing. We shall call this special gauge a reduced one in the rest of this letter.

5. Noether currents and gauge invariant quantities

Since the action (1) changes by only a total derivative term under the transformation (9), one can get the gauge currents by the Noether theorem. Now choosing the infinitesimal gauge transformation

$$\delta g\sim i\sigma_{1}g(x), \quad \delta A_{-}\sim i[\sigma_{1},A_{-}(x)], \quad \delta A_{+}=0\,,$$

and setting the on-shell and off-shell variation of the action to be equal, one finds the conservation law

Volume 266, number 3,4 PHYSICS LETTERS B 29 August 1991

$$\partial_+\mathcal{J}^{\rm L}_-(x)-\partial_-\mathcal{J}^{\rm L}_+(x)=([\partial_+gg^{-1}+gA_+g^{-1},A_-])^{(1)},$$
$$\mathcal{J}^{\rm L}_+(x)=-(\partial_+gg^{-1}+gA_+g^{-1})^{(1)},\quad \mathcal{J}^{\rm L}_-(x)=A^{(1)}_-, \tag{11}$$

where $A^{(1)}_-\equiv{\rm Tr}(\sigma_1A_-)$, etc. This conservation law correspond to the chiral $U_L(1)$ symmetry. There is another conservation law,

$$\partial_+\mathcal{J}^{\rm R}_-(x)-\partial_-\mathcal{J}^{\rm R}_+(x)=([A_+,g^{-1}\partial_-g+g^{-1}A_-g])^{(1)},$$
$$\mathcal{J}^{\rm R}_-(x)=-A^{(1)}_+,\quad \mathcal{J}^{\rm R}_-(x)=(g^{-1}\partial_-g+g^{-1}A_-g)^{(1)}, \tag{12}$$

corresponding to the infinitesimal chiral $U_R(1)$ transformation

$$\delta g\sim-{\rm i}g(x)\sigma_1,\quad \delta A_-=0,\quad \delta A_+\sim{\rm i}[\sigma_1,A_+(x)]\,.$$

Clearly, eqs. (11) and (12) are in fact the σ_1 components of the zero curvature equations (2) and (3). It is not difficult to prove that the right-hand side of eq. (11) (denoted by Q_L) and eq. (12) (denoted by Q_R) are invariant under the transformation (9) and thus are gauge independent. Using the result of section 2 one sees that

$$Q_{\rm L}=-\tfrac14{\rm i}\,\partial_+\cos\varphi(x),\quad Q_{\rm R}=-\tfrac14{\rm i}\,\partial_-\cos\varphi(x)\,;$$

therefore the only gauge independent field surviving in the reduced gauge is the sine-Gordon field $\varphi(x)$.

6. Hamiltonian reduction

In the reduced gauge, the action (1) can be written in terms of $\varphi(x)$ as

$$I(\varphi)=-\tfrac14\kappa\int{\rm d}^2x\,(\partial_\mu\varphi\,\partial^\mu\varphi-2\cos\varphi)\,, \tag{13}$$

so $\pi_\varphi=-\frac12\kappa\,\partial_0\varphi$ and the canonical equal-time Poisson bracket reads

$$\{\varphi(x),\partial_0\varphi(y)\}_{x_0=y_0}=-\frac{2}{\kappa}\delta(x_1-y_1)\,. \tag{14}$$

The Poisson bracket (14) can be derived from the WZNW model by hamiltonian reduction. The Poisson bracket for the standard WZNW model is given by the Kac–Moody-type chiral current algebra

$$\{J^{({\rm WZNW})a}_\pm(x_\pm),J^{({\rm WZNW})b}_\pm(y_\pm)\}=2{\rm i}\epsilon^{abc}J^{({\rm WZNW})c}_\pm(y_\pm)\delta(x_\pm-y_\pm)+\tfrac12\kappa\delta^{ab}\,\partial_\pm\delta(x_\pm-y_\pm)\,,$$
$$\{J^{({\rm WZNW})a}_+(x_+),J^{({\rm WZNW})b}_-(y_-)\}=0\,,$$

where

$$J^{({\rm WZNW})a}_+=\tfrac12\kappa\,{\rm Tr}(\sigma^a\,\partial_+gg^{-1}),\quad J^{({\rm WZNW})a}_-=\tfrac12\kappa\,{\rm Tr}(\sigma^ag^{-1}\,\partial_-g)\,.$$

For the restricted model (1), the currents $J^{({\rm WZNW})}_\pm$ are no longer chiral functions. Furthermore, the $\sigma_{1,2}$ components are restricted to zero. The only surviving components, $J^{({\rm WZNW})3}_\pm$ satisfy the following Poisson brackets:

$$\{J^{({\rm WZNW})3}_+(x),J^{({\rm WZNW})3}_+(y)\}_{x_-=y_-}=\tfrac12\kappa\,\partial_+\delta(x_+-y_+)\,,$$
$$\{J^{({\rm WZNW})3}_-(x),J^{({\rm WZNW})3}_-(y)\}_{x_+=y_+}=\tfrac12\kappa\,\partial_-\delta(x_--y_-)\,,$$
$$\{J^{({\rm WZNW})3}_+(x),J^{({\rm WZNW})3}_-(y)\}_{x_0=y_0}=0\,,$$

or, in terms of the sine-Gordon field $\varphi(x)$.

Volume 266, number 3,4 PHYSICS LETTERS B 29 August 1991

$$\{\partial_+\varphi(x), \partial_+\varphi(y)\}_{x_-=y_-} = \frac{2}{\kappa}\partial_+\delta(x_+-y_+), \quad \{\partial_-\varphi(x), \partial_-\varphi(y)\}_{x_+=y_+} = \frac{2}{\kappa}\partial_-\delta(x_--y_-),$$

which is equivalent to (14) in the usual Minkowski spacetime coordinates.

It is notable that there is a surviving discrete gauge freedom, say $\mathbb{Z}_2$ symmetry even in the reduced gauge. The elements of the surviving $\mathbb{Z}_2$ symmetry group read

$$\alpha=\beta=\exp(\tfrac{1}{2}in\pi\sigma_1), \quad n=0,1,$$

which transform $A_-\leftrightarrow A_+$, $g\leftrightarrow g^{-1}$, i.e., they transform the left-hand-side and the right-hand-side fields into each other. In terms of the sine-Gordon field $\varphi(x)$, this correspond to $\varphi\to-\varphi$ as studied in ref. [6].

7. Energy–momentum tensor and conformal symmetry breaking

The energy–momentum tensor of the action (1) reads

$$T^{\rho\tau}=T^{\rho\tau}_{\rm WZNW}+\kappa\,{\rm Tr}\{(\eta^{\rho\sigma}-\epsilon^{\rho\sigma})A_\sigma\,\partial^\tau gg^{-1}+(\eta^{\rho\sigma}+\epsilon^{\rho\sigma})g^{-1}\,\partial^\tau gA_\sigma$$
$$-\eta^{\rho\tau}[(\eta^{\lambda\sigma}+\epsilon^{\lambda\sigma})(A_\lambda\,\partial_\sigma gg^{-1}+g^{-1}\,\partial_\lambda gA_\sigma+A_\lambda gA_\sigma g^{-1})-A_0(\mu+\nu)+A_1(\mu-\nu)]\}, \qquad (15)$$

where

$$T^{\rho\tau}_{\rm WZNW}=\kappa\,{\rm Tr}(\eta^{\rho\sigma}\,\partial_\sigma gg^{-1}\,\partial^\tau gg^{-1}-\tfrac{1}{2}\eta^{\rho\tau}\,\partial_\sigma gg^{-1}\,\partial^\sigma gg^{-1}),$$

and $A_0\equiv\frac{1}{2}(A_++A_-)$, $A_1\equiv\frac{1}{2}(A_+-A_-)$.

In the reduced gauge, eq. (15) is reduced into

$$T^{\rho\tau}=T^{\rho\tau}_{\rm WZNW}+\eta^{\rho\tau}\kappa\,{\rm Tr}(A_-gA_+g^{-1}), \qquad (15r)$$

or in components

$$T^{++}\equiv\tfrac{1}{4}(T^{00}+T^{11}+2T^{01})=\tfrac{1}{4}\kappa\,{\rm Tr}(\partial_+gg^{-1}\,\partial_+gg^{-1})=\frac{1}{2\kappa}(J_+^{({\rm WZNW})3})^2, \qquad (16)$$

$$T^{--}\equiv\tfrac{1}{4}(T^{00}+T^{11}-2T^{10})=\tfrac{1}{4}\kappa\,{\rm Tr}(g^{-1}\,\partial_-gg^{-1}\,\partial_-g)=\frac{1}{2\kappa}(J_-^{({\rm WZNW})3})^2, \qquad (17)$$

$$T^{+-}=T^{-+}\equiv\tfrac{1}{4}(T^{00}-T^{11})=\tfrac{1}{2}\kappa\,{\rm Tr}(A_-gA_+g^{-1}). \qquad (18)$$

Remember that for the restricted model, $\partial_-J_+^{({\rm WZNW})}\neq0\neq\partial_+J_-^{({\rm WZNW})}$, so T^{++} and T^{--} are no longer holomorphic or antiholomorphic functions of light-cone variables. Using $T_\alpha{}^\alpha=\eta^{\alpha\beta}T_{\alpha\beta}$ one finds $T_\alpha{}^\alpha=T^{00}-T^{11}=4T^{+-}\neq0$. The non-zero trace of the energy–momentum tensor comes form the constraint terms. It explicitly breaks the scale invariance of the WZNW theory.

Substituting the result of section 2 into eq. (15r), one gets the sine-Gordon energy–momentum tensor

$$T^{\mu\nu}=-\tfrac{1}{2}\kappa\,\partial^\mu\varphi\,\partial^\nu\varphi+\tfrac{1}{4}\kappa\eta^{\mu\nu}\,\partial_\lambda\varphi\,\partial^\lambda\varphi-\tfrac{1}{2}\kappa\eta^{\mu\nu}\cos\varphi, \qquad (19)$$

which is in precise agreement with that which can be derived from the action (13).

8. Linearized equations and spectral parameter of the completely integrable system

In the reduced gauge, the zero curvature equations (2) and (3) can be regarded as the compatibility conditions of the following linear systems:

Volume 266, number 3,4 PHYSICS LETTERS B 29 August 1991

$$\partial_- U_L = -A_- U_L, \quad \partial_+ U_L = (\tfrac{1}{2}\mathrm{i}J^3_+(x)\sigma_3 + \mu) U_L, \tag{20}$$

$$\partial_- U_R = -(\tfrac{1}{2}\mathrm{i}J^3_-(x)\sigma_j + \nu) U_R, \quad \partial_+ U_R = A_+ U_R. \tag{21}$$

Choosing $m=1/n=\lambda_+=1/\lambda_-=\lambda$, eqs. (20) and (21) are reduces to the familiar form of sine-Gordon linearized equations

$$\partial_- U_L = -\tfrac{1}{2}\mathrm{i}\lambda^{-1}\begin{pmatrix} 0 & \exp(\mathrm{i}\varphi) \\ \exp(-\mathrm{i}\varphi) & 0 \end{pmatrix} U_L, \quad \partial_+ U_L = \tfrac{1}{2}\mathrm{i}\begin{pmatrix} \partial_+\varphi & \lambda \\ \lambda & -\partial_+\varphi \end{pmatrix} U_L, \tag{22}$$

$$\partial_- U_R = -\tfrac{1}{2}\mathrm{i}\begin{pmatrix} \partial_-\varphi & \lambda^{-1} \\ \lambda^{-1} & -\partial_-\varphi \end{pmatrix} U_R, \quad \partial_+ U_R = \tfrac{1}{2}\mathrm{i}\lambda\begin{pmatrix} 0 & \exp(-\mathrm{i}\varphi) \\ \exp(\mathrm{i}\varphi) & 0 \end{pmatrix} U_R, \tag{23}$$

in the asymptotic coordinate systems. The compatibility conditions give rise to the sine-Gordon equation independent of the choice of λ. This shows that the action (1) and the equations of motion (2)–(5) are invariant under

$$A_- \to \lambda^{-1} A_-, \quad A_+ \to \lambda A_+, \quad \mu \to \lambda\mu, \quad \nu \to \lambda^{-1}\nu$$

on the constraint surface. Thus one can regard λ as a spectral parameter which ensures the integrability. We further note that the linear systems (20) and (21) are the very Drinfeld–Sokolov linear systems, and that they can be transformed into each other by the restricted WZNW field $g(x)$. Now one sees what we mean by restricted WZNW field. It is the very field that relates the two asymptotic directions of the pseudosphere of the sine-Gordon equation.

Using the solutions U_L and U_R of the Drinfeld–Sokolov linear systems, the action (1) can be rewritten as

$$I(g, U_L, U_R) = S(U_L^{-1} g U_R) - S(U_L^{-1}) - S(U_R) - \kappa \int \mathrm{d}^2\xi\, \mathrm{Tr}(U_L \partial_- U_L^{-1}\mu + \partial_t U_R U_R^{-1}\nu)\,; \tag{24}$$

the gauge fixing condition for the reduced gauge now reads

$$\mathrm{Tr}(\sigma_3 U_L \partial_- U_R^{-1}) = \mathrm{Tr}(\sigma_3 \partial_+ U_R U_R^{-1}) = 0\,.$$

Furthermore, because of the similarity of the two Drinfeld–Sokolov linear systems, one immediately has

$$U_L^{-1} g U_R = 1, \quad \text{i.e.,} \quad U_L U_R^{-1} = g(x) \tag{25}$$

on the mass shell.

9. Non-linear sigma model

Following the last section, $U_L(x, \lambda)$ and $U_R(x, \lambda)$ are in general λ-dependent. Defining

$$N(x,\lambda) \equiv U_L^{-1}\sigma_3 U_L = U_R^{-1}\sigma_3 U_R\,,$$

one has the equation of motion [15]

$$[N, \partial_\mu \partial^\mu N] = 0$$

for the non-linear sigma model. Using $N(x, \lambda)$ one can construct various infinitely many conserved currents: notably, one can get the Backlund transformation and Ricatti equation [16,17] for the sine-Gordon model.

Volume 266, number 3,4 PHYSICS LETTERS B 29 August 1991

10. Sl(n, ℝ) restricted WZNW model and affine Toda equation

For $g(x)\in \mathrm{SL}(n,\mathbb{R})$, choosing

$$\mu=\tfrac{1}{2}\lambda\Lambda,\quad \nu=\tfrac{1}{2}\lambda^{-1}\Lambda^{\mathrm{T}},\quad \Lambda=\begin{pmatrix}0&0&\cdots&0&1\\1&0&&0&0\\0&1&\ddots&\vdots&\vdots\\\vdots&&\ddots&0&0\\0&&&1&0\end{pmatrix},\tag{26}$$

and fixing the gauge such that $A_\pm$ take zero components in the Cartan subalgebra, the constraint equations become

$$\mathrm{Tr}[E_\alpha(\partial_+ gg^{-1}+gA_+g^{-1}-\mu)]=0\,,\tag{27}$$

$$\mathrm{Tr}[E_\alpha(g^{-1}\,\partial_- g+g^{-1}A_-g-\nu)]=0,\quad \alpha\in\Phi\,.\tag{28}$$

Taking the Gauss decomposition of $g\in\mathrm{SL}(n,\mathbb{R})$, and using the fact that $g(x)$ is independent of λ, it leads to

$gA_+g^{-1}=\mu,\quad g^{-1}A_-g=\nu$ and $g(x)$ is diagonal .

Now choosing the fields

$$g=\begin{pmatrix}\exp(\frac{1}{2}\varphi_1)&&&0\\&\exp(\frac{1}{2}\varphi_2)&&\\&&\ddots&\\0&&&\exp(\frac{1}{2}\varphi_n)\end{pmatrix},\quad \varphi_1+\varphi_2+\ldots+\varphi_n=0\,,\tag{29}$$

one has

$$A_+=\tfrac{1}{2}\lambda\begin{pmatrix}0&0&\cdots&0&\exp[\frac{1}{2}(\varphi_n-\varphi_1)]\\\exp[\frac{1}{2}(\varphi_1-\varphi_2)]&0&&0&0\\0&\exp[\frac{1}{2}(\varphi_2-\varphi_3)]&&\vdots&\vdots\\\vdots&&\ddots&0&0\\0&0&\cdots&\exp[\frac{1}{2}(\varphi_{n-1}-\varphi_n)]&0\end{pmatrix},\tag{30}$$

$$A_-=\tfrac{1}{2}\lambda^{-1}\begin{pmatrix}0&\exp[\frac{1}{2}(\varphi_1-\varphi_2)]&0&\cdots&0\\0&0&\exp[\frac{1}{2}(\varphi_2-\varphi_3)]&&\vdots\\\vdots&\vdots&0&\ddots&0\\0&0&\vdots&&\exp[\frac{1}{2}(\varphi_{n-1}-\varphi_n)]\\\exp[\frac{1}{2}(\varphi_n-\varphi_1)]&0&0&\vdots&0\end{pmatrix},\tag{31}$$

$$\partial_+gg^{-1}+gA_+g^{-1}=\frac{1}{2}\begin{pmatrix}\partial_+\varphi_1&0&\cdots&0&\lambda\\\lambda&\partial_+\varphi_2&&&0\\0&\lambda&\ddots&&\vdots\\\vdots&&\ddots&\partial_+\varphi_{n-1}&0\\0&0&\cdots&\lambda&\partial_+\varphi_n\end{pmatrix},\tag{32}$$

Volume 266, number 3,4 PHYSICS LETTERS B 29 August 1991

$$g^{-1}\partial_- g+g^{-1}A_- g=\frac{1}{2}\begin{pmatrix} \partial_-\varphi_1 & \lambda^{-1} & 0 & \dots & 0 \\ 0 & \partial_-\varphi_2 & \lambda^{-1} & & \\ \vdots & & \partial_-\varphi_3 & \ddots & \\ 0 & & & \ddots & \lambda^{-1} \\ \lambda^{-1} & 0 & 0 & \dots & \partial_-\varphi_n \end{pmatrix}. \tag{33}$$

The zero curvature equations (2) and (3) now become the integrability conditions for the affine Toda equation

$$\partial_+\partial_-\varphi_i+\tfrac{1}{2}\{\exp[\tfrac{1}{2}(\varphi_i-\varphi_{i+1})]-\exp[\tfrac{1}{2}(\varphi_{i-1}-\varphi_i)]\}=0,\quad \varphi_1+\varphi_2+\dots+\varphi_n=0,\quad \varphi_{n+1}=\varphi_1,\quad \varphi_0=\varphi_n\,. \tag{34}$$

There remains a $\mathbb{Z}_n$ symmetry in the affine Toda equation (34), say, invariant under $\varphi_i\to\varphi_{i+1}$. The results of the former sections for the sine-Gordon model can be rightly generalized to the affine Toda case; for example, the gauge invariant quantities are related to the affine Toda fields, the Poisson brackets for φ_i are derived from the WZNW model using hamiltonian reduction, the non-tracelessness of the energy–momentum tensor is brought about by constraints, the Drinfeld–Sokolov linear systems are transformed into each other by $g(x)$, etc. The generalization to other groups is also available. For all classical groups A_n, B_n, C_n, D_n and the exceptional groups G_2, F_4, E_6, E_7 and E_8, the affine Toda equations can be derived, the only differences being that the exponents for each group may disappear or repeat accordingly [8].

11. Loop SL(2, ℝ), affine SL(2, ℝ) and conformal affine Toda fields

In the SL(2, ℝ) case, if one chooses

$$\mu=\sigma_+ +\lambda\sigma_-,\quad \nu=\sigma_- +\lambda^{-1}\sigma_+\,,$$

one can obtain, following similar considerations, the sinh-Gordon equation. This procedure can also be regarded as the values of $A_\pm$ being extended to the sl(2) loop algebra. $A_\pm$ can even be extended to be elements of the affine sl(2) algebra. In that case, one introduces

$$\mu=E_{\alpha_1}+E_{\alpha_0}\,, \tag{35}$$

$$\nu=E_{-\alpha_1}+E_{-\alpha_0}\,, \tag{36}$$

where α_1 and α_0 are simple roots of the affine sl(2) algebra, E_{α_1} and E_{α_0} are the corresponding root vectors. Following similar considerations to those in the former sections, one fixes the gauge so that $g(x)$ is diagonal,

$$g=\exp(\Phi),\quad \Phi=\tfrac{1}{2}gH+\eta d+\tfrac{1}{2}\xi c\,, \tag{37}$$

where H, d, c form the Cartan subalgebra of the affine sl(2). In this fixed gauge,

$$A_+=\exp(-\mathrm{ad}\,\Phi)\,\mu\,, \tag{38}$$

$$A_-=\exp(\mathrm{ad}\,\Phi)\,\nu\,, \tag{39}$$

and the zero curvature equations give rise to the conformal affine Toda equation [18]

$$\partial_+\partial_-\varphi=\exp(2\varphi)-\exp(2\eta-2\varphi)\,, \tag{40}$$

$$\partial_+\partial_-\eta=0\,, \tag{41}$$

$$\partial_+{}'gv_-\xi=\exp(2\eta-2\varphi)\,. \tag{42}$$

In ref. [18], the zero curvature condition for the conformal affine Toda equations are introduced a priori as a definition. Here we have derived it naturally from the constrained WZNW action. The action in ref. [18] can

Volume 266, number 3,4 PHYSICS LETTERS B 29 August 1991

also be derived from our action (1) in the reduced gauge. In addition, the ratio of the left and right monodromy matrix $U_L U_R^{-1}$ is just equal to the restricted WZNW field $g(x)$.

Now by eq. (41), one has $\eta(x)=\eta_+(x_+)+\eta_-(x_-)$. Using $\exp(-\eta_\pm d)$ to make a similar transformation for the zero curvature equations, we obtain

$$A_+ \to \begin{pmatrix} 0 & \exp(-\varphi) \\ \lambda\exp(-\eta_+)\exp(\varphi) & 0 \end{pmatrix}, \tag{43}$$

$$A_- \to \begin{pmatrix} 0 & \lambda^{-1}\exp(-\eta_-)\exp(\varphi) \\ \exp(-\varphi) & 0 \end{pmatrix}, \tag{44}$$

$$\partial_+ g g^{-1} + g A_+ g^{-1} \to \begin{pmatrix} \frac{1}{2}\partial_+\varphi & 1 \\ \lambda\exp(-\eta_+) & -\frac{1}{2}\partial_+\varphi \end{pmatrix} + \tfrac{1}{2}\partial_+\xi c, \tag{45}$$

$$g^{-1}\partial_- g + g^{-1} A_- g \to \begin{pmatrix} \frac{1}{2}\partial_-\varphi & \lambda^{-1}\exp(-\eta_-) \\ 1 & -\frac{1}{2}\partial_-\varphi \end{pmatrix} + \tfrac{1}{2}\partial_-\xi c, \tag{46}$$

where the terms containing $\lambda^{\pm 1}$ correspond to the root vecters $E_{\pm\alpha_0}$. In the case $c=0$, the affine sl(2) algebra reduces to the loop sl(2) algebra, and the zero curvature condition gives $\partial_+\partial_-\tilde{\varphi}+4\exp(-\frac{1}{2}\eta)\sinh\tilde{\varphi}=0$, which can be reduced to the usual form of the sinh-Gordon equation by a conformal change of coordinates, $x_+ \to \int^{x_+} dx'_+ \exp[-\frac{1}{2}\eta_+(x'_+)]$ and $x_- \to \int^{x_-} dx'_- \exp[-\frac{1}{2}\eta_-(x'_-)]$. In the non-linear sigma model, this is just the scale of $|\partial_+ N|$ and $|\partial_- N|$ as shown in refs. [16,15].

In this letter, we realized several completely integrable systems by imposing constraints on the WZNW model, mainly the sine-Gordon and affine Toda equations. As discussed in the introduction, our constraint procedure for the WZNW theory is different from the reduction of the principal chiral model of ref. [13]. The differences not only lie in the motivation but also in the details of reduction procedures. In ref. [13], the sine-Gordon and affine Toda equations are obtained by simply fixing some particular solutions of the zero curvature equations of the principal chiral model. There the $\mathbb{Z}_2$ or $\mathbb{Z}_n$ symmetries of the Lax pair are needed a priori for the reduction and the relations between the sine-Gordon fields φ or affine Toda fields φ^i and the principal chiral field $\tilde{g}(x)$ is not clearly understood. In our case such symmetries arise naturally as the consequences of the constraints, and the sine-Gordon field and affine Toda fields are made clear just to be the constrained WZNW field $g(x)$.

With the procedure carried through in this letter, one can use the well known method for the quantization of the WZNW model and its constrained case to quantize the resulting systems, and then study the relationship between these theories and conformal or off-critical conformal quantum field theories. On the other hand, because the WZNW model is known to be related to the three-dimensional Chern–Simons theory, one can transform there integrable systems into three spacetime dimensions and investigate the Yang–Baxter properties of the factorizable monodromy matrices. It is also interesting to study the parallel supersymmetric case.

Acknowledgement

The authors would like to thank Professor J. Balog, Professor L. Fehér, Professor L. O'Raifeartaigh, Professor P. Forgacs and Professor A. Wipf for mailing us ref. [8] and Professor O. Babelon and Professor L. Bonora for ref. [18]. Two of the authors (L.C. and H.-X.Y.) would also like to thank Mr. Yang Zhong-Xia for helpful discussions. This work is supported in part by the National Natural Science Foundation of China through the Nankai Institute of Mathematics.

Volume 266, number 3,4 PHYSICS LETTERS B 29 August 1991

Note added

While writing the manuscript, we read the interesting paper in ref. [19]. Different from that paper, the main purpose of our work is to relate the conformally invariant WZNW model to the off-critical sine-Gordon equation. So we start from an action with a constraint which explicitly breaks the conformal invariance. Thus instead of imposing constraints on certain components of WZNW currents such as $\mathrm{Tr}[\sigma^+ J_+(x)] = m$, which is only a special solution of the equations(s) $\partial_- J_+(x) = 0$, we proved that the only possible solution to these equations are those, such as $J_+(x) \sim$ const., with a constraint carefully imposed on the action. The relation of the two Drinfeld–Sokolov linear systems by $g(x)$ is completely new to us.

References

[1] A.B. Zamolodchikov, Theor. Math. Phys. 63 (1985) 1205;
V.A. Fateev and A.B. Zamolodchikov, Nucl. Phys. B 280 (1987) 644.
[2] V.A. Fateev and S.L. Luk'yanov, Intern. J. Mod. Phys. A 3 (1988) 507.
[3] A.N. Leznov and M.V. Savaliev, Lett. Math. Phys. 3 (1979) 489.
[4] A. Bilal and J.-L. Gervais, Phys. Lett. B 206 (1988) 412; Nucl. Phys. B 314 (1989) 646; B 318 (1989) 579.
[5] A.B. Zamolodchikov, Intern. J. Mod. Phys. A 4 (1989) 4235.
[6] T. Eguchi and S.K. Yang, Phys. Lett. B 224 (1989) 373; B 235 (1990) 282.
[7] H.W. Braden, E. Corrigan, P.E. Dorey and R. Sasaki, Nucl. Phys. B 318 (1990) 689.
[8] J. Balog, L. Fehér, L. O'Raifeartaigh, P. Forgacs and A. Wipf, Phys. Lett. B 227 (1989) 214; Dublin/Zurich preprints DIAS-STP-89-31, DIAS-STP-90-2.
[9] M. Bershadsky and H. Ooguri, Commun. Math. Phys. 126 (1989) 49.
[10] Q.-H. Park, Nucl. Phys. B 333 (1990) 267.
[11] A. Alekseev and S. Shatashvili, Nucl. Phys. B 323 (1989) 719.
[12] A.M. Polyakov, Mod. Phys. Lett. A 2 (1987) 893;
V.G. Knizhnik, A.M. Polyakov and A.B. Zamolodchikov, Mod. Phys. Lett. A 3 (1988) 819.
[13] A. Mikhailov, Physica D 3 (1981) 73; Pisma JETP 32 (1980) 187;
A. Mikhailov, M. Olshanetsky and A. Perelomov, Commun. Math. Phys. 79 (1981) 473.
[14] V. Drinfeld and S. Sokolov, J. Sov. Math. Phys. 30 (1984) 1975.
[15] B.-Y. Hou, B.-Y. Hou and P. Wang, J. Phys. A 18 (1985) 165.
[16] K. Pohlmeyer, Commun. Math. Phys. 46 (1976) 207.
[17] B.-Y. Hou, J. Math. Phys. 25 (1984) 2325.
[18] O. Babelon and L. Bonora, Phys. Lett. B 244 (1990) 220.
[19] H. Aratyn, L.A. Ferreira, J.F. Gomes and A.H. Zimerman, preprint IFT/P-24/90.

J. Phys. A: Math. Gen. 26 (1993) 4951–4965. Printed in the UK

Cyclic representation and function difference representation of the Z_n Sklyanin algebra

Bo-yu Hou†, Kang-jie Shi†‡ and Zhong-xia Yang†

† Institute of Modern Physics, Northwest University, Xian 710069, People's Republic of China

‡ Center of Theoretical Physics, CCAST (World Laboratory), P.O. Box 8730, Beijing 100080, People's Republic of China

Received 16 October 1992

Abstract. We obtain the cyclic representation of Z_n Sklyanin algebra. From this we derive its function difference representation. When $n=2$, it coincides with the known result.

1. Introduction

Recently, there has been considerable attention and intensive study on quantum group theory. From the physics point of view, the quantum group can represent both the exchange relation symmetry of vertex operators in conformal field theory and the symmetry of the six-vertex model and other exactly solvable statistical models, which have triangular functions as their Boltzmann weights.

It is expected that Sklyanin algebra [1] plays a similar role in integrable-massive field theory [2], which has its exchange relations expressed by elliptic functions, and in some exactly solvable statistical models [3–8], which have their Boltzmann weights expressed in elliptic functions. They are reduced to the corresponding triangular models when the modular parameter approaches infinity.

Starting from the eight-vertex model [3], Sklyanin [7] derived the sl(n) Sklyanin algebra for $n=2$. He subsequently constructed the single variable function representation of the algebra, and consequently the classification of the representations. Using the limit of these representations, he was the first to give the highest weight representation and the minimal cyclic representation of the quantum group $\mathrm{sl}_q(2)$.

Starting from $n\otimes n$ [3, 9] model, Cherednik [10] generalized the Sklyanin algebra to the generic sl(n). Wei *et al* [11, 12] worked out the structure constants for the algebra. The authors of this paper gave explicitly [13] the process by which Sklyanin algebra is reduced to $U_q(\hat{\mathrm{sl}}(n))$. Zhou *et al* [14, 15] constructed the tensor product representation of the Sklyanin algebra through fusion. However, the explicit expressions of the cyclic representation have yet to be found.

Hasegawa and Yamada [16] managed to construct the Yang–Baxter operator $L(u)$ for the eight-vertex model by using the cyclic representation [1] of Sklyanin algebra. It was found that the operator $L(u)$ could be factorized, which is similar to the triangular case in [18, 19]. They further derived the broken Z_n model [7, 8]. Bazhanov *et al* [19] point out that the factorizability of $L(u)$ means that we could construct the IRF model from the corresponding vertex model. Recently, making use of this approach, Quano

0305-4470/93/194951+15$07.50

[20] worked out the cyclic representation of the $L(u)$ operator for the Sklyanin algebra. (He also gave a new type of $A_n^{(1)}$ Kashiwara–Miwa model [6, 7].)

In this paper, we give an explicit cyclic representation of the Sklyanin operator S_α, and we find that it has $2n-1$ parameters, which is the same number as for the minimal cyclic representation of quantum group $U_q(\mathrm{sl}(n))$. In the mean time, the number of parameters for the cyclic representation of the $L(u)$ operator is also increased. On the other hand, the difference expression of the Sklyanin algebra operator, which is highly lauded by Smith [22], is the foundation [1] of representation theory for the $n=2$ case. We can make further study of this expression and compare it with its triangular limit, the quantum algebra.

1.1. The intertwiner for $A_{n-1}^{(1)}$ IRF model and Z_n symmetric vertex model and its factorized YBE operator $L(z)$

1.1.1. The Z_n symmetric Belavin R-matrix and Sklyanin algebra (SA). For a given positive integer n, we define $n\otimes n$ matrices g, h, I_α:

$$g_{jk}=\omega^j\delta_{jk} \qquad h_{jk}=\delta_{j+1,k} \qquad \omega=\exp\left(\frac{2\pi i}{n}\right) \tag{1}$$

$$I_\alpha=I_{(\alpha_1,\alpha_2)}=g^{\alpha_2}h^{\alpha_1}.$$

Let $I_\alpha^{(j)}=I\otimes\ldots\otimes I_\alpha\otimes\ldots$, I_α is at the jth space

$$W_\alpha(z)=\theta\begin{bmatrix}\frac{1}{2}+\frac{\alpha_2}{n}\\ \frac{1}{2}+\frac{\alpha_1}{n}\end{bmatrix}\left(z+\frac{w}{n},\tau\right)\Big/\theta\begin{bmatrix}\frac{1}{2}+\frac{\alpha_2}{n}\\ \frac{1}{2}+\frac{\alpha_1}{n}\end{bmatrix}\left(\frac{w}{n},\tau\right)\equiv\frac{\sigma_\alpha\left(z+\frac{w}{n}\right)}{\sigma_\alpha\left(\frac{w}{n}\right)} \tag{2}$$

$$\theta\begin{bmatrix}a\\ b\end{bmatrix}(z,\tau)=\sum_{m\in Z}\exp\{i\pi(m+a)[(m+a)\tau+2(z+b)]\}$$

$$\sigma_\alpha(z)\equiv\theta\begin{bmatrix}\frac{1}{2}+\frac{\alpha_2}{n}\\ \frac{1}{2}+\frac{\alpha_1}{n}\end{bmatrix}(z,\tau)$$

then the Z_n symmetric Belavin R-matrix is written as

$$R_{jk}(z)=\sum_{\alpha\in Z_n^2}W_\alpha(z)I_\alpha^{(j)}(I_\alpha^{-1})^{(k)}. \tag{3}$$

They satisfy the YBE

$$R_{12}(z_1-z_2)R_{13}(z_1)R_{23}(z_2)=R_{23}(z_2)R_{13}(z_1)R_{12}(z_1-z_2).$$

The operator representation of YBE is the $n\otimes n$ matrix $L_j(z)$ satisfying the following equation:

$$R_{12}(z_1-z_2)L_1(z_1)L_2(z_2)=L_2(z_2)L_1(z_1)R_{12}(z_1-z_2) \tag{4}$$

where $L_1(z_1)=L(z_1)\otimes I$, $L_2(z_1)=I\otimes L(z_2)$.

If the $L_j(z)$ could be expressed as

$$L_j(z)=\sum_{\alpha\in Z_n^2} W_\alpha(z+c)I_\alpha^{(j)}S_\alpha \tag{5}$$

then the operator S_α satisfies Z_n SA

$$\sum_\gamma C_{\alpha\beta\gamma}(w,\tau)S_{\alpha-\gamma}S_\gamma=0 \tag{6}$$

where $\alpha, \beta, \gamma\in Z_n^2$, structure constant $C_{\alpha\beta\gamma}$, operator S_α is independent of the spectrum parameter z. On the other hand, we can find a solution $L_j(z)$ of (4), if we have a representation of S_α satisfying (6),

1.1.2. Cyclic representation state and the Boltzmann weight of the IRF model. According to Jimbo *et al* [6], we have

$$W=\bigotimes_{j=0}^{n-1} V_j \qquad V_j\cong C^N.$$

We choose a set of canonical bases such that

$$Zu_j=u_{j-1} \qquad Xu_j=q^ju_j \qquad q=\exp\left(\frac{2\pi i}{N}\right)$$

then we have the cyclic representation space W^0

$$W^0=\{w\in W \mid Z_0\ldots Z_{n-1}w=w\}.$$

This could be constructed from the base vector

$$w_m=\sum_{k=0}^{N-1} u_{m_0+k}\otimes\ldots\otimes u_{m_{n-1}+k} \qquad m=(m_0\ldots m_{n-1})\in Z_N^n \tag{7a}$$

obviously

$$w_{(m_0,\ldots,m_{n-1})}=w_{(m_0+k,\ldots,m_{n-1}+k)}$$

hence w_m can be uniquely described by the element in $\{Q\}$

$$Q=Z_N^n \bmod Z_N(1,\ldots,1). \tag{7b}$$

We can choose from the equivalent $m\in Z_N^n$ one with $m_0=0$

$$\{0, m_1', m_2'\ldots m_{n-1}'\}$$

where $m_j'=m_j-m_0 \bmod N$.

Define

$$[m]=(0,\ldots,m_{n-1}') \tag{8}$$

then a base vector is determined by the $n-1$ numbers $m_j'\in Z_N$. We may obtain $[a]$ from a given vector $a\in\{w_m\}$; the opposite is also true, so that W^0 is N^{n-1} dimensional.

Given any two states $a, b\in\{w_m\}$, they are called admissible if they satisfy

$$[a]-[b]=[e_j] \bmod Z_N^n \qquad \text{with } 0\leqslant j\leqslant n-1 \tag{9}$$

where $e_j=(0,\ldots,1,0\ldots)$, 1 is at the jth place.

We write $b=a-e_j$, for a given state a there are n states b satisfying this relation. Define

$$W_z\begin{bmatrix} m & n-e_j \\ m-e_j & m-2e_j \end{bmatrix}=\frac{\sigma_0\left(z+\frac{N'}{N}\right)}{\sigma_0\left(\frac{N'}{N}\right)}$$

$$W_z\begin{bmatrix} m & m-e_j \\ m-e_j & m-e_j-e_k \end{bmatrix}=\frac{\sigma_0\left(z+m_{jk}\frac{N'}{N}\right)}{\sigma_0\left(m_{jk}\frac{N'}{N}\right)} \qquad (10)$$

$$W_z\begin{bmatrix} m & m-e_j \\ m-e_k & m-e_j-e_k \end{bmatrix}=\frac{\sigma_0(z)\sigma_0\left(m_{jk}\frac{N'}{N}-\frac{N'}{N}\right)}{\sigma_0\left(\frac{N'}{N}\right)\sigma_0\left(m_{jk}\frac{N'}{N}\right)}$$

where m, e_j, $e_k\in Z_N^n$, N, N' coprime to each other

$$m_{jk}=\bar{m}_j-\bar{m}_k \qquad \bar{m}_j=m_j-\frac{1}{n}\sum_{j'=0}^{n-1} m_{j'}+w_j \qquad (11)$$

$$w_j\neq w_k \bmod N\wedge \qquad \text{if } j\neq k.$$

We may also define Boltzmann weight

$$W^{\text{cyc}}\begin{bmatrix} a & b \\ d & c \end{bmatrix}$$

for a state configuration

$$\begin{bmatrix} a & b \\ c & d \end{bmatrix}$$

round a face by (a, b, c, $d\in\{w_m\}$)

$$W_z^{\text{cyc}}\begin{bmatrix} a & b \\ c & d \end{bmatrix}=\begin{cases} W_z\begin{bmatrix} [a] & [b] \\ [c] & [d] \end{bmatrix} & \text{if } (a,b), (b,c), (a,d) \text{ and } (d,c) \text{ are admissible} \\ 0 & \text{otherwise.} \end{cases} \qquad (12)$$

It can be shown that the non-vanishing W_z^{cyc} is one of these in (10), furthermore it is single-valued with respect to Z_N^n (the right-hand side of (10) is invariant under $m_j\to m_j+N$).

1.1.3. The intertwiner of face-vertex models and factorized L(z). Jimbo *et al* [21, 20] introduced the following intertwiner

$$\varphi_{ml}(z)={}^t(\varphi_{ml}^{(0)}(z),\ \ldots,\ \varphi_{ml}^{(n-1)}(z))$$

$$\varphi_{ml}^{(k)}(z)=\begin{cases}\theta^{(k)}(z+nw\bar{m}_j) & \text{if } [m]-[l]=e_j\\ 0 & \text{otherwise}\end{cases} \tag{13a}$$

$$\theta^{(j)}(u)\equiv\theta\begin{bmatrix}\frac{1}{2}-\frac{j}{n}\\ \frac{1}{2}\end{bmatrix}(u,\ n\tau).$$

This is a single-valued (with respect to m, l) n-dimensional vector. For a given state m, there are n states such that φ is non-vanishing. If different sets of numbers $(m_0,\ldots)\in Z_N^n$, $(m_0'\ldots)\in Z_N^n$ represent the same state, then (13) gives an identical vector $\varphi_m(z)$.

Jimbo *et al* [21] showed that, if $w=N'/N$, the following is true:

$$R(z^1-z_2)\varphi_{ab}(z_1)\otimes\varphi_{bc}(z_2)=\sum_d W_{z_1-z_2}^{\rm cyc}\begin{bmatrix}a & b\\ d & c\end{bmatrix}\varphi_{dc}(z_1)\otimes\varphi_{ab}(z_2). \tag{13b}$$

According to Bazhanov *et al* [17], if we could find a row vector

$$\bar{\varphi}_{ab}(z_1)(z)=(\bar{\varphi}_{ab}^{(0)}(z_1),\ \ldots,\ \bar{\varphi}_{ab}^{(n-1)}(z_1)$$

such that

$$\sum_k \bar{\varphi}_{ab}^{(k)}\varphi_{ac}^{(k)}=\begin{cases}\delta_{bc} & (a,b),\ (a,c)\quad \text{admissible}\\ 0 & \text{otherwise.}\end{cases} \tag{14a}$$

Consequently we have n b, c^s, furthermore, we have

$$\sum_b \varphi_{ab}^{(j)}\bar{\varphi}_{ab}^{(k)}=\delta_{jk} \tag{14b}$$

$$\bar{\varphi}_{dc}(z_1)\otimes\varphi_{ad}(z_2)R(z_1-z_2)=\sum_b W_{z_1-z_2}^{\rm cyc}\begin{bmatrix}a & b\\ d & c\end{bmatrix}\bar{\varphi}_{ab}(z_1)\otimes\bar{\varphi}_{bc}(z_2). \tag{15}$$

Proof. From (13)

$$R(z_1-z_2)\varphi_{ab}(z_1)\otimes\varphi_{bc}(z_2)=\sum_d W_{z_1-z_2}\begin{bmatrix}a & b\\ d & c\end{bmatrix}\varphi_{dc}(z_1)\otimes\varphi_{ab}(z_2).$$

Multiplying both sides from the right by $\bar{\varphi}_{ab}(z_1)\otimes\bar{\varphi}_{bc}(z_2)$ and summing over b, c we have

$$\begin{aligned}&\sum_b\sum_c R(z_1-z_2)\varphi_{ab}(z_1)\bar{\varphi}_{ab}(z_1)\otimes\varphi_{bc}(z_2)\bar{\varphi}_{bc}(z_2)\\ &=\sum_b R(z_1-z_2)\varphi_{ab}(z_1)\bar{\varphi}_{ab}(z_1)\otimes I\\ &=R(z_1-z_2)I\otimes I\\ &=\sum_{bc}\sum_d W_{z_1-z_2}\begin{bmatrix}a & b\\ d & c\end{bmatrix}\varphi_{dc}(z_1)\bar{\varphi}_{ab}(z_1)\otimes\varphi_{ad}(z_2)\bar{\varphi}_{bc}(z_2).\end{aligned}$$

Then we multiply both sides from the left by $\bar{\varphi}_{d'c'}(z_1)\otimes\bar{\varphi}_{ad'}(z_2)$ to give

$$\bar{\varphi}_{d'c'}(z_1)\otimes\bar{\varphi}_{ad'}(z_2)R(z_1-z_2)$$

$$=\sum_{bc}\sum_{d}W_{z_1-z_2}\begin{bmatrix}a & b\\ d & c\end{bmatrix}\bar{\varphi}_{d'c'}(z_1)\varphi_{dc}(z_1)\bar{\varphi}_{ab}(z_1)\otimes\delta_{dd'}\bar{\varphi}_{bc}(z_2)$$

$$=\sum_{bc}W_{z_1-z_2}\begin{bmatrix}a & b\\ d' & c\end{bmatrix}\delta_{c'c}\bar{\varphi}_{ab}(z_1)\otimes\bar{\varphi}_{bc}(z_2)$$

$$=\sum_{b}W_{z_1-z_2}\begin{bmatrix}a & b\\ d' & c'\end{bmatrix}\bar{\varphi}_{ab}(z_1)\otimes\bar{\varphi}_{bc'}(z_2).$$

Thus we have

$$\bar{\varphi}_{dc}(z_1)\otimes\bar{\varphi}_{ad}(z_2)R(z_1-z_2)=\sum_{b}W_{z_1-z_2}\begin{bmatrix}a & b\\ d & c\end{bmatrix}\bar{\varphi}_{ab}(z_1)\otimes\bar{\varphi}_{bc}(z_2).$$

Let

$$[L(z)]_{ab}=\varphi_{ab}(z)\bar{\varphi}_{ab}(z) \tag{16}$$

then $L(z)$ satisfies (5).

Proof. As

$$[R(z_1-z_2)L_1(z)L_1(z)]_{ac}=\sum_{b}R(z_1-z_2)\varphi_{ab}(z_1)\bar{\varphi}_{ab}(z_1)\otimes\varphi_{bc}(z_2)\bar{\varphi}_{bc}(z_2) \tag{16a}$$

from (14), the right-hand side of (16*a*) is

$$\sum_{b}\sum_{d}W_{z_1-z_2}\begin{bmatrix}a & b\\ d & c\end{bmatrix}\varphi_{dc}(z_1)\bar{\varphi}_{ab}(z_1)\otimes\varphi_{ad}(z_2)\bar{\varphi}_{bc}(z_2)$$

$$=\sum_{d}\varphi_{dc}(z_1)\otimes\varphi_{ad}(z_2)\sum_{b}W_{z_1-z_2}\begin{bmatrix}a & b\\ d & c\end{bmatrix}\bar{\varphi}_{ab}(z_1)\bar{\varphi}_{bc}(z_2).$$

From (15), the left-hand side of (16*a*) is

$$\sum_{d}\varphi_{dc}(z_1)\otimes\varphi_{ad}(z_2)\bar{\varphi}_{dc}(z_1)\otimes\bar{\varphi}_{ad}(z_2)R(z_1-z_2)$$

$$=\sum_{d}\varphi_{dc}(z_1)\bar{\varphi}_{dc}(z_1)\otimes\varphi_{ad}(z_2)\bar{\varphi}_{ad}(z_2)R(z_1-z_2)$$

$$=[L_2(z_2)L_1(z_1)R(z_1-z_2)]_{ac}.$$ □

This is the cyclic $L(z)$ obtained by Quano [20]. It generalizes the results of Hasegawa *et al* [16].

Remark. We can generalize the $L(z)$ in (16) as follows:
Put $\varphi_{ab}(z+\zeta)$ instead of $\varphi_{ab}(z)$ in (16), then the resulting

$$[L(z)]_{ml}=\varphi_{ml}(z+\zeta)\bar{\varphi}_{ml}(z) \tag{16b}$$

also satisfies YBE (5).

The $\bar{\varphi}^{(i)}_{a,a-e_l}(z)$ which satisfies (14*a*) is the matrix element of the inverse of $N\otimes n$ matrix A, where $A_{ij}=\varphi^{(i)}_{a,a-e_l}$, i.e.

$$A=\begin{bmatrix}\theta^{(0)}(nz_0) & \cdots & \theta^{(0)}(nz_{n-1})\\ \vdots & \cdots & \vdots \\ \theta^{(n-1)}(nz_0) & \cdots & \theta^{(n-1)}(nz_{n-1})\end{bmatrix} \tag{17}$$

where $nz_j=z+nw\bar{m}_j$. Thus

$$\bar{\varphi}^{(i)}_{a,a-e_l}(z)=B_{ij}(z)/\mathrm{Det}\,A \tag{18}$$

where B_{ij} is the cofactor matrix of A_{ij}. It can be shown [20] that

$$\mathrm{Det}\,A=C\sigma_0\left(\sum_{j=0}^{n-1}z_j-p_n\right)\prod_{l<k}\sigma_0(z_l-z_k) \tag{19}$$

where C is a z_i-independent constant and $p_n=(n-1)/2$.

2. Cyclic and function difference representations of SA

2.1. Cyclic representation of SA

We now look for the S_α in (6) which corresponds to $L(z)$ in (16). As I_α is invertible, and

$$\mathrm{tr}\,I_\alpha(I_\beta)^{-1}=n\delta_{\alpha_1\beta_1}\delta_{\alpha_2\beta_2} \tag{20}$$

it is easy to show that $\{I_\alpha\}$ forms a complete set of bases. Let

$$L(z)=\sum_\alpha I_\alpha U_\alpha(z) \qquad U_\alpha:\ \text{operator}$$

then

$$U_\alpha(z)=\frac{1}{n}\,\mathrm{tr}\,L(z)(I_\alpha)^{-1}.$$

If $U_\alpha(z)=W_\alpha(z+\mu,w,\tau)S_\alpha$, where W_α is given by (2), then S_α is a representation of SA and satisfies (7). From (16)

$$\begin{aligned} L_{ml}(z)&=\varphi_{ml}\bar{\varphi}_{ml}(z+\zeta) \\ n[U^\alpha]_{ml}&=\mathrm{tr}[L_{ml}(z)(I_\alpha)^{-1}] \\ &=\sum_{ik}[\varphi^{(i)}_{ml}(z)][\bar{\varphi}^{(k)}_{ml}](I_\alpha)^{-1}_{kl} \\ &=\sum_{ik}\bar{\varphi}^{(k)}_{ml}(z)(I_\alpha)^{-1}_{ki}\varphi^{(i)}_{ml}(z+\zeta) \\ &=\bar{\varphi}_{ml}(z)(I_\alpha)^{-1}\varphi_{ml}(z+\zeta) \\ &=\begin{cases}\bar{\varphi}_{m,m-e_j}(z)\psi_{m,m-e_j}(z+\zeta) & \text{if } [m]-[l]=e_j \\ 0 & \text{if } (m,l) \text{ not admissible}\end{cases}\end{aligned} \tag{21}$$

where $\psi_{m,m-e_j}(z)=(I_\alpha)^{-1}\varphi_{m,m-e_j}(z)$.

Consequently

$$n[U_a]_{ml}=\begin{cases}\sum_i \dfrac{B_{ij}(z)}{\text{Det } A(z)}\times \psi^{(l)}_{m,m-e_j}(z) & [m]-[l]=e_j\\ 0 \qquad (m,\, l) \text{ not admissible}\end{cases}$$

$$n[U_a]_{m,m-e_j}=\frac{\text{Det}\begin{bmatrix}\theta^0(nz_0) & \dots & \psi^0(nz_j) & \dots & \theta^0(nz_{n-1})\\ \vdots & \vdots & \vdots & \vdots & \vdots\\ \theta^{n-1}(nz_0) & \dots & \psi^0(nz_j) & \dots & \theta^{n-1}(nz_{n-1})\end{bmatrix}}{\text{Det}\begin{bmatrix}\theta^0(nz_0) & \dots & \theta^0(nz_j) & \dots & \theta^0(nz_{n-1})\\ \vdots & \vdots & \vdots & \vdots & \vdots\\ \theta^{n-1}(nz_0) & \dots & \theta^0(nz_j) & \dots & \theta^{n-1}(nz_{n-1})\end{bmatrix}} \tag{22}$$

From (13), it can be shown that

$$\begin{aligned}
&g^{-1}\varphi_{m,m-e_j}(z)=(-1)\varphi_{m,m-e_j}(z+1)\\
&h^{-1}\varphi_{m,m-e_j}(z)=\exp\left[\frac{2\pi\mathrm{i}}{n}\left(z+n\bar{m}_jw+\frac{1}{2}+\frac{\tau}{2}\right)\right]\varphi_{m,m-e_j}(z+\tau)\\
&\psi(z)=(I_a)^{-1}\varphi_{m,m-e_j}(z)=h^{-\alpha_2}g^{-\alpha_1}\varphi_{m,m-e_j}(z+\tau)\\
&\qquad=\exp\left[\frac{2\pi\mathrm{i}}{n}\left(\alpha_2 z+n\frac{\alpha_1}{2}+\alpha_1\alpha_2+\frac{\alpha_2^2\tau}{2}+n\alpha_2\bar{m}_jw+\frac{1}{2}\right)\right]\\
&\qquad\quad\times\varphi_{m,m-e_j}(z+\alpha_1+\alpha_2\tau)
\end{aligned} \tag{23}$$

$$\begin{aligned}
n[U^{\alpha}]_{ml}&=\bar{\varphi}_{mm-e_j}(z)(I_\alpha)^{-1}\varphi_{mm-e^j}(z)\\
&=\text{Det}\begin{bmatrix}\theta^0(nz_0) & \dots & \theta^0(nz'_j) & \dots & \theta^0(nz_{n-1})\\ \vdots & \vdots & \vdots & \vdots & \vdots\\ \theta^{n-1}(nz_0) & \dots & \theta^{n-1}(nz'_j) & \dots & \theta^{n-1}(nz_{n-1})\end{bmatrix}\\
&\quad\times\text{Det}\begin{bmatrix}\theta^0(nz_0) & \dots & \theta^0(nz_j) & \dots & \theta^0(nz_{n-1})\\ \vdots & \vdots & \vdots & \vdots & \vdots\\ \theta^{n-1}(nz_0) & \dots & \theta^{n-1}(nz_j) & \dots & \theta^{n-1}(nz_{n-1})\end{bmatrix}^{-1}\times\text{Factor}
\end{aligned} \tag{24}$$

where

$$z_i=\frac{1}{n}(z+nm_iw)\qquad z'_j=\frac{1}{n}(\alpha_1+\alpha_2\tau+\zeta)+z_j\qquad i,j=0,\dots,n-1 \tag{25}$$

$$\text{Factor}=\exp\left\{\frac{2\pi\mathrm{i}}{n}\left[\alpha_2(z+\zeta)+n\frac{\alpha_1}{2}+\alpha_1\alpha_2+\frac{\alpha_2^2\tau}{2}+n\alpha_2\bar{m}_jw+\frac{\alpha_2}{2}\right]\right\}.$$

Let

$$\mathrm{Det}\begin{bmatrix} \theta^0(nz_0) & \cdots & \cdots \\ \vdots & \vdots & \vdots \\ \cdots & \cdots & \theta^{n-1}(nz_{n-1}) \end{bmatrix} = f(z_0). \tag{26}$$

As

$$\begin{aligned} &\theta\begin{bmatrix} a \\ b \end{bmatrix}(z+\tau,\ \tau) = \exp(-\pi\mathrm{i}\tau - 2\pi\mathrm{i}(z+b))\theta\begin{bmatrix} a \\ b \end{bmatrix}(z,\ \tau) \\ &\theta\begin{bmatrix} a \\ b \end{bmatrix}(z+1,\ \tau) = \exp(2\pi\mathrm{i}a)\theta\begin{bmatrix} a \\ b \end{bmatrix}(z,\ \tau) \end{aligned} \tag{27}$$

$$\begin{aligned} &f(z_0+\tau) = \exp(-\pi\mathrm{i}\tau - 2\pi\mathrm{i}(nz_0+\tfrac{1}{2}))f(z_0) \\ &f(z_0+1) = \exp(n\pi\mathrm{i}a)f(z_0). \end{aligned} \tag{28}$$

Thus $f(z_0)$ has n zeros in $\wedge_\tau$, and the sum of the zeros is

$$\sum_{\wedge_\tau} \text{zeros} = \frac{n-1}{2}.$$

As a function of z_0, $f(z_0)$ has $n-1$ obvious zeros, $z_0 = z_1, z_2, \ldots, z_{n-1}$, so the last zero is at $z_0 = -z_1 - z_2 - \ldots - z_{n-1}$. It is easy to check that the function

$$g(z_0) = \sigma_0\left(z_0 + z_1 + \ldots + z_{n-1} - \frac{n-1}{2}\right) \prod_{k \neq 0} \sigma_0(z_0 - z_k)$$

has the same zeros as $f(z_0)$. As the function $f(z_0)/g(z_0)$ is a double periodic pure function, it can only be a z_0-independent constant, i.e.

$$\frac{f(z_0)}{g(z_0)} = C.$$

The same analysis is also true for z_i, $i = 1, \ldots, n-1$, so we finally have

$$f(z_0) = \sigma_0\left(\sum_{i=0}^{n-1} z_i - \frac{n-1}{2}\right) \prod_{i<k} \sigma_0(z_i - z_k). \tag{29}$$

Applying this to (24) we have

$$n[U_a]_{m,m-e_j} = \frac{\sigma_0\left(\sum_{i\neq j} z_i + z_j' - \frac{n-1}{2}\right) \prod_{k\neq j} \sigma_0(z_j' - z_k)}{\sigma_0\left(\sum_{i=0}^{n-1} z_i - \frac{n-1}{2}\right) \prod_{k\neq j} \sigma_0(z_j - z_k)} \times \text{Factor}$$

where z_i, z'_j are given by (25).

$$n[U_\alpha]_{m,m-e_j}=\frac{\sigma_0\left(z+w\delta+\frac{\alpha_1+\alpha_2\tau+\zeta}{n}-\frac{n-1}{2}\right)}{\sigma_0\left(z+\delta-\frac{n-1}{2}\right)}$$

$$\times\prod_{k\neq j}\frac{\sigma_0\left[\frac{\alpha_1+\alpha_2\tau+\zeta}{n}+(\bar{m}_j-\bar{m}_k)w\right]}{\sigma_0[(\bar{m}_j-\bar{m}_k)w]}\times\text{Factor}.$$

$$\delta=\sum_j\bar{m}_j=\sum_j w_j.$$

As

$$\sigma_0\left(z+\frac{\alpha_1+\alpha_2\tau}{n}\right)=\exp\left[\frac{-\pi i}{n^2}(2n\alpha_2 z+2\alpha_1\alpha_2+\alpha_2^2\tau+n\alpha_2)\right]\times\sigma_\alpha(z)$$

so

$$[U_\alpha]_{m,m-e_j}=\frac{\sigma_\alpha\left(z+\frac{\zeta}{n}+w\delta-\frac{n-1}{2}\right)}{n\sigma_0\left(z+w\delta-\frac{n-1}{2}\right)}\prod_{k\neq j}\frac{\sigma_\alpha\left[\frac{\zeta}{n}+(\bar{m}_j-\bar{m}_k)w\right]}{\sigma_0[(\bar{m}_j-\bar{m}_k)w]}$$

$$\times\exp\left(-\frac{2\pi i}{n}\left[\alpha_2\left(z+\frac{\alpha_2\tau+2\alpha_1+n}{2}+\xi+w\delta-\frac{n-1}{2}+n\bar{m}_jw-\sum\bar{m}_kw\right)\right]\right)$$

$$\times\exp\left(\frac{2\pi i}{n}\left[\alpha_2(z+\xi)+\alpha_2^2\tau+\alpha_1\alpha_2+\frac{n\alpha_1}{2}+n\alpha_2\bar{m}_jw+\frac{\alpha_2}{2}\right]\right)$$

$$=e^{\pi i\alpha_1}\frac{\sigma_\alpha\left(z+\frac{\zeta}{n}+w\delta-\frac{n-1}{2}\right)}{n\sigma_0\left(z+w\delta-\frac{n-1}{2}\right)}\prod_{k\neq 1}\frac{\sigma_\alpha\left[\frac{\zeta}{n}+(m_j+w_j-m_kw_k)w\right]}{\sigma_0[(m_j+w_j-m_k-w_k)w]}. \tag{30}$$

Comparing this with (2) and (5) we can see that, aside from some insignificant terms such as $[n\sigma_0(z+w\delta-(n-1)/2)]^{-1}$, $L(z)$ in (16*b*) has the desired form of (6); we can consequently write the matrix element of the cyclic representation of the SA

$$(S_\alpha)_{m,m-e_j}-(-1)^{\alpha_1}\sigma_\alpha\left(\frac{w}{n}\right)\prod_{k\neq j}\frac{\sigma_\alpha\left[\frac{\zeta}{n}+(m_j+w_j-m_k-w_k)w\right]}{\sigma_0[(m_j+w_j-m_k-w_k)w]} \tag{31}$$

$$L(z)=\frac{\sum_\alpha W_\alpha\left(z-\frac{w}{n}+\frac{\zeta}{n}+w\delta-\frac{n-1}{2}\right)I_\alpha S_\alpha}{n\sigma_0\left(z+w\delta-\frac{n-1}{2}\right)}. \tag{32}$$

Equation (32) is invariant under $\alpha_i \to \alpha_i + n$ because

$$\sigma_{\alpha+(nq,np)}(z,\tau) = (-1)^q \exp(2\pi i q\alpha_2/n)\sigma_\alpha(z,\tau) \qquad p, q \in Z$$

$$S_\alpha \to S_\alpha \qquad \text{as } \alpha_2 \to \alpha_2 + n$$

$$S_\alpha \to (-1)^n \times \exp\left(2\pi i\left(\frac{1}{2}+\frac{\alpha_2}{n}\right)\times n\right) = S_\alpha \qquad \text{as } \alpha_1 \to \alpha_1 \to n.$$

Remark. S_α, $L(z)$ are operators acting on $W^{(0)}$ (see (7)), and their matrix elements are well defined on the standard bases w_m given in (7*a*) (i.e. they are invariant under $m_i \to m_i + N$, $m \succ (1, 1, \ldots, 1) + m$).

2.2. *Cocycle coefficient of the cyclic representation*

We can expand the SA in the following way:

As Sklyanin algebra (6) is defined by second-order homogeneous equations with respect to S_α, and the sum of the lower indices of two Ss are the same (mod $n \times n$). There are only n states $|m - e_i\rangle$, $i = 0, 1, \ldots, n-1$, from which we can get to $|m\rangle$ by acting S_α on them, and there are n^2 states $|m - e_i - e[k]\rangle$, $i, k = 0, 1, \ldots, n-1$, from which we can get to $|m\rangle$ by acting $S_\alpha S_\beta$ on them. The possible routes from $|m\rangle$ to $|m - e_i - e_k\rangle$ are

(a) $i = k$

$$|m - e_i - e_i\rangle \to |m - e_i\rangle \to |m\rangle$$

(b) $i \neq k$

(i) $|m_i - e_k\rangle \to |m - e_i\rangle \to |m\rangle$

(ii) $|m - e_i - e_k\rangle \to |m - e_k\rangle \to |m\rangle$.

We propose to expand S_α as

$$(S_\alpha)_{m,m-e_j} \to e^{A(m,j,\alpha)}(S_\alpha)_{m,m-e_j}. \tag{33}$$

If

$$A(m, i, \alpha - \gamma) + A(m - e_i, k, \gamma) = B(m, i, k, \alpha) \tag{34}$$

is independent of γ, $m - e_i$, then (6) is still true, we require $e^{A(m,j,\alpha)}$ to be well defined with respect to m, $\alpha \in Z_n^2$, the simplest non-trivial choice is

$$A(m, j, \alpha) = \delta_j + \frac{2\pi i}{n}(l_1\alpha_1 + l_2\alpha_2)$$

where $\delta_j \in C$, $l_1, l_2 \in Z$.

The second term in our choice of $A(m, j, \alpha)$ is actually the part isomorphic to Sklyanin algebra. Finally we have

$$(S_\alpha)_{m,m-e_j} = \frac{\exp\left(\delta_j + 2\pi i\left(\frac{\alpha_1}{2} + l_1\frac{\alpha_1}{n} + l_2\frac{\alpha_2}{n}\right)\right)}{\sigma_\alpha\left(\frac{w}{n}\right)} \prod_{k \neq j} \frac{\sigma_\alpha\left[\frac{\zeta}{n} + (m_j + w_j - m_k - w_k)w\right]}{\sigma_0[(m_j + w_j - m_k - w_k)w]} \tag{35}$$

in which we have the parameters δ_j, ζ, w_j-w_k. As, when $\delta_j\to\delta_j+v$, S''_α is essentially unchanged, we choose $\sum\delta_j=0$, so we have $2n-1$ independent parameters in the representation.

It is remarkable that we cannot get rid of the coefficient

$$\exp\left(\delta_j+2\pi i\left(\frac{\alpha_1}{2}+l_1\frac{\alpha_1}{n}+l_2\frac{\alpha_2}{n}\right)\right)$$

through a similar transformation, because as we take the sequence $|m-Ne_i\rangle\to\ldots\to|m\rangle$, the product of these coefficients does not give one. As the initial and the final states in the sequence are identical, similar transformation should give a trivial coefficient.

$$\exp\left(\delta_j+2\pi i\left(\frac{\alpha_1}{2}+l_1\frac{\alpha_1}{n}+l_2\frac{\alpha_2}{n}\right)\right)$$

is actually a cocycle coefficient on n-torus.

2.3. *Function difference representation of* SA

So far in our treatment we have restricted ourselves to the case where $w=N'/N$ is rational, and we find that the representation is well defined on $\{Q\}$. We can ease our restrictions such that we associate each bases vector with (m), $m=\{m_0,\ldots,m_{n-1}\}\in Z^n$. The admissible condition is still $(m)-(1)=(e_j)$, i.e. $m_i-l_i=\delta_{ij}$. And we take (35) as $(\bar{S}_\alpha)_{(m)(m-e_j)}$. Obviously we have

$$\sum_{\gamma m'}C_{\alpha\beta\gamma}(\bar{S}_{\alpha-\gamma})_{(m)(m')}(\bar{S}_\gamma)_{(m')(m'')}=0. \tag{36}$$

According to Jimbo *et al* [8], equation (36) holds irrespective of whether w is a rational number or not.

Let $(m_i-w_i)w\equiv u_i$, then (35) becomes

$$(S_\alpha)_{m,m-e_j}(w)\equiv(S_\alpha)_j(u,w)$$

$$=\exp\left(\delta_j+2\pi i\left(\frac{\alpha_1}{2}+l_1\frac{\alpha_1}{n}+l_2\frac{\alpha_2}{n}\right)\right)\times\sigma_\alpha\left(\frac{w}{n}\right)\times\prod_{k\neq j}\frac{\sigma_\alpha\left[\frac{\zeta}{n}+(u_j-u_k)w\right]}{\sigma_0[(u_j-u_k)w]}. \tag{37}$$

Then (6) becomes

(a) $m''=m-2e_i$ $\qquad\sum_\gamma C_{\alpha\beta\gamma}(S_{\alpha-\gamma})_i(u)(S_\gamma)_i(u_0,\ldots,u_i-w,\ldots)=0$

(b) $m''=m-e_j-e_k$ $\qquad j\neq k;$

$$\sum_\gamma C_{\alpha\beta\gamma}\{(S_{\alpha-\gamma})_j(u)(S_\gamma)_k(u_0,\ldots,u_j-w,\ldots)$$
$$+(S_{\alpha-\gamma})_k(u)(S_\gamma)_j(u_0,\ldots,u_k-w,\ldots)\}=0. \tag{38}$$

As w_j is generic, so is $u\equiv\{u_j\}$, and it is easy to see that (38) is true for generic u.

Next we consider the function difference representation of SA. Define the operator $\hat{S}_\alpha$ acting on function $f(u)=f(u_0,\ldots,u_{n-1})$ as

$$\hat{S}_\alpha f(u)=\sum_j(S_\alpha)_j(u,w)f(u_0,\ldots,u_{n-1})$$

then

$$\sum_{\gamma} C_{\alpha\beta\gamma}\hat{S}_{\alpha-\gamma}\hat{S}_{\gamma}f(u)$$

$$=\sum_{\gamma} C_{\alpha\beta\gamma}\hat{S}_{\alpha-\gamma}\sum_{k}(S_{\gamma})_k f(u_0,\ldots,u_k-w,\ldots)$$

$$=\sum_{\gamma} C_{\alpha\beta\gamma}\sum_{j}(S_{\alpha-\gamma})_j(u)$$

$$\times\Big[\sum_{k\neq j}(S_{\gamma})_k(u_0,\ldots,u_j-w,\ldots)$$

$$\times f(u_0,\ldots,u_j-w,\ldots,u_k-w,\ldots)$$

$$+(S_{\gamma})_j(u_0,\ldots,u_j-w,\ldots)f(u_0,\ldots,u_j-2w,\ldots)\Big]$$

$$=\sum_{\gamma} C_{\alpha\beta\gamma}\sum_{j<k}[(S_{\alpha-\gamma})_j(u)(S_{\gamma})_k(u_0,\ldots,u_j-w,\ldots)$$

$$+(S_{\alpha-\gamma})_k(u)(S_{\gamma})_j(u_0,\ldots,u_k-w,\ldots)]$$

$$\times f(u_0,\ldots,u_j-w,\ldots,u_k-w,\ldots)$$

$$+\sum_{\gamma} C_{\alpha\beta\gamma}\sum_{j}(S_{\alpha-\gamma})_j(u)(S_{\gamma})_j(u_0,\ldots,u_j-w,\ldots)$$

$$\times f(u_0,\ldots,u_j-2w,\ldots).$$

So (39) guarantees the identity

$$\sum_{\gamma} C_{\alpha\beta\gamma}\hat{S}_{\alpha-\gamma}\hat{S}_{\gamma}f(u)=0.$$

This is the function difference representation of SA.

We now compare our results with those given by Sklyanin at $n=2$. At $n=2$, we consider the function as $f(u_0,u_1)=f(2u)$, where $u_0-u_1=2u$, $u_0+u_1=2v$. Let $\delta_0=\delta_1=\pi i$, $l_0=0$, $l_2=1$ in (37). We then have

$$(S_\alpha)_0(u)=\sigma_\alpha\left(\frac{w}{2}\right)\frac{\sigma_\alpha\left(\frac{\zeta}{n}+2u\right)}{\sigma_0(2u)}-(-1)^{\alpha_1+\alpha_2-1}$$

$$(S_\alpha)_1(u)=\sigma_\alpha\left(\frac{w}{2}\right)\frac{\sigma_\alpha\left(\frac{\zeta}{n}-2u\right)}{\sigma_0(-2u)}(-1)^{\alpha_1+\alpha_2-1}.$$

So

$$S_\alpha f(2u)=\left[\frac{\sigma_\alpha\left(\frac{w}{2}\right)\sigma_\alpha\left(\frac{\zeta}{n}+2u\right)}{\sigma_0(2u)}f(2u-w)-\frac{\sigma_\alpha\left(\frac{w}{2}\right)\sigma_\alpha\left(\frac{\zeta}{n}-2u\right)}{\sigma_0(2u)}f(2u+w)\right]\times(-1)^{\alpha_1+\alpha_2-1}$$

where $f(2u)=F(u)$, $\zeta/n=lw$.

$$S_\alpha F(u)=\left[\frac{\sigma_\alpha\left(\frac{w}{2}\right)\sigma_\alpha(lw+2u)}{\sigma_0(2u)}F\left(u-\frac{w}{2}\right)-\frac{\sigma_\alpha\left(\frac{w}{2}\right)\sigma_\alpha(lw-2u)}{\sigma_0(2u)}F\left(u+\frac{w}{2}\right)\right]\times(-1)^{\alpha_1+\alpha_2-1}$$

$$\equiv S_\alpha\left(u+\frac{lw}{2}\right)F\left(u-\frac{w}{2}\right)-S_\alpha\left(-u+\frac{lw}{2}\right)F\left(u+\frac{w}{2}\right)$$

where $f(2u)=F(u)$, $\zeta/n=lw$.

Let $w/2=-\eta$, remember that $\sigma_{00}(-\eta)=0_{00}(-\eta)=-\sigma_{00}(\eta)$, $\sigma_i(-\eta)=\sigma_i(\eta)$, $i=\alpha_2, \alpha_1=1,0;\ 1,1;\ 01$. We have

$$S_\alpha F(u)=\frac{S_\alpha(u-l\eta)F(u+\eta)-S_\alpha(-u-l\eta)F(u-n)}{0_{11}(2u)}$$

where S_α and the corresponding I_α, S_α are

$$S_\alpha=\begin{cases}S_{00}=\theta_{11}(\eta)\theta_{11}(2u)\\ S_{01}=\theta_{10}(\eta)\theta_{10}(2u)\\ S_{10}=\theta_{01}(\eta)\theta_{01}(2u)\\ S_{11}=-\theta_{00}(\eta)\theta_{00}(2u)\end{cases}\qquad I_\alpha=\begin{cases}I=I_{00}\\ \sigma_z=I_{01}\\ \sigma_x=I_{10}\\ \mathrm{i}\sigma_y=I_{11}\end{cases}\qquad S_\alpha=\begin{cases}S_0\\ S_3\\ S_1\\ S_2\end{cases}$$

where $\sigma_x, \sigma_y, \sigma_z$ are the Pauli matrices, and S_0, S_1, S_2, S_3 are the corresponding Sklyanin operators. From the automorphism of Sklyanin $(S_0, S_1, S_2, S_3)\to(S_0, S_3, -S_2, S_1)$, we have

$$S_0'=\theta_{11}(\eta)\theta_{11}(2u)\qquad S_1'=\theta_{10}(\eta)\theta_{10}(2u)$$

$$S_2'=\theta_{00}(\eta)\theta_{00}(2u)\qquad S_3'=\theta_{01}(\eta)\theta_{01}(2u).$$

Comparing S_α' with S_α'' used by Sklyanin

$$L(z)=\sum W_\alpha(z)\sigma_\alpha S_\alpha''$$

we see that $\mathrm{i}S_2'=S_2''$, $S_j'=S_j''$, $j=0, 1, 3$.

Acknowledgments

The authors would like to thank M Jimbo, T Miwa for inspiring discussions. We would also like to thank Y Quano for his valuable preprint. This work is supported in part by the Natural Science Fund of China.

References

[1] Sklyanin E K 1982 *Funct. Anal. Appl.* **16** 263
[2] Frenkel I B and Reshetikhin N Yu 1991 *Yale Preprint*
[3] Baxter R J 1972 *Ann. Phys. (USA)* **79**
[4] Belavin A A 1981 *Nucl. Phys.* B **180** 189
[5] Andrews G E, Baxter R J and Forrester P J 1984 *J. Stat. Phys.* **35** 193
[6] Jimbo M, Miwa T and Okado M 1988 *Commun. Math. Phys.* **116** 507
[7] Kashiwara M and Miwa T 1986 *Nucl. Phys.* B **275** 121

[8] Jimbo M, Miwa T and Okado M 1986 *Nucl. Phys.* B **275** 517
[9] Richey M P and Tracy C A 1986 *J. Stat. Phys.* **42** 311
Cherednik I V 1983 *Sov. J. Nucl. Phys.* **36** 320
[10] Cherednik I V 1985 *Funct. Anal. Appl.* **19** 77
[11] Hou B Y and Wei H *J. Math. Phys.* **30** 2750
[12] Quano Y H and Fuji A 1991 *Mod. Phys. Lett.* A **6** 3635
[13] Hou B Y, Shi K J and Yang Z X 1991 *Infinite Analysis* eds Tsuchiya A and Jimbo M (Singapore: World Scientific)
[14] Hou B Y and Zhou Y K 1990 *J. Phys. A: Math. Gen.* **23** 1147
[15] Cherednik I V 1986 *Sov. Math. Dokl.* **33** 507
[16] Hsaegawa K and Yamada Y *Phys. Lett.* **146A** 387
[17] Bazhanov V V and Stroganov Yu G 1990 *J. Stat. Phys.* **59** 799
[18] Bernard B and Pasquier V *Preprint* Spht/89-204
[19] Bazhanov V V, Kashaev R M, Mangazeev V V and Stroganov Yu G 1991 *Commun. Math. Phys.* **138** 391
[20] Quano Y H 1992 *Preprint* University of Tokyo UT-613
[21] Jimbo M, Miwa T and Okado M 1988 *Nucl. Phys.* B **300** 74
—— 1981 *Lett. Math. Phys.* **14** 123
[22] Smith S P and Staniszkis J M 1991 *Preprint* University of Washington

Physics Letters A 178 (1993) 73–80
North-Holland

PHYSICS LETTERS A

Function space representation of the Z_n Sklyanin algebra

Bo-yu Hou [a], Kang-jie Shi [b,a], Wen-li Yang [a] and Zhong-xia Yang [b,a]

[a] *Institute of Modern Physics, Northwest University, Xi'an 710069, China*
[b] *Center of Theoretical Physics, CCAST (World Laboratory), P.O. Box 8730, Beijing 100080, China*

Received 20 January 1993; revised manuscript received 6 April 1993; accepted for publication 27 April 1993
Communicated by A.R. Bishop

We give a direct proof of the function space representation for the Z_n Sklyanin algebra (SA). We show that a class of finite representation for the Z_n SA can indeed be obtained by properly choosing the base functions in the function space representation. In particular, we give the base functions which can yield the well-known fundamental representation of the Z_n SA.

1. Introduction

With the rapid development in the studies of quantum group theory [1], its ever present importance in both the conformal field theory and some exactly solvable statistical models [2] has been quickly and thoroughly appreciated and understood.

The Sklyanin algebra (SA) [3,4] may be regarded as an extension of the quantum group. First derived from the eight-vertex model, the Z_2 SA is fully studied. With its generalization [5–8] to Z_n ($n\geqslant 3$), some progress has been made in the study of the Z_n SA [9], but this was only at the beginning. Since the Z_n SA is expected to play a similar role in massive CFTs and in certain exactly solvable statistical models [9] as the quantum group in massless CFTs and in some exactly solvable statistical models, a full investigation is worthwhile.

Recently, inspired by Bazhanov et al. [10] and based on the intertwiner of the Z_n elliptic vertex model and the $A_n^{(1)}$ face model by Jimbo et al. [11,12], Quano and Fujii [13] constructed the factorized $L(u)$ operator for the Z_n symmetric R matrix. One may extend the $L(u)$ definition to include a parameter δ. With this $L(u, \delta)$ we have worked out the cyclic representation of the Z_n SA and further derived the one parameter function space representation of the Z_n SA in ref. [14].

In this paper, we directly prove this function space representation by using the Riemann identity and the symmetry properties given by Richey and Tracy [15]. This is equivalent to a direct proof that the factorized $L(u, \delta)$ satisfies the Yang–Baxter equation. As a simple illustration, we construct the base functions corresponding to the fundamental representation ($\rho(S_\alpha)=I_\alpha^{-1}$), which constitute the building blocks for fusing higher representations.

2. Z_n SA

We construct the Z_n symmetric Belavin R matrix and the corresponding A_n SA as follows [5–8,11,13,15,16]. Let g, h, I_α be $n\times n$ matrices with

$$g_{jk}=\omega^j\delta_{jk}\,,\qquad h_{jk}=\delta_{j+1,k}\,,\qquad I_\alpha=I_{(\alpha_1,\alpha_2)}=g^{\alpha_2}h^{\alpha_1}\,,\qquad I_0=I\,,$$
$$i,j,k,\alpha_1,\alpha_2\in Z_n\,,\qquad \omega=\exp(2\pi i/n)\,. \tag{2.1}$$

Volume 178, number 1,2 PHYSICS LETTERS A 5 July 1993

Define $I_\alpha^{(j)} \equiv I\otimes ... \otimes I_\alpha \otimes I \otimes ...$, I_α is at the jth space, and

$$W_\alpha(z) \equiv \frac{\sigma_\alpha(z+\eta)}{\sigma_\alpha(\eta)}, \tag{2.2}$$

where

$$\sigma_\alpha(z) \equiv \theta\begin{pmatrix} \frac{1}{2}+\alpha_1/n \\ \frac{1}{2}+\alpha_2/n \end{pmatrix}(z,\tau), \tag{2.3}$$

$$\theta\begin{pmatrix} a \\ b \end{pmatrix}(z,\tau) \equiv \sum_{m\in Z} \exp[i\pi(m+a)^2\tau + 2\pi i(m+a)(z+b)],$$

$$\sigma_\alpha(z) = c\sigma_{-\alpha}(-z) = c'\sigma_{\alpha+n\beta}(z), \quad \sigma_0(0) = \sigma_{n\beta}(0) = 0, \tag{2.4}$$

with $C(\alpha)$, $C'(\alpha, \beta)$ independent of z. The Z_n symmetric Belavin R matrix elements are written as

$$R_{jk}(z) = \sum_{\alpha\in Z_n^2} W_\alpha(z) I_\alpha^{(j)} (I_\alpha^{-1})^{(k)}. \tag{2.5}$$

They satisfy the Yang–Baxter equation (YBE),

$$R_{12}(z_1-z_2)R_{13}(z_1)R_{23}(z_2) = R_{23}(z_2)R_{13}(z_1)R_{12}(z_1-z_2). \tag{2.6}$$

The $L(z)$ operator of an R matrix is an $n\times n$ matrix with operator elements satisfying the following equation,

$$R_{12}(z_1-z_2)L_1(z_1)L_2(z_2) = L_2(z_2)L_1(z_1)R_{12}(z_1-z_2), \tag{2.7}$$

where $L_1(z_1) = L(z_1)\otimes I$, $L_2(z_2) = I\otimes L(z_2)$.

If the $L_j(z)$ can be expressed as

$$L(z) = \sum_{\alpha\in Z_n^2} W_\alpha(z+c) I_\alpha S_\alpha, \tag{2.8}$$

where c is an arbitrary parameter, then the S_α are the SA [5,7,13] generators. From (2.7), it is shown that S_α satisfies

$$\sum_{\gamma\in Z_n^2} C_{\alpha\beta\gamma} S_\gamma S_{\alpha+\beta-\gamma} = 0, \tag{2.9}$$

where $\alpha, \beta \in Z_n^2$, and the constants

$$C_{\alpha\beta\gamma} = \frac{\sigma_\beta(2\eta)\sigma_{\alpha+\beta-2\gamma}(0)\omega^{\gamma_1-\alpha_1}}{\sigma_{\gamma-\alpha}(\eta)\sigma_{\alpha+\beta-\gamma}(\eta)\sigma_\gamma(\eta)\sigma_{\beta-\gamma}(\eta)}\omega^{(\beta_1-\gamma_1)(\gamma_2-\alpha_2)}$$
$$= \frac{\omega^{(\beta_1-\gamma_1)(\gamma_2-\alpha_2)}\sigma_\beta(2\eta)}{\sigma_0(u-v)\sigma_\beta(u+2\eta)\sigma_\alpha(v)}[W_{\gamma-\alpha}(u-v)W_{\alpha+\beta-\gamma}(u)W_\gamma(v) - W_{\beta-\gamma}(u-v)W_{\alpha+\beta-\gamma}(v)W_\gamma(u)]|_{u,v} \tag{2.10}$$

(u and v can take any values) are independent of the choice of u, v [7].

3. Function space representation

3.1. Definition of the representation

Let $\{H\}$ be the space of n-variable complex functions $F = F(u_1, ..., u_n) \in \{H\}$. We define [14]

Volume 178, number 1,2 PHYSICS LETTERS A 5 July 1993

$$\hat{S}_\alpha F=\sum_j S_\alpha^{(j)}F(u_1,u_2,\dots,u_j-\eta,u_{j+1},\dots,u_n)=\sum_j S_\alpha^{(j)}F(u_j-\eta)\,, \tag{3.1.}$$

where

$$S_\alpha^{(j)}(u)=\exp(\pi i\alpha_2)\sigma_\alpha(\eta)\prod_{k\neq j}\frac{\sigma_\alpha(nu_j-nu_k+\delta)}{\sigma_0(nu_j-nu_k)}\,. \tag{3.2}$$

δ is an arbitrary parameter. We constructed and proved this representation in ref. [14]. In the following, we give an alternative proof, i.e., a direct proof.

In order to prove that (3.1) truly defines the Z_n SA representation, we need to show it satisfies (2.9), i.e.,

$$\begin{aligned}T&=\sum_\gamma C_{\alpha\beta\gamma}\hat{S}_\gamma\hat{S}_{\alpha+\beta-\gamma}F=\sum_\gamma C_{\alpha\beta\gamma}\hat{S}_\gamma\Big(\sum_{j1}S^{(j1)}_{\alpha+\beta-\gamma}F(u_{j1}-\eta)\Big)\\
&=\sum_{\gamma,j1,j2}C_{\alpha\beta\gamma}S_\gamma^{(j2)}S^{(j1)}_{\alpha+\beta-\gamma}(u_{j2}-\eta)F(u_{j1}-\eta,u_{j2}-\eta)\\
&=\sum_{\gamma,j}C_{\alpha\beta\gamma}S_\gamma^{(j)}S^{(j)}_{\alpha+\beta-\gamma}(u_j-\eta)F(u_j-2\eta)+\sum_{\gamma,j1\neq j2}C_{\alpha\beta\gamma}S_\gamma^{(j2)}S^{(j1)}_{\alpha+\beta-\gamma}(u_{j2}-\eta)F(u_{j1}-\eta,u_{j2}-\eta)\\
&\equiv\sum_j S_jF(u_j-2\eta)+\sum_{j1<j2}D_{j1j2}F(u_{j1}-\eta,u_{j2}-\eta)=0\end{aligned} \tag{3.3}$$

Thus we need only to show that $S_j=0$, $D_{j1,j2}=0$, for j, $j_1<j_2=1,\dots,n$. Without losing generality, we prove that $S_1=0$, $D_{12}=0$ in the following.

3.2. The proof of $S_1=0$, $D_{12}=0$

First we introduce two essential identities [15,17,18].

$$(1)\ n\omega^{-\eta_1\eta_2}\prod_{i=1}^{2n}\sigma_{\eta-\rho(i)}(u_i)=\sum_{\epsilon\in Z_n^2}\omega^{\langle\epsilon,\eta\rangle-\epsilon_1\epsilon_2}\prod_{i=1}^{2n}\sigma_{\epsilon+\rho(i)}(v_i)\,, \tag{R.I.}$$

where $\rho(1)+\rho(2)+\dots+\rho(2n)=(0,0)$, $\langle\epsilon,\eta\rangle=\epsilon_1\eta_2-\epsilon_2\eta_1$,

$$v_i=-u_i+\frac{1}{n}\sum_{j=1}^{2n}u_j\,. \tag{3.4}$$

This is the Riemann identity, to which we refer simply as (R.I.) later.

(2) For a, $b\in Z_n$, we define

$$S^{ab}(z)=\sum_{\alpha\in Z_n}W_{(-a,\alpha)}(z)\omega^{-b\alpha}\,.$$

According to Richey and Tracy [15], for $P=\pm1$, the following identity holds,

$$S^{ab}(Pn\eta)=PS^{-Pb,-Pa}(Pn\eta)\,, \tag{R.T.}$$

which is referred to as (R.T.) later. Considering $\gamma=(\gamma_1,\gamma_2)\in Z_n^2$ we can derive

$$\sum_{\gamma\in Z_n^2}W_\gamma(Pn\eta)\omega^{\langle\theta,\gamma\rangle}=PnW_{-P\theta}(Pn\eta) \tag{3.5}$$

from (R.T.). Combining (R.I.) and (3.5) we have for $\rho(i)$, u_i, v_i satisfying (3.4)

Volume 178, number 1,2 PHYSICS LETTERS A 5 July 1993

$$\sum_\gamma W'_{\pm(\gamma-\alpha)}(Pn\eta)\omega^{\langle\beta,\gamma\rangle-\gamma_1\gamma_2}\prod_i\sigma_{\gamma-\rho(i)}(u_i)\overset{(\mathrm{R.I.})}{=}\sum_\gamma W'_{\pm(\gamma-\alpha)}(Pn\eta)\omega^{\langle\beta,\gamma\rangle}\frac{1}{n}\sum_\epsilon\omega^{\langle\epsilon,\gamma\rangle-\epsilon_1\epsilon_2}\prod_i\sigma_{\epsilon+\rho(i)}(v_i)$$

$$\overset{\gamma'=\pm(\gamma-\alpha)}{=}\frac{1}{n}\sum_{\gamma',\epsilon}W'_{\gamma'}(Pn\eta)\omega^{\pm\langle\epsilon+\beta,\gamma'\rangle+\langle\epsilon+\beta,\alpha\rangle-\epsilon_1\epsilon_2}\prod_i\sigma_{\epsilon+\rho(i)}(v_i)$$

$$\overset{(3.5)}{=}P\sum_\epsilon W'_{\mp P(\epsilon+\beta)}(Pn\eta)\omega^{\langle\epsilon+\beta,\alpha\rangle-\epsilon_1\epsilon_2}\prod_i\sigma_{\epsilon+\rho(i)}(u_i)\,. \tag{3.6}$$

We prove next that $S_1=0$ and $D_{12}=0$.

(I) According to (3.3),

$$S_1=\sum_\gamma C_{\alpha\beta\gamma}S_\gamma^{(i)}(u)S_{\alpha+\beta-\gamma}^{(j)}(u_j-\eta)\,. \tag{3.7}$$

Let $nu_1-nu_k+\delta=x_k$, $k=2,\ldots,n$ and take the values of $C_{\alpha\beta\gamma}$ from (2.10), then (3.7) is written as

$$S_1=\frac{A}{\prod_{k=2}^n\sigma_0(x_k-\delta)\sigma_0(x_k-\delta-n\eta)}\,, \tag{3.8}$$

where

$$A=\frac{\sigma_\beta(2\eta)\exp[\pi\mathrm{i}(\alpha_2+\beta_2)](-1)^n}{\sigma_0(u-v)\sigma_\beta(u+2\eta)\sigma_\alpha(v)}(A_1-A_2)\,,$$

$$A_1=\sum_\gamma W_{\gamma-\alpha}(u-v)\sigma_{\gamma-\alpha-\beta}(-u-\eta)\sigma_\gamma(v+\eta)\omega^{(\beta_1-\gamma_1)(\gamma_2-\alpha_2)}\prod_{k=2}^n\sigma_\gamma(x_k)\sigma_{\gamma-\alpha-\beta}(-x_k+n\eta)\,,$$

$$A_2=\sum_\gamma W_{\beta-\gamma}(u-v)\sigma_{\gamma-\alpha-\beta}(-v-\eta)\sigma_\gamma(u+\eta)\omega^{(\beta_1-\gamma_1)(\gamma_2-\alpha_2)}\prod_{k=2}^n\sigma_\gamma(x_k)\sigma_{\gamma-\alpha-\beta}(-x_k+n\eta)\,,$$

where the identity $\sigma_\alpha(x)=-\exp(2\pi\mathrm{i}\alpha_1/n)\sigma_{-\alpha}(-x)$ is used.

Take $u=-\alpha_1\tau/n-\alpha_2/n\equiv-\alpha$, $v=-\alpha+n\eta$, and recall that $\sigma_{\alpha+\beta}(z)=\sigma_{\alpha-(n-1)\beta}(z)(-1)^{\beta_2}\omega^{\beta_2(\alpha_1+\beta_1)}$, then A_1 becomes

$$A_1=\sum_\gamma W'_{\gamma-\alpha}(-n\eta)\sigma_{\gamma+(n-1)\alpha}(n\eta+\eta)\sigma_{\gamma+(n-1)\beta}(-\eta)\omega^{\langle\beta,\gamma\rangle-\gamma_1\gamma_2}a_1\prod_{k=2}^n\sigma_\gamma(x_k)\sigma_{\gamma-\alpha-\beta}(-x_k+n\eta)\,, \tag{3.9}$$

where a_1 is independent of γ. We have from (3.6)

$$A_1=-\sum_\epsilon W'_{\beta+\epsilon}(-n\eta)\omega^{\langle\beta+\epsilon,\alpha\rangle-\epsilon_1\epsilon_2}\sigma_{\epsilon-(n-1)\alpha}(-\eta)\sigma_{\epsilon-(n-1)\beta}(n\eta+\eta)a_1\prod_{k=2}^n\sigma_\epsilon(n\eta-x_k)\sigma_{\epsilon+\alpha+\beta}(x_k)$$

$$\overset{\epsilon\to\gamma-\alpha-\beta}{=}-\sum_\gamma W'_{\gamma-\alpha}(-n\eta)\sigma_{\gamma+(n-1)\alpha}(n\eta+\eta)\sigma_{\gamma+(n-1)\beta}(-\eta)\omega^{\langle\beta,\gamma\rangle-\gamma_1\gamma_2}a_1\prod_{k=2}^n\sigma_\gamma(x_k)\sigma_{\gamma-\alpha-\beta}(-x_k+n\eta)$$

$$=-A_1\quad\Rightarrow\quad A_1=0\,,$$

and A_2 is given by

$$A_2=\sum_\gamma W_{\beta-\gamma}(-n\eta)\sigma_{\gamma+(n-1)\beta}(-n\eta-\eta)\sigma_{\gamma+(n-1)\alpha}(\eta)\omega^{\langle\beta,\gamma\rangle-\gamma_1\gamma_2}a_1\prod_{k=2}^n\sigma_\gamma(x_k)\sigma_{\gamma-\alpha-\beta}(-x_k+n\eta)\,. \tag{3.10}$$

It is easy to show that

$$W_{\beta-\gamma}(-n\eta)\sigma_{\gamma+(n-1)\beta}(-n\eta-\eta)=W_{\beta-\gamma}(n\eta)\sigma_{\gamma+(n-1)\beta}(n\eta-\eta)\,. \tag{3.11}$$

Volume 178, number 1,2 PHYSICS LETTERS A 5 July 1993

Thus (3.10) becomes

$$A_2=\sum_{\gamma} W_{\beta-\gamma}(n\eta)\sigma_{\gamma+(n-1)\beta}(n\eta-\eta)\sigma_{\gamma+(n-1)\alpha}(\eta)\omega^{\langle\beta,\gamma\rangle-\gamma_1\gamma_2}a_1\prod_{k=2}^{n}\sigma_\gamma(x_k)\sigma_{\gamma-\alpha-\beta}(-x_k+n\eta)\,. \tag{3.12}$$

From (3.6) and considering $W_{\epsilon+\beta}=W_{\epsilon-(n-1)\beta}$, we have

$$A_2=\sum_{\epsilon}\sigma_{\epsilon-(n-1)\beta}(n\eta+\eta)\sigma_{\epsilon-(n-1)\alpha}(-\eta+n\eta)\omega^{\langle\epsilon,\beta\rangle-\epsilon_1\epsilon_2}a_1\prod_{k=2}^{n}\sigma_\epsilon(-x_k+n\eta)\sigma_{\epsilon+\alpha+\beta}(x_k)$$

$$\stackrel{\text{(R.I.)}}{=} n\sigma_{n\beta}(0)\sigma_{\beta+(n-1)\alpha}(2\eta)a_1\omega^{-\beta_1\beta_2}\prod_{k=2}^{n}\sigma_\beta(x_k+\eta)\sigma_{-\alpha}(n\eta+\eta-x_k)=0 \tag{3.13}$$

As $A_1=0$, $A_2=0$ from (3.7), we have

$$A=0 \;\Rightarrow\; S_1=0 \qquad \text{from (3.8)}\,. \tag{3.14}$$

(II) We turn to the proof of $D_{12}=0$. From (3.3) we have

$$D_{12}=\sum_{\gamma} C_{\alpha\beta\gamma}[S_\gamma^{(1)}(u)S^{(2)}_{\alpha+\beta-\gamma}(u_1-\eta)+S_\gamma^{(2)}(u)S^{(1)}_{\alpha+\beta-\gamma}(u_2-\eta)]\,. \tag{3.15}$$

Let $nu_i=x_i$, $i=1,\ldots,n$, and take $u=-\eta+\delta$, $v=(n-1)\eta+\delta$ in the expression of $C_{\alpha\beta\gamma}$, we can derive from (2.10) and (3.2)

$$D_{12}=\frac{\sigma_\beta(2\eta)(-1)^{n+\alpha_2+\beta_2}B}{\sigma_0(u-v)\sigma_\beta(u+2\eta)\sigma_\alpha(v)\sigma_0(x_1-x_2)\prod_{k=3}^{n}\sigma_0(x_1-x_k)\sigma_0(x_2-x_k)}\,, \tag{3.16}$$

$$B=\frac{B_1(b_1)-B_2(b_1)}{\sigma_0(x_2-x_1+n\eta)}-\frac{B_1(b_2)-B_2(b_2)}{\sigma_0(x_1-x_2+n\eta)}\,, \tag{3.17}$$

$$B_1(b_1)=\sum_{\gamma} W_{\gamma-\alpha}(-n\eta)\sigma_{\gamma+(n-1)(\alpha+\beta)}(-\delta)\sigma_\gamma(n\eta+\delta)\omega^{\langle\beta,\gamma\rangle-\gamma_1\gamma_2}$$

$$\times\sigma_\gamma(x_1-x_2+\delta)\sigma_{\gamma-\alpha-\beta}(x_1-x_2-\delta-n\eta)a_2\prod_{k=3}^{n}\sigma_\gamma(x_1-x_k+\delta)\sigma_{\gamma-\alpha-\beta}(x_k-x_2-\delta)\,, \tag{3.18}$$

where $a_2=(-1)^{\alpha_2+\beta_2}\omega^{\alpha_1\alpha_2+\beta_1\beta_2+\alpha_1\beta_2}$. $B_1(b_2)$ is obtained by $x_1\leftrightarrow x_2$ in (3.18),

$$B_2(b_1)=\sum_{\gamma} W_{\beta-\gamma}(-n\eta)\sigma_{\gamma+(n-1)(\alpha+\beta)}(-n\eta-\delta)\sigma_\gamma(\delta)\omega^{\langle\beta,\gamma\rangle-\gamma_1\gamma_2}$$

$$\times\sigma_\gamma(x_1-x_2+\delta)\sigma_{\gamma-\alpha-\beta}(x_1-x_2-\delta-n\eta)a_2\prod_{k=3}^{n}\sigma_\gamma(x_1-x_k+\delta)\sigma_{\gamma-\alpha-\beta}(x_k-x_2-\delta)\,. \tag{3.19}$$

$B_2(b_2)$ is obtained by $x_1\leftrightarrow x_2$ in (3.19).

Similar to (3.9), it is easy to check that (3.18) vanishes,

$$B_1(b_1)=0 \qquad \text{which implies } B_1(b_2)=0\,. \tag{3.20}$$

Let $x_3=x_1+\delta+\beta+\eta$, so that the factor in (3.19), $\sigma_\gamma(x_1-x_3+\delta)\sim\sigma_{\gamma-\beta}(-\eta)\sim\sigma_{\beta-\gamma}(\eta)$, which eliminates the denominator of $W_{\beta-\gamma}$. We then have

Volume 178, number 1,2 PHYSICS LETTERS A 5 July 1993

$$B_2(b_1)=a_3\sum_\gamma \sigma_{\gamma-\beta}(n\eta-\eta)\sigma_{\gamma+(n-1)(\alpha+\beta)}(-n\eta-\delta)\sigma_\gamma(\delta)\sigma_\gamma(x_1-x_2+\delta)\sigma_{\gamma-\alpha-\beta}(x_1-x_2-\delta-n\eta)$$

$$\times\omega^{\langle\beta,\gamma\rangle-\gamma_1\gamma_2}\sigma_{\gamma-\alpha}(x_1-x_2+\eta)\prod_{k=4}^{n}\sigma_\gamma(x_1-x_k+\delta)\sigma_{\gamma-\alpha-\beta}(x_k-x_2-\delta)$$

$$\overset{\text{(R.I.)}}{=} na_3\sigma_0(x_1-x_2-n\eta)\sigma_{-\beta-(n-1)(\alpha+\beta)}(x_1-x_2+n\eta+\delta-\eta)\sigma_{-\beta}(x_1-x_2-\delta-\eta)\sigma_{-\beta}(-\delta-\eta)$$

$$\times\sigma_\alpha(\delta+n\eta-\eta)\omega^{-\beta_1\beta_2}\sigma_{-\beta+\alpha}(-2\eta)\prod_{k=4}^{n}\sigma_{-\beta}(x_k-x_2-\delta-\eta)\sigma_\alpha(x_1-x_k+\delta-\eta)\,, \tag{3.21}$$

where

$$a_3=a_2\exp\left[-\frac{2\pi i\beta_1^2}{n^2}\tau-\frac{2\pi i\beta_1}{n}\left(2\eta+x_1-x_2+\frac{\beta_2-\alpha_2}{n}\right)\right].$$

On the other hand, for $x_3=x_1+\delta+\beta+\eta$ we have from $x_1\leftrightarrow x_2$ of (3.19),

$$B_2(b_2)=a_3\sum_\gamma W_{\beta-\gamma}(-n\eta)\omega^{\langle\beta,\gamma\rangle-\gamma_1\gamma_2}\sigma_{\gamma+(n-1)(\alpha+\beta)}(-n\eta-\delta)\sigma_\gamma(\delta)\sigma_\gamma(x_2-x_1+\delta)\sigma_{\gamma-\alpha-\beta}(x_2-x_1-\delta-n\eta)$$

$$\times\sigma_{\gamma-\beta}(x_2-x_1-\eta)\sigma_{\gamma-\alpha}(\eta)\prod_{k=4}^{n}\sigma_\gamma(x_2-x_k+\delta)\sigma_{\gamma-\alpha-\beta}(x_k-x_1-\delta)$$

$$\overset{(3.6)}{=} -\sum_\epsilon a_3W_{-\beta-\epsilon}(-n\eta)\omega^{\langle\epsilon,\beta\rangle-\epsilon_1\epsilon_2}\sigma_{\epsilon-(n-1)(\alpha+\beta)}(x_2-x_1+\delta+n\eta-2\eta)\sigma_\epsilon(x_2-x_1-2\eta-\delta)\sigma_\epsilon(-2\eta-\delta)$$

$$\times\sigma_{\epsilon+\beta}(-\eta)\sigma_{\epsilon+\alpha+\beta}(-2\eta+\delta+n\eta)\sigma_{\epsilon+\alpha}(x_2-x_1-3\eta)\prod_{k=4}^{n}\sigma_\epsilon(x_k-x_1-2\eta-\delta)\sigma_{\epsilon+\alpha+\beta}(x_2-x_k-2\eta+\delta)\,.$$

Thus $\sigma_{\epsilon+\beta}(-\eta)\sim\sigma_{-\epsilon-\beta}(\eta)$ eliminates the denominator of $W_{-\beta-\epsilon}$ and we can use (R.I.) to obtain

$$B_2(b_2)=-a_3n\sigma_0(x_2-x_1-n\eta)\sigma_{\beta+(n-1)(\alpha+\beta)}(\eta-\delta-n\eta)\sigma_\beta(\eta+\delta)\sigma_\beta(x_2-x_1+\eta+\delta)\sigma_{-\alpha}(x_2-x_1+\eta-\delta-n\eta)$$

$$\times\sigma_{\beta-\alpha}(2\eta)\prod_{k=4}^{n}\sigma_\beta(x_2-x_k+\eta+\delta)\sigma_{-\alpha}(x_k-x_1+\eta-\delta)\omega^{-\beta_1\beta_2}\,. \tag{3.22}$$

Consequently

$$f(x_3)\equiv\frac{B_2(b_1)}{\sigma_0(x_2-x_1+n\eta)}-\frac{B_2(b_2)}{\sigma_0(x_1-x_2+n\eta)}=0 \tag{3.23a}$$

at $x_3=x_1+\delta+\beta+\eta$. Similarly we can show that

$$f(x_3)=0 \quad \text{at } x_3=x_1+\delta+\alpha-\eta\,. \tag{3.23b}$$

It is easy to find that $f(x_3)$ as a quasiperiodic function of x_3 has two zeros, and the sum of the positions of these two zeros is [15,18] Σ zeros$=x_1+x_2+\alpha+\beta+2\delta$. As this is incompatible with (3.23), we could only have

$$f(x_3)\equiv 0\,. \tag{3.24}$$

With (3.24), (3.20), we have $B=0\Rightarrow D_{12}=0$.

This completes the proof for $n\geqslant 3$. The $n=2$ case can be treated straightforwardly. We thus conclude that the difference operator $\hat{S}_\alpha$ defined in (3.1), (3.2) satisfies the SA (2.9).

Volume 178, number 1,2 PHYSICS LETTERS A 5 July 1993

4. Fundamental representations of Z_n SA on the function space

Take a set of functions

$$F_\gamma = \exp(\pi i n\gamma_2) \prod_{l=1}^{n} \sigma_\gamma\left(nu_l - \sum u_k\right), \qquad \gamma \in Z_n^2 \tag{4.1}$$

and $\delta=\eta$ in $S_\alpha^{(j)}$. Then according to (3.1) and (3.2), we have

$$\hat{S}_0 F_0 = \sigma_0(\eta) \sum_j \left(\prod_{k\neq j} \frac{a_0(nu_j - nu_k + \eta)}{\sigma_0(nu_j - nu_k)} \right) \sigma_0\left(nu_j - \sum u_k - (n-1)\eta\right) \prod_{j\neq 0} \sigma_0\left(nu_j - \sum u_k + \eta\right). \tag{4.2}$$

Let $\hat{S}_0 F_0 \equiv G(\eta, \{u_i\})$.

Lemma.

$$G(\eta, \{u_i\}) = 0\,, \qquad \text{for } \eta = \frac{\alpha}{n} \Lambda_\tau = \frac{1}{n}(\alpha_1 \tau + \alpha_2)\,, \qquad \alpha_1, \alpha_2 \in Z_n\,. \tag{4.3}$$

Proof. For $\alpha=(0, 0)$, because of $\sigma_0(\eta)$ on the r.h.s. of (4.2), obviously $G(0, \{u_i\})=0$. For $\eta=(\alpha_1\tau+\alpha_2)/n\neq 0$, we have

$$G\left(\frac{\alpha}{n}\Lambda_\tau, \{u_i\}\right) = \sigma_\alpha(0) \left(\sum_j \prod_{k\neq j} \frac{\sigma_\alpha(nu_j - nu_k)}{\sigma_0(nu_j - nu_k)} \right) F_\alpha a_4\,, \tag{4.4}$$

where $a_4 = \exp(-\pi i \alpha_1^2 \tau - \pi i \alpha_1)$. Let

$$G_1(\{x_i\}) = \sum_j \prod_{k\neq j} \frac{\sigma_\alpha(x_j - x_k)}{\sigma_0(x_j - x_k)} \tag{4.5}$$

be a function of x_1,

$$G_1(x_1+1) = \exp(-2\pi i \alpha_1/n) G_1(x_1)\,, \qquad G_1(x_1+\tau) = \exp(2\pi i \alpha_2/n) G_1(x_1)\,. \tag{4.6}$$

It is easy to see that for $G_1(x_1)$ as a function of x_1, the only singular points are poles located at $x_1 = x_2, \dots, x_n \bmod \Lambda_\tau$. One can check that $G_1(x_1)$ is actually regular at $x_1=x_2$. That is, $x_1=x_2$ is not a pole of $G_1(x_1)$. In the same way $x_1=x_3, \dots, x_n$ are not poles either. So as a quasiperiodic function $G_1(x_1)$ has no poles. The entire function satisfying (4.6) can only be zero. We conclude that

$$G_1(x_1) \equiv 0 \quad \Rightarrow \quad G(\eta, \{u_i\}) = 0\,, \qquad \eta = \frac{1}{n}(\alpha_1\tau + \alpha_2)_\gamma\,, \qquad \alpha \in Z_n^2\,. \tag{4.7}$$

In this way, we have located n^2 zeros of $G(\eta, \{u_i\})$ as a function of η. But as a quasiperiodic function of η, it has only n^2 zeros, and the zeros are the same as $\sigma_0(n\eta)$. From the quasiperiodic properties of $G(\eta)$ and $\sigma_0(n\eta)$ we have

$$G(\eta, \{u_i\}) = \frac{1}{n} \sigma_0(n\eta) Q(\{u_i\})\,, \tag{4.8}$$

where $Q(\{u_i\})$ is independent of η. Taking $\eta \to 0$, we have

$$Q(\{u_i\}) = \prod_j \sigma_0\left(nu_j - \sum u_k\right) = F_0\,. \tag{4.9}$$

Volume 178, number 1,2 PHYSICS LETTERS A 5 July 1993

Equation (4.8) is thus

$$\hat{S}_0 f_0 = \frac{1}{n}\sigma_0(n\eta)F_0 . \qquad (4.10)$$

Let $u_1 \to u_1 - (\gamma_1\tau + \gamma_2)/n$, in both F_0 and $S_0^{(j)}$ in (4.10), we have

$$\hat{S}_0 F_\gamma = \frac{1}{n}\sigma_0(n\eta)F_\gamma . \qquad (4.11)$$

In (4.11), let $\eta \to \eta + (\alpha_1\tau + \alpha_2)/n$, we have

$$\hat{S}_\alpha F_{\gamma+\alpha} = \frac{1}{n}\exp[2\pi i\alpha_2(\gamma_1 + \alpha_1)/n]\sigma_0(n\eta)F_\gamma . \qquad (4.12)$$

Redefining $\gamma \to \gamma - \alpha$, we have

$$\hat{S}_\alpha F_\gamma = \frac{1}{n}\sigma_0(n\eta)\omega^{\alpha_2\gamma_1}F_{\gamma-\alpha} . \qquad (4.13)$$

We choose a basis set V_l, $l=0, ..., n-1, V_l = \sum_m F_{-l,m}$ and then

$$\hat{S}_\alpha V_l = \frac{1}{n}\sigma_0(n\eta)\omega^{-l\alpha_2}V_{l+\alpha_1} = \frac{1}{n}\sigma_0(n\eta)\sum_{l'}(I_\alpha^{-1})_{l'l}V_{l'} \qquad (4.14)$$

Since $\sigma_0(n\eta)/n$ is an irrelevant overall factor, (4.14) indeed constitutes the fundamental representation of the Z_n SA: $\rho(\hat{S}_\alpha) \sim I_\alpha^{-1}$ [7].

Acknowledgement

This work was supported in part by the National Natural Science Foundation of China.

References

[1] V.G. Drinfel'd, Sov. Math. Dokl. 32 (1985) 254;
M. Jimbo, Lett. Math. Phys. 10 (1985) 63.
[2] R.J. Baxter, Exactly solved models in statistical mechanics (Academic Press, New York, 1982).
[3] E.K. Sklyanin, Funct. Anal. Appl. 16 (1982) 263.
[4] E.K. Sklyanin, Funct. Anal. Appl. 17 (1983) 273.
[5] I.V. Cherednik, Sov. J. Nucl. Phys. 36 (1983) 320.
[6] I.V. Cherednik, Funct. Anal. Appl. 19 (1985) 77.
[7] B.Y. Hou and H. Wei, J. Math. Phys. 30 (1989) 2750.
[8] Y.-H. Quano and A. Fujii, Mod. Phys. Lett. A 6 (1991) 3635.
[9] B.Y. Hou and Y.K. Zhou, J. Phys. A 23 (1990) 1147.
[10] V.V. Bazhanov, R.M. Kashaev, V.V. Mangazeev and Yu.G. Stroganov, Commun. Math. Phys. 138 (1991) 391.
[11] M. Jimbo, T. Miwa and M. Okado, Nucl. Phys. B 300 (1988) 74.
[12] K. Hasegawa and Y. Yamada, Phys. Lett. A 146 (1990) 387.
[13] Y.H. Quano and A. Fujii, Preprint UT-603, Tokyo University (1991);
Y.H. Quano, preprint UT-613, Tokyo University (1992).
[14] B.Y. Hou, K.J. Shi and Z.X. Yang, preprint NWU-IMP-92 0914, Northwest University.
[15] M.P. Richey and C.A. Tracy, J. Stat. Phys. 42 (1986) 311.
[16] A.A. Belavin, Nucl. Phys. B 180 (1981) 189.
[17] C. Tracy, Physica D 16 (1985) 203.
[18] D. Mumford, Tata lectures on theta (Birkhauser, Basel, 1983).

Physics Letters B 298 (1993) 103–110
North-Holland

PHYSICS LETTERS B

Hamiltonian reductions of self-dual Yang–Mills theory

Bo-yu Hou and Liu Chao
Institute of Modern Physics, Northwest University, Xian 710069, China

Received 5 July 1992

The hamiltonian reductions of self-dual Yang–Mills theory associated with left–right dual regular constant constraints are analysed. It is shown that the reduced theory is precisely the four dimensional analogue of the nonabelian Toda theory. In the special case of principal gradation with constraints of the first degree this theory becomes the usual Toda theory in four dimensions. The effective action and the linear systems are all obtained from the reduction procedure.

1. Introduction

Establishing a nonperturbative approach to classical and quantum field theories has long been the dreams of theoretical and mathematical physicists. Over the last decade or so, much progress has been made toward the complete integrabilities of the models in two dimensions, where the existance of infinite dimensional symmetry algebras or alternatively Yang–Baxter algebras played the central role. As the results of investigations in two dimensions, the conformal field theories and quantum group theories have respectively found their places in 2D nonperturbative field theories, and later it turned out that they are not completely independent objects: it is now farely well understood that the fusing matrices of the conformal blocks are exactly the quantum $6j$ symbols of the corresponding quantum group.

Among various progresses in two dimensions, the methods for constructing exactly integrable field theoretical models are always of interest, since in the process of gaining more integrable systems the common properties giding or characterising their integrabilities may be found. In the present days the hamiltonian reduction method of WZNW theory [1–8] has been proven to be a powerful method in constructing integrable systems in two dimensions. This method has also been found to have close relations with the reductions of the chiral symmetry algebras of the WZNW model [1,4–7].

Comparing to the case of 2D field theories, much less work has been done toward the nonperturbative treatment of the 4D real physical world. The central difficulty lies in that in four dimensions it is very difficult to find an infinite set of symmetry algebras, or, in other words, infinitely many conservation laws. However, there exists a nice theory which has long been known to be integrable in four dimensions, namely the self-dual Yang–Mills theory (SDYM) [9]. This theory has a similar hamiltonian structure with respect to the 2D WZNW model [10]. Therefore, using the experiences gained in the 2D cases, we can explore various reduced cases of SDYM phase space. We can see that the reduced theories are integrable by construction. Thus a large class of 4D integrable systems can be obtained from this procedure, and exploiting the common algebraic properties of these models is hoped to be helpful in understanding the integrability conditions in four dimensions.

The present work is just a first attempt toward the hamiltonian reductions of SDYM theory. Here we consider only the case of left–right dual constant constraints. Other cases are planned to be treated elsewhere.

2. Description of the SDYM theory

Let us first give a description of the 4D SDYM theory. We adopt the J-gauge convention [9], in which the field equations read

Volume 298, number 1,2 PHYSICS LETTERS B 7 January 1993

$$\partial_y B_{\bar y} + \partial_z B_{\bar z} = 0 \text{ (left moving)} \quad (1)$$

or

$$\partial_{\bar y} B_y + \partial_{\bar z} B_z = 0 \text{ (right moving)}\,, \quad (2)$$

where

$$B_y = \partial_y J J^{-1}, \quad B_{\bar y} = J^{-1}\partial_{\bar y} J\,,$$
$$B_z = \partial_z J J^{-1}, \quad B_{\bar z} = J^{-1}\partial_{\bar z} J\,; \quad (3)$$

the field J belongs to some real noncompact gauge group G, the four-space–time (x^0, x^1, x^2, x^3) are chosen to have the metric diag$(+, -, +, -)$ and here we are using the "lightcone variables"

$$y = x^0 - x^1, \quad \bar y = x^0 + x^1\,,$$
$$z = x^2 - x^3, \quad \bar z = x^2 + x^3\,. \quad (4)$$

The four-space–time with the given metric can be regarded as the product space $S^2 \times S^2$ (or the product of any two closed riemannian surfaces) which enables us to write the effective action of our SDYM equations (1) and (2) as follows [10]:

$$I_{\mathrm{SDYM}}[J] = \kappa\left(\int d^2z\, S_y[J] + \int d^2y\, S_z[J]\right), \quad (4)$$

where κ is a dimensional coupling constant,

$$S_y[J] = \tfrac{1}{2}\int d^2y \langle \partial_\mu J J^{-1} \partial^\mu J J^{-1}\rangle$$
$$- \tfrac{1}{3}\int d^3y\, \epsilon^{ijk}\langle \partial_i J J^{-1}\partial_j J J^{-1}\partial_k J J^{-1}\rangle\,,$$
$$\mu = (x^0, x^1), \quad i, j, k = (x^0, x^1, u) \quad (5)$$

is the WZNW action in the y-sphere (where u denotes the topological dimension in the solid y-sphere), and similarly $S_z[J]$ is the WZNW action in the z-sphere.

Owing to the resemblance of the SDYM action to the WZNW action, it is desirable that many of the properties of WZNW theory can also be found in SDYM theory. This is indeed the case. For example, the well known Polyakov–Wiegmann formula for the WZNW action,

$$S[gh] = S[g] + S[h] + \kappa\int d^2x \langle g^{-1}\partial_- g\, \partial_+ h h^{-1}\rangle\,,$$

can be analogously written for the SDYM theory as

$$I_{\mathrm{SDYM}}[JK] = I_{\mathrm{SDYM}}[J] + I_{\mathrm{SDYM}}[K]$$
$$+ \kappa\int d^2y\, d^2z \langle J^{-1}\partial_{\bar y} J \partial_y K K^{-1} + J^{-1}\partial_{\bar z} J \partial_z K K^{-1}\rangle. \quad (6)$$

Thus the chiral invariance

$$S[h(x_+) g \bar h(x_-)] = S[g]$$

of WZNW theory can be easily generalized to the following symmetries of the SDYM theory:

$$I_{\mathrm{SDYM}}[K(y, z) J \bar K(\bar y, \bar z)] = I_{\mathrm{SDYM}}[J]\,. \quad (7)$$

In fact, the left and right SDYM equations (1) and (2) can be regarded as the conservation laws with respect to these symmetry transformations [11]. Furthermore, we have the "conserved charges"

$$\bar j(\bar y, \bar z) = \tfrac{1}{2}\kappa\int dz\, J^{-1}\partial_{\bar y} J\,,$$

$$j(y, z) = \tfrac{1}{2}\kappa\int d\bar z\, \partial_y J J^{-1}\,, \quad (8)$$

with

$$\partial_{\bar y} j(y, z) = 0, \quad \partial_y \bar j(\bar y, \bar z) = 0\,. \quad (9)$$

The canonical Poisson brackets of the SDYM theory are also very similar to those of the WZNW theory [12]. Choosing x^0 as time variable, the equal-time Poisson brackets between the "charges" $j(y, z)$ and $\bar j(\bar y, \bar z)$ are

$$\{j_a(y, z), j_b(y', z')\} = j_{[a,b]}(y, z)\delta(y-y')\delta(z-z')$$
$$+ \tfrac{1}{2}\kappa\langle a, b\rangle \partial_y \delta(y-y')\delta(z-z')\,, \quad (10a)$$
$$\{\bar j_a(\bar y, \bar z), \bar j_b(\bar y', \bar z')\} = \bar j_{[a,b]}(\bar y, \bar z)\delta(\bar y-\bar y')\delta(\bar z-\bar z')$$
$$+ \tfrac{1}{2}\kappa\langle a, b\rangle \partial_{\bar y}\delta(\bar y-\bar y')\delta(\bar z-\bar z')\,, \quad (10b)$$
$$\{j_a(y, z), \bar j_b(\bar y', \bar z')\} = 0\,, \quad (10c)$$

where a, b etc. are elements of the Lie algebra g of the group G. This Poisson algebra is nothing but the 4D analogue of the Kac–Moody algebra. If we choose another coordinate as time variable, we could arive at other sets of 4D Kac–Moody algebras.

At the end of this section let us introduce the linear systems of the 4D SDYM theory. They are [13]

$$\partial_z \chi = \lambda(\partial_{\bar y} + J^{-1}\partial_{\bar y} J)\chi\,,$$
$$\partial_y \chi = -\lambda(\partial_{\bar z} + J^{-1}\partial_{\bar z} J)\chi\,, \quad (11)$$

Volume 298, number 1,2 PHYSICS LETTERS B 7 January 1993

and

$$\partial_{\bar{z}}\bar{\chi} = -\frac{1}{\lambda}(\partial_y - \partial_y J J^{-1})\bar{\chi},$$

$$\partial_{\bar{y}}\bar{\chi} = \frac{1}{\lambda}(\partial_z - \partial_z J J^{-1})\bar{\chi}, \qquad (12)$$

where λ is an arbitrary complex-valued spectral parameter which ensures the existance of infinite many nonlocal conserved currents. The SDYM equations (1) and (2) can easily be obtained as the compatibility conditions of the linear systems (11) and (12), respectively.

3. The left–right dual constant constraints

As described in section 2, the 4D SDYM theory resembles in many aspects to the 2D WZNW theory. Thus we naturally expect that the method for reducing the phase space of WZNW theory can also be applied to SDYM theory.

In order to reduce the SDYM phase space, let us recall that in the conventional hamiltonian reductions of the WZNW model, the constraints are imposed on the KM currents with a finite degree under some integral gradation of the underlying Lie algebra. To have a precise description, let us assume that g is graded with some grading operator H [1] in the Cartan subalgebra,

$$g = \bigoplus_{n=-m}^{m} g^n. \qquad (13)$$

Denoting by g_d (g_{-d}) the nilpotent subalgebras consisting of the elements of degree $\geqslant d$ ($\leqslant -d$),

$$g_d = \bigoplus_{n\geqslant d} g^n, \quad g_{-d} = \bigoplus_{n\leqslant -d} g^{-n}, \qquad (14)$$

we have the linear space decomposition

$$g = g_{-d} \oplus g_{(-d,d)} \oplus g_d, \qquad (15)$$

where

$$g_{(-d,d)} = \bigoplus_{n=-d+1}^{d-1} g^n \qquad (16)$$

is the complement of $g_{-d}\oplus g_d$ in g which is generally not a subalgebra but merely a linear subspace of g and is orthogonal to $g_{-d}\oplus g_d$ under the Killing form of g.

In what follows we shall only consider the left–right dual regular constant constraints of finite degree. *A regular element* μ of degree d is an element in g^d such that

$$\mathrm{Ker}(\mathrm{ad}\, u) \cap g_{-d} = \{0\}. \qquad (17)$$

We shall assume without proof that such elements exist.

Now choose two arbitrary constant regular elements μ and ν in g_d, and let $\bar{\mu}$ and $\bar{\nu}$ be the images of μ and ν respectively under the Cartan involution of g. The constraints we shall impose on the SDYM theory can be described as follows:

$$\langle g_{-d}, B_y - \mu\rangle = \langle g_{-d}, B_z - \nu\rangle = 0,$$

$$\langle g_d, B_{\bar{y}} - \bar{\mu}\rangle = \langle g_d, B_{\bar{z}} - \bar{\nu}\rangle = 0. \qquad (18)$$

To compare with the constraints imposed on the KM currents in the 2D WZNW cases, let us now make a few remarks.

First, the regular, constant constraints of finite degree were imposed on the KM currents of the 2D WZNW model in order to preserve the conformal symmetry, or in other words, the Virasoro symmetry in two dimensions. Now in our 4D case, we also have the (4D analogues of) Virasoro symmetries generated by the "Sugawara constructions",

$$L = \frac{2}{\kappa}\langle j(y,z) j(y,z)\rangle,$$

$$\bar{L} = \frac{2}{\kappa}\langle \bar{j}(\bar{y},\bar{z})\bar{j}(\bar{y},\bar{z})\rangle. \qquad (19)$$

The 4D Virasoro algebra has the generating relation

$$\{L(y,z), L(y',z')\} = 2L(y,z)\delta'(y-y')\delta^2(z-z)$$
$$+\partial_y L(y,z)\delta(y-y')\delta^2(z-z'), \qquad (20)$$

and a similar one for $\bar{L}(\bar{y},\bar{z})$. Following a parallel discussion as in the 2D cases we can show that our constraints (18) also preserve the Virasoro symmetry. But now there is no direct relation between the conformal transformations and Virasoro symmetries, since the Virasoro elements $L(y,z)$ and $\bar{L}(\bar{y},\bar{z})$ are defined in a nonlocal way in terms of the local field J.

Second, one may wonder why we do not impose the constraints directly on the "KM charges" $j(y,z)$ and $\bar{j}(\bar{y},\bar{z})$. The reason is again the nonlocality of the charges $j(y,z)$ and $\bar{j}(\bar{y},\bar{z})$. This nonlocality will prevent us from getting a reduced theory of local fields if we impose the constraints on the level of charges. Thus our choice of constraints is considered to be a more convenient one in order to obtain the desired 4D reduced theories.

4. Realization of the reduced theory

Now we turn to the realization of the reduced theory associated to the constraints described above.

According to the linear space decomposition (15) of g, and because of the noncompactness of G, we can write the SDYM field J in the neighborhood of identity as follows:

$$J=XYZ\,,$$
$$X\in\exp(g_{-d}),\quad Y\in\exp(g_{(-d,d)}),\quad Z\in\exp(g_d)\,. \tag{21}$$

Therefore, by eq. (3), we have

$$B_y=\partial_y XX^{-1}+X\partial_y YY^{-1}X^{-1}+XY\partial_y ZZ^{-1}Y^{-1}X^{-1}\,, \tag{22a}$$
$$B_{\bar y}=Z^{-1}\partial_{\bar y}Z+Z^{-1}Y^{-1}\partial_{\bar y}YZ+Z^{-1}Y^{-1}X^{-1}\partial_{\bar y}XYZ\,, \tag{22b}$$

and two other ones with $y\rightleftharpoons z$, $\bar y\rightleftharpoons \bar z$.

Take B_y as an example. According to the graded decomposition of g, the first two terms in eq. (22a) lie entirely in the linear subspace of g with degree $n<d$. Thus the constraints imposed on B_y as in eq. (18) can be equivalently expressed as

$$(XY\partial_y ZZ^{-1}Y^{-1}X^{-1})^{\geqslant d}=\mu\,. \tag{23}$$

Following an induction procedure as made in ref. [7], we can show that

$$\partial_y ZZ^{-1}=Y^{-1}\mu Y\,. \tag{24a}$$

Similarly we have

$$\partial_z ZZ^{-1}=Y^{-1}\nu Y\,, \tag{24b}$$
$$X^{-1}\partial_{\bar y}X=Y\bar\mu Y^{-1}\,, \tag{24c}$$
$$X^{-1}\partial_{\bar z}X=Y\bar\nu Y^{-1}\,. \tag{24d}$$

Now let us return to the SDYM equations (1) and (2). By conjugations with Z and X^{-1} respectively, we have

$$\partial_y(ZB_{\bar y}Z^{-1})-[\partial_y ZZ^{-1},ZB_{\bar y}Z^{-1}]+\partial_z(ZB_{\bar z}Z^{-1})-[\partial_z ZZ^{-1},ZB_{\bar z}Z^{-1}]=0\,,$$
$$\partial_{\bar y}(X^{-1}B_yX)+[X^{-1}\partial_{\bar y}X,X^{-1}B_yX]+\partial_{\bar z}(X^{-1}B_zX)+[X^{-1}\partial_{\bar z}X,X^{-1}B_zX]=0\,. \tag{25}$$

Substituting eqs. (22) and (24) into these two equations we have, in the $g_{(-d,d)}$ sectors, that

$$\partial_y(Y^{-1}\partial_{\bar y}Y)+\partial_z(Y^{-1}\partial_{\bar z}Y)-[Y^{-1}\mu Y,\bar\mu]-[Y^{-1}\nu Y,\bar\nu]=0\,, \tag{26}$$
$$\partial_{\bar y}(\partial_y YY^{-1})+\partial_{\bar z}(\partial_z YY^{-1})+[Y\bar\mu Y^{-1},\mu]+[Y\bar\nu Y^{-1},\nu]=0\,. \tag{27}$$

These two equations are nothing but the 4D analogues of the equations of motion of the nonabelian Toda field theories obtained from the hamiltonian reductions of the 2D WZNW model. We thus see that not only the complete structure of the 4D SDYM theory is very similar to that of the WZNW model, their reduced cases are also very similar. These similarities may shed some light on the understanding of the 4D integrabilities.

5. Gauged approach

In the previous section we realized the equations of motions for the reduced 4D theories by directly solving the constraints (18) imposed on the SDYM equations (1) and (2). In this section, we shall establish a lagrangian realization of the constraints, i.e. imposing the constraints on some gauged effective action of SDYM fields.

The idea comes from the 4D analogous identity (6) of the Polyakov–Wiegmann formula. By repeated use of this formula, it is easy to write down the equation

Volume 298, number 1,2 PHYSICS LETTERS B 7 January 1993

$$I_{\rm SDYM}[HJK]=I_{\rm SDYM}[H]+I_{\rm SDYM}[J]+I_{\rm SDYM}[K]$$
$$+\kappa\int d^2y\,d^2z\langle(H^{-1}\partial_{\bar y}H)(\partial_y JJ^{-1})$$
$$+(J^{-1}\partial_{\bar y}J)(\partial_y KK^{-1})+(H^{-1}\partial_{\bar z}H)(\partial_z JJ^{-1})$$
$$+(J^{-1}\partial_{\bar z}J)(\partial_z KK^{-1})$$
$$+(H^{-1}\partial_{\bar y}H)J(\partial_y KK^{-1})J^{-1}$$
$$+(H^{-1}\partial_{\bar z}H)J(\partial_z KK^{-1})J^{-1}\rangle\,, \qquad (28)$$

where H, J, K are all elements of the group G.

Now let us consider the action

$$(I_{\rm SDYM})^{\rm gauged}[HJK]=I_{\rm SDYM}[HJK]$$
$$-\kappa\int d^2y\,d^2z\langle H^{-1}\partial_{\bar y}H\mu+H^{-1}\partial_{\bar z}H\nu$$
$$+\partial_y KK^{-1}\bar\mu+\partial_z KK^{-1}\bar\nu\rangle\,, \qquad (29)$$

where H, K are chosen to belong to the subgroup generated by the subalgebras g_{-d} and g_d respectively, and the constant elements μ, ν etc. are just the regular constraints described in section 3.

The first term of (29) is explicitly invariant under the gauge transformations

$$H\to H\alpha^{-1},\quad J\to\alpha J\beta^{-1},\quad K\to\beta K\,,$$
$$\alpha\in\exp(g_{-d}),\quad \beta\in\exp(g_d)\,. \qquad (30)$$

With a little algebra we can see that the remaining terms of (29) are also invariant under the above transformation. Therefore eq. (29) gives a gauged SDYM action with gauge groups $\exp(g_{-d})_{\rm L}\otimes\exp(g_d)_{\rm R}$.

Using eq. (28) and noticing that the fields H, K are chosen to lie in the nilpotent subgroups $\exp(g_{-d})$ and $\exp(g_d)$ respectively, we can rewrite the action (29) as follows:

$$(I_{\rm SDYM})^{\rm gauged}[J,A_y,A_z,A_{\bar y},A_{\bar z}]$$
$$=I_{\rm SDYM}[J]+\kappa\int d^2y\,d^2z\langle A_{\bar y}\partial_y JJ^{-1}+A_{\bar z}\partial_z JJ^{-1}$$
$$+A_yJ^{-1}\partial_{\bar y}J+A_zJ^{-1}\partial_{\bar z}J+A_{\bar y}JA_yJ^{-1}$$
$$+A_{\bar z}JA_zJ^{-1}-A_{\bar y}\mu-A_{\bar z}\nu-A_y\bar\mu-A_z\bar\nu\rangle\,, \qquad (31)$$

where

$$A_y=\partial_y KK^{-1},\quad A_z=\partial_z KK^{-1}\,,$$
$$A_{\bar y}=H^{-1}\partial_{\bar y}H,\quad A_{\bar z}=H^{-1}\partial_{\bar z}H\,, \qquad (32)$$

and the gauge transformations (30) can be rewritten as

$$J\to\alpha J,\quad J\to J\beta^{-1}\,,$$
$$A_{\bar y}\to\alpha A_{\bar y}\alpha^{-1}+\alpha\partial_{\bar y}\alpha^{-1},\quad A_y\to\beta A_y\beta^{-1}+\partial_y\beta\beta^{-1}\,,$$
$$A_{\bar z}\to\alpha A_{\bar z}\alpha^{-1}+\alpha\partial_{\bar z}\alpha^{-1},\quad A_z\to\beta A_z\beta^{-1}+\partial_z\beta\beta^{-1}\,. \qquad (33)$$

Following the standard procedure, we have the following Euler–Lagrange equations for the action (31):

$$\partial_{\bar y}(\partial_y JJ^{-1}+JA_yJ^{-1})+[A_{\bar y},\partial_y JJ^{-1}+JA_yJ^{-1}]$$
$$+\partial_y A_{\bar y}+\partial_{\bar z}(\partial_z JJ^{-1}+JA_zJ^{-1})$$
$$+[A_{\bar z},\partial_z JJ^{-1}+JA_zJ^{-1}]+\partial_z A_{\bar z}$$
$$=0\,, \qquad (34a)$$

$$\partial_y(J^{-1}\partial_{\bar y}J+J^{-1}A_{\bar y}J)+[J^{-1}\partial_{\bar y}J+J^{-1}A_{\bar y}J,A_y]$$
$$+\partial_{\bar y}A_y+\partial_z(J^{-1}\partial_{\bar z}J+J^{-1}A_{\bar z}J)$$
$$+[J^{-1}\partial_{\bar z}J+J^{-1}A_{\bar z}J,A_z]+\partial_{\bar z}A_z$$
$$=0\,, \qquad (34b)$$

$$\langle g_{-d},\partial_y JJ^{-1}+JA_yJ^{-1}-\mu\rangle=0\,, \qquad (34c)$$
$$\langle g_{-d},\partial_z JJ^{-1}+JA_zJ^{-1}-\nu\rangle=0\,, \qquad (34d)$$
$$\langle g_d,J^{-1}\partial_{\bar y}J+J^{-1}A_{\bar y}J-\bar\mu\rangle=0\,, \qquad (34e)$$
$$\langle g_d,J^{-1}\partial_{\bar z}J+J^{-1}A_{\bar z}J-\bar\nu\rangle=0\,. \qquad (34f)$$

Now we can again write the field J as in eq. (21). Then by considering the gauge symmetry (33) of the action (32), we can fix the gauge by setting

$$J=Y\,. \qquad (35)$$

From this unique condition and eqs. (34c)–(34f), we have

$$A_y=Y^{-1}\mu Y,\quad A_z=Y^{-1}\nu Y\,, \qquad (36a)$$
$$A_{\bar y}=Y\bar\mu Y^{-1},\quad A_{\bar z}=Y\bar\nu Y^{-1}\,. \qquad (36b)$$

Substituting eqs. (35) and (36) into (34a), (34b), the equations of motion (26) and (27) for the reduced theory follow.

At the end of this section, let us mention that the effective action for the reduced theory (26) and (27) can also be obtained from the gauge fixing condition (35). Substituting (35) and (36) into the action (31), it follows that

Volume 298, number 1,2 PHYSICS LETTERS B 7 January 1993

$$I_{\text{eff}}[Y]=I_{\text{SDYM}}[Y]$$
$$-\kappa\int d^2y\, d^2z\langle Y^{-1}\mu Y\bar{\mu}+Y^{-1}\nu Y\bar{\nu}\rangle\,. \tag{37}$$

6. Linear systems

This section is devoted to the construction of the linear systems for the reduced systems (26) and (27). Let us recall that the same objects for the unreduced SDYM fields are given by eqs. (11) and (12). Furthermore, we use again the form of J as in eq. (21). Now define that

$$T=Z\chi,\quad \bar{T}=X^{-1}\bar{\chi}\,. \tag{38}$$

By direct calculations, we can show, for example, that

$$\begin{aligned}\partial_z T&=\partial_z(Z\chi)=(\partial_z ZZ^{-1})(Z\chi)+Z\partial_z\chi\\&=(\partial_z ZZ^{-1})(Z\chi)+\lambda Z(\partial_{\bar{y}}+J^{-1}\partial_{\bar{y}}J)\chi\\&=(\partial_z ZZ^{-1})(Z\chi)+\lambda[\partial_{\bar{y}}(Z\chi)\\&\quad-(\partial_{\bar{y}}ZZ^{-1})(Z\chi)+Z(J^{-1}\partial_{\bar{y}}J)Z^{-1}(Z\chi)]\\&=(Y^{-1}\nu Y)(Z\chi)+\lambda(\partial_{\bar{y}}+Y^{-1}\partial_{\bar{y}}Y+\bar{\mu})(Z\chi)\\&=[Y^{-1}\nu Y+\lambda(\partial_{\bar{y}}+Y^{-1}\partial_{\bar{y}}Y+\bar{\mu})]T\,,\end{aligned} \tag{38'}$$

where we have used eqs. (11) and (21). Following the similar procedure we have, respectively, the following linear systems for the system (26) and (27):

$$\begin{aligned}\partial_z T&=[Y^{-1}\nu Y+\lambda(\partial_{\bar{y}}+Y^{-1}\partial_{\bar{y}}Y+\bar{\mu})]T\,,\\\partial_y T&=[Y^{-1}\mu Y-\lambda(\partial_{\bar{z}}+Y^{-1}\partial_{\bar{z}}Y+\bar{\nu})]T\,,\end{aligned} \tag{39}$$

$$\begin{aligned}\partial_{\bar{z}}\bar{T}&=-\left(Y\bar{\nu}Y^{-1}+\frac{1}{\lambda}(\partial_y-\partial_y YY^{-1}-\mu)\right)\bar{T}\,,\\\partial_{\bar{y}}\bar{T}&=-\left(Y\bar{\mu}Y^{-1}-\frac{1}{\lambda}(\partial_z-\partial_z YY^{-1}-\nu)\right)\bar{T}\,.\end{aligned} \tag{40}$$

Here we can see one different point of the 4D theories from the 2D ones; there is always a spectral parameter in the linear systems of 4D reduced theories, while in the 2D cases, the linear systems of the nonabelian Toda systems contain no spectral parameter. This difference might be one of the main characteristics which indicates the different meanings of 2D and 4D integrabilities.

7. An example: the case of principal gradation with $d=1$

So far we have been treating the problem of hamiltonian reductions of the SDYM theory associated with left–right dual regular constant constraints of arbitrary degree d in an arbitrary integral gradation of the underlying Lie algebra g. In order that the readers have a more intuitive understanding of the above abstract constructions, we now consider a special case in which the integral gradation is chosen as the principal one and the degree of the constraints is fixed to be equal to 1. It should be remarked here that in this concrete case, the regular elements do exist for all classical and exceptional Lie algebras; they are simply given by the linear combinations of the simple roots,

$$\mu=\sum_{\alpha\ \text{simple}}\mu^\alpha E_\alpha,\quad \nu=\sum_{\alpha\ \text{simple}}\nu^\alpha E_\alpha\,. \tag{41}$$

Their images under the Cartan involution of g are given by

$$\bar{\mu}=\sum_{\alpha\ \text{simple}}\mu^\alpha F_\alpha,\quad \bar{\nu}=\sum_{\alpha\ \text{simple}}\nu^\alpha F_\alpha\,. \tag{42}$$

Notice that in the present case, the decomposition (15) of g is nothing but the Cartan decomposition

$$\mathrm{g}=\mathrm{g}_-\oplus\mathrm{h}\oplus\mathrm{g}_+\,, \tag{43}$$

and correspondingly the decomposition (21) of J is precisely the Gauss decomposition. Thus after solving the constraints, we arrive at an effective theory with only the Cartan degrees of freedom. Eqs. (26) and (27) now read

$$\begin{aligned}&(\partial_y\partial_{\bar{y}}+\partial_z\partial_{\bar{z}})\Phi\\&\quad+\sum_{\alpha\ \text{simple}}[(\mu^\alpha)^2+(\nu^\alpha)^2]\exp(\alpha\cdot\Phi)H_\alpha\\&\quad=0\,,\end{aligned} \tag{44}$$

where

$$Y=\exp(-\Phi)\equiv\exp\left(-\sum\Phi^\alpha H_\alpha\right)\,. \tag{45}$$

We see that eq. (44) is just the 4D Toda equation. The effective action and linear systems for the 4D Toda system can all be obtained following the same

Volume 298, number 1,2 PHYSICS LETTERS B 7 January 1993

line as shown in section 5 and section 6. The effective action reads

$$I_{\text{Toda}}[\Phi]=\kappa\int d^2y\,d^2z \sum_{\alpha\ \text{simple}} \{\partial_y\Phi^\alpha\partial_{\bar y}\Phi^\alpha + \partial_z\Phi^\alpha\partial_{\bar z}\Phi^\alpha-[(\mu^\alpha)^2+(\nu^\alpha)^2]\exp(\alpha\cdot\Phi)\}\,, \tag{46}$$

and the linear systems are

$$\partial_z T=[\exp(\text{ad}\,\Phi)\nu+\lambda(\partial_{\bar y}-\partial_{\bar y}\Phi+\bar\mu)]T\,,$$
$$\partial_y T=[\exp(\text{ad}\,\Phi)\mu-\lambda(\partial_{\bar z}-\partial_{\bar z}\Phi+\bar\nu)]T\,, \tag{47}$$

$$\partial_{\bar z}\bar T=-\left(\exp(-\text{ad}\,\Phi)\bar\nu+\frac{1}{\lambda}(\partial_y+\partial_y\Phi-\mu)\right)\bar T\,,$$
$$\partial_{\bar y}\bar T=-\left(\exp(-\text{ad}\,\Phi)\bar\mu-\frac{1}{\lambda}(\partial_z+\partial_z\Phi-\nu)\right)\bar T\,. \tag{48}$$

In this paper we discussed the hamiltonian reductions of SDYM theory associated with left–right dual regular constant constraints. It is shown that the reduced theory is equivalent to a 4D generalization of the nonabelian Toda theory obtained previously as the reductions of WZNW theory. As is stressed for many times in the context, the constraints considered here are relatively only some special choices. It is hoped that all of the rich structures of the reductions of the WZNW model can find their correspondances here in the reductions of SDYM theory, for example, the nondual constraints (which lead to the KdV systems in the 2D cases) [6], the space–time dependent constraints [8], and the cases in which there is no grading operator H (in two dimensions such cases correspond to the conformal breaking reductions).

It is interesting to notice that the reductions of SDYM theory considered in the present work is Virasoro preserving as in the 2D cases. But now since the Virasoro symmetry of SDYM theory only corresponds to some special kinds of nonlocal transformations, it seems necessary to consider more about the effect of the Virasoro preserving nature on the integrabilities of the reduced theory. Furthermore, it is in question that whether there is a W algebra survived as the reduction of the KM symmetry as in 2D cases. If such algebras exist, what effect would they have toward the 4D integrabilities? We hope to answer these questions in some other works.

At the end of this paper let us mention that there have already been numerous works considering various reductions of the SDYM theory [14,15]. But we stress that to our knowledge, most of the literature considers only the dimensional reductions which reduces the space–time dimensions and lead to 2D reduced theories. But now we consider the hamiltonian reductions and there are no changes in the structure of space–time throughout the reduction. So both kinds of reductions cannot be confused.

8. Final remarks

We are working on the nonstandard self-dual Yang–Mills theory in this article. By "nonstandard" we mean the fact that in the usual case, the J field of the SDYM theory takes values on a symmetric space such as SL(N, ℂ)SU(N), while in this article, we extended the range of values of J onto the group G as was done in ref. [12]. The relation between the nonstandard SDYM theory and the standard one is very similar to that between the nonlinear sigma model and the principal model in two dimensions, where the lagrangian of both systems can be written in the same form but with the values of the field taken on different spaces.

Although we work only on the nonstandard SDYM theory in the context, the similar reduction procedure works even for the standard theories, at least in the case of principal gradation, where the role of the Gauss decomposition (21) is played by the so-called R-gauge condition (cf. refs. [9,10]). The main difference between the nonstandard and the standard SDYM theory is that there might not be a complete Kac–Moody structure in the phase space of the latter, which makes the reduction problem somewhat more complicated.

References

[1] J. Balog, L. Feher, L. O'Raifeartaigh, P. Forgacs and A. Wipf, Phys. Lett. B 227 (1989) 214; Ann. Phys. 203 (1991) 76; L. O'Raifeartaigh and A. Wipf, preprint DIAS-STP-90-19; L. O'Raifeartaigh, preprints DIAS-STP-90-43, DIAS-STP-90-45;

Volume 298, number 1,2 PHYSICS LETTERS B 7 January 1993

L. O'Raifeartaigh, P. Ruelle, I. Tsutsui and A. Wipf, prepint DIAS-STP-91-03;
L. Feher, L. O'Raifeartaigh, P. Ruelle, I. Tsutsui and A. Wipf, preprints DIAS-STP-91-17, DIAS-STP-91-29;
L. Feher, preprint DIAS-STP-91-22.

[2] H. Aratyn, L.A. Ferreira, J.F. Gomes and A.H. Zimerman, preprint IFT/P-24/90.

[3] T. Inami, prepint KUNS 1038, HE(TH)90/14.

[4] J. Fuchs, Phys. Lett. B 262 (1991) 249.

[5] F.A. Bais, T. Tjin and P. Van Driel, Nucl. Phys. B 357 (1991) 632.

[6] M. Bershadsky and O. Ooguri, CMP (1990);
M. Bershadsky, preprint IASSNS-HEP-90/44 (CMP 1991).

[7] B.-Y. Hou and L. Chao, preprints NWU/IMP/911115, 911117, 920327.

[8] B.-Y. Hou and L. Chao, NWU/IMP preprint;
B.-Y. Hou and H.-X. Yang, in preparation.

[9] C.N. Yang, Phys. Rev. Lett. 38 (1977) 1377;
S. Ward, Phys. Lett. A 61 (1977) 81.

[10] B.-Y. Hou and X.-C. Song, Stony brook preprint (1983), unpublished;
V.P. Nair and J. Schiff, Phys. Lett. B 246 (1990) 423.

[11] M.K. Prasad, A. Sinka and L.-L. Chau Wang, Phys. Rev. 23 (1981) 2321.

[12] L.-L. Chau and I. Yamanaka, preprint UCDPHYS-PUB-91-6.

[13] L.L. Chau, in: Integrable systems, Nankai lectures on Mathematical physics, ed. X.-C. Song (World Scientific, Singapore, 1987).

[14] I.J. Mason and G.A.J. Sparling, Phys. Lett. A 137 (1989) 29;
R.S. Ward, Phil. Trans. R. Soc. London A 315 (1985) 451.

[15] I. Bakas, Mod. Phys. Lett. A 6 (1991) 1561, preprint UMD-PP91-211.

NUCLEAR PHYSICS B

ELSEVIER Nuclear Physics B 436 (1995) 638–658

Heterotic Toda fields

Liu Chao, Bo-Yu Hou

CCAST (World Laboratory), P.O.Box 8730, Beijing 100080, China
and Institute of Modern Physics, Northwest Univ., Xian 710069, China

Received 21 December 1993; revised 5 July 1994; accepted 17 November 1994

Abstract

A non-left–right symmetric conformal integrable Toda field theory is constructed. It is found that the conformal algebra for this model is the product of a left chiral W_{r+1} algebra and a right chiral $W_{r+1}^{(2)}$ algebra. The general classical solution is constructed out of the chiral vectors satisfying the so-called classical exchange algebra. In addition, we derived an explicit Wronskian type solution in relation to the constrained WZNW theory. We also showed that the A_∞ limit of this model is precisely the (B_2, C_1) flow of the standard Toda lattice hierarchy.

1. Introduction

Of all the conformal invariant integrable models Toda field theories are the most interesting and extensively studied ones. Within the framework of Toda field theories, one finds the conformal Toda (CT), loop Toda (or affine Toda, denoted as LT for short), conformal affine Toda (CAT) [1] and their various extensions, especially the "2-extensions" [2,3,13] studied by the authors some time earlier. One of the major reasons why conformal invariant Toda fields are so attractive is their nice property of yielding the W-algebra symmetries. Nowadays it is becoming a common practice to treat the conformally noninvariant Toda theories as the result of appropriate deformations [4,5] of the corresponding conformal invariant ones. In view of this, the integrability of all Toda type field theories is governed by the conformal (W-algebra) symmetries of the undeformed Toda fields.

In the study of conformal invariant field theories, people are used to treat only one half of the complete model, namely one of the two chiral sectors. This is because of the left-right symmetry of the model under consideration. The left–right symmetry is in fact some kind of "parity" (P) or "charge-parity" (CP) invariance, which causes an indistinguishability between both chiral sectors. To be specific, let us consider the CT

SSDI 0550-3213(94)00530-3

and the 2-extended (2-E) CT models. The equations of motion for these two models can be written respectively as

$$\partial_+\partial_-\varphi^i - \exp\left(-\sum_j K_{ij}\varphi^j\right) = 0\,, \qquad i,j = 1,2,\ldots,r$$

and

$$\partial_+\partial_-\varphi^j - \sum_{i,j} \mathrm{sign}(i-j)\mathrm{sign}(k-j)\psi_+^i K_{ij}\psi_-^k K_{kj}\omega^j + \sum_{i,(i\neq j)} \omega^i\omega^j K_{ij} = 0\,,$$

$$\partial_+\psi_-^j = \sum_i \mathrm{sign}(i-j)\psi_+^i K_{ij}\omega^j\,,$$

$$\partial_-\psi_+^j = \sum_i \mathrm{sign}(i-j)\psi_-^i K_{ij}\omega^j\,,$$

$$\omega^i \equiv \exp\left(\sum_j K_{ij}\varphi^j\right) \qquad (i,j,k = 1,2,\ldots,r)\,,$$

where K_{ij} are entries of the Cartan matrix of the corresponding rank-r finite-dimensional Lie algebra (now chosen as simply laced), and the signature function $\mathrm{sign}(i-j)$ is defined such that it takes the value "zero" at equal arguments. One sees that the left–right symmetry of CT is actually the P-invariance $x_+ \leftrightarrow x_-$, but for the 2-ECT case, it is represented by the following "CP invariance":

$$P : x_+ \leftrightarrow x_-\,, \qquad C : \psi_+^i \leftrightarrow \psi_-^i\,. \tag{1}$$

It is not difficult to check that the fields $\psi_\pm^i$ have the conformal weights $(1/2,0)$ and $(0,1/2)$, respectively, therefore the "charge conjugation" in Eq. (1) is to be understood as the conjugation of conformal charges instead of the normal (electronic) charges.

Despite the elegant properties of the left–right symmetric models like CT and 2-ECT mentioned above, there exists something in nature which is *not* left–right symmetric (such as neutrinos). Therefore, exploiting a conformal invariant model having no left–right symmetry might be interesting. Such theories already exist in superstring theories, i.e. the heterotic string theory [6]. But now we concern ourselves with a Toda type integrable theory which is also "heterotic" (which we call heterotic Toda field theory, denoted HTFT for short). We shall construct such a model by defining explicitly its Lax pair representation, discuss its conformal properties (which is represented by the product of a left chiral W_{r+1} algebra and a right chiral $W_{r+1}^{(2)}$ algebra—a mixture of CT and 2-ECT theories), and consider the chiral exchange algebra, classical solution and the relations to WZNW and Toda lattice hierarchies [8,5] as well.

2. The heterotic Toda model

Let us start by constructing explicitly the HTFT mentioned above. Stressing its integrability, we begin with the Lax pair representation of the model. The Lax pair is defined in the cylindrical space-time as follows:

$$\partial_+ T = \left[\frac{1}{2}\partial_+\Phi + \exp\left(-\frac{1}{2}\mathrm{ad}\,\Phi\right)(\bar{\Psi}_+ + \mu)\right] T,$$
$$\partial_- T = -\left[\frac{1}{2}\partial_-\Phi + \exp\left(\frac{1}{2}\mathrm{ad}\,\Phi\right)\nu\right] T, \qquad (2)$$

where $x_\pm \equiv t \pm x$, $\partial_\pm \equiv \partial_{x_\pm}$, and

$$\mu = \frac{1}{2}\sum_{i,j=1}^{r}\mathrm{sign}(i-j)[E_i, E_j], \qquad \nu = \sum_{i=1}^{r} F_i,$$
$$\Phi = \sum_{i=1}^{r}\varphi^i H_i, \qquad \Psi_+ = \sum_{i=1}^{r}\psi_+^i F_i, \qquad \bar{\Psi}_+ = [\mu, \Psi_+].$$

In the above definitions, $\{H_i, E_i, F_i\}$ denote the Chevalley generators of the rank-r finite-dimensional Lie algebra g (which will be restricted to be the classical A_r algebra for simplicity), and all the component fields, φ^i, ψ_+^i, are assumed to be periodic in the spatial coordinate, x.

The equations of motion for the HTFT follow from the compatibility conditions of the Lax pair (2). They turn out to be

$$\partial_+\partial_-\Phi + [\nu, \exp(-\mathrm{ad}\,\Phi)\bar{\Psi}_+] = 0,$$
$$\partial_-\Psi_+ - \exp(\mathrm{ad}\,\Phi)\nu = 0. \qquad (3)$$

or, in component form,

$$\partial_+\partial_-\varphi^i - \sum_j \mathrm{sign}(j-i)K_{ji}\psi_+^j\omega^i = 0,$$
$$\partial_-\psi_+^i - \omega^i = 0, \quad \omega^i = \exp(-\sum_j K_{ji}\varphi^j). \qquad (4)$$

One may wonder how we can imagine such a complicated model by direct construction. Indeed, the model (3), (4) appears not to be very simple at first glance, but from the point of view of hamiltonian reductions of WZNW theories [9]—which is now a quite common way of constructing extended Toda models—the origin of the model (4) can be made very clear: it is nothing but the constrained WZNW theory under the following constraints:

$$\langle \mathbf{g}_-, \partial_+ g g^{-1} - \mu\rangle = 0, \qquad \langle \mathbf{g}_+, g^{-1}\partial_- g - \nu\rangle = 0, \qquad (5)$$

where $\mathbf{g}_\pm$ are the maximal nilpotent subalgebras consisted of positive and negative step operators, respectively, and $\langle\ ,\ \rangle$ is the standard Killing form. Notice that the above constraint equations are not of the left–right dual style, this is why the present model is heterotic. Actually, if one performs the CP transform (1) on the model (4), one would arrive at a *different* but *dual* model, with the roles of the left and right chiral sectors interchanged.

There are two ways to identify the conformal invariance of the HTFT. The first way is to study the conformal-preserving nature of the WZNW $\to$ HTFT reduction. In this way one can show that the conformal algebra of the theory (4) is nothing but the product of a left chiral (the x_--depending sector) W_{r+1} algebra and a right chiral $W^{(2)}_{r+1}$ algebra. The second way of exploiting the conformal invariance of the model (4) is to show explicitly the behaviors of the equations of motion under conformal change of spacetime variables. One can easily check that under the conformal transformations of the coordinates

$$x_+ \to \tilde{x}_+ = f(x_+), \qquad x_- \to \tilde{x}_- = h(x_-),$$

the equations of motion (4) is left invariant provided the fields φ^i, ψ^i_+ transform as follows:

$$\varphi^i \to \tilde{\varphi}^i = \varphi^i + \sum_j \left(K^{-1}\right)^{ji} \ln(f')^{1/2}h' \Rightarrow \omega^i \to \tilde{\omega}^i = (f')^{-1/2}(h')^{-1}\omega^i,$$

$$\psi^i_+ \to \tilde{\psi}^i_+ = (f')^{-1/2}\psi^i_+.$$

This shows that the fields ψ^i_+ and ω^i are primary fields of conformal weights $(0, 1/2)$ and $(1, 1/2)$, respectively. One may therefore expect that the conformal spectrum of the left chiral sector consists of only integers, and that of the right chiral sector may consist of integers and half-integers. This observation is in agreement with the above-mentioned $(W_{r+1})_L \otimes (W^{(2)}_{r+1})_R$ symmetry.

Now let us write down the effective action for the model (4). It reads

$$\begin{aligned} I[\Phi, \Psi_+] &= \frac{1}{2}\int \mathrm{d}^2x \langle \partial_+\Phi\partial_-\Phi + \bar{\Psi}_+\partial_-\Psi_+ - 2\left(\exp(-\mathrm{ad}\,\Phi)\bar{\Psi}_+\right)\nu\rangle \\ &= \frac{1}{2}\int \mathrm{d}^2x \sum_{ij}\Big[\partial_+\varphi^i K_{ij}\partial_-\varphi^j + \mathrm{sign}(i-j)\psi^i_+ K_{ij}\partial_-\psi^j_+ \\ &\quad -2\,\mathrm{sign}(i-j)\psi^i_+ K_{ij}\exp(-\sum \varphi^l K_{lj})\Big]. \end{aligned} \tag{6}$$

We see that the form of the kinematic terms in Eq. (6) are very similar to that of the heterotic string theory [6]. The difference lies in the fact that in the heterotic string theory the ψ fields are fermionic and the full theory has a heterotic supersymmetry, whilst in the present case, ψ^i_+ are bosonic fields and therefore the theory has no supersymmetry (the curious similarity between bosonic conformal algebras having the integer half-integer conformal spectrum and the real superconformal algebras is still an open area for further study, at least in the authors' point of view).

Given the effective action (6), we are now ready to define the conjugate momenta and the canonical Poisson brackets in the usual way,

$$\pi(\varphi_i) = \sum_j K_{ij}\partial_0\varphi^j, \qquad \pi(\psi^i_+) = \frac{1}{2}\sum_j \mathrm{sign}(j-i)\psi^j_+ K_{ji},$$

$$\{\pi(\varphi^i)(x), \varphi^j(y)\} = \delta^{ij}\delta(x-y), \qquad \{\pi(\psi^i_+)(x), \psi^j_+(y)\} = \delta^{ij}\delta(x-y),$$

where of course the δ-functions are also assumed to be periodic due to the periodicity of the fundamental fields. However, since the fields ψ_+^i are of the first order in derivatives in the action (6), one should treat the definitions of the canonical momenta of these fields as primary constraints and replace the corresponding naive Poisson brackets by Dirac Poisson brackets. It follows that provided the rank r of the underlying Lie algebra A_r is even, all the above constraints are of second class and there are no further constraints in the theory. The final Dirac brackets for the ψ-fields read (here we denote the Dirac brackets again by $\{\ ,\ \}$)

$$\{\psi_+^i(x),\ \psi_+^j(y)\} = (M^{-1})_{ji}\delta(x-y),$$

where $M_{ij} \equiv \mathrm{sign}(i-j)K_{ij}$, which is invertible provided r is even [1]. From the above Poisson brackets we can calculate the fundamental Poisson relation (FPR) for the spatial component of the Lax connection. Then, upon integration with the initial value $T(0) = 1$, the FPR for the ultralocal transport matrix T follows [7]

$$\{T(x) \otimes_, T(x)\} = [r_\pm,\ T(x) \otimes T(x)], \tag{7}$$

where $r_\pm$ are the so-called classical r-matrices which are well known in the standard CT theory,

$$r_+ = \frac{1}{2}\left\{\sum (K^{-1})^{ij} H_i \otimes H_j + 2\sum_{\alpha>0} E_\alpha \otimes F_\alpha\right\},$$

$$r_- = -\frac{1}{2}\left\{\sum (K^{-1})^{ij} H_i \otimes H_j + 2\sum_{\alpha>0} F_\alpha \otimes E_\alpha\right\}. \tag{8}$$

We mention that the FPR (7) together with the r-matrices (8) are the characterizing properties for the integrabilities of all the Toda type field theories. They are also the starting points for studying the exchange algebras [7,3] and dressing symmetries [10,3] for such theories. But here shall not consider these issues in detail. Instead, we briefly discuss the origin of the chiral exchange algebra in HTFT with a specific emphasis on its relations to the classical solutions and the W-algebra symmetries of the model. These discussions are part of the content of the next section.

3. Exchange algebra and Leznov–Saveliev analysis

In this section we shall discuss some aspects connected with the chiral vectors in the model. First let us show how there are chiral vectors embedded in the present theory.

3.1. Existence of chiral vectors

As is well known, the Lax pair representation for integrable systems admits a gauge freedom. In other words, the compatibility conditions for a Lax pair are left invariant

[1] In the following context, as far as Poisson brackets are concerned, we shall assume that r is even. But the other results such as the Wronskian type solutions, etc., hold true without this restriction.

while the transport matrix T is shifted from the left by a gauge group element. Therefore, one can choose different gauges for the Lax pair to obtain an optimized form for the current usage. In the present case, we can choose the following convenient gauges:

$$\partial_+ T_L = \left\{\partial_+\Phi + \bar{\Psi}_+ + \mu\right\} T_L,$$
$$\partial_- T_L = -\left\{\exp(\mathrm{ad}\,\Phi)\nu\right\} T_L; \tag{9}$$

and

$$\partial_+ T_R = \left\{\exp(-\mathrm{ad}\,\Phi)(\bar{\Psi}_+ + \mu)\right\} T_R,$$
$$\partial_- T_R = -\left\{\partial_-\Phi + \nu\right\} T_R, \tag{10}$$

where

$$T_L = \exp\left(\frac{1}{2}\Phi\right) T, \qquad T_R = \exp\left(-\frac{1}{2}\Phi\right) T. \tag{11}$$

Assuming that $|\lambda^i\rangle$ and $\langle\lambda^i|$ are the highest weight and dual highest weight vectors in ith fundamental representation of g, respectively, it follows from Eqs. (9) and (10) that the vectors

$$\xi^{(i)}(x_+) \equiv \langle\lambda^i|T_L, \qquad \bar{\xi}^{(i)}(x_-) \equiv T_R^{-1}|\lambda^i\rangle \tag{12}$$

are chiral,

$$\partial_- \xi^{(i)}(x_+) = 0, \qquad \partial_+ \bar{\xi}^{(i)}(x_-) = 0.$$

Moreover, performing another gauge transformation $T_L \longrightarrow \tilde{T}_L = \exp(\Psi_+)T_L$, we can show that the vectors

$$\zeta^{(i)}(x_+) \equiv \langle\lambda^i - \alpha^i|\tilde{T}_L = \langle\lambda - \alpha^i|\exp(\Psi_+)T_L \tag{13}$$

are also chiral (α^i being the ith simple root)

$$\partial_- \zeta^{(i)}(x_+) = 0.$$

These chiral vectors play a central role in the rest of this article.

3.2. *Exchange algebra for the chiral vectors*

The chiral vectors obtained in the last subsection obey a very nice exchange algebra. The method for obtaining such exchange algebras is now quite familiar in the CT and 2-ECT theories. In the present model, one can first calculate the Poisson brackets between $\exp(\Phi)$, $\exp(\Psi_+)$ and T, then, using the definitions of the chiral vectors (12) and (13) and by straightforward calculations, one obtains the following exchange relations (throughout this article, all the chiral quantities are assumed to be evaluated at equal time $t = t_0$):

$$\{\xi^{(i)}(x)\otimes_, \xi^{(j)}(y)\}=\xi^{(i)}(x)\otimes\xi^{(j)}(y)\left(r_+\theta(x-y)+r_-\theta(y-x)\right), \tag{14}$$

$$\{\xi^{(i)}(x)\otimes_, \bar{\xi}^{(j)}(y)\}=-\left(\xi^{(i)}(x)\otimes 1\right)r_-\left(1\otimes\bar{\xi}^{(j)}(y)\right), \tag{15}$$

$$\{\bar{\xi}^{(i)}(x)\otimes_, \xi^{(j)}(y)\}=-\left(1\otimes\xi^{(j)}(y)\right)r_+\left(\bar{\xi}^{(i)}(x)\otimes 1\right), \tag{16}$$

$$\{\bar{\xi}^{(i)}(x)\otimes_, \bar{\xi}^{(j)}(y)\}=\left(r_-\theta(x-y)+r_+\theta(y-x)\right)\bar{\xi}^{(i)}(x)\otimes\bar{\xi}^{(j)}(y), \tag{17}$$

$$\begin{aligned}\{\zeta^{(i)}(x)\otimes_, \zeta^{(j)}(y)\}=&\zeta^{(i)}(x)\otimes\zeta^{(j)}(y)\left(r_+\theta(x-y)+r_-\theta(y-x)\right)\\&+(M^{-1})_{ji}\delta(x-y)\xi^{(i)}(x)\otimes\xi^{(j)}(y),\end{aligned} \tag{18}$$

$$\{\xi^{(i)}(x)\otimes_, \zeta^{(j)}(y)\}=\xi^{(i)}(x)\otimes\zeta^{(j)}(y)\left(r_+\theta(x-y)+r_-\theta(y-x)\right), \tag{19}$$

$$\{\zeta^{(i)}(x)\otimes_, \xi^{(j)}(y)\}=\zeta^{(i)}(x)\otimes\xi^{(j)}(y)\left(r_+\theta(x-y)+r_-\theta(y-x)\right), \tag{20}$$

$$\{\zeta^{(i)}(x)\otimes_, \bar{\xi}^{(j)}(y)\}=-\left(\zeta^{(i)}(x)\otimes 1\right)r_-\left(1\otimes\bar{\xi}^{(j)}(y)\right), \tag{21}$$

$$\{\bar{\xi}^{(i)}(x)\otimes_, \zeta^{(j)}(y)\}=-\left(1\otimes\zeta^{(j)}(y)\right)r_+\left(\bar{\xi}^{(i)}(x)\otimes 1\right). \tag{22}$$

Recalling that the fields Φ, Ψ_+ are periodic, we have the following monodromy properties for the chiral vectors:

$$\begin{aligned}\xi^{(i)}(x+2\pi)&=\xi^{(i)}(x)T,\\ \zeta^{(i)}(x+2\pi)&=\zeta^{(i)}(x)T,\\ \bar{\xi}^{(i)}(x+2\pi)&=T^{-1}\bar{\xi}^{(i)}(x),\end{aligned}$$

where $T\equiv T(2\pi)$. Moreover, there is a set of nontrivial Poisson brackets between the above chiral vectors and the monodromy matrix,

$$\begin{aligned}\{T\otimes_, T\}&=\left[r_\pm, T\otimes T\right],\\ \{\xi^{(i)}(x)\otimes_, T\}&=-\left(\xi^{(i)}(x)\otimes 1\right)r_-,\\ \{\zeta^{(i)}(x)\otimes_, T\}&=-\left(\zeta^{(i)}(x)\otimes 1\right)r_-,\\ \{\bar{\xi}^{(i)}(x)\otimes_, T\}&=(1\otimes T)\,r_+\left(\bar{\xi}^{(i)}(x)\otimes 1\right).\end{aligned}$$

Notice that although the objects $\xi, \zeta, \bar{\xi}$ are chiral, there is a nontrivial coupling among them. This can happen only through the zero modes and the corresponding conjugate variables. Actually, just as in the usual Toda context, the degrees of freedom corresponding to the zero modes are contained in the diagonal part of the monodromy matrix. Therefore in order to choose an appropriate basis in which the chiralities are completely split we have to first diagonalize the monodromy matrix. This in practice is connected to the so-called Drinfeld–Sokolov linear systems, and we shall leave the task for sketching such procedures to Subsection 3.4.

3.3. Leznov–Saveliev analysis

Let us now study the relations between the chiral vectors (12), (13) and the general solution of the model. As will be shown below, the general solution of the model can be represented by the products of these chiral vectors, which can be rewritten in terms of appropriate matrix elements in the fundamental representations of the underlying Lie algebra. Such analysis were first carried out by Leznov and Saveliev in the case of standard Toda theories, which is why the current subsection is titled as above.

From Eqs. (12) and (13) we have

$$\exp(\varphi^i) = \langle\lambda^i|\exp(\Phi)|\lambda^i\rangle = \langle\lambda^i|T_L T_R^{-1}|\lambda^i\rangle = \xi^{(i)}(x_+)\bar{\xi}^{(i)}(x_-), \tag{23}$$

$$\psi_+^i = \frac{\langle\lambda^i - \alpha^i|\exp(\Psi_+)T_L T_R^{-1}|\lambda^i\rangle}{\langle\lambda^i|T_L T_R^{-1}|\lambda^i\rangle} = \frac{\zeta^{(i)}(x_+)\bar{\xi}^{(i)}(x_-)}{\xi^{(i)}(x_+)\bar{\xi}^{(i)}(x_-)}. \tag{24}$$

Now recalling that the matrices T_L and T_R differ from each other only by a diagonal part from the left (see Eq. (11)), we can make the following Gauss decompositions:

$$T_L = \mathrm{e}^{K_+}N_-M_+, \qquad T_R = \mathrm{e}^{K_-}N_+M_-, \tag{25}$$

where $K_\pm$ are respectively the diagonal parts of T_L and T_R, N_+, M_+ and N_-, M_- are upper and lower triangular matrices with entries on the diagonal equal to one. These upper and lower triangular matrices are intrinsically related by the Gauss decompositions of the original transport matrix T,

$$T = \mathrm{e}^{H_+}N_-M_+ = \mathrm{e}^{H_-}N_+M_-,$$

where $H_\pm$ are the diagonal part of T under each Gauss decomposition. Now substituting the Gauss decompositions (25) into the definitions (12) and (13), it follows that

$$\xi^{(i)}(x_+) = \langle\lambda^i|\mathrm{e}^{K_+}M_+, \qquad \bar{\xi}^{(i)}(x_-) = M_-^{-1}\mathrm{e}^{-K_-}|\lambda^i\rangle, \tag{26}$$

$$\zeta^{(i)}(x_+) = \langle\lambda^i - \alpha^i|\mathrm{e}^{K_+}\left(\mathrm{e}^{-\mathrm{ad}\,K_+}\mathrm{e}^{\Psi_+}\right)N_-M_+. \tag{27}$$

Expanding the matrix N_- into the form

$$N_- = \exp(\chi^{(-1)})\exp(\chi^{(-2)})\ldots \tag{28}$$

with $\chi^{(-i)}$ being a lower triangular matrix with nonzero entries only in the ith lower-diagonal, we can rewrite Eq. (27) into the form

$$\zeta^{(i)}(x_+) = \langle\lambda^i - \alpha^i|\mathrm{e}^{K_+}\left\{1 + \mathrm{e}^{-\mathrm{ad}\,K_+}P_+\right\}M_+,$$
$$P_+ \equiv \Psi_+ + \mathrm{e}^{\mathrm{ad}\,K_+}\chi^{(-1)}. \tag{29}$$

The chirality of $\xi^{(i)}$ and $\bar{\xi}^{(i)}$ then implies that the matrices $K_\pm$ and $M_\pm$ in Eq. (26) are chiral,

$$\partial_\pm K_\mp = \partial_\pm M_\mp = 0.$$

Consequently the chirality of $\zeta^{(i)}$ implies the similar property of P_+,

$$\partial_- P_+ = 0.$$

Substituting Eqs. (26) and (29) into Eqs. (23), (24), it turns out that the general solution of the model (4) is completely determined by the chiral quantities $K_\pm, M_\pm$ and P_+,

$$\exp(\varphi^i) = \langle\lambda^i|\mathrm{e}^{K_+} M_+ M_-^{-1} \mathrm{e}^{-K_-}|\lambda^i\rangle,$$
$$\psi^i_+ = \frac{\langle\lambda^i - \alpha^i|\mathrm{e}^{K_+}\left(1+\mathrm{e}^{-\mathrm{ad}\,K_+}P_+\right) M_+ M_-^{-1}\mathrm{e}^{-K_-}|\lambda^i\rangle}{\langle\lambda^i|\mathrm{e}^{K_+} M_+ M_-^{-1}\mathrm{e}^{-K_-}|\lambda^i\rangle}. \tag{30}$$

We have to point out that the chiral quantities $K_\pm, M_\pm$ and P_+ are not all independent: they have to obey some linear partial differential equations. To specify what differential equations are satisfied by these objects, let us recall that the gauge-transformed transport matrices T_L and T_R satisfy the following equations:

$$\partial_+ T_L T_L^{-1} = \partial_+\Phi + \bar{\Psi}_+ + \mu, \qquad \partial_- T_R T_R^{-1} = -(\partial_-\Phi + \nu). \tag{31}$$

Substituting the Gauss decompositions (25) into Eq. (31) and projecting onto the upper- and lower-triangular parts respectively, we have

$$\left[N_-\left(\partial_+ M_+ M_+^{-1}\right) N_-^{-1}\right]_+ = \mathrm{e}^{-\mathrm{ad}\,K_+}(\bar{\Psi}_+ + \mu),$$
$$\left[N_+\left(M_-\partial_- M_-^{-1}\right) N_+^{-1}\right]_- = \mathrm{e}^{-\mathrm{ad}\,K_-}\nu.$$

Considering the fact that μ has nonvanishing entries only on the second upper diagonal, ν has nonvanishing entries only on the first lower diagonal, and also recalling the further decomposition (28) of N_- in terms $\chi^{(-i)}$, we finally get

$$\partial_+ M_+ M_+^{-1} = \mathrm{e}^{-\mathrm{ad}\,K_+}(\bar{P}_+ + \mu), \qquad M_-\partial_- M_-^{-1} = \mathrm{e}^{-\mathrm{ad}\,K_-}\nu, \tag{32}$$

with

$$\bar{P}_+ \equiv [\mu,\ P_+]. \tag{33}$$

Eq.(30) together with (32), (33) constitute the general solution of the model (4).

Remark. The construction we made in this subsection is in some sense a little formal because of the nontrivial couplings between both chiralities. In order to reformulate the general solution of the model in terms of free fields—which decouples from each other—we again need to diagonalize the monodromy matrix. This additional issue also ensures that the general solution obtained in this way is single-valued (i.e. periodic in x) and local (i.e. Poisson commute).

3.4. Sketch for a free-field representation

Let us now give a brief sketch for the free-field representation of the general solution. The construction is based on the following Drinfeld–Sokolov (DS) linear systems:

$$\partial_+ Q_+ = (\partial_+ K_+ + \bar{P}_+ + \mu) Q_+, \qquad \partial_- Q_- = Q_-(\partial_- K_- + \nu),$$

where $\partial_\pm K_\pm$ and $\bar{P}_+$ are the same as in the last subsection and are assumed to be periodic. Since in the above DS systems everything is chiral, we introduce the chiral vectors

$$\sigma^{(i)}(x) = \langle \lambda^i | Q_+(x),$$
$$\bar{\sigma}^{(i)}(x) = Q_-(x)\,|\lambda^i\rangle,$$
$$\varsigma^{(i)}(x) = \langle \lambda^i - \alpha^i | \mathrm{e}^{P_+} Q_+(x),$$

where the DS solutions $Q_\pm$ are normalized as $Q_\pm(0) = 1$.

It is obvious that these chiral vectors have the following monodromy properties:

$$\sigma^{(i)}(x+2\pi) = \sigma^{(i)}(x)S,$$
$$\varsigma^{(i)}(x+2\pi) = \varsigma^{(i)}(x)S,$$
$$\bar{\sigma}^{(i)}(x+2\pi) = \bar{S}\bar{\sigma}^{(i)}(x),$$

where $S \equiv Q_+(2\pi)$ and $\bar{S} \equiv Q_-(2\pi)$, which are respectively upper and lower triangular.

Now introducing the Poisson brackets

$$\{\partial_\pm K_\pm(x) \otimes, \partial_\pm K_\pm(y)\} = \mp(\partial_x - \partial_y)\delta(x-y)\sum_{i,j}(K^{-1})^{ij} H_i \otimes H_j,$$
$$\{\bar{P}_+(x) \otimes, P_+(y)\} = \delta(x-y)\sum_i E_i \otimes F_i,$$

it can be proved that σ, ς and $\bar{\sigma}$ satisfy the following exchange relations:

$$\{\sigma^{(i)}(x) \otimes,\ \sigma^{(j)}(y)\} = \sigma^{(i)}(x) \otimes \sigma^{(j)}(y)\left(r_+\theta(x-y) + r_-\theta(y-x)\right),$$
$$\{\varsigma^{(i)}(x) \otimes,\ \varsigma^{(j)}(y)\} = \varsigma^{(i)}(x) \otimes \varsigma^{(j)}(y)\left(r_+\theta(x-y) + r_-\theta(y-x)\right),$$
$$+(M^{-1})_{ji}\delta(x-y)\sigma^{(i)}(x) \otimes \sigma^{(j)}(y),$$
$$\{\sigma^{(i)}(x) \otimes,\ \varsigma^{(j)}(y)\} = \sigma^{(i)}(x) \otimes \varsigma^{(j)}(y)\left(r_+\theta(x-y) + r_-\theta(y-x)\right),$$
$$\{\bar{\sigma}^{(i)}(x) \otimes,\ \bar{\sigma}^{(j)}(y)\} = \left(r_-\theta(x-y) + r_+\theta(y-x)\right)\bar{\sigma}^{(i)}(x) \otimes \bar{\sigma}^{(j)}(y),$$
$$\{\sigma^{(i)}(x) \otimes,\ \bar{\sigma}^{(j)}(y)\} = \{\varsigma^{(i)}(x) \otimes,\ \bar{\sigma}^{(j)}(y)\} = 0.$$

Moreover, it can be checked that the following expressions are solutions of the equations of motion:

$$\exp(\varphi^i) = \sigma^{(i)} U \bar{\sigma}^{(i)},$$
$$\psi_+^i = \frac{\varsigma^{(i)} U \bar{\sigma}^{(i)}}{\sigma^{(i)} U \bar{\sigma}^{(i)}},$$

where U is any constant matrix acting on the space of the ith fundamental representation of A_r. The problems which still need to be solved are that the above solution has to be periodic and local. These two requirements drastically reduce the degrees of freedom in choosing the constant matrix U, as will be shown below.

Let us consider the periodicity. Inserting the monodromy properties into the above solutions and letting the fields φ^i and ψ^i_+ be periodic, we have

$$S U \bar{S} = U. \tag{34}$$

This requirement can be fulfilled as follows. Since the monodromy matrix S is upper triangular, it can be diagonalized by a unique strictly upper triangular matrix g,

$$S = g\kappa g^{-1}, \qquad \kappa = \mathrm{e}^{2\pi\partial_+ K_+(0)},$$

where $\partial_+ K_+(0)$ means $\partial_+ K_+(x)|_{x=0}$, which is obviously diagonal. Similarly we can diagonalize the monodromy matrix $\bar{S}$ of the other chirality by a strictly lower triangular matrix $\bar{g}$,

$$\bar{S} = \bar{g}\bar{\kappa}\bar{g}^{-1}, \qquad \bar{\kappa} = \mathrm{e}^{-2\pi\partial_- K_-(0)},$$

with the notation $\partial_- K_-(0) = \partial_- K_-(x)|_{x=0}$. It is then evident that the periodicity condition (34) is satisfied if the constant matrix U takes the form

$$U = g\mathcal{D}\bar{g}^{-1},$$

together with a constraint condition imposed on the positive and negative zero modes,

$$\kappa\bar{\kappa} = 1 \quad \text{or} \quad \partial_+ K_+(0) = \partial_- K_-(0).$$

Actually, the above conditions simply imply that the left and right monodromies and the constant matrix U can be diagonalized simultaneously and the diagonal parts of the left and right monodromies must be equal.

The remaining problem—the problem of locality of the general solution—is far more difficult to prove. However, the general principle for this proof is rather simple [24]. Starting from the Poisson brackets for $\partial_\pm K_\pm$ and $\bar{P}_+$, one can obtain well-defined exchange relations between each pair of the normalized chiral fields $\sigma^{(i)}$, $\varsigma^{(i)}$, $\bar{\sigma}^{(i)}$ and the monodromy matrices S and $\bar{S}$. Then by direct calculations using the general solution formula, one can prove that the only admissible diagonal matrix $\mathcal{D}$ satisfying the locality condition is the one of the form

$$\mathcal{D} = \Theta\bar{\Theta}, \qquad \Theta = \mathrm{e}^{\Pi - \partial_+ K_+(0)}, \qquad \bar{\Theta} = \mathrm{e}^{\bar{\Pi} + \partial_- K_-(0)},$$

where Π and $\bar{\Pi}$ are the conjugate variables of the zero modes $\partial_+ K_+(0)$ and $\partial_- K_-(0)$ respectively. As the explicit calculations are considerably long and tedious, we prefer to publish them in a separate publication rather than present them in the present article.

Readers who are smart enough might already have noticed that Eq. (32), when appropriately gauge transformed, yields the following DS-*like* linear systems:

$$\partial_+ V_+ = (\partial_+ K_+ + \bar{P}_+ + \mu)V_+, \qquad \partial_- V_- = V_-(\partial_- K_- + \nu),$$

where $V_+ \equiv \mathrm{e}^{K_+} M_+$, $V_- \equiv M_-^{-1}\mathrm{e}^{-K_-}$. However, as the normalizations of $V_\pm$ and $Q_\pm$ are different, we cannot identify the last equations with the standard DS systems. In fact, these two objects differ from each other by a right-shift with a constant matrix which results in different monodromy properties of the corresponding chiral vectors. To be more explicit, the chiral vectors $\xi^{(i)}$, $\zeta^{(i)}$ and $\bar{\xi}^{(i)}$ described in Subsection 3.1 are related to the chiral vectors $\sigma^{(i)}$, $\varsigma^{(i)}$ and $\bar{\sigma}^{(i)}$ as follows:

$$\sigma^{(i)} = \xi^{(i)} V_+^{-1}(0) = \xi^{(i)} M_+^{-1}(0) \mathrm{e}^{-K_+(0)},$$
$$\varsigma^{(i)} = \zeta^{(i)} V_+^{-1}(0) = \xi^{(i)} M_+^{-1}(0) \mathrm{e}^{-K_+(0)},$$
$$\bar{\sigma}^{(i)} = V_-^{-1}(0) \bar{\xi}^{(i)} = \mathrm{e}^{K_-(0)} M_-(0) \bar{\xi}^{(i)}.$$

Therefore, we can relate the general solution described in terms of the chiral vectors $\xi^{(i)}$, $\zeta^{(i)}$ and $\bar{\xi}^{(i)}$ to the one described by $\sigma^{(i)}$, $\varsigma^{(i)}$ and $\bar{\sigma}^{(i)}$ as

$$U = V_+(0) V_-(0) = \mathrm{e}^{K_+(0)} M_+(0) M_-^{-1}(0) \mathrm{e}^{-K_-(0)}.$$

So to ensure that the solution given by Eqs. (23), (24) be periodic and local, all that is needed is to choose appropriate initial values for the fields $K_\pm$ and $M_\pm$.

Remark. In this subsection we considered only the zero modes of the fields $K_\pm$. There are, however, zero mode problems for the fields P_+ and $\bar{P}_+$. It is these zero modes that make the chiral vectors $\sigma^{(i)}$ and $\varsigma^{(i)}$ interact nontrivially. Since the zero modes for the fields P_+ and $\bar{P}_+$ *do not* lie in the diagonal, it is much more difficult to disentangle them than disentangling the zero modes of $\partial_\pm K_\pm$. We hope to come back to this point later.

4. Wronskian type special solution and connections with WZNW model

The general solution obtained in the last section involves matrix elements in all the fundamental representations. Such a solution is very useful when studying the symmetries of the model. But in practice, one is often interested in the explicit space-time behaviors of the fundamental fields. In this case, the general solution given above may appear not to be very helpful. In this section, we make another attempt in deriving the explicit solution of the model.

4.1. Wronskian type special solution

Let us start from the chiral embedding of the light cone coordinates x_+, x_- into the product space $V = \mathbf{R}^{r+1} \otimes \mathbf{R}^{2(r+1)}$. Such embedding are expressed by the identifications

$$\bar{X}^i = \bar{\xi}^i(x_-), \quad X^i = \xi^i(x_+), \quad Y^i = \zeta^i(x_+), \qquad i = 1, 2, \ldots, r+1 \tag{35}$$

where $\bar{X}^i$ and X^i, Y^i are coordinates of the left $\mathbf{R}^{r+1}$ and the right $\mathbf{R}^{2(r+1)}$ spaces, respectively, $\bar{\xi}^i, \xi^i$ and ζ^i are arbitrary functions of the arguments (*these functions must obey some fixed monodromy properties in order to maintain the dynamics since they are exactly the components of the chiral vectors $\xi^{(1)}, \zeta^{(1)}$ and $\bar{\xi}^{(1)}$ as we shall show later. However, given the discussions in the last subsection, we are left with no doubt in the fact that there* exist *physically meaningful solutions for our model, and that is enough* *for the discussions that follow. Therefore we do not care about any explicit monodromy behaviors of the above chiral embedding functions).*

Now define two sets of $(r+1)$ column vectors $\bar{f}_a$ and row vectors f_a as follows:

$$\bar{f}^i_a(x_-) = \partial_-^{a-1}\bar{\xi}^i(x_-),$$
$$f^i_{2a-1}(x_+) = \partial_+^{a-1}\xi^i(x_+),$$
$$f^i_{2a}(x_+) = \partial_+^{a-1}\zeta^i(x_+), \qquad i = 1,2,\dots,r+1.$$

From these vectors we can construct the $(r+1)\times(r+1)$ matrix of inner products

$$g_{ab}(x_-,\ x_+) \equiv f_a\bar{f}_b = \sum_{i=1}^{r+1} f^i_a(x_+)\bar{f}^i_b(x_-), \quad a,b = 1,2,\dots,r+1.$$

We also introduce the following notation:

$$\Delta_a \equiv \det\begin{pmatrix} g_{11} & \cdots & g_{1a} \\ \vdots & & \vdots \\ g_{a1} & \cdots & g_{aa}\end{pmatrix}, \quad a = 1,\ 2,\ \dots, r+1,$$
$$\Delta_0 \equiv 1.$$

Using these notations, we now define two sets of new vectors e_a and $\bar{e}_a$,

$$e^i_a \equiv \frac{1}{\sqrt{\Delta_{a-1}\Delta_a}}\begin{vmatrix} g_{11} & \cdots & g_{1,a-1} & f^i_1 \\ \vdots & & \vdots & \vdots \\ g_{a1} & \cdots & g_{a,a-1} & f^i_a\end{vmatrix}, \quad \bar{e}^i_a \equiv \frac{1}{\sqrt{\Delta_{a-1}\Delta_a}}\begin{vmatrix} g_{11} & \cdots & g_{1,a} \\ \vdots & & \vdots \\ g_{a-1,1} & \cdots & g_{a-1,a} \\ \bar{f}^i_1 & \cdots & \bar{f}^i_a\end{vmatrix}, \tag{36}$$

where e_a are $(r+1)$ row vectors, and $\bar{e}_a$ are $(r+1)$ column vectors. It can be easily proved that the vectors e_a and $\bar{e}_a$ are orthogonal to each other,

$$(e_a,\ \bar{e}_b) = \sum_{i=1}^{r+1} e^i_a\bar{e}^i_b = \delta_{ab}. \tag{37}$$

This is due to the following Laplacian expansions of the definition (36):

$$e^i_a = \sqrt{\frac{\Delta_{a-1}}{\Delta_a}}\sum_{l=1}^{a}\frac{\Delta_a(l,\ a)}{\Delta_{a-1}}f^i_l(x_+) \equiv \sqrt{\frac{\Delta_{a-1}}{\Delta_a}}\sum_{l=1}^{a}(A^{-1})_{al}f^i_l(x_+),$$
$$\bar{e}^i_a = \sqrt{\frac{\Delta_{a-1}}{\Delta_a}}\sum_{l=1}^{a}\frac{\Delta_a(a,\ l)}{\Delta_{a-1}}\bar{f}^i_l(x_-) \equiv \sum_{l=1}^{a}\bar{f}^i_l(x_-)(C^{-1})_{la}\sqrt{\frac{\Delta_{a-1}}{\Delta_a}}, \tag{38}$$

where $\Delta_a(i,j)$ are algebraic cominors of the entry (i,j). According to these Laplacian expansions, we can express the derivatives of e_a and $\bar{e}_a$ in terms of their linear combinations,

$$\partial_+ e_a = (\omega_+)_a^b e_b, \qquad \partial_- e_a = (\omega_-)_a^b e_b,$$
$$\partial_+ \bar{e}_a = \bar{e}_b(\bar{\omega}_+)_a^b, \qquad \partial_- \bar{e}_a = \bar{e}_b(\bar{\omega}_-)_a^b,$$

where as usual, the subscripts $\pm$ specifies the upper or lower triangularities of the corresponding ω matrices. The orthogonality condition (37) now implies that $\omega_\pm$ and $\bar{\omega}_\pm$ are not independent objects,

$$(\omega_\pm)_a^b = -(\bar{\omega}_\pm)_a^b.$$

Furthermore, straightforward calculations show that only a few of the matrix elements in $\omega_\pm$ are nonvanishing, the nonvanishing elements being positioned on the main diagonal and the first and the second upper/lower diagonals. Explicitly, we have

$$\begin{aligned}
\partial_+ e_a &= \frac{1}{2}\partial_+ \ln\left(\frac{\Delta_a}{\Delta_{a-1}}\right) e_a + \sqrt{\frac{\Delta_{a-1}\Delta_{a+1}}{\Delta_a^2}}\left(\frac{\Delta_a(a-1,\ a)}{\Delta_{a-1}} - \frac{\Delta_{a+2}(a+1,\ a+2)}{\Delta_{a+1}}\right) e_{a+1} \\
&\quad + \sqrt{\frac{\Delta_{a-1}\Delta_{a+1}}{\Delta_a^2}}\sqrt{\frac{\Delta_a\Delta_{a+2}}{\Delta_{a+1}^2}} e_{a+2}, \\
\partial_- e_a &= -\frac{1}{2}\partial_- \ln\left(\frac{\Delta_a}{\Delta_{a-1}}\right) e_a - \sqrt{\frac{\Delta_{a-2}\Delta_a}{\Delta_{a-1}^2}} e_{a-1}, \quad a = 1, 2, \ldots, r+1, \\
e_{-1} &= e_0 = e_{r+2} = e_{r+3} = 0.
\end{aligned} \tag{39}$$

Denoting

$$\varphi^a = \ln \Delta_a, \qquad \psi_+^a = -\frac{\Delta_{a+1}(a,\ a+1)}{\Delta_a}, \tag{40}$$

Eq. (39) becomes

$$\begin{aligned}
\partial_+ e_a &= \frac{1}{2}\partial_+(\varphi^a - \varphi^{a-1}) e_a + (w^a)^{1/2}(\psi_+^{a-1} - \psi_+^{a+1}) e_{a+1} + (w^a w^{a+1})^{1/2} e_{a+2}, \\
\partial_- e_a &= -\frac{1}{2}\partial_-(\varphi^a - \varphi^{a-1}) e_a - (w^{a-1})^{1/2} e_{a-1}, \\
w^a &\equiv \exp(\varphi^{a-1} + \varphi^{a+1} - 2\varphi^a).
\end{aligned} \tag{41}$$

The compatibility condition $[\partial_+,\ \partial_-] e_a = 0$ of Eq. (41) then yields

$$\begin{aligned}
\partial_+\partial_-\varphi^j - (\psi_+^{j-1} - \psi_+^{j+1}) w^j &= 0, \qquad \partial_-\psi_+^j - w^j = 0, \quad j = 1, \ldots, r, \\
\partial_+\partial_-\varphi^{r+1} &= 0, \qquad \partial_-\psi_+^{r+1} = 0.
\end{aligned}$$

Except for the nonvanishing functions φ^{r+1} and ψ_+^{r+1}, the above equations are very similar to the equations of motion (4) of the HTFT. So if we can somehow fix the values of φ^{r+1} and ψ_+^{r+1} to zero, then the above construction will really result in an explicit solution to the system (4).

In order to fix the values of φ^{r+1} and ψ_+^{r+1} to zero, let us mention that the Δ symbols Δ_{r+1} and $\Delta_{r+2}(r+1,\ r+2)$ are simply products of chiral objects,

$$\Delta_{r+1} = \det g = U\bar{U}, \qquad \Delta_{r+2}(r+1,\ r+2) = V\bar{U},$$

where U, $\bar{U}$ are respectively the determinants of the matrices consisting of the functions f_a^i and $\bar{f}_a^i$, and V is defined as follows:

$$V = \begin{vmatrix} f_1^1 & f_1^2 & \cdots & f_1^{r+1} \\ \vdots & \vdots & & \vdots \\ f_r^1 & f_r^2 & \cdots & f_r^{r+1} \\ f_{r+2}^1 & f_{r+2}^2 & \cdots & f_{r+2}^{r+1} \end{vmatrix}.$$

Now it is clear from Eq. (40) that if we set $\varphi^{r+1} = \psi_+^{r+1} = 0$, then the following conditions on U and V have to be satisfied:

$$U = \bar{U} = 1, \qquad V = 0. \tag{42}$$

In terms of the original embedding functions, these constraints are nothing but linear differential equations for the $3(r+1)$ arbitrary functions. Such equations can be easily solved for any three of the embedding functions, and this finishes the construction of the Wronskian type solution.

4.2. Connections with WZNW theory

Readers who are familiar with the classical W-geometrical theory of Gervais et al. [12] may have already recognized that the constructions made above are almost entirely based on the similar construction in the conventional CT theory. That *is* the point. In the case of standard CT theory, the Wronskian type solution is closely related to the Drinfeld–Sokolov gauges of the corresponding $(W_{r+1})_L \otimes (W_{r+1})_R$ symmetries of the model. Now we come to show that the conformal algebra for the HTFT is $(W_{r+1})_L \otimes (W_{r+1}^{(2)})_R$ in contrast to the CT $((W_{r+1})_L \otimes (W_{r+1})_R$ algebra) and 2-ECT $((W_{r+1}^{(2)})_L \otimes (W_{r+1}^{(2)})_R$ algebra) cases. The difference in conformal algebras in the left and the right chiral sectors is the most crucial property of the present model.

Remembering the definitions of the matrix elements g_{ab}, we have

$$\partial_+ g_{ab} = g_{a+2,b}, \quad \partial_- g_{ab} = g_{a,b+1}, \quad a = 1,2,\ldots,r-1, \quad b = 1,2,\ldots, r.$$

In matrix form, we have

$$\partial_+ g = J_+ g, \qquad \partial_- g = g J_-, \tag{43}$$

where

$$J_- = \begin{pmatrix} 0 & & & & * \\ 1 & 0 & & & * \\ & 1 & \ddots & & \vdots \\ & & \ddots & 0 & * \\ & & & 1 & * \end{pmatrix}, \qquad J_+ = \begin{pmatrix} 0 & 0 & 1 & & & \\ & 0 & 0 & 1 & & \\ & & \ddots & \ddots & \ddots & \\ & & & 0 & 0 & 1 \\ * & * & \cdots & * & * & * \\ * & * & \cdots & * & * & * \end{pmatrix}, \tag{44}$$

and the non-zero non-constant entries "$*$" in (44) are to be determined. Eq. (43) is nothing but a constrained version of the well known WZNW equations, with $J_\pm$

being the constrained WZNW currents. We recognize that this form of the constrained WZNW currents is written in the well-known Drinfeld–Sokolov gauges of the $(W_{r+1})_L$ and $(W^{(2)}_{r+1})_R$ algebras, where the "$*$" entries are just the W-algebra generators. It is straightforward to check that the vectors $\boldsymbol{f}_a$ and $\bar{\boldsymbol{f}}_a$ also solve Eq. (43),

$$\partial_+ \boldsymbol{f}_a = \sum_{b=1}^{r+1} (J_+)_{ab} \boldsymbol{f}_b, \qquad \partial_- \bar{\boldsymbol{f}}_a = \sum_{b=1}^{r+1} \bar{\boldsymbol{f}}_b (J_-)_{ba}. \tag{45}$$

Hence the matrices consisting of the functions f^i_a and $\bar{f}^i_a$ are the chiral components of the WZNW field g, respectively, and there is no problem in concluding that all the "$*$" in Eq. (44) are chiral objects.

Now let us go one step further to determine the relations between the W-algebra generators and the embedding functions (35).

Substituting the relations $\partial_+ \boldsymbol{f}_a = \boldsymbol{f}_{a+2}$, $\partial_- \bar{\boldsymbol{f}}_b = \boldsymbol{f}_{b+1}$ into Eq. (45), we have

$$\sum_{c=1}^{r+1} (J_+)_{ac} \boldsymbol{f}_c = \boldsymbol{f}_{a+2}, \qquad \sum_{c=1}^{r+1} \bar{\boldsymbol{f}}_c (J_-)_{cb} = \bar{\boldsymbol{f}}_{b+1}. \tag{46}$$

Solving the above linear system of equations for the variables $(J_+)_{ac}$ and $(J_-)_{cb}$ with $a = r, r+1$ and $b = r+1$, it results in

$$\begin{aligned} (J_+)_{r,c} &= -\frac{R_c(c,\ r+2)}{U}, \\ (J_+)_{r+1,c} &= -\frac{S_c(c,\ r+2)}{U}, \\ (J_-)_{r+1,c} &= -\frac{\bar{R}(c,\ r+2)}{U}, \end{aligned} \tag{47}$$

where R_c, $\bar{R}_c$ and S_c are given as follows:

$$R_c = \begin{vmatrix} f^1_1 & \cdots & f^{r+1}_1 & f^c_1 \\ \vdots & & \vdots & \vdots \\ f^1_{r+1} & \cdots & f^{r+1}_{r+1} & f^c_{r+1} \\ f^1_{r+2} & \cdots & f^{r+1}_{r+2} & f^c_{r+2} \end{vmatrix} = 0, \qquad S_c = \begin{vmatrix} f^1_1 & \cdots & f^{r+1}_1 & f^c_1 \\ \vdots & & \vdots & \vdots \\ f^1_{r+1} & \cdots & f^{r+1}_{r+1} & f^c_{r+1} \\ f^1_{r+3} & \cdots & f^{r+1}_{r+3} & f^c_{r+3} \end{vmatrix} = 0,$$

$$\bar{R}_c = R_c(\text{with } f^i_a \leftrightarrow \bar{f}^i_a).$$

Eq. (47) determines the W-algebra generators completely in terms of the embedding functions (35). It is interesting to notice that the constraint conditions in (42) imply that

$$(J_+)_{r,r+1} = 0, \quad (J_+)_{r,r} + (J_+)_{r+1,r+1} = 0, \quad (J_-)_{r+1,r+1} = 0.$$

This not only ensures the tracelessness of the A_r WZNW currents (44) but also makes $J_\pm$ fit in the framework of Eq. (5).

Now let us give some brief remarks on the aspects which have not been mentioned above.

(i) Using the standard method, we can also reduce Eq. (45) into the Gelfand–Dickey equations for the W-algebras in the left and right chiral sectors. The one for the $(W_{r+1})_L$ is a scalar differential equation, and the one for $(W^{(2)}_{r+1})_R$ is a matrix one. The matrix Gelfand–Dickey equation corresponding to the $W^{(2)}_{r+1}$ algebra was first given in Ref. [13].

(ii) The matrices A, C defined in Eq. (38) have very profound meanings in the WZNW setting of the model. In fact, from (38), we can write

$$f_a = \sum_{l=1}^{a} A_{al}\sqrt{\frac{\Delta_l}{\Delta_{l-1}}}e_l, \qquad \bar{f}_a = \sum_{l=1}^{a} \bar{e}_l\sqrt{\frac{\Delta_l}{\Delta_{l-1}}}C_{la}, \tag{48}$$

from which we have

$$g_{ab} = f_a\bar{f}_b = ABC, \quad (B)_{ab} \equiv \left[\exp(\Phi)\right]_{ab} = \frac{\Delta_a}{\Delta_{a-1}}\delta_{ab}.$$

The last equation shows that the matrices A, C are exactly the lower- and upper-triangular parts in the Gauss decomposition of the constrained WZNW field g. Moreover, it can be shown that A, C satisfy the following equations:

$$A^{-1}\partial_- A = \exp(\mathrm{ad}\,\Phi)\nu, \qquad \partial_+ CC^{-1} = \exp(-\mathrm{ad}\,\Phi)(\bar{\Psi}_+ + \mu).$$

These equations are precisely the constrained WZNW equations re-expressed in terms of the Gauss components A and C.

(iii) The chiral embedding functions (35) are just the components of the chiral vectors $\bar{\xi}^{(1)}, \xi^{(1)}$ and $\zeta^{(1)}$ corresponding to the defining representation of the Lie algebra A_r. This statement makes the discussions in this and the last sections finally unified, and it follows that the W-algebras $(W_{r+1})_L$ and $(W^{(2)}_{r+1})_R$ are related to the simple exchange algebra (14)–(22) by Eq. (47). So it seems that the exchange algebra (14)–(22) is more fundamental than the W-symmetry algebras.

Now let us spend some more words on the identification of chiral embedding functions with the chiral vectors (12) and (13). Notice that while the constraints (42) are imposed, Eq. (41) will become exactly the Lax pair (2) of A_r HTFT, with the transport matrix T defined as

$$T_{ai} \equiv e^i_{\ a}, \qquad T^{-1}_{ia} \equiv \bar{e}^i_{\ a}.$$

This observation enables us to rewrite Eq. (48) as

$$f^i_a = \sum_{l=1}^{a} A_{al}\left[\exp\left(\frac{1}{2}\Phi\right)\right]_{ll} T_{li}, \qquad \bar{f}^i_a = \sum_{l=1}^{a} (T^{-1})_{il}\left[\exp\left(\frac{1}{2}\Phi\right)\right]_{ll} C_{la}. \tag{49}$$

On the other hand, the definitions of A and C imply that

$$A_{aa} = 1, \qquad A_{a+1,a} = \psi^a_+, \qquad C_{aa} = 1. \tag{50}$$

Substituting Eq. (50) into (49) and remembering that $\xi^i = f^i_1, \zeta^i = f^i_2, \bar{\xi}^i = \bar{f}^i_1$, we finally get

$$\bar{\xi}^i = (T^{-1})_{i1}\left[\exp\left(\frac{1}{2}\Phi\right)\right]_{11} C_{11} = \left\{T^{-1}\exp\left(\frac{1}{2}\Phi\right)\right\}_{i1}, \tag{51}$$

$$\xi^i = A_{11}\left[\exp\left(\frac{1}{2}\Phi\right)\right]_{11}(T)_{1i} = \left\{\exp\left(\frac{1}{2}\Phi\right)T\right\}_{1i}, \tag{52}$$

$$\zeta^i = \left\{\exp(\Psi_+)\exp\left(\frac{1}{2}\Phi\right)T\right\}_{2i}. \tag{53}$$

Eqs. (51)–(53) are exactly the definitions (12) and (13) rewritten in the defining representation of the Lie algebra A_r.

To end the present section, let us mention that the Wronskian type solution (40) is only a special solution. The key point in relating this special solution to the general solution given in the last section is to perform *chiral* gauge transformations starting from the Drinfeld–Sokolov type gauges (44). In terms of the embedding functions, such gauge transformations are expressed by replacing these chiral functions by their arbitrary linear combinations, and that is all about the W-geometrical picture about the HTFT.

5. Connections with Toda lattice hierarchy

In this section we are going to consider the connections between HTFT and the well-known Toda lattice hierarchy (TLH). This is a totally different view point from the discussions made in the previous sections.

First let us briefly recall the usual description of TLH (here we are using the notations of Ref. [5]). Let $\mathcal{O}$ be the space of the unitary hermitian states $\{|n\rangle\}$ equipped with the metric

$$\langle m|n\rangle = \delta_{mn}.$$

Let Λ be the shifting operator acting on $\mathcal{O}$,

$$\Lambda|n\rangle = |n+1\rangle, \qquad \langle n|\Lambda = \langle n-1|.$$

It follows that any operator W acting on the space $\mathcal{O}$ can be expressed as

$$W = \sum_{j\in\mathbb{Z}} W_j\Lambda^j, \qquad W_j = \sum_n |n\rangle W_j(n)\langle n|.$$

Moreover, each operator W admits a unique factorization

$$W = W_+ + W_-, \qquad W_+ = \sum_{j\geq 0} W_j\Lambda^j, \qquad W_- = \sum_{j<0} W_j\Lambda^j.$$

Now consider two special operators of the form

$$L = \Lambda + \sum_{n=0}^{\infty} u_n \Lambda^{-n}, \qquad M = \sum_{n=-1}^{\infty} v_n \Lambda^n$$

and define

$$B_n \equiv (L^n)_+, \qquad C_n \equiv (M^n)_-;$$

the TLH is then determined by the following evolution equations for the operators L and M:

$$\begin{aligned} \partial_n L &= [B_n, L], & \bar{\partial}_n L &= [C_n, L], \\ \partial_n M &= [B_n, M], & \bar{\partial}_n M &= [C_n, M], \end{aligned}$$

or, equivalently, by the compatibility conditions

$$\begin{aligned} &[\partial_n - B_n, \partial_m - B_m] = 0, \\ &\left[\bar{\partial}_n - C_n, \bar{\partial}_m - C_m\right] = 0, \\ &\left[\partial_n - B_n, \bar{\partial}_m - C_m\right] = 0. \end{aligned} \tag{54}$$

The claim of this section is that the A_∞ limit of the HTFT (4) is just the (B_2, C_1) flow in the hierarchy (54). To justify this claim let us recall that B_2 and C_1 must be of the form

$$\begin{aligned} B_2 &= \sum_s \left(|s\rangle\langle s|\Lambda^2 + |s\rangle b_1(s)\langle s|\Lambda + |s\rangle b_0(s)\langle s|\right), \\ C_1 &= \sum_s |s\rangle c_1(s)\langle s|\Lambda^{-1}. \end{aligned} \tag{55}$$

Substituting Eq. (55) into the second line of (54), it follows that

$$\begin{aligned} \partial_+ c_1(a) + c_1(a) b_0(a+1) - b_0(a) c_1(a) &= 0, \\ \partial_- b_1(a) - c_1(a) + c_1(a-2) &= 0, \\ \partial_- b_0(a) - c_1(a) b_1(a+1) + b_1(a) c_1(a-1) &= 0, \end{aligned} \tag{56}$$

where $\partial_+ \equiv \partial_2$, $\partial_- \equiv \bar{\partial}_1$. Changing the variables $b_0(a) \to \partial_+(\varphi^{a-1} - \varphi^a)$, $b_1(a) \to \psi_+^a - \psi_+^{a-2}$, Eq. (56) becomes

$$\begin{aligned} \partial_+\partial_-\varphi^a - (\psi_+^{a-1} - \psi_+^{a+1}) \exp(\varphi^{a-1} + \varphi^{a+1} - 2\varphi^a) &= 0, \\ \partial_-\psi_+^a - \exp(\varphi^{a-1} + \varphi^{a+1} - 2\varphi^a) &= 0, \quad a = -\infty, \dots, \infty. \end{aligned} \tag{57}$$

We see that Eq. (57) is just the A_∞ version of Eq. (4). Therefore, the model we are considering is really a Toda type theory.

Recently, it is of increasing interest to study the so-called continuous Toda field theories [15–19]. Such theories have intimate relations with the W-infinity algebras and self-dual gravities. So it seems worthwhile to remark here that the continuous limit of the present model will become a (1+2)-dimensional heterotic Toda theory. Such a theory is expected to be related to the heterotic W-infinity gravities. We hope to come back to this point later.

6. Summary and discussions

Let us now summarize the whole article and discuss some more points.

In this article we constructed a heterotic Toda field theory, studied its chiral exchange algebra and gave the classical general solution and a Wronskian type special solution. We also showed that the Wronskian type solution is closely related to the W-algebra symmetries of the model. The chiral vectors $\xi^{(i)}, \bar{\xi}^{(i)}$ and $\zeta^{(i)}$ are the cross-point of all these different aspects. Besides all this, we showed that the HTFT is really a member of the TLH. This final statement may be of interest in the theories of W-infinity algebras and self-dual gravities. It can be expected that in the continuous limit the HTFT yields two sets of W-infinity algebras, one of them the usual $(w_\infty)_L$ algebra, and the other a generalized $(w_\infty)_L$ algebra which may be denoted $(w_\infty^{(2)})_R$. The possibility for the existence of $(w_\infty^{(2)})_R$ type algebras was first proposed by the authors of Ref. [3]. A detailed study on the W-infinity algebras and W-infinity gravities in the framework of HTFT will be presented elsewhere.

It is also worth mentioning that here we only worked on the Lie algebra A_r. We can also construct similar models based on other finite-dimensional Lie algebras as well as loop and affine Lie algebras. In the latter cases, we shall arrive at heterotic analogies of LT and CAT theories. Then the soliton behaviors in these heterotic loop Toda and conformal affine Toda theories may become interesting subjects of further study. Moreover, as the complex affine (loop) Toda theories are receiving considerable attention [20–23] because they give physically meaningful (real) energy-momentum tensors, we also hope to study the case of complex coupling constants of these heterotic models. All the above-mentioned studies are now being carried out.

Acknowledgements

The authors are very grateful to Professor J.B. Zuber for helpful discussions and comments. They also would like to express their gratitude toward the referee for useful suggestions and comments on the monodromy aspects of the chiral vectors.

References

[1] O. Babelon and L. Bonora, Phys. Lett. B 244 (1990) 220.
[2] L. Chao, Commum. Theor. Phys. 15 (1993) 221.
[3] B.-Y. Hou and L. Chao, Int. J. Mod. Phys. A 8 (1993) 1105; A 8 (1993) 3773.
[4] V.A. Fatteev and A.B. Zamolodchikov, Nucl. Phys. B 280 (1987) 644;
T. Eguchi and S.K. Yang, Phys. Lett. B 224 (1989) 373; B 235 (1990) 282.
[5] M. Fukuma and T. Takebe, Mod. Phys. Lett. A 5 (1990) 509.
[6] A.M. Polyakov, Gauge field and strings (Harwood Academic, Publishers, 1987) p. 269.
[7] O. Babelon, Phys. Lett. B 215 (1988) 523.
[8] K. Ueno and K. Takasaki, Toda lattice hierarchy, Advanced Studies in Pure and Applied Mathematics 4, pp. 1 (1984).
[9] For reviews, see L. Feher, L. O'Raifeartaigh, P. Ruelle, I. Ttsutsui and A. Wipf, Phys. Rep. 222 (1993) 1;
see also B.-Y. Hou and L. Chao, Int. J. Mod. Phys. A 7 (1992) 7015.

[10] O. Babelon and D. Bernard, Phys. Lett. B 260 (1991) 81; Commun. Math. Phys. 149 (1992) 279.
[11] A.N. Leznov and M.V. Saveliev, Phys. Lett. B 79 (1978) 294; Lett. Math. Phys. 3 (1979) 207; 3 (1979) 489; Commun. Math. Phys. 74 (1980) 111; Lett. Math. Phys. 6 (1982) 505; Commun. Math. Phys. 89 (1983) 59.
[12] J.-L. Gervais and Y. Matsuo, Commun. Math. Phys. 152 (1993) 317.
[13] L. Chao and B.-Y. Hou, Ann. Phys. (in press).
[14] K. Takasaki and T. Takebe, Lett. Math. Phys. 23 (1991) 205; T. Takebe, Commun. Math. Phys. 129 (1990) 281.
[15] Q.-H. Park, Phys. Lett. B 236 (1990) 429.
[16] J. Avan, Phys. Lett. A 168 (1992) 263.
[17] M.V. Saveliev and S.A. Savelieva, Phys. Lett. B 313 (1993) 55.
[18] K. Takasaki, Kyoto Univ. Preprint KUCP-0057/93.
[19] S.-Y. Lou, Ningbo Normal College preprint 1993.
[20] T. Hollowood, Nucl. Phys. B 384 (1992) 523.
[21] H.-C. Liao, D. Olive and N. Turok, Phys. Lett. B 298 (1993) 95.
[22] D. Olive, N. Turok and J.W. Underwood, preprints Imperial/TP/91-92/35, Imperial/TP/92-93/29.
[23] M.A.C. Kneipp and D. Olive, Swansea preprint SWAT/92-93/6.
[24] O. Babelon, L. Bonora and F. Toppan, Commun. Math. Phys. 140 (1990) 93;
E. Aldrovandi, L. Bonora, V. Bonservizi and R. Paunov, Int. J. Mod. Phys. A 9 (1994) 57.

ELSEVIER

Nuclear Physics B 478 (1996) 723–757

NUCLEAR PHYSICS B

Algebraic Bethe ansatz for the eight-vertex model with general open boundary conditions

Heng Fan [a,b], Bo-yu Hou [b], Kang-jie Shi [a,b], Zhong-xia Yang [c,b]

[a] *CCAST (World Laboratory), P.O. Box 8730, Beijing 100080, China*

[b] *Institute of Modern Physics, Northwest University, P.O. Box 105, Xian 710069, China*

[c] *Graduate School of Academia Sinica, P.O. Box 3908, Beijing 100039, China*

Received 11 April 1996; accepted 25 July 1996

Abstract

By using the intertwiner and face–vertex correspondence relation, we obtain the Bethe ansatz equation of the eight-vertex model with open boundary conditions in the framework of algebraic Bethe ansatz method. The open boundary condition under consideration is the general solution of the reflection equation for the eight-vertex model with only one restriction on the free parameters of the right side reflecting boundary matrix. The reflecting boundary matrices used in this paper thus may have off-diagonal elements. Our construction can also be used for the Bethe ansatz of SOS model with reflection boundaries.

PACS: 75.10.J; 05.20; 05.30

Keywords: Eight-vertex model; Face Boltzmann weights; Bethe ansatz; Reflection equation

1. Introduction

One of the most important goals of exactly solvable lattice models is to find the eigenvalues and eigenvectors of the transfer matrix of a system, and then to obtain the thermodynamic limit of this system.

Bethe examined the completely isotropic case of the *XXX* model and found the eigenvalues and eigenvectors of its Hamiltonian [1]. After Bethe's work, Yang and Yang analyzed the anisotropic *XXZ* model by means of the Bethe ansatz [2]. Then, Baxter in his remarkable papers, gave a solution for the completely anisotropic *XYZ* model [3]. He discovered a relation between the quantum *XYZ* model and the eight-vertex model, which is one of the two-dimensional exactly solvable lattice models. Faddeev and Takhtajan simplified Baxter's formulae and proposed the quantum inverse

PII S0550-3213(96)00398-7

scattering method or algebraic Bethe ansatz method to solve the six-vertex and eight-vertex models, whose spin-chain equivalent are the XXZ spin model and the XYZ spin model, respectively [4]. After Yang–Baxter–Faddeev–Takhtajan's work, many exactly solvable models have been solved by the algebraic Bethe ansatz [5,6], the functional Bethe ansatz [7,8], the coordinate Bethe ansatz [9], etc.

Typically, the two-dimensional exactly solvable lattice models are solved by imposing periodic boundary conditions in which the Yang–Baxter equation provides a sufficient condition for the integrability of the models,

$$R_{12}(z_1-z_2)R_{13}(z_1-z_3)R_{23}(z_2-z_3)=R_{23}(z_2-z_3)R_{13}(z_1-z_3)R_{12}(z_1-z_2), \tag{1}$$

where the R-matrix is the Boltzmann weight for the vertex models in two-dimensional statistical mechanics. As usual, $R_{12}(z), R_{13}(z)$ and $R_{23}(z)$ act in $\mathbb{C}^n\otimes\mathbb{C}^n\otimes\mathbb{C}^n$ with $R_{12}(z)=R(z)\otimes 1, R_{23}(z)=1\otimes R(z)$, etc.

The exactly solvable models with non-periodic boundary conditions have been studied earlier in Refs. [10–14]. For periodic boundary conditions, the left and right boundaries are "closed" to form a cylinder, so we would like to call the non-periodic boundary condition the "open boundary" condition for simplicity, and such lattice systems are usually equivalent to the open spin chains. Recently, integrable models with open boundary conditions have attracted a great deal of interest. This was initiated by Cherednik [15] and Sklyanin [16], who proposed a systematic approach to handle the open boundary condition problems involving the so-called reflection equation (RE)

$$\begin{aligned}&R_{12}(z_1-z_2)K_1(z_1)R_{21}(z_1+z_2)K_2(z_2)\\&\quad=K_2(z_2)R_{12}(z_1+z_2)K_1(z_1)R_{21}(z_1-z_2)\,.\end{aligned} \tag{2}$$

The open boundary conditions are determined by the boundary reflecting matrix K satisfying the RE.

By using a non-trivial generalization of the quantum inverse scattering method, Sklyanin obtained the Bethe ansatz equation of the six-vertex model with open boundary conditions by the algebraic Bethe ansatz method [16]. The transfer matrix with a particular choice of boundary conditions is the quantum group $U_q(sl(2))$ invariant [17,18]. After Sklyanin's pioneering work, many exactly solvable lattice models with open boundary conditions have been solved. Mezincescu and Nepomechie solved the $A_1^{(1)}$ and $A_2^{(2)}$ vertex models by using the fusion procedure [19]. Foerster and Karowski solved the $spl_q(2,1)$ invariant Hamiltonian which contains a non-trivial boundary term by using the nested algebraic Bethe ansatz [20]. Using the same method, de Vega and Gonzalez-Ruiz solved the A_n vertex model, and also analyzed the thermodynamic limit of this system [21]. Yue, Fan and Hou solved the general $SU_q(n|m)$ vertex model [22]. For other progress on open boundary conditions along this direction, see e.g. Refs. [23–32].

For Baxter's eight-vertex model with open boundary conditions, some progress has been made. Jimbo et al. obtained the difference equation of the n-point function for a semi-infinite *XYZ* chain [33]. Zhou has recently studied the fused eight-vertex model and found the functional relations for the eight-vertex model with open boundary conditions [34], in which the reflecting K matrix which satisfies the vertex RE, Eq. (2), is diagonal. Using this K matrix, Batchelor et al. have obtained the surface free energy for the eight-vertex model [29].

In this paper, we will use the algebraic Bethe ansatz method to solve the eight-vertex model with open boundary conditions. It is known that the K matrix is a solution of RE. The general solution of RE for the six-vertex model was obtained by de Vega [35]. In Refs. [36,37], solutions of RE for the eight-vertex model have been found. The general solution of RE for the eight-vertex model which has three free parameters was found by two groups [38,39]. In our approach we use the general solution for left and right boundaries. The reflecting boundary K matrices thus may have off-diagonal elements. We need only to impose one relation on the free parameters of the right K matrix, if the free parameters of the left K matrix are arbitrarily given.

It is known that in Baxter's original work, in stead of the Yang–Baxter relation, the star–triangle relation plays the key role, which can be obtained from Yang–Baxter relation by using the intertwiners. This was later generalized to the $\mathbb{Z}_n$ Baxter–Belavin model [40,41] to describe the interaction-round-a-face model by Jimbo, Miwa and Okado [42]. This intertwiner method was also used to solve the Bethe ansatz problem for similar cases [43,44]. In order to get algebraic Bethe ansatz equation of the eight-vertex model with open boundary conditions, we need to describe the exchange relations of the monodromy matrix in the "face language". Thus in our paper, we need to convert our boundary conditions of the vertex model to that of a face model. Our approach is equivalent to an SOS model with open boundaries satisfying the face RE. The face RE was first proposed by Behrend, Pearce and O'Brien [45]. In Ref. [45], they also find a diagonal solution of the face RE for the ABF [46] model. By using the intertwiners, the face RE is derived directly from the vertex RE and the general solution of the face RE for the eight-vertex SOS model was found by other groups [47,34]. In this paper we actually use a diagonal solution of the face RE at the left boundary and an upper triangular solution at the right boundary for the eight-vertex SOS model. These open boundary conditions are different from the case discussed by Zhou in Ref. [34]. Since a diagonal matrix is a special case of the upper triangular matrix, our approach can be applied to an SOS model with boundaries proposed by Behrend et al. We can prove that by taking a special case, the Bethe ansatz equation for the RSOS model, the ABF model, can also be obtained.

The outline of this paper is as follows. In Section 2 we first review the eight-vertex model and reflecting open boundary conditions; the model under consideration in this paper will also be constructed. In Section 3, by using the correspondence of the face and vertex models, we derive the face RE. As mentioned above, the face RE will play a key role in the algebraic Bethe ansatz method for the eight-vertex model instead of the vertex RE. The transfer matrix with boundary conditions will also be constructed by using the face weight. In Section 4 we will find the local vacuum for the eight-vertex model with a boundary, which is the same as the vacuum state given by Baxter [3,7] in the Bethe ansatz of the eight-vertex model with periodic boundary conditions. The face boundary matrix derived directly from the vertex boundary matrix is obtained. In Section 5, the Bethe ansatz problem is solved for the eight-vertex model with open boundary conditions. Section 6 contains some discussions and future work.

2. Description of the model

2.1. The R-matrix

We first start from the R-matrix of the eight-vertex model. Denote $\alpha = (\alpha_1, \alpha_2)$, $\alpha_1, \alpha_2 = 0, 1$. Let g and h be 2×2 matrices with elements $g_{ii'} = (-1)^i \delta_{ii'}, h_{ii'} = \delta_{i+1,i'}, i, i' = 0, 1$.

Define 2×2 matrices $I_\alpha = I_{(\alpha_1,\alpha_2)} = h^{\alpha_1} g^{\alpha_2}, I_0 = I = identity$, and define $I_\alpha^{(j)} = I \otimes I \otimes \ldots I_\alpha \otimes \ldots \otimes I, I_\alpha$ is at jth space. As usual, $I_\alpha^{(j)}$ act in $V = V_1 \otimes \ldots \otimes V_l$, where the space V is consisted of l two-dimensional spaces. We then introduce some notations used in this paper. They are

$$\theta \begin{bmatrix} a \\ b \end{bmatrix} (z, \tau) \equiv \sum_{m \in \mathbb{Z}} \exp \left\{ \pi \sqrt{-1} (m + a) [(m + a)\tau + 2(z + b)] \right\}, \tag{3}$$

$$\sigma_\alpha(z) \equiv \theta \begin{bmatrix} \frac{1}{2} + \frac{\alpha_1}{2} \\ \frac{1}{2} + \frac{\alpha_2}{2} \end{bmatrix} (z, \tau), \tag{4}$$

$$h(z) \equiv \sigma_{(0,0)}(z), \tag{5}$$

$$\theta^{(i)}(z) \equiv \theta \begin{bmatrix} \frac{1}{2} - \frac{i}{2} \\ \frac{1}{2} \end{bmatrix} (z, 2\tau), \quad i = 0, 1, \tag{6}$$

$$W_\alpha(z) = \frac{1}{2} \frac{\sigma_\alpha(z + \frac{w}{2})}{\sigma_\alpha(\frac{w}{2})}. \tag{7}$$

The R-matrix of the eight-vertex model takes the form (Fig. 1a)

$$R_{jk}(z) = \sum_\alpha W_\alpha(z) I_\alpha^{(j)} (I_\alpha^{-1})^{(k)} \tag{8}$$

which satisfies the Yang–Baxter equation (Fig. 2a)

$$\begin{aligned} &R_{ij}(z_i - z_j) R_{ik}(z_i - z_k) R_{jk}(z_j - z_k) \\ &= R_{jk}(z_j - z_k) R_{ik}(z_i - z_k) R_{ij}(z_i - z_j). \end{aligned} \tag{9}$$

It can be proved that the R-matrix of the eight-vertex model satisfies the following unitarity and cross-unitarity conditions:

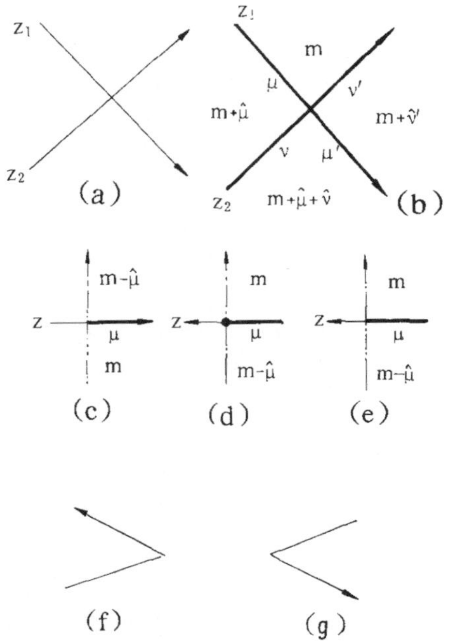

Fig. 1. Graphic representations. The dashed line is the boundary between the "vertex world" and the "face world", while solid lines are boundaries faces with different face weights. (a) $R_{12}(z_1-z_2)$, (b) $W(m|z_1-z_2)^{\mu'\nu'}_{\mu\nu}$, (c) $\phi_{m,\mu}(z)$, (d) $\bar{\phi}_{m,\mu}(z)$, (e) $\tilde{\phi}_{m,\mu}(z)$, (f) $K(z)$, (g) $\tilde{K}(z)$.

$$\text{unitarity}: \quad R_{ij}(z)R_{ji}(-z) = \rho(z)\cdot\text{id}, \tag{10}$$

$$\text{cross-unitarity}: \quad R^{t_i}_{ij}(z)R^{t_i}_{ji}(-z-2w) = \rho'(z)\cdot\text{id}, \tag{11}$$

where id is the identity and t_i denotes transposition in the ith space. $\rho(z)$ and $\rho'(z)$ are scalars satisfying

$$\rho(z) = \rho(-z), \tag{12}$$

$$\rho'(z) = \rho'(-z-2w). \tag{13}$$

In Eqs. (8)–(11), the indices take the value $i,j,k = 1,\ldots,l$.

2.2. Reflection equation and reflecting boundary conditions

We now deal with an exactly solvable lattice model with reflecting boundary conditions. The R-matrix defined above is the Boltzmann weights for this lattice model. In order to construct the transfer matrix of this system, we must introduce a reflecting boundary matrix $K(z)$ which is a 2×2 matrix and satisfies the reflection equation proposed by Cherednik [15] and Sklyanin [16] (Fig. 2b),

$$\begin{aligned} &R_{12}(z_1-z_2)K_1(z_1)R_{21}(z_1+z_2)K_2(z_2) \\ &\quad = K_2(z_2)R_{12}(z_1+z_2)K_1(z_1)R_{21}(z_1-z_2)\,. \end{aligned} \tag{14}$$

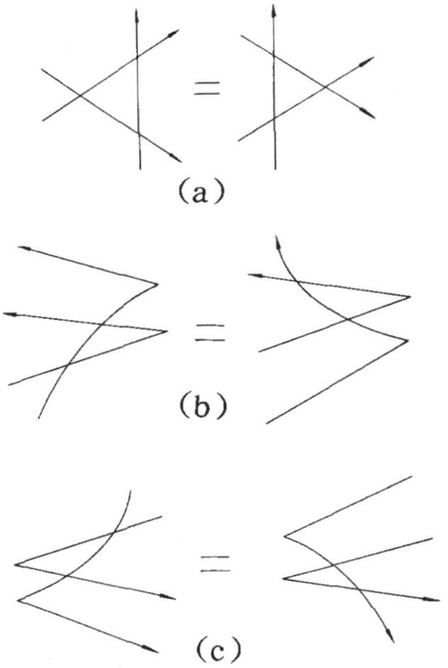

Fig. 2. YBE and RE. (a) YBE (9), (b) RE (14), (c) dual RE (16).

As mentioned in the introduction, two groups have independently obtained the general solution K of this reflection equation for the eight-vertex model. Here we take the solution $K(z)$ as [39]

$$K(z) = \sum_{\alpha} C_{\alpha} \frac{I_{\alpha}}{\sigma_{\alpha}(-z)}, \tag{15}$$

where C_{α} are arbitrary parameters. Correspondingly, we have the dual reflection equation which is necessary in the rest of this paper (Fig. 2c):

$$\begin{aligned} &R_{12}(z_2 - z_1)\tilde{K}_1(z_1)R_{21}(-z_1 - z_2 - 2w)\tilde{K}_2(z_2) \\ &= \tilde{K}_2(z_2)R_{12}(-z_1 - z_2 - 2w)\tilde{K}_1(z_1)R_{21}(z_2 - z_1). \end{aligned} \tag{16}$$

We take the solution of this reflection equation as

$$\tilde{K}(z) = \sum_{\alpha} \tilde{C}_{\alpha} \frac{I_{\alpha}}{\sigma_{\alpha}(z + w)}, \tag{17}$$

where $\tilde{C}_{\alpha}$ are also arbitrary parameters. Usually, we also call the reflecting boundary K and $\tilde{K}$ matrices the right and left boundary matrices, respectively.

In order to deal with the systems with open boundary conditions, let us define two forms of standard "row-to-row" monodromy matrices $S_1(z_1)$ and $T_1(z_1)$ which act in the space $V = V_1 \otimes V_2 \otimes \ldots \otimes V_l$ by

$$S_1(z_1) = R_{l1}(u_l + z_1)R_{l-1,1}(u_{l-1} + z_1)\ldots R_{31}(u_3 + z_1),$$

$$T_1(z_1) = R_{13}(z_1 - u_3) R_{14}(z_1 - u_4) \ldots R_{1l}(z_1 - u_l), \tag{18}$$

where $(u_l, u_{l-1}, \ldots u_3) \equiv \{u_i\}$ are arbitrary parameters. $S_2(z_2)$ and $T_2(z_2)$ can similarly be defined.

Considering that $\{u_i\}$ are the same for $S_1(z_1)$ and $S_2(z_2)$, and noticing that two R-matrices acting on four different spaces commute with each other, we find

$$R_{21}(z_1 - z_2) S_1(z_1) S_2(z_2) = R_{21}(z_1 - z_2) R_{l1}(u_l + z_1) R_{l2}(u_l + z_2) \ldots \tag{19}$$

Using the Yang–Baxter equation repeatedly, we have

$$R_{21}(z_1 - z_2) S_1(z_1) S_2(z_2) = S_2(z_2) S_1(z_1) R_{21}(z_1 - z_2). \tag{20}$$

Similarly, we can obtain

$$T_1(z_1) R_{12}(z_1 + z_2) S_2(z_2) = S_2(z_2) R_{12}(z_1 + z_2) T_1(z_1), \tag{21}$$

$$T_2(z_2) T_1(z_1) R_{12}(z_1 - z_2) = R_{12}(z_1 - z_2) T_1(z_1) T_2(z_2). \tag{22}$$

For the periodic boundary condition cases that have been studied extensively before, the transfer matrix is defined as the trace of the standard "row-to-row" monodromy matrix. But for the open boundary conditions cases, instead of the standard "row-to-row" monodromy matrix, we should define the "double-row" monodromy matrices which take the form

$$\begin{aligned} k_1(z_1) &= T_1(z_1) K_1(z_1) S_1(z_1), \\ k_2(z_2) &= T_2(z_2) K_2(z_2) S_2(z_2). \end{aligned} \tag{23}$$

Using the relations listed above, we can prove that $k_i(z_i)$ satisfies the reflection equation

$$\begin{aligned} &R_{12}(z_1 - z_2) k_1(z_1) R_{21}(z_1 + z_2) k_2(z_2) \\ &\quad = k_2(z_2) R_{12}(z_1 + z_2) k_1(z_1) R_{21}(z_1 - z_2). \end{aligned} \tag{24}$$

As commonly used in the framework of the quantum inverse scattering method, $k_i(z_i)$ are 2×2 matrices with elements defined as operators acting in the space $V' = V_3 \otimes V_4 \otimes \ldots \otimes V_l$, which is the so-called quantum space; the spaces V_1 and V_2 are auxiliary spaces. Eqs. (14) and (24) show that $k(z)$ is the co-module of $K(z)$.

2.3. The transfer matrix

Now, let us formulate the transfer matrix with open boundary conditions:

$$\begin{aligned} t(z_i) &= \mathrm{Tr}_{V_i} \left\{ \tilde{K}_i(z_i) k_i(z_i) \right\} \\ &= \sum_{kl} \tilde{K}(z_i)_{kl} k(z_i)_{lk}, \end{aligned} \tag{25}$$

with $i = 1, 2$. Since the transfer matrices are defined as the trace over the auxiliary spaces $V_i, i = 1, 2$, they should be independent of V_1 and V_2, and are represented as operators acting in the quantum space $V_3 \otimes \ldots \otimes V_l$. With the help of the unitarity and cross-unitarity relations of the R-matrix, the Yang–Baxter relation, the reflection equation and its dual reflection equation, we can prove that the transfer matrices with different spectrum commute with each other [48],

$$t(z_1)t(z_2) = t(z_2)t(z_1). \tag{26}$$

This ensures the integrability of this system.

The aim of this paper is to find the eigenvalues and eigenvectors of the transfer matrix which defines the Hamiltonian of the system under consideration. We will use the algebraic Bethe ansatz method to solve this problem. The transfer matrix is defined as a linear function of the elements of the "double-row" monodromy matrix. So, it is necessary to find the proper linear combinations of the elements of the "double-row" monodromy matrix whose commutation relations are suitable for the algebraic Bethe ansatz. Besides this, we also need to find a "vacuum" state which is independent of the spectrum z. It is well known that this "vacuum" state can be obtained easily for the six-vertex model with periodic or open boundary conditions. For the eight-vertex model, it is not a trivial problem. We will study the commutation relations and the "vacuum" state problems in the following sections.

3. Commutation relations

It is known that for the six-vertex model and other trigonometric vertex models we can obtain the necessary commutation relations directly from the reflection equation in which $k(z)$ is the "double-row" monodromy matrix. But for the eight-vertex model, whose R-matrix has eight non-zero elements, we can not obtain such relations directly from the reflection equation. We have to use the vertex–face correspondence to solve this problem. This means we should properly combine the elements of $k(z)$ so that we can find simple commutation relations that can be dealt with by the algebraic Bethe ansatz method.

3.1. Face–vertex correspondence

We first define a two-element column vector $\phi_{m,\mu}(z), \mu = 0,1, m \in \mathbb{Z}$, whose kth element is [7,42,52] (Fig. 1c)

$$\phi^k_{m,\mu} = \theta^{(k)}(z + (-1)^\mu wa + w\beta), \tag{27}$$

where $a = m+\gamma, \gamma, \beta \in \mathbb{C}, k = 0,1$. We call m the face weight, μ the face index, which takes the values $0,1$. ϕ is usually called the three-spin operator. It can be proved that we can find row vectors $\bar{\phi}, \tilde{\phi}$ satisfying the following conditions for generic w, β, γ:

$$\begin{aligned} \tilde{\phi}_{m+\hat{\mu},\mu}(z)\phi_{m+\hat{\nu},\nu}(z) &= \delta_{\mu\nu}, \\ \bar{\phi}_{m,\mu}(z)\phi_{m,\nu}(z) &= \delta_{\mu\nu}, \end{aligned} \tag{28}$$

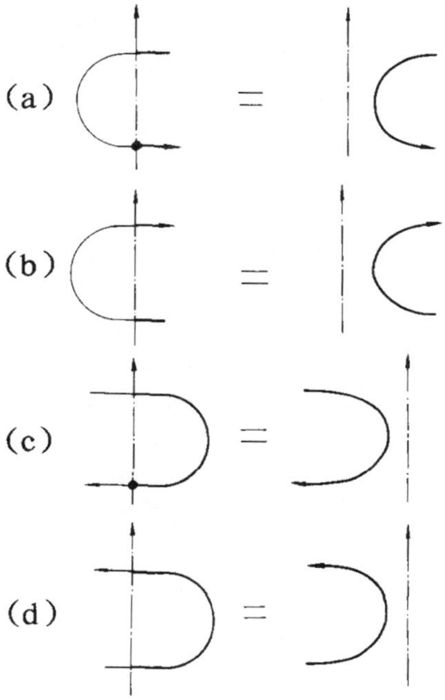

Fig. 3. Orthogonal relations. Note the differences between (a) and (b), and between (c) and (d). (a) First equation of (28), (b) Second equation of (28), (c) First equation of (29), (d) Second equation of (29).

where

$$\hat{\mu} \equiv (-1)^{\mu}, \qquad \hat{\nu} \equiv (-1)^{\nu}.$$

The above relations can also be written in another form

$$\sum_{\nu=0}^{1} \phi_{m+\hat{\nu},\nu}(z)\tilde{\phi}_{m+\hat{\nu},\nu}(z) = I,$$

$$\sum_{\mu=0}^{1} \phi_{m,\mu}(z)\bar{\phi}_{m,\mu}(z) = I. \tag{29}$$

As usual, I is the 2×2 unit matrix (Fig. 3).

We define the face Boltzmann weights for the interaction-round-a-face model (IRF) as follows [7,42,46,52]:

$$W(m|z)^{\mu\mu}_{\mu\mu} = \frac{h(z+w)}{h(w)},$$

$$W(m|z)^{\nu\mu}_{\mu\nu} = \frac{h(w(m+\gamma)-(-1)^{\mu}z)}{h(w(m+\gamma))}, \quad \mu \neq \nu,$$

$$W(m|z)^{\mu\nu}_{\mu\nu} = \frac{h(z)h(w(m+\gamma)-(-1)^{\mu}w)}{h(w)h(w(m+\gamma))}, \quad \mu \neq \nu, \tag{30}$$

where the face indices μ, ν take the values 0, 1. The other face Boltzmann weights are defined as zeroes, so we can see explicitly that for a given face weight m, we only have six non-zero face Boltzmann weights in total. Traditionally, the face Boltzmann weights for the eight-vertex SOS model are denoted as $W_z\begin{bmatrix} a & b \\ c & d \end{bmatrix}$. Its relation with the notations used in this paper is (Fig. 1b):

$$W(m|z)_{\mu\nu}^{\mu'\nu'} = W_z\begin{bmatrix} m+\hat{\mu}+\hat{\nu} & m+\hat{\nu}' \\ m+\hat{\mu} & m \end{bmatrix}. \tag{31}$$

The reason that we use notations (30) is that not many zero face Boltzmann weights will appear in our calculation. It is also convenient to compare the results of the eight-vertex model case with the six-vertex model case.

The face Boltzmann weights of the IRF model defined above have a relation with the R-matrix of the eight-vertex model, which is usually called the face–vertex correspondence [3,7,42],

$$\begin{aligned} &R_{12}(z_1-z_2)\phi^{(1)}_{m+\hat{\mu}+\hat{\nu},\mu}(z_1)\phi^{(2)}_{m+\hat{\nu},\nu}(z_2) \\ &= \sum_{\mu',\nu'} W(m|z_1-z_2)^{\mu\nu}_{\mu',\nu'}\phi^{(2)}_{m+\hat{\mu}'+\hat{\nu}',\nu'}(z_2)\phi^{(1)}_{m+\hat{\mu}',\mu'}(z_1), \end{aligned} \tag{32}$$

where $\phi^{(i)}$ denote that it act in the ith space $(i=1,2)$.

With the help of the properties of $\phi, \tilde{\phi}, \bar{\phi}$ (28), (29), we can derive the following relations from the above face–vertex correspondence relation:

$$\begin{aligned} &\tilde{\phi}^{(1)}_{m+\hat{\mu},\mu}(z_1)R_{12}(z_1-z_2)\phi^{(2)}_{m+\hat{\nu},\nu}(z_2) \\ &= \sum_{\mu'\nu'} W(m|z_1-z_2)^{\mu'\nu}_{\mu\nu'}\tilde{\phi}^{(1)}_{m+\hat{\mu}'+\hat{\nu},\mu'}(z_1)\phi^{(2)}_{m+\hat{\mu}+\hat{\nu}',\nu'}(z_2), \end{aligned} \tag{33}$$

$$\begin{aligned} &\tilde{\phi}^{(2)}_{m+\hat{\mu}+\hat{\nu},\nu}(z_2)\tilde{\phi}^{(1)}_{m+\hat{\mu},\mu}(z_1)R_{12}(z_1-z_2) \\ &= \sum_{\mu'\nu'} W(m|z_1-z_2)^{\mu'\nu'}_{\mu\nu}\tilde{\phi}^{(1)}_{m+\hat{\mu}'+\hat{\nu}',\mu'}(z_1)\tilde{\phi}^{(2)}_{m+\hat{\nu}',\nu'}(z_2), \end{aligned} \tag{34}$$

$$\begin{aligned} &\bar{\phi}^{(2)}_{m,\nu}(z_2)R_{12}(z_1-z_2)\phi^{(1)}_{m,\mu}(z_1) \\ &= \sum_{\nu'\mu'} W(m-\hat{\mu}-\hat{\nu}'|z_1-z_2)^{\mu\nu'}_{\mu'\nu}\phi^{(1)}_{m-\hat{\nu},\mu'}(z_1)\bar{\phi}^{(2)}_{m-\hat{\mu},\nu'}(z_2), \end{aligned} \tag{35}$$

$$\begin{aligned} &\bar{\phi}^{(2)}_{m+\hat{\mu}+\hat{\nu},\nu}(z_2)\bar{\phi}^{(1)}_{m+\hat{\mu},\mu}(z_1)R_{12}(z_1-z_2) \\ &= \sum_{\mu'\nu'} W(m|z_1-z_2)^{\mu'\nu'}_{\mu\nu}\bar{\phi}^{(1)}_{m+\hat{\mu}'+\hat{\nu}',\mu'}(z_1)\bar{\phi}^{(2)}_{m+\hat{\nu}',\nu'}(z_2). \end{aligned} \tag{36}$$

All of the relations obtained above have describe the correspondence between face and vertex models. Usually, we call ϕ the intertwiner of the face–vertex correspondence (Fig. 4).

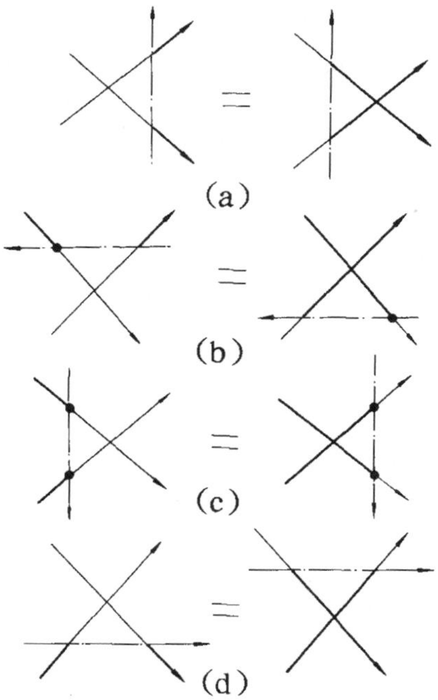

Fig. 4. Face–vertex correspondences. Note the differences between (b) and (d). (a) Eq. (32), (b) Eq. (33), (c) Eq. (34), (d) Eq. (35).

3.2. Commutation relations for elements of the face boundary reflecting k matrix

As mentioned above, in order to obtain comparatively simple commutation relations that can be dealt with by the algebraic Bethe ansatz method, we should change the "vertex" reflection equations to "face" reflection equations, since there the new "R-matrix" face Boltzmann weights have only six non-zero elements instead of eight non-zero elements in the "vertex" case. For this purpose, by using the three-spin operator ϕ, we change the vertex boundary reflecting matrix to the face boundary reflecting matrix, and find the commutation relations between the elements of the "face" type monodromy matrix, which are useful for the quantum inverse scattering method.

We first change the matrix $\tilde{K}(z)$ defined in Eqs. (16), (17) to the face boundary reflecting matrix $\tilde{K}(m|z)^{\nu}_{\mu}$, using the unitarity properties of the intertwiner (29). We find

$$\begin{aligned}\tilde{K}(z) &= \sum_{\mu}\left\{\phi_{m,\mu}(-z)\bar{\phi}_{m,\mu}(-z)\tilde{K}(z)\sum_{\nu}[\phi_{m-\hat{\mu}+\hat{\nu},\nu}(z)\tilde{\phi}_{m-\hat{\mu}+\hat{\nu},\nu}(z)]\right\}\\ &= \sum_{\mu\nu}\phi_{m,\mu}(-z)\tilde{\phi}_{m-\hat{\mu}+\hat{\nu},\nu}(z)\tilde{K}(m|z)^{\nu}_{\mu},\end{aligned}$$

where

$$\tilde{K}(m|z)^{\nu}_{\mu} \equiv \bar{\phi}_{m,\mu}(-z)\tilde{K}(z)\phi_{m-\hat{\mu}+\hat{\nu},\nu}(z) . \tag{37}$$

Thus, the transfer matrix of the eight-vertex model with open boundary conditions can be rewritten as

$$\begin{aligned} t(z) &= \mathrm{Tr}\left(\tilde{K}(z)k(z)\right) \\ &= \mathrm{Tr}\left\{\sum_{\mu\nu}\phi_{m,\mu}(-z)\tilde{\phi}_{m-\hat{\mu}+\hat{\nu},\nu}(z)\tilde{K}(m|z)^{\nu}_{\mu}k(z)\right\} \\ &= \sum_{\mu\nu}[\tilde{\phi}_{m-\hat{\mu}+\hat{\nu},\nu}(z)k(z)\phi_{m,\mu}(-z)]\tilde{K}(m|z)^{\nu}_{\mu} \\ &\equiv \sum_{\mu\nu}k(m|z)^{\mu}_{\nu}\tilde{K}(m|z)^{\nu}_{\mu}, \end{aligned} \tag{38}$$

which is true for arbitrary m. Here we also introduce the definition of the face boundary reflecting k matrix as

$$k(m|z)^{\mu}_{\nu} = \tilde{\phi}_{m-\hat{\mu}+\hat{\nu},\nu}(z)k(z)\phi_{m,\mu}(-z). \tag{39}$$

We call m and $m-\hat{\mu}+\hat{\nu}$ the initial and final weight of $k(m|z)^{\mu}_{\nu}$, respectively. Thus, we have written out the transfer matrix by using the face form of the model.

Next, we will derive the face reflection equation directly from the vertex reflection equation by using the intertwiner. Multiply both sides of Eq. (14) from the left by $\tilde{\phi}^{(1)}_{m+\hat{\mu}_0,\mu_0}(z_1)\tilde{\phi}^{(2)}_{m+\hat{\mu}_0+\hat{\nu}_0,\nu_0}(z_2)$, from the right by $\phi^{(1)}_{m+\hat{\mu}_3,\mu_3}(-z_1)\ \phi^{(2)}_{m+\hat{\mu}_3+\hat{\nu}_3,\nu_3}(-z_2)$. Notice the properties such as $\tilde{\phi}^{(2)}$ commutes with $\tilde{\phi}^{(1)}$ and $k_1(z_1)$ and use the face–vertex correspondence relations. We then get the face reflection equation (see Appendix A and Figs. 5 and 6) [45,47,34],

$$\begin{aligned} & k(m+\hat{\mu}_2+\hat{\nu}_1|z_1)^{\mu_2}_{\mu_1}k(m+\hat{\mu}_3+\hat{\nu}_3|z_2)^{\nu_3}_{\nu_2} \\ &\quad \times W(m|z_1+z_2)^{\nu_2\mu_3}_{\nu_1\mu_2}W(m|z_1-z_2)^{\mu_1\nu_1}_{\mu_0\nu_0} \\ &\quad = k(m+\hat{\mu}_1+\hat{\nu}_2|z_2)^{\nu_1}_{\nu_0}k(m+\hat{\mu}_3+\hat{\nu}_3|z_1)^{\mu_2}_{\mu_1} \\ &\quad \times W(m|z_1+z_2)^{\mu_1\nu_2}_{\mu_0\nu_1}W(m|z_1-z_2)^{\nu_3\mu_3}_{\nu_2\mu_2}. \end{aligned} \tag{40}$$

Here and below summation over repeated indices is assumed. One can find that this equation is true for arbitrary μ_0,ν_0,μ_3,ν_3. We should point out here that this face reflection equation is different from the one proposed by Behrend et al. in Ref. [45], if the cross-inversion relation of the face Boltzmann weights [42] is applied, the two equations are equivalent.

Now, we will let the indices in the above relation take special values so that we obtain the necessary commutation relations. When $\mu_0=\nu_0=0, \mu_3=\nu_3=1$, we get

$$\begin{aligned} & k(m|z_1)^1_0k(m-2|z_2)^1_0W(m|z_1+z_2)^{01}_{01}W(m|z_1-z_2)^{00}_{00} \\ &\quad = k(m|z_2)^1_0k(m-2|z_1)^1_0W(m|z_1+z_2)^{01}_{01}W(m|z_1-z_2)^{11}_{11}. \end{aligned} \tag{41}$$

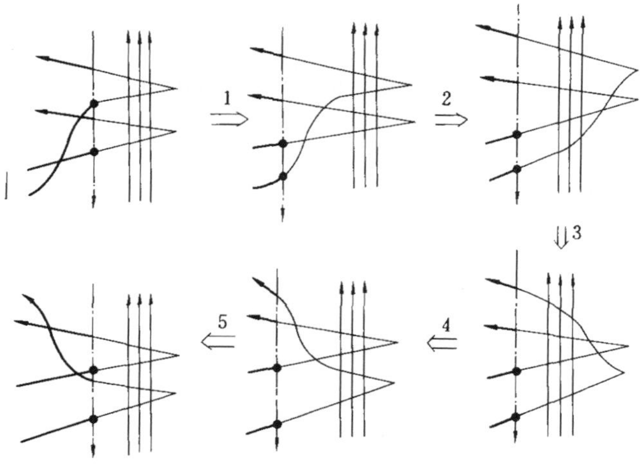

Fig. 5. Derivation of Eq. (40) by graphic representation. Steps 1 and 5 by the face–vertex correspondence in Fig. 4; steps 2 and 4 by YBE in Fig. 2a; step 3 by RE in Fig. 2b.

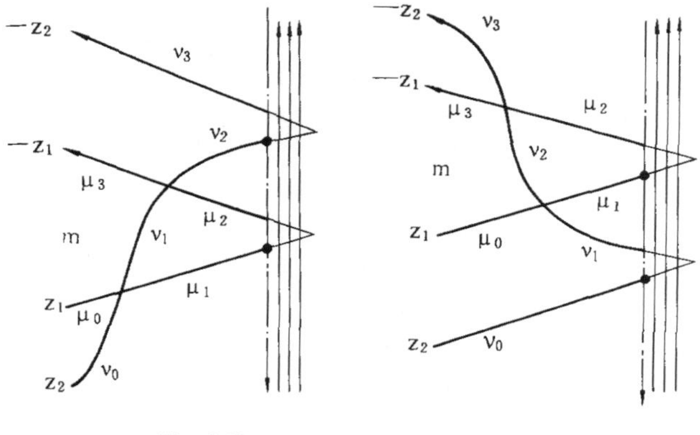

Fig. 6. Eq. (40) with face indices.

From the definition of the face Boltzmann weights, we know that $W(m|z)^{00}_{00} = W(m|z)^{11}_{11} = \frac{h(z+w)}{h(w)}$, so we find

$$k(m|z_1)^1_0 k(m-2|z_2)^1_0 = k(m|z_2)^1_0 k(m-2|z_1)^1_0. \tag{42}$$

This means that the positions of z_1 and z_2 can be exchanged in this form.

Let $\mu_0 = \nu_0 = \nu_3 = 0, \mu_3 = 1$. We obtain

$$\begin{aligned} & k(m+2|z_2)^0_0 k(m|z_1)^1_0 W(m|z_1+z_2)^{00}_{00} W(m|z_1-z_2)^{01}_{01} \\ & = k(m|z_1)^1_0 k(m|z_2)^0_0 W(m|z_1+z_2)^{01}_{01} W(m|z_1-z_2)^{00}_{00} \\ & \quad - k(m|z_2)^1_0 k(m|z_1)^0_0 W(m|z_1+z_2)^{01}_{01} W(m|z_1-z_2)^{01}_{10} \\ & \quad - k(m|z_2)^1_0 k(m|z_1)^1_1 W(m|z_1+z_2)^{10}_{01} W(m|z_1-z_2)^{01}_{01}. \end{aligned} \tag{43}$$

Denote

$$\begin{aligned}
A(m|z) &\equiv k(m|z)_0^0, \\
D(m|z) &\equiv k(m|z)_1^1, \\
B(m|z) &\equiv k(m|z)_0^1, \\
C(m|z) &\equiv k(m|z)_1^0.
\end{aligned} \tag{44}$$

So, the matrix $\begin{pmatrix} A & B \\ C & D \end{pmatrix}$ is the boundary k matrix in the "face" form. Relations (42), (43) give the commutation relations of BB and AB, respectively. In order to use the algebraic Bethe ansatz method, we must also have the commutation relation of DB.

Let $\mu_0 = \mu_3 = \nu_3 = 1, \nu_0 = 0$. Exchanging z_1 and z_2, we get one relation; exchanging z_1 and z_2 in Eq. (43), we find another relation. Combining the two relations and using Eq. (43) we find the commutation relation of DB:

$$\begin{aligned}
& k(m+2|z_2)_1^1 k(m|z_1)_0^1 W(m+2|z_1+z_2)_{01}^{01} W(m+2|z_1-z_2)_{10}^{10} \\
& \quad \times W(m|z_1+z_2)_{00}^{00} W(m|z_1-z_2)_{01}^{01} \\
& \quad = k(m|z_2)_0^1 k(m|z_1)_0^0 W(m+2|z_1+z_2)_{10}^{01} W(m|z_1+z_2)_{01}^{01} \\
& \qquad \times \{ W(m+2|z_2-z_1)_{11}^{11} W(m|z_2-z_1)_{00}^{00} \\
& \qquad -W(m+2|z_2-z_1)_{10}^{01} W(m|z_1-z_2)_{10}^{01} \} \\
& \qquad +k(m|z_2)_0^1 k(m|z_1)_1^1 W(m+2|z_2-z_1)_{10}^{01} W(m|z_2-z_1)_{01}^{01} \\
& \qquad \times \{ -W(m+2|z_1+z_2)_{11}^{11} W(m|z_1+z_2)_{00}^{00} \\
& \qquad +W(m+2|z_1+z_2)_{10}^{01} W(m|z_1+z_2)_{01}^{10} \} \\
& \qquad +k(m|z_1)_0^1 k(m|z_2)_0^0 W(m+2|z_1+z_2)_{10}^{01} W(m|z_1+z_2)_{01}^{01} \\
& \qquad \times \{ -W(m+2|z_2-z_1)_{11}^{11} W(m|z_2-z_1)_{10}^{01} \\
& \qquad +W(m+2|z_2-z_1)_{10}^{01} W(m|z_1-z_2)_{00}^{00} \} \\
& \qquad +k(m|z_1)_0^1 k(m|z_2)_1^1 W(m+2|z_2-z_1)_{11}^{11} W(m|z_2-z_1)_{01}^{01} \\
& \qquad \times \{ W(m+2|z_1+z_2)_{11}^{11} W(m|z_1+z_2)_{00}^{00} \\
& \qquad -W(m+2|z_1+z_2)_{10}^{01} W(m|z_1+z_2)_{01}^{10} \} \, .
\end{aligned} \tag{45}$$

In the right-hand side of this equation, behind $k(m|z_2)_0^1$, we find not only term $k(m|z_1)_1^1$ but also term $k(m|z_1)_0^0$. This will cause trouble in the proceeding of the algebraic Bethe ansatz method, especially in the case where the thermodynamic limit of this system is taken. In order to solve this problem, we should reformulate A and D as A and $\tilde{D}$ so that when we commute $\tilde{D}$ with B, only the term $\tilde{D}$ exists behind $B(z_2)$, which is the notation of $k(m|z_2)_0^1$.

It is not easy to find $\tilde{D}$ by direct calculation, but the work of Sklyanin [16] gives us a hint to formulate the term $\tilde{D}$. Sklyanin pointed out in his paper that the term $\tilde{D}$ in the six-vertex model case is one of the elements of the inverse matrix of the monodromy matrix. Now we will study the inverse of the face form monodromy matrix for the eight-vertex model case.

For simplicity, abusing the face weights m, etc., we can write the face reflection equation as

$$\begin{aligned} & W_{12}(z_1 - z_2) k_1(z_1) W_{21}(z_1 + z_2) k_2(z_2) \\ & \quad = k_2(z_2) W_{12}(z_1 + z_2) k_1(z_1) W_{21}(z_1 - z_2). \end{aligned} \tag{46}$$

It is just the same as the vertex reflection equation. If the inverse of $k(z)$ exists, we have

$$\begin{aligned} & k_2^{-1}(z_2) W_{12}(z_1 - z_2) k_1(z_1) W_{21}(z_1 + z_2) \\ & \quad = W_{12}(z_1 + z_2) k_1(z_1) W_{21}(z_1 - z_2) k_2^{-1}(z_2). \end{aligned} \tag{47}$$

This means that $k_2^{-1}(-z_2)$ and $k_2(z_2)$ has a similar exchange relation with $k_1(z_1)$ in the formalism. We have known that the commutation relation of $k(m+2|z_2)_0^0$ with $k(m|z_1)_0^1$ is comparatively simple, which is also useful for the algebraic Bethe ansatz. From the above equation we assume that the commutation relation of $k^{-1}(m+2|z_2)_0^0$ with $k(m|z_1)_0^1$ should have similar properties in the formalism. We know that $k_2^{-1}(z_2)_0^0$ is a linear combination of A and D in the six-vertex model. We hope that this is also true for the case of the eight-vertex model. Fortunately, we have obtained the expected result.

Defining

$$\begin{aligned} Q(m|z-w)_{\mu_3}^{\nu_2} &= k(m+\hat{\bar{\mu}}_3+\hat{\nu}_2|z-w)_{\mu_1}^{\mu_2} W(m|2z-w)_{\nu_1\mu_2}^{\nu_2\bar{\mu}_3} W(m|-w)_{01}^{\mu_1\nu_1}, \\ Q'(m|z-w)_{\nu_1}^{\mu_0} &= k(m|z-w)_{\mu_1}^{\mu_2} W(m|2z-w)_{\bar{\mu}_0\nu_1}^{\mu_1\nu_2} W(m|-w)_{\nu_2\mu_2}^{01}, \end{aligned} \tag{48}$$

where, as usual, summation over repeated indices is assumed and $\bar{0}=1, \bar{1}=0$, we can show (Appendix B) that

$$\begin{aligned} Q(m|z-w)_\nu^{\nu'} k(m+\hat{\bar{\nu}}+\hat{\nu}''|z)_{\nu'}^{\nu''} &= \rho(m,\nu|z)\delta_{\nu\nu''}, \\ k(m+\hat{\bar{\mu}}''+\hat{\mu}'|z)_\mu^{\mu'} Q'(m|z-w)_{\mu'}^{\mu''} &= \rho'(m,\mu|z)\delta_{\mu\mu''}, \end{aligned} \tag{49}$$

where $\rho(m,\nu|z)$, and $\rho'(m,\mu|z)$ are scalars of the "quantum space". We can rescale Q and Q' such that $\rho, \rho' \to 1$. But this is not necessary in the following derivation. Multiply (40) by $Q(m+\hat{\nu}+\hat{\mu}_0|z_2-w)_\nu^{\nu_0}$ from left and by $Q'(m+\hat{\mu}_3+\hat{\nu}'|z_2-w)_{\nu_3}^{\nu'}$ from the right. Summation over ν_0, ν_3 gives

$$\begin{aligned} & Q(m+\hat{\nu}+\hat{\mu}_0|z_2-w)_\nu^{\nu_0} k(m+\hat{\mu}_3+\hat{\nu}_2|z_1)_{\mu_1}^{\mu_2} W(m|z_1+z_2)_{\nu_1\mu_2}^{\nu_2\mu_3} \\ & \quad \times W(m|z_1-z_2)_{\mu_0\nu_0}^{\mu_1\nu_1} \rho'(m+\hat{\mu}_3+\hat{\nu}', \nu'|z_2)\delta_{\nu_2\nu'} \\ & \quad = \rho(m+\hat{\nu}+\hat{\mu}_0, \nu|z_2)\delta_{\nu\nu_1} k(m+\hat{\mu}_3+\hat{\nu}_3|z_1)_{\mu_1}^{\mu_2} Q'(m+\hat{\mu}_3+\hat{\nu}'|z_2-w)_{\nu_3}^{\nu'} \\ & \qquad \times W(m|z_1+z_2)_{\mu_0\nu_1}^{\mu_1\nu_2} W(m|z_1-z_2)_{\nu_2\mu_2}^{\nu_3\mu_3}. \end{aligned} \tag{50}$$

In the derivation of LHS, we use the fact that $W(m|z_1+z_2)_{\nu_1\mu_2}^{\nu_2\mu_3} \neq 0$ only if $\hat{\nu}_1+\hat{\mu}_2=\hat{\nu}_2+\hat{\mu}_3$. Eq. (50) becomes

$$Q(m+\hat{\nu}+\hat{\mu}_0|z_2-w)_{\nu}^{\nu_0}k(m+\hat{\mu}_3+\hat{\nu}'|z_1)_{\mu_1}^{\mu_2}$$
$$\times W(m|z_1+z_2)_{\nu_1\mu_2}^{\nu'\mu_3}W(m|z_1-z_2)_{\mu_0\nu_0}^{\mu_1\nu_1}$$
$$=k(m+\hat{\mu}_3+\hat{\nu}_3|z_1)_{\mu_1}^{\mu_2}Q'(m+\hat{\mu}_3+\hat{\nu}'|z_2-w)_{\nu_3}^{\nu'}$$
$$\times W(m|z_1+z_2)_{\mu_0\nu}^{\mu_1\nu_2}W(m|z_1-z_2)_{\nu_2\mu_2}^{\nu_3\mu_3}\frac{\rho(m+\hat{\nu}+\hat{\mu}_0,\nu|z_2)}{\rho'(m+\hat{\mu}_3+\hat{\nu}',\nu'|z_2)}, \tag{51}$$

which is similar to (40). Put $\mu_0=\nu=\nu'=0,\mu_3=1$, and notice $m+\hat{0}+\hat{0}=m+2, m+\hat{0}+\hat{1}=m$. We then have

$$Q(m+2|z_2-w)_0^1k(m|z_1)_0^0W(m|z_1+z_2)_{10}^{01}W(m|z_1-z_2)_{01}^{01}$$
$$+Q(m+2|z_2-w)_0^0k(m|z_1)_0^1W(m|z_1+z_2)_{01}^{01}W(m|z_1-z_2)_{00}^{00}$$
$$+Q(m+2|z_2-w)_0^1k(m|z_1)_1^1W(m|z_1+z_2)_{01}^{01}W(m|z_1-z_2)_{01}^{10}$$
$$=k(m|z_1)_0^1Q'(m|z_2-w)_0^0W(m|z_1+z_2)_{00}^{00}W(m|z_1-z_2)_{01}^{01}\frac{\rho(m+2,0|z_2)}{\rho'(m,0|z_2)}. \tag{52}$$

Using the same derivation as that in Appendix B, we can calculate

$$\frac{\rho(m+2,0|z_2)}{\rho'(m,0|z_2)}=\frac{(-1)h(wa)h(w(a+1))}{h(w(a-1))h(w(a+2))}, \tag{53}$$

where $a\equiv m+\gamma$ and γ is a parameter in defining ϕ. From (48) and considering the explicit form of W, we can write Q,Q' in components of $k(z)$,

$$Q(m|z-w)_0^1=k(m-2|z-w)_0^1W(m|2z-w)_{11}^{11}W(m|-w)_{01}^{01}, \tag{54}$$

which is proportional to $B(m-2|z-w)$, and

$$Q(m|z-w)_0^0=-\left[k(m|z-w)_1^1W(m|2z-w)_{10}^{10}\right.$$
$$\left.-k(m|z-w)_0^0W(m|2z-w)_{10}^{01}\right]W(m|-w)_{01}^{01}$$
$$=-Q'(m|z-w)_0^0$$
$$\equiv-\tilde{D}(m|z-w)W(m|-w)_{01}^{01}, \tag{55}$$

where

$$\tilde{D}(m|z)\equiv-k(m|z)_0^0W(m|2z+w)_{10}^{01}+k(m|z)_1^1W(m|2z+w)_{10}^{10}$$
$$=\frac{-Q(m|z)}{W(m|-w)_{01}^{01}} \tag{56}$$

is indeed a linear combination of $A(m|z)\equiv k(m|z)_0^0$ and $D(m|z)\equiv k(m|z)_1^1$. Substitute (53)–(55) into (52) and write $k(m|z_2-w)_1^1$ in the equation in terms of $\tilde{D}(m|z_2-w)$ and $A(m|z_2-w)$. We then change the parameter z_2-w to z_2, giving the expected equation:

$$
\begin{aligned}
&\tilde{D}(m+2|z_2)B(m|z_1)\\
&\quad = B(m|z_1)\tilde{D}(m|z_2)\frac{h(z_1-z_2-w)h(z_1+z_2+2w)}{h(z_1+z_2+w)h(z_1-z_2)}\\
&\quad +B(m|z_2)A(m|z_1)\frac{h(2z_1)h(z_1+z_2+w(a+2))h(2z_2+2w)}{h(2z_1+w)h(z_1+z_2+w)h(w(a+1))}\\
&\quad +B(m|z_2)\tilde{D}(m|z_1)\frac{h(z_2-z_1+w(a+1))h(2z_2+2w)h(w)}{h(2z_1+w)h(z_1-z_2)h(w(a+1))},
\end{aligned}
\tag{57}
$$

where $h(z)\equiv\sigma_0(z)$. In the derivation we have used a formula of θ function [7]

$$
\begin{aligned}
&h(u+x)h(u-x)h(v+y)h(v-y)-h(u+y)h(u-y)h(v+x)h(v-x)\\
&\quad = h(u+v)h(u-v)h(x+y)h(x-y).
\end{aligned}
\tag{58}
$$

We need also to change D in the exchange relation of A and B by a linear combination of $\tilde{D}$ and A. Thus (43) is rewritten as

$$
\begin{aligned}
&A(m+2|z_2)B(m|z_1)\\
&\quad = B(m|z_1)A(m|z_2)\frac{h(z_1+z_2)h(z_1-z_2+w)}{h(z_1+z_2+w)h(z_1-z_2)}\\
&\quad -B(m|z_2)A(m|z_1)\frac{h(2z_1)h(z_1-z_2+w(a+1))h(w)}{h(2z_1+w)h(z_1-z_2)h(w(a+1))}\\
&\quad -B(m|z_2)\tilde{D}(m|z_1)\frac{h(-z_1-z_2+wa)[h(w)]^2}{h(z_1+z_2+w)h(2z_1+w)h(w(a+1))}.
\end{aligned}
\tag{59}
$$

We have also used (58) in the derivation. With the permutation relations of AB, $\tilde{D}B$ and BB (57), (59), (42), we obtain the algebraic Bethe ansatz equations provided

(1) the transfer matrix $t(z)$ is a linear combination of A and $\tilde{D}$;

(2) there is a "vacuum state" of the quantum space which is an eigenstate of A and $\tilde{D}$ but not an eigenstate of B.

We will study these problems in Section 4.

4. Vacuum state and boundary conditions

4.1. Vacuum state

The algebraic Bethe ansatz requires the construction of a state of the "quantum" space, which is an eigenstate of the operators A and D all with spectrum parameter z. This state is called a vacuum state, which is also an eigenstate of $\tilde{D}$. Before introducing the vacuum state, let us make some preparations. We first express S and T (see Eq. (18)) in terms of the "face language", change their auxiliary indices to face indices, and express $k(m|z)^{\nu}_{\mu}$ by such expressions of S and T. From (23), (39), the operator $k(z)$ with "face" indices can be written as

$$
k(m|z)^{\nu'}_{\mu'}=\tilde{\phi}_{m+\hat{\mu}'-\hat{\nu}',\mu'}(z)T(z)K(z)S(z)\phi_{m,\nu'}(-z). \tag{60}
$$

From (28), (29) we have

$$K(z)=\sum_{\mu\nu}\left\{\phi_{m_0+\hat{\mu}-\hat{\nu},\mu}(z)\bar{\phi}_{m_0+\hat{\mu}-\hat{\nu},\mu}(z)K(z)\phi_{m_0,\nu}(-z)\bar{\phi}_{m_0,\nu}(-z)\right\}. \tag{61}$$

Combining these two equations gives

$$k(m|z)_{\mu'}^{\nu'}=\sum_{\mu\nu}K(m_0|z)_{\mu}^{\nu}T(m-\hat{\nu}',m_0-\hat{\nu}|z)_{\mu'\mu}S(m,m_0|z)_{\nu'\nu}, \tag{62}$$

where we define

$$K(m_0|z)_{\mu}^{\nu}\equiv\tilde{\phi}_{m_0+\hat{\mu}-\hat{\nu},\mu}(z)K(z)\phi_{m_0,\nu}(-z), \tag{63}$$

$$T(m-\hat{\nu}',m_0-\hat{\nu}|z)_{\mu'\mu}\equiv\tilde{\phi}_{m+\hat{\mu}'-\hat{\nu}',\mu'}(z)T(z)\phi_{m_0+\hat{\mu}-\hat{\nu},\mu}(z), \tag{64}$$

$$S(m,m_0|z)_{\nu'\nu}\equiv\bar{\phi}_{m_0,\nu}(-z)S(z)\phi_{m,\nu'}(-z). \tag{65}$$

Eq. (62) is true for all m and m_0.

Next, assume that we can properly choose the parameters of the right reflecting matrix such that for given m_0 and all z, $K(m_0|z)_1^0=0$. This is possible. We will study this problem in the second part of this section. Actually, this requirement constitute the only restriction on the boundary matrices in our approach.

Then, multiply $\tilde{\phi}_{m,1}^{(1)}(z)\bar{\phi}_{m_0+1,0}^{(2)}(-z)$ from the left and multiply $\phi_{m_0+1,0}^{(1)}(z)\phi_{m,1}^{(2)}(-z)$ from the right to Eq. (21). Using the face-vertex correspondence relations (32)–(36) we can prove the following exchange relation of S and T with face indices:

$$\begin{aligned}&W(m+1|2z)_{11}^{11}S(m-1,m_0+1|z)_{10}T(m,m_0|z)_{10}\\&\quad+W(m+1|2z)_{10}^{01}S(m+1,m_0+1|z)_{00}T(m,m_0|z)_{00}\\&\quad=W(m_0+1|2z)_{10}^{01}T(m+1,m_0+1|z)_{11}S(m,m_0|z)_{11}\\&\qquad+W(m_0-1|2z)_{00}^{00}T(m+1,m_0-1|z)_{10}S(mm_0|z)_{10},\end{aligned} \tag{66}$$

which will be useful in our derivation. For later convenience we rewrite S,T as

$$\begin{aligned}S(z)&=R_{l0}(u_l+z)R_{l-1,0}(u_{l-1}+z)\dots R_{10}(u_1+z),\\T(z)&=R_{01}(z-u_1)R_{02}(z-u_2)\dots R_{0l}(z-u_l).\end{aligned} \tag{67}$$

They are acting on the space $V_0\otimes V_1\otimes\ldots\otimes V_l$, where V_0 is the auxiliary space. The final preparation is the following observation. In Eqs. (32)–(36), the summation over face indices has at most two terms in the eight-vertex model. In the following cases, due to the non-zero condition of $W(m|z)_{\mu\nu}^{\mu'\nu'}$, there is actually only one term in the RHS of the equations,

$$\begin{aligned}&(1)\quad \text{Eq. (32) when } \mu=\nu\\&(2)\quad \text{Eq. (33) when } \mu\neq\nu\\&(3)\quad \text{Eq. (35) when } \mu\neq\nu.\end{aligned} \tag{68}$$

Thus we have (Fig. 7)

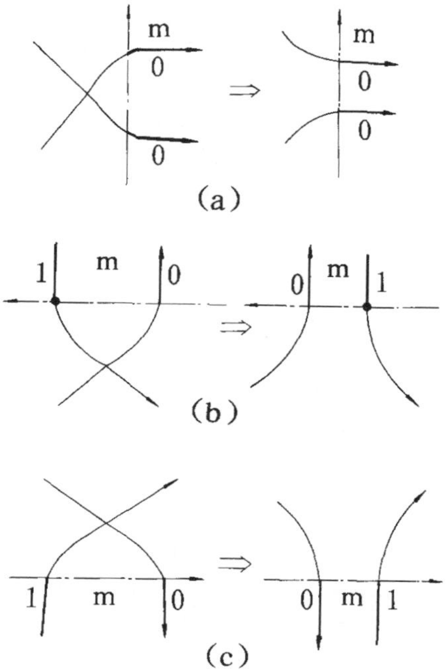

Fig. 7. Special cases in Fig. 4 where the summation over face indices is actually only one term. Note the difference between (b) and (c). (a) Eq. (69), (b) Eq. (70), (c) Eq. (71).

$$R_{12}(z_1-z_2)\phi^{(1)}_{m+2,0}(z_1)\phi^{(2)}_{m+1,0}(z_2)=W(|z_1-z_2)^{00}_{00}\phi^{(2)}_{m+2,0}(z_2)\phi^{(1)}_{m+1,0}(z_1), \quad (69)$$

$$\bar{\phi}^{(1)}_{m-1,1}(z_1)R_{12}(z_1-z_2)\phi^{(2)}_{m+1,0}(z_2)=W(m|z_1-z_2)^{10}_{10}\bar{\phi}^{(1)}_{m,1}(z_1)\phi^{(2)}_{m,0}(z_2), \quad (70)$$

$$\bar{\phi}^{(2)}_{m,1}(z_2)R_{12}(z_1-z_2)\phi^{(1)}_{m,0}(z_1)=W(m|z_1-z_2)^{01}_{01}\phi^{(1)}_{m+1,0}(z_1)\bar{\phi}^{(2)}_{m-1,1}(z_2), \quad (71)$$

which will be repeatedly used in the proof of the vacuum state.

The above are preparations for introducing the vacuum state. Now, define the vacuum state as

$$|0\rangle^m_{m_0}\equiv\phi^{(l)}_{m_0,0}(u_l)\phi^{(l-1)}_{m_0-1,0}(u_{l-1})\dots\phi^{(2)}_{m_0-(l-2),0}(u_2)\phi^{(1)}_{m_0-(l-1),0}(u_1)\,, \quad (72)$$

where $m=m_0-l$. This is precisely the same vacuum state introduced by Baxter in the original work of the Bethe ansatz for the eight-vertex model with periodic boundary conditions [3]. For the vacuum state defined in (72) we can show that $S(m,m_0|z)_{00}$ and $T(m,m_0|z)_{11}$ change $|0\rangle^m_{m_0}$ to $|0\rangle^{m-1}_{m_0-1}$, while $S(m,m_0|z)_{11}$ and $T(m,m_0|z)_{00}$ change $|0\rangle^m_{m_0}$ to $|0\rangle^{m+1}_{m_0+1}$ (with some coefficients). The operators $S(m,m_0|z)_{01}$ and $T(m,m_0|z)_{10}$ change $|0\rangle^m_{m_0}$ to zero. The proof goes as follows.

We have

$$\begin{aligned}S(m,m_0|z)_{00}&=\bar{\phi}^{(0)}_{m_0,0}(-z)S(z)\phi^{(0)}_{m,0}(-z) \quad (73)\\&=\bar{\phi}^{(0)}_{m_0,0}(-z)R_{l0}(u_l+z)R_{l-1,0}(u_{l-1}+z)\dots R_{1,0}(u_1+z)\phi^{(0)}_{m,0}(-z).\end{aligned}$$

Since vectors and operators belonging to different spaces may change their positions in an equation, we obtain

$$S(m,m_0|z)_{00}|0\rangle^m_{m_0} = \bar{\phi}^{(0)}_{m_0,0}(-z)R_{l0}(u_l+z)\phi^{(l)}_{m_0,0}(u_l)R_{l-1,0}(u_{l-1}+z)\phi^{(l-1)}_{m_0-1,0}(u_{l-1}) \\ \ldots R_{20}(u_2+z)\phi^{(2)}_{m_0-(l-2),0}(u_2)R_{10}(u_1+z)\phi^{(1)}_{m+1,0}(u_1)\phi^{(0)}_{m,0}(-z). \tag{74}$$

By using (69) we move $\phi^{(0)}_{,0}(-z)$ towards the left step by step. At each step we eliminate an R-matrix getting a W factor, and change the face weight of $\phi^{(0)}_{,0}(-z)$ and that of the $\phi^{(i)}_{,0}$ encountered. Thus we have

$$\begin{aligned} S(m,m_0|z)_{00}|0\rangle^m_{m_0} &= \ldots R_{20}(u_2+z)\phi^{(2)}_{m+2,0}(u_2)\phi^{(0)}_{m+1,0}(-z)\phi^{(1)}_{m,0}(u_1)W(|u_1+z)^{00}_{00} \\ &= \ldots \\ &= \bar{\phi}^{(0)}_{m_0,0}(-z)\phi^{(0)}_{m_0,0}(-z)\phi^{(l)}_{m_0-1,0}(u_l)\ldots\phi^{(1)}_{m,0}(u_1) \\ &\quad\times\prod_{i=1}^{l} W(|u_i+z)^{00}_{00}. \end{aligned} \tag{75}$$

Due to the orthogonal relation (28), this becomes

$$\begin{aligned} S(m,m_0|z)_{00}|0\rangle^m_{m_0} &= \prod_{i=1}^{l} W(|u_i+z)^{00}_{00}|0\rangle^{m-1}_{m_0-1} \\ &\equiv s_{00}(z)|0\rangle^{m-1}_{m_0-1}. \end{aligned} \tag{76}$$

Similarly, one can show

$$\begin{aligned} S(m,m_0|z)_{01}|0\rangle^m_{m_0} &= \bar{\phi}^{(0)}_{m_0,1}(-z)\phi^{(0)}_{m_0,0}(-z)\ldots \\ &= 0. \end{aligned} \tag{77}$$

For the action of S_{11}, from Eqs. (65), (67) and (72), we write

$$\begin{aligned} &S(m,m_0|z)_{11}|0\rangle^m_{m_0} \\ &= \bar{\phi}^{(0)}_{m_0,1}(-z)R_{l0}(u_l+z)\phi^{(l)}_{m_0,0}(u_l)R_{l-1,0}(u_{l-1}+z)\phi^{(l-1)}_{m_0-1,0}(u_{l-1}) \\ &\quad\times\ldots R_{10}(u_1+z)\phi^{(0)}_{m+1,0}(u_1)\phi^{(0)}_{m,1}(-z). \end{aligned} \tag{78}$$

Using (71) we move $\bar{\phi}^{(0)}_{,1}(-z)$ towards the right step by step. At each step we eliminate an R-matrix getting a W factor, and change the face weight of the $\bar{\phi}^{(0)}_{,1}(-z)$ and $\phi^{(i)}_{,0}$ encountered. We then have (Fig. 8)

$$\begin{aligned} S(m,m_0|z)_{11}|0\rangle^m_{m_0} &= \prod_{i=1}^{l} W(m_0-(i-1)|u_i+z)^{01}_{01}\bar{\phi}^{(0)}_{m_0-l,1}(-z) \\ &\quad\times\phi^{(0)}_{m,1}(-z)\phi^{(l)}_{m_0+1,0}(u_l)\ldots\phi^{(1)}_{m+2,0}(u_1) \end{aligned}$$

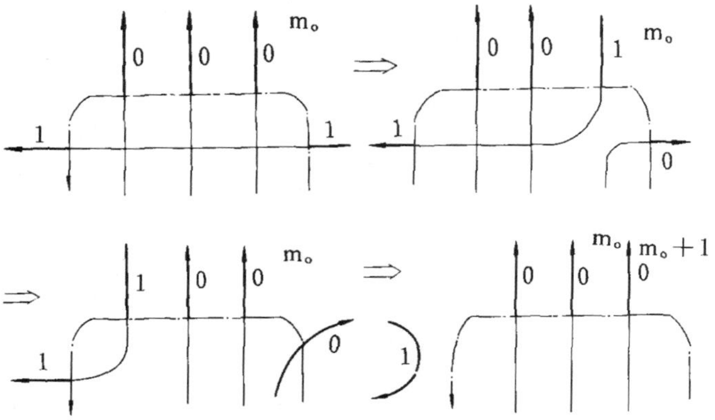

Fig. 8. Action of S_{11} on the vacuum state, Eq. (79) (derivation by graphic representation, using Figs. 7c and 3d).

$$= \prod_{i=1}^{l} W(m_0 - (i-1)|u_i + z)_{01}^{01} |0\rangle_{m_0+1}^{m+1}$$

$$\equiv s_{11}(m|z)|0\rangle_{m_0+1}^{m+1} . \tag{79}$$

Here $m_0 - l = m$. A similar derivation can be performed for T, giving

$$\begin{aligned} T(m, m_0|z)_{00}|0\rangle_{m_0}^{m} &= \prod_{i=1}^{l} W(|z - u_i)_{00}^{00}|0\rangle_{m_0+1}^{m+1} \\ &\equiv t_{00}(z)|0\rangle_{m_0+1}^{m+1}, \\ T(m, m_0|z)_{10}|0\rangle_{m_0}^{m} &= 0, \end{aligned} \tag{80}$$

$$\begin{aligned} T(m, m_0|z)_{11}|0\rangle_{m_0}^{m} &= \prod_{i=1}^{l} W(m + (i-1)|z - u_i)_{10}^{10}|0\rangle_{m_0-1}^{m-1} \\ &\equiv t_{11}(m|z)|0\rangle_{m_0-1}^{m-1}, \end{aligned} \tag{81}$$

which completes our proof.

We now assume that in (62), $K(m_0|z)_1^0 = 0$, and consider $\hat{\mu} \equiv (-1)^{\mu}$, obtaining

$$\begin{aligned} k(m|z)_0^0 = &K(m_0|z)_0^0 T(m-1, m_0-1|z)_{00} S(m, m_0|z)_{00} \\ &+ K(m_0|z)_0^1 T(m-1, m_0+1|z)_{00} S(m, m_0|z)_{01} \\ &+ K(m_0|z)_1^1 T(m-1, m_0+1|z)_{01} S(m, m_0|z)_{01}, \end{aligned} \tag{82}$$

$$\begin{aligned} k(m|z)_1^1 = &K(m_0|z)_0^0 T(m+1, m_0-1|z)_{10} S(m, m_0|z)_{10} \\ &+ K(m_0|z)_0^1 T(m+1, m_0+1|z)_{10} S(m, m_0|z)_{11} \\ &+ K(m_0|z)_1^1 T(m+1, m_0+1|z)_{11} S(m, m_0|z)_{11}. \end{aligned} \tag{83}$$

By (66), $k(m|z)_1^1$ can be written as

$$k(m|z)_1^1 = \frac{K(m_0|z)_0^0}{W(m_0-1|2z)_{00}^{00}}\{W(m+1|2z)_{11}^{11}S(m-1,m_0+1|z)_{10}T(m,m_0|z)_{10}$$
$$+W(m+1|2z)_{10}^{01}S(m+1,m_0+1|z)_{00}T(m,m_0|z)_{00}\}$$
$$+K(m_0|z)_0^1T(m+1,m_0+1|z)_{10}S(m,m_0|z)_{11}$$
$$+\left\{K(m_0|z)_1^1 - \frac{K(m_0|z)_0^0W(m_0+1|2z)_{10}^{01}}{W(m_0-1|2z)_{00}^{00}}\right\}$$
$$\times T(m+1,m_0+1|z)_{11}S(m,m_0|z)_{11}. \quad (84)$$

Acting on the vacuum state $|0\rangle_{m_0}^m$ and by Eqs. (76), (79)–(81), these two operators yield

$$k(m|z)_0^0|0\rangle_{m_0}^m = K(m_0|z)_0^0T(m-1,m_0-1|z)_{00}s_{00}(z)|0\rangle_{m_0-1}^{m-1} + 0 + 0$$
$$= K(m_0|z)_0^0t_{00}(z)s_{00}(z)|0\rangle_{m_0}^m$$
$$\equiv \tau(m|z)_0^0|0\rangle_{m_0}^m, \quad (85)$$

and

$$k(m|z)_1^1|0\rangle_{m_0}^m = \frac{K(m_0|z)_0^0W(m+1|2z)_{10}^{01}}{W(m_0-1|2z)_{00}^{00}}s_{00}(z)t_{00}(z)|0\rangle_{m_0}^m + 0$$
$$+\left\{K(m_0|z)_1^1 - \frac{K(m_0|z)_0^0W(m_0+1|2z)_{10}^{01}}{W(m_0-1|2z)_{00}^{00}}\right\}$$
$$\times t_{11}(m+1|z)s_{11}(m|z)|0\rangle_{m_0}^m$$
$$\equiv \tau(m|z)_1^1|0\rangle_{m_0}^m. \quad (86)$$

From Eqs. (85) and (86) we see that when $K(m_0|z)_1^0 = 0$, the vacuum state $|0\rangle_{m_0}^m$ is indeed an eigenstate of $A = k(m|z)_0^0$ and $D = k(m|z)_1^1$. Thus it is also an eigenstate of $\tilde{D}$. The eigenvalues depend on $m, \{u_i\}, l$ and z, the spectrum of A and D.

4.2. Boundary conditions

The algebraic Bethe ansatz requires a vacuum state, which needs the right boundary to satisfy

$$K(m_0|z)_1^0 = 0. \quad (87)$$

Also, the transfer matrix $t(z)$ must be a linear combination of A and $\tilde{D}$, which imposes the left boundary to satisfy

$$\tilde{K}(m^0|z)_0^1 = 0 \quad (88)$$

and

$$\tilde{K}(m^0|z)_1^0 = 0, \quad (89)$$

for m^0, which we will specify in the next section. We will see that m^0 is constrained with m, m_0 of the vacuum state by

$$m + l\hat{0} = m + l = m_0,$$
$$m + (-\hat{1} + \hat{0})l' = m + 2l' = m^0. \tag{90}$$

In Eq. (90), l is the column number of the lattice and l' is a positive integer, which will be the number of B operators in constructing the eigenstates of $t(z)$ (see (105)).

From the general solution of RE (15), (17),

$$K(z) = \sum_\alpha C_\alpha \frac{I_\alpha}{\sigma_\alpha(-z)}, \qquad \tilde{K}(z) = \sum_\alpha \tilde{C}_\alpha \frac{I_\alpha}{\sigma_\alpha(z+w)},$$

and the definitions

$$K(m_0|z)^\nu_\mu \equiv \tilde{\phi}_{m_0+\hat{\mu}-\hat{\nu},\mu}(z) K(z) \phi_{m_0,\nu}(-z),$$
$$\tilde{K}(m|z)^\nu_\mu \equiv \bar{\phi}_{m,\mu}(-z) \tilde{K}(z) \phi_{m-\hat{\mu}+\hat{\nu},\nu}(z), \tag{91}$$

we can derive the following results (see Appendix C):

$$K(m|z)^0_1 = (-1)\frac{\sum_\alpha C_\alpha (-1)^{\alpha_2}\sigma_\alpha(wa + w\beta - \frac{1}{2})}{h(z + w\beta + w - \frac{1}{2})h(wa - w)}, \tag{92}$$

$$\tilde{K}(m|z)^0_1 = \frac{\sum_\alpha \tilde{C}_\alpha (-1)^{\alpha_2}\sigma_\alpha(-w(a-1) + w\beta - \frac{1}{2})}{h(-z + w\beta - \frac{1}{2})h(wa)}, \tag{93}$$

$$\tilde{K}(m|z)^1_0 = (-1)\frac{\sum_\alpha \tilde{C}_\alpha (-1)^{\alpha_2}\sigma_\alpha(w(a+1) + w\beta - \frac{1}{2})}{h(-z + w\beta - \frac{1}{2})h(wa)}, \tag{94}$$

where $a \equiv m + \gamma$. We can easily see from (92)–(94) that the conditions (87)–(89) are actually independent of z, and depend only on $\{C_\alpha\}, \{\tilde{C}_\alpha\}$ and the parameters β, γ, w, τ, etc. For any given generic $\{\tilde{C}_\alpha\}$, we may solve the equation

$$\sum_\alpha \tilde{C}_\alpha (-1)^{\alpha_2}\sigma_\alpha(2\eta) = 0. \tag{95}$$

The LHS of (95) is a doubly quasi-periodic holomorphic function of η. From the quasi-periodicity [7,41,42] we see that it has four zeros in $\Lambda_\tau : \eta \to \eta + 1, \eta \to \eta + \tau$. Assuming η_1, η_2 to be two different zeros, we obtain α and β by solving

$$-w(a-1) + w\beta - \tfrac{1}{2} = 2\eta_1,$$
$$w(a+1) + w\beta - \tfrac{1}{2} = 2\eta_2. \tag{96}$$

Then from $a = m^0 + \gamma$ we obtain γ according to a given m^0. Using such β, γ from generic $\{\tilde{C}_\alpha\}$, one may construct $\phi, \bar{\phi}$ and $\tilde{\phi}$ with which Eqs. (88) and (89) are satisfied. Since β and γ are completely determined, Eq. (87) is a constraint for $\{C_\alpha\}$ at the right boundary. This is the only restriction we must impose for our approach. There are three free parameters at the right boundary for the general solution (15), (an overall scalar of K is not important). With the constraint (87), there are still two free parameters left. If we further require

$$K(m_0|z)_0^1 = 0, \tag{97}$$

then only one free parameter at the right boundary survives. We can show that

$$\frac{K(m_0|z)_1^1}{K(m_0|z)_0^0} = \frac{h(\xi - z)h(aw + \xi + z)}{h(\xi + z)h(aw + \xi - z)},$$

where ξ is the right boundary free parameter introduced in [45], and similarly for the left boundary. In this case the left and right boundaries are equivalent to those of the SOS model (except for a symmetric factor) introduced by Behrend, Pearce and O'Brien in [45], where they derived the solutions directly from the face reflection equation. This implies that our approach can be used for the Bethe ansatz of the SOS model with such boundary conditions.

5. Bethe ansatz

We see that for a given m^0, by properly choosing β and γ, we can ensure that

$$\tilde{K}(m^0|z)_0^1 = \tilde{K}(m^0|z)_1^0 = 0 \tag{98}$$

at the left boundary. We need the parameters $\{C_\alpha\}$ at the right boundary to satisfy

$$\sum_\alpha C_\alpha(-1)^{\alpha_2}\sigma_\alpha(wa + w\beta - \tfrac{1}{2}) = 0 \tag{99}$$

for an integer m_0, where $a \equiv m_0 + \gamma$, to ensure that $K(m_0|z)_1^0 = 0$. Assume that $2l' = m^0 - (m_0 - l)$ is a positive even integer, where l is the number of column of the lattice. We can proceed the standard algebraic Bethe ansatz [3,4,7,16] as follows.

We have the transfer matrix from Eq. (38),

$$t(z) = \tilde{K}(m^0|z)_0^0 k(m^0|z)_0^0 + \tilde{K}(m^0|z)_1^1 k(m^0|z)_1^1. \tag{100}$$

Due to (44) and (56), $t(z)$ can be rewritten as

$$t(z) = \mu_0(z)A(m^0|z) + \mu_1(z)\tilde{D}(m^0|z). \tag{101}$$

We also have the exchange relations of $A, \tilde{D}$ with B, Eqs. (57) and (59). They can be compactly written as

$$\begin{aligned} A(m|u)B(m-2|v) &= a_{00}(m,u,v)B(m-2|v)A(m-2|u) \\ &\quad + b_{00}(m,u,v)B(m-2|u)A(m-2|v) \\ &\quad + b_{01}(m,u,v)B(m-2|u)\tilde{D}(m-2|v), \end{aligned} \tag{102}$$

$$\begin{aligned} \tilde{D}(m|u)B(m-2|v) &= a_{11}(m,u,v)B(m-2|v)\tilde{D}(m-2|u) \\ &\quad + b_{10}(m,u,v)B(m-2|u)A(m-2|v) \\ &\quad + b_{11}(m,u,v)B(m-2|u)\tilde{D}(m-2|v). \end{aligned} \tag{103}$$

Besides, we have (42), the exchange relation of the B's,

$$B(m|u)B(m-2|v)=B(m|v)B(m-2|u). \tag{104}$$

Consider a vector of the quantum space

$$\Phi=B(m^0-2|z_1)B(m^0-4|z_2)\dots B(m^0-2l'|z_{l'})|0\rangle^m_{m_0}, \tag{105}$$

where the number of B is l', $m^0-2l'=m$. Due to (104), it is symmetric for the z_i's. We will show that for properly chosen $z_1,\dots,z_{l'}$ satisfying the so-called Bethe ansatz equations, Φ is an eigenvector (eigenstate) of $t(z)$.

We have

$$\begin{aligned}
t(z)\Phi=&\left\{\mu_0(z)A(m^0|z)+\mu_1(z)\tilde{D}(m^0|z)\right\}\Phi\\
=&\left\{\mu_0(z)A(m^0|z)+\mu_1(z)\tilde{D}(m^0|z)\right\}\\
&\times B(m^0-2|z_1)\dots B(m|z_{l'})|0\rangle^m_{m_0}\\
=&\left\{B(m^0-2|z_1)\left[\mu_0(z)a'_{00}A(m^0-2|z)+\mu_1(z)a'_{11}\tilde{D}(m^0-2|z)\right]\right.\\
&+B(m^0-2|z)\left[\mu_0(z)b'_{00}A(m^0-2|z_1)+\mu_0(z)b'_{01}\tilde{D}(m^0-2|z_1)\right.\\
&\left.\left.+\mu_1(z)b'_{10}A(m^0-2|z_1)+\mu_1(z)b'_{11}\tilde{D}(m^0-2|z_1)\right]\right\}\\
&\times B(m^0-4|z_2)\dots|0\rangle^m_{m_0}\\
=&\left\{B(m^0-2|z_1)\left[(\mu a')_0A(m^0-2|z)+(\mu a')_1\tilde{D}(m^0-2|z)\right]\right.\\
&\left.+B(m^0-2|z)\left[(\mu b')_0A(m^0-2|z_1)+(\mu b')_1\tilde{D}(m^0-2|z_1)\right]\right\}\\
&\times B(m^0-4|z_2)\dots|0\rangle^m_{m_0}\\
=&\left\{B(m^0-2|z_1)B(m^0-4|z_2)\left[(\mu a'a'')_0A(m^0-4|z)\right.\right.\\
&\left.+(\mu a'a'')_1\tilde{D}(m^0-4|z)\right]\\
&+B(m^0-2|z)B(m^0-4|z_2)\left[(\mu b'a''_1)_0A(m^0-4|z_1)\right.\\
&\left.+(\mu b'a''_1)\tilde{D}(m^0-4|z_1)\right]\\
&+B(m^0-2|z)B(m^0-4|z_1)\left[(\mu b'b'')_0A(m^0-4|z_2)\right.\\
&\left.\left.+(\mu b'b'')_1\tilde{D}(m^0-4|z_2)\right]\right\}\\
&\times B(m^0-6|z_3)\dots|0\rangle^m_{m_0},
\end{aligned} \tag{106}$$

where a',a'',a''_1,b',b'' are 2×2 matrices. The matrices a',a'',a''_1 are diagonal. The elements of these matrices are determined by m^0,z and z_i via Eqs. (102), (103). The notations of these matrices are

$$\begin{aligned}
a(m^0,z,z_1)&\equiv a',\\
a(m^0-2,z,z_2)&\equiv a'',\\
a(m^0-2,z_1,z_2)&\equiv a''_1,\\
b(m^0,z,z_1)&\equiv b',\\
b(m^0-2,z,z_2)&\equiv b''
\end{aligned} \tag{107}$$

for short. The notations $(\mu\ldots)_i$ represents the ith component of the product of the row vector μ with matrix "...". In the following derivation, the notations are similar. Repeatedly using Eqs. (102), (103) to move A and $\tilde{D}$ to the right of all B's, we obtain

$$\begin{aligned} t(z)\Phi = & B(m^0-2|z_1)\ldots B(m^0-2l'|z_{l'}) \\ & \times\left[(\mu a'a''\ldots a^{(l')})_0 A(m^0-2l'|z)+(\mu a'a''\ldots a^{(l')})_1\tilde{D}(m^0-2l'|z)\right]|0\rangle^m_{m_0} \\ & +B(m^0-2|z)B(m^0-4|z_2)\ldots B(m^0-2l'|z_{l'}) \\ & \times\left[(\mu b'a''_1\ldots a^{(l')}_1)_0 A(m^0-2l'|z_1)\right. \\ & \left.+(\mu b'a''_1\ldots a^{(l')}_1)_1\tilde{D}(m^0-2l'|z_1)\right]|0\rangle^m_{m_0} \\ & +B(m^0-2|z)B(m^0-4|z_1)B(m^0-6|z_3)\ldots \\ & \times\left[\ldots A(m^0-2l'|z_2)+\ldots\tilde{D}(m^0-2l'|z_2)\right]|0\rangle^m_{m_0}+\ldots \end{aligned} \tag{108}$$

The vector $|0\rangle^m_{m_0}$ is an eigenvector of $A(m|z)$ and $D(m|z)$. The eigenvalues are given by (85), (86). Thus we have

$$\begin{aligned} A(m|z)|0\rangle^m_{m_0} &= \lambda_0(z)|0\rangle^m_{m_0}, \\ \tilde{D}(m|z)|0\rangle^m_{m_0} &= \lambda_1(z)|0\rangle^m_{m_0}. \end{aligned} \tag{109}$$

Therefore, noticing $m^0-2l'=m$, we can write $t(z)\Phi$ as

$$\begin{aligned} t(z) = & \left[(\mu a'a''\ldots a^{(l')}_0)_0\lambda_0(z)+(\mu a'\ldots a^{(l')})_1\lambda_1(z)\right] \\ & \times B(m^0-2|z_1)\ldots B(m|z_{l'})|0\rangle^m_{m_0} \\ & +\left[(\mu b'a''_1\ldots a^{(l')}_1)_0\lambda_0(z_1)+(\mu b'a''_1\ldots a^{(l')}_1)_1\lambda_1(z_1)\right] \\ & \times B(m^0-2|z)B(m^0-4|z_2)\ldots B(m|z_{l'})|0\rangle^m_{m_0} \\ & +\left[(\ldots)_0\lambda_0(z_2)+(\ldots)_1\lambda_1(z_2)\right]B(m^0-2|z)B(m^0-4|z_1) \\ & \times B(m^0-6|z_3)\ldots|0\rangle^m_{m_0}+\ldots \end{aligned} \tag{110}$$

Because of (102)–(104) we see that $t(z)\Phi$ must be a linear combination of

$$\begin{aligned} B(m^0-2|z_1)\ldots B(m|z_{l'})|0\rangle^m_{m_0} &\equiv \Psi_0=\Phi, \\ B(m^0-2|z)B(m^0-4|z_2)\ldots|0\rangle^m_{m_0} &\equiv \Psi_1, \\ &\vdots \\ B(m^0-2|z)B(m^0-4|z_1)\ldots B(\ldots|z_{i-1})B(\ldots|z_{i+1})\ldots|0\rangle^m_{m_0} &\equiv \Psi_i, \\ &\vdots \\ B(m^0-2|z)B(m^0-4|z_1)\ldots B(m|z_{l'-1})|0\rangle^m_{m_0} &\equiv \Psi_{l'}. \end{aligned} \tag{111}$$

This is because we can always change the order of the B's such that the z_i inside the B's are arranged according to the order of i. So we have

$$t(z)\Phi \equiv C_0^1\Psi_0 + C_1^1\Psi_1 + C_2^1\Psi_2 + \ldots + C_{l'}^1\Psi_{l'}. \tag{112}$$

The problem now is that although the forms of C_0^1 and C_1^1 are simple and clear, C_i^1 for $i \geqslant 2$ are represented by a complicated summation. However, using the fact that Φ is a symmetric function of z_i, we can greatly simplify the calculation. Let us exchange z_i and z_1 in Φ. This does not change Φ. Then we can use the above standard procedure to obtain

$$t(z)\Phi = C_0^i\Psi + C_1^i\Psi_1 + \ldots + C_i^i\Psi_i + C_{i+1}^i\Psi_{i+1} + \ldots, \tag{113}$$

where C_0^i and C_i^i can be obtained by exchanging z_i and z_1 in C_0^1 and C_1^1. Assume that $\Psi_0, \Psi_1, \ldots, \Psi_{l'}$ are linearly independent vectors. Then each coefficient for the linear decomposition of $t(z)\Phi$ by $\{\Psi_i\}$ is unique. Thus we have $C_i^1 = C_i^i$. Putting all $C_i^1 = 0$ for $i \neq 0$ in (112) we get

$$t(z)\Phi = C_0^1\Phi \equiv \tau(z)\Phi, \tag{114}$$

i.e. Φ is an eigenstate of the transfer matrix $t(z)$ with eigenvalue

$$\tau(z) = \mu(z)a'a''\ldots a^{(l')}\lambda(z). \tag{115}$$

The spectrum parameters z_i are determined by the l' conditions $C_i^1 = 0, i = 1, \ldots, l'$. The first condition is

$$C_1^1 = \mu(z)b'a_1''\ldots a_1^{(l')}\lambda(z_1) = 0. \tag{116}$$

The other $l' - 1$ conditions can be obtained by exchanging z_1 and z_i in (116). These are the Bethe ansatz equations. Using the explicit form of (101)–(103) (i.e. Eqs. (56), (100), (57) and (59)), we can prove that these equations are actually independent of the spectrum parameter z. This implies that Φ is an eigenstate of all transfer matrices with arbitrary spectrum.

To end this section, we present here some results. The left boundary matrix is diagonal, we can show that its diagonal elements can be explicitly written as

$$\tilde{K}(m^0|z)_0^0 = h((a^0 - 1)w)h(\tilde{\xi} - z - w)h((a^0 + 1)w + \tilde{\xi} + z)F(z),$$
$$\tilde{K}(m^0|z)_1^1 = h((a^0 + 1)w)h(z + \tilde{\xi} + w)h((a^0 - 1)w + \tilde{\xi} - z)F(z),$$

where $\tilde{\xi}$ is the left boundary free parameter. We notice that this solution is identified with the solution given in Ref. [45]. $F(z)$ is a function of z depending on the scale of the left boundary matrix, which is not essential. So we get

$$\mu_0(z) = h(2z + 2w)h(\tilde{\xi} - z)h(z + \tilde{\xi} + a^0 w)F(z),$$
$$\mu_1(z) = h(w)h(z + \tilde{\xi} + w)h(\tilde{\xi} - z + (a^0 - 1)w)F(z).$$

The eigenvalue of the transfer matrix of the eight-vertex model with open boundary conditions is

$$\tau(z)=\mu_0(z)\lambda_0(z)\prod_{i=1}^{l'}\frac{h(z_i+z)h(z_i-z+w)}{h(z_i+z+w)h(z_i-z)}$$
$$+\mu_1(z)\lambda_1(z)\prod_{i=1}^{l'}\frac{h(z_i-z-w)h(z_i+z+2w)}{h(z_i+z+w)h(z_i-z)}. \qquad (117)$$

Here $\{z_i\}$ should satisfy the Bethe ansatz equations

$$\frac{\lambda_0(z_i)}{\lambda_1(z_i)}\prod_{j=1,j\neq i}^{l'}\frac{h(z_i+z_j)h(z_i-z_j+w)}{h(z_i-z_j-w)h(z_i+z_j+2w)}$$
$$=\frac{h(w)h(\tilde{\xi}-z_i+w(a^0-1))h(\tilde{\xi}+z_i+w)}{h(2z_i)h(\tilde{\xi}+z_i+wa^0)h(\tilde{\xi}-z_i)} \qquad (118)$$

for $i=1,\ldots,l'$. From (85), (86) and the definitions of $A,\tilde{D}$, we have

$$\lambda_0(z)=K(m_0|z)_0^0\prod_{i=1}^{l}\left[\frac{h(z+u_i+w)h(z-u_i+w)}{[h(w)]^2}\right],$$
$$\lambda_1(z)=\left[K(m_0|z)_1^1-K(m_0|z)_0^0\frac{h(2z+w(a_0+1))h(w)}{h(w(a_0+1))h(2z+w)}\right]$$
$$\times\frac{h(2z+w)h(w(a_0+1))}{h(w)h(wa_0)}\prod_{i=1}^{l}\left[\frac{h(z+u_i)h(z-u_i)}{[h(w)]^2}\right],$$

where $a^0\equiv m^0+\gamma=a+2l'$, $a_0\equiv m_0+\gamma\equiv a+l$, $a\equiv m+\gamma$.

6. Discussions

We can obtain our trigonometric limit as follows. The intertwiner (or three-spin operator) defined in (27) is

$$\phi^k_{m,\mu}(z)=\theta\begin{bmatrix}\frac{1}{2}-\frac{k}{2}\\ \frac{1}{2}\end{bmatrix}(z+(-1)^\mu w(m+\gamma)+w\beta,2\tau). \qquad (119)$$

Define $\gamma'=\gamma+\frac{\tau}{2w}$, $\beta'=\beta+\frac{\tau}{2w}$ and note $\mu=0,1$. Eq. (27) reads

$$\phi^k_{m,\mu}(z)=\theta\begin{bmatrix}\frac{1}{2}-\frac{k}{2}\\ \frac{1}{2}\end{bmatrix}(z+(-1)^\mu w(m+\gamma')+w\beta'+(\mu-1)\tau,2\tau)$$
$$=\xi(\mu)\theta\begin{bmatrix}\frac{\mu-k}{2}\\ \frac{1}{2}\end{bmatrix}(z+(-1)^\mu w(m+\gamma')+w\beta',2\tau), \qquad (120)$$

where

$$\xi(\mu) = \exp\left[-2\pi i\left(\frac{\mu-1}{2}\right)\left(\frac{\mu-1}{2}\tau + z + (-1)^{\mu}w(m+\gamma') + w\beta' + \tfrac{1}{2}\right)\right] \tag{121}$$

is independent of k. We then rescale ϕ to $\phi' = \xi^{-1}\phi$. At the same time, we must perform a gauge transformation which changes W to W' to ensure the face–vertex correspondence. When $\tau \to i\infty, \phi'^{k}_{m,\mu}(z) \to \delta_{\mu k}$. W' goes to a trigonometric R-matrix, which is different from the R-matrix in Ref. [16] only by a constant factor. The boundary condition $K(m|z)^0_1 = 0$, $\tilde{K}(m|z)^0_1 = \tilde{K}(m|z)^1_0 = 0$ (if we add $K(m|z)^1_0 = 0$) also approaches that of Ref. [16]. Thus we can show that the trigonometric limit of our model is that of Ref. [16]. It is a six-vertex model with integrable reflection boundaries. In such a model, the number of B's (l') in the Bethe ansatz state Φ is arbitrary. It is reasonable that the eigenstate of transfer matrix with maximum absolute value of the eigenvalue (ETMM) is a Bethe ansatz state. Each such six-vertex model can be obtained as the limit of a sequence of eight-vertex models with reflection boundaries. In this limit procedure, the Bethe ansatz equations, vacuum states and operators A, B, C, D are all approaching those of the six-vertex model. Thus we can reasonably assume in this sequence of eight-vertex models that there is a sequence of Bethe ansatz states that approach the ETMM of the six-vertex model. Thus it is quite possible that these Bethe ansatz states are ETMM of the eight-vertex models, especially when they are very close to the limiting six-vertex model since the eigenvalues of the transfer matrix $t(z)$ are discrete [7]. The fact whether a Bethe ansatz state is with the maximum absolute value of eigenvalue should not depend on continuous parameters if there is no phase transition in the procedure, also since the eigenvalues are discrete. For the above six-vertex model, when l is given (i.e. the column number), the true discrete variable is l'. The above sequence of Bethe ansatz states should have the same l'. Thus, it is reasonable that in our approach, when the left and right boundary conditions determine a proper l', the Bethe ansatz state has an eigenvalue of the maximum absolute value, which is the most important state in thermodynamics.

Johnson, Krinsky and McCoy calculated the energy of excitations of the *XYZ* model after Baxter obtained the Bethe ansatz of the eight-vertex model with periodic boundary conditions [48]. Now, using the results presented in this paper, we may also calculate the energy of the excitations of the *XYZ* model with boundaries, which is expected to be similar to the former results. However, we may further study the surface energy and other physical quantities that may be affected by the boundary conditions. For the trigonometric case, these have been studied by de Vega and Batchelor et al. [50,51]. By analyzing the eigenvalues of the transfer matrix, one may also get the boundary free energy, thermodynamic limit and finite size corrections. Other physical phenomena are also worth studying such as surface critical exponents and scaling, the central charges in conformal field theory, etc. It is well known that the eight-vertex model is equivalent to the SOS model; if one impose some restrictions on the SOS model, we can obtain the restricted SOS model (ABF model). So, if we impose some restrictions on the Bethe ansatz of the eight-vertex model with boundaries, we should find the Bethe ansatz for the ABF model with open boundary conditions. We may assume that these restrictions are similar to those Destri and de Vega found in obtaining the Bethe ansatz for the trigonometric ABF model [18]. All of this is worth being studied in the future.

Acknowledgements

We wish to thank Professors P.A. Pearce and Y.K. Zhou for their very helpful discussions and their valuable preprints. One of us, B-y.H. would like to thank Prof. Pearce for his hospitality in Melbourne.

Appendix A. Quadratic relation of the components of the face type $k(z)$

The left-hand side of Eq. (24) is $R_{12}(z_1-z_2)k_1(z_1)R_{21}(z_1+z_2)k_2(z_2)$. We multiply it from the left by $\tilde{\phi}^{(1)}_{m+\hat{\mu}_0,\mu_0}(z_1)\tilde{\phi}^{(2)}_{m+\hat{\mu}_0+\hat{\nu}_0,\nu_0}(z_2)$ and multiply it from the right by $\phi^{(1)}_{m+\hat{\mu}_3,\mu_3}(-z_1)\phi^{(2)}_{m+\hat{\mu}_3+\hat{\nu}_3,\nu_3}(-z_2)$, obtaining

$$\mathrm{LHS}=\tilde{\phi}^{(1)}_{m+\hat{\mu}_0,\mu_0}(z_1)\tilde{\phi}^{(2)}_{m+\hat{\mu}_0+\hat{\nu}_0,\nu_0}(z_2)R_{12}(z_1-z_2)\ldots \tag{A.1}$$

Using (34) we can eliminate R_{12} to get

$$\begin{aligned}\mathrm{LHS}=&W(m|z_1-z_2)^{\mu_1\nu_1}_{\mu_0\nu_0}\tilde{\phi}^{(1)}_{m+\hat{\mu}_1+\hat{\nu}_1,\mu_1}(z_1)\tilde{\phi}^{(2)}_{m+\hat{\nu}_1,\nu_1}(z_2)k_1(z_1)\\&\times R_{21}(z_1+z_2)k_2(z_2)\phi^{(1)}_{m+\hat{\mu}_3,\mu_3}(-z_1)\phi^{(2)}_{m+\hat{\mu}_3+\hat{\nu}_3,\nu_3}(-z_2).\end{aligned} \tag{A.2}$$

Move $\tilde{\phi}^{(2)}$ over $k_1(z_1)$ and move $\phi^{(1)}$ over $k_2(z_2)$, so that they are at the left and right side of R_{21}, respectively. LHS becomes

$$\begin{aligned}\mathrm{LHS}=&\ldots\tilde{\phi}^{(2)}_{m+\hat{\nu}_1,\nu_1}(z_2)R_{21}(z_1+z_2)\phi^{(1)}_{m+\hat{\mu}_3,\mu_3}(-z_1)\ldots\\=&\ldots W(m|z_1+z_2)^{\nu_2\mu_3}_{\nu_1\mu_2}\phi^{(1)}_{m+\hat{\mu}_2+\hat{\nu}_1,\mu_2}(-z_1)\tilde{\phi}^{(2)}_{m+\hat{\nu}_1+\hat{\mu}_2,\nu_2}(z_2)\ldots\end{aligned} \tag{A.3}$$

Since the Boltzmann weight W is non-zero only if $\hat{\nu}_2+\hat{\mu}_3=\hat{\nu}_1+\hat{\mu}_2$, we have

$$\mathrm{LHS}=\ldots\phi^{(1)}_{m+\hat{\mu}_2+\hat{\nu}_1,\mu_2}(-z_1)W(m|z_1+z_2)^{\nu_2\mu_3}_{\nu_1\mu_2}\tilde{\phi}^{(2)}_{m+\hat{\mu}_3+\hat{\nu}_2,\nu_2}(z_2)\ldots \tag{A.4}$$

Now $k_1(z_1)$ and $k_2(z_2)$ all have their "own" $\tilde{\phi}$ and ϕ at the left and right sides. By definition one concludes that

$$\begin{aligned}\mathrm{LHS}=&k(m+\hat{\mu}_2+\hat{\nu}_1|z_1)^{\mu_2}_{\mu_1}k(m+\hat{\mu}_3+\hat{\nu}_3|z_2)^{\nu_3}_{\nu_2}\\&\times W(m|z_1-z_2)^{\mu_1\nu_1}_{\mu_0\nu_0}W(m|z_1+z_2)^{\nu_2\mu_3}_{\nu_1\mu_2}.\end{aligned} \tag{A.5}$$

The derivation of RHS is similar.

Appendix B. Left and right inverse matrices of $k(z)$

When $n = 2$, we have

$$R_{12}(z) = R_{21}(z) = R_{12}(z)^{t_1 t_2}. \tag{B.1}$$

When $z = -w$, we have

$$P_-(12)R_{12}(-w) = R_{12}(-w)P_-(12) = R_{12}(-w), \tag{B.2}$$

where $P_-(12)$ is the anti-symmetric operator of the space $V_1 \otimes V_2$ satisfying $P_-(12)^2 = P_-(12)$. On the other hand, the R-matrix has the property

$$I_\alpha^{(i)} I_\alpha^{(j)} R_{ij}(z)[I_\alpha^{(i)}]^{-1}[I_\alpha^{(j)}]^{-1} = R_{ij}(z). \tag{B.3}$$

Thus

$$\begin{aligned} &I_\alpha^{(3)} P_-(12)R_{32}(z_2)R_{31}(z_1)P_-(12)[I_\alpha^{(3)}]^{-1} \\ &= P_-(12)I_\alpha^{(3)} R_{32}(z_2)[I_\alpha^{(3)}]^{-1} I_\alpha^{(3)} R_{31}(z_1)[I_\alpha^{(3)}]^{-1} P_-(12) \\ &= P_-(12)[I_\alpha^{(2)}]^{-1} R_{32}(z_2) I_\alpha^{(2)} [I_\alpha^{(1)}]^{-1} R_{31}(z_1) I_\alpha^{(1)} P_-(12) \\ &= P_-(12)[I_\alpha^{(2)}]^{-1}[I_\alpha^{(1)}]^{-1} R_{32}(z_2)R_{31}(z_1) I_\alpha^{(2)} I_\alpha^{(1)} P_-(12). \end{aligned} \tag{B.4}$$

It is not difficult to show that

$$I_\alpha^{(1)} I_\alpha^{(2)} P_-(12) = (-1)^{\alpha_1+\alpha_2} P_-(12) = P_-(12)[I_\alpha^{(1)}]^{-1}[I_\alpha^{(2)}]^{-1}. \tag{B.5}$$

Thus

$$\begin{aligned} &I_\alpha^{(3)} P_-(12)R_{32}(z_2)R_{31}(z_1)P_-(12)[I_\alpha^{(3)}]^{-1} \\ &= P_-(12)R_{32}(z_2)R_{31}(z_1)P_-(12) \equiv U. \end{aligned} \tag{B.6}$$

The operator U acting on $V_1 \otimes V_2 \otimes V_3$ is invariant under a similar transformation by $I_\alpha^{(3)}$, which implies that U is equivalent to the unit operator on V_3. From (B.1) we see that this is also true for $U' = P_-(12)R_{23}(z_2)R_{13}(z_1)P_-(12)$. We then consider the RE (14) for the case $z_1 = z - w, z_2 = z$, and have

$$\begin{aligned} G &= R_{12}(-w)k_1(z-w)R_{21}(2z-w)k_2(z) \\ &= R_{12}(-w)T_1(z-w)K_1(z-w)S_1(z-w)R_{21}(2z-w)T_2(z)K_2(z)S_2(z). \end{aligned}$$

Due to (21),

$$\begin{aligned} G &= R_{12}(-w)T_1(z-w)K_1(z-w)T_2(z)R_{21}(2z-w)S_1(z-w)K_2(z)S_2(z) \\ &= R_{12}(-w)T_1(z-w)T_2(z)K_1(z-w)R_{21}(2z-w)K_2(z)S_1(z-w)S_2(z). \end{aligned} \tag{B.7}$$

Rewrite $R_{12}(-w)$ as $P_-(12)R_{12}(-w)$, and move $R_{12}(-w)$ towards the right. Each time when it goes over a pair of $R_{1i}R_{2i}$ in T_1T_2 by the Yang–Baxter equation (9), we rewrite $R_{12}(-w)$ as $P_-^2(12)R_{12}(-w)$ and leave P_-^2. Then we obtain

$$G = [P_-(12)R_{23}R_{13}P_-(12)R_{24}R_{14}P_-(12)\dots P_-(12)R_{2l}R_{1l}P_-(12)]$$
$$\times R_{12}(-w)K_1(z-w)R_{21}(2z-w)K_2(z)S_1(z-w)S_2(z)$$
$$\equiv [M]\times\dots \tag{B.8}$$

Move $R_{12}(-w)$ to the left side of S_1 by the RE. One has

$$G = [M]K_2(z)R_{12}(2z-w)K_1(z-w)R_{21}(-w)S_1(z-w)S_2(z). \tag{B.9}$$

Similarly, we move $R_{21}(-w)$ step by step to the right side of S_1S_2, obtaining

$$G = [M]\dots[N]R_{21}(-w)$$
$$[N] = [P_-(12)R_{l2}R_{l1}P_-(12)P_-(12)R_{l-1,2}R_{l-1,1}P_-(12)\dots$$
$$\times P_-(12)R_{32}R_{31}P_-(12)]. \tag{B.10}$$

From the properties of U and U', we see that G is proportional to the identity operator in the "quantum" space $V' = V_3\otimes\dots\otimes V_l$. Using a similar derivation as in Appendix A, we multiply $\tilde\phi^{(1)}\tilde\phi^{(2)}$ from the left and multiply $\phi^{(1)}\phi^{(2)}$ from the right of G, and conclude that when $z_1 = z-w, z_2 = z$, both LHS and RHS of (40) are proportional to the identity operator in the quantum space V'. Properly choosing indices and noticing that $W(m|-w)_{\mu\nu}^{\mu'\nu'} = \dots\delta_{\mu\bar\nu}\delta_{\mu'\bar\nu'}$, we arrive at Eq. (49).

From the above derivation, we also see that G is anti-symmetric in the classical (auxiliary) indices of the space $V_1\otimes V_2$ and is independent of m. Thus

$$\rho'(m,\mu|z) = \tilde\phi^i_{m+\hat\mu,\bar\mu}(z-w)\tilde\phi^j_{m+\hat\mu+\hat\mu,\mu}(z)G_{ij}^{i'j'}\phi^{i'}_{m+\hat1,1}(-z+w)\phi^{j'}_{m+\hat1+\hat0,0}(-z)$$
$$= \left\{\tilde\phi^1_{\dots}(z-w)\tilde\phi^0_{\dots}(z) - \tilde\phi^0_{\dots}(z-w)\tilde\phi^1_{\dots}(z)\right\}$$
$$\times\left\{\phi^1_{\dots}(-z+w)\phi^0_{\dots}(-z) - \phi^0_{\dots}(-z+w)\phi^1_{\dots}(-z)\right\}G_{10}^{10}. \tag{B.11}$$

Similarly, $\rho(m,\nu|z)$ can also be expressed as G_{10}^{10} multiplied by a factor which depends only on $\phi,\tilde\phi$. Therefore the ratio $\rho(m+2|z)_0/\rho'(m|z)_0$ is independent of G_{10}^{10}. It is completely determined by $\phi,\tilde\phi$. Direct calculation gives (53).

Appendix C. Derivation of the boundary condition

From the definition we have

$$\phi^k_{m,\mu}(z) = \theta\begin{bmatrix}\frac12-\frac{k}{2}\\ \frac12\end{bmatrix}(z+(-1)^\mu wa + w\beta, 2\tau)$$
$$\equiv \theta^{(k)}(z+w((-1)^\mu a+\beta))$$
$$\equiv \theta^{(k)}(z+\chi), \tag{C.1}$$

where $a\equiv m+\gamma$. When α_1,α_2 are integers, by the expression of θ function, we have

$$\theta^{(k)}(z+\chi+\alpha_1\tau+\alpha_2)$$
$$=\exp\left[-2\pi i\left(\frac{\alpha_1}{2}\right)\left(\frac{\alpha_1}{2}\tau+z+\chi+\tfrac{1}{2}\right)+2\pi i\left(\tfrac{1}{2}-\frac{k}{2}\right)\alpha_2\right]\theta^{(k-\alpha_1)}(z+\chi), \tag{C.2}$$

giving

$$h^{\alpha_1}g^{\alpha_2}\phi_{m,\mu}(z)=(-1)^{\alpha_2}\exp\left[2\pi i\left(\frac{\alpha_1}{2}\right)\left(\frac{\alpha_1}{2}\tau+z+\chi+\tfrac{1}{2}+\alpha_2\right)\right]$$
$$\times\phi_{m,\mu}(z+\alpha_1\tau+\alpha_2). \tag{C.3}$$

On the other hand, from the property of zeros of doubly quasi-periodic holomorphic function, one can show [52] that

$$\mathrm{Det}\begin{bmatrix}\theta^{(0)}(z_1) & \theta^{(0)}(z_2)\\ \theta^{(1)}(z_1) & \theta^{(1)}(z_2)\end{bmatrix}=C\times h\left(\frac{z_1+z_2-1}{2}\right)h\left(\frac{z_1-z_2}{2}\right), \tag{C.4}$$

where C is independent of z_1 and z_2. If we write

$$A\equiv\begin{bmatrix}\phi^0_{m+\hat{0},0}(z) & \phi^0_{m+\hat{1},1}(z)\\ \phi^1_{m+\hat{0},0}(z) & \phi^1_{m+\hat{1},1}(z)\end{bmatrix}, \tag{C.5}$$

then $\tilde{\phi}^k_{m+\hat{\mu},\mu}(z)$ are elements of its inverse matrix. Thus for any column vector Ψ, the quantity $B\equiv\sum_k\tilde{\phi}^k_{m+\hat{1},1}(z)\Psi^k$ can be written as

$$B=\frac{1}{\mathrm{Det}\,A}\mathrm{Det}\begin{bmatrix}\phi_{m+\hat{0},0}(z) & \Psi^0\\ \phi_{m+\hat{0},0}(z) & \Psi^1\end{bmatrix}. \tag{C.6}$$

Combining (C.3) and (C.6) gives

$$B'_\alpha\equiv\tilde{\phi}_{m-2,1}(z)h^{\alpha_1}g^{\alpha_2}\phi_{m,0}(-z)$$
$$=\left(\frac{1}{\mathrm{Det}\,A'}\right)\mathrm{Det}\begin{bmatrix}\theta^{(0)}(z+wa+w\beta) & \theta^{(0)}(-z+wa+w\beta+\alpha_1\tau+\alpha_2)\\ \theta^{(1)}(z+wa+w\beta) & \theta^{(1)}(-z+wa+w\beta+\alpha_1\tau+\alpha_2)\end{bmatrix}$$
$$\times\exp\left[2\pi i\left(\frac{\alpha_1}{2}\right)\left(\frac{\alpha_1\tau}{2}-z+wa+w\beta+\tfrac{1}{2}+\alpha_2\right)\right]\times(-1)^{\alpha_2}. \tag{C.7}$$

The two determinants can be obtained from (141), thus

$$B'_\alpha=-\frac{\left[\sigma_\alpha(wa+w\beta-\frac{1}{2})\sigma_\alpha(-z)(-1)^{\alpha_2}\right]}{\left[h(z+w\beta+w-\frac{1}{2})h(wa-w)\right]}. \tag{C.8}$$

Substituting the definition of $K(m|z)^0_1$ and the expression of $K(z)$ (15), (63), we have

$$K(m|z)^0_1=(-1)\frac{\sum_\alpha C_\alpha(-1)^{\alpha_2}\sigma_\alpha(wa+w\beta-\frac{1}{2})}{h(z+w\beta+w-\frac{1}{2})h(wa-w)}. \tag{C.9}$$

The other elements of $K(m|z)$ can be similarly obtained. They are

$$K(m|z)_0^1 = a \to -a \quad \text{in RHS of the above equation;}$$

$$K(m|z)_0^0 = \frac{1}{h(z + w\beta + w - \frac{1}{2})h(wa - w)} \times \sum_\alpha C_\alpha (-1)^{\alpha_2} \frac{\sigma_\alpha(w\beta + w - \frac{1}{2})\sigma_\alpha(-z + wa - w)}{\sigma_\alpha(-z)}; \tag{C.10}$$

$$K(m|z)_1^1 = a \to -a \quad \text{in RHS of Eq. (147);}$$

$$\tilde{K}(m|z)_1^0 = \frac{1}{h(-z + w\beta - \frac{1}{2})h(wa)} \sum_\alpha \tilde{C}_\alpha (-1)^{\alpha_2} \sigma_\alpha(-wa + w + w\beta - \tfrac{1}{2}); \tag{C.11}$$

$$\tilde{K}(m|z)_0^1 = a \to -a \quad \text{in RHS of Eq. (148).}$$

$$\tilde{K}(m|z)_0^0 = \frac{1}{h(-z + w\beta - \frac{1}{2})h(wa)} \sum_\alpha \tilde{C}_\alpha (-1)^{\alpha_2} \frac{\sigma_\alpha(w\beta - \frac{1}{2})\sigma_\alpha(z + wa)}{\sigma_\alpha(z + w)}; \tag{C.12}$$

$$\tilde{K}(m|z)_1^1 = a \to -a \quad \text{in RHS of Eq. (149).}$$

References

[1] H.A. Bethe, Z. Phys. 49 (1931) 205.
[2] C.N. Yang and C.P. Yang, Phys. Rev. 105 (1966) 321, 322; 151 (1966) 258.
[3] R.J. Baxter, Phys. Rev. Lett. 26 (1971) 832, 834; Ann. Phys. 80 (1972) 193, 323.
[4] L.A. Takhtajan and L.D. Faddeev, Russ. Math. Surv. 34 (1979) 11.
[5] O. Babelon, H.J. de Vega and C.M. Viallet, Nucl. Phys. B 200 [FS 4] (1982) 266.
[6] C.L. Shultz, Physica A 122 (1983) 71.
[7] R.J. Baxter, Exactly Sloved Models in Statistical Mechanics (Academic Press, London, 1982).
[8] V.V. Bazhanov and N.Yu. Reshetikhin, Int. J. Mod. Phys. B 4 (1989) 115.
[9] E.H. Lieb and F.Y. Wu, Phase Transitions and Critical Phenomena, ed. C. Domb and M.S. Green (Academic Press, London) 1 (1982) 61.
[10] B.M. McCoy and T.T. Wu, Phys. Rev. 162 (1967) 436; 174 (1968) 546.
[11] R.Z. Bariev, Theor. Math. Phys. 40 (1979) 623.
[12] M. Gaudin, Phys. Rev. A 4 (1971) 386.
[13] J.L. Cardy, Nucl. Phys. B 275 (1986) 200.
[14] F.C. Alcaraz, M.N. Barber, M.T. Batchelor, R.J. Baxter and G.R.W. Quispel, J. Phys. A 20 (1987) 6397.
[15] I.V. Cherednik, Theor. Math. Phys. 17 (1983) 77; 61 (1984) 911.
[16] E.K. Sklyanin, J. Phys. A 21 (1988) 2375.
[17] P.P. Kulish and E.K. Sklyanin, J. Phys. A 24 (1991) L435.
[18] C. Destri and H.J.de Vega, Nucl. Phys. B 361 (1992) 361; B 374 (1992) 692.
[19] L. Mezincescu and R.I. Nepomechie, J. Phys. A 24 (1991) L19; Mod. Phys. Lett. A 6 (1991) 2497; Nucl. Phys. B 372 (1992) 597.
[20] A. Foerster and M. Karowski, Nucl. Phys. B 408 (1994) 512.
[21] H. de Vega and A. Gonzalez-Ruiz, Nucl. Phys. B 417 (1994) 553; Mod. Phys. Lett. A 9 (1994) 2207.
[22] R. Yue, H. Fan and B. Hou, Nucl. Phys. B 462 (1996) 167.
[23] S. Ghoshal and A. Zamolodchikov, Int. J. Mod. Phys. A 9 (1994) 3841.
[24] C.M. Yung and M.T. Batchelor, Nucl. Phys. B 435 (1995) 430.
[25] M. Jimbo, K. Kedem, T. Kojima, H. Konno and T. Miwa, Nucl. Phys. B 441 (1995) 437.
[26] P. Fendley, S. Saleur and N.P. Warner, Nucl. Phys. B 430 (1995) 577; B 428 (1994) 681.
[27] Y.K. Zhou, Nucl. Phys. B 453 (1995) 619.

[28] C. Ahn and W.M. Koo, hep-th/9508080.
[29] M.T. Batchelor and Y.K. Zhou, Phys. Rev. Lett. 76 (1996) 2826.
[30] T. Inami, S. Odake and Y.Z. Zhang, Phys. Lett. B 359 (1995) 118.
[31] Y.K. Zhou, preprint MRR 079-95.
[32] H.J. de Vega, Int. J. Mod. Phys. A 4 (1989) 2371.
[33] M. Jimbo, K. Kedem, T. Kojima, H. Konno and T. Miwa, Nucl. Phys. B 448 (1995) 429.
[34] Y.K. Zhou, Nucl. Phys. B 458 (1996) 504.
[35] H.J. de Vega and A. Gonzalez-Ruiz, J. Phys. A 26 (1993) L519.
[36] B.Y. Hou and R.H. Yue, Phys. Lett. A 183 (1993) 169.
[37] H.J. de Vega and A. Gonzalez-Ruiz, J. Phys. A 27 (1994) 6129.
[38] T. Inami and H. Konno, J. Phys. A 27 (1994) L913.
[39] B.Y. Hou, K.J. Shi, H. Fan and Z.X. Yang, Comm. Theor. Phys. 23 (1995) 163.
[40] A.A. Belavin, Nucl. Phys. B 180 [FS 2] (1981) 189.
[41] M.P. Richey and C.A. Tracy, J. Stat. Phys. 42 (1986) 311.
[42] M. Jimbo, T. Miwa and M. Okado, Nucl. Phys. B 300 [FS 22] (1988) 74.
[43] B.Y. Hou, M.L. Yan and Y.K. Zhou, Nucl. Phys. B 324 (1989) 715.
[44] T. Takebe, UTMS95-25(1995).
[45] R.E. Behrend, P.A. Pearce and D.L. O'Brien, hep-th/9507118.
[46] G.E. Andrews, R.J. Baxter and P.J. Forrester, J. Stat. Phys. 35 (1984) 193.
[47] H. Fan, B.Y. Hou and K.J. Shi, J. Phys. A 28 (1995) 4743.
[48] H. Fan, B.Y. Hou, K.J. Shi and Z.X. Yang, Phys. Lett. A 200 (1995) 109.
[49] J.D. Johnson, S. Krinsky and B.M. McCoy, Phys. Rev. Lett. 29 (1972) 492; Phys. Rev. A 8 (1973) 2526.
[50] H.J.de Vega and F. Woynarovich, Nucl. Phys. B 251 [FS 13] (1985) 439.
[51] M.T. Batchelor and C.J. Hamer, J. Phys. A 5 (1990) 761.
[52] Y.H. Quano, University of Tokyo preprint UT-613 (1992).

J. Phys. A: Math. Gen. **30** (1997) 6131–6145. Printed in the UK PII: S0305-4470(97)82002-1

An $\hbar$-deformation of the W_N algebra and its vertex operators

Bo-yu Hou† and Wen-li Yang†‡

† Institute of Modern Physics, Northwest University, Xian 710069, People's Republic of China§

‡ CCAST (World Laboratory), PO Box 8730, Beijing 100080, People's Republic of China

Received 19 February 1997

Abstract. In this paper, we derive an $\hbar$-deformation of the W_N algebra and its quantum Miura transformation. The vertex operators for this $\hbar$-deformed W_N algebra and its commutation relations are also obtained.

1. Introduction

Recently, the studies of q-deformation of some infinite-dimensional algebra—q-deformed affine algebra [4, 6, 11], q-deformed Virasoro [2, 21] algebra and W_N-algebra [2, 3, 9, 10]—have attracted much attention from physicist and mathematicians. The q-deformed affine algebra and its vertex operators provide a powerful method to study the state space and the correlation function of the solvable lattice model both in the bulk case [16] and the boundary case [14]. However, the symmetry of q-deformed affine algebra only corresponds to the current algebra (affine Lie algebra) symmetry, not to Virasoro-and W-algebra-type symmetry, in conformal field theory (CFT). The q-deformation of Virasoro and W algebra, which it is thought would play the role of symmetry algebra for the solvable lattice model, has been expected for a long time. Awata *et al* [2] also constructed the q-deformed W_N algebra (including Virasoro algebra) and the associated Miura transformation from a study of the Macdonald symmetrical functions. On the other hand, Frenkel and Reshetikhin [10] succeeded in constructing the q-deformed classical W_N algebra and the corresponding Miura transformation in an analysis of the $U_q(\hat{sl}_N)$ algebra at the critical level. Feigin and Frenkel [9] then obtained the quantum version of this q-deformed classical W_N algebra, i.e. the q-deformed W_N algebra. The q-deformed Virasoro algebra has also been given by Lukyanov and Pugai [21] in studying the bosonization for the ABF (Andrews–Baxter–Forrester) model. The bosonization for vertex operators of q-deformed Virasoro [17, 21] and W_N algebra [1, 2] have been constructed.

However, there exists another important deformation of infinite-dimensional algebra, which plays an important role in completely integrable field theories (in order to make a comparison with q-deformation, we call it $\hbar$-deformation). This deformation for affine algebra was created by Drinfeld [7] in studies of the Yangian. It has been shown that the Yangian ($DY(\hat{sl}_2)$) is the dynamical non-Abelian symmetry algebra for the SU(2)-invariant Thirring model [13, 18, 19, 23]. Naively, the $\hbar$-deformed affine algebra (or Yangian) would be expected to play the same role in integrable field theories as the q-deformed affine algebra

§ Mailing address.

in the solvable lattice model. Naturally, the $\hbar$-deformed Virasoro and W algebra, which would play the role of symmetry algebra of some integrable field model, are expected. We have succeeded in constructing the $\hbar$-deformed Virasoro algebra in [12] and shown that this $\hbar$-deformed Virasoro algebra is the dynamical symmetry algebra of the restricted sine–Gordon model. In this paper, we construct the $\hbar$-deformed W_N algebra (including the $\hbar$-deformed Virasoro algebra as a special case of $N = 2$), the corresponding quantum Miura transformation and its vertex operators. The $\hbar$-deformed W_N algebra becomes the usual non-deformed W_N algebra [8] with some centre charge which is related to parameter ξ, when $\hbar \longrightarrow 0$ and ξ and β are fixed.

This paper is arranged as follows. In section 2, we define the $\hbar$-deformed W_N algebra and its Miura transformation. The screening currents and vertex operators are derived in sections 3 and 4.

2. $\hbar$-deformation of W_N algebra

In this section, we start by defining an $\hbar$-deformed W_N algebra via the quantum Miura transformation.

2.1. $A^{(1)}_{N-1}$-type weight

In this subsection, we shall give some notation about the $A^{(1)}_{N-1}$-type weight which will be used in the following parts of this paper. Let $\epsilon_\mu (1 \leqslant \mu \leqslant N)$ be the orthonormal basis in $\mathbb{R}^N$, which is supplied with the inner product $\langle \epsilon_\mu, \epsilon_\nu \rangle = \delta_{\mu\nu}$. Set

$$\bar{\epsilon}_\mu = \epsilon_\mu - \epsilon \qquad \epsilon = \frac{1}{N}\sum_{\mu=1}^{N} \epsilon_\mu. \tag{1}$$

The $A^{(1)}_{N-1}$ type weight lattice is the linear space of

$$P = \sum_{\mu=1}^{N} Z\bar{\epsilon}_\mu.$$

Note that $\sum_{\mu=1}^{N} \bar{\epsilon}_\mu = 0$. Let $\omega_\mu (1 \leqslant \mu \leqslant N-1)$ be the fundamental weights

$$\omega_\mu = \sum_{\nu=1}^{\mu} \bar{\epsilon}_\nu$$

and α_μ the simple roots $(1 \leqslant \mu \leqslant N-1)$

$$\alpha_\mu = \bar{\epsilon}_\mu - \bar{\epsilon}_{\mu+1} = \epsilon_\mu - \epsilon_{\mu+1}. \tag{2}$$

An ordered pair $(b, a) \in \mathbb{P}^2$ is called admissible if only if there exists $\mu \in (1 \leqslant \mu \leqslant N-1)$ such that

$$b - a = \bar{\epsilon}_\mu.$$

An ordered set of four weights $\begin{pmatrix} c & d \\ b & a \end{pmatrix} \in \mathbb{P}^4$ is called an admissible configuration around a face if and only if the pairs (b, a), (c, b), (d, a) and (c, d) are all admissible pairs. To each admissible configuration around a face we shall associate a Boltzmann weight in section 4.

2.2. Quantum Miura transformation

Let us consider free bosons $\lambda_i(t)$ $(i = 1, \ldots, N)$ with a continuous parameter $t \in \{\mathbb{R} - 0)\}$ which satisfy

$$[\lambda_i(t), \lambda_i(t')] = \frac{4\mathrm{sh}((N-1)\hbar t/2)\mathrm{sh}(\hbar\xi t/2)\mathrm{sh}(\hbar(\xi+1)t/2)}{t\mathrm{sh}(N\hbar t/2)}\delta(t+t') \tag{3}$$

$$[\lambda_i(t), \lambda_j(t')] = -\frac{4\mathrm{sh}(\hbar t/2)\mathrm{sh}(\hbar\xi t/2)\mathrm{sh}(\hbar(\xi+1)t/2)\,\mathrm{e}^{\mathrm{sign}(j-i)N\hbar t/2}}{t\mathrm{sh}(N\hbar t/2)}\delta(t+t') \qquad i \neq j \tag{4}$$

with the deformed parameter $\hbar$ and a generic parameter ξ, where $\lambda_i(t)$ is subject to the following condition:

$$\sum_{l=1}^{N} \lambda_l(t)\,\mathrm{e}^{l\hbar t} = 0. \tag{5}$$

One can check that the restricted condition is compatable with equations (3) and (4).

Remark. The free bosons with continuous parameter in the case of $N = 2$, were first introduced by Jimbo *et al* [15] when studying the massless XXZ mode. This kind of bosons could be used to construct the bosonization of a Yangian double with centre $DY(\hat{sl}_N)$.

Let us define the fundamental operators $\Lambda_i(\beta)$ and the $\hbar$-deformed W_N algebra generators $T_i(\beta)$ for $i = 1, \ldots, N$ as follows,

$$\Lambda_i(\beta) =: \exp\left\{-\int_{-\infty}^{\infty} \lambda_i(t)\,\mathrm{e}^{\mathrm{i}\beta t}\,\mathrm{d}t\right\}: \tag{6}$$

$$T_l(\beta) = \sum_{1\leqslant j_1<j_2<\cdots<j_l\leqslant N} :\Lambda_{j_1}\left(\beta + \mathrm{i}\frac{l-1}{2}\hbar\right)\Lambda_{j_2}\left(\beta + \mathrm{i}\frac{l-3}{2}\hbar\right)\ldots\Lambda_{j_l}\left(\beta - \mathrm{i}\frac{l-1}{2}\hbar\right): \tag{7}$$

and $T_0(\beta) = 1$. Here $: O :$ stands for the usual bosonic normal ordering of some operator O such that the bosons $\lambda_i(t)$ with non-negative mode $t > 0$ are in the right. The restricted condition for bosons $\lambda_i(t)$ in equation (5) results in $T_N(\beta) = 1$. Actually, the generators $T_i(\beta)$ are obtained by the following quantum Miura transformation:

$$\begin{aligned} &:(\mathrm{e}^{\mathrm{i}\hbar\partial_\beta} - \Lambda_1(\beta))(\mathrm{e}^{\mathrm{i}\hbar\partial_\beta} - \Lambda_2(\beta - \mathrm{i}\hbar))\ldots(\mathrm{e}^{\mathrm{i}\hbar\partial_\beta} - \Lambda_N(\beta - \mathrm{i}(N-1)\hbar)): \\ &\quad = \sum_{l=0}^{N}(-1)^l T_l\left(\beta - \mathrm{i}\frac{l-1}{2}\hbar\right)\mathrm{e}^{\mathrm{i}(N-l)\hbar\partial_\beta}. \end{aligned} \tag{8}$$

Remark. $\mathrm{e}^{\mathrm{i}\hbar\partial_\beta}$ is the $\hbar$-shift operator such that

$$\mathrm{e}^{\mathrm{i}\hbar\partial_\beta} f(\beta) = f(\beta + \mathrm{i}\hbar).$$

If we take the limit of $\xi \longrightarrow -1$, the above generators $T_l(\beta)$ reduce to the classical version of the $\hbar$-deformed W_N algebra, which can be obtained by studying the Yangian double with centre $DY(\hat{sl}_N)$ at the critical level (i.e. $l = -N$). For the case of $N = 2$, the corresponding classical $\hbar$-deformed W_2 (Virasoro) algebra has been given by Ding *et al* [5]. Moreover, for the general case of $2 \leqslant N$, the corresponding classical $\hbar$-deformed W_N algebra has been obtained by Hou and Yang [24]. Thus, we call the limit ($\xi \longrightarrow -1$ with $\hbar$ and β fixed) the classical limit.

Let us consider another limit: $\hbar \longrightarrow 0$ with fixed ξ. Then we have $\Lambda_i(\beta) = 1 + i\hbar\chi_i(\beta) + o(\hbar)$ and $e^{i\hbar\partial_\beta} = 1 + i\hbar\partial_\beta + o(\hbar)$. Hence the right-hand side of (8) in this limit becomes

$$: (i\hbar)^N(\partial_\beta - \chi_1(\beta))(\partial_\beta - \chi_2(\beta))\ldots(\partial_\beta - \chi_N(\beta)) : + o(\hbar^N) \tag{9}$$

and we obtain the normally ordered Miura transformation corresponding to the non-deformed W_N algebra introduced by Fateev and Lukyanov [8]. Therefore, the non-deformed W_N algebra (the ordinary one) with the centre charge $(N-1) - (N(N+1)/\xi(1+\xi))$ can be obtained by taking this kind of limit. In this sense, we call this limit ($\hbar \longrightarrow 0$ with fixed ξ and β) the conformal limit.

2.3. Relations of the $\hbar$-deformed W_N algebra

In order to obtain the commutation relations for bosonic operators, we should make a comment about regularization. When one computes the exchange relation of bosonic operators, one often encounters an integral

$$\int_0^\infty F(t)\,dt$$

which is divergent at $t = 0$. Hence we adopt the regularization given by Jimbo *et al* [15]. Namely, the above integral should be understood as the contour integral

$$\int_C F(t)\frac{\log(-t)}{2i\pi}\,dt \tag{10}$$

where the contour C is chosen as the same as that in [15]. From the definition of fundamental operators $\Lambda_i(\beta)$ and the commutation relations of bosons $\lambda_i(t)$, we can derive the following OPEs (operator product equations):

$$\Lambda_i(\beta_1)\Lambda_i(\beta_2) = \phi_{i=i}(\beta_2 - \beta_1) : \Lambda_i(\beta_1)\Lambda_i(\beta_2) : \tag{11}$$

$$\Lambda_i(\beta_1)\Lambda_j(\beta_2) = \phi_{i<j}(\beta_2 - \beta_1) : \Lambda_i(\beta_1)\Lambda_i(\beta_2) : \qquad i < j \tag{12}$$

$$\Lambda_i(\beta_1)\Lambda_j(\beta_2) = \phi_{i>j}(\beta_2 - \beta_1) : \Lambda_i(\beta_1)\Lambda_j(\beta_2) : \qquad i > j \tag{13}$$

$$\begin{aligned}
\phi_{i=i}(\beta) &= \left[\Gamma\left(\frac{i\beta}{N\hbar} - \frac{\xi}{N}\right)\Gamma\left(\frac{i\beta}{N\hbar} + 1 - \frac{1}{N}\right)\Gamma\left(\frac{i\beta}{N\hbar} + \frac{1+\xi}{N}\right)\Gamma\left(\frac{i\beta}{N\hbar} + 1\right)\right] \\
&\quad \times \left[\Gamma\left(\frac{i\beta}{N\hbar}\right)\Gamma\left(\frac{i\beta}{N\hbar} - \frac{1+\xi}{N} + 1\right)\Gamma\left(\frac{i\beta}{N\hbar} + \frac{1}{N}\right)\Gamma\left(\frac{i\beta}{N\hbar} + 1 + \frac{\xi}{N}\right)\right]^{-1} \\
\phi_{i<j}(\beta) &= \left[\Gamma\left(\frac{i\beta}{N\hbar} - \frac{1}{N}\right)\Gamma\left(\frac{i\beta}{N\hbar} - \frac{\xi}{N}\right)\Gamma\left(\frac{i\beta}{N\hbar} + \frac{1+\xi}{N}\right)\right] \\
&\quad \times \left[\Gamma\left(\frac{i\beta}{N\hbar} - \frac{1+\xi}{N}\right)\Gamma\left(\frac{i\beta}{N\hbar} + \frac{\xi}{N}\right)\Gamma\left(\frac{i\beta}{N\hbar} + \frac{1}{N}\right)\right]^{-1} \\
\phi_{i>j}(\beta) &= \left[\Gamma\left(\frac{i\beta}{N\hbar} + 1 - \frac{1}{N}\right)\Gamma\left(\frac{i\beta}{N\hbar} + 1 - \frac{\xi}{N}\right)\Gamma\left(\frac{i\beta}{N\hbar} + 1 + \frac{1+\xi}{N}\right)\right] \\
&\quad \times \left[\Gamma\left(\frac{i\beta}{N\hbar} + 1 - \frac{1+\xi}{N}\right)\Gamma\left(\frac{i\beta}{N\hbar} + 1 + \frac{\xi}{N}\right)\Gamma\left(\frac{i\beta}{N\hbar} + 1 + \frac{1}{N}\right)\right]^{-1}.
\end{aligned} \tag{14}$$

To calculate the general OPEs, the integral representation for the Γ-function is very useful:

$$\Gamma(z) = \exp\left\{\int_0^\infty \left(\frac{e^{-zt} - e^{-t}}{1 - e^{-t}} + (z-1)\,e^{-t}\right)\frac{dt}{t}\right\} \qquad \mathrm{Re}(z) > 0. \tag{15}$$

Remark. The above OPEs can be considered as the operator scaling limit of q-deformed $\Lambda_i(z)$ given by Awata *et al* [2] or by Feigin and Frenkel [9] in studying the q-deformed W_N algebra. The scaling limit is taken as as follows:

$$z = p^{-i\beta/\bar{h}} \qquad q = p^{-\xi} \qquad \Lambda_i(\beta) = \lim_{p\longrightarrow 1} \Lambda_i(z) \equiv \lim_{p\longrightarrow 1} \Lambda_i(p^{(-i\beta/\bar{h})}). \tag{16}$$

Theorem 1. The generators $T_1(\beta)$ and $T_m(\beta)$ of the $\bar{h}$-deformed W_N algebra satisfy the following relations,

$$\begin{aligned} & f_{1m}^{-1}(\beta_2-\beta_1)T_1(\beta_1)T_m(\beta_2) - f_{1m}^{-1}(\beta_1-\beta_2)T_m(\beta_2)T_1(\beta_1) \\ &\quad = -2\mathrm{i}\pi\left\{\mathrm{i}\bar{h}\xi(\xi+1)\left(T_{m+1}\left(\beta_2+\frac{\mathrm{i}\bar{h}}{2}\right)\delta\left(\beta_1-\beta_2-\mathrm{i}\frac{m+1}{2}\bar{h}\right)\right.\right. \\ &\quad \left.\left. -T_{m+1}\left(\beta_2-\frac{\mathrm{i}\bar{h}}{2}\right)\delta\left(\beta_1-\beta_2+\mathrm{i}\frac{m+1}{2}\bar{h}\right)\right)\right\} \end{aligned} \tag{17}$$

where

$$\begin{aligned} f_{1m}(\beta) = & \left[\Gamma\left(\frac{\mathrm{i}\beta}{N\bar{h}}+1-\frac{1+m}{2N}\right)\Gamma\left(\frac{\mathrm{i}\beta}{N\bar{h}}+1+\frac{1-m}{2N}\right)\right. \\ & \left.\times\Gamma\left(\frac{\mathrm{i}\beta}{N\bar{h}}-\frac{\xi}{N}-\frac{1-m}{2N}\right)\Gamma\left(\frac{\mathrm{i}\beta}{N\bar{h}}+\frac{\xi}{N}+\frac{1+m}{2N}\right)\right] \\ & \times\left[\Gamma\left(\frac{\mathrm{i}\beta}{N\bar{h}}+1-\frac{\xi}{N}-\frac{1+m}{2N}\right)\Gamma\left(\frac{\mathrm{i}\beta}{N\bar{h}}+1+\frac{\xi}{N}+\frac{1-m}{2N}\right)\right. \\ & \left.\times\Gamma\left(\frac{\mathrm{i}\beta}{N\bar{h}}-\frac{1-m}{2N}\right)\Gamma\left(\frac{\mathrm{i}\beta}{N\bar{h}}+\frac{1+m}{2N}\right)\right]^{-1}. \end{aligned} \tag{18}$$

Proof. Using the OPEs. (11)–(13), we obtain that when $\mathrm{Im}\beta_2 \ll \mathrm{Im}\beta_1$

$$\Lambda_l(\beta_1) : \Lambda_{j_1}\left(\beta_2+\mathrm{i}\frac{m-1}{2}\bar{h}\right)\dots\Lambda_{j_m}\left(\beta_2-\mathrm{i}\frac{m-1}{2}\bar{h}\right):$$

is equal to

$$f_{1m}(\beta_2-\beta_1) : \Lambda_l(\beta_1)\Lambda_{j_1}\left(\beta_2+\mathrm{i}\frac{m-1}{2}\bar{h}\right)\dots\Lambda_{j_m}\left(\beta_2-\mathrm{i}\frac{m-1}{2}\bar{h}\right):$$

if $l = j_k$ for some $k \in \{1,\dots,m\}$; and

$$\begin{aligned} f_{1m}(\beta_2-\beta_1) & \left[\left(\mathrm{i}\frac{\beta_2-\beta_1}{N\bar{h}}-\frac{\xi}{N}-\frac{1}{2N}+\frac{2k-m}{2N}\right)\left(\mathrm{i}\frac{\beta_2-\beta_1}{N\bar{h}}+\frac{\xi}{N}+\frac{1}{2N}+\frac{2k-m}{2N}\right)\right] \\ & \times\left[\left(\mathrm{i}\frac{\beta_2-\beta_1}{N\bar{h}}-\frac{1}{2N}+\frac{2k-m}{2N}\right)\left(\mathrm{i}\frac{\beta_2-\beta_1}{N\bar{h}}+\frac{1}{2N}+\frac{2k-m}{2N}\right)\right]^{-1} \\ & \times : \Lambda_l(\beta_1)\Lambda_{j_1}\left(\beta_2+\mathrm{i}\frac{m-1}{2}\bar{h}\right)\dots\Lambda_{j_m}\left(\beta_2-\mathrm{i}\frac{m-1}{2}\bar{h}\right): \end{aligned}$$

if $j_k < l < j_{k+1}$. Here and in the following case $l < j_1$ corresponds to $k = 0$ and the case $l > j_m$ corresponds to $k = m$. On the other hand, when $\mathrm{Im}\beta_2 \gg \mathrm{Im}\beta_1$,

$$: \Lambda_{j_1}\left(\beta_2+\mathrm{i}\frac{m-1}{2}\bar{h}\right)\dots\Lambda_{j_m}\left(\beta_2-\mathrm{i}\frac{m-1}{2}\bar{h}\right) : \Lambda_l(\beta_1)$$

is equal to

$$f_{1m}(\beta_1-\beta_2) : \Lambda_{j_1}\left(\beta_2+\mathrm{i}\frac{m-1}{2}\bar{h}\right)\dots\Lambda_{j_m}\left(\beta_2-\mathrm{i}\frac{m-1}{2}\bar{h}\right)\Lambda_l(\beta_1):$$

if $l = j_k$ for some $k \in \{1, \ldots, m\}$; and

$$f_{1m}(\beta_1-\beta_2)\left[\left(\mathrm{i}\frac{\beta_1-\beta_2}{N\hbar}-\frac{\xi}{N}-\frac{1}{2N}-\frac{2k-m}{2N}\right)\left(\mathrm{i}\frac{\beta_1-\beta_2}{N\hbar}+\frac{\xi}{N}+\frac{1}{2N}-\frac{2k-m}{2N}\right)\right]$$
$$\times\left[\left(\mathrm{i}\frac{\beta_1-\beta_2}{N\hbar}-\frac{1}{2N}-\frac{2k-m}{2N}\right)\left(\mathrm{i}\frac{\beta_1-\beta_2}{N\hbar}+\frac{1}{2N}-\frac{2k-m}{2N}\right)\right]^{-1}$$
$$\times:\Lambda_{j_1}\left(\beta_2+\mathrm{i}\frac{m-1}{2}\hbar\right)\ldots\Lambda_{j_m}\left(\beta_2-\mathrm{i}\frac{m-1}{2}\hbar\right)\Lambda_l(\beta_1):$$

if $j_k < l < j_{k+1}$. Noting that

$$\lim_{\epsilon\longrightarrow 0+}\left(\frac{1}{x+\mathrm{i}\epsilon}-\frac{1}{x-\mathrm{i}\epsilon}\right)=-2\mathrm{i}\pi\delta(x) \tag{19}$$

we can obtain the commutation relations (17) for $T_1(\beta_1)$ and $T_m(\beta_2)$ after some straightforward calculations. Therefore, we complete the proof of theorem 1. □

In fact, the commutation relations for the generators of the $\hbar$-deformed W_N algebra have already been defined from the commutation relations of the fundamental operators $\Lambda_i(\beta)$ in equation (6) and the corresponding quantum Miura transformation in (7). So, one can also derive some similar commutation relations between $T_i(\beta)$ and $T_j(\beta)$ with $i, j > 1$ using the same method as that in the proof of theorem 1. These commutation relations are quadratic, and involve products of $T_{i-r}(\beta)$ and $T_{j+r}(\beta)$ with $r = 1, \ldots, \min(i, j) - 1$.

In the case of $N = 2$, this $\hbar$-deformed W_2 algebra becomes an $\hbar$-deformed Virasoro algebra, which we have studied in [12]. Here, we give an example for the case $N = 3$. The generators of this case are

$$T_1(\beta)=\Lambda_1(\beta)+\Lambda_2(\beta)+\Lambda_3(\beta) \tag{20}$$

$$T_2(\beta)=:\Lambda_1\left(\beta+\frac{\mathrm{i}\hbar}{2}\right)\Lambda_2\left(\beta-\frac{\mathrm{i}\hbar}{2}\right):+:\Lambda_1\left(\beta+\frac{\mathrm{i}\hbar}{2}\right)\Lambda_3\left(\beta-\frac{\mathrm{i}\hbar}{2}\right):$$
$$+:\Lambda_2\left(\beta+\frac{\mathrm{i}\hbar}{2}\right)\Lambda_3\left(\beta-\frac{\mathrm{i}\hbar}{2}\right):. \tag{21}$$

The commutation relations for these two generators are

$$f_{11}^{-1}(\beta_2-\beta_1)T_1(\beta_1)T_1(\beta_2)-f_{11}^{-1}(\beta_1-\beta_2)T_1(\beta_2)T_1(\beta_1)$$
$$=-2\mathrm{i}\pi\left\{\mathrm{i}\hbar\xi(\xi+1)\left(T_2\left(\beta_2+\frac{\mathrm{i}\hbar}{2}\right)\delta(\beta_1-\beta_2-\mathrm{i}\hbar)\right.\right.$$
$$\left.\left.-T_2\left(\beta_2-\frac{\mathrm{i}\hbar}{2}\right)\delta(\beta_1-\beta_2+\mathrm{i}\hbar)\right)\right\} \tag{22}$$

$$f_{12}^{-1}(\beta_2-\beta_1)T_1(\beta_1)T_2(\beta_2)-f_{12}^{-1}(\beta_1-\beta_2)T_2(\beta_2)T_1(\beta_1)$$
$$=-2\mathrm{i}\pi\left\{\mathrm{i}\hbar\xi(\xi+1)\left(\delta\left(\beta_1-\beta_2-\mathrm{i}\frac{3}{2}\hbar\right)-\delta\left(\beta_1-\beta_2+\mathrm{i}\frac{3}{2}\hbar\right)\right)\right\} \tag{23}$$

$$f_{22}^{-1}(\beta_2-\beta_1)T_2(\beta_1)T_2(\beta_2)-f_{22}^{-1}(\beta_1-\beta_2)T_2(\beta_2)T_2(\beta_1)$$
$$=-2\mathrm{i}\pi\left\{\mathrm{i}\hbar\xi(\xi+1)\left(T_1\left(\beta_2+\frac{\mathrm{i}\hbar}{2}\right)\delta(\beta_1-\beta_2-\mathrm{i}\hbar)\right.\right.$$
$$\left.\left.-T_1\left(\beta_2-\frac{\mathrm{i}\hbar}{2}\right)\delta(\beta_1-\beta_2+\mathrm{i}\hbar)\right)\right\} \tag{24}$$

where the coefficient function $f_{ij}(\beta)$ are

$$f_{11}(\beta)=\left[\Gamma\left(\frac{i\beta}{N\hbar}+1-\frac{1}{N}\right)\Gamma\left(\frac{i\beta}{N\hbar}+1\right)\Gamma\left(\frac{i\beta}{N\hbar}-\frac{\xi}{N}\right)\Gamma\left(\frac{i\beta}{N\hbar}+\frac{\xi}{N}+\frac{1}{N}\right)\right]$$
$$\times\left[\Gamma\left(\frac{i\beta}{N\hbar}+1-\frac{\xi}{N}-\frac{1}{N}\right)\Gamma\left(\frac{i\beta}{N\hbar}+1+\frac{\xi}{N}\right)\Gamma\left(\frac{i\beta}{N\hbar}\right)\Gamma\left(\frac{i\beta}{N\hbar}+\frac{1}{N}\right)\right]^{-1}$$

$$f_{12}(\beta)=\left[\Gamma\left(\frac{i\beta}{N\hbar}+1-\frac{3}{2N}\right)\Gamma\left(\frac{i\beta}{N\hbar}+1-\frac{1}{2N}\right)\right.$$
$$\left.\times\Gamma\left(\frac{i\beta}{N\hbar}-\frac{\xi}{N}+\frac{1}{2N}\right)\Gamma\left(\frac{i\beta}{N\hbar}+\frac{\xi}{N}+\frac{3}{2N}\right)\right]$$
$$\times\left[\Gamma\left(\frac{i\beta}{N\hbar}+1-\frac{\xi}{N}-\frac{3}{2N}\right)\Gamma\left(\frac{i\beta}{N\hbar}+1+\frac{\xi}{N}-\frac{1}{2N}\right)\right.$$
$$\left.\times\Gamma\left(\frac{i\beta}{N\hbar}+\frac{1}{2N}\right)\Gamma\left(\frac{i\beta}{N\hbar}+\frac{3}{2N}\right)\right]^{-1}$$

$$f_{11}(\beta)=f_{22}(\beta).$$

3. Screening currents

In this section, we will consider the screening currents for the $\hbar$-deformed W_N algebra. First, we introduce some zero mode operators. To each vector $\alpha\in\mathbb{P}$ (the $A^{(1)}_{N-1}$-type weight lattice defined in section 2.1), we associate operators P_α and Q_α which satisfy

$$[iP_\alpha, Q_\beta]=\langle\alpha,\beta\rangle \qquad (\alpha,\beta\in\mathbb{P}). \tag{25}$$

We shall deal with the bosonic Fock spaces $F_{l,k}(l,k\in\mathbb{P})$ generated by $\lambda_i(-t)(t>0)$ over the vacuum states $|l,k\rangle$ .The vacuum states $|l,k\rangle$ are defined by

$$\lambda_i(t)|l,k\rangle=0 \qquad \text{if } t>0$$
$$P_\beta|l,k\rangle=\langle\beta,\alpha_+l+\alpha_-k\rangle|l,k\rangle$$
$$|l,k\rangle=e^{i\alpha_+Q_l+i\alpha_-Q_k}|0,0\rangle$$

where $\alpha_\pm$ are some parameters related to ξ

$$\alpha_+=-\sqrt{\frac{1+\xi}{\xi}} \qquad \alpha_-=\sqrt{\frac{\xi}{1+\xi}} \tag{26}$$

and we also introduce α_0,

$$\alpha_0=\frac{1}{\sqrt{\xi(1+\xi)}}. \tag{27}$$

To each simple root α_j $(j=1,\ldots,N-1)$, let us introduce two series bosons $s_j^\pm(t)$ which are defined by

$$s_j^+(t)=\frac{e^{(j\hbar t/2)}}{2\mathrm{sh}(\xi\hbar t/2)}(\lambda_j(t)-\lambda_{j+1}(t)) \tag{28}$$
$$s_j^-(t)=\frac{e^{(j\hbar t/2)}}{2\mathrm{sh}((1+\xi)\hbar t/2)}(\lambda_j(t)-\lambda_{j+1}(t)). \tag{29}$$

By these simple root bosons, we can define the screening currents as follows:

$$S_j^+(\beta) =: \exp\left\{-\int_{-\infty}^{\infty} s_j^+(t)\, e^{i\beta t}\, dt\right\} : e^{-i\alpha_+ Q_{\alpha_j}} \tag{30}$$

$$S_j^-(\beta) =: \exp\left\{\int_{-\infty}^{\infty} s_j^-(t)\, e^{i\beta t}\, dt\right\} : e^{-i\alpha_- Q_{\alpha_j}}. \tag{31}$$

Then we have:

Theorem 2. The screening currents $S_j^+(\beta)$ satisfy

$$\begin{aligned}
&[: (e^{i\hbar\partial_\beta} - \Lambda_1(\beta))(e^{i\hbar\partial_\beta} - \Lambda_2(\beta - i\hbar)) \dots (e^{i\hbar\partial_\beta} - \Lambda_N(\beta - i(N-1)\hbar)) :, S_j^+(\sigma)] \\
&\quad = i\hbar(1+\xi)\{: (e^{i\hbar\partial_\beta} - \Lambda_1(\beta)) \dots (e^{i\hbar\partial_\beta} - \Lambda_{j-1}(\beta - i(j-2)\hbar)) D_{\sigma, i\hbar\xi} \\
&\quad \times \left(2\pi i \delta\left(\sigma - \beta - i\frac{j+\xi}{2}\hbar\right) A_j^+(\sigma)\right) \\
&\quad \times e^{i\hbar\partial_\beta}(e^{i\hbar\partial_\beta} - \Lambda_{j+2}(\beta - i(j+1)\hbar)) \dots (e^{i\hbar\partial_\beta} - \Lambda_N(\beta - i(N-1)\hbar)) :
\end{aligned} \tag{32}$$

with

$$A_j^+(\sigma) =: \Lambda_j\left(\sigma - i\frac{j+\xi}{2}\hbar\right) S_j^+(\sigma) :$$

and operator $D_{\sigma, i\hbar\xi}$ being a difference operator with variable σ:

$$D_{\sigma,\eta} f(\sigma) \equiv f(\sigma) - f(\sigma + \eta).$$

Proof. From formulae (28), we obtain the following commutation relations:

$$[\lambda_j(t), s_j^+(t')] = -\frac{2\, e^{t(1-j)\hbar/2} \mathrm{sh}((1+\xi)\hbar t/2)}{t} \delta(t+t')$$

$$[\lambda_{j+1}(t), s_j^+(t')] = \frac{2\, e^{-t(1+j)\hbar/2} \mathrm{sh}((1+\xi)\hbar t/2)}{t} \delta(t+t')$$

$$[\lambda_j(t), s_l^+(t')] = 0 \qquad \text{if } |j-l| > 1.$$

From these commutation relations and formula (15), we can derive the following OPEs:

$$\Lambda_j(\beta_1) S_j^+(\beta_2) = f_{jj}^+(\beta_2 - \beta_1) : \Lambda_j(\beta_1) S_j^+(\beta_2) : \tag{33}$$

$$S_j^+(\beta_1) \Lambda_j(\beta_2) = f_{jj}^+(\beta_2 - \beta_1) : S_j^+(\beta_1) \Lambda_j(\beta_2) : \tag{34}$$

$$\Lambda_{j+1}(\beta_1) S_j^+(\beta_2) = f_{j+1j}^+(\beta_2 - \beta_1) : \Lambda_{j+1}(\beta_1) S_j^+(\beta_2) : \tag{35}$$

$$S_j^+(\beta_1) \Lambda_{j+1}(\beta_2) = f_{j+1j}^+(\beta_2 - \beta_1) : S_j^+(\beta_1) \Lambda_{j+1}(\beta_2) : \tag{36}$$

$$S_j^+(\beta_1) \Lambda_l(\beta_2) =: S_j^+(\beta_1) \Lambda_l(\beta_2) :=: \Lambda_l(\beta_2) S_j^+(\beta_1) : \qquad \text{if } |j-l| > 1 \tag{37}$$

and

$$f_{jj}^+(\beta) = \left(\frac{i\beta}{N\hbar} - \frac{\xi}{2N} - \frac{1}{N} + \frac{j}{2N}\right) \Big/ \left(\frac{i\beta}{N\hbar} - \frac{\xi}{2N} + \frac{j}{2N} + \frac{\xi}{N}\right)$$

$$f_{j+1j}^+(\beta) = \left(\frac{i\beta}{N\hbar} - \frac{\xi}{2N} + \frac{1+\xi}{N} + \frac{j}{2N}\right) \Big/ \left(\frac{i\beta}{N\hbar} - \frac{\xi}{2N} + \frac{j}{2N}\right).$$

Formula (37) implies that the left-hand side of (32) equals

$$\begin{aligned}
&: (e^{i\hbar\partial_\beta} - \Lambda_1(\beta)) \dots (e^{i\hbar\partial_\beta} - \Lambda_{j-1}(\beta - i(j-2)\hbar)) \\
&\quad \times [(e^{i\hbar\partial_\beta} - \Lambda_j(\beta - i(j-1)\hbar))(e^{i\hbar\partial_\beta} - \Lambda_{j+1}(\beta - i(j)\hbar)), S_j^+(\sigma)] \\
&\quad \times (e^{i\hbar\partial_\beta} - \Lambda_{j+2}(\beta - i(j+1)\hbar)) \dots (e^{i\hbar\partial_\beta} - \Lambda_N(\beta - i(N-1)\hbar)) : .
\end{aligned} \tag{38}$$

Therefore, it is sufficient to consider the commutation relation

$$[:(e^{i\hbar\partial_\beta}-\Lambda_j(\beta-i(j-1)\hbar))(e^{i\hbar\partial_\beta}-\Lambda_{j+1}(\beta-i(j)\hbar)):,S_j^+(\sigma)].$$

According to the OPEs (33)–(36), we can derive

$$[:\Lambda_j(\beta-i(j-1)\hbar)\Lambda_{j+1}(\beta-i(j)\hbar):,S_j^+(\sigma)]=0.$$

Now we only need to consider the commutation relations between the term $\Lambda_j(\beta-i(j-1)\hbar)+\Lambda_{j+1}(\beta-i(j-1)\hbar)$ and the screening current $S_j^+(\sigma)$. From the OPEs (33)–(36), formula (19), noting that

$$:\Lambda_j\left(\beta-i\frac{j-\xi}{2}\hbar\right)S_j^+(\beta+i\xi\hbar):=:\Lambda_{j+1}\left(\beta-i\frac{j-\xi}{2}\hbar\right)S_j^+(\beta):$$

and using the same method as that in the proof of theorem 1, we have the following commutation relation:

$$\begin{aligned}&[\Lambda_j(\beta-i(j-1)\hbar)+\Lambda_{j+1}(\beta-i(j-1)\hbar),S_j^+(\sigma)]\\&\quad=i\hbar(1+\xi)D_{\sigma,i\hbar\xi}\left(2\pi i\delta\left(\sigma-\beta-i\frac{j+\xi}{2}\hbar\right):A_j^+(\sigma)\right):.\end{aligned}$$

Therefore, equation (32) has been obtained. □

Using the same method, we have:

Theorem 3. The second series screening currents $S_j^-(\sigma)$ satisfy

$$\begin{aligned}&[:(e^{i\hbar\partial_\beta}-\Lambda_1(\beta))(e^{i\hbar\partial_\beta}-\Lambda_2(\beta-i\hbar))\dots(e^{i\hbar\partial_\beta}-\Lambda_N(\beta-i(N-1)\hbar)):,S_j^-(\sigma)]\\&\quad=i\hbar\xi\{:(e^{i\hbar\partial_\beta}-\Lambda_1(\beta))\dots(e^{i\hbar\partial_\beta}-\Lambda_{j-1}(\beta-i(j-2)\hbar))D_{\sigma,-i\hbar(1+\xi)}\\&\quad\times\left(2\pi i\delta\left(\sigma-\beta+i\frac{-j+\xi+1}{2}\hbar\right)A_j^-(\sigma)\right)\\&\quad\times e^{i\hbar\partial_\beta}(e^{i\hbar\partial_\beta}-\Lambda_{j+2}(\beta-i(j+1)\hbar))\dots(e^{i\hbar\partial_\beta}-\Lambda_N(\beta-i(N-1)\hbar)):\end{aligned}\tag{39}$$

with

$$A_j^-(\sigma)=:\Lambda_j\left(\sigma+i\frac{-j+\xi+1}{2}\hbar\right)S_j^-(\sigma):.$$

Therefore, the screening currents $S_j^{\pm}(\beta)$ commute with any $\hbar$-deformed W_N algebra generators up to total difference.

Remark. In taking the conformal limit ($\hbar\longrightarrow 0$ and with ξ and β fixed), the screening currents $S_j^{\pm}(\beta)$ will become the ordinary screening current [8].

Theorem 2 and 3 imply that one can construct the intertwining operators (namely, vertex operators) for the $\hbar$-deformed W_N algebra using the screening currents $S_j^{\pm}(\beta)$. In the next section we shall construct the vertex operators for the $\hbar$-deformed W_N algebra.

4. The vertex operators and their exchange relations

In this section, we construct the type I and type II vertex operators for this $\bar{h}$-deformed W_N algebra through the two series screening currents $S_j^{\pm}(\beta)$. First, we set

$$\hat{\pi}_\mu = \alpha_0^{-1} P_{\bar{\epsilon}_\mu} \qquad \hat{\pi}_{\mu\nu} = \hat{\pi}_\mu - \hat{\pi}_\nu \tag{40}$$

and

$$\hat{\pi}_{\mu\nu} F_{l,k} = \langle \epsilon_\mu - \epsilon_\nu, -(1+\xi)l + \xi k \rangle F_{l,k}. \tag{41}$$

Note that

$$\mathrm{e}^{-\mathrm{i}\alpha_\pm Q_\gamma} \hat{\pi}_\sigma \, \mathrm{e}^{\mathrm{i}\alpha_\pm Q_\gamma} = \hat{\pi}_\sigma + \alpha_0^{-1}\alpha_\pm \langle \sigma, \gamma \rangle \tag{42}$$

and this formula is very useful for calculating commutation relations of vertex operators. To each fundamental weight of ω_j $(j = 1, \ldots, N-1)$, let us introduce two series bosons $a_j(t)$ and $a_j'(t)$ which are defined by

$$a_j = \sum_{k=1}^{j} \frac{\mathrm{e}^{-\bar{h}(j-2k+1)t/2}}{2\mathrm{sh}(\bar{h}\xi t/2)} \lambda_k(t) \qquad a_j' = \sum_{k=1}^{j} \frac{\mathrm{e}^{-\bar{h}(j-2k+1)t/2}}{2\mathrm{sh}(\bar{h}(1+\xi)t/2} \lambda_k(t) \tag{43}$$

and also define

$$U_{\omega_j}(\beta) =: \exp\left\{ \int_{-\infty}^{\infty} a_j(t)\, \mathrm{e}^{\mathrm{i}\beta t}\, \mathrm{d}t \right\} : \mathrm{e}^{\mathrm{i}\alpha_+ Q_{\omega_j}}$$
$$U'_{\omega_j}(\beta) =: \exp\left\{ \int_{-\infty}^{\infty} -a_j'(t)\, \mathrm{e}^{\mathrm{i}\beta t}\, \mathrm{d}t \right\} : \mathrm{e}^{\mathrm{i}\alpha_- Q_{\omega_j}}. \tag{44}$$

Because the vertex operators associated with each fundamental weight $\omega_j (= 2, \ldots, N-1)$ can be constructed from the skew-symmetric fusion of the basic ones $U_{\omega_1}(\beta)$ and $U'_{\omega_1}(\beta)$ [1], it is sufficient to only deal with the vertex operators corresponding to the fundamental weight ω_1. In order to calculate the exchange relations of the vertex operators, we first derive the following commutation relations:

$$[a_j(t), s_l^+(t')] = -\frac{\mathrm{sh}(\bar{h}(1+\xi)t/2)}{t\mathrm{sh}(\bar{h}\xi t/2)} \delta_{j,l}\delta(t+t') \qquad [a_j(t), s_l^-(t')] = -\frac{1}{t}\delta_{j,l}\delta(t+t')$$

$$[a_j'(t), s_l^-(t')] = -\frac{\mathrm{sh}(\bar{h}\xi t/2)}{t\mathrm{sh}(\bar{h}(1+\xi)t/2)} \delta_{j,l}\delta(t+t') \qquad [a_j'(t), s_l^+(t')] = -\frac{1}{t}\delta_{j,l}\delta(t+t')$$

$$[a_1(t), a_1(t')] = -\frac{\mathrm{sh}((N-1)\bar{h}t/2)\mathrm{sh}((1+\xi)\bar{h}t/2)}{t\mathrm{sh}(N\bar{h}t/2)\mathrm{sh}(\xi\bar{h}t/2)}\delta(t+t')$$

$$[a_1(t), a_1'(t')] = -\frac{\mathrm{sh}((N-1)\bar{h}t/2)}{t\mathrm{sh}(N\bar{h}t/2)}\delta(t+t')$$

$$[a_1'(t), a_1'(t')] = -\frac{\mathrm{sh}((N-1)\bar{h}t/2)\mathrm{sh}(\xi\bar{h}t/2)}{t\mathrm{sh}(N\bar{h}t/2)\mathrm{sh}((1+\xi)\bar{h}t/2)}\delta(t+t').$$

From the above relations, and taking the regularization in section 2.3, we can derive the following exchange relations,

$$U_{\omega_1}(\beta_1)U_{\omega_1}(\beta_2) = r_1(\beta_1-\beta_2)U_{\omega_1}(\beta_2)U_{\omega_1}(\beta_1)$$
$$U'_{\omega_1}(\beta_1)U'_{\omega_1}(\beta_2) = r_1'(\beta_1-\beta_2)U'_{\omega_1}(\beta_2)U'_{\omega_1}(\beta_1)$$
$$U_{\omega_1}(\beta_1)U'_{\omega_1}(\beta_2) = \tau_1(\beta_1-\beta_2)U'_{\omega_1}(\beta_2)U_{\omega_1}(\beta_1)$$
$$S_j^+(\beta_1)S_{j+1}^+(\beta_2) = -f(\beta_1-\beta_2, 0)S_{j+1}^+(\beta_2)S_j^+(\beta_1)$$
$$S_j^-(\beta_1)S_{j+1}^-(\beta_2) = -f'(\beta_1-\beta_2, 0)S_{j+1}^-(\beta_2)S_j^-(\beta_1)$$

$$U_{\omega_1}(\beta_1)S_1^+(\beta_2) = -f(\beta_1-\beta_2,0)S_1^+(\beta_2)U_{\omega_1}(\beta_1)$$
$$U'_{\omega_1}(\beta_1)S_1^- = -f'(\beta_1-\beta_2,0)S_1^-(\beta_2)U'_{\omega_1}(\beta_1)$$
$$U_{\omega_1}(\beta_1)S_1^-(\beta_2) = -U_{\omega_1}(\beta_1)S_1^-(\beta_2)$$
$$U'_{\omega_1}(\beta_1)S_1^+(\beta_2) = -U'_{\omega_1}(\beta_1)S_1^+(\beta_2) \tag{45}$$

where the fundamental function $f(\beta,w)$ and $f'(\beta,w)$ (which play a very important role in constructing the vertex operators) are defined by

$$f(\beta,w) = \sin\pi\left(\frac{\mathrm{i}\beta}{\overline{h}\xi}-\frac{1}{2\xi}-\frac{w}{\xi}\right)\Big/\sin\pi\left(\frac{\mathrm{i}\beta}{\overline{h}\xi}+\frac{1}{2\xi}\right)$$
$$f'(\beta,w) = \sin\pi\left(\frac{\mathrm{i}\beta}{\overline{h}(1+\xi)}+\frac{1}{2(1+\xi)}+\frac{w}{1+\xi}\right)\Big/\sin\pi\left(\frac{\mathrm{i}\beta}{\overline{h}(1+\xi)}-\frac{1}{2(1+\xi)}\right) \tag{46}$$

and

$$r(\beta) = \exp\left\{-\int_0^\infty \frac{2\mathrm{sh}((N-1)\overline{h}t/2)\mathrm{sh}((1+\xi)\overline{h}t/2)\mathrm{sh}(\mathrm{i}\beta t)}{t\mathrm{sh}(N\overline{h}t/2)\mathrm{sh}(\xi\overline{h}t/2)}\,\mathrm{d}t\right\} \tag{47}$$
$$r'(\beta) = \exp\left\{-\int_0^\infty \frac{2\mathrm{sh}((N-1)\overline{h}t/2)\mathrm{sh}(\xi\overline{h}t/2)\mathrm{sh}(\mathrm{i}\beta t)}{t\mathrm{sh}(N\overline{h}t/2)\mathrm{sh}((1+\xi)\overline{h}t/2)}\,\mathrm{d}t\right\} \tag{48}$$
$$\tau(\beta) = \sin\pi\left(\frac{1}{2N}-\frac{\mathrm{i}\beta}{N\overline{h}}\right)\Big/\sin\pi\left(\frac{1}{2N}+\frac{\mathrm{i}\beta}{N\overline{h}}\right). \tag{49}$$

Now let us define the type I vertex operators $Z'_\mu(\beta)$ and the type II vertex operators $Z_\mu(\beta)(\mu=1,\ldots,N)$

$$Z_\mu(\beta) = \int_{C_1}\prod_{j=1}^{\mu-1}\mathrm{d}\eta_j\, U'_{\omega_1}(\beta)S_1^-(\eta_1)S_2^-(\eta_2)\ldots S_{\mu-1}^-(\eta_{\mu-1})\prod_{j=1}^{\mu-1}f'(\eta_j-\eta_{j-1},\hat{\pi}_{j\mu}) \tag{50}$$

$$Z'_\mu(\beta) = \int_{C_2}\prod_{j=1}^{\mu-1}\mathrm{d}\eta_j\, U_{\omega_1}(\beta)S_1^+(\eta_1)S_2^+(\eta_2)\ldots S_{\mu-1}^+(\eta_{\mu-1})\prod_{j=1}^{\mu-1}f(\eta_j-\eta_{j-1},\hat{\pi}_{j\mu}). \tag{51}$$

It is easy to see that the vertex operators $Z_\mu(\beta)$ and $Z'_\mu(\beta)$ are some bosonic operators intertwing the Fock spaces $F_{l,k}$

$$Z_\mu(\beta): F_{l,k}\longrightarrow F_{l,k+\overline{\epsilon}_\mu} \qquad Z'_\mu(\beta): F_{l,k}\longrightarrow F_{l+\overline{\epsilon}_\mu,k}. \tag{52}$$

Here we set $\eta_0=\beta$, the integration contour C_1 is chosen as the contour corresponding to the integration variable η_j enclosing the poles $\eta_{j-1}+\mathrm{i}\frac{1}{2}\overline{h}-\mathrm{i}\overline{h}\xi n(0\leqslant n)$, and the other integration contour C_2 is chosen as the contour corresponding to the integration variable η_j enclosing the poles $\eta_{j-1}-\mathrm{i}\frac{1}{2}\overline{h}-\mathrm{i}(1+\xi)\overline{h}n(0\leqslant n)$.

The constructure form of our type I (type II) vertex operators seems to be similar to that of the vertex operators for the $A_{N-1}^{(1)}$ face model given by Asai *et al* [1], but with different bosonic operators and 'coefficient parts' function $f(\beta,w)$ $(f'(\beta,w))$. Thus the same trick [1] can be used to calculate the commutation relations for our vertex operators. Using the method which was presented by Asai *et al* in appendix B of the [1], we can derive the commutation relations for vertex operators $Z_\mu(\beta)$ and $Z'_\mu(\beta)$,

$$Z'_\mu(\beta_1)Z'_\nu(\beta_2) = \sum_{\mu'\nu'}^{\overline{\epsilon}_\mu+\overline{\epsilon}_\nu=\overline{\epsilon}_{\mu'}+\overline{\epsilon}_{\nu'}} Z'_{\mu'}(\beta_2)Z'_{\nu'}(\beta_1)\hat{W}'\begin{pmatrix}\hat{\pi}+\overline{\epsilon}_\mu+\overline{\epsilon}_\nu & \hat{\pi}+\overline{\epsilon}_{\nu'} \\ \hat{\pi}+\overline{\epsilon}_\nu & \hat{\pi}\end{pmatrix}|\beta_1-\beta_2\Big) \tag{53}$$

$$Z_\mu(\beta_1)Z_\nu(\beta_2) = \sum_{\mu'\nu'}^{\overline{\epsilon}_\mu+\overline{\epsilon}_\nu=\overline{\epsilon}_{\mu'}+\overline{\epsilon}_{\nu'}} Z_{\mu'}(\beta_2)Z_{\nu'}(\beta_1)\hat{W}\begin{pmatrix}\hat{\pi}+\overline{\epsilon}_\mu+\overline{\epsilon}_\nu & \hat{\pi}+\overline{\epsilon}_{\nu'} \\ \hat{\pi}+\overline{\epsilon}_\nu & \hat{\pi}\end{pmatrix}|\beta_1-\beta_2\Big) \tag{54}$$

$$Z_\mu(\beta_1)Z'_\nu(\beta_2) = Z'_\nu(\beta_2)Z'_\mu(\beta_1)\tau(\beta_1-\beta_2) \tag{55}$$

and the braid matrices (connection matrices) $\hat{W}\begin{pmatrix} \hat{\pi}+\bar{\epsilon}_\mu+\bar{\epsilon}_\nu & \hat{\pi}+\bar{\epsilon}_{\nu'} \\ \hat{\pi}+\bar{\epsilon}_\nu & \hat{\pi} \end{pmatrix}|\beta)$ and $\hat{W}'\begin{pmatrix} \hat{\pi}+\bar{\epsilon}_\mu+\bar{\epsilon}_\nu & \hat{\pi}+\bar{\epsilon}_{\nu'} \\ \hat{\pi}+\bar{\epsilon}_\nu & \hat{\pi} \end{pmatrix}|\beta)$ are some functions taking values on operators $\hat{\pi}_{\mu\nu}$ such as

$$\hat{W}'\left(\begin{matrix} \hat{\pi}+2\bar{\epsilon}_\mu & \hat{\pi}+\bar{\epsilon}_\mu \\ \hat{\pi}+\bar{\epsilon}_\mu & \hat{\pi} \end{matrix}|\beta\right) = r(\beta) \tag{56}$$

$$\hat{W}'\left(\begin{matrix} \hat{\pi}+\bar{\epsilon}_\mu+\bar{\epsilon}_\nu & \hat{\pi}+\bar{\epsilon}_\nu \\ \hat{\pi}+\bar{\epsilon}_\nu & \hat{\pi} \end{matrix}|\beta\right) = -r(\beta)\left[\sin\frac{\pi}{\xi}\sin\pi\left(\frac{i\beta}{\hbar\xi}-\frac{\hat{\pi}_{\mu\nu}}{\xi}\right)\Big/\sin\pi\left(\frac{\hat{\pi}_{\mu\nu}}{\xi}\right)\sin\pi\left(\frac{i\beta}{\hbar\xi}+\frac{1}{\xi}\right)\right] \tag{57}$$

$$\hat{W}'\left(\begin{matrix} \hat{\pi}+\bar{\epsilon}_\mu+\bar{\epsilon}_\nu & \hat{\pi}+\bar{\epsilon}_\mu \\ \hat{\pi}+\bar{\epsilon}_\nu & \hat{\pi} \end{matrix}|\beta\right) = r(\beta)\left[\sin\pi\left(\frac{i\beta}{\hbar\xi}\right)\sin\pi\left(\frac{1}{\xi}+\frac{\hat{\pi}_{\mu\nu}}{\xi}\right)\Big/\sin\pi\left(\frac{\hat{\pi}_{\mu\nu}}{\xi}\right)\sin\pi\left(\frac{i\beta}{\hbar\xi}+\frac{1}{\xi}\right)\right] \tag{58}$$

$$\hat{W}\left(\begin{matrix} \hat{\pi}+2\bar{\epsilon}_\mu & \hat{\pi}+\bar{\epsilon}_\mu \\ \hat{\pi}+\bar{\epsilon}_\mu & \hat{\pi} \end{matrix}|\beta\right) = r'(\beta) \tag{59}$$

$$\hat{W}\left(\begin{matrix} \hat{\pi}+\bar{\epsilon}_\mu+\bar{\epsilon}_\nu & \hat{\pi}+\bar{\epsilon}_\nu \\ \hat{\pi}+\bar{\epsilon}_\nu & \hat{\pi} \end{matrix}|\beta\right) = -r'(\beta)\left\{\left[\sin\frac{\pi}{1+\xi}\sin\pi\left(\frac{i\beta}{\hbar(1+\xi)}+\frac{\hat{\pi}_{\mu\nu}}{1+\xi}\right)\right\}\right.$$
$$\left.\times\left\{\sin\pi\left(\frac{\hat{\pi}_{\mu\nu}}{1+\xi}\right)\sin\pi\left(\frac{i\beta}{\hbar(1+\xi)}-\frac{1}{1+\xi}\right)\right]\right\}^{-1} \tag{60}$$

$$\hat{W}\left(\begin{matrix} \hat{\pi}+\bar{\epsilon}_\mu+\bar{\epsilon}_\nu & \hat{\pi}+\bar{\epsilon}_\mu \\ \hat{\pi}+\bar{\epsilon}_\nu & \hat{\pi} \end{matrix}|\beta\right) = r'(\beta)\left\{\left[\sin\pi\left(\frac{i\beta}{\hbar(1+\xi)}\right)\sin\pi\left(\frac{1}{1+\xi}+\frac{\hat{\pi}_{\mu\nu}}{1+\xi}\right)\right\}\right.$$
$$\left.\times\left\{\sin\pi\left(\frac{\hat{\pi}_{\mu\nu}}{1+\xi}\right)\sin\pi\left(\frac{i\beta}{\hbar(1+\xi)}-\frac{1}{1+\xi}\right)\right]\right\}^{-1}. \tag{61}$$

Therefore, these connection matrices do not commute with the vertex operators and the exchange relations should be written in the same order as in equations (53) and (54). Noting that

$$r(-\beta)|_{\xi\longrightarrow 1+\xi} = r'(\beta)\Delta_N(\beta)$$
$$\Delta_N(\beta) = \sin\pi\left(\frac{i\beta}{N\hbar}+\frac{1}{N}\right)\Big/\sin\pi\left(\frac{i\beta}{N\hbar}-\frac{1}{N}\right)$$

we find that the matrices $\hat{W}\begin{pmatrix} \hat{\pi}+\bar{\epsilon}_\mu+\bar{\epsilon}_\nu & \hat{\pi}+\bar{\epsilon}_{\nu'} \\ \hat{\pi}+\bar{\epsilon}_\nu & \hat{\pi} \end{pmatrix}|\beta)$ and $\hat{W}'\begin{pmatrix} \hat{\pi}+\bar{\epsilon}_\mu+\bar{\epsilon}_\nu & \hat{\pi}+\bar{\epsilon}_{\nu'} \\ \hat{\pi}+\bar{\epsilon}_\nu & \hat{\pi} \end{pmatrix}|\beta)$ are related to each other as follows:

$$\hat{W}'\left(\begin{matrix} \hat{\pi}+\bar{\epsilon}_\mu+\bar{\epsilon}_\nu & \hat{\pi}+\bar{\epsilon}_{\nu'} \\ \hat{\pi}+\bar{\epsilon}_\nu & \hat{\pi} \end{matrix}|-\beta\right)|_{\xi\longrightarrow 1+\xi} = \Delta_N(\beta)\hat{W}\left(\begin{matrix} \hat{\pi}+\bar{\epsilon}_\mu+\bar{\epsilon}_\nu & \hat{\pi}+\bar{\epsilon}_{\nu'} \\ \hat{\pi}+\bar{\epsilon}_\nu & \hat{\pi} \end{matrix}|\beta\right).$$

When $N=2$, the factor $\Delta_N(\beta)=-1$, as occurred when studying the $\hbar$-deformed Virasoro algebra [12]. If both sides of equations (53) and (54) are acted on the special Fock space $F_{l,k}$, noting that $l_{\mu\nu}$ and $k_{\mu\nu}$

$$l_{\mu\nu} = \langle\bar{\epsilon}_\mu-\bar{\epsilon}_\nu, l\rangle \qquad k_{\mu\nu} = \langle\bar{\epsilon}_\mu-\bar{\epsilon}_\nu, k\rangle$$

are all integer, we have

$$Z'_\mu(\beta_1)Z'_\nu(\beta_2)|_{F_{l,k}} = \sum_{\mu'\nu'}^{\bar{\epsilon}_\mu+\bar{\epsilon}_\nu=\bar{\epsilon}_{\mu'}+\bar{\epsilon}_{\nu'}} Z'_{\mu'}(\beta_2)Z'_{\nu'}(\beta_1)|_{F_{l,k}}W'\left(\begin{matrix} l+\bar{\epsilon}_\mu+\bar{\epsilon}_\nu & l+\bar{\epsilon}_{\nu'} \\ l+\bar{\epsilon}_\nu & l \end{matrix}|\beta_1-\beta_2\right) \tag{62}$$

$$Z_\mu(\beta_1)Z_\nu(\beta_2)|_{F_{l,k}} = \sum_{\mu'\nu'}^{\bar\epsilon_\mu+\bar\epsilon_\nu=\bar\epsilon_{\mu'}+\bar\epsilon_{\nu'}} Z_{\mu'}(\beta_2)Z_{\nu'}(\beta_1)|_{F_{l,k}} W\begin{pmatrix} k+\bar\epsilon_\mu+\bar\epsilon_\nu & k+\bar\epsilon_{\nu'} \\ k+\bar\epsilon_\nu & k \end{pmatrix}|\beta_1-\beta_2\Big) \tag{63}$$

and

$$W'\begin{pmatrix} l+2\bar\epsilon_\mu & l+\bar\epsilon_\mu \\ l+\bar\epsilon_\mu & l \end{pmatrix}|\beta\Big) = r(\beta)$$

$$W'\begin{pmatrix} l+\bar\epsilon_\mu+\bar\epsilon_\nu & l+\bar\epsilon_\nu \\ l+\bar\epsilon_\nu & l \end{pmatrix}|\beta\Big) = r(\beta)\left[\sin\frac{\pi}{\xi}\sin\pi\left(\frac{i\beta}{\hbar\xi}+\frac{l_{\mu\nu}}{\xi}\right)\Big/\sin\pi\left(\frac{l_{\mu\nu}}{\xi}\right)\sin\pi\left(\frac{i\beta}{\hbar\xi}+\frac{1}{\xi}\right)\right]$$

$$W'\begin{pmatrix} l+\bar\epsilon_\mu+\bar\epsilon_\nu & l+\bar\epsilon_\mu \\ l+\bar\epsilon_\nu & l \end{pmatrix}|\beta\Big) = r(\beta)\left[\sin\pi\left(\frac{i\beta}{\hbar\xi}\right)\sin\pi\left(-\frac{1}{\xi}+\frac{l_{\mu\nu}}{\xi}\right)\Big/\sin\pi\left(\frac{l_{\mu\nu}}{\xi}\right)\sin\pi\left(\frac{i\beta}{\hbar\xi}+\frac{1}{\xi}\right)\right]$$

$$W\begin{pmatrix} k+2\bar\epsilon_\mu & k+\bar\epsilon_\mu \\ k+\bar\epsilon_\mu & k \end{pmatrix}|\beta\Big) = r'(\beta)$$

$$\begin{aligned} W\begin{pmatrix} k+\bar\epsilon_\mu+\bar\epsilon_\nu & k+\bar\epsilon_\nu \\ k+\bar\epsilon_\nu & k \end{pmatrix}|\beta\Big) &= r'(\beta)\left\{\left[\sin\frac{\pi}{1+\xi}\sin\pi\left(\frac{i\beta}{\hbar(1+\xi)}-\frac{k_{\mu\nu}}{1+\xi}\right)\right\}\right. \\ &\quad\times\left\{\sin\pi\left(\frac{k_{\mu\nu}}{1+\xi}\right)\sin\pi\left(\frac{i\beta}{\hbar(1+\xi)}-\frac{1}{1+\xi}\right)\right]\Big\}^{-1} \end{aligned}$$

$$\begin{aligned} W\begin{pmatrix} k+\bar\epsilon_\mu+\bar\epsilon_\nu & k+\bar\epsilon_\mu \\ k+\bar\epsilon_\nu & k \end{pmatrix}|\beta\Big) &= r'(\beta)\left\{\left[\sin\pi\left(\frac{i\beta}{\hbar(1+\xi)}\right)\sin\pi\left(-\frac{1}{1+\xi}+\frac{k_{\mu\nu}}{1+\xi}\right)\right\}\right. \\ &\quad\times\left\{\sin\pi\left(\frac{k_{\mu\nu}}{1+\xi}\right)\sin\pi\left(\frac{i\beta}{\hbar(1+\xi)}-\frac{1}{1+\xi}\right)\right]\Big\}^{-1}. \end{aligned}$$

It can be checked that the Boltzmann weights $W\begin{pmatrix} c & d \\ b & a \end{pmatrix}|\beta\Big)$ and $W\begin{pmatrix} c & d \\ b & a \end{pmatrix}|\beta\Big)$ satisfy the star-triangle equations (or Yang–Baxter equation)

$$\begin{aligned} &\sum_g W\begin{pmatrix} d & e \\ c & g \end{pmatrix}|\beta_1\Big) W\begin{pmatrix} c & g \\ b & a \end{pmatrix}|\beta_2\Big) W\begin{pmatrix} e & f \\ g & a \end{pmatrix}|\beta_1-\beta_2\Big) \\ &= \sum_g W\begin{pmatrix} g & f \\ b & a \end{pmatrix}|\beta_1\Big) W\begin{pmatrix} d & e \\ g & f \end{pmatrix}|\beta_2\Big) W\begin{pmatrix} d & g \\ c & b \end{pmatrix}|\beta_1-\beta_2\Big) \end{aligned} \tag{64}$$

and unitary relation

$$\sum_g W\begin{pmatrix} c & g \\ b & a \end{pmatrix}|-\beta\Big) W\begin{pmatrix} c & d \\ g & a \end{pmatrix}|\beta\Big) = \delta_{bd}. \tag{65}$$

In fact, the Yang–Baxter equation (64) and the unitary relation (65) are direct results of the exchange relation of the vertex operators in equations (62) and (63) (the associativity of algebra $Z_\mu(\beta)$ and $Z'_\mu(\beta)$).

Remark. In fact, we have constructed the bosonization of the vertex operators for the trigonometric SOS (solid-on-solid) model of $A_{N-1}^{(1)}$ type.

5. Discussions

We have constructed an $\bar{h}$-deformed W_N algebra and its quantum Miura transformation. The $\bar{h}$-deformation of the W_N algebra can be obtained by two ways: one can first derive the classical version of the $\bar{h}$-deformed W_N algebra by studying the Yangian double with centre $DY(\hat{sl}_N)$ at the critical level, following Frenkel and Reshetikhin in their study of the $U_q(\hat{sl}_N)$ at the critical level [10], and then construct the (quantum) $\bar{h}$-deformed W_N algebra by quantizing the classical one; another way to construct the $\bar{h}$-deformed W_N algebra is by taking some scaling limit of the q-deformed W_N algebra such as equation (16). In fact, the same phenomena also occur in studying the Yangian double with centre $DY(\hat{sl}_N)$: the $DY(\hat{sl}_N)$ can be considered as some scaling limit of the $U_q(\hat{sl}_N)$ algebra.

We have only considered the $\bar{h}$-deformed W_N algebra for generic ξ. When ξ is some rational number ($\xi = p/q$, p and q are two coprime integers), the realization of the $\bar{h}$-deformed W_N algebra in the Fock space $F_{l,k}$ would be highly reducible and we have to throw out some states from the Fock space $F_{l,k}$ to obtain the irreducible component $H_{l,k}$. (Here we choose the same symbols as in [1]). For $N = 2$, the irreducible space $H_{l,k}$ (i.e. $L_{l,k}$ in [12]) can be obtained by some BRST cohomology [12]. Unfortunately, the constructure of the BRST complex and the calculation of cohomology for $3 \leqslant N$ is still an open problem.

We have also constructed the vertex operators of type I and type II. These vertex operators satisfy some Fadeev–Zamolodchikov algebra with face type Boltzmann weight as its constructure constant. In order to obtain the correlation functions as the traces of products of these vertex operators, we need to introduce a boost operator H,

$$H = \sum_{j=1}^{N-1} \int_0^{\infty} \frac{t^2 \mathrm{sh}(\xi \bar{h} t/2)}{\mathrm{sh}((1+\xi)\bar{h}t/2)} a_j(-t) s_j^{+}(t)\, \mathrm{d}t \tag{66}$$

which enjoys the property

$$\mathrm{e}^{2\bar{h}H} Z_\mu(t)\, \mathrm{e}^{-2\bar{h}H} = Z_\mu(t - 2\mathrm{i}\bar{h}) \qquad \mathrm{e}^{2\bar{h}H} Z'_\mu(t)\, \mathrm{e}^{-2\bar{h}H} = Z'_\mu(t - 2\mathrm{i}\bar{h}).$$

Moreover, using the skew-symmetric fusion of N vertex operators, one can obtain some invertibility for our vertex operators of the form such as (3.19) and (c.20) in [1]. Then the correlation function can be described by the following trace function:

$$G(\beta_1, \ldots, \beta_{Nn})_{\mu_1, \ldots, \mu_{Nn}} = \frac{\mathrm{tr}(\mathrm{e}^{-2\bar{h}H} Z'_{\mu_1}(\beta_1) \ldots Z'_{\mu_{Nn}}(\beta_{Nn}))}{\mathrm{tr}(\mathrm{e}^{-2\bar{h}H})}. \tag{67}$$

References

[1] Asai Y, Jimbo M, Miwa T and Pugai Y 1996 Bosonization of vertex operators for $A_{n-1}^{(1)}$ face model, RIMS-1082

[2] Awata H, Kubo H, Odake S and Shiraishi J 1996 *Comm. Math. Phys.* **179** 401
Awata H, Kubo H, Odake S and Shiraishi J 1995 A quantum deformation of the Virasoro algebra and the Macdonald symmetric functions *Preprint* q-alg/9507034, YITP/V-95-30
Awata H, Kubo H, Odake S and Shiraishi J Virasoro type symmetry in solvable model.

[3] Awata H, Matsuo Y H, Odake S and Shiraishi J 1995 *Phys. Lett.* **347B** 49

[4] Chari V and Pressley A 1991 *Comm. Maths. Phys.* **142** 261

[5] Ding X M, Hou B Y and Zhao L 1997 $\bar{h}$-(Yangian) deformation Miura Map and Virasoro algebra *Preprint* q-alg/9701014

[6] Drinfeld V G 1987 Quantum groups *Proc. Int. Congress of Mathematicians (Berkeley)* p 798

[7] Drinfeld V G 1988 A new realization of Yangians and quantized affine algebras *Sov. Math. Dokl.* **32** 212

[8] Fateev V and Lukyanov S 1988 *Int. J. Mod. Phys.* A **3** 507
Fateev V and Lukyanov S 1990 Additional symmetries and exactly solvable models in two-dimensional field theory *Sov. Sci. Rev.* **15** 1

[9] Feigin B and Frenkel E 1996 *Comm. Math. Phys.* **178** 653

[10] Frenkel E and Reshetikhin N 1996 *Comm. Math. Phys.* **178** 237

[11] Frenkel I and Reshetikhin N 1992 *Comm. Math. Phys.* **146** 1

[12] Hou B Y and Yang W L 1996 A $\bar{h}$-deformed Virasoro algebra as a hidden symmetry of the Restricted sin-Gordon model *Preprint* hep-th/9612235
Hou B Y and Yang W L 1996 Boundary $A_1^{(1)}$ face model *Comm. Theor. Phys.* to be published

[13] Iohara K and Kohno M 1996 *Lett. Math. Phys.* **37** 319

[14] Jimbo M, Kedem R, Kojima T, Konno H and Miwa T 1995 *Nucl. Phys.* **441B** 437

[15] Jimbo M, Konno H and Miwa T 1996 Massless XXZ model and degeneration of the Elliptic algebra $A_{Q,P}(\hat{sl}_2)$ *Preprint* RIMS-1105

[16] Jimbo M and Miwa T 1994 *Analysis of Solvable Lattice Models (CBMS Regional Conference Series in Mathematics 85)* (AMS)

[17] Kadeishvili A A 1996 Vertex operators for deformed Virasoro algebra *Preprint* hep-th/9604153

[18] Khoroshkin S 1996 Central extension of the Yangian double *Preprint* q-alg/9602031

[19] Khoroshkin S, Lebedev D and Pakuliak S 1996 Traces of intertwining operators for the Yangian Double *Preprint* q-alg/9605039, ITEP-TH-8/96, BONN-TH-96-03

[20] Lukyanov S 1995 *Comm. Math. Phys.* **167** 183

[21] Lukyanov S and Pugai Y 1996 *Nucl. Phys.* B **473** 631
Lukyanov S and Pugai Y 1994 Bosonization of ZF algebra: Direction toward deformed Virasoro algebra *Preprint* hep-th/9412128, RU-94-41

[22] Miwa T and Weston R 1996 Boundary ABF models *Preprint* hep-th/9610094

[23] Smirnov F A 1992 *J. Mod. Phys.* A **7** (suppl 1B) 813
Smirnov F A 1992 *J. Mod. Phys.* A **7** (suppl 1B) 839

[24] Hou B Y and Yang W L The classical $\bar{h}$-deformed W_N algebra, in preparation

Commun. Theor. Phys. **29** (1998) pp. 443–446
Vol. 29, No. 3, April 30, 1998

Lagrangian Form of the Self-Dual Equations for SU(N) Gauge Fields on Four-Dimensional Euclidean Space*

HOU Boyu† and SONG Xingchang‡

Institute for Theoretical Physics, State University of New York at Stony Brook, Stony Brook, New York 11794, USA

(Received September 1, 1997)

Abstract *By compactifying the four-dimensional Euclidean space into $S_2 \times S_2$ manifold and introducing two topological relevant Wess–Zumino terms to $H_n \equiv SL(n,c)/SU(n)$ nonlinear sigma model, we construct a Lagrangian form for $SU(n)$ self-dual Yang–Mills field, from which the self-dual equations follow as the Euler–Lagrange equations.*

PACS numbers: 11.10.Ef, 11.15.Ha

Key words: self-dual Yang–Mills theory, topological field

It is well known that the self-dual Yang–Mills fields (SDYM)[1,2] in four-dimensional Euclidean space play an important role in understanding the properties of whole non-Abelian gauge theories, e.g., the existence of the instanton solutions and the static monopole solutions.[3] Quite a lot of efforts have been made to investigate their properties and to find the solutions for SDYM.[4–6] However, to our knowledge, up to now the self-dual conditions have not been able to put into the Lagrangian form except in the special gauge, the so-called R-gauge.

On the other hand, there has been considerable progress in the investigation of the two-dimensional nonlinear sigma models, or chiral fields, within recent years. The main interest comes from the great similarity of these models to four-dimensional gauge fields. Both theories are renormalizable and asymptotic free, possess some kind of soliton-like solutions, Backlund transformations, infinite number of conservation laws and the linear system of equations, etc.[7] As to the direct connection between these two theories, we only know that in some planes the four-dimensional SDYM fields reduce to some two-dimensional nonlinear sigma models and we can pass from one to another by exerting some special conditions.[6,8] Some time ago one of the present authors proposed a new nonlinear sigma model, called the H_n sigma model, taking values in the symmetric space SL$\,(n,c)/\,$SU(n), and pointed out that this model has the same kinematical properties as SDYM field but yields a different form of dynamical equation.

More recently, Witten[9] has put forward the chiral model by introducing the Wess–Zumino term[10] associated with the topological properties of the group in higher dimensions. The same procedure has been applied to the two-dimensional nonlinear sigma model.[9] In this note by a similar consideration to introduce appropriate topological terms and choose the parameters properly, we show that the H_n sigma model proposed earlier[8] can be used to describe the SDYM field correctly.

First we start from the SU(n) gauge field. Consider the gauge potential A_μ and the field strength $F_{\mu\nu}$ defined by

$$A_\mu \equiv A^a_\mu \lambda^a/2\mathrm{i}\,, \qquad F_{\mu\nu} \equiv F^a_{\mu\nu}\lambda^a/2\mathrm{i} = \partial_\mu A_\nu - \partial_\nu A_\mu + [A_\mu, A_\nu]\,, \tag{1}$$

*The project supported in part by the NSF Contract No. PHY-81-09110-A-01. One of the authors (X.C. SONG) was supported by a Fung King-Hey Fellowship through the Committee for Educational Exchange with China

†Permanent address: Department of Physics, Northwest University, Xi'an 710069, China
‡Permanent address: Institute of Theoretical Physics, Peking University, Beijing 100871, China

where λ^a are Hermitian generators of SU(n) group. Following Yang,[2] we introduce four new complex coordinate variables $y, \bar{y}, z$ and $\bar{z}$ defined as

$$\sqrt{2}\,y = y_1 + \mathrm{i}\,y_2 \equiv x_1 + \mathrm{i}\,x_2\,, \qquad \sqrt{2}\,\bar{y} = y_1 - \mathrm{i}\,y_2 \equiv x_1 - \mathrm{i}\,x_2\,;$$
$$\sqrt{2}\,z = z_1 + \mathrm{i}\,z_2 \equiv x_3 + \mathrm{i}\,x_4\,, \qquad \sqrt{2}\,\bar{z} = z_1 - \mathrm{i}\,z_2 \equiv x_3 - \mathrm{i}\,x_4\,. \tag{2}$$

It is simple to check that the self-dual conditions

$$F_{\mu\nu} = \frac{1}{2}\varepsilon_{\mu\nu\alpha\beta}\,F_{\alpha\beta} \tag{3}$$

with $\varepsilon_{1234} = +1$ now take the form

$$F_{yz} = F_{\bar{y}\bar{z}} = 0\,, \qquad F_{y\bar{y}} + F_{z\bar{z}} = 0\,. \tag{4}$$

Then the first two of the above conditions can be immediately integrated to give

$$A_u = D^{-1}\,D_u\,, \qquad A_{\bar{u}} = \bar{D}^{-1}\,\bar{D}_{\bar{u}}\,, \tag{5}$$

where $u = y, z, D$ and $\bar{D}$ are arbitrary SL(n,c) matrices depending on $y, \bar{y}, z$ and $\bar{z}$. For real gauge fields[2] $A^a_\mu \doteq$ real for real x, then

$$\bar{D} \doteq (D^\dagger)^{-1}\,. \tag{6}$$

A gauge transformation is characterized by the replacement

$$D \longrightarrow D\,h\,, \qquad \bar{D} \longrightarrow \bar{D}\,h \tag{7}$$

with $h = h(y,\bar{y},z,\bar{z}) \in$ SL$(n,c), h^\dagger\,h \doteq I =$ unit $n\times n$ matrix, under which the gauge potential A_μ and the field strength $F_{\mu\nu}$ transform as

$$A_\mu \longrightarrow h^{-1}(\,A_\mu + \partial_\mu\,)\,h\,, \qquad F_{\mu\nu} \longrightarrow h^{-1}\,F_{\mu\nu}\,h\,. \tag{8}$$

respectively. Introduce a Hermitian matrix[5]

$$J = D\,\bar{D}^{-1} \doteq D\,D^\dagger\,, \tag{9}$$

which has the very important property of being invariant under the gauge transformation (7). Then the last of the self-dual conditions (4) becomes a "quasi-conservation" equation, i.e.,

$$(\,J^{-1}\,J_y\,)_{\bar{y}} + (\,J^{-1}\,J_z\,)_{\bar{z}} = 0\,. \tag{10}$$

Now we collect D and $\bar{D}$ to form an element of the group SL$(n,c) \times$ SL(n,c)

$$\Phi = \begin{pmatrix} D & \\ & \bar{D} \end{pmatrix}. \tag{11}$$

The symmetric space SL$(n,c) \times$ SL$(n,c)/$SL$(n,c)_{\mathrm{diag}}$ is defined by introducing an involution operator n

$$n = \begin{pmatrix} & I \\ I & \end{pmatrix}, \qquad n^2 = \boldsymbol{I} = \begin{pmatrix} I & \\ & I \end{pmatrix}. \tag{12}$$

The elements of the diagonal subgroup SL$(n,c)_{\mathrm{diag}}$ are those being invariant under the action of the operator n

$$\boldsymbol{h} = \begin{pmatrix} h & \\ & \bar{h} \end{pmatrix}, \qquad n\,\boldsymbol{h}\,n = \boldsymbol{h} \quad (\text{i.e.}, \bar{h} = h)\,. \tag{13}$$

The canonical variable[8] of the principal SL(n,c) chiral field is

$$\boldsymbol{N} = \Phi\,n\,\Phi^{-1} = \begin{pmatrix} & D\,\bar{D}^{-1} \\ \bar{D}\,D^{-1} & \end{pmatrix} = \begin{pmatrix} & J \\ J^{-1} & \end{pmatrix}, \qquad \boldsymbol{N}^2 = \boldsymbol{I}\,. \tag{14}$$

The Lagrangian for the principal SL (n,c) field is

$$\mathcal{L}_0 = -\frac{\lambda^2}{4}\,\mathrm{Tr}\,\{\partial_\mu N \partial_\mu N\} = -\frac{\lambda^2}{2}\mathrm{tr}\{\partial_\mu J\,\partial_\mu J^{-1}\}\,, \tag{15}$$

where λ^2 is the dimensional parameter, tr denotes the trace of the $n\times n$ matrix, and Tr the trace of the double matrix defined as in Eq. (11). For real x, SL (n,c) matrices D and $\bar{D}$ are relevant to each other, $\bar{D} \doteq (D^\dagger)^{-1}$ and $h=\bar{h}\doteq(h^\dagger)^{-1}$ is unitary. In this case our SL (n,c) principal chiral field reduces to the $H_n \equiv \mathrm{SL}\,(n,c)/\,\mathrm{SU}(n)$ model.[8] Notice the minus sign in front of Eq. (15), which is necessary since, for real x_μ, J is Hermitian instead of unitary (as in the case of SU(n) principal chiral field). The Euler–Lagrange equation for the principal SL (n,c) field described by this Lagrangian (15) is easily obtained

$$\partial_\mu\,(\,J^{-1}\partial_\mu J\,) = 0\,, \tag{16}$$

which is a conservation (continuous) equation rather than the "quasi-conservation" form as given in Eq. (10).

Now imagine the four-dimensional Euclidean space R^4 being compactified to an $S_2\times S_2$ manifold, i.e., the direct product of a y-sphere and a z-sphere. Wess–Zumino-like terms can be constructed by a similar treatment as in the case of two- or four-dimensional models.[9] The chiral field $J(x)=J(y,\bar{y},z,\bar{z})$ is a mapping of space manifold $S_2\times S_2$ into the group manifold $G=\mathrm{SL}\,(n,c)$. Since we know for any group manifold G, the second homotopy class $\pi_2(G)=0$, and $\pi_3\,(\mathrm{SL}\,(n,c))=\pi_3(\mathrm{SU}(n))=Z$, the mapping from $S_2\times S_2$ into SL (n,c) manifold can be extended to a mapping of $B\times B$ to SL (n,c), where B's are solid balls whose boundaries are imagines of two S_2 into SL (n,c) . If y_i (z_i, $i=1,2,3$) are coordinates for ball B_y (B_z), we may introduce Wess–Zumino terms as follows:

$$\Gamma_y = \frac{\mathrm{i}}{12\pi}\varepsilon^{i,j,k}\int_{B\times S_2}\mathrm{tr}\Big(J^{-1}\frac{\partial J}{\partial y^i}J^{-1}\frac{\partial J}{\partial y^j}J^{-1}\frac{\partial J}{\partial y^k}\Big)\,\mathrm{d}^3y\,\mathrm{d}^2z\,,$$

$$\Gamma_z = \frac{\mathrm{i}}{12\pi}\varepsilon^{i,j,k}\int_{S_2\times B}\mathrm{tr}\Big(J^{-1}\frac{\partial J}{\partial z^i}J^{-1}\frac{\partial J}{\partial z^j}J^{-1}\frac{\partial J}{\partial z^k}\Big)\,\mathrm{d}^2y\mathrm{d}^3z\,. \tag{17}$$

Γ_y and Γ_z are well-defined modulo a constant $2\pi\,\mathrm{i}$, as is required by the fact that the Euclidean action S_E appears in the path integral in the expression $\exp(-S_E)$. Now the whole action is

$$S_E = S_0 + n\,\Gamma_y + n'\,\Gamma_z\,,\qquad S_0=\int \mathrm{d}^4x\,\mathcal{L}_0 = \int \mathrm{d}^2y\,\mathrm{d}^2z\,\mathcal{L}_0\,. \tag{18}$$

The Euler–Lagrange equation is[9]

$$\lambda^2\frac{\partial}{\partial x_\mu}\Big(J^{-1}\frac{\partial J}{\partial x_\mu}\Big) - \frac{\mathrm{i}\,n}{4\pi}\Big[\frac{\partial}{\partial x_1}\Big(J^{-1}\frac{\partial J}{\partial x_2}\Big) - \frac{\partial}{\partial x_2}\Big(J^{-1}\frac{\partial J}{\partial x_1}\Big)\Big] - \frac{\mathrm{i}\,n'}{4\pi}\Big[\frac{\partial}{\partial x_3}\Big(J^{-1}\frac{\partial J}{\partial x_4}\Big) - \frac{\partial}{\partial x_4}\Big(J^{-1}\frac{\partial J}{\partial x_3}\Big)\Big] = 0\,. \tag{19}$$

Therefore we see that if

$$\lambda^2 = \frac{n}{4\pi} = -\frac{n'}{4\pi}\,, \tag{20}$$

equation (19) becomes

$$\frac{\partial}{\partial \bar{y}}\Big(J^{-1}\frac{\partial J}{\partial y}\Big) + \frac{\partial}{\partial \bar{z}}\Big(J^{-1}\frac{\partial J}{\partial z}\Big) = 0\,, \tag{21}$$

which is just the "quasi-conservation" equation (10).

In the triangular gauge[2] (R-gauge), $D\,(\bar{D})$ is chosen as the lower (upper) triangular matrix with real diagonal elements. Further[5] we can set $D=LF$ and $\bar{D}^{-1}=F\bar{L}$, where $L\,(\bar{L})$ is a lower (upper) triangular matrix with ones along the diagonal and F is a purely diagonal real matrix with the determinant one. L^{-1} ($\bar{L}^{-1}$) has the similar property, and for SU(n) group and real space we have $\bar{L} = L^{\dagger}$. It is not difficult to check that, in this special gauge, not only the variation of Γ_u, ($u = y, z$) but also the functionals Γ_u themselves are total divergence. Therefore the total action S_E in Eq. (18) can be written as an integral over the ordinary space $S_2 \times S_2$

$$S_E = \int \mathrm{d}^4x\mathcal{L}$$

with

$$\mathcal{L} = 2\lambda^2\{2\,\mathrm{tr}\,(F_u F_{\bar{u}} F^{-2}) + \mathrm{tr}\,(F^{-2}L^{-1}L_u\,F^2\,\bar{L}_{\bar{u}}\,\bar{L}^{-1})\}\,, \tag{22}$$

which coincides with that given in Ref. [5]. Furthermore, for the case of SU(2) gauge field, our H_2 model turns out to be Pohlmeyer's SO(1,3)/SO(3) nonlinear sigma model,[6] and equation (22) reduces to the Lagrangian given by Pohlmeyer.

Acknowledgment

The authors would like to express their thanks to Professor C.N. Yang for his hospitality and inspiration.

Note added

This note was an unpublished Stony Brook preprint # ITP-SB-83-66, which appeared in October 1983. Later on similar subject was rediscussed by several authors.[11] More recently it caused a lot of interests in investigating the integrability of SDYM and relevant topics.[12] Some of our friends suggested us to publish this material, and so we present it here for these needs.

References

[1] R.S. Ward, Phys. Lett. **A61** (1977) 81; M.F. Atiyah and R.S. Ward, Commun. Math. Phys. **55** (1977) 117.

[2] C.N. Yang, Phys. Rev. Lett. **38** (1977) 1377.

[3] For review, see for example, A. Actor Rev. Mod. Phys. **51** (1979) 461; M.K. Prasad, Physica **1D** (1980) 167.

[4] E. Corrigan, D.B. Fairlie, P. Goddard and R.G. Yates, Phys. Lett. **72B** (1978) 354; Commun. Math. Phys. **58** (1978) 223; M.K. Prasad, A. Sinha and L.L. Chau, Phys. Rev. Lett. **43** (1979) 750; Phys. Rev. **D23** (1981) 2321.

[5] Y. Brihaye, D.E. Fairlie, J. Nuyts and R.G. Yates, J. Math. Phys. **19** (1978) 2528.

[6] K. Pohlmeyer, Commun. Math. Phys. **72** (1980) 37; S. Takeno, Prog. Theor. Phys. **66** (1981) 1250.

[7] For a review, see L.L. Chau, in *Proc. CIFMO School & Workshop at Oaxtepec*, Mexico, 1982, *Lecture Notes in Physics*, Vol. 189, eds H. Araki *et al.*, Springer–Verlag (1983) p. 111.

[8] K.C. CHOU and X.C. SONG, Commun. Theor. Phys. (Beijng, China) **1** (1982) 69; ibid. **1** (1982) 185.

[9] E. Witten, Nucl. Phys. **B223** (1983) 422, Princeton University preprint.

[10] J. Wess and B. Zuminò, Phys. Lett. **B37** (1971) 95.

[11] S. Donaldson, Proc. Lond. Math. Soc. **50** (1985) 1; V.P. Nair and J. Schiff, Phys. Lett. **B246** (1990) 423; Nucl. Phys. **B371** (1992) 329.

[12] See for example, L.L. Chau and I. Yamanaka, Phys. Rev. Lett. **70** (1993) 1916; W.A. Bardeen, Prog. Theor. Phys. Suppl. **123** (1996) 1; T. Inami, H Kanno, T. Ueno and C.S. Xiong, Phys. Lett. **B399** (1997) 104, and the references therein.

International Journal of Modern Physics A, Vol. 13, No. 7 (1998) 1129–1144

$\hbar$ (YANGIAN) DEFORMATION OF THE MIURA MAP AND VIRASORO ALGEBRA

XIANG-MAO DING
Institute of Theoretical Physics, Academy of China, Beijing 100080, China

BO-YU HOU and LIU ZHAO
Institute of Modern Physics, Northwest University, Xian 710069, China

Received 26 May 1997

An $\hbar$-deformed Virasoro Poisson algebra is obtained using the Wakimoto realization of the Sugawara operator for the Yangian double $DY_\hbar(\mathrm{sl}_2)_c$ at the critical level $c = -2$.

1. Introduction

In this work we construct the Yangian-deformed Miura map and the corresponding (deformed) Virasoro algebra. Since Drinfeld's proposal in his works,[8,9] the Yangian algebra $Y_\hbar(g)$ has been known as a kind of Hopf algebra associated with rational solutions of the quantum Yang–Baxter equation characterized by an additive parameter $\hbar$, and as a deformation of the universal enveloping algebra of the loop algebra $g[u, u^{-1}]$ associated with the simple Lie algebra g. Since the rational solution of the Yang–Baxter equation is characterized by an additive parameter $\hbar$, we call the corresponding deformed algebras $\hbar$ deformations. Many faces of the Yangian algebra have been thoroughly studied in the literature with emphasis both on algebraic aspects and on physical applications. Algebraically, the development of studies of Yangian algebra is almost parallel to the development of studies of the so-called quantum algebra associated with trigonometric solutions of the Yang–Baxter equation and characterized by a multiplicative parameter q (hence the name q-deformed algebras[10]), and such parallelism became even more perfect after the work of Khoroshkin *et al.*, who successfully constructed the quantum double for Yangian algebra (the Yangian double[18]) and made a central extension for the Yangian double.[19,20,21,17] Physically the Yangian algebra [actually its quantum double $DY_\hbar(g)$] is found to be the dynamical symmetry algebra of many massive quantum integrable field theories, among which we would like to mention the Thirring model, principal chiral model and nonlinear sigma model.[30,6,23] Generally speaking the field-theoretical models yielding Yangian symmetries often correspond to certain scaling limits of some lattice statistical models away from criticality. In this sense the Yangian algebra is in a position somewhere between the usual Kac–Moody Lie algebra and the quantum algebra.

In the study of q-deformed Lie algebras there was a long-standing problem which was not resolved until recently, i.e. the construction of a q deformation of Virasoro algebra. After several attempts by different authors (see e.g. Ref. 26), in Ref. 15 Frenkel and Reshetikhin obtained a version of q-deformed Virasoro and W algebras as q-deformed Gelfand–Dickey Poisson algebras. Later on, the quantum version of their algebras was also obtained by several groups; see Refs. 28, 24, 2, 3 and 13. Vertex operators connected with such algebras have also been studied, in Refs. 25, 29 and 1. The central idea of Ref. 15 can be briefly summarized as follows. In the undeformed case, the (quantum) Virasoro algebra can be constructed from the Kac–Moody algebra, e.g. $\widehat{\mathrm{sl}}_2$, by means of Sugawara construction, and by an appropriate renormalization, the Virasoro generating function becomes a center in the formal completion of the universal enveloping algebra of $\widehat{\mathrm{sl}}_2$ at the critical level $k=-2$ (the dual Coexeter number with a minus sign). The Poisson brackets for the Virasoro algebra can then be obtained from the Wakimoto realization of the $\widehat{\mathrm{sl}}_2$ Kac–Moody current in the limit $k+2 \to 0$, and this construction has a natural connection with the famous Miura transformation (actually the Virasoro Poisson brackets have to be obtained via this transformation). In the q-deformed case, Frenkel and Reshetikhin[15] successfully made a parallel development. Using the Ding–Frenkel equivalence[7] of Drinfeld currents[11] and the Reshetikhin–Semenov-Tian-Shansky realization[27] of q affine algebras, they obtained the center of the formal completion of the q algebra $U_q(\widehat{\mathrm{sl}}_2)_k$. Then using the q-deformed Wakimoto realization[4] of the Drinfeld currents they showed that that center is nothing but a q deformation of the Miura map. Finally, they obtained the Poisson bracket algebra for q-deformed Virasoro algebra using the q Miura map at the critical level $k=-2$. The q-deformed W algebras are obtained in a similar spirit.

It is interesting to ask whether the constructions that worked in undeformed and q-deformed cases also work in the $\hbar$-deformed case. The answer is yes, but the construction is in some sense not straightforward, as will be shown in the main text of this paper. The $\hbar$ deformation of Virasoro algebra is our central object, and we feel that this algebra is important enough to be studied in detail, because it is the cornerstone of several important algebraic objects: (i) it is a deformation of the conventional Virasoro algebra; (ii) it is the scaling limit of the q-deformed Virasoro algebra obtained in Ref. 15; (iii) its connection with Yangian algebra with center is precisely the sort of connection between the q-deformed Virasoro and q affine algebras; (iv) it is the classical counterpart of the quantum $\hbar$-deformed Virasoro algebra obtained from the quantum q Virasoro algebra by taking the scaling limit.[16] Moreover, the connection with the $\hbar$-deformed Miura map is also an important problem because, while extended to algebras of higher rank, this may reveal a new kind of deformed Gelfand–Dickey equation and may also play some role in an $\hbar$-deformed Drinfeld–Sokolov reduction scheme.[12]

The outline of this work is as follows. In Sec. 2 we shall collect necessary backgrounds and formulas by a brief review of the Yangian doubles $DY(\mathrm{gl}_2)_c$ and $DY_\hbar(\mathrm{sl}_2)_c$. Section 3 is devoted to the construction of the $\hbar$-deformed Sugawara operator. Then, using the Wakimoto realization given in Ref. 22, we derive the $\hbar$-deformed Miura map in Sec. 4. In Sec. 5 we present the Poisson bracket for the $\hbar$-deformed Virasoro algebra, and Sec. 6 is devoted to some discussions and the outlook.

2. The Yangian Algebras $DY_\hbar(\mathrm{gl}_2)_c$ and $DY_\hbar(\mathrm{sl}_2)_c$

The Yangian algebra we shall make use of is actually the central extension $DY_\hbar(\mathrm{gl}_2)_c$ and $DY_\hbar(\mathrm{sl}_2)_c$ of the quantum doubles of $Y_\hbar(\mathrm{gl}_2)$ and $Y_\hbar(\mathrm{sl}_2)$ respectively, with central element c. These algebras can be realized in three equivalent ways, namely using the Chevalley generators, Drinfeld currents and Reshetkhin–Semenov–Tian–Shansky formalism. In this section we shall collect the necessary formulas by making a brief review of the algebra $DY_\hbar(\mathrm{gl}_2)_c$ and treating the algebra $DY_\hbar(\mathrm{sl}_2)_c$ as the subalgebra of $DY_\hbar(\mathrm{sl}_2)_c$ modulo a Heisenberg subalgebra.

In terms of Drinfeld currents, the algebra $DY_\hbar(\mathrm{gl}_2)_c$ can be regarded as the formal completion of the algebra generated by the currents $k_i^\pm(u)$ $(i = 1, 2)$, $e^\pm(u)$, $f^\pm(u)$ together with a derivative d and a center element c with the generating relations,[17]

$$[d, e(u)] = \frac{d}{du}\, e(u)\,,$$

$$[d, f(u)] = \frac{d}{du}\, f(u)\,,$$

$$[d, k_i^\pm(u)] = \frac{d}{du}\, k_i^\pm(u)\,, \quad i = 1, 2\,,$$

$$k_i^\pm(u)k_j^\pm(v) = k_j^\pm(v)k_i^\pm(u)\,, \quad i, j = 1, 2\,,$$

$$\rho(u_- - v_+)k_i^+(u)k_i^-(v) = k_i^-(v)k_i^+(u)\rho(u_+ - v_-)\,, \quad i = 1, 2\,,$$

$$\rho(u_+ - v_- - \hbar)k_2^+(u)k_1^-(v) = k_1^-(v)k_2^+(u)\rho(u_- - v_+ - \hbar)\,,$$

$$\rho(u_+ - v_- + \hbar)k_1^+(u)k_2^-(v) = k_2^-(v)k_1^+(u)\rho(u_- - v_+ + \hbar)\,,$$

$$e(u)e(v) = \frac{u - v + \hbar}{u - v - \hbar}\, e(v)e(u)\,,$$

$$f(u)f(v) = \frac{u - v - \hbar}{u - v + \hbar}\, f(v)f(u)\,,$$

$$k_1^\pm(u)e(v) = \frac{u_\pm - v}{u_\pm - v + \hbar}\, e(v)k_1^\pm(u)\,,$$

$$k_2^{\pm}(u)e(v) = \frac{u_{\pm} - v}{u_{\pm} - v - \hbar} e(v)k_2^{\pm}(u),$$

$$k_1^{\pm}(u)f(v) = \frac{u_{\mp} - v + \hbar}{u_{\mp} - v} f(v)k_1^{\pm}(u),$$

$$k_2^{\pm}(u)f(v) = \frac{u_{\mp} - v - \hbar}{u_{\mp} - v} f(v)k_2^{\pm}(u),$$

$$[e(u), f(v)] = \frac{1}{h}\left(\delta(u_- - v_+)k_2^+(u_-)k_1^+(u_-)^{-1} - \delta(u_+ - v_-)k_2^-(v_-)k_1^-(v_-)^{-1}\right), \tag{1}$$

where

$$\delta(u-v) = \sum_{n+m=-1} u^n v^m, \qquad \delta(u-v)g(u) = \delta(u-v)g(v), \qquad u_{\pm} = u \pm \frac{1}{4}\hbar c,$$

and the function $\rho(u)$ is to be specified in due course. The equivalence of the Drinfeld currents to the Chevalley generators is manifest in the following Laurent mode expansions of the currents:

$$e^{\pm}(u) = \pm \sum_{\substack{l \geq 0 \\ l < 0}} e[l]u^{-l-1},$$

$$f^{\pm}(u) = \pm \sum_{\substack{l \geq 0 \\ l < 0}} f[l]u^{-l-1},$$

$$k_i^{\pm}(u) = 1 \pm \hbar \sum_{\substack{l \geq 0 \\ l < 0}} k_i[l]u^{-l-1}.$$

In the main text of this paper, we shall actually need the equivalence between the Drinfeld currents and Reshetikhin–Semenov–Tian–Shansky realization; the latter is given by the Yang–Baxter type relations,[27]

$$\begin{aligned} R^{\pm}(u-v)L_1^{\pm}(u)L_2^{\pm}(v) &= L_2^{\pm}(v)L_1^{\pm}(u)R^{\pm}(u-v), \\ R^{+}(u_- - v_+)L_1^{+}(u)L_2^{-}(v) &= L_2^{-}(v)L_1^{+}(u)R^{+}(u_+ - v_-), \end{aligned} \tag{2}$$

where

$$R^{\pm}(u) = \rho^{\pm}(u)\begin{pmatrix} 1 & & & \\ & \frac{u}{u+\hbar} & \frac{\hbar}{u+\hbar} & \\ & \frac{\hbar}{u+\hbar} & \frac{u}{u+\hbar} & \\ & & & 1 \end{pmatrix},$$

$$\rho^{\pm}(u) = \left(\frac{\Gamma^2\left(\frac{1}{2} \mp \frac{u}{2\hbar}\right)}{\Gamma\left(\mp\frac{u}{2\hbar}\right)\Gamma\left(1 \mp \frac{u}{2\hbar}\right)}\right)^{\pm 1},$$

and the function $\rho(u)$ appearing in (1) is precisely $\rho^+(u)$. Notice that the scalar functions $\rho^\pm(u)$ in the R matrices $R^\pm(u)$ are chosen such that the unitarity and crossing symmetry for the R matrices hold, i.e.

$$R^+(u)R^-(-u) = 1\,, \qquad (C \otimes \mathrm{id})R^\pm(u)(C \otimes \mathrm{id})^{-1} = R^\mp(-u-\hbar)^{t_1}\,,$$

where t_1 refers to the transpose in the first component space, and C is the charge conjugation given in matrix form in

$$C = \begin{pmatrix} & -1 \\ 1 & \end{pmatrix}$$

The equivalence between the two realizations (1) and (2) is an analog of the well-known Ding–Frenkel equivalence[7] of two similar realizations of q affine algebras; see also Ref. 20 in the case of the Yangian double $DY_\hbar(\mathrm{sl}_2)_c$. In our case, the key point is that, in Eq. (2), $L^\pm(u)$ can be given a Gauss decomposition:

$$L^\pm(u) = \begin{pmatrix} 1 & 0 \\ \hbar f^\pm(u_\mp) & 1 \end{pmatrix} \begin{pmatrix} k_1^\pm(u) & 0 \\ 0 & k_2^\pm(u) \end{pmatrix} \begin{pmatrix} 1 & \hbar e^\pm(u_\pm) \\ 0 & 1 \end{pmatrix}, \tag{3}$$

where the diagonal entries $k_i^\pm(u)$ are identified with the Drinfeld currents $k_i^\pm(u)$ in (1), and the off-diagonal entries $e^\pm(u)$ and $f^\pm(u)$ are related to the Drinfeld currents $e(u)$ and $f(u)$ by

$$e(u) = e^+(u) - e^-(u), \qquad f(u) = f^+(u) - f^-(u)\,.$$

The algebra $DY_\hbar(\mathrm{gl}_2)_c$ can be split into two subalgebras: the Yangian double $DY_\hbar(\mathrm{sl}_2)_c$ and a Heisenberg subalgebra. The Heisenberg subalgebra is generated by the currents

$$K^\pm(u) \equiv k_2^\pm(u+\hbar)k_1^\pm(u) - 1\,. \tag{4}$$

It is an easy practice to show that $K^\pm(u)$ actually commute with all generating functions of $DY_\hbar(\mathrm{gl}_2)_c$ and thus generate a central subalgebra. The Yangian double $DY_\hbar(\mathrm{sl}_2)_c$ is thus obtained from $DY_\hbar(\mathrm{gl}_2)_c$ by taking the quotient with respect to this center. The resulting generating relations differ from that of $DY_\hbar(\mathrm{gl}_2)_c$ only in those involving the currents $k_i^\pm(u)$,

$$\begin{aligned}
[d, h^\pm(u)] &= \frac{d}{du} h^\pm(u)\,, \\
[h^\pm(u), h^\pm(v)] &= 0\,, \\
h^\pm(u)e(v) &= \frac{u_\pm - v + \hbar}{u_\pm - v - \hbar} e(v)h^\pm(u)\,, \\
h^\pm(u)f(v) &= \frac{u_\mp - v - \hbar}{u_\mp - v + \hbar} f(v)h^\pm(u)\,, \\
h^+(u)h^-(v) &= \frac{u_+ - v_- + \hbar}{u_- - v_+ + \hbar} \cdot \frac{u_- - v_+ - \hbar}{u_+ - v_- - \hbar} h^-(v)h^+(u)\,,
\end{aligned}$$

where $h^{\pm}(u)$ is defined as

$$h^{\pm}(u) = k_2^{\pm}(u)k_1^{\pm}(u)^{-1}\,. \tag{5}$$

In ending this section let us note that one can recover the original currents $k_{1,2}^{\pm}(u)$ from Eqs. (4) and (5) in the forms,

$$\begin{aligned} k_1^+(u) &= \prod_{l\geq 0} \frac{h^+(u-(2l+1)\hbar)}{h^+(u-2l\hbar)}\,, \quad k_2^+(u) = \prod_{l\geq 0} \frac{h^+(u-(2l+1)\hbar)}{h^+(u-(2l+2)\hbar)}\,,\\ k_1^-(u) &= \prod_{l\geq 0} \frac{h^-(u+(2l+2)\hbar)}{h^-(u+(2l+1)\hbar)}\,, \quad k_2^-(u) = \prod_{l\geq 0} \frac{h^-(u+2l\hbar)}{h^-(u+(2l+1)\hbar)}\,. \end{aligned} \tag{6}$$

These formulas will be used in Sec. 3.

3. ħ-Deformed Sugawara Construction

Let

$$L(u) = L^-\left(u-\frac{\hbar}{2}\right)L^+\left(u+\frac{\hbar}{2}\right)^{-1}\,. \tag{7}$$

The trace

$$l(u) = \mathrm{tr}\, L(u) = L_{11}(u) + L_{22}(u)\,,$$

as formal power series, would then lie in the formal completion of $DY_\hbar(\mathrm{gl}_2)_c$. Following Ref. 27 we may conclude that at $c=-2$ the coefficients of $l(u)$ are central elements of the formal completion of $DY_\hbar(\mathrm{gl}_2)_c$.

To express $l(u)$ in terms of the equivalence between the Drinfeld currents and the Reshetikhin–Semenov-Tian-Shansky formalism, we will now set $c=-2$. It follows from (7) and (3) that

$$\begin{aligned} L_{11}(u) &= k_1^-\left(u-\frac{\hbar}{2}\right)k_1^+\left(u+\frac{\hbar}{2}\right)^{-1}\\ &\quad + \hbar^2 k_1^-\left(u-\frac{\hbar}{2}\right)e^+(u)k_2^+\left(u+\frac{\hbar}{2}\right)^{-1} f^+(u+\hbar)\\ &\quad - \hbar^2 k_1^-\left(u-\frac{\hbar}{2}\right)e^-(u)k_2^+\left(u+\frac{\hbar}{2}\right)^{-1} f^+(u+\hbar)\,,\\ L_{22}(u) &= k_2^-\left(u-\frac{\hbar}{2}\right)k_2^+\left(u+\frac{\hbar}{2}\right)^{-1}\\ &\quad - \hbar^2 f^-(u-\hbar)k_1^-\left(u-\frac{\hbar}{2}\right)e^+(u)k_2^+\left(u+\tfrac{\hbar}{2}\right)^{-1}\\ &\quad + \hbar^2 f^-(u-\hbar)k_1^-\left(u-\frac{\hbar}{2}\right)e^-(u)k_2^+\left(u+\frac{\hbar}{2}\right)^{-1}\,. \end{aligned} \tag{8}$$

The last two equations have not yet been written purely in terms of Drinfeld currents. In order to do so, we have to combine $e^{\pm}(u)$ into $e(u)$ and $f^{\pm}(u)$ into $f(u)$. The first step will of course be moving the $f^{+}(u+\hbar)$ from the right of $k_2^{+}\left(u+\frac{\hbar}{2}\right)^{-1}$ to the left in $L_{11}(u)$, and $f^{-}(u-\hbar)$ from the left of $k_1^{-}\left(u-\frac{\hbar}{2}\right)^{-1}$ to the right in $L_{22}(u)$. To achieve this we have to use the commutation relations for $f^{\pm}(u)$ and $k_i^{\pm}(v)$. The required relations read

$$
\begin{aligned}
f^{-}(v_{+})k_1^{-}(u) &= \frac{u-v}{u-v+\hbar}\,k_1^{-}(u)f^{-}(v_{+}) + \frac{\hbar}{u-v+\hbar}\,f^{-}(u_{+})k_1^{-}(u)\,,\\
k_2^{+}(v)^{-1}f^{+}(u_{-}) &= \frac{u-v}{u-v+\hbar}\,f^{+}(u_{-})k_2^{+}(v)^{-1} + \frac{\hbar}{u-v+\hbar}\,k_2^{+}(v)^{-1}f^{+}(v_{-})\,.
\end{aligned}
$$

Multiplying by $u-v+\hbar$, we can see that, at points $v=u+\hbar$, the last equations become

$$
\begin{aligned}
f^{-}(u_{+})k_1^{-}(u) &= k_1^{-}(u)f^{-}(u_{+}+\hbar)\,,\\
k_2^{+}(u)^{-1}f^{+}(u_{-}) &= f^{+}(u_{-}-\hbar)k_2^{+}(u)^{-1}\,.
\end{aligned}
$$

Notice that when $c=-2$, we have $u_{\pm}=u\mp\frac{1}{2}\hbar$. Therefore, the above equations can be written as

$$
\begin{aligned}
f^{-}(u-\hbar)k_1^{-}\left(u-\frac{1}{2}\hbar\right) &= k_1^{-}\left(u-\frac{1}{2}\hbar\right)f^{-}(u)\,,\\
k_2^{+}\left(u+\frac{1}{2}\hbar\right)^{-1}f^{+}(u+\hbar) &= f^{+}(u)k_2^{+}\left(u+\frac{1}{2}\hbar\right)^{-1}\,.
\end{aligned}
\tag{9}
$$

Substituting (9) into (8), we are led to

$$
\begin{aligned}
L_{11}(u) &= k_1^{-}\left(u-\frac{\hbar}{2}\right)k_1^{+}\left(u+\frac{\hbar}{2}\right)^{-1}\\
&\quad + \hbar^2 k_1^{-}\left(u-\frac{\hbar}{2}\right)[e^{+}(u)-e^{-}(u)]f^{+}(u)k_2^{+}\left(u+\frac{\hbar}{2}\right)^{-1}\,,\\
L_{22}(u) &= k_2^{-}\left(u-\frac{\hbar}{2}\right)k_2^{+}\left(u+\frac{\hbar}{2}\right)^{-1}\\
&\quad - \hbar^2 k_1^{-}\left(u-\frac{\hbar}{2}\right)f^{-}(u)[e^{+}(u)-e^{-}(u)]k_2^{+}\left(u+\frac{\hbar}{2}\right)^{-1}\,.
\end{aligned}
$$

Finally, we have the expression for $l(u)$,

$$
\begin{aligned}
l(u) &= k_1^{-}\left(u-\frac{\hbar}{2}\right)k_1^{+}\left(u+\frac{\hbar}{2}\right)^{-1} + k_2^{-}\left(u-\frac{\hbar}{2}\right)k_2^{+}\left(u+\frac{\hbar}{2}\right)^{-1}\\
&\quad + \hbar^2 k_1^{-}\left(u-\frac{\hbar}{2}\right):e(u)f(u):k_2^{+}\left(u+\frac{\hbar}{2}\right)^{-1}\,,
\end{aligned}
\tag{10}
$$

where

$$:e(u)f(u): = e(u)f^+(u) - f^-(u)e(u)\,. \tag{11}$$

The final equation (10) is just the required ℏ-deformed Sugawara operator.

4. Wakimoto Module of $DY_\hbar(\mathrm{sl}_2)_c$ and ℏ-Deformed Miura Map

Given the ℏ-deformed Sugawara construction, our next goal is to show its connection with the corresponding (deformed) Miura map. This can be fulfilled by making use of the Wakimoto module of the Yangian double.

We shall adopt the Wakimoto module of $DY_\hbar(\mathrm{sl}_2)_c$ given by Konno.[22,a]

Introduce the following three sets of Heisenberg algebras with generators λ_n, b_n, c_n, $n \in Z - \{0\}$, $\exp(\pm q_\lambda)$, $\exp(\pm q_b)$, $\exp(\pm q_c)$, p_λ, p_b and p_c:

$$\begin{aligned} [\lambda_m, \lambda_n] &= \frac{k+2}{2} m\delta_{m+n,0}\,, & [p_\lambda, q_\lambda] &= \frac{k+2}{2}\,, \\ [b_m, b_n] &= -m\delta_{m+n,0}\,, & [p_b, q_b] &= -1\,, \\ [c_m, c_n] &= m\delta_{m+n,0}\,, & [p_c, q_c] &= 1\,. \end{aligned} \tag{12}$$

For $X = \lambda$, b, c, define

$$X(u; A, B) = \sum_{n>0} \frac{X_{-n}}{n}(u + A\hbar)^n - \sum_{n>0} \frac{X_n}{n}(u + B\hbar)^{-n} + \log(u + B\hbar)p_X + q_X\,,$$

and together, $X(u; A) = X(u; A, A)$. We also use the abbreviations

$$X^+(u; B) = -\sum_{n>0} \frac{X_n}{n}(u + B\hbar)^{-n}\,,$$

$$X^-(u; A) = \sum_{n>0} \frac{X_{-n}}{n}(u + A\hbar)^n\,.$$

The triple mode Fock space is defined as follows. Let $|0\rangle$ be a vector satisfying

$$X_n|0\rangle = 0\,, \qquad n > 0\,; \qquad p_X|0\rangle = 0\,.$$

Then $|l, s, t\rangle \equiv \exp\left(\frac{l}{k+2}q_\lambda + sq_b + tq_c\right)|0\rangle$ is a vacuum state with λ, b, c charges l, $-s$, t respectively. The Fock space is generated by the action of λ_{-n}, b_{-n}, $c_{-n} (n > 0)$ on $|l, s, t\rangle$:

$$\mathcal{F}_{l,s,t} = \left\{ \prod_{n>0} \lambda_{-n} \prod_{n'>0} b_{-n'} \prod_{n''>0} \right\} c_{-n''}|l, s, t\rangle\,.$$

On $\mathcal{F}_{l,s,t}$, the normal ordering of $\exp(X(u; A, B))$ is defined as

$$:\exp(X(u; A, B)): = \exp(X^-(u; A)) \exp(q_X)(u + B\hbar)^{p_X} \exp(X^+(u; B))\,.$$

[a]Our notation here differs from that of Konno in Ref. 22 in the following way: the boson λ corresponds to a_Φ of Konno, and b and c correspond to a_ϕ and a_χ respectively, and there is a shift of spectral parameters in the bosonization formulas of Drinfeld current because our starting definition of the Yangian double $DY_\hbar(\mathrm{sl}_2)_c$ differs from that of Konno by such a shift.

Now we are ready to define the Wakimoto module for the Yangian double $DY_\hbar(\mathrm{sl}_2)_c$. This is nothing but a homomorphism from the above-defined Heisenberg algebras to $DY_\hbar(\mathrm{sl}_2)_c$ under which the action of the Drinfeld currents on $\mathcal{F}_{l,s,t}$ is given by[22]

$$\begin{aligned}
c &= k\,,\\
d &= d_\lambda + d_b + d_c\,,\\
d_\lambda &= \frac{2}{k+2}\left(\lambda_{-1}p_\lambda + \sum_{n>0}\lambda_{-(n+1)}\lambda_n\right)\,,\\
d_b &= -b_{-1}p_b - \sum_{n>0} b_{-(n+1)}b_n\,,\\
d_c &= c_{-1}p_c + \sum_{n>0} c_{-(n+1)}c_n\,,\\
h^+(u) &= \exp\left[\lambda^+\left(u;-\frac{3}{4}k\right) - \lambda^+\left(u;-\left(\frac{3}{4}k+2\right)\right)\right.\\
&\quad\left. + b^+\left(u;-\frac{3}{4}k\right) - b^+\left(u;-\left(\frac{3}{4}k+2\right)\right)\right]\left(\frac{u-\frac{3}{4}k\hbar}{u-\left(\frac{3}{4}k+2\right)\hbar}\right)^{p_\lambda+p_b}\,,\\
h^-(u) &= \exp\left[\frac{2}{k+2}\left(\lambda^-\left(u;-\left(\frac{5}{4}k+3\right)\right) - \lambda^-\left(u;-\left(\frac{1}{4}k+1\right)\right)\right)\right]\\
&\quad\times\exp\left[b^-\left(u;-\left(\frac{5}{4}k+3\right)\right) - b^-\left(u;-\left(\frac{1}{4}k+1\right)\right)\right]\,,\\
e(u) &= -\frac{1}{\hbar} : (\exp(-c(u;-(k+1))) - \exp(-c(u;-(k+2))))\\
&\quad\times\exp(-b(u;-(k+1),-(k+2))):\,,\\
f(u) &= \frac{1}{\hbar} : \left(\exp\left[\lambda^+\left(u;-\frac{1}{2}k\right) - \lambda^+\left(u;-\left(\frac{1}{2}k+2\right)\right)\right]\left(\frac{u-\frac{1}{2}k\hbar}{u-\left(\frac{1}{2}k+2\right)\hbar}\right)^{p_\lambda}\right.\\
&\quad\times\exp\left[b\left(u;-\left(\frac{1}{2}k+1\right),-\frac{1}{2}k\right) + c\left(u;-\left(\frac{1}{2}k+1\right)\right)\right]\\
&\quad-\exp\left[\frac{2}{k+2}\left(\lambda^-\left(u;-\left(\frac{3}{2}k+3\right)\right) - \lambda^-\left(u;-\left(\frac{1}{2}k+1\right)\right)\right)\right]\\
&\quad\left.\times\exp\left[b\left(u;-\left(\frac{3}{2}k+3\right),\,-\left(\frac{3}{2}k+2\right)\right) + c\left(u;-\left(\frac{3}{2}k+2\right)\right)\right]\right): \,.
\end{aligned}$$

Introducing the notation

$$\begin{aligned}
\Upsilon^+(u) &= \left(u-\left(\frac{1}{2}k+1\right)\hbar\right)^{p_\lambda}\exp\left(\lambda^+\left(u;-\left(\frac{1}{2}k+1\right)\right)\right)\,,\\
\Upsilon^-(u) &= \exp\left(\frac{2}{k+2}\lambda^-(u;-(k+2))\right)\,,
\end{aligned}$$

we can recast the expressions for $h^{\pm}(u)$ and $f(u)$ into shorter forms:

$$h^{+}(u)=\Upsilon^{+}(u_{-}+\hbar)\Upsilon^{+}(u_{-}-\hbar)^{-1}\left(\frac{u-\frac{3}{4}k\hbar}{u-(\frac{3}{4}k+2)\hbar}\right)^{p_b}$$
$$\times\exp\left[b^{+}\left(u;-\frac{3}{4}k\right)-b^{+}\left(u;-\left(\frac{3}{4}k+2\right)\right)\right], \tag{13}$$

$$h^{-}(u)=\Upsilon^{-}\left(u_{+}-\frac{k+2}{2}\hbar\right)\Upsilon^{-}\left(u_{+}+\frac{k+2}{2}\hbar\right)^{-1}$$
$$\times\exp\left[b^{-}\left(u;-\left(\frac{5}{4}k+3\right)\right)-b^{-}\left(u;-\left(\frac{1}{4}k+1\right)\right)\right], \tag{14}$$

$$f(u)=\frac{1}{\hbar}:\Upsilon^{+}(u+\hbar)\Upsilon^{+}(u-\hbar)^{-1}$$
$$\times\exp\left[b\left(u;-\left(\frac{1}{2}k+1\right),-\frac{1}{2}k\right)+c\left(u;-\left(\frac{1}{2}k+1\right)\right)\right]$$
$$-\Upsilon^{-}\left(u-\frac{k+2}{2}\hbar\right)\Upsilon^{-}\left(u+\frac{k+2}{2}\hbar\right)^{-1}$$
$$\times\exp\left[b\left(u;-\left(\frac{3}{2}k+3\right),-\left(\frac{3}{2}k+2\right)\right)+c\left(u;-\left(\frac{3}{2}k+2\right)\right)\right]:.$$

In order to obtain the $\hbar$-deformed Miura map, we need to express $l(u)$ in terms of Laurent modes of only one of the three bosons, λ. To achieve this goal, we need a bosonic expression for $k_i^{\pm}(u)$, which can be obtained by substituting (13) and (14) into (6):

$$\begin{aligned}
k_1^{+}(u)&=\frac{\Upsilon^{+}\left(u-\frac{k}{2}\hbar\right)}{\Upsilon^{+}\left(u-\frac{k}{2}\hbar+\hbar\right)}\left(\frac{u-(k+1)\hbar}{u-k\hbar}\right)^{p_b}\\
&\quad\times\exp\left(b^{+}(u;-(k+1))-b^{+}(u;-k)\right),\\
k_1^{-}(u)&=\prod_{l\geq 0}\frac{\Upsilon^{-}(u+k\hbar+(2l+2)\hbar)\Upsilon^{-}(u+(2l+1)\hbar)}{\Upsilon^{-}(u+k\hbar+(2l+3)\hbar)\Upsilon^{-}(u+2l)\hbar)}\\
&\quad\times\exp\left(b^{-}(u;-(k+1))-b^{-}(u;-(k+2))\right),\\
k_2^{+}(u)&=\frac{\Upsilon^{+}\left(u-\frac{k}{2}\hbar\right)}{\Upsilon^{+}\left(u-\frac{k}{2}\hbar-\hbar\right)}\left(\frac{u-(k+1)\hbar}{u-(k+2)\hbar}\right)^{p_b}\\
&\quad\times\exp\left(b^{+}(u;-(k+1))-b^{+}(u;-(k+2))\right),\\
k_2^{-}(u)&=\prod_{l\geq 0}\frac{\Upsilon^{-}(u+k\hbar+(2l+2)\hbar)\Upsilon^{-}(u+(2l-1)\hbar)}{\Upsilon^{-}(u+k\hbar+(2l+1)\hbar)\Upsilon^{-}(u+2l)\hbar)}\\
&\quad\times\exp\left(b^{-}(u;-(k+3))-b^{-}(u;-(k+2))\right).
\end{aligned} \tag{15}$$

We have to note that, though in the form of Eq. (18) the $\hbar$-deformed Miura map looks very similar to the q-deformed version, the actual way of mapping from the bosonic field λ to $l(u)$ is much more complicated than the q-deformed case. The complexity comes about in the infinite product structure in the expression for $\Lambda^-(u)$. However, despite such complexities the operator product between Λ^+ and Λ^- is rather simple. It reads

$$\Lambda^+(u)\Lambda^-(v) = \frac{\rho(u-v-(k+2)\hbar)}{\rho(u-v)}\Lambda^-(v)\Lambda^+(u)\,.$$

Another remark concerns with the fact that the Sugawara operator $l(u)$ is associated with the two-dimensional representation of sl_2. To obtain analogous operators associated with higher-dimensional representations of sl_2, one can apply the fusing procedure which is similar to the q-deformed case, i.e. the operator $l^{(n)}(u)$ associated with the $(n+1)$-dimensional representation of sl_2 can be obtained from the iterative relation

$$l^{(1)}(u-n\hbar)l^{(n)}(u) = l^{(n+1)}(u) + l^{(n-1)}(u)\,,$$
$$l^{(1)}(u) = l(u)\,, \qquad l^{(0)}(u) = 1\,.$$

The explicit form for $l^{(n)}(u)$ reads

$$\begin{aligned}
l^{(n)}(u) = {} & \Lambda\left(u-\frac{\hbar}{2}\right)\Lambda\left(u-\frac{3\hbar}{2}\right)\Lambda\left(u-\frac{5\hbar}{2}\right)\cdots\Lambda\left(u-\frac{(2n-1)\hbar}{2}\right)\\
& + \Lambda\left(u+\frac{\hbar}{2}\right)^{-1}\Lambda\left(u-\frac{3\hbar}{2}\right)\Lambda\left(u-\frac{5\hbar}{2}\right)\cdots\Lambda\left(u-\frac{(2n-1)\hbar}{2}\right)\\
& + \Lambda\left(u+\frac{\hbar}{2}\right)^{-1}\Lambda\left(u-\frac{\hbar}{2}\right)^{-1}\Lambda\left(u-\frac{5\hbar}{2}\right)\cdots\Lambda\left(u-\frac{(2n-1)\hbar}{2}\right)\\
& + \Lambda\left(u+\frac{\hbar}{2}\right)^{-1}\Lambda\left(u-\frac{\hbar}{2}\right)^{-1}\Lambda\left(u-\frac{3\hbar}{2}\right)^{-1}\cdots\Lambda\left(u-\frac{(2n-1)\hbar}{2}\right)\\
& + \cdots\cdots\\
& + \Lambda\left(u+\frac{\hbar}{2}\right)^{-1}\Lambda\left(u-\frac{\hbar}{2}\right)^{-1}\Lambda\left(u-\frac{3\hbar}{2}\right)^{-1}\cdots\Lambda\left(u-\frac{(2n-3)\hbar}{2}\right)^{-1}.
\end{aligned}$$

5. Poisson Brackets for the $\hbar$-Deformed Virasoro Algebra

Let us recall that in the undeformed case, the Miura map provides a free field representation of the classical (i.e. Poisson bracket) Virasoro algebra. In the q-deformed case, such a map also gives a q-deformed Virasoro Poisson algebra from the Poisson bracket of a q-defomed bosonic field. In this section we shall show how we can obtain an analogous $\hbar$-deformed algebra.

First let us explain how a Poisson bracket could arise from a purely quantum theory. The key point is as follows. When $k+2$ approaches zero, the commutation relations in (12), divided by $k+2$, naturally induce a Poisson bracket structure,

$$\{\lambda_m, \lambda_n\} = \frac{1}{2} m\delta_{m+n,0}, \qquad \{p_\lambda, q_\lambda\} = \frac{1}{2}. \tag{20}$$

This Poisson structure turn the quantum $\hbar$-deformed free field $\lambda(u;A,B)$ into a classical object. In the meantime, all functions of the quantum field λ are also turned into classical ones, i.e. the noncommuting objects now become commutative and the original commutation relations are turned into Poisson brackets.

To obtain the Poisson bracket for the $\hbar$-deformed Virasoro algebra, we need to take the limit of $l(u)$ as $k+2 \to 0$. Simply substituting $k+2=0$ into the expressions of $\Upsilon^{\pm}(u)$ does not make sense because $\Upsilon^{-}(u)$ is not a well-defined object as $k+2 \to 0$. However, at level of $\Lambda^{\pm}(u)$, well-defined limits could be obtained. The limits of $\Lambda^{\pm}(u)$ read

$$\Lambda^{\pm}(u) = A^{\pm}\left(u - \frac{1}{2}\hbar\right) A^{\pm}\left(u + \frac{1}{2}\hbar\right)^{-1},$$

$$A^{+}(u) = u^{-p_\lambda} \exp\left\{\sum_{l>0} \frac{\lambda_n}{n} u^{-n}\right\},$$

$$A^{-}(u) = \prod_{l\geq 0} B^{-}(u + (2l+1)\hbar),$$

$$B^{-}(u) = \exp\left\{\sum_{n>0} 2\hbar\lambda_{-n} u^{n-1}\right\}.$$

Using the Poisson brackets (20), we can easily calculate

$$\begin{aligned}
\{A^{+}(u), B^{-}(v)\} &= \hbar \frac{\partial}{\partial v} \sum_{n>0} \frac{1}{n}\left(\frac{v}{u}\right)^n A^{+}(u)B^{-}(v) \\
&= -\hbar \frac{\partial}{\partial v} \log\left(1 - \frac{v}{u}\right) A^{+}(u)B^{-}(v) \qquad (|u| > |v|), \\
\{B^{-}(u), A^{+}(v)\} &= \hbar \frac{\partial}{\partial u} \log\left(1 - \frac{u}{v}\right) B^{-}(u)A^{+}(v) \qquad (|v| > |u|),
\end{aligned}$$

from which we obtain

$$\begin{aligned}
\{A^{+}(u), A^{-}(v)\} &= -\hbar \frac{\partial}{\partial v} \sum_{l\geq 0} \log\left(1 - \frac{v + (2l+1)\hbar}{u}\right) A^{+}(u)A^{-}(v) \\
&= -\hbar \frac{\partial}{\partial v} \log\left[\prod_{l\geq 0}\left(1 - \frac{v + (2l+1)\hbar}{u}\right)\right] A^{+}(u)A^{-}(v), \\
\{A^{-}(u), A^{+}(v)\} &= \hbar \frac{\partial}{\partial u} \log\left[\prod_{l\geq 0}\left(1 - \frac{u + (2l+1)\hbar}{v}\right)\right] A^{-}(u)A^{+}(v),
\end{aligned}$$

and further,

$$\{\Lambda^+(u),\Lambda^-(v)\} = \hbar\frac{\partial}{\partial v}\log\rho(u-v)\Lambda^+(u)\Lambda^-(v)\,,$$

$$\{\Lambda^-(u),\Lambda^-(u)\} = -\hbar\frac{\partial}{\partial u}\log\rho(v-u)\Lambda^-(u)\Lambda^+(v)\,.$$

Remembering the definition (19) of $\Lambda(u)$, we have

$$\{\Lambda(u),\Lambda(v)\} = \hbar\left[\frac{\partial}{\partial v}\log\rho(u-v) - \frac{\partial}{\partial u}\log\rho(v-u)\right]\Lambda(u)\Lambda(v)\,.$$

Finally, we have for

$$l(u)\to s(u) = \Lambda\left(u-\frac{1}{2}\hbar\right) + \Lambda\left(u+\frac{1}{2}\hbar\right)^{-1}$$

the Poisson bracket

$$\{s(u),s(v)\} = \hbar\left[\frac{\partial}{\partial v}\log\rho(u-v) - \frac{\partial}{\partial u}\log\rho(v-u)\right]s(u)s(v) + \hbar\delta(u-v-\hbar) - \hbar\delta(u-v+\hbar)\,,$$

where the δ functions are defined in the same way as in Eq. (1).

Before concluding this paper let us consider the connections between the $\hbar$-deformed Miura map and some $\hbar$ difference equations. These are analogs of the well-known Gelfand–Dickey equations written in the simplest case: the classical Miura map

$$\partial^2 - q(t) = \left(\partial - \frac{1}{2}\chi(t)\right)\left(\partial + \frac{1}{2}\chi(t)\right)\,. \tag{21}$$

Let $\mathcal{D}_\hbar$ be the $\hbar$ "derivative" defined as

$$\mathcal{D}_\hbar Q(u) = Q(u-\hbar)\,.$$

Then the $\hbar$-deformed version of Eq. (21) can be written as

$$\left(\Lambda\left(u-\frac{\hbar}{2}\right)\mathcal{D}_\hbar - 1\right)\left(\Lambda\left(u+\frac{\hbar}{2}\right)^{-1}\mathcal{D}_\hbar - 1\right) = \mathcal{D}_\hbar^2 - s(u)\mathcal{D}_\hbar + 1\,.$$

In particular, if $Q(u)$ is a solution of the $\hbar$ difference equation

$$\left(\mathcal{D}_\hbar^2 - s(u)\mathcal{D}_\hbar + 1\right)Q(u+\hbar) = 0\,;$$

then $s(u)$ can be expressed in the form

$$s(u) = \frac{Q(u-\hbar)}{Q(u)} + \frac{Q(u+\hbar)}{Q(u)}\,.$$

This equation is completely in analogy with the q-deformed version given in Ref. 15, which was first used by Baxter[5] in studying the eight-vertex model. Equations of this type have close connections with Bethe ansatz equations.

6. Concluding Remarks and Out-Looking

In this paper we have constructed the $\hbar$-deformed Miura map and the corresponding Virasoro algebra. This algebra can be viewed as either a Yangian deformation of the classical Virasoro algebra or a scaling limit of the q-deformed Virasoro algebra obtained by Frenkel and Reshetikhin in Ref. 15. However, the construction in this paper is much more complicated than the q-deformed case, because the $\hbar$-deformed Sugawara operator involves an infinite product structure in terms of the bosonic expression $\Upsilon^{-}(u)$. As a classical (Poisson bracket) algebra, our algebra is the classical limit of the quantum $\hbar$-deformed Virasoro algebra obtained from the quantum q-deformed Virasoro algebra by taking the proper limit, and may also be viewed as a deformation of the conventional quantum Virasoro algebra governing the conformal field theory.

It is desirable that there exists an infinite family of $\hbar$-deformed algebras (call them $\hbar$-deformed W algebras), each corresponding to a Yangian double of different underlying Lie algebras. The q-deformed W algebras have already been constructed in Ref. 15 and Refs. 28, 24, 2, 3, 13, in both classical and quantum form. However the $\hbar$ deformation of W algebras is not known yet and we hope to work out this problem in our next publication.

Perhaps the most important application of the conventional classical W algebras (i.e. Gelfand–Dickey Poisson algebras) is in the Hamiltonian description of integrable hierarchies, such as the KdV hierarchy. For q-deformed W algebras the corresponding differential–difference systems as deformed integrable hierarchies were obtained in Ref. 14. It is interesting to perform the analogous constructions for $\hbar$-deformed W algebras.

In the conventional (undeformed) cases, W algebra generators are connected with principal cominors of certain Wronskian determinants. It is quite interesting to ask the question whether q and/or $\hbar$ analogs exist. Such analogies are important if we want to identify the existence of the q and/or $\hbar$ versions of nonstandard W algebras, namely W algebras beyond the standard W_n series, $W_3^{(2)}$ for instance. We also hope to consider these problems in our future studies.

Acknowledgments

X.-M. Ding would like to thank Profs. Ke Wu, Shi-Kun Wang and Zhong-Yuang Zhu for valuable discussions.

References

1. Y. Asai, M. Jimbo, T. Miwa and Y. Pugai, *J. Phys.* **A29**, 6595 (1996).
2. H. Awata, H. Kubo, S. Odake and J. Shiraishi, *Commun. Math. Phys.* **179**, 401 (1996).

3. H. Awata, H. Kubo, S. Odake and J. Shiraishi, "Quantum deformation of W_n algebras," preprint q-alg/9612001.
4. H. Awata, S. Odake and J. Shiraishi, *Commun. Math. Phys.* **162**, 61 (1994).
5. R. J. Baxter, *Ann. Phys.* **70**, 193 (1972); R. J. Baxter, *Exactly Solved Models of Statistical Mechanics*, (Academic, 1982).
6. D. Bernard and A. LeClair, *Nucl. Phys.* **B399**, 709 (1993).
7. J. Ding and I. B. Frenkel, *Commun. Math. Phys.* **156**, 277 (1993).
8. V. G. Drinfeld, *Sov. Math. Dokl.* **283**, 1060 (1985).
9. V. G. Drinfeld, *Sov. Math. Dokl.* **32**, 212 (1988).
10. V. G. Drinfeld, "Quantum groups," in *Proc. Int. Cong. of Mathematicians* (Berkeley, 1987), pp. 798–820.
11. V. G. Drinfeld, *Sov. Math. Dokl.* **36**, 212 (1987).
12. V. G. Drinfeld and V. V. Sokolov, *Sov. Math. Dokl.* **23**, 457 (1981); *J. Sov. Math.* **30**, 1975 (1985).
13. B. Feigin and E. Frenkel, *Commun. Math. Phys.* **178**, 653 (1996).
14. E. Frenkel, *Int. Math. Res. Notices* **2**, 55 (1996).
15. E. Frenkel and N. Reshetikhin, *Commun. Math. Phys.* **178**, 237 (1996).
16. B.-Y. Hou and W.-L. Yang, "$\hbar$-deformed Virasoro algebra as hidden symmetry of the restricted sine–Gordon model," preprint hep-th/9612235.
17. K. Iohara and M. Konno, *Lett. Math. Phys.* **37**, 319 (1996).
18. S. Khoroshkin and V. Tolstoy, *Lett. Math. Phys.* **36**, 373 (1996).
19. S. Khoroshkin, "Central extension of the Yangian double." in *Collection SMF, Colloque "Septièmes Rencontres du Contact Franco-Belge en Algèbre"* (June 1995, Reins), preprint q-alg/9602031.
20. S. Khoroshkin and D. Lebedev, "Intertwining operators for the central extension of the Yangian double," preprint q-alg/9602030.
21. S. Khoroshkin, D. Lebedev and S. Pakuliak, *Lett. Math. Phys.* **41**, 31 (1997).
22. H. Konno, *Lett. Math. Phys.* **40**, 321 (1997).
23. A. LeClair and F. Smirnov, *Int. J. Mod. Phys.* **A7**, 2997 (1992).
24. S. Lukyanov, *Phys. Lett.* **B367**, 121 (1996).
25. S. Lukyanov and Ya. Pugai, *Nucl. Phys.* **B473**, 631 (1996)
26. S. Lukyanov and Ya. Pugai, *J. Exp. Theor. Phys.* **82**, 1021 (1996).
27. N. Yu. Reshetikhin and M. A. Semenov-Tyan-Shansky, *Lett. Math. Phys.* **19**, 133 (1990).
28. J. Shiraishi, H. Kubo, H. Awata and S. Odake, *Lett. Math. Phys.* **38**, 33 (1996).
29. J. Shiraishi, H. Kubo, Y. Morita, H. Awata and S. Odake, *Lett. Math. Phys.* **41**, 65 (1997).
30. F. A. Smirnov, *Int. J. Mod. Phys.* **A7** (Suppl. 1B), 813, 839 (1992).

ELSEVIER

Nuclear Physics B 541 (1999) 483–505

NUCLEAR PHYSICS B

Algebraic Bethe ansatz for the supersymmetric t–J model with reflecting boundary conditions

Heng Fan [a], Bo-yu Hou [b], Kang-jie Shi [a,b]

[a] *CCAST (World Laboratory), P.O. Box 8730, Beijing 100080, China*

[b] *Institute of Modern Physics, P.O. Box 105, Northwest University, Xian, 710069, China* [1]

Received 17 June 1998; revised 7 October 1998; accepted 10 November 1998

Abstract

In the framework of the graded quantum inverse scattering method (QISM), we obtain the eigenvalues and eigenvectors of the supersymmetric t–J model with reflecting boundary conditions in FFB background. The corresponding Bethe ansatz equations are obtained. © 1999 Elsevier Science B.V.

PACS: 75.10.Jm; 05.20; 05.30
Keywords: Supersymmetric t–J model; Algebraic Bethe ansatz; Reflection equation

1. Introduction

One-dimensional highly correlated electron models, such as the t–J model, were the subject of numerous studies in the context of high-T_c superconductivity. The Hamiltonian of the t–J model includes the near-neighbour hopping (t) and antiferromagnetic exchange (J) [1,2],

$$H=\sum_{j=1}^{L}\left\{-t\mathcal{P}\sum_{\sigma=\pm 1}(c_{j,\sigma}^{\dagger}c_{j+1,\sigma}+\text{H.c.})\mathcal{P}+J(\mathbf{S}_j\mathbf{S}_{j+1}-\tfrac{1}{4}n_n n_{j+1})\right\}. \tag{1}$$

It was stated that this model is supersymmetric and integrable for $J=\pm 2t$ [3,4], and the supersymmetric t–J model was also studied in Refs. [5–7], for a review see Ref. [8]. Essler and Korepin et al. show that the one-dimensional Hamiltonian can be obtained

[1] Mailing address.

from the transfer matrix of the two-dimensional supersymmetric exactly solvable lattice model [7]. They used the graded QISM [9,10] and obtained the eigenvalues and eigenvectors for the supersymmetric t–J model with periodic boundary conditions in three different backgrounds, for related work see, for example, Ref. [11].

We know that the exactly solvable models are generally solved by imposing periodic boundary conditions. Recently, solvable models with reflecting boundary conditions have been attracting a great deal of interest [12–33]. Besides the original Yang–Baxter equation [34,35], the reflection equations also play a key role to prove the commuting of the transfer matrices with reflecting boundary conditions [12,13]. The Hamiltonian includes non-trivial boundary terms which are determined by the boundary K-matrices. In the present paper, we will use the algebraic Bethe ansatz method to solve the eigenvalue and eigenvector problems of the supersymmetric t–J model with reflecting boundary conditions in the framework of the graded QISM (FFB grading), and the Bethe ansatz equations are also obtained.

The graded reflection equation was introduced in [17], and later applied to fermionic models [18,19]. In this paper, we will use the graded reflection equation to study the supersymmetric t–J model. For the supersymmetric t–J model, the spin of the electrons and the charge "hole" degrees of freedom play a very similar role forming a graded superalgebra with two fermions and one boson (FFB). The holes obey boson commutation relations, while the spinons are fermions, see Ref. [8] and the references therein. The graded approach has the advantage of making real distinction between bosonic and fermionic degrees of freedom. So, it is interesting to study the supersymmetric t–J model with reflecting boundary conditions by the graded algebraic Bethe ansatz method. We should mention that the trigonometric R-matrix related to the supersymmetric t–J model with reflecting boundary conditions was studied in [20,21] by using the usual reflection equation, the results have also been extended to other cases [22,23]; some physical quantities were calculated in Ref. [24].

The paper is organized as follows. In Section 2 we will introduce the model and the notations. In Section 3 we will prove the integrability of model with reflecting boundary conditions in the graded sense. The general solution of the reflection equation is also presented in this section. In Section 4 we use the algebraic Bethe ansatz method to obtain the eigenvalues and eigenvectors of the supersymmetric t–J model. Section 5 includes a brief summary and some discussions.

2. Description of the model

We first give a brief review of the graded version of the QISM. For convenience we take the notations used by Essler and Korepin [7], and what we consider in this paper is the FFB grading, that means the grading is fermionic, fermionic and bosonic. In terms of the Grassmann parities this means that $\epsilon_1 = \epsilon_2 = 1$ and $\epsilon_3 = 0$. The R is defined as

$$\hat{R}(\lambda) = b(\lambda)I + a(\lambda)\Pi, \tag{2}$$

where

$$a(\lambda)=\frac{\lambda}{\lambda+i},\qquad b(\lambda)=\frac{i}{\lambda+i},\tag{3}$$

and the identity operator is given by $I_{a_1a_2}^{b_1b_2}=\delta_{a_1b_1}\delta_{a_2b_2}$, the matrix Π permutes the individual linear spaces in the tensor product space,

$$\Pi_{a_1a_2}^{b_1b_2}=\delta_{a_1b_2}\delta_{a_2b_1}(-1)^{\epsilon_{b_1}\epsilon_{b_2}}.\tag{4}$$

Explicitly, we can write the R-matrix as

$$\hat{R}(\lambda)=\begin{pmatrix}
b(\lambda)-a(\lambda) & 0 & 0 & 0 & 0 & 0 & 0 & 0 & 0\\
0 & b(\lambda) & 0 & -a(\lambda) & 0 & 0 & 0 & 0 & 0\\
0 & 0 & b(\lambda) & 0 & 0 & 0 & a(\lambda) & 0 & 0\\
0 & -a(\lambda) & 0 & b(\lambda) & 0 & 0 & 0 & 0 & 0\\
0 & 0 & 0 & 0 & b(\lambda)-a(\lambda) & 0 & 0 & 0 & 0\\
0 & 0 & 0 & 0 & 0 & b(\lambda) & 0 & a(\lambda) & 0\\
0 & 0 & a(\lambda) & 0 & 0 & 0 & b(\lambda) & 0 & 0\\
0 & 0 & 0 & 0 & 0 & a(\lambda) & 0 & b(\lambda) & 0\\
0 & 0 & 0 & 0 & 0 & 0 & 0 & 0 & 1
\end{pmatrix}.\tag{5}$$

As is well known the Yang–Baxter [34,35] relation plays a key role for integrable models with periodic boundary conditions, which takes the form

$$\hat{R}(\lambda-\mu)L_n(\lambda)\otimes L_n(\mu)=L_n(\mu)\otimes L_n(\lambda)\hat{R}(\lambda-\mu),\tag{6}$$

where the tensor product is in the graded sense

$$(F\otimes G)_{ac}^{bd}=F_{ab}G_{cd}(-1)^{\epsilon_c(\epsilon_a+\epsilon_b)}.\tag{7}$$

In the following, all tensor products are in this graded sense. The n means the nth quantum space which is standard in QISM [9,10]. We can also write the Yang–Baxter relation explicitly as

$$\begin{aligned}&\hat{R}(\lambda-\mu)_{a_1a_2}^{c_1c_2}L_n(\lambda)_{c_1\alpha_n}^{b_1\gamma_n}L_n(\mu)_{c_2\gamma_n}^{b_2\beta_n}(-1)^{\epsilon_{c_2}(\epsilon_{c_1}+\epsilon_{b_1})}\\
&=L_n(\mu)_{a_1\alpha_n}^{c_1\gamma_n}L_n(\lambda)_{a_2\gamma_n}^{c_2\beta_n}(-1)^{\epsilon_{a_2}(\epsilon_{a_1}+\epsilon_{c_1})}\hat{R}(\lambda-\mu)_{c_1c_2}^{b_1b_2}.\end{aligned}\tag{8}$$

The L-operator can be constructed from the R-matrix

$$L_n(\lambda)_{a\alpha}^{b\beta}=\Pi_{a\alpha}^{c\gamma}\hat{R}(\lambda)_{c\gamma}^{b\beta}=[b(\lambda)\Pi+a(\lambda)I]_{a\alpha}^{b\beta}.\tag{9}$$

So, the L-operator is of the form

$$L_n(\lambda)=\begin{pmatrix}
a(\lambda)-b(\lambda)e_n^{11} & -b(\lambda)e_n^{21} & b(\lambda)e_n^{31}\\
-b(\lambda)e_n^{12} & a(\lambda-b(\lambda)e_n^{22} & b(\lambda)e_n^{32}\\
b(\lambda)e_n^{13} & b(\lambda)e_n^{23} & a(\lambda)+b(\lambda)e_n^{33}
\end{pmatrix},\tag{10}$$

where e_n^{ab} are quantum operators acting in the nth quantum space with matrix representation $(e_n^{ab})_{\alpha\beta} = \delta_{a\alpha}\delta_{b\beta}$. The monodromy matrix $T_L(\lambda)$ is defined as the matrix product over the L-operators on all sites of the lattice,

$$T_L(\lambda) = L_L(\lambda) L_{L-1}(\lambda) \ldots L_1(\lambda), \tag{11}$$

where the tensor product is still in the graded sense, and we will not point it out in the following:

$$\begin{aligned} &\{[T_L(\lambda)]^{ab}\}_{\beta_1 \ldots \beta_L}^{\alpha_1 \ldots \alpha_L} \\ &\quad = L_L(\lambda)_{a\alpha_L}^{c_L\beta_L} L_{L-1}(\lambda)_{c_L\alpha_{L-1}}^{c_{L-1}\beta_{L-1}} \ldots L_1(\lambda)_{c_2\alpha_1}^{b\beta_1} (-1)^{\sum_{j=2}^{L}(\epsilon_{\alpha_j}+\epsilon_{\beta_j})\sum_{i=1}^{j-1}\epsilon_{\alpha_i}}. \end{aligned} \tag{12}$$

By repeatedly using the Yang–Baxter relation (6), one can easily prove that the monodromy matrix also satisfies the Yang–Baxter relation

$$\check{R}(\lambda-\mu)[T_L(\lambda) \otimes T_L(\mu)] = T_L(\mu) \otimes T_L(\lambda) \check{R}(\lambda-\mu). \tag{13}$$

The transfer matrix $\tau_{\text{peri}}(\lambda)$ of this model is defined as the supertrace of the monodromy matrix in the auxiliary space. It is defined as the following in the general case:

$$\tau_{\text{peri}}(\lambda) = \text{str}[T_L(\lambda)] = \sum (-1)^{\epsilon_a} [T_L(\lambda)]^{aa}. \tag{14}$$

For the case considered in this paper, if we represent

$$T_L(\lambda) = \begin{pmatrix} A_{11}(\lambda) & A_{12}(\lambda) & B_1(\lambda) \\ A_{21}(\lambda) & A_{22}(\lambda) & B_2(\lambda) \\ C_1(\lambda) & C_2(\lambda) & D(\lambda) \end{pmatrix}, \tag{15}$$

the transfer matrix is then given as

$$\tau(\lambda)_{\text{peri}} = -A_{11}(\lambda) - A_{22}(\lambda) + D(\lambda). \tag{16}$$

As a consequence of the Yang–Baxter relation (13), we can prove that the transfer matrices commute with each other for different spectrum parameters,

$$[\tau_{\text{peri}}(\lambda), \tau_{\text{peri}}(\mu)] = 0. \tag{17}$$

It has been proved that the Hamiltonian obtained by taking the first logarithmic derivative at zero spectral parameter

$$H_{(2)} = -i \left. \frac{d \ln[\tau(\lambda)]}{d\lambda} \right|_{\lambda=0} = -\sum_{k=1}^{L} \Pi^{k,k+1} \tag{18}$$

is equivalent to the Hamiltonian of the supersymmetric t–J model [7]. Here we have omitted the identities.

What is mentioned above is for periodic boundary conditions. We will study the case of the reflecting boundary conditions for the supersymmetric t–J model. For convenience, we change the braided R-matrix $\hat{R}$ (2), (5) to the non-braided R-matrix

$$R(\lambda)=b(\lambda)\Pi+a(\lambda)I$$
$$=\begin{pmatrix} a(\lambda)-b(\lambda) & 0 & 0 & 0 & 0 & 0 & 0 & 0 & 0\\ 0 & a(\lambda) & 0 & -b(\lambda) & 0 & 0 & 0 & 0 & 0\\ 0 & 0 & a(\lambda) & 0 & 0 & 0 & b(\lambda) & 0 & 0\\ 0 & -b(\lambda) & 0 & a(\lambda) & 0 & 0 & 0 & 0 & 0\\ 0 & 0 & 0 & 0 & a(\lambda)-b(\lambda) & 0 & 0 & 0 & 0\\ 0 & 0 & 0 & 0 & 0 & a(\lambda) & 0 & b(\lambda) & 0\\ 0 & 0 & b(\lambda) & 0 & 0 & 0 & a(\lambda) & 0 & 0\\ 0 & 0 & 0 & 0 & 0 & b(\lambda) & 0 & a(\lambda) & 0\\ 0 & 0 & 0 & 0 & 0 & 0 & 0 & 0 & 1 \end{pmatrix}. \tag{19}$$

So, we change the Yang–Baxter relation (13) as

$$R(\lambda-\mu)T_1(\lambda)T_2(\mu)=T_2(\mu)T_1(\lambda)R(\lambda-\mu). \tag{20}$$

Here subscripts 1, 2 mean the auxiliary space. The definition for monodromy matrix T remain the same as before.

3. Integrability of the supersymmetric t–J model with reflecting boundary conditions

As we know the Yang–Baxter relation is enough to prove the integrability of the exactly solvable model with periodic boundary conditions. For the reflecting boundary conditions, besides the Yang–Baxter relation, we also need the reflection equation and the dual reflection equation to prove the integrability of the solvable model. The reflection equation was first proposed by Cherednik [13]. In order to prove the integrability of solvable models with reflecting boundary conditions, Sklyanin proposed the dual reflection equation [12]. Generally, the dual reflection equation, which depends on the unitarity and cross-unitarity relations of the R-matrix, takes different forms for different models.

For the R-matrix considered in this paper (19), one can prove that the R-matrix satisfies the unitarity relation

$$R_{12}(\lambda)R_{21}(-\lambda)=1, \tag{21}$$

where $R_{21}=\Pi R_{12}\Pi$. We can also find that this R-matrix has a symmetry $R_{21}=R_{12}$.

We define the supertransposition 'st' as

$$(A^{\mathrm{st}})_{ij}=A_{ji}(-1)^{(\epsilon_i+1)\epsilon_j}. \tag{22}$$

For the case considered in this paper $\epsilon_1 = \epsilon_2 = 1, \epsilon_3 = 0$, we can rewrite the above relation explicitly as

$$\begin{pmatrix} A_{11} & A_{12} & B_1 \\ A_{21} & A_{22} & B_2 \\ C_1 & C_2 & D \end{pmatrix}^{st} = \begin{pmatrix} A_{11} & A_{21} & C_1 \\ A_{12} & A_{22} & C_2 \\ -B_1 & -B_2 & D \end{pmatrix}. \tag{23}$$

Here for convenience, we also define the inverse of the supertransposition $\bar{st}$ as

$$\{A^{st}\}^{\bar{st}} = A.$$

Considering the R-matrix presented above, we find that the R-matrix satisfies the following cross-unitarity relation:

$$R_{12}^{st_1}(i-\lambda)R_{21}^{st_1}(\lambda) = \rho(\lambda), \qquad \rho(\lambda) = \frac{(i-\lambda)\lambda}{(\lambda+i)(2i-\lambda)}. \tag{24}$$

Here st_1 means supertransposition in the first space.

Next, we introduce the graded version of the reflection equation as

$$R_{12}(\lambda-\mu)K_1(\lambda)R_{21}(\lambda+\mu)K_2(\mu) = K_2(\mu)R_{12}(\lambda+\mu)K_1(\lambda)R_{21}(\lambda-\mu). \tag{25}$$

It can also be rewritten as

$$\begin{aligned} &R(\lambda-\mu)_{a_1a_2}^{b_1b_2}K(\lambda)_{b_1}^{c_1}R(\lambda+\mu)_{b_2c_1}^{c_2d_1}K(\mu)_{c_2}^{d_2}(-1)^{(\epsilon_{b_1}+\epsilon_{c_1})\epsilon_{b_2}} \\ &= K(\mu)_{a_2}^{b_2}R(\lambda+\mu)_{a_1b_2}^{b_1c_2}K(\lambda)_{b_1}^{c_1}R(\lambda-\mu)_{c_2c_1}^{d_2d_1}(-1)^{(\epsilon_{b_1}+\epsilon_{c_1})\epsilon_{c_2}}. \end{aligned} \tag{26}$$

Here the reflecting K is the solution of the reflection equation. We will just consider the diagonal K matrix in this paper, so we suppose

$$K(\lambda)_a^b = \delta_{ab}k_a(\lambda). \tag{27}$$

Substituting this condition into the reflection equation (26), we find the only non-trivial relation is

$$\begin{aligned} &R(\lambda-\mu)_{a_1a_2}^{a_1a_2}R(\lambda+\mu)_{a_2a_1}^{a_1a_2}k(\lambda)_{a_1}k(\mu)_{a_1} \\ &\quad +R(\lambda-\mu)_{a_1a_2}^{a_2a_1}R(\lambda+\mu)_{a_1a_2}^{a_1a_2}k(\lambda)_{a_2}k(\mu)_{a_1} \\ &= R(\lambda+\mu)_{a_1a_2}^{a_1a_2}R(\lambda-\mu)_{a_2a_1}^{a_1a_2}k(\mu)_{a_2}k(\lambda)_{a_1} \\ &\quad +R(\lambda+\mu)_{a_1a_2}^{a_2a_1}R(\lambda-\mu)_{a_1a_2}^{a_2a_1}k(\mu)_{a_2}k(\lambda)_{a_2}. \end{aligned} \tag{28}$$

Solving this relation, we can find two different types of solutions to the graded reflection equation

$$K_I(\lambda) = \begin{pmatrix} \xi+\lambda & & \\ & \xi+\lambda & \\ & & \xi-\lambda \end{pmatrix}, \tag{29}$$

$$K_{II}(\lambda) = \begin{pmatrix} \xi + \lambda & & \\ & \xi - \lambda & \\ & & \xi - \lambda \end{pmatrix}, \tag{30}$$

where ξ is an arbitrary parameter. According to the form of the cross-unitarity relation of the R-matrix (24), we propose the graded dual reflection equation

$$\begin{aligned} &R_{12}(\mu - \lambda) K_1^+(\lambda) R_{21}(i - \lambda - \mu) K_2^+(\mu) \\ &\quad = K_2^+(\mu) R_{12}(i - \lambda - \mu) K_1^+(\lambda) R_{21}(\mu - \lambda). \end{aligned} \tag{31}$$

One can find that there is an isomorphism between the reflection equation (25) and dual reflection equation (31). Given a solution of the reflection equation (25), we can also find a solution of the dual reflection equation (31). But in the sense of the commuting transfer matrix, the reflection equation and the dual reflection equation are independent of each other. We have two types of the solutions of the dual reflection equation

$$K_I^+(\lambda) = \begin{pmatrix} \xi^+ - \lambda & & \\ & \xi^+ - \lambda & \\ & & \xi^+ - i + \lambda \end{pmatrix}, \tag{32}$$

$$K_{II}^+(\lambda) = \begin{pmatrix} \xi^+ - \lambda & & \\ & \xi^+ - i + \lambda & \\ & & \xi^+ - i + \lambda \end{pmatrix}. \tag{33}$$

Here ξ^+ is an arbitrary parameter which is independent of ξ.

Note here that there are two different types of solutions K and K^+, respectively, and we know that K and K^+ are independent of each other in the sense of the transfer matrix, so there are four different transfer matrices altogether corresponding to those K and K^+: $\{K_I^+, K_I\}$; $\{K_I^+, K_{II}\}$; $\{K_{II}^+, K_I\}$; $\{K_{II}^+, K_{II}\}$.

Following the method of Sklyanin [12], we define the double-row monodromy matrix for the case of reflecting boundary conditions

$$\mathcal{T}(\lambda) = T(\lambda) K(\lambda) T^{-1}(-\lambda). \tag{34}$$

Using the Yang–Baxter relation (20), one can easily prove that this double-row monodromy matrix also satisfies the reflection equation (25),

$$R_{12}(\lambda - \mu) \mathcal{T}_1(\lambda) R_{21}(\lambda + \mu) \mathcal{T}_2(\mu) = \mathcal{T}_2(\mu) R_{12}(\lambda + \mu) \mathcal{T}_1(\lambda) R_{21}(\lambda - \mu). \tag{35}$$

We thus define the transfer matrix with open boundary conditions as

$$t(\lambda) = \operatorname{str} K^+(\lambda) \mathcal{T}(\lambda). \tag{36}$$

As before str means supertrace. Next, we will prove that the defined transfer matrices with different spectral parameters commute with each other. Generally, in this sense, we mean that the model is integrable.

We first take supertransposition in the first space.

$$t(\lambda)t(\mu)=\mathrm{str}_1 K_1^+(\lambda)\mathcal{T}_1(\lambda)\mathrm{str}_2 K_2^+(\mu)\mathcal{T}_2(\mu)$$
$$=\mathrm{str}_{12}K_1^+(\lambda)^{st_1}K_2^+(\mu)\mathcal{T}_1^{st_1}(\lambda)\mathcal{T}_2(\mu).$$

Now we insert the cross-unitarity relation (24) of the R-matrix, and take inverse of supertransposition in the first space, we have

$$\ldots=\frac{1}{\rho(\lambda+\mu)}\mathrm{str}_{12}K_1^+(\lambda)^{st_1}K_2^+(\mu)R_{12}^{st_1}(i-\lambda-\mu)R_{21}^{st_1}(\lambda+\mu)\mathcal{T}_1^{st_1}(\lambda)\mathcal{T}_2(\mu),$$
$$=\frac{1}{\rho(\lambda+\mu)}\mathrm{str}_{12}\{K_1^+(\lambda)^{st_1}K_2^+(\mu)R_{12}^{st_1}(i-\lambda-\mu)\}^{\bar{st_1}}$$
$$\times\{\mathcal{T}_1(\lambda)R_{21}(\lambda+\mu)\mathcal{T}_2(\mu)\},$$
$$=\frac{1}{\rho(\lambda+\mu)}\mathrm{str}_{12}\{K_2^+(\mu)R_{12}(i-\lambda-\mu)K_1^+(\lambda)\}\{\mathcal{T}_1(\lambda)R_{21}(\lambda+\mu)\mathcal{T}_2(\mu)\}.$$

Here the fact that $K_1^+(\lambda)^{st}$ and $K_2^+(\mu)$ commute is used. This is because K and K^+ are supermatrices. By inserting the unitarity relation of the R-matrix (21), and using the RE (35) and the dual RE (31), we have

$$\ldots=\frac{1}{\rho(\lambda+\mu)}\mathrm{str}_{12}\{K_2^+(\mu)R_{12}(i-\lambda-\mu)K_1^+(\lambda)R_{21}(\mu-\lambda)\}$$
$$\times\{R_{12}(\lambda-\mu)\mathcal{T}_1(\lambda)R_{21}(\lambda+\mu)\mathcal{T}_2(\mu)\}.$$
$$=\frac{1}{\rho(\lambda+\mu)}\mathrm{str}_{12}\{R_{12}(\mu-\lambda)K_1^+(\lambda)R_{21}(i-\mu-\lambda)K_2^+(\mu)\}$$
$$\times\{\mathcal{T}_2(\mu)R_{12}(\lambda+\mu)\mathcal{T}_1(\lambda)R_{21}(\lambda-\mu)\}. \tag{37}$$

By applying almost the same procedure as before, and again using the unitarity relation (21) and the cross-unitarity relation (24), we have

$$\ldots=\frac{1}{\rho(\lambda+\mu)}\mathrm{str}_{12}\{K_1^+(\lambda)R_{21}(i-\mu-\lambda)K_2^+(\mu)\}\{\mathcal{T}_2(\mu)R_{12}(\lambda+\mu)\mathcal{T}_1(\lambda)\}.$$
$$=\frac{1}{\rho(\lambda+\mu)}\mathrm{str}_{12}\{R_{21}^{st_1}(i-\mu-\lambda)K_1^+(\lambda)^{st_1}K_2^+(\mu)\}$$
$$\times\{\mathcal{T}_2(\mu)\mathcal{T}_1^{st_1}(\lambda)R_{12}^{st_1}(\lambda+\mu)\}.$$
$$=\mathrm{str}_2 K_2^+(\mu)\mathcal{T}_2(\mu)\mathrm{str}_1 K_1^+(\lambda)\mathcal{T}_1(\lambda)$$
$$=t(\mu)t(\lambda). \tag{38}$$

Thus we have proved that the transfer matrix constitutes a commuting family, which gives an infinite set of conserved quantities.

Corresponding to this transfer matrix, we can also obtain the Hamiltonian

$$H_{(2)}^{\mathrm{Bound.}}=-\frac{i}{2}\frac{d\ln[t(\lambda)]}{d\lambda}\bigg|_{\lambda=0}=-\sum_{k=1}^{L-1}\Pi^{k,k+1}-\frac{i}{2}K_1'(0)-\frac{\mathrm{str}_1 K_1^+(0)\Pi^{L,1}}{\mathrm{str}\,K^+(0)}.$$

The boundary terms are determined by the reflecting K-matrices.

4. Algebraic Bethe ansatz method

4.1. Transfer matrix and the vacuum state

According to the definition of the monodromy matrix (11), we can write the inverse of the monodromy matrix as

$$T^{-1}(-\lambda)=L_1^{-1}(-\lambda)L_2^{-1}(-\lambda)\dots L_L^{-1}(-\lambda). \tag{39}$$

With the help of the definition relation (34), we can rewrite the double-row monodromy matrix explicitly as

$$\begin{aligned}\mathcal{T}(\lambda)&=T(\lambda)K(\lambda)T^{-1}(-\lambda)\\&=\begin{pmatrix}A_{11}(\lambda)&A_{12}(\lambda)&B_1(\lambda)\\A_{21}(\lambda)&A_{22}(\lambda)&B_2(\lambda)\\C_1(\lambda)&C_2(\lambda)&D(\lambda)\end{pmatrix}\times\begin{pmatrix}k_1(\lambda)&0&0\\0&k_2(\lambda)&0\\0&0&k_3(\lambda)\end{pmatrix}\\&\quad\times\begin{pmatrix}\bar{A}_{11}(-\lambda)&\bar{A}_{12}(-\lambda)&\bar{B}_1(-\lambda)\\\bar{A}_{21}(-\lambda)&\bar{A}_{22}(-\lambda)&\bar{B}_2(-\lambda)\\\bar{C}_1(-\lambda)&\bar{C}_2(-\lambda)&\bar{D}(-\lambda)\end{pmatrix}\\&=\begin{pmatrix}\mathcal{A}_{11}(\lambda)&\mathcal{A}_{12}(\lambda)&\mathcal{B}_1(\lambda)\\\mathcal{A}_{21}(\lambda)&\mathcal{A}_{22}(\lambda)&\mathcal{B}_2(\lambda)\\\mathcal{C}_1(\lambda)&\mathcal{C}_2(\lambda)&\mathcal{D}(\lambda)\end{pmatrix}.\end{aligned} \tag{40}$$

For the periodic boundary conditions, Essler and Korepin choose the reference state in the kth quantum space and the vacuum $|0\rangle$ as [7]

$$|0\rangle_n=\begin{pmatrix}0\\0\\1\end{pmatrix},\qquad |0\rangle=\otimes_{k=1}^{L}|0\rangle_k. \tag{41}$$

What we study in this paper is the case of the reflecting boundary conditions, we assume the vacuum state remains the same as the case of periodic boundary conditions. That means the above state $|0\rangle$ is still the vacuum state for the reflecting boundary conditions. According to the definition of the monodromy matrix $T(\lambda)$ and the inverse of the monodromy matrix $T^{-1}(\lambda)$, we have the following results:

$$\begin{aligned}T(\lambda)|0\rangle&=\begin{pmatrix}[a(\lambda)]^L&0&0\\0&[a(\lambda)]^L&0\\C_1(\lambda)&C_2(\lambda)&1\end{pmatrix}|0\rangle,\\T^{-1}(-\lambda)|0\rangle&=\begin{pmatrix}[a(\lambda)]^L&0&0\\0&[a(\lambda)]^L&0\\\bar{C}_1(-\lambda)&\bar{C}_2(-\lambda)&1\end{pmatrix}|0\rangle.\end{aligned} \tag{42}$$

Now let us see the values of the double-row monodromy matrix $\mathcal{T}$ acting on the vacuum state. One can easily obtain

$$\mathcal{D}(\lambda)|0\rangle = k_3(\lambda)D(\lambda)\bar{D}(-\lambda)|0\rangle = k_3(\lambda)|0\rangle, \tag{43}$$

$$\mathcal{B}_1(\lambda)|0\rangle = 0, \qquad \mathcal{B}_2(\lambda)|0\rangle = 0, \tag{44}$$

$$\mathcal{C}_1(\lambda)|0\rangle \neq 0, \qquad \mathcal{C}_2(\lambda)|0\rangle \neq 0. \tag{45}$$

It is non-trivial for the other elements,

$$\mathcal{A}_{12}(\lambda)|0\rangle = k_3(\lambda)B_1(\lambda)\bar{C}_2(-\lambda)|0\rangle,$$

$$\mathcal{A}_{21}(\lambda)|0\rangle = k_3(\lambda)B_2(\lambda)\bar{C}_1(-\lambda)|0\rangle, \tag{46}$$

$$\mathcal{A}_{22}(\lambda)|0\rangle = [k_2(\lambda)A_{22}(\lambda)\bar{A}_{22}(-\lambda) + k_3(\lambda)B_2(\lambda)\bar{C}_2(-\lambda)]|0\rangle, \tag{47}$$

$$\mathcal{A}_{11}(\lambda)|0\rangle = [k_1(\lambda)A_{11}(\lambda)\bar{A}_{11}(-\lambda) + k_3(\lambda)B_1(\lambda)\bar{C}_1(-\lambda)]|0\rangle. \tag{48}$$

In order to obtain the results of the above relations, we should use the graded Yang–Baxter relation. From relation (20), we can find the following explicit relation:

$$[T^{-1}(-\lambda)]_{a_2}^{b_2}R(2\lambda)_{a_1b_2}^{b_1c_2}T(\lambda)_{b_1}^{c_1}(-1)^{(\epsilon_{b_1}+\epsilon_{c_1})\epsilon_{c_2}}$$
$$= T(\lambda)_{a_1}^{b_1}R(2\lambda)_{b_1a_2}^{c_1b_2}[T^{-1}(-\lambda)]_{b_2}^{c_2}(-1)^{(\epsilon_{a_1}+\epsilon_{b_1})\epsilon_{a_2}}. \tag{49}$$

Acting with the two sides of this relation on the vacuum state, and taking special values for the indices, for cases $a_1 = 1, a_2 = 3, c_1 = 3, c_2 = 2$ and $a_1 = 2, a_2 = 3, c_1 = 3, c_2 = 1$, with the help of relation (42), we have the results

$$\mathcal{A}_{12}(\lambda)|0\rangle = 0, \qquad \mathcal{A}_{21}(\lambda)|0\rangle = 0. \tag{50}$$

For cases $a_1 = 2, a_2 = 3, c_1 = 3, c_2 = 2$, we have

$$B_2(\lambda)\bar{C}_2(-\lambda)|0\rangle = b(2\lambda)\bar{D}(-\lambda)D(\lambda)|0\rangle - b(2\lambda)A_{22}(\lambda)\bar{A}_{22}(-\lambda)|0\rangle. \tag{51}$$

Substituting this relation into relation (47), we find

$$\mathcal{A}_{22}(\lambda)|0\rangle = \{[k_2(\lambda) - k_3(\lambda)b(2\lambda)]a^{2L}(\lambda) + b(2\lambda)k_3(\lambda)\}|0\rangle. \tag{52}$$

Here for convenience, we introduce a transformation

$$\mathcal{A}_{22}(\lambda) = \tilde{\mathcal{A}}_{22}(\lambda) + b(2\lambda)\mathcal{D}(\lambda). \tag{53}$$

Thus we can find the value of the element $\tilde{\mathcal{A}}_{22}$ acting on the vacuum state

$$\tilde{\mathcal{A}}_{22}(\lambda)|0\rangle = [k_2(\lambda) - k_3(\lambda)b(2\lambda)]a^{2L}(\lambda)|0\rangle. \tag{54}$$

The above transformation is very important in the later algebraic Bethe ansatz method. Instead of $\mathcal{A}_{22}(\lambda)$, we use $\tilde{\mathcal{A}}_{22}(\lambda)$ acting on the assumed eigenvectors, so we find that there is only one wanted term which is necessary for the algebraic Bethe ansatz method. Similarly, we have the relation

$$B_1(\lambda)\bar{C}_1(-\lambda)|0\rangle = b(2\lambda)\bar{D}(-\lambda)D(\lambda)|0\rangle - b(2\lambda)A_{11}(\lambda)\bar{A}_{11}(-\lambda)|0\rangle. \tag{55}$$

Introducing a similar transformation

$$\mathcal{A}_{11}(\lambda) = \tilde{\mathcal{A}}_{11}(\lambda) + b(2\lambda)\mathcal{D}(\lambda), \tag{56}$$

we have

$$\tilde{\mathcal{A}}_{11}(\lambda)|0\rangle = [k_1(\lambda) - k_3(\lambda)b(2\lambda)]a^{2L}(\lambda)|0\rangle. \tag{57}$$

We summarize the above results as

$$\begin{aligned}\tilde{\mathcal{A}}_{11}(\lambda)|0\rangle &= W_1(\lambda)a^{2L}(\lambda)|0\rangle,\\ \tilde{\mathcal{A}}_{22}(\lambda)|0\rangle &= W_2(\lambda)a^{2L}(\lambda)|0\rangle,\\ \mathcal{D}(\lambda)|0\rangle &= U_3(\lambda)|0\rangle.\end{aligned} \tag{58}$$

Corresponding to two different types of solutions K of the reflection equation, $W_j(\lambda), j = 1, 2$, and $U_3(\lambda)$ take the following values:

For $K_I(\lambda)$:

$$\begin{aligned}W_1(\lambda) &= \frac{2\lambda(\lambda+\xi+i)}{2\lambda+i},\\ W_2(\lambda) &= \frac{2\lambda(\lambda+\xi+i)}{2\lambda+i},\\ U_3(\lambda) &= (\xi-\lambda).\end{aligned} \tag{59}$$

For $K_{II}(\lambda)$:

$$\begin{aligned}W_1(\lambda) &= \frac{2\lambda(\lambda+\xi+i)}{2\lambda+i},\\ W_2(\lambda) &= \frac{2\lambda(\xi-\lambda)}{2\lambda+i},\\ U_3(\lambda) &= (\xi-\lambda).\end{aligned} \tag{60}$$

Considering the transformation (53), (56) and definition of the transfer matrix with reflecting boundary conditions, we can rewrite the transfer matrix as

$$\begin{aligned}t(\lambda) = \mathrm{str}\,K^+(\lambda)\mathcal{T}(\lambda) &= -k_1^+(\lambda)\mathcal{A}_{11}(\lambda) - k_2^+(\lambda)\mathcal{A}_{22}(\lambda) + k_3^+(\lambda)\mathcal{D}(\lambda)\\ &= -W_1^+(\lambda)\tilde{\mathcal{A}}_{11}(\lambda) - W_2^+(\lambda)\tilde{\mathcal{A}}_{22}(\lambda) + U_3^+(\lambda)\mathcal{D}(\lambda).\end{aligned} \tag{61}$$

Here $W_j^+, j = 1, 2$ and U_3^+ take the following form:

For $K_I^+(\lambda)$:

$$\begin{aligned}W_1^+(\lambda) &= \xi^+ - \lambda,\\ W_2^+(\lambda) &= \xi^+ - \lambda,\\ U_3^+(\lambda) &= \frac{(2\lambda-i)(\xi^+ + \lambda + i)}{2\lambda+i}.\end{aligned} \tag{62}$$

For $K^{+}_{II}(\lambda)$:

$$W_1^{+}(\lambda)=\xi^{+}-\lambda,$$
$$W_2^{+}(\lambda)=\xi^{+}-i+\lambda,$$
$$U_3^{+}(\lambda)=\frac{(2\lambda-i)(\xi^{+}+\lambda)}{2\lambda+i}. \tag{63}$$

4.2. Commutation relations and the first step of the nested algebraic Bethe ansatz method

For the algebraic Bethe ansatz method, we should obtain the commutation relations between the elements of $\mathcal{T}$. In the case of the reflecting boundary condition, instead of the Yang–Baxter relation, we need the reflection equation (35) to obtain the necessary commutation relations. First of all, we introduce a transformation

$$\mathcal{A}_{ab}(\lambda)=\tilde{\mathcal{A}}_{ab}(\lambda)+\delta_{ab}b(2\lambda)\mathcal{D}(\lambda), \tag{64}$$

which is consistent with the former transformations (53), (56). Next we intend to find the commutation relations between $\tilde{\mathcal{A}}_{aa}$, $\mathcal{D}$ and $\mathcal{C}_b$. The commutation relation between $\mathcal{A}_{aa}$ and $\mathcal{C}_b$ is not necessary, because two or more wanted terms will appear, which can not be handled for the algebraic Bethe ansatz method. For convenience, we write the reflection equation explicitly as

$$\begin{aligned}&R(\lambda-\mu)^{b_1b_2}_{a_1a_2}\mathcal{T}(\lambda)^{c_1}_{b_1}R_{21}(\lambda+\mu)^{d_1c_2}_{c_1b_2}\mathcal{T}(\mu)^{d_2}_{c_2}(-1)^{(\epsilon_{b_1}+\epsilon_{c_1})\epsilon_{b_2}}\\&=\mathcal{T}(\mu)^{b_2}_{a_2}R(\lambda+\mu)^{b_1c_2}_{a_1b_2}\mathcal{T}(\lambda)^{c_1}_{b_1}R_{21}(\lambda-\mu)^{d_1d_2}_{c_1c_2}(-1)^{(\epsilon_{b_1}+\epsilon_{c_1})\epsilon_{c_2}}.\end{aligned} \tag{65}$$

Taking special values for the indices of this reflection equation, for the case $a_1=a_2=3, d_1, d_2\neq 3$, we find

$$\mathcal{C}_{d_1}(\lambda)\mathcal{C}_{d_2}(\mu)=-\mathcal{C}_{c_2}(\mu)\mathcal{C}_{c_1}(\lambda)R(\lambda-\mu)^{d_2d_1}_{c_2c_1}. \tag{66}$$

We will use this property to construct the eigenvector of the transfer matrix. Note here all indices in the commutation relation take values 1, 2. For other commutation relations, this is also true and we will not point it out later. For the case $a_1=a_2=d_2=3, d_1\neq 3$, and considering the transformation (61), we have the commutation relation between $\mathcal{D}$ and $\mathcal{C}$,

$$\begin{aligned}\mathcal{D}(\lambda)\mathcal{C}_d(\mu)=&\frac{(\lambda+\mu)(\lambda-\mu-i)}{(\lambda+\mu+i)(\lambda-\mu)}\mathcal{C}_d(\mu)\mathcal{D}(\lambda)\\&+\frac{2i\mu}{(\lambda-\mu)(2\mu+i)}\mathcal{C}_d(\lambda)\mathcal{D}(\mu)\\&-\frac{i}{\lambda+\mu+i}\mathcal{C}_b(\lambda)\tilde{\mathcal{A}}_{bd}(\mu).\end{aligned} \tag{67}$$

To obtain the commutation relation between $\tilde{\mathcal{A}}$ and $\mathcal{C}$, the calculation is very complicated and tedious. Here we just give a sketch of it. Taking the indices $a_2=3$, a_1, d_1, $d_2\neq 3$, we have

$$
\begin{aligned}
&a(\lambda-\mu)a(\lambda+\mu)\mathcal{A}_{a_1d_1}(\lambda)\mathcal{C}_{d_2}(\mu)\\
&\quad-(1-\delta_{a_1d_1})b(\lambda-\mu)a(\lambda+\mu)\mathcal{C}_{d_1}(\lambda)\mathcal{A}_{a_1d_2}(\mu)\\
&\quad+\delta_{a_1d_1}\{b(\lambda-\mu)b(\lambda+\mu)\mathcal{D}(\lambda)\mathcal{C}_{d_2}(\mu)\\
&\quad-b(\lambda-\mu)R_{21}(\lambda+\mu)^{a_1c_1}_{c_1a_1}\mathcal{C}_{c_1}(\lambda)\mathcal{A}_{c_1d_2}(\mu)\}\\
&=-\mathcal{T}(\mu)^3_3R(\lambda+\mu)^{3a_1}_{a_13}\mathcal{T}(\lambda)^{c_1}_3R_{21}(\lambda-\mu)^{d_1d_2}_{c_1a_1}\\
&\quad+\mathcal{T}(\mu)^{b_2}_3R(\lambda+\mu)^{b_1c_2}_{a_1b_2}\mathcal{T}(\lambda)^{c_1}_{b_1}R_{21}(\lambda-\mu)^{d_1d_2}_{c_1c_2}.
\end{aligned}
\tag{68}
$$

Substitute the transformation (61) into this relation and consider it for three cases:

- Case I: $a_1 \neq d_1, d_1 = d_2 or d_1 \neq d_2$;
- Case II: $a_1 = d_1 = d_2$;
- Case III: $a_1 = d_1 \neq d_2$.

The results of the above relation can be calculated. However, it is still too complicated to be handled for the algebraic Bethe ansatz method. Fortunately, we can summarize all of these results to a much concise relation:

$$
\begin{aligned}
\tilde{\mathcal{A}}_{a_1d_1}(\lambda)\mathcal{C}_{d_2}(\mu)=&\frac{(\lambda-\mu+i)(\lambda+\mu+2i)}{(\lambda-\mu)(\lambda+\mu+i)}\\
&\times r_{12}(\lambda+\mu+i)^{c_1b_2}_{a_1c_2}r_{21}(\lambda-\mu)^{d_1d_2}_{b_1b_2}\mathcal{C}_{c_2}(\mu)\tilde{\mathcal{A}}_{c_1b_1}(\lambda)\\
&+\frac{2i(\lambda+i)}{(\lambda-\mu)(2\lambda+i)}r(2\lambda+i)^{b_2d_1}_{a_1b_1}\mathcal{C}_{b_1}(\lambda)\tilde{\mathcal{A}}_{b_2d_2}(\mu)\\
&-\frac{4i\mu(\lambda+i)}{(2\lambda+i)(2\mu+i)(\lambda+\mu+2i)}r(2\lambda+i)^{d_2d_1}_{a_1b_2}\mathcal{C}_{b_2}(\lambda)\mathcal{D}(\mu).
\end{aligned}
\tag{69}
$$

Where the elements of the r-matrix are defined as the elements of the original R-matrix for the case all of its indices just take values 1,2,

$$
r(\lambda)^{bd}_{ac}=a(\lambda)\delta_{ab}\delta_{cd}-b(\lambda)\delta_{ac}\delta_{bd}=a(\lambda)I+b(\lambda)\Pi^{(1)},
\tag{70}
$$

where $\Pi^{(1)}$ is the 4×4 permutation matrix corresponding to the grading $\epsilon_1=\epsilon_2=1$. We can write the r-matrix as

$$
r(\lambda)=\begin{pmatrix}a(\lambda)-b(\lambda)&&&\\&a(\lambda)&-b(\lambda)&\\&-b(\lambda)&a(\lambda)&\\&&&a(\lambda)-b(\lambda)\end{pmatrix}.
\tag{71}
$$

Similar to the periodic boundary condition case, we construct a set of eigenvectors of the transfer matrix with reflecting boundary conditions as

$$
\mathcal{C}_{d_1}(\mu_1)\mathcal{C}_{d_2}(\mu_2)\dots\mathcal{C}_{d_n}(\mu_n)|0\rangle F^{d_1\dots d_n}.
\tag{72}
$$

Here $F^{d_1\dots d_n}$ is a function of the spectral parameters μ_j. This technique is standard for the algebraic Bethe ansatz method. Acting with the transfer matrix on this eigenvectors, we should find the eigenvalues $\Lambda(\lambda)$ of the transfer matrix $t(\lambda)$ and a set of Bethe ansatz equations. Act first with $\mathcal{D}$ on the eigenvector defined above, and use next the commutation relation (67), consider the value of $\mathcal{D}$ acting on the vacuum state (58), we have

$$\begin{aligned}&\mathcal{D}(\lambda)\mathcal{C}_{d_1}(\mu_1)\mathcal{C}_{d_2}(\mu_2)\dots\mathcal{C}_{d_n}(\mu_n)|0\rangle F^{d_1\dots d_n}\\&=U_3(\lambda)\prod_{i=1}^{n}\frac{(\lambda+\mu_i)(\lambda-\mu_i-i)}{(\lambda+\mu_i+i)(\lambda-\mu_i)}\\&\quad\times\mathcal{C}_{d_1}(\mu_1)\mathcal{C}_{d_2}(\mu_2)\dots\mathcal{C}_{d_n}(\mu_n)|0\rangle F^{d_1\dots d_n}+\text{u.t.},\end{aligned}\tag{73}$$

where we have omitted the unwanted terms u.t.

Then we act $\tilde{\mathcal{A}}_{aa}(\lambda)$ on the assumed eigenvector, using repeatedly the commutation relations (69), we have

$$\begin{aligned}&\tilde{\mathcal{A}}_{aa}(\lambda)\mathcal{C}_{d_1}(\mu_1)\mathcal{C}_{d_2}(\mu_2)\dots\mathcal{C}_{d_n}(\mu_n)|0\rangle F^{d_1\dots d_n}\\&=\prod_{i=1}^{n}\frac{(\lambda-\mu_i+i)(\lambda+\mu_i+2i)}{(\lambda-\mu_i)(\lambda+\mu_i+i)}r_{12}(\lambda+\mu_1+i)_{ac_1}^{a_1e_1}r_{21}(\lambda-\mu_1)_{b_1e_1}^{ad_1}\\&\quad\times r_{12}(\lambda+\mu_2+i)_{a_1c_2}^{a_2e_2}r_{21}(\lambda-\mu_2)_{b_2e_2}^{b_1d_2}\dots r_{12}(\lambda+\mu_n+i)_{a_{n-1}c_n}^{a_ne_n}r_{21}(\lambda-\mu_n)_{b_ne_n}^{b_{n-1}d_n}\\&\quad\times\mathcal{C}_{c_1}(\mu_1)\dots\mathcal{C}_{c_n}(\mu_n)\tilde{\mathcal{A}}_{a_nb_n}(\lambda)|0\rangle F^{d_1\dots d_n}+\text{u.t.}\end{aligned}\tag{74}$$

Summarizing relations (50), (58), we know that

$$\mathcal{A}_{a_nb_n}(\lambda)|0\rangle=\delta_{a_nb_n}W_{a_n}(\lambda)a^{2L}(\lambda)|0\rangle.\tag{75}$$

We can rewrite the transfer matrix as

$$\begin{aligned}t(\lambda)&=-W_1^+(\lambda)\tilde{\mathcal{A}}_{11}(\lambda)-W_2^+(\lambda)\tilde{\mathcal{A}}_{22}(\lambda)+U_3^+(\lambda)\mathcal{D}(\lambda)\\&=-W_a^+(\lambda)\tilde{\mathcal{A}}_{aa}(\lambda)+U_3^+(\lambda)\mathcal{D}(\lambda).\end{aligned}\tag{76}$$

Thus the eigenvalue of the transfer matrix with reflecting boundary condition is written as

$$\begin{aligned}&t(\lambda)\mathcal{C}_{d_1}(\mu_1)\mathcal{C}_{d_2}(\mu_2)\dots\mathcal{C}_{d_n}(\mu_n)|0\rangle F^{d_1\dots d_n}\\&=U_3^+(\lambda)U_3(\lambda)\prod_{i=1}^{n}\frac{(\lambda+\mu_i)(\lambda-\mu_i-i)}{(\lambda+\mu_i+i)(\lambda-\mu_i)}\mathcal{C}_{d_1}(\mu_1)\dots\mathcal{C}_{d_n}(\mu_n)|0\rangle F^{d_1\dots d_n}\\&\quad+a^{2L}(\lambda)\prod_{i=1}^{n}\frac{(\lambda-\mu_i+i)(\lambda+\mu_i+2i)}{(\lambda-\mu_i)(\lambda+\mu_i+i)}\\&\quad\times\mathcal{C}_{c_1}(\mu_1)\dots\mathcal{C}_{c_n}(\mu_n)|0\rangle t^{(1)}(\lambda)_{d_1\dots d_n}^{c_1\dots c_n}F^{d_1\dots d_n}+\text{u.t.},\end{aligned}\tag{77}$$

where $t^{(1)}(\lambda)$ is the so called nested transfer matrix, and with the help of the relation (74), it can be defined as

$$t^{(1)}(\lambda)^{c_1\ldots c_n}_{d_1\ldots d_n} = -W_a^+(\lambda)\left\{r(\lambda+\mu_1+i)^{a_1e_1}_{ac_1}r(\lambda+\mu_2+i)^{a_2e_2}_{a_1c_2}\ldots r(\lambda+\mu_1+i)^{a_ne_n}_{a_{n-1}c_n}\right\}$$
$$\times\delta_{a_nb_n}W_{a_n}(\lambda)\left\{r_{21}(\lambda-\mu_n)^{b_{n-1}d_n}_{b_ne_n}\ldots r_{21}(\lambda-\mu_2)^{b_1d_2}_{b_2e_2}r_{21}(\lambda-\mu_1)^{ad_1}_{b_1e_1}\right\}. \tag{78}$$

We find that this nested transfer matrix can be defined as a transfer matrix with reflecting boundary conditions corresponding to the anisotropic case

$$t^{(1)}(\lambda) = \mathrm{str}\,K^{(1)+}(\tilde\lambda)T^{(1)}(\tilde\lambda,\{\tilde\mu_i\})K^{(1)}(\tilde\lambda)T^{(1)^{-1}}(-\tilde\lambda,\{\tilde\mu_i\}) \tag{79}$$

with the grading $\epsilon_1=\epsilon_2=1$, where we denote $\tilde\lambda=\lambda+\frac{i}{2},\tilde\xi=\xi+\frac{i}{2},\tilde\xi^+=\xi^+-\frac{i}{2}$, we will also denote $\tilde\mu=\mu+\frac{i}{2}$ later. Explicitly we have $K^{(1)}{}^+_I(\tilde\lambda)=$ id. up to a whole factor $\tilde\xi^++i-\tilde\lambda$, and

$$K^{(1)}{}^+_{II}(\tilde\lambda)=\begin{pmatrix}W_1^+(\tilde\lambda) & \\ & W_2^+(\tilde\lambda)\end{pmatrix}=\begin{pmatrix}\tilde\xi^++i-\tilde\lambda & \\ & \tilde\xi^+-i+\tilde\lambda\end{pmatrix} \tag{80}$$

corresponding to K_I^+ and K_{II}^+ respectively. We also have $K_I^{(1)}(\tilde\lambda)=$ id. up to a whole factor $(2\tilde\lambda-i)(\tilde\lambda+\tilde\xi)/2\tilde\lambda$, and

$$K_{II}^{(1)}(\tilde\lambda)=\begin{pmatrix}W_1(\tilde\lambda) & \\ & W_2(\tilde\lambda)\end{pmatrix}=\frac{(2\tilde\lambda-i)}{2\tilde\lambda}\begin{pmatrix}\tilde\xi+\tilde\lambda & \\ & \tilde\xi-\tilde\lambda\end{pmatrix} \tag{81}$$

corresponding to K_I and K_{II}. The row-to-row monodromy matrix $T^{(1)}(\tilde\lambda,\{\tilde\mu_i\})$ (corresponding to the periodic boundary condition) is defined as

$$T^{(1)}_{aa_n}(\tilde\lambda,\{\tilde\mu_i\})^{e_1\ldots e_n}_{c_1\ldots c_n}=r(\tilde\lambda+\tilde\mu_1)^{a_1e_1}_{ac_1}r(\tilde\lambda+\tilde\mu_2)^{a_2e_2}_{a_1c_2}\ldots r(\tilde\lambda+\tilde\mu_1)^{a_ne_n}_{a_{n-1}c_n}$$
$$=L_1^{(1)}(\tilde\lambda+\tilde\mu_1)L_2^{(1)}(\tilde\lambda+\tilde\mu_2)\ldots L_n^{(1)}(\tilde\lambda+\tilde\mu_1). \tag{82}$$

The L-operator takes the form

$$L_k^{(1)}(\tilde\lambda)=\begin{pmatrix}a(\tilde\lambda)-b(\tilde\lambda)e_k^{11} & -b(\tilde\lambda)e_k^{21}\\ -b(\tilde\lambda)e_k^{12} & a(\tilde\lambda)-b(\tilde\lambda)e_k^{22}\end{pmatrix}, \tag{83}$$

and we also have

$$T^{(1)^{-1}}(-\tilde\lambda,\{\tilde\mu_i\})=r_{21}(\tilde\lambda-\tilde\mu_n)^{b_{n-1}d_n}_{b_ne_n}\ldots r_{21}(\tilde\lambda-\tilde\mu_2)^{b_1d_2}_{b_2e_2}r_{21}(\tilde\lambda-\tilde\mu_1)^{ad_1}_{b_1e_1}$$
$$=L_n^{(1)^{-1}}(-\tilde\lambda+\tilde\mu_n)\ldots L_2^{(1)^{-1}}(-\tilde\lambda+\tilde\mu_2)L_1^{(1)^{-1}}(-\tilde\lambda+\tilde\mu_1), \tag{84}$$

where we have used the unitarity relation of the r-matrix $r_{12}(\lambda)r_{21}(-\lambda)=1$.

In this section, we show that the problem to find the eigenvalue of the original transfer matrix $t(\lambda)$ become the problem to find the eigenvalue of the nested transfer matrix $t^{(1)}(\lambda)$. In relation (77), one can see that besides the wanted term which is dedicated to the eigenvalue, we also have the unwanted terms which must be canceled so that the assumed eigenvector is indeed the eigenvector of the transfer matrix. With the help of the symmetry property (66) of the assumed eigenvector (72), using the standard algebraic Bethe ansatz method, we find that if $\mu_1, \ldots, \mu_n$ satisfies the following Bethe ansatz equations, the unwanted terms will vanish:

$$a^{-2L}(\mu_j)U_3(\mu_j)U_3^+(\mu_j)\frac{\mu_j}{\mu_j+i}\prod_{i=1,i\neq j}^{n}\frac{(\mu_j+\mu_i)(\mu_j-\mu_i-i)}{(\mu_j-\mu_i+i)(\mu_j+\mu_i+2i)}$$
$$=\Lambda^{(1)}(\mu_j),\quad j=1,2,\ldots,n. \tag{85}$$

Here we have used the notation $\Lambda^{(1)}$ to denote the eigenvalue of the nested transfer matrix $t^{(1)}(\lambda)$.

Thus what we should do next is to find the eigenvalue of the nested transfer matrix $t^{(1)}$.

4.3. The nested algebraic Bethe ansatz method

We hope that the eigenvalue of the nested transfer matrix can be solved similarly to that of the original transfer matrix. It seems that there is a logical error to call $t^{(1)}$ the transfer matrix, because we did not show that $t^{(1)}$ commute with each other for different spectral parameters. On the other hand, $K^{(1)}$ and $K^{(1)+}$ have already been defined explicitly in the above section, they are not obtained from, for example, the reflection equation and the dual reflection equation. In this section, we will show that all of these problems can be solved in the framework of the reflecting boundary condition case.

We know that the following graded Yang–Baxter relation with grading $\epsilon_1=\epsilon_2=1$ is satisfied:

$$r(\lambda-\mu)L_1^{(1)}(\lambda)L_2^{(1)}(\mu)=L_2^{(1)}(\mu)L_1^{(1)}(\lambda)r(\lambda-\mu). \tag{86}$$

So, we also have the Yang–Baxter relation for the row-to-row monodromy matrix

$$r(\lambda-\mu)T_1^{(1)}(\lambda,\{\mu_i\})T_2^{(1)}(\mu,\{\mu_i\})=T_2^{(1)}(\mu,\{\mu_i\})T_1^{(1)}(\lambda,\{\mu_i\})r(\lambda-\mu). \tag{87}$$

We propose the reflection equation

$$r_{12}(\lambda-\mu)K_1^{(1)}(\lambda)r_{21}(\lambda+\mu)K_2^{(1)}(\mu)$$
$$=K_2^{(1)}(\mu)r_{12}(\lambda+\mu)K_1^{(1)}(\lambda)r_{21}(\lambda-\mu). \tag{88}$$

Solving this reflection equation directly, we find that

$$K^{(1)}(\lambda)=\begin{pmatrix}\xi+\lambda & \\ & \xi-\lambda\end{pmatrix} \tag{89}$$

is a solution of the reflection equation and it is easy to show that $K^{(1)}=\text{id.}$ is also a solution. These solutions of the reflection equation are consistent with the results defined in the above subsection. So, we can show that the nested double-row monodromy matrix

$$\mathcal{T}^{(1)}(\lambda,\{\mu_i\}) \equiv T^{(1)}(\lambda,\{\mu_i\})K^{(1)}(\lambda)T^{(1)^{-1}}(-\lambda,\{\mu_i\}) \tag{90}$$

also satisfies the reflection equation

$$\begin{aligned} & r_{12}(\lambda-\mu)\mathcal{T}_1^{(1)}(\lambda,\{\mu_i\})r_{21}(\lambda+\mu)\mathcal{T}_2^{(1)}(\mu,\{\mu_i\}) \\ &\quad = \mathcal{T}_2^{(1)}(\nu,\{\mu_i\})r_{12}(\lambda+\mu)\mathcal{T}_1^{(1)}(\lambda,\{\mu_i\})r_{21}(\lambda-\mu). \end{aligned} \tag{91}$$

One can prove that the r-matrix satisfies a unitarity relation

$$r_{12}^{st_1}(\lambda)r_{21}^{st_1}(2i-\lambda) = a(\lambda)a(2i-\lambda)\cdot \mathrm{id}. \tag{92}$$

According to this relation and the unitarity relation of the r-matrix $r_{12}(\lambda)r_{21}(-\lambda) = 1\cdot\mathrm{id.}$, we propose the following dual reflection equation:

$$\begin{aligned} & r_{12}(\mu-\lambda)K_1^{(1)+}(\lambda)r_{21}(\lambda+\mu+2i)K_2^{(1)+}(\mu) \\ &\quad = K_2^{(1)+}(\mu)r_{12}(\lambda+\mu+2i)K_1^{(1)+}(\lambda)r_{21}(\mu-\lambda). \end{aligned} \tag{93}$$

We can also find that the solution of the dual reflection equation is consistent with the $K^{(1)+}$ results (80) presented in the above subsection. A similar procedure as that presented in Section 3 can also be applied now, we find that the defined nested transfer matrix indeed constitutes a commuting family for different spectral parameters.

Now, let us again use the algebraic Bethe ansatz method to obtain the eigenvalue $\Lambda^{(1)}(\lambda)$ of the nested transfer matrix $t^{(1)}(\lambda)$. We write the nested double-row monodromy matrix as

$$\begin{aligned} \mathcal{T}^{(1)}(\lambda,\{\mu_i\}) &= \begin{pmatrix} \mathcal{A}^{(1)}(\lambda) & \mathcal{B}^{(1)}(\lambda) \\ \mathcal{C}^{(1)}(\lambda) & \mathcal{D}^{(1)}(\lambda) \end{pmatrix} \\ &= T^{(1)}(\lambda,\{\mu_i\})K^{(1)}(\lambda)T^{(1)^{-1}}(-\lambda,\{\mu_i\}) \\ &= \begin{pmatrix} A^{(1)}(\lambda) & B^{(1)}(\lambda) \\ C^{(1)}(\lambda) & D^{(1)}(\lambda) \end{pmatrix} \begin{pmatrix} k_1^{(1)}(\lambda) & \\ & k_2^{(1)}(\lambda) \end{pmatrix} \begin{pmatrix} \bar{A}^{(1)}(-\lambda) & \bar{B}^{(1)}(-\lambda) \\ \bar{C}^{(1)}(-\lambda) & \bar{D}^{(1)}(-\lambda) \end{pmatrix}. \end{aligned} \tag{94}$$

For convenience, we again introduce a transformation

$$\mathcal{A}^{(1)}(\lambda) = \tilde{\mathcal{A}}^{(1)}(\lambda) - \frac{i}{2\lambda - i}\mathcal{D}^{(1)}(\lambda). \tag{95}$$

Because the nested double-row monodromy matrix satisfies the reflection equation (91), we can find the following commutation relations:

$$\begin{aligned} \mathcal{D}^{(1)}(\lambda)\mathcal{C}^{(1)}(\mu) &= \frac{(\lambda-\mu+i)(\lambda+\mu)}{(\lambda-\mu)(\lambda+\mu-i)}\mathcal{C}^{(1)}(\mu)\mathcal{D}^{(1)}(\lambda) \\ &\quad - \frac{2i\mu}{(\lambda-\mu)(2\mu-i)}\mathcal{C}^{(1)}(\lambda)\mathcal{D}^{(1)}(\mu) \\ &\quad + \frac{i}{\lambda+\mu-i}\mathcal{C}^{(1)}(\lambda)\tilde{\mathcal{A}}^{(1)}(\mu), \end{aligned} \tag{96}$$

$$\tilde{\mathcal{A}}^{(1)}(\lambda)\mathcal{C}^{(1)}(\mu)=\frac{(\lambda-\mu-i)(\lambda+\mu-2i)}{(\lambda-\mu)(\lambda+\mu-i)}\mathcal{C}^{(1)}(\mu)\tilde{\mathcal{A}}^{(1)}(\lambda)$$
$$+\frac{2i(\lambda-i)}{(\lambda-\mu)(2\lambda-i)}\mathcal{C}^{(1)}(\lambda)\tilde{\mathcal{A}}^{(1)}(\mu)$$
$$-\frac{4i\mu(\lambda-i)}{(\lambda+\mu-i)(2\lambda-i)(2\mu-i)}\mathcal{C}^{(1)}(\lambda)\mathcal{D}^{(1)}(\mu), \qquad (97)$$

$$\mathcal{C}^{(1)}(\lambda)\mathcal{C}^{(1)}(\mu)=\mathcal{C}^{(1)}(\mu)\mathcal{C}^{(1)}(\lambda). \qquad (98)$$

As the reference states, for the nesting we pick

$$|0\rangle_k^{(1)}=\begin{pmatrix}0\\1\end{pmatrix}, \qquad |0\rangle^{(1)}=\otimes_{k=1}^{n}|0\rangle_k^{(1)}. \qquad (99)$$

With the help of the definition (82), (84), we know the results of the nested monodromy matrix and the inverse of the monodromy matrix acting on the reference state

$$T^{(1)}(\lambda,\{\mu_i\})|0\rangle^{(1)}=\begin{pmatrix}A^{(1)}(\tilde{\lambda}) & B^{(1)}(\tilde{\lambda})\\ C^{(1)}(\tilde{\lambda}) & D^{(1)}(\tilde{\lambda})\end{pmatrix}|0\rangle^{(1)}$$
$$=\begin{pmatrix}\prod_{i=1}^{n}a(\tilde{\lambda}+\tilde{\mu}_i) & 0\\ C^{(1)}(\tilde{\lambda}) & \prod_{i=1}^{n}[a(\tilde{\lambda}+\tilde{\mu}_i)-b(\tilde{\lambda}+\tilde{\mu})]\end{pmatrix}|0\rangle^{(1)}, \qquad (100)$$

$$T^{(1)^{-1}}(-\lambda,\{\mu_i\})|0\rangle^{(1)}=\begin{pmatrix}\bar{A}^{(1)}(\tilde{\lambda}) & \bar{B}^{(1)}(\tilde{\lambda})\\ \bar{C}^{(1)}(\tilde{\lambda}) & \bar{D}^{(1)}(\tilde{\lambda})\end{pmatrix}|0\rangle^{(1)}$$
$$=\begin{pmatrix}\prod_{i=1}^{n}a(\tilde{\lambda}-\tilde{\mu}_i) & 0\\ C^{(1)}(\tilde{\lambda}) & \prod_{i=1}^{n}[a(\tilde{\lambda}-\tilde{\mu}_i)-b(\tilde{\lambda}-\tilde{\mu})]\end{pmatrix}|0\rangle^{(1)}. \qquad (101)$$

Substituting those relations into relation (94), we can obtain the results of the nested double-row monodromy matrix acting on the nested vacuum state $|0\rangle^{(1)}$,

$$\mathcal{B}^{(1)}(\tilde{\lambda})|0\rangle^{(1)}=0, \qquad \mathcal{C}^{(1)}(\tilde{\lambda})|0\rangle^{(1)}\neq 0, \qquad (102)$$

$$\mathcal{D}^{(1)}(\tilde{\lambda})|0\rangle^{(1)}=U_2(\tilde{\lambda})\prod_{i=1}^{n}\Big\{[a(\tilde{\lambda}+\tilde{\mu}_i)-b(\tilde{\lambda}+\tilde{\mu}_i)]$$
$$\times[a(\tilde{\lambda}-\tilde{\mu}_i)-b(\tilde{\lambda}-\tilde{\mu}_i)]\Big\}|0\rangle^{(1)}. \qquad (103)$$

Here we use the notation $U_2=k_2^{(1)}$, $U_2(\tilde{\lambda})=(2\tilde{\lambda}-i)(\tilde{\lambda}+\tilde{\xi})/2\tilde{\lambda}$ for the K_I case, and $U_2(\tilde{\lambda})=(2\tilde{\lambda}-i)(\tilde{\xi}-\tilde{\lambda})/2\tilde{\lambda}$ for the K_{II} case.

The result of element $\tilde{\mathcal{A}}^{(1)}$ is not so direct. We know that

$$\mathcal{A}^{(1)}(\tilde{\lambda})|0\rangle^{(1)} = k_1^{(1)}(\tilde{\lambda})A^{(1)}(\tilde{\lambda})\bar{A}^{(1)}(-\tilde{\lambda})|0\rangle^{(1)} + k_2^{(1)}(\tilde{\lambda})B^{(1)}(\tilde{\lambda})\bar{C}^{(1)}(-\tilde{\lambda})|0\rangle^{(1)}. \quad (104)$$

Using the Yang–Baxter relation (87), the above relation becomes

$$\begin{aligned}\mathcal{A}^{(1)}(\tilde{\lambda})|0\rangle^{(1)} &= k_1^{(1)}(\tilde{\lambda})A^{(1)}(\tilde{\lambda})\bar{A}^{(1)}(-\tilde{\lambda})|0\rangle^{(1)} + k_2^{(1)}(\tilde{\lambda})\frac{b(2\tilde{\lambda})}{a(2\tilde{\lambda})-b(2\tilde{\lambda})} \\ &\quad \times [A^{(1)}(\tilde{\lambda})\bar{A}^{(1)}(-\tilde{\lambda}) - \bar{D}^{(1)}(-\tilde{\lambda})D^{(1)}(\tilde{\lambda})]|0\rangle^{(1)} \\ &= \left[k_1^{(1)}(\tilde{\lambda}) + k_2^{(1)}(\tilde{\lambda})\frac{b(2\tilde{\lambda})}{a(2\tilde{\lambda})-b(2\tilde{\lambda})}\right] \cdot \\ &\quad \times \prod_{i=1}^{n}[a(\tilde{\lambda}+\tilde{\mu}_i)a(\tilde{\lambda}-\tilde{\mu}_i)]|0\rangle^{(1)} - \frac{i}{2\tilde{\lambda}-i}\mathcal{D}^{(1)}(\tilde{\lambda})|0\rangle^{(1)}.\end{aligned} \quad (105)$$

With the help of the transformation (95), we find

$$\tilde{\mathcal{A}}^{(1)}(\tilde{\lambda})|0\rangle^{(1)} = U_1(\tilde{\lambda})\prod_{i=1}^{n}[a(\tilde{\lambda}+\tilde{\mu}_i)a(\tilde{\lambda}-\tilde{\mu}_i)]|0\rangle^{(1)}, \quad (106)$$

where we denote

$$U_1(\tilde{\lambda}) = k_1^{(1)}(\tilde{\lambda}) + k_2^{(1)}(\tilde{\lambda})\frac{b(2\tilde{\lambda})}{a(2\tilde{\lambda})-b(2\tilde{\lambda})},$$

and U_1 takes the following form explicitly.

For $K_I(\lambda)$,

$$U_1(\tilde{\lambda}) = \tilde{\lambda} + \tilde{\xi}. \quad (107)$$

For $K_{II}(\lambda)$,

$$U_1(\tilde{\lambda}) = \tilde{\lambda} + \tilde{\xi} - i. \quad (108)$$

The nested transfer matrix takes the form

$$\begin{aligned} t^{(1)}(\tilde{\lambda}) &= \mathrm{str}\, K^{(1)}(\tilde{\lambda})\mathcal{T}^{(1)}(\tilde{\lambda}) \\ &= -k_1^{(1)+}(\tilde{\lambda})\mathcal{A}^{(1)}(\tilde{\lambda}) - k_2^{(1)+}(\tilde{\lambda})\mathcal{D}^{(1)}(\tilde{\lambda}) \\ &= -U_1^+(\tilde{\lambda})\tilde{\mathcal{A}}^{(1)}(\tilde{\lambda}) - U_2^+(\tilde{\lambda})\mathcal{D}^{(1)}(\tilde{\lambda}), \end{aligned} \quad (109)$$

where we denote

$$U_1^+ = k_1^{(1)+}, \qquad U_2^+(\lambda) = k_2^{(1)+}(\lambda) - \frac{i}{2\lambda - i}k_1^{(1)+}(\lambda),$$

that means the following.

For the K_I^+ case,

$$U_1^+(\tilde{\lambda}) = \tilde{\xi}^+ + i - \tilde{\lambda}, \tag{110}$$

$$U_2^+(\tilde{\lambda}) = \frac{2(\tilde{\lambda} - i)(\tilde{\xi}^+ + i - \tilde{\lambda})}{2\tilde{\lambda} - i}. \tag{111}$$

For the K_{II}^+ case,

$$U_1^+(\tilde{\lambda}) = \tilde{\xi}^+ + i - \tilde{\lambda},$$
$$U_2^+(\tilde{\lambda}) = \frac{2(\tilde{\lambda} + \tilde{\xi}^+)(\tilde{\lambda} - i)}{2\tilde{\lambda} - i}. \tag{112}$$

Using the standard algebraic Bethe ansatz method, assume that the eigenvector of the nested transfer matrix is constructed as $C(\tilde{\mu}_1^{(1)})C(\tilde{\mu}_2^{(1)})\ldots C(\tilde{\mu}_m^{(1)})|0\rangle^{(1)}$. Acting with the nested transfer matrix on this eigenvector, using repeatedly the commutation relations (96), (97), we have the eigenvalue

$$\begin{aligned}
\Lambda^{(1)}(\tilde{\lambda}) = {} & -U_1^+(\tilde{\lambda})U_1(\tilde{\lambda}) \prod_{i=1}^{n} [a(\tilde{\lambda} + \tilde{\mu}_i)a(\tilde{\lambda} - \tilde{\mu}_i)] \\
& \times \prod_{l=1}^{m} \left\{ \frac{(\tilde{\lambda} - \tilde{\mu}_l^{(1)} + i)(\tilde{\lambda} + \tilde{\mu}_l^{(1)})}{(\tilde{\lambda} - \tilde{\mu}_l^{(1)})(\tilde{\lambda} + \tilde{\mu}_l^{(1)} - i)} \right\} - U_2^+(\tilde{\lambda})U_2(\tilde{\lambda}) \\
& \times \prod_{i=1}^{n} \{[a(\tilde{\lambda} + \tilde{\mu}_i) - b(\tilde{\lambda} + \tilde{\mu}_i)][a(\tilde{\lambda} - \tilde{\mu}_i) - b(\tilde{\lambda} - \tilde{\mu}_i)]\} \\
& \times \prod_{l=1}^{m} \left\{ \frac{(\tilde{\lambda} - \tilde{\mu}_l^{(1)} - i)(\tilde{\lambda} + \tilde{\mu}_l^{(1)} - 2i)}{(\tilde{\lambda} - \tilde{\mu}_l^{(1)})(\tilde{\lambda} + \tilde{\mu}_l^{(1)} - i)} \right\},
\end{aligned} \tag{113}$$

where $\tilde{\mu}_1^{(1)}, \ldots, \tilde{\mu}_m^{(1)}$ should satisfy the following Bethe ansatz equations:

$$\begin{aligned}
& \frac{U_1^+(\tilde{\mu}_j^{(1)})U_1(\tilde{\mu}_j^{(1)})}{U_2^+(\tilde{\mu}_j^{(1)})U_2(\tilde{\mu}_j^{(1)})} \frac{\tilde{\mu}_j^{(1)}}{(\tilde{\mu}_j^{(1)} - i)} \prod_{i=1}^{n} \frac{(\tilde{\mu}_j^{(1)} + \tilde{\mu}_i)(\tilde{\mu}_j^{(1)} - \tilde{\mu}_i)}{(\tilde{\mu}_j^{(1)} + \tilde{\mu}_i - i)(\tilde{\mu}_j^{(1)} - \tilde{\mu}_i - i)} \\
& = \prod_{l=1, \neq j}^{m} \frac{(\tilde{\mu}_j^{(1)} - \tilde{\mu}_l^{(1)} - i)(\tilde{\mu}_j^{(1)} + \tilde{\mu}_l^{(1)} - 2i)}{(\tilde{\mu}_j^{(1)} - \tilde{\mu}_l^{(1)} + i)(\tilde{\mu}_j^{(1)} + \tilde{\mu}_l^{(1)})}, \quad j = 1, \ldots, m.
\end{aligned} \tag{114}$$

Thus, the eigenvalue of the transfer matrix $t(\lambda)$ with a reflecting boundary condition (36) is obtained as

$$\begin{aligned}
\Lambda(\lambda) = {} & U_3^+(\lambda)U_3(\lambda) \prod_{i=1}^{n} \frac{(\lambda + \mu_i)(\lambda - \mu_i - i)}{(\lambda + \mu_i + i)(\lambda - \mu_i)} \\
& + a^{2L}(\lambda) \prod_{i=1}^{n} \frac{(\lambda - \mu_i + i)(\lambda + \mu_i + 2i)}{(\lambda - \mu_i)(\lambda + \mu_i + i)} \Lambda^{(1)}(\tilde{\lambda}).
\end{aligned} \tag{115}$$

Here for convenience, we give a summary of the values U and U^+.

Case I:

$$U_1^+(\tilde{\lambda}) = \tilde{\xi}^+ + i - \tilde{\lambda},$$
$$U_2^+(\tilde{\lambda}) = \frac{2(\tilde{\lambda} - i)(\tilde{\xi}^+ + i - \tilde{\lambda})}{2\tilde{\lambda} - i},$$
$$U_3^+(\lambda) = \frac{(2\lambda - i)(\xi^+ + \lambda + i)}{2\lambda + i}. \tag{116}$$

Case II:

$$U_1^+(\tilde{\lambda}) = \tilde{\xi}^+ + i - \tilde{\lambda},$$
$$U_2^+(\tilde{\lambda}) = \frac{2(\tilde{\lambda} + \tilde{\xi}^+)(\tilde{\lambda} - i)}{2\tilde{\lambda} - i},$$
$$U_3^+(\lambda) = \frac{(2\lambda - i)(\xi^+ + \lambda)}{2\lambda + i}. \tag{117}$$

Case I:

$$U_1(\tilde{\lambda}) = \tilde{\lambda} + \tilde{\xi},$$
$$U_2(\tilde{\lambda}) = \frac{(2\tilde{\lambda} - i)(\tilde{\lambda} + \tilde{\xi})}{2\tilde{\lambda}},$$
$$U_3(\lambda) = (\xi - \lambda). \tag{118}$$

Case II:

$$U_1(\tilde{\lambda}) = \tilde{\lambda} + \tilde{\xi} - i,$$
$$U_2(\tilde{\lambda}) = \frac{(2\tilde{\lambda} - i)(\tilde{\xi} - \tilde{\lambda})}{2\tilde{\lambda}},$$
$$U_3(\lambda) = (\xi - \lambda). \tag{119}$$

We know that U and U^+ are independent of each other, so there are four combinations for $\{U, U^+\}$ such as $\{I, I\}$, $\{I, II\}$, $\{II, I\}$ and $\{II, II\}$.

5. Summary and discussions

In this paper, we study the supersymmetric t–J model with reflecting boundary conditions. We first studied the unitarity relation and the cross-unitarity relation of the R-matrix. According to these relations, we proposed the reflection equation and the dual reflection equation for this supersymmetric t–J model. Solving the reflection equation and the dual reflection equation we give two types of solutions to them, respectively. The transfer matrix for the supersymmetric t–J model is then constructed, and we proved that the transfer matrix constitutes a commuting family. Using the nested algebraic Bethe ansatz method, we obtain the eigenvalues of the transfer matrix.

We discussed all of the above in the FFB grading. For the periodic boundary conditions, the supersymmetric t–J model was studied in three different background gradings [7]. We can also study the FBF and BFF grading for the supersymmetric t–J model with reflecting boundary conditions. The integrability can be proved to be similar to that found in this paper. We have found that the unitarity and cross-unitarity relations of the R-matrix take the same form as that of FFB grading. We also have similar solutions for the reflection equation. Using the similar nested algebraic Bethe ansatz method, we can find the eigenvalues of the transfer matrix with reflecting boundary conditions with grading FBF and BFF.

The R-matrix of the supersymmetric t–J model are rational R-matrix, there are trigonometric R-matrix corresponding to a generalized supersymmetric t–J model. We can also study this generalized supersymmetric t–J model with reflecting boundary conditions in the graded sense.

It is of interest to continue to study the thermodynamic limit of the result obtained in this paper. Thus we can find some physical quantities such as free energy, surface free energy and interfacial tension etc.

We can also extend the supersymmetric t–J model to a more general supersymmetric case. The R-matrix is equivalent to the R-matrix of the Perk–Shultz model [36].

Acknowledgements

This work is supported in part by the National Science Foundation of China, H.F. is partly supported by the Science Fund of NWU.

References

[1] P.W. Anderson, Science 235 (1987) 1196; Phys. Rev. Lett. 65 (1990) 2306.
[2] F.C. Zhang and T.M. Rice, Phys. Rev. B 37 (1988) 3759.
[3] C.K. Lai, J. Math. Phys. 15 (1974) 167.
[4] B. Sutherland, Phys. Rev. B 12 (1975) 3795.
[5] P. Schlottmann, Phys. Rev. B 12 (1987) 5177.
[6] P.A. Bares and G. Blatter, Phys. Rev. Lett. 64 (1990) 2567.
[7] F.H.L. Essler and V.E. Korepin, Phys. Rev B 46 (1992) 9147.
[8] P. Schlottmann, Int. J. Mod. Phys. B 11 (1997) 355.
[9] L.A. Takhtajan and L.D. Faddeev, Russ. Math. Surv. 34 (1979) 11.
[10] V.E. Korepin, G. Izergin and N.M. Bogoliubov, Quantum Inverse Scattering Method, Correlation Functions and Algebraic Bethe Ansatz (Cambridge Univ. Press, Cambridge, 1992).
[11] A. Foerster and M. Karowski, Nucl. Phys. B 396 (1993) 611.
[12] E.K. Sklyanin, J. Phys. A 21 (1988) 2375.
[13] I.V. Cherednik, Theor. Math. Phys. 17 (1983) 77; 61 (1984) 911.
[14] H.J. de Vega, Int. J. Mod. Phys. A 4 (1989) 2371.
[15] L. Mezincescue and R.I. Nepomechie, J. Phys. A 24 (1991) L19; Mod. Phys. Lett. A 6 (1991) 2497.
[16] C. Destri, H.J. de Vega, Nucl. Phys. B 361 (1992) 361; Nucl. Phys. B 374 (1992) 692.
[17] L. Mezincescu and R.I. Nepomechie, Quantum Field Theory, Statistical Mechanics, Quantum Groups and Topology, eds. T. Curtright et al. (World Scientific, Singapore, 1992) p. 200.
[18] H.Q. Zhou, J. Phys. A 30 (1997) 711; Phys. Lett. A 228 (1997) 48.

[19] M. Shiroishi and M. Wadati, J. Phys. Soc. Jpn. 66 (1997) 2288.
[20] A. Foerster and M. Karowski, Nucl. Phys. B 408 (1994) 512.
[21] A. Gonzalez-Ruiz, Nucl. Phys. B 424 (1994) [FS] 468.
[22] H. de Vega and A. Gonzalez-Ruiz, Nucl. Phys. B 417 (1994) 553; Mod. Phys. Lett. A 9 (1994) 2207.
[23] R.H. Yue, H. Fan and B.Y. Hou, Nucl. Phys. B 462 (1996) 167.
[24] F.H.L. Essler, J. Phys. A 29 (1996) 6183.
[25] E. Corrigen, P.E. Dorey, R.H. Rietdijk and R. Sasaki, Phys. Lett. B 333 (1994) 83.
[26] P. Fendley, S. Saleur and N.P. Warner, Nucl. Phys. B 430 (1995) 577; B 428 (1994) 681.
[27] A. Leclair, G. Mussardo, H. Saleur and S. Skorik, Nucl. Phys. B 453 [FS] (1995) 581.
[28] M.T. Batchelor and Y.K. Zhou, Phys. Rev. Lett. 76 (1996) 2826.
[29] R.E. Behrend and P.A. Pearce, J. Phys. A 29 (1996) 7828.
[30] S. Ghoshal and A. Zamolodchikov, Int. J. Mod. Phys. A 9 (1994) 3841.
[31] M. Jimbo, K. Kedem, T. Kojima, H. Konno and T. Miwa, Nucl. Phys. B 441 (1995) 437.
[32] H. Fan, B.Y. Hou, K.J. Shi and Z.X. Yang, Nucl. Phys. B 478 (1996) 723;
H. Fan, Nucl. Phys. B 488 (1997) 409;
H. Fan, B.Y. Hou and K.J. Shi, Nucl. Phys. B 496 [PM] (1997) 551.
[33] M. Shiroishi and M. Wadati, J. Phys. Soc. Jpn. 66, No.1 (1997) 1.
[34] C.N. Yang, Phys. Rev. Lett. 19 (1967) 1312.
[35] R.J. Baxter, Exactly Sloved Models in Statistical Mechanics (Academic Press, London, 1982).
[36] J.H. Perk and C.L. Shultz, Phys. Lett. A 84 (1981) 3759.

ELSEVIER

Nuclear Physics B 556 [FS] (1999) 485–504

NUCLEAR PHYSICS B

www.elsevier.nl/locate/npe

The twisted quantum affine algebra $U_q(A_2^{(2)})$ and correlation functions of the Izergin–Korepin model

Bo-Yu Hou [a], Wen-Li Yang [a,b], Yao-Zhong Zhang [b]

[a] *Institute of Modern Physics, Northwest University, Xian 710069, China*

[b] *Department of Mathematics, University of Queensland, Brisbane, Qld 4072, Australia*

Received 7 April 1999; accepted 4 June 1999

Abstract

We derive the exchange relations of the vertex operators of $U_q(A_2^{(2)})$ and show that these vertex operators give the bosonization of the Izergin–Korepin model. We give an integral expression of the correlation functions of the Izergin–Korepin model and derive the difference equations which they satisfy. © 1999 Elsevier Science B.V. All rights reserved.

1. Introduction

It is well known that quantum affine algebras play an essential role in the studies of low-dimensional massive integrable models such as quantum spin chains, since they provide the symmetry algebras of these models. Based on the works of q-vertex operators [1,2] and level-one highest weight representations of the quantum affine algebra $U_q(\widehat{sl}_2)$, the Kyoto group [3,4] developed a new method which enabled the group to diagonalize the XXZ spin-$\frac{1}{2}$ chain directly in the thermodynamic limit and moreover compute the correlation functions of the spin operators. This approach was later generalized to higher spin XXZ chains [5–7], vertex models with $U_q(\widehat{sl}_n)$ symmetries [8], the face type statistical mechanics models [9,10], and more recently to integrable models with quantum affine superalgebra symmetry [11].

In this paper, we will extend the above programme further to the Izergin–Korepin 19-vertex model [12–16] which has twisted quantum affine algebra $U_q(A_2^{(2)})$ as its non-abelian symmetry. The bosonic realization of $U_q(A_2^{(2)})$ and its level-one vertex operators were constructed recently [17]. In Section 3 we use the results of Ref. [17] to calculate the exchange relations of the q-vertex operators. We show that these vertex operators satisfy the Faddeev–Zamolodchikov algebra with the R-matrix of $U_q(A_2^{(2)})$ as

0550-3213/99/$ - see frontmatter
PII S0550-3213(99)00348-X

its constructure constant. Miki's construction of $U_q(A_2^{(2)})$ is also given. In Section 4, generalizing the Kyoto group's work [3] to the case of twisted quantum affine algebra $U_q(A_2^{(2)})$, we give the bosonization of the Izergin–Korepin model. In Section 5 we compute the correlation functions of the local operators (such as the spin operator S_z) and give an integral expression of the correlation functions. A set of difference equations satisfied by the correlation functions have also been derived.

2. Preliminaries

In this section, we briefly review the bosonization of the twisted quantum affine algebra $U_q(A_2^{(2)})$ at level one [17].

2.1. *Quantum affine algebra $U_q(A_2^{(2)})$*

The symmetric Cartan matrix of the twisted affine Lie algebra $A_2^{(2)}$ is

$$(a_{ij}) = \begin{pmatrix} 8 & -4 \\ -4 & 2 \end{pmatrix},$$

where $i, j = 0, 1$. Quantum affine algebra $U_q(A_2^{(2)})$ is a q-analogue of the universal enveloping algebra of $A_2^{(2)}$ generated by the Chevalley generators $\{e_i, f_i, t_i^{\pm 1}, d | i = 0, 1\}$, where d is the usual derivation operator. The defining relations are [18,19]

$$t_i t_j = t_j t_i, \qquad t_i d = d t_i, \qquad [d, e_i] = \delta_{i,0} e_i, \qquad [d, f_i] = -\delta_{i,0} f_i,$$

$$t_i e_j t_i^{-1} = (q^{\frac{1}{2}})^{a_{ij}} e_j, \qquad t_i f_j t_i^{-1} = (q^{\frac{1}{2}})^{-a_{ij}} f_j,$$

$$[e_i, f_j] = \delta_{ij} \frac{t_i - t_i^{-1}}{q_i - q_i^{-1}},$$

$$\sum_{r=0}^{1-a_{ij}} (-1)^r \begin{bmatrix} 1 - a_{ij} \\ r \end{bmatrix}_{q_i} (e_i)^r e_j (e_i)^{1-a_{ij}-r} = 0, \quad \text{if} \quad i \neq j,$$

$$\sum_{r=0}^{1-a_{ij}} (-1)^r \begin{bmatrix} 1 - a_{ij} \\ r \end{bmatrix}_{q_i} (f_i)^r f_j (f_i)^{1-a_{ij}-r} = 0, \quad \text{if} \quad i \neq j,$$

where $q_1 = q^{\frac{1}{2}}$, $q_0 = q^2$, $t_i = q_i^{h_i}$, and

$$[n]_{q_i} = \frac{q_i^n - q_i^{-n}}{q_i - q_i^{-1}}, \qquad [n]_{q_i}! = [n]_{q_i}[n-1]_{q_i} \cdots [1]_{q_i},$$

$$\begin{bmatrix} n \\ r \end{bmatrix}_{q_i} = \frac{[n]_{q_i}!}{[n-r]_{q_i}![r]_{q_i}!}.$$

$U_q(A_2^{(2)})$ is a quasi-triangular Hopf algebra endowed with Hopf algebra structure,

$$\Delta(t_i)=t_i\otimes t_i,\quad \Delta(e_i)=e_i\otimes 1+t_i\otimes e_i,\quad \Delta(f_i)=f_i\otimes t_i^{-1}+1\otimes f_i, \tag{2.1}$$

$$\epsilon(t_i)=1,\quad \epsilon(e_i)=\epsilon(f_i)=0, \tag{2.2}$$

$$S(e_i)=-t_i^{-1}e_i,\quad S(f_i)=-f_it_i,\quad S(t_i^{\pm1})=t_i^{\mp1},\quad S(d)=-d. \tag{2.3}$$

$U_q(A_2^{(2)})$ can also be realized by the Drinfeld generators [17,20] $\{d,\ X_m^{\pm},\ a_n,\ K^{\pm1},\ \gamma^{\pm1/2}|m\in\mathbb{Z},\ n\in\mathbb{Z}_{\neq0}\}$. The relations read

$$\gamma \text{ is central},\quad [K,a_n]=0,\quad [d,K]=0,\quad [d,a_n]=na_n, \tag{2.4}$$

$$[a_m,a_n]=\delta_{m+n,0}\frac{\{[4n]_{q_1}-(-1)^n[2n]_{q_1}\}(\gamma^m-\gamma^{-m})}{m(q_1-q_1^{-1})}, \tag{2.5}$$

$$KX_m^{\pm}=q^{\pm1}X_m^{\pm}K,\quad [d,X_m^{\pm}]=mX_m^{\pm}, \tag{2.6}$$

$$[a_m,X_n^{\pm}]=\pm\frac{\{[4m]_{q_1}-(-1)^m[2m]_{q_1}\}}{m}\gamma^{\mp|m|/2}X_{n+m}^{\pm}, \tag{2.7}$$

$$[X_m^+,X_n^-]=\frac{1}{q_1-q_1^{-1}}(\gamma^{(m-n)/2}\psi_{m+n}^+-\gamma^{-(m-n)/2}\psi_{m+n}^-), \tag{2.8}$$

$$(z-wq^{\pm2})(z+wq^{\mp1})X^{\pm}(z)X^{\pm}(w)=(zq^{\pm2}-w)(zq^{\mp1}+w)X^{\pm}(w)X^{\pm}(z), \tag{2.9}$$

where the corresponding Drinfeld currents $\psi^{\pm}(z)$ and $X^{\pm}(z)$ are defined by

$$\psi^+(z)=\sum_{m=0}^{\infty}\psi_m^+z^{-m}=K\exp\left\{(q_1-q_1^{-1})\sum_{k=1}^{\infty}a_kz^{-k}\right\},$$

$$\psi^-(z)=\sum_{m=0}^{\infty}\psi_{-m}^-z^{m}=K^{-1}\exp\left\{-(q_1-q_1^{-1})\sum_{k=1}^{\infty}a_{-k}z^{k}\right\},$$

$$X^{\pm}(z)=\sum_{n\in Z}X_m^{\pm}z^{-m}.$$

The Chevalley generators are related to the Drinfeld generators by the formulae

$$t_1=K,\quad e_1=X_0^+,\quad t_0=\gamma K^{-2},\quad f_1=X_0^-, \tag{2.10}$$

$$e_0=K^{-2}[X_0^-,X_1^-]_q,\quad f_0=\frac{1}{[4]_{q_1}^2}[X_{-1}^+,X_0^+]_{q^{-1}}K^2. \tag{2.11}$$

2.2. *Bosonization of $U_q(A_2^{(2)})$ at level one*

Let us introduce the bosonic q-oscillators [17] $\{a_n, Q, P|n\in Z-\{0\}\}$ which satisfy the commutation relations

$$[a_m,a_n]=\delta_{m+n,0}\frac{1}{m}\{[4m]_{q_1}-(-1)^m[2m]_{q_1}\}[2m]_{q_1},$$

$$[P,a_m]=[Q,a_m]=0,\quad [P,Q]=1.$$

Set $F_1=\oplus_{n\in Z}C[a_{-1},a_{-2},\dots]\,e^{\frac{Q}{2}+nQ}|0\rangle$, where the Fock vacuum vector $|0\rangle$ is defined by

$$a_n|0\rangle = 0 \quad \text{for} \quad n > 0, \quad P|0\rangle = 0.$$

Then, on F_1 the Drinfeld currents of $U_q(A_2^{(2)})$ at level-one are realized by the free boson fields as [17]

$$\begin{aligned}
&\gamma = q, \qquad K = q^P, \\
&\psi^+(z) = q^P \exp\left\{(q_1 - q_1^{-1})\sum_{n=1}^{\infty} a_n z^{-n}\right\}, \\
&\psi^-(z) = q^{-P} \exp\left\{-(q_1 - q_1^{-1})\sum_{n=1}^{\infty} a_{-n} z^{n}\right\}, \\
&X^{\pm}(z) = \exp\left\{\pm\sum_{n=1}^{\infty}\frac{a_{-n}}{[2n]_{q_1}} q^{\mp\frac{n}{2}} z^n\right\}\exp\left\{\mp\sum_{n=1}^{\infty}\frac{a_n}{[2n]_{q_1}} q^{\mp\frac{n}{2}} z^{-n}\right\} e^{\pm Q} z^{\pm P+\frac{1}{2}}.
\end{aligned} \tag{2.12}$$

By the relations between Chevalley basis and Drinfeld basis (2.10), (2.11), one can check that

$$\begin{aligned}
&t_1\, e^{\frac{Q}{2}}|0\rangle = q^{\frac{1}{2}}\, e^{\frac{Q}{2}}|0\rangle = (q^{\frac{1}{2}})^{\langle\alpha_1,\Lambda_1\rangle}\, e^{\frac{Q}{2}}|0\rangle, \\
&t_0\, e^{\frac{Q}{2}}|0\rangle = e^{\frac{Q}{2}}|0\rangle = (q^{\frac{1}{2}})^{\langle\alpha_0,\Lambda_1\rangle}\, e^{\frac{Q}{2}}|0\rangle, \\
&e_i\, e^{\frac{Q}{2}}|0\rangle = 0, \quad i = 0, 1,
\end{aligned}$$

where α_1, $\alpha_0 = -2\alpha_1 + \delta$ are the simple roots of $A_2^{(2)}$, Λ_1 is one basic fundamental weight, and δ is the imaginary root. The Fock space F_1 coincides with the level-one irreducible highest weight module $V(\Lambda_1)$ with the highest weight vector given by $|\Lambda_1\rangle = e^{\frac{Q}{2}}|0\rangle$.

2.3. Level-one vertex operators

Let V be the three-dimensional evaluation representation of $U_q(A_2^{(2)})$,$\{v_1, v_0, v_{-1}\}$ be the basis vectors of V, and $E_{i,j}$ be the 3×3 matrix whose (i,j) element is unity and zero otherwise. Then the three-dimensional level-0 representation V_z of $U_q(A_2^{(2)})$ is given by

$$\begin{aligned}
&e_1 = \alpha^{-1}[2]_{q_1}E_{1,0} + \alpha E_{0,-1}, \qquad e_0 = zE_{-1,1}, \\
&f_1 = \alpha^{-1}[2]_{q_1}E_{-1,0} + \alpha E_{0,1}, \qquad f_0 = z^{-1}E_{1,-1}, \\
&t_1 = q_1^2 E_{1,1} + q_1^{-2}E_{-1,-1} + E_{0,0}, \qquad t_0 = q_0 E_{-1,-1} + q_0^{-1}E_{1,1} + E_{0,0},
\end{aligned}$$

where $\alpha = \{[2]_{q_1} q^{-\frac{1}{2}}\}^{\frac{1}{2}}$.

We define the dual modules V_z^{*b} of V_z by $\pi_{V^{*b}}(a) = \pi_V(b(a))^t$, $\forall a \in U_q(A_2^{(2)})$, where t is the transposition operation. b is the anti-automorphism defined by

$$b(x) = (-q)^{\hat{\rho}} S(x) (-q)^{-\hat{\rho}}$$

$$\stackrel{\text{def}}{=} (-q)^{3d}(-q^{\frac{1}{2}})^{\frac{1}{2}h_1} S(x)(-q)^{-3d}(-q^{\frac{1}{2}})^{-\frac{1}{2}h_1}, \quad \forall x \in U_q(A_2^{(2)}), \tag{2.13}$$

where $2\hat{\rho} = 6d + \frac{1}{2}h_1$. A convenient feature of b is that $b^2 = \text{id}$ (since $S^2(x) = q^{-2\hat{\rho}} x q^{2\hat{\rho}}$) and that the following isomorphism holds:

$$C : V_z \longrightarrow V_z^{*b}, \qquad v_i \otimes z^n \longrightarrow v_{-i}^* \otimes z^n. \tag{2.14}$$

Throughout, we denote by $V(\lambda)$ a level-one irreducible highest weight $U_q(A_2^{(2)})$ module with highest weight λ. Consider the following intertwiners of $U_q(A_2^{(2)})$ modules:

$$\Phi_\lambda^{\mu V}(z) : V(\lambda) \longrightarrow V(\mu) \otimes V_z, \qquad \Phi_\lambda^{\mu V^*}(z) : V(\lambda) \longrightarrow V(\mu) \otimes V_z^{*b},$$
$$\Psi_\lambda^{V\mu}(z) : V(\lambda) \longrightarrow V_z \otimes V(\mu), \qquad \Psi_\lambda^{V^*\mu}(z) : V(\lambda) \longrightarrow V_z^{*b} \otimes V(\mu).$$

They are intertwiners in the sense that for any $x \in U_q(A_2^{(2)})$,

$$\Theta(z) \cdot x = \Delta(x) \cdot \Theta(z), \qquad \Theta(z) = \Phi(z), \Phi^*(z), \Psi(z), \Psi^*(z). \tag{2.15}$$

$\Phi(z)$ ($\Phi^*(z)$) is called a type I (dual) vertex operator and $\Psi(z)$ ($\Psi^*(z)$) a type II (dual) vertex operator.

We expand the vertex operators as

$$\Phi(z) = \sum_{j=1,0,-1} \Phi_j(z) \otimes v_j, \qquad \Phi^*(z) = \sum_{j=1,0,-1} \Phi_j^*(z) \otimes v_j^*, \tag{2.16}$$

$$\Psi(z) = \sum_{j=1,0,-1} v_j \otimes \Psi_j(z), \qquad \Psi^*(z) = \sum_{j=1,0,-1} v_j^* \otimes \Psi_j^*(z). \tag{2.17}$$

Define the operators $\Phi_j(z), \Phi_j^*(z), \Psi_j(z)$ and $\Psi_j^*(z)$ ($j = 1, 0, -1$) acting on the Fock space F_1 by

$$\Phi_{-1}(z) = \exp\left\{\sum_{n=1}^{\infty} \frac{[2n]_{q_1}}{n} q^{\frac{9}{2}n} \omega_{-n} (-z)^n\right\}$$
$$\times \exp\left\{\sum_{n=1}^{\infty} \frac{[2n]_{q_1}}{n} q^{-\frac{7}{2}n} \omega_n (-z)^{-n}\right\} e^{Q} (zq^4)^{P+\frac{1}{2}}, \tag{2.18}$$

$$\Phi_0(z) = \left\{\frac{q^{-\frac{1}{2}}}{[2]_{q_1}}\right\}^{\frac{1}{2}} [\Phi_{-1}(z), f_1]_q,$$
$$\Phi_1(z) = \left\{\frac{q^{\frac{1}{2}}}{[2]_{q_1}}\right\}^{\frac{1}{2}} [\Phi_0(z), f_1], \tag{2.19}$$

$$\Psi_1(z) = \exp\left\{-\sum_{n=1}^{\infty} \frac{[2n]_{q_1}}{n} q^{\frac{1}{2}n} \omega_{-n} z^n\right\}$$
$$\times \exp\left\{-\sum_{n=1}^{\infty} \frac{[2n]_{q_1}}{n} q^{-\frac{3}{2}n} \omega_n z^{-n}\right\} e^{-Q} (-zq)^{-P+\frac{1}{2}}, \tag{2.20}$$

$$\Psi_0(z)=\left\{\frac{q^{\frac{-1}{2}}}{[2]_{q_1}}\right\}^{\frac{1}{2}}[\Psi_1(z),e_1]_q,$$

$$\Psi_1(z)=\left\{\frac{q^{\frac{1}{2}}}{[2]_{q_1}}\right\}^{\frac{1}{2}}[\Psi_0(z),e_1], \tag{2.21}$$

$$\Phi_i^*(z)=\Phi_{-i}(-zq^{-3}), \qquad \Psi_i^*(z)=\Psi_{-i}(-zq^3), \tag{2.22}$$

where $[a,b]_x=ab-xba$ and the bosonic q-oscillators ω_m are defined by

$$\omega_m=-\frac{m}{\{[4m]_{q_1}-(-1)^m[2m]_{q_1}\}[2m]_{q_1}}a_m,$$

such that $[a_n,\omega_m]=\delta_{m+n,0}$.

According to Ref. [17], the operators $\Phi(z)$, $\Phi^*(z)$, $\Psi(z)$ and $\Psi^*(z)$ defined in (2.18)–(2.22) are the only vertex operators of $U_q(A_2^{(2)})$ which intertwine the level-one irreducible highest weight $U_q(A_2^{(2)})$ modules,

$$\begin{aligned}\Phi(z):V(\Lambda_1)&\longrightarrow V(\Lambda_1)\otimes V_z, &\qquad \Phi^*(z):V(\Lambda_1)&\longrightarrow V(\Lambda_1)\otimes V_z^{*b},\\ \Psi(z):V(\Lambda_1)&\longrightarrow V_z\otimes V(\Lambda_1), &\qquad \Psi^*(z):V(\Lambda_1)&\longrightarrow V_z^{*b}\otimes V(\Lambda_1).\end{aligned} \tag{2.23}$$

3. Exchange relations of vertex operators

In this section, we derive the exchange relations of the type I and type II vertex operators of $U_q(A_2^{(2)})$.

3.1. The R-matrix

We introduce some abbreviations:

$$(z;p_1,p_2,\ldots,p_m)=\prod_{\{l_1,l_2,\ldots,l_m\}=0}^{\infty}(1-zp_1^{l_1}p_2^{l_2}\cdots p_m^{l_m}),$$

$$\Theta_p(z)=(z;p)(pz^{-1};p)(p;p), \qquad \{z\}=(z;q^6,q^6).$$

Let $\overline{R}(z)\in\mathrm{End}(V\otimes V)$ be the R-matrix of $U_q(A_2^{(2)})$:

$$\overline{R}(z)(v_i\otimes v_j)=\sum_{k,l}\overline{R}_{kl}^{i,j}(z)v_k\otimes v_l, \qquad \forall v_i,v_j,v_k,v_l\in V, \tag{3.1}$$

where the matrix elements are given by

$$\overline{R}_{1,1}^{1,1}\left(\frac{z_1}{z_2}\right)=\overline{R}_{-1,-1}^{-1,-1}\left(\frac{z_1}{z_2}\right)=1, \qquad \overline{R}_{0,-1}^{-1,0}\left(\frac{z_1}{z_2}\right)=\overline{R}_{1,0}^{0,1}\left(\frac{z_1}{z_2}\right)=\frac{(q-q^{-1})z_2}{z_1q-z_2q^{-1}},$$

$$\overline{R}_{-1,0}^{-1,0}\left(\frac{z_1}{z_2}\right)=\overline{R}_{0,-1}^{0,-1}\left(\frac{z_1}{z_2}\right)=\overline{R}_{1,0}^{1,0}\left(\frac{z_1}{z_2}\right)=\overline{R}_{0,1}^{0,1}\left(\frac{z_1}{z_2}\right)=\frac{z_1-z_2}{z_1q-z_2q^{-1}},$$

$$\overline{R}^{0,-1}_{-1,0}\left(\frac{z_1}{z_2}\right)=\overline{R}^{1,0}_{0,1}\left(\frac{z_1}{z_2}\right)=\frac{(q-q^{-1})z_1}{z_1q-z_2q^{-1}},$$
$$\overline{R}^{-1,1}_{-1,1}\left(\frac{z_1}{z_2}\right)=\overline{R}^{1,-1}_{1,-1}\left(\frac{z_1}{z_2}\right)=\frac{(z_1-z_2)q^2(z_1q+z_2)}{(z_1q-z_2q^{-1})(z_1q^4+z_2q)},$$
$$\overline{R}^{1,-1}_{-1,1}\left(\frac{z_1}{z_2}\right)=\frac{(q^2-1)z_1\{z_2+z_2q^2+z_1q^3-z_1q^2\}}{(z_1q-z_2q^{-1})(z_1q^4+z_2q)},$$
$$\overline{R}^{-1,1}_{1,-1}\left(\frac{z_1}{z_2}\right)=\frac{(q^2-1)z_2\{z_2-z_2q+z_1q^3+z_1q\}}{(z_1q-z_2q^{-1})(z_1q^4+z_2q)},$$
$$\overline{R}^{1,-1}_{0,0}\left(\frac{z_1}{z_2}\right)=\overline{R}^{0,0}_{-1,1}\left(\frac{z_1}{z_2}\right)=\frac{(q-q^{-1})q^{\frac{7}{2}}z_1(z_1-z_2)}{(z_1q-z_2q^{-1})(z_1q^4+z_2q)},$$
$$\overline{R}^{-1,1}_{0,0}\left(\frac{z_1}{z_2}\right)=\overline{R}^{0,0}_{1,-1}\left(\frac{z_1}{z_2}\right)=\frac{(q^{-1}-q)q^{\frac{3}{2}}z_2(z_1-z_2)}{(z_1q-z_2q^{-1})(z_1q^4+z_2q)},$$
$$\overline{R}^{0,0}_{0,0}\left(\frac{z_1}{z_2}\right)=\frac{z_1-z_2}{z_1q-z_2q^{-1}}+\frac{(q-q^{-1})(q+q^4)z_2z_1}{(z_1q-z_2q^{-1})(z_1q^4+z_2q)},$$
$$R^{ij}_{kl}=0,\quad \text{otherwise}.$$

Define the R-matrices $R^I(z)$, $R^{II}(z)$ and $R^U(z)$,

$$R^I(z)=r(z)\overline{R}(z),\qquad R^{II}(z)=\overline{r}(z)\overline{R}(z),\qquad R^U(z)=\rho(z)\overline{R}(z),\tag{3.2}$$

where

$$r(z)=z^{-1}\frac{(-q^3z;q^6)(q^2z;q^6)(q^6z^{-1};q^6)(-q^5z^{-1};q^6)}{(-q^3z^{-1};q^6)(q^2z^{-1};q^6)(q^6z;q^6)(-q^5z;q^6)},$$
$$\overline{r}(z)=\frac{(q^4z;q^6)(-q^3z;q^6)(-qz^{-1};q^6)(q^6z^{-1};q^6)}{(-q^3z^{-1};q^6)(q^4z^{-1};q^6)(q^6z;q^6)(-qz;q^6)},$$
$$\rho(z)=\frac{r(z)}{\tau(zq^{-1})},$$
$$\tau(z)=z^{-1}\frac{\Theta_{q^6}(qz)\Theta_{q^6}(-q^2z)}{\Theta_{q^6}(q^5z)\Theta_{q^6}(-q^4z)}.$$

Then the R-matrices satisfy Yang–Baxter equation (YBE) on $V\otimes V\otimes V$

$$R^i_{12}(z)R^i_{13}(zw)R^i_{23}(w)=R^i_{23}(w)R^i_{13}(zw)R^i_{12}(z),\quad \text{for}\quad i=I,II,U,$$

and moreover enjoys: (i) the initial condition, $R^i(1)=P$ for $i=I,II,U$, with P being the permutation operator; (ii) the unitarity condition,

$$R^i_{12}\left(\frac{z}{w}\right)R^i_{21}\left(\frac{w}{z}\right)=1,\quad \text{for}\quad i=I,II,U$$

where $R^i_{21}(z)=PR^i_{12}(z)P$; and (iii) the crossing relations

$$(R^i)^{k,l}_{m,n}(z)=(-q^{\frac{1}{2}})^{-\rho_l}(R^{(i)})^{-n,k}_{-l,m}(-z^{-1}q^{-3})(-q^{\frac{1}{2}})^{\rho_n},\quad \text{for}\quad i=I,II.$$

Here and throughout,

$$\rho_1 = -1, \qquad \rho_0 = 0, \qquad \rho_{-1} = 1. \tag{3.3}$$

3.2. The Faddeev–Zamolodchikov algebra

Now, we are in the position to calculate the exchange relations of the type I and type II vertex operators of $U_q(A_2^{(2)})$ given in (2.18)–(2.22).

Firstly we bosonize the derivation operator d:

$$d = \sum_{n=1}^{\infty} n a_{-n}\omega_n - \frac{P^2}{2}. \tag{3.4}$$

One can easily check that this d operator obeys the following commutation relations:

$$\begin{aligned} q^d X^{\pm}(z) q^{-d} &= X^{\pm}(zq^{-1}), \\ q^d \Phi_i(z) q^{-d} &= \Phi_i(zq^{-1}), \\ q^d \Psi_i(z) q^{-d} &= \Psi_i(zq^{-1}), \end{aligned} \tag{3.5}$$

as required.

Define

$$\oint dz\, f(z) = \mathrm{Res}(f) = f_{-1},$$

for formal series function

$$f(z) = \sum_{n\in Z} f_n z^n.$$

Then the Chevalley generators of $U_q(A_2^{(2)})$ can be expressed by the integrals

$$\begin{aligned} e_1 &= \oint z^{-1} dz\, X^+(z), \\ f_1 &= \oint z^{-1} dz\, X^-(z), \\ e_0 &= K^{-2} \oint\oint dz\, w^{-1} dw [X^-(w), X^-(z)]_q, \\ f_0 &= \frac{1}{[4]_{q_1}} \oint\oint z^{-2} dz\, w^{-1} dw [X^+(z), X^+(w)]_{q^{-1}}. \end{aligned}$$

From the normal order relations in Appendix A, one can also obtain the integral expression of the vertex operators defined in (2.18)–(2.22)

$$\Phi_0(z) = \left\{ \frac{q^{-\frac{1}{2}}}{[2]_{q_1}} \right\}^{\frac{1}{2}} \oint \frac{dw}{w} \frac{(q^{-1}-q)}{zq^3(1+\frac{zq^5}{w})(1+\frac{w}{zq^3})} :\Phi_{-1}(z) X^-(w):, \tag{3.6}$$

$$\Phi_1(z)=\frac{1-q^2}{[2]_{q_1}}\oint\frac{dw_1}{w_1}\oint\frac{dw}{w}$$
$$\times\frac{(1-\frac{w}{w_1})\{w_1^2(1-\frac{w}{w_1q^2})(1+\frac{wq}{w_1})(1+\frac{zq^5}{w_1})-zwq^2(1+\frac{w_1}{zq^3})(1-\frac{wq^2}{w_1})(1+\frac{w_1q}{w})\}}{z^3q^{12}(1+\frac{zq^5}{w_1})(1+\frac{wq}{w_1})(1+\frac{w_1}{zq^3})(1+\frac{w_1q^5}{w})(1+\frac{w}{zq^5})(1+\frac{w}{zq^3})}$$
$$\times:\Phi_{-1}(z)X^-(w)X^-(w_1):,\tag{3.7}$$

$$\Psi_0(z)=\left\{\frac{q^{-\frac{1}{2}}}{[2]_{q_1}}\right\}^{\frac{1}{2}}\oint\frac{dw}{w}\frac{(1-q^2)}{wq(1-\frac{w}{zq^2})(1-\frac{z}{w})}:\Psi_1(z)X^+(w):,\tag{3.8}$$

$$\Psi_{-1}(z)=\frac{1-q^2}{[2]_{q_1}}\oint\frac{dw_1}{w_1}$$
$$\times\oint\frac{dw}{w}\frac{(1-\frac{w}{w_1})\{(1-\frac{w_1}{wq^2})(1-\frac{z}{w_1})(1+\frac{w}{w_1q})w_1-zq(1-\frac{w}{w_1q^2})(1-\frac{w_1}{zq^2})(1+\frac{w_1}{wq})\}}{zwq(1-\frac{w}{zq^2})(1-\frac{z}{w})(1-\frac{w_1}{zq^2})(1+\frac{w_1}{wq})(1-\frac{z}{w_1})(1+\frac{w}{w_1q})}$$
$$\times:\Psi_1(z)X^+(w)X^+(w_1):.\tag{3.9}$$

By the technique proposed in [10], using the above integral expressions and the relations given in Appendices A and B, and after tedious calculations, we can show that the bosonic vertex operators defined in (2.18)–(2.22) satisfy the Faddeev–Zamolodchikov (ZF) algebra,

$$\Phi_j(z_2)\Phi_i(z_1)=\sum_{kl}R^I\left(\frac{z_1}{z_2}\right)_{i,j}^{k,l}\Phi_k(z_1)\Phi_l(z_2),\tag{3.10}$$

$$\Psi_i^*(z_1)\Psi_j^*(z_2)=-\sum_{kl}R^{II}\left(\frac{z_1}{z_2}\right)_{kl}^{ij}\Psi_l^*(z_2)\Psi_k^*(z_1),\tag{3.11}$$

$$\Psi_i^*(z_1)\Phi_j(z_2)=\tau\left(\frac{z_1}{z_2}\right)\Phi_j(z_2)\Psi_i^*(z_1),\tag{3.12}$$

and moreover, the bosonic vertex operators defined in (2.18)–(2.22) have the following invertibility relations:

$$\Phi_i(z)\Phi_j^*(z)=g\delta_{ij}(-q^{\frac{1}{2}})^{\rho_i},\qquad g=\frac{(q^6;q^6)(-q^5;q^6)}{(-q^3;q^6)(q^2;q^6)},\tag{3.13}$$

$$\sum_k(-q^{\frac{1}{2}})^{-\rho_k}\Phi_k^*(z)\Phi_k(z)=g,\tag{3.14}$$

and

$$\Psi_i(z_1)\Psi_j^*(z_2)=\frac{\bar{g}\delta_{ij}(-q^{\frac{1}{2}})^{-\rho_i}}{1-\frac{z_2}{z_1}}+\text{regular term}\quad\text{when } z_1\longrightarrow z_2,\tag{3.15}$$
$$\bar{g}=\frac{q(q^4;q^6)(-q^3;q^6)}{(-q;q^6)(q^6;q^6)}.$$

In the derivation of the above relations the following fact is helpful:

$$:\Phi_{-1}(z)\Phi_{-1}(-zq^{-3})X^{-}(zq^{2})X^{-}(-zq^{3})(zq^{2})^{-1}(-zq^{3})^{-1}:=\mathrm{id},$$
$$:\Psi_{1}(z)\Psi_{1}(-zq^{3})X^{+}(zq^{2})X^{+}(-zq^{3})(zq^{2})^{-1}(-zq^{3})^{-1}:=\mathrm{id}.$$

3.3. Miki's construction of $U_q(A_2^{(2)})$

We generalize Miki's construction to the twisted quantum affine algebra $U_q(A_2^{(2)})$ case.

Define

$$(L^{+}(z))_i^j=\Phi_i(zq^{\frac{1}{2}})\Psi_j^*(zq^{-\frac{1}{2}}),$$
$$(L^{-}(z))_i^j=\Phi_i(zq^{-\frac{1}{2}})\Psi_j^*(zq^{\frac{1}{2}}).$$

Then from the Faddeev–Zamolodchikov algebra in (3.10)–(3.12) and the identity

$$-\frac{r(z)}{\bar{r}(z)}=\frac{\tau(zq^{-1})}{\tau(z^{-1}q^{-1})}=\frac{\tau(zq)}{\tau(z^{-1}q)},$$

we can verify by straightforward computations that the L-operators $L^{\pm}(z)$ give a realization of the Reshetikhin–Semenov-Tian-Shansky (RS) algebra [21] at level one in the twisted quantum affine algebra $U_q(A_2^{(2)})$,

$$R^U\left(\frac{z}{w}\right)L_1^{\pm}(z)L_2^{\pm}(w)=L_2^{\pm}(w)L_1^{\pm}(z)R^U\left(\frac{z}{w}\right),$$
$$R^U\left(\frac{z^+}{w^-}\right)L_1^{+}(z)L_2^{-}(w)=L_2^{-}(w)L_1^{+}(z)R^U\left(\frac{z^-}{w^+}\right),$$

where $L_1^{\pm}(z)=L^{\pm}(z)\otimes 1$, $L_2^{\pm}(z)=1\otimes L^{\pm}(z)$ and $z^{\pm}=zq^{\pm\frac{1}{2}}$. We remark that the Drinfeld bases (2.4)–(2.9) of the twisted quantum affine algebra $U_q(A_2^{(2)})$ can also be derived by using the twisted RS algebra and the corresponding Gauss decomposition [22].

4. The Izergin–Korepin model

In this section, we give a mathematical definition of the Izergin–Korepin model on an infinite lattice.

4.1. Space of states

By means of the R-matrix (3.1) of $U_q(A_2^{(2)})$, one can define the Izergin–Korepin model on the infinite lattice $\ldots\otimes V\otimes V\otimes V\ldots$. Let h be the operator on $V\otimes V$ such that

$$PR\left(\frac{z_1}{z_2}\right)=1+uh+\ldots,\qquad u\longrightarrow 0,$$

$$P: \text{the permutation operator}, \qquad e^u \equiv \frac{z_1}{z_2}.$$

The Hamiltonian H of the Izergin–Korepin model is defined by [14]

$$H = \sum_{l \in Z} h_{l+1,l}. \tag{4.1}$$

H acts formally on the infinite tensor product,

$$\ldots V \otimes V \otimes V \ldots. \tag{4.2}$$

It can be easily checked that

$$[U_q'(A_2^{(2)}), H] = 0,$$

where $U_q'(A_2^{(2)})$ is the subalgebra of $U_q(A_2^{(2)})$ with the derivation operator d being dropped. So $U_q'(A_2^{(2)})$ plays the role of infinite dimensional *non-abelian symmetries* of the Izergin–Korepin model on the infinite lattice. Since the level-one vertex operators only exist between $V(\Lambda_1)$ and itself, following Ref. [3], we can replace the infinite tensor product (4.2) by the level-0 $U_q(A_2^{(2)})$ module,

$$\mathbf{H} = \text{Hom}(V(\Lambda_1), V(\Lambda_1)) \cong V(\Lambda_1) \otimes V(\Lambda_1)^{*b},$$

where $V(\Lambda_1)$ is level-one irreducible highest weight $U_q(A_2^{(2)})$ module and $V(\Lambda_1)^{*b}$ is the dual module of $V(\Lambda_1)$. By Theorem 2, this homomorphism can be realized by applying the type I vertex operators repeatedly. So we shall make the (hypothetical) identification:

$$\text{“the space of physical states”} = V(\Lambda_1) \otimes V(\Lambda_1)^{*b}.$$

Namely, we take

$$\mathbf{H} \equiv \text{End}(\text{V}(\Lambda_1))$$

as the space of states of the Izergin–Korepin model on the infinite lattice. The left action of $U_q(A_2^{(2)})$ on $\mathbf{H}$ is defined by

$$x.f = \sum x_{(1)} \circ f \circ b(x_{(2)}), \quad \forall x \in U_q(A_2^{(2)}), \quad f \in \mathbf{H},$$

where we have used notation $\Delta(x) = \sum x_{(1)} \otimes x_{(2)}$. A linear operator of the form $O = A \otimes B$ ($A \in \text{End}(\text{V}(\Lambda_1))$ and $B \in \text{End}(\text{V}(\Lambda_1)^{*\text{b}})$) operate on a state f as $f \longrightarrow A \circ f \circ B'$.

Note that $\mathbf{H}$ has the unique canonical element which is referred as to the physical vacuum [4] and denoted by $|\text{vac}\rangle$

$$|\text{vac}\rangle = \chi^{-\frac{1}{2}}(-\text{q})^{-\hat{\rho}}, \tag{4.3}$$

where χ coincide with the character of $U_q(A_2^{(2)})$ module $V(\Lambda_1)$

$$\chi = \mathrm{tr}_{V(\Lambda_1)}(q^{-2\hat{\rho}}) = q^{\frac{1}{4}}(1+q)\prod_{n=1}^{\infty}(1+q^{6n-1})(1+q^{6n+1}).$$

The proof of the above character formula is given in Appendix C.

4.2. Local structure and local operators

Following Jimbo and Miwa [4], we use the type I vertex operators and their variants to incorporate the local structure into the space of physical states $\mathbf{H}$, that is to formulate the action of local operators of the Izergin–Korepin model on the infinite tensor product (4.2) in terms of their actions on $\mathbf{H}$.

Using the isomorphisms (cf. Theorem 2)

$$\begin{aligned} \Phi(z) &: V(\Lambda_1) \longrightarrow V(\Lambda_1)\otimes V_z, \\ \Phi^{*,t}(-zq^3) &: V_z \otimes V(\Lambda_1)^{*b} \longrightarrow V(\Lambda_1)^{*b}, \end{aligned} \tag{4.4}$$

were t is the transposition on the quantum space, we have the following identification:

$$V(\Lambda_1)\otimes V(\Lambda_1)^{*b} \xrightarrow{\sim} V(\Lambda_1)\otimes V_z \otimes V(\Lambda_1)^{*b} \xrightarrow{\sim} V(\Lambda_1)\otimes V(\Lambda_1)^{*b}.$$

The resulting isomorphism can be identified with the translation (or shift) operator defined by

$$T = g^{-1}\sum_i \Phi_i(1)\otimes \Phi_i^{*,t}(-q^3).$$

Its inverse is given by

$$T^{-1} = g^{-1}\sum_i \Phi_i^{*}(1)\otimes \Phi_i^{t}(-q^{-3}).$$

Thus we can define the local operators on V as operators on $\mathbf{H}$ [4]. Let us label the tensor components from the middle as $1,2,\ldots$ for the left half and as $0,-1,-2,\ldots$ for the right half. The operators acting on the site 1 are defined by

$$E_{i,j} \stackrel{\mathrm{def}}{=} E_{i,j}^{(1)} = g^{-1}(-q^{-\frac{1}{2}})^{-\rho_i}\Phi_i^{*}(1)\Phi_j(1)\otimes \mathrm{id}. \tag{4.5}$$

In particular, we have spin operator S_z

$$S_z = g^{-1}\{\Phi_1^{*}(1)\Phi_1(1)(-q^{-\frac{1}{2}}) + \Phi_0^{*}(1)\Phi_0(1) - \Phi_{-1}^{*}(1)\Phi_{-1}(1)(-q^{-\frac{1}{2}})^{-1}\}\otimes \mathrm{id}.$$

More generally we set

$$E_{i,j}^{(n)} = T^{-(n-1)}E_{i,j}T^{n-1} \quad (n\in Z). \tag{4.6}$$

Then, from the invertibility relations of the type I vertex operators of $U_q(A_2^{(2)})$, we can show that the local operators $E_{ij}^{(n)}$ acting on $\mathbf{H}$ satisfy the following relations:

$$E_{i,j}^{(m)}E_{k,l}^{(n)} = \begin{cases} \delta_{jk}E_{i,l}^{(n)} & \text{if} \quad m=n, \\ E_{k,l}^{(n)}E_{i,j}^{(m)} & \text{if} \quad m\neq n. \end{cases} \tag{4.7}$$

Moreover, we can define transfer matrix of the Izergin–Korepin model on the infinite lattice as

$$T(z)=g^{-1}\sum_i \Phi_i(z)\otimes\Phi_i^{*,t}(-zq^3). \tag{4.8}$$

The commutativity of the transfer matrix, $[T(z),T(w)]=0$, follows from the ZF algebraic relations (3.10). The translation operator is $T=T(1)$, as expected. Comparing with the definition of the Hamiltonian of the Izergin–Korepin model (4.1), the action of such a Hamiltonian on $\mathbf{H}$ can be given by

$$H=z\frac{d}{dz}\{\ln T(z)\}|_{z=1}. \tag{4.9}$$

As is expected from the physical point of view, the vacuum vector $|\mathrm{vac}\rangle$ is translationally invariant and singlet (i.e. belong to the trivial representation of $U_q(A_2^{(2)})$)

$$\begin{aligned}&T|\mathrm{vac}\rangle=|\mathrm{vac}\rangle,\\&x.|\mathrm{vac}\rangle=\epsilon(x)|\mathrm{vac}\rangle.\end{aligned} \tag{4.10}$$

This is proved as follows:

$$\begin{aligned}T(z)|\mathrm{vac}\rangle &= \chi^{-\frac{1}{2}}g^{-1}\sum_i \Phi_i(z)(-q)^{-\hat{\rho}}\Phi_i^*(-zq^3)\\
&\stackrel{\mathrm{def}}{=}\chi^{-\frac{1}{2}}g^{-1}\sum_i \Phi_i(z)(-q)^{-3d}(-q^{\frac{1}{2}})^{-\frac{1}{2}h_1}\Phi_i^*(-zq^3)\\
&=\chi^{-\frac{1}{2}}g^{-1}\sum_i \Phi_i(z)\Phi_i^*(zq^6)(-q^{\frac{1}{2}})^{\rho_i}(-q)^{-\hat{\rho}}\\
&\stackrel{\mathrm{def}}{=}\chi^{-\frac{1}{2}}g^{-1}\sum_i \Phi_{-i}^*(-zq^3)\Phi_{-i}(-zq^3)(-q^{\frac{1}{2}})^{-\rho_{-i}}(-q)^{-\hat{\rho}}\\
&=\chi^{-\frac{1}{2}}(-q)^{-\hat{\rho}}=|\mathrm{vac}\rangle,\end{aligned}$$

where we have used the fact $\rho_i=-\rho_{-i}$, (2.13) and (3.13). Similarly,

$$\begin{aligned}x.|\mathrm{vac}\rangle&=\chi^{-\frac{1}{2}}\sum x_{(1)}(-q)^{\hat{\rho}}b(x_{(2)})\\
&=\chi^{-\frac{1}{2}}\sum x_{(1)}S(x_{(2)})(-q)^{\hat{\rho}}\\
&=\epsilon(x)|\mathrm{vac}\rangle.\end{aligned}$$

In the third line we have used the axioms of Hopf algebra [4]

$$m\circ(\mathrm{id}\otimes S)\circ\Delta=\epsilon.$$

This completes the proof.

For any local operator O on $\mathbf{H}$ defined in (4.6), its vacuum expectation value is given by

$$\langle \mathrm{vac}|\mathrm{O}|\mathrm{vac}\rangle \overset{\mathrm{def}}{=} \langle \mathrm{O}\rangle = \frac{\mathrm{tr}_{V(\Lambda_1)}(q^{-2\hat{\rho}}\mathrm{O})}{\mathrm{tr}_{V(\Lambda_1)}(q^{-2\hat{\rho}})} = \frac{\mathrm{tr}_{V(\Lambda_1)}(q^{-6d-\frac{1}{2}h_1}\mathrm{O})}{\mathrm{tr}_{V(\Lambda_1)}(q^{-6d-\frac{1}{2}h_1})},$$

where the normalization $\langle \mathrm{vac}|\mathrm{vac}\rangle = 1$ has been chosen. We shall denote the correlator $\langle \mathrm{vac}|\mathrm{O}|\mathrm{vac}\rangle$ by $\langle O\rangle$.

By Proposition 4 and the definition of the local operators $E_{i,j}^{(n)}$ (4.6), we have

$$\langle E_{i,j}^{(n)}\rangle = \langle \mathrm{vac}|T^{-(n-1)}E_{i,j}T^{(n-1)}|\mathrm{vac}\rangle = \langle \mathrm{vac}|E_{i,j}|\mathrm{vac}\rangle = \langle E_{i,j}\rangle$$

4.3. The n-particle states and form factors

In order to construct the general eigenstates of the transfer matrix of the Izergin–Korepin model (4.8), we employ the type II vertex operators. Define the n-particle states

$$\begin{aligned}|\xi_n,\xi_{n-1},\cdots,\xi_1\rangle_{i_n,i_{n-1},\cdots,i_1} &= \bar{g}^{-\frac{n}{2}}\Psi_{i_n}^*(\xi_n)\Psi_{i_{n-1}}^*(\xi_{n-1})\cdots\Psi_{i_1}^*(\xi_1)|\mathrm{vac}\rangle \\ &= \bar{g}^{-\frac{n}{2}}\chi^{-\frac{1}{2}}\Psi_{i_n}^*(\xi_n)\Psi_{i_{n-1}}^*(\xi_{n-1})\cdots\Psi_{i_1}^*(\xi_1)(-q)^{-\hat{\rho}},\end{aligned} \tag{4.11}$$

and its dual states

$${}_{i_n,i_{n-1},\cdots,i_1}\langle \xi_n,\xi_{n-1},\cdots,\xi_1| = \bar{g}^{-\frac{n}{2}}\chi^{-\frac{1}{2}}(-q)^{-\hat{\rho}}\Psi_{i_1}(\xi_1)\Psi_{i_2}(\xi_2)\cdots\Psi_{i_n}(\xi_n). \tag{4.12}$$

Using the commutation relation of the type I vertex operators,(3.12), one can verify that

$$\begin{aligned}&T(z)|\xi_n,\cdots,\xi_1\rangle_{i_n,i_{n-1},\cdots,i_1} \\ &= g^{-1}\bar{g}^{-\frac{n}{2}}\chi^{-\frac{1}{2}}\sum_i \Phi_i(z)\Psi_{i_n}^*(\xi_n)\Psi_{i_{n-1}}^*(\xi_{n-1})\cdots\Psi_{i_1}^*(\xi_1)(-q)^{-\hat{\rho}}\Phi_i^*(-zq^3) \\ &= g^{-1}\bar{g}^{-\frac{n}{2}}\chi^{-\frac{1}{2}}\sum_i \Phi_i(z)\Psi_{i_n}^*(\xi_n)\Psi_{i_{n-1}}^*(\xi_{n-1})\cdots\Psi_{i_1}^*(\xi_1)\Phi_i^*(zq^6)(-q^{\frac{1}{2}})^{\rho_i}(-q)^{-\hat{\rho}} \\ &= g^{-1}\bar{g}^{-\frac{n}{2}}\chi^{-\frac{1}{2}}\sum_i\prod_{j=1}^n \tau(\xi_j/z)\Phi_i(z)\Phi_i^*(zq^6)(-q^{\frac{1}{2}})^{\rho_i}\Psi_{i_n}^*(\xi_n)\Psi_{i_{n-1}}^*(\xi_{n-1}) \\ &\quad\times\cdots\Psi_{i_1}^*(\xi_1)(-q)^{-\hat{\rho}} \\ &= \prod_{j=1}^n \tau(\xi_j/z)|\xi_n,\cdots,\xi_1\rangle_{i_n,i_{n-1},\cdots,i_1}.\end{aligned}$$

Here we have used $\tau(zq^6) = \tau(z)$. Therefore, the n-particle states (4.11) are the eigenstates of transfer matrix $T(z)$. Likewise for its dual states (4.12). Hence, the Hamiltonian H (4.9) on the n-particle states are given by

$$H|\xi_n, \cdots, \xi_1\rangle_{i_n, i_{n-1}, \cdots, i_1} = \sum_{j=1}^{n} \epsilon(\xi_j)|\xi_n, \cdots, \xi_1\rangle_{i_n, i_{n-1}, \cdots, i_1},$$

where

$$\epsilon(z) = -z\frac{d}{dz}\ln \tau(z).$$

These results coincide with the energy of the elementary excitations derived from the Bethe ansatz method [23].

In much the same way as the correlation functions, the form factors of a local operator O of the form $O = A \otimes B$ can be given by

$$\langle \mathrm{vac}|\mathrm{O}|\xi_n, \cdots, \xi_1\rangle_{i_n, i_{n-1}, \cdots, i_1}$$
$$= \frac{\overline{g}^{-\frac{n}{2}} \mathrm{tr}_{V(\Lambda_1)}((-q)^{-\hat{\rho}} A \Psi^*_{i_n}(\xi_n)\Psi^*_{i_{n-1}}(\xi_{n-1}) \cdots \Psi^*_{i_1}(\xi_1)(-q)^{-\hat{\rho}} B')}{\mathrm{tr}_{V(\Lambda_1)}(q^{-2\hat{\rho}})}$$

In particular, for the spin operator S_z, we have

$$\langle \mathrm{vac}|\mathrm{S_z}|\xi_n, \cdots, \xi_1\rangle_{i_n, i_{n-1}, \cdots, i_1}$$
$$= g^{-1}[\overline{g}^{-\frac{n}{2}} \mathrm{tr}_{V(\Lambda_1)}(q^{-2\hat{\rho}}\{\Phi^*_1(1)\Phi_1(1)(-q^{\frac{1}{2}}) + \Phi^*_0(1)\Phi_0(1)$$
$$-\Phi^*_{-1}(1)\Phi_{-1}(1)(-q^{\frac{1}{2}})^{-1}\}\Psi^*_{i_n}(\xi_n)\Psi^*_{i_{n-1}}(\xi_{n-1}) \cdots \Psi^*_{i_1}(\xi_1))]$$
$$\times [\mathrm{tr}_{V(\Lambda_1)}(q^{-2\hat{\rho}})]^{-1}.$$

5. Correlation functions

The aim of this section is to calculate $\langle E_{mn}\rangle$. The generalization to the calculation of the multi-point functions and the form factors is straightforward.

Set

$$P^m_n(z_1, z_2|q) = \frac{\mathrm{tr}_{V(\Lambda_1)}(q^{-2\hat{\rho}}\Phi^*_m(z_1)\Phi_n(z_2))}{\mathrm{tr}_{V(\Lambda_1)}(q^{-2\hat{\rho}})},$$

then $\langle E_{mn}\rangle = g^{-1}P^m_n(z, z|q)(-q^{\frac{1}{2}})^{-\rho_n}$. Using the Clavelli–Shapiro technique [24] which we will present in Appendix C, we get

$$P^m_n(z_1, z_2|q) \stackrel{\mathrm{def}}{=} \delta_{mn}P_m(z_1, z_2|q) = \frac{\delta_{mn}(C_{-1})^2(C^{(-)})^2}{\chi}G\left(-\frac{z_2 q^3}{z_1}\right)$$
$$\times \oint \frac{dw}{w}\oint \frac{dw_1}{w_1}\theta\left(\frac{-z_1 z_2 q^5}{ww_1}\right)I_m(z_1, z_2; w, w_1), \tag{5.1}$$

where $\theta(z)$ is the elliptic function,

$$\theta(z) \stackrel{\text{def}}{=} \sum_{n\in Z} q^{3(n+\frac{1}{2})^2} z^n,$$

$$\begin{aligned}
&I_1(z_1,z_2;w,w_1)\\
&=\left(\frac{-z_1z_2q^5}{ww_1}\right)^{\frac{1}{2}} W_1\left(-\frac{wq^3}{z_1}\right)W_1\left(-\frac{w_1q^3}{z_1}\right)\\
&\quad\times\left\{\frac{w}{z_2q^4}W_1\left(\frac{w}{z_2}\right)W_1\left(\frac{w_1}{z_2}\right)H\left(\frac{w_1}{w}\right)+\frac{z_2q^5}{w}W_2\left(\frac{z_2}{w_1}\right)W_2\left(\frac{z_2}{w}\right)H\left(\frac{w}{w_1}\right)\right.\\
&\quad\left.-qW_2\left(\frac{z_2}{w}\right)W_1\left(\frac{w_1}{z_2}\right)H\left(\frac{w_1}{w}\right)-W_2\left(\frac{z_2}{w_1}\right)W_1\left(\frac{w}{z_2}\right)H\left(\frac{w}{w_1}\right)\right\},\\
&I_0(z_1,z_2;w,w_1)\\
&=\left(\frac{-z_1z_2q^5}{ww_1}\right)^{\frac{1}{2}} W_1\left(-\frac{wq^3}{z_1}\right)W_2\left(\frac{z_2}{w}\right)H\left(\frac{w_1}{w}\right)\\
&\quad\times\left\{\frac{z_1q^2}{w}W_2\left(-\frac{z_1}{wq^3}\right)W_1\left(\frac{w_1}{z_2}\right)-\frac{z_2q^5}{w_1}W_1\left(-\frac{w_1q^3}{z_1}\right)W_2\left(\frac{z_2}{w_1}\right)\right.\\
&\quad\left.+W_1\left(-\frac{wq^3}{z_1}\right)W_1\left(\frac{w_1}{z_2}\right)+q^2\left(\frac{-z_1z_2q^5}{ww_1}\right)^{\frac{1}{2}}W_2\left(-\frac{z_1}{wq^3}\right)W_2\left(\frac{z_2}{w_1}\right)\right\},\\
&I_{-1}(z_1,z_2;w,w_1)\\
&=-\left(\frac{-z_1z_2q^5}{ww_1}\right)^{\frac{3}{2}} W_2\left(\frac{z_2}{w}\right)W_2\left(\frac{z_2}{w_1}\right)\\
&\quad\times\left\{qW_2\left(-\frac{z_1}{wq^3}\right)W_1\left(-\frac{w_1q^3}{z_1}\right)H\left(\frac{w_1}{w}\right)\right.\\
&\quad+W_2\left(-\frac{z_1}{w_1q^3}\right)W_1\left(-\frac{wq^3}{z_1}\right)H\left(\frac{w}{w_1}\right)\\
&\quad+\frac{w}{z_1q}W_1\left(-\frac{wq^3}{z_1}\right)W_1\left(-\frac{w_1q^3}{z_1}\right)H\left(\frac{w_1}{w}\right)\\
&\quad\left.+\frac{z_1q^2}{w}W_2\left(-\frac{z_1}{w_1q^3}\right)W_2\left(-\frac{z_1}{wq^3}\right)H\left(\frac{w}{w_1}\right)\right\}
\end{aligned}$$

and the functions $G(z)$, $W_1(z)$, $W_2(z)$, $H(z)$ and the constant C_{-1}, $C^{(-)}$ are given in Appendix D (C.1)–(C.4).

In particular, the correlation function of spin operator S_z is given by

$$\langle \mathrm{vac}|\mathrm{S_z}|\mathrm{vac}\rangle = \mathrm{P_1(z,z|q)(-q^{-\frac{1}{2}}) + P_0(z,z|q) - P_{-1}(z,z|q)(-q^{-\frac{1}{2}})^{-1}}, \quad (5.2)$$

where $P_m(z_1,z_2|q)$ are defined in (5.1).

We now derive the difference equations satisfied by these one-point functions. Noting (3.5) and using the cyclicity of trace, we get the difference equations of $P_m(z_1,z_2|q)$

$$P_m(z_1,z_2|q) = \chi^{-1}\mathrm{tr}_{V(\Lambda_1)}(q^{-6d-\frac{1}{2}h_1}\Phi_m^*(z_1)\Phi_m(z_2))$$

$$
\begin{aligned}
&= \chi^{-1}\mathrm{tr}_{V(\Lambda_1)}(q^{-6d-\frac{1}{2}h_1}\Phi_{-m}(-z_1q^{-3})\Phi_m(z_2))\\
&= \chi^{-1}\sum_n R^l(z_2,-z_1q^{-3})^{n,-n}_{m,-m}\mathrm{tr}_{V(\Lambda_1)}(q^{-6d-\frac{1}{2}h_1}\Phi_n(z_2)\Phi^*_n(z_1))\\
&= \chi^{-1}\sum_n q^{-\rho_n}R^l(z_2,-z_1q^{-3})^{n,-n}_{m,-m}\mathrm{tr}_{V(\Lambda_1)}(\Phi_n(z_2q^6)q^{-6d-\frac{1}{2}h_1}\Phi^*_n(z_1))\\
&= \sum_n q^{-\rho_n}R^l(z_2,-z_1q^{-3})^{n,-n}_{m,-m}P_n(z_1,z_2q^6|q)\,.
\end{aligned}
$$

Acknowledgements

W.L. Yang thanks Prof. Fan for fruitful discussions. Y.-Z. Zhang would like to thank the Australian Research Council IREX programme for an Asia–Pacific Link Award and Institute of Modern Physics of Northwest University for hospitality. The financial support from the National Natural Science Foundation of China, and the Australian Research Council large, small and QEII fellowship grants is also gratefully acknowledged.

Appendix A

In this appendix, we give the normal order relations of fundamental bosonic fields:

$$
\begin{aligned}
&\Phi_{-1}(z)\Phi_{-1}(w) = zq^4 g(w/z) : \Phi_{-1}(z)\Phi_{-1}(w) :,\\
&\Phi_{-1}(z)X^-(w) = \frac{1}{zq^4(1+\frac{w}{zq^3})} : \Phi_{-1}(z)X^-(w) :,\\
&X^-(w)\Phi_{-1}(z) = \frac{1}{w(1+\frac{zq^5}{w})} : \Phi_{-1}(z)X^-(w) :,\\
&\Phi_{-1}(z)X^+(w) = (zq^4+w) : \Phi_{-1}(z)X^+(w) : X^+(w)\Phi_{-1}(z),\\
&\Psi_1(z)\Psi_1(w) = -zq\bar{g}(w/z) : \Psi_1(z)\Psi_1(w) :,\\
&\Psi_1(z)\Phi_{-1}(w) = -(zq)^{-1}h_1(w/z) : \Psi_1(z)\Phi_{-1}(w) :,\\
&\Phi_{-1}(z)\Psi_1(w) = (zq^4)^{-1}h_2(w/z) : \Phi_{-1}(z)\Psi_1(w) :,\\
&\Psi_1(z)X^-(w) = (w-zq) : \Phi_{-1}(z)X^-(w) :, \qquad X^-(w)\Psi_1(z),\\
&X^+(w)\Psi_1(z) = \frac{1}{w-z} : \Psi_1(z)X^+(w) :,\\
&\Psi_1(z)X^+(w) = -\frac{1}{zq(1-\frac{w}{zq^2})} : \Psi_1(z)X^+(w) :,
\end{aligned}
$$

where

$$
g(z) = \frac{(-q^3z;q^6)(q^2z;q^6)}{(q^6z;q^6)(-q^5z;q^6)}, \qquad \bar{g}(z) = \frac{(-qz;q^6)(z;q^6)}{(q^4z;q^6)(-q^3z;q^6)},
$$

$$h_1(z)=\frac{(-q^8z;q^6)(q^7z;q^6)}{(q^5z;q^6)(-q^4z;q^6)},\qquad h_2(z)=\frac{(-q^2z;q^6)(qz;q^6)}{(q^{-1}z;q^6)(-q^{-2}z;q^6)},$$

and

$$X^+(z)X^+(w)=\frac{(z-w)(z-wq^{-2})}{z+wq^{-1}}:X^+(z)X^+(w):,$$

$$X^-(z)X^-(w)=\frac{(z-w)(z-wq^2)}{z+wq}:X^-(z)X^-(w):,$$

$$X^+(z)X^-(w)=\frac{z+w}{(z-wq)(z-wq^{-1})}:X^+(z)X^+(w):,$$

$$X^-(w)X^+(z)=\frac{z+w}{(w-zq)(w-zq^{-1})}:X^+(z)X^+(w):.$$

Appendix B

By means of the bosonic realization of $U_q(A_2^{(2)})$, the integral expressions of the vertex operators and the technique given in Ref. [10], one can check the following relations:

- For the type I vertex operators

$$[\Phi_1(z),f_1]_{q^{-1}}=0,\qquad \Phi_1(z)=\frac{1}{\alpha}[\Phi_0(z),f_1],$$

$$\Phi_0(z)=\frac{\alpha}{[2]_{q_1}}[\Phi_{-1}(z),f_1]_q,$$

$$[\Phi_1(z),e_1]=\frac{[2]_{q_1}}{\alpha}t_1\Phi_0(z),\qquad [\Phi_0(z),e_1]=\alpha t_1\Phi_{-1}(z),$$

$$[\Phi_{-1}(z),e_1]=0,$$

$$\Phi_1(z)t_1=qt_1\Phi_1(z),\qquad \Phi_0(z)t_1=t_1\Phi_0(z),$$

$$\Phi_{-1}(z)t_1=q^{-1}t_1\Phi_{-1}(z),$$

where we take $\alpha=\{[2]_{q_1}q^{-\frac{1}{2}}\}^{\frac{1}{2}}$.

- For the type II vertex operators

$$\Psi_{-1}(z)=\frac{1}{\alpha}[\Psi_0(z),e_1],\qquad \Psi_0(z)=\frac{\alpha}{[2]_{q_1}}[\Psi_1(z),e_1]_q,$$

$$[\Psi_{-1}(z),e_1]_{q^{-1}}=0,$$

$$[\Psi_1(z),f_1]=0,\qquad [\Psi_0(z),f_1]=\alpha t_1^{-1}\Psi_1(z),$$

$$[\Psi_{-1}(z),f_1]=\frac{[2]_{q_1}}{\alpha}t_1^{-1}\Psi_0(z),$$

$$\Psi_1(z)t_1=qt_1\Psi_1(z),\qquad \Psi_0(z)t_1=t_1\Psi_0(z),$$

$$\Psi_{-1}(z)t_1=q^{-1}t_1\Psi_{-1}(z).$$

Appendix C

In computation of the correlation functions, one encounters the trace of the form

$$\mathrm{tr}\left[x^{-d}\exp\left(\sum_{m=1}^{\infty}A_m a_{-m}\right)\exp\left(\sum_{m=1}^{\infty}B_m a_m\right)f^P\right],$$

where A_m,B_m and f are all some coefficients. We can calculate the contributions from the oscillators modes and the zero modes *separately*. The trace over the oscillator modes can be carried out as follows by using the Clavelli–Shapiro technique [24]. Let us introduce the extra oscillators a'_m which commutate with a_m. a'_m satisfy the same commutation relations as those satisfied by a_m. Introduce the operators

$$H_m=\frac{a_m\otimes 1}{1-x^m}+1\otimes a'_{-m},\quad m>0,$$

$$H_m=a_m\otimes 1+\frac{1\otimes a'_{-m}}{x^m-1},\quad m<0,$$

which act on the space of the tensor Fock spaces of $\{a_m\}$ and $\{a'_m\}$. Then for any bosonic operator $O(a_m)$, one can show

$$\mathrm{tr}(x^{-d}O(a_m))=\frac{\langle 0|O(H_m)|0\rangle}{\prod_{n=1}(1-x^n)}$$

providing that d satisfies the derivation properties (2.4). We write $\langle 0|O(H_m)|0\rangle\equiv\langle\langle O(a_m)\rangle\rangle$. Then by Wick's theorem, one obtains

$$\langle\langle\Phi_{-1}(z)\Phi_{-1}(w)\rangle\rangle=C_{-1}C_{-1}G(w/z),$$

$$\langle\langle\Phi_{-1}(z)X^-(w)\rangle\rangle=C_{-1}C^{(-)}W_1(w/z),$$

$$\langle\langle X^-(z)\Phi_{-1}(w)\rangle\rangle=C_{-1}C^{(-)}W_2(w/z),$$

$$\langle\langle X^-(z)X^-(w)\rangle\rangle=C^{(-)}C^{(-)}H(w/z),$$

where

$$C_{-1}=\frac{\{-q^9\}\{q^8\}}{\{q^{12}\}\{-q^{11}\}},\qquad C^{(-)}=\frac{(q^6;q^6)(q^8;q^6)}{(-q^7;q^6)},\tag{C.1}$$

$$G(z)=\frac{\{-q^3z\}\{q^2z\}\{-q^9z^{-1}\}\{q^8z^{-1}\}}{\{q^6z\}\{-q^5z\}\{q^{12}z^{-1}\}\{-q^{11}z^{-1}\}},\tag{C.2}$$

$$W_1(z)=\frac{(-q^{11}z^{-1};q^6)}{(-q^{-3}z;q^6)},\qquad W_2(z)=\frac{(-q^3z^{-1};q^6)}{(-q^5z;q^6)},\tag{C.3}$$

$$H(z)=\frac{(q^2z;q^6)(z;q^6)(q^8z^{-1};q^6)(q^6z^{-1};q^6)}{(-qz;q^6)(-q^7z^{-1};q^6)}.\tag{C.4}$$

Now, we use the above technique to calculate the character of $U_q(A_2^{(2)})$ module $V(\Lambda_1)$

$$\chi = \mathrm{tr}_{V(\Lambda_1)}(q^{-2\hat{\rho}}) = \mathrm{tr}_{F_1}(q^{-6d}q^{-\frac{1}{2}h_1}) = \mathrm{tr}_{F_1}(q^{-6d}q^{-P})$$
$$= \frac{1}{\prod_{n=1}^{\infty}(1-q^{6n})}\sum_{n\in Z} q^{3(n+\frac{1}{2})^2}q^{-(n+\frac{1}{2})}$$
$$= q^{\frac{1}{4}}(1+q)\prod_{n=1}^{\infty}(1+q^{6n-1})(1+q^{6n+1}) .$$

We have used the Jacobi tripe product identity [25]

$$\sum_{n\in Z} q^{(n+v-\frac{1}{2})^2}t^n = q^{\frac{1}{12}}q^{v^2-v+\frac{1}{6}}\prod_{n=1}^{\infty}(1-q^{2n})(1+q^{2(n+v-1)}t)(1+q^{2(n-v)}t^{-1})$$

References

[1] I.B. Frenkel and N. Yu. Reshetikhin, Commun. Math. Phys. 146 (1992) 1.
[2] I.B. Frenkel and N. Jing, Proc. Nat'l. Acad. Sci. USA 85 (1988) 9373.
[3] B. Davies, O. Foda, M. Jimbo, T. Miwa and A. Nakayashiki, Commun. Math. Phys.151 (1993) 89.
[4] M. Jimbo and T. Miwa, Algebraic Analysis of Solvable Lattice Models, CBMS Regional Conference Series in Mathematics, Vol. 85, AMS, 1994.
[5] M. Idzumi, Int. J. Mod. Phys. A 9 (1994) 449.
[6] H. Bougurzi and R. Weston, Nucl. Phys. B 417 (1994) 439.
[7] J. Hong, S.J. Kang, T. Miwa and R. Weston, J. Phys. A 31 (1998) L515.
[8] Y. Koyama, Commun. Math. Phys. 164(1994) 277.
[9] S. Lukyanov and Y. Pugai, Nucl. Phys. B 473 (1996) 631.
[10] Y. Asai, M. Jimbo,T. Miwa and Y. Pugai, J. Phys. A 29 (1996) 6595.
[11] W.L. Yang and Y.-Z. Zhang, Nucl. Phys. B 547 (1999) 599.
[12] A.G. Izergin and V.E. Korepin, Commun. Math. Phys. 79 (1981) 303.
[13] N. Yu. Reshetikhin, Lett. Math. Phys. 7 (1993) 205.
[14] L. Mezincescu and R.I. Nepomechie, Nucl. Phys. 372 (1992) 597; Int. J. Mod. Phys. A 6 (1991) 5231.
[15] A. Kuniba, Nucl. Phys. B 335 (1991) 801.
[16] M.J. Martins, Nucl. Phys. B 450 (1995) 768.
[17] N.H. Jing and K.C. Misra, Vertex operators for twisted quantum affine algebras, e-print q-alg/9701034.
[18] V. Chari and A. Pressley, Commun. Math. Phys. 196 (1998) 461.
[19] G.W. Delius, M.D. Gould and Y.-Z. Zhang, Int. J. Mod. Phys. A 11 (1996) 3415.
[20] V.G. Drinfeld, Sov. Math. Dokl. 36 (1988) 212.
[21] N. Yu. Reshetikhin, M.A. Semenov-Tian-Shansky, Lett. Math. Phys. 19 (1990) 133.
[22] W.L. Yang and Y.-Z. Zhang, The twisted quantum affine algebra $U_q(A_2^{(2)})$ and its bosonization at any level., in preparation.
[23] V.I. Vichirko and N. Yu. Reshetikhin, Thero. Math. Phys. Vol. 56, No. 2 (1983) 260.
[24] L. Clavelli and J.A. Shapiro, Nucl. Phys. B 57 (1973) 490.
[25] D. Mumford, Tata Lectures on the Theta Function (Birhauser, Basel, 1983).

ELSEVIER

28 June 1999

PHYSICS LETTERS A

Physics Letters A 257 (1999) 189–194

Impurity of arbitrary spin embed in the 1-D Hubbard model with open boundary conditions

Bo-yu Hou, Xiao-qiang Xi, Rui-hong Yue

Institute of Modern Physics, Northwest University P.O. Box 105, Xian 710069, China

Received 18 January 1999; received in revised form 11 May 1999; accepted 14 May 1999
Communicated by A.R. Bishop

Abstract

In this letter we study the exact solution of the 1-D Hubbard model with arbitrary spin impurity under the open boundary conditions by the coordinate Bethe ansatz method, then derive the eigenvalue and the corresponding Bethe ansatz equations © 1999 Published by Elsevier Science B.V. All rights reserved.

PACS: 75.30.Hx; 71.10.Fd; 71.27.+a

In recent years, strongly correlated systems have been of great interest because of the study in the field of high-temperature superconductors. In [1,2], we find that their importance as the most fundamental systems related to the theory of high-temperature superconductors, besides this, these electron systems pose many theoretical challenges due to their essentially non-perturbative nature. The Hubbard chain and t–J model are frequently invoked as the basic models of the 1-D strongly correlated electron systems. Especially, when the repulsion energy U for two electrons located on the same atom is much larger than the bandwidth of the electrons, the Hubbard model reduces to the t–J model with an occupation of one electron per site.

Recently great progress has been made for the 1-D systems [3]. In particular for the quantum impurity problem such as the Kondo problem and tunneling in quantum wires. The Kondo effect [4] and its study told us that the impurities play an important role in the strongly correlated electron compounds. Even a small amount of defects may change the properties of the electron system.

Exact results even for a simplified model exploring just some features of the system are always useful and provide a testing ground for approaches intended for the full problem of higher order of complexity. Therefore it is of interest to construct integrable systems of the strongly correlated systems including impurities. Indeed, there has been much

PII: S0375-9601(99)00324-2

development in studying the effects of the impurities in the many-body quantum systems within the framework of the integrable systems [5–9], although the impurity usually destroys the integrability of the model when it is introduced into an exactly solved model. Andrei and Johanesson incorporated a magnetic impurity of the arbitrary spin into the isotropic spin-$\frac{1}{2}$ Heisenberg chain with integrability preserved, the results were extended to the Babujian–Takhtajan spin-chain in Refs. [10]. Wang and Voit give the exact result of the single impurity Kondo problem in the 1D δ-potential Fermi gas [11]. In Refs. [12], the integrable model with the impurity coupled with periodic t–J chain was solved. The supersymmetric t–J model [13] with impurity was studied in Refs. [14–17]. In Refs. [18–20], the open boundary t–J model with magnetic impurities was discussed. Zvyagin and Schlottmann studied the magnetic impurities embedded in the Hubbard model [21]. In Refs. [22], Bares and Grzegorczyk have studied periodic Hubbard chain with a Kondo impurity.

In this letter, we construct a Hamiltonian of the impurity embedded in the Hubbard model with open boundary conditions [23]. The impurity is coupled to the strongly correlated electron system and has arbitrary spin. The boundary R matrices are similar to those in Refs. [23] and satisfy the reflection equation. The electron-impurity scattering matrix is given explicitly and satisfies the Yang–Baxter equation. The two conditions ensured the integrability of our impurity model. The interaction parameters of the impurity with the electron are parameterized by the constants which are related to the potential of the impurity in order to contain the integrability of the system. The Hamiltonian with the impurity is diagonalized exactly by using the coordinate Bethe ansatz method. The Bethe ansatz equations are obtained explicitly by comparing the transfer matrix with the pioneering work of Sklyanin [24].

The 1-D Hubbard model is one of the most important solvable models in condensed matter physics. It describes interacting electrons on 1-D lattice. The low-excitation spectra depends on the parameters of the model such as the Coulomb repulsion U, the band-width t, the chemical potential μ and the magnetic field h [25–27]. After the impurity embedded in the Hubbard model with general open boundary conditions, we introduce the Hamiltonian in as follows:

$$\begin{aligned}\mathscr{H} = & -t \sum_{j=-L}^{L-1} \sum_{\sigma=\uparrow\downarrow} \left(c_{j\sigma}^{+} c_{j+1\sigma} + c_{j+1\sigma}^{+} c_{j\sigma} \right) \\ & + U \sum_{j=-L}^{L} n_{j\uparrow} n_{j\downarrow} + \mu \sum_{j=-L}^{L} (n_{j\uparrow} + n_{j\downarrow}) \\ & - \frac{h}{2} \sum_{j=-L}^{L} (n_{j\uparrow} - n_{j\downarrow}) \\ & + \sum_{\sigma=\uparrow\downarrow} (P_{-L\sigma} n_{-L\sigma} + P_{L\sigma} n_{L\sigma}) \\ & + V \sum_{\sigma=\uparrow\downarrow} n_{0\sigma} + J \boldsymbol{\sigma} \cdot \mathbf{S}_{\rm im}, \end{aligned} \tag{1}$$

where $c_{i\sigma}^{+}$ creates an electron of spin σ at site i and $n_{i\sigma} = c_{i\sigma}^{+} c_{i\sigma}$. $P_{-L\sigma}$ and $P_{L\sigma}$ are the free parameters describing the boundary external fields, J the exchange coupling the localized spin-S impurity to the conduction electrons and V a local charge interaction. $\boldsymbol{\sigma}$ stands for the spin of electron defined by $(c_{0\uparrow}^{+}, c_{0\downarrow}^{+})\ \boldsymbol{\sigma} (c_{0\uparrow}, c_{0\downarrow})^{T}$. Hereafter we shall assume $t = 1$.

We know that when impurity is embedded in the 1-D integrable model, the integrability usually is destroyed. Without the integrability, we can not use the standard Bethe ansatz method [24] to derive the eigenvalue and the Bethe ansatz equations. The following is what we do to ensure the integrability.

To solve the eigenvalue problem associated with the Hamiltonian of Eq. (1), we make the following eigenstate with N electrons and M down-spin electrons

$$\Psi_{NM} = \sum_{\{x_j\}} f(x_1, \cdots, x_N) c_{x_1\sigma_1}^{+} \cdots c_{x_N\sigma_N}^{+} |\mathrm{vac}\rangle, \tag{2}$$

here, the x_j and σ_j denote the position and the spin of electrons respectively. We note that the wave function f also depends on the spin variables $\{\sigma_1, \cdots, \sigma_N\}$. The wave function f takes the form

$$\begin{aligned} f(x_1, \cdots, x_N) & = \sum_{Q,P} \epsilon_P \Theta_0(x_{\tilde{Q}}) \Theta(x_Q) A_{\sigma_{Q_1} \cdots \sigma_{Q_N}} \\ & \times \left(\tilde{Q} | k_{p_1}, \cdots, k_{p_N} \right) \exp\left\{ i \sum_{j=1}^{N} k_{p_j} x_{q_j} \right\}, \end{aligned} \tag{3}$$

where Q runs over S_N, the permutation group of N particles, and P over all the permutations and the ways of negations of k's. There are $N!\times 2^N$ possibilities for P, while $N!$ for Q. We recall that ϵ_P denotes the sign of P. If the permutation is even, P makes $\epsilon_P=-1$ for odd number of k's negative and $\epsilon_P=1$ for even number of k's negative. $\Theta(x_Q)=\Theta(x_{Q_1}<\cdots<x_{Q_N})=1$ when the ordering associated to the permutation Q of the coordinates is respected and otherwise zero. $\Theta_0(x_{\tilde{Q}})=\prod_j^N\Theta(\nu_j x_j)$, where $\nu_j(\tilde{Q})=\pm 1$ according to whether the coordinate is to the left or to the right of the impurity as specified by the permutation $\tilde{Q}$ of $N+1$ objects, and where $\Theta(x)=1$ if $x>0$ and otherwise zero with the prescription $\Theta(0)=\frac{1}{2}$.

First, let us consider the case of one-particle state. The eigenstate can be assumed to be

$$\Psi_1=\sum_{x=-L}^{L} f(x)c_{x\sigma}^{+}|\mathrm{vac}\rangle. \tag{4}$$

Substituting this ansatz the into Schrödinger equation $H\Psi=E\Psi$, one can get

$$\begin{aligned}
Ef(x)&=-f(x+1)-f(x-1)\\
&\quad+f(x)\left[\mu-\frac{h}{2}(1-2M)\right],\\
&0<|x|\le L-1,\\
Ef(-L)&=-f(-L+1)\\
&\quad+f(-L)\left[\mu-\frac{h}{2}(1-2M)+P_{-L\sigma}\right],\\
Ef(L)&=-f(L-1)\\
&\quad+f(L)\left[\mu-\frac{h}{2}(1-2M)+P_{L\sigma}\right],\\
Ef(0)&=-f(1)-f(-1)\\
&\quad+f(0)\left[\mu-\frac{h}{2}(1-2M)+(\tilde{V}+JP^{js})\right],
\end{aligned} \tag{5}$$

where $\tilde{V}=V-\frac{J}{2}$, $P^{js}=\frac{1}{2}+\boldsymbol{\sigma}\cdot\mathbf{S}_{\mathrm{im}}$. We assume the wave function $f(x)$ to be

$$f(x)=\begin{cases}A_\sigma(+|k)e^{ikx}-A_\sigma(+|-k)e^{-ikx} & x>0\\ A_\sigma(-|k)e^{ikx}-A_\sigma(-|-k)e^{-ikx} & x<0.\\ \frac{1}{2}[f(0^+)+f(0^-)] & x=0\end{cases} \tag{6}$$

From this ansatz and Eq. (5), we have

$$\begin{aligned}
&A_\sigma(-|k)\alpha_\sigma(-k)=A_\sigma(-|-k)\alpha_\sigma(k),\\
&A_\sigma(+|k)\beta_\sigma(k)=A_\sigma(+|-k)\beta_\sigma(-k),\\
&\alpha_\sigma(k)=\left(P_{-L\sigma}+e^{ik}\right)e^{ikL},\\
&\beta_\sigma(k)=\left(P_{L\sigma}+e^{ik}\right)e^{ikL},\\
&A_\sigma(+|k)\left[2i\sin(k)-(\tilde{V}+JP^{js})\right]\\
&\quad=A_\sigma(-|k)\left[2i\sin(k)+(\tilde{V}+JP^{js})\right],\\
&A_\sigma(+|-k)\left[2i\sin(k)+(\tilde{V}+JP^{js})\right]\\
&\quad=A_\sigma(-|-k)\left[2i\sin(k)-(\tilde{V}+JP^{js})\right].
\end{aligned} \tag{7}$$

Next let us consider the general eigenstates (2). Substituting the ansatz (2) and (3) into the Schrödinger equation, we can get the energy

$$E=N\mu-\tfrac{1}{2}h(N-2M)-2\sum_{j=1}^{N}\cos k_j. \tag{8}$$

and the following relations [1]:

$$\begin{aligned}
&A_{\cdots\sigma_j\sigma_{j+1}\cdots}\left(\pm|\cdots,k_{p_j},k_{p_{j+1}},\cdots\right)\\
&\quad=\sum Y_{k_{p_j}k_{p_{j+1}}}^{\sigma_j'\sigma_{j+1}'}A_{\cdots\sigma_j'\sigma_{j+1}'\cdots}\\
&\qquad\times\left(\pm|\cdots,k_{p_{j+1}},k_{p_j},\cdots\right),
\end{aligned} \tag{9}$$

$$\begin{aligned}
&A_{\cdots\sigma_j\cdots}\left(\pm|\cdots,k_{p_j},\cdots\right)\\
&\quad=R_{js}^{\pm}\left(k_{p_j}\right)A_{\cdots\sigma_j\cdots}\left(\mp|\cdots,k_{p_j},\cdots\right),
\end{aligned} \tag{10}$$

$$\begin{aligned}
&A_{\sigma_{q_1}\cdots}\left(-|k_{p_1},\cdots\right)\\
&\quad=U_{\sigma_{q_1}}\left(k_{p_1}\right)A_{\sigma_{p_1}\cdots}\left(-|-k_{p_1},\cdots\right),
\end{aligned} \tag{11}$$

$$\begin{aligned}
&A_{\cdots\sigma_{q_N}}\left(+|\cdots,k_{p_N}\right)\\
&\quad=V_{\sigma_{q_N}}\left(-k_{p_N}\right)A_{\cdots\sigma_N}\left(+|\cdots,-k_{p_N}\right),
\end{aligned} \tag{12}$$

[1] In the following formula, for the sake of convenience, we only sign the place of k_{p_i} (other k's place is unchanged). No specific note later.

where the operator Y is defined by

$$Y^{\sigma_j\sigma_{j+1}}_{k_{p_j}k_{p_{j+1}}} = \frac{iU/2}{\sin k_{p_{j+1}} - \sin k_{p_j} + iU/2}\mathrm{I} + \frac{\sin k_{p_{j+1}} - \sin k_{p_j}}{\sin k_{p_{j+1}} - \sin k_{p_j} + iU/2}P^{\sigma_j\sigma_{j+1}}, \tag{13}$$

here $P^{\sigma_j\sigma_{j+1}}$ denotes the permutation operator acting on the spin variables σ_j's, and $R^{\pm}_{js}$ is

$$R^{\pm}_{js}(\sin k) = \frac{2i\sin k \pm (\tilde{V} + JP^{js})}{2i\sin k \mp (\tilde{V} + JP^{js})}, \tag{14}$$

the other two operators $U_\sigma(k) = \frac{\alpha_\sigma(k)}{\alpha_\sigma(-k)}, V_\sigma(k) = \frac{\beta_\sigma(k)}{\beta_\sigma(-k)}$. From (9)–(12) we get

$$A_{\sigma_1\cdots\sigma_N}(\pm|k_{p_1},\cdots,k_{p_N}) = \sum_{\sigma_1'\cdots\sigma_N'} \{U(k_{p_1})X_{\hat{1}2}X_{\hat{1}3}\cdots X_{\hat{1}s}\cdots X_{\hat{1}N}V(k_{p_1}) \times X_{N1}\cdots X_{s1}\cdots X_{31}X_{21}\}^{\sigma_1'\cdots\sigma_N'}_{\sigma_1\cdots\sigma_N} A_{\sigma_1'\cdots\sigma_N'} \times(\pm|k_{p_1},\cdots,k_{p_N}), \tag{15}$$

where

$$X_{ij} = \frac{iU/2}{\sin k_{p_i} - \sin k_{p_j} + iU/2}P_{\sigma_i\sigma_j} + \frac{\sin k_{p_i} - \sin k_{p_j}}{\sin k_{p_i} - \sin k_{p_j} + iU/2}\mathrm{I},$$

$$X_{\hat{i}j} = \frac{iU/2}{-\sin k_{p_i} - \sin k_{p_j} + iU/2}P_{\sigma_i\sigma_j} + \frac{-\sin k_{p_i} - \sin k_{p_j}}{-\sin k_{p_i} - \sin k_{p_j} + iU/2}\mathrm{I},$$

$$U(k_{p_j}) = \mathrm{diag.}\left(\frac{\alpha_\uparrow(k)}{\alpha_\uparrow(-k)}, \frac{\alpha_\downarrow(k)}{\alpha_\downarrow(-k)}\right),$$

$$V(k_{p_j}) = \mathrm{diag.}\left(\frac{\beta_\uparrow(k)}{\beta_\uparrow(-k)}, \frac{\beta_\downarrow(k)}{\beta_\downarrow(-k)}\right),$$

$$X_{\hat{i}s} = \frac{-2i\sin k_i + (\tilde{V} + JP^{js})}{-2i\sin k_i - (\tilde{V} + JP^{js})},$$

$$X_{si} = \frac{2i\sin k_i - (\tilde{V} + JP^{js})}{2i\sin k_i + (\tilde{V} + JP^{js})}. \tag{16}$$

Let us diagonalize Eq. (15). We want to determine the matrices $U(k)$'s and $V(k)$'s, the parameters $\hat{V}$ and J so that the system is integrable, i.e., Eq. (15) can be diagonalized for arbitrary sets of values of k's. The key observation is that the form of Eq. (15) is similar to that of the transfer matrix of the inhomogeneous XXX model under the open boundary condition which was discussed by Sklyanin [24], tracking his ways, we will introduce the operator T acting on the function I A later.

The first step is to determine the parameters $\tilde{V}$ and J. We know that if we want to solve the Hubbard chain with impurity in the model, the electron-impurity scattering matrix R_{js} must satisfy the Yang–Baxter equation: $R_{ij}R_{is}R_{js} = R_{js}R_{is}R_{ij}$. Then we get $J = -\frac{U}{2}$, $\tilde{V} = \pm\frac{U}{2}(s+\frac{1}{2})$. Substituting those results into R_{js}, one can get

$$R_{js}(\sin k) = \frac{\sin k}{\sin k - i\tilde{V}}\mathrm{I} - \frac{iU/2}{\sin k - i\tilde{V}}P_{js}. \tag{17}$$

Now let us show the operator T:

$$\begin{aligned}T(\sin k) = \mathrm{tr}_0 K_0^+(\sin k)L_{01}(-\sin k,\sin k_1) &\times L_{02}(-\sin k,\sin k_2)\cdots \\ &\times L_{0s}^-(-\sin k)\cdots \\ &\times L_{0N}(-\sin k,\sin k_N)K_0^-(\sin k) \\ &\times L_{0N}(\sin k,\sin k_N)\cdots \\ &\times L_{0s}^+(\sin k)\cdots \\ &\times L_{02}(\sin k,\sin k_2) \\ &\times L_{01}(\sin k,\sin k_1),\end{aligned} \tag{18}$$

where

$$L_{0j}(\sin k,\sin k_j) = \frac{\sin k - \sin k_j}{\sin k - \sin k_j + iU/2}\mathrm{I} + \frac{iU/2}{\sin k - \sin k_j + iU/2}P_{0j},$$

$$L_{0s}^{\pm}(\sin k) = \frac{\sin k}{\sin k \mp i\tilde{V}}\mathrm{I} \pm \frac{iU/2}{\sin k \mp i\tilde{V}}P_{0s},$$

$$\begin{aligned}K_0^+(\sin k) &= \frac{2\sin k + iU/2}{2\sin k(2\sin k + iU)} \\ &\quad\times \mathrm{diag.}\big((2\sin k + iU/2)U_\uparrow(k) \\ &\quad - i(U/2)U_\downarrow(k)(2\sin k + iU/2)U_\downarrow(k) \\ &\quad - i(U/2)U_\uparrow(k)\big),\end{aligned}$$

$$K_0^-(\sin k) = \mathrm{diag.}\big(V_\uparrow(k), V_\downarrow(k)\big). \tag{19}$$

In terms of the operator $T(\mu)$ Eq. (15) is given by the form

$$T(\sin k_{p_1})A(\pm|k_{p_1},\cdots,k_{p_N}) = A(\pm|k_{p_1},\cdots,k_{p_N}), \tag{20}$$

where the eigenvalue is given by 1. We note that if $T(\mu)T(\nu)=T(\mu)T(\nu)$ for any μ and ν, then the transfer matrix can be diagonalized, i.e., the model is integrable.

From the solution of the reflection equation of the XXX model [24] we find that $T(\mu)T(\nu)=T(\nu)T(\mu)$ if the $U(k)$ and $V(k)$ satisfy the following relations:

$$V_\uparrow(k)V_\downarrow(-k)=\frac{\zeta_-+\sin k}{\zeta_--\sin k},$$

$$U_\uparrow(k)U_\downarrow(-k)=\frac{\zeta_++\sin k}{\zeta_+-\sin k}. \tag{21}$$

Making use of the expression of the eigenvalue of the inhomogeneous transfer matrix for the XXX model with open boundary condition [24], we get the eigenvalue $\Lambda(\sin k)$ of $T(\sin k)$.

$$\Lambda(\sin k) = \frac{(2\sin k+iU/2)U_\downarrow(k)V_\downarrow(k)}{(2\sin k+iU)(\zeta_+-\sin k)(\zeta_--\sin k)}\tilde{\Lambda}(\sin k)$$

$$\begin{aligned}\tilde{\Lambda}(\sin k) &= \frac{2\sin k+iU}{2\sin k+iU/2}(\zeta_++\sin k)\Delta_+(\sin k+iU/4)\\ &\times\prod_{m=1}^{M}\frac{(\sin k-v_m-iU/4)(\sin k+v_m-iU/4)}{(\sin k-v_m+iU/4)(\sin k+v_m+iU/4)}\\ &-\frac{1}{2\sin k+iU/2}(\sin k-\zeta_++iU/4)\Delta_-(\sin k+iU/4)\\ &\times\prod_{m=1}^{M}\frac{(\sin k-v_m+i3U/4)(\sin k+v_m+i3U/4)}{(\sin k-v_m+iU/4)(\sin k+v_m+iU/4)},\end{aligned} \tag{22}$$

where

$$\begin{aligned}\Delta_+(x) &= (\zeta_-+x-iU/4)\times\delta_+(x)\delta_-(-x)\phi(x-iU/4),\\ \Delta_-(x) &= (\zeta_--x-iU/4)\delta_+(-x)\delta_-(x)\times\phi(x-iU/4)(2x-iU/2),\\ \delta_+(x) &= [x-(iU/2)s]\prod_{j=1}^{N}(x-\sin k_j+iU/4),\\ \delta_-(x) &= [x+(iU/2)s]\prod_{j=1}^{N}(x-\sin k_j-iU/4),\\ \phi^{-1}(x) &= (x-i\tilde{V})^2\prod_{j=1}^{N}(x-\sin k_j+iU/2)\times(-x-\sin k_j-iU/2).\end{aligned} \tag{23}$$

From the condition that $\Lambda(\sin k_{p_j})=1$ and the Bethe ansatz equations for the XXX model with open boundary, we obtain the following Bethe ansatz equations:

$$\begin{aligned}&\frac{(P_{-L\uparrow}e^{-ik_j}+1)(P_{L\uparrow}+e^{ik_j})}{(P_{-L\uparrow}e^{ik_j}+1)(P_{L\uparrow}+e^{-ik_j})}e^{i2k_j(2L+1)}\\ &=\frac{[\sin k_j\pm i(U/2)(s+1/2)]^2}{[\sin k_j+i(U/2)(s+1/2)]^2}\times\prod_{m=1}^{M}\frac{(\sin k_j-v_m+iU/4)(\sin k_j+v_m+iU/4)}{(\sin k_j-v_m-iU/4)(\sin k_j+v_m-iU/4)},\end{aligned} \tag{24}$$

$$\begin{aligned}&\frac{(\zeta_+-v_m-iU/4)(\zeta_--v_m-iU/4)}{(\zeta_++v_m-iU/4)(\zeta_-+v_m-iU/4)}\times\prod_{n\neq m}^{M}\frac{(v_m-v_n+iU/2)(v_m+v_n+iU/2)}{(v_m-v_n-iU/2)(v_m+v_n-iU/2)}\\ &=\frac{[v_m-i(U/2)s]^2}{[v_m+i(U/2)s]^2}\times\prod_{j=1}^{N}\frac{(v_m-\sin k_j+iU/4)(v_m+\sin k_j+iU/4)}{(v_m-\sin k_j-iU/4)(v_m+\sin k_j-iU/4)}.\end{aligned} \tag{25}$$

where

$$\zeta_+=\begin{cases}\infty & \text{for } P_{-L\uparrow}=P_{-L\downarrow}\\ -\dfrac{1-P_{-L\uparrow}^2}{2iP_{-L\uparrow}} & \text{for } P_{-L\uparrow}=-P_{-L\downarrow}\end{cases},$$

$$\zeta_-=\begin{cases}\infty & \text{for } P_{L\uparrow}=P_{L\downarrow}\\ -\dfrac{1-P_{L\uparrow}^2}{2iP_{L\uparrow}} & \text{for } P_{L\uparrow}=-P_{L\downarrow}\end{cases}.$$

Before ending this letter, let us give a short conclusion and discussion. In this letter, we introduced impurity of arbitrary spin to the Hubbard model with open boundary conditions. Then discussed the case of the integrability of the system, there are four kinds of boundary conditions and two kinds of impurity. i.e. there are eight cases ensured the integrability of the system. The main results of this letter are Eq. (8), the energy of the system, and Eqs. (24) and (25), the Bethe ansatz equation of k_j. From the Bethe ansatz equations one can see that there is a shift of the energy after embedding the impurity, and the shift is connected with the spin of the impurity. In a future paper, we will discuss how the impurity influences the energy in detail and consider the thermodynamic limit of this impurity system, such as the magnetic susceptibility, the specific heat etc.

References

[1] P.W. Anderson, Science 235 (1987) 1196.
[2] F.C. Zhang, T.M. Rice, Phys. Rev. B 37 (1988) 3759.
[3] P. Schlottmann, Int. J. Mod. Phys. B 11 (1997) 355.
[4] J. Kondo, Prog. Theor. Phys. 32 (1964) 37; Sol. State Phys. 23 (1969) 183.
[5] N. Andrei, H. Johnesson, Phys. Lett. A 100 (1984) 108.
[6] H.J. de Vega, L. Mezincescu, R. Nepomechie, Phys. Rev. B 49 (1994) 13223; Int. J. Mod. Phys. B 8 (1994) 3473.
[7] A.M. Tsvelick, P.B. Wiegmann, Adv. Phys. 32 (1983) 453.
[8] N. Andrei, K. Furuya, J.H. Lowenstein, Rev. Mod. Phys. 55 (1983) 331; N. Andrei, A. Jerez, Phys. Rev. Lett. 74 (1995) 4507.
[9] H. Frahm, A.A. Zvyagin, J. Phys. Cond. Matt. 9 (1997) 9939.
[10] K. Lee, P. Schlottmann, Phys. Rev. B 37 (1983) 379; P. Schlottmann, J. Phys. Cond. Matt. 3 (1991) 6617.
[11] Y. P Wang, J. Voit, Phys. Rev. Lett. 77 (1996) 4934.
[12] G. Bedurfig, F.H.L. Eßler, H. Frahm, Phys. Rev. Lett. 69 (1996) 5098; Nucl. Phys. B 498 (1997) 697.
[13] A. Foerster, J. Phys. A 29 (1996) 7625.
[14] P.-A. Bares, Exact results for a one-dimensional $t-J$ model with impurity, Cond-mat/9412011.
[15] J. Links, A. Foerster, J. Phys. A 32 (1999) 147.
[16] P. Schlottmann, A.A. Zvyagin, Phys. Rev. B 55 (1997) 5027; A.A. Zvyagin, P. Schlottmann, J. Phys. Cond. Matt. 9 (1997) 3543; E 9 (1997) 6479.
[17] P. Schlottmann, A.A. Zvyagin, Nucl. Phys. B 501 (1997) 728; A.A. Zvyagin, Phys. Rev. Lett. 79 (1997) 4661.
[18] Zhan–Ning Hu, Fu–Cho Pu, Two magnetic impurities with arbitrary spins in open boundary $t-J$ model, preprint.
[19] G. Bedürftig, H. Frahm, Open $t-J$ chain with boundary impurities, Cond-mat/9903202.
[20] H.Q. Zhou, X.Y. Ge, J. Links, M.D. Gould, Graded reflection equation algebras and integrable Kondo impurities in the one-dimensional $t-J$ model, Cond-mat/9809056.
[21] A.A. Zvyagin, P. Schlottmann, Phys. Rev. B 56 (1997) 300.
[22] P.A. Bares, K. Grzegorczyk, Hubbard Chain with a Kondo Impurity, Cond-mat/9810310.
[23] Tetsuo Deguchi, Ruihong Yue, Exact Solutions of 1-D Hubbard model with open boundary conduction and the conformal dimensions under boundary magnetic fields, Cond-mat/9704138.
[24] E.K. Sklyanin, J. Phys. A 21 (1988) 2375.
[25] E. Lieb, F.Y. Wu, Phys. Rev. Lett. 20 (1968) 1445.
[26] A.A. Ovchinnikov, Sov. Phys. JETP 30 (1997) 1160.
[27] F. Woynarovich, J. Phys. C 15 (1982) 85, 97; C 16 (1983) 5213, 6593.

Progress of Theoretical Physics Supplement No. 135, 1999 149

Elliptic Ruijsenaars-Schneider and Calogero-Moser Models Represented by Sklyanin Algebra and $sl(n)$ Gaudin Algebra

Kai CHEN,[a] Heng FAN,[b,a] Bo-yu HOU,[a] Kang-jie SHI,[a] Wen-li YANG[a] and Rui-hong YUE[a]

[a] *Institute of Modern Physics, Northwest University P. O. Box 105, Xian 710069, China*
[b] *Department of Physics, Graduate School of Science University of Tokyo, Tokyo 113-0033, Japan*

(Received May 28, 1999)

The relationship between Elliptic Ruijsenaars-Schneider (RS) and Calogero-Moser (CM) models with Sklyanin algebra is presented. Lax pair representations of the Elliptic RS and CM are reviewed. For $n = 2$ case, the eigenvalue and eigenfunction for Lamé equation are found by using the result of the Bethe ansatz method.

§1. Introduction

A general description of classical completely integrable models of n one-dimensional particles with two-body interactions $V(q_i - q_j)$ was given in Ref. 1). To each simple Lie algbra and choice of interaction, one can associate a classically completely integrable system[1)-4)] such as a rational, hyperbolic, trigonometric or elliptic CM model.

The Lax pair representation (Lax representation) of a system is a direct method of showing its integrability, and the complete set of integrals of motion can also be constructed easily. The Lax representation and its corresponding r-matrix for rational, hyperbolic and trigonometric A_{n-1} CM models was constructed by Avan el al.[2)] The Lax representation for the elliptic CM models was constructed by Krichever[5)] and the corresponding r-matrix was given by Sklyanin[6)] and Braden et al.[7)] There exists a specific feature in that the r-matrices of the Lax representations for these models turn out to be dynamical (i.e., they depend on the dynamical variables) and satisfy dynamical Yang-Baxter equations.[8),7),9),6)]

For the dynamical r-matrix, the fundamental Poisson algebra of the Lax operator, whose structural constants are given by a dynamical r-matrix, is generally no longer closed. The quantization problem and its geometrical interpretation are also difficult. Considering all of these, a non-dynamical r-matrix is found for these systems.[10),11)] The trigonometric limit of these results can be found in Ref. 42). We know the Lax representation for a completely integrable model is not unique. The different Lax representations of an integrable system are conjugate to each other (for the field system they are related by a 'gauge' transformation). The corresponding r-matrices are related by a 'gauge' transformation which is the classical dynamical twisting relation[12)] between those r-matrices.

The RS model is a relativistic generalization of a CM model. It describes a completely integrable system of n one-dimensional interacting relativistic particles. It can be related to the dynamics of solitons in some integrable relativistic field theory.[8),13),9),14)] Its discrete-time version has been connected with the Bethe ansatz equation of the solvable statistical model.[15)] Recent developments have shown that it can be obtained by a Hamiltonian reduction of the cotangent bundle of some Lie group,[16)] and can be considered as the gauged WZW theory.[17)] The Lax representation and its corresponding r-matrix for rational, hyperbolic and trigonometric A_{n-1} type RS models was constructed by Avan et al.[2)] The Lax representation for the elliptic RS models was constructed by Reuijsenaars,[18)] and the corresponding r-matrix was given by Nijhoff[15)] and Suris.[19)] The main difference between the r-matrices of the relativistic (RS) and non-relativistic (CM) models is that the latter is given in terms of a linear Poisson-Lie bracket, whereas the former (RS model) is given in terms of a quadratic Poisson-Lie bracket. In contrast with the dynamical Yang-Baxter equation of the r-matrix for the CM model,[16)] the generalized Yang-Baxter relation for the quadratic Poisson-Lie bracket (RS model) with a dynamical r-matrix is still an open problem.[20)] Moreover, the Poisson bracket of the Lax operator is no longer closed, and consequently the quantum version of the classical L-operator has not been constructed. However, as for the CM model, a different Lax representation which is conjugated to the original one can be found. The corresponding r-matrix changes by a 'gauge' transformation. The resulting r-matrix may be non-dynamical. Such a transformation may be called the classical dynamical twisting of the associated linear Poisson-Lie bracket. Due to the quadratic Poisson-Lie bracket of the RS model, there exist dynamical twisting relations between the r-matrices of Lax operators related by gauge transformations. Such dynamical twisting is the semi-classical limit of the quantum dynamical twisting of the R-matrix in Ref. 12). For recent progress in the study of CM models, see, for example, Refs. 21)-24).

The paper is organized as follows: In §2, we present some general formulae for dynamical systems. In §3, we review some results for the elliptic RS and CM models. The non-dynamical r-matrices for the integrable elliptic systems are then presented. Their quantization conditions correspond to the quantum Yang-Baxter relation, and the R-matrix is simply the Z_n-symmetric Belavin model.[28)] In §4, we will present the relationship between the Sklyanin algebra[6),32)] and the integrable systems. In §5, we will obtain the eigenvalue and eigenfunction for the Lamé equation. The Lamé operator is equivalent to the Hamiltonian of the elliptic CM model. Section 6 has some brief summary.

§2. The dynamical twisting of classical r-matrix

A Lax pair (L, M) consists of two functions on the phase space of the system with values in some Lie algebra g, such that the evolution equations may be written in the following form

$$\frac{dL}{dt} = [L, M], \tag{1}$$

where [,] denotes the bracket in the Lie algebra g. If we have formulated the Lax pair relation, the conserved quantities (integrals of motion) can be constructed easily. It follows that the adjoint-invariant quantities $\mathrm{tr}L^l$ $(l = 1, \cdots, n)$ are the integrals of the motion. In order to implement the Liouville theorem onto this set of possible action variables we need them to be Poisson-commuting. As shown in Ref. 25), the commutativity of the integrals $\mathrm{tr}L^l$ follows if the Lax operator can be deduced from the fundamental Poisson bracket

$$\{L_1(u), L_2(v)\} = [r_{12}(u,v), L_1(u)] - [r_{21}(v,u), L_2(v)] \tag{2}$$

or quadratic form [19]

$$\begin{aligned}\{L_1(u), L_2(v)\} = &L_1(u)L_2(v)r_{12}^-(u,v) - r_{21}^+(v,u)L_1(u)L_2(v)\\ &+L_1(u)s^+(u,v)L_2(v) - L_2(v)s^-(u,v)L_1(u),\end{aligned} \tag{3}$$

where we use the notation

$$L_1 = L \otimes 1, \qquad L_2 = 1 \otimes L, \qquad a_{21} = Pa_{12}P, \tag{4}$$

and P is the permutation operator such that $Px \otimes y = y \otimes x$.

For the above relations to define a consistent Poisson bracket, one should impose some constraints on the r-matrices. The skew-symmetry of the Poisson bracket requires that

$$r_{21}^{\pm}(v,u) = -r_{12}^{\pm}(u,v), \qquad s_{21}^{+}(v,u) = s_{12}^{-}(u,v), \tag{5}$$

$$r_{12}^{+}(u,v) - s_{12}^{+}(u,v) = r_{12}^{-}(u,v) - s_{12}^{-}(u,v). \tag{6}$$

For the numerical r-matrices $r^{\pm}(u,v), s^{\pm}(u,v)$, some constraint conditions (sufficient conditions) are imposed on the r-matrix in order to make it satisfy the Jacobi identity. [26] However, generally speaking, the r-matrices $r^{\pm}(u,v), s^{\pm}(u,v)$ depend on dynamical variables, and the Jacobi identity which implies an algebraic constraint for the r-matrices takes a very complicated form

$$[L_1, [r_{12}, r_{13}] + [r_{12}, r_{23}] + [r_{32}, r_{13}] + \{L_2, r_{13}\} - \{L_3, r_{12}\}] + \text{cyc.perm} = 0. \tag{7}$$

It should be remarked that for a given integrable system, we can choose different Lax formulations. The r-matrices corresponding to different Lax formulations are generally different. So, in some cases, we can transform a dynamical r-matrix into a non-dynamical r-matrix. [10),11)] The different Lax representations of a system are conjugate to each other: if $(\tilde{L}, \tilde{M})$ is another Lax pair of the same dynamical system conjugate to with the old one (L, M), it means that

$$\begin{aligned}&\frac{d\tilde{L}}{dt} = [\tilde{L}, \tilde{M}],\\ &\tilde{L}(u) = g(u)L(u)g^{-1}(u),\\ &\tilde{M}(u) = g(u)M(u)g^{-1}(u) - \left(\frac{d}{dt}g(u)\right)g^{-1}(u),\end{aligned} \tag{8}$$

where $g(u) \in G$ whose Lie algebra is g. Then we have
Proposition: The Lax pair $(\tilde{L}, \tilde{M})$ has the following r-matrix structure

$$\{\tilde{L}_1(u), \tilde{L}_2(v)\} = [\tilde{r}_{12}(u,v), \tilde{L}_1(u)] - [\tilde{r}_{21}(v,u), \tilde{L}_2(v)], \tag{9}$$

where

$$\begin{aligned}\tilde{r}_{12}(u,v) &= g_1(u)g_2(v)r_{12}(u,v)g_1^{-1}(u)g_2^{-1}(v) + g_2(v)\{g_1(u), L_2(v)\}g_1^{-1}(u)g_2^{-1}(v)\\ &+\frac{1}{2}[\{g_1(u), g_2(v)\}g_1^{-1}(u)g_2^{-1}(v), g_2(v)L_2(v)g_2^{-1}(v)].\end{aligned} \tag{10}$$

For a given Lax pair (L, M) and the corresponding r-matrix, if there exists a g such that

$$\begin{aligned}h_{12} &= \{g_1(u)g_2(v)r_{12}(u,v)g_1^{-1}(u)g_2^{-1}(v) + g_2(v)\{g_1(u), L_2(v)\}g_1^{-1}(u)g_2^{-1}(v)\\ &+\frac{1}{2}[\{g_1(u), g_2(v)\}g_1^{-1}(u)g_2^{-1}(v), g_2(v)L_2(v)g_2^{-1}(v)]\end{aligned} \tag{11}$$

and

$$\partial_{q_i} h_{12} = \partial_{p_j} h_{12} = 0 \tag{12}$$

then a non-dynamical Lax representation of the system exists.

By a straightforward calculation, we can also find that the twisted Lax pair $(\tilde{L}, \tilde{M})$ and the corresponding r-matrix $\tilde{r}_{12}$ satisfy

$$[\tilde{L}_1, [\tilde{r}_{12}, \tilde{r}_{13}] + [\tilde{r}_{12}, \tilde{r}_{23}] + [\tilde{r}_{32}, \tilde{r}_{13}] + \{\tilde{L}_2, \tilde{r}_{13}\} - \{\tilde{L}_3, \tilde{r}_{12}\}] + \text{cycl.perm} = 0. \tag{13}$$

Similarly, for the quadratic form, the Lax pair $(\tilde{L}, \tilde{M})$ has the following r-matrix structure

$$\begin{aligned}\{\tilde{L}_1(u), \tilde{L}_2(v)\} &= \tilde{L}_1(u)\tilde{L}_2(v)\tilde{r}_{12}^-(u,v) - \tilde{r}_{12}^+(u,v)\tilde{L}_1(u)\tilde{L}_2(v)\\ &+\tilde{L}_1(u)\tilde{s}_{12}^+(u,v)\tilde{L}_2(v) - \tilde{L}_2(v)\tilde{s}_{12}^-(u,v)\tilde{L}_1(u),\end{aligned} \tag{14}$$

where

$$\begin{aligned}\tilde{r}_{12}^-(u,v) &= g_1(u)g_2(v)r_{12}^-(u,v)g_1^{-1}(u)g_2^{-1}(v) - \tilde{\Delta}_{12}(u,v) + \tilde{\Delta}_{21}(v,u),\\ \tilde{r}_{12}^+(u,v) &= g_1(u)g_2(v)r_{12}^+(u,v)g_1^{-1}(u)g_2^{-1}(v) - \tilde{\Delta}_{12}^{(1)}(u,v) + \tilde{\Delta}_{21}^{(1)}(v,u),\\ \tilde{s}_{12}^+(u,v) &= g_1(u)g_2(v)s_{12}^+(u,v)g_1^{-1}(u)g_2^{-1}(v) - \tilde{\Delta}_{21}(v,u) - \tilde{\Delta}_{12}^{(1)}(u,v),\\ \tilde{s}_{12}^-(u,v) &= g_1(u)g_2(v)s_{12}^-(u,v)g_1^{-1}(u)g_2^{-1}(v) - \tilde{\Delta}_{12}(u,v) - \tilde{\Delta}_{21}^{(1)}(v,u),\\ \tilde{\Delta}_{12}(u,v) &= \tilde{L}_2^{-1}(v)\Delta_{12}(u,v), \quad \tilde{\Delta}_{12}^{(1)}(u,v) = \Delta_{12}(u,v)\tilde{L}_2^{-1}(v),\\ \Delta_{12}(u,v) &= g_2(v)\{g_1(u), L_2(v)\}g_1^{-1}(u)g_2^{-1}(v)\\ &+\frac{1}{2}[\{g_1(u), g_2(v)\}g_1^{-1}(u)g_2^{-1}(v), g_2(v)L_2(v)g_2^{-1}(v)]\end{aligned} \tag{15}$$

There are still relations:

$$\tilde{r}^{\pm}_{21}(v,u) = -\tilde{r}^{\pm}_{12}(u,v), \qquad \tilde{s}^{+}_{21}(v,u) = \tilde{s}^{-}_{12}(u,v),$$
$$\tilde{r}^{+}_{12}(u,v) - \tilde{s}^{+}_{12}(u,v) = \tilde{r}^{-}_{12}(u,v) - \tilde{s}^{-}_{12}(u,v). \tag{16}$$

And also we have, for given Lax pair (L, M) and the corresponding r-matrices, if there exists a g such that

$$g_1(u)g_2(v)s^{+}_{12}(u,v)g_1^{-1}(u)g_2^{-1}(v) - \tilde{\varDelta}_{21}(v,u) - \tilde{\varDelta}^{(1)}_{12}(u,v) = 0,$$
$$g_1(u)g_2(v)s^{-}_{12}(u,v)g_1^{-1}(u)g_2^{-1}(v) - \tilde{\varDelta}_{12}(u,v) - \tilde{\varDelta}^{(1)}_{21}(v,u) = 0,$$
$$h_{12}(u,v) = g_1(u)g_2(v)r^{-}_{12}(u,v)g_1^{-1}(u)g_2^{-1}(v) - \tilde{\varDelta}_{12}(u,v) + \tilde{\varDelta}_{21}(v,u)$$
$$= g_1(u)g_2(v)r^{+}_{12}(u,v)g_1^{-1}(u)g_2^{-1}(v) - \tilde{\varDelta}^{(1)}_{12}(u,v) + \tilde{\varDelta}^{(1)}_{21}(v,u) \tag{17}$$

and

$$\partial_{q_i}h_{12} = \partial_{p_j}h_{12} = 0, \tag{18}$$

then a non-dynamical Lax representation with Sklyanin Poisson-Lie bracket for the system exists.

§3. Lax pair for elliptic RS and CM models

We first define some elliptic functions:

$$\theta^{(j)}(u) = \theta\begin{bmatrix} \frac{1}{2} - \frac{j}{n} \\ \frac{1}{2} \end{bmatrix}(u, n\tau), \qquad \sigma(u) = \theta\begin{bmatrix} \frac{1}{2} \\ \frac{1}{2} \end{bmatrix}(u,\tau),$$
$$\theta\begin{bmatrix} a \\ b \end{bmatrix}(u,\tau) = \sum_{n=-\infty}^{\infty} \exp\{i\pi[(m+a)^2\tau + 2(m+a)(z+b)]\},$$
$$\theta'^{(j)}(u) = \partial_u\{\theta^{(j)}(u)\}, \sigma'(u) = \partial_u\{\sigma(u)\}, \xi(u) = \partial_u\{\ln\sigma(u)\}, \tag{19}$$

where τ is a complex number with $\mathrm{Im}(\tau) > 0$.

The Ruijsenaars-Schneider model is a system of n one-dimensional relativistical particles interacting by a two-body potential. In terms of the canonical variables p_i, q_i $(i = 1, \cdots, n)$ enjoying the canonical Poisson bracket

$$\{p_i, p_j\} = 0, \qquad \{q_i, q_j\} = 0, \qquad \{q_i, p_j\} = \delta_{ij}, \tag{20}$$

the Hamiltonian of the system is expressed as [18)]

$$H = mc^2\sum_{j=1}^{n}\cosh\left(p_j\prod_{k\neq j}\left\{\frac{\sigma(q_{jk}+\gamma)\sigma(q_{jk}-\gamma)}{\sigma^2(q_{jk})}\right\}^{\frac{1}{2}}\right), \tag{21}$$

where $q_{jk} = q_j - q_k$, m denotes the particle mass, c the speed of light, and γ is the coupling constant. The above defined Hamiltonian is known to be completely integrable.[18),27)] As we mentioned above, the Lax representation (Lax operator of the classical L-operator) is one of the most effective ways to show that the system is integrable. One Lax representation for the elliptic RS model was first formulated by Ruijsenaars:[18)]

$$L_R(u)^i_j = \frac{e^{p_j}\sigma(q_{ji}+u+\gamma)}{\sigma(\gamma+q_{ji})\sigma(u)}\prod_{k\neq j}^{n}\left\{\frac{\sigma(q_{jk}+\gamma)\sigma(q_{jk}-\gamma)}{\sigma^2(q_{jk})}\right\}^{\frac{1}{2}}, \quad i,j=1,\cdots,n. \tag{22}$$

Here, we use another Lax representation $\tilde{L}_R$[20)]

$$\tilde{L}_R(u)^i_j = \frac{e^{p_j}\sigma(u+q_{ji}+\gamma)}{\sigma(u)\sigma(q_{ji}+\gamma)}\prod_{k\neq j}\frac{\sigma(\gamma+q_{jk})}{\sigma(q_{jk})}. \tag{23}$$

$\tilde{L}_R$ can be obtained from the standard Ruijsenaars' $L_R(u)$ by using a Poisson map

$$q_i \longrightarrow q_i, \quad p_i \longrightarrow p_i + \frac{1}{2}\ln\prod_{k\neq i}\frac{\sigma(q_{ik}+\gamma)}{\sigma(q_{ik}-\gamma)}. \tag{24}$$

Following the work of Nijhoff et al.,[20)] the fundamental Poisson bracket of the Lax operator $\tilde{L}_R(u)$ can be given in the following quadratic r-matrix form with dynamical r-matrices

$$\begin{aligned}\{\tilde{L}_R(u)_1, \tilde{L}_R(v)_2\} = {} & \tilde{L}_R(u)_1\tilde{L}_R(v)_2 r^-_{12}(u,v) - r^+_{12}(u,v)\tilde{L}_R(u)_1\tilde{L}_R(v)_2\\ & + \tilde{L}_R(u)_1 s^+_{12}(u,v)\tilde{L}_R(v)_2 - \tilde{L}_R(v)_2 s^-_{12}(u,v)\tilde{L}_R(u)_1,\end{aligned} \tag{25}$$

where

$$\begin{aligned}r^-_{12}(u,v) = a_{12}(u,v) - s_{12}(u) + s_{21}(v), \quad & r^+_{12}(u,v) = a_{12}(u,v) + u^+_{12} + u^-_{12},\\ s^+_{12}(u,v) = s_{12}(u) + u^+_{12}, \quad & s^-_{12}(u,v) = s_{21}(v) - u^-_{12}\end{aligned} \tag{26}$$

and

$$\begin{aligned}& u^{\pm}_{12} = \sum_{ij}\xi(q_{ji}\pm\gamma)e_{ii}\otimes e_{jj},\\ & a_{12}(u,v) = r^0_{12}(u,v) + \sum_{i=1}\xi(u-v)e_{ii}\otimes e_{ii} + \sum_{i\neq j}\xi(q_{ij})e_{ii}\otimes e_{jj},\\ & r^0_{12}(u,v) = \sum_{i\neq j}\frac{\sigma(q_{ij}+u-v)}{\sigma(q_{ij})\sigma(u-v)}e_{ij}\otimes e_{ji}, \quad s_{12}(u) = \sum_{i,j}\left(\tilde{L}_R(u)\partial_\gamma\tilde{L}_R(u)\right)^i_j e_{ij}\otimes e_{jj}.\end{aligned} \tag{27}$$

The following properties are satisfied:

$$\begin{aligned}r^{\pm}_{21}(v,u) = -r^{\pm}_{12}(u,v), \quad & s^+_{21}(v,u) = s^-_{12}(u,v),\\ r^+_{12}(u,v) - s^+_{12}(u,v) = {} & r^-_{12}(u,v) - s^-_{12}(u,v).\end{aligned} \tag{28}$$

Here we would like to reformulate the Lax formulation for the RS model. Define an $n \otimes n$ matrix $A(u;q)$ as:

$$A(u;q)^i_j \equiv A(u, q_1, \cdots, q_n)^i_j = \theta^{(i)}\left(u + nq_j - \sum_{k=1}^{n} q_k + \frac{n-1}{2}\right). \quad (29)$$

Here we should point out that $A(u;q)^i_j$ corresponds to the intertwiner function $\phi^{(i)}_j$ between the Z_n-symmmetric Belavin R-matrix [28), 29)] and the $A^{(1)}_{n-1}$ face model. [30), 31)]

Define

$$g(u) = A(u;q)\Lambda(q), \quad \Lambda(q)^i_j = h_i(q)\delta^i_j,$$
$$h_i(q) \equiv h_i(q_1, \cdots, q_n) = \frac{1}{\prod_{l\neq i}\sigma(q_{il})}. \quad (30)$$

We can construct the new Lax operator $L(u)$ as

$$L(u) = g(u)\tilde{L}_R(u)g^{-1}(u). \quad (31)$$

More explicitly, it can be written as:

$$L(u)^i_j = \sum_{k=1}^{n} \frac{1}{\sigma(\gamma)} A(u + n\gamma; q)^i_k A^{-1}(u;q)^k_j e^{p_k}, \quad i,j = 1, 2, \cdots, n. \quad (32)$$

It can be proved that the fundamental Poisson bracket of $L(u)$ can be given in the quadratic Poisson-Lie form with a nondynamical r-matrix:

$$\{L_1(u), L_2(v)\} = [r_{12}(u-v), L_1(u)L_2(v)]. \quad (33)$$

Here the numerical r-matrix is the classical Z_n-symmetric r-matrix.[32)] It takes the form

$$r^{lk}_{ij}(v) =$$
$$\begin{cases} (1-\delta^l_i)\frac{\theta'^{(0)}(0)\theta^{(i-j)}(v)}{\theta^{(l-j)}(v)\theta^{(i-l)}(0)} + \delta^l_i\delta^k_j\left(\frac{\theta'^{(i-j)}(v)}{\theta^{(i-j)}}(v) - \frac{\sigma'(v)}{\sigma(v)}\right) & \text{if } i+j = l+l \mod n \\ 0 & \text{otherwise} \end{cases}. \quad (34)$$

We know the Z_n symmetric r-matrix satisfies the nondynamical classical Yang-Baxter equation

$$[r_{12}(v_1 - v_2), r_{13}(v_1 - v_3)] + [r_{12}(v_1 - v_2), r_{23}(v_2 - v_3)]$$
$$+ [r_{13}(v_1 - v_3), r_{23}(v_2 - v_3)] = 0. \quad (35)$$

We also know that this r-matrix is antisymmetric and Z_n symmetric:

$$\text{Antisymmetry}: \quad -r_{21}(-v) = r_{12}(v),$$
$$Z_n \otimes Z_n\text{Symmetry}: r_{12}(v) = (a \otimes a)r_{12}(v)(a \otimes a)^{-1}, \quad (36)$$

where $a = g, h$, and g, h are $n \otimes n$ matrices defined as:

$$h_{ij} = \delta_{i+1,j \bmod n}, \quad g_{ij} = \omega^i \delta_{i,j}. \tag{37}$$

For convenience, we also define another $n \otimes n$ matrix

$$I_\alpha \equiv I_{\alpha_1,\alpha_2} \equiv g^{\alpha_2} h^{\alpha_1}, \tag{38}$$

where $\alpha_1, \alpha_2 \in Z_n$ and $\omega = \exp(2\pi i/n)$.

Next, we will consider the non-relativistic limit of the Lax operator $L(u)$. First rescale the momenta $\{p_i\}$, the coupling constant γ and the Lax operator as follows:[20]

$$p_i := -\beta p'_i, \quad n\gamma := \beta s, \quad L(u) := \sigma\left(\frac{\beta s}{n}\right) L'(u). \tag{39}$$

Here notation L' is introduced. The non-relativistic limit is then obtained by taking the limit $\beta \to 0$. We have the following asymptotic properties

$$L'(u)^i_j = \delta^i_j - \beta\left(\sum_k \{A(u;q)^i_k A(u;q)^k_j p'_k - s\partial_s(A(u;q)^i_k)A^{-1}(u;q)^k_j\}\right) + O(\beta^2). \tag{40}$$

If we make the canonical transformation

$$p'_i \to p'_i - \frac{s}{n}\frac{\partial}{\partial q_i}\ln M(q), \quad M(q) = \prod_{i<j}\sigma(q_{ij}), \tag{41}$$

we finally obtain the Lax operator of the elliptic A_{n-1} CM model[10]

$$L_{\rm CM}(u)^i_j = -\lim_{\beta\to 0}\frac{L'(u)^i_j - \delta^i_j}{\beta}\Bigg|_{p'_i \to p'_i - \frac{s}{n}\frac{\partial}{\partial q_i}\ln M(q)}. \tag{42}$$

Here we have

$$\{L_{\rm CM}(u)_1, L_{\rm CM}(v)_2\} = [r_{12}(u-v), L_{\rm CM}(u)_1 + L_{\rm CM}(v)_2]. \tag{43}$$

For the newly constructed Lax representation $L(u)$, the quantization becomes no longer difficult. Define the Z_n-symmetric Belavin's R-matrix as:

$$R^{lk}_{ij}(u) = \begin{cases} \frac{\theta'^{(0)}(0)\sigma(u)\sigma(\sqrt{-1}\hbar)}{\sigma'(0)\theta^{(0)}(v)\sigma(v+\sqrt{-1}\hbar)}\frac{\theta^{(0)}(v)\theta^{(i-j)}(v+\sqrt{-1}\hbar)}{\theta^{(i-l)}(\sqrt{-1}\hbar)\theta^{(l-j)}(v)} & \text{if } i+j = l+k \mod n \\ 0 & \text{otherwise.} \end{cases} \tag{44}$$

We know this R-matrix satisfies the quantum Yang-Baxter equation

$$R_{12}(u_1-u_2)R_{13}(u_1-u_3)R_{23}(u_2-u_3) = R_{23}(u_2-u_3)R_{13}(u_1-u_3)R_{12}(u_1-u_2). \tag{45}$$

The R-matrix is Z_n symmetric in the sense that

$$R_{12}(u) = (a \otimes a) R_{12}(u) (a \otimes a)^{-1}, \quad a = g, h. \tag{46}$$

By taking the special limit $\hbar \to 0$, we can obtain the classical Z_n symmetric r-matrix

$$R_{12}(u)|_{\hbar\to 0} = 1 \otimes 1 + \sqrt{-1}\hbar r_{12}(u) + o(\hbar^2). \tag{47}$$

Now let us study the quantum L operator, using the usual canonical quantization procedure

$$p_j \to \hat{p}_j = -\sqrt{-1}\hbar \frac{\partial}{\partial q_j}, \quad q_j \to q_j, \quad j = 1, \cdots, n. \tag{48}$$

The corresponding quantum L-operator can be formulated as:

$$\begin{aligned} \hat{L}(u)_l^m &= \frac{1}{\sigma(\gamma)} \sum_{k=1}^{n} A(u + n\gamma; q)_k^m A^{-1}(u; q)_l^k e^{\hat{p}_k} \\ &= \frac{1}{\sigma(\gamma)} \sum_{k=1}^{n} A(u + n\gamma; q)_k^m A^{-1}(u; q)_l^k e^{-\sqrt{-1}\hbar \frac{\partial}{\partial q_k}}. \end{aligned} \tag{49}$$

It should be remarked that this quantum L-operator is simply the factorised difference representation for the elliptic L-operator.[31),33)] The above defined quantum L-operator satisfies the quantum Yang-Baxter relation

$$R_{12}(u - v) \hat{L}_1(u) \hat{L}_2(v) = \hat{L}_2(v) \hat{L}_1(u) R_{12}(u - v). \tag{50}$$

The proof can be found in Refs. 31), 34), 33) and 35).

§4. RS and CM models related with Sklyanin algebra

We introduce here some notation for elliptic functions:

$$\begin{aligned} \sigma_\alpha(u) &= \theta \begin{bmatrix} \frac{1}{2} + \frac{\alpha_1}{n} \\ \frac{1}{2} + \frac{\alpha_2}{n} \end{bmatrix} (u, \tau), \\ W_\alpha(u) &= \frac{\sigma_\alpha(u + \sqrt{-1}\hbar)}{\sigma_\alpha(\sqrt{-1}\hbar)} \frac{\sigma_0(\sqrt{-1}\hbar)}{\sigma_0(u + \sqrt{-1}\hbar)}. \end{aligned} \tag{51}$$

The above mentioned quantum R-matrix can be rewritten as following up to a scale:

$$R(u) = \sum_\alpha W_\alpha(u) I_\alpha \otimes I_\alpha^{-1}, \tag{52}$$

as before $\alpha \equiv \alpha_1, \alpha_2$ and $\alpha_i \in Z_n, i = 1, 2$.

The quantum L-operator $\hat{L}$ obtained in the last section can be represented by the generators of Sklyanin algebra S_α:

$$\hat{L}(u) = \sum_\alpha V_\alpha(u) I_\alpha S_\alpha, \tag{53}$$

where

$$V_\alpha(u) = \frac{\sigma_\alpha(u' + \sqrt{-1}\hbar\xi)}{n\sigma_0(u')\sigma_\alpha(\sqrt{-1}\hbar)}, \quad u' = u + n\sqrt{-1}\hbar\delta - \frac{n-1}{2}, \tag{54}$$

where δ is a constant.

The quantum Yang-Baxter relation (49) gives the defining relations of the Sklyanin algebra:[32),6)]

$$\sum_\gamma \frac{\omega^{\gamma_1-\alpha_1+(\beta_1-\gamma_1)(\gamma_1-\alpha_2)}\sigma_{\alpha+\beta-2\gamma}(0)\sigma_\beta(2\sqrt{-1}\hbar)}{\sigma_{\gamma-\alpha}(\sqrt{-1}\hbar)\sigma_{\alpha+\beta-\gamma}(\sqrt{-1}\hbar)\sigma_\gamma(\sqrt{-1}\hbar)\sigma_{\beta-\gamma}(\sqrt{-1}\hbar)} S_{\alpha+\beta-\gamma}S_\gamma = 0, \tag{55}$$

with $\alpha_i, \beta_i, \gamma_i \in Z_n, i = 1, 2$.

We can give a realization of the generators of Sklyanin algebra as:

$$S_\alpha = \sum_j S_{j\alpha} e^{-n\sqrt{-1}\hbar\frac{\partial}{\partial q_j}}. \tag{56}$$

Here we introduce the symbol $S_{j\alpha}$ for the elliptic function

$$S_{j\alpha} = (-1)^{\alpha_1}\sigma_\alpha(\sqrt{-1}\hbar)\prod_{k\neq j}\frac{\sigma_\alpha(\sqrt{-1}\hbar\xi + q_{jk})}{\sigma_0(q_{jk})}. \tag{57}$$

Next, we will consider the classical limit of the above defining relations. Letting $\hbar \to 0$, the quantum R-matrix become the classical r-matrix, and we explicitly have the elements of the r-matrix in (34) presented in the last section, here we use another notation

$$R(u) = 1 + \sqrt{-1}\hbar r(u) + O(\hbar^2). \tag{58}$$

The classical r-matrix is written as:

$$r(u) = \sum_\alpha w_\alpha(u) I_\alpha \otimes I_\alpha^{-1}, \tag{59}$$

where

$$\begin{aligned} w_0(u) &= 0, \\ w_\alpha(u) &= \frac{\sigma_\alpha(u)\sigma_0'(0)}{\sigma_\alpha(0)\sigma_0(u)}, \quad \alpha \neq 0. \end{aligned} \tag{60}$$

In order to consider the classical limit of $\hat{L}$, we first present the classical limit of $V_\alpha(u)$:

$$V_0(u) = \frac{\sigma_0(u') + \sqrt{-1}\hbar\xi\sigma_0'(u') + O(\hbar^2)}{n\sigma_0(u')\sigma_0(\sqrt{-1}\hbar)}, \tag{61}$$

$$V_\alpha(u) = \frac{\sigma_\alpha(u')}{n\sigma_\alpha(0)\sigma_0(u')} + \frac{\sqrt{-1}\hbar}{n\sigma_0(u')}\left[\frac{\xi\sigma_\alpha'(u')\sigma_\alpha(0) - \sigma_\alpha(u')\sigma_\alpha'(0)}{\sigma_\alpha^2(0)}\right] + O(\hbar^2),$$
$$\alpha \neq 0. \tag{62}$$

From the definition of the operator S_α, we easily have

$$S_\alpha = \sum_j S_{j\alpha}\left(1 - n\sqrt{-1}\hbar\frac{\partial}{\partial q_j} + O(\hbar^2)\right). \tag{63}$$

In the limit $\hbar \to 0$, the elliptic functions $S_{j\alpha}$ take the forms:

$$S_{j0} = \sigma(\sqrt{-1}\hbar)\left[1 + \sqrt{-1}\hbar\xi\sum_{k\neq j}\frac{\sigma_0'(q_{jk})}{\sigma_0(q_{jk})} + O(\hbar^2)\right], \tag{64}$$

$$S_{j\alpha} = (-1)^{\alpha_1}\sigma_\alpha(0)\prod_{k\neq j}\frac{\sigma_\alpha(q_{jk})}{\sigma_0(q_{jk})}\left[1 + \sqrt{-1}\hbar\left(\frac{\sigma_\alpha'(0)}{\sigma_\alpha(0)} + \xi\sum_{k\neq j}\frac{\sigma_\alpha'(q_{jk})}{\sigma_\alpha(q_{jk})}\right) + O(\hbar^2)\right], \quad \alpha \neq 0. \tag{65}$$

So, we have

$$\begin{aligned} V_0(u)S_0 &= 1 + \sqrt{-1}\hbar\left[\xi\frac{\sigma_0'(u')}{\sigma_0(u')}\right] + \frac{\sqrt{-1}\hbar}{n}\sum_j\left[\xi\sum_{k\neq j}\frac{\sigma_0'(q_{jk})}{\sigma_0(q_{jk})} - n\frac{\partial}{\partial q_j}\right] + O(\hbar^2), \\ &= 1 + \sqrt{-1}\hbar\left[\xi\frac{\sigma_0'(u')}{\sigma_0(u')}\right] + \frac{\sqrt{-1}\hbar}{n}\sum_j\left[-n\frac{\partial}{\partial q_j}\right] + O(\hbar^2), \end{aligned} \tag{66}$$

$$V_\alpha(u)S_\alpha = \sqrt{-1}\hbar(-1)^{\alpha_1}\frac{\sigma_\alpha(u')}{n\sigma_0(u')}\sum_j\prod_{k\neq j}\frac{\sigma_\alpha(q_{jk})}{\sigma_0(q_{jk})}\left[\xi\sum_{k\neq j}\frac{\sigma_\alpha'(q_{jk})}{\sigma_\alpha(q_{jk})} - n\frac{\partial}{\partial q_j}\right] + O(\hbar^2), \quad \alpha \neq 0. \tag{67}$$

here we have $\sum_j\sum_{k\neq j}\frac{\sigma_0'(q_{jk})}{\sigma_0(q_{jk})} = 0$, because $\frac{\sigma_0'(q_{jk})}{\sigma_0(q_{jk})}$ is an odd function.

We can finally expand the quantum $\hat{L}$ operator in the order of $\hbar$ when we take a limit $\hbar \to 0$. However, we first introduce some notation

$$\hat{L}(u) = \sum_\alpha V_\alpha(u)S_\alpha I_\alpha = 1 + \sqrt{-1}\hbar l(u) + O(\hbar^2), \tag{68}$$

where $l(u)$ is the classical l operator. We may represent $l(u)$ in terms of generators of the "classical" Sklyanin algebra s_α:

$$l(u) = \frac{\xi}{n}\frac{\sigma_0'(u')}{\sigma_0(u')} - \sum_\alpha v_\alpha(u)s_\alpha. \tag{69}$$

The function $v_\alpha(u)$ is defined as:

$$v_0(u) = \frac{1}{n}, \tag{70}$$

$$v_\alpha(u) = \frac{1}{n}\frac{\sigma_\alpha(u')\sigma_\alpha(0)}{\sigma_0(u')}, \quad \alpha \neq 0. \tag{71}$$

From the above obtained results, we can realize the generators of the "classical" Sklyanin algebra in the following forms:

$$s_0 = \sum_j n \frac{\partial}{\partial q_j}, \tag{72}$$

$$s_\alpha = (-1)^{\alpha_1} \sigma_\alpha(0) \sum_j \prod_{k \neq j} \frac{\sigma_\alpha(q_{jk})}{\sigma_0(q_{jk})} \left[n \frac{\partial}{\partial q_j} - \xi \sum_{k \neq j} \frac{\sigma'_\alpha(q_{jk})}{\sigma_\alpha(q_{jk})} \right], \quad \alpha \neq 0. \tag{73}$$

On the other hand, here we can say we give a definition of the generators of the "classical" Sklyanin algebra.

In the classical limit, the quantum Yang-Baxter relation becomes the following

$$[l_1(u), l_2(v)] = [r_{12}(u-v), l_1(u) + l_2(v)]. \tag{74}$$

Substitute the l-operator with the generators of the "classical" Sklyanin algebra, and through tedious calculation, we have

$$[s_\alpha, s_\gamma] = (\omega^{\alpha_1 \gamma_2} - \omega^{\alpha_2 \gamma_1}) \left(\frac{\sigma'_0(0)}{n} \right) s_{\alpha+\gamma}. \tag{75}$$

On the other hand, we find that $\{I_\alpha\}$ satisfy a similar relation

$$[I_\alpha, I_\gamma] = (\omega^{\alpha_1 \gamma_2} - \omega^{\alpha_2 \gamma_1}) I_{\alpha+\gamma}. \tag{76}$$

So, after rescaling s_α, we find $\{s_\alpha\}$ and $\{I_\alpha\}$ satisfy the same algebra.

Finally we should point out that if we substitute $\frac{\partial}{\partial q_k}$ by the corresponding canonical variable p_k, the l-operator will become a T-operator, and the commutative bracket on the left-hand side of the above relation (74) will change to the standard Poisson-Lie bracket. Here we rewrite as:

$$\{T_1(u), T_2(v)\} = [r_{12}(u-v), T_1(u) + T_2(v)]. \tag{77}$$

§5. CM model, Gaudin model, Lamé-equation and the Bethe ansatz

For the difference factorized operator $\hat{L}$, we can find some commuting families which are related to conserved operators. By using the fusion procedure, the commuting family take the form

$$D_m = \text{tr}[\hat{L}(u) \otimes \cdots \otimes \hat{L}(u) P_-^m],$$

there are m $\hat{L}$'s above, P_-^m is the anti-symmetric projector. In the classical limit, we also have a similar commuting family

$$a_m(u) = \sum_{\alpha_i \neq 0} v_{\alpha_1(u)} \cdots v_{\alpha_m}(u) s_{\alpha_1} \cdots s_{\alpha_m} \text{tr}[I_{\alpha_1} \otimes \cdots I_{\alpha_m} P_-^m], \tag{78}$$

where $\alpha_i \in Z_n^2, i = 1, \cdots, m$. Let $u' = 0$, so $u = u_0 = \frac{n-1}{2} - n\hbar\xi$, and after rescaling $a_l(u)$, we have

$$a_m(u_0) = \sum_{\alpha_i \neq 0} s_{\alpha_1} \cdots s_{\alpha_m} \text{tr}[I_{\alpha_1} \otimes \cdots \otimes I_{\alpha_m} P_-^m]. \tag{79}$$

We will discuss a special case $n = 2$, explicitly we have

$$s_{01} = 2\sigma_{01}(0)\left\{\xi\frac{\sigma'_{01}(q_{12})}{\sigma_0(q_{12})} - \frac{\sigma_{01}(q_{12})}{\sigma_0(q_{12})}\left(\frac{\partial}{\partial q_1} - \frac{\partial}{\partial q_2}\right)\right\},$$
$$s_{10} = 2\sigma_{10}(0)\left\{\xi\frac{\sigma'_{10}(q_{12})}{\sigma_0(q_{12})} - \frac{\sigma_{10}(q_{12})}{\sigma_0(q_{12})}\left(\frac{\partial}{\partial q_1} - \frac{\partial}{\partial q_2}\right)\right\},$$
$$s_{11} = 2\sigma_{11}(0)\left\{\xi\frac{\sigma'_{11}(q_{12})}{\sigma_0(q_{12})} - \frac{\sigma_{11}(q_{12})}{\sigma_0(q_{12})}\left(\frac{\partial}{\partial q_1} - \frac{\partial}{\partial q_2}\right)\right\}. \quad (80)$$

We will calculate the non-trivial conserved operator

$$4a_2(u) = \sum_{\alpha\neq 0}\frac{\sigma_\alpha(u')\sigma_{-\alpha}(u')}{\sigma_\alpha(0)\sigma_{-\alpha}(0)}s_\alpha s_{-\alpha}\omega^{-\alpha_1\alpha_2}. \quad (81)$$

After some tedious calculations, we have

$$a_2 = -\xi^2\frac{\sigma''_0(q)}{\sigma_0(q)} + 2\xi\frac{\sigma'_0(q)}{\sigma_0(q)}\left(\frac{\partial}{\partial q_1} - \frac{\partial}{\partial q_2}\right) + 4\frac{\partial^2}{\partial q_1\partial q_2}$$
$$-\left(\frac{\partial}{\partial q_1} + \frac{\partial}{\partial q_2}\right)^2 - (\xi^2 + 2\xi)\left[\frac{\sigma'_0(u')^2}{\sigma_0(u')^2} - \frac{\sigma''_0(u')}{\sigma_0(u')}\right], \quad (82)$$

where $q = q_1 - q_2$. This relation is just the same as that obtained by Hasegawa.[36)]

Since 1 and $\frac{\partial}{\partial q_1} + \frac{\partial}{\partial q_2}$ are also conserved quantities defined above, after some tedious calculations we have another conserved operator

$$H = \frac{\partial^2}{\partial q^2} - \xi\frac{\sigma'_0(q)}{\sigma_0(q)}\frac{\partial}{\partial q} + \frac{\xi^2}{4}\frac{\sigma''_0(q)}{\sigma_0(q)}. \quad (83)$$

We can change it to a more familiar form. Let $\xi = 2\beta$, and suppose ψ and Λ are an eigenfunction and eigenvalue of the above Hamiltonian

$$H\psi = \Lambda\psi. \quad (84)$$

At the same time, we introduce a transformation of this eigenfunction $\psi = \tilde{\psi}[\sigma_0(q)]^\beta$, we thus have the following relations:

$$H\tilde{\psi}[\sigma_0(q)]^\beta = \left[\frac{d^2}{dq^2} - 2\beta\frac{\sigma'_0(q)}{\sigma_0(q)}\frac{d}{dq} + \beta^2\frac{\sigma''_0(q)}{\sigma_0(q)}\right]\tilde{\psi}[\sigma_0^\beta(q)]^\beta = \Lambda\tilde{\psi}[\sigma_0^\beta(q)]^\beta. \quad (85)$$

This means:

$$H'\tilde{\psi} \equiv \left[\frac{d^2}{dq^2} + \beta(\beta+1)\frac{d^2}{dq^2}\ln\sigma_0(q)\right]\tilde{\psi} = \Lambda\tilde{\psi}. \quad (86)$$

One finds that H' is simply the Hamiltonian of the CM model, see, for example, Refs. 3), 4), 37) and 38). It is also connected to the Lamé operator, see Ref. 39) and the references therein.

Next, we will calculate the eigenfunction and eigenvalue of the above defined Lamé operator H. Here we first review some of the results obtained by Felder and Varchenko.[39] The difference operator L which is equivalent to S_0, one generator of the Sklyanin algebra when $n = 2$, is given by

$$L\psi(q) = \frac{\sigma_0(q+2\hbar\beta)}{\sigma_0(q)}\psi(q-2\hbar) + \frac{\sigma_0(q-2\hbar\beta)}{\sigma_0(q)}\psi(q+2\hbar). \tag{87}$$

This difference operator is also called the q-deformed Lamé operator. In the framework of the quantum inverse scattering method, there is a result as follows:[39]

Let $(t_1, \cdots, t_m, c)$ be a solution of the Bethe ansatz equations:

$$\frac{\sigma_0(t_i - 2\hbar\beta)}{\sigma_0(t_i + 2\hbar 2\beta)} \prod_{j\neq i} \frac{\sigma_0(t_j - t_i - 2\hbar)}{\sigma_0(t_j - t_i + 2\hbar)} = e^{4\hbar c}, \quad i = 1, \cdots, \beta, \tag{88}$$

such that $t_i \neq t_j \bmod Z + \tau Z$ if $i \neq j$. Then

$$\psi(q) = e^{cq}\prod_{j}^{\beta}\sigma_0(q+t_j), \tag{89}$$

is a solution of the q-deformed Lamé equation $L\psi = \epsilon\psi$ with eigenvalue

$$\epsilon = e^{-2\hbar c}\frac{\sigma_0(4\hbar\beta)}{\sigma_0(2\hbar\beta)}\prod_{j=1}^{\beta}\frac{\sigma_0(t_j + (2\beta-2)\hbar)}{\sigma_0(t_j + 2\beta\hbar)}. \tag{90}$$

By taking the special limit $\hbar \to 0$, the difference operator becomes:

$$L = 2 + 4\hbar^2\left[\frac{d^2}{dq^2} - 2\beta\frac{\sigma_0'(q)}{\sigma_0(q)}\frac{d}{dq} + \beta^2\frac{\sigma_0''(q)}{\sigma_0(q)}\right] + O(\hbar^4). \tag{91}$$

We find the term with order $\hbar^2$ is exactly the Hamiltonian presented in relation (85)

$$H = \frac{d^2}{dq^2} - 2\beta\frac{\sigma_0'(q)}{\sigma_0(q)}\frac{d}{dq} + \beta^2\frac{\sigma_0''(q)}{\sigma_0(q)}. \tag{92}$$

Since we already know the eigenfunction of the difference operator L is $\psi(q) = e^{cq}\prod_{j=1}^{\beta}\sigma_0(q+t_j)$ which does not depend on $\hbar$, we need only expand the eigenvalue of L in the order of $\hbar$. We can obtain the eigenvalue of the Lamé operator H

$$\begin{aligned}\epsilon = 2 &- 4\hbar\left[c + \sum_{j=1}^{\beta}\frac{\sigma_0'(t_j)}{\sigma_0(t_j)}\right] + 8\hbar^2 c\left[c + \sum_{j=1}^{\beta}\frac{\sigma_0'(t_j)}{\sigma_0(t_j)}\right] \\ &-4\hbar^2c^2 + 4\beta^2\hbar^2\frac{\sigma_0'''(0)}{\sigma_0'(0)} + 4\hbar^2\left\{\sum_{j=1}^{\beta}\left[\frac{\sigma_0'(t_j)}{\sigma_0(t_j)}\right]^2 + 2\sum_{i>j}\frac{\sigma_0'(t_i)\sigma_0'(t_j)}{\sigma_0(t_i)\sigma_0(t_j)}\right\} \\ &+4\hbar^2(1-2\beta)\sum_{j=1}^{\beta}\left[\frac{\sigma_0''(t_j)}{\sigma_0(t_j)} - \left(\frac{\sigma_0'(t_j)}{\sigma_0(t_j)}\right)^2\right] + O(\hbar^4).\end{aligned} \tag{93}$$

At the same time we take the limit $\hbar \to 0$ for the Bethe ansatz equation, obtaining

$$c+\beta\frac{\sigma_0'(t_i)}{\sigma_0(t_i)}-\sum_{j,j\neq i}\frac{\sigma_0'(t_i-t_j)}{\sigma_0(t_i-t_j)}=0. \tag{94}$$

Considering this Bethe ansatz equation, we can finally find the eigenvalue of the Lamé operator Λ,

$$\Lambda=(1-2\beta)\left[\sum_{j=1}^{\beta}\frac{\sigma_0'(t_j)}{\sigma_0(t_j)}\right]'+\beta^2\frac{\sigma_0'''(0)}{\sigma_0'(0)}. \tag{95}$$

Here we have the results:

Let $(t_1,\cdots,t_m,c)$ be a solution of the Bethe ansatz equations:

$$c+\beta\frac{\sigma_0'(t_i)}{\sigma_0(t_i)}-\sum_{j,j\neq i}\frac{\sigma_0'(t_i-t_j)}{\sigma_0(t_i-t_j)}=0, \qquad i=1,\cdots,\beta, \tag{96}$$

such that $t_i\neq t_j \bmod Z+\tau Z$ if $i\neq j$. Then

$$\psi(q)=e^{cq}\prod_{j}^{\beta}\sigma_0(q+t_j) \tag{97}$$

is a solution of the equation

$$H\psi(q)=\left[\frac{d^2}{dq^2}-2\beta\frac{\sigma_0'(q)}{\sigma_0(q)}\frac{d}{dq}+\beta^2\frac{\sigma_0''(q)}{\sigma_0(q)}\right]\psi(q)=\Lambda\psi(q), \tag{98}$$

with eigenvalue

$$\Lambda=(1-2\beta)\left[\sum_{j=1}^{\beta}\frac{\sigma_0'(t_j)}{\sigma_0(t_j)}\right]'+\beta^2\frac{\sigma_0'''(0)}{\sigma_0'(0)}. \tag{99}$$

We can also obtain these results by directly using the algebraic Bethe ansatz method for the $A_1^{(1)}$ Gaudin model,[40)] just like the algebraic Bethe ansatz for the XYZ Gaudin model.[41)]

§6. Summary

We review some developments concerning the non-dynamical structure of the elliptic RS and CM models. We also give a solution to the Lamé equation. The eigenfunction and eigenvalue for the Lamé operator are found through the results of the Bethe ansatz.

The results of the last sections are only for the $n=2$ case. For general n, we can also obtain the eigenvalue and eigenfunction for the generalized Lamé operator by using the algebraic Bethe ansatz method for the $sl(n)$ elliptic Gaudin model. The conserved quantities also correspond to the Hamiltonian of the elliptic CM model.

Acknowlegements

H. Fan is supported by the Japan Society for the Promotion of Science. He would like to thank the hospitality of Professor Wadati's group in the Department of Physics, University of Tokyo. B. Y. Hou would like to thank the hospitality of Professor Sasaki and the Yukawa Institute in Kyoto University. This work is supported in part by the NSFC, Clibming project, NWU teachers' fund and Shaanxi province's fund.

References

1) M. A. Olshanetsky and A. M. Perelomov, Phys. Rep. **71** (1981), 313.
2) J. Avan and T. Talon, Phys. Lett. **B303** (1993), 33.
3) F. Calogero, Lett. Nuovo Cim. **13** (1975), 411; **16** (1976), 77.
4) J. Moser, Adv. Math. **16** (1975), 1.
5) I. M. Krichever, Func. Anal. Appl. **14** (1980), 282.
6) E. K. Sklyanin, Func. Anal. Appl. **16** (1982), 263; **17** (1983), 320; Commun. Math. Phys. **150** (1992), 181; hep-th/9308060.
7) H. W. Braden and T. Suzuki, Lett. Math. Phys. **30** (1994), 147.
8) O. Babelon and D. Bernard, Phys. Lett. **B317** (1993), 363.
9) H. W. Braden, Andrew and N. W. Hone, Phys. Lett. **B380** (1996), 296.
10) B. Y. Hou and W. L. Yang, Lett. Math. Phys. **44** (1998), 35; J. of Phys. **A32** (1999), 1475.
11) B. Y. Hou and W. L. Yang, IMPNWU-971219; math.QA/9802104.
12) J. Avan, O. Babelon and E. Billey, Commun. Math. Phys. **178** (1996), 281.
13) H. W. Braden and R. Sasaki, Prog. Theor. Phys. **97** (1997), 1003.
14) H. W. Braden, E. Corrigan, P. E. Dorey and R. Sasaki, Nucl. Phys. **B338** (1990), 689; **B356** (1991), 469.
15) F. W. Nijhoff, O. Ragnisco and V. B. Kuznetsov, Commun. Math. Phys. **176** (1996), 681.
16) G. E. Arutyunov, S. A. Frolov and P. B. Medredev, J. of Phys. **A30** (1997), 5051; J. Math. Phys. **38** (1997), 5682.
17) A. Gorsky and N. Nekrosov, Nucl. Phys. **B414** (1994), 213; **B436** (1995), 582.
18) S. N. M. Ruijsenaars, Commun. Math. Phys. **115** (1988), 127.
19) Yuri B. Suris, hep-th/9603011.
20) F. W. Nijhoff, V. B. Kuznetrov, E. K. Sklyanin and O. Ragnisco, J. of Phys. **A29** (1996), L333.
21) A. J. Bordner, E. Corrigan and R. Sasaki, Prog. Theor. Phys. **100** (1998), 1107.
22) A. J. Bordner, R. Sasaki and K. Takasaki, Prog. Theor. Phys. **101** (1999), 487.
23) A. J. Bordner and R. Sasaki, Prog. Theor. Phys. **101** (1999), 799.
24) A. J. Bordner, E. Corrigan and R. Sasaki, Prog. Theor. Phys. **102** (1999), 499; hep-th/9905011.
25) O. Babelon and D. Bernard, Phys. Lett. **B317** (1993), 363.
26) L. Freidel and J. M. Maillet, Phys. Lett. **B262** (1991), 278.
27) S. N. M. Ruijsenaars and H. Schneider, Ann. of Phys. **170** (1986), 370.
28) A. A. Belavin, Nucl. Phys. **B180** (1980), 109.
29) M. P. Richey and C. A. Tracy, J. Stat. Phys. **42** (1986), 311.
30) M. Jimbo, T. Miwa and M. Okado, Nucl. Phys. **B300** (1988), 74.
31) B. Y. Hou, K. J. Shi and W. L. Yang, J. of Phys. **A26** (1993), 4951.
32) B. Y. Hou and H. Wei, J. Math. Phys. **30** (1989), 2750.
33) K. Hasegawa, J. of Phys. **A26** (1993), 3211; J. Math. Phys. **35** (1994), 6158.
34) Y. H. Quano and A. Fujii, Mod. Phys. Lett. **A6** (1991), 3635.
35) B. Y. Hou, K. J. Shi, W. L. Yang and Z. X. Yang, Phys. Lett. **A178** (1993), 73.
36) K. Hasegawa, q-alg/9512029; Commun. Math. Phys. **187** (1997), 289.
37) E. D'Hoker and D. H. Phong, hep-th/9808156.
38) K. Hikami and Y. Komori, J. Phys. Soc. Jpn. **67** (1998), 4037.
39) G. Felder and A. Varchenko, q-alg/9605024; Nucl. Phys. **B480** (1996), 485.

40) R. H. Yue et al., Preprint (NWU-IMP9906).
41) E. K. Sklyanin and T. Takebe, Phys. Lett. **A219** (1996), 217; solv-int/9807008.
42) K. Chen, B. Y. Hou, W. L. Yang and Y. Zhen, Chinese. Phys. Lett. **16** (1999), 1.

JOURNAL OF MATHEMATICAL PHYSICS VOLUME 41, NUMBER 1 JANUARY 2000

The dynamical twisting and nondynamical *r*-matrix structure of the elliptic Ruijsenaars–Schneider model

Bo-yu Hou and Wen-Li Yang[a)]
CCAST (World Laboratory), P.O. Box 8730, Beijing 100080, People's Republic of China

(Received 8 December 1998; accepted for publication 26 July 1999)

From the dynamical twisting of the classical r-matrix, we obtain a new Lax operator for the elliptic Ruijsenaars–Schneider model (cf. Ruijsenaars). The corresponding r-matrix is shown to be the classical Z_n-symmetric elliptic r-matrix, which is the same as that obtained in the study of the nonrelativistic version—the A_{n-1} Calogero–Moser model. © *2000 American Institute of Physics.* [S0022-2488(00)02801-2]

I. INTRODUCTION

Following the successes of the Calogero–Moser (CM) models,[1,2] a relativistic generalization of the CM models—the so-called Ruijsenaars–Schneider (RS) models have been proposed,[3] which the integrability has been conserved. The RS model describes a completely integrable system of n one-dimensional interacting relativistic particles. Its importance lies in the fact that it is related to the dynamics of solitons in some integrable relativistic field theories[4,5] and its discrete-time version has been connected with the Bethe anstaz equation of the solvable lattice statistical model.[6] A recent development demonstrated that it can be obtained by a Hamiltonian reduction of the contangent bundle of some Lie group,[7] and can also be considered as the gauged WZW theory.[8] The study of the RS model would play a universal role in study of completely integrable multiparticle systems. Among all types of RS models, the elliptic RS model is the most general one and other types such as the rational, hyperbolic, and trigonometric types are just various degenerations of the elliptic one. In this paper, we shall study the elliptic A_{n-1} type RS model with generic $n(n>2)$.

The Lax representation and its corresponding r-matrix structure for rational, hyperbolic, and trigonometric A_{n-1} type RS models were constructed by Avan and Talon.[9] The Lax representation for the elliptic RS models was constructed by Ruijsenaars,[10] and the corresponding r-matrix structure was given by Nijhoff, Kuznetsov, and Sklyanin[11] and Suris.[12] It turns out that the r-matrix structure of the RS model is given in terms of a quadratic Poisson–Lie bracket with dynamical r-matrices (i.e., the r-matrix depends upon the dynamical variables). Particularly, in contrast to the dynamical Yang–Baxter equation of the r-matrix structure of the CM model, the generalized Yang–Baxter relations for the quadratic Poisson–Lie bracket with a dynamical r-matrix is still an open problem.[11] Since the Poisson bracket of the Lax operator is no longer closed, the quantum version of such classical L-operator has not been able to be constructed.

It is well known that the Lax representation for completely integrable models is not unique. It has been recognized[13,14] that the r-matrix of a model can be changed drastically by the choice of Lax representation. In our former work,[14] we succeeded in constructing a new Lax operator (cf. Krichever's[15]) for the elliptic A_{n-1} CM model and showing that the corresponding r-matrix is a nondynamical one, which is the classical Z_n-symmetric elliptic r-matrix.[16,14] Very recently, we found a "good" Lax operator for the elliptic RS model with a very special case $n=2$.[17] In the present paper, extending our former work in Ref. 17, we construct a "good" Lax operator (in the

a)Author to whom all correspondence should be addressed: Institute of Modern Physics, Northwest University, Xian 710069, China; electronic mail: wlyang@phy.nwu.edu.cn

0022-2488/2000/41(1)/357/13/$17.00

sense that it has a nondynamical r-matrix structure) for the elliptic RS model with a general $n(n>2)$.

The paper is organized as follows. In Sec. II, we construct the dynamical twisting relations of the classical r-matrix for the quadratic Poisson–Lie bracket. The condition that a "good" Lax representation could exist is found. In Sec. III, we provide brief reviews of the work of Nijhoff and co-workers on the dynamical r-matrix of the elliptic RS model. In Sec. IV, we construct a "good" Lax representation for an elliptic RS model with generic n, and obtain the corresponding nondynamical r-matrix structure. The quantum version L-operator of the Lax operator is constructed in Sec. V. Finally, we give a summary and discussions. The Appendix contains some detailed calculations.

II. THE DYNAMICAL TWISTING OF CLASSICAL r-MATRIX

In this section we will give some general theories of the completely integrable finite particles systems.

A Lax pair (L, M) consists of two functions on the phase space of the system with values in some Lie algebra g, such that the evolution equations may be written in the following form

$$\frac{dL}{dt}=[L,M], \tag{II.1}$$

where [ ] denotes the bracket in the Lie algebra g. The interest in the existence of such a pair lies in the fact that it allows for an easy construction of conserved quantities (integrals of motion). It follows that the adjoint-invariant quantities $\mathrm{tr}\,L^l$ $(l=1,...,n)$ are the integrals of the motion. In order to implement Liouville theorem onto this set of possible action variables we need them to be Poisson commuting. As shown in Ref. 13, the commutativity of the integrals $\mathrm{tr}\,L^l$ of the Lax operator can be deduced from the fact that the fundamental Poisson bracket $\{L_1(u),L_2(v)\}$ could be represented in the linear commutator form

$$\{L_1(u),L_2(v)\}=[r_{12}(u,v),L_1(u)]-[r_{21}(v,u),L_2(v)], \tag{II.2}$$

or quadratic form[18]

$$\{L_1(u),L_2(v)\}=L_1(u)L_2(v)r^-_{12}(u,v)-r^+_{12}(v,u)L_1(u)L_2(v)$$
$$+L_1(u)s^+_{12}(u,v)L_2(v)-L_2(v)s^-_{12}(u,v)L_1(u), \tag{II.3}$$

where we have used the notation

$$L_1\equiv L\otimes 1,\quad L_2\equiv 1\otimes L,\quad a_{21}=Pa_{12}P,$$

and P is the permutation operator such that $Px\otimes y=y\otimes x$.

The dynamical twisting of the linear Poisson–Lie bracket (II.2) was studied in Ref. 14 (we refer therein) and also studied by Babelon and Viallet.[13] We are to investigate the general dynamical twisting of the quadratic Poisson–Lie bracket (II.3).

In order to define a consistent Poisson bracket, one should impose some constraints on the r-matrices. The skew-symmetry of the Poisson bracket requires that

$$r^{\pm}_{21}(v,u)=-r^{\pm}_{12}(u,v),\quad s^+_{21}(v,u)=s^-_{12}(u,v), \tag{II.4}$$

$$r^+_{12}(u,v)-s^+_{12}(u,v)=r^-_{12}(u,v)-s^-_{12}(u,v). \tag{II.5}$$

As for the numerical r-matrices $r^{\pm}(u,v)$, $s^{\pm}(u,v)$ case, some constraint conditions (sufficient condition) imposed on the r-matrices to satisfy the Jacobi identity were given by Freidel and Maillet.[18] However, generally speaking, the Jacobi identity for the dynamical r-matrices $r^{\pm}(u,v)$, $s^{\pm}(u,v)$ would take a very complicated form.

It should be remarked that such a classification (from dynamical and nondynamical r-matrix structure) is by no means unique, which drastically depends on the Lax representation which one chooses for a system. Therefore, there is no one-to-one correspondence between a given dynamical system and a defined r-matrix. The same dynamical system may have several Lax representations and several r-matrix.[14] The different Lax representations of a system are conjugated with each other. Namely, if $(\tilde{L},\tilde{M})$ is one of other Lax pair of the same dynamical system conjugated with the old one (L, M), it means that

$$\tilde{L}(u)=g(u)L(u)g^{-1}(u), \quad \tilde{M}(u)=g(u)M(u)g^{-1}(u)-\left(\frac{d}{dt}g(u)\right)g^{-1}(u), \tag{II.6}$$

where $g(u)\in G$ whose Lie algebra is g. Then, we have

Proposition 1: The Lax pair $(\tilde{L},\tilde{M})$ has the following r-matrix structure:

$$\begin{aligned}\{\tilde{L}_1(u),\tilde{L}_2(v)\}=&\tilde{L}_1(u)\tilde{L}_2(v)\tilde{r}^-_{12}(u,v)-\tilde{r}^+_{12}(v,u)\tilde{L}_1(u)\tilde{L}_2(v)\\&+\tilde{L}_1(u)\tilde{s}^+_{12}(u,v)\tilde{L}_2(v)-\tilde{L}_2(v)\tilde{s}^-_{12}(u,v)\tilde{L}_1(u),\end{aligned} \tag{II.7}$$

where

$$\tilde{r}^-_{12}(u,v)=g_1(u)g_2(v)r^-_{12}(u,v)g_1^{-1}(u)g_2^{-1}(v)-\tilde{\Delta}_{12}(u,v)+\tilde{\Delta}_{21}(v,u),$$

$$\tilde{r}^+_{12}(u,v)=g_1(u)g_2(v)r^+_{12}(u,v)g_1^{-1}(u)g_2^{-1}(v)-\tilde{\Delta}^{(1)}_{12}(u,v)+\tilde{\Delta}^{(1)}_{21}(v,u),$$

$$\tilde{s}^+_{12}(u,v)=g_1(u)g_2(v)s^+_{12}(u,v)g_1^{-1}(u)g_2^{-1}(v)-\tilde{\Delta}_{21}(v,u)-\tilde{\Delta}^{(1)}_{12}(u,v),$$

$$\tilde{s}^-_{12}(u,v)=g_1(u)g_2(v)s^-_{12}(u,v)g_1^{-1}(u)g_2^{-1}(v)-\tilde{\Delta}_{12}(u,v)-\tilde{\Delta}^{(1)}_{21}(v,u),$$

$$\tilde{\Delta}_{12}(u,v)=\tilde{L}_2^{-1}(v)\Delta_{12}(u,v), \quad \tilde{\Delta}^{(1)}_{12}(u,v)=\Delta_{12}(u,v)\tilde{L}_2^{-1}(v),$$

$$\begin{aligned}\Delta_{12}(u,v)=&+\tfrac{1}{2}[\{g_1(u),g_2(v)\}g_1^{-1}(u)g_2^{-1}(v),g_2(v)L_2(v)g_2^{-1}(v)]\\&\times g_2(v)\{g_1(u),L_2(v)\}g_1^{-1}(u)g_2^{-1}(v)\end{aligned}$$

and the properties of (II.4) and (II.5) are conserved

$$\tilde{r}^{\pm}_{21}(v,u)=-\tilde{r}^{\pm}_{12}(u,v), \quad \tilde{s}^+_{21}(v,u)=\tilde{s}^-_{12}(u,v),$$

$$\tilde{r}^+_{12}(u,v)-\tilde{s}^+_{12}(u,v)=\tilde{r}^-_{12}(u,v)-\tilde{s}^-_{12}(u,v).$$

Proof: The proof is directly substituting (II.6) into the fundamental Poisson bracket (II.3) and use the following identity:

$$[[a_{12},L_1],L_2]=[[a_{12},L_2],L_1],$$

where a_{12} is any matrix on $g\otimes g$. □

It can be seen that: (I) The Lax operator $L(u)$ is transferred as a similarity transformation from the different Lax representation; (II) the corresponding M has undergone the usual gauge transformation; (III) the r-matrices are transferred as some generalized gauge transformation, which can be considered as the generalized classical version of the dynamically twisting relation of the quantum R-matrix.[19] Therefore, it is of great value to find a "good" Lax representation for a system if it exists, in which the corresponding r-matrices are all nondynamical ones and

$r^{+}_{12}(u,v)=r^{-}_{12}(u,v)$, $s^{\pm}_{12}(u,v)=0$. In this special case, the corresponding Poisson–Lie bracket becomes the Sklyanin bracket and the well-studied theories[20,21] can be directly applied in the system.

Corollary 1: For given Lax pair (L, M) and the corresponding r-matrices, if there exist $g(u)$ *satisfied*

$$g_1(u)g_2(v)s^{+}_{12}(u,v)g_1^{-1}(u)g_2^{-1}(v)-\tilde{\Delta}_{21}(v,u)-\tilde{\Delta}^{(1)}_{12}(u,v)=0,$$
$$g_1(u)g_2(v)s^{-}_{12}(u,v)g_1^{-1}(u)g_2^{-1}(v)-\tilde{\Delta}_{12}(u,v)-\tilde{\Delta}^{(1)}_{21}(v,u)=0, \tag{II.8}$$

$$\partial_{g_i}h_{12}=\partial_{p_j}h_{12}=0, \tag{II.9}$$

where

$$h_{12}(u,v)=g_1(u)g_2(v)r^{-}_{12}(u,v)g_1^{-1}(u)g_2^{-1}(v)-\tilde{\Delta}^{(1)}_{12}(u,v)+\tilde{\Delta}_{21}(v,u)$$
$$\equiv g_1(u)g_2(v)r^{+}_{12}(u,v)g_1^{-1}(u)g_2^{-1}(v)-\tilde{\Delta}^{(1)}_{12}(u,v)+\tilde{\Delta}^{(1)}_{21}(v,u), \tag{II.10}$$

the nondynamical Lax representation with Sklyanin Poisson–Lie Bracket of the system would exist.

The main purpose of this paper is to find a "good" Lax representation for the elliptic RS model with generic $n(n>2)$.

III. REVIEW OF THE ELLIPTIC RS MODEL

We first define some elliptic functions

$$\theta^{(j)}(u)=\theta\begin{bmatrix}\frac{1}{2}-\frac{j}{n}\\ \frac{1}{2}\end{bmatrix}(u,n\tau),\qquad \sigma(u)=\theta\begin{bmatrix}\frac{1}{2}\\ \frac{1}{2}\end{bmatrix}(u,\tau), \tag{III.1}$$

$$\theta\begin{bmatrix}a\\ b\end{bmatrix}(u,\tau)=\sum_{m=-\infty}^{\infty}\exp\{\sqrt{-1}\pi[(m+a)^2\tau+2(m+a)(z+b)]\},$$
$$\theta'^{(j)}(u)=\partial_u\{\theta^{(j)}(u)\},\quad \sigma'(u)=\partial_u\{\sigma(u)\},\quad \xi(u)=\partial_u\{\ln\sigma(u)\}, \tag{III.2}$$

where τ is a complex number with $\mathrm{Im}(\tau)>0$.

The Ruijsenaars–Schneider model is the system of n one-dimensional relativistic particles interacting by the two-body potential. In terms of the canonical variables p_i, q_i $(i=1,\ldots,n)$ enjoying in the canonical Poisson bracket

$$\{p_i,p_j\}=0,\quad \{q_i,q_j\}=0,\quad \{q_i,p_j\}=\delta_{ij},$$

the Hamiltonian of the system is expressed as[10]

$$H=mc^2\sum_{j=1}^{n}\cosh p_j\prod_{k\neq j}\left\{\frac{\sigma(q_{jk}+\gamma)\sigma(q_{jk}-\gamma)}{\sigma^2(q_{jk})}\right\}^{1/2},\quad q_{jk}=q_j-q_k. \tag{III.3}$$

Here, m denotes the particle mass, c denotes the speed of light, γ is the coupling constant. The Hamiltonian (III.3) is known to be completely integrable. The most effective way to show its integrability is to construct the Lax representation for the system (namely, to find the classical Lax operator). One L-operator for the elliptic RS model was given by Ruijsenaars[10]

$$L_R(u)^i_j=\frac{e^{p_j}\sigma(\gamma+u+q_{ji})}{\sigma(\gamma+q_{ji})\sigma(u)}\prod_{k\neq j}^{n}\left\{\frac{\sigma(q_{jk}+\gamma)\sigma(q_{jk}-\gamma)}{\sigma^2(q_{jk})}\right\}^{1/2},\quad i,j=1,\dots,n. \tag{III.4}$$

Alternatively, we adopt another Lax operator $\tilde{L}_R$, which is similar to that of Nijhoff Kutznetsov, and Sklyanin in Ref. 11,

$$\tilde{L}_R(u)^i_j=\frac{e^{p_j}\sigma(\gamma+u+q_{ji})}{\sigma(u)\sigma(\gamma+q_{ji}}\prod_{k\neq j}\frac{\sigma(\gamma+q_{jk})}{\sigma(q_{jk})}. \tag{III.5}$$

The relation of $\tilde{L}_R$ with the standard Ruijsenaars' $L_R(u)$ can be obtained from a Poisson map (or a canonical transformation)

$$q_i\rightarrow q_i,\quad p_i\rightarrow p_i+\frac{1}{2}\ln\prod_{k\neq i}\frac{\sigma(q_{ik}+\gamma)}{\sigma(q_{ik}-\gamma)}. \tag{III.6}$$

Proposition 2: The map defined in (III.6) is a Poisson map.
Proof: Proposition 2 can be proven by considering the symplectic two-form

$$\begin{aligned}\sum_i d\left(p_i+\frac{1}{2}\ln\prod_{k\neq i}\frac{\sigma(q_{ik}+\gamma)}{\sigma(q_{ik}-\gamma)}\right)\wedge dq_i&=\sum_i dp_i\wedge dq_i-\frac{1}{2}\sum_{k\neq i}\left(\frac{\sigma'(q_{ik}+\gamma)}{\sigma(q_{ik}+\gamma)}-\frac{\sigma(q_{ik}-\gamma)}{\sigma(q_{ik}-\gamma)}\right)dq_k\wedge dq_i\\&=\sum_i dp_i\wedge dq_i-\frac{1}{2}\sum_{k<i}\left\{\left(\frac{\sigma'(q_{ik}+\gamma)}{\sigma(q_{ik}+\gamma)}+\frac{\sigma'(q_{ik}-\gamma)}{\sigma(q_{ik}-\gamma)}\right)\right.\\&\quad\left.-\left(\frac{\sigma'(q_{ik}-\gamma)}{\sigma(q_{ik}-\gamma)}+\frac{\sigma'(q_{ik}+\gamma)}{\sigma(q_{ik}+\gamma)}\right)\right\}dq_k\wedge dq_i=\sum_i dp_i\wedge dq_i,\end{aligned}$$

where we have used the property that the elliptic function $\sigma(u)$ is an odd function with regard to argument u. □

It is well known that the Poisson bracket is invariant under the Poisson map. Hence the study of the r-matrix structure for the standard Ruijsenaars Lax operator $L_R(u)$ is equivalent to that of Lax operator $\tilde{L}_R(u)$.

Following the work of Nijhoff, Kuznetsov, and Sklyanin,[11] the fundamental Poisson bracket of the Lax operator $\tilde{L}_R(u)$ can be given in the following quadratic r-matrix form with a dynamical r-matrices:

$$\begin{aligned}\{\tilde{L}_R(u)_1,\tilde{L}_R(v)_2\}&=\tilde{L}_R(u)_1\tilde{L}_R(v)_2r^-_{12}(u,v)-r^+_{12}(v,u)\tilde{L}_R(u)_1\tilde{L}_R(v)_2\\&\quad+\tilde{L}_R(u)_1s^+_{12}(u,v)\tilde{L}_R(v)_2-\tilde{L}_R(v)_2s^-_{12}(u,v)\tilde{L}_R(u)_1,\end{aligned} \tag{III.7}$$

where

$$r^-_{12}(u,v)=a_{12}(u,v)-s_{12}(u)+s_{21}(v),\quad r^+_{12}(u,v)=a_{12}(u,v)+u^+_{12}+u^-_{12},$$

$$s^+_{12}(u,v)=s_{12}(u)+u^+_{12},\quad s^-_{12}(u,v)=s_{21}(v)-u^-_{12},$$

and

$$a_{12}(u,v)=r^0_{12}(u,v)+\sum_{i=1}\xi(u-v)e_{ii}\otimes e_{ii}+\sum_{i\neq j}\xi(q_{ij})e_{ii}\otimes e_{jj},$$

$$r^0_{12}(u,v)=\sum_{i\neq j}\frac{\sigma(q_{ij}+u-v)}{\sigma(q_{ij})\sigma(u-v)}e_{ij}\otimes e_{ji},\quad s_{12}(u)=\sum_{i,j}(\tilde{L}_R(u)\partial_\gamma\tilde{L}_R(u))^i_je_{ij}\otimes e_{jj},$$

$$u_{12}^{\pm}=\sum_{i,j}\xi(q_{ji}\pm\gamma)e_{ii}\otimes e_{jj}.$$

The matrix element of e_{ij} is equal to $(e_{ij})l/k=\delta_{il}\delta_{jk}$. It can be checked that the following symmetric condition holds for the r-matrices $r_{12}^{\pm}(u,v)$ and $s_{12}^{\pm}(u,v)$:

$$r_{21}^{\pm}(v,u)=-r_{12}^{\pm}(u,v),\quad s_{21}^{+}(v,u)=s_{12}^{-}(u,v), \tag{III.8}$$

$$r_{12}^{+}(u,v)-s_{12}^{+}(u,v)=r_{12}^{-}(u,v)-s_{12}^{-}(u,v). \tag{III.9}$$

The classical r-matrices $r_{12}^{\pm}(u,v)$, $s_{12}^{\pm}(u,v)$ are of dynamical ones (i.e., the matrix element depends upon the dynamical variables q_i). The quadratic Poisson bracket (III.7) and the symmetric conditions of (III.8) and (III.9) lead to the evolution integrals $\mathrm{tr}(\tilde{L}_R(u))^l$.

Due to the r-matrices depending on the dynamical variables, the Poisson bracket of $\tilde{L}_R(u)$ is no longer closed. Due to the complexity of the r-matrices (III.7) it is still an open problem to check the generalized Yang–Baxter relations for the RS model. Moreover, the quantum version of the algebraic relation (III.7) is still not found. The same situation also occurs for the standard Lax operator $L_R(u)$, and the corresponding r-matrices were given by Suris.[12]

IV. THE "GOOD" LAX REPRESENTATION OF THE ELLIPTIC RS MODEL AND ITS r-MATRIX

The L-operator of the elliptic RS model given by Ruijsenaars $L_R(u)$ in (III.4) [or its Poisson equivalent counterpart $\tilde{L}_R(u)$ in (III.5)] and corresponding r-matrix $r_{12}(u,v)$ given by Suris,[12] (or given by Nijhoff, Kuznetov, and Sklyanin[11]) leads to some difficulties in the investigation of the RS model. This motivates us to find a "good" Lax representation of the RS model. As see from Proposition 1 and Corollary 1 in Sec. II, this means finding $g(u)$ which satisfies (II.8) and (II.9). In our previous work,[17] we succeeded in finding such a $g(u)$ for the elliptic RS model with a special case $n=2$. Fortunately, we could also find such a $g(u)$ for the elliptic RS model with a generic $n(n>2)$ (This kind of L-operator does not always exist for a general completely integrable system). The fundamental Poisson bracket of this new L-operator $L(u)$ would be expressed in the Sklyanin Poisson–Lie bracket form with a numeric r-matrix. The corresponding r-matrix is the classical Z_n-symmetric r-matrix in Ref. 14. Namely, the elliptic RS and the corresponding nonrelativistic version—the elliptic A_{n-1} CM model,[14] are governed by the exact same r-matrix (cf. Ref. 12) in some gauge. In order to compare with the L-operator given by Ruijsenaars $L_R(u)$ and its Poisson equivalence $\tilde{L}_R(u)$, we call this L-operator the new Lax operator (alternatively, a "good" Lax operator).

Set an $n\otimes n$ matrix $A(u;q)$,

$$A(u;q)_j^i\equiv A(u;q_1,q_2,\ldots,q_n)_j^i=\theta^{(i)}\left(u+nq_j-\sum_{k=1}^{n}q_k+\frac{n-1}{2}\right). \tag{IV.1}$$

We remark that $A(u,q)_j^i$ corresponds to the interwiner function $\varphi_j^{(i)}$ between the Z_n-symmetric Belavin model and the $A_{n-1}^{(1)}$ face model[22] in Ref. 23

Define

$$g(u)=A(u;q)\Lambda(q),\quad \Lambda(q)_j^i=h_i(q)\delta_j^i,$$

$$h_i(q)\equiv h_i(q_1,\ldots,q_n)=\frac{1}{\Pi_{l\neq i}\sigma(q_{il})}.$$

Let us construct the new Lax operator $L(u)$,

$$L(u)=g(u)\tilde{L}_R(u)g^{-1}(u). \tag{IV.2}$$

It will turn out that such a Lax operator $L(u)$ gives a "good" Lax representation for the elliptic RS model. This is our main result of this paper. To recover this, let us express the "good" Lax operator $L(u)$ more explicitly.

Proposition 3: The Lax operator $L(u)$ can be rewritten in the factorized form

$$L(u)^i_j=\sum_{k=1}^{n}\frac{1}{\sigma(\gamma)}A(u+n\gamma;q)^i_kA^{-1}(u;q)^k_je^{p_k},\quad i,j=1,2,...,n. \tag{IV.3}$$

Proof: First, let us introduce a matrix $T(u)$ with matrix elements

$$T(u)^i_j=\sum_k e^{p_j}A^{-1}(u;q)^i_kA(u+n\gamma;q)^k_j.$$

From the definition of $A(u;q)^i_j$ and the determinant formula of Vandermonde type[23]

$$\det[\theta^{(j)}(u_k)]=\text{const.}\times\sigma\left(\frac{1}{n}\sum_k u_k-\frac{n-1}{2}\right)\prod_{1\le j<k\le n}\sigma\left(\frac{u_k-u_j}{n}\right), \tag{IV.4}$$

where the const. does not depend upon $\{u_k\}$, we have

$$\sum_k A^{-1}(u;q)^i_kA(u+n\gamma;q)^k_j=\frac{\sigma(\gamma+u+q_{ji})}{\sigma(u)}\prod_{k\neq i}\frac{\sigma(\gamma+q_{jk})}{\sigma(q_{ik})}.$$

Namely,

$$T(u)^i_j=\frac{e^{p_j}\sigma(\gamma+u+q_{ji})}{\sigma(u)}\prod_{k\neq i}\frac{\sigma(\gamma+q_{jk})}{\sigma(q_{ik})}$$
$$=\frac{1}{\prod_{k\neq i}\sigma(q_{ik})}\left\{\frac{e^{p_j}\sigma(\gamma+u+q_{ji})\sigma(\gamma)}{\sigma(u)\sigma(\gamma+q_{ji})}\prod_{k\neq j}\frac{\sigma(\gamma+q_{jk})}{\sigma(q_{jk})}\right\}\prod_{k\neq j}\sigma(q_{jk}).$$

Then, we obtain

$$\frac{1}{\sigma(\gamma)}\sum_k A(u+n\gamma;q)^i_kA^{-1}(u;q)^k_je^{p_k}$$
$$=\frac{1}{\sigma(\gamma)}\sum_{m,l}A(u;q)^i_mT^m_l(u)A^{-1}(u;q)^l_j$$
$$=\sum_{m,l}\frac{A(u;q)^i_m}{\prod_{k\neq m}\sigma(q_{mk})}\left\{\frac{e^{-p_l}\sigma(\gamma+u+q_{lm})}{\sigma(u)\sigma(\gamma+q_{lm})}\prod_{k\neq l}\frac{\sigma(\gamma+q_{lk})}{\sigma(q_{lk})}\right\}A^{-1}(u;q)^l_j\prod_{k\neq l}\sigma(q_{lk})$$
$$=\sum_{m,l}g(u)^i_m\tilde{L}_R(u)^m_lg^{-1}(u)^l_j\equiv L(u)^i_j.$$

□

Let us consider the nonrelativistic limit of our Lax operator $L(u)$. First, rescale the momenta $\{p_i\}$, the coupling constant γ, and the Lax operator $L(u)$ as follows:[11]

$$p_i:=-\beta p_i',\quad n\gamma:=\beta s,\quad L(u):=\sigma\left(\frac{\beta s}{n}\right)L'(u), \tag{IV.5}$$

where p_i' is the conjugated momenta of q_i in the CM model.

Then the nonrelativistic limit is obtained by taking $\beta\to 0$, we have the following asymptotic properties:

$$L'(u)^i_j=\delta^i_j-\beta\left(\sum_k \{A(u;q)^i_k A^{-1}(u;q)^k_j p'_k - s\partial_u(A(u;q)^i_k)A^{-1}(u;q)^k_j\}\right)+0(\beta^2).$$

If we make the canonical transformation

$$p_i'\to p_i'-\frac{s}{n}\frac{\partial}{\partial q_i}\ln M(q),\quad M(q)=\prod_{i<j}\sigma(q_{ij}),$$

we obtain the "good" Lax operator of the elliptic A_{n-1} CM model in Ref. 14,

$$L_{\mathrm{CM}}(u)^i_j=-\lim_{\beta\to 0}\frac{L'(u)^i_j-\delta^i_j}{\beta}\bigg|_{p_i'\to p_i'-(s/n)(\partial/\partial q_i)\ln M(q)}. \tag{IV.6}$$

Now, we have a position to calculate the r-matrix structure of the "good" Lax operator $L(u)$ for the elliptic RS model. From Proposition 3 and through the straightforward calculation, we have the main theorem of this paper:

Theorem 1: *(Main Theorem) The fundamental Poisson bracket of $L(u)$ can be given in the quadratic Poisson–Lie form with a nondynamical r-matrix (or Sklyanin bracket)*

$$\{L_1(u),L_2(v)\}=[r_{12}(u-v),L_1(u)L_2(v)], \tag{IV.7}$$

where the numeric r-matrix $r_{12}(u)$ is the classical Z_n-symmetric r-matrix[14]

$$r^{lk}_{ij}(v)=\begin{cases}(1-\delta^l_i)\dfrac{\theta'^{(0)}(0)\theta^{(i-j)}(v)}{\theta^{(i-j)}(v)\theta^{(i-l)}(0)}+\delta^l_i\delta^k_j\left(\dfrac{\theta'^{(i-j)}(v)}{\theta^{(i-j)}(v)}-\dfrac{\sigma'(v)}{\sigma(v)}\right) & \text{if } i+j=l+k \bmod n\\ 0 \quad \text{otherwise.}\end{cases} \tag{IV.8}$$

Remark: I. The elliptic RS and CM model are governed by the exact same nondynamical r-matrix in the special Lax representation.

II. It was shown in Ref. 14 that such a Z_n-symmetric r-matrix satisfies the nondynamical classical Yang–Baxter equation

$$[r_{12}(v_1-v_2),r_{13}(v_1-v_3)]+[r_{12}(v_1-v_2),r_{23}(v_2-v_3)]+[r_{13}(v_1-v_3),r_{23}(v_2-v_3)]=0,$$

and enjoys in the antisymmetric properties

$$-r_{21}(-v)=r_{12}(v). \tag{IV.9}$$

Moreover, the r-matrix $r_{12}(u)$ also enjoys in the $Z_n\otimes Z_n$ symmetry

$$r_{12}(v)=(a\otimes a)r_{12}(v)(a\otimes a)^{-1}\quad \text{for } a=g,h, \tag{IV.10}$$

where the $n\times n$ matrices h, g are defined in Sec. V.

Corollary 2: The Lax operator $L_{\mathrm{CM}}(u)$ of the elliptic A_{n-1} CM model in (IV.6) satisfies the nondynamical linear Poisson–Lie bracket

$$\{L_{CM}(u)_1,L_{\mathrm{CM}}(v)_2\}=[r_{12}(u-v),L_{\mathrm{CM}}(u)_1+L_{\mathrm{CM}}(v)_2]. \tag{IV.11}$$

The direct proof that such a "good" (classical) Lax operator $L_{\mathrm{CM}}(u)$ of the elliptic A_{n-1} CM model satisfies (IV.11) was given in Ref. 14.

V. THE QUANTUM L-OPERATOR FOR THE ELLIPTIC QUANTUM RS MODEL

In this section, we will construct the quantum L-operator for the quantum elliptic RS model, which satisfies the nondynamical "$RLL=LLR$" relation.

We first introduce the elliptic Z_n-symmetric quantum R-matrix related to Z_n-symmetric Belavin model, which is the quantum version of the classical Z_n-symmetric r-matrix defined in (IV.8).

We define $n\times n$ matrices h, g, and I_α by

$$h_{ij}=\delta_{i+1,j \bmod n}, \quad g_{ij}=\omega^i\delta_{ij}, \quad I_{\alpha_1,\alpha_2}\equiv I_\alpha=g^{\alpha_2}h^{\alpha_1},$$

where $\alpha_1,\ \alpha_2\in Z_n$ and $\omega=\exp(2\pi(\sqrt{-1}/n))$. Define the Z_n-symmetric Belavin's R-matrix[24,22,23]

$$R_{ij}^{lk}(v)=\begin{cases}\dfrac{\theta'^{(0)}(0)\sigma(v)\sigma(\sqrt{-1}\hbar)}{\sigma'(0)\theta^{(0)}(v)\sigma(v+\sqrt{-1}\hbar)}\ \dfrac{\theta^{(0)}(v)\theta^{(i-j)}(v+\sqrt{-1}\hbar}{\theta^{(i-l)}(\sqrt{-1}\hbar)\theta^{(i-j)}(v)} & \text{if } i+j=l+k \bmod n\\ 0 \quad \text{otherwise,}\end{cases}\tag{V.1}$$

where $\hbar$ is the Planck's constant and $\sqrt{-1}\hbar$ is usually called the crossing parameter of the R-matrix. We remark that our R-matrix coincides with the usual one in Ref. 23 up to a scalar factor

$$\frac{\theta'^{(0)}(0)\sigma(v)}{\sigma'(0)\theta^{(0)}(v)}\prod_{j=1}^{n-1}\frac{\theta^{(j)}(v)}{\theta^{(j)}(0)},$$

which is to satisfy (V.4). The R-matrix satisfies the quantum Yang–Baxter equation (QYBE)

$$R_{12}(v_1-v_2)R_{13}(v_1-v_3)R_{23}(v_2-v_3)=R_{23}(v_2-v_3)R_{13}(v_1-v_3)R_{12}(v_1-v_2).\tag{V.2}$$

Moreover, the R-matrix enjoys in following $Z_n\otimes Z_n$ symmetric properties

$$R_{12}(v)=(a\otimes a)R_{12}(v)(a\otimes a)^{-1} \quad \text{for } a=g,h.\tag{V.3}$$

The Z_n-symmetric r-matrix has the following relation with its quantum counterpart:

$$\begin{aligned}&R_{12}(v)|_{\hbar=0}=1\otimes 1,\\ &R_{12}(v)=1\otimes 1+\sqrt{-1}\hbar r_{12}(v)+0(\hbar^2) \quad \text{when } \hbar\to 0.\end{aligned}\tag{V.4}$$

Now we construct the quantum version of L-operator $L(u)$. The usual canonical quantization procedure reads

$$p_j\to\hat{p}_j=-\sqrt{-1}\hbar\frac{\partial}{\partial q_j}, \quad q_j\to q_i, \quad j=1,\dots,n.$$

Then, the corresponding quantum L-operator $\hat{L}(u)$ consequently reads

$$\hat{L}(u)_l^m=\frac{1}{\sigma(\gamma)}\sum_{k=1}^{n}A(u+n\gamma;q)_k^m A^{-1}(u;q)_l^k e^{\hat{p}_k}=\frac{1}{\sigma(\gamma)}\sum_{k=1}^{n}A(u+n\gamma;q)_k^m A^{-1}(u;q)_l^k e^{-\sqrt{-1}\hbar(\partial/\partial q_k}.\tag{V.5}$$

It should be remarked that such a quantum L-operator is just the factorized difference representation for the elliptic L-operator.[23] So, we have

Theorem 2: *(References 23, 26 and 25) The quantum L-operator $\hat{L}(u)$ defined in (V.5) satisfies*

$$R_{12}(u-v)\hat{L}_1(u)\hat{L}_2(v)=\hat{L}_2(v)\hat{L}_1(u)R_{12}(u-v), \tag{V.6}$$

and $R_{12}(u)$ is the Z_n-symmrtric R-matrix.

The proof of Theorem 2 was given by Hou, Shi, and Yang in Ref. 23, by Quano and Fujii in Ref. 26, by Hasegawa in Ref. 25, through the face-vertex corresponding relations independently. The direct proof was also given in Ref. 27.

From the quantum L-operator $\hat{L}(u)$ and the fundamental relation $RLL-LLR$, Hasegawa constructed the skew-symmetric fusion of $\hat{L}(u)$ and succeeded in relating them with the elliptic type Macdonald operator in Ref. 25, which is actually equivalent to the quantum Ruijsenaar's operators.

VI. DISCUSSION

In this paper, we only consider the most general RS model—the elliptic RS model. Such a nondynamical r-matrix structure should exist for the degenerated case: the rational, hyperbolic, and trigonometric RS model.

From the results of Ref. 25 and 28, when the coupling constant $\gamma/\sqrt{-1}\hbar=$non-negative integer, the corresponding quantum L-operator $\hat{L}(u)$ has finite dimensional representation. This means that the states of the quantum RS model should degenerate in this special case.

ACKNOWLEDGMENTS

This work has been financially supported by National Natural Science Foundation of China. We would like to thank Heng Fan for a careful reading of the manuscript and many helpful comments. W.L.Y. was also partially supported by a grant from Northwest University.

APPENDIX: THE PROOF OF THEOREM 1

In this appendix, we give the proof of Theorem 1, which is the main result of this paper.

Lemma 1: The classical L-operator $L(u)$ for the elliptic RS model satisfies the following algebraic relations:

$$[r_{12}(u-v),L_1(u)L_2(v)]_{\alpha\beta}^{\rho\delta}=\sum_{i,j}\left\{A(u+n\gamma;q)_i^{\rho}A^{-1}(v;q)_{\alpha}^{i}e^{p_i}\frac{\partial}{\partial q_i}(A(v+n\gamma;q)_j^{\partial}A^{-1}(v;q)_{\beta}^{j})e^{p_j}\right.$$
$$\left.-A(v+n\gamma;q)_i^{\delta}A^{-1}(v;q)_{\beta}^{i}e^{p_i}\frac{\partial}{\partial q_i}(A(u+n\gamma;q)_j^{\rho}A^{-1}(u;q)_{\alpha}^{j})e^{p_j}\right\}.$$

Proof: Let us introduce the difference operators $\{\hat{D}_j\}$,

$$\hat{D}_j=\exp\left(-\sqrt{-1}\hbar\frac{\partial}{\partial q_j}\right)\quad\text{and}\quad\hat{D}_jf(q)=f(q_1,\ldots,q_{j-1},q_j-\sqrt{-1}\hbar,q_{j+1},\ldots,q_n).$$

Define

$$T(i,j)_{\alpha\beta}^{\rho\delta}=\begin{cases}\Sigma_{\rho',\delta'}R(u-v)_{\rho'\delta'}^{\rho\delta}A(u+n\gamma;q)_i^{\rho'}A^{-1}(u;q)_{\alpha}^{i}\hat{D}_i(A(v+n\gamma;q)_i^{\delta'}A^{-1}(v;q)_{\beta}^{i}) & \text{if } i=j\\ \Sigma_{\rho',\delta'}R(u-v)_{\rho',\delta'}^{\rho\delta}\{A(u+n\gamma;q)_i^{\rho'}A^{-1}(u;q)_{\alpha}^{i}\hat{D}_i(A(v+n\gamma;q)_j^{\delta'}A^{-1}(v;q)_{\beta}^{j}) & \\ \quad+A(u+n\gamma;q)_j^{\rho'}A^{-1}(u;q)_{\alpha}^{j}\hat{D}_j(A(v+n\gamma;q)_i^{\delta'}A^{-1}(v;q)_{\beta}^{i})\} & \text{if } i\neq j\end{cases},$$

and

$$G(i,j)^{\rho\delta}_{\alpha\beta}$$

$$=\begin{cases} \Sigma_{\rho',\delta'}R(u-v)^{\rho'\delta'}_{\alpha\beta}A(v+n\gamma;q)^{\delta}_{i}A^{-1}(v;q)^{i}_{\delta'}\hat{D}_i(A(u+n\gamma;q)^{\rho}_{i}A^{-1}(u;q)^{i}_{\rho'}) & \text{if } i=j \\ \Sigma_{\rho',\delta'}R(u-v)^{\rho'\delta'}_{\alpha\beta}\{A(v+n\gamma;q)^{\delta}_{i}A^{-1}(v;q)^{i}_{\delta'}\hat{D}_i(A(u+n\gamma;q)^{\rho}_{j}A^{-1}(u;q)^{j}_{\rho'}) \\ \quad +A(v+n\gamma;q)^{\delta}_{j}A^{-1}(v;q)^{j}_{\delta'}\hat{D}_j(A(u+n\gamma;q)^{\rho}_{i}A^{-1}(u;q)^{i}_{\rho'}) & \text{if } i\neq j. \end{cases}$$

The quantum L-operator $\hat{L}(u)$ satisfying the "$RLL=LLR$" relation results in

$$T(i,j)^{\rho\delta}_{\alpha\beta}=G(i,j)^{\rho\delta}_{\alpha\beta}. \tag{A1}$$

Considering the asymptotic properties when $\hbar\to 0$,

$$R_{12}(u)=1+\sqrt{-1}\hbar r_{12}(u)+0(\hbar^2),$$

$$\hat{D}_j=1-\sqrt{-1}\hbar\frac{\partial}{\partial q_j}+0(\hbar^2),$$

we have the following.

(I) If $i=j$:

$$\begin{aligned} T(i,j)^{\rho\delta}_{\alpha\beta}&\equiv T^{(0)}(i,j)^{\rho\delta}_{\alpha\beta}+\sqrt{-1}\hbar T^{(1)}(i,j)^{\rho\delta}_{\alpha\beta}+0(\hbar^2) \\ &=A(u+n\gamma;q)^{\rho}_{i}A^{-1}(u;q)^{i}_{\alpha}A(v+n\gamma;q)^{\delta}_{i}A^{-1}(v;q)^{i}_{\beta}+\sqrt{-1}\hbar\sum_{\rho',\delta'}r(u-v)^{\rho\delta}_{\rho'\delta'}A(u \\ &\quad +n\gamma;q)^{\rho'}_{i}A^{-1}(u;q)^{i}_{\alpha}A(v+n\gamma;q)^{\delta'}_{i}A^{-1}(v;q)^{i}_{\beta}-\sqrt{-1}\hbar A(u+n\gamma;q)^{\rho}_{i}A^{-1}(u;q)^{i}_{\alpha} \\ &\quad \times\frac{\partial}{\partial q_i}(A(v+n\gamma;q)^{\delta}_{i}A^{-1}(v;q)^{i}_{\beta})+0(\hbar^2). \end{aligned}$$

(II) If $i\neq j$:

$$\begin{aligned} T(i,j)^{\rho\delta}_{\alpha\beta}&\equiv T^{(0)}(i,j)^{\rho\delta}_{\alpha\beta}+\sqrt{-1}\hbar T^{(1)}(i,j)^{\rho\delta}_{\alpha\beta}+0(\hbar^2) \\ &=A(u+n\gamma;q)^{\rho}_{i}A^{-1}(u;q)^{i}_{\alpha}A(v+n\gamma;q)^{\delta}_{j}A^{-1}(v;q)^{j}_{\beta}+A(u+n\gamma;q)^{\rho}_{j}A^{-1}(u;q)^{j}_{\alpha} \\ &\quad \times A(v+n\gamma;q)^{\delta}_{i}A^{-1}(v;q)^{i}_{\beta}+\sqrt{-1}\hbar\sum_{\rho',\delta'}r(u-v)^{\rho\delta}_{\rho'\delta'}\{A(u+n\gamma;q)^{\rho'}_{i}A^{-1}(u;q)^{i}_{\alpha} \\ &\quad \times A(v+n\gamma;q)^{\delta'}_{j}A^{-1}(v;q)^{j}_{\beta}+A(u+n\gamma;q)^{\rho'}_{j}A^{-1}(u;q)^{j}_{\alpha}A(v+n\gamma;q)^{\delta'}_{i}A^{-1}(u;q)^{i}_{\beta}\} \\ &\quad -\sqrt{-1}\hbar\left\{A(u+n\gamma;q)^{\rho}_{i}A^{-1}(u;q)^{i}_{\alpha}\frac{\partial}{\partial q_i}(A(v+n\gamma;q)^{\delta}_{j}A^{-1}(v;q)^{j}_{\beta})\right. \\ &\quad \left.+A(u+n\gamma;q)^{\rho}_{j}A^{-1}(u;q)\frac{\partial}{\partial q_j}(A(v+n\gamma;q)^{\delta}_{i}A^{-1}(v;q)^{i}_{\beta})\right\}+0(\hbar^2). \end{aligned}$$

(III) If $i=j$:

$$G(i,j)^{\rho\delta}_{\alpha\beta}\equiv G^{(0)}(i,j)^{\rho\delta}_{\alpha\beta}+\sqrt{-1}\hbar G^{(1)}(i,j)^{\rho\delta}_{\alpha\beta}+0(\hbar^2)$$

$$=A(u+n\gamma;q)^{\rho}_{i}A^{-1}(u;q)^{i}_{\alpha}A(v+n\gamma;q)^{\delta}_{i}A^{-1}(v;q)^{i}_{\beta}+\sqrt{-1}\hbar\sum_{\rho',\delta'}r(u-v)^{\rho'\delta'}_{\alpha\beta}$$

$$\times A(v+n\gamma;q)^{\delta}_{i}A^{-1}(v;q)^{i}_{\delta'}A(u+n\gamma;q)^{\rho}_{i}A^{-1}(u;q)^{i}_{\rho'}$$

$$-\sqrt{-1}\hbar A(v+n\gamma;q)^{\delta}_{i}A^{-1}(v;q)^{i}_{\beta}\frac{\partial}{\partial q_i}(A(u+n\gamma;q)^{\rho}_{i}A^{-1}(u;q)^{i}_{\alpha}+0(\hbar^2).$$

(IV) If $i\neq j$:

$$G(i,j)^{\rho\delta}_{\alpha\beta}\equiv G^{(0)}(i,j)^{\rho\delta}_{\alpha\beta}+\sqrt{-1}\hbar G^{(1)}(i,j)^{\rho\delta}_{\alpha\beta}+0(\hbar^2)$$

$$=A(u+n\gamma;q)^{\rho}_{i}A^{-1}(u;q)^{i}_{\alpha}A(v+n\gamma;q)^{\delta}_{j}A^{-1}(v;q)^{j}_{\beta}+A(u+n\gamma;q)^{\rho}_{j}A^{-1}(u;q)^{j}_{\alpha}$$

$$\times A(v+n\gamma;q)^{\delta}_{i}A^{-1}(v;q)^{i}_{\beta}+\sqrt{-1}\hbar\sum_{\rho',\delta'}r(u-v)^{\rho'\delta'}_{\alpha\beta}\{A(v+n\gamma;q)^{\delta}_{i}A^{-1}(v;q)^{i}_{\delta'}$$

$$\times A(u+n\gamma;q)^{\rho}_{j}A^{-1}(u;q)^{j}_{\rho'}+A(v+n\gamma;q)^{\delta}_{j}A^{-1}(v;q)^{j}_{\delta'}A(u+n\gamma;q)^{\rho}_{i}A^{-1}(u;q)^{i}_{\rho'}\}$$

$$-\sqrt{-1}\hbar\left\{A(v+n\gamma;q)^{\delta}_{i}A^{-1}(v;q)^{i}_{\beta}\frac{\partial}{\partial q_i}(A(u+n\gamma;q)^{\rho}_{j}A^{-1}(u;q)^{j}_{\alpha})\right.$$

$$\left.+A(v+n\gamma;q)^{\delta}_{j}A^{-1}(v;q)^{j}_{\beta}\frac{\partial}{\partial q_j}(A(u+n\gamma;q)^{\rho}_{i}A^{-1}(u;q)^{i}_{\alpha})\right\}+0(\hbar^2).$$

Noting (A1) and considering the term of the first order with regard to $\hbar$, we have

$$T^{(1)}(i,j)^{\rho\delta}_{\alpha\beta}=G^{(1)}(i,j)^{\rho\delta}_{\alpha\beta}. \tag{A2}$$

Multiplying by $e^{p_i+p_j}$ from both sides of (A.2) and summing up for i and j, we have

$$\sum_{i,j}T^{(1)}(i,j)^{\rho\delta}_{\alpha\beta}e^{p_i}e^{p_j}=\sum_{i,j}G^{(1)}(i,j)^{\rho\delta}_{\alpha\beta}e^{p_i}e^{p_j}.$$

Due to the commutativity of $\{e^{p_j}\}$, we obtain

$$\sum_{\rho',\delta' i,j}\{r(u-v)^{\rho\delta}_{\rho'\delta'}\{A(u+n\gamma;q)^{\rho'}_{i}A^{-1}(u;q)^{i}_{\alpha}e^{p_i}A(v+n\gamma;q)^{\delta'}_{j}A^{-1}(v;q)^{j}_{\beta}e^{p_j}-r(u-v)^{\rho'\delta'}_{\alpha\beta}$$

$$\times A(v+n\gamma;q)^{\delta}_{i}A^{-1}(v;q)^{i}_{\delta'}e^{p_i}A(u+n\gamma;q)^{\rho}_{j}A^{-1}(u;q)^{j}_{\rho'}e^{p_j}\}$$

$$=\sum_{i,j}\left\{A(u+n\gamma;q)^{\rho}_{i}A^{-1}(v;q)^{i}_{\alpha}e^{p_i}\frac{\partial}{\partial q_i}(A(v+n\gamma;q)^{\delta}_{j}A^{-1}(v;q)^{j}_{\beta})e^{p_j}\right.$$

$$\left.-A(v+n\gamma;q)^{\delta}_{i}A^{-1}(v;q)^{i}_{\beta}e^{p_i}\frac{\partial}{\partial q_i}(A(u+n\gamma;q)^{\rho}_{j}A^{-1}(u;q)^{j}_{\alpha})e^{p_j}\right\}.$$

Namely, we have

$$[r_{12}(u-v),L_1(u)L_2(v)]_{\alpha\beta}^{\rho\delta}=\sum_{i,j}\left\{A(u+n\gamma;q)_i^{\rho}A^{-1}(v;q)_{\alpha}^{i}e^{p_i}\frac{\partial}{\partial q_i}(A(v+n\gamma;q)_j^{\delta}A^{-1}(v;q)_{\beta}^{j})e^{p_j}\right.$$
$$\left.-A(v+n\gamma;q)_i^{\delta}A^{-1}(v;q)_{\beta}^{i}e^{p_i}\frac{\partial}{\partial q_i}(A(u+n\gamma;q)_j^{\rho}A^{-1}(u;q)_{\alpha}^{j})e^{p_j}\right\}.$$

□

Now, we have a position to calculate the fundamental Poisson bracket of $L(u)$,

$$\{L_1(u),L_2(v)\}_{\alpha\beta}^{\rho\delta}=\{L(u)_{\alpha}^{\rho},L(v)_{\beta}^{\delta}\}$$
$$=\left\{\sum_i A(u+n\gamma;q)_i^{\rho}A^{-1}(u;q)_{\alpha}^{i}e^{p_i},\sum_j A(v+n\gamma;q)_j^{\delta}A^{-1}(v;q)_{\beta}^{j}e^{p_j}\right\}$$
$$=\sum_{i,j}\left\{A(u+n\gamma;q)_i^{\rho}A^{-1}(u;q)_{\alpha}^{i}e^{p_i}\frac{\partial}{\partial q_i}(A(v+n\gamma;q)_j^{\delta}A^{-1}(v;q)_{\beta}^{j})e^{p_j}\right.$$
$$\left.-A(v+n\gamma;q)_i^{\delta}A^{-1}(v;q)_{\beta}^{i}e^{p_i}\frac{\partial}{\partial q_i}(A(u+n\gamma;q)_j^{\rho}A^{-1}(u;q)_{\alpha}^{j})e^{p_j}\right\}$$
$$=[r_{12}(u-v),L_1(u)L_2(v)]_{\alpha\beta}^{\rho\delta}.$$

We have used the Lemma 1 in the last equation. Thus, we have

$$\{L_1(u),L_2(v)\}=[r_{12}(u-v),L_1(u)L_2(v)].$$

[1] F. Calogero, Lett. Nuovo Cimento **13**, 411 (1975); **16**, 77 (1976).
[2] J. Moser, Adv. Math. **16**, 1 (1975).
[3] S. N. M. Ruijsenaars and H. Schneider, Ann. Phys. (N.Y.) **170**, 370 (1986).
[4] O. Babelon and D. Bernard, Phys. Lett. B **317**, 363 (1993).
[5] H. W. Braden and R. Sasaki, Prog. Theor. Phys. **97**, 1003 (1997).
[6] F. W. Nijhoff, O. Ragnisco, and V. B. Kuznetsov, Commun. Math. Phys. **176**, 681 (1996).
[7] G. E. Arutyunov, S. A. Frolov, and P. B. Medredev, e-print hep-th/9607170; hep-th/9608013.
[8] A. Gorsky and N. Nekrasov, Nucl. Phys. B **414**, 213 (1994); **436**, 582 (1995).
[9] J. Avan and T. Talon, Phys. Lett. B **303**, 33 (1993).
[10] S. N. M. Ruijsenaars, Commun. Math. Phys. **110**, 191 (1987).
[11] F. W. Nijhoff, V. B. Kuznetsov, and E. K. Sklyanin, e-print solv-int/9603006.
[12] Y. B. Suris, Phys. Lett. A **225**, 253 (1997).
[13] O. Babelon and C. M. Viallet, Phys. Lett. B **237**, 411 (1989).
[14] B. Y. Hou and W. L. Yang, Lett. Math. Phys. **44**, 35 (1998); J. Phys. A **32**, 1475 (1999).
[15] I. M. Krichever, Funct. Anal. Appl. **14**, 282 (1980).
[16] B. Y. Hou and H. Wei, J. Math. Phys. **30**, 2750 (1989).
[17] B. Y. Hou and W.-L. Yang, e-print solv-int/9802015, Comm. Theor. Phys. (in press).
[18] L. Freidel and J.-M. Maillet, Phys. Lett. B **262**, 278 (1991).
[19] J. Avan, O. Babelon, and E. Billey, Commun. Math. Phys. **178**, 281 (1996).
[20] A. A. Belavin and V. G. Drinfeld, Soviet Sci. Rev. Sect. C **4**, 93 (1984).
[21] L. D. Faddeev and L. Takhtajan, *Hamiltonian Methods in the Theory of Solitons* (Springer, Berlin, 1987).
[22] M. Jimbo, T. Miwa, and M. Okado, Nucl. Phys. B **300**, 74 (1988).
[23] B. Y. Hou, K. J. Shi, and Z. X. Yang, J. Phys. A **26**, 4951 (1993).
[24] M. P. Richey and C. A. Tracy, J. Stat. Phys. **42**, 311 (1986).
[25] K. Hasegawa, Commun. Math. Phys. **187**, 289 (1997); J. Math. Phys. **35**, 6158 (1994).
[26] Y. H. Quano and A. Fujii, Mod. Phys. Lett. A **8**, 1585 (1993).
[27] B. Y. Hou, K. J. Shi, W. L. Yang, and Z. X. Yang, Phys. Lett. A **178**, 73 (1993).
[28] B. Y. Hou, K. J. Shi, W. L. Yang, Z. X. Yang, and S. Y. Zhou, Int. J. Mod. Phys. A **12**, 2927 (1997).

JOURNAL OF MATHEMATICAL PHYSICS VOLUME 42, NUMBER 10 OCTOBER 2001

The Lax pairs for elliptic C_n and BC_n Ruijsenaars–Schneider models and their spectral curves

Kai Chen[a)] and Bo-yu Hou[b)]
Institute of Modern Physics, Northwest University, Xi'an 710069, People's Republic of China

Wen-li Yang[c)]
Physikalisches Institut der Universitat Bonn, Nussallee 12, 53115 Bonn, Germany

(Received 1 March 2001; accepted for publication 17 May 2001)

We study the elliptic C_n and BC_n Ruijsenaars–Schneider models which are elliptic generalization of systems given in previous paper by the present authors [Chen *et al.*, J. Math. Phys. **41**, 8132 (2000)]. The Lax pairs for these models are constructed by Hamiltonian reduction technology. We show that the spectral curves can be parametrized by the involutive integrals of motion for these models. Taking nonrelativistic limit and scaling limit, we verify that they lead to the systems corresponding to Calogero–Moser and Toda types. © *2001 American Institute of Physics.* [DOI: 10.1063/1.1389091]

I. INTRODUCTION

The Ruijsenaars–Schneider (RS) and Calogero–Moser (CM) models, as integrable many-body models, recently have attracted remarkable attention and have been extensively studied. They describe one-dimensional N-particle systems with pairwise interaction. Their importance lies in various fields ranging from lattice models in statistics physics,[1,2] to the field theory and gauge theory,[3,4] to the Seiberg–Witten theory,[5] etc. In particular, the study of the RS model is of great importance since it is the integrable relativistic generalization of the corresponding CM model.[6,7]

The Lax pairs for the CM models in various root systems have been constructed by Olshanetsky, and Perelomov[8] using reduction on symmetric space, and are further given by Inozemtsev in Ref. 9 without spectral parameter. It was almost 20 years until D'Hoker and Phong[10] constructed the Lax pairs with a spectral parameter for each of the finite dimensional Lie algebras, and the untwisted and twisted Calogero–Moser systems were introduced. Subsequently, Bordner *et al.*[11–13] succeeded in giving two types of Lax pairs associated to all of the Lie algebra: the root type and the minimal type, with and without spectral parameters. Even for all of the Coxeter group, the construction has been obtained in Ref. 14. In Ref. 15, Hurtubise and Markman utilized a so-called "structure group," which combines a semisimple group and Weyl group, to construct CM systems associated with the Hitchin system, which in some degree generalizes the result of Refs. 10–14. Furthermore, the quantum version of the generalization has been developed in Refs. 16 and 17 at least for degenerate potentials of trigonometry after the works of Olshanetsky and Perelomov.[18]

So far as for the RS model, only the Lax pair of the A_{N-1} type RS model was obtained[6,2,19–22] and succeeded in recovering it by applying the Hamiltonian reduction procedure on a two-dimensional current group.[23] Although the commutative operators for the RS model based on various type Lie algebras have been given by Komori and co-workers,[24,25] Diejen,[26,27] and Hasegawa *et al.*,[1,28] the Lax integrability (or Lax pair representation) of the other type of RS model is still an open problem[5] except for a few degenerate cases.[27,30]

[a)]Electronic mail: kai@phy.nwu.edu.cn
[b)]Electronic mail: byhou@phy.nwu.edu.cn
[c)]Electronic mail: wlyang@th.physik.uni-bonn.de

In Refs. 29 and 30, we succeeded in constructing the Lax pair for C_n and BC_n RS systems only with the degenerate case (without spectral parameters). The r-matrix structure for them have been derived by Avan *et al.*[31] In this paper, we study the Lax pair for the most general C_n and BC_n RS models—the elliptic C_n and BC_n RS models. We shall give the explicit forms of Lax pairs for these systems by Hamiltonian reduction. We calculate the spectral curves for these systems, which are shown to be parametrized by a set of involutive integrals of motion. In particular, taking their nonrelativistic limit and scaling limit, we shall recover the systems of corresponding CM and Toda types, respectively. The other various degenerate cases are also be discussed and the connection between the Lax pair with a spectral parameter and the one without the spectral parameter is commented on.

The paper is organized as follows. The basic materials of the A_{N-1} RS model are reviewed in Sec. II, where we propose a Lax pair associating with the Hamiltonian which has a reflection symmetry with respect to the particles in the origin. This includes construction of a Lax pair for the A_{N-1} RS system together with its symmetry analysis. The main results are shown in Secs. III and IV. In Sec. III, we present the Lax pairs for the elliptic C_n and BC_n RS models by reducing the A_{N-1} RS model. The explicit forms for the Lax pairs are given in Sec. IV. Section V is devoted to deriving the spectral curves for these systems and their nonrelativistic counterpart, the Calogero–Moser model and scaling limit of the Toda model. Section VI shows the various degenerate limits: the trigonometric, hyperbolic, and rational cases. The last section is a brief summary and discussion.

II. THE A_{N-1}-TYPE RUIJSENAARS–SCHNEIDER MODEL

As a relativistic-invariant generalization of the A_{N-1}-type nonrelativistic Calogero–Moser model, the A_{N-1}-type Ruijsenaars–Schneider systems are completely integrable. The system's integrability was first shownd by Ruijsenaars.[6,7] The Lax pair for this model has been constructed in Refs. 6, 2, 19–22. Recent progress has shown that the compactification of higher dimension SUSY Yang–Mills theory and Seiberg–Witten theory can be described by this model.[5] Instanton correction of the prepotential associated with the sl_2 RS system has been calculated in Ref. 32.

A. Model and equations of motion

Let us briefly give the basics of this model. In terms of the canonical variables p_i, $x_i(i,j=1,\ldots,N)$ enjoying the canonical Poisson bracket

$$\{p_i,p_j\}=\{x_i,x_j\}=0, \quad \{x_i,p_j\}=\delta_{ij}, \tag{II.1}$$

the Hamiltonian of the A_{N-1} RS system reads

$$\mathcal{H}_{A_{N-1}}=\sum_{i=1}^{N}\left(e^{p_i}\prod_{k\neq i}f(x_i-x_k)+e^{-p_i}\prod_{k\neq i}g(x_i-x_k)\right), \tag{II.2}$$

where

$$\begin{aligned} f(x)&:=\frac{\sigma(x-\gamma)}{\sigma(x)}, \\ g(x)&:=f(x)|_{\gamma\to-\gamma}, \quad x_{ik}:=x_i-x_k, \end{aligned} \tag{II.3}$$

and γ denotes the coupling constant. Here, $\sigma(x)$ is the Weierstrass σ function which is an entire, odd and quasiperiodic function with a fixed pair of the primitive quasiperiods $2\omega_1$ and $2\omega_3$. It can be defined as the infinite product

$$\sigma(x)=x\prod_{w\in\Gamma\setminus\{0\}}\left(1-\frac{x}{w}\right)\exp\left[\frac{x}{w}+\frac{1}{2}\left(\frac{x}{w}\right)^2\right],$$

where $\Gamma=2\omega_1\mathbb{Z}+2\omega_3\mathbb{Z}$ is the corresponding period lattice. Defining a third dependent quasiperiod $2\omega_2=-2\omega_1-2\omega_3$, one has

$$\sigma(x+2\omega_k)=-\sigma(x)e^{2\eta_k(x+\omega_k)},\quad \zeta(x+2\omega_k)=\zeta(x)+2\eta_k,\quad k=1,2,3,$$

where

$$\zeta(x)=\frac{\sigma'(x)}{\sigma(x)},\quad \wp(x)=-\zeta'(x),$$

and $\eta_k=\zeta(\omega_k)$ satisfy $\eta_1\omega_3-\eta_3\omega_1=\pi i/2$.

Notice that in Ref. 6 Ruijsenaars used another "gauge" of the momenta such that two are connected by the following canonical transformation:

$$x_i\to x_i,\quad p_i\to p_i+\frac{1}{2}\ln\prod_{j\neq i}^{N}\frac{f(x_{ij})}{g(x_{ij})}. \tag{II.4}$$

The Lax matrix for this model has the form(for the general elliptic case)

$$L(\lambda)=\sum_{i,j=1}^{N}\frac{\Phi(x_i-x_j+\gamma,\lambda)}{\Phi(\gamma,\lambda)}\exp(p_j)b_jE_{ij}, \tag{II.5}$$

where

$$\Phi(x,\lambda):=\frac{\sigma(x+\lambda)}{\sigma(x)\sigma(\lambda)},\quad b_j:=\prod_{k\neq j}f(x_j-x_k),\quad (E_{ij})_{kl}=\delta_{ik}\delta_{jl} \tag{II.6}$$

and λ is the spectral parameter. It is shown in Refs. 21, 33, 34 that the Lax operator satisfies the quadratic fundamental Poisson bracket

$$\{L_1,L_2\}=L_1L_2a_1-a_2L_1L_2+L_2s_1L_1-L_1s_2L_2, \tag{II.7}$$

where $L_1=L_{A_{N-1}}\otimes Id, L_2=Id\otimes L_{A_{N-1}}$ and the four matrices read

$$\begin{aligned} a_1&=a+w,\quad s_1=s-w,\\ a_2&=a+s-s^*-w,\quad s_2=s^*+w. \end{aligned} \tag{II.8}$$

The forms of a,s,w are

$$a(\lambda,\mu)=-\zeta(\lambda-\mu)\sum_{k=1}^{N}E_{kk}\otimes E_{kk}-\sum_{k\neq j}\Phi(x_j-x_k,\lambda-\mu)E_{jk}\otimes E_{kj},$$

$$s(\lambda)=\zeta(\lambda)\sum_{k=1}^{N}E_{kk}\otimes E_{kk}+\sum_{k\neq j}\Phi(x_j-x_k,\lambda)E_{jk}\otimes E_{kk}, \tag{II.9}$$

$$w=\sum_{k\neq j}\zeta(x_k-x_j)E_{kk}\otimes E_{jj}.$$

The asterisk means

$$r^*=\Pi r\Pi\ \text{ with }\ \Pi=\sum_{k,j=1}^{N}E_{kj}\otimes E_{jk}. \tag{II.10}$$

Noticing that

$$L(\lambda)^{-1}{}_{ij}=\frac{\sigma(\gamma+\lambda)\sigma(\lambda+(N-1)\gamma)}{\sigma(\lambda)\sigma(\lambda+N\gamma)}\times\sum_{i,j=1}^{N}\frac{\Phi(x_i-x_j-\gamma,\lambda+N\gamma)}{\Phi(-\gamma,\lambda+N\gamma)}\exp(-p_i)b_j'E_{ij}, \tag{II.11}$$

$$b_j':=\prod_{k\neq j}g(x_j-x_k) \tag{II.12}$$

(the proof of the above-given identity is sketched in the Appendix) one can get the characteristic polynomials of $L_{A_{N-1}}$ (Refs. 35 and 34),

$$\det(L(\lambda)-v\cdot Id)=\sum_{j=0}^{N}\Phi(\gamma,\lambda)^{-j}(-v)^{N-j}\frac{\mathcal{H}_j^{+}}{\sigma^j(\gamma)}\times\frac{\sigma(\lambda+j\gamma)}{\sigma(\lambda)}, \tag{II.13}$$

and that of $L_{A_{N-1}}^{-1}$ by using formula given in Eq. (A8),

$$\det\left(\frac{\sigma(\lambda)\sigma(\lambda-N\gamma)}{\sigma(\lambda-\gamma)\sigma(\lambda-(N-1)\gamma)}\times L(\lambda-N\gamma)^{-1}-v\cdot Id\right)$$
$$=\sum_{j=0}^{N}\Phi(-\gamma,\lambda)^{-j}(-v)^{N-j}\times\frac{(\mathcal{H}_j^{-})}{\sigma^j(-\gamma)}\frac{\sigma(\lambda-j\gamma)}{\sigma(\lambda)}, \tag{II.14}$$

where $(\mathcal{H}_0^{\pm})_{A_{N-1}}=(\mathcal{H}_N^{\pm})_{A_{N-1}}=1$, and

$$(\mathcal{H}_i^{+})_{A_{N-1}}=\sum_{\substack{J\subset\{1,\ldots,N\}\\|J|=i}}\exp\left(\sum_{j\in J}p_j\right)\prod_{\substack{j\in J\\k\in\{1,\ldots,N\}\setminus J}}f(x_j-x_k), \tag{II.15}$$

$$(\mathcal{H}_i^{-})_{A_{N-1}}=\sum_{\substack{J\subset\{1,\ldots,N\}\\|J|=i}}\exp\left(\sum_{j\in J}-p_j\right)\prod_{\substack{j\in J\\k\in\{1,\ldots,N\}\setminus J}}g(x_j-x_k). \tag{II.16}$$

Defining

$$(\mathcal{H}_i)_{A_{N-1}}=(\mathcal{H}_i^{+})_{A_{N-1}}+(\mathcal{H}_i^{-})_{A_{N-1}}, \tag{II.17}$$

from the fundamental Poisson bracket Eq. (II.7), we can verify that

$$\{(\mathcal{H}_i)_{A_{N-1}},(\mathcal{H}_j)_{A_{N-1}}\}=\{(\mathcal{H}_i^{\varepsilon})_{A_{N-1}},(\mathcal{H}_j^{\varepsilon'})_{A_{N-1}}\}=0,\quad \varepsilon,\varepsilon'=\pm,\quad i,j=1,\ldots,N. \tag{II.18}$$

In particular, the Hamiltonian Eq. (II.2) can be rewritten as

$$\mathcal{H}_{A_{N-1}}\equiv\mathcal{H}_1=(\mathcal{H}_1^{+})_{A_{N-1}}+(\mathcal{H}_1^{-})_{A_{N-1}}$$
$$=\sum_{j=1}^{N}(e^{p_j}b_j+e^{-p_j}b_j')$$
$$=\mathrm{Tr}\left(L(\lambda)+\frac{\sigma(\lambda)\sigma(\lambda+N\gamma)}{\sigma(\gamma+\lambda)\,\sigma(\lambda+(N-1)\gamma)}L(\lambda)^{-1}\right). \tag{II.19}$$

It should be remarked that the set of integrals of motion Eq. (II.17) has a reflection symmetry which is the key property for the later reduction to C_n and BC_n cases, i.e., if we set

$$p_i \leftrightarrow -p_i, \quad x_i \leftrightarrow -x_i, \tag{II.20}$$

then the Hamiltonians flows $(\mathcal{H}_i)_{A_{N-1}}$ are invariant with respect to this symmetry.

The canonical equations of motion associated with the Hamiltonian flows $\mathcal{H}_1^+$ in its generic (elliptic) form read

$$\ddot{x}_i = \sum_{j \neq i} \dot{x}_i \dot{x}_j (V(x_{ij}) - V(x_{ji})), \quad i = 1, \ldots, N, \tag{II.21}$$

where the potential $V(x)$ is given by

$$V(x) = \zeta(x) - \zeta(x+\lambda), \tag{II.22}$$

in which $\zeta(x) = \sigma'(x)/\sigma(x)$. Here, $x_i = x_i(t)$, $p_i = p_i(t)$, and the superimposed dot denotes t differentiation.

B. The construction of Lax pair for the A_{N-1} RS model

As for the A_{N-1} RS model, a generalized Lax pair has been given in Refs. 6, 2, and 19–22. But there is a common character that the time evolution of the Lax matrix $L_{A_{N-1}}$ is associated with the Hamiltonian $(\mathcal{H}_1^+)_{A_{N-1}}$. We will see in Sec. III that the Lax pair cannot reduce from that kind of forms directly. Instead, we give a new Lax pair in which the evolution of $L_{A_{N-1}}$ is associated with the Hamiltonian $\mathcal{H}_{A_{N-1}}$,

$$\dot{L}_{A_{N-1}} = \{L_{A_{N-1}}, \mathcal{H}_{A_{N-1}}\} = [M_{A_{N-1}}, L_{A_{N-1}}], \tag{II.23}$$

where $M_{A_{N-1}}$ can be constructed with the help of (r,s) matrices as follows:

$$M_{A_{N-1}} = \mathrm{Tr}_2\left((s_1 - a_2)\left(1 \otimes \left(L(\lambda) - \frac{\sigma(\lambda)\sigma(\lambda+N\gamma)}{\sigma(\gamma+\lambda)\,\sigma(\lambda+(N-1)\gamma)} L(\lambda)^{-1} \right)\right)\right). \tag{II.24}$$

The explicit expression of entries for $M_{A_{N-1}}$ is

$$M_{ij} = \Phi(x_{ij}, \lambda) e^{p_j} b_j - \Phi(x_{ij}, \lambda+N\gamma) e^{-p_i} b_j', \quad i \neq j, \tag{II.25}$$

$$M_{ii} = (\zeta(\lambda) + \zeta(\gamma)) e^{p_i} b_i - (\zeta(\lambda+\gamma) - \zeta(\gamma)) e^{-p_i} b_i' \tag{II.26}$$

$$+ \sum_{j \neq i} ((\zeta(x_{ij}+\gamma) - \zeta(x_{ij})) e^{p_j} b_j$$

$$+ \frac{\Phi(x_{ji}+\gamma, \lambda)}{\Phi(\gamma, \lambda)} \Phi(x_{ij}, \lambda+N\gamma) e^{-p_i} b_j'). \tag{II.27}$$

III. HAMILTONIAN REDUCTION OF C_n AND BC_n RS MODELS FROM A_{N-1}-TYPE MODELS

Let us first mention some results about the integrability of Hamiltonian (II.2). In Ref. 7 Ruijsenaars demonstrated that the symplectic structure of the C_n- and BC_n-types of RS systems can be proved integrable by embedding their phase space to a submanifold of the A_{2n-1} and A_{2n} type RS ones, respectively, while in Refs. 26, 27, and 25, Diejen and Komori, respectively, gave a series of commuting difference operators which led to their quantum integrability. However, there are not any results about their Lax representations so far except for the special degenerate case.[29,30] In this section, we concentrate our treatment on the exhibition of the explicit forms for general C_n and BC_n RS systems.

For the convenience of analysis of symmetry, let us first give vector representation of A_{N-1} Lie algebra. Introducing an N-dimensional orthonormal basis of $\mathbb{R}^N$,

$$e_j \cdot e_k = \delta_{j,k}, \quad j,k=1,\ldots,N. \tag{III.1}$$

Then the sets of roots and vector weights of A_{N-1} type are,

$$\Delta = \{e_j - e_k : j,k=1,\ldots,N\}, \tag{III.2}$$

$$\Lambda = \{e_j : j=1,\ldots,N\}. \tag{III.3}$$

The dynamical variables are canonical coordinates $\{x_j\}$ and their canonical conjugate momenta $\{p_j\}$ with the Poisson brackets of Eq. (II.1). In a general sense, we denote them by N-dimensional vectors x and p,

$$x=(x_1,\ldots,x_N)\in \mathbb{R}^N, \quad p=(p_1,\ldots,p_N)\in \mathbb{R}^N,$$

so that the scalar products of x and p with the roots $\alpha \cdot x$, $p\cdot\beta$, etc., can be defined. The Hamiltonian of Eq. (II.2) can be rewritten as

$$\mathcal{H}_{A_{N-1}} = \sum_{\mu\in\Lambda} \left(\exp(\mu\cdot p) \prod_{\Delta\ni\beta=\mu-\nu} f(\beta\cdot x) + \exp(-\mu\cdot p) \prod_{\Delta\ni\beta=-\mu+\nu} g(\beta\cdot x) \right). \tag{III.4}$$

Here, the condition $\Delta\ni\beta=\mu-\nu$ means that the summation is over roots β such that for $\exists\nu \in\Lambda$,

$$\mu-\nu=\beta\in\Delta. \tag{III.5}$$

So does for $\Delta\ni\beta=-\mu+\nu$.

A. The C_n model

The set of C_n roots consists of two parts, long roots and short roots:

$$\Delta_{C_n} = \Delta_L \cup \Delta_S, \tag{III.6}$$

in which the roots are conveniently expressed in terms of an orthonormal basis of $\mathbb{R}^n$:

$$\begin{aligned} \Delta_L &= \{\pm 2e_j : \quad j=1,\ldots,n\}, \\ \Delta_S &= \{\pm e_j \pm e_k, : \quad j,k=1,\ldots,n\}. \end{aligned} \tag{III.7}$$

In the vector representation, vector weights Λ are

$$\Lambda_{C_n} = \{e_j, -e_j : \quad j=1,\ldots,n\}. \tag{III.8}$$

The Hamiltonian of the C_n model is given by

$$\mathcal{H}_{C_n} = \frac{1}{2}\sum_{\mu\in\Lambda_{C_n}} \left(\exp(\mu\cdot p) \prod_{\Delta_{C_n}\ni\beta=\mu-\nu} f(\beta\cdot x) + \exp(-\mu\cdot p) \prod_{\Delta_{C_n}\ni\beta=-\mu+\nu} g(\beta\cdot x) \right). \tag{III.9}$$

From the above-mentioned data, we notice that either for A_{N-1} or C_n Lie algebra, any root $\alpha \in\Delta$ can be constructed in terms with vector weights as $\alpha=\mu-\nu$ where $\mu,\nu\in\Lambda$. By simple comparison of representation between A_{N-1} and C_n, one can find that if replacing e_{j+n} with $-e_j$

in the vector weights of A_{2n-1} algebra, we can obtain the vector weights of C_n algebra. This also holds for the corresponding roots. This gives us a hint that it is possible to get the C_n model by this kind of reduction.

For the A_{2n-1} model let us set restrictions on the vector weights with

$$e_{j+n}+e_j=0, \quad \text{for } j=1,\ldots,n, \tag{III.10}$$

which correspond to the following constraints on the phase space of the A_{2n-1}-type RS model with

$$\begin{aligned} G_i &\equiv (e_{i+n}+e_i)\cdot x = x_i + x_{i+n} = 0, \\ G_{i+n} &\equiv (e_{i+n}+e_i)\cdot p = p_i + p_{i+n} = 0, \quad i=1,\ldots,n. \end{aligned} \tag{III.11}$$

Following Dirac's method,[36] we can show

$$\{G_i,\mathcal{H}_{A_{2n-1}}\}\simeq 0, \quad \text{for } \forall i \in \{1,\ldots,2n\}, \tag{III.12}$$

i.e., $\mathcal{H}_{A_{2n-1}}$ is the first class Hamiltonian corresponding to the constraints in Eq. (III.11). Here the "weak equal" symbol$\simeq$represents that only after calculating the result of the left-hand side of the identity could we use the conditions of constraints. It should be pointed out that the most necessary condition ensuring Eq. (III.12) is the symmetry property of formula (II.20) for the Hamiltonian Eq. (II.2). So for an arbitrary dynamical variable A, we have

$$\begin{aligned} \dot{A} = \{A,\mathcal{H}_{A_{2n-1}}\}_D &= \{A,\mathcal{H}_{A_{2n-1}}\} - \{A,G_i\}\Delta_{ij}^{-1}\{G_j,\mathcal{H}_{A_{2n-1}}\} \\ &\simeq \{A,\mathcal{H}_{A_{2n-1}}\}, \quad i,j=1,\ldots,2n, \end{aligned} \tag{III.13}$$

where

$$\Delta_{ij} = \{G_i,G_j\} = 2\begin{pmatrix} 0 & Id \\ -Id & 0 \end{pmatrix}, \tag{III.14}$$

and $\{,\}_D$ denotes the Dirac bracket. By straightforward calculation, we have the nonzero Dirac brackets of

$$\begin{aligned} \{x_i,p_j\}_D &= \{x_{i+n},p_{j+n}\}_D = \tfrac{1}{2}\delta_{i,j}, \\ \{x_i,p_{j+n}\}_D &= \{x_{i+n},p_j\}_D = -\tfrac{1}{2}\delta_{i,j}. \end{aligned} \tag{III.15}$$

Using the above-mentioned data together with the fact that $\mathcal{H}_{A_{N-1}}$ is the first class Hamiltonian-[see Eq. (III.12)], we can directly obtain a Lax representation of the C_n RS model by imposing constraints G_k on Eq. (II.23),

$$\begin{aligned} \{L_{A_{2n-1}},\mathcal{H}_{A_{2n-1}}\}_D &= \{L_{A_{2n-1}},\mathcal{H}_{A_{2n-1}}\}|_{G_k,k=1,\ldots,2n}, \\ &= [M_{A_{2n-1}},L_{A_{2n-1}}]|_{G_k,k=1,\ldots,2n} = [M_{C_n},L_{C_n}], \end{aligned} \tag{III.16}$$

$$\{L_{A_{2n-1}},\mathcal{H}_{A_{2n-1}}\}_D = \{L_{C_n},\mathcal{H}_{C_n}\}, \tag{III.17}$$

where

$$\mathcal{H}_{C_n} = \tfrac{1}{2}\mathcal{H}_{A_{2n-1}}|_{G_k,k=1,\ldots,2n},$$

$$L_{C_n}=L_{A_{2n-1}}|_{G_k,k=1,\ldots,2n},$$
$$M_{C_n}=M_{A_{2n-1}}|_{G_k,k=1,\ldots,2n}, \tag{III.18}$$

so that

$$\dot{L}_{C_n}=\{L_{C_n},\mathcal{H}_{C_n}\}=[M_{C_n},L_{C_n}]. \tag{III.19}$$

Nevertheless, the $(\mathcal{H}_1^+)_{A_{N-1}}$ is not the first class Hamiltonian, so the Lax pair given by many authors previously cannot reduce to the C_n case directly in this way.

B. The BC_n model

The BC_n root system consists of three parts: long, middle, and short roots:

$$\Delta_{BC_n}=\Delta_L\cup\Delta\cup\Delta_S, \tag{III.20}$$

in which the roots are conveniently expressed in terms of an orthonormal basis of $\mathbb{R}^n$:

$$\Delta_L=\{\pm 2e_j:\quad j=1,\ldots,n\},$$
$$\Delta=\{\pm e_j\pm e_k:\quad j,k=1,\ldots,n\}, \tag{III.21}$$
$$\Delta_S=\{\pm e_j:\quad j=1,\ldots,n\}.$$

In the vector representation, vector weights Λ can be

$$\Lambda_{BC_n}=\{e_j,-e_j,0:\quad j=1,\ldots,n\}. \tag{III.22}$$

The Hamiltonian of the BC_n model is given by

$$\mathcal{H}_{BC_n}=\frac{1}{2}\sum_{\mu\in\Lambda_{BC_n}}\left(\exp(\mu\cdot p)\prod_{\Delta_{BC_n}\ni\beta=\mu-\nu}f(\beta\cdot x)+\exp(-\mu\cdot p)\prod_{\Delta_{BC_n}\ni\beta=-\mu+\nu}g(\beta\cdot x)\right). \tag{III.23}$$

By similar comparison of representations between A_{N-1} and BC_n, one can find that if replacing e_{j+n} with $-e_j$ and e_{2n+1} with 0 in the vector weights of the A_{2n} Lie algebra, we can obtain the vector weights of the BC_n one. The same holds for the corresponding roots. So by the same procedure as the C_n model, we could get the Lax representation of the BC_n model.

For the A_{2n} model, we set restrictions on the vector weights with

$$e_{j+n}+e_j=0\quad\text{for}\quad j=1,\ldots,n,$$
$$e_{2n+1}=0, \tag{III.24}$$

which correspond to the following constraints on the phase space of the A_{2n}-type RS model with

$$G_i'\equiv(e_{i+n}+e_i)\cdot x=x_i+x_{i+n}=0,$$
$$G_{i+n}'\equiv(e_{i+n}+e_i)\cdot p=p_i+p_{i+n}=0,\quad i=1,\ldots,n,$$
$$G_{2n+1}'\equiv e_{2n+1}\cdot x=x_{2n+1}=0, \tag{III.25}$$
$$G_{2n+2}'\equiv e_{2n+1}\cdot p=p_{2n+1}=0.$$

Similarly, we can show

$$\{G_i,\mathcal{H}_{A_{2n}}\}\simeq 0,\quad \text{for}\quad \forall i\in\{1,\ldots,2n+1,2n+2\},\tag{III.26}$$

i.e., $\mathcal{H}_{A_{2n}}$ is the first class Hamiltonian corresponding to the above-mentioned constraints Eq. (III.25). So L_{BC_n} and M_{BC_n} can be constructed as follows:

$$\begin{aligned}L_{BC_n}&=L_{A_{2n}}|_{G'_k,k=1,\ldots,2n+2},\\ M_{BC_n}&=M_{A_{2n}}|_{G'_k,k=1,\ldots,2n+2},\end{aligned}\tag{III.27}$$

while $\mathcal{H}_{BC_n}$ is

$$\mathcal{H}_{BC_n}=\tfrac{1}{2}\mathcal{H}_{A_{2n}}|_{G_k,k=1,\ldots,2n+2},\tag{III.28}$$

due to the similar derivation of Eqs. (III.13)–(III.19).

IV. LAX REPRESENTATIONS OF THE C_n AND BC_n RS MODELS

A. The C_n model

The Hamiltonian of the C_n RS system is Eq. (III.9), so the canonical equations of motion are

$$\dot{x}_i=\{x_i,\mathcal{H}\}=e^{p_i}b_i-e^{-p_i}b'_i,\tag{IV.1}$$

$$\begin{aligned}\dot{p}_i=\{p_i,\mathcal{H}\}=&\sum_{j\neq i}^{n}\left(e^{p_j}b_j(h(x_{ji})-h(x_j+x_i))+e^{-p_j}b'_j(\hat{h}(x_{ji})-\hat{h}(x_j+x_i))\right)\\ &-e^{p_i}b_i\left(2h(2x_i)+\sum_{j\neq i}^{n}(h(x_{ij})+h(x_i+x_j))\right)\\ &-e^{-p_i}b'_i\left(2\hat{h}(2x_i)+\sum_{j\neq i}^{n}(\hat{h}(x_{ij})+\hat{h}(x_i+x_j))\right),\end{aligned}\tag{IV.2}$$

where

$$\begin{aligned}&h(x):=\frac{d\ln f(x)}{dx},\quad \hat{h}(x):=\frac{d\ln g(x)}{dx},\\ &b_i=f(2x_i)\prod_{k\neq i}^{n}(f(x_i-x_k)f(x_i+x_k)),\\ &b'_i=g(2x_i)\prod_{k\neq i}^{n}(g(x_i-x_k)g(x_i+x_k)).\end{aligned}\tag{IV.3}$$

The Lax matrix for the C_n RS model can be written in the following form:

$$(L_{C_n})_{\mu\nu}=e^{\nu\cdot p}b_\nu\frac{\Phi((\mu-\nu)\cdot x+\gamma,\lambda)}{\Phi(\gamma,\lambda)},\tag{IV.4}$$

which is a $2n\times 2n$ matrix whose indices are labeled by the vector weights, denoted by $\mu,\nu\in\Lambda_{C_n}$, M_{C_n} can be written as

$$M_{C_n}=D+Y, \tag{IV.5}$$

where D denotes the diagonal part and Y denotes the off-diagonal part

$$Y_{\mu\nu}=e^{\nu\cdot p}b_\nu\Phi(x_{\mu\nu},\lambda)+e^{-\mu\cdot p}b'_\nu\Phi(x_{\mu\nu},\lambda+N\gamma), \tag{IV.6}$$

$$D_{\mu\mu}=(\zeta(\lambda)+\zeta(\gamma))e^{\mu\cdot p}b_\mu-(\zeta(\lambda+\gamma)-\zeta(\gamma))e^{-\mu\cdot p}b'_\mu$$
$$+\sum_{\nu\neq\mu}\left((\zeta(x_{\mu\nu}+\gamma)-\zeta(x_{\mu\nu}))e^{\nu\cdot p}b_\nu+\frac{\Phi(x_{\nu\mu}+\gamma,\lambda)}{\Phi(\gamma,\lambda)}\Phi(x_{\mu\nu},\lambda+N\gamma)e^{-\mu\cdot p}b'_\nu\right) \tag{IV.7}$$

and

$$b_\mu=\prod_{\Delta_{C_n}\ni\beta=\mu-\nu}f(\beta\cdot x),$$

$$b'_\mu=\prod_{\Delta_{C_n}\ni\beta=\mu-\nu}g(\beta\cdot x), \tag{IV.8}$$

$$x_{\mu\nu}:=(\mu-\nu)\cdot x.$$

The L_{C_n}, M_{C_n} satisfies the Lax equation

$$\dot{L}_{C_n}=\{L_{C_n},\mathcal{H}_{C_n}\}=[M_{C_n},L_{C_n}], \tag{IV.9}$$

which is equivalent to the equations of motion (IV.1) and (IV.2). The Hamiltonian $\mathcal{H}_{C_n}$ can be rewritten as the trace of L_{C_n},

$$\mathcal{H}_{C_n}=\operatorname{tr}L_{C_n}=\frac{1}{2}\sum_{\mu\in\Lambda_{C_n}}(e^{\mu\cdot p}b_\mu+e^{-\mu\cdot p}b'_\mu). \tag{IV.10}$$

B. The BC_n model

The Hamiltonian of the BC_n model is expressed in Eq. (III.23), so the canonical equations of motion are

$$\dot{x}_i=\{x_i,\mathcal{H}\}=e^{p_i}b_i-e^{-p_i}b'_i, \tag{IV.11}$$

$$\dot{p}_i=\{p_i,\mathcal{H}\}=\sum_{j\neq i}^{n}(e^{p_j}b_j(h(x_{ji})-h(x_j+x_i))+e^{-p_j}b'_j(\hat{h}(x_{ji})-\hat{h}(x_j+x_i)))$$
$$-e^{p_i}b_i\left(h(x_i)+2h(2x_i)+\sum_{j\neq i}^{n}(h(x_{ij})+h(x_i+x_j))\right)$$
$$-e^{-p_i}b'_i\left(\hat{h}(x_i)+2\hat{h}(2x_i)+\sum_{j\neq i}^{n}(\hat{h}(x_{ij})+\hat{h}(x_i+x_j))\right)-b_0(h(x_i)+\hat{h}(x_i)), \tag{IV.12}$$

where

$$b_i = f(x_i) f(2x_i) \prod_{k \neq i}^{n} (f(x_i - x_k) f(x_i + x_k)),$$

$$b'_i g(x_i) g(2x_i) \prod_{k \neq i}^{n} (g(x_i - x_k) g(x_i + x_k)), \tag{IV.13}$$

$$b_0 = \prod_{i=1}^{n} f(x_i) g(x_i).$$

The Lax pair for the BC_n RS model can be constructed as the form of Eqs. (IV.4)–(IV.8) where one should replace the matrices labels with $\mu, \nu \in \Lambda_{BC_n}$, and roots with $\beta \in \Delta_{BC_n}$.

The Hamiltonian $\mathcal{H}_{BC_n}$ can also be rewritten as the trace of L_{BC_n},

$$\mathcal{H}_{BC_n} = \mathrm{tr}\, L_{BC_n} = \frac{1}{2} \sum_{\mu \in \Lambda_{BC_n}} (e^{\mu \cdot p} b_\mu + e^{-\mu \cdot p} b'_\mu). \tag{IV.14}$$

V. SPECTRAL CURVES OF THE C_n AND BC_n RS SYSTEMS

It has recently been pointed out in Refs. 4, 5, 37 and 38, that the SU(N) RS model is related to five-dimensional gauge theories. In the context of Seiberg–Witten theory, the elliptic RS integrable system can be linked with the relevant low energy effective action when a compactification from five dimension to four dimension is imposed with all of the fields belonging to the adjoint representation of the SU(N) gauge group.[5] More evidence for this correspondence between the SYM and RS models is depicted by calculating instanton correction of prepotential for SU(2) Seiberg-Witten theory in Ref. 32.

As for the spectral curve and its relation to the Seiberg–Witten spectral curve, much progress has been made in the correspondence of "Calogero-Moser integrable theories and gauge theories." See the recent reviews in Refs. 39 and 40, and references therein. In the following we will give the spectral curves for C_n and BC_n systems, which are shown to be parametrized by the integrals of motion of the corresponding system. We will also see that the elliptic Calogero–Moser, Toda (affine and nonaffine) ones are particular limits of these systems.

A. Spectral curve of the C_n RS system

Given the Lax operator with spectral parameter for the Calogero–Moser system and of the RS system associated with Lie algebras $\mathcal{G}$, the spectral curve for the given system is defined as

$$\Gamma : R(v,\lambda) = \det(L(\lambda) - v \cdot Id) \equiv 0. \tag{V.1}$$

It is natural that the function $R(v,z)$ is invariant under time evolution,

$$\frac{\mathrm{d}}{\mathrm{d}t} R(v,\lambda) = \{\mathcal{H}, R(v,\lambda)\} = 0. \tag{V.2}$$

Thus, $R(v,\lambda)$ must be a function of only the n independent integrals of motion, which in super-Yang–Mills theory play the role of *moduli*, parametrizing the supersymmetric vacua of the gauge theory. This has been confirmed in the case of the elliptic Calogero–Moser system for general Lie algebra in Refs. 41 and 42 and in the case of the elliptic SU(N) RS system for the perturbative limit and some nonperturbative special point.[5]

As for the C_n RS system, the spectral curve can be generated by the Lax matrix $L(\lambda)_{C_n}$ as follows:

$$\det(L(\lambda)_{C_n}-v\cdot Id)=\sum_{j=0}^{2n}\frac{(\sigma(\lambda))^{(j-1)}\sigma(\lambda+j\gamma)}{(\sigma(\gamma+\lambda))^j}(-v)^{2n-j}(H_j)_{C_n}=0, \tag{V.3}$$

where $(\mathcal{H}_0)_{C_n}=(\mathcal{H}_{2n})_{C_n}=1$ and $(\mathcal{H}_i)_{C_n}=(\mathcal{H}_{2n-i})_{C_n}$ who Poisson commute

$$\{(\mathcal{H}_i)_{C_n},(\mathcal{H}_j)_{C_n}\}=0,\quad i,j=1,\ldots,n. \tag{V.4}$$

This can be deduced by verbose but straightforward calculation to verify that the $(\mathcal{H}_i)_{A_{2n-1}}$, $i=1,\ldots,2n$ is the first class Hamiltonian with respect to the constraints (III.11), using Eqs. (II.18), (III.13) and the first formula of Eq. (III.18).

The explicit form of $(\mathcal{H}_l)_{C_n}$ is

$$(\mathcal{H}_l)_{C_n}=\sum_{\substack{J\subset\{1,\ldots,n\},\ |J|\leq l\\ \varepsilon_j=\pm 1, j\in J}}\exp(p_{\varepsilon J})\,F_{\varepsilon J;J^c}\,U_{J^c,l-|J|},\quad l=1,\ldots,n \tag{V.5}$$

with

$$p_{\varepsilon J}=\sum_{j\in J}\varepsilon_j p_j,$$

$$F_{\varepsilon J;\,K}=\prod_{\substack{j,j'\in J\\ j<j'}}f^2(\varepsilon_j x_j+\varepsilon_{j'} x_{j'})\prod_{j\in J k\in K}f(\varepsilon_j x_j+x_k)f(\varepsilon_j x_j-x_k)\prod_{j\in J}f(2\varepsilon_j x_j), \tag{V.6}$$

$$U_{I,p}=\sum_{\substack{I'\subset I\\ |I'|=[p/2]}}\prod_{\substack{j\in I'\\ k\in I\backslash I'}}f(x_{jk})f(x_j+x_k)g(x_{jk})g(x_j+x_k)\begin{cases}0, & (p\ \text{odd})\\ 1, & (p\ \text{even})\end{cases}.$$

Here, $[p/2]$ denotes the integer part of $p/2$. As an example, for the C_2 RS model, the independent Hamiltonian flows $(\mathcal{H}_1)_{C_2}$ and $(\mathcal{H}_2)_{C_2}$ generated by the Lax matrix L_{C_2} are[29]

$$\begin{aligned}(\mathcal{H}_1)_{C_2}=\mathcal{H}_{C_2}=&e^{p_1}f(2x_1)\,f(x_{12})f(x_1+x_2)+e^{-p_1}g(2x_1)\,g(x_{12})g(x_1+x_2)\\ &+e^{p_2}f(2x_2)\,f(x_{21})f(x_2+x_1)+e^{-p_2}g(2x_2)\,g(x_{21})g(x_2+x_1),\end{aligned} \tag{V.7}$$

$$\begin{aligned}(\mathcal{H}_2)_{C_2}=&e^{p_1+p_2}f(2x_1)\,(f(x_1+x_2))^2f(2x_2)+e^{-p_1-p_2}g(2x_1)\,(g(x_1+x_2))^2g(2x_2)\\ &+e^{p_1-p_2}f(2x_1)\,(f(x_{12}))^2f(-2x_2)+e^{p_2-p_1}g(2x_1)\,(g(x_{12}))^2g(-2x_2)\\ &+2f(x_{12})\,g(x_{12})\,f(x_1+x_2)g(x_1+x_2).\end{aligned} \tag{V.8}$$

Similar to the form of "gauge" transformation of Eq. (II.4), we can check that the involutive Hamiltonians of Eq. (V.5) are identical to the one given by Diejen in Ref. 26 with the following transformation

$$x_i\to x_i,\quad p_i\to p_i+\frac{1}{2}\ln\left(\frac{f(2x_i)}{g(2x_i)}\prod_{j\neq i}^{n}\frac{f(x_{ij})f(x_i+x_j)}{g(x_{ij})g(x_i+x_j)}\right). \tag{V.9}$$

B. Spectral curve of the BC_n model

Similar to the calculation of the C_n case, the spectral curve of the BC_n RS system can be generated by Lax matrix $L(\lambda)_{BC_n}$ as follows

$$\det(L(\lambda)_{BC_n}-v\cdot Id)=0. \tag{V.10}$$

The explicit form of the spectral curve is

$$\det(L(\lambda)_{BC_n}-v\cdot Id)=\sum_{j=0}^{2n+1}\frac{(\sigma(\lambda))^{(j-1)}\sigma(\lambda+j\gamma)}{(\sigma(\gamma+\lambda))^{j}}(-v)^{2n+1-j}(\mathcal{H}_j)_{BC_n}=0, \tag{V.11}$$

where $(\mathcal{H}_0)_{BC_n}=(\mathcal{H}_{2n})_{BC_n}=1$ and $(\mathcal{H}_i)_{BC_n}=(\mathcal{H}_{2n+1-i})_{BC_n}$ Poisson commute

$$\{(\mathcal{H}_i)_{BC_n},(\mathcal{H}_j)_{BC_n}\}=0, \quad \forall i,j\in\{1,\ldots,n\}. \tag{V.12}$$

This can be deduced similarly to the C_n case to verify that $(\mathcal{H}_i)_{A_{2n}}$, $i=1,\ldots,2n$ is the first class Hamiltonian with respect to the constraints (III.25).

The explicit forms of $(\mathcal{H}_l)_{BC_n}$ are

$$(\mathcal{H}_l)_{BC_n}=\sum_{\substack{J\subset\{1,\ldots,n\},\ |J|\le l\\ \varepsilon_j=\pm 1,\ j\in J}}\exp(p_{\varepsilon J})\,F_{\varepsilon J;J^c}U_{J^c,l-|J|}, \quad l=1,\ldots,n \tag{V.13}$$

with

$$p_{\varepsilon J}=\sum_{j\in J}\varepsilon_j p_j,$$

$$F_{\varepsilon J;K}=\prod_{\substack{j,j'\in J\\ j<j'}}f^2(\varepsilon_j x_j+\varepsilon_{j'}x_{j'})\prod_{\substack{j\in J\\ k\in K}}f(\varepsilon_j x_j+x_k)f(\varepsilon_j x_j-x_k)\prod_{j\in J}f(2\varepsilon_j x_j)\prod_{j\in J}f(\varepsilon_j x_j), \tag{V.14}$$

$$U_{I,p}=\sum_{\substack{I'\subset I\\ |I'|=[p/2]}}\prod_{\substack{j\in I'\\ k\in I\setminus I'}}f(x_{jk})f(x_j+x_k)g(x_{jk})g(x_j+x_k)\begin{cases}\prod_{i\in I\setminus I'}f(x_i)g(x_i), & (p\ \text{odd})\\ \prod_{i\in I}f(x_i)g(x_i), & (p\ \text{even})\end{cases}.$$

It is similar to the C_n case for the relation between $(\mathcal{H}_l)_{BC_n}$ with the one given in Ref. 26:

$$x_i\to x_i, \quad p_i\to p_i+\tfrac{1}{2}\ln\left(\frac{f(x_i)}{g(x_i)}\frac{f(2x_i)}{g(2x_i)}\prod_{j\ne i}^{n}\frac{f(x_{ij})f(x_i+x_j)}{g(x_{ij})g(x_i+x_j)}\right). \tag{V.15}$$

Remarks: So far we have Lax matrices with the spectral parameter of Eq. (IV.4) for the C_n and BC_n RS models, and the corresponding spectral curve equation of Eqs. (V.3) and (V.11). It is expected that they will be useful to study the relation between the $5d$ SUSY gauge theory and these integrable models which have been pointed out in Ref. 5. More exactly, it is expected that these spectral curves would be identical to the complex curve in the context of SUSY gauge theory associated with the corresponding gauge group. On the other hand, these nonsimple laced models may be potential candidates which are connected with orientifold in brane theory, corresponding to the fact that the A_{n-1} RS model is connected with orbifold. This exact correspondence in these directions is missed and certainly desires further investigation.

C. Limit to the Calogero–Moser system and Toda system

The Calogero–Moser system can be achieved by taking the so-called "nonrelativistic limit." The procedure is by rescaling $p_\mu \mapsto \beta p_\mu$, $\gamma \mapsto \beta\gamma$ and letting $\beta \mapsto 0$, followed by making a canonical transformation

$$p_\mu \mapsto p_\mu + \gamma \sum_{\Delta \ni \eta = \mu - \nu} \zeta(\eta \cdot x). \tag{V.16}$$

Here $p_\mu = \mu \cdot p$, such that

$$L \mapsto Id + \beta L_{\mathrm{CM}} + O(\beta^2), \tag{V.17}$$

and

$$\mathcal{H} \mapsto N + 2\beta^2 \mathcal{H}_{\mathrm{CM}} + O(\beta^2), \tag{V.18}$$

where $N = 2n$ for the C_n model and $N = 2n+1$ for BC_n model.

L_{CM} can be expressed as

$$L_{\mathrm{CM}} = p \cdot H + X, \tag{V.19}$$

where

$$H_{\mu\nu} = \mu \delta_{\mu\nu}, \quad X_{\mu\nu} = \gamma \Phi(x_{\mu\nu}, \lambda)(1 - \delta_{\mu\nu}). \tag{V.20}$$

The Hamiltonian $\mathcal{H}_{\mathrm{CM}}$ of the CM model can be given by

$$\mathcal{H}_{\mathrm{CM}} = \frac{1}{2} p^2 - \frac{\gamma^2}{2} \sum_{\alpha \in \Delta} \wp(\alpha \cdot x) = \frac{1}{4} \mathrm{tr}\, L^2 + \mathrm{const}, \tag{V.21}$$

where

$$\mathrm{const} = -\frac{N(N-1)\gamma^2}{4} \wp(\lambda).$$

All of the above-mentioned results are identified with the results of Refs. 8, 10, 12–15 up to a suitable choice of coupling parameters. Now the degenerate RS spectral curve reduces to

$$\Gamma: \ R(v, \lambda) = \det(L(\lambda)_{\mathrm{CM}} - v \cdot Id) \equiv 0, \tag{V.22}$$

which is exactly identified with the spectral curve analyzed in Refs. 39 and 41.

Starting from the CM system to the Toda system is more directly due to the progress that the limit to Toda for the general Lie algebra has been studied extensively in Refs. 43–45. The main idea is making a suitable scaling limit with the following parametrization:

$$\omega_1 = -i\pi, \quad \omega_3 \in \mathbf{R}_+, \quad \tau \equiv \frac{\omega_3}{\omega_1} = i\omega_3/\pi, \tag{V.23}$$

and shifting the dynamical variable x,

$$\begin{aligned} x &\to Q - 2\omega_3 \delta \rho^\vee, \quad p \to P, \\ \lambda &\to \log Z - \omega_3, \quad Z \in \mathbf{R}_+, \end{aligned} \tag{V.24}$$

TABLE I. Root system of A_{n-1}, C_n and BC_n types.

$\mathcal{G}$	All roots	Simple roots Π	$h_{\mathcal{G}}$	Dual Weyl vector $\rho^\vee$	Vector weights
A_{n-1}	$\pm e_i \pm e_j$, $1 \leq i, j \leq n$, $i \neq j$	$e_i - e_{i+1}$, $i=1,\ldots,n-1$	n	$\Sigma_{j=1}^n (n-j) e_j$	e_i, $i=1,\ldots,n$
C_n	$\pm e_i \pm e_j, \pm 2e_i$, $1 \leq i, j \leq n$, $i \neq j$	$e_i - e_{i+1}, 2e_n$, $i=1,\ldots,n-1$	$2n$	$\Sigma_{j=1}^n (n+\frac{1}{2}-j) e_j$	$e_i, -e_i$, $i=1,\ldots,n$
BC_n	$\pm e_i \pm e_j, \pm 2e_i, \pm e_i$ $1 \leq i, j \leq n$, $i \neq j$	$e_i - e_{i+1}, e_n$, $i=1,\ldots,n-1$	$2n+1$	$\Sigma_{j=1}^n (n+1-j) e_j$ (define)	$e_i, -e_i, 0$, $i=1,\ldots,n$

in which $h_{\mathcal{G}}$ is the Coxeter number for the corresponding root system $\mathcal{G}$, $\rho^\vee$ the dual of the Weyl vector defined as $\rho^\vee = \frac{1}{2}\Sigma_{\alpha \in \Delta_+} 2\alpha/\alpha^2$, and δ satisfies $\delta \leq 1/h_{\mathcal{G}}$.

For convenience, we give the basics of these root systems as shown in Table I.

As for the C_n model, selecting $\rho^\vee = \rho^\vee_{C_n}$, $\gamma = im\, e^{\omega_3 \delta}$, one has the nonaffine C_n Toda model from the Hamiltonian of the CM model Eq. (V.21),

$$\mathcal{H}^{\text{Toda}}_{C_n} = \frac{1}{2} P^2 + m^2 \sum_{j=1}^{n-1} e^{Q_j - Q_{j+1}} + m^2\, e^{2Q_n}, \tag{V.25}$$

for $\delta < 1/h_{C_n}$ and $C_n^{(1)}$ Toda model

$$\mathcal{H}^{\text{Toda}}_{C_n^{(1)}} = \frac{1}{2} P^2 + m^2 e^{-2Q_1} + m^2 \sum_{j=1}^{n-1} e^{Q_j - Q_{j+1}} + m^2 e^{2Q_n} \tag{V.26}$$

for $\delta = 1/h_{C_n}$.

The same holds for the BC_n model. Selecting $\rho^\vee = \rho^\vee_{BC_n}$, $\gamma = im\, e^{\omega_3 \delta}$, one has the nonaffine B_n Toda model from the Hamiltonian of the CM model Eq. (V.21),

$$\mathcal{H}^{\text{Toda}}_{B_n} = \frac{1}{2} P^2 + m^2 \sum_{j=1}^{n-1} e^{Q_j - Q_{j+1}} + m^2\, e^{Q_n} \tag{V.27}$$

for $\delta < 1/h_{BC_n}$ and BC_n Toda model

$$\mathcal{H}^{\text{Toda}}_{BC_n} = \frac{1}{2} P^2 + m^2 e^{-2Q_1} + m^2 \sum_{j=1}^{n-1} e^{Q_j - Q_{j+1}} + m^2 e^{Q_n} \tag{V.28}$$

for $\delta = 1/h_{BC_n}$.

If we use the following gauge for $\Phi(x,\lambda)$:[46]

$$\Phi(x,\lambda) \to \frac{\sigma(x+\lambda)}{\sigma(\lambda)\sigma(x)} \exp(\zeta(\lambda)x), \tag{V.29}$$

it does not destroy the validity[45] for the Lax pair, We have the following limit for $\gamma\Phi(\alpha\cdot x,\lambda)$:

$$
\begin{aligned}
\gamma\Phi(\alpha\cdot x,\lambda) &\to -m\exp\left(\frac{\alpha\cdot Q}{2}\right) \quad \text{for } \alpha\in\Pi \ \ (\delta\leqslant 1/h_{\mathcal{G}}) \\
&\to mZ\exp\left(-\frac{\alpha\cdot Q}{2}\right) \quad \text{for } \alpha=\alpha_h \ \ (\delta=1/h_{\mathcal{G}}) \\
&\to 0 \quad \text{otherwise,} \\
\gamma\Phi(-\alpha\cdot x,\lambda) &\to m\exp\left(\frac{\alpha\cdot Q}{2}\right) \quad \text{for } \alpha\in\Pi(\delta\leqslant 1/h_{\mathcal{G}}) \\
&\to -\frac{m}{Z}\exp\left(-\frac{\alpha\cdot Q}{2}\right), \quad \text{for } \alpha=\alpha_h(\delta=1/h_{\mathcal{G}}) \\
&\to 0 \quad \text{otherwise.}
\end{aligned} \tag{V.30}
$$

So the Lax operator now reads

$$
L_{\text{Toda}}=P\cdot H-im\sum_{\alpha\in\Pi}\exp\left(\frac{\alpha\cdot Q}{2}\right)[E(\alpha)-E(-\alpha)]+im\exp\left(\frac{\alpha_0\cdot Q}{2}\right)[ZE(-\alpha_0)-Z^{-1}E(\alpha_0)], \tag{V.31}
$$

where $E(\alpha)_{\mu\nu}=\delta_{\mu-\nu,\alpha}$. This Lax operator holds for all the root systems of $A_{n-1}(A_{n-1}^{(1)})$, $C_n(C_n^{(1)})$, $B_n(BC_n)$ and coincides with the standard form given in Ref. 8. It is not difficult to find that the parameter Z now plays the role of a spectral parameter for the affine Toda model based on $\mathcal{G}^{(1)}$. When we refer to the Toda models based on a finite Lie algebra $\mathcal{G}$, we should only drop the terms containing the affine root α_0.

So the degenerate spectral curve for the Toda $A_{n-1}^{(1)}$, $C_n^{(1)}$, and $BC_n(A_{2n}^{(2)})$ systems can be defined

$$
\Gamma: R(v,\lambda)=\det(L(\lambda)_{\text{Toda}}-v\cdot Id)\equiv 0, \tag{V.32}
$$

which is identical to the one given in Refs. 47 and 48.

VI. DEGENERATE CASES

Let us now consider the other various special degenerate cases. As is well known, if one or both of the periods of the Weierstrass sigma function $\sigma(x)$ become infinite, there will occur three degenerate cases associated with trigonometric, hyperbolic, and rational systems. The degenerate limits of the functions $\Phi(x,\lambda)$, $\sigma(x)$, and $\zeta(x)$ will give corresponding Lax pairs which include spectral parameter. Moreover, when the spectral parameter value is on a certain limit, the Lax pairs without spectral parameter will be derived.

A. Trigonometric limit

The limit can be obtained by sending ω_3 to $i\infty$ with $\omega_1=\pi/2$, so that

$$
\begin{aligned}
\sigma(x)&\to e^{(1/6)x^2}\sin x, \\
\zeta(x)&\to\cot x+\tfrac{1}{3}x,
\end{aligned} \tag{VI.1}
$$

and the function

$$
\Phi(x,\lambda)\equiv\frac{\sigma(x+\lambda)}{\sigma(x)\sigma(\lambda)}
$$

reduces to

$$\Phi(x,\lambda)\rightarrow(\cot\lambda-\cot x)e^{(1/3)xu}. \tag{VI.2}$$

By replacing the corresponding functions $\Phi(x,\lambda)$, $\sigma(x)$, and $\zeta(x)$ with the above-given form for the Lax pairs, we will get the corresponding spectral parameter dependent Lax pairs. For simplicity, we notice that the exponential part of the above-mentioned functions can be removed by applying suitable "gauge" transformation of the Lax matrix on which condition the functions can be valued as follows:

$$\begin{aligned}\sigma(x)&\rightarrow\sin x,\\ \zeta(x)&\rightarrow\cot x,\\ \Phi(x,\lambda)&\rightarrow(\cot\lambda-\cot x).\end{aligned} \tag{VI.3}$$

As for the spectral parameter independent Lax pair, furthermore, we can take the limit $\lambda\rightarrow i\infty$, so the function

$$\Phi(x,\lambda)\rightarrow\frac{1}{\sin x}, \tag{VI.4}$$

while the corresponding Lax matrix becomes

$$L_{\mu\nu}=e^{\nu\cdot p}b_\nu\frac{\sin\gamma}{\sin((\mu-\nu)\cdot x+\gamma)}, \tag{VI.5}$$

which is exactly the same as the spectral parameter independent Lax matrix given in Ref. 30.

B. Hyperbolic limit

In this case, the periods can be chosen by sending ω_1 to $i\infty$ with $\omega_3=\pi/2$, so following all the procedures in achieving the result of the trigonometric case, we can find the hyperbolic Lax pairs by simple replacement of the functions appearing in the trigonometric Lax pair as follows:

$$\begin{aligned}\sin x&\rightarrow\sinh x,\\ \cos x&\rightarrow\cosh x,\\ \cot x&\rightarrow\coth x.\end{aligned} \tag{VI.6}$$

The same as for the trigonometric case, we can get the Lax pairs with and without spectral parameter.

C. Rational limit

As far as the form of the Lax pair for the rational-type system is concerned, we can achieve it by making the following substitutions:

$$\begin{aligned}\sigma(x)&\rightarrow x,\\ \zeta(x)&\rightarrow\frac{1}{x},\\ \Phi(x,\lambda)&\rightarrow\frac{1}{x}+\frac{1}{\lambda}\end{aligned} \tag{VI.7}$$

for the spectral parameter dependent Lax pair, while furthermore, taking the limit $\lambda \rightarrow i\infty$, we can obtain the spectral parameter independent Lax pair. The explicit form of Lax matrix without spectral parameter is

$$L_{\mu\nu}=e^{\nu\cdot p}b_{\nu}\frac{\gamma}{(\mu-\nu)\cdot x+\gamma}. \tag{VI.8}$$

which completely coincides with the spectral parameter independent Lax matrix given in Ref. 30.

Remark: As for the various degenerate cases for the CM and Toda systems, one can follow the same procedure as for the RS model [please refer to Eqs. (VI.1)–(VI.8)].

VII. CONCLUDING REMARKS

In this paper, we have proposed the Lax pairs for elliptic C_n and BC_n RS models. The spectral parameter dependent and independent Lax pairs for the trigonometric, hyperbolic, and rational systems can be derived as the degenerate limits of the elliptic potential case. The spectral curves of these systems are given and shown depicted by the complete sets of involutive constant integrals of motion. They are expected be related to the five-dimensional gauge theory[4,5] and even to brane theory, which desires further study. In the nonrelativistic limit(scaling limit), these systems lead to CM(Toda) systems associated with the root systems of C_n and BC_n. There are still many open problems. For example, it seems to be a challenging subject to carry out a Lax pair with as many independent coupling constants as independent Weyl orbits in the set of roots, as done for the Calogero–Moser systems(see Refs. 8, 11–15). What is also interesting is to generalize the results obtained in this paper to the systems associated with all other Lie Algebras even to those associated with all the finite reflection groups.[14] Moreover, the issue of getting the r-matrix structure for these systems is deserved due to the success of calculating for the trigonometric BC_n RS system by Avan *et al.* in Ref. 31.

ACKNOWLEDGMENTS

We would like to thank Professor K. J. Shi and Professor L. Zhao for useful and stimulating discussions. This work has been supported financially by the National Natural Science Foundation of China. W.-L.Y. has been supported by the Alexander von Humboldt Foundation.

APPENDIX

In this appendix we prove the identity equation (II.11) and then derive the relation between the Lax operator $L(\lambda)$ and its inverse of $L(\lambda)^{-1}$.

Using the result given in Ref. 6 of Eq. (B5), we have the following conclusion:

Let

$$C_{ij}=\frac{\sigma(q_i-r_j+\lambda)}{\sigma(q_i-r_j+\mu)}, \quad i,j=1,\ldots,N, \tag{A1}$$

then one has

$$\det(C)=\sigma(\lambda-\mu)^{N-1}\sigma(\lambda+(N-1)\mu+\Sigma)\times\prod_{i<j}\sigma(q_i-q_j)\sigma(r_j-r_i)\prod_{i,j}\frac{1}{\sigma(q_i-r_j+\mu)}, \tag{A2}$$

where

$$\Sigma=\sum_{i=1}^{N}(q_i-r_j). \tag{A3}$$

So it is straightforward to compute the inverse of matrix C,

$$(C^{-1})_{ij}=\text{the cofactor of } C \text{ with respect to } C_{ji}$$
$$=\frac{\sigma(\lambda+(N-2)\mu+q_i-q_j)}{\sigma(\lambda-\mu)\sigma(\lambda+(N-1)\mu)\sigma(q_i-q_j-\mu)}$$
$$\times\frac{\Pi_l\sigma(q_j-q_l+\mu)\Pi_l\sigma(q_i-q_l-\mu)}{\Pi_{k\neq i}\sigma(q_i-q_k)\Pi_{k\neq j}\sigma(q_j-q_k)}. \tag{A4}$$

From Eq. (II.5), we have

$$L(\lambda)=\sum_{i,j=1}^{N}\frac{\Phi(x_{ij}+\gamma,\lambda)}{\Phi(\gamma,\lambda)}\exp(p_j)b_jE_{ij}$$
$$=\frac{1}{\Phi(\gamma,\lambda)}\sum_{i,j=1}^{N}\frac{\sigma(x_{ij}+\gamma+\lambda)}{\sigma(x_{ij}+\gamma)\sigma(\lambda)}\exp(p_j)b_j$$
$$=\frac{1}{\Phi(\gamma,\lambda)}\sum_{i,j=1}^{N}G_{ij}\exp(p_j)b_j, \tag{A5}$$

where

$$G_{ij}:=\Phi(x_{ij}+\gamma,\lambda)=\frac{\sigma(x_{ij}+\gamma+\lambda)}{\sigma(x_{ij}+\gamma)\sigma(\lambda)},$$

with the help of Eq. (A4), one has

$$(G^{-1})_{ij}=\frac{\sigma(\lambda+(N-1)\gamma+x_{ij})}{\sigma(\lambda+N\gamma)\sigma(x_{ij}-\gamma)}\times\frac{\Pi_k\sigma(x_{jk}+\gamma)\Pi_k\sigma(x_{ik}-\gamma)}{\Pi_{k\neq i}\sigma(x_{ik})\Pi_{k\neq j}\sigma(x_{jk})}, \tag{A6}$$

so that

$$L(\lambda)^{-1}{}_{ij}=\Phi(\gamma,\lambda)(G^{-1})_{ij}b_j^{-1}\exp(-p_i))E_{ij}$$
$$=\frac{-\sigma(\gamma)^2\,\sigma(\lambda+\gamma)\sigma(\lambda+(N-1)\gamma+x_{ij})}{\sigma(\lambda)\sigma(\gamma)\sigma(\lambda+N\gamma)\sigma(x_{ij}-\gamma)}\times\exp(-p_i)\prod_{k\neq j}\frac{\sigma(x_{jk}+\gamma)}{\sigma(x_{jk})}$$
$$=\frac{\sigma(\gamma+\lambda)\,\sigma(\lambda+(N-1)\gamma)}{\sigma(\lambda)\sigma(\lambda+N\gamma)}\times\frac{\Phi(x_{ij}-\gamma,\lambda+N\gamma)}{\Phi(-\gamma,\lambda+N\gamma)}\exp(-p_i)b_j'E_{ij}. \tag{A7}$$

By comparing the forms of $L(\lambda)$ and $L(\lambda)^{-1}{}_{ij}$, we find $L(\lambda)^{-1}{}_{ij}$ can be expressed with $L(\lambda)$ as

$$L(\lambda)^{-1}{}_{ij}=L(\lambda)_{ij}|_{\gamma\to-\gamma,\lambda\to\lambda+N\gamma}\times\frac{\sigma(\gamma+\lambda)\,\sigma(\lambda+(N-1)\gamma)}{\sigma(\lambda)\sigma(\lambda+N\gamma)}\exp(-p_i-p_j). \tag{A8}$$

[1] K. Hasegawa, "Ruijsenaars' commuting difference operators as commuting transfer matrices," Commun. Math. Phys. **187**, 289 (1997).
[2] F.W. Nijhoff, V.B. Kuznetsov, E.K. Sklyanin, and O. Ragnisco, "Dynamical r-matrix for the elliptic Ruijsenaars–Schneider system," J. Phys. A **29**, L333 (1996).
[3] A. Gorsky and A. Marshakov, "Towards effective topological gauge theories on spectral curves," Phys. Lett. B **375**, 127 (1996).
[4] N. Nekrasov, "Five-dimension gauge theories and relativistic integrable systems," hep-th/9609219; Nucl. Phys. B **531**, 323 (1998).
[5] H.W. Braden, A. Marshakov, A. Mironov, and A. Morozov, "The Ruijsenaars–Schneider model in the context of Seiberg–Witten theory," hep-th/9902205; Nucl. Phys. B **558**, 371 (1999).
[6] S.N.M. Ruijsenaars, "Complete integrability of relativistic Calogero–Moser systems and elliptic function identities," Commun. Math. Phys. **110**, 191 (1987).
[7] S.N.M. Ruijsenaars, "Action-angle maps and scattering theory for some finite-dimensional integrable systems," Commun. Math. Phys. **115**, 127 (1988).
[8] M.A. Olshanetsky and A.M. Perelomov, "Classical integrable finite-dimensional systems related to Lie algebras," Phys. Rep. **71**, 314 (1981); A.M. Perelomov, *Integrable Systems of Classical Mechanics and Lie Algebras* (Birkhäuser, Boston, 1990).
[9] VI. Inozemtsev, "Lax representation with spectral parameter on a torus for integrable particle systems," Lett. Math. Phys. **17**, 11 (1989).
[10] E. D'Hoker and D.H. Phong, "Calogero–Moser Lax pairs with spectral parameter for general Lie algebras," hep-th/9804124; Nucl. Phys. B **530**, 537 (1998).
[11] A.J. Bordner, E. Corrigan, and R. Sasaki, "Calogero–Moser Models. I. A new formulation," hep-th/9805106; Prog. Theor. Phys. **100**, 1107 (1998).
[12] A.J. Bordner, R. Sasaki, and K. Takasaki, "Calogero–Moser Models. II. Symmetries and foldings," hep-th/9809068; Prog. Theor. Phys. **101**, 487 (1999).
[13] A.J. Bordner and R. Sasaki, "Calogero–Moser Models III: Elliptic potentials and twisting," hep-th/9812232; Prog. Theor. Phys. **101**, 799 (1999).
[14] A.J. Bordner, E. Corrigan, and R. Sasaki, "Generalized Calogero–Moser models and universal Lax pair operators," hep-th/9905011; Prog. Theor. Phys. **102**, 499 (1999).
[15] J.C. Hurtubise and E. Markman, "Calogero–Moser systems and Hitchin systems," math/9912161.
[16] A.J. Bordner, N.S. Manton, and R. Sasaki, "Calogero–Moser models. V. Supersymmetry and quantum Lax pair," hep-th/9910033; Prog. Theor. Phys. **103**, 463 (2000).
[17] S.P. Khastgir, A.J. Pocklington, and R. Sasaki, "Quantum Calogero–Moser models: Integrability for all root systems," math/0005277.
[18] M.A. Olshanetsky and A.M. Perelomov, "Quantum integrable systems related to Lie algebras," Phys. Rep. **94**, 313 (1983).
[19] M. Bruschi and F. Calogero, "The Lax pair representation for an integrable class of relativistic dynamical systems," Commun. Math. Phys. **109**, 481 (1987).
[20] I. Krichever and A. Zabrodin, "Spin generalization of the Ruijsenaars–Schneider model, the non-Abelian 2D Toda chain, and representations of the Sklyanin algebra," Usp. Mat. Nauk **50**, 3 (1995).
[21] Y.B. Suris, "Why are the rational and hyperbolic Ruijsenaars–Schneider hierarchies governed by the same R-operators as the Calogero–Moser ones?," hep-th/9602160.
[22] Y.B. Suris, "Elliptic Ruijsenaars–Schneider and Calogero–Moser hierarchies are governed by the same r-matrix," solv-int/9603011; Phys. Lett. A **225**, 253 (1997).
[23] G.E. Arutyunov, S.A. Frolov, and P.B. Medvedev, "Elliptic Ruijsenaars–Schneider model from the cotangent bundle over the two-dimensional current group," hep-th/9608013; J. Math. Phys. **38**, 5682 (1997).
[24] Y. Komori and K. Hikami, "Conserved operators of the generalized elliptic Ruijsenaars models," J. Math. Phys. **39**, 6175 (1998).
[25] Y. Komori, "Theta functions associated with the affine root systems and the elliptic Ruijsenaars operators," math.QA/9910003.
[26] J.F. van Diejen, "Integrability of difference Calogero–Moser systems," J. Math. Phys. **35**, 2983 (1994).
[27] J.F. van Diejen, "Commuting difference operators with polynomial eigenfunctions," Compositio. Math. **95**, 183 (1995).
[28] K. Hasegawa, T. Ikeda, and T. Kikuchi, "Commuting difference operators arising from the elliptic $C_2^{(1)}$-face model," J. Math. Phys. **40**, 4549 (1999).
[29] K. Chen, B.Y. Hou, and W.-L. Yang, "The Lax pair for C_2-type Ruijsenaars–Schneider model," hep-th/0004006; Chin. Phys. **10**, 550 (2001).
[30] K. Chen, B.Y. Hou, and W.-L. Yang, "Integrability of the C_n and BC_n Ruijsenaars–Schneider models," hep-th/0006004; J. Math. Phys. **41**, 8132 (2000).
[31] J. Avan and G. Rollet, "BC_n Ruijsenaars–Schneider models: R-matrix structure and Hamiltonians," hep-th/0008174.
[32] Y. Ohta, "Instanton correction of prepotential in Ruijsenaars model associated with $N=2$ SU(2) Seiberg–Witten theory," hep-th/9909196; J. Math. Phys. **41**, 4541 (2000).
[33] J. Avan, "Classical dynamical r-matrices for Calogero–Moser systems and their generalizations," q-alg/9706024.
[34] V.B. Kuznetsov, F.W. Nijhoff, and E.K. Sklyanin, "Separation of variables for the Ruijsenaars system," Commun. Math. Phys. **189**, 855 (1997).
[35] S.N.M. Ruijsenaars and H. Schneider, "A new class of integrable systems and its relation to solitons," Ann. Phys. (Leipzig) **170**, 370 (1986).

[36] Paul A.M. Dirac, *Lectures on Quantum Physics* (Yeshiva University Press, New York, 1964).
[37] A. Mironov and Morozov, "Double elliptic systems: Problems and perspectives," hep-th/0001168.
[38] A. Mironov, "Seiberg–Witten theory and duality in integrable systems," hep-th/0011093.
[39] E. D'Hoker and D.H. Phong, "Lectures on supersymmetric Yang–Mills theory and integrable systems," hep-th/9912271.
[40] A. Marshakov, *Seiberg–Witten Theory and Integrable Systems* (World Scientific, Singapore, 1998).
[41] E. D'Hoker and D.H. Phong, "Spectral curves for Super-Yang–Mills with adjoint hypermultiplet for general Lie algebras," hep-th/9804126; Nucl. Phys. B **534**, 697 (1998).
[42] E. D'Hoker and D.H. Phong, "Lax pairs and spectral curves for Calogero–Moser and spin Calogero–Moser systems," hep-th/9903002.
[43] V. I. Inozemtsev, "The finite Toda lattices," Commun. Math. Phys. **121**, 628 (1989).
[44] E. D'Hoker and D.H. Phong, "Calogero–Moser and Toda systems for twisted and untwisted affine Lie Algebras," hep-th/9804125; Nucl. Phys. B **530**, 611 (1998).
[45] S.P. Khastgir, R. Sasaki, and K. Takasaki, "Calogero–Moser Models. IV. Limits to Toda theory," hep-th/9907102; Prog. Theor. Phys. **102**, 749 (1999).
[46] I.M. Krichever, "Elliptic solutions of the Kadomtsev–Petviashvili equation and integrable systems of particles," Funct. Anal. Appl. **14**, 282 (1980).
[47] E. Martinec and N. Warner, "Integrable Systems and supersymmetric gauge theory," hep-th/9509161; Nucl. Phys. B **459**, 97 (1996).
[48] K. Takasaki, "Whitham deformation of Seiberg–Witten curves for classical gauge groups," hep-th/9901120; Int. J. Mod. Phys. A **15**, 3635 (2000).

Letters in Mathematical Physics **61:** 205–220, 2002.

Solitons on Noncommutative Orbifold T^2/Z_N

BO-YU HOU, KANG-JIE SHI, and ZHAN-YING YANG
Institute of Modern Physics, Northwest University, Xi'an, 710069, P.R. China.
e-mail: {byhou, kjshi, yzz}@phy.nwu.edu.cn

(Received: 23 April 2002; revised version: 25 July 2002)

Abstract. Following the construction of the projection operators on T^2 presented by Gopakumar, Headrick and Spradlin, we construct a set of projection operators on the integral noncommutative orbifold $T^2/G(G = Z_N, N = 2, 3, 4, 6)$ which correspond to a set of solitons on T^2/Z_N in noncommutative field theory. In this way, we derive an explicit form of projector on T^2/Z_6 as an example. We also construct a complete set of projectors on T^2/Z_N by series expansions for integral case.

Mathematics Subject Classifications (2000). 35Q51, 81T75.

Key words. noncommutative orbifold, projection operators, soliton.

1. Introduction

Noncommutative geometry is an interesting topic in mathematics [1–3]. During the last few years, noncommutative field theories have renewed the physist's interest primarily due to the discovery that noncommutative gauge theories naturally arise from the low energy dynamics of D-branes in the presence of a background B field and as various limits of M-theory compactification [4–6]. Quantum field theory on a noncommutative space is useful in understanding various physical phenomena, such as string behaviors and D-brane dynamics. They also appear as theories describing the behavior of electron gas in the presence of a strong, external magnetic field, the quantum Hall effect [7]. Recently, Susskind and Hu, Zhang [8] proposed that noncommutative Chern–Simons theory on the plane may provide a description of (fractionally filled) quantum Hall fluid. Being nonlocal, noncommutative field theory may help to understand nonlocality at a short distant in quantum gravity.

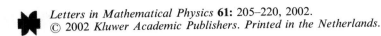

After the connection between string theory and noncommutative field theories was unraveled, the study of solitons in noncommutative space attracted much attention [9–13]. Soliton solutions in field theory and string theory often shed light on the nonperturbative and strong coupling behavior of the theory, thus it is interesting to investigate these solutions in noncommutative field theories. Gopakumar *et al.* found that soliton solution of noncommutative flat space can be exactly given in terms of projection operators [12]. Harvey *et al.* set up a new method to investigate the soliton solution and M. Hamanaka and S. Terashima generalised this 'solution generating technique' to a BPS monopole in $3+1$ dimension [13]. Martinec and Moore discussed how D-branes on orbifolds fit naturally into the algebraic framework as described by projection operators [10]. Thus the study of projection operators in various noncommutative spaces are important in string field theory. Rieffel has presented a general formula for the projection operators on noncommutative torus [14]. Boca further described the projection operators on orbifold T^2/G and discussed the relation between their trace and the commutator q of the operators U and V on a noncommutative torus ($UV = VUe^{2i\pi q}$). He proved the existence of nontrivial projection operators and explicitly presented an example with trace $1/q$ for q an integer in the Z_4 case [15]. Boca expressed the projection operators in terms of the θ function depending, respectively, on U and V. Konechny *et al.* [16, 17] have also given some Z_2, Z_4 invariant projection operators. Martinec and Moore pointed out that no explicit expressions for the projection operators in Z_3, Z_6 case have been found. Gopakumar *et al.* [9] succeeded in constructing the projection operator on a noncommutative integral torus ($q = \mathrm{A} =$ integer) with generic τ. Here the integer A measures the quantums of magnetic flux passing through the torus, its ratio with the area servers as the noncommutative parameter. We find that if the vacuum state $|0\rangle$ in their paper is replaced by any state vector $|\phi\rangle$, their construction still works. We notice that if the state vector $|\phi\rangle$ has some symmetries, the operators are just the projection operators on the orbifold T^2/G ($G = Z_N$ is a symmetry group). Then we give a set of projection operators in T^2/Z_N. As an example, an explicit projection operator in T^2/Z_6 is obtained by this approach.

This Letter is organized as following. We introduce the noncommutative orbifold T^2/Z_N in Section 2 and in the next section we review the construction proposed by Gopakumar *et al.* on the integral torus T^2. We show how this approach can be used to construct the projection operators which are invariant under the transformation group Z_N in Section 4. In Section 5, we presented an explicit form of projector on T^2/Z_6 as an example, using theta functions with $\hat{y}_1$ and $\hat{y}_2$ as variables. In Section 6, we provide a complete set of projectors on T^2/Z_N by series expansions for the integral case.

2. Noncommutative Orbifold T^2/Z_N

In this section, we introduce operators on the noncommutative orbifold T^2/Z_N. First we introduce two operators $\hat{y}_1$ and $\hat{y}_2$ on noncommutative R^2 which satisfy the commutation relation

$$[\hat{y}_1, \hat{y}_2] = i. \tag{1}$$

Define the operators

$$U_1 = e^{-il\hat{y}_2}, \qquad U_2 = e^{il(\tau_2\hat{y}_1 - \tau_1\hat{y}_2)}, \tag{2}$$

where l, τ_1, τ_2 are all real numbers and l, $\tau_2 > 0$. All operators on R^2 which commute with U_1 and U_2 constitute the operators defined on noncommutative torus T^2. This torus is formed as a manifold which identifies two points $(\hat{y}_1, \hat{y}_2) \sim (\hat{y}_1, \hat{y}_2) + \mathbf{r}$ with

$\mathbf{r} = m\mathbf{l_1} + n\mathbf{l_2}$ on the noncommutative plane R^2, where $\mathbf{l_1} = (l, 0)$, $\mathbf{l_2} = (l\tau_1, l\tau_2)$. Thus, we have

$$\begin{aligned} U_1^{-1}\hat{y}_1 U_1 &= \hat{y}_1 + l, \qquad & U_2^{-1}\hat{y}_1 U_2 &= \hat{y}_1 + l\tau_1, \\ U_1^{-1}\hat{y}_2 U_1 &= \hat{y}_2, \qquad & U_2^{-1}\hat{y}_2 U_2 &= \hat{y}_2 + l\tau_2. \end{aligned} \tag{3}$$

The operators U_1 and U_2 are two different wrapping operators around the noncommutative torus and their commutation relation is $U_1U_2 = U_2U_1 e^{-2\pi i\frac{l^2\tau_2}{2\pi}}$. When $A = l^2\tau_2/2\pi$ is an integer, we call the torus integral. Next, we introduce a linear transformation R, which gives

$$R^{-1}\hat{y}_1 R = a\hat{y}_1 + b\hat{y}_2, \qquad R^{-1}\hat{y}_2 R = c\hat{y}_1 + d\hat{y}_2. \tag{4}$$

Setting $U_1 = U$, $U_2 = V$, under R, if U and V change as in [10]*, then

$$\begin{aligned} Z_2 &: U \to U^{-1}, \qquad & V &\to V^{-1}, \\ Z_3 &: U \to V, \qquad & V &\to U^{-1}V^{-1}, \\ Z_4 &: U \to V, \qquad & V &\to U^{-1}, \\ Z_6 &: U \to V, \qquad & V &\to U^{-1}V, \end{aligned} \tag{5}$$

then we refer to R as a Z_N symmetry rotation of the torus T^2. The operators on noncommutative orbifold T^2/Z_N are the operators of T^2 which are invariant under transformation R.

If we define operators $\hat{y}'_1$ and $\hat{y}'_2$ as

$$\begin{aligned} &\hat{y}_1 = a\hat{y}'_1 + b\hat{y}'_2, \qquad \hat{y}_2 = \frac{1}{a}\hat{y}'_2, \\ &a = \sqrt{\frac{\tau'_2}{\tau_2}}, \qquad b = \frac{-\tau'_1 + \tau_1}{\sqrt{\tau_2\tau'_2}}, \qquad l' = \frac{l}{a}, \end{aligned} \tag{6}$$

then we get

$$[\hat{y}'_1, \hat{y}'_2] = i, \qquad U_1 = e^{-il'\hat{y}'_2}, \qquad U_2 = e^{il'(\tau'_2\hat{y}'_1 - \tau'_1\hat{y}'_2)}. \tag{7}$$

From the above result, we notice that the noncommutative torus is invariant after taking a suitable module parameter $\tau = \tau_1 + i\tau_2$. Now we consider the rotations in symmetric orbifolds $T^2/Z_N(Z_N = Z_2, Z_3, Z_4, Z_6)$. Let $\tau = e^{\frac{2\pi i}{N}} = e^{i\theta}$, then the transformation

$$R^{-1}\hat{y}_1 R = \cos\theta\hat{y}_1 + \sin\theta\hat{y}_2, \qquad R^{-1}\hat{y}_2 R = \cos\theta\hat{y}_2 - \sin\theta\hat{y}_1. \tag{8}$$

will give the corresponding transformation of the operators U_1 and U_2 as in Equation (5). Such R can be realized by

$$R = e^{-i\theta\frac{\hat{y}_1^2 + \hat{y}_2^2}{2} + i\frac{\theta}{2}}. \tag{9}$$

*These relations may include some phase factors. It will not change the derivation in the text, see Equation (10).

Next, we take Z_6 and Z_4 as examples.

(1) $\theta = \pi/3$:

$$\begin{aligned}&\tau_1 = \tfrac{1}{2}, \qquad \tau_2 = \frac{\sqrt{3}}{2}, \quad N = 6,\\ &U_1 = e^{-il\hat{y}_2}, \qquad U_2 = e^{il\left(\frac{\sqrt{3}}{2}\hat{y}_1 - \frac{1}{2}\hat{y}_2\right)},\\ &R^{-1}U_1R = U_2, \qquad R^{-1}U_2R = e^{-\pi iA}U_1^{-1}U_2.\end{aligned} \tag{10}$$

From the above result we find that the lattice remain invariant under rotation.

(2) $\theta = \pi/2$:

$$\begin{aligned}&\tau_1 = 0, \qquad \tau_2 = 1, \quad N = 4,\\ &U_1 = e^{-il\hat{y}_2}, \qquad U_2 = e^{il\hat{y}_1},\\ &R^{-1}U_1R = U_2, \qquad R^{-1}U_2R = U_1^{-1}.\end{aligned} \tag{11}$$

This shows that the whole lattice remains invariant. We can realize the operators $\hat{y}_1$ and $\hat{y}_2$ as the operators in Fock space. Introducing

$$a = \frac{\hat{y}_2 - i\hat{y}_1}{\sqrt{2}}, \qquad a^+ = \frac{\hat{y}_2 + i\hat{y}_1}{\sqrt{2}}, \tag{12}$$

we have $[a, a^+] = 1$. The rotation R can be expressed by a, a^+ via

$$R = e^{-i\theta a^+ a}. \tag{13}$$

3. Review GHS Construction for Soliton

In this section, we review the results in paper [9]. A noncommutative space R^2 has been orbifolded to a torus T^2 with double periods l and τl. The generators are

$$U_1 = e^{-il\hat{y}_2}, \qquad U_2 = e^{il(\tau_2\hat{y}_1 - \tau_1\hat{y}_2)}, \tag{14}$$

where $[\hat{y}_1, \hat{y}_2] = i$, here we just consider the case when $A = \tau_2 l^2/2\pi$ is an integer. Introduce a state vector

$$|\psi\rangle = \sum_{j_1, j_2} C_{j_1, j_2} U_1^{j_1} U_2^{j_2} |\Omega\rangle \quad (j_1, j_2 \in Z) \tag{15}$$

that satisfies

$$\langle\psi| U_1^{j_1} U_2^{j_2} |\psi\rangle = \delta_{j_1,0}\delta_{j_2,0}. \tag{16}$$

The state $|\Omega\rangle$ will be specified later. Then a projection operator on T^2 can be constructed as

$$P = \sum_{j_1, j_2} U_1^{j_1} U_2^{j_2} |\psi\rangle\langle\psi| U_2^{-j_2} U_1^{-j_1}. \tag{17}$$

The power series of $\hat{y}_1$ and $\hat{y}_2$ can be made of the power series of a and a^+. Moreover, the formula $|0\rangle\langle 0| =: \mathrm{e}^{-a^+a}:$ indicates that any $|\psi\rangle\langle\psi|$ can be constituted by the power series of a and a^+. The projection operator is therefore spanned by the operators $\hat{y}_1$ and $\hat{y}_2$. It is easy to check that $P^2 = P$ and $U_i^{-1}PU_i = P$. So P is a projection operator on noncommutative T^2. The kq representation [18, 19] provides a basis for the common eigenstate of U_1 and U_2:

$$|k, q\rangle = \sqrt{\frac{l}{2\pi}}\mathrm{e}^{-i\tau_1\hat{y}_2^2/2\tau_2}\sum_j \mathrm{e}^{ijkl}|q + jl\rangle, \tag{18}$$

where the ket on the right is a $\hat{y}_1$ eigenstate. We have

$$\begin{aligned} &U_1|k, q\rangle = \mathrm{e}^{-ilk}|k, q\rangle, \qquad U_2|k, q\rangle = \mathrm{e}^{il\tau_2 q}|k, q\rangle, \\ &\mathrm{id} = \int_a^{\frac{2\pi}{l}+a}\mathrm{d}k\int_b^{l+b}\mathrm{d}q|k, q\rangle\langle k, q|, \end{aligned} \tag{19}$$

where a and b are real numbers and

$$|k, q\rangle = |k + \frac{2\pi}{l}, q\rangle = \mathrm{e}^{ilk}|k, q + l\rangle$$

In terms of wave functions in the kq representation, $|\psi\rangle$ becomes

$$C_\psi(k, q) = \langle k, q|\psi\rangle = \sum_{j_1, j_2} C_{j_1, j_2}\mathrm{e}^{-ij_1lk+ij_2l\tau_2 q}\langle k, q|\Omega\rangle = \tilde{c}(k, q)C_0(k, q), \tag{20}$$

where

$$\tilde{c}(k, q) = \sum_{j_1, j_2} C_{j_1, j_2}\mathrm{e}^{-ij_1lk+ij_2l\tau_2 q}, \; C_0(k, q) = \langle k, q|\Omega\rangle.$$

Note that $\tilde{c}(k, q)$ is doubly periodic:

$$\tilde{c}\left(k + \frac{2\pi}{l}, q\right) = \tilde{c}\left(k, q + \frac{l}{A}\right) = \tilde{c}(k, q).$$

The orthonormality condition (16) becomes

$$\delta_{j_1,0}\delta_{j_2,0} = \int_0^{\frac{2\pi}{l}}\mathrm{d}k\int_0^l \mathrm{d}q\,\mathrm{e}^{-ij_1lk+ij_2l\tau_2 q}|\tilde{c}(k, q)|^2|C_0(k, q)|^2. \tag{21}$$

The coefficient C_{j_1, j_2} can be obtained if and only if $\tilde{c}(k, q)$ is a double periodic function with periods $2\pi/l$ and l/A for k and q, respectively. So we rewrite the above equation as

$$\delta_{j_1,0}\delta_{j_2,0} = \int_0^{\frac{2\pi}{l}}\mathrm{d}k\int_0^{l/A}\mathrm{d}q\,\mathrm{e}^{-ij_1lk+ij_2l\tau_2 q}|\tilde{c}(k, q)|^2\sum_{n=0}^{A-1}\left|C_0\left(k, q + \frac{ln}{A}\right)\right|^2. \tag{22}$$

This holds for any j_1 and j_2 if and only if

$$|\tilde{c}(k, q)|^2\sum_{n=0}^{A-1}\left|C_0\left(k, q + \frac{ln}{A}\right)\right|^2 = \frac{A}{2\pi}.$$

Then we have

$$|\tilde{c}(k,q)| = \frac{1}{\sqrt{\frac{2\pi}{A}\sum_{n=0}^{A-1}\left|C_0\left(k,q+\frac{\ln}{A}\right)\right|^2}}. \tag{23}$$

Setting $e^{i\beta}$ as a phase factor of $\tilde{c}(k,q)$, we have

$$C_\psi(k,q) = \frac{C_0(k,q)e^{i\beta}}{\sqrt{\frac{2\pi}{A}\sum_{n=0}^{A-1}\left|C_0\left(k,q+\frac{\ln}{A}\right)\right|^2}}. \tag{24}$$

4. The Projection Operator on T^2/Z_N

In the previous section, we reviewed how to construct projection operators on a noncommutative torus. In this section, we will discuss how to construct a projection operator on the noncommutative orbifold T^2/Z_N following the result of the previous section. Recall the projection operator

$$P = \sum_{j_1,j_2} U_1^{j_1}U_2^{j_2}|\psi\rangle\langle\psi|U_2^{-j_2}U_1^{-j_1} \tag{25}$$

and transform it by rotation R:

$$R^{-1}PR = \sum_{j_1,j_2}(U_1')^{j_1}(U_2')^{j_2}R^{-1}|\psi\rangle\langle\psi|R(U_2')^{-j_2}(U_1')^{-j_1}, \tag{26}$$

where $U_i' = R^{-1}U_iR$. Considering the transformation group $G = Z_N$, we get

$$R^{-1}PR = \sum_{j_1',j_2'} U_1^{j_1'}U_2^{j_2'}R^{-1}|\psi\rangle\langle\psi|RU_2^{-j_2'}U_1^{-j_1'}, \tag{27}$$

where

$$\begin{array}{lll} j_1' = -j_1, & j_2' = -j_2 & \text{for } Z_2 \text{ case,} \\ j_1' = -j_2, & j_2' = j_1 - j_2 & \text{for } Z_3 \text{ case,} \\ j_1' = -j_2, & j_2' = j_1 & \text{for } Z_4 \text{ case,} \\ j_1' = -j_2, & j_2' = j_1 + j_2 & \text{for } Z_6 \text{ case.} \end{array}$$

Then we can obtain $R^{-1}PR = P$ as long as

$$R|\psi\rangle = e^{i\alpha}|\psi\rangle. \tag{28}$$

We can show that this can be satisfied if $R|\Omega\rangle = e^{i\alpha}|\Omega\rangle$. In the next step, we take $G = Z_6$ as an example to prove this (it is easy to generalize this to other cases). Assume

$$|\psi\rangle = \sum_{j_1,j_2} C_{j_1,j_2}U_1^{j_1}U_2^{j_2}|\Omega\rangle(j_1,j_2 \in Z) \tag{29}$$

satisfies

$$\langle\psi|U_1^{j_1}U_2^{j_2}|\psi\rangle = \delta_{j_1,0}\delta_{j_2,0}. \tag{30}$$

Setting $R|\Omega\rangle = e^{i\alpha}|\Omega\rangle$, we have

$$R|\psi\rangle = \sum_{j_1,j_2} C_{j_1,j_2} U_1^{-j_2} U_2^{j_1+j_2} R|\Omega\rangle$$
$$= \sum_{j_1,j_2} C'_{j_1,j_2} U_1^{j_1} U_2^{j_2}|\Omega\rangle, \tag{31}$$

where

$$C'_{j_1,j_2} = C_{j_1+j_2,-j_1} e^{i\alpha} \tag{32}$$

and

$$\langle\psi|R^{-1}U_1^{j_1}U_2^{j_2}R|\psi\rangle = \langle\psi|U_2^{j_1}(U_1^{-j_2}U_2^{j_2})|\psi\rangle$$
$$= \delta_{j_1+j_2,0}\delta_{-j_2,0} = \delta_{j_1,0}\delta_{j_2,0}. \tag{33}$$

Adding the condition $C^*_{j_1,j_2} = C_{-j_1,-j_2}e^{-2i\beta}$, we have

$$\tilde{c}^*(k,q) = \tilde{c}(k,q)e^{-2i\beta}. \tag{34}$$

The unique solution satisfying the above equations is

$$\tilde{c}(k,q) = \frac{e^{i\beta}}{\sqrt{\frac{2\pi}{A}\sum_{n=0}^{A-1}\left|C_0\left(k, q+\frac{\ln}{A}\right)\right|^2}}. \tag{35}$$

In brief, the C_{j_1,j_2} which satisfies conditions (29), (30), (34) is uniquely determined, as demonstrated in Section 3. On the other hand, from Equation (32) we find that

$$(C'_{j_1,j_2})^* = C'_{-j_1,-j_2}e^{-2i(\alpha+\beta)}, \tag{36}$$

giving

$$\tilde{c}'^*(k,q) = \tilde{c}'(k,q)e^{-2i(\alpha+\beta)}. \tag{37}$$

Equations (31), (33) and (37) also uniquely determine $\tilde{c}'(k,q)$ as

$$\tilde{c}'(k,q) = \frac{e^{i(\alpha+\beta)}}{\sqrt{\frac{2\pi}{A}\sum_{n=0}^{A-1}\left|C_0\left(k, q+\frac{\ln}{A}\right)\right|^2}} = \tilde{c}(k,q)e^{i\alpha} \tag{38}$$

for $R|\psi\rangle$. We then have $|\psi'\rangle = R|\psi\rangle = e^{i\alpha}|\psi\rangle$. In conclusion, if the vector $|\Omega\rangle$ satisfies $R|\Omega\rangle = e^{i\alpha}|\Omega\rangle$, the state $|\psi\rangle$ will satisfy Equation (28) and the projection operator by GHS construction will be a projection operator on noncommutative orbifold T^2/G.

In the above discussion, we know that the crucial point is that the state vector $|\Omega\rangle$ must be invariant under the rotation R, namely $R|\Omega\rangle = e^{i\alpha}|\Omega\rangle$. We now show how to construct such a state vector. The operator for rotation is

$$R = e^{-i\theta a^+ a}. \tag{39}$$

We can set

$$|\Omega\rangle = \sum_{j=0}^{N-1} R^j \mathrm{e}^{\frac{2\pi i}{N} js} |\phi\rangle, \tag{40}$$

where $|\phi\rangle$ is an arbitrary state vector, $R = \mathrm{e}^{-\frac{2\pi i}{N} a^+ a}$, $s = 0, \ldots, N-1$. Since

$$R^N |\phi\rangle = \mathrm{e}^{-2\pi i a^+ a} \left(\sum_n c_n (a^+)^n \right) |0\rangle = \sum_n \mathrm{e}^{-2\pi i n} c_n (a^+)^n |0\rangle = |\phi\rangle, \tag{41}$$

thus

$$\begin{aligned} R|\Omega\rangle &= \sum_{j=0}^{N-1} R^{j+1} \mathrm{e}^{i\frac{2\pi}{N} js} |\phi\rangle \\ &= \mathrm{e}^{-is\frac{2\pi}{N}} |\Omega\rangle \\ &= \mathrm{e}^{i\alpha} |\Omega\rangle. \end{aligned} \tag{42}$$

If we obtain the expression for $C_0(k, q)$, it is easy to write the expression for the field configuration for the projection operator by Equation (24) ([9]) or get the Fourier expansion by Equations (55), (56). In next step, we take an example to show how to construct $C_0(k, q) = \langle k, q|\Omega\rangle$. Introduce the coherent states

$$|z\rangle = \mathrm{e}^{-\frac{1}{2} z\bar{z}} \mathrm{e}^{a^+ z} |0\rangle, \tag{43}$$

where $z = x + iy$, $\bar{z} = x - iy$, which satisfies

$$\frac{1}{\pi} \int_{-\infty}^{\infty} \mathrm{d}x\, \mathrm{d}y |z\rangle\langle z| = \text{identity}. \tag{44}$$

Thus, from Equation (39), we get

$$\begin{aligned} R|z\rangle &= \mathrm{e}^{-\frac{1}{2} z\bar{z}} \mathrm{e}^{a^+ \omega z} |0\rangle \\ &= |\omega z\rangle \end{aligned} \tag{45}$$

where $\omega = \mathrm{e}^{-i\frac{2\pi}{N}}$ and we can employ $\langle y_2|0\rangle = \frac{1}{\pi^{1/4}} \mathrm{e}^{-y_2^2/2}$, $|y_2\rangle$ as the eigenstate of the operator $\hat{y}_2$, to obtain

$$\langle y_2|z\rangle = \frac{1}{\pi^{1/4}} \mathrm{e}^{-z^2/2 - z\bar{z}/2}\, \mathrm{e}^{-y_2^2/2 + \sqrt{2} z y_2}. \tag{46}$$

We have

$$\begin{aligned} \langle k, q|z\rangle &= \int \langle k, q|y_2\rangle\langle y_2|z\rangle \mathrm{d}y_2 \\ &= \frac{1}{\sqrt{l}\pi^{1/4}} \theta\begin{bmatrix} 0 \\ 0 \end{bmatrix} \left(\frac{q + \frac{\tau}{\tau_2} k - i\sqrt{2} z}{l}, \frac{\tau}{A} \right) \mathrm{e}^{-\frac{\tau}{2i\tau_2} k^2 + ikq + \sqrt{2} kz - (z^2 + z\bar{z})/2}. \end{aligned} \tag{47}$$

Letting

$$|\phi\rangle = \frac{1}{\pi} \int_{-\infty}^{\infty} \mathrm{d}x\, \mathrm{d}y |z\rangle\langle z|\phi\rangle = \int_{-\infty}^{\infty} \mathrm{d}x\, \mathrm{d}y F(z) |z\rangle, \tag{48}$$

we can obtain:

$$C_0(k,q) = \langle k,q|\Omega\rangle = \frac{1}{\pi}\int \left\langle k,q\right| \sum_{j=0}^{N-1} R^j e^{i\frac{2\pi}{N}js} \left|z\right\rangle\left\langle z\right|\phi\big\rangle \mathrm{d}x\,\mathrm{d}y$$

$$= \int\left[\sum_{j=0}^{N-1} e^{i\frac{2\pi}{N}js}\langle k,q|\omega^j z\rangle\right] F(z)\mathrm{d}x\,\mathrm{d}y, \tag{49}$$

where $F(z)$ is an arbitrary function. After obtaining $C_0(k,q)$, we can then compute $\langle k,q|\psi\rangle$ by Equation (24) and the Fourier coefficient for the projection operator P on the noncommutative orbifold T^2/Z_N by Equation (56).

5. An Example for the T^2/Z_6 Case

In this section, we give an example for T^2/Z_6 case. We first review the Weyl–Moyal transformation on a torus and then present the explicit expression for the projection operator by the Fourier series of the operators $\hat{y}_1$ and $\hat{y}_2$. Define

$$A(\hat{p}) = \sum_{j_1,j_2} U_1^{j_1} U_2^{j_2} b(\hat{p}) U_2^{-j_2} U_1^{-j_1} (j_1,j_2 \in Z), \tag{50}$$

where $U_1 = e^{is_2\hat{p}_1}$, $U_2 = e^{is_1\hat{p}_2}$ and $\hat{p}_1,\hat{p}_2$ are linear combinations of $\hat{y}_1$ and $\hat{y}_2$, $[\hat{p}_1,\hat{p}_2] = i$. It is easy to see that $U_i^{-1}A(\hat{p})U_i = A(\hat{p})$, namely $A(\hat{p})$ is an operator on the noncommutative torus T^2. The field configuration for $A(\hat{p})$ is

$$\Phi_A(p) = \frac{(2\pi)^2}{s_1 s_2}\sum_{mn} \mathrm{tr}\{e^{2\pi i[(\hat{p}_1-p_1)\frac{m}{s_1}+(\hat{p}_2-p_2)\frac{n}{s_2}]} b(\hat{p})\}. \tag{51}$$

We can also reobtain $A(\hat{p})$ by the Weyl–Moyal transformation from $\Phi_A(p)$,

$$A(\hat{p}) = \sum_{mn}\frac{1}{2\pi s_1 s_2}\int_0^{s_1}\mathrm{d}p_1\int_0^{s_2}\mathrm{d}p_2 \Phi_A(p) e^{2\pi i[(\hat{p}_1-p_1)\frac{m}{s_1}+(\hat{p}_2-p_2)\frac{n}{s_2}]}. \tag{52}$$

We now set

$$\hat{p}_1 = -\hat{y}_2, \qquad \hat{p}_2 = \hat{y}_1 - \frac{\tau_1}{\tau_2}\hat{y}_2, \qquad s_1 = l\tau_2, \qquad s_2 = l, \qquad b(\hat{p}) = |\psi\rangle\langle\psi|,$$

and have

$$U_1 = e^{-il\hat{y}_2}, \qquad U_2 = e^{il(\tau_2\hat{y}_1 - \tau_1\hat{y}_2)}. \tag{53}$$

Then the operator $A(\hat{p})$ becomes the projection operator P. The field configuration for the projection operator P is

$$\Phi_p(y) = \frac{(2\pi)^2}{l^2\tau_2}\sum_{j_1 j_2}\langle\psi| e^{\frac{2\pi i}{l}(j_1(\hat{y}_1-y_1)+\frac{j_2-\tau_1 j_1}{\tau_2}(\hat{y}_2-y_2))}|\psi\rangle. \tag{54}$$

The Fourier expansion for the projection operator is obtained by Weyl–Moyal transformation (52),

$$P=\sum_{j_1j_2} D_{j_1j_2}\mathrm{e}^{-\frac{2\pi i}{l}(j_1\hat{y}_1+\frac{j_2-\tau_1 j_1}{\tau_2}\hat{y}_2)}, \tag{55}$$

$$\begin{aligned} D_{j_1j_2} &= \frac{1}{A}\langle\psi|\mathrm{e}^{\frac{2\pi i}{l}(j_1\hat{y}_1+\frac{j_2-\tau_1 j_1}{\tau_2}\hat{y}_2)}|\psi\rangle \\ &= \frac{1}{A}\int_0^{\frac{2\pi}{l}}\mathrm{d}k\int_0^l \mathrm{d}q\langle k,q|\psi\rangle\left\langle\psi|k,q-\frac{ls}{A}\right\rangle \mathrm{e}^{2\pi i j_1(q/l-s/A)}\mathrm{e}^{imlk}\mathrm{e}^{\pi i j_1 j_2/A}, \end{aligned} \tag{56}$$

where $j_2 = mA + s$, $s = 0, \dots, A-1$.

Rewrite the above equations as

$$\begin{aligned} P &= \sum_{j_1j_2} D_{j_1j_2}\mathrm{e}^{-2\pi i j_2\frac{\hat{y}_2}{l\tau_2}}\mathrm{e}^{-2\pi i j_1\frac{\hat{y}_1-\frac{\tau_1}{\tau_2}\hat{y}_2}{l}}\mathrm{e}^{-\pi i j_1 j_2/A} \\ &= \sum_s\left(\mathrm{e}^{-2\pi i\frac{\hat{y}_2}{l\tau_2}}\right)^s\sum_{j_1m}\left(\mathrm{e}^{-2\pi i A\frac{\hat{y}_2}{l\tau_2}}\right)^m\left(\mathrm{e}^{-2\pi i\frac{\hat{y}_1-\frac{\tau_1}{\tau_2}\hat{y}_2}{l}}\right)^{j_1}D_{j_1ms} \\ &= \sum_s(u_1)^s\sum_{j_1m}D_{j_1ms}(u_1^A)^m(u_2)^{j_1} \end{aligned} \tag{57}$$

where

$$u_1=\mathrm{e}^{-2\pi i\frac{\hat{y}_2}{l\tau_2}},\qquad u_2=\mathrm{e}^{-2\pi i\frac{\hat{y}_1-\frac{\tau_1}{\tau_2}\hat{y}_2}{l}}$$

and

$$\begin{aligned} D_{j_1ms} &= D_{j_1j_2}\mathrm{e}^{-\pi i j_1 j_2/A}|_{j_2=mA+s} \\ &= \frac{1}{A}\int_0^{\frac{2\pi}{l}}\mathrm{d}k\int_0^l \mathrm{d}q'\left\langle k,q'+\frac{ls}{A}\middle|\psi\right\rangle\langle\psi|k,q'\rangle\mathrm{e}^{2\pi i j_1\left(\frac{q'}{l}\right)}\mathrm{e}^{2\pi i m\left(\frac{lk}{2\pi}\right)}, \end{aligned} \tag{58}$$

where $q' = q-(ls/A)$. This is a calculation for the Fourier coefficient of periodic (in k and q) function $\langle k,q+\frac{s}{A}|\psi\rangle\langle\psi|k,q\rangle = f_s$, that is

$$f_s=\frac{A}{2\pi}\sum_{j_1m}D_{j_1ms}\mathrm{e}^{-2\pi i j_1\left(\frac{q}{l}\right)}\mathrm{e}^{-2\pi i m\left(\frac{lk}{2\pi}\right)}. \tag{59}$$

Notice that in Equation (57), u_1^A and u_2 commute with each other. Thus, when we obtain an explicit expression of $f_s(K,Q)$ in terms of $Q=\mathrm{e}^{-2\pi i\left(\frac{q}{l}\right)}$ and $K=\mathrm{e}^{-2\pi i\left(\frac{lk}{2\pi}\right)}$ for real q/l and $lk/2\pi$, we can immediately write P as

$$P=\frac{2\pi}{A}\sum_s u_1^s f_s(u_1^A,u_2). \tag{60}$$

We calculate a projector P as an example in the following. Let $|\Omega\rangle = |0\rangle$, which is obviously R invariant. We have

$$\langle kq|\psi\rangle = C_\psi(k,q)=\frac{C_0(k,q)}{\sqrt{\frac{2\pi}{A}\sum_{n=0}^{A-1}\left|C_0\left(k,q+\frac{\ln}{A}\right)\right|^2}} \tag{61}$$

with $C_0(k,q) = \langle k, q|0\rangle$ ([9]). Then the corresponding f_s is

$$f_s = \frac{C_0\big(k, q+\frac{ls}{A}\big)C_0^*(k,q)}{\frac{2\pi}{A}\sum_{n=0}^{A-1} C_0\big(k, q+\frac{\mathrm{ln}}{A}\big)C_0^*\big(k, q+\frac{\mathrm{ln}}{A}\big)}, \tag{62}$$

where

$$\begin{aligned} C_0(k,q) &= \frac{1}{\sqrt{l}\pi^{1/4}}\theta\begin{bmatrix}0\\0\end{bmatrix}\left(\frac{q}{l}+\frac{\tau k}{l\tau_2}, \frac{\tau}{A}\right)\mathrm{e}^{-\frac{\tau}{2i\tau_2}k^2+ikq} \\ &= \sqrt{\frac{Ai}{l\tau\sqrt{\pi}}}\theta\begin{bmatrix}0\\0\end{bmatrix}\left(\frac{lk}{2\pi}+\frac{Aq}{l\tau}, -\frac{A}{\tau}\right)\mathrm{e}^{-\pi i\frac{Aq^2}{\tau l^2}}. \end{aligned} \tag{63}$$

Thus, for real $q/l = u$ and $lk/2\pi = v$, we have

$$\begin{aligned} g(u,v)_{ss'} &= C_0\left(k, q+\frac{ls}{A}\right)C_0^*\left(k, q+\frac{ls'}{A}\right) \\ &= \frac{1}{l\sqrt{\pi}}\theta\begin{bmatrix}0\\0\end{bmatrix}\left(u+\frac{s}{A}+\frac{\tau v}{A}, \frac{\tau}{A}\right)\theta\begin{bmatrix}0\\0\end{bmatrix}\left(u+\frac{s'}{A}+\frac{\tau^* v}{A}, \frac{-\tau^*}{A}\right)\times \\ &\quad \times \mathrm{e}^{\pi i\frac{\tau-\tau^*}{A}v^2+2\pi i\frac{s-s'}{A}v}, \end{aligned} \tag{64}$$

$$\begin{aligned} &= \frac{A}{l|\tau|\sqrt{\pi}}\theta\begin{bmatrix}0\\0\end{bmatrix}\left(v+\frac{A}{\tau}\left(u+\frac{s}{A}\right), -\frac{A}{\tau}\right)\theta\begin{bmatrix}0\\0\end{bmatrix}\left(v+\frac{A}{\tau^*}\left(u+\frac{s'}{A}\right), \frac{A}{\tau^*}\right)\times \\ &\quad \times \mathrm{e}^{-\pi i\frac{A}{\tau}\left(u+\frac{s}{A}\right)^2+\pi i\frac{A}{\tau^*}\left(u+\frac{s'}{A}\right)^2} \end{aligned} \tag{65}$$

due to

$$\theta\begin{bmatrix}0\\0\end{bmatrix}(z,\tau)^* = \theta\begin{bmatrix}0\\0\end{bmatrix}(z*, -\tau*). \tag{66}$$

In the T^2/Z_6 case, $\tau = \mathrm{e}^{\frac{i\pi}{3}}$, so we have

$$g(u+1, v) = g(u, v+1) = g(u,v), \tag{67}$$

$$g(u+\tau, v) = \mathrm{e}^{-2\pi i A(2u+\frac{v}{A}+x)}g(u,v), \tag{68}$$

$$g(u, v+A\tau) = \mathrm{e}^{-2\pi i(2v+Au+y)}g(u,v) \tag{69}$$

where

$$x = \tau - \frac{1}{2} + \frac{s+s'}{A} \quad \text{and} \quad y = A\tau - \frac{A}{2} + \frac{s}{\tau} + \frac{s'}{\tau^*}.$$

From the above equation one can prove

$$\begin{aligned} &g(u,v) \\ &= \Bigg\{\sum_{i=1}^{2}\theta\begin{bmatrix}0\\0\end{bmatrix}(Au+v-v_i+\tfrac{1}{2}, \tau A)\theta\begin{bmatrix}0\\0\end{bmatrix}\left(v-v_i+\frac{A\tau}{2}, \tau A\right)\times \\ &\quad \times\theta\begin{bmatrix}0\\0\end{bmatrix}\left(u+\frac{s}{A}+\frac{\tau}{A}v_i, \frac{\tau}{A}\right)\theta\begin{bmatrix}0\\0\end{bmatrix}\left(u+\frac{s'}{A}+\frac{\tau^* v_i}{A}, -\frac{\tau^*}{A}\right), \\ &\quad X(v_i)\Bigg\}\Bigg/\left[\theta\begin{bmatrix}0\\0\end{bmatrix}(Au+\tfrac{1}{2}, \tau A)\theta\begin{bmatrix}0\\0\end{bmatrix}\left(\frac{\tau A}{2}, \tau A\right)\right\} \equiv \{G_{ss'}(u,v)\}/[r(u)]\Bigg], \end{aligned} \tag{70}$$

where

$$v_1 = \frac{1}{2}\left(\frac{A}{2} + \frac{A\tau}{2} + \frac{1}{2} - \frac{As}{\tau} - \frac{As'}{\tau^*}\right), \tag{71}$$

$$v_2 = \frac{1}{2}\left(\frac{A}{2} + \frac{A\tau}{2} - \frac{1}{2} - \frac{As}{\tau} - \frac{As'}{\tau^*}\right), \tag{72}$$

$$X(v) = \frac{1}{l\sqrt{\pi}} \mathrm{e}^{-\frac{2\pi\tau_2}{A}v^2 + 2\pi i v\left(\frac{s}{A} - \frac{s'}{A}\right)}. \tag{73}$$

Thus

$$f_s = \frac{G_{s0}(u, v)}{\frac{2\pi}{A}\sum_{n=0}^{A-1} G_{nn}(u, v)}. \tag{74}$$

The proof is as follows. Because of Equations (67) and (69), the entire function g of v belongs to a two-dimensional function space of v. We properly choose two functions as a base. Then fix the coefficients (they are functions of u) at two special values of v (v_1 and v_2), we finally obtain Equation (70). Let

$$\theta\begin{bmatrix} 0 \\ 0 \end{bmatrix}(z, \tau) = \Theta(Z, \tau) = \sum_m \mathrm{e}^{\pi i m^2 \tau} Z^m, \qquad Z = \mathrm{e}^{-2\pi i z}. \tag{75}$$

Then

$$\begin{aligned} &G_{ss'}(u, v) \\ &\quad = \sum_{i=1}^{2} \Theta(KQ^A \mathrm{e}^{2\pi i\left(v_i - \frac{1}{2}\right)}, A\tau)\Theta(K\mathrm{e}^{2\pi i\left(v_i - \frac{A\tau}{2}\right)}, A\tau) \times \\ &\qquad \times \Theta\left(Q\mathrm{e}^{-2\pi i\left(\frac{\tau}{A}v_i + \frac{s}{A}\right)}, \frac{\tau}{A}\right)\Theta\left(Q\mathrm{e}^{-2\pi i\left(\frac{\tau^*}{A}v_i + \frac{s'}{A}\right)}, -\frac{\tau^*}{A}\right) \times \\ &\qquad \times X(v_i) \equiv \Phi_{ss'}(K, Q), \end{aligned} \tag{76}$$

giving

$$P = \sum_{s=0}^{A-1} u_1^s \frac{\Phi_{s0}(u_1^A, u_2)}{\sum_{n=0}^{A-1} \Phi_{nn}(u_1^A, u_2)}. \tag{77}$$

This is an explicit T^2/Z_6 projector.

6. Complete Set of Projections

We assume that all operators in T^2 can be expressed as

$$\hat{B} = \sum_{j_1 j_2} U_1^{j_1} U_2^{j_2} \hat{b} U_2^{-j_2} U_1^{-j_1} \tag{78}$$

for some operators $\hat{b}$ in noncommutative space R^2. Reorganize the complete set $|k, q\rangle$ as

$$\{|k, q\rangle\} = \{|k, q_0, s\rangle\}, \quad q = q_0 + \frac{l}{A}s,$$

where $k \in [0, 2\pi/l)$, $q_0 \in [0, l/A)$, $s = 0, \ldots, A-1$. Equation (19) becomes

$$\mathrm{id} = \sum_{s=0}^{A-1} \int_0^{\frac{2\pi}{l}} \mathrm{d}k \int_0^{\frac{l}{A}} \mathrm{d}q_0 |k, q_0, s\rangle\langle k, q_0, s|. \tag{79}$$

Combining the above equation and

$$\sum_j \mathrm{e}^{ijx} = \sum_m 2\pi\delta(x + 2\pi m), \tag{80}$$

we can obtain

$$\begin{aligned}&\left\langle k, q_0 + \frac{l}{A} n |\hat{B}| k', q_0' + \frac{l}{A} n'\right\rangle \\ &\quad = \frac{(2\pi)}{A}\delta(k'-k)\delta(q_0' - q_0)\left\langle k, q_0 + \frac{l}{A} n |\hat{b}| k, q_0 + \frac{l}{A} n'\right\rangle.\end{aligned} \tag{81}$$

So we have

$$\begin{aligned}&\left\langle k, q_0 + \frac{l}{A} n |\hat{A}\hat{B}| k', q_0' + \frac{l}{A} n'\right\rangle \\ &\quad = \left\langle k, q_0 + \frac{l}{A} n |\hat{A}(\mathrm{id})\hat{B}| k', q_0' + \frac{l}{A} n'\right\rangle \\ &\quad = \int_0^{\frac{2\pi}{l}}\int_0^{\frac{l}{A}} \sum_{n''} \left\langle k, q_0 + \frac{l}{A} n |\hat{a}| k'', q_0'' + \frac{l}{A} n''\right\rangle\left\langle k'', q_0'' + \frac{l}{A} n'' |\hat{b}| k', q_0' + \frac{l}{A} n'\right\rangle \mathrm{d}k''\mathrm{d}q_0'' \\ &\quad = \int_0^{\frac{2\pi}{l}}\int_0^{\frac{l}{A}} \frac{(2\pi)^2}{A^2}\sum_{n''} \delta(k''-k)\delta(k'-k'')\delta(q_0' - q_0'')\delta(q_0'' - q_0), \\ &\left\langle k, q_0 + \frac{l}{A} n |\hat{a}| k, q_0 + \frac{l}{A} n''\right\rangle\left\langle k'', q_0'' + \frac{l}{A} n'' |\hat{b}| k'', q_0'' + \frac{l}{A} n'\right\rangle \mathrm{d}k''\mathrm{d}q_0'' \\ &\quad = \frac{(2\pi)^2}{A^2}\sum_{n''}\delta(k'-k)\delta(q_0' - q_0)\left\langle k, q_0 + \frac{l}{A} n |\hat{a}| k, q_0 + \frac{l}{A} n''\right\rangle \times \\ &\quad\quad \times \left\langle k, q_0 + \frac{l}{A} n'' |\hat{b}| k, q_0 + \frac{l}{A} n'\right\rangle.\end{aligned} \tag{82}$$

From Equations (81) and (82), one concludes that the necessary and sufficient condition of a projection operator in T^2 is

$$\sum_{n''} M_b(k, q_0)_{nn''} M_b(k, q_0)_{n''n'} = M_b(k, q_0)_{nn'} \tag{83}$$

at each point (k, q_0), where

$$M_b(k, q_0)_{n'n''} = \frac{2\pi}{A}\left\langle k, q_0 + \frac{l}{A} n' |\hat{b}| k, q_0 + \frac{l}{A} n''\right\rangle,$$

The matrix $M(k, q_0)$ satisfying the above equation is always diagonizable. Thus

$$M_b(k, q_0) = S(k, q_0)^{-1}\bar{M}_b(k, q_0)S(k, q_0) \tag{84}$$

gives the complete set of projections of the form (78) in T^2, where S is an arbitrary invertible matrix and $\bar{M}$ is a diagonalized matrix with entry 0 and 1, that is to say that

$$\bar{M}_b(k, q_0)_{nn'} = \delta_{nn'}\epsilon_n(k, q_0), \tag{85}$$

$$\epsilon_n(k, q_0) = \{0, 1\}. \tag{86}$$

Next, we study the conditions for P being invariant under rotation R,

$$R^{-1}PR = P. \tag{87}$$

First notice that from Equation (5), the common eigenvectors $|k, q_0 + (l/A)s\rangle$ of U_1 and U_2 are still eigenvectors of

$$U_1' = R^{-1}U_1R, \qquad U_2' = R^{-1}U_2R. \tag{88}$$

We have

$$U_1'\left|k, q_0 + \frac{l}{A}s\right\rangle = \lambda_1\left|k, q_0 + \frac{l}{A}s\right\rangle, \tag{89}$$

$$U_2'\left|k, q_0 + \frac{l}{A}s\right\rangle = \lambda_2\left|k, q_0 + \frac{l}{A}s\right\rangle, \tag{90}$$

giving

$$U_1R\left|k, q_0 + \frac{l}{A}s\right\rangle = \lambda_1R\left|k, q_0 + \frac{l}{A}s\right\rangle, \tag{91}$$

$$U_2R\left|k, q_0 + \frac{l}{A}s\right\rangle = \lambda_2R\left|k, q_0 + \frac{l}{A}s\right\rangle. \tag{92}$$

Thus $R|k, q_0 + (l/A)s\rangle$ is still a common eigenvector of U_1 and U_2. Since the eigenvalue of U_2 is A fold degenerate, we conclude from Equation (19) that

$$R\left|k, q_0 + \frac{l}{A}n\right\rangle = \sum_{n'} A(k, q_0)_n^{n'}\left|k', q_0' + \frac{l}{A}n'\right\rangle \tag{93}$$

for a definite (k', q_0'). One can derive explicit relations for (k', q_0') and (k, q_0) for all Z_N cases, which is essentially the linear relations W: $(k, q_0) \to (k', q_0')$ and $(W)^N = \text{id}$. Since R is unitary, the matrix $A(k, q_0)$ is also unitary, namely

$$A^*(k, q_0)_{nn'} = A^{-1}(k, q_0)_{n'n}. \tag{94}$$

One can show that the map W is an area preserving map, thus

$$\delta(k_1 - k_2)\delta(q_{01} - q_{02}) = \delta(k_1' - k_2')\delta(q_{01}' - q_{02}'). \tag{95}$$

Then we have

$$\begin{aligned}
&\left\langle k_1, q_{01}+\frac{l}{A}n_1\right|R^{-1}PR\left|k_2, q_{02}+\frac{l}{A}n_2\right\rangle \\
&\quad = \delta(k_1'-k_2')\delta(q_{01}'-q_{02}')\left\langle k_1', q_{01}'+\frac{l}{A}n_1'\right|\hat{b}\left|k_2', q_{02}'+\frac{l}{A}n_2'\right\rangle\times \\
&\qquad \times A^*(k_1, q_{01})_{n_1n_1'}A(k_2, q_{02})_{n_2n_2'} \\
&\quad = \delta(k_1-k_2)\delta(q_{01}-q_{02})\left\langle k_1, q_{01}+\frac{l}{A}n_1\right|\hat{b}\left|k_2, q_{02}+\frac{l}{A}n_2\right\rangle.
\end{aligned} \tag{96}$$

Finally, we obtain

$$M(kq_0)_{n_1n_2} = \sum_{n_1'n_2'} A^*(k, q_0)_{n_1n_1'}M(k'q_0')_{n_1'n_2'}A(k, q_0)_{n_2n_2'} \tag{97}$$

$$M(k'q_0') = A^t(k, q_0)M(kq_0)A^*(k, q_0). \tag{98}$$

That is, from the matrix M of a giving point (k, q_0), we can get a definite M for the point $(k', q_0') = W(k, q_0)$, if the corresponding operator is R rotation invariant. From the explicit expression of W, we can show that one can always divide the area σ $(k \in [0, 2\pi/l), q_0 \in [0, l/A))$ into N pieces $\sigma_1, \sigma_2, \ldots, \sigma_N$ for T^2/Z_N, where W: $\sigma_i \to \sigma_{i+1}, \sigma_N \to \sigma_1$. We can arbitrarily choose $M(k, q_0)$ in σ_1 by Equation (84) and get $M(k, q_0)$ in $\sigma_2, \ldots, \sigma_N$ by Equation (98). In such way, we obtain all projective operators in T^2/Z_N by series expansion

$$\begin{aligned}
P &= \sum_{j_1j_2} D_{j_1j_2} e^{-\frac{2\pi i}{l}(j_1\hat{y}_1 + \frac{j_2-\tau_1j_1}{\tau_2}\hat{y}_2)} \\
D_{j_1j_2} &= \frac{1}{A}\mathrm{tr}\left[\hat{b}e^{\frac{2\pi i}{l}(j_1(\hat{y}_1-\frac{\tau_1}{\tau_2}\hat{y}_2)+j_2\frac{\hat{y}_2}{\tau_2})}\right] \\
&= \frac{1}{A}\sum_{nn'}\int dk\, dq_0\, dk'\, dq_0' \left\langle k, q_0+\frac{l}{A}n\right|\hat{b}\left|k', q_0'+\frac{l}{A}n'\right\rangle\times \\
&\qquad \times \left\langle k', q_0'+\frac{l}{A}n'\right|e^{\frac{2\pi i}{l}\left(j_1\left(\hat{y}_1-\frac{\tau_1}{\tau_2}\hat{y}_2\right)+j_2\frac{\hat{y}_2}{\tau_2}\right)}\left|k, q_0+\frac{l}{A}n\right\rangle \\
&= \frac{1}{A}\sum_{nn'}\int dk\, dq_0\, dk'\, dq_0' \left\langle k, q_0+\frac{l}{A}n\right|\hat{b}\left|k', q_0'+\frac{l}{A}n'\right\rangle\times \\
&\qquad \times \delta(k-k')\delta(q_0-q_0')E(k, q_0, j_1, j_2)_{n'n} \\
&= \sum_{nn'}\int dk\, dq_0 (2\pi)^{-1}M(k, q_0)_{nn'}E(k, q_0, j_1, j_2)_{n'n} \\
&= \int dk\, dq_0 (2\pi)^{-1}\mathrm{tr}M(k, q_0)E(k, q_0, j_1, j_2),
\end{aligned} \tag{99}$$

where

$$E(k, q_0, j_1, j_2)_{n'n} = e^{2\pi i j_1(q_0+\frac{l}{A}n')/l+\frac{\pi i j_1j_2}{A}+i(j_2+n'-n)\frac{lk}{A}}\sum_j \delta_{j_2+n'-n, jA}. \tag{100}$$

References

1. Connes, A.: *Non-commutative Geometry*, Academic Press, New York, 1994.
2. Landi, G.: An introduction to non-commutative space and their geometry, hep-th/9701078; Varilly, J.: An introduction to non-commutative geometry, physics/9709045.
3. Madore, J.: *An Introduction to Non-commutative Differential Geometry and its Physical Applications*, 2nd edn, Cambridge Univ. Press, 1999.
4. Connes, A., Douglas, M. and Schwartz, A.: Matrix theory compactification on Tori, *J. High Energy Phys.* **9802** (1998) 003, hep-th/9711162; M. Dougals, C. Hull, JHEP 9802 (1998) 008, hep-th/9711165.
5. Seiberg, N. and Witten, E.: String theory and non-commutative geometry, *J. High Energy Phys.* **9909** (1999), 032, hep-th/9908142; Schomerus, V.: D-branes and deformation quantization, *J. High Energy Phys.* **9906** (1999), 030.
6. Witten, E. Noncommutative geometry and string field theory, *Nuclear Phys. B* **268** (1986), 253.
7. Laughlin, R. B.: In: R. Prange and S. Girvin (eds), *The quantum Hall Effect*, Springer, New York, 1987, p233.
8. Susskind, L.: hep-th/0101029; Hu, J. P. and Zhang, S. C.: cond-mat/0112432.
9. Gopakumar, R., Headrick, M. and Spradin, M.: On noncommutative multi-solitons, hep-th/0103256.
10. Martinec, E. J. and Moore, G.: Noncommutative solitons on orbifolds, hep-th/0101199.
11. Gross, D. J. and Nekrasov, N. A.: Solitons in noncommutative gauge theory, hep-th/0010090; Douglas, M. R. and Nekrasov, N. A.: Noncommutative field theory, hep-th/0106048.
12. Gopakumar, R., Minwalla, S. and Strominger, A.: Noncommutative soliton, *J. High Energy Phys.* **005** (2000), 048, hep-th/0003160.
13. Harvey, J.: Komaba lectures on noncommutative solitons and D-branes, hep-th/0102076; Harvey, J. A., Kraus, P. and Larsen, F.: *J. High Energy Phys.* **0012** (200), 024, hep-th/0010060; Hamanaka, M. and Terashima, S.: On exact noncommutative BPS solitons, *J. High Energy Phys.* **0103** (2001), 034, hep-th/0010221.
14. Rieffel, M.: *Pacific J. Math.* **93** (1981), 415.
15. Boca, F. P.: *Comm. Math. Phys.* **202** (1999), 325.
16. Konechny, A. and Schwartz, A.: Compactification of M(atrix) theory on noncommutative toroidal orbifolds *Nuclear Phys. B* **591**, 667 (2000), hep-th/9912185; Module spaces of maximally supersymmetric solutions on noncommutative tori and noncommutative orbifolds, *J. High Energy Phys.* **0009** (2000), 005, hep-th/0005167.
17. Walters, S.: Projective modules over noncommutative sphere, *J. London Math. Soc.* **51** (1995), 589; Chern characters of Fourier modules, *Canad. J. Math.* **52** (2000), 633.
18. Bacry, H., Grossman, A. and Zak, J.: *Phys. Rev. B* **12** (1975), 1118.
19. Zak, J.: In H. Ehrenreich, F. Seitz and D. Turnbull (eds). *Solid State Physics*, Academic Press, New York, 1972.

ELSEVIER

Nuclear Physics B 638 (2002) 220–242

NUCLEAR PHYSICS B

www.elsevier.com/locate/npe

Non-commutative algebra of functions of 4-dimensional quantum Hall droplet

Yi-Xin Chen [a], Bo-Yu Hou [b], Bo-Yuan Hou [c]

[a] *Zhejiang Institute of Modern Physics, Zhejiang University, Hangzhou 310027, PR China*
[b] *Institute of Modern Physics, Northwestern University, Xi'an 710069, PR China*
[c] *Graduate School, Chinese Academy of Science, Beijing 100039, PR China*

Received 11 March 2002; received in revised form 13 May 2002; accepted 18 June 2002

Abstract

We develop the description of non-commutative geometry of the 4-dimensional quantum Hall fluid's theory proposed recently by Zhang and Hu. The non-commutative structure of fuzzy S^4, which is the base of the bundle S^7 obtained by the second Hopf fibration, i.e., $S^7/S^3 = S^4$, appears naturally in this theory. The fuzzy monopole harmonics, which are the essential elements in the non-commutative algebra of functions on S^4, are explicitly constructed and their obeying the matrix algebra is obtained. This matrix algebra is associative. We also propose a fusion scheme of the fuzzy monopole harmonics of the coupling system from those of the subsystems, and determine the fusion rule in such fusion scheme. By products, we provide some essential ingredients of the theory of $SO(5)$ angular momentum. In particular, the explicit expression of the coupling coefficients, in the theory of $SO(5)$ angular momentum, are given. We also discuss some possible applications of our results to the 4-dimensional quantum Hall system and the matrix brane construction in M-theory.

Keywords: 4-dimensional quantum Hall system; Fuzzy monopole harmonics; Non-commutative geometry

1. Introduction

The planar coordinates of quantum particles in the lowest Landau level of a constant magnetic field provide a well-known and natural realization of non-commutative space [1]. The physics of electrons in the lowest Landau level exhibits many fascinating properties. In particular, when the electron density lies in at certain rational fractions of the density

E-mail addresses: yxchen@zimp.zju.edu.cn (Y.-X. Chen), byhou@phy.nwu.edu.cn (B.-Y. Hou).

0550-3213/02/$ – see front matter
PII: S0550-3213(02)00499-6

corresponding to a fully filled lowest Landau level, the electrons condensed into special incompressible fluid states whose excitations exhibit such unusual phenomena as fractional charge and fractional statistics. For the filling fractions $1/m$, the physics of these states is accurately described by certain wave functions proposed by Laughlin [2], and more general wave functions may be used to describing the various types of excitations about the Laughlin states.

There has recently appeared an interesting connection between quantum Hall effect and non-commutative field theory. In particular, Susskind [3] proposed that non-commutative Chern–Simons theory on the plane may provide a description of the (fractional) quantum Hall fluid, and specifically of the Laughlin states. Susskind's non-commutative Chern–Simons theory on the plane describes a spatially infinite quantum Hall system. It, i.e., does the Laughlin states at filling fractions ν for a system of an infinite number of electrons confined in the lowest Landau level. The fields of this theory are infinite matrices which act on an infinite Hilbert space, appropriate to account for an infinite number of electrons. Subsequently, Polychronakos [4] proposed a matrix regularized version of Susskind's non-commutative Chern–Simons theory in an effort to describe finite systems with a finite number of electrons in the limited spatial extent. This matrix model was shown to reproduce the basic properties of the quantum Hall droplets and two special types of excitations of them. The first type of excitations is arbitrary area-preserving boundary excitations of the droplet. The another type of excitations are the analogs of quasi-particle and quasi-hole states. These quasi-particle and quasi-hole states can be regarded as non-perturbative boundary excitations of the droplet. Furthermore, it was shown that there exists a complete minimal basis of exact wave functions for the matrix regularized version of non-commutative Chern–Simons theory at arbitrary level ν^{-1} and rank N, and that those are one to one correspondence with Laughlin wave functions describing excitations of a quantum Hall droplet composed of N electrons at filling fraction ν [5]. It is believed that the matrix regularized version of non-commutative Chern–Simons theory is precisely equivalent to the theory of composite fermions in the lowest Landau level, and should provide an accurate description of fractional quantum Hall state. However, it does appear an interesting conclusion that they are agreement on the long distance behavior, but the short distance behavior is different [6].

In the matrix regularized version of non-commutative Chern–Simons theory, a confining harmonic potential must be added to the action of this matrix model to keep the particles near the origin. In fact, there has been a translationally invariant version of Laughlin quantum Hall fluid for the two-dimensional electron gas in which it is not necessary to add any confining potential. Such model is Haldane' that of fractional quantum Hall effect based on the spherical geometry [7]. Haldane's model is set up by a two-dimensional electron gas of N particles on a spherical surface in radial monopole magnetic field. A Dirac's monopole is at the center of two-dimensional sphere. This compact sphere space can be mapped to the flat Euclidean space by standard stereographical mapping. In fixed limit, the connection between this model and non-commutative Chern–Simons theory can be exhibited clearly. Exactly, the non-commutative property of particle's coordinates in Haldane's model should be described in terms of fuzzy two-sphere.

In this paper, we do not plan to discuss such description of Haldane's model in detail, but want to exhibit the character of non-commutative geometry of 4-dimensional generalization of Haldane's model, proposed recently by Zhang and Hu [8]. The 4-dimensional generalization of the quantum Hall system is composed of many particles moving in four-dimensional space under a $SU(2)$ gauge field. Instead of the two-sphere geometry in Haldane's model, Zhang and Hu considered particles on a four-sphere surface in radial Yang's $SU(2)$ monopole gauge field [9], which replaces the Dirac's monopole field of Haldane's model. This Yang's $SU(2)$ monopole gauge potential defined on four-sphere can be transformed to the instanton potential of the $SU(2)$ Yang–Mills theory [10] upon a conformal transformation from four-sphere to the 4-dimensional Euclidean space. Zhang and Hu had shown that at appropriate integer and fractional filling fractions the generalization of system forms an incompressible quantum fluid. They [12] also investigated collective excitations at the boundary of the 4-dimensional quantum Hall droplet proposed by them. In their discussion, an non-commutative algebraic relation between the coordinates of particle moving on four-sphere plays the key role. According to our understanding about Haldane's and Zhang et al. works, we think that fuzzy sphere structures in their models is the geometrical origin of non-commutative algebraic relations of the particle coordinates. We shall clarify this idea in this paper.

Non-commutative spheres have found a variety of physical applications [13–20]. The description of fuzzy two-sphere [14] was discovered in early attempts to quantize the super-membrane [16]. The fuzzy four-sphere appeared in [15,17]. The connection of non-commutative second Hopf bundle with the fuzzy four-sphere has been investigated [19] from quantum group. The fuzzy four-sphere was used [21] in the context of the matrix theory of BFSS [22] to described time-dependent 4-brane solutions constructed from zero-brane degrees of freedom. Furthermore, the non-commutative descriptions of spheres also arise in various contexts in the physics of D-branes. The descriptions of them, e.g., were used to exhibit the non-commutative properties and dielectric effects of D-branes [23]. Recently, Ho and Ramgoolam [24,25] had studied the matrix descriptions of higher-dimensional fuzzy spherical branes in the matrix theory. They have found that the finite matrix algebras associated with the various fuzzy spheres have a natural basis which falls in correspondence with tensor constructions of irreducible representations of the corresponding orthogonal groups. In their formalism, they gave the connection between various fuzzy spheres and matrix algebras by introducing a projection from matrix algebra to fuzzy spherical harmonics. Their fuzzy spheres obey non-associative algebras because of the non-associativity induced by the projected multiplication. The complication of projection makes their constructions of fuzzy spherical harmonics formal.

The goal of this paper is to explore the character of non-commutative geometry of 4-dimensional quantum Hall system proposed recently by Zhang and Hu. Recently, Fabinger [26] had pointed that there exists a connection of the fuzzy S^4 with Zhang and Hu's quantum Hall model of S^4. The string theory and brane matrix theory related with such non-commutative structure of fuzzy S^4 are discussed by [26–28]. However, the structure of non-commutative algebra of functions on S^4 is not still clear. It is known that the key idea of non-commutative geometry is in replacement of commutative algebra of functions on a smooth manifold by a non-commutative deformation of it [20]. We shall explore the structure the character of non-commutative geometry of 4-dimensional quantum Hall system to find non-commutative algebra of functions on S^4.

We should emphasize that the non-commutative structure of fuzzy S^4 of Zhang and Hu's quantum Hall model is different with those commonly considered by people. In fact, the S^4 of Zhang and Hu's quantum Hall model is the base of the Hopf bundle S^7 obtained by the second Hopf fibration, i.e., $S^7/S^3 = S^4$, from the connection between the second Hopf map and Yang's $SU(2)$ [29,30]. Equivalently, $\frac{SO(5)}{SU(2)\times SU(2)} = \frac{S^7}{SU(2)=S^3} = S^4$ since $S^7 = \frac{SO(5)}{SU(2)}$. The bundle $S^7 = \frac{SO(5)}{SU(2)}$ can be parametered $\theta, \alpha, \beta, \gamma, \alpha_I, \beta_I, \gamma_I$ (see Section 2 in detail). By further smearing out the common $U(1)$ gauge symmetry parametered by γ_I, we can obtain the bundle $\frac{SO(5)}{SU(2)\times U(1)} = \frac{SO(5)}{U(2)} = S^6$ and its fibre $\frac{SU(2)}{U(1)} = \frac{S^3}{S^1} = S^2$. Furthermore, we can find the sections of the bundle S^7 or S^6 by solving the eigenfunctions of Zhang and Hu's model. The eigenfunctions of the LLL consist of the space on which non-commutative algebra of functions on S^4 act.

Since at special filling factors, the quantum disordered ground state of 4-dimensional quantum Hall effect is separated from all excited states by a finite energy gap, the lowest energy excitations are quasi-particle or quasi-hole excitations near the lowest Landau level. The quantum disordered ground state of 4-dimensional quantum Hall effect is the state composed coupling by particles lying in the lowest Landau level state. At appropriate integer and fractional filling fractions, the system forms an incompressible quantum liquid, which is called as a 4-dimensional quantum Hall droplet. In fact, the spherical harmonic operators for fuzzy four-sphere are related with quasi-particle's or quasi-hole's creators of 4-dimensional quantum Hall effect. These operators is composed of a complete set of the matrices with the fixed dimensionality. Focusing on the space of particle's lowest Landau level state, we shall construct explicitly these fuzzy spherical harmonics, also called the fuzzy monopole harmonics, and discuss the nontrivial algebraic relation between them. Furthermore, we shall clarify the physical implications of them.

This paper is organized as follows. Section 2 introduces the 4-dimensional quantum Hall model proposed by Zhang and Hu, and analyzes the property of Hilbert space and the symmetrical structures of this 4-dimensional quantum Hall system. We shall emphasize the intrinsic properties of the Yang's $SU(2)$ monopole included in this system, and give the explicit forms of normalized wave functions of this system, which is given by (8) in our paper. The wave functions corresponding to the irreducible representation $R = [r_1 = I, r_2 = I]$ of $SO(5)$ are those in the LLL. They are the sections of the bundle $S^7 = SO(5)/SU(2)$ over S^4 with fibre $SU(2) = S^3$ parametered by $\alpha_I, \beta_I, \gamma_I$. These degeneracy wave functions consist of the LLL Hilbert space which the non-commutative algebra of functions on S^4 acts on. The elements of this algebra are constructed in the following section. Section 3 describes the elements of non-commutative algebra of functions on S^4 which is related with the fuzzy four-sphere from the geometrical and symmetrical structures of 4-dimensional quantum Hall droplet. We find the matrix forms and the symbols of the elements, given by (15) and (13), respectively. We give the explicit construction of complete set of this matrix algebra, determined by (19). In Section 4, we find the system of algebraic equations satisfied by the generators of the matrix algebra. The results are given by Eqs. (30) and (31). For the matrix forms of the elements, this non-commutative algebra should be understood as the algebraic relation of matrix mulitiplication, and for the symbols of the elements, it should be done as that of Moyal product. It can be seen from these results and the complete set (19) that the non-commutative algebra is closed. The associativity of this algebra is shown by the relation (33). Furthermore, a fusion scheme of the fuzzy monopole harmonics of the coupling system from those of the subsystems, and its fusion rule are established in this section, which are given by the relations (40) and (41). Section 5 includes discussions about the physical interpretations of the results and remarks on some physical applications of them.

2. The Hilbert space of 4-dimensional quantum Hall system

The 4-dimensional quantum Hall system is composed of many particles moving in four-dimensional space under a $SU(2)$ gauge field. The Hamiltonian of a single particle moving on four-sphere S^4 is read as

$$\mathcal{H}=\frac{\hbar^2}{2MR^2}\sum_{a<b}\Lambda_{ab}^2, \tag{1}$$

where M is the inertia mass and R the radius of S^4. The symmetry group of S^4 is $SO(5)$. Because the particle is coupling with a $SU(2)$ gauge field A_a, Λ_{ab} in Eq. (1) is the dynamical angular momentum given by $\Lambda_{ab}=-i(x_aD_b-x_bD_a)$. From the covariant derivative $D_a=\partial_a+A_a$, one can calculate the gauge field strength from the definition $f_{ab}=[D_a,D_b]$. Λ_{ab} does not satisfy the commutation relations of $SO(5)$ generators. Similar to in Dirac's monopole field, the angular momentum of a particle in Yang's $SU(2)$ monopole field can be defined as $L_{ab}=\Lambda_{ab}-if_{ab}$, which indeed obey the $SO(5)$ commutation relations. Yang [11] proved that L_{ab} can generate all $SO(5)$ irreducible representations.

In general, the representations of $SO(5)$ can be put in one-to-one correspondence with Young diagrams, labelled by the row lengths $[r_1,r_2]$, which obey the constraints $0\leqslant r_2\leqslant r_1$. For such a representation, the eigenvalue of Casimir operator is given by $A(r_1,r_2)=\sum_{a<b}L_{ab}^2=r_1^2+r_2^2+3r_1+r_2$, and its dimensionality is

$$D(r_1,r_2)=\frac{1}{6}(1+r_1-r_2)(1+2r_2)(2+r_1+r_2)(3+2r_1). \tag{2}$$

The $SU(2)$ gauge field is valued in the $SU(2)$ Lie algebra $[I_i,I_j]=i\varepsilon_{ijk}I_k$. The value of this $SU(2)$ Casimir operator $\sum_i I_i^2=I(I+1)$ specifies the dimension of the $SU(2)$ representation in the monopole potential. I is an important parameter of L_{ab} generating all $SO(5)$ irreducible representations and the Hamiltonian Eq. (1). In fact, for a given I, if one deals with the eigenvalues and eigenfunctions the operator of angular momentum $\sum_{a<b}L_{ab}^2$, it can be found that the $SO(5)$ irreducible representations, which the eigenfunctions called by Yang [11] as $SU(2)$ monopole harmonics belong to, are restricted. Such $SO(5)$ irreducible representations are labelled by the integers $[r_1,r_2=I]$, and $r_1\geqslant I$. Based on the expressions of $SU(2)$ monopole potentials given by Yang [11], or by Zhang and Hu [8], one can show that $\sum_{a<b}\Lambda_{ab}^2=\sum_{a<b}L_{ab}^2-\sum_i I_i^2$ by straightforwardly evaluating. This implies that the eigenvalues and eigenfunctions of the Hamiltonian Eq. (1) can be read off from those of the operator $\sum_{a<b}L_{ab}^2$. Hence, for a

given I, the energy eigenvalues of the Hamiltonian Eq. (1) are read as

$$E_{[r_1,r_2=I]}=\frac{\hbar^2}{2MR^2}\big[A(r_1,r_2=I)-2I(I+1)\big]. \tag{3}$$

The degeneracy of energy level is given by the dimensionality of the corresponding irreducible representation $D(r_1,r_2=I)$.

The ground state of the Hamiltonian Eq. (1) plays a key role in the procedure of construction of many-body wave function and the discussion of incompressibility of 4-dimensional quantum Hall system. This ground state, also called the lowest Landau level (LLL) state, is described by the least admissible irreducible representation of $SO(5)$, i.e., labelled by $[r_1=I,r_2=I]$ for a given I. The LLL state is $D(r_1=I,r_2=I)=\frac{1}{6}(2I+1)(2I+2)(2I+3)$ fold degenerate, and its energy eigenvalue is $\frac{\hbar^2}{2MR^2}2I$. Zhang and Hu [8] found the explicit form of the ground state wave function in the spinor coordinates. This wave function is read as

$$\langle x_a,\mathbf{n}_i|m_1,m_2,m_3,m_4\rangle=\sqrt{\frac{p!}{m_1!m_2!m_3!m_4!}}\Psi_1^{m_1}\Psi_2^{m_2}\Psi_3^{m_3}\Psi_4^{m_4}, \tag{4}$$

with integers $m_1+m_2+m_3+m_4=p=2I$. The orbital coordinate x_a, which is defined by the coordinate point of the 4-dimensional sphere $X_a=Rx_a$, is related with the spinor coordinates Ψ_α with $\alpha=1,2,3,4$ by the relations $x_a=\overline{\Psi}\Gamma_a\Psi$ and $\sum_\alpha\overline{\Psi}_\alpha\Psi_\alpha=1$. The five 4×4 Dirac matrices Γ_a with $a=1,2,3,4,5$ satisfy the Clifford algebra $\{\Gamma_a,\Gamma_b\}=2\delta_{ab}$. The isospin coordinates $\mathbf{n}_i=\bar{u}\sigma_i u$ with $i=1,2,3$ are given by an arbitrary two-component complex spinor (u_1,u_2) satisfying $\sum_\sigma\bar{u}_\sigma u_\sigma=1$. Zhang and Hu gave the explicit solution of the spinor coordinate with respect to the orbital coordinate as following

$$\begin{pmatrix}\Psi_1\\ \Psi_2\end{pmatrix}=\sqrt{\frac{1+x_5}{2}}\begin{pmatrix}u_1\\ u_2\end{pmatrix},\qquad \begin{pmatrix}\Psi_3\\ \Psi_4\end{pmatrix}=\sqrt{\frac{1}{2(1+x_5)}}(x_4-ix_i\sigma_i)\begin{pmatrix}u_1\\ u_2\end{pmatrix}. \tag{5}$$

By computing the geometric connection, one can get a non-Abelian gauge potential A_a, which is just the $SU(2)$ gauge potential of a Yang monopole defined on 4-dimensional sphere S^4 [8,9]. Since we do not need the explicit form of it here, we do not write out that of it.

The description of the 4-dimensional quantum Hall liquid involves the quantum many-body problem of N particle's moving on the 4-dimensional sphere S^4 in the Yang's $SU(2)$ monopole field lying in center of the sphere S^4. The wave functions of many particles can be constructed by the nontrivial product of the single particle wavefunctions, among which every single particle wave function is given by the LLL wavefunction Eq. (4). In the case of integer filling, the many-particle wave function is simply the Slater determent composed of N single-particle wave functions. For the fractional filling fractions, the many particle wave function cannot be expressed as the Laughlin form of a single product. But the amplitude of the many-particle wave function can also be interpreted as the Boltzmann weight for a classical fluid. One can see that it describes an incompressible liquid by means of plasma analogy. Therefore, at the integer or fractional filling fractions, the 4-dimensional system of the generalizing quantum Hall effect forms an incompressible quantum liquid [8]. We shall call this 4-dimensional system as a 4-dimensional quantum Hall droplet.

The space of the degenerate states in the LLL is very important not only for the description of the 4-dimensional quantum Hall droplet but also for that of edge excitations and quasi-particle or quasi-hole excitations of the droplet. In fact, this space of the degenerate states is the space which we shall construct the matrix algebra acting on in the following section. In order to construct the complete set of matrix algebra of fuzzy S^4, we need to know the explicit forms of the wave functions associated with all irreducible representations $[r_1, r_2 = I]$ of $SO(5)$. Although Yang had found the wave functions for all the $[r_1, r_2 = I]$ states, the form of his parameterizing the four sphere S^4 is not convenient for our purpose. Following Hu and Zhang [12], we can parameterize the four sphere S^4 by the following coordinate system

$$\begin{aligned}
x_1 &= \sin\theta \sin\frac{\beta}{2}\sin(\alpha - \gamma),\\
x_2 &= -\sin\theta \sin\frac{\beta}{2}\cos(\alpha - \gamma),\\
x_3 &= -\sin\theta \cos\frac{\beta}{2}\sin(\alpha + \gamma),\\
x_4 &= \sin\theta \cos\frac{\beta}{2}\cos(\alpha + \gamma),\\
x_5 &= \cos\theta,
\end{aligned} \tag{6}$$

where $\theta, \beta \in [0, \pi)$ and $\alpha, \gamma \in [0, 2\pi)$. The direction of the isospin is specified by α_I, β_I and γ_I.

As the above explanation, we can get the eigenfunctions of the Hamiltonian Eq. (1) from the eigenfunctions of the operator $\sum_{a<b} L_{ab}^2$. The angular momentum operators L_{ab} consists of an orbital part $L_{\mu\nu}^{(0)} = -i(x_\mu \partial_\nu - x_\nu \partial_\mu)$ with $\mu, \nu = 1, 2, 3, 4$ and an isospin part involving the $SU(2)$ monopole field. The angular momentum operators $L_{\mu\nu}^{(0)}$ generate the rotation in the subspace (x_1, x_2, x_3, x_4), and satisfy the commutation relations of $SO(4)$ generators. They can be decomposed into two $SU(2)$ algebras: $\hat{J}_{1i}^{(0)} = \frac{1}{2}(\frac{1}{2}\varepsilon_{ijk}L_{jk}^{(0)} + L_{4i})$ and $\hat{J}_{2i}^{(0)} = \frac{1}{2}(\frac{1}{2}\varepsilon_{ijk}L_{jk}^{(0)} - L_{4i})$. They satisfy the identity of operator $\sum_i \hat{J}_{1i}^{(0)2} = \sum_i \hat{J}_{2i}^{(0)2}$. Therefore, if one would use the operators $\hat{J}_{1i}^{(0)}$ and $\hat{J}_{2i}^{(0)}$ to generate the $SO(4)$ blocks of the $SO(5)$ irreducible representations, he cannot obtain all irreducible representations of $SO(5)$. However, because of the coupling to the Yang's $SU(2)$ monopole potential, these orbital $SO(4)$ generators are modified into $L_{\mu\nu}$, which are decomposed into $\hat{J}_{1i} = \hat{J}_{1i}^{(0)}$ and $\hat{J}_{2i} = \hat{J}_{2i}^{(0)} + \hat{I}_i$. Indeed, the $SO(4)$ generators $\hat{J}_{1i}$ and $\hat{J}_{2i}$ can be used to generate the $SO(4)$ block states of all $SO(5)$ irreducible representations [11]. Such $SO(4)$ block states can be labelled by the $SO(4)$ quantum numbers $J \equiv \left(\begin{smallmatrix} j_1 & j_2 \\ j_{1z} & j_{2z} \end{smallmatrix}\right)$, where $\sum_i \hat{J}_{1i}^2 = j_1(j_1 + 1)$ and $\sum_i \hat{J}_{2i}^2 = j_2(j_2 + 1)$. The j_{1z} and j_{2z} are the magnetic quantum numbers of two $SU(2)$ algebras.

Applying the $SO(4)$ operators to the quantum states described by the Hamiltonian Eq. (1) in the irreducible representations of $SO(5)$, one can see that the complete set of quantum observables of the system is composed of the operators $\sum_{a<b} L_{ab}^2$, $\hat{J}_1^2$, $\hat{J}_2^2$, $\hat{J}_{1z}$ and $\hat{J}_{2z}$. Thus, there exist the simultaneous eigenfunctions of those operators, which are just the wave functions of energy eigenvalue being $E_{[r_1, r_2 = I]}$. Noticing that the isospin

operators $\hat{I}_i$ are coupling with the operator $\hat{J}_{2i}^{(0)}$ into the angular momentum operators $\hat{J}_{2i}$, we should also introduce the quantum number labelling the isospin parameter I, which is written as $I^{(0)} \equiv \binom{0\ I}{0\ I_z}$. The parameters of group $(\theta, \alpha, \beta, \gamma, \alpha_I, \beta_I, \gamma_I)$ are abbreviated to (Ω). The wave functions corresponding to the irreducible representation $R \equiv [r_1, r_2 = I]$ of $SO(5)$ are denoted as $\mathcal{D}_{JI^{(0)}}^{(R)}(\Omega)$. By using the equation of parameterizing S^4 and the parameters of the isospin direction, and following Yang [11], one can obtain the wave functions $\mathcal{D}_{JI^{(0)}}^{(R)}(\Omega)$ obeying the system of equations as following

$$
\begin{aligned}
&\hat{J}_1^2\mathcal{D}_{JI^{(0)}}^{(R)}(\Omega) = j_1(j_1+1)\mathcal{D}_{JI^{(0)}}^{(R)}(\Omega), \qquad \hat{J}_{1z}\mathcal{D}_{JI^{(0)}}^{(R)}(\Omega) = j_{1z}\mathcal{D}_{JI^{(0)}}^{(R)}(\Omega),\\
&\hat{I}^2\mathcal{D}_{JI^{(0)}}^{(R)}(\Omega) = I(I+1)\mathcal{D}_{JI^{(0)}}^{(R)}(\Omega),\\
&\hat{J}_2^2\mathcal{D}_{JI^{(0)}}^{(R)}(\Omega) = j_2(j_2+1)\mathcal{D}_{JI^{(0)}}^{(R)}(\Omega), \qquad \hat{J}_{2z}\mathcal{D}_{JI^{(0)}}^{(R)}(\Omega) = j_{2z}\mathcal{D}_{JI^{(0)}}^{(R)}(\Omega),\\
&\left\{\frac{1}{\sin^3\theta}\frac{\partial}{\partial\theta}\left(\sin^3\theta\frac{\partial}{\partial\theta}\right) - \frac{4\hat{J}_1^2}{\sin^2\theta} - \frac{2(1-\cos\theta)}{\sin^2\theta}\left(\hat{J}_2^2 - \hat{J}_1^2 - \hat{I}^2\right)\right.\\
&\quad \left. - \frac{(1-\cos\theta)(3+\cos\theta)}{\sin^2\theta}\hat{I}^2\right\}\mathcal{D}_{JI^{(0)}}^{(R)}(\Omega) = -A(r_1, r_2 = I)\mathcal{D}_{JI^{(0)}}^{(R)}(\Omega).
\end{aligned} \tag{7}
$$

The first line and second line of the equations tell us that we can use two $SU(2)$ D-functions to realize the wave function with respect to the dependence of the group parameters α, β, γ and $\alpha_I, \beta_I, \gamma_I$. Furthermore, we also should consider the coupling relations $\hat{J}_{2i} = \hat{J}_{2i}^{(0)} + \hat{I}_i$, $\hat{J}_{1i} = \hat{J}_{1i}^{(0)}$ and $\sum_i \hat{J}_{1i}^{(0)2} = \sum_i \hat{J}_{2i}^{(0)2}$ to make the wave function obey the third line of the equations. The final line is the equation to determine the θ dependence of the wave function, which had been solved by Yang [11]. Now, we can write the explicit solution form of the normalized wavefunction as

$$
\begin{aligned}
\mathcal{D}_{JI^{(0)}}^{(R)}(\Omega) = &\sqrt{\frac{2r_1+3}{2}}(\sin\theta)^{-1}d_{j_1-j_2,-j_1-j_2-1}^{(r_1+1)}(\theta)\frac{\sqrt{(2I+1)(2j_1+1)}}{8\pi^2}\\
&\times \sum_{j_{2z}^{(0)}, I_z'} D_{j_{1z}j_{2z}^{(0)}}^{j_1}(\alpha, \beta, \gamma)\langle j_2^{(0)} = j_1, j_{2z}^{(0)}; I, I_z' | j_2, j_{2z}\rangle D_{I_z' I_z}^{(I)}(\alpha_I, \beta_I, \gamma_I),
\end{aligned} \tag{8}
$$

where

$$
\begin{aligned}
d_{j_1-j_2,-j_1-j_2-1}^{(r_1+1)}(\theta) = &\sqrt{\frac{2^{-2(j_1-j_2)}(r_1+j_1-j_2+1)!(r_1-j_1+j_2+1)!}{(r_1+j_1+j_2+2)!(r_1-j_1-j_2)!}}\\
&\times (1-\cos\theta)^{\frac{2j_1+1}{2}}(1+\cos\theta)^{-\frac{2j_2+1}{2}}P_{r_1-j_1+j_2+1}^{(2j_1+1,-2j_2-1)}(\cos\theta),
\end{aligned} \tag{9}
$$

and $P_n^{(\alpha,\beta)}$ is the Jacobi polynomial.

Although $d_{j_1-j_2,-j_1-j_2-1}^{(r_1+1)}(\theta)$ is the d-function of $SO(3)$ rotation group, it should be emphasized that here the quantum numbers of the subgroup $SO(4)$ of $SO(5)$ have replaced the usual magnetic quantum number of $SO(3)$. Practically, $\sqrt{\frac{2r_1+3}{2}}(\sin\theta)^{-1}d_{j_1-j_2,-j_1-j_2-1}^{(r_1+1)}(\theta)$ is the d-function of $SO(5)$ rotation group in the special case. The D-function $D_{j_{1z}j_{2z}^{(0)}}^{j_1}(\alpha, \beta,$

γ) is the standard $SU(2)$ representation matrix for the Euler angles α, β, γ. They are the $SU(2)$ rotation matrix elements generated rotationally by the operators $\hat{J}_{1i}$. Similarly, $D^{(I)}_{I'_z I_z}(\alpha_I, \beta_I, \gamma_I)$ are those generated by the isospin operators $\hat{I}_i$. The coupling coefficients $\langle j_2^{(0)} = j_1, j_{2z}^{(0)}; I, I'_z | j_2, j_{2z}\rangle$, i.e., the Clebsch–Gordon coefficients, show the coupling behavior of the $SU(2)$ angular momentums $\hat{J}_{1i}$, $\hat{I}_i$ and $\hat{J}_{2i}$ by means of $\hat{J}^{(0)}_{2i}$.

The isospin direction can be normally specified by two angles α_I and β_I. However, the wave function $\mathcal{D}^{(R)}_{JI^{(0)}}(\Omega)$ depends on the Eular angles α_I, β_I and γ_I of the isospin space. In fact, in the $(\hat{I}^2, \hat{I}_z)$ picture, the γ_I dependence of $\mathcal{D}^{(R)}_{JI^{(0)}}(\Omega)$ is given by the $U(1)$ phase factor $\exp\{-iI_z\gamma_I\}$, where I_z is simply an $U(1)$ gauge index. Therefore, different values of I_z correspond to the same physical state. One can smear the γ_I dependence of the wave function by the gauge choice. Such gauge choice can be fixed by taking $I_z = I$ or $I_z = -I$. If such gauge choice is taken at every step in all calculations, we call this choice as taking the physical gauge. Analogous to doing usually in the field theory, there exists another gauge choice, which such gauge choice is taken at the end of calculations. The latter gauge choice are called as taking the covariant gauge. We shall take the covariant gauge in this paper, which this thick was also used in the reference [12].

The wave functions $\mathcal{D}^{(R)}_{JI^{(0)}}(\Omega)$ are the $SU(2)$ monopole harmonics. Exactly, they are the spherical harmonics on the coset space $SO(5)/SU(2)$, which is locally isomorphic to the sphere $S^4 \times S^3$. By smearing the $U(1)$ degree of freedom of the γ_I dependence, i.e., taking the physical gauge, they can be viewed as the spherical harmonics on the coset space $SO(5)/U(2)$, which is locally isomorphic to the sphere $S^4 \times S^2$. Globally, $SO(5)/U(2)$ is a bundle over the S^4 with fibre S^2. The wave functions are the cross sections in this nontrivial fibre bundle. This implies that there exists a stabilizer group of the wave function solutions $U(2)$, and the action of $SO(5)$ on the state generates a space of the wave function solutions which is the space of cross sections in $SO(5)/U(2)$. If we take the covariant gauge, the $SU(2)$ monopole harmonics should be regarded as the cross sections in the nontrivial fibre bundle $SO(5)/SU(2)$. Then, the stabilizer group of the wave function solutions is $SU(2)$. $SO(5)$ doing the state generates that of cross sections in $SO(5)/SU(2)$. The procedure of parameterizing the four sphere S^4 Eq. (6) and building up the isomorphic relation of the $SU(2)$ group manifold and the three sphere S^3 is just that of smearing the stabilizer subgroup $SU(2)$. Of course, if one want to take the physical gauge, he can smear the $U(2)$ by further smearing the $U(1)$ gauge subgroup. Such globally geometrical structure and symmetrical structure can guide us to develop some techniques which we need in this paper.

Let us introduce the state vector $\left|{R \atop J}\right\rangle$, which belongs to the Hilbert space $\mathcal{H}^{(R)}$ composed of the $SU(2)$ monopole harmonics $\mathcal{D}^{(R)}_{JI^{(0)}}(\Omega)$. Of course, $\left|{R \atop I^{(0)}}\right\rangle$ is an element of the Hilbert space $\mathcal{H}^{(R)}$. In general, taking a fixed vector in the Hilbert space, one can use the unitary irreducible representation of an arbitrary Lie group acting in the Hilbert space to produce the coherent state for this Lie group. Now, we are interesting to the coherent state corresponding to the coset space $SO(5)/SU(2)$. The $SU(2)$ is the isotropic subgroup of $SO(5)$ for the state $\left|{R \atop I^{(0)}}\right\rangle$ since $SU(2)$ is the stabilizer group of the wave function solutions. If we use the unitary irreducible representations of $SO(5)$ acting on the state $\left|{R \atop I^{(0)}}\right\rangle$ to produce the coherent states, the coherent state vectors belonging to a left coset class of

$SO(5)$ with respect to the subgroup $SU(2)$ differ only in a phase factor and so determine the same state. Consequently, the coherent state vectors depend only on the group parameters parameterizing the coset space $SO(5)/SU(2)$. Thus, we can now introduce the following coherent state vector

$$|\Omega, R\rangle = \sum_{J, I_z} \mathcal{D}^{*(R)}_{J I^{(0)}}(\Omega) \left| \begin{matrix} R \\ J \end{matrix} \right\rangle |I, I_z\rangle, \tag{10}$$

where the star stands for the complex conjugate. In order to realize the covariant gauge, we have added the isospin frame $|I, I_z\rangle$ to the monopole harmonics $\mathcal{D}^{*(R)}_{J I^{(0)}}(\Omega)$. In fact, we can use the finite rotation $\widehat{\mathcal{R}}(\Omega) = \exp\{i\alpha \hat{J}_{1z}\} \exp\{i\beta \hat{J}_{1y}\} \exp\{i\gamma \hat{J}_{1z}\} \exp\{i\theta \hat{L}_{54}\} \exp\{i\alpha_I \hat{I}_z\} \times \exp\{i\beta_I \hat{I}_y\} \exp\{i\gamma_I \hat{I}_z\}$ acting on the state vector $\left| \begin{smallmatrix} R \\ I^{(0)} \end{smallmatrix} \right\rangle$ to realize the general finite rotation of $SO(5)$ acting on the state vector $\left| \begin{smallmatrix} R \\ I^{(0)} \end{smallmatrix} \right\rangle$ up to a phase factor, and to generate the above coherent state vector.

The wave functions in the coherent state picture are given by

$$\left\langle \Omega, R \middle| \begin{matrix} R \\ J \end{matrix} \right\rangle = \sum_{I_z} \langle I, I_z | \mathcal{D}^{(R)}_{J I^{(0)}}(\Omega). \tag{11}$$

The l.h.s. of Eq. (11) is not with the label I since it naturelly appears in the label $R = [r_1, r_2 = I]$, which corresponds to the $SU(2)$ monopole harmonics. It should be emphasized that the above wave functions become the wave functions in the physical gauge only if one smears the isospin frame of them and projects back to the α_I and β_I angles. In the sense of the finite rotation of $SO(5)$, the explicit forms of wave function solutions given here are the wave function solutions of the 4-dimensional spherically symmetrical top with the $SU(2)$ self-rotating in the isospin direction. The Yang's $SU(2)$ monopole harmonics [11] can be interpret as the wave functions in the coherent state picture in the physical gauge.

Based on the orthogonality and completeness of the state vectors $\left| \begin{smallmatrix} R \\ J \end{smallmatrix} \right\rangle$ belonging to an irreducible representation $R = [r_1, r_2 = I]$ of $SO(5)$, we can give the completeness condition of the coherent state vectors

$$\frac{D(r_1, r_2 = I)}{A(\Omega)} \int d\Omega\, |\Omega, R\rangle\langle\Omega, R| = 1, \tag{12}$$

where $A(\Omega) = \mathrm{Area}(S^4 \times S^3) = \frac{1}{12}(8\pi^2)^2$. Every irreducible representation of $SO(5)$ corresponds to a complete set of the coherent state vectors. The coherent states of the different irreducible representations are orthogonal each other. The coherent state corresponding to the LLL states is very important for the description of non-commutative geometry of 4-dimensional quantum Hall droplet.

In order to avoid the label of the irreducible representation of the LLL states confusing with the parameter I of the model, we denote the irreducible representation $R = [r_1 = I, r_2 = I]$ as $\frac{P}{2} \equiv [r_1 = I, r_2 = I]$. The LLL degeneracy states consist of the Hilbert space $\mathcal{H}^{(P/2)}$. Because of the LLL states are the lowest energy states of particle's living in, for a given I, the smallest admissible irreducible representation of $SO(5)$ is $R = [r_1 = I, r_2 = I] = P/2$. Therefore, the irreducible representations of $SO(5)$ are truncated since there exists an Yang's $SU(2)$ monopole. If we focus on the Hilbert space of the LLL states, we can determine the matrix forms of tensor operator for the $SU(2)$ monopole harmonics, which are the $D(r_1 = I, r_2 = I) \times D(r_1 = I, r_2 = I)$ matrices. The coupling relation between the tensor operators provides a truncated parameter for the tensor operator for the $SU(2)$ monopole harmonics. The number of independent operators is $(D(r_1 = I, r_2 = I))^2$, which we shall explain in more detail in the next section. Therefore, we should replace the functions by the $D(r_1 = I, r_2 = I) \times D(r_1 = I, r_2 = I)$ matrices on the fuzzy S^4. Exactly, the cross sections in the fibre bundle $SO(5)/SU(2)$, which is a bundle over S^4 with fibre S^3, are replaced by the $D(r_1 = I, r_2 = I) \times D(r_1 = I, r_2 = I)$ matrices on the fuzzy S^4. Thus, the algebra on the fuzzy S^4 becomes non-commutative. The direct product of N single-particle Hilbert spaces $\mathcal{H}^{(\frac{P}{2})}$ can be used to build up the Hilbert space of 4-dimensional quantum Hall droplet. Hence, the fuzzy S^4 appears naturally in the description of particle's moving on the S^4 with the Yang's $SU(2)$ monopole at the center of the sphere. Consequently, the fuzzy S^4 is the description of non-commutative geometry of 4-dimensional quantum Hall droplet.

3. Fuzzy monopole harmonics and Matrix operators of fuzzy S^4

The construction of fuzzy S^4 is to replace the functions on S^4 by the non-commutative algebra taken in the irreducible representations of $SO(5)$. This is a full matrix algebra which is generated by the fuzzy monopole harmonics. These fuzzy monopole harmonics consist of a complete basis of the matrix space. In this section, we shall find the explicit forms of such fuzzy monopole harmonics by means of the expressions of the $SU(2)$ monopole harmonics given in the previous section. Our main task of this section is to construct the operators corresponding to the $SU(2)$ monopole harmonics, and to give the matrix elements of these operators acting on the LLL states.

Although for a given I, the $SU(2)$ monopole harmonics of the smallest admissible irreducible representations of $SO(5)$ can be regarded as the single-particle wave functions of the LLL in 4-dimensional quantum Hall system, the $SU(2)$ monopole harmonics smaller than the smallest admissible irreducible representations of $SO(5)$ are useful for us to construct the fuzzy monopole harmonics on $SO(5)/SU(2)$. Such $SU(2)$ monopole harmonics can be read off from the expressions obtained in the previous section by the changing of the parameter I. If we replace I by J, the $SO(5)$ irreducible representation of the monopole harmonics becomes $R = [r_1, r_2 = J]$. Equivalently, we can use $R = [r_1, r_2]$ and $J^{(0)} \equiv \binom{0\ r_2}{0\ r_{2z}}$ to label them. Thus, the $SU(2)$ monopole harmonics are generally expressed as $\mathcal{D}^{(R)}_{J J^{(0)}}(\Omega)$, which can be obtained by replacing $I^{(0)}$ with $J^{(0)}$ in the Eq. (8). We can find their corresponding coherent states by the same replacement.

By using the standard techniques of the generalized coherent state [31], we can now construct the operator $\widehat{\mathcal{Y}}^R_J$ corresponding to the $SU(2)$ monopole harmonics $\mathcal{D}^{(R)}_{J J^{(0)}}(\Omega)$. Noticing that this operator is an operator of acting in the LLL Hilbert space and $\mathcal{D}^{(R)}_{J J^{(0)}}(\Omega)$

is regarded as the basic function, we can express it as the following form

$$\widehat{\mathcal{Y}}_J^R = \frac{D(r_1 = I, r_2 = I)}{A(\Omega)} \int d\Omega \sum_{r_{2z}} \langle r_2, r_{2z} | \mathcal{D}^{(R)}_{J\,J^{(0)}}(\Omega) \left| \Omega, \frac{P}{2} \right\rangle \left\langle \Omega, \frac{P}{2} \right|. \tag{13}$$

The matrix elements of this operator in the LLL Hilbert space are read as

$$\begin{aligned} \left\langle \begin{matrix} \frac{P}{2} \\ K^1 \end{matrix} \middle| \widehat{\mathcal{Y}}_J^R \middle| \begin{matrix} \frac{P}{2} \\ K^2 \end{matrix} \right\rangle &= \frac{D(r_1 = I, r_2 = I)}{A(\Omega)} \\ &\times \int d\Omega \sum_{I_{1z}, r_{2z}, I_{2z}} \mathcal{D}^{*(P/2)}_{K^1 I^{(0)}}(\Omega) \mathcal{D}^{(R)}_{J\,J^{(0)}}(\Omega) \mathcal{D}^{(P/2)}_{K^2 I^{(0)}}(\Omega) \\ &\times \langle I, I_{1z} | r_2, r_{2z} \rangle | I, I_{2z} \rangle. \end{aligned} \tag{14}$$

The above integral can be analytically performed by making use of the integral formulae about the Jacobi polynomial and the product of three $SU(2)$ D-functions, and the properties of $SU(2)$ D-function. The result is

$$\begin{aligned} \left\langle \begin{matrix} \frac{P}{2} \\ K^1 \end{matrix} \middle| \widehat{\mathcal{Y}}_J^R \middle| \begin{matrix} \frac{P}{2} \\ K^2 \end{matrix} \right\rangle &= \left\langle \begin{matrix} \frac{P}{2} \\ K^1 \end{matrix} \middle\| \begin{matrix} R \\ J \end{matrix} \middle\| \begin{matrix} \frac{P}{2} \\ K^2 \end{matrix} \right\rangle \begin{Bmatrix} k_1^1 & j_1 & k_1^2 \\ I & r_2 & I \\ k_2^1 & j_2 & k_2^2 \end{Bmatrix} \\ &\times \langle k_1^1, k_{1z}^1 | j_1, j_{1z}; k_1^2, k_{1z}^2 \rangle \langle k_2^1, k_{2z}^1 | j_2, j_{2z}; k_2^2, k_{2z}^2 \rangle, \end{aligned} \tag{15}$$

where $\left\langle \begin{smallmatrix} P/2 \\ K^1 \end{smallmatrix} \middle\| \begin{smallmatrix} R \\ J \end{smallmatrix} \middle\| \begin{smallmatrix} P/2 \\ K^2 \end{smallmatrix} \right\rangle$ is independent of the $SU(2)$ magnetic quantum numbers, and is composed of two parts, i.e., it is equal to $\Phi(k_1^1, k_2^1, k_1^2, k_2^2, j_1, j_2, I, r_2)\Theta(k_1^1, k_2^1, k_1^2, k_2^2, j_1, j_2, I, r_1, r_2)$. The part of Φ is from the contribution of the integral with respect to the variables $\alpha, \beta, \gamma, \alpha_I, \beta_I$ and γ_I, which is provided by the normalized coefficients and the arisen factors when we expressed the sum over all $SU(2)$ magnetic quantum numbers of six $SU(2)$ coupling coefficients as the 9-j symbol. Its expression is given by

$$\begin{aligned} \Phi &= (-1)^{j_1 + j_2 + j_1 + k_1^1 + k_1^2} \\ &\times \sqrt{\frac{(2I+1)^3 (2k_2^1+1)^2 (2k_2^2+1)(2k_1^2+1)(2r_2+1)(2j_1+1)(2j_2+1)}{(8\pi^2)^2}}. \end{aligned} \tag{16}$$

The another part Θ is given by the integrated part of θ. It is read as

$$\begin{aligned} \Theta &= \sqrt{\frac{(2I+1)^2(2r_1+1)}{8}} \int_0^\pi d\theta\, d^{(I+1)}_{k_1^1 - k_2^1, -I-1}(\theta)\, d^{(I+1)}_{k_1^2 - k_2^2, -I-1}(\theta)\, d^{(r_1+1)}_{j_1 - j_2, -j_1 - j_2 - 1}(\theta) \\ &= \left\{ 2^{2(k_1^1 + k_1^2 + j_1) - 2(k_2^1 + k_2^2 + j_2) + 3} \right\}^{-\frac{1}{2}} \\ &\times \left\{ \frac{(I + k_1^1 - k_2^1 + 1)!(I + k_1^2 - k_2^2 + 1)!(I + k_2^1 - k_1^1 + 1)!}{(2I+3)^2(2r_1+3)((2I+2)!)^2} \right. \\ &\times \left. (I + k_2^2 - k_1^2 + 1)! \right\}^{-\frac{1}{2}} \end{aligned}$$

$$\times \left\{ \frac{(r_1+j_1+j_2+2)!(r_1-j_1-j_2)!}{(r_1+1+j_2-j_1)!(r_1+1-j_2+j_1)!} \right\}^{-\frac{1}{2}} \frac{D(r_1=I, r_2=I)}{A(\Omega)}$$
$$\times \frac{2^{2I+j_1-j_2-2k_2^1-2k_2^2}\Gamma(j_1+k_1^1+k_1^2+2)\Gamma(k_2^1+k_2^2-j_2+1)}{(r_1-j_1+j_2+1)!\Gamma(2I+3+j_1-j_2)\Gamma(2j_1+2)}$$
$$\times \Gamma(r_1+j_1+j_2+3)$$
$$\times {}_3F_2(-r_1+j_1-j_2-1, r_1+j_1-j_2+2, j_1+k_1^1+k_1^2+2; 2j_1+2, 2I+3+j_1-j_2; 1), \tag{17}$$

where ${}_3F_2$ is the hyper-geometric function.

Although the 9-j symbol of $SO(3)$ is also independent of the $SU(2)$ magnetic quantum numbers, Φ, Θ and it are very important for the matrix form of the operator since they all depend on the $SO(4)$ subgroup quantum numbers of $SO(5)$. Hence, they may not be smeared out by re-scaling. In particular, the 9-j symbol together with two $SU(2)$ coupling coefficients will provide the selection rules of the matrix elements of the operator $\widehat{\mathcal{Y}}_J^R$. From the coupling relations of 9-j symbol's elements and the triangle relations of $SU(2)$ angular momentum coupling, we find that r_2 can be evaluated in the range zero to $2I$. Furthermore, because of $k_1^1+k_2^1=k_1^2+k_2^2=I$, the highest values of these k's is I. We know that the maximums of j_1 and j_2 both are $2I$ from the coupling relations of two $SU(2)$ coupling coefficients. The scheme of the $SO(5)$ irreducible representation containing the $SO(4)$ blocks [11] tells us that the largest admissible value of r_1 is $2I$. Acting in the LLL Hilbert space, the operators $\widehat{\mathcal{Y}}_J^R$ producing the nonzero contributions are those belonging to the irreducible representations of $SO(5)$ of $0 \leqslant r_2 \leqslant r_1 \leqslant 2I$.

The number of such operators can be calculated by means of the dimension formula of irreducible representation of $SO(5)$. From Eq. (2), we have

$$\sum_{0\leqslant r_2\leqslant r_1\leqslant 2I} D(r_1,r_2) = \frac{1}{6}\sum_{0\leqslant r_2\leqslant r_1\leqslant 2I}(1+r_1-r_2)(1+2r_2)(2+r_1+r_2)(3+2r_1)$$
$$= \frac{1}{36}(2I+1)^2(2I+2)^2(2I+3)^2. \tag{18}$$

This is exactly equal to the square of the dimensionality of the LLL Hilbert space. This show that the operators

$$\widehat{\mathcal{Y}}_J^R, \quad 0\leqslant r_2\leqslant r_1\leqslant 2I \tag{19}$$

constitute a complete set of irreducible representations of $SO(5)$ in the matrix algebra with the square of the dimension of the LLL Hilbert space. We also call them as the fuzzy monopole harmonics.

Subsequently, we shall discuss the coupling of the $SO(5)$ angular momentums. It is helpful to explore the Hilbert space structure of many-body system corresponding to the system (1), in special, of the 4-dimensional quantum Hall system. The Kronecker product of two unitary irreducible representations of a semi-simple group is completely reducible, but it is generally not simply reducible. Hence, a given representation may appear more than once in the decomposition of the Kronecker product. However, in the case of $SO(5)$ angular momentum, every admissible irreducible representation only appears once in the decomposition of the Kronecker product [11]. This conclusion is a corollary of Theorems 2, 4 and 8 in Ref. [11]. Based on this fact, we have the following coupling relation of the state vectors

$$\left|\begin{matrix} R \\ J \end{matrix}\right\rangle = \sum_{J_1,J_2} \left\langle \begin{matrix} R_1 & R_2 \\ J_1 & J_2 \end{matrix} \middle| \begin{matrix} R \\ J \end{matrix} \right\rangle \left| \begin{matrix} R_1 & R_2 \\ J_1 & J_2 \end{matrix} \right\rangle. \tag{20}$$

Then, for the case of $SO(5)$ angular momentum, the decomposition of the Kronecker product of two $SO(5)$ irreducible representations is accomplished by the $SO(5)$ coupling coefficients $\langle \begin{smallmatrix} R_1 & R_2 \\ J_1 & J_2 \end{smallmatrix} | \begin{smallmatrix} R \\ J \end{smallmatrix} \rangle$. These coupling coefficients obey the usual orthogonality relations. We again emphasize that here all state vectors $|\begin{smallmatrix} R \\ J \end{smallmatrix}\rangle$ belong to the Hilbert space $\mathcal{H}^{(R)}$, which is composed of the monopole harmonics $\mathcal{D}^{(R)}_{J\,J^{(0)}}(\Omega)$.

In the previous section, we have explained that $\sum_{r_{2z}} \mathcal{D}^{*(R)}_{J\,J^{(0)}}(\Omega)|r_2, r_{2z}\rangle$ can be regarded as the wave function with the $U(1)$ gauge degree of freedom generated by the finite rotation of $SO(5)$ acting on the fixed state vector in the Hilbert space, of course, $\sum_{r_{2z}} \langle r_2, r_{2z}|\mathcal{D}^{(R)}_{J\,J^{(0)}}(\Omega)$ also can be done. By using this property of the wave functions and the orthogonality relations of the coupling coefficients, we find that

$$\begin{aligned} &\sum_{R,J,r_{2z},r^1_{2z},r^2_{2z}} \langle r_2, r_{2z}|\mathcal{D}^{(R)}_{J\,J^{(0)}}(\Omega) \left\langle \begin{matrix} R \\ J^{(0)} \end{matrix} \middle| \begin{matrix} R_1 & R_2 \\ J^{(0)}_1 & J^{(0)}_2 \end{matrix} \right\rangle \left\langle \begin{matrix} R_1 & R_2 \\ J_1 & J_2 \end{matrix} \middle| \begin{matrix} R \\ J \end{matrix} \right\rangle \\ &\quad = \sum_{r^1_{2z}} \langle r^1_2, r^1_{2z}|\mathcal{D}^{(R_1)}_{J_1 J^{(0)}_1}(\Omega) \sum_{r^2_{2z}} \langle r^2_2, r^2_{2z}|\mathcal{D}^{(R_2)}_{J_2 J^{(0)}_2}(\Omega), \end{aligned} \tag{21}$$

where $R_i = [r^i_1, r^i_2]$, and $J^{(0)}_i = \begin{pmatrix} 0 & r^i_2 \\ 0 & r^i_{2z} \end{pmatrix}$ for $i = 1, 2, 3$ (see below). By means of the orthogonality and normalized condition of the $SU(2)$ monopole harmonic, we can obtain the explicit expression of the coupling coefficients of the $SO(5)$ angular momentum

$$\begin{aligned} \left\langle \begin{matrix} R_2 & R_3 \\ J_2 & J_3 \end{matrix} \middle| \begin{matrix} R_1 \\ J_1 \end{matrix} \right\rangle &= \langle R_1 \| R_2 \| R_3 \rangle^{-1} \\ &\quad \times \int d\Omega \sum_{r^1_{2z},r^2_{2z},r^3_{2z}} \mathcal{D}^{*(R_1)}_{J_1 J^{(0)}_1}(\Omega) \mathcal{D}^{(R_2)}_{J_2 J^{(0)}_2}(\Omega) \mathcal{D}^{(R_3)}_{J_3 J^{(0)}_3}(\Omega) \\ &\quad \times \langle r^1_2, r^1_{2z}|r^2_2, r^2_{2z}\rangle |r^3_2, r^3_{2z}\rangle, \end{aligned} \tag{22}$$

where $\langle R_1 \| R_2 \| R_3 \rangle = \sum_{r^1_{2z},r^2_{2z},r^3_{2z}} \langle \begin{smallmatrix} R_1 \\ J^{(0)}_1 \end{smallmatrix} | \begin{smallmatrix} R_2 & R_3 \\ J^{(0)}_2 & J^{(0)}_3 \end{smallmatrix} \rangle$ is invariant under the rotation transformation of $SO(5)$, hence, is a pure scale factor.

Performing the calculation of the above integral, we get

$$\begin{aligned} \left\langle \begin{matrix} R_2 & R_3 \\ J_2 & J_3 \end{matrix} \middle| \begin{matrix} R_1 \\ J_1 \end{matrix} \right\rangle &= \frac{\langle \begin{smallmatrix} R_1 \\ J_1 \end{smallmatrix} \| \begin{smallmatrix} R_2 \\ J_2 \end{smallmatrix} \| \begin{smallmatrix} R_3 \\ J_3 \end{smallmatrix} \rangle}{\langle R_1 \| R_2 \| R_3 \rangle} \begin{Bmatrix} j^1_1 & j^2_1 & j^3_1 \\ r^1_2 & r^2_2 & r^3_2 \\ j^1_2 & j^2_2 & j^3_2 \end{Bmatrix} \\ &\quad \times \langle j^1_1, j^1_{1z}|j^2_1, j^2_{1z}; j^3_1, j^3_{1z}\rangle \langle j^1_2, j^1_{2z}|j^2_2, j^2_{2z}; j^3_2, j^3_{2z}\rangle. \end{aligned} \tag{23}$$

Similar to the matrix elements of the fuzzy monopole harmonics, $\left\langle {R_1 \atop J_1} \middle\| {R_2 \atop J_2} \middle\| {R_3 \atop J_3} \right\rangle = \widetilde{\Theta}\widetilde{\Phi}$. $\widetilde{\Theta}$ and $\widetilde{\Phi}$ are simply similar to Θ and Φ respectively. In $\widetilde{\Phi}$, it is only different of the replacing the quantum numbers in Φ with the corresponding quantum number at present. The expression of $\widetilde{\Phi}$ is given by

$$\widetilde{\Phi} = (-1)^{j_1^2 - j_1^3 + j_2^2 - j_2^3 + r_2^2 - r_2^3} \sqrt{\frac{(2r_2^1 + 1)(2j_2^1 + 1)}{(8\pi^2)^2(2j_1^1 + 1)}} \times \prod_{i=1}^{3} (-1)^{j_1^i - r_2^i} \sqrt{(2r_2^i + 1)(2j_1^i + 1)(2j_2^i + 1)}. \tag{24}$$

The expression of $\widetilde{\Theta}$ is given by the following integration

$$\widetilde{\Theta} = \int_{-1}^{1} dx \left(1 - x^2\right)^{-\frac{1}{2}} \prod_{i=1}^{3} \left[\frac{2^{2j_1^i - 2j_2^i + 1}(r_1^i + j_1^i + j_2^i + 2)!(r_1^i - j_1^i - j_2^i)!}{(2r_1^i + 3)(r_1^i - j_1^i + j_2^i + 1)!(r_1^i + j_1^i - j_2^i + 1)!} \right]^{-\frac{1}{2}} \times (1 - x)^{\frac{2j_1^i + 1}{2}} (1 + x)^{-\frac{2j_2^i + 1}{2}} P_{r_1 - j_1^i + j_2^i + 1}^{(2j_1^i + 1, -2j_2^i - 1)}(x). \tag{25}$$

Performing the analysis parallel to the selection rules of the matrix elements, we find that the coupling coefficient vanishes unless the r_1^i and r_2^i satisfy the generalized triangular conditions as following

$$|r_2^1 - r_2^2| \leqslant r_2^3 \leqslant r_1^3 \leqslant r_1^1 + r_1^2, \tag{26}$$

etc. The selection rules of the $SO(5)$ coupling coefficients about the $SO(4)$ subgroup quantum numbers are provided by the 9-j symbol and two $SU(2)$ coupling coefficients.

Usually, the evaluation of a coupling coefficient involving a set of basis states labeled by the irreducible representations of a chain of nested subgroup for a semi-simple group. In this approach, the most important ingredient is Racah's factorization lemma, which ensure the coupling coefficient to factorize into the coupling coefficients of the subgroup. Then, the evaluation of a coupling coefficient becomes the calculations of the re-coupling coefficients, also called the isoscalar factors. In fact, the calculation of the isoscalar factor is very difficult. The advantage of this method is that it can overcome the trouble with a semi-simple group being not generally simply reducible. However, for the case of $SO(5)$ angular momentum, the group $SO(5)$ is simply reducible in the $SO(5)$ level, i.e., every irreducible representation of $SO(5)$ only appears once in the decomposition of the direct product of two $SO(5)$ irreducible representations. Our deriving the explicit expression of the coupling coefficients of $SO(5)$ is just based on this fact. Since the expression of the coupling coefficients given here is analytically exact, it is not necessary further to calculate the isoscalar factors appearing in the factorization of the coupling coefficients of the chain of nested subgroup.

In fact, $\widetilde{\Theta}$ and $\widetilde{\Phi}$ together with the 9-j symbol of $SO(3)$ in the Eq. (23) provide the isoscalar factor for $SO(5) \supset SU(2) \times SU(2)$ in the level of $SO(5)$ angular momentum. Here, $SU(2) \times SU(2)$ is the relatively direct product, of which the rotation transformation is generated by one part $\hat{J}_{1i}$ and another relative part $\hat{J}_{2i} = \hat{J}_{2i}^{(0)} + \hat{I}_i$ by means of

$\sum_i \hat{J}_{1i}^2 = \sum_i \hat{J}_{2i}^{(0)2}$. Furthermore, we give here the analytically exact expressions of all isoscalar factors of $SO(5)$ in the $SO(5)$ angular momentum level. It is emphasized worthily that all irreducible representations of $SO(5)$ discussed by us here are those belonging to the $SO(5)$ angular momentum, exactly, those corresponding to the Hilbert spaces which are composed of the $SU(2)$ monopole harmonics.

We can construct many-particle's states from the single-particle states by using of the expression of the coupling coefficients. In the procedure of many-particle state construction, the re-coupling coefficients can be obtained by performing summation over products of the coupling coefficients. Therefore, the expression of the coupling coefficients given by us here is important for the study of the 4-dimensional quantum Hall system and the physical system with the rotation symmetry of $SO(5)$. In fact, the calculation of the matrix elements of fuzzy monopole harmonics is that of the coupling coefficients.

On the other hand, the matrix operators $\widehat{\mathcal{Y}}_J^R, 0 \leqslant r_2 \leqslant r_1 \leqslant 2I$ are the single-body operators acting in the LLL Hilbert space since the matrix forms of them are provided by the matrix elements between the LLL states of single particle system. By determining the ground states, i.e., the LLL states, which can be obtained by the linear combinations of these operators acting on the vacuum state, we can use these operators to generate all possible single particle states including the quasi-particle and quasi-hole states near the LLL states. In this sense, the fuzzy monopole harmonics given by us are the generatores of the general wave functions of single particle system. In order to explore the detail structure of this system's Hilbert space and the properties of quantum states and operators in the 4-dimensional quantum Hall droplet, we need further to discuss the algebraical structure of these operators, i.e., the commutation relations between them.

4. Matrix algebra of fuzzy S^4 and non-commutative geometry from the system

Now, we turn to the discussion of the matrix algebra of the fuzzy monopole harmonics. It is clear that the matrix operators $\widehat{\mathcal{Y}}_J^R, 0 \leqslant r_2 \leqslant r_1 \leqslant 2I$ belong to the irreducible representations of $SO(5)$. All operators belong to the irreducible representations of $SO(5)$ must satisfy the following relation of the tensor products of the operators

$$\widehat{\mathcal{T}}_{J_1}^{R_1} \widehat{\mathcal{T}}_{J_2}^{R_2} = \sum_{R,J} \left\langle \begin{matrix} R \\ J \end{matrix} \middle| \begin{matrix} R_1 & R_2 \\ J_1 & J_2 \end{matrix} \right\rangle \widehat{\mathcal{T}}_J^R. \tag{27}$$

Another useful relation is the Wigner–Eckart theorem for the theory of $SO(5)$ angular momentum. It is read as

$$\left\langle \begin{matrix} R_1 \\ J_1 \end{matrix} \middle| \widehat{\mathcal{T}}_J^R \middle| \begin{matrix} R_2 \\ J_2 \end{matrix} \right\rangle = \left\langle \begin{matrix} R_1 & R \\ J_1 & J \end{matrix} \middle| \begin{matrix} R_2 \\ J_2 \end{matrix} \right\rangle \langle R_1 \| \widehat{\mathcal{T}}^R \| R_2 \rangle, \tag{28}$$

where $\langle R_1 \| \widehat{\mathcal{T}}^R \| R_2 \rangle$ is independent of the subgroup $SO(4)$ quantum numbers of $SO(5)$, which is called as the reduced matrix elements of the operator $\widehat{\mathcal{T}}_J^R$. This theorem can be derived by using of the standard method of the group representation [32] and the fact that every irreducible representation of $SO(5)$ appears once in the decomposition of the Kronecker product of two irreducible representations of $SO(5)$ in the case of $SO(5)$ angular momentum.

We can obtain the relation between the reduced matrix elements by calculating the matrix elements of the tensor product's relation of the operators. Furthermore, we scale the operators $\widehat{\mathcal{Y}}_J^R$ into

$$\widetilde{\widetilde{\mathcal{Y}}}_J^R = \frac{\widehat{\mathcal{Y}}_J^R}{\langle \frac{P}{2} \| \widehat{\mathcal{Y}}^R \| \frac{P}{2} \rangle}. \tag{29}$$

By means of the relation between the reduced matrix elements, we can now read off the matrix algebraic relation of the operators $\widetilde{\widetilde{\mathcal{Y}}}_J^R$ from the operator product relation (27). This relation is

$$\widetilde{\widetilde{\mathcal{Y}}}_{J_1}^{R_1} \widetilde{\widetilde{\mathcal{Y}}}_{J_2}^{R_2} = \sum_{R,J} \left\langle \begin{matrix} R \\ J \end{matrix} \middle| \begin{matrix} R_1 & R_2 \\ J_1 & J_2 \end{matrix} \right\rangle \left\{ \begin{matrix} R & R_1 & R_2 \\ \frac{P}{2} & \frac{P}{2} & \frac{P}{2} \end{matrix} \right\} \left[D\left(\frac{P}{2}\right) \right]^{-1} \widetilde{\widetilde{\mathcal{Y}}}_J^R, \tag{30}$$

where we have define that $D(R) = D(r_1, r_2)$ and $D(\frac{P}{2}) = D(r_1 = I, r_2 = I)$. The operator of the l.h.s. of Eq. (30) is given by the matrix multiplication between the operators $\widetilde{\widetilde{\mathcal{Y}}}_{J_1}^{R_1}$ and $\widetilde{\widetilde{\mathcal{Y}}}_{J_2}^{R_2}$, which act in the LLL Hilbert space. The 6-J symbol of $SO(5)$ is given by

$$\begin{aligned} \left\{ \begin{matrix} R & R_1 & R_2 \\ \frac{P}{2} & \frac{P}{2} & \frac{P}{2} \end{matrix} \right\} = \sum_{J,J_1,J_2,K,K_1,K_2} & \left\langle \begin{matrix} R_1 & R_2 \\ J_1 & J_2 \end{matrix} \middle| \begin{matrix} R \\ J \end{matrix} \right\rangle \left\langle \begin{matrix} \frac{P}{2} \\ K_2 \end{matrix} \middle| \begin{matrix} R & \frac{P}{2} \\ J & K_1 \end{matrix} \right\rangle \\ & \times \left\langle \begin{matrix} \frac{P}{2} & R_1 \\ K_1 & J_1 \end{matrix} \middle| \begin{matrix} \frac{P}{2} \\ K \end{matrix} \right\rangle \left\langle \begin{matrix} \frac{P}{2} & R_2 \\ K & J_2 \end{matrix} \middle| \begin{matrix} \frac{P}{2} \\ K_2 \end{matrix} \right\rangle. \end{aligned} \tag{31}$$

It can be seen easily that the 6-J symbol of $SO(5)$ is possessed of the properties very similar to the 6-j symbol of $SO(3)$.

The matrix algebra (30) of the fuzzy S^4 is formally analogous to that of the fuzzy S^2 [33–35]. Because of the operators $\widetilde{\widetilde{\mathcal{Y}}}_J^R, 0 \leqslant r_2 \leqslant r_1 \leqslant 2I$ consist of a complete basis of the $D(P/2) \times D(P/2)$ matrix algebra, any operator acting in the LLL Hilbert space can be expressed as a linear combination of the matrices $\widetilde{\widetilde{\mathcal{Y}}}_J^R, 0 \leqslant r_2 \leqslant r_1 \leqslant 2I$. The product of such operators again becomes the linear combination of the matrices due to the matrix algebraic relation (30). In the other words, the matrix algebra is a fundamental relation to determine the algebraic relations of all operators acting in the LLL Hilbert space. The origin of the matrix operators and their algebra appearing in the 4-dimensional quantum Hall system is due to the existence of the Yang's $SU(2)$ monopole in the system. Its appearance makes the coordinate space S^4, which the particles live in, become non-commutative. The description of this non-commutative geometry can be obtained by replacing the algebra of the functions with the matrix operators, i.e., the fuzzy monopole harmonics. The procedure of our finding these matrix operators and their matrix algebra above is just establishing the description of this non-commutative geometry. In this sense, the matrix algebra given here is the algebra of fuzzy S^4, and describes the non-commutativity of the coordinate space S^4. When the strength of the Yang's $SU(2)$ monopole vanishes, i.e., $I = 0$, the dimension of the matrix becomes one-dimensional and trivial, and then the matrix algebra does the algebra of the functions. Consequently, the fuzzy S^4 becomes a classical S^4 when $I = 0$.

The matrix algebra given by us is an associative algebra. The simple interpretation of the associativity of the matrix algebra is that since these matrices are the operators acting in the LLL Hilbert space, they can be expressed as

$$\widetilde{\widetilde{\mathcal{Y}}}{}^R_J = \sum_{K^1,K^2} \left\langle \begin{matrix} \frac{P}{2} \\ K^1 \end{matrix} \middle| \widetilde{\widetilde{\mathcal{Y}}}{}^R_J \middle| \begin{matrix} \frac{P}{2} \\ K^2 \end{matrix} \right\rangle \left| \begin{matrix} \frac{P}{2} \\ K^1 \end{matrix} \right\rangle \left\langle \begin{matrix} \frac{P}{2} \\ K^2 \end{matrix} \right|. \tag{32}$$

Thus, the products of three operators become those of three matrices. The associativity of the matrix algebra is equivalently described by the identity of the matrices

$$\begin{aligned} &\sum_{K',K''} \left\langle \begin{matrix} \frac{P}{2} \\ K^1 \end{matrix} \middle| \widetilde{\widetilde{\mathcal{Y}}}{}^{R_1}_{J_1} \middle| \begin{matrix} \frac{P}{2} \\ K' \end{matrix} \right\rangle \left\langle \begin{matrix} \frac{P}{2} \\ K' \end{matrix} \middle| \widetilde{\widetilde{\mathcal{Y}}}{}^{R_2}_{J_2} \middle| \begin{matrix} \frac{P}{2} \\ K'' \end{matrix} \right\rangle \left\langle \begin{matrix} \frac{P}{2} \\ K'' \end{matrix} \middle| \widetilde{\widetilde{\mathcal{Y}}}{}^{R_3}_{J_3} \middle| \begin{matrix} \frac{P}{2} \\ K^2 \end{matrix} \right\rangle \\ &\quad = \sum_{K',K''} \left\langle \begin{matrix} \frac{P}{2} \\ K^1 \end{matrix} \middle| \widetilde{\widetilde{\mathcal{Y}}}{}^{R_1}_{J_1} \middle| \begin{matrix} \frac{P}{2} \\ K'' \end{matrix} \right\rangle \left\langle \begin{matrix} \frac{P}{2} \\ K'' \end{matrix} \middle| \widetilde{\widetilde{\mathcal{Y}}}{}^{R_2}_{J_2} \middle| \begin{matrix} \frac{P}{2} \\ K' \end{matrix} \right\rangle \left\langle \begin{matrix} \frac{P}{2} \\ K' \end{matrix} \middle| \widetilde{\widetilde{\mathcal{Y}}}{}^{R_3}_{J_3} \middle| \begin{matrix} \frac{P}{2} \\ K^2 \end{matrix} \right\rangle. \end{aligned} \tag{33}$$

Obviously, the results of the above summations are same because there does not exist any singularity to make the summarizing sequences change the results of the summations. Of course, one can straightforwardly show the associativity of the operator algebra of the fuzzy S^4 in the similar manner of the proof of the associativity for the case of fuzzy S^2 [33]. This is a more complicate and more technical procedure in which the generalized Biedenharn–Elliott relation for the 6-J symbols of $SO(5)$ should be established. Although this generalized relation is very useful, we do not discuss it here.

From the above discussions, one can see that the matrix algebra (30) is the multiplicative relation of the matrices produced by the fuzzy monopole harmonics acting in the LLL Hilbert space of single particle system. These matrices consist of a complete set of all operators belonging to the LLL Hilbert space of single particle system. It is well known that the Hilbert space of the many particles can be constructed from the single particle's Hilbert spaces by the coupling of some few of single-particles. The most simplest way of fuzzy monopole harmonics'construction of the system of N particles is to construct first the most elementary fuzzy monopole harmonics of the system of N particles by making of the symmetrical direct sum of N single particle fuzzy monopole harmonics. Then, one can produce all fuzzy monopole harmonics of the system of N particles by using of the matrix algebra (30). This way cannot provide the generally truncated rule of the irreducible representations of $SO(5)$ corresponding to the fuzzy monopole harmonics, which are composed of the LLL Hilbert space of the system of N particles. On the other hand, it is not obvious that the connection between the construction of the fuzzy monopole harmonics in this way and the wave functions of quasi-particle or quasi-hole excitations in the Laughlin's and Haldane's forms. We shall give another scheme of the construction of fuzzy monopole harmonics of the system of N particles in the rest of this section, and the most simplest way is a special case of the following scheme.

Subsequently, we shall discuss how the fuzzy monopole harmonics of the coupling system are constructed from the subsystems composed of the coupling system. Since the particles spread over the four-sphere S^4, this study is very important for the description of non-commutative geometry of the 4-dimensional quantum Hall droplet. Our starting point is the tensor product relation of two tensor operators belonging to the irreducible representations of $SO(5)$ which correspond to the Yang's $SU(2)$ monopole harmonics. The tensor operators $\widehat{T}_{J^1}^{R^1}(1)$ and $\widehat{T}_{J^2}^{R^2}(2)$ are supposed to work on the subsystem 1 and the subsystem 2, respectively, of a system in order to distinguish them labelled by adding to $1, 2$. Because of these operators belonging to the irreducible representations of $SO(5)$, they must satisfy the tensor product relation of the operators

$$\widehat{T}_J^R = \sum_{J^1, J^2} \left\langle \begin{matrix} R^1 & R^2 \\ J^1 & J^2 \end{matrix} \middle| \begin{matrix} R \\ J \end{matrix} \right\rangle \widehat{T}_{J^1}^{R^1}(1) \widehat{T}_{J^2}^{R^2}(2), \tag{34}$$

where $\widehat{T}_J^R$ is the tensor operator of the coupling system. In order to calculate the matrix elements, we must choose the coupling scheme of the subsystem in the system. Suppose that we want to calculate the operator matrix elements between the left vectors belonging to the irreducible representation of $R^{1\prime}$ and $R^{2\prime}$ coupling into R' and the right vectors doing that of $R^{1\prime\prime}$ and $R^{2\prime\prime}$ coupling into R''.

We can first make use the tensor product relation (34) and the Wigner–Eckart theorem (28) to obtain an expression for the reduced matrix element in such coupling scheme. The result is

$$\begin{aligned} \langle R' \| \widehat{T}^R \| R'' \rangle = [D(R'')]^{-1} & \begin{Bmatrix} R' & R & R'' \\ R^{1\prime} & R^1 & R^{1\prime\prime} \\ R^{2\prime} & R^2 & R^{2\prime\prime} \end{Bmatrix} \\ & \times \langle R^{1\prime} \| \widehat{T}^{R^1}(1) \| R^{1\prime\prime} \rangle \langle R^{2\prime} \| \widehat{T}^{R^2}(2) \| R^{2\prime\prime} \rangle, \end{aligned} \tag{35}$$

where we have defined the 9-J symbol of the $SO(5)$ angular momentum, which is given by

$$\begin{aligned} & \begin{Bmatrix} R' & R & R'' \\ R^{1\prime} & R^1 & R^{1\prime\prime} \\ R^{2\prime} & R^2 & R^{2\prime\prime} \end{Bmatrix} \\ & = \sum_{\text{all of J's}} \left\langle \begin{matrix} R^1 & R^2 \\ J^1 & J^2 \end{matrix} \middle| \begin{matrix} R \\ J \end{matrix} \right\rangle \left\langle \begin{matrix} R^{1\prime\prime} & R^{2\prime\prime} \\ J^{1\prime\prime} & J^{2\prime\prime} \end{matrix} \middle| \begin{matrix} R'' \\ J'' \end{matrix} \right\rangle \left\langle \begin{matrix} R^{1\prime} & R^1 \\ J^{1\prime} & J^1 \end{matrix} \middle| \begin{matrix} R^{1\prime\prime} \\ J^{1\prime\prime} \end{matrix} \right\rangle \\ & \quad \times \left\langle \begin{matrix} R^{2\prime} & R^2 \\ J^{2\prime} & J^2 \end{matrix} \middle| \begin{matrix} R^{2\prime\prime} \\ J^{2\prime\prime} \end{matrix} \right\rangle \left\langle \begin{matrix} R'' \\ J'' \end{matrix} \middle| \begin{matrix} R' & R \\ J' & J \end{matrix} \right\rangle \left\langle \begin{matrix} R' \\ J' \end{matrix} \middle| \begin{matrix} R^{1\prime} & R^{2\prime} \\ J^{1\prime} & J^{2\prime} \end{matrix} \right\rangle. \end{aligned} \tag{36}$$

Then, we scale the tensor operator $\widehat{T}_J^R$ as

$$\widehat{\widetilde{T}}_J^R = \frac{\widehat{T}_J^R}{\langle R' \| \widehat{T}^R \| R'' \rangle}. \tag{37}$$

Finally, from the tensor product relation of operators we can read off a fundamental formula of the coupling system's operators constructed by means of the tensor product of the subsystem's operators, which is

$$\begin{aligned} \widehat{\widetilde{T}}_J^R = \sum_{R^2, R^{2\prime}, R^{2\prime\prime}, J^1, J^2} & [D(R^{2\prime\prime}) D(R^{1\prime\prime})]^{-1} \left\langle \begin{matrix} R^1 & R^2 \\ J^1 & J^2 \end{matrix} \middle| \begin{matrix} R \\ J \end{matrix} \right\rangle \\ & \times \begin{Bmatrix} R^1 & R^2 & R \\ R^{1\prime} & R^{2\prime} & R' \\ R^{1\prime\prime} & R^{2\prime\prime} & R'' \end{Bmatrix} \widehat{\widetilde{T}}_{J^1}^{R^1}(1) \widehat{\widetilde{T}}_{J^2}^{R^2}(2). \end{aligned} \tag{38}$$

It should be pointed that in the procedure of our obtaining the above formula, we have used the orthogonality relation of the 9-J symbol of the $SO(5)$ angular momentum. In fact, the inverse relation of the above formula also is important since it can be regarded as the operator product expansion of two operators. In order to make the transfer from one to another among them become convenient, here we write out this orthogonality relation

$$\sum_{R,R',R''}\left[D(R'')\right]^{-1}\begin{Bmatrix} R^1 & R^2 & R \\ R^{1\prime} & R^{2\prime} & R' \\ R^{1\prime\prime} & R^{2\prime\prime} & R'' \end{Bmatrix}\begin{Bmatrix} R' & R & R'' \\ R^{1\prime} & R^1 & R^{1\prime\prime} \\ R^{2\prime} & R^2 & R^{2\prime\prime} \end{Bmatrix}=D\left(R^{2\prime\prime}\right)D\left(R^{1\prime\prime}\right). \tag{39}$$

Now, we return to our considering the 4-dimensional quantum Hall system. At special filling factors, the quantum ground state of the 4-dimensional quantum Hall effect is separated from all excited states by a finite energy gap. This quantum ground state of the system of many particles is build up by some few of the particles lying in their LLL states in the coupling manner corresponding to a special filling factor. This quantum ground state is the LLL state of the system of many particles. The lowest energy excitations are the collective excitations of the system of many particles. Such collective excitations also can be built up by the lowest energy excitations of these particles in the fixed coupling manner. In fact, the matrix operators given in the above section are related with the generators of the lowest energy excitations. The present goal is to establish a scheme that the generators of matrix algebra corresponding to the collective excitations are constructed from the generators of the lowest energy excitations of the single particle.

Because of the existence of the finite energy gap, all matrix operators act in their LLL Hilbert spaces. Generally, for the system including an Yang's $SU(2)$ monopole, its LLL Hilbert space is described by the irreducible representation of $SO(5)$ $R=[r_1=r_2,r_2]$. This implies that for our considering system all irreducible representations labelling the matrix elements of the operators should belong to such irreducible representations of $SO(5)$ as $R=[r_1=r_2,r_2]$, e.g., R', $R^{1\prime}$, $R^{2\prime}$, R'', $R^{1\prime\prime}$ and $R^{2\prime\prime}$ in Eq. (38) should be this kind of the irreducible representations of $SO(5)$. Furthermore, we have that $R'=R''$, $R^{1\prime}=R^{1\prime\prime}$ and $R^{2\prime}=R^{2\prime\prime}$ because the operators are realized by the matrices. Now, we can establish a fusion scheme of the matrices of the coupling system from the matrices of the subsystems based on the fundamental formula (38), which is

$$\tilde{\tilde{\mathcal{Y}}}^R_J=\sum_{R^2,J^1,J^2}\left[D(R^{2\prime})D(R^{1\prime})\right]^{-1}\left\langle\begin{matrix} R^1 & R^2 \\ J^1 & J^2 \end{matrix}\middle|\begin{matrix} R \\ J \end{matrix}\right\rangle\begin{Bmatrix} R^1 & R^2 & R \\ R^{1\prime} & R^{2\prime} & R' \\ R^{1\prime} & R^{2\prime} & R' \end{Bmatrix}\tilde{\tilde{\mathcal{Y}}}^{R^1}_{J^1}(1)\tilde{\tilde{\mathcal{Y}}}^{R^2}_{J^2}(2), \tag{40}$$

where $R'=[r_1'=r_2',r_2']$, $R^{1\prime}=[r_1^{1\prime}=r_2^{1\prime},r_2^{1\prime}]$ and $R^{2\prime}=[r_1^{2\prime}=r_2^{2\prime},r_2^{2\prime}]$.

By using of the generalized triangular relation of the $SO(5)$ coupling coefficients and the property of the 9-J symbol of $SO(5)$, a fusion rule of this fusion scheme can be read off as following

$$\begin{aligned} &\left|r_2^1-r_2^2\right|\leqslant r_2\leqslant r_1\leqslant r_1^1+r_1^2, && 0\leqslant r_2\leqslant r_1\leqslant 2r_2', \\ &\left|r_2^{1\prime}-r_2^{2\prime}\right|\leqslant r_2'\leqslant r_1'\leqslant r_2^{1\prime}+r_2^{2\prime}, && 0\leqslant r_2^2\leqslant r_1^2\leqslant 2r_2^{2\prime}, \\ & && 0\leqslant r_2^1\leqslant r_1^1\leqslant 2r_2^{1\prime}. \end{aligned} \tag{41}$$

For the 4-dimensional quantum Hall system composed of N particles, one can repeatedly use the fusion formula and its fusion rule for $N-1$ times to obtain the elements of the matrix algebra of this system. However, the matrix operators obtained in the finals are certainly those of the fuzzy monopole harmonics of the system. That is, if the LLL Hilbert space is the space that corresponds to the irreducible representation of $SO(5)$ $Q=[q_1=q_2,q_2]$, the dimension of this Hilbert space is $D(Q)$ and the operators obtained in the finals are the $D(Q)\times D(Q)$ matrices. These matrix operators consist of the complete set of irreducible representations of $SO(5)$ in $D(Q)\times D(Q)$ matrices, and their number is $D(Q)\times D(Q)$. The non-commutativity of these operators is described by the matrix algebra (30) of replacing $\frac{P}{2}$ with Q. This matrix algebra is universal for the 4-dimensional quantum Hall fluids. Hence, the matrix algebra is the description of the coordinate non-commutativity of particle's moving on S^4 in the 4-dimensional quantum Hall system. Exactly, this matrix algebra should be viewed as the description of non-commutative geometry of the 4-dimensional quantum Hall droplet since these matrix operators act in the LLL Hilbert space of our considering system.

Although both Eqs. (30) and (40) describe the operator product's relations of the fuzzy monopole harmonics, the significances of them are different. The former should be view as the matrix multiplicater relation of the fuzzy monopole harmonics of the same system, and the latter as the operator tensor product relation of the fuzzy monopole harmonics of the different subsystems. Because the 4-dimensional quantum Hall droplet is a system of many-body, they determine all operators, which act in the LLL Hilbert space of the 4-dimensional quantum Hall droplet, to obey the rules. As our explaining in Section 3, the fuzzy monopole harmonics can be considered as the generatores of the general wave functions of single particle system. Hence, the matrix algebra (30) and the operator tensor product (40) can be used the construction of the wave functions of the LLL and its collective excitations of the 4-dimensional quantum Hall system. We shall discuss this construction elsewhere.

5. Summary and outlook

Similar to particle's motion on a plane in a constant magnetic field, the existence of Yang's $SU(2)$ monopole in the 4-dimensional quantum Hall system make the coordinates of particle's moving on the four-sphere become non-commutative. The appearance of such monopole also results in the irreducible representations of $SO(5)$ belonging to the Hilbert space of the system to be truncated. The similar phenomenon occurs in the description of fuzzy two-sphere. This clues us to the description of non-commutative geometry of the 4-dimensional quantum Hall system. Here we found that the fuzzy S^4 describes the non-commutative geometry of the 4-dimensional quantum Hall droplet. By determining the explicit forms of fuzzy monopole harmonics and their matrix algebra, we established the description of non-commutative geometry of the 4-dimensional quantum Hall droplet.

The $SU(2)$ monopole harmonics with the isospin rotating frame can be interpreted as the wave functions of a 4-dimensional symmetrical top under the rotation transformation of $SO(5)$. Based on this view, we given the explicit expression of coupling coefficients of the $SO(5)$ angular momentum. The theory of angular momentum of $SO(5)$ is surprisingly simple and excellent, which is very similar to that of $SO(3)$. Many relations, paralleled the relations appearing in the $SO(3)$ theory, can be explicitly written off in $SO(5)$'s that. The expression of coupling coefficients given here is essential to the theory of angular momentum of $SO(5)$. These results are useful for the physics of atoms and molecules with the higher symmetry, i.e., $SO(5)$ symmetry.

Following the proof of equivalence of two-dimensional quantum Hall physics and non-commutative field theory given recently by Hellerman and Raamsdonk [5], we can clarify the physical implication of the fuzzy monopole harmonics here. The second-quantized field theory description of the quantum Hall fluid for various filling fractions should involve some non-commutative field theory. On the 2-dimensional plane, such non-commutative field theory is the regularized matrix version of the non-commutative $U(1)$ Chern–Simons theory. Our constructing the fuzzy monopole harmonics are a complete set of matrix version of non-commutative field theory corresponding to the 4-dimensional quantum Hall fluid. Because the creation and annihilation operators in the second-quantized field theory description of the quantum Hall fluid can be built up by this complete set. In this sense, these fuzzy monopole harmonics can be viewed as the creation and annihilation operators in the second-quantized field theory description of the quantum Hall fluid. The matrix algebra obeyed by these fuzzy monopole harmonics can be interpreted as the non-commutative relations satisfied by the creation and annihilation operators. The construction of the fuzzy monopole harmonics and their matrix algebra given by us is only the first step to establish the second-quantized field theory description of the 4-dimensional quantum Hall fluid, but it is an essential step. The fusion scheme of the fuzzy monopole harmonics and its inverse relation provide an approach for the calculation of correlation functions in the non-commutative field theory. In fact, the methods and some results present here can be straightforwardly used for the study of the 4-dimensional quantum Hall system.

On the other hand, it is interesting to relate the matrix algebra of fuzzy S^4 given here with some applications in D-brane dynamics in string theory and M-theory. The ability to construct the higher-dimensional brane configurations using D0-branes is essential for a success of the Matrix theory of BFSS [22], where for example arbitrary membrane configurations in M-theory must be described in terms of the low energy degrees of freedom of the D0-branes. Myers [23] found that D0-branes expand into spherical D2-branes in the constant background RR fields. From the matrix model construction of Kabat and Taylor [36], the non-commutative solution for such spherical D2-brane actually represents the bound state of a spherical D2-brane with some D0-branes. The fuzzy S^4 was used in the context of the Matrix theory of BFSS to describe time-dependent 4-brane solutions constructed from the D0-brane degrees of freedom. In this sense, the fuzzy monopole harmonics $\widetilde{\widetilde{\mathcal{Y}}}_J^R$ can be used to describe classical solutions of the corresponding matrix brane model. However, the matrix algebra of fuzzy S^4 present here is different with those of fuzzy S^4 in the references [21,24,25,37]. The study of the matrix brane construction associated with the matrix algebra of fuzzy S^4 present here and the relation between it and the matrix brane constructions of the above references is an interesting topic. The work of this aspect is in progress.

Acknowledgements

We would like to thank K.J. Shi and L. Zhao for many valuable discussions. Y.X. Chen thanks the Institute of Modern Physics of Northwestern University in Xi'an for hospitality during his staying in the Institute. The work was partly supported by the NNSF of China and by the Foundation of Ph.D's program of Education Ministry of China.

References

[1] G.V. Dunne, R. Jackiw, C.A. Trugenberger, Phys. Rev. D 41 (1990) 661.
[2] R. Laughlin, Phys. Rev. Lett. 50 (1983) 1395.
[3] L. Susskind, The quantum Hall fluid and non-commutative Chern–Simons theory, hep-th/0101029.
[4] A.P. Polychronakos, Quantum Hall states as matrix Chern–Simons theory, hep-th/0103013.
[5] S. Hellerman, M.V. Raamsdonk, Quantum Hall physics equals non-commutative field theory, hep-th/0103179.
[6] D. Karabali, B. Sakita, Chern–Simons matrix model: coherent states and relation to Laughlin wave functions, hep-th/0106016.
[7] F.D.M. Haldane, Phys. Rev. Lett. 51 (1983) 605.
[8] S.C. Zhang, J.P. Hu, Science 294 (2001) 823.
[9] C.N. Yang, J. Math. Phys. 19 (1978) 320.
[10] A. Belavin, A. Polyakov, A. Schwartz, Y. Tyupkin, Phys. Lett. B 59 (1975) 85.
[11] C.N. Yang, J. Math. Phys. 19 (1978) 2622.
[12] J.P. Hu, S.C. Zhang, Collective excitations at the boundary of a 4D quantum Hall droplet, cond-mat/0112432.
[13] A. Connes, Non-Commutative Geometry, Academic Press, 1994.
[14] J. Madore, Class. Quantum Grav. 9 (1992) 69.
[15] H. Grosse, C. Klimcik, P. Presnajder, Commun. Math. Phys. 180 (1996) 429.
[16] B. de Wit, J. Hoppe, H. Nicolai, Nucl. Phys. B 305 (1988) 545.
[17] H. Grosse, A. Strohmaier, Lett. Math. Phys. 48 (1999) 163.
[18] A. Connes, G. Landi, Commun. Math. Phys. 221 (2001) 141.
[19] F. Bonechi, N. Ciccoli, M. Tarlini, Non-commutative instantons on the 4-sphere from quantum group, math.QA/0012236.
[20] A. Konechny, A. Schwarz, Phys. Rep. 360 (2002) 353.
[21] J. Castellino, S. Lee, W. Taylor, Nucl. Phys. B 526 (1998) 334.
[22] T. Banks, W. Fischler, S. Shenker, L. Susskind, Phys. Rev. D 55 (1997) 5112.
[23] R. Myers, JHEP 9911 (1999) 037, hep-th/991053.
[24] S. Ramgoolam, On spherical harmonics for fuzzy spheres in diverse dimensions, hep-th/0105006.
[25] P.M. Ho, S. Ramgoolam, Higher-dimensional geometries from matrix brane constructions, hep-th/0111278.
[26] M. Fabinger, Higher-dimensional quantum Hall effect in string theory, hep-th/0201016.
[27] D. Karabali, V.P. Nair, Quantum Hall effect in higher dimensions, hep-th/0203264.
[28] Y. Kimura, Non-commutative gauge theory on fuzzy four-sphere and matrix model, hep-th/0204256.
[29] M. Minami, Prog. Theor. Phys. 63 (1980) 303.
[30] E. Demler, S.C. Zhang, Ann. Phys. 271 (1999) 83.
[31] A.M. Peremolov, Generalized Coherent State and their Applications, Springer-Verlag, Berlin, 1986.
[32] B.G. Wybourne, Classical Groups for Physicists, Wiley, 1974.
[33] A. Alekseev, A. Recknagel, V. Schomerus, JHEP 9909 (1999) 023, hep-th/9908040.
[34] C. Chan, C. Chen, F. Lin, H. Yang, CP^n model on fuzzy sphere, hep-th/0105087.
[35] B.Y. Hou, B.Y. Hou, R.H. Yue, Fuzzy sphere bimodule, ABS construction to the exact soliton solution, hep-th/0109091.
[36] D. Kabat, W. Taylor, Adv. Theor. Math. Phys. 2 (1998) 181.
[37] N. Constable, R. Myers, O. Tafjord, JHEP 0106 (2001) 023, hep-th/0102080.

ELSEVIER

Available online at www.sciencedirect.com
SCIENCE DIRECT

NUCLEAR PHYSICS B

Nuclear Physics B 663 [FS] (2003) 467–486

www.elsevier.com/locate/npe

Algebraic Bethe ansatz for the elliptic quantum group $E_{\tau,\eta}(sl_n)$ and its applications

B.Y. Hou [a], R. Sasaki [b], W.-L. Yang [a,b]

[a] *Institute of Modern Physics, Northwest University, Xian 710069, PR China*
[b] *Yukawa Institute for Theoretical Physics, Kyoto University, Kyoto 606-8502, Japan*

Received 10 March 2003; received in revised form 17 April 2003; accepted 2 May 2003

Abstract

We study the tensor product of the *higher spin representations* (see the definition in Section 2.2) of the elliptic quantum group $E_{\tau,\eta}(sl_n)$. The transfer matrices associated with the $E_{\tau,\eta}(sl_n)$-module are exactly diagonalized by the nested Bethe ansatz method. Some special cases of the construction give the exact solution for the $\mathbb{Z}_n$ Belavin model and for the elliptic A_{n-1} Ruijsenaars–Schneider model.

PACS: 03.65.Fd; 05.30.-d

Keywords: Integrable models; Elliptic quantum group; Bethe ansatz; $\mathbb{Z}_n$ Belavin model; Ruijsenaars–Schneider model

1. Introduction

Bethe ansatz method has proved to be the most powerful and (probably) unified method to construct the common eigenvectors of commuting families of operators (usually called *transfer matrices*) in two-dimensional integrable models [1–3]. Faddeev et al. [1] reformulated the Bethe ansatz method in a representation theory form: *transfer matrices* are associated with representations of certain algebras with quadratic relations (now called quantum groups). The eigenvectors are constructed by acting certain algebra elements on the "highest weight vectors". This type of Bethe ansatz is known as the algebraic Bethe ansatz.

E-mail address: wlyang@yukawa.kyoto-u.ac.jp (W.-L. Yang).

0550-3213/03/$ – see front matter
doi:10.1016/S0550-3213(03)00390-0

Whereas this construction has been very successful in rational and trigonometric integrable models [4,5], its extension to elliptic models had been problematic due to the fact that for the underlying algebras $-\mathbb{Z}_n$ Sklyanin algebras [6,7] the highest weight representations were not properly defined. Therefore, the algebraic Bethe ansatz method had not be applied to the elliptic integrable models directly.

Recently, a definition of elliptic quantum groups $E_{\tau,\eta}(\mathfrak{g})$ associated with any simple classical Lie algebra $\mathfrak{g}$ was given [8]. The highest weight representations of the elliptic quantum groups (cf. $\mathbb{Z}_n$ Sklyanin algebras) are now well-defined [8,9]. This enabled Felder et al. [10] to apply successfully the algebraic Bethe ansatz method for constructing the eigenvectors of the *transfer matrices* of the elliptic integrable models associated with modules over $E_{\tau,\eta}(sl_2)$, and Billey [11] to apply the algebraic nested Bethe ansatz method for constructing the eigenvalues of the *transfer matrices* associated with a special module over $E_{\tau,\eta}(sl_n)$ (see Section 5).

In this paper, we will extend the above construction of Bethe ansatz method further to the elliptic quantum group $E_{\tau,\eta}(sl_n)$ on a generic *higer spin* module $W = V_{\Lambda^{(l_1)}}(z_1) \otimes V_{\Lambda^{(l_2)}}(z_2) \otimes \cdots \otimes V_{\Lambda^{(l_m)}}(z_m)$. After briefly reviewing the definition of the elliptic quantum group $E_{\tau,\eta}(sl_n)$, we study one parameter series of highest weight representations (*higher spin representations*) of $E_{\tau,\eta}(sl_n)$, and introduce the associated operator algebra and the *transfer matrices* corresponding to the $E_{\tau,\eta}(sl_n)$-module in Section 2. In Section 3, we describe the algebraic Bethe ansatz for $E_{\tau,\eta}(sl_n)$. Finally, we give some applications of our construction to the $\mathbb{Z}_n$ Belavin models in Section 4, and the elliptic Ruijsenaars–Schneider model associated to A_{n-1} root system in Section 5.

2. The elliptic quantum group and its modules associated to A_{n-1}

2.1. *The elliptic quantum group associated to A_{n-1}*

We first review the definition of the elliptic quantum group $E_{\tau,\eta}(sl_n)$ associated to A_{n-1} [8]. Let $\{\epsilon_i \mid i = 1, 2, \ldots, n\}$ be the orthonormal basis of the vector space $\mathbb{C}^n$ such that $\langle \epsilon_i, \epsilon_j \rangle = \delta_{ij}$. The A_{n-1} simple roots are $\{\alpha_i = \epsilon_i - \epsilon_{i+1} \mid i = 1, \ldots, n-1\}$ and the fundamental weights $\{\Lambda_i \mid i = 1, \ldots, n-1\}$ satisfying $\langle \Lambda_i, \alpha_j \rangle = \delta_{ij}$ are given by

$$\Lambda_i = \sum_{k=1}^{i} \epsilon_k - \frac{i}{n} \sum_{k=1}^{n} \epsilon_k.$$

Set

$$\hat{i} = \epsilon_i - \bar{\epsilon}, \quad \bar{\epsilon} = \frac{1}{n} \sum_{k=1}^{n} \epsilon_k, \; i = 1, \ldots, n, \quad \text{then} \sum_{i=1}^{n} \hat{i} = 0. \tag{2.1}$$

For each dominant weight $\Lambda = \sum_{i=1}^{n-1} a_i \Lambda_i$, $a_i \in \mathbb{Z}^+$, there exists an irreducible highest weight finite-dimensional representation V_Λ of A_{n-1} with the highest vector $|\Lambda\rangle$. For example, the fundamental vector representation is V_{Λ_1}. In this paper, we only consider the symmetric tensor-product representation of $\overbrace{V_{\Lambda_1} \otimes V_{\Lambda_1} \otimes \cdots \otimes V_{\Lambda_1}}^{l}$ (or, the *higher spin-l representation* of A_{n-1}), namely, the one parameter series of highest weight representations $V_{\Lambda^{(l)}}$, with

$$\Lambda^{(l)} = l\Lambda_1, \quad l \in \mathbb{Z} \quad \text{and} \quad l > 0. \tag{2.2}$$

This corresponds to the Young diagram $\overbrace{\square\square\square\cdots\square}^{l}$.

Let $\mathfrak{h}$ be the Cartan subalgebra of A_{n-1} and $\mathfrak{h}^*$ be its dual. A finite-dimensional diagonalizable $\mathfrak{h}$-module is a complex finite-dimensional vector space W with a weight decomposition $W = \bigoplus_{\mu\in\mathfrak{h}^*} W[\mu]$, so that $\mathfrak{h}$ acts on $W[\mu]$ by $x\,v = \mu(x)\,v$, $(x \in \mathfrak{h}, v \in W[\mu])$. For example, the fundamental vector representation $V_{\Lambda_1} = \mathbb{C}^n$, the non-zero weight spaces are $W[\hat{i}] = \mathbb{C}\epsilon_i$, $i = 1, \ldots, n$.

Let us fix τ such that $\mathrm{Im}(\tau) > 0$ and a generic complex number η. For convenience, we introduce another parameter $w = n\eta$ related to η. Let us introduce the following elliptic functions

$$\sigma(u) = \theta\begin{bmatrix} \frac{1}{2} \\ \frac{1}{2} \end{bmatrix}(u, \tau), \qquad \theta^{(j)}(u) = \theta\begin{bmatrix} \frac{1}{2} - \frac{j}{n} \\ \frac{1}{2} \end{bmatrix}(u, n\tau), \tag{2.3}$$

$$\theta\begin{bmatrix} a \\ b \end{bmatrix}(u, \tau) = \sum_{m=-\infty}^{\infty} \exp\{\sqrt{-1}\pi[(m+a)^2\tau + 2(m+a)(u+b)]\}. \tag{2.4}$$

For a generic complex weight $\xi \in \sum_{i=1}^{n-1} \mathbb{C}\Lambda_i$, we introduce a weight lattice with a shift by ξ, $\mathbb{P}_\xi$ as follows

$$\mathbb{P}_\xi = \xi + \sum_{i=1}^{n} m_i \hat{i}, \quad m_i \in \mathbb{Z}. \tag{2.5}$$

For $\lambda \in \mathbb{P}_\xi$, define

$$\lambda_i = \langle\lambda, \epsilon_i\rangle, \qquad \lambda_{ij} = \lambda_i - \lambda_j = \langle\lambda, \epsilon_i - \epsilon_j\rangle, \quad i, j = 1, \ldots, n. \tag{2.6}$$

In this paper, we restrict $\lambda \in \mathbb{P}_\xi$ so that the inverse of the matrix $S(z, \lambda)$ in (4.8) does exist. Let $R(z, \lambda) \in \mathrm{End}(\mathbb{C}^n \otimes \mathbb{C}^n)$ be the R-matrix given by

$$\begin{aligned} R(z, \lambda) = {} & \sum_{i=1}^{n} R_{ii}^{ii}(z, \lambda) E_{ii} \otimes E_{ii} \\ & + \sum_{i\neq j} \{R_{ij}^{ij}(z, \lambda) E_{ii} \otimes E_{jj} + R_{ij}^{ji}(z, \lambda) E_{ji} \otimes E_{ij}\}, \end{aligned} \tag{2.7}$$

in which E_{ij} is the matrix with elements $(E_{ij})_k^l = \delta_{jk}\delta_{il}$. The coefficient functions are

$$R_{ii}^{ii}(z, \lambda) = 1, \qquad R_{ij}^{ij}(z, \lambda) = \frac{\sigma(z)\sigma(\lambda_{ij}w + w)}{\sigma(z + w)\sigma(\lambda_{ij}w)}, \tag{2.8}$$

$$R_{ij}^{ji}(z, \lambda) = \frac{\sigma(w)\sigma(z + \lambda_{ij}w)}{\sigma(z + w)\sigma(\lambda_{ij}w)}, \tag{2.9}$$

and λ_{ij} is defined in (2.6). The R-matrix satisfies the dynamical (modified) quantum Yang–Baxter equation

$$R_{12}\big(z_1 - z_2, \lambda - h^{(3)}\big)R_{13}(z_1 - z_3, \lambda)R_{23}\big(z_2 - z_3, \lambda - h^{(1)}\big)$$
$$= R_{23}(z_2 - z_3, \lambda)R_{13}\big(z_1 - z_3, \lambda - h^{(2)}\big)R_{12}(z_1 - z_2, \lambda), \tag{2.10}$$

with the initial condition

$$R^{kl}_{ij}(0, \lambda) = \delta^l_i\, \delta^k_j. \tag{2.11}$$

We adopt the notation: $R_{12}(z, \lambda - h^{(3)})$ acts on a tensor $v_1 \otimes v_2 \otimes v_3$ as $R(z; \lambda - \mu) \otimes \mathrm{Id}$ if $v_3 \in W[\mu]$.

A representation of the elliptic quantum group $E_{\tau,\eta}(sl_n)$ (an $E_{\tau,\eta}(sl_n)$-module) is by definition a pair (W, L) where W is a diagonalizable $\mathfrak{h}$-module and $L(z, \lambda)$ is a meromorphic function of λ and the spectral parameter $z \in \mathbb{C}$, with values in $\mathrm{End}_{\mathfrak{h}}(\mathbb{C}^n \otimes W)$ (the endomorphism commuting with the action of $\mathfrak{h}$). It obeys the so-called "RLL" relation

$$R_{12}\big(z_1 - z_2, \lambda - h^{(3)}\big)L_{13}(z_1, \lambda)L_{23}\big(z_2, \lambda - h^{(1)}\big)$$
$$= L_{23}(z_2, \lambda)L_{13}\big(z_1, \lambda - h^{(2)}\big)R_{12}(z_1 - z_2, \lambda), \tag{2.12}$$

where the first and second space are auxiliary spaces ($\mathbb{C}^n$) and the third space plays the role of quantum space (W). The total weight conservation condition for the L-operator reads

$$\big[h^{(1)} + h^{(3)}, L_{13}(z, \lambda)\big] = 0.$$

In terms of the elements of the L-operator defined by

$$L(z, \lambda)(e_i \otimes v) = \sum_{j=1}^{n} e_j \otimes L^j_i(z, \lambda)v, \quad v \in W, \tag{2.13}$$

the above condition can be expressed equivalently as

$$f(h)L^j_i(z, \lambda) = L^j_i(z, \lambda)f(h + \hat{\imath} - \hat{\jmath}), \tag{2.14}$$

in which $f(h)$ is any meromorphic function of h and h measures the weight of the quantum space (W).

2.2. *Modules over $E_{\tau,\eta}(sl_n)$ and the associated operator algebra*

The basic example of an $E_{\tau,\eta}(sl_n)$-module is $(\mathbb{C}^n, L)$ with $L(z, \lambda) = R(z - z_1, \lambda)$, which is called the fundamental vector representation $V_{\Lambda_1}(z_1)$ with the evaluation point z_1. It is obvious that "RLL" relation is satisfied as a consequence of the dynamical Yang–Baxter equation (2.10). Other modules can be obtained by taking tensor products: if $(W_1, L^{(1)})$ and $(W_2, L^{(2)})$ are $E_{\tau,\eta}(sl_n)$-modules, where $L^{(j)}$ acts on $(\mathbb{C}^n \otimes W_j)$, then also $(W_1 \otimes W_2, L)$ with

$$L(z, \lambda) = L^{(1)}\big(z, \lambda - h^{(2)}\big)L^{(2)}(z, \lambda) \quad \text{acting on } \mathbb{C}^n \otimes W_1 \otimes W_2. \tag{2.15}$$

An $E_{\tau,\eta}(sl_n)$-submodule of an $E_{\tau,\eta}(sl_n)$-module (W, L) is a pair (W_1, L_1) where W_1 is an $\mathfrak{h}$-submodule of W such that $\mathbb{C}^n \otimes W_1$ is invariant under the action of all the $L(z, \lambda)$, and

$L_1(z,\lambda)$ is the restriction to this invariant subspace. Namely, the $E_{\tau,\eta}(sl_n)$-submodules are $E_{\tau,\eta}(sl_n)$-modules.

Using the fusion rule of $E_{\tau,\eta}(sl_n)$ (2.15) one can construct the symmetric $E_{\tau,\eta}(sl_n)$-submodule of l-tensors of fundamental vector representations:

$$V_{\Lambda^{(l)}}(z_1) = \text{symmetric subspace of } V_{\Lambda_1}(z_1)\otimes V_{\Lambda_1}(z_1-w)\otimes\cdots \otimes V_{\Lambda_1}\big(z_1-(l-1)w\big),$$

where $\Lambda^{(l)}$ is defined by (2.2). We call such an $E_{\tau,\eta}(sl_n)$-module the *higher spin-l representation* with evaluation point z_1. These series of representations in the case of $\mathbb{Z}_n$ Sklyanin algebra have been studied in [6] for $n=2$ case and in [12,13] for generic n case. From direct calculation, we find that the $E_{\tau,\eta}(sl_n)$-module $V_{\Lambda^{(l)}}(z)$ is an irreducible highest weight module of $E_{\tau,\eta}(sl_n)$ with the highest vector $|\Lambda^{(l)}\rangle \in W[l\Lambda_1]=\mathbb{C}|\Lambda^{(l)}\rangle$. It satisfies the following highest weight conditions

$$L^1_1(u,\lambda)\big|\Lambda^{(l)}\big\rangle = \big|\Lambda^{(l)}\big\rangle,\qquad L^i_1(u,\lambda)\big|\Lambda^{(l)}\big\rangle = 0,\quad i=2,\ldots,n, \tag{2.16}$$

$$L^i_j(u,\lambda)\big|\Lambda^{(l)}\big\rangle = \delta^i_j\frac{\sigma(u-z)\sigma(\lambda_{i1}w+lw)}{\sigma(u-z+lw)\sigma(\lambda_{i1}w)}\big|\Lambda^{(l)}\big\rangle,\quad i,j=2,\ldots,n, \tag{2.17}$$

$$f(h)\big|\Lambda^{(l)}\big\rangle = f(l\hat{1})\big|\Lambda^{(l)}\big\rangle, \tag{2.18}$$

where $f(h)$ is any meromorphic function of h, which measures the weight of the quantum space W.

For any $E_{\tau,\eta}(sl_n)$-module, as in [10] one can define an associated operator algebra of difference operators on the space $\mathrm{Fun}(W)$ of meromorphic functions of $\lambda\in\mathbb{P}_\xi$ with values in W. The algebra is generated by h and the operators $\tilde{L}(z)\in\mathrm{End}(\mathbb{C}^n\otimes\mathrm{Fun}(W))$ acting as

$$\tilde{L}(z)(e_i\otimes f)(\lambda) = \sum_{j=1}^{n} e_j\otimes L^j_i(z,\lambda)f(\lambda-\hat{\imath}). \tag{2.19}$$

One can derive the following exchange relation of the difference operator $\tilde{L}(z)$ from the "RLL" relation (2.12), the weight conservation condition of $L^j_i(z,\lambda)$ (2.14) and the fact $[h^{(1)}+h^{(2)},\ R_{12}(z,\lambda)]=0$,

$$R_{12}(z_1-z_2,\lambda-h)\tilde{L}_{13}(z_1)\tilde{L}_{23}(z_2) = \tilde{L}_{23}(z_2)\tilde{L}_{13}(z_1)R_{12}(z_1-z_2,\lambda), \tag{2.20}$$

$$f(h)\tilde{L}^j_i(z) = \tilde{L}^j_i(z)f(h+\hat{\imath}-\hat{\jmath}), \tag{2.21}$$

where $f(h)$ is any meromorphic function of h and h measures the weight of the quantum space W.

The *transfer matrices* associated to an $E_{\tau,\eta}(sl_n)$-module (W,L) [10] is a difference operator acting on the space $\mathrm{Fun}(W)[0]$ of meromorphic functions of λ with values in the zero-weight space of W. It is defined by

$$T(u)f(\lambda) = \sum_{i=1}^{n}\tilde{L}^i_i(u)f(\lambda) = \sum_{i=1}^{n}L^i_i(u,\lambda)f(\lambda-\hat{\imath}). \tag{2.22}$$

The exchange relation of $\tilde{L}$-operators (2.20) and (2.21) imply that, for any $E_{\tau,\eta}(sl_n)$-module, the above transfer matrices preserve the space $H = \mathrm{Fun}(W)[0]$ of functions with values in the zero weight space $W[0]$. Moreover, they commute pairwise on H:

$$[T(u), T(v)]\big|_H = 0.$$

In this paper we will study the tensor product $E_{\tau,\eta}(sl_n)$-module $W = V_{\Lambda^{(l_1)}}(z_1) \otimes V_{\Lambda^{(l_2)}}(z_2) \otimes \cdots \otimes V_{\Lambda^{(l_m)}}(z_m)$. With the generic evaluation points $\{z_i\}$, the module is an irreducible highest weight $E_{\tau,\eta}(sl_n)$-module [8,9]. Let $\Lambda = \Lambda^{(l_1)} + \cdots + \Lambda^{(l_m)}$, then $W[\Lambda] = \mathbb{C}|\Lambda\rangle$ with

$$|\Lambda\rangle \equiv \big|\Lambda^{(l_1)}, \Lambda^{(l_2)}, \ldots, \Lambda^{(l_m)}\big\rangle \equiv \big|\Lambda^{(l_1)}\big\rangle \otimes \big|\Lambda^{(l_2)}\big\rangle \otimes \cdots \otimes \big|\Lambda^{(l_m)}\big\rangle. \tag{2.23}$$

The vector $|\Lambda\rangle$, viewed as a constant function in $\mathrm{Fun}(W)$, obeys the following highest weight conditions:

$$\tilde{L}_1^1(z)|\Lambda\rangle = A(z,\lambda)|\Lambda\rangle, \qquad \tilde{L}_1^i(z)|\Lambda\rangle = 0, \quad i = 2,\ldots,n,$$
$$\tilde{L}_j^i(z)|\Lambda\rangle = \delta_j^i D_i(z,\lambda)|\Lambda\rangle, \quad i,j = 2,\ldots,n, \qquad f(h)|\Lambda\rangle = f(N\hat{1})|\Lambda\rangle.$$

The highest weight functions read

$$A(z,\lambda) = 1, \qquad D_i(z,\lambda) = \left\{\prod_{k=1}^{m} \frac{\sigma(z-p_k)}{\sigma(z-q_k)}\right\} \frac{\sigma(\lambda_{i1}w + Nw)}{\sigma(\lambda_{i1}w)}, \qquad i = 2,\ldots,n, \tag{2.24}$$

where

$$p_k = z_k, \qquad q_k = z_k - l_k w, \qquad N = \sum_{k=1}^{m} l_k, \quad k = 1,\ldots,m. \tag{2.25}$$

3. Algebraic Bethe ansatz for $E_{\tau,\eta}(sl_n)$

In this section we fix a highest weight $E_{\tau,\eta}(sl_n)$-module W of weight Λ, the functions $A(z,\lambda)$, $D_i(z,\lambda)$ (2.24), with the highest vector $|\Lambda\rangle$. We assume that $N = \sum_{k=1}^{m} l_k = n \times l$ with l being an integer, so that the zero-weight space $W[0]$ can be non-trivial and so that the algebraic Bethe ansatz method can be applied as in [10,14–16].

Let us adopt the standard notation for convenience:

$$\mathcal{A}(u) = \tilde{L}_1^1(u), \qquad \mathcal{B}_i(u) = \tilde{L}_i^1(u), \quad i = 2,\ldots,n, \tag{3.1}$$
$$\mathcal{C}_i(u) = \tilde{L}_1^i(u), \qquad \mathcal{D}_i^j(u) = \tilde{L}_i^j(u), \quad i,j = 2,\ldots,n. \tag{3.2}$$

The transfer matrices $T(u)$ become

$$T(u) = \mathcal{A}(u) + \sum_{i=2}^{n} \mathcal{D}_i^i(u). \tag{3.3}$$

Any non-zero vector $|\Omega\rangle \in \mathrm{Fun}(W)[\Lambda]$ is of form $|\Omega\rangle = g(\lambda)|\Lambda\rangle$, for some meromorphic function $g \neq 0$. When $N = n \times l$, the weight Λ can be written in the form

$$\Lambda = nl\Lambda_1 = l\sum_{k=1}^{n-1}(\epsilon_1 - \epsilon_{k+1}). \tag{3.4}$$

Noting (2.21), the zero-weight vector space is spanned by the vectors of the following form

$$\mathcal{B}_{i_{N_1}}(v_{N_1})\mathcal{B}_{i_{N_1-1}}(v_{N_1-1})\cdots\mathcal{B}_{i_1}(v_1)|\Omega\rangle, \tag{3.5}$$

where $N_1 = (n-1) \times l$ and among the indices $\{i_k \mid k = 1, \dots, N_1\}$, the number of $i_k = j$, denoted by $\#(j)$, should be

$$\#(j) = l, \quad j = 2, \dots, n. \tag{3.6}$$

The above states (3.5) actually belong to the zero-weight space $W[0]$, because

$$\begin{aligned} & f(h)\mathcal{B}_{i_{N_1}}(v_{N_1})\mathcal{B}_{i_{N_1-1}}(v_{N_1-1})\cdots\mathcal{B}_{i_1}(v_1)|\Omega\rangle \\ &= \mathcal{B}_{i_{N_1}}(v_{N_1})\mathcal{B}_{i_{N_1-1}}(v_{N_1-1})\cdots\mathcal{B}_{i_1}(v_1)f\left(h + \sum_{k=1}^{N_1}\hat{i}_k - N_1\hat{1}\right)|\Omega\rangle \\ &= f\left(l\sum_{i=2}^{n}\hat{i} - (n-1)l\hat{1} + nl\hat{1}\right)\mathcal{B}_{i_{N_1}}(v_{N_1})\mathcal{B}_{i_{N_1-1}}(v_{N_1-1})\cdots\mathcal{B}_{i_1}(v_1)|\Omega\rangle \\ &= f(0)\mathcal{B}_{i_{N_1}}(v_{N_1})\mathcal{B}_{i_{N_1-1}}(v_{N_1-1})\cdots\mathcal{B}_{i_1}(v_1)|\Omega\rangle, \end{aligned}$$

for any meromorphic function f. We will seek the common eigenvectors of the *transfer matrices* $T(u)$ in the form

$$\big|\lambda; \{v_k\}\big\rangle = F^{i_1,i_2,\dots,i_{N_1}}\big(\lambda; \{v_k\}\big)\mathcal{B}_{i_{N_1}}(v_{N_1})\mathcal{B}_{i_{N_1-1}}(v_{N_1-1})\cdots\mathcal{B}_{i_1}(v_1)|\Omega\rangle, \tag{3.7}$$

with the restriction condition (3.6) and the parameters $\{v_k\}$ will be specified later by the Bethe ansatz equations (3.45). We adopt here the convention that the repeated indices imply summation over $2, 3, \dots, n$, and the notation that

$$v_k = v_k^{(0)}, \quad k = 1, 2, \dots, N_1. \tag{3.8}$$

For convenience, let us introduce the following set of integers

$$N_i = (n-i) \times l, \quad i = 1, 2, \dots, n-1,$$

and $(n-1)$ complex parameters $\{\alpha^{(i)} \mid i = 1, 2, \dots, n-1\}$ and set $\alpha^{(n)} = -\sum_{k=1}^{n-1}\alpha^{(k)}$. Associated with $\{\alpha^{(i)}\}$, let us introduce

$$\bar{\alpha}^{(i)} = \frac{1}{n-i-1}\left\{\alpha^{(i+1)} - \frac{1}{n-1}\sum_{k=i+1}^{n}\alpha^{(k)}\right\}, \quad i = 0, 1, \dots, n-2. \tag{3.9}$$

From the exchange relations (2.20) and (2.21), one can derive the commutation relations among $\mathcal{A}(u)$, $\mathcal{D}_i^j(u)$ and $\mathcal{B}_i(u)$ $(i, j = 2, \dots, n)$ (for details, see Appendix A). The relevant commutation relations are

$$\mathcal{A}(u)\mathcal{B}_i(v) = r(v-u, \lambda_{i1})\mathcal{B}_i(v)\mathcal{A}(u) + s(u-v, \lambda_{1i})\mathcal{B}_i(u)A(v), \tag{3.10}$$

$$\mathcal{D}_i^j(u)\mathcal{B}_l(v) = r(u-v, \lambda_{j1} - h_{j1}) \sum_{\alpha,\beta=2}^{n} R_{i\ l}^{\alpha\beta}(u-v, \lambda)\mathcal{B}_\beta(v)\mathcal{D}_\alpha^j(u)$$
$$- s(u-v, \lambda_{1j} - h_{1j})\mathcal{B}_i(u)\mathcal{D}_l^j(v), \tag{3.11}$$

$$\mathcal{B}_i(u)\mathcal{B}_j(v) = \sum_{\alpha,\beta=2}^{n} R_{i\ j}^{\alpha\beta}(u-v, \lambda)\mathcal{B}_\beta(v)\mathcal{B}_\alpha(u), \tag{3.12}$$

$$f(h)\mathcal{A}(u) = \mathcal{A}(u)f(h), \quad f(h)\mathcal{B}_i(u) = \mathcal{B}_i(u)f(h+\hat{\imath}-\hat{1}), \tag{3.13}$$

$$f(h)\mathcal{D}_i^j(u) = \mathcal{D}_i^j(u)f(h+\hat{\imath}-\hat{\jmath}), \tag{3.14}$$

where the function r, s are defined by

$$r(u,\nu) = \frac{\sigma(u+w)\sigma(\nu w)}{\sigma(u)\sigma(\nu w + w)}, \qquad s(u,\nu) = \frac{\sigma(w)\sigma(u+\nu w)}{\sigma(u)\sigma(\nu w - w)}, \quad \nu \in \mathbb{C}.$$

Because of the simple poles in the functions r, s at $u \in \mathbb{Z} + \tau\mathbb{Z}$, let us assume for convenience that all $\{v_k\}$ are distinct *modulo* $\mathbb{Z} + \tau\mathbb{Z}$ and consider u at a generic point. Take

$$|\Omega\rangle = g(\lambda)|\Lambda\rangle, \tag{3.15}$$

as the so-called quasi-vacuum satisfying the following conditions

$$\mathcal{A}(u)|\Omega\rangle = \frac{g(\lambda-\hat{1})}{g(\lambda)}|\Omega\rangle, \qquad \mathcal{C}_i(u)|\Omega\rangle = 0, \quad i = 2, \ldots, n, \tag{3.16}$$

$$\mathcal{D}_i^j(u)|\Omega\rangle = \delta_j^i D_i(u,\lambda)\frac{g(\lambda-\hat{\imath})}{g(\lambda)}|\Omega\rangle, \quad i,j = 2,\ldots,n, \tag{3.17}$$

$$f(h)|\Omega\rangle = f(N\hat{1})|\Omega\rangle, \qquad \mathcal{B}_i(u)|\Omega\rangle \neq 0, \quad i = 2,\ldots,n, \tag{3.18}$$

where $D_i(u,\lambda)$, p_k and q_k are given in (2.24) and (2.25). We will specify the function $g(\lambda)$ later (3.22).

Now, let us evaluate the action of $\mathcal{A}(u)$ on $|\lambda; \{v_k\}\rangle$. Many terms will appear when we move $\mathcal{A}(u)$ from the left to the right of $\mathcal{B}_i(v_k)$'s. They can be classified into two types: wanted and unwanted terms. The wanted terms in $\mathcal{A}(u)|\lambda; \{v_k\}\rangle$ can be obtained by retaining the first term in the commutation relation (3.10). The unwanted terms arising from the second term of (3.10), have some $B(v_k)$ replaced by $B(u)$. One unwanted term where $B(v_{N1})$ is replaced by $B(u)$ can be obtained by using firstly the second terms of (3.10), then repeatedly using the first term of (3.10). Thanks to the commutation relation (3.12), one can easily obtain *the other unwanted terms* where $B(v_k)$ is replaced by $B(u)$. So, we can find the action of $A(u)$ on $|\lambda; \{v_k\}\rangle$

$$\mathcal{A}(u)F^{i_1,i_2,\ldots,i_{N_1}}(\lambda; \{v_k\})\mathcal{B}_{i_{N_1}}(v_{N_1})\mathcal{B}_{i_{N_1-1}}(v_{N_1-1})\cdots\mathcal{B}_{i_1}(v_1)|\Omega\rangle$$

$$
\begin{aligned}
&= F^{i_1,i_2,\ldots,i_{N_1}}(\lambda-\hat{1};\{v_k\})\left\{\prod_{k=1}^{N_1} r\left(v_k-u,\left\langle\lambda-\sum_{s=k+1}^{N_1}\hat{i}_s,\epsilon_{i_k}-\epsilon_1\right\rangle\right)\right\}\\
&\quad\times B_{i_{N_1}}(v_{N_1})\cdots B_{i_1}(v_1)\frac{g(\lambda-\hat{1})}{g(\lambda)}|\Omega\rangle\\
&\quad+F^{i_1,i_2,\ldots,i_{N_1}}(\lambda-\hat{1};\{v_k\})s(u-v_{N_1},\lambda_{1i_{N_1}})\\
&\quad\quad\times\left\{\prod_{k=1}^{N_1-1} r\left(v_k-v_{N_1},\left\langle\lambda-\sum_{s=k+1}^{N_1}\hat{i}_s,\epsilon_{i_k}-\epsilon_1\right\rangle\right)\right\}\\
&\quad\quad\times B_{i_{N_1}}(u)B_{i_{N_1-1}}(v_{N_1-1})\cdots B_{i_1}(v_1)\frac{g(\lambda-\hat{1})}{g(\lambda)}|\Omega\rangle\\
&\quad+\text{o.u.t.},
\end{aligned}
\tag{3.19}
$$

where o.u.t. stands for *the other unwanted terms*. Due to the restriction condition (3.6) and $N_1=(n-1)\times l$, we have

$$
\begin{aligned}
&\left\{\prod_{k=1}^{N_1} r\left(v_k-u,\left\langle\lambda-\sum_{s=k+1}^{N_1}\hat{i}_s,\epsilon_{i_k}-\epsilon_1\right\rangle\right)\right\}\\
&\quad=\left\{\prod_{k=1}^{N_1}\frac{\sigma(v_k-u+w)}{\sigma(v_k-u)}\right\}\left\{\prod_{j=2}^{n}\frac{\sigma(\lambda_{j1}w-lw+w)}{\sigma(\lambda_{j1}w+w)}\right\},
\end{aligned}
\tag{3.20}
$$

$$
\sum_{k=1}^{N_1}\hat{i}_k=l\sum_{i=2}^{n}\hat{i}=-l\hat{1}.
\tag{3.21}
$$

In order that the Bethe ansatz equation (3.45) and the eigenvalues (3.30) should be independent of λ, one needs to choose (cf. [10])

$$
g(\lambda)=e^{\langle\alpha^{(1)}\epsilon_1,\lambda w\rangle}\prod_{j=2}^{n}\left\{\prod_{k=1}^{l}\frac{\sigma(\lambda_{j1}w+kw)}{\sigma(w)}\right\},
\tag{3.22}
$$

where α is an arbitrary complex number. The action of $\mathcal{A}(u)$ on $|\lambda;\{v_k\}\rangle$ becomes

$$
\begin{aligned}
&\mathcal{A}(u)F^{i_1,i_2,\ldots,i_{N_1}}(\lambda;\{v_k\})\mathcal{B}_{i_{N_1}}(v_{N_1})\mathcal{B}_{i_{N_1-1}}(v_{N_1-1})\cdots\mathcal{B}_{i_1}(v_1)|\Omega\rangle\\
&\quad=e^{\frac{1-n}{n}\alpha^{(1)}w}\prod_{k=1}^{N_1}\frac{\sigma(v_k-u+w)}{\sigma(v_k-u)}F^{i_1,i_2,\ldots,i_{N_1}}(\lambda-\hat{1};\{v_k\})B_{i_{N_1}}(v_{N_1})\cdots B_{i_1}(v_1)|\Omega\rangle\\
&\quad\quad+s(u-v_{N_1},\lambda_{1i_{N_1}})\frac{\sigma(\lambda_{i_{N_1}1}w+w)}{\sigma(\lambda_{i_{N_1}1}w)}e^{\frac{1-n}{n}\alpha^{(1)}w}\prod_{k=1}^{N_1-1}\frac{\sigma(v_k-v_{N_1}+w)}{\sigma(v_k-v_{N_1})}\\
&\quad\quad\times F^{i_1,i_2,\ldots,i_{N_1}}(\lambda-\hat{1};\{v_k\})B_{i_{N_1}}(u)B_{i_{N_1-1}}(v_{N_1-1})\cdots B_{i_1}(v_1)|\Omega\rangle\\
&\quad\quad+\text{o.u.t.}
\end{aligned}
\tag{3.23}
$$

Next, we evaluate the action of $\mathcal{D}_i^i(u)$ on $|\lambda;\{v_k\}\rangle$:

$$\begin{aligned}
&\mathcal{D}_i^i(u)F^{i_1,i_2,\ldots,i_{N_1}}\big(\lambda;\{v_k\}\big)\mathcal{B}_{i_{N_1}}(v_{N_1})\mathcal{B}_{i_{N_1-1}}(v_{N_1-1})\cdots\mathcal{B}_{i_1}(v_1)|\Omega\rangle\\
&\quad=\left\{\prod_{k=1}^{N_1}r\big(u-v_k,\big\langle\lambda-h-(N_1-k)\hat{1},\epsilon_i-\epsilon_1\big\rangle\big)\right\}\\
&\qquad\times\left\{\prod_{k=1}^{m}\frac{\sigma(u-p_k)}{\sigma(u-q_k)}\right\}\frac{\sigma(\lambda_{i1}w+(N-l)w)}{\sigma(\lambda_{i1}w-lw)}\frac{g(\lambda-\hat{i}+l\hat{1})}{g(\lambda+l\hat{1})}\\
&\qquad\times\big\{L^{(1)i}{}_i(u,\lambda)\big\}_{i_1,i_2,\ldots,i_{N_1}}^{i'_1,i'_2,\ldots,i'_{N_1}}F^{i_1,i_2,\ldots,i_{N_1}}\big(\lambda-\hat{i};\{v_k\}\big)\\
&\qquad\times\mathcal{B}_{i'_{N_1}}(v_{N_1})\mathcal{B}_{i'_{N_1-1}}(v_{N_1-1})\cdots\mathcal{B}_{i'_1}(v_1)|\Omega\rangle\\
&\qquad+\text{u.t.},
\end{aligned}\tag{3.24}$$

where u.t. stands for the all *unwanted terms*. We have introduced the convenient notation

$$\begin{aligned}
&\big\{L^{(1)j}{}_i(u,\lambda)\big\}_{i_1,i_2,\ldots,i_{N_1}}^{i'_1,i'_2,\ldots,i'_{N_1}}\\
&\quad=R^{j,\ i'_1}_{\gamma_{N_1-1},\ i_1}\left(u-v_1;\lambda-\sum_{k=2}^{N_1}\hat{i}'_k\right)R^{\gamma_{N_1-1},\ i'_2}_{\gamma_{N_1-2},\ i_2}\left(u-v_2;\lambda-\sum_{k=3}^{N_1}\hat{i}'_k\right)\times\cdots\\
&\qquad\times R^{\gamma_1,\ i'_{N_1}}_{i,\ \ i_{N_1}}(u-v_{N_1};\lambda),
\end{aligned}\tag{3.25}$$

and all the indices take the values $2,\ldots,n$. Noting the explicit form of the function $g(\lambda)$ given in (3.22) and the zero-weight of the state $|\lambda;\{v_k\}\rangle$, the action of $\mathcal{D}_i^i(u)$ becomes

$$\begin{aligned}
&\mathcal{D}_i^i(u)\big|\lambda;\{v_k\}\big\rangle\\
&\quad=e^{\frac{\alpha^{(1)}w}{n}}\left\{\prod_{k=1}^{N_1}\frac{\sigma(u-v_k+w)}{\sigma(u-v_k)}\right\}\left\{\prod_{k=1}^{m}\frac{\sigma(u-p_k)}{\sigma(u-q_k)}\right\}\\
&\qquad\times\big\{L^{(1)i}{}_i(u,\lambda)\big\}_{i_1,i_2,\ldots,i_{N_1}}^{i'_1,i'_2,\ldots,i'_{N_1}}F^{i_1,i_2,\ldots,i_{N_1}}\big(\lambda-\hat{i};\{v_k\}\big)\\
&\qquad\times\mathcal{B}_{i'_{N_1}}(v_{N_1})\mathcal{B}_{i'_{N_1-1}}(v_{N_1-1})\cdots\mathcal{B}_{i'_1}(v_1)|\Omega\rangle\\
&\qquad-e^{\frac{\alpha^{(1)}w}{n}}s(u-v_{N_1},\lambda_{1i})\frac{\sigma(\lambda_{i1}w+w)}{\sigma(\lambda_{i1}w)}\\
&\qquad\times\left\{\prod_{k=1}^{N_1-1}\frac{\sigma(v_{N_1}-v_k+w)}{\sigma(v_{N_1}-v_k)}\right\}\left\{\prod_{k=1}^{m}\frac{\sigma(v_{N_1}-p_k)}{\sigma(v_{N_1}-q_k)}\right\}\\
&\qquad\times\big\{L^{(1)i}{}_i(v_{N_1},\lambda)\big\}_{i_1,i_2,\ldots,i_{N_1}}^{i'_1,i'_2,\ldots,i'_{N_1}}F^{i_1,i_2,\ldots,i_{N_1}}\big(\lambda-\hat{i};\{v_k\}\big)\\
&\qquad\times\mathcal{B}_{i'_{N_1}}(u)\mathcal{B}_{i'_{N_1-1}}(v_{N_1-1})\cdots\mathcal{B}_{i'_1}(v_1)|\Omega\rangle\\
&\qquad+\text{o.u.t.}
\end{aligned}\tag{3.26}$$

Keeping the initial condition of R-matrix (2.11) and the notation (3.25) in mind, we can find the action of the transfer matrices on the state $|\lambda;\{v_k\}\rangle$:

$$\begin{aligned}
&T(u)\big|\lambda;\{v_k\}\big\rangle\\
&=\left(\mathcal{A}(u)+\sum_{i=2}^{n}\mathcal{D}_i^i(u)\right)\big|\lambda;\{v_k\}\big\rangle\\
&=e^{\frac{1-n}{n}\alpha^{(1)}w}\prod_{k=1}^{N_1}\frac{\sigma(v_k-u+w)}{\sigma(v_k-u)}F^{i_1,i_2,\ldots,i_{N_1}}\big(\lambda-\hat{1};\{v_k\}\big)B_{i_{N_1}}(v_{N_1})\cdots B_{i_1}(v_1)|\Omega\rangle\\
&\quad+e^{\frac{\alpha^{(1)}w}{n}}\left\{\prod_{k=1}^{N_1}\frac{\sigma(u-v_k+w)}{\sigma(u-v_k)}\right\}\left\{\prod_{k=1}^{m}\frac{\sigma(u-p_k)}{\sigma(u-q_k)}\right\}\\
&\quad\times\big\{T^{(1)}(u)\big\}_{i_1,i_2,\ldots,i_{N_1}}^{i'_1,i'_2,\ldots,i'_{N_1}}F^{i_1,i_2,\ldots,i_{N_1}}\big(\lambda;\{v_k\}\big)\\
&\quad\times\mathcal{B}_{i'_{N_1}}(v_{N_1})\mathcal{B}_{i'_{N_1-1}}(v_{N_1-1})\cdots\mathcal{B}_{i'_1}(v_1)|\Omega\rangle\\
&\quad+s(u-v_{N_1},\lambda_{1i_{N_1}})\frac{\sigma(\lambda_{i_{N_1}1}w+w)}{\sigma(\lambda_{i_{N_1}1}w)}e^{\frac{1-n}{n}\alpha^{(1)}w}\prod_{k=1}^{N_1-1}\frac{\sigma(v_k-v_{N_1}+w)}{\sigma(v_k-v_{N_1})}\\
&\quad\times F^{i_1,i_2,\ldots,i_{N_1}}\big(\lambda-\hat{1};\{v_k\}\big)B_{i_{N_1}}(u)B_{i_{N_1-1}}(v_{N_1-1})\cdots B_{i_1}(v_1)|\Omega\rangle\\
&\quad-\left(\sum_{i=2}^{n}e^{\frac{\alpha^{(1)}w}{n}}s(u-v_{N_1},\lambda_{1i})\frac{\sigma(\lambda_{i1}w+w)}{\sigma(\lambda_{i1}w)}\right.\\
&\qquad\times\left\{\prod_{k=1}^{N_1-1}\frac{\sigma(v_{N_1}-v_k+w)}{\sigma(v_{N_1}-v_k)}\right\}\left\{\prod_{k=1}^{m}\frac{\sigma(v_{N_1}-p_k)}{\sigma(v_{N_1}-q_k)}\right\}\\
&\qquad\times\delta_{i'_{N_1}}^{i}\big\{T^{(1)}(v_{N_1})\big\}_{i_1,i_2,\ldots,i_{N_1}}^{i'_1,i'_2,\ldots,i'_{N_1}}F^{i_1,i_2,\ldots,i_{N_1}}\big(\lambda;\{v_k\}\big)\\
&\qquad\left.\times\mathcal{B}_{i'_{N_1}}(u)\mathcal{B}_{i'_{N_1-1}}(v_{N_1-1})\cdots\mathcal{B}_{i'_1}(v_1)|\Omega\rangle\right)\\
&\quad+\text{o.u.t.},
\end{aligned}\tag{3.27}$$

where we have introduced the reduced transfer matrices $T^{(1)}(u)$

$$\begin{aligned}
&T^{(1)}(u)F^{i'_1,i'_2,\ldots,i'_{N_1}}\big(\lambda;\{v_k\}\big)\\
&=\sum_{i=2}^{n}L^{(1)^i}{}_i(u,\lambda)_{i_1,i_2,\ldots,i_{N_1}}^{i'_1,i'_2,\ldots,i'_{N_1}}F^{i_1,i_2,\ldots,i_{N_1}}\big(\lambda-\hat{i};\{v_k\}\big).
\end{aligned}\tag{3.28}$$

Thanks to the commutation relations (3.12), one can obtain the explicit expressions of all the o.u.t. Eq. (3.27) tells that the state $|\lambda;\{v_k\}\rangle$ is not an eigenvector of the transfer matrices $T(u)$ *unless* F*'s* are the eigenvectors of the reduced transfer matrices $T^{(1)}(u)$. The condition that the third and fourth terms in the above equation should cancel each other and also the all o.u.t. terms vanish, will give a restriction on the N_1 parameters $\{v_k\}$, the so-called Bethe ansatz equations. Hence we arrive at the final results:

$$T(u)\big|\lambda;\{v_k\}\big\rangle = t\big(u;\{v_k\}\big)\big|\lambda;\{v_k\}\big\rangle. \tag{3.29}$$

The eigenvalue reads

$$t\big(u;\{v_k\}\big) = e^{(1-n)\bar{\alpha}^{(0)}w}\left\{\prod_{k=1}^{N_1}\frac{\sigma(v_k-u+w)}{\sigma(v_k-u)}\right\} + e^{\bar{\alpha}^{(0)}w}\left\{\prod_{k=1}^{N_1}\frac{\sigma(u-v_k+w)}{\sigma(u-v_k)}\right\}\left\{\prod_{k=1}^{m}\frac{\sigma(u-p_k)}{\sigma(u-q_k)}\right\}t^{(1)}\big(u;\{v_k^{(1)}\}\big), \tag{3.30}$$

in which

$$T^{(1)}(u)F^{i'_1,i'_2,\ldots,i'_{N_1}}\big(\lambda;\{v_k\}\big) = e^{-\frac{1}{n(n-1)}\sum_{i=2}^{n}\alpha^{(i)}w}\,t^{(1)}\big(u;\{v_k^{(1)}\}\big)F^{i'_1,i'_2,\ldots,i'_{N_1}}\big(\lambda;\{v_k\}\big), \tag{3.31}$$

$$F^{i'_1,i'_2,\ldots,i'_{N_1}}\big(\lambda-\hat{1};\{v_k\}\big) = e^{\frac{1}{n}\sum_{i=2}^{n}\alpha^{(i)}w}F^{i'_1,i'_2,\ldots,i'_{N_1}}\big(\lambda;\{v_k\}\big), \tag{3.32}$$

and $\{v_k \mid k=1,2,\ldots,N_1\}$ satisfy

$$e^{-n\bar{\alpha}^{(0)}w}\left\{\prod_{k=1,k\neq s}^{N_1}\frac{\sigma(v_k-v_s+w)}{\sigma(v_k-v_s-w)}\right\} = \left\{\prod_{k=1}^{m}\frac{\sigma(v_s-p_k)}{\sigma(v_s-q_k)}\right\}t^{(1)}\big(v_s;\{v_k^{(1)}\}\big). \tag{3.33}$$

The parameters $\{v_k^{(1)} \mid k=1,2,\ldots,N_2\}$ will be specified later by the Bethe ansatz equations (3.45).

The diagonalization of the transfer matrices $T(u)$ is now reduced to diagonalization of the reduced transfer matrices $T^{(1)}(u)$ in (3.31). The explicit expression of $T^{(1)}(u)$ given in (3.28) implies that $T^{(1)}(u)$ can be considered as the transfer matrices of an $E_{\tau,\eta}(sl_{n-1})$-module $W^{(1)}$ (or reduced space): N_1 tensor product of fundamental representations of $E_{\tau,\eta}(sl_{n-1})$ with evaluation points $\{v_k\}$. We can use the same method to find the eigenvalue of $T^{(1)}(u)$ as we have done for the diagonalization of $T(u)$. Similarly to (3.1), (3.2), we introduce

$$\mathcal{A}^{(1)}(u) = \{\tilde{L}^{(1)}\}_2^2(u), \qquad \mathcal{B}_i^{(1)}(u) = \{\tilde{L}^{(1)}\}_i^2(u), \qquad i=3,\ldots,n, \tag{3.34}$$

$$\mathcal{C}_i^{(1)}(u) = \{\tilde{L}^{(1)}\}_2^i(u), \qquad \mathcal{D}^{(1)j}{}_i(u) = \{\tilde{L}^{(1)}\}_i^j(u), \qquad i,j=3,\ldots,n. \tag{3.35}$$

Then the reduced transfer matrices (3.28) can be rewritten

$$T^{(1)}(u) = \mathcal{A}^{(1)}(u) + \sum_{i=3}^{n}\mathcal{D}^{(1)i}{}_i(u). \tag{3.36}$$

We seek the common eigenvectors of the reduced transfer matrices $T^{(1)}(u)$ in an analogous form to (3.7):

$$|\lambda; \{v_k^{(1)}\}\rangle^{(1)}$$
$$= \sum_{i_1,\ldots,i_{N_2}=3}^{n} F^{(1)^{i_1,i_2,\ldots,i_{N_2}}}(\lambda; \{v_k^{(1)}\}) \mathcal{B}_{i_{N_2}}^{(1)}(v_{N_2}^{(1)}) \mathcal{B}_{i_{N_2-1}}^{(1)}(v_{N_2-1}^{(1)}) \cdots \mathcal{B}_{i_1}^{(1)}(v_1^{(1)}) |\Omega^{(1)}\rangle, \tag{3.37}$$

with the restriction of zero-weight conditions similar to (3.6). The quasi-vacuum $|\Omega^{(1)}\rangle$ is the corresponding highest weight vector in the reduced space $W^{(1)}$. We can also derive the relevant commutation relations among $\mathcal{A}^{(1)}(u)$, $\mathcal{D}^{(1)}{}_i^j$ and $B_i^{(1)}(u)$ $(i, j = 3, 4, \ldots, n)$ similar to (3.10)–(3.14). Using the same method as we have done for the diagonalization of $T(u)$, we obtain the eigenvalue of the reduced transfer matrices $T^{(1)}(u)$:

$$T^{(1)}(u)|\lambda; \{v_k^{(1)}\}\rangle^{(1)} = t^{(1)}(u; \{v_k^{(1)}\})|\lambda; \{v_k^{(1)}\}\rangle^{(1)}. \tag{3.38}$$

Though very complicated, the coefficients $F^{i_1,i_2,\ldots,i_{N_1}}(\lambda; \{v_k\})$ could be extracted from the eigenvectors $|\lambda; \{v_k^{(1)}\}\rangle^{(1)}$, in principle. The eigenvalue $t^{(1)}(u; \{v_k^{(1)}\})$ is given by

$$t^{(1)}(u; \{v_k^{(1)}\})$$
$$= e^{(2-n)\bar{\alpha}^{(1)}w} \left\{ \prod_{k=1}^{N_2} \frac{\sigma(v_k^{(1)} - u + w)}{\sigma(v_k^{(1)} - u)} \right\}$$
$$+ e^{\bar{\alpha}^{(1)}w} \left\{ \prod_{k=1}^{N_2} \frac{\sigma(u - v_k^{(1)} + w)}{\sigma(u - v_k^{(1)})} \right\} \left\{ \prod_{k=1}^{N_1} \frac{\sigma(u - v_k)}{\sigma(u - v_k + w)} \right\} t^{(2)}(u; \{v_k^{(2)}\}). \tag{3.39}$$

The parameters $\{v_k^{(2)} | \ k = 1, 2, \ldots, N_3\}$ will be specified later by the Bethe ansatz equations (3.45). The function $t^{(2)}(u; \{v_k^{(2)}\})$ is the eigenvalue of the second reduced transfer matrices $T^{(2)}(u)$

$$T^{(2)}(u) F^{(1)^{i'_1,\ldots,i'_{N_2}}}(\lambda; \{v_k^{(1)}\})$$
$$= e^{-\frac{1}{(n-1)(n-2)} \sum_{i=3}^{n} \alpha^{(i)} w} t^{(2)}(u; \{v_k^{(2)}\}) F^{(1)^{i'_1,\ldots,i'_{N_2}}}(\lambda; \{v_k^{(1)}\}), \tag{3.40}$$

and $T^{(2)}(u)$ is given by (3.25) and (3.28) with all indices taking values over $3, \ldots, n$ and depending on $\{v_k^{(1)}\}$ instead of $\{v_k\}$. Hence, the diagonalization of $T(u)$ is further reduced to the diagonalization of $T^{(2)}(u)$ in (3.40). This is so-called nested Bethe ansatz. Repeating the above procedure further $n-2$ times, one can reduce to the last reduced transfer matrices $T^{(n-1)}(u)$ which is trivial to get the eigenvalues. At the same time we need to introduce the $\frac{n(n-1)}{2} l$ parameters $\{\{v_k^{(i)} | \ k = 1, 2, \ldots, N_{i+1}\}, i = 0, 1, \ldots, n-2\}$ to specify the eigenvectors of the corresponding reduced transfer matrices $T^{(i)}(u)$ (including the original one $T(u) = T^{(0)}(u)$), and $n-1$ arbitrary complex numbers $\{\alpha^{(i)} \mid i = 0, 1, \ldots, n-2\}$ to specify the quasi-vacuum $|\Omega^{(i)}\rangle$ of each step as in (3.15) and (3.22). Finally, we obtain all the eigenvalues of the reduced transfer matrices $T^{(i)}(u)$ with the eigenvalue $t^{(i)}(u; \{v_k^{(i)}\})$ in a recurrence form

$$t^{(i)}\big(u;\{v_k^{(i)}\}\big) = e^{(i+1-n)\bar{\alpha}^{(i)}w}\left\{\prod_{k=1}^{N_{i+1}}\frac{\sigma(v_k^{(i)}-u+w)}{\sigma(v_k^{(i)}-u)}\right\} + e^{\bar{\alpha}^{(i)}w}\left\{\prod_{k=1}^{N_{i+1}}\frac{\sigma(u-v_k^{(i)}+w)}{\sigma(u-v_k^{(i)})}\right\}\left\{\prod_{k=1}^{N_i}\frac{\sigma(u-p_k^{(i)})}{\sigma(u-q_k^{(i)})}\right\}t^{(i+1)}\big(u;\{v_k^{(i+1)}\}\big),$$

$$i=0,1,\ldots,n-2, \tag{3.41}$$

$$t^{(0)}\big(u;\{v_k^{(0)}\}\big)=t\big(u;\{v_k\}\big),\qquad t^{(n-1)}(u)=1, \tag{3.42}$$

where $\bar{\alpha}^{(i)}$, $i=0,1,\ldots,n-2$, are given by (3.9), $\bar{\alpha}=\bar{\alpha}^{(0)}$, $N_0=m$ and

$$p_k^{(0)}=p_k=z_k,\qquad q_k^{(0)}=q_k=z_k-l_kw,\quad k=1,2,\ldots,m, \tag{3.43}$$

$$p_k^{(i)}=v_k^{(i-1)},\qquad q_k^{(i)}=v_k^{(i-1)}-w,\quad i=1,2,\ldots,n-2,\ k=1,2,\ldots,N_i. \tag{3.44}$$

The $\{\{v_k^{(i)}\}\}$ satisfy the following Bethe ansatz equations

$$e^{(i-n)\bar{\alpha}^{(i)}w}\left\{\prod_{k=1,k\neq s}^{N_{i+1}}\frac{\sigma(v_k^{(i)}-v_s^{(i)}+w)}{\sigma(v_k^{(i)}-v_s^{(i)}-w)}\right\} = \left\{\prod_{k=1}^{N_i}\frac{\sigma(v_s^{(i)}-p_k^{(i)})}{\sigma(v_s^{(i)}-q_k^{(i)})}\right\}t^{(i+1)}\big(v_s^{(i)};\{v_k^{(i+1)}\}\big). \tag{3.45}$$

We conclude this section with some remarks on analytic properties of the functions $t^{(i)}(u;\{v_k^{(i)}\})$. By construction, the eigenvalue functions $t^{(i)}(u;\{v_k^{(i)}\})$ of the transfer matrices should not be singular at $u=v_k^{(i)}$ (*modulo* $\mathbb{Z}+\tau\mathbb{Z}$) with $0\leqslant i\leqslant n-2$, $1\leqslant k\leqslant N_{i+1}$. On the other hand, the Bethe ansatz equations (3.33) and (3.45) are derived from the requirement that the unwanted terms should vanish. It is interesting to note that these constraints could be understood from a different point of view: from Eq. (3.30), we know that $u=v_k^{(0)}=v_k$ (*modulo* $\mathbb{Z}+\tau\mathbb{Z}$) is a possible simple pole position of $t(u;\{v_k\})$. However, the constraints on $\{v_k\}$, the Bethe ansatz equations (3.33) simply tell that the *residue* of $t(u;\{v_k\})$ at $u=v_k$ (*modulo* $\mathbb{Z}+\tau\mathbb{Z}$) is vanishing. Hence, $t(u;\{v_k\})$ is analytic at $u=v_k$ (*modulo* $\mathbb{Z}+\tau\mathbb{Z}$). Similarly, the Bethe ansatz equations (3.45) ensure the analyticity of $t^{(i)}(u;\{v_k^{(i)}\})$ at all $u=v_k^{(i)}$ (*modulo* $\mathbb{Z}+\tau\mathbb{Z}$). Therefore, the eigenvalues of transfer matrices $T(u)$ are analytic functions of u.

4. The $\mathbb{Z}_n$ Belavin model

We shall show in this section how to obtain the eigenvectors and the corresponding eigenvalues of the transfer matrices for $\mathbb{Z}_n$ Belavin model with periodic boundary condition from our results in Section 3.

First, we construct the $\mathbb{Z}_n$ Belavin R-matrix [17]. Let g, h, I_α be $n\times n$ matrices with the elements

$$g_{ik}=\omega^i\delta_{ik},\qquad h_{ik}=\delta_{i+1,k}, \tag{4.1}$$

$$I_\alpha=I_{(\alpha_1,\alpha_2)}=g^{\alpha_2}h^{\alpha_1}, \tag{4.2}$$

where $i,k,\alpha_1,\alpha_2\in\mathbb{Z}_n$ and $\omega=\exp\{\frac{2\pi\sqrt{-1}}{n}\}$. We introduce the σ-functions

$$\sigma_\alpha(u)=\theta\begin{bmatrix}\frac{1}{2}+\frac{\alpha_1}{n}\\ \frac{1}{2}+\frac{\alpha_2}{n}\end{bmatrix}(u,\tau),\qquad \sigma_{0,0}(u)=\sigma(u). \tag{4.3}$$

The $\mathbb{Z}_n$ Belavin model is based on the following R-matrix

$$R_B(u)=\frac{\sigma(w)}{n\sigma(u+w)}\sum_{\alpha\in Z_n^2}\frac{\sigma_\alpha(u+\frac{w}{n})}{\sigma_\alpha(\frac{w}{n})}I_\alpha\otimes I_\alpha^{-1}, \tag{4.4}$$

which satisfies the quantum Yang–Baxter equation

$$\begin{aligned}&R_{12}(u_1-u_2)R_{13}(u_1-u_3)R_{23}(u_2-u_3)\\&\quad=R_{23}(u_2-u_3)R_{13}(u_1-u_3)R_{12}(u_1-u_2),\end{aligned} \tag{4.5}$$

and the initial condition: $R_{ij}^{kl}(0)=\delta_i^l\delta_j^k$. For generic evaluation points $z_1,\ldots,z_m$, one then defines the commuting transfer matrices

$$T_B(u)=\mathrm{tr}_0\,L_B(u), \tag{4.6}$$

$$L_B(u)=\{R_B\}_{01}(u-z_1)\cdots\{R_B\}_{0m}(u-z_m), \tag{4.7}$$

where the L-operator $L_B(u)$ acts on $\mathbb{C}^n\otimes(\mathbb{C}^n)^{\otimes m}$ and the transfer matrices $T_B(u)$ act on $(\mathbb{C}^n)^{\otimes m}$.

Using the intertwiner introduced by Jimbo et al. [18], we can define the matrix $S(z,\lambda)$ with the elements $S_j^i(z,\lambda)=\frac{\theta^{(i)}(z+nw\lambda_j)}{\prod_{k\neq j}\sigma(\lambda_{kj}w)}$. When $\lambda\in\mathbb{P}_\xi$, the inverse matrix of $S(z,\lambda)$ exists. Then, one finds the following *twisting relation* holds:

$$\begin{aligned}&\{R_B\}_{12}(u_1-u_2)S(u_1,\lambda)^{(1)}S\big(u_2,\lambda-h^{(1)}\big)^{(2)}\\&\quad=S(u_2,\lambda)^{(2)}S\big(u_1,\lambda-h^{(2)}\big)^{(1)}R_{12}(u_1-u_2,\lambda),\end{aligned} \tag{4.8}$$

in which $R(u,\lambda)$ is defined by (2.7), (2.8) and (2.9). The modified quantum Yang–Baxter equation (2.10) and the quantum Yang–Baxter equation (4.5) are equivalent to each other due to the above *twisting relation* (4.8). Let us introduce

$$\begin{aligned}&S_m(z_1,\ldots,z_m;\lambda)\\&\quad=S(z_m,\lambda)^{(m)}S\big(z_{m-1},\lambda-h^{(m)}\big)^{(m-1)}\cdots S\left(z_1,\lambda-\sum_{k=2}^m h^{(k)}\right)^{(1)}.\end{aligned} \tag{4.9}$$

For the special case of $l_1=l_2=\cdots=l_m=1$, we find that the L-operator $L(u,\lambda)$ corresponding to the $E_{\tau,\eta}(sl_n)$-module W is equivalent to the L-operator $L_B(u)$ of $\mathbb{Z}_n$ Belavin model (4.7) by the *twisting relation*

$$\begin{aligned}&S_m(z_1,\ldots,z_m;\lambda)S(u,\lambda-h)^{(0)}L(u,\lambda)\\&\quad=L_B(u)S(u,\lambda)^{(0)}S_m\big(z_1,\ldots,z_m;\lambda-h^{(0)}\big).\end{aligned} \tag{4.10}$$

In this formula $S_m(z_1,\ldots,z_m;\lambda)$ acts on the factors from 1 to m, and $h=h^{(1)}+\cdots+h^{(m)}$. Acting the both sides of (4.10) on a state $e_i\otimes v$ with $v\in W[0]$, replacing λ by $\lambda+\hat{i}$, and noting (2.14), we find

$$\sum_{j=1}^{n}\big(S(u,\lambda+\hat{j})e_j\big)\otimes\big(S_m(z_1,\ldots,z_m;\lambda+\hat{i})L_i^j(u,\lambda+\hat{i})v\big)$$
$$=L_B(u)\big\{\big(S(u,\lambda+\hat{i})e_i\big)\otimes\big(S_m(z_1,\ldots,z_m;\lambda)v\big)\big\}. \tag{4.11}$$

Therefore, one can derive

$$T_B(u)S_m(z_1,\ldots,z_m;\lambda)|_H=\sum_{i=1}^{n}S_m(z_1,\ldots,z_m;\lambda+\hat{i})L_i^i(u,\lambda+\hat{i})|_H, \tag{4.12}$$

where $H=\mathrm{Fun}(W)[0]$ is preserved by the transfer matrices.

Let $f\mapsto\int f(\lambda)$ be a linear functional on the space $\mathcal{F}$ of functions of $\lambda\in\mathbb{C}^n$, such that $\int f(\lambda+\hat{i})=\int f(\lambda)$, $i=1,\ldots,n$, for all $f\in\mathcal{F}$. Extend $\int$ to vector-valued functional by acting component-wise. Then for each eigenfunction $\Psi(\lambda)$ of $T(u)$ with eigenvalue $t(u)$, the transfer matrices of $\mathbb{Z}_n$ Belavin model $T_B(u)$ act on the vector $\int S_m(z_1,\ldots,z_m;\lambda)\Psi(\lambda)$

$$T_B(u)\int S_m(z_1,\ldots,z_m;\lambda)\Psi(\lambda)$$
$$=\sum_{i=1}^{n}\int S_m(z_1,\ldots,z_m;\lambda+\hat{i})L_i^i(u,\lambda+\hat{i})\Psi(\lambda)$$
$$=\sum_{i=1}^{n}\int S_m(z_1,\ldots,z_m;\lambda)L_i^i(u,\lambda)\Psi(\lambda-\hat{i})$$
$$=\int S_m(z_1,\ldots,z_m;\lambda)T(u)\Psi(\lambda)=t(u)\int S_m(z_1,\ldots,z_m;\lambda)\Psi(\lambda). \tag{4.13}$$

Namely, the vector $\int S_m(z_1,\ldots,z_m;\lambda)\Psi(\lambda)$ is an eigenvector of $T_B(u)$ with the same eigenvalue. Fortunately, such a linear functional on $\mathcal{F}$ has been given in [10,14] for XYZ spin chain ($\mathbb{Z}_2$-Belavin model) and its higher spin generalizations [16], for the generic $\mathbb{Z}_n$ Belavin model [15]. And our results recover those of [10] for $n=2$ case and coincide with those of [15] for $\mathbb{Z}_n$ Belavin model.

As in [16], one can construct the higher spin generalization of $\mathbb{Z}_n$ Belavin model by putting at each site a local L-operator

$$L_k(u;l_k)=\sum_{\alpha\in Z_n^2}\frac{\sigma_\alpha(u-z_k+\frac{w}{n})}{\sigma_\alpha(\frac{w}{n})}I_\alpha\otimes\rho_{l_k}\{S_\alpha\},\quad l_k\in\mathbb{Z}^+,\ S_\alpha\in SK, \tag{4.14}$$

where ρ_{l_k} is the spin-l_k representation of $\mathbb{Z}_n$ Sklyanin algebra SK with dimension $\frac{(n+l_k-1)!}{(n-1)!l_k!}$ [12]. The corresponding transfer matrices of the model are given by

$$T_B\big(u;\{l_k\}\big)=\mathrm{tr}_0\big\{L_1(u;l_1)L_2(u;l_2)\cdots L_m(u;l_m)\big\}. \tag{4.15}$$

Then, our construction in Section 3 actually results in the eigenvalues formulation (3.30) of the transfer matrices (4.15), with the Bethe ansatz equations (3.45). In particular, our result recovers that of [16] when $n = 2$.

5. The elliptic A_{n-1} Ruijsenaars–Schneider model

If we take $E_{\tau,\eta}(sl_n)$-module W for the special case[1]

$$W = V_{\Lambda^{(l_1)}}(z_1), \quad \Lambda^{(l_1)} = l_1\Lambda_1 = N\Lambda_1 = n \times l\Lambda_1, \tag{5.1}$$

the zero-weight space of this module is one-dimensional and the associated *transfer matrices* can be given by [9]

$$T(u) = \frac{\sigma(u - z_1 + lw)}{\sigma(u - z_1 + nlw)} M. \tag{5.2}$$

The operator M is independent of u and is given by

$$M = \sum_{i=1}^{n} \left\{ \prod_{j \neq i} \frac{\sigma(\lambda_{ij} w + lw)}{\sigma(\lambda_{ij} w)} \Gamma_i \right\}. \tag{5.3}$$

Here $\{\Gamma_i\}$ are elementary difference operators: $\Gamma_i f(\lambda) = f(\lambda - \hat{\imath})$. In fact, for a special choice of the parameters [19], this difference operator M is the Hamiltonian of elliptic A_{n-1} type Ruijsenaars–Schneider model [20] with the special *coupling constant* $\gamma = lw$, up to conjugation by a function [13,21]. Now, we consider the spectrum of M. The results of Section 3, enable us to obtain the spectrum of the Hamiltonian of the elliptic A_{n-1} Ruijsenaars–Schneider model as well as the eigenfunctions, in terms of the associated transfer matrices (5.2). Since M is independent of u, we can evaluate the eigenvalue of $T(u)$ at a special value of u, $u = z_1$. Then the expression of the eigenvalue $t(u; \{v_k\})$ simplifies drastically, for the second term in the right-hand side of (3.30) (the one depending on the eigenvalue of the reduced transfer matrices $t^{(1)}(u; \{v_k^{(1)}\})$) vanishes because $u - p_1^{(0)} = 0$ in the case (5.8). Finally, we obtain the eigenvalues of M:

$$e^{(1-n)\bar{\alpha}w} \frac{\sigma(nlw)}{\sigma(lw)} \left\{ \prod_{k=1}^{(n-1)\times l} \frac{\sigma(v_k - z_1 + w)}{\sigma(v_k - z_1)} \right\}, \tag{5.4}$$

where $\{v_k\}$ and $\{\{v_k^{(i)}\}\}$ are determined by the nested Bethe ansatz equations

$$e^{(i-n)\bar{\alpha}^{(i)}w} \left\{ \prod_{k=1, k\neq s}^{N_{i+1}} \frac{\sigma(v_k^{(i)} - v_s^{(i)} + w)}{\sigma(v_k^{(i)} - v_s^{(i)} - w)} \right\}$$
$$= \left\{ \prod_{k=1}^{N_i} \frac{\sigma(v_s^{(i)} - p_k^{(i)})}{\sigma(v_s^{(i)} - q_k^{(i)})} \right\} t^{(i+1)}\left(v_s^{(i)}; \{v_k^{(i+1)}\}\right), \quad i = 0, 1, \ldots, n-2. \tag{5.5}$$

---

[1] Billey used algebraic Bethe ansatz method over the $E_{\tau,\eta}(sl_n)$-module $W = V_{\Lambda^{(1)}}(z_1) \otimes \cdots \otimes V_{\Lambda^{(1)}}(z_1 - (nl-1)w)$ to obtain eigenvalues of elliptic N-body Ruijsenaars operator [11].

The functions $t^{(i)}(u;\{v_k^{(i)}\})$ appearing in (5.5) are defined by the following recurrence relations

$$\begin{aligned} &t^{(i)}\big(u;\{v_k^{(i)}\}\big) \\ &\quad = e^{(i+1-n)\bar{\alpha}^{(i)}w}\left\{\prod_{k=1}^{N_{i+1}}\frac{\sigma(v_k^{(i)}-u+w)}{\sigma(v_k^{(i)}-u)}\right\} \\ &\quad\quad + e^{\bar{\alpha}^{(i)}w}\left\{\prod_{k=1}^{N_{i+1}}\frac{\sigma(u-v_k^{(i)}+w)}{\sigma(u-v_k^{(i)})}\right\}\left\{\prod_{k=1}^{N_i}\frac{\sigma(u-p_k^{(i)})}{\sigma(u-q_k^{(i)})}\right\}t^{(i+1)}\big(u;\{v_k^{(i+1)}\}\big), \\ &\quad i=1,\ldots,n-2, \end{aligned} \tag{5.6}$$

$$t^{(n-1)}(u)=1, \tag{5.7}$$

where $\bar{\alpha}^{(i)}$, $i=0,1,\ldots,n-2$, are given by (3.9), $\bar{\alpha}=\bar{\alpha}^{(0)}$, $N_0=1$ and

$$p_1^{(0)}=z_1,\qquad q_1^{(0)}=z_1-nlw, \tag{5.8}$$

$$p_k^{(i)}=v_k^{(i-1)},\qquad q_k^{(i)}=v_k^{(i-1)}-w,\quad i=1,2,\ldots,n-2,\ k=1,2,\ldots,N_i. \tag{5.9}$$

6. Conclusions

We have studied some modules over the elliptic quantum group associated with A_{n-1} type root systems. There are elliptic deformation of symmetric tensor product of the fundamental vector representation of A_{n-1} (we have called it *higher spin representation*). Using the nested Bethe ansatz method, we exactly diagonalize the commuting transfer matrices associated with the tensor product of the $E_{\tau,\eta}(sl_n)$-modules with generic evaluation points. The eigenvalues of the transfer matrices and associated Bethe ansatz equation are given by (3.30) and (3.45). For the special case of $n=2$, our result recovers that of Felder et al. [10].

We also take the applications of our construction to some integrable models. For $l_1=l_2=\cdots=1$ case, our result gives the spectrum problem of $\mathbb{Z}_n$ Belavin model with periodic boundary condition, which coincides with the result [15]. For the special case $W=V_{\Lambda^{(nl)}}(z_1)$, or equivalently $l_1=\cdots=l_{nl}=1$ and $z_i=z_1-(i-1)w$, our result gives the spectrum problem of the elliptic A_{n-1} type Ruijsenaars operators, which is "relativistic" generalization [22] of the Calogero–Moser type integrable differential operators, with the special *coupling constant* $\gamma=lw$, which coincides with the result [11]. Moreover, for the generic case, our result gives eigenvalues of the *transfer matrices* given by (4.15) associated with higher spin generalization of $\mathbb{Z}_n$ Belavin model with periodic boundary condition, and it further enables us to construct the eigenvalues of all types of Ruijsenaars operators associated with the A_{n-1} root system in the Bethe ansatz form [19].

Acknowledgements

We thank professors A. Belavin, S. Odake and K.J. Shi for their useful discussion. This work is supported in part by Grant-in-Aid for Scientific Research from the Ministry of Education, Culture, Sports, Science and Technology, No. 12640261. W.-L. Yang is supported by the Japan Society for the Promotion of Science.

Appendix A. The commutation relations

The starting point for deriving the commutation relations among $\mathcal{A}(u)$, $\mathcal{D}_i^j(u)$ and $\mathcal{B}_i(u)$ $(i, j = 2, \ldots, n)$ is the exchange relations (2.20) and (2.21). We rewrite (2.20) in the component form

$$\sum_{i',j'=1}^{n} R_{i'\ j'}^{a\ b}(u-v, \lambda-h)\tilde{L}_c^{i'}(u)\tilde{L}_d^{j'}(v) = \sum_{i',j'=1}^{n} \tilde{L}_{j'}^{b}(v)\tilde{L}_{i'}^{a}(u)R_{c\ d}^{i'j'}(u-v, \lambda). \tag{A.1}$$

For $a = b = d = 1, c = i \neq 1$, we obtain

$$\begin{aligned}\mathcal{A}(v)\mathcal{B}_i(u) &= \frac{1}{R_{i1}^{i1}(u-v, \lambda-\hat{1}-\hat{i})}\mathcal{B}_i(u)\mathcal{A}(v) - \frac{R_{i1}^{1i}(u-v, \lambda-\hat{1}-\hat{i})}{R_{i1}^{i1}(u-v, \lambda-\hat{1}-\hat{i})}\mathcal{B}_i(v)\mathcal{A}(u)\\ &= \frac{1}{R_{i1}^{i1}(u-v, \lambda)}\mathcal{B}_i(u)\mathcal{A}(v) - \frac{R_{i1}^{1i}(u-v, \lambda)}{R_{i1}^{i1}(u-v, \lambda)}\mathcal{B}_i(v)\mathcal{A}(u),\end{aligned} \tag{A.2}$$

by noting the definitions (2.8) and (2.9). The commutation relation (3.10) is a simple consequence of the above equation.

Similarly, for $a = j \neq 1, b = 1, c = i \neq 1, d = l \neq 1$, we obtain

$$\begin{aligned}\mathcal{D}_i^j(u)\mathcal{B}_l(v) &= \sum_{\alpha,\beta=2}^{n} \left\{ \frac{R_{i\ l}^{\alpha\beta}(u-v, \lambda)}{R_{j1}^{j1}(u-v, \lambda-h)}\mathcal{B}_\beta(v)\mathcal{D}_\alpha^j(v) \right\}\\ &\quad - \frac{R_{1j}^{j1}(u-v, \lambda-h)}{R_{j1}^{j1}(u-v, \lambda-h)}\mathcal{B}_i(u)\mathcal{D}_l^j(v),\end{aligned} \tag{A.3}$$

from which follows the commutation relation (3.11). For $a = b = 1, c = i \neq 1, d = j \neq 1$, we obtain

$$\begin{aligned}\mathcal{B}_i(u)\mathcal{B}_j(v) &= \sum_{\alpha,\beta=2}^{n} \mathcal{B}_\beta(v)\mathcal{B}_\alpha(u)R_{ij}^{\alpha\beta}(u-v, \lambda)\\ &= \sum_{\alpha,\beta=2}^{n} R_{ij}^{\alpha\beta}(u-v, \lambda-\hat{\alpha}-\hat{\beta})\mathcal{B}_\beta(v)\mathcal{B}_\alpha(u)\\ &= \sum_{\alpha,\beta=2}^{n} R_{ij}^{\alpha\beta}(u-v, \lambda)\mathcal{B}_\beta(v)\mathcal{B}_\alpha(u),\end{aligned} \tag{A.4}$$

which leads to the commutation relation (3.12). The other commutation relations (3.13), (3.14) are derived from (2.21).

References

[1] L.A. Takhtajan, L.D. Faddeev, Russian Math. Surveys 34 (1979) 11.
[2] H.B. Thacker, Rev. Mod. Phys. 53 (1982) 253.
[3] V.E. Korepin, N.M. Bogoliubov, A.G. Izergin, Quantum Inverse Scattering Method and Correlation Functions, Cambridge Univ. Press, Cambridge, 1993.
[4] O. Babelon, H.J. de Vega, C.M. Viallet, Nucl. Phys. B 200 (1982) 266.
[5] C.L. Schultz, Physica A 122 (1983) 71.
[6] E.K. Sklyanin, Func. Anal. Appl. 17 (1983) 273.
[7] B.Y. Hou, H. Wei, J. Math. Phys. 30 (1989) 2750.
[8] G. Felder, Elliptic quantum groups, in: Proceedings of the International Congress of Mathematical Physics, Paris, 1994, International Press, 1995, p. 211.
[9] G. Felder, A. Varchenko, Commun. Math. Phys. 181 (1996) 741;
G. Felder, A. Varchenko, J. Stat. Phys. 89 (1997) 963.
[10] G. Felder, A. Varchenko, Nucl. Phys. B 480 (1996) 485.
[11] E. Billey, Algebraic nested Bethe ansatz for the elliptic Ruijsenaars model, math.QA/9806068.
[12] B.Y. Hou, K.J. Shi, W.-L. Yang, Z.X. Yang, S.Y. Zhou, Int. J. Mod. Phys. A 12 (1997) 2927.
[13] K. Hasegawa, Commun. Math. Phys. 187 (1997) 289.
[14] R.J. Baxter, Ann. Phys. (N.Y.) 76 (1973) 1;
R.J. Baxter, Ann. Phys. (N.Y.) 76 (1973) 25;
R.J. Baxter, Ann. Phys. (N.Y.) 76 (1973) 48.
[15] B.Y. Hou, M.L. Yan, Y.K. Zhou, Nucl. Phys. B 324 (1989) 715.
[16] T. Takebe, J. Phys. A 25 (1992) 1071.
[17] A.A. Belavin, Nucl. Phys. B 180 (1981) 189.
[18] M. Jimbo, T. Miwa, M. Okado, Lett. Math. Phys. 14 (1987) 123;
M. Jimbo, T. Miwa, M. Okado, Nucl. Phys. B 300 (1988) 74.
[19] B.Y. Hou, R. Sasaki, W.-L. Yang, Eigenvalues of Ruijsenaars–Schneider model associated with A_{n-1} root system in the Bethe ansatz formalism, YITP-03-13.
[20] S.N.M. Ruijsenaars, H. Schneider, Ann. Phys. (N.Y.) 170 (1986) 370;
S.N.M. Ruijsenaars, H. Schneider, Commun. Math. Phys. 110 (1987) 191.
[21] B.Y. Hou, W.-L. Yang, Phys. Lett. 261 (1999) 259;
B.Y. Hou, W.-L. Yang, J. Math. Phys. 41 (2000) 357.
[22] H. Braden, R. Sasaki, Prog. Theor. Phys. 97 (1997) 1003.

Commun. Theor. Phys. (Beijing, China) **49** (2008) pp. 439–450
© Chinese Physical Society
Vol. 49, No. 2, February 15, 2008

Affine $A_3^{(1)}$ $N = 2$ Monopole as the D Module and Affine ADHMN Sheaf*

HOU Bo-Yu[1,†] and HOU Bo-Yuan[2]

[1]Institute of Modern Physics, Northwest University, Xi'an 710069, China

[2]Graduate School, the Chinese Academy of Sciences, Beijing 100049, China

(Received January 12, 2007)

Abstract *A Higgs–Yang–Mills monopole scattering spherical symmetrically along light cones is given. The left incoming anti-self-dual α plane fields are holomorphic, but the right outgoing SD β plane fields are antiholomorphic, meanwhile the diffeomorphism symmetry is preserved with mutual inverse affine rapidity parameters μ and μ^{-1}. The Dirac wave function scattering in this background also factorized respectively into the (anti)holomorphic amplitudes. The holomorphic anomaly is realized by the center term of a quasi Hopf algebra corresponding to an integrable conformal affine massive field. We find explicit Nahm transformation matrix (Fourier–Mukai transformation) between the Higgs YM BPS (flat) bundles (D modules) and the affinized blow up ADHMN twistors (perverse sheafs). Thus we establish the algebra for the 't Hooft–Hecke operators in the Hecke correspondence of the geometric Langlands program.*

PACS numbers: 14.80.Hv
Key words: affine BPS monopole, affinized ADHMN sheaf, affinized Nahm transformation, 't Hooft–Hecke operator, geometric Langlands program

1 Introduction

The 't Hooft BPS monopole always plays an important role in the gauge field theory, for example, for the Seiberg–Witten monopole condensation[1] and for the 't Hooft operator.[2] In this paper, we first give an affine monopole, which affinizes the minitwistor. The key point lies in that, when we do the transformation from the spacetime fixed frame to the comoving frame, in fact we transform to the static κ symmetric Killing gauge in Green–Schwarz theory. Meanwhile, we start from the BPS Higgs–Yang–Mills bundle, which is flat over the self-dual plane α and the anti-self-dual plane β, to the affinized ADHMN sheaf (i.e. the twistor sheaf).

Then we solve the Dirac equation in this background. Here we covariantly transform to the same comoving frame. This will manifest the Nahm transformation to the twistor space. These are realized as the soliton solution of the conformal affine Toda field.[3,4] Furthermore, this will give the elementary building block for the moduli space of exactly solvable worldsheet theory (Bena, Polchinski, Roiban,[5] and Refs. [6] and [7]). Our work shows the explicit links between Yang–Mills theory, quantum massive integrable field and quasi-Hopf (Drinfeld[8]) twistors for the scaled elliptic algebra and the scaled W algebra.[9–14] So it generalizes the algebraic formulation in the Langlands program by Frenkel *et al.*[15–17] to the four-dimensional case.

2 Affine Monopole Solution

Generalizing Kapustin and Witten's discussion,[2] we consider an affine family of the $N = 2$ supersymmetric gauge field in the 4d-Minkowski space $\mathbb{M}$, which has been twisted relevant to the geometric Langlands program. We pick a homotopic homomorphism $\mathfrak{K}$ from the spacetime symmetry $\mathrm{Spin}(4)_S$, which is the universal cover SU(4) of the conformal SO(4,2), to the R symmetry $\mathrm{Spin}(6)_R$, i.e. $\mathfrak{K}$: $\mathrm{Spin}(4)_S \to \mathrm{Spin}(4)_R \subset \mathrm{Spin}(6)_R$. By the choice of these Bochner–Martinelli kernel homotopy operator $\mathfrak{K}$, we get a family of $N = 2$ loop supersymmetry $\widetilde{\mathrm{SU}}(4)$ with affine parameter $t \in \mathbb{CP}_1$. To establish the Hecke correspondence, it should be further centrally extended to $A_3^{(1)}$. The time reverse and orientation reversal symmetry described by Ref. [2] implies, and the global Riemann–Hecke correspondence requires that the coordinates x_μ ($\mu = 0, 1, 2, 3$) of $\mathbb{M}$ be complexifield, embedded into $\mathbb{C}^4$ by analytically continuing to the upper and lower complex planes respectively, such that the connection of the D module becomes flat in the SD planes α and the ASD planes β,

$$[D_{i+}, D_{\perp i+}] + *[D_{i+}, D_{\perp i+}] = 0\,, \qquad (\alpha) \qquad (1)$$

$$[D_{\bar{i}-}, D_{\perp\bar{i}-}] - *[D_{\bar{i}-}, D_{\perp\bar{i}-}] = 0\,. \qquad (\beta) \qquad (2)$$

Remark The flat connection introduced by Yang[18] is the $i = 1$ case of Eqs. (1) and (5).

Here in the covariant derivative

$$D_\mu = \partial_\mu + A_\mu\,, \qquad (3)$$

we have

$$A_\mu = A_\mu^{\hat{a}} T^{\hat{a}}\,, \quad \mu = 0, 1, 2, 3\,, \qquad (4)$$

where $T^{\hat{a}}$ is the $A_3^{(1)}$ generators, the $*$ denotes the 4-dimensional Hodge dual, and the tangent vectors ∂ in the **fixed null frame** for the α planes are

$$\partial_{i_+} \equiv \frac{1}{\sqrt{2}}\Big(\frac{\partial}{\partial x^i} + \frac{\partial}{\partial x^0}\Big)\,, \quad \partial_{\perp i_+} \equiv \frac{1}{\sqrt{2}}\Big(\frac{\partial}{\partial x^j} + \mathrm{i}\frac{\partial}{\partial x^k}\Big)\,, (5)$$

*The project supported by National Natural Science Foundation of China under Grant No. 90403019

†E-mail: byhou@nwu.edu.cn

$$\partial_{i_-} \equiv \frac{1}{\sqrt{2}}\Big(\frac{\partial}{\partial x^i} - \frac{\partial}{\partial x^0}\Big), \quad \partial_{\perp i_-} \equiv \frac{1}{\sqrt{2}}\Big(\frac{\partial}{\partial x^j} - \mathrm{i}\frac{\partial}{\partial x^k}\Big), \quad (6)$$

while the tangent vectors $\bar\partial$ for the β planes are

$$\bar\partial_{i_-} \equiv \frac{1}{\sqrt{2}}\Big(\frac{\partial}{\partial \bar x^i} - \frac{\partial}{\partial \bar x^0}\Big), \quad \bar\partial_{\perp i_-} \equiv \frac{1}{\sqrt{2}}\Big(\frac{\partial}{\partial \bar x^j} - \mathrm{i}\frac{\partial}{\partial \bar x^k}\Big), \quad (7)$$

$$\bar\partial_{i_+} \equiv \frac{1}{\sqrt{2}}\Big(\frac{\partial}{\partial \bar x^i} + \frac{\partial}{\partial \bar x^0}\Big), \quad \bar\partial_{\perp i_+} \equiv \frac{1}{\sqrt{2}}\Big(\frac{\partial}{\partial \bar x^j} + \mathrm{i}\frac{\partial}{\partial \bar x^k}\Big). \quad (8)$$

The x_μ ($\bar x_\mu$) are the complexified Cartesian coordinate of $\mathbb{M}$, analytical continued upperwise (lowerwise) respectively. We use $\perp$ to denote perpendicular in α (β) plane, i,j,k are the cyclic permutation of $1,2,3$. The tangential vectors ∂_{i_+} and $\partial_{\perp i_+}$ are left null, while ∂_{i_-} and $\partial_{\perp i_-}$ are right null. To establish the Hecke correspondence, we should have left (right) D operators flat on α (β) plane and act on left (right) Hilbert space $\mathcal{H}_{L,R}$ respectively (c.f. Sec. 3). In this paper we consider the level one case, i.e. $\mathcal{H}_{L,R}$ are $|\Lambda_i\rangle$, $\langle\bar\Lambda_i|$ ($i=0,1,2,3$) respectively (c.f. Sec. 4). Over the α plane (1), the gauge field $\mathcal{F}_{\mu\nu}$ of the left bundle is anti-self dual, while over the β plane (2), $\mathcal{F}_{\bar\mu\bar\nu}$ is self-dual. Here $\mu\nu$ or $\bar\mu\bar\nu$ implies that x_μ approaches the real slice from upper or lower half complex planes of x_μ. We will find the monopole solution separately for the α (β) null planes, such that it is incoming (outgoing) along the left (right) real null lines $\partial_{i_+}(\bar\partial_{i_-})$ spherical symmetrically, i.e., we boost the static BPS monopole (Appendix A) along the incoming (outgoing) real null lines. The interaction between incoming and outgoing waves at all the scattering points yields the central extension (c.f. Sec. 4).

The *level one*, i.e. the grade one in principle gradation of the affine algebra, Lorentz-covariant spherical symmetric ansatz of the nonzero components of **incoming** $A_\mu^{\hat a}$ in the **tensor product** form of spherical **comoving** spacetime and gauge **frames** are (cf. Appendix A)

$$K_{T^+} = \frac{\mathrm{i}F(\mathrm{r})}{\sqrt{2}\mathrm{r}}\rho^{-1}E^{-1}, \quad (9)$$

$$K_{T^-} = -\frac{\mathrm{i}F(\mathrm{r})}{\sqrt{2}\mathrm{r}}\rho^{-1}E, \quad (10)$$

$$H_{T^+} = \frac{-\partial\gamma/\partial\varphi + \cos\theta}{\mathrm{r}\sin\theta}\rho^{-2}, \quad (11)$$

$$H_{T^-} = \frac{\partial\gamma/\partial\varphi + \cos\theta}{\mathrm{r}\sin\theta}\rho^{-2}, \quad (12)$$

$$K_{r^-} = A_{r^-} = G(\mathrm{r})\rho^{-2}, \quad (13)$$

$$A^c_{r^+} = \zeta(\mathrm{r}), \quad (14)$$

$$A^d_\mu = 0, \quad (15)$$

$$D_\mu = \partial_\mu + A_\mu, \quad A_\mu = H_\mu + K_\mu,$$

$$D_\mu^{(H)} = \partial_\mu + H_\mu,$$

where the cyclic element

$$E = \begin{pmatrix} 0 & 1 & & \\ & 0 & 1 & \\ & & 0 & 1 \\ 1 & & & 0 \end{pmatrix}, \quad (16)$$

and the Coxeter ρ lies in the center of U(4)

$$\rho = \begin{pmatrix} 1 & & & \\ & i & & \\ & & -1 & \\ & & & -i \end{pmatrix}, \quad (17)$$

i.e., $(\rho^{-1})_{i,i} = (\rho^{-1}E)_{i,i+1} = (-\omega)^i$, $(\rho^{-1}E^{-1})_{i+1,i} = (-\omega)^i$, $\omega \equiv \mathrm{e}^{2\pi\mathrm{i}/4}$, here ρE ($E\rho = \omega E\rho$) are the generators of the affine Heisenberg algebra (group), used for the construction of the vertex operators in the principle realization of the affine algebra.[19] The **spherical incoming space time null frames**

$$\mathbf{e}^{T^+} = -\frac{1}{\sqrt{2}}|\mathrm{r}|\,\mathrm{e}^{-\mathrm{i}\gamma}(\mathrm{d}\theta + \mathrm{i}\sin\theta\mathrm{d}\varphi), \quad (18)$$

$$\mathbf{e}^{T^-} = \frac{1}{\sqrt{2}}|\mathrm{r}|\,\mathrm{e}^{\mathrm{i}\gamma}(\mathrm{d}\theta - \mathrm{i}\sin\theta\mathrm{d}\varphi), \quad (19)$$

$$\mathbf{e}^{r^\pm} = \frac{1}{\sqrt{2}}\mu^{\pm1}(\mathrm{d}r \pm \mathrm{d}t),$$

$$r^2 = x_1^2 + x_2^2 + x_3^2, \quad (20)$$

where $\mathrm{r} = \mu r^+ + \bar\mu^{-1}\bar r^-$, $r^\pm = (r\pm t)/2$, $\gamma(\theta,\varphi) = \mp\varphi$ in north (south) Wu–Yang gauge. We have decomposed the connection A_μ into the gauge connection H_μ, which lies in the center U(1) of U(4), and the covariant constant independent of (θ,φ) components K_μ. The connection H_μ turns to be the U(1) Dirac monopole component for the 't Hooft monopole. Now the K_μ in Eqs. (9), (10), and (13) are not connections, but are tensors with the basis elements $A_{\beta,j}$ (given by ρ, E) in Lemma (14.6) of Ref. [19]. For the level one representation $|\Lambda_i\rangle$, in the expression of the principle realization in affine algebra for the vertex operator Γ, only the level one $j=1$ term in the infinite sum of the following equation, i.e. Eq. (14.6.7) of Ref. [19] contributes,

$$\Gamma^\beta = \langle\Lambda_0, A_{\beta,0}\rangle \exp\Big(\sum_{j=1}^\infty \lambda_{\beta j'} z^{b_j} x_j\Big) \times \exp\Big(-\sum_{j=1}^\infty \lambda^{b_j}_{\beta,N+1-j'} b_j^{-1} z^{-b_j}\frac{\partial}{\partial x_j}\Big). \quad (14.6.7)$$

The $j=1$ term is usually written as $E_{ij}^{(1)}$, with the grade "1" corresponding to the exponent "1" of the loop parameter z in the above equation.

The explicit expression of the central term ζ will be given later, which will contribute the holomorphic anomaly. However it is irrelevant for the field strength over $\alpha(\beta)$ plane here. Actually, the incoming and outgoing $K_\mu\gamma^\mu$ in reduced cup product form (Sec. 3) turns to be the Affine–Toda–Lax connection generated by the grade $1(-1)$ part of the soliton generating vertex operator (Sec. 4) in the principle realization.[3,4]

Our ansatz for level -1 nonzero component $A_{\bar\mu}^{\hat a}$ in **the outgoing comoving null frames** over the β planes (2) are

$$K_{\bar T^+} = \frac{\mathrm{i}F(\mathrm{r})}{\sqrt{2}\mathrm{r}}\rho E, \quad (21)$$

$$K_{\bar{T}^-} = \frac{\mathrm{i}F(\mathbb{r})}{\sqrt{2}\mathbb{r}}\rho E^{-1}\,, \tag{22}$$

$$H_{\bar{T}^+} = \frac{\partial\bar{\gamma}/\partial\bar{\varphi} - \cos\bar{\theta}}{\mathbb{r}\sin\bar{\bar{\theta}}}\rho^2\,, \tag{23}$$

$$H_{\bar{T}^-} = \frac{-\partial\bar{\gamma}/\partial\bar{\varphi} - \cos\bar{\theta}}{\mathbb{r}\sin\theta}\rho^2\,, \tag{24}$$

$$K_{\bar{r}^+} = G(\mathbb{r})\rho^2\,, \tag{25}$$

$$A^c_{\bar{r}^-} = \zeta(\mathbb{r})\,, \tag{26}$$

$$A^d_{\bar{\mu}} = 0\,. \tag{27}$$

As in the case of the static BPS monopole (Eqs. (A13) and (A14)), to obtain the field strength from the comoving frame ansatz (9) $\sim$ (15) for the incoming potential becomes very simple, by using the generalized Gauss–Codazzi equations (A43) and (A44). Now, only the Dirac potential component is dependent on (θ,φ) to get its $F^{(H)}_{\mu\nu}$ involving differential calculation, all other K_μ terms become (θ,φ) independent tensor matrices, its contribution to the field strength is just the tensor products and r directional covariant derivatives. The holomorphic ASD fields strength over α plane are

$$F_{T^+,r^-} = \frac{-F(\mathbb{r})G(\mathbb{r})}{\mathbb{r}}\rho E^{-1}\,, \tag{28}$$

$$F_{T^-,r^-} = \frac{F(\mathbb{r})G(\mathbb{r})}{\mathbb{r}}\rho E\,, \tag{29}$$

$$F_{T^+,T^-} = -\frac{F^2(\mathbb{r})}{\mathbb{r}^2}\rho^{-2} - \frac{1}{\mathbb{r}^2}\rho^{-2}\,, \tag{30}$$

$$F_{T^+,r^+} = -\frac{F'(\mathbb{r})}{\mathbb{r}}\rho^{-1}E^{-1}\,, \tag{31}$$

$$F_{T^-,r^+} = -\frac{F'(\mathbb{r})}{\mathbb{r}}\rho^{-1}E\,, \tag{32}$$

$$F_{r^+,r^-} = G'(\mathbb{r})\rho^{-2}\,, \tag{33}$$

where $G'(\mathbb{r}) = (\mathrm{d}/\mathrm{d}\mathbb{r})G(\mathbb{r})$, $F'(\mathbb{r}) = (\mathrm{d}/\mathrm{d}\mathbb{r})F(\mathbb{r})$.

The anti-self-dual equation becomes

$$F'(\mathbb{r}) = G(\mathbb{r})F(\mathbb{r})\,, \tag{34}$$

$$G'(\mathbb{r}) = \frac{F^2(\mathbb{r})-1}{\mathbb{r}^2}\,. \tag{35}$$

Remark the seemingly mismatch factor ρ^2 is due to the opposite chirality in the $\gamma_\mu D^{(H)}_\mu$ and $\gamma_\mu K^\mu$ (c.f. Sec. 3) under homotopy, which includes a $\rho^2 \in \mathbb{Z}_2$ factor in the Hodge duality in the target space, moduli space, as the opposite rotation of δA_z, $\delta\phi_z$ under the action of the complex structure (e.g. Kapustin and Witten[2] (4.3)).

The unique "normalizable" solution is

$$F(\mathbb{r}) = \frac{\mathbb{r}}{\operatorname{sh}\mathbb{r}}\,, \tag{36}$$

$$G(\mathbb{r}) = \frac{1}{\mathbb{r}} - \mu\operatorname{cth}\mathbb{r}\,. \tag{37}$$

This solution is the unique solution such that the field strength has appropriate asymptotical property both at the infinity and at the origin. When we calculate the left (right) part of our solution, we keep $\bar{r}_-(r_+)$ fixed. This implies as in Sec. 4, pushing forward to the worldsheet by the static (with respect to "time" $r_-(\bar{r}_+)$) gauge, the dependence on worldsheet coordinates $z = r_-(\bar{z} = \bar{r}_+)$ turns to be (anti)holomorphic, respectively.

From the outgoing comoving null frames Eqs. (21) $\sim$ (27), the antiholomorphic SD field strength equals the Hermitian conjugate of ASD part, only the sign of $\bar{H}_{T^+}$, $\bar{H}_{T^-}$ changes, since the orientation of the basis is reverse,

$$\bar{\mathfrak{e}}^{T^+} = -\frac{1}{\sqrt{2}}\mathrm{e}^{-\mathrm{i}\bar{\gamma}}\mathbb{r}(\mathrm{d}\bar{\theta} - \mathrm{i}\sin\bar{\theta}\mathrm{d}\bar{\varphi})\,,$$

$$\bar{\mathfrak{e}}^{r^\pm} = \frac{1}{\sqrt{2}}\bar{\mu}^{\mp 1}(\mathrm{d}\bar{r} \pm \mathrm{d}\bar{t})\,,$$

$$\bar{\mathfrak{e}}^{T^-} = \frac{1}{\sqrt{2}}\mathrm{e}^{\mathrm{i}\bar{\gamma}}\mathbb{r}(\mathrm{d}\bar{\theta} + \mathrm{i}\sin\bar{\theta}\mathrm{d}\bar{\varphi})\,.$$

3 Dirac Equation in Affine Monopole Background

Now we turn to the Dirac equation in this affine monopole background

$$\text{For the right } D \text{ bundle } \gamma^\mu D_\mu|\psi\rangle = 0\,, \tag{38}$$

$$\text{For the left } D \text{ bundle } \langle\psi|(\gamma^{\bar{\mu}}D_{\bar{\mu}})^\dagger = 0\,. \tag{39}$$

As in the last section, c.f. Appendix B, we decompose $A = A^{\hat{a}}_\mu \mathbf{T}^{\hat{a}}\mathrm{d}x^\mu$ into the potential H_μ of Dirac monopole and the vector boson K_μ covariant with respect to both the spin S and the "R-spin" T,

$$A_\mu = H_\mu + K_\mu\,. \tag{40}$$

Let

$$D_\mu = D^{(H)}_\mu + K_\mu\,.$$

For the right D module, the outgoing spherical waves propagate along r^+ with $r^- =$ constant. We introduce

$$\kappa \equiv -\mathrm{i}\epsilon_{ijk}\sigma_i\hat{r}_j D^{(H)}_k + 1\,, \tag{41}$$

where $i,j,k = T^+, T^-, r^+$, then one can prove that

$$\gamma_i D^{(H)}_i = \gamma_{r^+}\Big(\frac{\partial}{\partial r^+} + \frac{1}{r^+}\Big) - \frac{1}{r^+}\gamma_{r^+}\kappa\,, \tag{42}$$

where $\gamma_{r^+} = \Sigma^2 \otimes \sigma_{r^+}$, $\sigma_{r^+} = \sum_i \sigma_i\hat{r}^+_i$.

Let the fixed frame wave function in the tensor product form of "R-spin" $|I\rangle^\beta_\nu$ and the space time spin $|S\rangle^\alpha_\lambda$ be factorized into (t,r) and (φ,θ) dependent part‡

$$|\psi\rangle = \psi^{\alpha\beta}_{\lambda\nu}|S\rangle^\alpha_\lambda|I\rangle^\beta_\nu\,, \quad \alpha,\beta = +,-;$$

$$\lambda,\nu = \pm\frac{1}{2}\,. \tag{43}$$

Then the incoming Dirac equation (38) becomes

$$\gamma_\mu D^{(H)}_\mu|\psi\rangle + \sum_{a,j=1,2}\epsilon_{r^+aj}(\Sigma^2\otimes\sigma_j)^{\alpha\alpha'}_{\lambda\lambda'}\mathbb{F}(\mathbb{r})^{\alpha'\alpha'';\beta'\beta''}_{\lambda'\lambda'';\nu'\nu''}(\Sigma^2\otimes\sigma_a)^{\beta\beta'}_{\nu\nu'}\psi^{\alpha''\beta''}_{\lambda''\nu''}|S\rangle^{\alpha''}_{\lambda''}|I\rangle^{\beta''}_{\nu''}$$
$$+\,\eta_{r^-r^+}(\Sigma^{r^+}\otimes\sigma_{r^-})^{\alpha\alpha'}_{\lambda\lambda'}\mathbb{G}(\mathbb{r})^{\alpha'\alpha'';\beta'\beta''}_{\lambda'\lambda'';\nu'\nu''}(\Sigma^{r^-}\otimes\sigma_{r^+})^{\beta\beta'}_{\nu\nu'}\psi^{\alpha''\beta''}_{\lambda''\nu''}|S\rangle^{\alpha''}_{\lambda''}|I\rangle^{\beta''}_{\nu''} = 0\,, \tag{44}$$

‡We adopt the convention $|v\rangle = (v_1,\cdots,v_n)^{\mathrm{T}} = \sum_{i=1}^n v_i|e\rangle_i$, $|e\rangle_1 = (1,0,\cdots,0)^{\mathrm{T}},\cdots$

where

$$\mathbb{F}^{\alpha\alpha';\beta\beta'}_{\lambda\lambda';\nu\nu'} \equiv \Big((\gamma_+)^{\alpha\alpha'}_{\lambda\lambda'} \underset{ST}{\otimes} (\rho)^{\beta\beta'}_{\nu\nu'}\Big)\frac{F(\mathbb{r})}{\mathbb{r}}\,, \quad \mathbb{G}^{\alpha\alpha';\beta\beta'}_{\lambda\lambda';\nu\nu'} \equiv \Big((\gamma_0)^{\alpha\alpha'}_{\lambda\lambda'} \underset{ST}{\otimes} (\rho)^{\beta\beta'}_{\nu\nu'}\Big)G(\mathbb{r})\,; \tag{45}$$

and the γ_μ matrices is decomposed as the direct product $\otimes$ of the 4d-chirality Σ and the spin σ,

$$\gamma_i = \Sigma^2 \otimes \sigma_i\,, \quad \gamma_0 = \Sigma^1 \otimes \sigma_0\,, \quad \gamma_5 = \Sigma^3 \otimes \mathbf{1}\,,$$

$$\sigma_1 = \begin{pmatrix} 0 & 1 \\ 1 & 0 \end{pmatrix}, \quad \sigma_2 = \begin{pmatrix} 0 & -\mathrm{i} \\ \mathrm{i} & 0 \end{pmatrix}, \tag{46}$$

$$\sigma_{r^+} = \sigma_3 = \begin{pmatrix} 1 & 0 \\ 0 & -1 \end{pmatrix}, \quad \sigma_{r^-} = \sigma_0 = -\begin{pmatrix} \mathrm{i} & 0 \\ 0 & \mathrm{i} \end{pmatrix}, \tag{47}$$

$$\Sigma^{r^+} \equiv \Sigma^1 = \begin{pmatrix} 0 & 1 \\ 1 & 0 \end{pmatrix}, \quad \Sigma^{r^-} \equiv \Sigma^2 = \begin{pmatrix} 0 & -\mathrm{i} \\ \mathrm{i} & 0 \end{pmatrix},$$

$$\Gamma \equiv \Sigma^3 = \begin{pmatrix} 1 & 0 \\ 0 & -1 \end{pmatrix}, \quad \Gamma^{\pm} = \frac{1}{2}(1 \pm \Gamma)\,. \tag{48}$$

and $\eta^{r^- r^+}$ is the light-cone metric in real null direction.

We use the D function to transform from the fixed frame basis $|S\rangle = \begin{pmatrix} |S\rangle^+ \\ |S\rangle^- \end{pmatrix}$, $|I\rangle = \begin{pmatrix} |I\rangle^+ \\ |I\rangle^- \end{pmatrix}$ to the comoving frame $|s\rangle^{\pm}_{\lambda}$, $|i\rangle^{\pm}_{\nu}$,

$$|s\rangle^{\pm}_{\lambda} = \sum_{\rho} D^{\pm S}_{\rho\lambda}(\varphi,\theta,\gamma)|S\rangle^{\pm}_{\rho}\,,$$

$$|S\rangle_{\frac{1}{2}} = \begin{pmatrix} 1 \\ 0 \end{pmatrix}, \quad |S\rangle_{-\frac{1}{2}} = \begin{pmatrix} 0 \\ 1 \end{pmatrix}, \tag{49}$$

$$|i\rangle^{\pm}_{\lambda} = \sum_{\rho} D^{\pm T}_{\rho\lambda}(\varphi,\theta,\gamma)|I\rangle_{\rho}\,,$$

$$|I\rangle_{\frac{1}{2}} = \begin{pmatrix} 1 \\ 0 \end{pmatrix}, \quad |I\rangle_{-\frac{1}{2}} = \begin{pmatrix} 0 \\ 1 \end{pmatrix}, \tag{50}$$

which satisfy the spin part: $\sigma_r|s\rangle^{\pm}_{\lambda} = 2\lambda|s\rangle^{\pm}_{\lambda}$, $\sigma_r|i\rangle^{\pm}_{\nu} = 2\nu|i\rangle^{\pm}_{\nu}$, for the chirality part: $\Gamma|i\rangle^{\pm} = \pm|i\rangle^{\pm}$, $\Gamma|s\rangle^{\pm} = \pm|s\rangle^{\pm}$. The superscript S and T denote the finite rotation matrix function $D^{1/2}_{\lambda\rho}(\varphi,\theta,\gamma)$ acting in the spin space and the isospin space, respectively, with $S=1/2$, $T=1/2$.

Then the fixed frame equations (6) and (7) are transformed to the following comoving frame equation,

$$\Big(\gamma_{r^+}\Big(\frac{\partial}{\partial r^+} + \frac{1}{r^+}\Big)|\psi\rangle\Big) \tag{51}$$

$$+\Big\{\sum_{a,j=1,2} \epsilon_{3aj}(\hat{\Sigma}^2(x)^{\alpha\alpha'} \otimes \hat{\sigma}_j(x)_{\lambda\lambda'})(D^{\alpha S}_{\lambda''\lambda'}) \underset{ST}{\otimes} \mathbb{F}^{\alpha'\beta';\alpha''\beta''}_{\lambda''\lambda''';\nu''\nu'''}(\mathbb{r})(D^{\beta T}_{\nu''\nu'})(\hat{\Sigma}^2(x)^{\beta\beta'} \otimes \hat{\sigma}_a(x)_{\nu\nu'})$$

$$+(\hat{\Sigma}^{r^+}(x)^{\alpha\alpha'} \otimes \hat{\sigma}_{r^-}(x)_{\lambda\lambda'})(D^{\alpha S}_{\lambda''\lambda'})\eta_{r^- r^+} \underset{ST}{\otimes} (D^{\beta T}_{\nu''\nu'})\mathbb{G}^{\alpha'\beta';\alpha''\beta''}_{\lambda''\lambda''';\nu''\nu'''}(\mathbb{r})(\hat{\Sigma}^{r^-}(x)^{\beta\beta'} \otimes \hat{\sigma}_{r^+}(x)_{\nu\nu'})\Big\} f^{\alpha''\beta''}_{\lambda'''\nu'''}|s\rangle^{\alpha''}_{\lambda'''}|i\rangle^{\beta''}_{\nu'''}$$

$$=0\,, \tag{52}$$

$$\hat{\psi}^{\alpha\beta}_{\lambda\nu} = f^{\alpha\beta}_{\lambda\nu}(r,t)\mathbb{D}^{\alpha\beta}_{\lambda\nu}(\varphi,\theta,\gamma)\,. \tag{53}$$

Here the $\hat{\sigma}(x)$ and $\hat{\Sigma}(x)$ matrices acted on the comoving frame basis $|i\rangle^{\pm}_{\nu}$, $|s\rangle^{\pm}_{\lambda}$ become the constant matrices in Eqs. (46) and (47) (c.f. Eqs. (A24) $\sim$ (A28)). Then, since the κ operator in Eq. (42) turns to be zero as shown after Eq. (57), similar to Eq. (A29), this Dirac equation turns to be simply a "tangent vector equation along the left radial direction".

We note that in the second term of Eq. (51) the space spin S and the R spin T are coupled into

$$\epsilon_{3aj}(\sigma_j)_{\lambda'\lambda''}D^{\alpha S}_{\lambda\lambda'}|S\rangle^{\alpha}_{\lambda''} \underset{ST}{\otimes} (\sigma_a)_{\nu'\nu''}D^{\beta T}_{\nu\nu'}|I\rangle^{\beta}_{\nu''} = \epsilon_{3aj}(\hat{\sigma}_j)_{\lambda\lambda'}|s\rangle^{\alpha}_{\lambda'} \underset{ST}{\otimes} (\hat{\sigma}_a)_{\nu\nu'}|i\rangle^{\beta}_{\nu'}(-1)^{\lambda-1/2}(-\delta_{\alpha,\beta}\delta_{\lambda,\nu} + \delta_{\alpha,-\beta}\delta_{\lambda,-\nu}) \tag{54}$$

with nonvanishing (φ,θ) functions $\mathbb{D}$

$$\mathbb{D}^{\alpha\beta}_{\lambda\nu}(\varphi,\theta,\gamma) \equiv (-1)^{2\lambda+1}(-\delta_{\lambda,\nu}\delta_{\alpha,\beta} + \delta_{\lambda,-\nu}\delta_{\alpha,-\beta})|s\rangle^{\alpha}_{\lambda}|i\rangle^{\beta}_{\nu}\,, \tag{55}$$

by using

$$\frac{1}{4\pi}\sum_{\lambda',\nu'} C^{S,T,0}_{\lambda',\nu',0}D^{\alpha S}_{\lambda'\lambda}(\varphi,\theta,\gamma)D^{-\alpha T}_{\nu'\nu}(\varphi,\theta,\gamma) = \frac{1}{\sqrt{2\pi}}D^{0}_{0,\lambda+\nu}(\varphi,\theta,\gamma)C^{S,T,0}_{\lambda,\nu,0} = (-1)^{\lambda-1/2}\delta_{\lambda,-\nu}\,, \tag{56}$$

$$\frac{1}{4\pi}\sum_{\lambda',\nu'} C^{S,T,0}_{\lambda',\nu',0}D^{\alpha S}_{\lambda'\lambda}(\varphi,\theta,\gamma)D^{\alpha T}_{\nu'\nu}(\varphi,\theta,\gamma) = \frac{1}{\sqrt{2\pi}}D^{0}_{0,\lambda-\nu}(\varphi,\theta,\gamma)C^{S,T,0}_{\lambda,-\nu,0} = -(-1)^{\lambda-1/2}\delta_{\lambda,\nu}\,, \tag{57}$$

where the finite rotation matrices $D^{S}_{\mu'\mu}$ and $D^{T}_{\nu'\nu}$ from the fixed frames to the comoving frames are coupled to a singlet expressed by the well-known Kroneker matrix $(-1)^{\mu-1/2}\delta_{\mu,\mp\nu}$, and we use ϵ^{3aj} as the $C^{\frac{1}{2}\frac{1}{2}0}_{\lambda\,\nu\,0}$ to couple D^S and D^T to obtain $D^{J}_{m,\mu+\nu}s_\mu i_\nu$, this gives $J=0$, $\delta_{\mu,\pm\nu}$ and so $\kappa=0$ in Eq. (51) implicitly. Since the S T chirality have opposite helicity, so there are two terms in Eq. (54), with opposite sign.

We also note that for the last term of Eq. (51), S and T are coupled as

$$\eta_{r^- r^+}\sigma_{r^-}D^{\alpha S}|S\rangle \otimes \sigma_{r^+}D^{\beta T}|I\rangle = \eta^{r^- r^+}(\sigma^{r^-}|s\rangle^{\alpha}_{\lambda} \underset{ST}{\otimes} \sigma^{r^+}|i\rangle^{\beta}_{\nu})(-1)^{\lambda-1/2}(-\delta_{\alpha,\beta}\delta_{\lambda,\nu} + \delta_{\alpha,-\beta}\delta_{\lambda,-\nu})\,, \tag{58}$$

Thus, the D function has been simply coupled into the Kronecker matrix.

From the constraint for the zero mode sections in Eq. (50), the original $16 \otimes 16$ tensor product space $|s\rangle_\lambda^\alpha |i\rangle_\nu^\beta$ is reduced to 8 dimensions, including $|s\rangle_\lambda^\alpha |i\rangle_\lambda^\alpha$ and $|s\rangle_\lambda^\alpha |i\rangle_{-\lambda}^{-\alpha}$ only. Then after factorizing out the φ, θ dependent basis $|i\rangle$, $|s\rangle$, we can show as Eqs. (A32) $\sim$ (A34) that since for the zero mode section we have further constraint $f_{\lambda\nu}^{\alpha\beta} = -f_{-\lambda-\nu}^{\alpha\beta} \equiv f^{\alpha\beta}$. So only 4 independent components remain only.

Remind that, as has been proved by Hitchin[20] only the left imaginary null $\boldsymbol{K}_{T^-}$ and real null $\boldsymbol{K}_{r^-}$ contribute on α plane. The Dirac equation in the cup product frame turns to be

$$\Big(\Big(\frac{\partial}{\partial r^+} + \frac{1}{r^+}\Big)I + \boldsymbol{K}_{T^-} + \boldsymbol{K}_{r^-}\Big)|\xi\rangle \equiv \Big(\Big(\frac{\partial}{\partial r^+} + \frac{1}{r^+}\Big)I + \mathbb{K}_+\Big)|\xi\rangle = 0\,, \quad |\xi\rangle \equiv \left|\begin{matrix} f^{++} \\ -f^{+-} \\ f^{--} \\ -f^{-+} \end{matrix}\right\rangle, \tag{59}$$

where $\boldsymbol{K}_\mu$ are

$$\boldsymbol{K}_{T^-} = \sum_{i=1}^{4} \frac{F^i(\mathbb{r})}{\mathbb{r}} \boldsymbol{E}_{i,i+1}\,,$$

$$\boldsymbol{K}_{T^+} = -\sum_{i=1}^{4} \frac{F^i(\mathbb{r})}{\mathbb{r}} \boldsymbol{E}_{i+1,i}\,,$$

$$\boldsymbol{K}_{r^-} = \frac{1}{2}\sum_{i=1}^{4} G^i(\mathbb{r})(\boldsymbol{E}_{i,i} + \boldsymbol{E}_{i+1,i+1})\,,$$

$$F^i(\mathbb{r}) = \omega^i F(\mathbb{r})\,, \quad G^i(\mathbb{r}) = \omega^i G(\mathbb{r})\,.$$

Here we add a subscript $+$ to $\mathbb{K} = \boldsymbol{K}_{T^-} + \boldsymbol{K}_{r^-}$, to denote that it is the whole $\mathbb{K}$ along r^+. Further, we write $\mathbb{A}_+$ instead of $\mathbf{K}_+$, since the $D_\mu^{(H)}$ becomes simply $\partial/\partial r + 1/r$ in this gauge, by the vanishing of κ. Equation (59) in each i-th $2 \otimes 2$ block generated by the $\boldsymbol{E}_{i+1,i}$, $\boldsymbol{E}_{i,i+1}$ and $\boldsymbol{E}_{i,i}$, $\boldsymbol{E}_{i+1,i+1}$ is obtained in the same way as Appendix B. Here we generalize the cup product $\underline{2} \otimes \underline{2} \to \underline{1} \otimes \underline{3}$ into $(2,1) \otimes ((2,1) \oplus (1,2)) \to (1,1) \oplus (3,1)$ and conjugate. That is the same decomposition of the quaternion product for the kernel of Dirac operator in SDYM. The outgoing Dirac equation is manipulated in the same way,

$$\langle\xi|\Big(\Big(\frac{\partial}{\partial \bar{r}^-} + \frac{1}{\bar{r}^-}\Big)I + \mathbb{K}_-\Big) = 0\,. \tag{60}$$

For the consistence of this equation with Eq. (59) at the intersection points of the outgoing and incoming waves, we should affinize the algebra by the inclusion of the $\boldsymbol{c}$ and the $\boldsymbol{d}$ terms (see next section). The explicit form of $f^{\alpha\beta}(r^+, r^-)$ will be found in the next section by solving the scattering equation of the incoming and outgoing waves.

4 Conformal Affine Massive Models, Affine ADHMN Construction and 't Hooft Hecke Operator

In this section we will sketch how these zero mode solutions in the cup product form (Sec. 3) in the background of the flat connection of the D bundles (Sec. 2) are transformed to the conformal **affine** system by the affine Nahm construction, then turn to the 't Hooft–Hecke operator.

In the previous section, by transforming from the fixed frame to the tensor product form of the spherical comoving frame, the dependence of θ, φ, $\gamma(\theta,\varphi)$ disappear by "Fourier transformation", i.e. integrate with the Green function for these variable. Meanwhile we choose the geodesics coordinate $\gamma = \pm\varphi$ in the north (south) patch of S^2 ($\sim \mathbb{C}P_1$). But the dependence on r^+r^- remains, so we call it intermediate comoving frame. Now to get the final comoving frame without r^+r^- dependence. we adopt the following step. We start from the cup product form in Eq. (59), then, firstly, we rotate the $|\xi\rangle$ in Eq. (59) by an angle ϕ such that $\mathrm{e}^{\phi/2} = \coth(\mathbb{r})$ (c.f. Eq. (62)), so that D_{r^+} in Eq. (59) for each G^i, F^i term of the cup product becomes

$$\Big(\frac{\partial}{\partial \mathbb{r}} + \frac{F^i(\mathbb{r})}{\mathbb{r}}\begin{pmatrix} 0 & 1 \\ 1 & 0 \end{pmatrix} + G^i(\mathbb{r}) - \frac{1}{\mathbb{r}}\Big)\,.$$

By noticing

$$\Big(\frac{\mathrm{d}}{\mathrm{d}\mathbb{r}} + \Big(G^i(\mathbb{r}) - \frac{1}{\mathbb{r}}\Big)\Big)\frac{F^i(\mathbb{r})}{\mathbb{r}} = 0\,,$$

i.e., the diagonal functions are the logarithm derivative of the off-diagonal function. In these geodesic r, t fixed frames, we have

$$\mathbb{A}_\pm^i = \begin{pmatrix} G^i(\mathbb{r}) - \frac{1}{\mathbb{r}} & \frac{F^i(\mathbb{r})}{\mathbb{r}} \\ \frac{F^i(\mathbb{r})}{\mathbb{r}} & G^i(\mathbb{r}) - \frac{1}{\mathbb{r}} \end{pmatrix} \pm \zeta \boldsymbol{c}\,.$$

Then for the outgoing and incoming waves scattering along r_+ and r_- directions. The fixed frame $\mathbb{A}_\pm$ is given by integrating the light-cone comoving frame $a_\pm$ along the spectral line,

$$\mathbb{A}_\pm = \mathrm{P}\exp\int a_\pm(\mathbb{z};\mu)\,,$$

where

$$\begin{aligned} a_\pm(\mathbb{z};\mu) &= a_\pm(\mathbb{z}) \\ &= \frac{\partial_{z_\pm}(G - 1/\mathbb{z})}{G - 1/\mathbb{z}}\rho^{\mp 2} + \Big(\frac{\partial_{z_\pm}(F/\mathbb{z})}{F/\mathbb{z}}\Big)^{\pm 1}\rho^{\mp 1}E^{\pm 1} \\ &\quad + \zeta(\mathbb{z})c\,, \end{aligned} \tag{61}$$

where we have pushed forward $\mathbb{r} = \mu r_+ + \mu^{-1}\bar{r}_-$ to $z_+ = r_+, z_- = r_-$ on worldsheet. This $a_\pm(\mathbb{z})$ turns to be the Lax connection

$$a_\pm(\mathbb{z}) = \sum_{i=0}^{3} \partial_{z_\pm}\phi_i \omega^{\pm 2\mathrm{i}} E_{i,i} + \mathrm{e}^{\pm\phi_i}(E^{\pm 1})_{i,i\pm 1} + \zeta(\mathbb{Z})\boldsymbol{c}\,,$$

as we substitute it into the one soliton solution of the conformal affine Toda equation, by the following identification

$$\begin{aligned} &\frac{(F(\mathbb{z})/\mathbb{z})'}{F(\mathbb{z})/\mathbb{z}} = \mathrm{e}^{|\phi_{\mathrm{soliton}}|} = \frac{1 + \mathrm{e}^{2\mathbb{z}}}{1 - \mathrm{e}^{2\mathbb{z}}} = \coth\mathbb{z}\,, \\ &\phi_i = |\phi|\rho_i\,, \quad \eta = \frac{\pi\mathrm{i}}{2}\,, \quad \mathrm{d}\eta = 0\,, \\ &\zeta = \ln(1 - \mathrm{e}^{2\mathbb{z}}) \equiv \zeta_{\mathrm{sol}} - \zeta_{\mathrm{vac}}\,. \end{aligned} \tag{62}$$

The Lax equation implies the existence of the transport matrix U, $\partial_\pm U = a_\pm(\mathbb{z})U$. Let the monodromy matrix $\mathcal{T}_\pm$, the loop operator of a along $r_\pm = 0 \to \infty$, $\mathcal{T}_+ = \int_0^\infty a(\mathbb{z})\mathrm{d}z$, $\mathcal{T}_- = \int_0^\infty a(\mathbb{z})\mathrm{d}\bar{z}$, then $\mathcal{T}_\pm$ becomes independent of r, t and satisfy the **affine Nahm equation or the affine Donaldson's imaginary equation**

$$\lambda\frac{\mathrm{d}\mathcal{T}_+^\alpha(\lambda)}{\mathrm{d}\lambda} - [\mathcal{T}_+^h(\lambda), \mathcal{T}_+^\alpha(\lambda)] - \delta_+(s) = 0\,, \tag{63}$$

$$-\lambda\frac{\mathrm{d}\mathcal{T}_-^\alpha(\lambda)}{\mathrm{d}\lambda} - [\mathcal{T}_-^h(\lambda), \mathcal{T}_-^\alpha(\lambda)] - \delta_-(s) = 0\,, \tag{64}$$

and **real** equation,

$$\begin{aligned}&\lambda\frac{\mathrm{d}\mathcal{T}_+^\mathfrak{h}(\lambda)}{\mathrm{d}\lambda} - \frac{\mathrm{d}\mathcal{T}_-^\mathfrak{h}(\lambda)}{\mathrm{d}\lambda} + [\mathcal{T}_+^\mathfrak{h}(\lambda), \mathcal{T}_-^\mathfrak{h}(\lambda)] \\ &+ [\mathcal{T}_+^\alpha(\lambda), \mathcal{T}_-^\alpha(\lambda)] - \delta(s) \\ &= \zeta(\mathbb{r}_0)\,.\end{aligned} \tag{65}$$

which depend on $\lambda = \mathrm{e}^s$ only, here λ is the loop parameter of $\hat{g}$ and $\tilde{g}$, $\mathcal{T}_\pm(\lambda) = \mathcal{T}_-^\mathfrak{h}(\lambda)\rho + \mathcal{T}_\pm^\pm(\lambda)E^{\pm 1}$

$$\delta_\pm(s) = \frac{1}{2}\Big(\delta(s) \pm \frac{\mathrm{i}}{\pi}\mathcal{P}\frac{1}{s}\Big)\,.$$

Meanwhile we change r_+, r_- with respect to the scattering point (blow up point) $\mathbb{r}_0$ by $r_+ \to r_+ - r_{0+}$, $r_- \to r_- - r_{0-}$, $\mathbb{r}_0 \equiv \mu r_{0+} + \bar{\mu} r_{0-}$.

The residues of $\mathcal{T}$ around $\lambda = \mu$, $\bar{\lambda} = \bar{\mu}$ respectively are the EE^{-1} generators of the $A_3^{(1)}$. The parameter μ describes the central $\mathrm{U}(1)^c$ of $\mathrm{U}(4)$, i.e. the diffeomorphism, area preserving, dual twistor angle $\mathrm{U}(1)$ symmetry, which we affinize, and it happens that this serves also as the common dilation, rapidity shift parameter in Rindler coordinate. This dilation operators also rotate the fixed frame $|\Lambda_{\max}\rangle$ to the moving frame $|\xi_{\rm vac}\rangle$. After all these we have reach the affine twistor construction.

It is easy to check that the character of the blow up sheaf $\mathcal{M}_{n,r,k}$ (e.g. in Ref. [21]) is the Pontrjagin class $c_2 = n = 1$, rank $r = 4$, 1st Chern class $c_1 = k = 1$ case is the one soliton τ function of conformal affine Toda. Which can be factorized into $\langle\xi(\bar{\lambda})|\xi(\lambda)\rangle$.

The distributions in fact are given by the trace twisted by the density matrix $\tilde{\rho}$ (67)

$$\mathrm{Tr}|\xi\rangle\Gamma^\pm\langle\xi| = \delta_\pm(\lambda)\,,\quad \mathrm{Tr}|\xi\rangle\Gamma\langle\xi| = \delta(\lambda)\,,\quad \mathrm{Tr} = \mathrm{tr}\tilde{\rho}M$$

the occurrence of the density matrix, comes from that $|\xi\rangle$ as a state in massive integrable model is non-pure state.

To find the density matrix, besides the $\lambda\partial/\partial\lambda \equiv d$ and c, one should introduce the homotopy operator $\mathfrak{K}$. That is, we are dealing with a massive integrable field theory§ as the Unruh effect, the bare vacuum should be replaced by introducing the density matrix ρ,

$$\langle\xi_{\rm vac}|T(\lambda_2)T(\lambda_1)|\xi_{\rm vac}\rangle = \mathrm{tr}[\tilde{\rho}T(\lambda_1)T(\lambda_2)]\,, \tag{66}$$

$$\tilde{\rho} = \mathrm{e}^{2\pi\mathrm{i}\mathfrak{K}}\,, \tag{67}$$

now λ becomes the rapidity in the Rindler coordinates:

$$x = r\cosh\alpha\,,\quad t = r\sinh\alpha\,,$$

$$-\infty < \alpha < +\infty\,,\quad 0 < r < +\infty\,. \tag{68}$$

We can introduce the rapidity shift operator $\mathfrak{K}$, such that,

$$\mathrm{e}^{\alpha\mathfrak{K}}T(\lambda)\,\mathrm{e}^{-\alpha\mathfrak{K}} = T(\lambda - \alpha)\,. \tag{69}$$

where the shift operator, homotopy Bochner–Martinelli kernel operator $\mathfrak{K}$ is realized as the second term on the right-hand side of the following equation (70)

$$\begin{aligned}[\hat{X}, \hat{Y}] &= [\tilde{X}, \tilde{Y}]_\sim + \frac{1}{2}\oint\frac{\mathrm{d}\lambda}{2\mathrm{i}\pi}\,\mathrm{tr}(\partial_\lambda\tilde{X}(\lambda)\cdot\tilde{Y}(\lambda))C \\ &= [\tilde{X}, \tilde{Y}]_\sim + \mathfrak{K}\tilde{X}(\tilde{Y})C\,,\end{aligned} \tag{70}$$

where $\hat{X}, \hat{Y} \in \hat{\mathfrak{g}}$, $\tilde{X}, \tilde{Y} \in \tilde{\mathfrak{g}}$.

Actually, the parameter λ is introduced by following Nahm,[23] it is the Fourier transformation of time t originally restricted to the static time of BPS monopole in Ref. [23], but now generated to r_+, r_- for incoming and outgoing wave along r_+, r_-, which is further pushed down respectively to z_+, z_-, and further central extend upon scattering point $\mathbb{z}_0$. So this affine parameter is the rapidity, i.e. the Fourier transformation of the Rindler coordinates α. Together with the rotation matrix in the base space $D^{\rm S}(\theta,\varphi)$ and in the target space $D^{\rm T}(\theta,\varphi)$. The radial scattering function and the Rindler boost function expressed by the radial Bessel function give the Fourier transformation, which constitutes the affine monopole function for a blow up point.

Now as in Ref. [3], we rotate (dressing transform) the wavefunction $|\xi\rangle$ and $\langle\bar{\xi}|$ respectively from the $|\xi_{\rm vac}\rangle$ to $|\xi_{\rm sol}\rangle$, $\langle\bar{\xi}_{\rm vac}|$ to $\langle\bar{\xi}_{\rm sol}|$ at the same time changes $a_{\rm vac}$ to $a_{\rm sol}$ by conjugate by the vertex operator $V(z)$ ('t Hooft–Hecke operator). The positive (negative) frequency part $V_-(z)$, $(V_+(z))$ is determined by the Riemann–Hilbert transformation.

$$\begin{aligned}&\tilde{V}_- = V_-\,\mathrm{e}^{-\zeta(\mathbb{r})/2}\,, \\ &\tilde{V}_-^{-1}(\mu) = \frac{1}{2}\zeta^{-1}\frac{\partial}{\partial z_+}\zeta + \Big(\frac{P_+}{\lambda-\mu} + \frac{P_-}{\lambda+\mu}\Big) \\ &\equiv \begin{pmatrix}\sqrt{\frac{\mathrm{e}^{\mathbb{r}}+\mathrm{e}^{-\mathbb{r}}}{\mathrm{e}^{\mathbb{r}}-\mathrm{e}^{-\mathbb{r}}}}I & 0 \\ 0 & \sqrt{\frac{\mathrm{e}^{\mathbb{r}}-\mathrm{e}^{-\mathbb{r}}}{\mathrm{e}^{\mathbb{r}}+\mathrm{e}^{-\mathbb{r}}}}I\end{pmatrix} \\ &+ \frac{2\,\mathrm{e}^{2\mathbb{r}}}{\sqrt{1-\mathrm{e}^{4\mathbb{r}}}}\begin{pmatrix}\mu & \lambda \\ -\lambda & -\mu\end{pmatrix}\frac{\mu}{\lambda^2-\mu^2}\,,\end{aligned} \tag{71}$$

$$P_\pm \equiv \frac{\mu\,\mathrm{e}^{\mathbb{r}}}{\sqrt{1-\mathrm{e}^{2\mathbb{r}}}}\begin{pmatrix}\pm I & I \\ -I & \mp I\end{pmatrix}\,, \tag{72}$$

where all matrices are 4×4, the elements in each chiral block are 2×2 diagonal. $P_\pm$ is diagonal along the rotation axis for the operator V with rotation angle $\varphi_{\rm sol}/2 = \coth(\mathbb{z})$.

Remark The right-hand side of this equation is similar as the following operators in Hitchin's paper[24]

$$\mathrm{i}\frac{\mathrm{d}}{\mathrm{d}\lambda} + \Big(\frac{1}{\lambda-1} + \frac{1}{\lambda+1}\Big)T\,. \tag{73}$$

The Hirota equation of τ gives the background dependent holomorphic anomaly. After summing over various

§We follow the discussion for the quantum massive integrable field by Lukyanov in Ref. [22].

representation of $|\xi\rangle$, Nakajima[21] obtained the quantum τ. But the perturbative factor of the character (partition function) has been obtained by conjecture to fit the large N approximation. We have included the central extension with center $\boldsymbol{c}$ lying in the center $\mathbb{Z}$ of U(4), to construct the universal bundle. This extension by homotopy is the way of nonabelian localization as calculated by Beasley and Witten for the Seifert manifold.[25]

We may match our affine Nahm equation of the 't Hooft–Hecke operator with the Nahm equation of the surface operator given by Gukov and Witten,[26] since we have both 1st and 2nd Chern classes c_1 and c_2.

5 Discussion and Outlook

In fact equation (70) is the integral kernel of the Bochner–Martinelli formula. From this, the reside theorem will give the Poincare–Hopf localization index, i.e. the analytic index. After doing the Thom isomorphism, the representative of this index can be written as the integral of the A-roof genus by using the Bott reside formula. This is just the topological index.

In fact, the analytic index is just the anti-holomorphic function on the sheaf $T(x)$ and $\psi(x)$ can be expressed using the $T(x)$ in Ref. [27]. The one corresponding to the real null vector, $T_0 \pm T_1$, is just B_1, and the one corresponding to the imaginary vector, $T_2 \pm \mathrm{i}T_3$, is just B_2, and $\oint |\psi\rangle\langle\psi|\mathrm{d}\lambda$ is just $i^*j + j^*i$. The character for the one affine monopole solution is just the one in Ref. [27] with $r=2$, $n=4$, $k=1$.

If we use the Riemann–Hilbert transformation with N poles, we will get the N-soliton solution. The corresponding character is the character for the sheaf with the Young diagram that satisfies $|Y|=N$ in Ref. [27].

Now we turn to the problem of quantization. First we should use the Seiberg–Witten curve to determine the cut-off constant Λ, here the Seiberg–Witten curve is just the spectral curve of the Lax connections in the affine Toda system. There are four formalisms for quantization. The first one is the quantum group formalism, such as the massive integrable field theory (The one studied in Ref. [22]). The bosonic oscillator representation is given in Refs. [10] and [13] (here the relevant quantum group is $\hat{Sl}_4(p,q)$). In the second formalism, twistor and sheaf are used.[21,27] The third formalism is the field theoretic formalism, where the hyperkahler quotient plays an important role. In the last one, the spectral curve in the moduli space forms a Calabi–Yau 3-fold. Here, in the commutation relation, we should use ϕ_{quantum} to replace the ϕ_{vacuum}.[28]

In the quantum group formalism, beside expansing the ϕ_{vaccuum} by the q-oscillator, we also obtain the exact explicit commutation relation of the vertex operator. We can obtain the exact result in the thermodynamics limit which is not an approximate result in the large N limit. Moreover, the two kinds of vertex operators in the quantum group formalism are corresponding to the Wilson loop and the 't Hooft loop in Yang–Mills theory, respectively. Modular transformation will exchange both these two kinds of operators and the corresponding states characterize respectively the order parameter and disorder parameter. This is also called the level-rank duality.

The area-preserving transformation of the dual twist angle φ ($\lambda = \mathrm{e}^{\mathrm{i}\varphi}$) is just the diffeomorphism transformation. So it will connect the phase angle φ of the center term of the topological field theory and the attractor parameter μ of the black hole entropy.

The quasi-Hopf quantum double of the Drinfeld's quantum group not only realizes the almost factorization in Ref. [29], but also gives the center term of the holomorphic anomaly, which is obtained by solving the Hirota equation. In this situation, the polarization of the waving function can be used to explain the background independence.

Appendix A: BPS Monopole

The static spherical symmetric ansatz¶

$$A_i^a(x) = \epsilon_{iaj}\frac{F(r)-1}{r}\hat{r}^j\,, \tag{A1}$$

$$A_0^a(x) = \mathrm{i}G(r)\hat{r}^a(x)\,, \tag{A2}$$

where $r = (x_1^2 + x_2^2 + x_3^2)^{1/2}$, the unit radial vector $\hat{r}^j = r^j/r$, the space spin index $i,j=1,2,3$, the "isospin" SU(2) generator index $a=1,2,3$. This ansatz is U(1) symmetrical under the cooperative local rotation generated by both the space time spin $\hat{s}(x)=\hat{r}$ and the SU(2) isospin $\boldsymbol{T}^r(x) = \sum_a T^a\hat{r}^a$.

Decompose A as

$$\begin{aligned}&A_\mu = H_\mu + K_\mu\,, \quad H_i^a = -\varepsilon_{iaj}\frac{1}{r}\hat{r}^j\,,\\ &K_i^a = \epsilon_{iaj}\frac{F(r)}{r}\hat{r}^j\,, \quad K_0^a = \mathrm{i}G(r)\hat{r}^a\,,\\ &H_0^a = 0\,.\end{aligned} \tag{A3}$$

Let‖ $D_\mu = \partial_\mu - \mathrm{i}A_\mu = D_\mu^{(H)} - \mathrm{i}K_\mu$, herein

$$D_\mu^{(H)} = \partial_\mu - \mathrm{i}H_\mu\,. \tag{A4}$$

Then we have that, $\boldsymbol{T}^r(x)$ is covariant constant under the SU(2) "isospin" rotation; $D_i^{(H)}\boldsymbol{T}^r(x) = (\partial_i + \mathrm{i}H_i^aT^{[a]})(\sum_b T^{b]}\hat{r}^b) = \partial_i\hat{r}^aT^a - \mathrm{i}\epsilon_{iaj}\hat{r}^j\hat{r}^b[T^a,T^b] = 0$; $D_0^{(H)}\boldsymbol{T}^r(x)=0$; and meanwhile $\nabla_i\hat{r}^a = 0$ under the action of spin connection along the surface S^2 ($r=$ constant).

Transform from the spacetime fixed frames in "spherical basis"[28] $e^M (M=0,+1,-1)$:

$$e^0 = \mathrm{d}x^3\,, \quad e^{+1} \equiv -\frac{1}{\sqrt{2}}(\mathrm{d}x^1 + \mathrm{i}\mathrm{d}x^2)\,,$$

¶For the BPS monopole which is constituted by the Higgs–Yang–Mills field or spontaneously breaking Yang–Mills field, the A_0^a should be replaced by the Higgs field $\mathrm{i}\Phi^a$ and the ASD equation by the BPS equation.

Thus, we adopt the notation (A2) in view of further embedding into the SUSY affine monopole as the incoming and outgoing waves in Secs. 2 and 3. There, in case of $d=4$ SUYM, our ansatz A_μ turns to be the $\mathcal{A}_\mu$ of Ref. [2], $\mathcal{A}_\mu = A_\mu + \mathrm{i}\phi_{\mu+4}$ with $A_0 = \phi_5 = \phi_6 = \phi_7 = 0$.

‖In appendices, we adopt the physics convention with a Hermitian gauge field A_μ.

$$e^{-1} = \frac{1}{\sqrt{2}}(\mathrm{d}x^1 - \mathrm{i}\mathrm{d}x^2), \tag{A5}$$

to the local comoving frames on S^2:

$$\boldsymbol{e}^{T_+} = -\frac{1}{\sqrt{2}} r\,\mathrm{e}^{-\mathrm{i}\gamma}(\mathrm{d}\theta + \mathrm{i}\sin\theta\mathrm{d}\varphi)\,,$$
$$\boldsymbol{e}^{T_-} = \frac{1}{\sqrt{2}} r\,\mathrm{e}^{\mathrm{i}\gamma}(\mathrm{d}\theta - \mathrm{i}\sin\theta\mathrm{d}\varphi)\,, \tag{A6}$$

together with the normal vector $\boldsymbol{e}^r = \mathrm{d}r$ and time $\boldsymbol{e}^t = \mathrm{d}t$, $\gamma = \mp\varphi$ in the north (south) patch. In the spherical basis $\boldsymbol{e}^m$ $(m = 0, +1, -1)$, $\boldsymbol{e}^0 \equiv \boldsymbol{e}^r$, $\boldsymbol{e}^{\pm1} \equiv \boldsymbol{e}^{T_\pm}$, we have

$$\boldsymbol{e}^m = \sum_{M=0,\pm1} D^1_{Mm}(\varphi,\theta,\gamma)e^M\,, \tag{A7}$$

change the basis of the fixed frame SU(2) generator from the cartesian basis T^a $(a = 1, 2, 3)$ to that in the spherical basis T^M, then gauge transform to the comoving frame $\boldsymbol{T}^m(x)$

$$\boldsymbol{T}^m(x) = \sum_{M=0,\pm1} D^1_{Mm}(\varphi,\theta,\gamma)T^M \tag{A8}$$

e.g.

$$\boldsymbol{T}^r(x) = \boldsymbol{T}^0(x) = \sum_{a=1,2,3} \hat{r}^a T^a\,,$$

bold face $\boldsymbol{e}$ ($\boldsymbol{T}$) denotes the vector (generator) in comoving frame, here the letters from the beginning $(a, b, c = 1, 2, 3)$ and from the middle $(M, m = -1, 0, +1)$ of the alphabet denote the indices of orthogonal basis and spherical basis respectively. Capital letter $(M = -1, 0, +1)$ and lower-case $(m = -1, 0, +1)$ denote the indices of the fixed frame and the comoving frame respectively. Gauge transform $A_\mu(x)$ from the fixed gauge to the comoving gauge. The connection H_μ in Eq. (A3) gauge transform to the U(1) Dirac potential (A9) while matrix tensor K_μ transform covariantly to Eq. (A10),

$$\boldsymbol{H}^0_{T_\pm} = \boldsymbol{A}^0_{T_\pm} = \frac{\mp\partial\gamma/\partial\varphi + \cos\theta}{r\sin\theta}\,, \tag{A9}$$
$$-\boldsymbol{K}^-_{T_+}(x) = -\boldsymbol{A}^-_{T_+}(x) = \boldsymbol{K}^+_{T_-}(x) = \boldsymbol{A}^+_{T_-}(x) = -\frac{\mathrm{i}}{\sqrt{2}}\frac{F(r)}{r}\,,$$
$$\boldsymbol{K}^0_t(x) = \boldsymbol{A}^0_t(x) = G(r)\,, \tag{A10}$$
$$A(x) = \sum_{\substack{m=0,\pm1\\ n=T_+,T_-,t}} \boldsymbol{A}^m_n(x)\boldsymbol{e}^n(x)\boldsymbol{T}^m(x)\,. \tag{A11}$$

Separate $F_{\mu\nu}$ into the radial and the tangential electric and magnetic components

$$F^a_{i0} = \frac{E_r(r)x_i x^a}{r^2} + E_T(r)\Big(\delta^a_i - \frac{x_i x^a}{r^2}\Big)\,,$$
$$\varepsilon_{ijk}F^a_{ij} = M_r(r)\frac{x_k x^a}{r^2} + M_T(r)\Big(\delta^a_k - \frac{x_k x^a}{r^2}\Big)\,. \tag{A12}$$

From Eq. (A11) for $A(x)$, it is easy to see that the radial component $M_r(x)$ can be found by using the generalized Gauss equation (C9) for the reduced curvature,

$$M_r(x) = 2\boldsymbol{K}^-_+\boldsymbol{K}^+_- + F^{(H)}_{+-}.\boldsymbol{T}^0 = \frac{F^2(r)-1}{r^2}\,, \tag{A13}$$

where the $F^{(H)}_{+-} = (1/r^2)\boldsymbol{T}^0$ is the magnetic field of the Dirac monopole (A9) $\boldsymbol{T}^a.\boldsymbol{T}^b \equiv (1/2)\,\mathrm{Tr}\,\boldsymbol{T}^a.\boldsymbol{T}^b$; the radial $E_r(x)$ and the tangential $E_T(x)$, $M_T(x)$ can be found by using the generalized Codazzi equation (C10)

$$E_r(x) = G'(r)\,,\quad E_T(x) = \frac{G(r)F(r)}{r}\,,$$
$$M_T(x) = \frac{F'(r)}{r}\,. \tag{A14}$$

The normalizable solution of the anti-self-dual equation

$$F'(r) = G(r)F(r)\,,\quad G'(r) = \frac{F^2(r)-1}{r^2}\,, \tag{A15}$$

is

$$F(r) = \frac{\mu r}{\mathrm{sh}\,(\mu r)}\,,\quad G(r) = \frac{1}{r} - \mu\,\mathrm{cth}(\mu r)\,, \tag{A16}$$

which is holomorphic, and satisfies the condition $E_r = M_r \sim O(1) \Rightarrow F(r) \sim O(1)$, $G(r) \sim O(1/r)$ as $r \to 0$, and asymptotically $E_r = M_r \sim 1/r^2$, $E_T = M_T \sim O(1/r^2) \Rightarrow G \sim 1$, $F(r) \sim p(r)\,\mathrm{e}^{-\mu r}$, here $p(r)$ is a polynomial.

Appendix B: Zero Mode of Fermion in Static Spherical Symmetric BPS Monopole Background

Dirac equation

$$\gamma_\mu D^\mu \Psi(x) = 0\,, \tag{A17}$$

where $D_\mu = \partial_\mu - \mathrm{i}\sum_a T^a A^a_\mu$ with A_μ given by Eqs. (A1) and (A2); $\gamma_j = \begin{pmatrix} 0 & -\mathrm{i}\sigma_j \\ \mathrm{i}\sigma_j & 0\end{pmatrix}$, $\gamma_0 = -\mathrm{i}\begin{pmatrix} 0 & I \\ I & 0\end{pmatrix}$, $\gamma_5 = \begin{pmatrix} I & 0 \\ 0 & -I\end{pmatrix}$ in the γ_5 diagonal representation.

Let $\Psi(x) = \begin{pmatrix}\chi^+ \\ \chi^-\end{pmatrix}$ be static, then we get two decoupled 2-component equations with opposite chirality

$$(\sigma_j D_j \mp G(r)\boldsymbol{T}^r(x))\chi^\pm = 0\,. \tag{A18}$$

Introduce the spin shift operator κ by coupling the spin with the orbital momentum

$$\kappa = -\epsilon_{ijk}\sigma_i x_j D^{(H)}_k + 1\,, \tag{A19}$$

then

$$\sigma_i D^{(H)}_i = \sigma_r\Big(\frac{\partial}{\partial r} + \frac{1}{r}\Big) - \frac{1}{r}\sigma_r\kappa\,,\quad \sigma_r \equiv \sum_i \sigma_i\hat{r}_i\,. \tag{A20}$$

So in the Dirac operator besides the spin shift operator κ only the radial derivative remains

$$\Big[\sigma_r\Big(\frac{\partial}{\partial r} + \frac{1}{r}\Big) - \frac{1}{r}\sigma_r\kappa - \mathrm{i}\epsilon_{ijk}\frac{F(r)}{r}\hat{r}_i\sigma_j T_k \mp G(r)\boldsymbol{T}^r\Big]\chi^\pm = 0\,. \tag{A21}$$

In the spherical symmetric coordinate, let

$$\chi^\pm = \sum_J\sum_{\mu,\nu} f^{\pm J}_{\mu,\nu}(r)\mathbb{D}^J_{\mu,\nu}(\varphi,\theta,\gamma)\,, \tag{A22}$$

$$\mathbb{D}^J_{\mu,\nu}(\varphi,\theta,\gamma) \equiv \sqrt{\frac{2J+1}{4\pi}}D^J_{M,\mu+\nu}s_\mu i_\nu\,,$$
$$M = -(2J+1),\ldots,2J+1\,,$$
no summation for μ, ν. (A23)

$$
\begin{aligned}
&s_\mu(\varphi,\theta,\gamma)=\sum_{\mu'}\sqrt{\frac{1}{2\pi}}D^S_{\mu',\mu}(\varphi,\theta,\gamma)S_{\mu'}\,,\\
&\sigma_3 S_\mu = 2\mu S_\mu\,,\quad S=\frac{1}{2}\,, \qquad \text{(A24)}\\
&i_\nu(\varphi,\theta,\gamma)=\sum_{\nu'}\sqrt{\frac{1}{2\pi}}D^I_{\nu',\nu}(\varphi,\theta,\gamma)I_{\nu'}\,,\\
&T_3 I_\nu = 2\nu I_\nu\,,\quad I=\frac{1}{2}\,, \qquad \text{(A25)}
\end{aligned}
$$

and

$$
\begin{aligned}
&\sigma_r s_\mu = 2\mu s_\mu\,,\quad T_r i_\nu=\nu i_\nu\,, \qquad \text{(A26)}\\
&\sigma_m(x)=\sum_{M=0,\pm1}D^1_{Mm}(\varphi,\theta,\gamma)\sigma^M\,,\\
&\sigma^r(x)=\sigma^0(x)=\sum_{a=1,2,3}\hat{r}^a\sigma^a\,,\\
&\text{or}\ \ (\sigma^a(x))_{\lambda\nu}=D^{1/2}_{\lambda\mu}(\varphi,\theta,\gamma)(\sigma^a)_{\mu\rho}D^{*1/2}_{\rho\nu}(\varphi,\theta,\gamma)\,,
\end{aligned}
$$

where as in Eq. (A8) for the isospin generator $\boldsymbol{T}$, we introduce the spin generators $\sigma_1(x)$, $\sigma_2(x)$, $\sigma_3(x)$ in the comoving frames, later when we consider the spherical case, we always write σ_m simply, by omitting the argument x.

Then the $J=0,1$ components $\mathbb{D}^J_{\mu,\nu}(\varphi,\theta,\gamma)$ of the wave function χ satisfy

$$
\begin{aligned}
&J^2\mathbb{D}^J_{\mu,\nu}=J(J+1)\mathbb{D}^J_{\mu,\nu}\,,\quad J=0,1\,,\\
&J_3\mathbb{D}^J_{\mu,\nu}=M\mathbb{D}^J_{\mu,\nu}\,,\qquad M=-(2J+1),\ldots,2J+1\,,\\
&\sigma_r\mathbb{D}^J_{\mu,\nu}=2\mu\mathbb{D}^J_{\mu,\nu}\,,\qquad \mu,\nu=\pm\frac{1}{2}\,,\\
&T_r\mathbb{D}^J_{\mu,\nu}=\nu\mathbb{D}^J_{\mu,\nu}\,,\quad \kappa\mathbb{D}^J_{\mu,\nu}=\kappa_\mu\mathbb{D}^J_{-\mu,\nu}\,,\\
&\varepsilon_{ijk}\hat{r}_i\sigma_j\boldsymbol{T}_k\mathbb{D}^J_{\mu,\nu}=-\mathrm{i}2\mu\alpha^{1/2}_{\nu+\frac{1}{2}+\mu}\mathbb{D}^J_{-\mu,\nu+2\mu}\,,
\end{aligned}
$$

where

$$
\begin{aligned}
&\alpha^I_\nu=(I+\nu)^{1/2}(I-\nu+1)^{1/2}\,,\\
&\kappa_\nu=\alpha^J_{|\nu|+1/2}=\sqrt{\Big(J+\frac{1}{2}\Big)^2-\nu^2}\,. \qquad \text{(A27)}
\end{aligned}
$$

So the radial components $f^{\pm J}(r)$ satisfy

$$
\begin{aligned}
&\Big(\frac{\partial}{\partial r}+\frac{1}{r}\Big)f^{\pm J}_{\mu,\nu}(r)-\frac{\kappa_\mu}{r}f^{\pm J}_{-\mu,\nu}(r)\\
&+\alpha^I_{\nu+\frac{1}{2}+\mu}\frac{F(r)}{r}f^{\pm J}_{-\mu,\nu+2\mu}(r)\mp 2\mu\nu G(r)f^{\pm J}_{\mu,\nu}(r)\\
&=0\,. \qquad \text{(A28)}
\end{aligned}
$$

The convergence at $r\to\infty$, requires $\mp 2\mu\nu G(r)>0$ asymptotically. Thus for the $\gamma^5=1$ (-1) solutions $f^{+J}_{\mu,\nu}$ $(f^{-J}_{\mu,\nu})$, $\mu\nu$ always >0 (<0). But then the $(\kappa_\nu/r)f^{\pm J}_{-\mu,\nu}$ term disobeys this condition, so we should require $\kappa_\mu=0$. Thus J is restricted to be 0. The superscript J will be dropped later. This implies the well-known fact that for the zero mode, the total spin, which is contributed by the space spin and the isospin induced by the field, cancels. Based on the same reason in Sec. 3, we simply use the $J=0$ component in the tensor product of S and T. From $F(r)f^{\pm}_{-\mu,\nu+2\mu}$ term, $-\mu(\nu+2\mu)$ should be >0 (<0) for $f^+(f^-)$. Hence $\nu=-\mu$ and $f^+=0$, only f^- is convergent, it satisfies

$$
\begin{aligned}
&\Big(\frac{\partial}{\partial r}+\frac{1}{r}\Big)f^-_{\mu,-\mu}(r)-\frac{F(r)}{r}f^-_{-\mu,\mu}(r)-\frac{1}{2}G(r)f^-_{\mu,-\mu}(r)\\
&=0\,,
\end{aligned}
$$

i.e.

$$
\begin{aligned}
&\Big(\frac{\partial}{\partial r}+\frac{1}{r}\Big)f^-_{\frac{1}{2},-\frac{1}{2}}(r)-\frac{F(r)}{r}f^-_{-\frac{1}{2},\frac{1}{2}}(r)-\frac{1}{2}G(r)f^-_{\frac{1}{2},-\frac{1}{2}}(r)\\
&=0\,,\\
&\Big(\frac{\partial}{\partial r}+\frac{1}{r}\Big)f^-_{-\frac{1}{2},\frac{1}{2}}(r)-\frac{F(r)}{r}f^-_{\frac{1}{2},-\frac{1}{2}}(r)-\frac{1}{2}G(r)f^-_{-\frac{1}{2},\frac{1}{2}}(r)\\
&=0\,. \qquad \text{(A29)}
\end{aligned}
$$

The unique convergent solution is

$$
\begin{aligned}
f^-(r)&=f^-_{1/2,-1/2}(r)=-f^-_{-1/2,1/2}(r)\\
&\simeq r^{-1/2}\Big(\mathrm{sh}\frac{\beta r}{2}\Big)^{1/2}\Big(\mathrm{ch}\frac{\beta r}{2}\Big)^{-3/2}\,.
\end{aligned}
$$

which satisfies

$$
\Big(\frac{\partial}{\partial r}+\frac{1}{r}\Big)f^-(r)+\frac{F(r)}{r}f^-(r)-\frac{1}{2}G(r)f^-(r)=0\,. \qquad \text{(A30)}
$$

In fact, we have a local (comoving) homotopy isomorphism between the spin s_μ and isospin i_ν in the reduced cup product $\underline{2}\underset{ST}{\dot{\otimes}}\underline{2}\xRightarrow{\text{cup product}}\underline{1}$, in the tensor product $s_\mu\otimes i_\nu$.

$$
\underline{2}\underset{ST}{\otimes}\underline{2}=\underline{1}\oplus\underline{3}\,, \qquad \text{(A31)}
$$

the nondegenerate **zero mode** lies in $\underline{1}$ only. So we simply have

$$
\chi^-=\sum_{\mu\nu}f^-_{\mu\nu}s_\mu i_\nu(-1)^{\mu-1/2}\delta_{\mu,-\nu} \qquad \text{(A32)}
$$

by using

$$
\begin{aligned}
&\frac{1}{4\pi}\sum_{\mu',\nu'}C^{S,T,J}_{\mu',\nu',M}D^S_{\mu'\mu}(\varphi,\theta,\gamma)D^T_{\nu'\nu}(\varphi,\theta,\gamma)\\
&=\frac{1}{\sqrt{2\pi}}D^J_{M,\mu+\nu}(\varphi,\theta,\gamma)C^{S,T,J}_{\mu,\nu,0}\\
&=(-1)^{\mu-1/2}\delta_{\mu,-\nu}\,, \qquad \text{(A33)}\\
&S=T=\frac{1}{2}\,,\quad J=0\,.
\end{aligned}
$$

Here, the finite rotation matrices $D^S_{\mu'\mu}$ and $D^T_{\nu'\nu}$ from the fixed frames to the comoving frames are coupled to a singlet expressed by the well-known Kroneker matrix $(-1)^{\mu-1/2}\delta_{\mu,-\nu}$.

The fixed frame Dirac equation (A21) turns to be Eq. (A28) in the **tensor product comoving frame**, the last two terms come from Eq. (A28)

$$
\begin{aligned}
&\gamma_i K_i=\gamma_i K^a_i T^a=\gamma_i\epsilon_{ija}\frac{F(r)}{r}\hat{r}_j T_a\,,\\
&\gamma_0 K_0=\gamma_0\delta^{ab}K^a_0 T^b=\gamma_0\delta^{ab}iG(r)\hat{r}^a T^b\,.
\end{aligned}
$$

And further turn to be in the **cup product form**,

$$\chi^- \text{ part: } \Big(\mathrm{i}\epsilon_{3jk}\sigma_j \mathbf{T}_k \frac{F(r)}{r}\Big)\chi^- = \mathrm{i}(\sigma_1 \boldsymbol{T}_2 - \sigma_2 \boldsymbol{T}_1)\frac{F(r)}{r}\chi^- \xrightarrow{\text{cup product}} \frac{\mathrm{i}}{2}(\boldsymbol{E}_1^2 - \boldsymbol{E}_2^1)\frac{F(r)}{r}|f^-\rangle\,,$$

$$G(r)\hat{r}^a T^a \chi^- \xrightarrow{\text{cup product}} \frac{1}{2}G(r)(\boldsymbol{E}_1^1 - \boldsymbol{E}_2^2)|f^-\rangle\,,$$

where the cup product basis matrix $(\boldsymbol{E}_i^j)_{\alpha\beta} = \delta_{i\alpha}\delta_{j\beta}$. Notice that the radial gradient term has a σ_r, which turns to $(1/2)(\boldsymbol{E}_1^1 - \boldsymbol{E}_2^2)$ also, so we have at last the zero mode equation in cup product form (A29)

$$\Big(\Big(\frac{\partial}{\partial r} + \frac{1}{r}\Big)I - \frac{F(r)}{2r}\begin{pmatrix} 0 & 1 \\ 1 & 0 \end{pmatrix} - \frac{1}{2}G(r)I\Big)|f^-\rangle = 0\,,$$

$$|f^-\rangle = \frac{1}{2}\begin{pmatrix} 1 & -1 \\ -1 & 1 \end{pmatrix}\begin{pmatrix} f^-_{\frac{1}{2},-\frac{1}{2}}(r) \\ f^-_{-\frac{1}{2},\frac{1}{2}}(r) \end{pmatrix} = \frac{1}{2}\begin{pmatrix} f^-_{\frac{1}{2},-\frac{1}{2}}(r) - f^-_{-\frac{1}{2},\frac{1}{2}}(r) \\ -f^-_{\frac{1}{2},-\frac{1}{2}}(r) + f^-_{-\frac{1}{2},\frac{1}{2}}(r) \end{pmatrix}. \tag{A34}$$

Here we have factorized out the spherical dependence of the wave function in the reduced cup product frame

$$s_-^- i_+ \quad - \quad s_+^- i_-\,,$$

(the superscript $-$ of s denotes the chirality).

By Eq. (A10) we have

$$\frac{\mathrm{i}}{2}(\boldsymbol{E}_1^2 - \boldsymbol{E}_2^1)\frac{F(r)}{r} = \frac{1}{2}(\boldsymbol{E}_+^- - \boldsymbol{E}_-^+)\frac{F(r)}{r}$$
$$\Rightarrow \frac{1}{2}(\boldsymbol{K}_+^- - \boldsymbol{K}_-^+) = \boldsymbol{K}_+^-\,.$$

so at last equation (A34) turns to be

$$\Big[\Big(\frac{\partial}{\partial r} + \frac{1}{r}\Big) + \boldsymbol{K}_{T+} - \frac{1}{2}\boldsymbol{K}_t\Big]|f^-\rangle = 0\,,$$

i.e. Eq. (A30).

Remark Following Nahm[22] by Fourier transform to the momentum space, but different from Nahm, we adopt the light-cone spherical frame coordinates, then the $|f^-\rangle$ becomes the holomorphic sheaf, i.e., the mini-twistor.[26] Here, the geodesic flow along the spectral line, left real null line, $\nabla_r - (1/2)\boldsymbol{K}_t \sim \nabla_U - \mathrm{i}\Phi$,[20] and the $\boldsymbol{K}_{T+} \sim \nabla_x + \mathrm{i}\nabla_y = \bar{\partial}$,[20] spans the left null plane (α plane), as the natural flat connection on it.

Appendix C: Generalized Gauss Codazzi Equation

(i) *Condition of Reducibility*

The necessary and sufficient condition for the gauge field with group G defined on space-time manifold M to be reducible into gauge field with subgroup H, may be formulated as follows: A connection on the principal bundle $P(M,G)$ can be reduced into the connection of subbundle $Q(M,H)$, when and only when the associated bundle $E(M,G/H,G)$ has a section $\boldsymbol{n}(x) : M \to G/H$, which is invariant under parallel displacement. In order to employ this condition, we must find out proper expression for G/H, consequently we decompose the left invariant algebra g of group G canonically into $g = \mathfrak{h} + m$, where $\mathfrak{h}$ is the subalgebra corresponding to the stationary subgroup of the element on G/H. Observing the natural correspondence between G/H and the subspace spanned by $\mathfrak{h}$ in the left invariant Lie algebra, we may perform the reduction as follows.

(ii) Gauge field with group $G = \mathrm{SU}(2)$, which may be abelianized into $H = U(1)$. In this case, $\mathfrak{h}$ is one-dimensional everywhere, and its normalized base is taken as $\hat{\boldsymbol{n}}$, $\hat{\boldsymbol{n}}\cdot\hat{\boldsymbol{n}} \equiv -2\,\mathrm{tr}(\hat{\boldsymbol{n}}\cdot\hat{\boldsymbol{n}}) = 1$. Then the set of $\hat{\boldsymbol{n}}$ makes a unit sphere $S^2 \sim \mathrm{SU}(2)/\mathrm{U}(1)$ in the space of adjoint representation. The section $\hat{\boldsymbol{n}}(x)$ is the mapping of space-time M (except the singular point) onto S^2. This unit isospin field is invariant under the parallel displacement by gauge potential $\boldsymbol{A}_\mu(x)$,

$$\tilde{\nabla}\hat{\boldsymbol{n}}(x) \equiv \partial_\mu\hat{\boldsymbol{n}}(x) + e[\boldsymbol{A}_\mu(x), \hat{\boldsymbol{n}}(x)] = 0\,,$$
$$\boldsymbol{A}_\mu(x) \in g\,, \qquad \mu = 0,1,2,3\,, \tag{A35}$$

where under infinitesimal gauge transformation

$$\boldsymbol{A}'_\mu(x) = \boldsymbol{A}_\mu(x) + e[\boldsymbol{A}_\mu(x), \alpha(x)] + e\partial_\mu\alpha(x)\,,$$
$$\alpha(x) \in g\,. \tag{A36}$$

Using identity $\boldsymbol{V} = (\boldsymbol{V}\cdot\boldsymbol{N})\boldsymbol{N} + [\boldsymbol{N}, [\boldsymbol{V}, \boldsymbol{N}]]$, we can easily see that the necessary and sufficient condition of (A35) is

$$\boldsymbol{A}_\mu = (\boldsymbol{A}_\mu\cdot\hat{\boldsymbol{n}})\hat{\boldsymbol{n}} - \frac{1}{e}[\hat{\boldsymbol{n}}, \partial_\mu\hat{\boldsymbol{n}}] \tag{A37}$$

Here, the potential is SU(2) formally, but in reality it may be transformed at least locally into U(1) potential with a constant $\hat{\boldsymbol{n}}(x)$ as the generator, i.e. the gauge could be chosen to turn $\hat{\boldsymbol{n}}(x)$ into the same direction in some region of x, $\partial_\mu\hat{\boldsymbol{n}}(x) = 0$. Then $\boldsymbol{A}'_\mu(x)$ becomes explicit Abelian, i.e., it equals $(\boldsymbol{A}'\cdot\boldsymbol{n})\boldsymbol{n}$ in this region. If we fix the direction of $\hat{\boldsymbol{n}}(x)$, but rotate a gauge angle $\Gamma(x)$ around $\hat{\boldsymbol{n}}(x)$, then we obtain the U(1) transform generated by $e\hat{\boldsymbol{n}} : \boldsymbol{A}'_\mu = \boldsymbol{A}_\mu + e\hat{\boldsymbol{n}}\partial_\mu\Gamma$.

Substituting Eq. (A37) into

$$\boldsymbol{F}_{\mu\nu} = \partial_\mu\boldsymbol{A}_\nu - \partial_\nu\boldsymbol{A}_\mu + e[\boldsymbol{A}_\mu, \boldsymbol{A}_\nu]\,, \tag{A38}$$

we get

$$\boldsymbol{F}_{\mu\nu} = [\partial_\mu(\boldsymbol{A}_\nu\cdot\hat{\boldsymbol{n}}) - \partial_\nu(\boldsymbol{A}_\mu\cdot\hat{\boldsymbol{n}})]\hat{\boldsymbol{n}} - \frac{1}{e}[\partial_\mu\hat{\boldsymbol{n}}, \partial_\nu\hat{\boldsymbol{n}}]$$
$$= (\boldsymbol{F}_{\mu\nu}\cdot\hat{\boldsymbol{n}})\hat{\boldsymbol{n}}\,. \tag{A39}$$

In the region where $\hat{\boldsymbol{n}}(x)$ is well-defined, $\boldsymbol{F}_{\mu\nu}\cdot\hat{\boldsymbol{n}}$ satisfies

$$\partial^\mu({}^*\boldsymbol{F}_{\mu\nu}\cdot\hat{\boldsymbol{n}}) = \tilde{\nabla}^\mu({}^*\boldsymbol{F}_{\mu\nu}\cdot\hat{\boldsymbol{n}})$$
$$= (\tilde{\nabla}^{\mu*}\boldsymbol{F}_{\mu\nu})\cdot\hat{\boldsymbol{n}} + {}^*\boldsymbol{F}_{\mu\nu}\tilde{\nabla}^\mu\hat{\boldsymbol{n}}$$
$$= 0\,, \tag{A40}$$

where $*\boldsymbol{F}_{\mu\nu} \equiv \frac{1}{2}\epsilon_{\mu\nu\lambda\rho}\boldsymbol{F}^{\lambda\rho}$. Locally $\boldsymbol{F}_{\mu\nu}\cdot\hat{\boldsymbol{n}}$ is the same as the ordinary electromagnetic field without magnetic charge, and in explicit Abelian gauge it may be expressed by the U(1) potential $\boldsymbol{A}_\mu\cdot\hat{\boldsymbol{n}}$; $\boldsymbol{F}_{\mu\nu}\cdot\hat{\boldsymbol{n}}=\partial_\mu(\boldsymbol{A}_\nu\cdot\hat{\boldsymbol{n}})-\partial_\nu(\boldsymbol{A}_\mu\cdot\hat{\boldsymbol{n}})$. But, globally its magnetic flux through some two-dimensional space like close surface M' may be non-zero,

$$\frac{1}{2}\iint_{M'}\boldsymbol{F}_{\mu\nu}\cdot\hat{\boldsymbol{n}}\mathrm{d}x^\mu\wedge\mathrm{d}x^\nu$$
$$=\frac{l}{e}\iint_{S^2}\hat{\boldsymbol{n}}\cdot[\mathrm{d}\hat{\boldsymbol{n}},\delta\hat{\boldsymbol{n}}]=-\frac{4\pi l}{e}. \qquad (A41)$$

Here integer l is the times by which the surface M' covers the isospin sphere S^2 through the mapping $\hat{\boldsymbol{n}}(x)$. Physically it is the quantum number of the magnetic charge surrounded by the surface M'. If $l\neq 0$, it is impossible to turn $\hat{\boldsymbol{n}}(x)$ into one and the same direction globally by non-singular single-valued gauge transformation. Then there must be either singularity or overlapping regions with transition function, the corresponding Abelian potential being the Dirac–Schwinger potential with string or the Wu–Yang global potential. Above all, the characteristic $\pi_1(S^1)$ of bundle $Q(M,H)$ of U(1) gauge field corresponds one to one to $\pi_2(S^2)$ of the section $\boldsymbol{n}(x)$ on the associated coset bundle of SU(2), $\pi_1(S^1)\sim\pi_2(S^2)$. Their common characteristic number is determined physically by the dual charge. Mathematically $\hat{\boldsymbol{n}}(x)$ is the generator of the holonomy group of $P(M,G)$ under given connection; physically $e\hat{\boldsymbol{n}}(x)$ is the charge operator; abelianizable $\boldsymbol{A}_\mu$ is the potential of pure electromagnetic field $\boldsymbol{F}_{\mu\nu}$; and in the meantime, $\hat{\boldsymbol{n}}$ is the common isodirection of the six space-time components of $\boldsymbol{F}_{\mu\nu}$.

(iii) Non-abelianizable SU(2) potential $\boldsymbol{A}_\mu(x)$ and field strength $\boldsymbol{F}_{\mu\nu}(x)$. Now, the holonomy group is the whole SU(2), thus it is impossible to choose from its generators some $\hat{\boldsymbol{n}}(x)$ which remains invariant under parallel displacement. The noninvariant section $\hat{\boldsymbol{n}}(x)$ must be given otherwise. Physically, as the charge operator, $\hat{\boldsymbol{n}}(x)$ is the phase axis of wave functions of charged particles, or the isodirection of its current vector, or $\hat{\boldsymbol{n}}(x)=\phi(x)/|\phi(x)|$, where $\phi(x)$ is the Higgs particle. In pure gauge field without other particles, $\hat{\boldsymbol{n}}(x)$ may be the privileged direction determined by the intrinsic symmetry of the field, e.g. the generator of the stationary subgroup H for G invariant connection. (In the case of synchronous space spin and isospin spherical symmetry field as the BPS monopole, the privileged direction is synchronous with the vector radius $\hat{\boldsymbol{r}}$).

Once a section $\hat{\boldsymbol{n}}(x)$ is given, it determines a corresponding subbundle $Q(M,H)$, whose characteristic class is decided by the homotopic property of $\hat{\boldsymbol{n}}(x)$. Physically, as soon as the charge operator $\hat{\boldsymbol{n}}(x)$ is given, one can separate the U(1) Dirac component $\boldsymbol{H}_\mu(x)$ from the SU(2) potential $\boldsymbol{A}_\mu(x)$ as follows: Here $\boldsymbol{H}_\mu(x)$ is the U(1) gauge potential with $\hat{\boldsymbol{n}}(x)$ as the generator. Now from $\tilde{\nabla}_\mu\hat{\boldsymbol{n}}\equiv\partial_\mu\boldsymbol{n}+e[\boldsymbol{A}_\mu,\hat{\boldsymbol{n}}]$ which is not vanishing now, we get

$$\boldsymbol{A}_\mu=(\boldsymbol{A}_\mu\cdot\hat{\boldsymbol{n}})\hat{\boldsymbol{n}}-\frac{1}{e}[\hat{\boldsymbol{n}},\partial_\mu\hat{\boldsymbol{n}}]+\frac{1}{e}[\hat{\boldsymbol{n}},\nabla_\mu\hat{\boldsymbol{n}}]$$
$$\equiv\boldsymbol{H}_\mu+\boldsymbol{K}_\mu. \qquad (A42)$$

Here we have set $\boldsymbol{H}_\mu\equiv(\boldsymbol{A}_\mu\cdot\hat{\boldsymbol{n}})\hat{\boldsymbol{n}}-(1/e)[\hat{\boldsymbol{n}},\partial_\mu\hat{\boldsymbol{n}}]$. It is easy to prove that $\boldsymbol{H}_\mu$ satisfies Eqs. (A35) $\sim$ (A37), hence it is the U(1) part in $\boldsymbol{A}_\mu$, the remainder $(1/e)[\hat{\boldsymbol{n}},\nabla_\mu\hat{\boldsymbol{n}}]\equiv\boldsymbol{K}_\mu$ is gauge-covariant and represents the charged vector particles. Geometrically $e\boldsymbol{K}_\mu$ corresponds to the second fundamental form, e.g., $\nabla_\mu\hat{\boldsymbol{n}}=[e\boldsymbol{K}_\mu,\hat{\boldsymbol{n}}]$ is the generalized Weingarten formula (since $|\hat{\boldsymbol{n}}|=1$, the "normal" component is absent). Substituting Eq. (A42) into $\boldsymbol{F}_{\mu\nu}=\partial_\mu\boldsymbol{A}_\nu-\partial_\nu\boldsymbol{A}_\mu+e[\boldsymbol{A}_\mu,\boldsymbol{A}_\nu]$ and making comparison with Eq. (A39), we get

$$\boldsymbol{F}_{\mu\nu}\cdot\hat{\boldsymbol{n}}\hat{\boldsymbol{n}}-[\boldsymbol{K}_\mu,\boldsymbol{K}_\nu]$$
$$=F_{\mu\nu}\cdot\hat{\boldsymbol{n}}\hat{\boldsymbol{n}}-[\nabla_\mu\hat{\boldsymbol{n}},\nabla_\nu\hat{\boldsymbol{n}}]$$
$$=\partial_\mu(\boldsymbol{H}_\nu\cdot\hat{\boldsymbol{n}})\hat{\boldsymbol{n}}-\partial_\nu(\boldsymbol{H}_\mu\cdot\hat{\boldsymbol{n}})\hat{\boldsymbol{n}}-\frac{1}{e}[\partial_\mu\hat{\boldsymbol{n}},\partial_\nu\hat{\boldsymbol{n}}]$$
$$=\boldsymbol{F}_{\mu\nu}^{(H)}. \qquad (A43)$$

This is just the 't Hooft expression. Here $\boldsymbol{F}_{\mu\nu}^{(H)}$ is the U(1) field part in $\boldsymbol{F}_{\mu\nu}$ contributed by $\boldsymbol{H}_\mu$ (The Higgs particle does not contribute this U(1) field, only its isodirection coincides with that of charge operator). Geometrically equation (A43) is the generalized Gauss equation. $\boldsymbol{F}_{\mu\nu}^{(H)}\cdot\hat{\boldsymbol{n}}$, as the U(1) "subcurvature" of the total curvature $\boldsymbol{F}_{\mu\nu}$, satisfies the Bianchi identity on subbundle. At the same time we get the generalized Codazzi equation,

$$[\hat{\boldsymbol{n}},[\boldsymbol{F}_{\mu\nu},\hat{\boldsymbol{n}}]]=\tilde{\nabla}_\mu\boldsymbol{K}_\nu-\tilde{\nabla}_\nu\boldsymbol{K}_\mu. \qquad (A44)$$

Acknowledgments

We would like to thank X.C. Song, K.J. Shi, K. Wu, S. Hu, Y.X. Chen, Y.Z. Zhang, R.H. Yue, L. Zhao, X.M. Ding, C.H. Xiong, X.H. Wang, and S.M. Ke for their helpful discussions and would like to thank J.B. Wu for helpful discussions and for the help in writing the preliminary version of this paper; thanks Simonsyang and Binlijunwang for their helpful discussions on the version 1 and version 2 of this paper.

References

[1] N. Seiberg and E. Witten, Nucl. Phys. B **426** (1994) 19, arXiv:hep-th/9407087.

[2] A. Kapustin and E. Witten, arXiv:hep-th/0604151.

[3] O. Babelon and D. Bernard, Int. J. Mod. Phys. A **8** (1993) 507.

[4] D.I. Olive, N. Turok, and J.W.R. Underwood, Nucl. Phys. B **401** (1993) 663.

[5] I. Bena, J. Polchinski, and R. Roiban, Phys. Rev. D **69** (2004) 046002.

[6] B.Y. Hou, D.T. Peng, C.H. Xiong, and R.H. Yue, arXiv:hep-th/0406239.

[7] A.M. Polyakov, Mod. Phys. Lett. A **19** (2004) 1649.

[8] V.G. Drinfeld, Leningrad Math. J. **2** (1991) 829.

[9] B.Y. Hou and W.L. Yang, Commun. Theor. Phys. (Beijing, China) **31** (1999) 265.

[10] B.Y. Hou and W.L. Yang, J. Phys. A: Math. Gen. **30** (1997) 6131.

[11] S. Khoroshkin, D. Lebedev, and S. Pakuliak, arXiv:q-alg/9702002.

[12] S. Khoroshkin, D. Lebedev, S. Pakuliak, A. Stolin, and V. Tolstoy, arXiv:q-alg/9703043.

[13] B.Y. Hou and W.L. Yang, J. Phys. A: Math. Gen. **31** (1998) 5349.

[14] M. Jimbo, H. Konno, S. Odake, and J. Shiraishi, arXiv: q-alg/9712029.

[15] E. Frenkel, arXiv:hep-th/0512172.

[16] B. Feigin, E. Frenkel, and N. Reshetikhin, Commun. Math. Phys. **166** (1994) 27, arXiv:hep-th/9402022.

[17] B. Feigin and E. Frenkel, Commun. Math. Phys. **178** (1996) 653, arXiv:q-alg/9508009.

[18] C.N. Yang, Phys. Rev. Lett. **38** (1977) 1377.

[19] V. Kac, *Infinite Dimensional Lie Algebras*, Cambridge University Press, Cambridge (1990).

[20] N.J. Hitchin, Commun. Math. Phys. **83** (1982) 579.

[21] H. Nakajima and K. Yoshioka, arXiv:math.AG/0311058.

[22] S.L. Lukyanov, Phys. Lett. B **325** (1994) 409.

[23] W. Nahm, *The Construction of all Self-dual Multi-monopoles by the ADHM Method, in Monopoles in Quantum Field Theory*, eds. N.S. Craigie, P. Goddard, and W. Nahm, World Scientific, Singapore (1982).

[24] N.J. Hitchin, Commun. Math. Phys. **89** (1983) 145.

[25] C. Beasley and E. Witten, J. Diff. Geom. **70** (2005) 183, arXiv:hep-th/0503126.

[26] S. Gukov and E. Witten, arXiv:hep-th/0612073, p. 584.

[27] H. Nakajima and K. Yoshioka, arXiv:math.AG/0306198.

[28] M. Aganagic, R. Dijkgraaf, A. Klemm, M. Marino, and C. Vafa, Commun. Math. Phys. **261** (2006) 451.

[29] M. Aganagic, A. Neitzke, and C. Vafa, arXiv:hep-th/0504054.

PUBLISHED FOR SISSA BY SPRINGER

RECEIVED: *August 26, 2010*
REVISED: *November 4, 2010*
ACCEPTED: *December 10, 2010*
PUBLISHED: *January 3, 2011*

Determinant representations of scalar products for the open XXZ chain with non-diagonal boundary terms

Wen-Li Yang,[a] Xi Chen,[a] Jun Feng,[a] Kun Hao,[a] Bo-Yu Hou,[a] Kang-Jie Shi[a] and Yao-Zhong Zhang[b]

[a] *Institute of Modern Physics, Northwest University, Xian 710069, P.R. China*
[b] *The University of Queensland, School of Mathematics and Physics, Brisbane, QLD 4072, Australia*
E-mail: wlyang@nwu.edu.cn, chenxi0905@yahoo.cn, grammophon@163.com, hoke72@163.com, byhou@nwu.edu.cn, kjshi@nwu.edu.cn, yzz@maths.uq.edu.au

ABSTRACT: The determinant representation of the scalar products of the Bethe states of the open XXZ spin chain with non-diagonal boundary terms is studied. Using the vertex-face correspondence, we transfer the problem into the corresponding trigonometric solid-on-solid (SOS) model with diagonal boundary terms. With the help of the Drinfeld twist or factorizing F-matrix, we obtain the determinant representation of the scalar products of the Bethe states of the associated SOS model. By taking the on shell limit, we obtain the determinant representations (or Gaudin formula) of the norms of the Bethe states.

KEYWORDS: Bethe Ansatz, Lattice Integrable Models

ARXIV EPRINT: 1011.4719

doi:10.1007/JHEP01(2011)006

Contents

1 Introduction

The computation of correlation functions (or scalar products of Bethe states) is one of major challenging problems in the theory of quantum integrable models [1, 2]. There are two approaches in the literature for computing the correlation functions of a quantum integrable model. One is the vertex operator method (see e.g. [3–9]) which works only on an infinite lattice, and another one is based on the detailed analysis of the structure of the Bethe states [10, 11]. As for the second approach which usually works for models with finite size, it is well known that in the framework of quantum inverse scattering method (QISM) [2] Bethe states are obtained by applying pseudo-particle creation operators to reference state (pseudo-vacuum). However, the apparently simple action of creation operators is plagued with non-local effects arising from polarization clouds or compensating exchange terms on the level of local operators. This makes the direct calculation of correlation functions of models with finite size challenging.

Progress has recently been made on the second approach with the help of the Drinfeld twists or factorizing F-matrices [12]. Working in the F-basis provided by the F-matrices, the authors in [13, 14] managed to calculate the form factors and correlation functions of the XXX and XXZ chains with periodic boundary condition (or closed chains) analytically and expressed them in determinant forms. Then the determinant representation of the scalar products and correlation functions of the supersymmetric t-J model [15] (and references therein) and its q-deformed model [16] with periodic boundary condition was obtained within the corresponding F-basis given in [17, 18].

It was noticed [19, 20] that the F-matrices of the closed XXX and XXZ chains also make the pseudo-particle creation operators of the open XXX and XXZ chains with diagonal boundary terms polarization free. This is mainly due to the fact that the closed chain and the corresponding open chain with diagonal boundary terms share the same reference state [21]. However, the story for the open XXZ chain with non-diagonal boundary terms is quite different [22–44]. Firstly, the reference state (all spin up state) of the closed chain is no longer a reference state of the open chain with non-diagonal boundary terms [24–27, 35]. Secondly, at least two reference states (and thus two sets of Bethe states) are needed [45, 46] for the open XXZ chain with non-diagonal boundary terms in order to obtain its complete spectrum [47–50]. As a consequence, the F-matrix found in [13] is no longer the *desirable* F-matrix for the open XXZ chain with non-diagonal boundary terms.

The F-matrices for the face model was first introduced in [51]. Basing the work we have succeeded in obtaining the factorizing F-matrices for the open XXZ chain with integrable boundary conditions given by the non-diagonal K-matrices (2.9) and (2.11) [52]. In this paper, we shall investigate the determinant representations of the scalar products of the Bethe states of the open XXZ chain with non-diagonal boundary terms with the help of the associated F-matrices.

The paper is organized as follows. In section 2, we briefly describe the open XXZ chain with non-diagonal boundary terms, and introduce the pseudo-particle creation operators and the two sets of Bethe states of the model. In section 3, we introduce the face picture of the model and express the scalar products in terms of the operators in the face picture. In section 4, we present the F-matrix of the open XXZ chain in the face picture and give the completely symmetric and polarization free representations of the pseudo-particle creation/annihilation operators in the F-basis. In the face picture, with the help of the F-basis, we obtain the determinant representations of the scalar products of Bethe states in section 5. In section 6, we summarize our results and give some discussions. Some detailed technical proof is given in appendix A.

2 The inhomogeneous spin-$\frac{1}{2}$ XXZ open chain

Throughout, V denotes a two-dimensional linear space. The spin-$\frac{1}{2}$ XXZ chain can be constructed from the well-known six-vertex model R-matrix $\overline{R}(u) \in \mathrm{End}(V\otimes V)$ [2] given by

$$\overline{R}(u) = \begin{pmatrix} 1 & & & \\ & b(u) & c(u) & \\ & c(u) & b(u) & \\ & & & 1 \end{pmatrix}. \tag{2.1}$$

The coefficient functions read: $b(u) = \frac{\sin u}{\sin(u+\eta)}$, $c(u) = \frac{\sin \eta}{\sin(u+\eta)}$. Here we assume η is a generic complex number. The R-matrix satisfies the quantum Yang-Baxter equation (QYBE),

$$R_{1,2}(u_1 - u_2)R_{1,3}(u_1 - u_3)R_{2,3}(u_2 - u_3) = R_{2,3}(u_2 - u_3)R_{1,3}(u_1 - u_3)R_{1,2}(u_1 - u_2), \quad (2.2)$$

and the unitarity, crossing-unitarity and quasi-classical properties [35]. We adopt the standard notations: for any matrix $A \in \mathrm{End}(V)$, A_j (or A^j) is an embedding operator in the tensor space $V \otimes V \otimes \cdots$, which acts as A on the j-th space and as identity on the other factor spaces; $R_{i,j}(u)$ is an embedding operator of R-matrix in the tensor space, which acts as identity on the factor spaces except for the i-th and j-th ones.

One introduces the "row-to-row" (or one-row) monodromy matrix $T(u)$, which is an 2×2 matrix with elements being operators acting on $V^{\otimes N}$, where $N = 2M$ (M being a positive integer),

$$T_0(u) = \overline{R}_{0,N}(u - z_N)\overline{R}_{0,N-1}(u - z_{N-1}) \cdots \overline{R}_{0,1}(u - z_1). \quad (2.3)$$

Here $\{z_j | j = 1, \ldots, N\}$ are arbitrary free complex parameters which are usually called inhomogeneous parameters.

Integrable open chain can be constructed as follows [21]. Let us introduce a pair of K-matrices $K^-(u)$ and $K^+(u)$. The former satisfies the reflection equation (RE)

$$\begin{aligned} &\overline{R}_{1,2}(u_1 - u_2)K_1^-(u_1)\overline{R}_{2,1}(u_1 + u_2)K_2^-(u_2) \\ &\quad = K_2^-(u_2)\overline{R}_{1,2}(u_1 + u_2)K_1^-(u_1)\overline{R}_{2,1}(u_1 - u_2), \end{aligned} \quad (2.4)$$

and the latter satisfies the dual RE

$$\begin{aligned} &\overline{R}_{1,2}(u_2 - u_1)K_1^+(u_1)\overline{R}_{2,1}(-u_1 - u_2 - 2\eta)K_2^+(u_2) \\ &\quad = K_2^+(u_2)\overline{R}_{1,2}(-u_1 - u_2 - 2\eta)K_1^+(u_1)\overline{R}_{2,1}(u_2 - u_1). \end{aligned} \quad (2.5)$$

For open spin-chains, instead of the standard "row-to-row" monodromy matrix $T(u)$ (2.3), one needs to consider the "double-row" monodromy matrix $\mathbb{T}(u)$

$$\mathbb{T}(u) = T(u)K^-(u)\hat{T}(u), \quad \hat{T}(u) = T^{-1}(-u). \quad (2.6)$$

Then the double-row transfer matrix of the XXZ chain with open boundary (or the open XXZ chain) is given by

$$\tau(u) = tr(K^+(u)\mathbb{T}(u)). \quad (2.7)$$

The QYBE and (dual) REs lead to that the transfer matrices with different spectral parameters commute with each other [21]: $[\tau(u), \tau(v)] = 0$. This ensures the integrability of the open XXZ chain.

In this paper, we will consider the K-matrix $K^-(u)$ which is a generic solution to the RE (2.4) associated the six-vertex model R-matrix [53, 54]

$$K^-(u) = \begin{pmatrix} k_1^1(u) & k_2^1(u) \\ k_1^2(u) & k_2^2(u) \end{pmatrix} \equiv K(u). \quad (2.8)$$

The coefficient functions are

$$k_1^1(u) = \frac{\cos(\lambda_1 - \lambda_2) - \cos(\lambda_1 + \lambda_2 + 2\xi)e^{-2iu}}{2\sin(\lambda_1 + \xi + u)\sin(\lambda_2 + \xi + u)},$$
$$k_2^1(u) = \frac{-i\sin(2u)e^{-i(\lambda_1+\lambda_2)}e^{-iu}}{2\sin(\lambda_1 + \xi + u)\sin(\lambda_2 + \xi + u)},$$
$$k_1^2(u) = \frac{i\sin(2u)e^{i(\lambda_1+\lambda_2)}e^{-iu}}{2\sin(\lambda_1 + \xi + u)\sin(\lambda_2 + \xi + u)},$$
$$k_2^2(u) = \frac{\cos(\lambda_1 - \lambda_2)e^{-2iu} - \cos(\lambda_1 + \lambda_2 + 2\xi)}{2\sin(\lambda_1 + \xi + u)\sin(\lambda_2 + \xi + u)}. \tag{2.9}$$

At the same time, we introduce the corresponding *dual* K-matrix $K^+(u)$ which is a generic solution to the dual reflection equation (2.5) with a particular choice of the free boundary parameters:

$$K^+(u) = \begin{pmatrix} k^{+1}_{\ 1}(u) & k^{+1}_{\ 2}(u) \\ k^{+2}_{\ 1}(u) & k^{+2}_{\ 2}(u) \end{pmatrix} \tag{2.10}$$

with the matrix elements

$$k^{+1}_{\ 1}(u) = \frac{\cos(\lambda_1 - \lambda_2)e^{-i\eta} - \cos(\lambda_1 + \lambda_2 + 2\bar{\xi})e^{2iu+i\eta}}{2\sin(\lambda_1 + \bar{\xi} - u - \eta)\sin(\lambda_2 + \bar{\xi} - u - \eta)},$$
$$k^{+1}_{\ 2}(u) = \frac{i\sin(2u + 2\eta)e^{-i(\lambda_1+\lambda_2)}e^{iu-i\eta}}{2\sin(\lambda_1 + \bar{\xi} - u - \eta)\sin(\lambda_2 + \bar{\xi} - u - \eta)},$$
$$k^{+2}_{\ 1}(u) = \frac{-i\sin(2u + 2\eta)e^{i(\lambda_1+\lambda_2)}e^{iu+i\eta}}{2\sin(\lambda_1 + \bar{\xi} - u - \eta)\sin(\lambda_2 + \bar{\xi} - u - \eta)},$$
$$k^{+2}_{\ 2}(u) = \frac{\cos(\lambda_1 - \lambda_2)e^{2iu+i\eta} - \cos(\lambda_1 + \lambda_2 + 2\bar{\xi})e^{-i\eta}}{2\sin(\lambda_1 + \bar{\xi} - u - \eta)\sin(\lambda_2 + \bar{\xi} - u - \eta)}. \tag{2.11}$$

The K-matrices depend on four free boundary parameters $\{\lambda_1, \lambda_2, \xi, \bar{\xi}\}$. It is very convenient to introduce a vector $\lambda \in V$ associated with the boundary parameters $\{\lambda_i\}$,

$$\lambda = \sum_{k=1}^{2} \lambda_k \epsilon_k, \tag{2.12}$$

where $\{\epsilon_i, i = 1, 2\}$ form the orthonormal basis of V such that $\langle \epsilon_i, \epsilon_j \rangle = \delta_{ij}$.

2.1 Vertex-face correspondence

Let us briefly review the face-type R-matrix associated with the six-vertex model.

Set

$$\hat{\imath} = \epsilon_i - \bar{\epsilon}, \quad \bar{\epsilon} = \frac{1}{2}\sum_{k=1}^{2} \epsilon_k, \quad i = 1, 2, \qquad \text{then} \sum_{i=1}^{2} \hat{\imath} = 0. \tag{2.13}$$

Let $\mathfrak{h}$ be the Cartan subalgebra of A_1 and $\mathfrak{h}^*$ be its dual. A finite dimensional diagonalizable $\mathfrak{h}$-module is a complex finite dimensional vector space W with a weight decomposition

$W = \oplus_{\mu\in\mathfrak{h}^*} W[\mu]$, so that $\mathfrak{h}$ acts on $W[\mu]$ by $x\,v = \mu(x)\,v$, $(x \in \mathfrak{h},\ v \in W[\mu])$. For example, the non-zero weight spaces of the fundamental representation $V_{\Lambda_1} = \mathbb{C}^2 = V$ are

$$W[\hat{\imath}] = \mathbb{C}\epsilon_i,\ \ i = 1, 2. \tag{2.14}$$

For a generic $m \in V$, define

$$m_i = \langle m, \epsilon_i\rangle,\quad m_{ij} = m_i - m_j = \langle m, \epsilon_i - \epsilon_j\rangle,\quad i, j = 1, 2. \tag{2.15}$$

Let $R(u, m) \in \mathrm{End}(V \otimes V)$ be the R-matrix of the six-vertex SOS model, which is trigonometric limit of the eight-vertex SOS model [55] given by

$$R(u;m)=\sum_{i=1}^{2} R(u;m)_{ii}^{ii}E_{ii}\otimes E_{ii}+\sum_{i\neq j}\left\{R(u;m)_{ij}^{ij}E_{ii}\otimes E_{jj}+R(u;m)_{ij}^{ji}E_{ji}\otimes E_{ij}\right\}, \tag{2.16}$$

where E_{ij} is the matrix with elements $(E_{ij})_k^l = \delta_{jk}\delta_{il}$. The coefficient functions are

$$R(u;m)_{ii}^{ii} = 1, \qquad R(u;m)_{ij}^{ij} = \frac{\sin u \sin(m_{ij} - \eta)}{\sin(u + \eta)\sin(m_{ij})},\quad i \neq j, \tag{2.17}$$

$$R(u;m)_{ij}^{ji} = \frac{\sin\eta \sin(u + m_{ij})}{\sin(u + \eta)\sin(m_{ij})},\quad i \neq j, \tag{2.18}$$

and m_{ij} is defined in (2.15). The R-matrix satisfies the dynamical (modified) quantum Yang-Baxter equation (or the star-triangle relation) [55]

$$\begin{aligned} &R_{1,2}(u_1 - u_2; m - \eta h^{(3)})R_{1,3}(u_1 - u_3; m)R_{2,3}(u_2 - u_3; m - \eta h^{(1)}) \\ &\quad = R_{2,3}(u_2 - u_3; m)R_{1,3}(u_1 - u_3; m - \eta h^{(2)})R_{1,2}(u_1 - u_2; m). \end{aligned} \tag{2.19}$$

Here we have adopted

$$R_{1,2}(u, m - \eta h^{(3)})\, v_1 \otimes v_2 \otimes v_3 = (R(u, m - \eta\mu) \otimes \mathrm{id})\, v_1 \otimes v_2 \otimes v_3,\quad \text{if } v_3 \in W[\mu]. \tag{2.20}$$

Moreover, one may check that the R-matrix satisfies weight conservation condition,

$$\left[h^{(1)} + h^{(2)},\, R_{1,2}(u; m)\right] = 0, \tag{2.21}$$

unitary condition,

$$R_{1,2}(u; m)\, R_{2,1}(-u; m) = \mathrm{id} \otimes \mathrm{id}, \tag{2.22}$$

and crossing relation

$$R(u;m)_{ij}^{kl} = \varepsilon_l\, \varepsilon_j \frac{\sin(u)\sin((m - \eta\hat{\imath})_{21})}{\sin(u + \eta)\sin(m_{21})} R(-u - \eta; m - \eta\hat{\imath})_{\bar{l}\,i}^{\bar{j}\,k}, \tag{2.23}$$

where

$$\varepsilon_1 = 1,\ \varepsilon_2 = -1,\quad \text{and } \bar{1} = 2,\ \bar{2} = 1. \tag{2.24}$$

Define the following functions: $\theta^{(1)}(u) = e^{-iu}$, $\theta^{(2)}(u) = 1$. Let us introduce two intertwiners which are 2-component column vectors $\phi_{m,m-\eta\hat{\jmath}}(u)$ labelled by $\hat{1}$, $\hat{2}$. The k-th element of $\phi_{m,m-\eta\hat{\jmath}}(u)$ is given by

$$\phi^{(k)}_{m,m-\eta\hat{\jmath}}(u) = \theta^{(k)}(u + 2m_j). \tag{2.25}$$

Explicitly,

$$\phi_{m,m-\eta\hat{1}}(u) = \begin{pmatrix} e^{-i(u+2m_1)} \\ 1 \end{pmatrix}, \qquad \phi_{m,m-\eta\hat{2}}(u) = \begin{pmatrix} e^{-i(u+2m_2)} \\ 1 \end{pmatrix}. \tag{2.26}$$

Obviously, the two intertwiner vectors $\phi_{m,m-\eta\hat{\imath}}(u)$ are linearly *independent* for a generic $m \in V$.

Using the intertwiner vectors, one can derive the following face-vertex correspondence relation [24]

$$\begin{aligned}&\overline{R}_{1,2}(u_1 - u_2)\phi^1_{m,m-\eta\hat{\imath}}(u_1)\phi^2_{m-\eta\hat{\imath},m-\eta(\hat{\imath}+\hat{\jmath})}(u_2)\\ &\quad= \sum_{k,l} R(u_1 - u_2; m)^{kl}_{ij}\phi^1_{m-\eta\hat{l},m-\eta(\hat{l}+\hat{k})}(u_1)\phi^2_{m,m-\eta\hat{l}}(u_2).\end{aligned} \tag{2.27}$$

Then the QYBE (2.2) of the vertex-type R-matrix $\overline{R}(u)$ is equivalent to the dynamical Yang-Baxter equation (2.19) of the SOS R-matrix $R(u, m)$. For a generic m, we can introduce other types of intertwiners $\bar{\phi}$, $\tilde{\phi}$ which are both row vectors and satisfy the following conditions,

$$\bar{\phi}_{m,m-\eta\hat{\mu}}(u)\,\phi_{m,m-\eta\hat{\nu}}(u) = \delta_{\mu\nu}, \quad \tilde{\phi}_{m+\eta\hat{\mu},m}(u)\,\phi_{m+\eta\hat{\nu},m}(u) = \delta_{\mu\nu}, \tag{2.28}$$

from which one can derive the relations,

$$\sum_{\mu=1}^{2} \phi_{m,m-\eta\hat{\mu}}(u)\,\bar{\phi}_{m,m-\eta\hat{\mu}}(u) = \mathrm{id}, \tag{2.29}$$

$$\sum_{\mu=1}^{2} \phi_{m+\eta\hat{\mu},m}(u)\,\tilde{\phi}_{m+\eta\hat{\mu},m}(u) = \mathrm{id}. \tag{2.30}$$

One may verify that the K-matrices $K^{\pm}(u)$ given by (2.8) and (2.10) can be expressed in terms of the intertwiners and *diagonal* matrices $\mathcal{K}(\lambda|u)$ and $\tilde{\mathcal{K}}(\lambda|u)$ as follows

$$K^-(u)^s_t = \sum_{i,j} \phi^{(s)}_{\lambda-\eta(\hat{\imath}-\hat{\jmath}),\ \lambda-\eta\hat{\imath}}(u)\mathcal{K}(\lambda|u)^j_i\bar{\phi}^{(t)}_{\lambda,\ \lambda-\eta\hat{\imath}}(-u), \tag{2.31}$$

$$K^+(u)^s_t = \sum_{i,j} \phi^{(s)}_{\lambda,\ \lambda-\eta\hat{\jmath}}(-u)\tilde{\mathcal{K}}(\lambda|u)^j_i\tilde{\phi}^{(t)}_{\lambda-\eta(\hat{\jmath}-\hat{\imath}),\ \lambda-\eta\hat{\jmath}}(u). \tag{2.32}$$

Here the two *diagonal* matrices $\mathcal{K}(\lambda|u)$ and $\tilde{\mathcal{K}}(\lambda|u)$ are given by

$$\mathcal{K}(\lambda|u) \equiv \mathrm{Diag}(k(\lambda|u)_1,\, k(\lambda|u)_2) = \mathrm{Diag}\left(\frac{\sin(\lambda_1 + \xi - u)}{\sin(\lambda_1 + \xi + u)}, \frac{\sin(\lambda_2 + \xi - u)}{\sin(\lambda_2 + \xi + u)}\right), \tag{2.33}$$

$$\tilde{\mathcal{K}}(\lambda|u) \equiv \mathrm{Diag}(\tilde{k}(\lambda|u)_1,\, \tilde{k}(\lambda|u)_2)$$
$$= \mathrm{Diag}\left(\frac{\sin(\lambda_{12}-\eta)\sin(\lambda_1+\bar{\xi}+u+\eta)}{\sin\lambda_{12}\sin(\lambda_1+\bar{\xi}-u-\eta)}, \frac{\sin(\lambda_{12}+\eta)\sin(\lambda_2+\bar{\xi}+u+\eta)}{\sin\lambda_{12}\sin(\lambda_2+\bar{\xi}-u-\eta)}\right). \quad (2.34)$$

Although the vertex type K-matrices $K^{\pm}(u)$ given by (2.8) and (2.10) are generally non-diagonal, after the face-vertex transformations (2.31) and (2.32), the face type counterparts $\mathcal{K}(\lambda|u)$ and $\tilde{\mathcal{K}}(\lambda|u)$ become *simultaneously* diagonal. This fact enabled the authors to apply the generalized algebraic Bethe ansatz method developed in [25–27] for SOS type integrable models to diagonalize the transfer matrices $\tau(u)$ (2.7) [35, 45, 46].

2.2 Two sets of eigenstates

In order to construct the Bethe states of the open XXZ model with non-diagonal boundary terms specified by the K-matrices (2.9) and (2.11), we need to introduce the new double-row monodromy matrices $\mathcal{T}^{\pm}(m|u)$ [52]:

$$\mathcal{T}^{-}(m|u)^{\nu}_{\mu} = \tilde{\phi}^{0}_{m-\eta(\hat{\mu}-\hat{\nu}),m-\eta\hat{\mu}}(u)\ \mathbb{T}_0(u)\phi^{0}_{m,m-\eta\hat{\mu}}(-u), \quad (2.35)$$
$$\mathcal{T}^{+}(m|u)^{j}_{i} = \prod_{k\neq j}\frac{\sin(m_{jk})}{\sin(m_{jk}-\eta)}\,\phi^{t_0}_{m-\eta(\hat{j}-\hat{\imath}),m-\eta\hat{j}}(u)\left(\mathbb{T}^{+}(u)\right)^{t_0}\bar{\phi}^{t_0}_{m,m-\eta\hat{j}}(-u), \quad (2.36)$$

where t_0 denotes transposition in the 0-th space (i.e. auxiliary space) and $\mathbb{T}^{+}(u)$ is given by

$$\left(\mathbb{T}^{+}(u)\right)^{t_0} = T^{t_0}(u)\left(K^{+}(u)\right)^{t_0}\hat{T}^{t_0}(u). \quad (2.37)$$

These double-row monodromy matrices, in the face picture, can be expressed in terms of the face type R-matrix $R(u;m)$ (2.16) and K-matrices $\mathcal{K}(\lambda|u)$ (2.33) and $\tilde{\mathcal{K}}(\lambda|u)$ (2.34) (for the details see appendix A).

So far only two sets of Bethe states (i.e. eigenstates) of the transfer matrix for the models with non-diagonal boundary terms have been found [45, 46]. These two sets of states are [52]

$$|\{v^{(1)}_i\}\rangle^{(I)} = \mathcal{T}^{+}(\lambda+2\eta\hat{1}|v^{(1)}_1)^{1}_{2}\cdots\mathcal{T}^{+}(\lambda+2M\eta\hat{1}|v^{(1)}_M)^{1}_{2}|\Omega^{(I)}(\lambda)\rangle, \quad (2.38)$$
$$|\{v^{(2)}_i\}\rangle^{(II)} = \mathcal{T}^{-}(\lambda-2\eta\hat{2}|v^{(2)}_1)^{2}_{1}\cdots\mathcal{T}^{-}(\lambda-2M\eta\hat{2}|v^{(2)}_M)^{2}_{1}|\Omega^{(II)}(\lambda)\rangle, \quad (2.39)$$

where the vector λ is related to the boundary parameters (2.12). The associated reference states $|\Omega^{(I)}(\lambda)\rangle$ and $|\Omega^{(II)}(\lambda)\rangle$ are

$$|\Omega^{(I)}(\lambda)\rangle = \phi^{1}_{\lambda+N\eta\hat{1},\lambda+(N-1)\eta\hat{1}}(z_1)\phi^{2}_{\lambda+(N-1)\eta\hat{1},\lambda+(N-2)\eta\hat{1}}(z_2)\cdots\phi^{N}_{\lambda+\eta\hat{1},\lambda}(z_N), \quad (2.40)$$
$$|\Omega^{(II)}(\lambda)\rangle = \phi^{1}_{\lambda,\lambda-\eta\hat{2}}(z_1)\phi^{2}_{\lambda-\eta\hat{2},\lambda-2\eta\hat{2}}(z_2)\cdots\phi^{N}_{\lambda-(N-1)\eta\hat{2},\lambda-N\eta\hat{2}}(z_N). \quad (2.41)$$

It is remarked that $\phi^k = \mathrm{id}\otimes\mathrm{id}\cdots\otimes\overset{k-th}{\phi}\otimes\mathrm{id}\cdots$.

If the parameters $\{v^{(1)}_k\}$ satisfy the first set of Bethe ansatz equations given by

$$\frac{\sin(\lambda_2+\xi+v^{(1)}_\alpha)\sin(\lambda_2+\bar{\xi}-v^{(1)}_\alpha)\sin(\lambda_1+\bar{\xi}+v^{(1)}_\alpha)\sin(\lambda_1+\xi-v^{(1)}_\alpha)}{\sin(\lambda_2+\bar{\xi}+v^{(1)}_\alpha+\eta)\sin(\lambda_2+\xi-v^{(1)}_\alpha-\eta)\sin(\lambda_1+\xi+v^{(1)}_\alpha+\eta)\sin(\lambda_1+\bar{\xi}-v^{(1)}_\alpha-\eta)}$$

$$= \prod_{k\neq\alpha}^{M} \frac{\sin(v_\alpha^{(1)}+v_k^{(1)}+2\eta)\sin(v_\alpha^{(1)}-v_k^{(1)}+\eta)}{\sin(v_\alpha^{(1)}+v_k^{(1)})\sin(v_\alpha^{(1)}-v_k^{(1)}-\eta)} \times \prod_{k=1}^{2M} \frac{\sin(v_\alpha^{(1)}+z_k)\sin(v_\alpha^{(1)}-z_k)}{\sin(v_\alpha^{(1)}+z_k+\eta)\sin(v_\alpha^{(1)}-z_k+\eta)}, \quad \alpha=1,\ldots,M, \tag{2.42}$$

the Bethe state $|v_1^{(I)},\ldots,v_M^{(1)}\rangle^{(1)}$ becomes the eigenstate of the transfer matrix with eigenvalue $\Lambda^{(1)}(u)$ given by [52]

$$\begin{aligned}\Lambda^{(1)}(u) =& \frac{\sin(\lambda_2+\bar{\xi}-u)\sin(\lambda_1+\bar{\xi}+u)\sin(\lambda_1+\xi-u)\sin(2u+2\eta)}{\sin(\lambda_2+\bar{\xi}-u-\eta)\sin(\lambda_1+\bar{\xi}-u-\eta)\sin(\lambda_1+\xi+u)\sin(2u+\eta)} \\ &\times \prod_{k=1}^{M} \frac{\sin(u+v_k^{(1)})\sin(u-v_k^{(1)}-\eta)}{\sin(u+v_k^{(1)}+\eta)\sin(u-v_k^{(1)})} \\ &+\frac{\sin(\lambda_2+\bar{\xi}+u+\eta)\sin(\lambda_1+\xi+u+\eta)\sin(\lambda_2+\xi-u-\eta)\sin 2u}{\sin(\lambda_2+\bar{\xi}-u-\eta)\sin(\lambda_1+\xi+u)\sin(\lambda_2+\xi+u)\sin(2u+\eta)} \\ &\times \prod_{k=1}^{M} \frac{\sin(u+v_k^{(1)}+2\eta)\sin(u-v_k^{(1)}+\eta)}{\sin(u+v_k^{(1)}+\eta)\sin(u-v_k^{(1)})} \\ &\times \prod_{k=1}^{2M} \frac{\sin(u+z_k)\sin(u-z_k)}{\sin(u+z_k+\eta)\sin(u-z_k+\eta)}.\end{aligned} \tag{2.43}$$

If the parameters $\{v_k^{(2)}\}$ satisfy the second Bethe Ansatz equations

$$\begin{aligned}&\frac{\sin(\lambda_1+\xi+v_\alpha^{(2)})\sin(\lambda_1+\bar{\xi}-v_\alpha^{(2)})\sin(\lambda_2+\bar{\xi}+v_\alpha^{(2)})\sin(\lambda_2+\xi-v_\alpha^{(2)})}{\sin(\lambda_1+\bar{\xi}+v_\alpha^{(2)}+\eta)\sin(\lambda_1+\xi-v_\alpha^{(2)}-\eta)\sin(\lambda_2+\xi+v_\alpha^{(2)}+\eta)\sin(\lambda_2+\bar{\xi}-v_\alpha^{(2)}-\eta)} \\ &= \prod_{k\neq\alpha}^{M} \frac{\sin(v_\alpha^{(2)}+v_k^{(2)}+2\eta)\sin(v_\alpha^{(2)}-v_k^{(2)}+\eta)}{\sin(v_\alpha^{(2)}+v_k^{(2)})\sin(v_\alpha^{(2)}-v_k^{(2)}-\eta)} \\ &\times \prod_{k=1}^{2M} \frac{\sin(v_\alpha^{(2)}+z_k)\sin(v_\alpha^{(2)}-z_k)}{\sin(v_\alpha^{(2)}+z_k+\eta)\sin(v_\alpha^{(2)}-z_k+\eta)}, \quad \alpha=1,\ldots,M,\end{aligned} \tag{2.44}$$

the Bethe states $|v_1^{(2)},\ldots,v_M^{(2)}\rangle^{(II)}$ yield the second set of the eigenstates of the transfer matrix with the eigenvalues [45, 46],

$$\begin{aligned}\Lambda^{(2)}(u) =& \frac{\sin(2u+2\eta)\sin(\lambda_1+\bar{\xi}-u)\sin(\lambda_2+\bar{\xi}+u)\sin(\lambda_2+\xi-u)}{\sin(2u+\eta)\sin(\lambda_1+\bar{\xi}-u-\eta)\sin(\lambda_2+\bar{\xi}-u-\eta)\sin(\lambda_2+\xi+u)} \\ &\times \prod_{k=1}^{M} \frac{\sin(u+v_k^{(2)})\sin(u-v_k^{(2)}-\eta)}{\sin(u+v_k^{(2)}+\eta)\sin(u-v_k^{(2)})} \\ &+\frac{\sin(2u)\sin(\lambda_1+\bar{\xi}+u+\eta)\sin(\lambda_2+\xi+u+\eta)\sin(\lambda_1+\xi-u-\eta)}{\sin(2u+\eta)\sin(\lambda_1+\bar{\xi}-u-\eta)\sin(\lambda_2+\xi+u)\sin(\lambda_1+\xi+u)}\end{aligned}$$

$$\times \prod_{k=1}^{M} \frac{\sin(u+v_k^{(2)}+2\eta)\sin(u-v_k^{(2)}+\eta)}{\sin(u+v_k^{(2)}+\eta)\sin(u-v_k^{(2)})}$$
$$\times \prod_{k=1}^{2M} \frac{\sin(u+z_k)\sin(u-z_k)}{\sin(u+z_k+\eta)\sin(u-z_k+\eta)}. \tag{2.45}$$

3 Scalar products

It was shown that in order to compute correlation functions of the closed XXZ chain [2] and the open XXZ chain with diagonal boundary terms [19, 20], one suffices to calculate the scalar products of an on-shell Bethe state and a general state (an off-shell Bethe state). The aim of this paper is to give the explicit expressions of the following scalar products of the open XXZ chain with non-diagonal boundary terms:

$$S^{I,II}(\{u_\alpha\};\{v_i^{(2)}\}) = {}^{(I)}\langle\{u_\alpha\}|\{v_i^{(2)}\}\rangle^{(II)}, \quad S^{II,I}(\{u_\alpha\};\{v_i^{(1)}\}) = {}^{(II)}\langle\{u_\alpha\}|\{v_i^{(1)}\}\rangle^{(I)}, \tag{3.1}$$
$$S^{I,I}(\{u_\alpha\};\{v_i^{(1)}\}) = {}^{(I)}\langle\{u_\alpha\}|\{v_i^{(1)}\}\rangle^{(I)}, \quad S^{II,II}(\{u_\alpha\};\{v_i^{(2)}\}) = {}^{(II)}\langle\{u_\alpha\}|\{v_i^{(2)}\}\rangle^{(II)}, \tag{3.2}$$

where the dual states ${}^{(I)}\langle\{u_\alpha\}|$ and ${}^{(II)}\langle\{u_\alpha\}|$ are given by

$${}^{(I)}\langle\{u_\alpha\}| = \langle\Omega^{(I)}(\lambda)|\mathcal{T}^-(\lambda-2(M-1)\eta\hat{1}|u_M)^2_{\bar{1}}\ldots\mathcal{T}^-(\lambda|u_1)^2_{\bar{1}}, \tag{3.3}$$
$${}^{(II)}\langle\{u_\alpha\}| = \langle\Omega^{(II)}(\lambda)|\mathcal{T}^+(\lambda+2(M-1)\eta\hat{2}|u_M)^1_{\bar{2}}\ldots\mathcal{T}^+(\lambda|u_1)^1_{\bar{2}}, \tag{3.4}$$

and $\langle\Omega^{(I)}(\lambda)|$, $\langle\Omega^{(II)}(\lambda)|$ are

$$\langle\Omega^{(I)}(\lambda)| = \tilde{\phi}^1_{\lambda,\lambda-\eta\hat{1}}(z_1)\ldots\tilde{\phi}^N_{\lambda-(2M-1)\eta\hat{1},\lambda-2M\eta\hat{1}}(z_N), \tag{3.5}$$
$$\langle\Omega^{(II)}(\lambda)| = \tilde{\phi}^1_{\lambda+2M\eta\hat{2},\lambda+(2M-1)\eta\hat{2}}(z_1)\ldots\tilde{\phi}^N_{\lambda+\eta\hat{1},\lambda}(z_N). \tag{3.6}$$

Some remarks are in order. The parameters $\{u_\alpha\}$ in (3.1)–(3.2) are free parameters, namely, they do not need to satisfy the Bethe ansatz equations. In the face picture (see the subsection 3.1 below), the states corresponding to (3.3) and (3.4) are clearly the dual eigenstates for the associated face model provided that the parameters satisfy the associated Bethe ansatz equations. However in the spin chain picture they, in contrast with those of the open XXZ chain with diagonal boundary terms [20], are dual eigenstates for a different model with different boundary terms.[1]

The K-matrices $K^{\pm}(u)$ given by (2.8) and (2.10) are generally non-diagonal (in the vertex picture), after the face-vertex transformations (2.31) and (2.32), the face type counterparts $\mathcal{K}(\lambda|u)$ and $\tilde{\mathcal{K}}(\lambda|u)$ given by (2.33) and (2.34) *simultaneously* become diagonal. This fact suggests that it would be much simpler if one performs all calculations in the face picture.

3.1 Face picture

Let us introduce the face type one-row monodromy matrix (c.f (2.3))

$$T_F(l|u) \equiv T^F_{0,1\ldots N}(l|u)$$

[1]We thank the anonymous referee for his/her pointing out this issue.

$$= R_{0,N}(u - z_N; l - \eta \sum_{i=1}^{N-1} h^{(i)}) \ldots R_{0,2}(u - z_2; l - \eta h^{(1)}) R_{0,1}(u - z_1; l),$$

$$= \begin{pmatrix} T_F(l|u)_1^1 & T_F(l|u)_2^1 \\ T_F(l|u)_1^2 & T_F(l|u)_2^2 \end{pmatrix} \tag{3.7}$$

where l is a generic vector in V. The monodromy matrix satisfies the face type quadratic exchange relation [56–58]. Applying $T_F(l|u)_j^i$ to an arbitrary vector $|i_1, \ldots, i_N\rangle$ in the N-tensor product space $V^{\otimes N}$ given by

$$|i_1, \ldots, i_N\rangle = \epsilon_{i_1}^1 \ldots \epsilon_{i_N}^N, \tag{3.8}$$

we have

$$\begin{aligned} T_F(l|u)_j^i |i_1, \ldots, i_N\rangle &\equiv T_F(m; l|u)_j^i |i_1, \ldots, i_N\rangle \\ &= \sum_{\alpha_{N-1} \ldots \alpha_1} \sum_{i'_N \ldots i'_1} R(u - z_N; l - \eta \sum_{k=1}^{N-1} \hat{i'_k})_{\alpha_{N-1}\, i_N}^{i \quad\ i'_N} \cdots \\ &\quad \times R(u - z_2; l - \eta \hat{i'_1})_{\alpha_1\, i_2}^{\alpha_2\, i'_2} R(u - z_1; l)_{j\ \, i_1}^{\alpha_1\, i'_1} |i'_1, \ldots, i'_N\rangle, \end{aligned} \tag{3.9}$$

where $m = l - \eta \sum_{k=1}^{N} \hat{i}_k$. With the help of the crossing relation (2.23), the face-vertex correspondence relation (2.27) and the relations (2.28), following the method developed in [25–27, 52], we find that the scalar products (3.1)–(3.2) can be expressed in terms of the face-type double-row monodromy operators as follows:

$$\begin{aligned} S^{I,II}(\{u_\alpha\}; \{v_i^{(2)}\}) &= \langle 1, \ldots, 1 | \mathcal{T}_F^-(\lambda - 2(M-1)\eta\hat{1}, \lambda | u_M)_1^2 \ldots \mathcal{T}_F^-(\lambda, \lambda | u_1)_1^2 \\ &\quad \times \mathcal{T}_F^-(\lambda + 2\eta\hat{1}, \lambda | v_1^{(2)})_1^2 \ldots \mathcal{T}_F^-(\lambda + 2M\eta\hat{1}, \lambda | v_M^{(2)})_1^2 | 2, \ldots, 2\rangle, \quad (3.10) \\ S^{II,I}(\{u_\alpha\}; \{v_i^{(1)}\}) &= \langle 2, \ldots, 2 | \mathcal{T}_F^+(\lambda, \lambda + 2(M-1)\eta\hat{2} | u_M)_2^1 \ldots \mathcal{T}_F^+(\lambda, \lambda | u_1)_2^1 \\ &\quad \times \mathcal{T}_F^+(\lambda, \lambda - 2\eta\hat{2} | v_1^{(1)})_2^1 \ldots \mathcal{T}_F^+(\lambda, \lambda - 2M\eta\hat{2} | v_M^{(1)})_2^1 | 1, \ldots, 1\rangle, \quad (3.11) \\ S^{I,I}(\{u_\alpha\}; \{v_i^{(1)}\}) &= \langle 1, \ldots, 1 | \mathcal{T}_F^-(\lambda - 2(M-1)\eta\hat{1}, \lambda | u_M)_1^2 \ldots \mathcal{T}_F^-(\lambda, \lambda | u_1)_1^2 \\ &\quad \times \mathcal{T}_F^+(\lambda, \lambda + 2\eta\hat{1} | v_1^{(1)})_2^1 \ldots \mathcal{T}_F^+(\lambda, \lambda + 2M\eta\hat{1} | v_M^{(1)})_2^1 | 1, \ldots, 1\rangle, \quad (3.12) \\ S^{II,II}(\{u_\alpha\}; \{v_i^{(2)}\}) &= \langle 2, \ldots, 2 | \mathcal{T}_F^+(\lambda, \lambda + 2(M-1)\eta\hat{2} | u_M)_2^1 \ldots \mathcal{T}_F^+(\lambda, \lambda | u_1)_2^1 \\ &\quad \times \mathcal{T}_F^-(\lambda - 2\eta\hat{2}, \lambda | v_1^{(2)})_1^2 \ldots \mathcal{T}_F^-(\lambda - 2M\eta\hat{2}, \lambda | v_M^{(2)})_1^2 | 2, \ldots, 2\rangle. \quad (3.13) \end{aligned}$$

The above double-row monodromy matrix operators $\mathcal{T}_F^-(m, \lambda|u)_1^2$ and $\mathcal{T}_F^+(\lambda, m|u)_2^1$ are given in terms of the one-row monodromy matrix operator $T_F(m; l|u)_j^i$ [52]

$$\begin{aligned} \mathcal{T}_F^-(m, \lambda|u)_1^2 &= \frac{\sin(m_{21})}{\sin(\lambda_{21})} \prod_{k=1}^{N} \frac{\sin(u + z_k)}{\sin(u + z_k + \eta)} \\ &\times \left\{ \frac{\sin(\lambda_1 + \xi - u)}{\sin(\lambda_1 + \xi + u)} T_F(m, \lambda|u)_1^2 T_F(m + \eta\hat{2}, \lambda + \eta\hat{2}| - u - \eta)_2^2 \right. \\ &\left. - \frac{\sin(\lambda_2 + \xi - u)}{\sin(\lambda_2 + \xi + u)} T_F(m + 2\eta\hat{2}, \lambda|u)_2^2 T_F(m + \eta\hat{1}, \lambda + \eta\hat{1}| - u - \eta)_1^2 \right\}, \end{aligned} \tag{3.14}$$

$$\mathcal{T}_F^{+}(\lambda,m|u)_2^1=\prod_{k=1}^{N}\frac{\sin(u+z_k)}{\sin(u+z_k+\eta)}$$
$$\times\left\{\frac{\sin(\lambda_{12}-\eta)\sin(\lambda_1+\bar{\xi}+u+\eta)}{\sin(m_{12}-\eta)\sin(\lambda_1+\bar{\xi}-u-\eta)}T_F(\lambda+2\eta\hat{2},m+2\eta\hat{2}|u)_2^1T_F(\lambda+\eta\hat{2},m+\eta\hat{2}|-u-\eta)_2^2\right.$$
$$\left.-\frac{\sin(\lambda_{21}-\eta)\sin(\lambda_2+\bar{\xi}+u+\eta)}{\sin(m_{21}+\eta)\sin(\lambda_2+\bar{\xi}-u-\eta)}T_F(\lambda,m+2\eta\hat{2}|u)_2^2T_F(\lambda+\eta\hat{2},m+\eta\hat{2}|-u-\eta)_2^1\right\}.\quad(3.15)$$

In the next section we shall construct the Drinfeld twist (or factorizing F-matrix) in the face picture for the open XXZ chain with non-diagonal boundary terms. In the resulting F-basis, the two sets of pseudo-particle creation/annihilation operators $\mathcal{T}_F^{\pm}$ given by (3.14) and (3.15) take completely symmetric and polarization free forms simultaneously. These polarization free forms allow us to construct the explicit expressions of the scalar products (3.10)–(3.13).

4 F-basis

In this section, we construct the Drinfeld twist [12] (factorizing F-matrix) on the N-fold tensor product space $V^{\otimes N}$ (i.e. the quantum space of the open XXZ chain) and the associated representations of the pseudo-particle creation/annihilation operators in this basis.

4.1 Factorizing Drinfeld twist F

Let $\mathcal{S}_N$ be the permutation group over indices $1,\ldots,N$ and $\{\sigma_i|i=1,\ldots,N-1\}$ be the set of elementary permutations in $\mathcal{S}_N$. For each elementary permutation σ_i, we introduce the associated operator $R^{\sigma_i}_{1\ldots N}$ on the quantum space

$$R^{\sigma_i}_{1\ldots N}(l)\equiv R^{\sigma_i}(l)=R_{i,i+1}\left(z_i-z_{i+1}|l-\eta\sum_{k=1}^{i-1}h^{(k)}\right),\quad(4.1)$$

where l is a generic vector in V. For any σ, $\sigma'\in\mathcal{S}_N$, operator $R^{\sigma\sigma'}_{1\ldots N}$ associated with $\sigma\sigma'$ satisfies the following composition law [52](and references therein):

$$R^{\sigma\sigma'}_{1\ldots N}(l)=R^{\sigma'}_{\sigma(1\ldots N)}(l)\,R^{\sigma}_{1\ldots N}(l).\quad(4.2)$$

Let σ be decomposed in a minimal way in terms of elementary permutations,

$$\sigma=\sigma_{\beta_1}\ldots\sigma_{\beta_p},\quad(4.3)$$

where $\beta_i=1,\ldots,N-1$ and the positive integer p is the length of σ. The composition law (4.2) enables one to obtain operator $R^{\sigma}_{1\ldots N}$ associated with each $\sigma\in\mathcal{S}_N$. The dynamical quantum Yang-Baxter equation (2.19), weight conservation condition (2.21) and unitary condition (2.22) guarantee the uniqueness of $R^{\sigma}_{1\ldots N}$. Moreover, one may check that $R^{\sigma}_{1\ldots N}$ satisfies the following exchange relation with the face type one-row monodromy matrix (3.7)

$$R^{\sigma}_{1\ldots N}(l)T^F_{0,1\ldots N}(l|u)=T^F_{0,\sigma(1\ldots N)}(l|u)R^{\sigma}_{1\ldots N}(l-\eta h^{(0)}),\qquad\forall\sigma\in\mathcal{S}_N.\quad(4.4)$$

Now, we construct the face-type Drinfeld twist $F_{1\ldots N}(l) \equiv F_{1\ldots N}(l; z_1, \ldots, z_N)^2$ on the N-fold tensor product space $V^{\otimes N}$, which satisfies the following three properties:

$$\text{I. lower} - \text{triangularity}; \tag{4.5}$$

$$\text{II. non} - \text{degeneracy}; \tag{4.6}$$

$$\text{III. factorizing property}: \ R^{\sigma}_{1\ldots N}(l) = F^{-1}_{\sigma(1\ldots N)}(l) F_{1\ldots N}(l), \ \forall \sigma \in \mathcal{S}_N. \tag{4.7}$$

Substituting (4.7) into the exchange relation (4.4) yields the following relation

$$F^{-1}_{\sigma(1\ldots N)}(l) F_{1\ldots N}(l) T^F_{0,1\ldots N}(l|u) = T^F_{0,\sigma(1\ldots N)}(l|u) F^{-1}_{\sigma(1\ldots N)}(l - \eta h^{(0)}) F_{1\ldots N}(l - \eta h^{(0)}). \tag{4.8}$$

Equivalently,

$$F_{1\ldots N}(l) T^F_{0,1\ldots N}(l|u) F^{-1}_{1\ldots N}(l - \eta h^{(0)}) = F_{\sigma(1\ldots N)}(l) T^F_{0,\sigma(1\ldots N)}(l|u) F^{-1}_{\sigma(1\ldots N)}(l - \eta h^{(0)}). \tag{4.9}$$

Let us introduce the twisted monodromy matrix $\tilde{T}^F_{0,1\ldots N}(l|u)$ by

$$\begin{aligned} \tilde{T}^F_{0,1\ldots N}(l|u) &= F_{1\ldots N}(l) T^F_{0,1\ldots N}(l|u) F^{-1}_{1\ldots N}(l - \eta h^{(0)}) \\ &= \begin{pmatrix} \tilde{T}_F(l|u)^1_1 & \tilde{T}_F(l|u)^1_2 \\ \tilde{T}_F(l|u)^2_1 & \tilde{T}_F(l|u)^2_2 \end{pmatrix}. \end{aligned} \tag{4.10}$$

Then (4.9) implies that the twisted monodromy matrix is symmetric under $\mathcal{S}_N$, namely,

$$\tilde{T}^F_{0,1\ldots N}(l|u) = \tilde{T}^F_{0,\sigma(1\ldots N)}(l|u), \quad \forall \sigma \in \mathcal{S}_N. \tag{4.11}$$

Define the F-matrix:

$$F_{1\ldots N}(l) = \sum_{\sigma \in \mathcal{S}_N} \sum_{\{\alpha_j\}=1}^{2}{}^{*} \prod_{j=1}^{N} P^{\sigma(j)}_{\alpha_{\sigma(j)}} R^{\sigma}_{1\ldots N}(l), \tag{4.12}$$

where P^i_α is the embedding of the project operator P_α in the ith space with matric elements $(P_\alpha)_{kl} = \delta_{kl}\delta_{k\alpha}$. The sum $\sum^*$ in (4.12) is over all non-decreasing sequences of the labels $\alpha_{\sigma(i)}$:

$$\begin{aligned} \alpha_{\sigma(i+1)} &\geq \alpha_{\sigma(i)} \quad \text{if} \quad \sigma(i+1) > \sigma(i), \\ \alpha_{\sigma(i+1)} &> \alpha_{\sigma(i)} \quad \text{if} \quad \sigma(i+1) < \sigma(i). \end{aligned} \tag{4.13}$$

From (4.13), $F_{1\ldots N}(l)$ obviously is a lower-triangular matrix. Moreover, the F-matrix is non-degenerate because all its diagonal elements are non-zero. It was shown [52] that the F-matrix also satisfies the factorizing property (4.7). Hence, the F-matrix $F_{1\ldots N}(l)$ given by (4.12) is the desirable Drinfeld twist.

[2]In this paper, we adopt the convention: $F_{\sigma(1\ldots N)}(l) \equiv F_{\sigma(1\ldots N)}(l; z_{\sigma(1)}, \ldots, z_{\sigma(N)})$.

4.2 Completely symmetric representations

Direct calculation shows [52] that the twisted operators $\tilde{T}_F(l|u)_i^j$ defined by (4.10) indeed simultaneously have the following polarization free forms

$$\tilde{T}_F(l|u)_2^2 = \frac{\sin(l_{21}-\eta)}{\sin(l_{21}-\eta+\eta\langle H,\epsilon_1\rangle)} \otimes_i \begin{pmatrix} \frac{\sin(u-z_i)}{\sin(u-z_i+\eta)} & \\ & 1 \end{pmatrix}_{(i)}, \tag{4.14}$$

$$\tilde{T}_F(l|u)_1^2 = \sum_{i=1}^{N} \frac{\sin\eta\sin(u-z_i+l_{12})}{\sin(u-z_i+\eta)\sin l_{12}} E_{12}^i \otimes_{j\neq i} \begin{pmatrix} \frac{\sin(u-z_j)\sin(z_i-z_j+\eta)}{\sin(u-z_j+\eta)\sin(z_i-z_j)} & \\ & 1 \end{pmatrix}_{(j)}, \tag{4.15}$$

$$\tilde{T}_F(l|u)_2^1 = \frac{\sin(l_{21}-\eta)}{\sin(l_{21}+\eta\langle H,\epsilon_1-\epsilon_2\rangle)} \sum_{i=1}^{N} \frac{\sin\eta\sin(u-z_i+l_{21}+\eta+\eta\langle H,\epsilon_1-\epsilon_2\rangle)}{\sin(u-z_i+\eta)\sin(l_{21}+\eta+\eta\langle H,\epsilon_1-\epsilon_2\rangle)}$$
$$\times E_{21}^i \otimes_{j\neq i} \begin{pmatrix} \frac{\sin(u-z_j)}{\sin(u-z_j+\eta)} & \\ & \frac{\sin(z_j-z_i+\eta)}{\sin(z_j-z_i)} \end{pmatrix}_{(j)}, \tag{4.16}$$

where $H = \sum_{k=1}^{N} h^{(k)}$. Applying the above operators to the arbitrary state $|i_1,\ldots,i_N\rangle$ given by (3.8) leads to

$$\tilde{T}_F(m,l|u)_2^2 = \frac{\sin(l_{21}-\eta)}{\sin(l_2-m_1-\eta)} \otimes_i \begin{pmatrix} \frac{\sin(u-z_i)}{\sin(u-z_i+\eta)} & \\ & 1 \end{pmatrix}_{(i)}, \tag{4.17}$$

$$\tilde{T}_F(m,l|u)_1^2 = \sum_{i=1}^{N} \frac{\sin\eta\sin(u-z_i+l_{12})}{\sin(u-z_i+\eta)\sin l_{12}}$$
$$\times E_{12}^i \otimes_{j\neq i} \begin{pmatrix} \frac{\sin(u-z_j)\sin(z_i-z_j+\eta)}{\sin(u-z_j+\eta)\sin(z_i-z_j)} & \\ & 1 \end{pmatrix}_{(j)}, \tag{4.18}$$

$$\tilde{T}_F(m,l|u)_2^1 = \frac{\sin(l_{21}-\eta)}{\sin(m_{21}-2\eta)} \sum_{i=1}^{N} \frac{\sin\eta\sin(u-z_i+m_{21}-\eta)}{\sin(u-z_i+\eta)\sin(m_{21}-\eta)}$$
$$\times E_{21}^i \otimes_{j\neq i} \begin{pmatrix} \frac{\sin(u-z_j)}{\sin(u-z_j+\eta)} & \\ & \frac{\sin(z_j-z_i+\eta)}{\sin(z_j-z_i)} \end{pmatrix}_{(j)}. \tag{4.19}$$

It then follows that the two pseudo-particle creation operators (3.14) and (3.15) in the F-basis simultaneously have the following completely symmetric polarization free forms:

$$\tilde{\mathcal{T}}_F^-(m,\lambda|u)_1^2 = \frac{\sin m_{12}}{\sin(m_1-\lambda_2)} \prod_{k=1}^{N} \frac{\sin(u+z_k)}{\sin(u+z_k+\eta)}$$
$$\times \sum_{i=1}^{N} \frac{\sin(\lambda_1+\xi-z_i)\sin(\lambda_2+\xi+z_i)\sin 2u\sin\eta}{\sin(\lambda_1+\xi+u)\sin(\lambda_2+\xi+u)\sin(u-z_i+\eta)\sin(u+z_i)}$$
$$\times E_{12}^i \otimes_{j\neq i} \begin{pmatrix} \frac{\sin(u-z_j)\sin(u+z_j+\eta)\sin(z_i-z_j+\eta)}{\sin(u-z_j+\eta)\sin(u+z_j)\sin(z_i-z_j)} & \\ & 1 \end{pmatrix}_{(j)}, \tag{4.20}$$

$$\tilde{\mathcal{T}}_F^+(\lambda,m|u)_2^1 = \frac{\sin(m_{21}+\eta)}{\sin(m_2-\lambda_1)} \prod_{k=1}^{N} \frac{\sin(u+z_k)}{\sin(u+z_k+\eta)}$$

$$\times\sum_{i=1}^{N}\frac{\sin(\lambda_2+\bar{\xi}-z_i)\sin(\lambda_1+\bar{\xi}+z_i)\sin(2u+2\eta)\sin\eta}{\sin(\lambda_1+\bar{\xi}-u-\eta)\sin(\lambda_2+\bar{\xi}-u-\eta)\sin(u+z_i)\sin(u-z_i+\eta)}$$

$$\times E_{21}^{i}\otimes_{j\neq i}\begin{pmatrix}\frac{\sin(u-z_j)\sin(u+z_j+\eta)}{\sin(u-z_j+\eta)\sin(u+z_j)} & \\ & \frac{\sin(z_j-z_i+\eta)}{\sin(z_j-z_i)}\end{pmatrix}_{(j)}. \tag{4.21}$$

5 Determinant representations of the scalar products

Due to the fact that the states $|1,\ldots,1\rangle$, $|2,\ldots,2\rangle$ and their dual states $\langle 1,\ldots,1|$, $\langle 2,\ldots,2|$ are invariant under the action of the F-matrix $F_{1\ldots N}(l)$ (4.12), the calculation of the scalar products (3.10)–(3.13) can be performed in the F-basis. Namely,

$$S^{I,II}(\{u_\alpha\};\{v_i^{(2)}\})=\langle 1,\ldots,1|\tilde{\mathcal{T}}_F^-(\lambda-2(M-1)\eta\hat{1},\lambda|u_M)_{\hat{1}}^2\ldots\tilde{\mathcal{T}}_F^-(\lambda,\lambda|u_1)_{\hat{1}}^2$$
$$\times\tilde{\mathcal{T}}_F^-(\lambda+2\eta\hat{1},\lambda|v_1^{(2)})_{\hat{1}}^2\ldots\tilde{\mathcal{T}}_F^-(\lambda+2M\eta\hat{1},\lambda|v_M^{(2)})_{\hat{1}}^2|2,\ldots,2\rangle, \tag{5.1}$$

$$S^{II,I}(\{u_\alpha\};\{v_i^{(1)}\})=\langle 2,\ldots,2|\tilde{\mathcal{T}}_F^+(\lambda,\lambda+2(M-1)\eta\hat{2}|u_M)_{\hat{2}}^1\ldots\tilde{\mathcal{T}}_F^+(\lambda,\lambda|u_1)_{\hat{2}}^1$$
$$\times\tilde{\mathcal{T}}_F^+(\lambda,\lambda-2\eta\hat{2}|v_1^{(1)})_{\hat{2}}^1\ldots\tilde{\mathcal{T}}_F^+(\lambda,\lambda-2M\eta\hat{2}|v_M^{(1)})_{\hat{2}}^1|1,\ldots,1\rangle, \tag{5.2}$$

$$S^{I,I}(\{u_\alpha\};\{v_i^{(1)}\})=\langle 1,\ldots,1|\tilde{\mathcal{T}}_F^-(\lambda-2(M-1)\eta\hat{1},\lambda|u_M)_{\hat{1}}^2\ldots\tilde{\mathcal{T}}_F^-(\lambda,\lambda|u_1)_{\hat{1}}^2$$
$$\times\tilde{\mathcal{T}}_F^+(\lambda,\lambda+2\eta\hat{1}|v_1^{(1)})_{\hat{2}}^1\ldots\tilde{\mathcal{T}}_F^+(\lambda,\lambda+2M\eta\hat{1}|v_M^{(1)})_{\hat{2}}^1|1,\ldots,1\rangle, \tag{5.3}$$

$$S^{II,II}(\{u_\alpha\};\{v_i^{(2)}\})=\langle 2,\ldots,2|\tilde{\mathcal{T}}_F^+(\lambda,\lambda+2(M-1)\eta\hat{2}|u_M)_{\hat{2}}^1\ldots\tilde{\mathcal{T}}_F^+(\lambda,\lambda|u_1)_{\hat{2}}^1$$
$$\times\tilde{\mathcal{T}}_F^-(\lambda-2\eta\hat{2},\lambda|v_1^{(2)})_{\hat{1}}^2\ldots\tilde{\mathcal{T}}_F^-(\lambda-2M\eta\hat{2},\lambda|v_M^{(2)})_{\hat{1}}^2|2,\ldots,2\rangle. \tag{5.4}$$

In the above equations, we have used the identity: $\hat{1}=-\hat{2}$. Thanks to the polarization free representations (4.20) and (4.21) of the pseudo-particle creation/annihilation operators, we can obtain the determinant representations of the scalar products.

5.1 The scalar products $S^{I,II}$ and $S^{II,I}$

It was shown [60] that the scalar product $S^{I,II}(\{u_\alpha\};\{v_i^{(2)}\})$ (resp. $S^{II,I}(\{u_\alpha\};\{v_i^{(1)}\})$) can be expressed in terms of some determinant no matter the parameters $\{v_i^{(2)}\}$ (resp.$\{v_i^{(1)}\}$) satisfy the associated Bethe ansatz equations or not. In this subsection we do not require these parameters being the roots of the Bethe ansatz equations. Let us introduce two functions

$$\mathcal{Z}_N^{(I)}(\{\bar{u}_J\})\equiv S^{I,II}(\{u_\alpha\};\{v_i\})$$
$$=\langle 1,\ldots,1|\tilde{\mathcal{T}}_F^-(\lambda-2(M-1)\eta\hat{1},\lambda|\bar{u}_N)_{\hat{1}}^2\ldots\tilde{\mathcal{T}}_F^-(\lambda+2M\eta\hat{1},\lambda|\bar{u}_1)_{\hat{1}}^2|2,\ldots,2\rangle, \tag{5.5}$$

$$\mathcal{Z}_N^{(II)}(\{\bar{u}_J\})\equiv S^{II,I}(\{u_\alpha\};\{v_i\})$$
$$=\langle 2,\ldots,2|\tilde{\mathcal{T}}_F^+(\lambda,\lambda+2(M-1)\eta\hat{2}|\bar{u}_N)_{\hat{2}}^1\ldots\tilde{\mathcal{T}}_F^+(\lambda,\lambda-2M\eta\hat{2}|\bar{u}_1)_{\hat{2}}^1|1,\ldots,1\rangle, \tag{5.6}$$

where N free parameters $\{\bar{u}_J|J=1,\ldots N\}$ are given by

$$\bar{u}_i=u_i \text{ for } i=1,\ldots M,\qquad\text{and}\qquad \bar{u}_{M+i}=v_i \text{ for } i=1,\ldots M. \tag{5.7}$$

Note that these functions $\mathcal{Z}_N^{(I)}(\{\bar{u}_J\})$ and $\mathcal{Z}_N^{(II)}(\{\bar{u}_J\})$ correspond to the partition functions of the six-vertex model with domain wall boundary conditions and one reflecting end [59] specified by the non-diagonal K-matrices (2.8) and (2.10) respectively [60].

The polarization free representations (4.20) and (4.21) of the pseudo-particle creation/annihilation operators allowed ones [60] to express the above functions in terms of the determinants representations of some $N \times N$ matrices as follows:

$$\mathcal{Z}_N^{(I)}(\{\bar{u}_J\}) = \prod_{k=1}^{M} \frac{\sin(\lambda_{12}+2k\eta)\sin(\lambda_{12}-2k\eta+\eta)}{\sin(\lambda_{12}+k\eta)\sin(\lambda_{12}-k\eta+\eta)} \prod_{l=1}^{N}\prod_{i=1}^{N} \frac{\sin(\bar{u}_i+z_l)}{\sin(\bar{u}_i+z_l+\eta)}$$
$$\times \frac{\prod_{\alpha=1}^{N}\prod_{i=1}^{N}\sin(\bar{u}_\alpha-z_i)\sin(\bar{u}_\alpha+z_i+\eta)\det\mathcal{N}^{(I)}(\{\bar{u}_\alpha\};\{z_i\})}{\prod_{\alpha>\beta}\sin(\bar{u}_\alpha-\bar{u}_\beta)\sin(\bar{u}_\alpha+\bar{u}_\beta+\eta)\prod_{k<l}\sin(z_k-z_l)\sin(z_k+z_l)}, \quad (5.8)$$

$$\mathcal{Z}_N^{(II)}(\{\bar{u}_J\}) = \prod_{k=1}^{M} \frac{\sin(\lambda_{21}+\eta-2k\eta)\sin(\lambda_{21}-\eta+2k\eta)}{\sin(\lambda_{21}-k\eta)\sin(\lambda_{21}+k\eta-\eta)} \prod_{l=1}^{N}\prod_{i=1}^{N} \frac{\sin(\bar{u}_i+z_l)}{\sin(\bar{u}_i+z_l+\eta)}$$
$$\times \frac{\prod_{\alpha=1}^{N}\prod_{i=1}^{N}\sin(\bar{u}_\alpha+z_i)\sin(\bar{u}_\alpha-z_i+\eta)\det\mathcal{N}^{(II)}(\{\bar{u}_\alpha\};\{z_i\})}{\prod_{\alpha>\beta}\sin(\bar{u}_\alpha-\bar{u}_\beta)\sin(\bar{u}_\alpha+\bar{u}_\beta+\eta)\prod_{k<l}\sin(z_l-z_k)\sin(z_l+z_k)}, \quad (5.9)$$

where the $N \times N$ matrices $\mathcal{N}^{(I)}(\{\bar{u}_\alpha\};\{z_i\})$ and $\mathcal{N}^{(II)}(\{\bar{u}_\alpha\};\{z_i\})$ are given by

$$\mathcal{N}^{(I)}(\{\bar{u}_\alpha\};\{z_i\})_{\alpha,j} = \frac{\sin\eta\sin(\lambda_1+\xi-z_j)}{\sin(\bar{u}_\alpha-z_j)\sin(\bar{u}_\alpha+z_j+\eta)\sin(\lambda_1+\xi+\bar{u}_\alpha)}$$
$$\times\frac{\sin(\lambda_2+\xi+z_j)\sin(2\bar{u}_\alpha)}{\sin(\lambda_2+\xi+\bar{u}_\alpha)\sin(\bar{u}_\alpha-z_j+\eta)\sin(\bar{u}_\alpha+z_j)}, \quad (5.10)$$

$$\mathcal{N}^{(II)}(\{\bar{u}_\alpha\};\{z_i\})_{\alpha,j} = \frac{\sin\eta\sin(\lambda_2+\bar{\xi}-z_j)}{\sin(\bar{u}_\alpha-z_j)\sin(\bar{u}_\alpha+z_j+\eta)\sin(\lambda_2+\bar{\xi}-\bar{u}_\alpha-\eta)}$$
$$\times\frac{\sin(\lambda_1+\bar{\xi}+z_j)\sin(2\bar{u}_\alpha+2\eta)}{\sin(\lambda_1+\bar{\xi}-\bar{u}_\alpha-\eta)\sin(\bar{u}_\alpha-z_j+\eta)\sin(\bar{u}_\alpha+z_j)}. \quad (5.11)$$

The above determinant representations are crucial to construct the determinant representations of the remaining scalar products $S^{I,I}$ and $S^{II,II}$ in the next subsection.

5.2 The scalar products $S^{I,I}$ and $S^{II,II}$

Let us introduce two sets of functions $\{H_j^{(I)}(u;\{z_i\},\{v_i\})|j = 1,\ldots,M\}$ and $\{H_j^{(II)}(u;\{z_i\},\{v_i\})|j = 1,\ldots,M\}$

$$H_j^{(I)}(u;\{z_i\},\{v_i\}) = F_1(u)\prod_{l=1}^{N}\frac{\sin(u+z_l)}{\sin(u+z_l+\eta)}\frac{\prod_{k\neq j}\sin(u+v_k+2\eta)\sin(u-v_k+\eta)}{\sin(u-v_j)\sin(u+v_j+\eta)\sin(2u+\eta)}$$
$$-F_2(u)\prod_{l=1}^{N}\frac{\sin(u-z_l+\eta)}{\sin(u-z_l)}\frac{\prod_{k\neq j}\sin(u+v_k)\sin(u-v_k-\eta)}{\sin(u-v_j)\sin(u+v_j+\eta)\sin(2u+\eta)}, \quad (5.12)$$

$$H_j^{(II)}(u;\{z_i\},\{v_i\}) = F_3(u)\prod_{l=1}^{N}\frac{\sin(u-z_l)}{\sin(u-z_l+\eta)}\frac{\prod_{k\neq j}\sin(v_k+u+2\eta)\sin(v_k-u-\eta)}{\sin(u+v_j+\eta)\sin(u-v_j)\sin(2u+\eta)}$$
$$-F_4(u)\prod_{l=1}^{N}\frac{\sin(u+z_l+\eta)}{\sin(u+z_l)}\frac{\prod_{k\neq j}\sin(v_k+u)\sin(v_k-u+\eta)}{\sin(u+v_j+\eta)\sin(u-v_j)\sin(2u+\eta)}, \quad (5.13)$$

where the coefficients $\{F_i(u)|i = 1,2,3,4\}$ are

$$F_1(u) = \sin(\lambda_2+\bar{\xi}+u+\eta)\sin(\lambda_2+\xi-u-\eta)\sin(\lambda_1+\bar{\xi}-u-\eta)\sin(\lambda_1+\xi+u+\eta), \quad (5.14)$$

$$F_2(u) = \sin(\lambda_2 + \bar{\xi} - u)\sin(\lambda_2 + \xi + u)\sin(\lambda_1 + \bar{\xi} + u)\sin(\lambda_1 + \xi - u), \quad (5.15)$$

$$F_3(u) = \sin(\lambda_2+\bar{\xi}-u-\eta)\sin(\lambda_2+\xi+u+\eta)\sin(\lambda_1+\bar{\xi}+u+\eta)\sin(\lambda_1+\xi-u-\eta), \quad (5.16)$$

$$F_4(u) = \sin(\lambda_2 + \bar{\xi} + u)\sin(\lambda_2 + \xi - u)\sin(\lambda_1 + \bar{\xi} - u)\sin(\lambda_1 + \xi + u). \quad (5.17)$$

Let us consider the scalar product $S^{I,I}(\{u_\alpha\};\{v_i^{(1)}\})$ defined by (3.2). The expression (5.3) of $S^{I,I}(\{u_\alpha\};\{v_i^{(1)}\})$ under the F-basis and the polarization free representations (4.20) and (4.21) of the pseudo-particle creation/annihilation operators allow us to compute the scalar product following the similar procedure as that in [14] for the bulk case as follows. In front of each operators $\tilde{T}_F^-$ in (5.3), we insert a sum over the complete set of spin states $|j_1,\ldots,j_i\rangle\rangle$, where $|j_1,\ldots,j_i\rangle\rangle$ is the state with i spins being ϵ_2 in the sites $j_1,\ldots,j_i$ and $2M-i$ spins being ϵ_1 in the other sites. We are thus led to consider some intermediate functions of the form

$$G^{(i)}(u_1,\ldots,u_i|j_{i+1},\ldots,j_M;\{v_i^{(1)}\})=\langle\langle j_{i+1},\ldots,j_M|\tilde{T}_F^-(\lambda-2(i-1)\eta\hat{1},\lambda|u_i)_1^2\ldots\tilde{T}_F^-(\lambda,\lambda|u_1)_1^2$$
$$\times\tilde{T}_F^+(\lambda,\lambda+2\eta\hat{1}|v_1^{(1)})_2^1\ldots\tilde{T}_F^+(\lambda,\lambda+2M\eta\hat{1}|v_M^{(1)})_2^1|1,\ldots,1\rangle,$$
$$i=0,1,\ldots,M, \quad (5.18)$$

which satisfy the following recursive relation:

$$G^{(i)}(u_1,\ldots,u_i|j_{i+1},\ldots,j_M;\{v_i^{(1)}\})$$
$$=\sum_{j\neq j_{i+1},\ldots,j_M}\langle\langle j_{i+1},\ldots,j_M|\tilde{T}_F^-(\lambda-2(i-1)\eta\hat{1},\lambda|u_i)_1^2|j,j_{i+1},\ldots,j_M\rangle\rangle$$
$$\times G^{(i-1)}(u_1,\ldots,u_{i-1}|j,j_{i+1},\ldots,j_M;\{v_i^{(1)}\}), \quad i=1,\ldots,M. \quad (5.19)$$

Note that the last of these functions $\{G^{(i)}|i=0,\ldots,M\}$ is precisely the scalar product $S^{I,I}(\{u_\alpha\};\{v_i^{(1)}\})$, namely,

$$G^{(M)}(u_1,\ldots,u_M;\{v_i^{(1)}\})=S^{I,I}(\{u_\alpha\};\{v_i^{(1)}\}), \quad (5.20)$$

whereas the first one,

$$G^{(0)}(j_1,\ldots,j_M;\{v_i^{(1)}\})=\langle\langle j_1,\ldots,j_M|\tilde{T}_F^+(\lambda,\lambda+2\eta\hat{1}|v_1^{(1)})_2^1\ldots\tilde{T}_F^+(\lambda,\lambda+2M\eta\hat{1}|v_M^{(1)})_2^1|1,\ldots,1\rangle,$$

is closely related to the partition function computed in [60]. Solving the recursive relations (5.19), we find that if the parameters $\{v_k^{(1)}\}$ satisfy the first set of Bethe ansatz equations (2.42) the scalar product $S^{I,I}(\{u_\alpha\};\{v_i^{(1)}\})$ has the following determinant representation

$$S^{I,I}(\{u_\alpha\};\{v_i^{(1)}\})=\prod_{k=1}^{M}\left\{\frac{\sin(\lambda_{12}+2\eta-2k\eta)\sin(\lambda_{12}-\eta+2k\eta)}{\sin(\lambda_{12}-(k-1)\eta)\sin(\lambda_{12}+k\eta)}\prod_{l=1}^{N}\frac{\sin(u_k-z_l)\sin(v_k^{(1)}-z_l)}{\sin(u_k-z_l+\eta)\sin(v_k^{(1)}-z_l+\eta)}\right\}$$
$$\times\frac{\det\bar{\mathcal{N}}^{(I)}(\{u_\alpha\};\{v_i^{(1)}\})}{\prod_{\alpha<\beta}\sin(u_\alpha-u_\beta)\sin(u_\alpha+u_\beta+\eta)\prod_{k>l}\sin(v_k^{(1)}-v_l^{(1)})\sin(v_k^{(1)}+v_l^{(1)}+\eta)}, \quad (5.21)$$

where the $M\times M$ matrix $\bar{\mathcal{N}}^{(I)}(\{u_\alpha\};\{v_i^{(1)}\})$ is given by

$$\bar{\mathcal{N}}^{(I)}(\{u_\alpha\};\{v_i^{(1)}\})_{\alpha,j}=\frac{\sin\eta\sin(2u_\alpha)\sin(2v_j^{(1)}+2\eta)H_j^{(I)}(u_\alpha;\{z_i\},\{v_i^{(1)}\})}{\sin(\lambda_1+\xi+u_\alpha)\sin(\lambda_2+\xi+u_\alpha)\sin(\lambda_2+\bar{\xi}-v_j^{(1)}-\eta)\sin(\lambda_1+\bar{\xi}-v_j^{(1)}-\eta)}. \quad (5.22)$$

Using the similar method as above, we have that the scalar product $S^{II,II}(\{u_\alpha\};\{v_i^{(2)}\})$ has the following determinant representation provided that the parameters $\{v_k^{(2)}\}$ satisfy the second set of Bethe ansatz equations (2.44)

$$S^{II,II}(\{u_\alpha\};\{v_i^{(2)}\})=\prod_{k=1}^{M}\left\{\frac{\sin(\lambda_{12}+2k\eta)\sin(\lambda_{21}-\eta+2k\eta)}{\sin(\lambda_{12}+k\eta)\sin(\lambda_{21}+(k-1)\eta)}\prod_{l=1}^{N}\frac{\sin(u_k+z_l)\sin(v_k^{(2)}+z_l)}{\sin(u_k+z_l+\eta)\sin(v_k^{(2)}+z_l+\eta)}\right\}$$
$$\times\frac{\det\bar{\mathcal{N}}^{(II)}(\{u_\alpha\};\{v_i^{(2)}\})}{\prod_{\alpha<\beta}\sin(u_\alpha-u_\beta)\sin(u_\alpha+u_\beta+\eta)\prod_{k>l}\sin(v_k^{(2)}-v_l^{(2)})\sin(v_k^{(2)}+v_l^{(2)}+\eta)},\qquad(5.23)$$

where the $M\times M$ matrix $\bar{\mathcal{N}}^{(II)}(\{u_\alpha\};\{v_i^{(2)}\})$ is given by

$$\bar{\mathcal{N}}^{(II)}(\{u_\alpha\};\{v_i^{(2)}\})_{\alpha,j}=\frac{\sin\eta\sin(2u_\alpha+2\eta)\sin(2v_j^{(2)})H_j^{(II)}(u_\alpha;\{z_i\},\{v_i^{(2)}\})}{\sin(\lambda_2+\bar{\xi}-u_\alpha-\eta)\sin(\lambda_1+\bar{\xi}-u_\alpha-\eta)\sin(\lambda_2+\xi+v_j^{(2)})\sin(\lambda_1+\xi+v_j^{(2)})}.\qquad(5.24)$$

Now we are in position to compute the norms of the Bethe states which can be obtained by taking the limit $u_\alpha\to v_\alpha^{(i)}$, $\alpha=1,\dots M$. The norm of the first set of Bethe state (2.38) is

$$\begin{aligned}\mathbb{N}^{I,I}(\{v_\alpha^{(1)}\})&=\lim_{u_\alpha\to v_\alpha^{(1)}}S^{I,I}(\{u_\alpha\};\{v_i^{(1)}\})\\&=\prod_{k=1}^{M}\left\{\frac{\sin(\lambda_{12}+2\eta-2k\eta)\sin(\lambda_{12}-\eta+2k\eta)}{\sin(\lambda_{12}-(k-1)\eta)\sin(\lambda_{12}+k\eta)}\prod_{l=1}^{N}\frac{\sin^2(v_k^{(1)}-z_l)}{\sin^2(v_k^{(1)}-z_l+\eta)}\right\}\\&\quad\times\prod_{\alpha\neq\beta}\frac{\sin(v_\alpha^{(1)}+v_\beta^{(1)})\sin(v_\alpha^{(1)}-v_\beta^{(1)}-\eta)}{\sin(v_\alpha^{(1)}-v_\beta^{(1)})\sin(v_\alpha^{(1)}+v_\beta^{(1)}+\eta)}\det\Phi^{(I)}(\{v_\alpha^{(1)}\}),\end{aligned}\qquad(5.25)$$

where the matrix elements of $M\times M$ matrix $\Phi^{(I)}(\{v_\alpha\})$ are given by

$$\begin{aligned}\Phi_{\alpha,j}^{(I)}(\{v_\alpha\})=&\frac{\sin\eta\sin(\lambda_2+\bar{\xi}-v_\alpha)\sin(\lambda_1+\bar{\xi}+v_\alpha)\sin(\lambda_1+\xi-v_\alpha)}{\sin(\lambda_1+\xi+v_\alpha)\sin(\lambda_2+\bar{\xi}-v_j-\eta)\sin(\lambda_1+\bar{\xi}-v_j-\eta)}\\&\times\frac{\sin 2v_\alpha\sin(2v_j+2\eta)}{\sin(2v_\alpha+\eta)\sin(2v_j+\eta)}\prod_{l=1}^{N}\frac{\sin(v_\alpha-z_l+\eta)}{\sin(v_\alpha-z_l)}\\&\times\frac{\partial}{\partial v_\alpha}\ln\left\{\frac{\sin(\lambda_2+\bar{\xi}+v_j+\eta)\sin(\lambda_2+\xi-v_j-\eta)\sin(\lambda_1+\bar{\xi}-v_j-\eta)\sin(\lambda_1+\xi+v_j+\eta)}{\sin(\lambda_2+\bar{\xi}-v_j)\sin(\lambda_2+\xi+v_j)\sin(\lambda_1+\bar{\xi}+v_j)\sin(\lambda_1+\xi-v_j)}\right.\\&\left.\times\prod_{l=1}^{N}\frac{\sin(v_j+z_l)\sin(v_j-z_l)}{\sin(v_j+z_l+\eta)\sin(v_j-z_l+\eta)}\prod_{k\neq j}\frac{\sin(v_j+v_k+2\eta)\sin(v_j-v_k+\eta)}{\sin(v_j+v_k)\sin(v_j-v_k-\eta)}\right\},\end{aligned}\qquad(5.26)$$

the norm of the second set of Bethe state (2.39) is given by

$$\begin{aligned}\mathbb{N}^{II,II}(\{v_\alpha^{(2)}\})&=\lim_{u_\alpha\to v_\alpha^{(2)}}S^{II,II}(\{u_\alpha\};\{v_i^{(2)}\})\\&=\prod_{k=1}^{M}\left\{\frac{\sin(\lambda_{12}+2k\eta)\sin(\lambda_{21}-\eta+2k\eta)}{\sin(\lambda_{12}+k\eta)\sin(\lambda_{21}+(k-1)\eta)}\prod_{l=1}^{N}\frac{\sin^2(v_k^{(2)}+z_l)}{\sin^2(v_k^{(2)}+z_l+\eta)}\right\}\\&\quad\times\prod_{\alpha\neq\beta}\frac{\sin(v_\alpha^{(2)}+v_\beta^{(2)}+2\eta)\sin(v_\alpha^{(2)}-v_\beta^{(2)}-\eta)}{\sin(v_\alpha^{(2)}-v_\beta^{(2)})\sin(v_\alpha^{(2)}+v_\beta^{(2)}+\eta)}\det\Phi^{(II)}(\{v_\alpha^{(2)}\}),\end{aligned}\qquad(5.27)$$

where the matrix elements of $M \times M$ matrix $\Phi^{(II)}(\{v_\alpha\})$ are given by

$$\Phi^{(II)}_{\alpha,j}(\{v_\alpha\}) = \frac{\sin\eta\sin(\lambda_2+\xi+v_\alpha+\eta)\sin(\lambda_1+\bar{\xi}+v_\alpha+\eta)\sin(\lambda_1+\xi-v_\alpha-\eta)}{\sin(\lambda_1+\bar{\xi}-v_\alpha-\eta)\sin(\lambda_2+\xi+v_j)\sin(\lambda_1+\xi+v_j)}$$
$$\times\frac{\sin(2v_\alpha+2\eta)\sin 2v_j}{\sin(2v_\alpha+\eta)\sin(2v_j+\eta)}\prod_{l=1}^{N}\frac{\sin(v_\alpha-z_l)}{\sin(v_\alpha-z_l+\eta)}$$
$$\times\frac{\partial}{\partial v_\alpha}\ln\left\{\frac{\sin(\lambda_2+\bar{\xi}-v_j-\eta)\sin(\lambda_2+\xi+v_j+\eta)\sin(\lambda_1+\bar{\xi}+v_j+\eta)\sin(\lambda_1+\xi-v_j-\eta)}{\sin(\lambda_2+\bar{\xi}+v_j)\sin(\lambda_2+\xi-v_j)\sin(\lambda_1+\bar{\xi}-v_j)\sin(\lambda_1+\xi+v_j)}\right.$$
$$\left.\times\prod_{l=1}^{N}\frac{\sin(v_j+z_l)\sin(v_j-z_l)}{\sin(v_j+z_l+\eta)\sin(v_j-z_l+\eta)}\prod_{k\neq j}\frac{\sin(v_j+v_k+2\eta)\sin(v_j-v_k+\eta)}{\sin(v_j+v_k)\sin(v_j-v_k-\eta)}\right\}. \quad (5.28)$$

Moreover, one may check that if the parameters $\{u_\alpha\}$ satisfy the Bethe ansatz equations (i.e. on shell) but different from $\{v^{(i)}_\alpha\}$ the corresponding scalar products $S^{I,I}(\{u_\alpha\};\{v^{(1)}_i\})$ or $S^{II,II}(\{u_\alpha\};\{v^{(2)}_i\})$ vanishes, namely, the corresponding Bethe states are orthogonal.

6 Conclusions

We have studied scalar products between an on-shell Bethe state and a general state (or an off-shell Bethe state) of the open XXZ chain with non-diagonal boundary terms, where the non-diagonal K-matrices $K^{\pm}(u)$ are given by (2.9) and (2.11). In our calculation the factorizing F-matrix (4.12) in the face picture of the open XXZ chain, which leads to the polarization free representations (4.20) and (4.21) of the associated pseudo-particle creation/annihilation operators, has played an important role. It is found that the scalar products can be expressed in terms of the determinants (5.8), (5.9), (5.21) and (5.23). By taking the on shell limit, we obtain the determinant representations (or Gaudin formula) (5.25)–(5.26) and (5.27)–(5.28) of the norms of the Bethe states. However it should be emphasized that, in contrast with those of the open XXZ chain with diagonal boundary terms, the dual states (3.3) and (3.4) are generally no longer the eigenstates of the open chain with non-diagonal boundary terms even if the parameters satisfy the associated Bethe ansatz equations.

Acknowledgments

The financial supports from the National Natural Science Foundation of China (Grant Nos. 11075126 and 11031005), Australian Research Council and the NWU Graduate Cross-discipline Fund (08YJC24) are gratefully acknowledged. One of authors (W.L.Y.) would like to thank the school of Mathematics and Statistics of the University of Sydney where part of this work was done, especially Xin Liu, for hospitality. We also would like to thank the referee for his/her valuable suggestions to revise the manuscript.

A $\mathcal{T}^{\pm}(m|u)$ in the face picture

The K-matrices $K^{\pm}(u)$ given by (2.8) and (2.10) are generally non-diagonal (in the vertex picture), after the face-vertex transformations (2.31) and (2.32), the face type counterparts

$\mathcal{K}(\lambda|u)$ and $\tilde{\mathcal{K}}(\lambda|u)$ given by (2.33) and (2.34) *simultaneously* become diagonal. This fact suggests that it would be much simpler if one performs all calculations in the face picture.

Associated with the vertex type monodromy matrices $T(u)$ (2.3) and $\hat{T}(u)$ (2.6), we introduce the following operators

$$T(m,l|u)^j_\mu = \tilde{\phi}^0_{m+\eta\hat{j},m}(u)\,T_0(u)\,\phi^0_{l+\eta\hat{\mu},l}(u), \tag{A.1}$$

$$S(m,l|u)^\mu_i = \bar{\phi}^0_{l,l-\eta\hat{\mu}}(-u)\,\hat{T}_0(u)\,\phi^0_{m,m-\eta\hat{\imath}}(-u). \tag{A.2}$$

Moreover, for the case of $m = l - \eta\sum_{k=1}^N \hat{\imath}_k$, we introduce a generic state in the quantum space from the intertwiner vector (2.25)

$$|i_1,\ldots,i_N\rangle^m_l = \phi^1_{l,l-\eta\hat{\imath}_1}(z_1)\phi^2_{l-\eta\hat{\imath}_1,l-\eta(\hat{\imath}_1+\hat{\imath}_2)}(z_2)\ldots\phi^N_{l-\eta\sum_{k=1}^{N-1}\hat{\imath}_k,l-\eta\sum_{k=1}^N\hat{\imath}_k}(z_N). \tag{A.3}$$

We can evaluate the action of the operator $T(m,l|u)$ on the state $|i_1,\ldots,i_N\rangle^m_l$ from the face-vertex correspondence relation (2.27)

$$\begin{aligned}
T(m,l|u)^j_\mu|i_1,\ldots,i_N\rangle^m_l &= \tilde{\phi}^0_{m+\eta\hat{j},m}(u)\,T_0(u)\,\phi^0_{l+\eta\hat{\mu},l}(u)|i_1,\ldots,i_N\rangle^m_l\\
&= \tilde{\phi}^0_{m+\eta\hat{j},m}(u)\overline{R}_{0,N}(u-z_N)\ldots\overline{R}_{0,1}(u-z_1)\phi^0_{l+\eta\hat{\mu},l}(u)\phi^1_{l,l-\eta\hat{\imath}_1}(z_1)\ldots\\
&= \sum_{\alpha_1,i'_1} R(u-z_1;l+\eta\hat{\mu})^{\alpha_1 i'_1}_{\mu\, i_1}\phi^1_{l+\eta\hat{\mu},l+\eta\hat{\mu}-\eta\hat{\imath}'_1}(z_1)\tilde{\phi}^0_{m+\eta\hat{j},m}(u)\overline{R}_{0,N}(u-z_N)\ldots\\
&\quad\times\overline{R}_{0,2}(u-z_2)\phi^0_{l+\eta\hat{\mu}-\eta\hat{\imath}'_1,l-\eta\hat{\imath}_1}(u)\phi^2_{l-\eta\hat{\imath}_1,l-\eta(\hat{\imath}_1+\hat{\imath}_2)}(z_2)\ldots\\
&\vdots\\
&= \sum_{\alpha_1\ldots\alpha_{N-1}}\sum_{i'_1\ldots i'_N} R(u-z_N;l+\eta\hat{\mu}-\eta\sum_{k=1}^{N-1}\hat{\imath}'_k)^{j\quad i'_N}_{\alpha_{N-1}i_N}\ldots\\
&\quad\times R(u-z_1;l+\eta\hat{\mu})^{\alpha_1 i'_1}_{\mu\, i_1}|i'_1,\ldots,i'_N\rangle^{l+\eta\hat{\mu}-\eta\sum_{k=1}^N\hat{\imath}'_k}_{l+\eta\hat{\mu}}.
\end{aligned}\tag{A.4}$$

Comparing with (3.9), we have the following correspondence

$$T(m,l|u)^j_\mu|i_1,\ldots,i_N\rangle^m_l \longleftrightarrow T_F(m+\eta\hat{\mu};l+\eta\hat{\mu}|u)^j_\mu|i_1,\ldots,i_N\rangle, \tag{A.5}$$

where vector $|i_1,\ldots,i_N\rangle$ is given by (3.8). Hereafter, we will use O_F to denote the face version of operator O in the face picture.

Noting that

$$\hat{T}_0(u) = \overline{R}_{1,0}(u+z_1)\ldots\overline{R}_{N,0}(u+z_N),$$

we obtain the action of $S(m,l|u)^\mu_i$ on the state $|i_1,\ldots,i_N\rangle^m_l$

$$\begin{aligned}
S(m,l|u)^\mu_i|i_1,\ldots,i_N\rangle^m_l &= \sum_{\alpha_1\ldots\alpha_{N-1}}\sum_{i'_1\ldots i'_N} R(u+z_1;l)^{i'_1\,\mu}_{i_1\alpha_{N-1}}R(u+z_2;l-\eta\hat{\imath}_1)^{i'_2\,\alpha_{N-1}}_{i_2\alpha_{N-2}}\\
&\quad\times\ldots R(u+z_N;l-\eta\sum_{k=1}^{N-1}\hat{\imath}_k)^{i'_N\,\alpha_1}_{i_N\, i}|i'_1,\ldots,i'_N\rangle^{l-\eta\hat{\mu}-\eta\sum_{k=1}^N\hat{\imath}'_k}_{l-\eta\hat{\mu}}.
\end{aligned}\tag{A.6}$$

Then the crossing relation of the R-matrix (2.23) enables us to establish the following relation:

$$S(m,l|u)_i^{\mu}=\varepsilon_{\bar{i}}\varepsilon_{\bar{\mu}}\frac{\sin(m_{21})}{\sin(l_{21})}\prod_{k=1}^{N}\frac{\sin(u+z_k)}{\sin(u+z_k+\eta)}T(m,l|-u-\eta)_{\bar{\mu}}^{\bar{i}}, \tag{A.7}$$

where the parities are defined in (2.24) and m_{21} (or l_{21}) is defined in (2.15).

Now we are in the position to express $\mathcal{T}^{\pm}$ (2.35) and (2.36) in terms of $T(m,l)_j^i$ and $S(l,m)_j^i$ which both can be expressed in terms of the face type R-matrix (2.16). By (2.29) and (2.30), we have

$$\begin{aligned}\mathcal{T}^-(m|u)_i^j &= \tilde{\phi}^0_{m-\eta(\hat{\imath}-\hat{\jmath}),m-\eta\hat{\imath}}(u)\ \mathbb{T}(u)\ \phi^0_{m,m-\eta\hat{\imath}}(-u)\\ &= \tilde{\phi}^0_{m-\eta(\hat{\imath}-\hat{\jmath}),m-\eta\hat{\imath}}(u)T_0(u)K_0^-(u)\hat{T}_0(u)\phi^0_{m,m-\eta\hat{\imath}}(-u)\\ &=\sum_{\mu,\nu}\tilde{\phi}^0_{m-\eta(\hat{\imath}-\hat{\jmath}),m-\eta\hat{\imath}}(u)T_0(u)\phi^0_{l-\eta(\hat{\nu}-\hat{\mu}),l-\eta\hat{\nu}}(u)\tilde{\phi}^0_{l-\eta(\hat{\nu}-\hat{\mu}),l-\eta\hat{\nu}}(u)\\ &\qquad\times K_0^-(u)\phi^0_{l,l-\eta\hat{\nu}}(-u)\bar{\phi}^0_{l,l-\eta\hat{\nu}}(-u)\hat{T}_0(u)\phi^0_{m,m-\eta\hat{\imath}}(-u)\\ &=\sum_{\mu,\nu}T(m-\eta\hat{\imath},l-\eta\hat{\nu}|u)_{\mu}^{j}\mathcal{K}(l|u)_{\nu}^{\mu}S(m,l|u)_i^{\nu}\\ &\overset{\text{def}}{=}\mathcal{T}^-(m,l|u)_i^j,\end{aligned} \tag{A.8}$$

where the face-type K-matrix $\mathcal{K}(l|u)_\nu^\mu$ is given by

$$\mathcal{K}(l|u)_\nu^\mu=\tilde{\phi}^0_{l-\eta(\hat{\nu}-\hat{\mu}),l-\eta\hat{\nu}}(u)K_0^-(u)\phi^0_{l,l-\eta\hat{\nu}}(-u). \tag{A.9}$$

Similarly, we have

$$\begin{aligned}\mathcal{T}^+(m|u)_i^j &= \prod_{k\neq j}\frac{\sin m_{jk}}{\sin(m_{jk}-\eta)}\sum_{\mu,\nu}T(l-\eta\hat{\mu},m-\eta\hat{\jmath}|u)_i^{\nu}\tilde{\mathcal{K}}(l|u)_\nu^\mu S(l,m|u)_\mu^j\\ &\overset{\text{def}}{=}\mathcal{T}^+(l,m|u)_i^j\end{aligned} \tag{A.10}$$

with

$$\tilde{\mathcal{K}}(l|u)_\nu^\mu=\bar{\phi}^0_{l,l-\eta\hat{\mu}}(-u)K_0^+(u)\phi^0_{l-\eta(\hat{\mu}-\hat{\nu}),l-\eta\hat{\mu}}(u). \tag{A.11}$$

Thanks to the fact that when $l=\lambda$ the corresponding face-type K-matrices $\mathcal{K}(\lambda|u)$ (A.9) and $\tilde{\mathcal{K}}(\lambda|u)$ (A.11) become diagonal ones (2.33) and (2.34), we have

$$\mathcal{T}^-(m,\lambda|u)_i^j=\sum_{\mu}T(m-\eta\hat{\imath},\lambda-\eta\hat{\mu}|u)_\mu^j k(\lambda|u)_\mu S(m,\lambda|u)_i^\mu, \tag{A.12}$$

$$\mathcal{T}^+(\lambda,m|u)_i^j=\prod_{k\neq j}\frac{\sin m_{jk}}{\sin(m_{jk}-\eta)}\sum_{\mu}T(\lambda-\eta\hat{\mu},m-\eta\hat{\jmath}|u)_i^\mu\tilde{k}(\lambda|u)_\mu S(\lambda,m|u)_\mu^j, \tag{A.13}$$

where the functions $k(\lambda|u)_\mu$ and $\tilde{k}(\lambda|u)_\mu$ are given by (2.33) and (2.34) respectively. The relation (A.7) implies that one can further express $\mathcal{T}^{\pm}(m|u)_i^j$ in terms of only $T(m,l|u)_i^j$.

Here we present the results for the pseudo-particle creation operators $\mathcal{T}^-(m|u)^2_1$ in (2.39) and $\mathcal{T}^+(m|u)^1_2$ in (2.38):

$$\begin{aligned}
\mathcal{T}^-(m|u)^2_1 \equiv \mathcal{T}^-(m,\lambda|u)^2_1 = & \frac{\sin(m_{21})}{\sin(\lambda_{21})}\prod_{k=1}^{N}\frac{\sin(u+z_k)}{\sin(u+z_k+\eta)} \\
& \times\left\{\frac{\sin(\lambda_1+\xi-u)}{\sin(\lambda_1+\xi+u)}T(m+\eta\hat{2},\lambda+\eta\hat{2}|u)^2_1 T(m,\lambda|-u-\eta)^2_2\right. \\
& \left. -\frac{\sin(\lambda_2+\xi-u)}{\sin(\lambda_2+\xi+u)}T(m+\eta\hat{2},\lambda+\eta\hat{1}|u)^2_2 T(m,\lambda|-u-\eta)^2_1\right\},
\end{aligned} \tag{A.14}$$

$$\begin{aligned}
\mathcal{T}^+(m|u)^1_2 \equiv \mathcal{T}^+(\lambda,m|u)^1_2 = & \prod_{k=1}^{N}\frac{\sin(u+z_k)}{\sin(u+z_k+\eta)} \\
& \times\left\{\frac{\sin(\lambda_{12}-\eta)\sin(\lambda_1+\bar{\xi}+u+\eta)}{\sin(m_{12}-\eta)\sin(\lambda_1+\bar{\xi}-u-\eta)}T(\lambda+\eta\hat{2},m+\eta\hat{2}|u)^1_2 T(\lambda,m|-u-\eta)^2_2\right. \\
& \left. -\frac{\sin(\lambda_{21}-\eta)\sin(\lambda_2+\bar{\xi}+u+\eta)}{\sin(m_{21}+\eta)\sin(\lambda_2+\bar{\xi}-u-\eta)}T(\lambda+\eta\hat{1},m+\eta\hat{2}|u)^2_2 T(\lambda,m|-u-\eta)^1_2\right\}.
\end{aligned} \tag{A.15}$$

Similar to (A.5), we have the correspondence,

$$\mathcal{T}^-(m,l|u)^2_1|i_1,\ldots,i_N\rangle^m_l \longleftrightarrow \mathcal{T}^-_F(m,l|u)^2_1|i_1,\ldots,i_N\rangle, \tag{A.16}$$

$$\mathcal{T}^+(m,l|u)^1_2|i_1,\ldots,i_N\rangle^m_l \longleftrightarrow \mathcal{T}^+_F(m,l|u)^1_2|i_1,\ldots,i_N\rangle. \tag{A.17}$$

This gives rise the expressions of the operators $\mathcal{T}^\pm_F(m|u)$ given by (3.14) and (3.15).

Some remarks are in order. It follows from (A.4) that the action of the operator $T(m,l|u)$ on the state $|i_1,\ldots,i_N\rangle^m_l$ can be expressed in terms of the face type R-matrix (2.16). This implies that the corresponding actions of $\mathcal{T}^\pm(m|u)$ can also be expressed in terms of the R-matrix and the K-matrices (2.33) and (2.34). Moreover, the transfer matrix $t(u)$ (2.7) can be given as a linear combination of either $\mathcal{T}^-(m|u)^i_i$:

$$\begin{aligned}
\tau(u) &= tr(K^+(u)\mathbb{T}(u)) = \sum_{\mu,\nu} tr\left(K^+(u)\phi_{\lambda-\eta(\hat{\mu}-\hat{\nu}),\lambda-\eta\hat{\mu}}(u)\tilde{\phi}_{\lambda-\eta(\hat{\mu}-\hat{\nu}),\lambda-\eta\hat{\mu}}(u)\right. \\
&\qquad \left.\times\, \mathbb{T}(u)\phi_{\lambda,\lambda-\eta\hat{\mu}}(-u)\bar{\phi}_{\lambda,\lambda-\eta\hat{\mu}}(-u)\right) \\
&= \sum_{\mu,\nu}\tilde{\mathcal{K}}(\lambda|u)^\mu_\nu \mathcal{T}^-(\lambda|u)^\nu_\mu = \sum_{\mu}\tilde{k}(\lambda|u)_\mu \mathcal{T}^-(\lambda|u)^\mu_\mu,
\end{aligned} \tag{A.18}$$

or $\mathcal{T}^+(m|u)^i_i$:

$$\begin{aligned}
\tau(u) &= tr(K^+(u)\mathbb{T}(u)) = tr\left((\mathbb{T}^+(u))^{t_0}(K^-(u))^{t_0}\right) \\
&= \sum_{\mu,\nu} tr\Big((\mathbb{T}^+(u))^{t_0}(\bar{\phi}^0_{\lambda,\lambda-\eta\hat{\mu}}(-u))^{t_0}(\phi^0_{\lambda,\lambda-\eta\hat{\mu}}(-u))^{t_0}(K^-(u))^{t_0} \\
&\qquad \times(\tilde{\phi}^0_{\lambda-\eta(\hat{\mu}-\hat{\nu}),\lambda-\eta\hat{\mu}}(u))^{t_0}(\phi^0_{\lambda-\eta(\hat{\mu}-\hat{\nu}),\lambda-\eta\hat{\mu}}(u))^{t_0}\Big)
\end{aligned}$$

$$
\begin{aligned}
&= \sum_{\mu,\nu}\prod_{k\neq\mu}\frac{\sin(\lambda_{\mu k}-\eta)}{\sin(\lambda_{\mu k})}\mathcal{K}(\lambda|u)^{\nu}_{\mu}\mathcal{T}^{+}(\lambda|u)^{\mu}_{\nu} \\
&= \sum_{\mu}\prod_{k\neq\mu}\frac{\sin(\lambda_{\mu k}-\eta)}{\sin(\lambda_{\mu k})}k(\lambda|u)_{\mu}\mathcal{T}^{+}(\lambda|u)^{\mu}_{\mu}. \qquad (A.19)
\end{aligned}
$$

This implies that the transfer matrix, when acts on the subspace (2.38) (resp. (2.39)) with the parameters on shell or off-shell, can be expressed in terms of the operator $\mathcal{T}^{+}$ (resp. $\mathcal{T}^{-}$) as follows:

$$
\begin{aligned}
\tau(u)|\{v_i^{(1)}\}\rangle^{(I)} &= \sum_{\mu}\prod_{k\neq\mu}\frac{\sin(\lambda_{\mu k}-\eta)}{\sin(\lambda_{\mu k})}k(\lambda|u)_{\mu}\mathcal{T}^{+}(\lambda|u)^{\mu}_{\mu}|\{v_i^{(1)}\}\rangle^{(I)}, \\
\tau(u)|\{v_i^{(2)}\}\rangle^{(II)} &= \sum_{\mu}\tilde{k}(\lambda|u)_{\mu}\mathcal{T}^{-}(\lambda|u)^{\mu}_{\mu}|\{v_i^{(2)}\}\rangle^{(II)}. \qquad (A.20)
\end{aligned}
$$

It was shown that in ref. [52] the first set of Bethe states given by (2.38) generated by $\mathcal{T}^{+}(m|u)$ are the eigenstates of our transfer matrix (2.7) with the eigenvalue (2.43) if the parameters $\{v_i^{(1)}\}$ satisfy the Bethe ansatz equation (2.42), while in ref. [45, 46] the second ones are the eigenstates with the eigenvalue (2.45) provided that the corresponding parameters satisfy (2.44).

References

[1] F.A. Smirnov, *Form factors in completely integrable models of quantum field theory*, *Adv. Ser. Math. Phys.* **14** (1992) 1, World Scientific, Singapore (1992).

[2] V.E. Korepin, N.M. Bogoliubov and A.G. Izergin, *Quantum inverse scattering method and correlation functions*, Cambridge University Press, Cambridge U.K. (1993).

[3] I.B. Frenkel and N.Y. Reshetikhin, *Quantum affine algebras and holonomic difference equations*, *Commun. Math. Phys.* **146** (1992) 1 [SPIRES].

[4] B. Davies, O. Foda, M. Jimbo, T. Miwa and A. Nakayashiki, *Diagonalization of the XXZ Hamiltonian by vertex operators*, *Commun. Math. Phys.* **151** (1993) 89 [hep-th/9204064] [SPIRES].

[5] Y. Koyama, *Staggered polarization of vertex models with $U_q(\widehat{sl}(n))$-symmetry*, *Commun. Math. Phys.* **164** (1994) 277 [hep-th/9307197] [SPIRES].

[6] B.-Y. Hou, K.-J. Shi, Y.-S. Wang and W.-L. Yang, *Bosonization of quantum sine-Gordon field with boundary*, *Int. J. Mod. Phys.* **A 12** (1997) 1711 [hep-th/9905197] [SPIRES].

[7] W.-L. Yang and Y.-Z. Zhang, *Highest weight representations of $U_q(\widehat{sl}(2|1))$ and correlation functions of the q-deformed supersymmetric t-J model*, *Nucl. Phys.* **B 547** (1999) 599.

[8] W.-L. Yang and Y.-Z. Zhang, *Level-one highest weight representation of $U_q[sl(\widehat{N}|1)]$ and bosonization of the multicomponent super t-J model*, *J. Math. Phys.* **41** (2000) 5849.

[9] B.-Y. Hou, W.-L. Yang and Y.-Z. Zhang, *The twisted quantum affine algebra $U_q(A_2^{(2)})$ and correlation functions of the Izergin-Korepin model*, *Nucl. Phys.* **556** (1999) 485.

[10] V.E. Korepin, *Calculation of norms of Bethe wave functions*, *Commun. Math. Phys.* **86** (1982) 391 [SPIRES].

[11] A.G. Izergin, *Partition function of the six-vertex model in a finite volume*, *Sov. Phys. Dokl.* **32** (1987) 878.

[12] V.G. Drinfeld, *On constant quasiclassical solution of the QYBE*, *Sov. Math. Dokl.* **28** (1983) 667.

[13] J.M. Maillet and J. Sanchez de Santos, *Drinfeld twists and algebraic Bethe ansatz*, *Am. Math. Soc. Transl.* **201** (2000) 137 [q-alg/9612012] [SPIRES].

[14] N. Kitanine, J.M. Maillet and V. Terras, *Form factors of the XXZ Heisenberg spin-1/2 finite chain*, *Nucl. Phys.* **B 554** (1999) 647.

[15] S.-Y. Zhao, W.-L. Yang and Y.-Z. Zhang, *Determinant representation of correlation functions for the supersymmetric t-J model*, *Commun. Math. Phys.* **268** (2006) 505 [hep-th/0511028] [SPIRES].

[16] S.-Y. Zhao, W.-L. Yang and Y.-Z. Zhang, *On the construction of correlation functions for the integrable supersymmetric fermion models*, *Int. J. Mod. Phys.* **B 20** (2006) 505 [hep-th/0601065] [SPIRES].

[17] W.-L. Yang, Y.-Z. Zhang and S.-Y. Zhao, *Drinfeld twists and algebraic Bethe ansatz of the supersymmetric t-J model*, *JHEP* **12** (2004) 038 [cond-mat/0412182] [SPIRES].

[18] W.-L. Yang, Y.-Z. Zhang and S.-Y. Zhao, *Drinfeld twists and algebraic Bethe ansatz of the supersymmetric model associated with $U_q(gl(m|n))$*, *Commun. Math. Phys.* **264** (2006) 87 [hep-th/0503003] [SPIRES].

[19] Y.-S. Wang, *The scalar products and the norm of Bethe eigenstates for the boundary XXX Heisenberg spin-1/2 finite chain*, *Nucl. Phys.* **B 622** (2002) 633.

[20] N. Kitanine et al., *Correlation functions of the open XXZ chain I*, *J. Stat. Mech.* (2007) P10009 [arXiv:0707.1995] [SPIRES].

[21] E.K. Sklyanin, *Boundary conditions for integrable quantum systems*, *J. Phys.* **A 21** (1988) 2375 [SPIRES].

[22] R.I. Nepomechie, *Functional relations and Bethe ansatz for the XXZ chain*, *J. Stat. Phys.* **111** (2003) 1363 [hep-th/0211001] [SPIRES].

[23] R.I. Nepomechie, *Bethe ansatz solution of the open XXZ chain with nondiagonal boundary terms*, *J. Phys.* **A 37** (2004) 433 [hep-th/0304092] [SPIRES].

[24] J. Cao, H.-Q. Lin, K.-J. Shi and Y. Wang, *Exact solution of XXZ spin chain with unparallel boundary fields*, *Nucl. Phys.* **B 663** (2003) 487.

[25] W.L. Yang and R. Sasaki, *Exact solution of Z_n Belavin model with open boundary condition*, *Nucl. Phys.* **B 679** (2004) 495 [hep-th/0308127] [SPIRES].

[26] W.L. Yang and R. Sasaki, *Solution of the dual reflection equation for $A_{n-1}^{(1)}$ SOS model*, *J. Math. Phys.* **45** (2004) 4301 [hep-th/0308118] [SPIRES].

[27] W.-L. Yang, R. Sasaki and Y.-Z. Zhang, *Z_n elliptic Gaudin model with open boundaries*, *JHEP* **09** (2004) 046 [hep-th/0409002] [SPIRES].

[28] W. Galleas and M.J. Martins, *Solution of the SU(N) vertex model with non-diagonal open boundaries*, *Phys. Lett.* **A 335** (2005) 167.

[29] C.S. Melo, G.A.P. Ribeiro and M.J. Martins, *Bethe ansatz for the XXX-S chain with non-diagonal open boundaries*, *Nucl. Phys.* **B 711** (2005) 565.

[30] J. de Gier and P. Pyatov, *Bethe ansatz for the TemperleyLieb loop model with open boundaries*, *J. Stat. Mech.* (2004) P002.

[31] A. Nichols, V. Rittenberg and J. de Gier, *The effects of spatial constraints on the evolution of weighted complex networks*, *J. Stat. Mech.* (2005) P05003.

[32] J. de Gier, A. Nichols, P. Pyatov and V. Rittenberg, *Magic in the spectra of the XXZ quantum chain with boundaries at* $\Delta = 0$ *and* $\Delta = -1/2$, *Nucl. Phys.* **B 729** (2005) 387 [hep-th/0505062] [SPIRES].

[33] J. de Gier and F.H.L. Essler, *Bethe ansatz solution of the asymmetric exclusion process with open boundaries*, *Phys. Rev. Lett.* **95** (2005) 240601 [SPIRES].

[34] J. de Gier and F.H.L. Essler, *Exact spectral gaps of the asymmetric exclusion process with open boundaries*, *J. Stat. Mech.* (2006) P12011.

[35] W.-L. Yang, Y.-Z. Zhang and M. Gould, *Exact solution of the XXZ Gaudin model with generic open boundaries*, *Nucl. Phys.* **B 698** (2004) 503 [hep-th/0411048] [SPIRES].

[36] Z. Bajnok, *Equivalences between spin models induced by defects*, *J. Stat. Mech.* (2006) P06010 [hep-th/0601107] [SPIRES].

[37] W.-L. Yang and Y.-Z. Zhang, *Exact solution of the* $A_{n-1}^{(1)}$ *trigonometric vertex model with non-diagonal open boundaries*, *JHEP* **01** (2005) 021 [hep-th/0411190] [SPIRES].

[38] W.-L. Yang, Y.-Z. Zhang and R. Sasaki, A_{n-1} *Gaudin model with open boundaries*, *Nucl. Phys.* **729** (2005) 594 [hep-th/0507148] [SPIRES].

[39] A. Doikou and P.P. Martin, *On quantum group symmetry and Bethe ansatz for the asymmetric twin spin chain with integrable boundary*, *J. Stat. Mech.* (2006) P06004 [hep-th/0503019] [SPIRES].

[40] A. Dikou, *The open XXZ and associated models at q root of unity*, *J. Stat. Mech.* (2006) P09010.

[41] R. Murgan, R.I. Nepomechie and C. Shi, *Exact solution of the open XXZ chain with general integrable boundary terms at roots of unity*, *J. Stat. Mech.* (2006) P08006 [hep-th/0605223] [SPIRES].

[42] P. Baseilhac and K. Koizumi, *Exact spectrum of the XXZ open spin chain from the q-Onsager algebra representation theory*, *J. Stat. Mech.* (2007) P09006 [hep-th/0703106] [SPIRES].

[43] W. Galleas, *Functional relations from the Yang-Baxter algebra: eigenvalues of the XXZ model with non-diagonal twisted and open boundary conditions*, *Nucl. Phys.* **B 790** (2008) 524 [SPIRES].

[44] R. Murgan, *Bethe ansatz of the open spin-s XXZ chain with nondiagonal boundary terms*, *JHEP* **04** (2009) 076 [arXiv:0901.3558] [SPIRES].

[45] W.-L. Yang and Y.-Z. Zhang, *On the second reference state and complete eigenstates of the open XXZ chain*, *JHEP* **04** (2007) 044 [hep-th/0703222] [SPIRES].

[46] W.-L. Yang and Y.-Z. Zhang, *Multiple reference states and complete spectrum of the* Z_n *Belavin model with open boundaries*, *Nucl. Phys.* **B 789** (2008) 591 [arXiv:0706.0772] [SPIRES].

[47] R.I. Nepomechie and F. Ravanini, *Completeness of the Bethe ansatz solution of the open XXZ chain with nondiagonal boundary terms*, *J. Phys.* **A 36** (2003) 11391 [*Addendum ibid.* **A 37** (2004) 1945] [hep-th/0307095] [SPIRES].

[48] W.-L. Yang, R.I. Nepomechie and Y.-Z. Zhang, *Q-operator and T-Q relation from the fusion hierarchy*, *Phys. Lett.* **B 633** (2006) 664 [hep-th/0511134] [SPIRES].

[49] W.-L. Yang and Y.-Z. Zhang, *T-Q relation and exact solution for the XYZ chain with general nondiagonal boundary terms*, *Nucl. Phys.* **B 744** (2006) 312 [hep-th/0512154] [SPIRES].

[50] L. Frappat, R.I. Nepomechie and E. Ragoucy, *Out-of-equilibrium relaxation of the EdwardsWilkinson elastic line*, *J. Stat. Mech.* (2007) P09008.

[51] T.-D. Albert, H. Boos, R. Flume, R.H. Poghossian and K. Rulig, *An F-twisted XYZ model*, *Lett. Math. Phys.* **53** (2000) 201.

[52] W.-L. Yang and Y.-Z. Zhang, *Drinfeld twists of the open XXZ chain with non-diagonal boundary terms*, *Nucl. Phys.* **B 831** (2010) 408 [arXiv:1011.4120] [SPIRES].

[53] H.J. de Vega and A. Gonzalez Ruiz, *Boundary K matrices for the six vertex and the $n(2n-1)A_{n-1}$ vertex models*, *J. Phys.* **A 26** (1993) L519 [hep-th/9211114] [SPIRES].

[54] S. Ghoshal and A.B. Zamolodchikov, *Boundary S matrix and boundary state in two-dimensional integrable quantum field theory*, *Int. J. Mod. Phys.* **A 9** (1994) 3841 [*Erratum ibid.* **A 9** (1994) 4353] [hep-th/9306002] [SPIRES].

[55] R.J. Baxter, *Exactly solved models in statistical mechanics*, Academic Press, New York U.S.A. (1982).

[56] G. Felder and A. Varchenko, *Algebraic Bethe ansatz for the elliptic quantum group $E_{\tau,\eta}(sl_2)$*, *Nucl. Phys.* **B 480** (1996) 485.

[57] B.-Y. Hou, R. Sasaki and W.-L. Yang, *Algebraic Bethe ansatz for the elliptic quantum group $E_{\tau,\eta}(sl_n)$ and its applications*, *Nucl. Phys.* **B 663** (2003) 467 [hep-th/0303077] [SPIRES].

[58] B. Hou, R. Sasaki and W.-L. Yang, *Eigenvalues of Ruijsenaars-Schneider models associated with A_{n-1} root system in Bethe ansatz formalism*, *J. Math. Phys.* **45** (2004) 559 [hep-th/0309194] [SPIRES].

[59] O. Tsuchiya, *Determinant formula for the six-vertex model with reflecting end*, *J. Math. Phys.* **39** (1998) 5946.

[60] W.-L. Yang et al., *Determinant formula for the partition function of the six-vertex model with an non-diagonal reflecting end*, *Nucl. Phys.* **B 844** (2011) 289.

人世变而道长青*

——五十年治学路

侯伯宇

一、早年经历

侯伯宇出生于1930年9月11日,日寇侵占东北的20世纪30年代。侯伯宇的父亲为黄埔军校一期生,当时由周恩来介绍入中国共产党。上海工人第三次武装起义,其为主席团成员,培训工人纠察队,率队攻占南市警察局,后为南昌起义第一枪的教导团团长。因顾顺章叛变,寻找不到组织,辗转回到国民党部队。抗战时,侯伯宇随父于湘皖等地上小学,父亲提倡侯伯宇科学救国,做出更好的武器。侯伯宇最后到了精英荟萃的重庆南开中学,专长数学。原子弹迫使日本投降,使侯关注到物理学。1947年侯高二时考上清华,保留学籍。1948年,燕京保送考,侯全国第一,得到全额奖学金,交纳学费后,还留剩余可观的生活费。侯考虑上医学预科这门金饭碗,便与同学找仓孝龢老师(地下党员),问如何选择是好。仓说,如果中国老是这样,学什么都没用,但不会是这样的。于是,侯选定去清华学物理。1948年,随家去港,转台后入台大。侯觉得其教师水平、图书数量均远赶不上清华。侯得悉清华有同学仍去图书馆,就放弃出国的机会,回到北京。美国越过三八线,朝鲜战争爆发,侯报名参干,被录取学俄文。停战谈判后,侯向高教部部长助理写信申请得到回去学物理的机会。因鞍钢三大建设开始,缺翻译,侯含泪主动提出去鞍山。到了东北,无论政治、外文学习,或是翻译工作,侯均名列前茅。1955年"肃反",侯被鞍山市委列为"重点对象"。父亲找"肃反全国十人小组"的一位打了招呼,侯得以乘1956年"向科学进军号召"的东风,到全国唯一招插班生的西北大学,侯以自学为主,不久被全体同学选为班长。1958年,侯倦于"反右"的上纲,以及带领同学上屋顶轰麻雀,从粪坑捞蛆、数个数,故提请早日工作。交大欲成立加速器专业,西大将侯

---

* 本文原为侯伯宇先生应《二十世纪知名科学家学术成就概览》丛书《物理卷》之邀所写传文。草稿成时,先生已罹患重症,无力再行修改,以致此文最终无法付梓。但先生撰文时颇费心力,无论取材、文字皆斟酌再三,故借此集另附文题刊出,以为告慰。

分配给交大。虽侯学习成绩优异，但因去台问题无结论，政治表现又是中，虽侯是“反右核心小组成员”，交大还是将侯转为该校矿山机电系分出成立的西安矿业学院。这期间除“反右”运动外，侯伯宇抓紧一切时间钻研数理。1960 年他劳逸结合，恢复了周末与寒暑假，开始了科研工作。

二、分子链的相互作用 Green 函数

这是在 1960 年劳逸结合时期，有了星期日与寒暑假，侯挤出时间来钻研的处女作。西安矿业学院同事量化教师文振翼约侯论证唐敖庆在吉林大学的量化讨论班，涉及美国科学院院士 Hirshfelder 等未能求得交叠区同类项合并的式子。侯用回路积分，将各区统一为一个式子，各区用绕回路不同方向积出，均表为三 Bessel 函数积分。但交叠区的三 Bessel 函数，遍翻苏函数积分表，Erdeyli 高级超越函数等都没给出，只好将这三个 Bessel 函数按幂次级数递推表出。求和出现的项数，以几何级数方式增加，逐级得到简化的递推式。不幸并项得到的是条件收敛级数。又用边界条件的对称性，选定如何并项为偶次，于是得到收敛级数。

三、局域坐标系中的旋量球函数及算子，helicity 幅

考察从固定中心发出的辐射或在其上的散射，以及从质心系观察的散射与辐射时，既要考察到辐射源、散射中心的极化，又要考虑到辐射、散射波的极化依赖于方向角与径向距离的关系。通常采用 Clebsch-Gordan 系数 $C^{l,s,J}_{m,m_s,M}$，将固定坐标系的轨道矩 l、磁量子数 m、自旋 s、磁量子数 m_s，耦合为总动量矩 J、磁量子数 M，构成旋量球函数。侯伯宇发现，用从固定系标架到球面局域三足标架的转动矩阵，可直接表达旋量波函数，即所谓 helicity 幅。侯伯宇又注意到，将微分算子 ∇ 的横向分量作局域系 Hodge 对偶，得到轨道动量矩算子 $\vec{l}$，再将 $\vec{l}$ 表为可观察的总角动量与自旋的差，就可以利用 helicity 系的量子数直接表出梯度、旋度等公式。计算跃迁矩阵时，先表为始、末态旋量波函数及散射中心的多极旋量算子，再分离变量为角向和径向。角变量部分的三个 D 函数乘积的积分表示为两个 C-G 系数；径向部分表为与三者的量子数对应的三个 Bessel 型函数乘积的积分。将 Rossi 两本书中各径向、横向多极辐式的繁杂结果直截了当地表为用 C-G 系数耦

合到一起的已知的径向积分式。

评注：李杨 CP 宇称破缺奖，当时国际上各大加速器中心竟相研究粒子的碰撞反应。侯伯宇仔细推算了《物理学报》上各有关文章，发现将按沿径向极化等量子数写的波函数的各分量沿固定观察系投影，分类合并后，再按径向投影，甚为繁杂。采用始终按径向运算的方法。又发现可将微商 ∇ 的横向作 Hodge 对偶，化为代数运算，就可简洁写出结果。

20 世纪 60 年代，研究 Yang-Mills 丛，吴杨固定坐标规范，更加复杂。侯在径横向加上场的贡献，结果与 20 世纪 80 年代 Witten 拓扑场论用空间 $spinSO(2)_{左}$ 与 Yang-Mills 的 $SU(2)$ 合成的 $spin'(4)$ 相当。

四、SU_n 群不可约表示的正交归一基底与代数

Gelfand 和 Tseythn 于 1950 年给出 su(N) 代数表示式后十余年，其证明虽经多人努力，未能成功。侯伯宇 1965 年用玻色湮灭产生算子构成的负素根算子与钩算子，依次构成了从各 Gelfand 正则基降到与它相邻的各低阶相邻正则基的降算子多项式。同时又用正素根算子及钩算子构成升算子。再用这升算子递升得到各对偶正则基。侯伯宇利用玻色子与钩的对易法则成功地求得各正则基的归一化系数，从而构成了正交归一、完整的正则基。然后正确无误地将各正负素根算子表示式表为各子群内的钩因子及相邻子群间的不可约张量因子的乘积。改正了相角符号中以往文献从 Gelfand 沿袭下来的错误。侯伯宇的方法被用来构成 $su_q(N)$ 量子群与 Yangian 的正则基与代数。

评注：劳逸结合，同时给知识分子脱帽子，并提出向科研进军。侯报考中科院数学研究所张宗燧的研究生。张见到侯局部系文的稿子惊呼侯为 expert，而且侯的考分最高，张从绝不录取岁数大的改为录取岁数最大的侯。当时粒子物理研究弱电统一，用到 Yang-Mills。高秩 Lie 代数的表示是基本工具。这时 Weyl 基不正交归一，用降算子构成正交归一的 Gelfand 基，再用升算子作用在 Weyl 基上，得到共轭基。侯注意到相互共轭基的相角的差异，才如上得到正确的符号。Jimbo 将 Poisson 括号量子化为对易法则，得到量子群的 Gelfand 基。上野将振子改为 q ($\sim e^{\hbar}$) 振子，用侯的方法构成了量子群，这实际就是由普适包络代数得量子代数普适 R 矩阵的 Drinfeld-Hopf 代数方法。第九节侯用

quasi-twistedHopf 的 p、q 得到椭圆型 RSOS 及得其双标度极限 $\hbar$ 与交叉参数 η 的有质量场式。代替 Feigin 等采用的 Wakimoto 临界水平量子化。这也是第十节实现双 gerbal motivic 表示的关键。

五、各级微扰展开下自发对称破缺的规范无关性、幺正性及可重整性

规范场的自发对称破缺理论有两种相反而相成的方案。其中一种方案明显可重整,但因含非物理粒子,表观上似违反幺正性。另一种方案只含物理粒子,它明显地幺正,但是表观上却不可重整。't Hooft 证明了可重整方案的幺正性及两种方案的 S 矩阵元的等价。这些证明采用泛函积分研究全格林函数。但是研究量子色动力学各过程时,不可能求出全格林函数,必须用微扰论计算。1976 年,侯伯宇采用微扰论,逐步讨论可重整规范下内线非物理粒子的贡献在同阶图间的各种相消机构。明显地证明互相抵消后,除与环线对应有 $\delta^4(0)$ 项以外,只剩下与标粒子"外"线的质壳外动量互相依存的部分。当有些"外"线不在质壳上时,这些"外"线实际上是虚粒子内线,应与更外一层的线相连。如此类推,直到所有的外线都是质壳上的粒子线为止,于是非物理粒子的规范有关贡献正好全部抵消掉。发散项是整个被积式抵消掉,与正常化及重整化无关,也无须取极限 $\xi \to 0$,只剩下物理粒子的贡献,也就是幺正规范。这就明显地证明了物理过程的规范无关性及任意规范的幺正性。于是,量子电动力学 Abel 规范有关项按耦合常数幂次各级抵消为规范无关的 Ward-Takahashi 恒等式,被侯推广到 SU(2) Yang-Mills-Higgs 场。2003 年底,Witten 对极大 helicity 破坏辐作从动量空间到 Penrose 的 twistor(缠量)空间的 Fourier 变换,论证了变得的幅的支集在全纯曲线上,这相当于,生成侯推广的 Ward 恒等式的不变性为 Penrose 变换。

评注:1973 年,侯回到西北大学,教师们厌倦、抵制无休止的路线斗争。侯在给工农兵学员上课辅导之外的业余,科研直到深夜,得到默许,甚至于赞赏。这时,权威 S. Weinberg 等都纷纷考虑如何计算繁杂的 Feynman 图的同类项合并。侯的此文英文稿遭《中国科学》编委审稿的意见,是't Hooft 已解决的问题,不采用,但未退稿。亏得中文稿因《物理学报》所在单位与编辑部的理解,得以刊出。但缺英文稿中 s,f,u 各道求和式子。这些可用来具体证明 Coach、Witten 等的猜想。

六、规范场的拓扑非平庸的稳定解(磁涡流、单极)的整体宏观量子数及其规范协变量、电、磁荷流等的定域分布

1974 年't Hooft 提出由有质量 Higgs 场 $\vec{\phi}(x)$ 与 $SU(2)$ 规范场 $\vec{A}(x)$ 构成的单极。侯伯宇发现,$SU(2)$ 势 $\vec{A}(x)$ 可按照伴随丛 $\vec{\phi}(x)$ 的方向矢 $\hat{n}=\frac{\vec{\phi}(x)}{|\phi(x)|}$ 协变地分解为 $U(1)$ 规范场 $\vec{H}(x)$ 及协变矢量 $\vec{K}(x)$:

$$\vec{A}(x)-\vec{H}(x)+\vec{K}(x)$$

其中

$$\vec{H}(x)=(\vec{A}\cdot\hat{n})\hat{n}-\hat{n}\times d\hat{n}$$

它含与 n̂ 正交的第二项

$$\vec{K}(x)\equiv\hat{n}\times(d+\vec{A}\times)\hat{n}$$

方向矢 $\hat{n}(x)$ 局域在 U(1) 势 $\vec{H}(x)$ 平移下不变 $\partial_\mu\hat{n}(x)+[\vec{H}(x),\hat{n}(x)]=0$;整体在 $\int\vec{A}(x)dx$ 的和乐群下不变。$\hat{n}(x)$ 在无穷远边界上的映像对应的 Kronecker 指数也就是 $\hat{n}(x)$ 覆盖边界的次数,侯明显计算证明它等于 $U(1)$ 场强 $F(x)$ 穿过空间边界的通量 $\oiint_{r=\infty}F(x)$;这里 $F=d\vec{H}+\vec{H}\times\vec{H}$ 为 $U(1)$ 势 $\vec{H}$ 的曲率。侯伯宇与侯伯元发现 BPS 单极内自对偶电磁荷流、序参数的分布及其屏蔽穿透比,如二类超导。

评注:这时,政治运动愈演愈烈。侯将't Hooft 文打印稿和侯伯宇论证 $\hat{n}=\frac{\vec{\phi}(x)}{|\phi(x)|}$ 的定向相反对应相反的磁荷的《科学通报》文稿寄兰州段一士。段一士、葛墨林出差到西安时找到侯,介绍在原点得到 Jacobi 的 δ 函数型奇异。侯在春节最终算出,这是和乐群下不变的 $\hat{n}(x)$ 场在原点正好也就是单极支点,也是可去奇异,也就是现在 Witten 文的驯服奇异。这是 Floer 和乐理论作为连通和,分析 Gromov-Witten 不变量的出发点。2008 年,侯在京都报告中采用四维时空的胀开 $\hat{\mathbf{p}}^2$ 上的驯服分歧,见第十节。

七、模空间隐藏对称性的生成代数及由之产生的非定域守恒流

在用和乐群平移得到的活动标架中,将扁西方程下平庸守恒的拓扑磁荷流的生成

元，向与之Hodge对偶的、正交的、导致运动方程守恒的Noether电荷流的生成元扭转。将这二流矢按扭转角合成得到非定域流，其边界项的变更互补，其和为全散度，故守恒。此流生成以扭转角为loop参数的loop群。又将扭转算子对其复参数作微伸缩变换，展开得到Virasoro型变换。侯又与李卫构成引力度规扭势的仿射群与伸缩Virasoro群的半直接积。它升降多极矩，超出仅由Kinnerseley与Chitre的仿射型变换生成的Geroch群。从而实现可以由四维引力场的真空平庸解变到静轴或柱对称一切解的Geroch猜想。Geroch构造了有无穷多参数的Geroch群。Kinnerseley、Chitre变换为Geroch群的无穷小形式，但是它不能改变正则坐标，不能改变多极矩，从而并未实现Geroch猜想。而Hou-Li变换的确改变正则坐标，起到升降多极矩的作用。因而实现了Geroch猜想。变换的构成及各代数关系推导式被Julia引用于关键的扭曲自对偶约束及双谱参数，用来构成十一维伸缩超引力渐进几乎处处平的一切整体解。

评注：拓扑场论量子化的路径积分应包括将孤立子空间各sector求和，涉及孤立子空间的隐藏对称性。20世纪70年代初，孤立子的非局域对称性及相应守恒量为研究热点。我国周光召、宋行长、郭汉英、吴可与侯伯宇兄弟、王佩及吴咏时、葛墨林都做了大量的研究。侯用协变分解，得到Kac-Moody型变换，曾被一些同行称为“侯变换”，然后他们又称之为“H变换”，而H是Riemann-Hilbert中的H。侯又进一步做出了Virasoro型变换。J. Schwarz弦论大会总结，引用大量对称性变换。而实际上，Kac-Moody型是侯的，同一条类除Shaposnik文有关系，Dolan文源于侯耶鲁文外，其他文章根本不是Kac-Moody型。他引用Virasoro条，将陈文列为第一，而陈文是引用侯文而作的。又如不知道只限于self-dual-Yang-Mills而不是整个Yang-Mills的拉氏函数，这实际上侯与宋也做出来了。如果是犹太科学家或日本科学家的工作被如此乱引，一定会群起而纠正。关于侯、宋作的非定域变换亏得花工夫向乔玲丽介绍，请她协同写出，才被认知。而侯、宋的含Wess-Zumino-Witten项的self-dual拉氏量时间上前于Donaldson，却在投稿莫须有地退回后，束之高阁，我们未再力争，也是我们的不足。

八、量子化规范场的反常与其大范围拓扑性质

周光召在世纪之交谈到，他重新攀登物理事业，于20世纪80年代中，又达到国际最

前沿。当时他指导吴可,由北大宋行长、理论物理所郭汉英与侯伯宇等人合作的工作获得国家自然科学二等奖。侯伯宇的贡献为给出反常系列整体特征数的继承性以及3上同调链的提出与分析。Atiyah 用时空底流形曲率的 A-roof 多项式乘上规范场陈类构成的解析族指数来分析反常系列,Faddeev 用联络空间的规范群上同调链分析反常系列的拓扑,都没有给出整体拓扑指数。侯伯宇将流守恒反常、对易反常和 Jacobi 反常的局域微分继承式积分,然后用 Cech 双上同调理论建立了解析族指数反常系列与规范群上同调链的联系:他先用边缘同调算子 δ 与上同调解析微分算子∂构成双线性同伦算子 $K = \partial\delta + \delta\partial$,又将底流形上的 de Rham 外微分与 K 的和表为 Cartan 型全微分,再将之作用在规范群上同调链上,在底流形上积分。从而将族指数表为底流形普适覆盖交叠区上积分的交替和,由底流形覆盖网与联络空间上链的对偶,用 Cech 双上同调理论描述了反常递降方程的结构,给出两者统一的反常整体继承特征数。又将单形上同调链的递降方程推广到非三角剖分有非平庸相交边时的相对上同调,证明了联络 $A(x)$ 的仿射空间 $A(x)$ 与规范群 $\Gamma(x)$ 的陪集 $A(x)/\Gamma(x)$ 上 $\pi_k(A/G) = \pi_{k-1}(A/G)$。还将此陪集丛乘上 U(1)线丛,构成普适丛。选择规范 $D_\mu A = 0$ 使纤维与底正交。明显构造了微分上同调形式系列,积分得到整体特征的继承递降 $A(x)/\Gamma(x)$ 上 $\pi_k(A/G) = \pi_{k-1}(A/G) = \cdots$。侯伯宇与 Jackiw 同时分别提出3上同调链。但侯伯宇先发现结合律反常满足五角恒等式,且发现 Jacobi 反常的自洽条件是四重对易式,不是 Malcev 代数,澄清了 Jackiw 与 Zumino 的争论。

评注:在给出 Faddeev 规范群上同调的整体量子数的继承性时,侯伯宇、侯伯元发现了协变的 de Rham 上同调势,其实就是 Cartan 算子。进一步考虑到宇称对称,就是第十节的式子,此项工作的确如周光召所称的回到了国际最前列。周当时也说,在国际上要赶超就要按照他们的竞赛规则,注意真正的热点,也就是说弦论。实际上,就是不要闭关自守,政治挂帅,鼓吹自己的"层子"。但是,周光召因对我国科学事业的贡献和参与铀弹研制而被重用,离开了理论物理所,不再过问纯粹理论科研,而此项成果未见得到应有的赞许。与 Calabi-Yau 猜想的结合,造成了弦论的一次革命,而侯没有理解这在基础科学上根本、深远的意义,误以为该做有实效、可马上观测的理论,而转向了可积与量子群。

九、可积模型与量子群

侯伯宇等最先用库仑气玻色子方法的 Dotsenko 积分中的单值行为表出极小共形场共形块的辫子矩阵，又用积分直接推出 $SU(2)$ WZW 模型的高自旋共形块的聚合与辫子矩阵，证明它们分别用量子群的两种 q-$6j$ 系数表示。

评注：侯等证明了共形块为三角函数型，而其单值函数为椭圆函数型，比 Gepner、Qiu 早许多年。但执笔者只顾作推导，将其中辫子与聚合的五角等式，如何将通常角动量理论中 Biedenharn 的五角等式、六角等式推广为 q 的，这一重要结果淹没在大量的计算推导中。

（一）有限格点与多体长程作用

这里，椭圆函数型（如各向异性 XYZ 链）包含三角函数型，有理函数型（如 XXY，XXX 链）为其退化情况。

1982 年，Sklyanin 作出 $n=2$ 椭圆量子代数，Cherednik 于 1985 年试图推广到 $n>2$。侯伯宇等从顶角模型分离出谱，构成 $n>2$ 椭圆函数型经典及量子代数。创先构成其循环表示及差分算子表示。侯等用所发现的 $sl(\mathrm{n})$ 椭圆型差分算子表出 RSOS 模型杨 Baxter 表示矩阵，给出转移矩阵 T 的最高权矢 Λ 的权函数。求得 T 张量积的本征矢赝真空与 Λ 所差相因子，依赖于对 RSOS 高度 λ 与扭转参数 α 的关系。再构造高秩 nested Bethe Ansatz，使得 T 乘积不含不愿有项（unwanted term），得与 λ 无关的本征值。取一维零权空间得相对论多体模型。用聚合缠结算子变换得高自旋链。总之得到了 RSOS、长程多体及自旋链三种椭圆函数型模型的基本方程，即 BA 及其本征值、本征矢。

评注：侯此文在 Felder 的数学物理大会和数学大会报告之前就已做出。当时曾将结果投给 Communication in Mathematical Physics，审稿人要求做一些修改就可以考虑。撰写人杨仲侠认为是被否定，不肯再写。侯只好找到国内 Drinfeld-Manin 理论的专家来予以修改，不料他采用法国做椭圆量子代数的专家的一个不成功的方式将文章改得面目全非，无法采用，只好将之束之高阁。侯因为英文打字困难，多花了好多工夫，失去了这一好机会，未曾加以投出，侯也未坚持，以致大家所公认的为数年后 Moore 的结果。侯本应邀去首尔做这方面报告，因当时韩国领事馆工作人员的刁难，最后签证未能赶上飞机，失去了介绍的机会。原计算者得到了很好的结果，湮没在大量的推导中。侯将此结果拿给周光召看，周认为太数学了。这都是我们的缺陷。

（二）有质量量子代数的可积

侯伯宇等提出采用 Drinfeld 的扭曲 Hopf 代数可从 XYZ 格点的量子代数得到 RSOS 模型的量子代数，发现 $Z_n \times Z_n$ 面模型在无穷多格点极限，也就是热力学极限下，既有动力学扭曲又有中心扩张的量子代数，用动力学移动了的 Gauss 分解明显求出这流。将这种格点代数取双标度极限，或将临界 level 的 Yangian-double 量子化，二法得到同一种新代数，它描述有质量场的散射关系。并且 Miura 变换得有质量场的 W 代数。给出流代数、W 代数及其顶角算子的双畸变玻色振子式。

评注：Feigin、Frenkel 讨论了 $W_{p,q,\pi}$ 动力学 twisted 椭圆型格点 RSOS 代数，对应的是 Asai、Jimbo、Miwa、Pugai 文 $\alpha_+ = \sqrt{(r+1)/r}$、$\alpha_- = \sqrt{(r-1)/r}$，$\alpha_+$、$\alpha_-$ 与 π 即通常屏蔽荷的 double－scaling 极限，侯、杨做的有质量场代数，得到 $A_{\hbar,\eta,\hat{\pi}}(\hat{g}l_2)$ 代数。其中 $x^{2v} = p^{\frac{-i\beta}{\hbar}}$，$q = p^{-\frac{1}{\eta}}p \to 1$ 的 $\hbar$、η 是 p、q 的双 scaling 极限。

侯伯宇、侯伯元、马中骐独立于 Kirrilov 与 Reshetikhen，创先且较他们“明显而具体地推得各系数及其对称性与求和律”，将二子表示递降各顺序间所差 q 相角求和，得到由 q 系数构成的 Racah 型 C-G 系数。再改变求和指标得 $3j$ 对称的 Van der Wardaen 型 $3j$ 系数。进一步给出了 q-$3j$ 系数及其三个 j 顺序置换与 q 反转为 $1/q$ 时的关系。发现有聚合、辫子两种 q 拉卡系数及其关系。按照聚合顺序采用不同于原角动量理论的求和顺序成功推得聚合型系数的 q 乘积式，得 144 个系数的全部对称关系；两拉卡系数满足辫子群与五角结合律；与辫子矩阵满足 Skein 带图关系。特别是辫子和聚合的五角关系表达得极为清楚。求聚合时所用求和顺序加以改变，这是侯伯元在春节时算出来的。Gervais 说这正是他研究二维引力等聚合时所需的聚合型。

评注：马中骐在 John Hopkins 会报告介绍此文时，得到与会物理学家的赞扬。因为与大家熟悉的 C-G 系数算法相同，觉得比数学家写得易懂，杨振宁听说也很高兴。但有中国人说这是 Kirrilov 算出来的，而实际上，侯在京都见到 Kirrilov，他也承认，如前面正文所说，许多是我们具体算出来的。反常会后，侯伯宇去石溪。杨谈到 Kondo 效应可直接算出，值得注意，侯开始做可积模型、量子群。遂与西北大学石康杰、周玉魁、卫华、岳瑞宏、杨仲侠、范桁、杨文力、赵柳及侯伯元合作，做了以上所讲的切实可靠的工作，特别是椭圆型。但是在国内得不到有力支持，国外孤军奋战，全靠有些合作者在国外有机会与最前沿的专家交流讨论，得到认可。例如，周玉魁可与 Zuber 合作，杨文力除与 Sasaki 合作外，还与量子群创始人 Belavin 合作，使我们的工作在艰难地与京都学派竞争中，虽受到歧视，仍然得到量子群奠基人 Korepin、Semirnov 等一些俄国人的赞许。虽然对侯、侯、马之文，Reshetikhin 一度遇到来自中国人的障碍，但也给予了肯定。最近，在京都，见第十节，就推广到 $\hat{\mathbf{P}}^2$，这里要对侯、杨、赵、丁的有质量场 quasi-Hopf 动力学扭曲直接量子化，用到侯、杨的Ⅰ、Ⅱ型顶角算子的玻色振子表示式，及相应的 R 矩阵来表达量子 torus 上的 Donaldson-Thomas 不变量。

十、四维时空的胀开 $\hat{\mathbf{P}}^2$ 上的驯服分歧

这是我正与吴可等协作的项目。Witten 等未考虑宇称，从而是静止的 't Hoon 单极。它的支集为球心。联络的稳定性条件为有 level（水平）1 的循环（cyclic）矢量，取值在次对角线上。但是应该考虑宇称对称，如 Kontsevich 与 Soibelman。侯发现这样一来，求稳定联络时要考查 $\hat{\mathbf{P}}^2$ 如下。

$\hat{\mathbf{P}}^2$ 是 $\mathbf{P}^2$——即时空——在例外点上 C 的胀开，如下式：

$$\{([z_0:z_1:z_2],[z:\omega])\in\hat{\mathbf{P}}^2\times\mathbf{P}^1 \mid z_1\omega=z_2z\}$$

其例外点为：

$$l_\infty : z_0=0,\quad C: z_1=z_2=0$$

在双重彷射微分同胚 $(t_1,t_2)\in\mathbf{C}^*\times\mathbf{C}^*$ 下：

$$([z_0:z_1:z_2],[z:\omega])\mapsto([z_0:t_1z_1:t_2z_2],[t_1z:t_2\omega])$$

固定点为：

- 在 C 点，为 't Hooft 算子的支集：

$$p_1=([1:0:0],[1:0]),\quad p_2=([1:0:0],[0:1])$$

- 在无穷远线 l_∞：

$$q_1 = ([0:1:0],[1:0]),\quad q_2 = ([0:0:1],[0:1])$$

上,为标架。

侯伯宇提出的在 $\hat{\mathbf{P}}^2$ 的例外除子 C 有正规奇异性的稳定联络为:

$$A_+ = \partial_+ \Phi + e^{ad\Phi}\sum_{i=0}^{r} E_{+\alpha_i}$$

$$A_- = -\partial_- \Phi + e^{-ad\Phi}\sum_{i=0}^{r} E_{-\alpha_i} \tag{1}$$

其中 $\Phi = \phi\rho + \eta d + \xi c$ 式中:出射波 A_+、入射波 A_-,分别有支集 p_1、p_2,沿左右实零光线 $z_+ = \frac{z_1}{z_0}$、$z_- = \frac{z_2}{z_0}$ 传播。分别在包围出射(入射)谱线 $z_- = 0(z_+ = 0)$ 的球面 S^2 的 $\frac{z}{\omega}$、$\frac{\omega}{z}$ 点,有单位指向矢量。

在 $\hat{A}_4^{(1)}$ 的级(level)为 1 的表示,相容条件 $F = dA + A\wedge A = 0$ 的保面积($t_1t_2 = 1$)解为:

$$A_\pm(F) = \sum_{j=0}^{r-1}\left[\partial_\pm \phi_j\rho_j + e^{\pm\eta\phi_j}\lambda^{\pm1}(E)_{j,j+1}\right] + \partial_\pm \xi(F)c$$

$$(j\in \mathbf{Z}mod r,\ \mathrm{F} = \mu \mathrm{z}_+ + \mu^{-1}\mathrm{z}_-) \tag{2}$$

式中我们选 $\rho = diag(1,\omega,\cdots,\omega^r)(\omega = (-1)^{1/r})$,$d = \lambda\frac{d}{d\lambda}$。$\xi$ 在谱线相交处的全纯反常,取值于 Drinfeld 中心。

- 式(2)的解为:

$$\phi_j = \phi\rho_j,\quad \eta = \frac{\pi i}{r},\quad \xi = -2\ln\cosh\frac{\phi}{2}$$

也是共形仿射户田方程的解。

- 有驯服分歧的解为:

$$\phi = -\log\tanh F,\ \xi = -\log\sinh\frac{F}{2}$$

仿射缠量方程及 **monad** 构造

- 有全纯反常 $\partial_\pm \xi(F)c$ 时的孤子解的单值矩阵:

$$M_+ = \oint_{p1} A_+(F)\delta(z_-)dz_+dz_-,\quad M_- = \oint_{p2} A_-(F)\delta(z_+)dz_+dz_-$$

$$\begin{aligned} M_\pm(\lambda) &= M_\pm^h(\lambda)\rho + M_\pm^\alpha(\lambda)E^\alpha + M_\pm^c(\lambda)\mathrm{c} \\ &= M_\pm^h(\lambda)\rho + M_\pm^{\pm1}(\lambda)E^{\pm1} \pm Res_{[C]}\partial_\pm \xi(F)\theta(t)\mathrm{c} \end{aligned} \tag{3}$$

式中 $\lambda = e^t$。

- 复方程：

$$\lambda \frac{M_+^\alpha(\lambda)}{d\lambda} - [M_+^h(\lambda), M_+^\alpha(\lambda)] = 0 \tag{4}$$

$$-\lambda \frac{M_-^\alpha(\lambda)}{d\lambda} - [M_-^h(\lambda), M_-^\alpha(\lambda)] = 0 \tag{5}$$

- 实方程：

$$\lambda \frac{dM_+^h(\lambda)}{d\lambda} - \lambda \frac{dM_-^h(\lambda)}{d\lambda} + [M_+^h(\lambda), M_+^h(\lambda)] + [M_+^\alpha(\lambda), M_-^\alpha(\lambda)] = \mathrm{Res}_{[C]}\partial_+\xi(F)\delta_+(t)\mathrm{c} + \mathrm{Res}_{[C]}\partial_-\xi(F)\delta_-(t)\mathrm{c} \tag{6}$$

评注：复方程为仿射单极及反单极的仿射化小缠子(mini twistor)。

- 驯服情况：

$$\mathrm{Res}_{[C]}\partial_+\xi(F) = \mu, \ \mathrm{Res}_{[C]}\partial_-\xi(F) = \mu^{-1} \tag{7}$$

- Monad

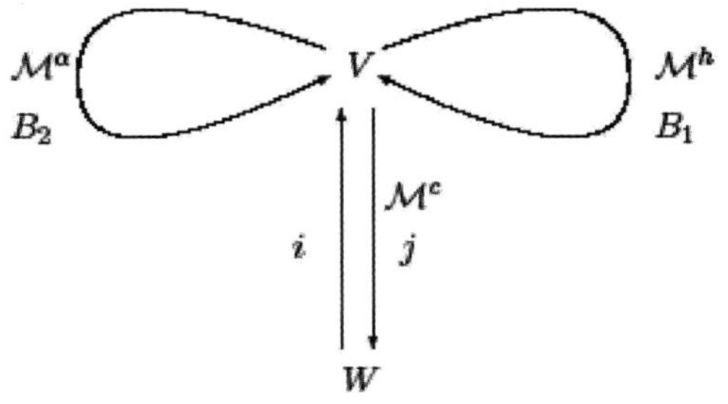

用仿射解实现了 Bryan 文中 King 的构造。即 Nakajima 和 Yoshioka 的 sheaf，但是要引入其 sheaf 的仿射化。

侯伯宇将这仿射 Monad 的代数结构明显表示为双 master 群 $GL(\infty,\infty)$ 的代数 $gl(\infty,\infty)$。Frenkel 与朱新文两年前就宣称要做这个代数，尚未见结果。为此先简单介绍熟知的 Kac 无穷维代数书中的复一维 $gl(\infty)$ 的情况。再介绍 Frenkel 与朱新文如何推广到复二维 $GL(\infty,\infty)$，最后写出侯伯宇给的 $gl(\infty,\infty)$ 的明显式。

$Gl(\infty)$ 是复一维数域 $K=C((t))$ 的连续内自同构 End(K) 的李代数。K 的紧开子空间称为格子(lattice)。例如 $O_K = C[[t]] \subset K$ 为格子。gl_∞^+ 为 O_K 的连续内自同构，Tr: gl_f为其离散内自同构的双边理想，即有开的核。

$$\tilde{g}l_\infty = \{(A,X) \in gl_\infty^+ \times gl_\infty \mid A - \pi(X) \in gl_f\}$$

式中各 π 由 $\pi: K \to O_K$ 诱导而得。$\tilde{g}l_\infty$ 给出普适中心扩张。

$$0 \to gl_f \xrightarrow{i} \tilde{g}l_\infty \xrightarrow{p} gl_\infty \to 0。$$

但由代数的正合系列只能推出对应的群的左正合。于是引进群的满同态 deg：$GL_\infty \to Z$

$$\deg(g) = \dim\left(\frac{O_K}{O_K \cap gO_K}\right) - \dim\left(\frac{gO_K}{O_K \cap gO_K}\right)$$

deg(g)的核为 $GL_\infty^0 = \ker(\deg)$。它的中心扩张上同调类给出中心扩张 $\hat{GL}_\infty$：

$$1 \to C^x \to G\hat{L}_\infty \to GL_\infty \to 1$$

又可以用几何方法构成 $\hat{GL}_\infty$，用场的行列式丛 $\mathbf{L}$ 如下：GL_∞ 的齐性空间为 Grassmannian $Gr(\mathbf{C})$，由 $Gr(\mathbf{R})$的二格子 L,L'确定 $Gr(\mathbf{R})$上的行列式丛 $\mathbf{L}$

$$\det(L \mid L') := \wedge^{top}(L/t^N(R \hat{\otimes} O_K)) \otimes (\wedge^{top}(L'/t^N(R \hat{\otimes} O_K))^{-1}$$

线性定域紧致的拓扑空间 V 上的行列式理论 Δ。对于 V 中任何格子对于 $L_1 \subset L_2$ 均有同构

$$\Delta_{L_1,L_2}: \Delta(L_1) \otimes \det(L_2/L_1) \to \Delta(L_2)$$

满足结合律的行列式。Grebe D_V 是这样一个范畴。它的对象是 V 上 Δ，而态射是行列式理论的同构。

令 $K = C((t))$，其对偶 $K^* = C((t))dt$，设 $CL_K = Cl(K \oplus K^*)$，它由基底 $\phi_n = t^n$，$\phi_n^* = t^n \dfrac{dt}{t}$ 产生。

$$[\phi_n, \phi_m]_+ = 0, \quad [\phi_n^*, \phi_m^*]_+ = 0, \quad [\phi_n, \phi_m^*] = \delta_{n,-m}$$

则可递进证明 Gr 上 $\mathbf{L}$ 丛的截面 $\Gamma(Gr, \mathbf{L}^*)^*$ 可表示为 Clifford 模如下：

$$\mathrm{Ind}_{\wedge(O_K \oplus O_K dt)}^{Cl_K}(C \mid 0\rangle_{O_K}) \cong \Gamma(Gr, \mathbf{L}^*) = U(Cl_K) \otimes_{U(Cl(K \oplus K^*) \oplus cI)} \mathbf{C}_K$$

式中真空矢 $\mathbf{C}_K$ 满足：$Cl(K \oplus K^*)C_K = 0$，$Cl(K \oplus K^*) = K \oplus K^*$ 的 level。

现在我们终于可构造仿射 monard 的代数 $gl(\infty, \infty)$。它是双复变量 s,t 的 Frenkel，朱新文要求的 level 1 的 Double erbil 表示 $g\hat{l}_{f,\infty}$。这时 Clifford 模的产生子为双指标 $m = (m_1, m_2)$，其对易律表为双变量 R 矩阵 $R(k_1,k_2)$，$k_1 = s + t$，$k_2 = s - t$。

$R(k_1,k_2)$ 为一分割的三维立体杨图，如 Okounkov、Reshetikhin 和 Vafa 的三维晶体的

双标度极限。但 Okounkov、Reshetikhin 和 Vafa 的生成元的自由振子式须改为侯伯宇、杨文力文中的关系式。这时中心的缺陷与边界条件给出无序，而 Miura 变换变 Monad 为 Monoid。

十一、结语

我的这些工作，除了与我的学生石康杰、杨仲侠、岳瑞宏、赵柳、丁祥茂、范桁、杨文力等，更是与侯伯元、马中骐、王佩、宋行长（通过行长又与周光召）、郭汉英、吴可、王世坤的合作是分不开的，特别是第八节中愉快而有成效的合作。我的工作也得到了一些朋友的理解与支持，例如《20 世纪中国知名科学家学术成就概览 · 物理学卷》主编陈佳洱、副主编于渌，以及化学界的知己老友孙家钟院士。同时，我这些年来的工作也离不开我的老伴儿曹淑霞始终不渝的支持。因此，我誓借几何 Langlands 可推广到 $\hat{\mathbf{P}}^2$ 这一机遇，以创新湮灭顽疾之苦痛，以创新来消灭死。誓以我有生之年，在国际竞相攀登科研高峰中顽强攀顶，以此作为最后的礼物。

2010 年 6 月

附录

（发表文章汇总）

1. 局部坐标系中的旋量球函数及算子，
侯伯宇，
物理学报 **19** (1963) 341—359
2. Green 函数及 δ 函数的三方向球函数展开式，
侯伯宇，
物理学报 **20**（1964）11—18
3. 散射矩阵的角分布不变变换群，
侯伯宇，
物理学报 **20**（1964）691—695
4. Three directional expansion of 1/R and δ function，
Bo-Yu Hou，
Scientia Sinica **14** (1965) 39—45
5. SU_3 群的多项式基底及其 Clebsch-Gordan 系数的明显表达式，
侯伯宇，
物理学报 **22**（1966）460—470
6. 从结构模型中重子流的独立结构因子看内部运动和对称性，
侯伯宇，
北京大学学报自然科学版 **2** (1966) 183—188
7. Orthonormal bases and infinitesimal operators of the irreducible representations of group U(N)，
Bo-Yu Hou，
Scientia Sinica **15** (1966) 763—772
8. 关于电磁作用与弱作用的统一理论，
侯伯宇，
西北大学学报自然科学版 **2**（1974）92—98
9. 电与磁对偶的双重协变规范场，
侯伯宇，
科学通报 **20** (1975) 273—276
10. 电磁对偶真空凝聚相内基本粒子谱的形成与统一作用的破缺，
侯伯宇，
西北大学学报自然科学版 Z **1**（1976）12—20
11. 静点磁荷系场的 SU (1)及 SU(2)规范势，
王永康，张高有，侯伯宇，
西北大学学报自然科学版 Z **1**（1976）21—27
12. 不可易规范场的对偶荷，
王永康，张高有，侯伯宇，
物理学报 **25** (1976) 514—520
13. 加速磁荷的推迟 SU(2)及 U(1)规范势与场强，

侯伯宇，
兰州大学学报 **2**（1977）37—40
14. 在磁单极附近荷电粒子的波函数，
冼鼎昌，侯伯宇，
科学通报 **22** (1977) 204—206
15. 不可易规范场的规范不变守恒流，
侯伯宇，
物理学报 **26** (1977) 433—435
16. 各级微扰展开下自发破缺的规范无关性、幺正性及可重整性，
侯伯宇，
物理学报 **26** (1977) 317—322
17. 关于 SU(2)规范场结构，
侯伯宇，
物理学报 **26** (1977) 83—86
18. 关于球对称的 SU(2) 规范场，
侯伯宇，谷超豪，胡和生，
复旦学报自然科学版 **01** (1977) 92—99
19. SU(n)规范场的可对易分量及带荷粒子分量，
侯伯宇，
复旦学报自然科学版 **03** (1977) 23—27
20. The decomposition and reduction of gauge field and dual charged solution of abelianizable field，
Bo-Yu Hou，Yi-Shi Duan，Mo-Lin Ge，
Scientia Sinica **21** (1978) 446—45
21. 具有同样 SU(2)场强与外源的不等价规范势，
侯伯宇，
高能物理与核物理 **2**（1978）61—66
22. SU(2)规范场的一系列“无源”解——静点磁荷系的 SU(2)势及 U(1)势，
王永康，张高有，侯伯宇，
高能物理与核物理 **2**（1978）368—370
23. 谈点非线性超对称与超引力，
侯伯宇，
郑州大学学报自然科学版 **2**（1979）37—48
24. 磁单极周围荷电粒子波函数在局部坐标系中的分离变量法，
侯伯宇，张高有，
西北大学学报自然科学版 **1**（1979）57—64
25. 静球对称无外源 SU(2)规范场解——其自对偶性、唯一性及自源荷流分析，
侯伯元，侯伯宇，

高能物理与核物理 **3** (1979) 255—265

26. SU(N)规范理论的倍步解，
王珮，侯伯宇，
高能物理与核物理 **3**（1979）555—571
27. 闵空间拓扑非平庸球对称场中零能费米子，
侯伯元，侯伯宇，
高能物理与核物理 **3** (1979) 697—707
28. 非阿贝尔规范场单极动量矩算子的构成，
侯伯宇，
科学通报 **24** (1979) 16—18
29. 规范场的分解约化及可 Abel 化场对偶荷解，
侯伯宇，段一士，葛墨林，
中国科学A **1**（1979）45—54
30. 用活动标架研究五维瞬子背景场中的波动方程、本征函数系与格林函数，
王珮，侯伯元，侯伯宇，
中国科学A **8**（1980）744—750
31. Non-selfdual instanton of two dimensional CP(N)(N>2) chiral theory，
Bo-Yu Hou，Bo-Yuan Hou，Pei Wang，Yu-Bin Wang，
Physics Letters B **93** (1980)，415—418
32. 单极动力学群的离散对称及阶梯算子，
侯伯宇，
西北大学学报自然科学版 **22**（1981）20—23
33. A discrete symmetry and the ladder operators for the dynamical group of monopole，
Bo-Yu Hou，
Annals of Physics **134** (1981) 373—375
34. Equations，eigenfunction and Green function in the field of O(5) instanton expressed by moving frames，
Bo-Yu Hou，Bo-Yuan Hou，Pei Wang，
Scientia Sinica. Series A **24** (1981) 342—355
35. Spherical symmetric static sourceless solution of SU(2) gauge field – an analysis of self duality，uniqueness，and self-induced charge current，
Bo-Yu Hou，Bo-Yuan Hou，
Chinese Physics **1** (1981) 624—634
36. Noether analysis for the hidden symmetry responsible for an infinite set of nonlocal currents，
Bo-Yu Hou，Mo-Lin Ge，Yong-Shi Wu，
Physical Review D **24** (1981) 2238—2244
37. 非局域流作为整体非局域变换产生的 Noether 流，

侯伯宇，
西北大学学报自然科学版 **36**（1982）19—28
38. 手征模型中的 H-变换、无穷多非定域守恒流的 Noether 分析与 Kac-Moody 代数结构及其推广，
乔玲丽，吴詠时，侯伯宇，葛墨林，
中国科学 A **10**（1982）907—918
39. Angular momentum operator in a nonabelian gauge field of an N-dimensional monopole and an SO(N) top operator，
Bo-Yu Hou，Bo-Yuan Hou，Pei Wang，
Journal of Mathematical Physics **23** (1982) 2488—2493
40. Nonlocal currents as Noether currents due to a nonlocally transformed global symmetry，
Bo-Yu Hou，
Communications in Theoretical Physics **1** (1982) 333—344
41. Soliton 方程的紧致代数结构旋量 AKNS 系统及其球面实现，
郭汉英，吴可，侯伯宇，向延育，王世坤，
数学物理学报 **3**（1983）241—247
42. CP(n)手征理论的非自偶解及相应的规范场，
侯伯宇，石康杰，王育邠，侯伯元，王珮，
高能物理与核物理 **7**（1983）150—159
43. H-transformation， Noether analysis of nonlocal infinitely conserved-currents and Kac-Moody algebraic structure in chiral model and their extensions，
Ling-Lie Chau，Yong-Shi Wu，Bo-Yu Hou，Mo-Lin Ge，
Scientia Sinica. Series A **26** (1983) 51—64
44. The gyroelectric ratio of supersymmetrical monopole determined by its structure，
Bo-Yu Hou，
Physics Letters B **125** (1983) 389—392
45. The stringless gauge potential of accelerated monopole，
Bo-Yu Hou，Kang-Jie Shi，Yu-Bin Wang，
IL Nuovo Cimento A **76** (1983) 556—568
46. A formula relating infinitesimal Backlünd-transformations to hierarchy generating operators，
Bo-Yu Hou，Gui-zhang Tu，
Journal of Physics A **16** (1983) 3955—3960
47. The non-self-dual solutions of the CP(n) chiral theory and the corresponding gauge-fields，
Bo-Yu Hou，Kang-Jie Shi，Yu-Bin Wang，Bo-Yuan Hou，Pei Wang，
Chinese Physics **3** (1983) 926—935
48. The Compact Algebra Structure Of Soliton Equations，Their Spinor AKNS System And Realization On Sphere，

Han-Ying Guo，Ke Wu，Bo-Yu Hou，Yan-Yu Hsiang，Shi-Kun Wang，
Acta Mathematica Scientia **3** (1983) 241—247
49. 手征模型的对偶性及其与 sine-Gordon 方程的几何关联，
王珮，侯伯元，侯伯宇，郭汉英，
物理学报 **33**（1984）294—301
50. Some series of infinitely many symmetry generators in symmetric space chiral models，
Bo-Yu Hou，
Journal of Mathematical Physics **25** (1984) 2325—2330
51. Gauge covariant formulation of symmetric space fields，
Ling-Li Chau，Bo-Yu Hou，
Physics Letters B **145** (1984) 374—352
52. A correspondence between the sigma model and the Liouville model，
H. Bohr，Bo-Yu Hou，S. Saito，
IL Nuovo Cimeno A **84** (1984) 237—248
53. O(3) nonlinear Sigma model and pseudospherical surface，
Bo-Yu Hou，Bo-Yuan Hou，Pei Wang，
Journal of Physics A **18** (1985) 165—185
54. Gauge covariant formulation of integrability properties of symmetric space fields，
Ling-Lie Chau，Bo-Yu Hou，Xing-Chang Song，
Physics Letters B **151** (1985) 421—427
55. Anomalies，cohomology and Chern-Simons，cochains，
Han-Yin Guo，Bo-Yu Hou，Shi-Kun Wang，Ke Wu，
Communications in Theoretical Physics **4** (1985) 145—155
56. Cohomology of gauge groups and characteristic of Chern-Simons type，
Han-Yin Guo，Bo-Yu Hou，Shi-Kun Wang，Ke Wu，
Communications in Theoretical Physics **4** (1985) 233—251
57. On the geometrical meaning of three dimensional topological term in 2+1 dimensional sigma model，
Bo-Yu Hou，Yu-Bin Wang，Li-Ning Zhang，
Communications in Theoretical Physics **4** (1985) 351—360
58. The third order cohomology cycle for gauge group and the anomalous associative law for gauge transformation，
Bo-Yu Hou，Bo-Yuan Hou，
Chinese Physics Letters **2** (1985) 49—52
59. Local symmetry and conserved-currents of principal chiral model with topological term，
Bo-Yu Hou，Yu-Bing Wang，
Communications in Theoretical Physics **5** (1986) 55—66
60. The global topological meaning of cocycles in a gauge group，

Bo-Yu Hou，Bo-Yuan Hou，Pei Wang，
Letters in Mathematical Physics **11** (1986) 179—187

61. How to eliminate the dilemma in 3 cocycle，
Bo-Yu Hou，Bo-Yuan Hou，Pei Wang，
Annals of Physics **143** (1986) 172—185

62. Integrability properties of symmetric space fields reduced from axially symmetric Einstein and Yang-Mills equation，
Ling-Lie Chau，Kuang-Chao Chou，Bo-Yu Hou，Xing-Chang Song，
Physical Review D **34** (1986) 1814—1823

63. The static axially symmetric self-dual Yang-Mills fields and surfaces of negative curvature，
Bo-Yu Hou，Bo-Yuan Hou，Pei Wang，
International Journal of Modern Physics A **1** (1986) 193—210

64. 平移群 3-上闭链与带膜波函数，
王佩，侯伯元，侯伯宇，
高能物理与核物理 **10** (1986)170—176

65. 反常、Chern-Simons 上链，
王世坤，吴可，侯伯宇，郭汉英，
物理学报 **35**（1986）89—93

66. 规范理论中反常的拓扑意义与规范群的 Cech-de Rham 上同调
王佩，侯伯元，侯伯宇，
物理学报 **35** (1986) 433—442

67. 平移群上同调与 Kronecker 映像，
侯伯宇，侯伯元，王佩，
物理学报 **35** (1986) 829—832

68. 联络空间上同调规范群上同调继承系和族指数定理，
侯伯宇，
物理学报 **35** (1986) 1662—1666

69. The Virasoro algebra in the solution space of Ernest's equation，
Bo-Yu Hou，Wei Lee，
Letters in Mathematical Physics **13** (1987) 1—6

70. The cohomology in translation group and Kronecker mapping，
Bo-Yu Hou，Bo-Yuan Hou，Pei Wang，
Chinese Physics **7** (1987) 87—90

71. Two dimensional angular momentum and Fermion in the field of magnetic flux tube，
Bo-Yu Hou，Bo-Yuan Hou，Wei-Xi Wang，Juan-Hu Yan，
Communications in Theoretical Physics **7** (1987) 49—69

72. 拓扑反常的递降继承，

侯伯宇，侯伯元，
高能物理与核物理 **11** (1987) 57—63
73. 引力场约化理论中的 Virasoro 代数，
侯伯宇，李卫，
高能物理与核物理 **11** (1987) 137—141
74. New Virasoro and Kac-Moody symmetries for the nonlinear $\sigma-$model，
Bo-Yu Hou，Wei Li，
Journal of Physics A **20** (1987) L897—L904
75. Cohomology in connection space，family index theorem and abelian gauge structure，
Bo-Yu Hou，Yao-Zhong Zhang，
Journal of Mathematical Physics **28** (1987) 1709—1715
76. The comments on constraint for anomalous Jacobi identity，
Bo-Yu Hou，Yao-Zhong Zhang，
Modern Physics Letters A **1** (1987) 103—110
77. 一种 Einstein 约化场方程和产生的新方法，
侯伯宇，李卫，
物理学报 **36** (1987) 930—934
78. Skyrme 模型的上同调分析和杂化手征袋模型的边效应，
侯伯宇，陈一新，张耀中，
中国科学 A **10** (1987) 1049—1057
79. 静轴对称 Einstein-Maxwell 引力场的新对称性，
侯伯宇，李卫，
高能物理与核物理 **12** (1988) 196—202
80. Cohomological Analysis in Skyrme Model and Boundary Effect of Chiral Bag Model，
Bo-Yu Hou，Yi-Xin Chen，Yao-Zhong Zhang，
Scientia Sinica.Series A **31** (1988)172—182
81. Some remarks on the extension of loop group，
Bo-Yu Hou，Yao-Zhong Zhang，
Journal of Physics A **21** (1988) 345—352
82. New method to generate the solutions of the reduced Einstein equations，
Bo-Yu Hou，Wei Li，
Chinese Physics **8** (1988) 343—346
83. Exact solution of the $Z_n \times Z_n$ symmetric generalization of Baxter 8-vertex model，
Zhou Yu-Kui，Yan Mu-Lin，Hou Bo-Yu，
Journal of Physics A **21** (1988) 1929—1934
84. Algebraic Bethe ansatz of Belavin's $Z_n \times Z_n$ symmetric model，
Zhou Yu-Kui，Yan Mu-Lin，Hou Bo-Yu，
Physics Letters A **133** (1988) 391—394

85. Generating functions with the new hidden symmetries for cylindrically Symmetric gravitational fields,
Bo-Yu Hou，Wei Li，
Classical and Quantum Gravity **6** (1989) 163—177
86. Algebras connected with the Z_n elliptic solution of Yang-Baxter equation，
Bo-Yu Hou，Hua Wei，
Journal of Mathematical Physics **30** (1989) 2750—2755
87. The Riemann-Hilbert transformation for an approach to a representation of the Virasoro group I，
Wei Li，Bo-Yu Hou，
Journal of Mathematical Physics **30** (1989) 1198—1204
88. On the fusion of face and vertex models，
Yu-Kui Zhou，Bo-Yu Hou，
Journal of Physics A **22** (1989) 5089—5096
89. Hamiltonian of a 1D quantum chain for Belavin's $Z_n \times Z_n$ model，
Hua Wei，Yu-Kui Zhou，Bo-Yu Hou，
Journal of Physics A **22** (1989) 1579—1586
90. Two And Three Cocycles In A Hybrid Bag Model: A Comment On The Bosonization Of Fermions，
Bo-Yu Hou ，Bo-Yuan Hou，Pei Wang，
Communications in Theoretical Physics **12** (1989) 109—116
91. Exact solution of the Belavin's $Z_n \times Z_n$ symmetric model，
Bo-Yu Hou，Mu-Lin Yan，Yu-Kui Zhou，
Nuclear Physics B **324**(1989) 715—728
92. Quantum group structure in the Unitary minimal model，
Bo-Yu Hou，Ding-Ping Lie，Rui-Hong Yue，
Physics Letters B **229** (1989) 45—50
93. Nonlocal Noether currents and conformal invariance for super chiral fields，
Wei Li，Bo-Yu Hou，San-Ru Hao，
Physical Review D **39** (1989) 1655—1658
94. 造就直达国际前沿研究阵地的理论物理研究生，
侯伯宇，
学位与研究生教育 **02**（1990）7—9
95. 利用 Riemann-Hilber 方法产生 Ernst 方程严格解的新途径，
侯伯宇，李卫，
应用数学学报 **13**（1990）168—175
96. SU(2)$_l$ WZW模型的基本问题，
侯伯宇，李康，

高能物理与核物理 **14** (1991)711—723

97. 幺正最小模型的聚合与辫子矩阵的明显表示，
侯伯宇，石康杰，岳瑞宏，
高能物理与核物理（1990）802—809
98. The Fiber bundle formulation of sigma model and SUSY sigma model，
Bo- Yu Hou，Pei Wang，Bo- Yuan Hou，Li- Ning Zhang，
Communications in Theoretical Physics **13** (1990) 41—48
99. The embedding of quantum algebra $\hat{U}(gl(n+1))$ in irreducible representations of $U(gl(n+1))$，
Bo-Yu Hou，Kang-Jie Shi，Song Xing-Chang，Pei Wang，
Communications in Theoretical Physics **13** (1990) 71—78
100. Clebsch-Gordan coefficients，Racah coefficients and Braiding fusion of quantum *sl*(2) enveloping algebra (I)，
Bo-Yu Hou，Bo-Yuan Hou，Zhong-Qi Ma，
Communications in Theoretical Physics **13** (1990) 181—198
101. Clebsch-Gordan coefficients，Racah coefficients and braiding fusion of quantum *sl*(2) enveloping algebra (II)，
Bo-Yu Hou，Bo-Yuan Hou，Zhong-Qi Ma，
Communications in Theoretical Physics **13** (1990) 341—354
102. Fusion procedure and Sklyanin algebra，
Bo-Yu Hou，Yu-Kui Zhou，
Journal of Physics A **23** (1990) 1147—1154
103. General solutions to fundamental problems of $SU(2)_k$ WZW models，
Bo-Yu Hou，Kang Li，Pei Wang，
Journal of Physics A **23** (1990) 3431—3446
104. Yang-Baxter equation，algebras and braid group for the Zn-symmetric statistical model，
Hua Wei，Bo-Yu Hou，
Chinese Physics Letters **7** (1990) 337—340
105. The Crossing matrices of WZW SU(2) model and minimal model with the quantum 6j symbols，
Bo-Yu Hou，Kang-Jie Shi，Pei Wang，Rui-Hong Yue，
Nuclear Physics B **345** (1990) 659—684
106. 仿射Toda场作为约束的WZNW模型，
侯伯宇，赵柳，
高能物理与核物理 **15** (1991) 701—710
107. XXZ模型及其对称代数的结构，

马中骐，侯伯元，侯伯宇，
高能物理与核物理 **15** (1991)812—821

108. Riemann-Hilbert transformation approach to a representation of Virasoro group II，
Wei Li，Bo-Yu Hou，
Journal of Mathematical Physics **32** (1991) 1328—1333

109. The XXZ model with Beraha values，
Bo-Yu Hou，Bo-Yuan Hou，Zhong-Qi Ma，
Journal of Physics A **24** (1991) 2847—2861

110. Modular invariance and the Feigin-Fuch representation of characters for $SU_k(2)$ WZW and minimal models，
Bo-Yu Hou，Rui-Hong Yue，
Journal of Physics A **24** (1991) 11—21

111. Solutions of Yang–Baxter equation in the vertex model and the face model for octet representation，
Bo-Yu Hou，Bo-Yuan Hou，Zhong-Qi Ma，Yu-Dong Yin，
Journal of Mathematical Physics **32** (1991) 2210—2218

112. Sine-Gordon and affine Toda fields as nonconformally constrained WZNW model，
Bo-Yu Hou，Liu Chao，Huan-Xiong Yang，
Physics Letters B **266** (1991) 353—362

113. Integrable quantum chain and the representation of the quantum group $SU_q(2)$，
Bo-Yu Hou，Kang-Jie Shi，Zhong-Xia Yang，Rui-Hong Yue，
Journal of Physics A **24** (1991) 3825—3836

114. *XXZ* 模型的Bethe Ansatz解与量子群的最高权态，
侯伯宇，杨仲侠，
西北大学学报自然科学版 **21** (1991) 7—11

115. 非幺正$SU_k(2)$ WZW模型的关联函数与交叉矩阵，
侯伯宇，石康杰，岳瑞宏，
高能物理与核物理 **16** (1992) 414—422

116. 快度取值超环的可因式化散射矩阵与对称性算子，
胡占宁，侯伯宇，
高能物理与核物理 **16** (1992)978—990

117. H_{XXZ}模型与量子$SU_q(2)$群的表示，
侯伯宇，石康杰，杨仲侠，岳瑞宏，
物理学报 **41** (1992)201—212

118. Zhong-Xia Yang，$sl_q(n)$ quantum algebra as a limit of the elliptic case，
Bo-Yu Hou，Kang-Jie Shi，
International Journal of Modern Physics A7 Supplement **1A** (1992) 391—403

119. Quantum Clebsch-Gordan coefficients for nongeneric q values，

Bo-Yu Hou，Bo-Yuan Hou，Zhong-Qi Ma，
Journal of Physics A **25** (1992)1211—1222
120. XXZ model and the structure of symmetric algebra，
Zhong-Qi Ma，Bo-Yuan Hou，Bo-Yu Hou，
Chinese Physics **12** (1992) 325—335
121. Some remarks on the eigenvalue problem for Z_n symmetric vertex and face models，
Yu-Kui Zhou，Bo-Yu Hou，
Physica A **187** (1992) 308—328
122. Conformally reduced WZNW theory，new extended chiral algebras and their associated Toda type integrable systems，
Bo-Yu Hou，Liu Chao，
International Journal of Modern Physics A **7** (1992) 391—403
123. A_N W algebras and classification，
Liu Chao，Bo-Yu Hou，
Modern Physics Letters A **7** (1992) 3419—3423
124. WZNW模型的手征Hamilton约化，
侯伯宇，赵柳，
中国科学A **23**（1993）487—492
125. 一个源于共形的SL（2，R）WZNW模型的非共形规范对称体系的研究，
杨焕雄，侯伯宇，
中国科学A **23**（1993）1283—1292
126. WZNW模型的共形约化，扩展的手征代数及相应的Toda类可积模型(Ⅰ)一般构架，
侯伯宇，赵柳，
高能物理与核物理 **17** (1993) 28—37
127. WZNW模型的共形约化，扩展的手征代数及相应的Toda类可积模型(Ⅱ)一个例子，
侯伯宇，赵柳，
高能物理与核物理 **17** (1993)241—251
128. WZNW模型的不同约化相应的W代数的联系，
侯伯宇，赵柳，
高能物理与核物理 **17** (1993)329—337
129. (广义)BFOFW模型的Hamilton程式及其量子化，
侯伯宇，杨焕雄，
高能物理与核物理 **17** (1993)524—533
130. Function space representation of the *Zn* Sklyanin algebra，
Bo-Yu Hou，Kang-Jie Shi，Wen-Li Yang，Zhong-Xia Yang，
Physics Letters A **178** (1993) 73—80
131. Cyclic representation and function difference representation of the Z_n Sklyanin algebra，
Bo-Yu Hou，Kang-Jie Shi，Zhong-Xia Yang，

Journal of Physics A **26** (1993) 4951—4965
132. General solution of the reflection equation for the eigh-vertex model,
Bo-Yu Hou，Rui-Hong Yue,
Physics Letters A **183** (1993) 169—174
133. Hamiltonian reductions of self-dual Yang-Mills theory,
Bo-Yu Hou，Liu Chao ，
Physics Letters B **298** (1993) 103—110
134. From integrability to conformal symmetry: Bosonic superconformal Toda theories,
Bo-Yu Hou，Liu Chao,
International Journal of Modern Physics A **8** (1993)1105—1124
135. Bosonic superconformal affine Toda theory: Exchange algebra and dressing symmetry,
Liu Chao，Bo-Yu Hou,
International Journal of Modern Physics A **8** (1993) 3773—3790
136. Hamiltonian reduction of WZNW model associated with chiral constraints,
Bo-Yu Hou，Liu Zhao,
Scientia Sinica. Series A **36** (1993) 1095—1102
137. 活动标架系下以O（3）非线性σ模型Lax－pair矩阵的Poisson－Lie结构，
丁祥茂，王延申，侯伯宇，
物理学报 **43**（1994）1—6
138. 一个经典完全可积的非共形约束的SL（2，R）Wess－Zumino－Novikov－Witten模型（Ⅰ），
王延申，杨文力，杨焕雄，侯伯宇，
物理学报 **43**（1994）175—184
139. 一个经典完全可积的非共形约束的SL（2，R）Wess－Zumino－Novikov－Witten模型（II），
王延申，杨文力，杨焕雄，侯伯宇，
物理学报 **43**（1994）185—190
140. On the solutions of two-extended principal conformal Toda theory ，
Liu Chao，Bo-Yu Hou,
Annals of Physics **230** (1994) 1—20
141. Hamiltonian formalism of two loop WZNW model under Chevalley basis,
Xiang-Mao Ding，Bo-Yu Hou，Huan-Xiong Yang,
International Journal of Modern Physics A **9** (1994) 341—364
142. The Dressing transformation of the conformal affine Toda,
Bo-Yu Hou，Wen-Li Yang,
International Journal of Modern Physics A **9** (1994) 2997—3006
143. Exchange relations of Lax pair for nonlinear sigma models,
Bo-Yu Hou，Xiang-Mao Ding，Yan-Shen Wang，Bo-Yuan Hou,

Modern Physics Letters A **9** (1994) 1521—1528

144. A conformally broken system as the constrained SL(2，R) WZNW model，its integrability and dressing symmetry，
Bo-Yu Hou，Huan-Xiong Yang，Wen-Li Yang，Yan-Shen Wang，
Journal of Physics A **27**(1994) 6525—6532

145. 活动标架系下O（n）非线性σ模型Laxpair矩阵的Poisson—Lie结构，
王延申，侯伯宇，
高能物理与核物理 **18**（1994）892—901

146. The Poisson-Lie structure of nonlinear O(N) sigma-model by using the moving-frame method，
Bo-Yu Hou，Bo-Yuan Hou，Yin-Wan Li，Biao Wu，
Journal of Physics A **27**(1994) 7209—7216

147. 三维精确可解统计模型——Baxter-Bazhanov模型的可积性条件，
胡占宁，侯伯宇，
高能物理与核物理 **19** (1995)123—130

148. Z_n格点模型的量子仿射变形代数，
石康杰，杨文力，侯伯宇，范桁，
高能物理与核物理 **19** (1995)229—240

149. Bosonic realization of boundary operators in SU(2) invariant Thirring model，
Liu Chao，Yan-Shen Wang，Wen-Li Yang，Bo-Yu Hou，Kang-Jie Shi，
International Journal of Modern Physics A **10** (1995) 4469—4482

150. Zamolodchikov - Faddeev algebra related to Z_n symmetric elliptical R-matrix，
Heng Fan，Bo-Yu Hou，Kang-Jie Shi，Wen-Li Yang，Zhong-Xia Yang，
Journal of Physics A **28** (1995)3157—3175

151. General solution of reflection equation for eight vertex SOS model，
Heng Fan，Bo-Yu Hou，Kang-Jie Shi，
Journal of Physics A **28** (1995) 4743—4749

152. q-deformed Chern class，Chern-Simons and cocycle hierarchy，
Bo-Yu Hou，Bo-Yuan Hou，Zhong-Qi Ma，
Journal of Physics A **28** (1995) 543—558

153. The Extended $W_3^{(2)}$ Toda system with conformal weights in 1/3 sequence，
Huan-Xiong Yang，Bo-Yu Hou，
Communications in Theoretical Physics **23** (1995) 73—82

154. Solution of reflection equation，
Bo-Yu Hou，Kang-Jie Shi，Heng Fan，Zhong-Xia Yang，
Communications in Theoretical Physics **23** (1995)163—166

155. Remarks on the star-triangle relation in the Baxter-Bazhanov model，

Zhan-Ning Hu，Bo-Yu Hou，
Journal of Statistical Physics **79** (1995) 759—764

156. Three-dimensional vertex model in statistical mechanics from Baxter-Bazhanov model，
Zhan-Ning Hu，Bo-Yu Hou，
Journal of Statistical Physics **82** (1995) 633—655

157. Heterotic Toda fields，
Liu Chao，Bo-Yu Hou，
Nuclear Physics B **436** (1995) 638—658

158. Integrable open-boundary conditions for the $Z_n \times Z_n$ Belavin model，
Heng Fan，Bo-Yu Hou，Kang-Jie Shi，Zhong-Xia Yang，
Physics Letters A **200** (1995) 109—114

159. Cyclic quantum dilogarithm，shift operator and star-square relation of the BB model，
Zhan-Ning Hu，Bo-Yu Hou，
Physics Letters A **201** (1995)151—155

160. Representations of q-oscillator algebra at a root of 1，
Bo-Yu Hou，Bo-Yuan Hou，Lian-Chao Xu，Shan-You Zhou，
Scientia Sinica.Series A **7** (1995)799—804

161. q - deformed Chern characters for quantum groups $SU_q(N)$，
Bo-Yu Hou，Bo-Yuan Hou，Zhong-Qi Ma，
Journal of Mathematical Physics **36** (1995) 5110—5138

162. SU(2) Thirring模型边界算子的Boson表达式，
王延申，石康杰，杨文力，赵柳，侯伯宇，
高能物理与核物理 **20** (1996)216—225

163. Z_n Belavin模型的反射矩阵，
陈敏，侯伯宇，石康杰，
高能物理与核物理 **20** (1996)794—800

164. Fusion procedure for Z_n Belavin model with open boundary conditions，
Heng Fan，Kang-Jie Shi，Bo-Yu Hou ，
Journal of Physics A **29** (1996) 5735—5743

165. Algebraic Bethe ansatz for the eight vertex model with general open-boundary conditions，
Heng Fan，Bo-Yu Hou，Kang-Jie Shi，Zhong-Xia Yang，
Nuclear Physics B **478** (1996)723—757

166. Exact diagonalization of the quantum supersymmetric $SU_q(n|m)$ model，
Rui-Hong Yue，Heng Fan，Bo-Yu Hou，
Nuclear Physics B **462** (1996) 167—191

167. 中心扩张的sl_2双重Yangian的Miura变换，
丁祥茂，侯伯宇，赵柳，
高能物理与核物理 **21** (1997) 697—705

168. 用玻色化顶点算子实现与Z_n对称的Belavin R矩阵相关的Z-F代数，
石康杰，杨文力，范桁，侯伯宇，
高能物理与核物理 **21** (1997)706—713

169. Boundary $A_1^{(1)}$ face model ，
Bo-Yu Hou，Wen-Li Yang，
Communications in Theoretical Physics **27** (1997) 257—262

170. Bosonization of Quantum sine-Gordon field with boundary，
Bo-Yu Hou，Kang-Jie Shi，Yan-Shen Wang，Wen-Li Yang，
International Journal of Modern Physics A **12** (1997) 1711—1741

171. Integrable boundary conditions associated with the $Z_n \times Z_n$ Belavin model and solutions of reflection equation，
Heng Fan，Kang-Jie Shi，Bo-Yu Hou，Zhong-Xia Yang，
International Journal of Modern Physics A **12** (1997) 2809—2823

172. Fusion for the function difference representation of Z_n Sklyanin algebra，
Bo-Yu Hou，Kang-Jie Shi，Wen-Li Yang，Zhong-Xia Yang，Shan-You Zhou，
International Journal of Modern Physics A **12** (1997) 2927—2945

173. Correlation functions of the SU(2)-Invariant Thirring model with a boundary，
Bo-Yu Hou，Kang-Jie Shi，Yan-Shen Wang，Wen-Li Yang，
Journal of Physics A **30** (1997) 251—263

174. Bosonization of vertex operators for the Z_n symmetry Belavin model，
Heng Fan，Bo-Yu Hou，Kang-Jie Shi，Wen-Li Yang，
Journal of Physics A **30** (1997) 5687—5696

175. An $\hbar$-deformation of the W_N algebra and its vertex operators，
Bo-Yu Hou，Wen-Li Yang，
Journal of Physics A **30** (1997) 6131—6145

176. Note on the algebra of screening currents for the quantum-deformed W-algebra，
Liu Zhao，Bo-Yu Hou，
Journal of Physics A **30** (1997) 7659—7666

177. Drinfeld constructions of the quantum affine superalgebra $U_q(gl(\hat{m/n}))$ ，
Heng Fan，Bo-Yu Hou，Kang-Jie Shi，
Journal of Mathematical Physics **38**(1997) 411—433

178. Properties of eigenstates of the six vertex model with twisted and open boundary conditions，
Heng Fan，Bo-Yu Hou，Kang-Jie Shi，
Journal of Mathematical Physics **38** (1997) 3446—3456

179. Representation of the boundary elliptic quantum group $BE_{\tau,\eta}(sl_2)$ and the Bethe ansatz，

Heng Fan, Bo-Yu Hou, Kang-Jie Shi,
Nuclear Physics B **496** (1997) 551—570

180. Integrable $A_{n-1}^{(1)}$ IRF model with reflecting boundary conditions,
Heng Fan, Bo-Yu Hou, Guang-Liang Li, Kang-Jie Shi,
Modern Physics Letters A **12** (1997) 1929—1942

181. Boundary K-operator for the elliptic R-operator acting on functional space,
Heng Fan, Bo-Yu Hou, Kang-Jie Shi,
Physics Letters A **230** (1997) 19—23

182. 由因式化L算子所构成的经典可积动力学体系,
石康杰, 李广良, 范桁, 侯伯宇,
高能物理与核物理 **22** (1998) 1100—1111

183. Drinfeld-Sokolov construction of heterotic Toda model,
Yan-Shen Wang, Liu Chao, Bo-Yu Hou,
Communications in Theoretical Physics **29** (1998)389—398

184. Lagrangian form of the self-dual equations for *SU*(*N*) gauge fields on four-dimensional Euclidean space,
Bo-Yu Hou, Xing-Chang Song,
Communications in Theoretical Physics **29** (1998)443—446

185. Bosonic realization for boundary states in quantum sine-Gordon field with boundary,
Bo-Yu Hou, Kang-Jie Shi, Yan-Shen Wang, Wen-Li Yang,
Communications in Theoretical Physics **29** (1998) 579—586

186. The general crossing relations for boundary reflection matrix,
Bo-Yu Hou, Kang-Jie Shi, Wen-Li Yang,
Communications in Theoretical Physics **30** (1998) 415—418

187. $\hbar$-(Yangian) deformation of the Miura map and Virasoro algebra,
Xiang-Mao Ding, Bo-Yu Hou, Liu Zhao,
International Journal of Modern Physics A **13** (1998) 1129—1144

188. Free boson representation of $DY_{\hbar}(gl_N)_k$ and $DY_{\hbar}(sl_N)_k$,
Xiang-Mao Ding, Bo-Yu Hou, Bo-Yuan Hou, Liu Zhao,
Journal of Mathematical Physics **39** (1998) 2273—2289

189. The elliptic quantum algebra $A_{q,p}(\hat{sl}_n)$ and its bosonization at level one,
Heng Fan, Bo-Yu Hou, Kang-Jie Shi, Wen-Li Yang,
Journal of Mathematical Physics **39** (1998) 4356—4368

190. A one-dimensional many-body integrable model from Z_n Belavin model with open boundary conditions,

Heng Fan，Bo-Yu Hou，Guang-Liang Li，Kang-Jie Shi，Yan-Shen Wang，
Journal of Mathematical Physics **39** (1998)4746—4758

191. The algebra $A_{\hbar,\eta}(\hat{g})$ and infinite Hopf family of algebras，
Bo-Yu Hou，Liu Zhao，Xiang-Mao Ding，
Journal of Geometry and Physics **27** (1998) 249—266

192. Dynamically twisted algebra $A_{q,p;\hat{\pi}}(\hat{gl_2})$ as current algebra generalizing screening currents of q-deformed Virasoro algebra，
Bo-Yu Hou，Wen-Li Yang，
Journal of Physics A **31** (1998) 5349—5369

193. Ruijsenaars–Macdonald-type difference operators from Zn Belavin model with open boundary conditions，
Bo-Yu Hou，Kang-Jie Shi，Yan-Shen Wang，Liu Zhao，
Journal of Physics A **31** (1998)5911—5923

194. The non-dynamical r-matrix structure of the elliptic Calogero - Moser model,
Bo-Yu Hou，Wen-Li Yang，
Letters in Mathematical Physics **44** (1998) 35—41

195. A new solution to the reflection equation for the Z_n symmetric Belavin model，
Heng Fan，Bo-Yu Hou，Guang-Liang Li，Kang-Jie Shi，
Physics Letters A **250** (1998) 79—87

196. Nondynamical *r*-matrix structure of the sl_2 trigonometric Ruijsenaars-Schneider model，
Kai Chen，Bo-Yu Hou，Wen-Li Yang，Yi Zhen，
Chinese Physics Letters **16** (1999) 1—3

197. An $\hbar$ -deformed Virasoro algebra as the hidden symmetry algebra of the Restricted sine-Gordon model，
Bo-Yu Hou，Wen-Li Yang，
Communications in Theoretical Physics **31** (1999) 265—270

198. The structure of affine Hecke algebra H_N for the XXZ model with boundary，
Bo-Yu Hou，Wen-Li Yang，
Communications in Theoretical Physics **32** (1999) 97—102

199. Z_n Belavin模型反射方程的多参数解，
石康杰，李广良，范桁，侯伯宇，
高能物理与核物理 **23** (1999) 425—435

200. 三角Calogero-Moser模型的非动力学r矩阵结构，
甄翼，杨文力，侯伯宇，陈凯，
高能物理与核物理 **23** (1999) 854—858

201. 在FBF背景下带反射边界条件的超对称t-J模型的代数Bethe ansatz方法，

惠小强，石康杰，侯伯宇，范桁，
高能物理与核物理 **23** (1999)961—979

202. $A_{n-1}^{(1)}$ 面模型反射方程的多参数解，

石康杰，李广良，范桁，侯伯宇，
高能物理与核物理 **23** (1999) 1163—1170

203. 经典 $\hbar$ -deformed W_N 代数，

杨文力，侯伯宇，
物理学报 **48** (1999) 1565—1570

204. 量子 $\hbar$ -deformed W_N 代数及其屏蔽流代数，

杨文力，侯伯宇，
物理学报 **48** (1999) 1571—1580

205. The nondynamical *r*-matrix structure for the elliptic A_{n-1} Calogero-Moser model，
Bo-Yu Hou，Wen-Li Yang，
Journal of Physics A **32** (1999) 1475—1486

206. Infinite Hopf family of elliptic algebras and bosonization，
Bo-Yu Hou，Liu Zhao，Xiang-Mao Ding，
Journal of Physics A **32** (1999)1951—1959

207. The general solutions to the reflection equation of the Izergin-Korepin model，
Heng Fan，Bo-Yu Hou，Guang-Liang Li，Kang-Jie Shi，Rui-Hong Yue，
Journal of Physics A **32** (1999) 6021—6032

208. Integrability of the Heisenberg chains with boundary impurities and their Bethe ansatz，
Bo-Yu Hou，Kang-Jie Shi，Rui-Hong Yue，Shao-You Zhao，
Journal of Physics A **32** (1999) 7623—7635

209. Algebraic Bethe Ansatz for the supersymmetric $t-J$ model with open boundary conditions，
Heng Fan，Bo-Yu Hou，Kang-Jie Shi，
Nuclear Physics B **541** (1999) 483—505

210. The twisted quantum affine algebra $U_q(A_2^{(2)})$ and correlation functions of

Izergin-Korepin model，
Bo-Yu Hou，Wen-Li Yang，Yao-Zhong Zhang，
Nuclear Physics B **556** (1999) 485—504

211. Impurity of arbitrary spin embedded in the 1-D Hubbard model with open boundary conditions，
Bo-Yu Hou，Xiao-Qiang Xi，Rui-Hong Yue，

Physics Letters A **257** (1999) 189—194

212. Elliptic Ruijsenaars-Schneider and Calogero-Moser models represented by Sklyanin algebra and *sln* Gaudin algebra,
Kai Chen, Heng Fan, Bo-Yu Hou, Kang-Jie Shi, Wen-Li Yang, Rui-Hong Yue,
Progress of Theoretical Physics. Supplement **135** (1999) 149—165

213. The nondynamical r-matrix structure for the Ruijsenaars - Schneider model and the classical Sklyanin algebra,
Bo-Yu Hou, Wen-Li Yang,
Physics Letters A **261** (1999) 259—264

214. 15顶角模型反射方程的常数解,
石康杰,李广良,范桁,侯伯宇,
高能物理与核物理 **24** (2000)11—20

215. 19顶角模型反射方程的常数解,
石康杰,李广良,范桁,岳瑞宏,侯伯宇,
高能物理与核物理 **24** (2000)21—34

216. 动力学椭圆代数 $A_{q,p,\hat{\pi}}(g\hat{l}_n)$ 的Drinfeld流,
侯伯宇,范桁,杨文力,曹俊鹏,
高能物理与核物理 **24** (2000)98—105

217. 同一个非动力学r矩阵所对应的有理Ruijsenaars-Schneider模型(n=2)及Calogero-Moser模型,
王美旭,杨文力,侯伯宇,
高能物理与核物理 **24** (2000)300—305

218. Integrability of the spin-1 XXX chain with boundary impurities and their Bethe ansatz,
Bo-Yu Hou, Kang-Jie Shi, Rui-Hong Yue, Shao-You Zhao,
Chinese Physics **9** (2000) 203—209.

219. Supersymmetric SU(1|2) Gaudin model,
Jun-Peng Cao, Bo-Yu Hou, Rui-Hong Yue,
Chinese Physics **10** (2000)103—108

220. Exact solution for extended Essler-Korepin-Schoutens model with open boundary conditions,
Guang-Liang Li, Rui-Hong Yue, Kang-Jie Shi, Bo-Yu Hou,
Chinese Physics **10** (2000)113—116

221. The nondynamical r-matrix structure of the elliptic Ruijsenaars -Schneider model with N=2,
Bo-Yu Hou, Wen-Li Yang,
Communications in Theoretical Physics **33** (2000) 371—376

222. Bethe Ansatz for the Spin-1 XXX Chain with Two Impurities,

Bo-Yu Hou，Kang-Jie Shi，Rui-Hong Yue，Shao-You Zhao，
Communications in Theoretical Physics **33** (2000) 559—564

223. Gauss decomposition of dynamical elliptic algebra $A_{q,p,\hat{\pi}}(g\hat{l}_n)$，
Bo-Yu Hou，Heng Fan，Wen-Li Yang，Jun-Peng Cao，
Communications in Theoretical Physics **34** (2000)77—82
224. The Lax pair and boundary K-matrices for XXC model，
Guang-Liang Li，Bo-Yu Hou，Kang-Jie Shi，Rui-Hong Yue，Heng Fan，
Communications in Theoretical Physics **34** (2000)635—642
225. The dynamical twisting and nondynamical *r*-matrix structure of the elliptic Ruijsenaars-Schneider model，
Bo-Yu Hou，Wen-Li Yang，
Journal of Mathematical Physics **41** (2000) 357—369
226. Integrability of the *Cn* and *BCn* Ruijsenaars-Schneider models，
Kai Chen，Bo-Yu Hou，Wen-Li Yang，
Journal of Mathematical Physics **41** (2000) 8132—8147
227. The exact solution of the SU(3) Hubbard model，
Bo-Yu Hou，Dan-Tao Peng，Rui-Hong Yue，
Nuclear Physics B **575** (2000) 561—578
228. Nested Bethe ansatz for Perk–Schultz model with open boundary conditions，
Guang-Liang Li，Rui-Hong Yue，Bo-Yu Hou，
Nuclear Physics B **586** (2000) 711—729
229. 椭圆量子群极小表示的系数代数，
石康杰，范桁，岳瑞宏，赵少游，侯伯宇，
高能物理与核物理 **25** (2001)628—635
230. Exact Solution for Perk-Schultz Model with Boundary Impurities，
Guang-Liang Li，Rui-Hong Yue，Kang-Jie Shi，Bo-Yu Hou，
Chinese Physics Letters **18** (2001)316—318
231. Rational SU(N) Gaudin Model，
Jun-Peng Cao，Bo-Yu Hou，Rui-Hong Yue，
Chinese Physics Letters **18** (2001)715—717
232. The Lax pair for C_2-type Ruijsenaars- Schneider model，
Kai Chen，Bo-Yu Hou，Wen-Li Yang，
Chinese Physics **10** (2001)550—554
233. Trigonometric SU(N) Gaudin model，
Jun-Peng Cao，Bo-Yu Hou，Rui-Hong Yue，
Chinese Physics **10** (2001)924—928
234. The thermodynamics of the anisotropic Heisenberg chain with boundary spin impurities，

Bo-Yu Hou，Kang-Jie Shi，Rui-Hong Yue，Shao-You Zhao，
Physics Letters A **285** (2001) 141—149

235. The Lax pair for the elliptic *Cn* and *BCn* Ruijsenaars- Schneider models and their spectral curves，
Kai Chen，Bo-Yu Hou，Wen-Li Yang，
Journal of Mathematical Physics **42** (2001) 4894—4914

236. Antother free Boson representation of Yangian double $DY_{\hbar}(sl_N)$ with Arbitary Level，
Bo-Yu Hou，Liu Zhao，Xiang-Mao Ding，
Communications in Theoretical Physics **35** (2001) 167—172

237. Free boson representation of $DY_{\hbar}(s\hat{l}(M+1|N+1))$ at level one，
Bo-Yu Hou，Wen-Li Yang，Yi Zhen，
Communications in Theoretical Physics **35** (2001)669—672

238. Rational SU(3) Gaudin Model，
Jun-Peng Cao，Bo-Yu Hou，Rui-Hong Yue，
Communications in Theoretical Physics **36** (2001)139—144

239. The Dn Ruijsenaars-Schneider model，
Kai Chen，Bo-Yu Hou，
Journal of Physics A **34** (2001) 7579—7590

240. Generalized $SU_q(1|2)$ Gaudin model，
Jun-Peng Cao，Bo-Yu Hou，Rui-Hong Yue，
Journal of Physics A **34** (2001) 3761—3768

241. The q-deformed supersymmetric t-J model with a boundary，
Bo-Yu Hou，Wen-Li Yang，Yao-Zhong Zhang，Yi Zhen，
Journal of Physics A **35** (2001) 2593—2608

242. Quantum Currents in the Coset Space SU(2)/U(1)，
Xiang-Mao Ding，Bo-Yu Hou，Liu Zhao，
Communications in Theoretical Physics **37** (2002)59—62

243. Accessible Information for Equally-Distant Partially-Entangled Alphabet State Resource，
San-Ru Hao，Bo-Yu Hou，Xiao-Qiang Xi，Rui-Hong Yue，
Communications in Theoretical Physics **37** (2002)149—154

244. The Geometric Construction of WZW Effective Action in Noncommutative Manifold，
Bo-Yu Hou，Yong-Qiang Wang，Zhan-Ying Yang，Rui-Hong Yue，
Communications in Theoretical Physics **37** (2002)349—352

245. Non-commutative Chiral QCD_2 Model，
Zhan-Ying Yang，Rui-Hong Yue，Bo-Yu Hou，Kang-Jie Shi，
Communications in Theoretical Physics **38** (2002)217—220

246. Nonadiabatic Geometric Quantum Computation with Asymmetric Superconducting Quantum Interference Device,
San-Ru Hao, Bo-Yu Hou, Xiao-Qiang Xi, Rui-Hong Yue,
Communications in Theoretical Physics **38** (2002)285—291
247. How to control a geometric quantum gate,
San-Ru Hao, Bo-Yu Hou, Xiao-Qiang Xi, Rui-Hong Yue,
Chinese Physics **11** (2002) 109—114
248. Average mutual information of the random block-message ensembles for many-letters,
San-Ru Hao, Bo-Yu Hou,
Chinese Physics **11** (2002)450—455
249. Elliptic Algebra and Integrable Models for Solitons on Noncommutative Torus T,
Bo-Yu Hou, Dan-Tao Peng,
International Journal of Modern Physics B **16** (2002) 2079—2087
250. Soliton on Noncommutative Orbifold T^2/Z_k,
Bo-Yu Hou, Kang-Jie Shi, Zhan-Ying Yang,
Letters in Mathematical Physics **61** (2002) 205—220
251. Fuzzy sphere bimodule, ABS construction to the exact soliton solutions,
Bo-Yu Hou, Bo-Yuan Hou, Rui-Hong Yue,
Journal of Physics A **35** (2002) 2937—2946
252. Non-commutative geometry of 4-dimensional quantum Hall droplet,
Yi-Xin Chen, Bo-Yu Hou, Bo-Yuan Hou,
Nuclear Physics B **638** (2002) 220—242
253. The Dynamical Elliptic Quantum Group gaudin models and their solutions,
Bo-Yu Hou, Kang-Jie Shi, Rui-Hong Yue, Shao-You Zhao,
Journal of Mathematical Physics **43** (2002) 4628—4640
254. Solitons on Noncommutative Torus as Elliptic Calogero Gaudin Models, Branes and Laughlin Wave Functions,
Bo-Yu Hou, Dan-Tao Peng, Kang-Jie Shi, Rui-Hong Yue,
International Journal of Modern Physics A **18** (2003) 2477—2500
255. Optimal Conclusive Teleportation of an Arbitrary d-Dimensional N-Particle Unknown State via a Partially Entangled Quantum Channel,
San-Ru Hao, Bo-Yu Hou, Xiao-Qiang Xi, Rui-Hong Yue,
Communications in Theoretical Physics **39** (2003)157—166
256. Quantum Standard Teleportation Based on the Generic Measurement Bases,
San-Ru Hao, Bo-Yu Hou, Xiao-Qiang Xi, Rui-Hong Yue,
Communications in Theoretical Physics **40** (2003)415—420
257. Open Superstring Star as a Continuous Moyal Product with B Field,
Xiao-Hui Wang, Bo-Yu Hou, Zhan-Ying Yang, Rui-Hong Yue,

Communications in Theoretical Physics **40** (2003) 451—456
258. Algebraic Bethe ansatz for the elliptic quantum group $E_{\tau,\ \eta}(sl_n)$ and its applications,
Bo-Yu Hou, Ryu Sasaki, Wen-Li Yang,
Nuclear Physics B **663** (2003) 467—486
259. The Dynamical Yang-Baxter Relation and the Minimal Representation of the Elliptic Quantum Group,
Heng Fan, Bo-Yu Hou, Kang-Jie Shi, Rui-Hong Yue, Shao-You Zhao,
Journal of Mathematical Physics **44** (2003) 1276—1296
260. Soliton Solutions on Noncommutative Orbifold T^2/Z_4,
Hui Deng, Bo-Yu Hou, Kang-Jie Shi, Zhan-Ying Yang, Rui-Hong Yue,
Journal of Mathematical Physics **45** (2004) 978—995
261. Eigenvalues of Ruijsenaars-Schneider models associated with A_{n-1} root system in Bethe ansatz formalism,
Bo-Yu Hou, Ryu Sasaki, Wen-Li Yang,
Journal of Mathematical Physics **45** (2004) 559—575
262. The manifest covariant soliton solutions on noncommutative orbifold T^2/Z_6 and T^2/Z_3,
Hui Deng, Bo-Yu Hou, Kang-Jie Shi, Zhan-Ying Yang, Rui-Hong Yue,
Reviews in Mathematical Physics **18** (2006) 255—284
263. Moduli Space of IIB Superstring and SUYM in $AdS_5 \times S^5$,
Bo-Yu Hou, Bo-Yuan Hou, Xiao-Hui Wang, Chuan-Hua Xiong, Rui-Hong Yue,
Communications in Theoretical Physics **45** (2006) 301—314
264. Flat Currents of Green-Schwarz Superstring in $AdS_2 \times S^2$ Background,
Xiao-Hui Wang, Zhan-Yun Wang, Xiao-Lin Cai, Pei Song, Bo-Yu Hou, Kang-Jie Shi,
Communications in Theoretical Physics **45** (2006)663—668
265. Affine $A_3^{(1)}$ N = 2 Monopole as the D Module and Affine ADHMN Sheaf,
Bo-Yu Hou, Bo-Yuan Hou,
Communications in Theoretical Physics **49** (2008) 439—450
266. Unitary Transformation in kq Representation,
Zhan-Ying Yang, Kai Zhang, Bo-Yu Hou, Kang-Jie Shi,
Communications in Theoretical Physics **52** (2009)103—107
267. Second Reference State and Complete Eigenstates of Open XYZ Chain,
Jun Feng, Xi Chen, Kun Hao, Bo-Yu Hou, Kang-Jie Shi, Cheng-Yi Sun, Wen-Li Yang,
Communications in Theoretical Physics **56** (2011)55—60
268. Projection Operator on Rational Noncommutative Orbifold T^2/Z_4,
Jun Feng, Zhan-Ying Yang, Kai Zhang, Bo-Yu Hou, Kang-Jie Shi,
Communications in Theoretical Physics **56** (2011)107—118
269. Determinant formula for the partition function of the six-vertex model with a non-diagonal

reflecting end,
Wen-Li Yang, Xi Chen, Jun Feng, Kun Hao, Bo-Yu Hou, Kang-Jie Shi, Yao-Zhong Zhang, *Nuclear Physics B* **844** (2011) 289—307

270. Determinant representations of scalar products for the open XXZ chain with non-diagonal boundary terms,
Wen-Li Yang, Xi Chen, Jun Feng, Kun Hao, Bo-Yu Hou, Kang-Jie Shi, Yao-Zhong Zhang, *Journal of High Energy Physics* **01** (2011) 006